Pulsars as Astrophysical Laboratories for Nuclear and Particle Physics

Studies in High Energy Physics, Cosmology and Gravitation

Other books in the series

Electron–Positron Physics at the Z
M G Green, S L Lloyd, P N Ratoff and D R Ward

Non-accelerator Particle Physics
Paperback edition
H V Klapdor-Kleingrothaus and A Staudt

Ideas and Methods of Supersymmetry and Supergravity
***or* A Walk Through Superspace**
Revised edition
I L Buchbinder and S M Kuzenko

Related titles

Electron–Positron Annihilation Physics
Edited by B Foster

Gauge Theories in Particle Physics
Second edition
I J R Aitchison and A J G Hey

Introduction to Gauge Field Theory
Revised edition
D Bailin and A Love

Supersymmetric Gauge Field Theory and String Theory
D Bailin and A Love

Hadron Interactions
P D B Collins and A D Martin

High Energy Phenomenology
Edited by K J Peach and L L J Vick

Pulsars as Astrophysical Laboratories for Nuclear and Particle Physics

F Weber

Institute for Theoretical Physics
Ludwig-Maximilians University
Munich, Germany

Institute of Physics Publishing
Bristol and Philadelphia

British Library Cataloguing-in-Publication Data

A catalogue record for this book is available from the British Library.

ISBN 0 7503 0332 8

Library of Congress Cataloging-in-Publication Data are available

Consultant Editor: Professor Dr Walter Greiner

Published by Institute of Physics Publishing, wholly owned by The Institute of Physics, London

Institute of Physics Publishing, Dirac House, Temple Back, Bristol BS1 6BE, UK

US Office: Institute of Physics Publishing, The Public Ledger Building, Suite 1035, 150 South Independence Mall West, Philadelphia, PA 19106, USA

Typeset in TEX using the IOP Bookmaker Macros

Contents

Preface

One of the most challenging but also most complicated problems of modern physics, both experimental and theoretical, consists of exploring the behavior of matter under extreme conditions of temperature and/or density and the determination of the equation of state associated with it. Knowing the properties of such matter is of key importance for laboratory physics, and for our understanding of the physics of the early Universe and its evolution in time to the present day, and of various astrophysical phenomena, and specifically *pulsars* which are generally accepted to be rotating neutron stars.

On the Earth, relativistic heavy-ion colliders provide the major tools by means of which hot and superdense matter can be created (for some 10^{-22} s) and its properties studied. On the other hand, it is known that nature has created two branches of stable and extremely dense stellar configurations that contain matter in one of the densest forms found in the Universe. These are white dwarfs, in which gravity is balanced by electron degeneracy pressure, and neutron stars (pulsars), held up by baryon degeneracy pressure. The enormous gravitational pull compresses most of the matter of neutron stars to densities that are up to an order of magnitude higher than those found in atomic nuclei, rendering neutron stars as unique astrophysical probes of cold (on the nuclear scale) superdense matter of practically 'infinite' lifetime.

Because of the high-pressure environment in the centers of neutron stars, numerous particle processes – ranging from hyperon population to quark deconfinement to the formation of boson condensates – may be competing with each other there. Moreover, there are theoretical suggestions of even more 'exotic' processes inside neutron stars, such as the formation of absolutely stable strange quark matter, a configuration of matter even more stable than the most stable atomic nucleus, ^{56}Fe. In the latter event, neutron stars would be largely composed of pure quark matter!

In contrast to most people's belief, neutron stars are anything but rare objects. In our Galaxy – the Milky Way – the number of neutron stars has been estimated to be possibly as large as 10^9, which implies that up to 1% of the mass of our Galaxy is confined in neutron stars. Enormous progress has been made in radio and X-ray astronomy in recent years, which has led to an unprecedented wealth of new observational data on neutron stars, and it now seems within reach for the first time ever to understand the properties of superdense matter, in the high-density, low-temperature portion of its phase diagram, from observed neutron star data. This research complements the high-temperature domain to be probed by the relativistic heavy-ion colliders, like those currently under construction at Brookhaven National Laboratory and at CERN.

Among those conferences on the properties of superdense matter that are being held regularly, the NATO Advanced Study Institute on Hot and Dense Nuclear Matter, organized by W Greiner, H Stöcker, and A Gallmann, has become institutional. It is one of those conference series that successfully creates an environment where physicists from all the different branches of physics quoted above come together to discuss the numerous facets of superdense matter, as related to their respective areas. The last meeting of this series took place in the fall of 1993 in Bodrum, Turkey. The idea of writing this book dates back to a conversation with Professor W Greiner during this meeting, when he asked me with a wink in his eye whether I would like to 'expand' the content of my lectures to the size of a book. Several years have passed since then, and so the book has evolved not only from my Bodrum lectures but from lecture notes for a full-year graduate course on relativistic many-body physics and relativistic astrophysics, which I taught at the Ludwig-Maximilians University of Munich, paired with an up-to-date account on the present status of research that aims to explore the behavior of superdense matter from the properties of neutron stars.

The successful completion of this book was dependent on the efforts of many people. Aside from Professor Greiner, whom I would like to thank very much for inviting me to write this text, I am particularly grateful to Professors M K Weigel (University of Munich) and N K Glendenning (Lawrence Berkeley National Laboratory, Berkeley) who, over the years, provided me with invaluable insight into numerous aspects of the physics at the interface between nuclear physics, particle physics, and relativistic astrophysics. It is practically impossible for me to thank all my other colleagues who gave me criticism, suggestions, and instructions. Those to whom I am most grateful are Professors Ch Alcock (LLNL), S Bastrukov (JINR), G Brown (SUNY), A Faessler (University of Tübingen), Dr Ch Kettner (University of Augsburg), Professor J Madsen (University of Aarhus), Dr B Myers (LBNL), Professor M Prakash (SUNY), Drs A Sedrakian (Universities of Rostock and Cornell), W Swiatecki (LBNL), Professors F K Thielemann (University of Basel), M Wiescher (University of Notre Dame), and my Munich colleagues Professor G Börner (MPA, Garching), Dr M Harlander (GeNUA), Dipl-Phys B Hermann (University of Munich), Professor W Hillebrandt (MPA, Garching), Dr H Huber (University of Munich), Drs T Janka and E Müller (MPA, Garching), Professor P Ring (Technical University of Munich), Dr Ch Schaab (University of Munich), and Professor W Weise (Technical University of Munich). Last, but not least, Professor J Lattimer (SUNY) deserves my very special thanks for carefully reading an earlier draft of the book and providing me with crucial feedback.

For assistance in the research that went into this book, I thank the Deutsche Forschungsgemeinschaft (DFG), the German Ministry for Research and Technology (BMFT), and the Max Kade Foundation of New York for research grants and fellowships.

Finally, I would like to add that there exist a considerable number of outstanding accompanying books and review articles that deal with some of the aspects of the subject matter developed here. I have tried to refer to several of them throughout the volume. Naturally, it is virtually impossible for me to give sufficient credit to all of them.

F Weber

September 1998

Chapter 1

Introduction

1.1 In search of the behavior of superdense matter

A leading area of research, both experimental and theoretical, concerns the exploration of the properties of matter under extreme conditions of temperature and density, and the determination of the equation of state – that is, the relation between pressure, temperature, and density of a physical system – associated with it [1, 2, 3]. Its knowledge is of key importance for our understanding of the physics of the early universe, its evolution in time to the present day, compact stars, various astrophysical phenomena, and laboratory physics [2, 3]. The properties of superdense matter in the high-temperature domain of its phase diagram is probed by relativistic heavy-ion colliders, as schematically illustrated in figure 1.1. On the other hand, it is well known that nature has created a large number of compact stellar objects, that is, white dwarfs and neutron stars, which contain matter in one of the densest forms found in the universe. The enormous gravitational pull that binds neutron stars compresses most of their matter to densities that are up to an order of magnitude higher than the density of atomic nuclei, making neutron stars unique probes of superdense matter. A recent quantitative determination of the number of neutron stars in our Galaxy has been performed in [5], using recently calculated models for massive stellar evolution and supernovae [6, 7, 8, 9] coupled to a model for Galactic chemical evolution. It was found that the number of neutron stars in our Galaxy is probably as large as 10^9 (plus a comparable number of black holes), rendering neutron stars as anything but rare stellar objects.

Neutron stars are associated with two classes of astrophysical galactic objects. The first class, *pulsars* [10, 11, 12, 13, 14] are generally accepted to be rotating neutron stars. Famous pulsar representatives are Crab, Vela, PSR 1937+21, and PSR 1957+20.[1] The fastest pulsars observed so far

[1] PSR stands for Pulsating Source of Radiation.

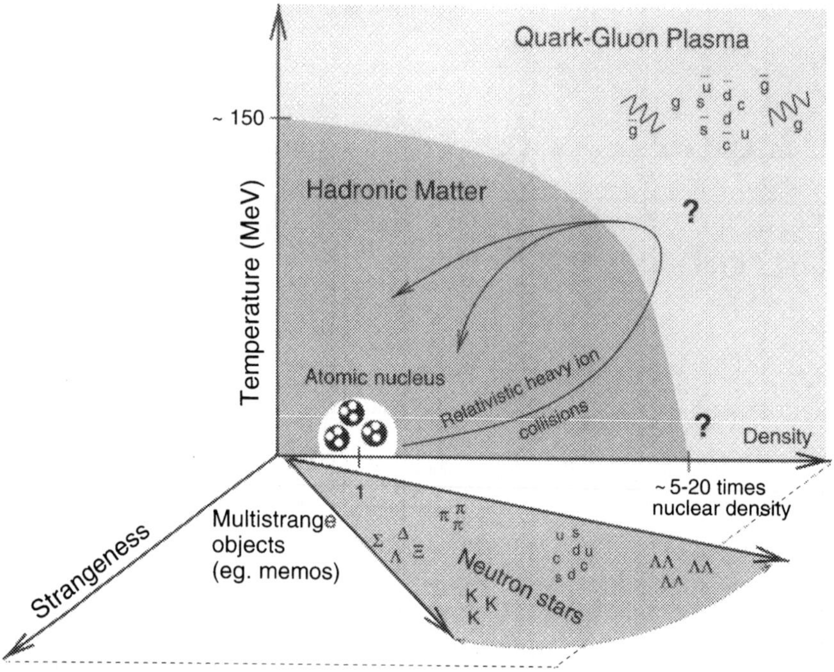

Figure 1.1. Phase diagram of superdense nuclear matter. Experimentally, relativistic heavy-ion collision experiments enable physicists to cast a brief glance at hot and superdense matter for times as short as about 10^{-22} s [4]. Complementary to this, the superdense matter in the low-temperature high-density region of the phase diagram forms a permanent component inside neutron stars.

have rotational periods of $P = 1.6$ ms, which corresponds to about 620 rotations per second! The second class are *compact X-ray sources* (e.g. Her X–1 and Vela X–1, MXB(1636–536)), certain of which are neutron stars in close binary orbits with an ordinary star [15]. Matter in their cores possess densities ranging from a few times the density of normal nuclear matter (2.5×10^{14} g cm^{-3}) to about an order of magnitude higher, depending on the star's mass. This circumstance links the quest of the behavior of superdense matter, which constitutes a nuclear physics and particle physics problem, to Einstein's theory of general relativity in the framework of which models of (rotating) neutron stars are to be constructed and their properties studied. In this respect, neutron stars together with black holes are examples where two of the frontiers of physics – the strong interaction and general relativity – come together. Paired with the unprecedented wealth of new observational neutron star data, it seems to be within reach for the first time ever to draw definitive conclusions about the behavior of superdense matter from observed neutron star data. The situation looked entirely different some 30 years ago when J A Wheeler wrote [16]

> So far no superdense star has been identified. Moreover, a cool superdense star – with radius of 10 km, with a surface temperature (after $\sim 10^6$ years of cooling) of 10^6 K, and at a distance of 10 pc, comparable to the distance of near-by stars – is fainter than the 19th magnitude and therefore hardly likely to be seen. The rapidity of cooling makes detection even more difficult.

1.2 Neutron stars as probes of superdense matter

Rotating neutron stars show a large range of extreme astrophysical phenomena [17, 18] and, as clocks, they show incredible rotation stability and promise to act as superb probes of superdense matter. All our information about such objects is contained in the directional distribution of their electromagnetic radiation and in its spectral composition (X-ray, radio, etc). To date about 800 pulsars are known, and the fastest of them have rotational periods in the millisecond range, e.g. pulsars 1937+21 and 1957+20 with periods of $P = 1.6$ ms. The discovery rate of new pulsars is rather high. It is important to note, however, that searches for rapidly rotating radio pulsars are biased, being least sensitive to short periods [19, 20, 21, 22]. Therefore the world's data of radio pulsars do not represent the true population of fast pulsars. This is particularly the case for pulsars of periods below about 1 ms. Most of the large surveys have had very poor sensitivity to millisecond pulsars with rotational periods below about 4 ms, thus presumably distorting the statistics on pulsar periods [19, 21, 22, 23]. The present cutoff in short periods at 1.6 ms is therefore possibly only an artifact of the search sensitivity. The growing number of pulsar discoveries with periods right down to this cutoff suggests that this is so and presents a special opportunity and challenge to radio astronomy [22, 23]. The high level of interest in the properties of millisecond pulsars [24], which – as we shall see in this volume – constitute particularly superb astrophysical laboratories for a wide range of physical studies, makes it clearly worthwhile to continue searching for the extremely weak signals emitted by these objects. In the next ten years or so, astronomers may have found out whether possibly even (exotic) sub-millisecond pulsars exist or not [21, 22, 23].

The realization that globular clusters are ideal environments for the formation of binaries including neutron star binaries, in which accretion from the companion spins up the compact star, promises many more data [25]. Indeed a considerable fraction of the presently known millisecond pulsars [24] are in globular clusters, as can be inferred from tables 1.1 and

1.2.[2] The surveys listed in these tables have been remarkably successful at detecting the extremely weak signals of millisecond pulsars. In the next 10 years or so, computing resources are expected to increase enough to allow astronomers to find out whether or not pulsars spinning at sub-millisecond periods exist. As we shall discuss in great detail in this volume, such exotic systems could seriously challenge their interpretation as extremely rapidly spinning neutron stars [26], made up of matter that stays either in the pure confined hadronic phase or exists in chemical equilibrium with deconfined quark matter. The theoretical lower limit on the rotational period of such stellar systems seems to be close to about 1 ms (i.e. 1000 Hz). Stars that spin below that limit begin to shed mass at the equator, by means of which stability is irrevocably lost. Besides that limit, the emission of gravity waves, driven by the development of aspherical perturbations that propagate along the surface of a rotating neutron star (similar to waves on the surface of the ocean), sets an even more stringent limit of fast rotation which may be as large as about 1.5 to 1.6 ms, depending on star mass. The only class of stars that could withstand stable rotation at periods considerably below 1 ms (say about 0.5 ms) would be bodies made up of self-bound matter, of which absolutely stable 3-flavor strange quark matter is the most plausible state. Such stellar objects are known as strange quark matter stars [27, 28, 29, 30]. So the future discovery of a single sub-millisecond pulsar, rotating at a period considerably below about 1 ms, would address the fundamental issue of the true ground state of strongly interacting matter and may provide strong evidence that we are witnessing a state of nuclear matter that is merely in a metastable phase.

Astrophysical constraints on the equation of state are imposed by several astrophysical sources. These are:

- Masses of about two dozen neutron stars [31, 32].
- Rotational periods, P, of rapidly spinning pulsars, as compiled in tables 1.1 and 1.2.
- Radio luminosities of pulsars.
- Pulse widths (duty cycles).
- First and second period derivative, $\dot{P}$ [12, 17, 33] and $\ddot{P}$.
- Sub-pulse multiplicity (if any).
- Radio-frequency spectral features (e.g. index, cutoff, etc).

[2] Information on radio pulsar resources is available from the World Wide Web at http://pulsar.princeton.edu/rpr.shtml. The addresses of a few selected homepages are:
Arecibo Observatory at http://aosun.naic.edu/,
European Pulsar Network at http://www.mpifr–bonn.mpg.de/pulsar/epn/epn.html,
Jodrell Bank Pulsar Group at http://www.jb.man.ac.uk/∼pulsar/index.html,
Max–Planck Institute for Radioastronomy, Bonn at http://www.mpifr-bonn.mpg.de/,
National Radio Astronomy Observatory Green Bank at http://www.gb.nrao.edu/,
Princeton Pulsar Group at http://pulsar.princeton.edu/, and
Parkes Radio Telescope at http://wwwpks.atnf.csiro.au/.

- Temperatures of neutron stars as determined by the X-ray observatories Einstein, EXOSAT, ROSAT, ASCA, Ginga [34, 35, 36, 37].
- Magnetic fields as inferred from cyclotron line features observed in binary X-ray pulsars like Her X–1, 4U 0115–63, 1E2259+586 [36, 38, 39, 40, 41]. The line feature in the latter X-ray pulsar was observed with the Ginga satellite, launched in 1987.
- Radius determination of nearby neutron star RXJ 185 635–3754 [42, 43, 44] as well as of other X-ray emitting neutron stars [45].
- Redshifted photons [46].
- Supernovae [7, 8, 47].

Of course, there are other morphological features, many of which, however, may be possessed by only a few pulsars [17]. Of the items listed just above, the theoretical and observational statuses of the masses, rotational periods, slowing-down rates, temperatures, and magnetic fields are currently most secure. For the second, for instance, the discovery of the first millisecond pulsar in 1982 by Backer *et al* [48] has stimulated considerable interest in the rotation of neutron stars. The physics of supernovae [7, 8, 47, 49, 50, 51, 52, 53] is known to involve many factors of comparable importance but high uncertainty so that it cannot be said to provide very severe constraints on the equation of state at the present time.

Some of the characteristic properties of neutron stars, which make them ideal probes for a wide range of physical studies,[3] are briefly reviewed in the following. The first property concerns the range of stable masses of neutron stars, which range from $0.1\,M_\odot$ to about $2\,M_\odot$. (The mass of the Sun is listed, together with other astrophysical constants, in table 1.3 of section 1.3. The other tables of this section, i.e. tables 1.4 to 1.7, contain useful conversion factors, and give an overview of the numerous physical and astrophysical quantities that shall be encountered frequently in the main body of this volume.) Being confined within a sphere of radius of about 10 km, the matter in the cores of neutron stars possess densities ranging from a few times ρ_0 to an order of magnitude higher. Here $\rho_0 = 0.15$ nucleons/fm^3 denotes the density of normal nuclear matter, which corresponds to a mass density of 2.5×10^{14} g cm^{-3}, or an energy density of 140 MeV fm^{-3} (see table 1.4). The number of baryons forming a neutron star is of the order of $A \sim 10^{57}$. A nucleon sitting at the surface of a neutron star is gravitationally bound by about 100 MeV. When neutron stars are formed they have interior temperatures of the order of $T \sim 10^{11}$ K (about 10 MeV). They cool by neutrino emission processes to interior temperatures of less than 10^{10} K within a few days. Throughout most of the active life of neutron stars as pulsars and X-ray sources, the temperature is between

[3] See, for instance, references [14, 54–67].

Table 1.1. Sample of millisecond pulsars in globular clusters [19, 22, 25].† The entries are: rotational period P, period derivative $\dot{P}$, and magnetic field at the surface, B.

Cluster	Pulsar	P (ms)	$\dot{P}$ (10^{-15})	$\log_{10} B$ (G)	Feature
47 Tuc	0021–72C	5.757	-4.98×10^{-5}		single
	0021–72D	5.358	-0.28×10^{-5}		single
	0021–72E	3.536			binary
	0021–72F	2.624			single
	0021–72G	4.040			single
	0021–72H	3.210			binary
	0021–72I	3.485			binary
	0021–72J ‡	2.101			binary
	0021–72M	3.677			binary
	0021–72N	3.075			single
M53	1310+18	33.163			binary
M5	1516+02A	5.553	6×10^{-5}	8.77	single
	1516+02B	7.947			binary
M4	1620–26	11.08	79.04×10^{-5}		binary
M13	1639+36A	10.378	0.00000		single
	1639+36B	3.528			binary
Terzan 5	1744–24A	11.56	-1.9×10^{-5}		binary
NGC 6624	1820–30A	5.440	0.00338	9.64	single
M28	1821–24	3.054	0.00161	9.35	single
NGC 6760	1908+00	3.6			binary
M15	2127+11D	4.65	-10.75×10^{-3}		single
	2127+11E	4.80	1.78×10^{-4}	8.96	single
	2127+11F	4.03	3.2×10^{-5}	8.56	single
	2127+11H	6.74	2.4×10^{-5}	8.61	single

† See also http://pulsar.princeton.edu/ftp/pub/catalog/.
‡ Possibly 4.201 ms pulsar with strong interpulse.

Table 1.2. Sample of millisecond pulsars discovered at Arecibo, Parkes, and Jodrell Bank (from [21, 22, 23]).† The entries are: rotational period P, period derivative $\dot{P}$, age τ, magnetic field at surface B, and transverse velocity V_T.

Pulsar	P (ms)	$\dot{P}$ (10^{-15})	$\log_{10}\tau$ (yr)	$\log_{10} B$ (G)	V_T (km s^{-1})
		Arecibo millisecond pulsars			
J0751+1807	3.48	8.0×10^{-6}	9.84	8.23	
J1025+10	16.45				
B1257+12	6.22	1143.34×10^{-7}	8.95	8.93	276
J1713+0747	4.57	8.52×10^{-6}	9.95	8.30	31
B1855+09	5.36	1783.63×10^{-8}	9.68	8.49	28
B1937+21	1.56	1051.19×10^{-7}	8.36	8.62	10
B1953+29	6.13	2.95×10^{-5}	9.52	8.63	
B1957+20	1.61	1685.15×10^{-8}	9.17	8.23	206
J2019+2425	3.93	7.02×10^{-6}	9.95	8.26	102
J2317+1439	3.44	2.42×10^{-6}	10.3	8.00	
J2322+2057	4.81	9.74×10^{-6}	8.89	8.34	94
		Parkes millisecond pulsars			
J0034–0534	1.88	6.7×10^{-6}	9.64	8.04	
J0437–4715	5.76	57.09×10^{-6}	9.20	8.76	72
J0613–0200	3.06	1.1×10^{-5}	9.69	8.26	
J0711–6830	5.49				
J1025–0709	5.16				
J1045–4509	7.47	1.9×10^{-5}	9.79	8.58	
J1455–3330	7.99	0.000 00	>10.3	<8.3	60
J1604–72	14.84				
J1643–1224	4.62	3.3×10^{-5}	9.34	8.60	
J1730–2304	8.12	1.9×10^{-5}	>9.6	<8.7	22
J1745–11	4.07				
J1804–2718	9 34				
J2052–08	4.51				
J2124–3358	4.93	0.000 00	>9.1	<8.7	
J2129–57	3.73				
J2145–0750	16.05	0.000 00	>10.1	<8.8	
		Jodrell Bank/Caltech millisecond pulsars			
J0218+4232	2.32	7.5×10^{-5}	9.58	8.18	
J1012+5307	5.26	1.46×10^{-5}		8.45	

† See also http://pulsar.princeton.edu/ftp/pub/catalog/.

10^7 and 10^9 K. Surface temperatures are an order of magnitude or more smaller. Measurements of the surface temperature of the famous Crab pulsar, for instance, have led to $T^\infty \approx 1.5 \times 10^6$ K (for the definition of T^∞, see table 1.5). Rotating magnetized neutron stars are surrounded by a plasma – the so-called magnetosphere – in which via plasma processes the electromagnetic radiation from pulsars is generated. The extremely strong magnetic fields near the surface of neutron stars in compact X-ray sources, which typically have values of $B \sim 10^{12}$ G (about thirteen orders of magnitude stronger than the Earth's magnetic field at the surface) play an important role in channeling the accreting matter onto the neutron star surface. Trümper *et al* have observed a feature in the pulsed hard X-ray spectrum of Her X–1 [38, 39], while Wheaton *et al* have observed a similar feature in 4U 0115–63 [40]. Interpreting the features as cyclotron lines implies $B \sim 4\text{–}6 \times 10^{12}$ G in Her X–1 and $B \sim 2 \times 10^{12}$ G in 4U 0115–63. It is interesting to note that such large values arise naturally from the collapse of a main sequence star with a typical frozen-in surface magnetic field of 100 G. The decrease in radius by a factor of $\sim 10^5$ leads to an increase of the magnetic field by a factor of $\sim 10^{10}$, which is in agreement with most of the theoretically derived surface fields for pulsars.

Figure 1.2 schematically illustrates the life of a neutron star born either in a supernova explosion or from a collapsing white dwarf that has become gravitationally unstable because of mass accretion from a companion star. A combination of the enormous magnetic field and the high rotational frequency, Ω, ignites electromagnetic beaming, by means of which pulsars register themselves in satellites and terrestrial radio telescopes, provided of course the instruments lie in the cone swept out by the electromagnetic beam. During the first $\sim 10^4$ years of its existence, the star's surface temperature falls to about 10^8 K is caused by the loss of thermal energy due to the neutrino emission. This epoch is followed by a continued cooling driven primarily by photon emission. The rotating neutron star is born with an enormous store of angular momentum and rotational energy, estimated to be $\sim 10^{55}$ MeV for the Crab pulsar, which it radiates slowly over millions of years by the weak processes of electromagnetic radiation and a wind of electron–positron pairs. Due to these radiative processes the star gradually spins down which eventually is interrupted by a sudden spin-ups, or glitches, as observed in the pulsar frequencies of Crab, $|\Delta\Omega|/\Omega \sim 10^{-8}$, and Vela, $|\Delta\Omega|/\Omega \sim 2 \times 10^{-6}$. Numerous models have been proposed to explain the origin of these sudden frequency spin-ups since they were first observed in 1969. These include starquakes, vortex pinning, magnetospheric instabilities, and instabilities in the motion of the superfluid neutrons.

Of fundamental importance for the theoretical analysis of the above listed properties and the physical phenomena of neutron stars is the equation of state – i.e. the functional dependence of pressure on total

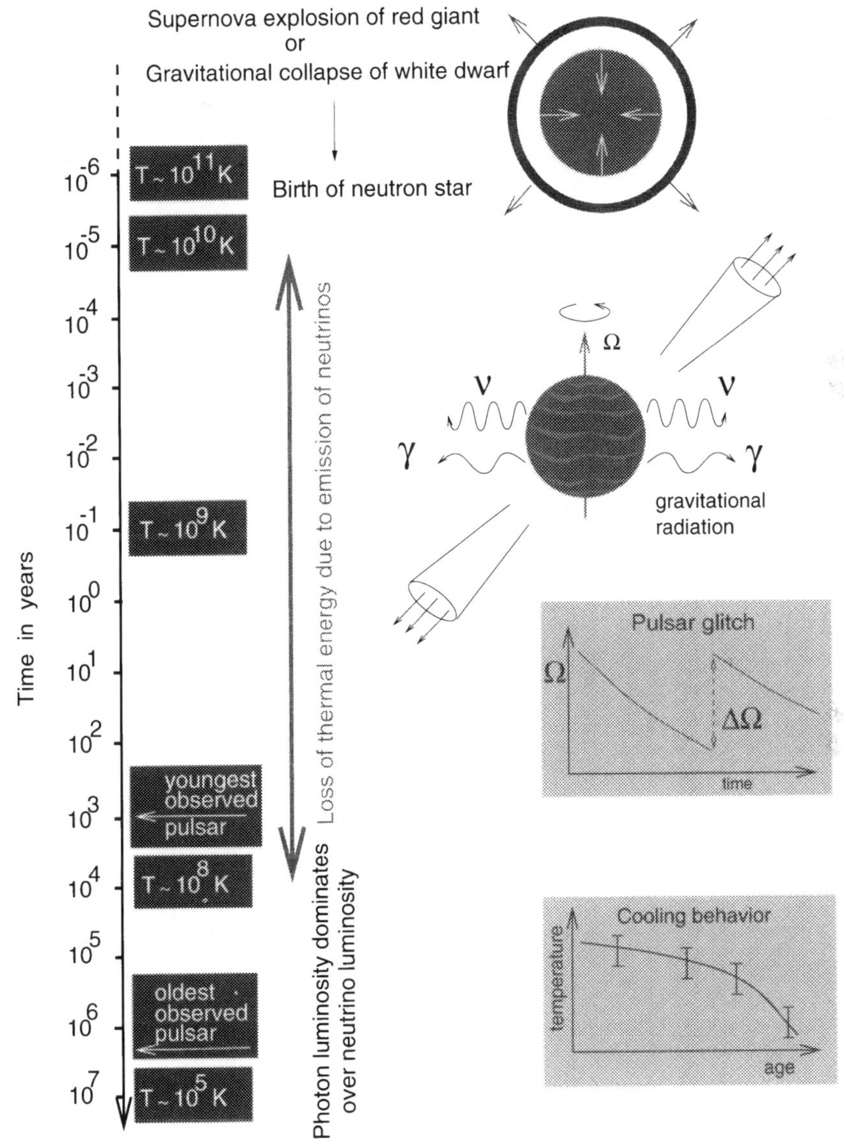

Figure 1.2. Life of a neutron star schematically illustrated.

energy density, $P(\epsilon)$ – of neutron star matter. Figure 1.3 illustrates the fundamental role of the equation of state for the construction of models of neutron stars. It forms the basic input quantity when solving Einstein's field equations for the structure of neutron stars [14, 58, 62, 64], as schematically

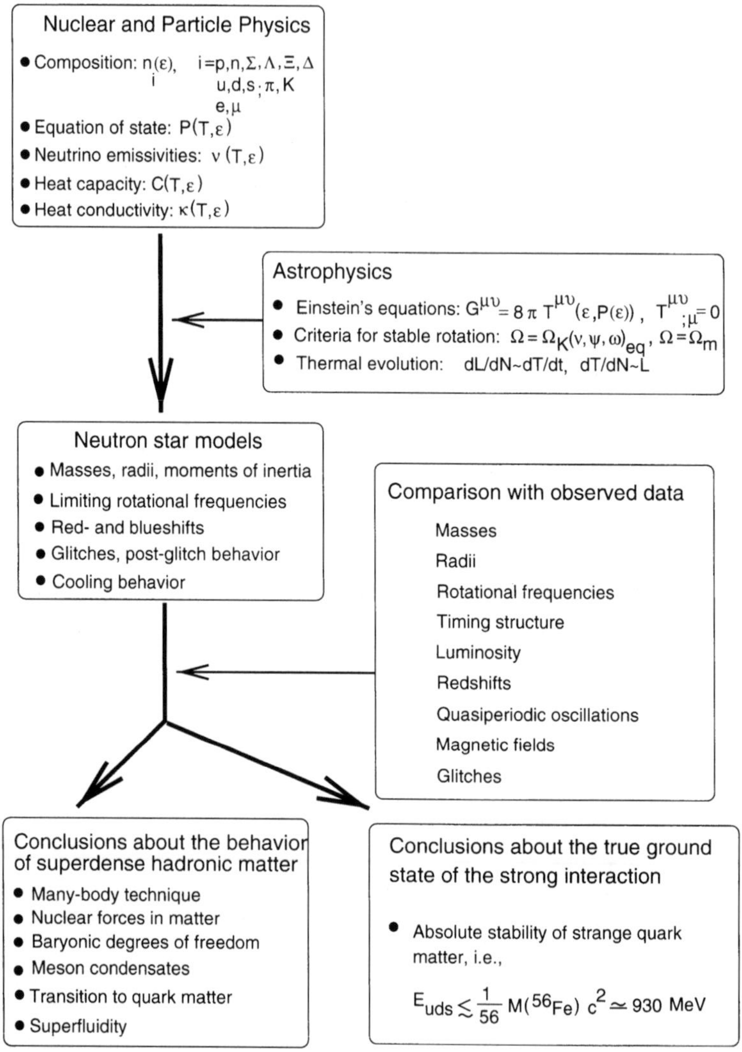

Figure 1.3. To find the properties of neutron stars, nuclear and particle physics are to be married with relativistic astrophysics. A comparison of the theoretically determined neutron star properties with the observed data enables one to shed light on various facets concerning the behavior of superdense matter.

illustrated in figure 1.3. For neutron stars, the density of matter spans an enormous range, from supernuclear densities of possibly more than 10 times the density of normal nuclear matter in the core down to the density of iron at the star's edge.

Most of the mass of the star is contributed by highly compressed matter at nuclear and supernuclear densities. The star's atmosphere is so thin, indeed so is the crust of all but the least massive stars, that these regions contribute negligibly to the bulk stellar properties like mass, radius, and moment of inertia [61]. For this reason the high-density part of the equation of state plays the decisive role in determining the properties of neutron stars heavier than about $1\,M_\odot$. The equations of state of neutron star matter up to neutron drip density [68, 69, 70], which occurs at about 4×10^{11} g cm^{-3}, and at densities above neutron drip but below the saturation density of nuclear matter [71, 72] is relatively well known. This is to a lesser extent the case for the equation of state in the vicinity of the saturation density of normal nuclear matter. Finally the physical properties of matter at still higher densities – referred to as the supernuclear density regime, for which model calculations predict besides neutrons and protons a significant number of hyperons ($\Sigma^{\pm,0}, \Lambda, \Xi^{0,-}$), possibly Δ's, and plausibly quarks (u, d, s-flavors) – are rather uncertain and the associated equation of state is therefore only poorly known. The models derived for it differ considerably with respect to the functional dependence of pressure on density. This has its origin in various sources such as: (1) the many-body technique used for the determination of the equation of state, (2) the model adopted for the nucleon–nucleon force, (3) the description of electrically charge neutral neutron star matter in terms of either (a) only neutrons, (b) neutrons and protons in chemical (β-) equilibrium with electrons and muons, or (c) nucleons, hyperons, and more massive baryon states in β-equilibrium with leptons, (4) the treatment of pion or kaon condensation, and (5) the inclusion of the possible phase transition of confined hadronic matter into deconfined quark matter.

The last item has attracted particular interest over the last 25 years, initiated with the early work by Fritzsch *et al* [73], Baym and Chin [74], Keister and Kisslinger [75], Chapline and Nauenberg [76], and Fechner and Joss [77]. It is generally agreed on that the high-pressure environment that exists in the cores of neutron stars constitutes an ideal physical environment where hadrons may transform into quark matter, forming a permanent component of matter in neutron stars. Nevertheless until recently no observational signal has even been proposed. This is so because whether or not the quark–hadron phase transition occurs in such stars makes only a small difference to their static properties, like the range of possible masses, radii, or even their limiting rotational periods. This however may be strikingly different for the timing structure of a pulsar, that is, its braking behavior in the course of stellar evolution, which deviates dramatically from the canonical behavior if a pulsar starts to convert some of its matter at the center to pure quark matter, as we shall see in this volume.

Neutron stars as well as their hypothetical counterparts – the strange quark matter stars – are objects of highly compressed matter so that the geometry of spacetime is changed considerably from flat space. Thus models of such stars are to be constructed in the framework of Einstein's general theory of relativity combined with theories of superdense many-body matter, each of which constitute a very teasing problem. The connection between both of these branches of physics is provided by Einstein's field equations (G denotes the gravitational constant, and μ, ν=0,1,2,3),

$$G^{\mu\nu} = 8\,\pi\,G\,T^{\mu\nu}(\epsilon, P(\epsilon))\,, \tag{1.1}$$

which couples the Einstein curvature tensor, $G^{\mu\nu}$, to the energy–momentum density tensor, $T^{\mu\nu}$, of the stellar matter. The tensor $T^{\mu\nu}$ contains the equation of state, $P(\epsilon)$, of the stellar matter, which is derivable from the stellar matter Lagrangian $\mathcal{L}_m(\{\chi\})$. In general, $\mathcal{L}_m$ is a complicated function of the numerous hadron and quark fields, χ, that plausibly acquire finite amplitudes up to the highest densities reached in the cores of compact stars. According to what has been said above, candidates for χ are the baryons of the SU(3) baryon octet, $p, n, \Sigma, \Lambda, \Xi, \Delta$, as well as the quark fields u, d, s [30, 78, 79]. Chemical equilibrium and electric charge neutrality of the stellar matter require the presence of leptons too, in which case $\chi = e^-, \mu^-$. Finally the interactions among baryons are described via the exchange of mesons with masses up to about 1 GeV, as schematically illustrated in figure 1.4. Hence, in this case $\chi = \sigma, \omega, \pi, \rho, \eta, \delta, \phi$. For a given $\mathcal{L}_m$, the models for the equation of state then follow as

$$\frac{\partial \mathcal{L}_m}{\partial \chi} - \partial_\mu \, \frac{\partial \mathcal{L}_m}{\partial(\partial_\mu \chi)} = 0 \quad \Longrightarrow \quad P(\epsilon)\,, \tag{1.2}$$

where χ stands for the various particle fields described just above.

In general, equations (1.1) and (1.2) were to be solved simultaneously since the baryons and quarks move in curved spacetime whose geometry, determined by Einstein's field equations, is coupled to the total mass energy density of the matter fields. In the case of neutron stars and quark matter stars, as for all astrophysical situations for which the long-range gravitational forces can be cleanly separated from the short-range forces, the deviation from flat spacetime over the length scale of the strong interaction, $\sim$ 1 fm, is however practically zero up to the highest densities reached in the cores of such stars (some 10^{15} g cm^{-3}). This is not to be confused with the global length scale of a neutron star, $\sim$ 10 km, for which $M/R \sim 0.3$, depending on the star's mass. That is to say, gravity curves spacetime only on a macroscopic scale. To achieve an appreciable curvature on a microscopic scale set by the strong interaction, mass densities greater than $\sim 10^{40}$ g cm^{-3} would be necessary [56, 80]! This circumstance divides the construction of models of compact stars into *two* distinct problems. Firstly, the effects of the short-range nuclear forces on the properties of matter are described in a comoving proper reference frame (local inertial

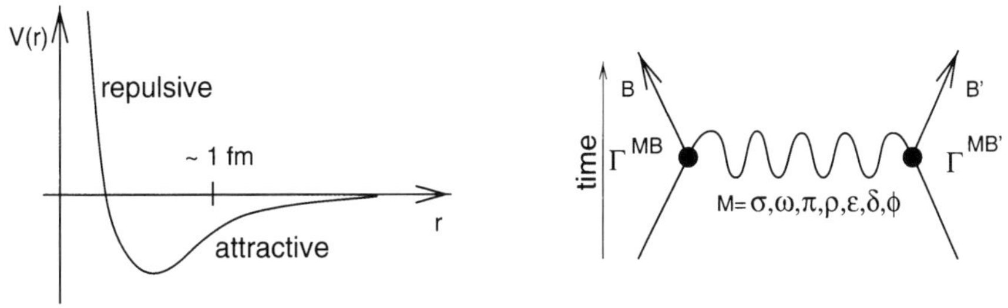

Figure 1.4. Schematic illustration of nucleon–nucleon interaction. Left: A nucleon–nucleon potential is used in non-relativistic Schrödinger-based many-body calculations. Right: In relativistic field theory, the interaction among two nucleons is described by the exchange of mesons, M, of masses up to about 1 GeV. The vertices Γ^{MB} describe the coupling between baryons and mesons.

frame), where spacetime is flat, by the parameters and laws of many-body physics. Secondly, the coupling between the long-range gravitational field and the matter is taken into account by solving Einstein's field equations for the gravitational field described by the general relativistic curvature of spacetime, leading to the global structure of the stellar configuration. While still rather complicated, this two-step analysis eases the problem considerably [81].

The nuclear equation of state of neutron star matter differs in several respects from the nuclear equation of state encountered in other branches of nuclear physics and astrophysics. For instance, normal nuclei are approximately isospin symmetric (that is, close equality in neutron and proton number) and are therefore bound by the strong force. The idealization of this behavior is realized in a system referred to as infinite, symmetric nuclear matter. In contrast to such a system, neutron stars consist of highly isospin-asymmetric nuclear matter, which is not bound by the strong interaction. Neutron stars are held together by the gravitational attraction only. (The situation would be different for the above mentioned class of hypothetical strange stars [27, 28, 29, 30] which are self-bound by confinement. Gravity makes them only a little bit more dense.) For that reason neutron star matter must be electrically charge neutral to very high precision [58, 61, 62], because the Coulomb force, like the gravitational, is long range, and any net charge on a star will reduce the binding that gravity provides. Another characteristic feature of neutron star matter has its origin in the violation of strangeness conservation by the weak interaction, which leads to the accumulation of a significant amount of net strangeness in neutron stars that is either [58, 61, 62, 82, 83, 84, 85, 86] confined in hyperons (up to 20% [61, 84, 85]), or exists in the form of unconfined strange quark matter that coexists in chemical equilibrium with the hyperons [87, 88, 89, 90]. One way or the other, neutron stars plausibly contain net strangeness as a permanent component of their matter, in sharp contrast to the dense and hot nuclear matter created in heavy-ion collisions which, due to the short time scales involved, has zero net strangeness. To sum this up, we stress that in a realistic determination of the equation of state of neutron star matter one *must* impose the two constraints of electric charge neutrality and chemical equilibrium on the equations of motion for the various baryon and quark fields. The pure neutron matter model, which still possesses some popularity in the literature, may be acceptable for a first-step analysis, but it ignores a lot of important physics. Trivially it fulfills the condition of electric charge neutrality, but not of chemical equilibrium. Therefore what is being described by this model is a state of matter that is highly excited relative to neutron star matter [61] which will quickly decay into the lower energy state.

Obviously determining the equation of state over such a huge density range as encountered in neutron stars is a formidable task. It rests mainly on theoretical physical arguments for which no direct experimental confirmation exists. Only by means of comparing the theoretically determined properties of neutron stars (pulsars) with the data base built up for these properties is one able to either confirm or refute competing ideas about the behavior of matter at supernuclear densities and thus constrain the equation of state associated with it. If one leaves the two outer crusts of a neutron star out of consideration, for which rather reliable equations of state are available in the literature, one is left with the dense nuclear regime, whose structure should naturally be studied in the framework of a relativistic field-theoretical theory. Relativistic theories of dense nuclear matter [62, 91, 92, 93] and finite nuclei [94, 95, 96, 97] have enjoyed a renaissance in recent years, and they have the virtue of describing nuclear matter at saturation, many features of finite nuclei, both spherical and deformed, and they extrapolate causally to high density.

One of the first modern attempts toward a realistic determination of the equation of state of superdense neutron star matter, utilizing the latest methods and theories that were accessible and practicable at that time, goes back to Glendenning [61]. He started off from a relativistic Lagrangian density, included all baryon states that become populated in superdense neutron star matter, and solved the equations of motion in the relativistic mean-field (i.e. relativistic Hartree) approximation. This level of many-body approximation however has its limitations. For instance, the π meson contributions are absent, the tensor coupling of the ρ meson is not included, and large σ-meson self-interactions are needed to achieve agreement with the incompressibility of nuclear matter. Some of these deficiencies were removed later in the framework of the relativistic Hartree–Fock scheme [62, 84, 85], where the π meson contributes to the exchange term, the ρ meson tensor coupling can be included, and somewhat smaller incompressibilities than for the linear mean-field model are obtained. The mathematical structure of the relativistic Hartree–Fock Lagrangians bear a close similarity to those Lagrangians that underly relativistic Brueckner–Hartree–Fock type calculations. In this latter case relativistic boson-exchange interactions – whose parameters are adjusted to the two-nucleon data and the properties of the deuteron – serve as input quantities. Because the latter two Lagrangians contain more degrees of freedom, the resulting equations of state are softer than the relativistic Hartree ones [98, 99, 100].

Both the relativistic Hartree and relativistic Hartree–Fock approximations have the advantage that the more massive baryons can be incorporated in a relatively easy fashion, but from a theoretical standpoint one is still dealing with a phenomenological theory since the coupling parameters are adjusted to the bulk properties of nuclear matter at saturation density. Only recently we have succeeded in performing calculations of symmetric and asymmetric nuclear matter within the relativistic Brueckner–Hartree–Fock scheme, solving the equations in the full Dirac space [101, 102] which avoids ambiguities in the determination of the self-energy that occur in the standard treatment by restriction to positive energy states only [101, 102, 103]. Using the modern Brockmann–Machleidt boson-exchange interactions [104] as an input, the outcome for the properties of the symmetric and asymmetric volume parts of the mass formula was encouraging, indicating that this 'parameter-free' method is applicable not only to symmetric nuclear matter but to asymmetric nuclear matter (specifically neutron star matter) too.

It is the main objective of this volume to provide a detailed overview of the present status of dense neutron star matter calculations, and to test several competing models derived for the equation of state of such matter with respect to their compatibility with the body of observed neutron star data. Of particular interest in this connection will be the analysis of the structure and stability of rapidly rotating neutron stars in the framework of Einstein's theory of general relativity.

Until the end of 1980 such investigations were very rare. Among the first who studied the properties of rapidly spinning neutron stars for arbitrary models for the equation of state were Friedman *et al* [105, 106, 107], Lattimer *et al* [108], and Weber *et al* [62, 109, 110, 111, 112, 113, 114]. The exact solution of Einstein's equations for massive rotating objects is known to be a cumbersome and complicated task [106]. An alternative treatment, which is easier to implement and has proven to be a practical tool for the construction of models of general relativistic, rapidly rotating neutron stars [111, 112], is Hartle's perturbative method [115, 116]. Within the latter a perturbative solution of the rotating stellar structure equations, based on the Schwarzschild metric, is developed. We shall perform our investigations in the framework of an improved version of this method [109].

The models for the equation of state of high-density neutron star matter will be derived in the framework of the relativistic, finite-density Green function formalism, solved at three different levels of approximation of which the relativistic Hartree approximation is the simplest one [78, 79]. It is followed by the more complex relativistic Hartree–Fock approximation [79]. The most complex and sophisticated approach we shall be dealing with is the relativistic ladder approximation to the **T**-matrix [92, 93, 117, 118], which accounts for 'dynamical' two-particle correlations calculated from summing repeated two-baryon scattering processes in dense matter. It is this approximation where relativistic meson-exchange interactions enter into the calculation. Besides the three recent interactions A, B, C of Brockmann and Machleidt [104], we shall be employing the 'Bonn' interaction of Machleidt, Holinde, and Elster [119] together with the 'HEA' potential of Holinde, Erkelenz, and Alzetta [120]. The relativistic ladder approximation is similar to the well-known relativistic Brueckner–Hartree–Fock approach [92, 93, 121, 122], which results as a byproduct from the ladder approximation. As already mentioned, the essential difference between the ladder approximation and the Hartree and Hartree–Fock approximations consists in the neglect of dynamical correlations in the latter two cases. The scattering matrix obeys an integral equation whose Born term is made up of a chosen meson-exchange interaction, with parameters determined by the nucleon–nucleon scattering phase shifts in free space and the properties of the deuteron. For that reason the ladder approximation is referred to as parameter-free. This is in sharp contrast to the relativistic Hartree and Hartree–Fock approximations whose parameters are to be adjusted to the bulk properties of infinite nuclear matter [100, 123, 124]. These are binding energy, symmetry energy, compression modulus, and effective nucleon mass, each at saturation density. The implementation of the boson-exchange models for the nucleon–nucleon interaction into Hartree and Hartree–Fock calculations fails because these do not saturate or, in some cases, do not even bind normal nuclear matter [117, 118, 125].

To summarize, the following open issues associated with the behavior of superdense matter will be explored:

- Applicability of different, competing many-body approximations like

  (a) relativistic Hartree,
  (b) relativistic Hartree–Fock,
  (c) relativistic Brueckner–Hartree–Fock,

  to superdense nuclear matter calculations.
- Relativistic, field-theoretical methods versus non-relativistic, Schrödinger-based treatments.
- Meson-exchange interactions (based on the exchange of $\sigma, \omega, \pi, \rho, \eta, \delta, \phi$ mesons) versus non-relativistic nucleon–nucleon potentials.

- Influence of dynamical two-particle correlations at intermediate nuclear densities calculated from the relativistic, ladder-approximated two-nucleon scattering **T**-matrix in matter.
- Concentration of hyperons ($\Sigma^{\pm,0}, \Lambda, \Xi^{0,-}$), more massive baryon states (i.e. the charged Δ resonance states), and quarks (u, d, s) in superdense neutron star matter.
- Importance of the causality condition, $\partial P / \partial \epsilon \leq 1$, which is automatically fulfilled by relativistic, field-theoretical equations of state at all densities, but appears to be violated by most non-relativistic equations of state.
- Implication of π^- and K^- Bose condensates on the properties of neutron stars.
- Superfluidity.
- Implications of the possible transition of confined hadronic matter into deconfined u, d, s-quark matter on neutron star properties. Specifically, does the transition register itself in an observational signal?

These issues are intimately connected to the bulk properties of neutron stars as well as to stellar features such as glitches, crustal heating, and the effect of neutron star atmospheres upon the stars' evolution and upon the interpretation of X-ray observations. The latter are not discussed here in the detail they deserve. The same is true for the non-relativistic equations of state, extrapolations from laboratory nuclei and ion-beam/heavy-ion experiments to extremely neutron-rich nuclei, the description of matter below nuclear densities, neutrino interactions in dense matter, the formation and early neutrino emission from proto-neutron stars (including the detection of neutrino signals and limits on neutrino and other particle properties from SN 1987A and future supernovae), the pulsar mechanism and the evolution of pulsars (e.g. magnetic field evolution, the death line, formation of millisecond pulsars), and the influence of intense magnetic fields upon both the equation of state and neutron star structure [14]. Since it is virtually impossible to cover all these aspects in too great a detail in a single volume, I contented myself with referring to the relevant literature insted.

1.3 Abbreviations, natural units, and conversion factors

The values of those astrophysical constants that are of relevance in this book are listed in table 1.3. It is customary, and convenient, to use what are called *natural units*. These arise from setting some of the fundamental constants equal to unity, which makes mathematical formulae look simpler and more readable. If not stated otherwise, we shall set $G = c = \hbar = 1$. This leads, for instance, for the mass of the Sun to have a value of $M_\odot = 1.476$ km,

Table 1.3. Physical and astrophysical quantities.

Quantity	Symbol	Value
Planck constant, reduced	$\hbar$	1.0546×10^{-34} J s
		$=6.5821 \times 10^{-22}$ MeV s
Speed of light	c	2.99792×10^{8} m s^{-1}
Gravitational constant	G	6.6726×10^{-11} m^3 kg^{-1} s^{-2}
Boltzmann constant	k_B	1.3807×10^{-23} J K^{-1}
Astronomical unit	AU	1.49598×10^{11} m
Parsec (1 AU/1 arc sec)	pc	3.08568×10^{16} m= 3.262 ly
Light year	ly	0.3066 pc = 0.9461×10^{16} m
Solar mass	$M_\odot$	1.99×10^{30} kg = 1.12×10^{60} MeV
Solar equatorial radius	$R_\odot$	6.96×10^{8} m
Solar luminosity	$L_\odot$	3.846×10^{26} W
Schwarzschild radius of Sun	$2M_\odot G/c^2$	2.953 km
Earth equatorial radius	$R_\oplus$	6.378×10^{6} m
Earth mass	$M_\oplus$	5.9737×10^{24} kg
Local disk density	$\epsilon_{\rm disk}$	3–12 $\times 10^{-24}$ g cm^{-3}
		$\approx$ 2–7 GeV/c^2 cm^{-3}
Local halo density	$\epsilon_{\rm halo}$	2–13 $\times 10^{-25}$ g cm^{-3}
		$\approx$ 0.1–0.7 GeV/c^2 cm^{-3}
Nuclear matter density	ϵ_0	2.5×10^{14} g cm^{-3}
		= 140 MeV fm^{-3}

and to the important conversion factor 1 = 197.329 MeV fm. This and other conversion factors are listed in table 1.4.

Throughout this volume we shall be dealing with a considerable number of physical variables and parameters. The meanings of those which will be encountered most frequently are summarized in tables 1.5 and 1.6. Finally, in table 1.7 we compile quantities, together with their mathematical

Table 1.4. Conversion factors.

Quantity	From units of	To	Multiplication factor
Momentum	fm^{-1}	MeV	197.329
Pressure	MeV fm^{-3}	dyne cm^{-2}	1.6022×10^{33}
Energy density	MeV fm^{-3}	g cm^{-3}	1.7827×10^{12}
Temperature	MeV	K	1.1604×10^{10}
Moment of inertia	g cm^2	km^3	7.4219×10^{-44}
Magnetic field	G	km^{-1}	2.8744×10^{-8}
Luminosity	erg s^{-1}	W	10^{7}

Table 1.5. Symbols for stellar quantities.

Quantity	Symbol	Definition †
Gravitational mass	M	
Star's baryon number	A	
Star's baryon mass	$M_{\rm A}$	$M_{\rm A} = m_n\, A$
Star's proper mass	$M_{\rm P}$	
Change of star mass due to rotation	ΔM	
Monopole mass perturbation function	m_0	
Quadrupole mass perturbation function	m_2	
Monopole stretching function	ξ_0	
Quadrupole stretching function	ξ_2	
Quadrupole moment	Π	
Star's angular velocity relative to infinity	$\Omega\ (\equiv \Omega^\infty)$	
Star's rotational period relative to infinity	$P\ (\equiv P^\infty)$	$P = 2\pi/\Omega$
Kepler (mass shedding) frequency	$\Omega_{\rm K}\ (\equiv \Omega_{\rm K}^\infty)$	
Kepler (mass shedding) period	$P_{\rm K}\ (\equiv P_{\rm K}^\infty)$	$P_{\rm K} = 2\pi/\Omega_{\rm K}$
Gravitational-radiation reaction instability frequency	$\Omega_{\rm GRR}$	
Equatorial velocity	$V_{\rm eq}$	
Star's angular velocity relative to a local inertial frame	$\bar\omega$	$\bar\omega = \Omega - \omega$
Moment of inertia	I	
Angular momentum	J	$J = I\,\Omega$
Rotational energy	T	
Total gravitational mass	W	$W = M_{\rm P} + T - M$
Stability parameter	t	$t = T/\vert W\vert$
Equatorial (polar) radius	$R_{\rm eq}\ (R_{\rm p})$	
Eccentricity	e	$e = (1 - (R_{\rm p}/R_{\rm eq})^2)^{1/2}$
Redshift	z	$z = (1 - 2M/R)^{-1/2} - 1$
Redshift in backward and forward direction	$z_{\rm B},\ z_{\rm F}$	
Luminosity at infinity	$L\ (\equiv L^\infty)$	$L^\infty = -g_{00} L_{\rm s} = {\rm e}^{2\Phi} L_{\rm s}$
Temperature at infinity	$T\ (\equiv T^\infty)$	$T^\infty = \sqrt{-g_{00}}\, T_{\rm s} = {\rm e}^{\Phi}\, T_{\rm s}$
Braking index of a pulsar	n	
Age of a pulsar	τ	$\tau = -\frac{1}{n-1}\,\Omega/\dot\Omega$

† No entry means that the definition (where it applies) will be given in the text. $L_{\rm s}$ and $T_{\rm s}$ denote surface luminosity and surface temperature, respectively.

Table 1.6. Symbols for quantities of many-body physics.

Quantity	Symbol
Baryon spectral function †	$\boldsymbol{\Xi}$, $\bar{\boldsymbol{\Xi}}$
Baryon self-energy (mass operator)	$\boldsymbol{\Sigma}$
Two-particle scattering matrix in matter	$\mathbf{T}$
Baryon propagator of intermediate scattering states	$\boldsymbol{\Lambda}$
Baryon–baryon interaction in free space (OBEP)	$\mathbf{V}$
Two-point baryon Green function	$\boldsymbol{g}_1^{BB'}$
Time-ordering operator	$\hat{T}$
Dirac matrices	$\gamma^\nu = (\gamma^0, \gamma^1, \gamma^2, \gamma^3)$
Pauli isospin matrices	$\boldsymbol{\tau} = (\tau^1, \tau^2, \tau^3)$
Pauli spin matrices	$\boldsymbol{\sigma} = (\sigma^1, \sigma^2, \sigma^3)$
Particle field operators	$\psi_B;\ \sigma, \omega, \pi, \rho, \eta, \delta, \phi$
Form factors, vertices, coupling constants	$F^{MB};\ \Gamma^{MB};\ g_{MB},\ f_{MB}$
Hyperon-to-nucleon coupling	$x_\sigma,\ x_\omega,\ x_\rho$
Meson propagators	$\Delta,\ \mathcal{D};\ \tilde{\Delta} \equiv F^2\Delta$
Mass of particle χ $(\chi = B; \sigma, \omega, \pi, \rho, \eta, \delta, \phi; u, d, c, s)$	m_χ
Effective particle mass	m_χ^*
Baryon number and electric charge of particle χ	$q_\chi,\ q_\chi^{\mathrm{el}}$
Cutoff masses	Λ_M
Fermi–Dirac distribution function	$f,\ \bar{f}$
Temperature	$T \quad (T_n \equiv T/10^n\ \mathrm{K})$
Many-body ground state	$\vert \boldsymbol{\Phi}_0 \rangle$
Single-particle energy	$\omega,\ \bar{\omega}$
Chemical potential	$\mu,\ \bar{\mu}$
Energy (mass) density	ϵ
Pressure	P
Baryon number	A
Baryon number density	$\rho \quad (\equiv A/V)$
Fermi momentum	$k_F,\ \bar{k}_F$
Bag constant	B
Energy per baryon	E/A
Internal energy density	$\mathcal{E}^{\mathrm{int}} \quad (\equiv \rho\, E/A)$
Incompressibility of nuclear matter	K
Symmetry energy	a_{sym}
Adiabatic index	Γ
Neutrino luminosity	ϵ_ν
Heat capacity (per particle)	$c_{\mathrm{v}}\ (\bar{c}_{\mathrm{v}})$
Thermal conductivity	κ_{T}
Total thermal conductivity	κ
Radiative and conductive absorption coefficients	$\kappa_{\mathrm{R}},\ \kappa_{\mathrm{c}}$
Superfluid energy gap	Δ_{F}
Relative particle fraction	$Y_\chi \quad (\equiv \rho^\chi/\rho)$

† Barred quantities refer to antiparticles.

Table 1.7. Symbols for quantities of general relativity theory.

Quantity	Symbol
Energy–momentum tensor	$T^{\mu\nu}$
Metric tensor	$g^{\mu\nu}$
Determinant of metric tensor	$g \equiv \det(g^{\mu\nu}) \equiv \|g^{\mu\nu}\|$
Covariant derivative	$\nabla_\mu = \partial/\partial x^\mu$
Euclidean metric tensor	g_{ij}
Euclidean covariant derivative	$\nabla_i = \partial/\partial x^i$
Christoffel symbol of the first kind	$[\lambda\mu, \nu]$
Christoffel symbol of the second kind	$\Gamma^\sigma_{\mu\nu}$
Riemann–Christoffel curvature tensor	$R^\tau{}_{\mu\nu\sigma}$
Ricci tensor	$R^{\mu\nu}$
Ricci scalar (scalar curvature)	R
Einstein tensor	$G^{\mu\nu}$
Observer's proper time	τ
Four-velocity	u^μ
Line element	$(\mathrm{d}s)^2$
Metric functions of a spherically symmetric star	Φ, Λ
Metric functions of a rotationally deformed star	ν, λ, μ, ψ

symbols, of Einstein's theory of general relativity. The standard sign convention for the metric tensor in *flat* spacetime $(+,-,-,-)$ adopted in the relativistic field-theoretical part of this book (chapters 4 through 12) is *not* the same as in the rest of this book (chapters 13 through 19), which deals with the properties of compact stellar models in *curved* spacetime. In the latter case we adopt the sign convention $(-,+,+,+)$, since this is the standard notation widely used in the literature on rotating general relativistic stars. While this may be considered as a nuisance by some readers, it will certainly be appreciated by those who want to familiarize themselves with the (original) literature that exists on compact rotating stars.

Moreover, to comply with the conventional textbook notation, it appears unavoidable to use several symbols, such as, for instance, τ, ω or Δ, for more than one quantity. However, this unfortunate circumstance is unlikely to confuse the reader, since all those symbols which have twofold meanings refer to entirely different physical or mathematical quantities, that is, they are either of astrophysical origin or enter in relativistic field theory or particle physics. All those quantities not listed in the tables here have their standard textbook meaning. In order to keep the notation to a minimum, Dirac spin and isospin matrix indices as well as the superscripts and subscripts carried by the two-point baryon Green function, $\boldsymbol{g}_1^{BB'}$, will be temporarily suppressed. This is well justified since it is generally a straightforward task to restore them again later, when calculating the final physical expression. The ground-state expectation value of an operator $\mathcal{O}$ is frequently abbreviated to $\langle \mathcal{O} \rangle \equiv \langle \boldsymbol{\Phi}_0 | \mathcal{O} | \boldsymbol{\Phi}_0 \rangle$. If not stated otherwise, the dot product signifies that all spacetime indices are contracted to create Lorentz scalars, and following Einstein's convention, repeated indices are generally summed or integrated over. Moreover, we shall employ quite often the usual subscripted notation to indicate derivatives, as for instance $\partial \Phi / \partial x \equiv \Phi_{,x}$, or, more generally, $\partial \Phi / \partial x^\mu \equiv \Phi_{,\mu}$. Occasionally, the subscripted notation is also used to abbreviate a quantity of the form $X/10^m$ to X_m, where X stands for pressure P, mass density ϵ, or temperature T, for instance. A temperature of say $T_9 = 1$ quoted in the text thus means 10^9 K. Finally, to save space, the step function $\Theta(k_{F_B} - |\boldsymbol{k}|)$ is abbreviated to $\Theta^B(\boldsymbol{k})$.

Chapter 2

Overview of relativistic stars

2.1 History and idea of neutron stars

In July or August of 1054, Chinese astronomers saw and recorded the spectacular explosion of a supernova, the violent death of a star that may have been as much as ten times more massive than our Sun. The star's flare up, taking place in the sky above the southern horn of the constellation Taurus (around 7000 light years away from Earth), the Chinese described as six times brighter than Venus and about as brilliant as the full Moon. This guest star, as the Chinese called it, was so bright that people saw it in the sky during the day for almost a month. During that time, the star was blazing with the light of about 400 million suns. The star remained visible in the evening sky for more than a year.

In the nine centuries since, astronomers have witnessed only two more comparable cataclysms in our Galaxy: the supernova explosions of 1572 (Tycho) and 1604 (Kepler). The supernova was forgotten for more than 600 years until the invention of telescopes, which revealed fainter celestial details than the human eye can detect. In 1731, the English physicist and amateur astronomer John Bevis observed the strings of gas and dust that form the nebula, which today has extended to a size of 10 by 15 light years. While hunting for comets in 1758, Charles Messier spotted the nebula, noting that it had no apparent motion. The nebula, christened as M 1, became the first in his famous *Catalogue of Nebulae and Star Clusters*, first published in 1774. Lord Rosse named the nebula the Crab in 1844 because its tentacle-like structure resembled the legs of the crustacean.

In the decades following Lord Rosse's work, astronomers continued to study the Crab because of their fascination for the strange object. In 1939, astronomer John Duncan concluded that the nebula was expanding and probably originated from a point source about 766 years earlier. Astronomer Walter Baade probed deeper into the nebula, observing in 1942 that a prominent star near the nebula's center might be related to its origin. Six years later, scientists discovered that the Crab was emitting among the strongest radio waves of any celestial object. Baade noticed in 1954 that the Crab possesses powerful magnetic fields, and in 1963, a high-altitude rocket detected X-ray energy from the nebula. But what was the radiation's origin? This was clarified in 1968 when an object in the nebula's center – Baade's prominent star – was discovered that emitted bursts of radio waves 33 times per second. Called the Crab pulsar (PSR 0531+21), it was among the first pulsars discovered. About one year later, scientists concluded that the pulsar was a highly magnetized neutron star rotating 33 times per second about its axis. Some historical events along the route that have led to the conception of neutron stars are summarized in the following:

- 1932: Discovery of neutron by Chadwick.
- 1934: Baade and Zwicky proposed the idea of neutron stars.
- 1939: Tolman and Oppenheimer and Volkoff performed first calculations of neutron star models.
- 1962: Giacconi launched a rocket carrying three Geiger counters which leads to the discovery that our Galaxy contains discrete X-ray sources (e.g. Sco X–1, Cyg X–1). By 1965 the number of known sources of celestial X-rays totaled 10. At this writing the number exceeds 10^5.
- 1964: Crab nebula identified as X-ray source.

This epoch was followed by events that ultimately confirmed that neutron stars really exist in nature:

- 1967: Discovery of first pulsar (Cambridge pulsar CP 1919) by Hewish *et al* [126]. At about the same time Pacini discussed the energy emission from a neutron star in connection with supernova remnants such as Crab [127, 128].
- 1968: Gold proposed that pulsars are rotating, highly magnetized neutron stars [129]. Strong additional support of this interpretation came soon afterwards from Gunn [130] and Ostriker and Gunn [131, 132].
- 1968: Existence of a pulsar in the Crab nebula.
- 1970: First astronomy satellite, Uhuru, launched by NASA. Before its demise in March of 1973, it had identified over 300 discrete X-ray sources (compact stellar objects accreting gas from a companion star).
- 1970s: Ten additional satellites were launched.
- 1975: Discovery of first binary radio pulsar, PSR 1913+16, by Hulse and Taylor [133].
- 1982: Discovery of first millisecond pulsar, PSR 1937+215, by Backer, Kulkarni, Heiles, Davis, and Goss [48].
- 1987: First 'nearby' supernova explosion in about 400 years, SN 1987A. The star that exploded was Sanduleak 202–69, an $18\,M_\odot$ blue supergiant, cataloged by Nick Sanduleak in 1969. Many of the

expectations of supernova explosions of large stars were fulfilled [6, 8], but up to the present day astronomers have waited in vain to observe the birth of a neutron star. One possible reason for that may be that it went into a black hole [134, 135].

- 1990s: Launch of a new generation of satellites like COMPTON,[1] ROSAT, ASCA, and the Hubble Space Telescope (HST), which have led to a flood of new observed pulsar data.

Some neutron stars, such as the Crab, emit radiation that appears to pulse on and off like a lighthouse beacon. Called pulsars [10, 11, 12, 13, 14, 136], they do not really turn radiation on and off – it just appears that way to the observers on Earth because the star is spinning. Astronomers pick up the radiation only when the pulsar's beam [137] sweeps across the Earth, which probably makes most pulsars invisible from Earth. The pulsation gives radio astronomers an immediate handle on the determination of the rotational frequency of a rotating neutron star. To date about 800 pulsars are known. The two fastest of them have rotational frequencies of 620 Hz, which makes them spin about 20 times faster than the neutron star in Crab, whose frequency is 33 Hz. Neutron stars, about 10 kilometers wide but with a mass greater than the mass of our Sun and thus a density of about a billion tons per teaspoonful, are the only stars that can rotate that rapidly without breaking apart. Acting as a celestial power station, the rotating neutron star in Crab generates enough energy to keep the entire nebula radiating over almost the whole electromagnetic spectrum and make it expand.

The distance to the Crab pulsar is known, since it is young enough (943 years) to expand perceptibly over a number of years, and the Doppler shift of radiating filaments moving toward and away from us can be measured. Thus these two pieces of information immediately give the distance to the nebula M 1, within some small uncertainty over its exact shape, which comes out to be about 2000 pc [138]. (The units are explained in table 1.3.) Other pulsars have distances hundreds to thousands of light years away from the Earth, and the inverse-distance-squared decline in apparent brightness favors observation of those at hundreds of light years. For pulsars that far from Earth to be readily detectable by existing radio telescopes they must have radio luminosities of about 10^{21} W (our Sun puts out 4×10^{26} W in sunlight, not radio) at a distance of 100 pc. For such a tiny object as a neutron star to be so luminous at radio frequencies indicates that it must have an important coupling to the electromagnetic field, and an intrinsic dipole magnetic field seems the most plausible assumption [137]. If

[1] COMPTON carries four different experimental setups. These are BATSE (Burst and Transient Source Experiment), OSSE (Oriented Scintillation Spectrometer Experiment), COMPTEL (Imaging Compton Telescope), and EGRET (Energetic Gamma Ray Experiment Telescope).

so, the field strength at the surface must be around 10^{12} G if the observed slowing down rates are to be matched (cf. section 3.5).

2.2 From neutron stars to strange stars to white dwarfs

Physicists know of three stable configurations for objects of stellar mass. These are normal stars, balanced in hydrostatic equilibrium between gravity and thermal pressure; white dwarfs, in which gravity is balanced by electron degeneracy pressure; and neutron stars, held up by neutron degeneracy pressure. According to model calculations, neutron stars are far from being composed of only neutrons but may contain a large fraction of strangeness-carrying hyperons which are possibly in phase equilibrium with u, d, s-quark matter. The very name 'neutron star' is therefore most likely to be a misnomer. Instead they should be termed *nucleon stars* because at densities above the critical density for kaon condensation the matter would be composed of nearly equal numbers of neutrons, protons, and K^- mesons [139], *hyperon stars* if hyperon states become populated in addition to the nucleons [140], or *hybrid stars* if the highly compressed matter in the cores of neutron stars transforms into its deconfined phase, quark matter [86]. Finally, there is the possibility that for a collection of more than a few hundred baryons, the ground state at zero pressure may be strange quark matter, made up of u, d, s quarks, instead of iron! If this is true, and if it is possible for neutron matter to tunnel to quark matter in at least some neutron stars, then it appears likely that all neutron stars would have to be *strange quark matter stars* [141, 142, 143, 144, 145]. In any event, hyperon stars as well as hybrid stars would be only metastable with respect to stars made up of *absolutely* stable (i.e. lower in energy than nuclear matter) 3-flavor strange quark matter [27, 28, 29, 146, 147, 148]. The structure of these classes of stars is schematically illustrated in figure 2.1. A number of aspects of absolutely stable strange quark matter for laboratory physics as well as astrophysics will be discussed immediately below in section 2.3. In contrast to the hybrid stars, which are made up of neutrons, protons, hyperons, and quarks in chemical equilibrium with leptons, strange quark matter stars consist of pure 3-flavor strange quark matter [20, 26, 149], eventually enveloped in a thin crust of low-density nuclear matter [148]. The properties of these three families of stars will be investigated in great detail in chapters 14 through 19.

Figure 2.2 shows the mass–radius relationship of the only two types of stable compact stars, neutron stars and white dwarfs, if strange quark matter is not absolutely stable with respect to atomic nuclei. One clearly notices the huge area in this plane that is void of compact stars. As we shall see in chapter 18, this would be dramatically different if strange quark matter were more stable than atomic nuclei.

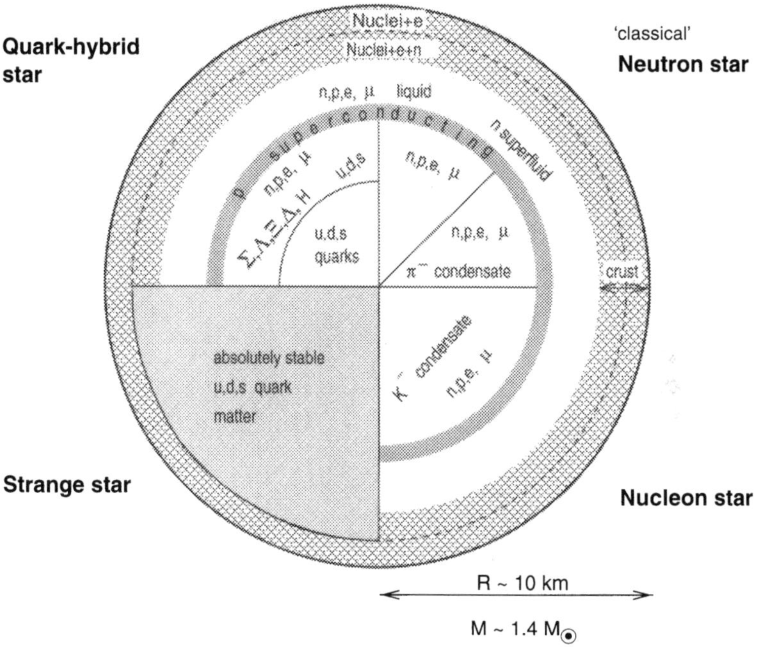

Figure 2.1. Cross section of a neutron star. Modern theories of superdense matter predict stellar compositions quite different from the classical conception of a neutron star's interior. These stretch from the existence of baryons with masses as high as the Δ resonance, a deconfinement transition to u, d, s-quark matter, and boson condensates, to the existence of absolutely stable strange quark matter.

2.3 Stars made up of absolutely stable strange quark matter

2.3.1 The strange quark matter hypothesis

The theoretical possibility that strange quark matter may constitute the *true* ground state of the strong interaction rather than ^{56}Fe was proposed by Bodmer [150], Witten [27], and Terazawa [151]. A schematic illustration of this so-called strange matter hypothesis is given in figure 2.3, which compares the energy per baryon of ^{56}Fe (that is, nuclear matter) with the energy per baryon of 2-flavor quark matter and 3-flavor strange quark matter. The latter two constitute color-singlet multi-quark matter made up of either u, d quarks or u, d, s quarks, respectively. 3-flavor strange quark matter, also referred to as strange matter, is always lower in energy than 2-flavor quark matter due to the extra Fermi well accessible to strange quarks. The vertical arrow roughly comprises the possible lower bound on

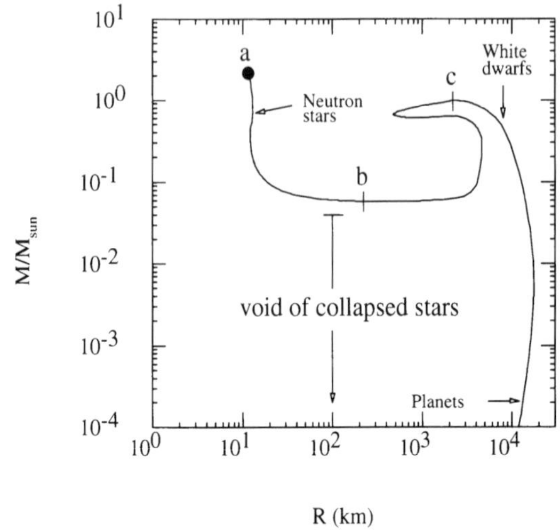

Figure 2.2. Mass versus radius of neutron stars, white dwarfs, and planets. Stars located between the lightest neutron star and the heaviest white dwarf, marked with vertical bars, are unstable against radial oscillations and thus cannot exist in nature. If strange matter is more stable than nuclear matter, the huge region void of collapsed stars should be populated with different categories of strange matter stars (cf. figure 18.32).

the energy per baryon of strange matter predicted by theory. Theoretically (see below), it may be even smaller than the energy per baryon of ^{56}Fe, in which case strange matter would be more stable than nuclear matter. This peculiar feature makes the strange matter hypothesis one of the most startling speculations of modern physics, which, if correct, would have implications of fundamental importance for our understanding of the early universe, its evolution in time to the present day, and astrophysical compact objects, as well as laboratory physics [28, 142, 145, 152] (cf. section 2.3.2). Unfortunately it seems unlikely that quantum chromodynamic (QCD) lattice calculations will be accurate enough in the foreseeable future to give a definitive prediction on the absolute stability of strange matter, and one is thus left with experiment [153, 154, 155, 156, 157, 158, 159] and astrophysical tests [144, 148, 160] to either confirm or reject the hypothesis.

In the following we present simple arguments that serve to formulate the strange matter hypothesis in more quantitative terms [142]. Being based on a perturbative treatment of QCD, final conclusions about the possible (meta) stability of strange matter can not be drawn from this analysis, of course. Nevertheless, one may argue that some of the qualitative features established from a perturbative analysis will have some correspondence in

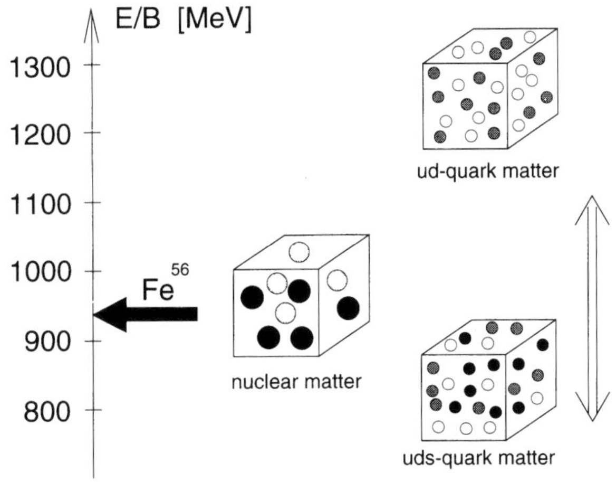

Figure 2.3. Comparison of the energy per baryon of ^{56}Fe (that is, nuclear matter) with that of 2-flavor (u, d quarks) and 3-flavor (u, d, s quarks) strange quark matter. Theoretically, the energy per baryon of strange quark matter may be below 930 MeV, rendering such matter more stable than nuclear matter.

the full non-perturbative regime [161].

Let us consider massless quarks inside a confining bag at zero temperature. For a massless quark flavor,[2] f, the Fermi momentum, p_{F_f}, equals the chemical potential, μ^f. The number densities are therefore given by $\rho^f = (\mu^f)^3/\pi^2$, the energy densities by $\epsilon^f = 3(\mu^f)^4/4\pi^2$, and the corresponding pressures $P^f = (\mu^f)^4/4\pi^2 = \epsilon^f/3$. At zero external bag pressure, the sum of the quark pressures is balanced by the confining bag pressure, B, i.e. $\sum_f P^f = B$. This gives for the total energy density inside the bag $\epsilon = \sum_f \epsilon^f + B$. The baryon number density is $\rho^f = \sum_f \rho^f/3$.

For a gas of massless u and d quarks electric charge neutrality (cf. section 4.2), $\frac{2}{3}\rho^u - \frac{1}{3}\rho^d = 0$, requires that $\rho^d = 2\rho^u$, or $\mu_2 \equiv \mu^u = \mu^d/2^{1/3}$. The corresponding 2-flavor quark pressure then follows as $P_2 \equiv P^u + P^d = (1 + 2^{4/3})\mu_2^4/4\pi^2 = B$. From the total energy density, $\epsilon_2 = 3P_2 + B = 4B$, and the baryon number density, $\rho_2 = (\rho^u + \rho^d)/3 = \mu_2^3/\pi^2$, one obtains for the energy per baryon of 2-flavor quark matter,

$$\left.\frac{E}{A}\right|_2 \equiv \frac{\epsilon_2}{\rho_2} = \frac{4B}{\rho_2} = 934\ \mathrm{MeV} \times B_{145}^{1/4}, \tag{2.1}$$

where $B_{145}^{1/4} \equiv B^{1/4}/145$ MeV. A value of 145 MeV is the lowest possible choice for $B^{1/4}$ for reasons discussed below (see also chapter 18).

[2] The more general case of finite quark masses as well as finite temperatures will be studied in chapter 18.

A 3-flavor quark gas is electrically neutral for $\rho^u = \rho^d = \rho^s$, i.e. $\mu_3 \equiv \mu^u = \mu^d = \mu^s$. For fixed bag constant, the 3-flavor quark gas should exert the same pressure as the 2-flavor gas, i.e., $P_3 = P_2$. This implies for the chemical potentials $\mu_3 = [(1 + 2^{4/3})/3]^{1/4} \mu_2$. Hence, the total baryon number in this case can be written as $\rho_3 = \mu_3^3/\pi^2 = [(1 + 2^{4/3})/3]^{3/4} \rho_2$. The energy per baryon is then given by

$$\left.\frac{E}{A}\right|_3 \equiv \frac{\epsilon_3}{\rho_3} = \frac{4B}{1.127\,\rho_2} = 829 \text{ MeV} \times B_{145}^{1/4}\,, \qquad (2.2)$$

since $\epsilon_3 = 3P_3 + B = 4B = \epsilon_2$. We thus see that the energy per baryon of a massless non-interacting 3-flavor quark gas is of order 100 MeV per baryon *lower* than in 2-flavor quark matter. The difference arises from $\rho_2/\rho_3 = [3/(1+2^{4/3})]^{3/4} \simeq 0.89$ which reflects that the baryon number can be packed more densely into 3-flavor quark matter (i.e. $\rho_3 > \rho_2$) due to the extra Fermi well accessible to the strange quarks.

The energy per baryon in a free gas of neutrons is the neutron mass, $E/A = m_n = 939.6$ MeV (cf. table 5.1). For an ^{56}Fe nucleus one has $E/A = (56 m_N - 56.6 \text{ MeV})/56 \simeq 930$ MeV, where m_N denotes the nucleon mass. Stability of u, d-quark matter relative to neutrons thus corresponds to $(E/A)_2 < m_n$, or $B^{1/4} < 145.9$ MeV ($B^{1/4} < 144.4$ MeV for stability relative to ^{56}Fe). The argument is often turned around because one observes neutrons and ^{56}Fe in nature, rather than u, d-quark matter. Hence $B^{1/4}$ must be *larger* than about 145 MeV.

Bulk u, d, s-quark matter is *absolutely* stable relative to ^{56}Fe for $B^{1/4} <$ 162.8 MeV, *metastable* relative to a neutron gas for $B^{1/4} < 164.4$ MeV, and relative to a gas of Λ particles for $B^{1/4} < 195.2$ MeV. These numbers are upper limits. A finite s-quark mass as well as a non-zero strong coupling constant decrease the limits on $B^{1/4}$ [142].

Finally we mention that the presence of ordinary nuclei in nature is not in contradiction to the possible absolute stability of strange matter, the reason being that conversion of an atomic nucleus of baryon number A into a lump of strange quark matter requires the *simultaneous* transformation of roughly A up and down quarks into strange quarks. The probability for this to happen involves a weak interaction coupling G_F to the power A (i.e. $\propto G_F^{2A}$), which makes nuclei with $A_{\rm crit} \gtrsim 6$, as required by the strange matter mass formula (2.3), stable for more than 10^{60} years! Note that sufficiently large baryon numbers, i.e. $A \gtrsim 6$ as in the above quoted example, are mandatory. Otherwise finite-size and shell effects will dominate over the volume term of strange matter according to [142, 162]

$$\frac{E}{A} \simeq \left(829 \text{ MeV} + 351 \text{ MeV } A^{-2/3}\right) B_{145}^{1/4}\,, \qquad (2.3)$$

rendering strange matter configurations with $A \lesssim 6$ metastable or completely unstable. An example of the latter is the $\Lambda = (u, d, s)$ particle. Possessing a baryon number of $A = 1$, its energy per baryon (i.e. its rest mass m_Λ) lies way above that of ^{56}Fe. Feeding the Λ with roughly equal numbers of u, d, s quarks reduces the energy per baryon. The actual A value for which the strange nugget may become absolutely stable depends on the poorly constrained surface energy of strange matter [163]. In the case of (2.3), strange matter becomes absolutely stable for $A > A_{\rm crit} = 6$.

It is understood that even if strange matter is the ground state, little or no such matter would have survived the high-temperature era of the very early universe because of evaporation to hadrons and the universe would have evolved essentially independent of which is the true ground state [152, 164, 165, 166, 167]. If the strange matter hypothesis is true, then a separate class of compact stars should exist, which are called strange (quark matter) stars. They form a distinct and disconnected branch of compact stars and are not a part of the continuum of equilibrium configurations that include white dwarfs and neutron stars. In principle both strange and neutron stars could exist. However if strange stars exist, the galaxy is likely to be contaminated by strange quark nuggets which would convert all neutron stars that they come into contact with to strange stars [20, 144, 145, 167].

At the present time there appears to be only one crucial astrophysical test of the strange quark matter hypothesis, and that is whether strange quark stars can give rise to the observed phenomena of pulsar glitches, schematically illustrated in figure 1.2. Glitches are sudden relatively small changes in the rotational frequency of a pulsar (cf. figure 2.4) which otherwise decreases very slowly with time due to the loss of rotational energy through the emission of electromagnetic dipole radiation and an electron–positron wind. They occur in various pulsars at intervals of days

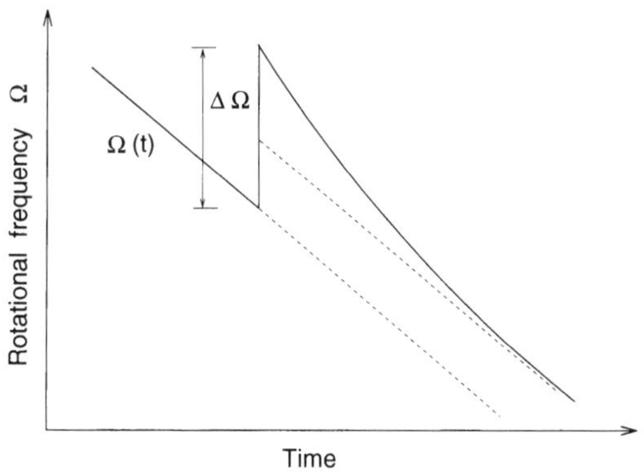

Figure 2.4. Schematic illustration of a glitch (discontinuous speedup) in a pulsar's rotational frequency followed by a subsequent relaxation epoch.

to months or years, and in some pulsars are small (Crab), and in others large (Vela) and infrequent ($\Delta\Omega/\Omega \sim 10^{-8} - 10^{-6}$, respectively). Glitches have been attributed variously to several causes related to the assumed structure of neutron stars. One such is the crust quake in which an oblate solid crust in its present shape slowly comes out of equilibrium with the forces acting on it as the period of rotation changes, and fractures when the built up stress exceeds the sheer strength of the crust material [168, 169]. The frequency and rate of change of frequency, Ω and $\dot{\Omega}$ respectively, slowly return to the trend preceding the glitch as the coupling between crust and core reestablishes their corotation. The existence of glitches may have a decisive impact on the question of whether or not strange matter is the ground state of strongly interacting matter.

The reason is that, provided the cores of the most massive of neutron stars have densities exceeding that required to convert hadrons to quark matter, of which strange matter is the lower energy state, and that such stars of sufficient mass can be made in natural processes, then almost certainly all other neutron stars have also been converted to strange stars [20, 144, 167]. This in turn would mean that the objects known to astronomers as pulsars would be rotating strange matter stars, not rotating neutron stars! The conversion would take place either as a result of neutron stars being bathed themselves by a flux of strange nuggets estimated to be as high as $0.1\,\mathrm{cm}^{-2}\,\mathrm{s}^{-1}$ produced in strange star collisions in the galaxy [20], or because of neutron star formation in a supernova from a progenitor star that already contains strange seeds, either acquired after formation of the progenitor star or as a result of the gaseous material out of which the progenitor star was formed already containing seeds of strange matter produced in the same way either within our galaxy or in the neighboring cluster of galaxies [144]. In any case the nugget will grow to consume the entire neutron star [170, 171, 172]. Therefore, since under the above proviso neutron stars have already been converted, strange stars must be capable of producing pulsar glitches else the hypothesis of strange matter fails. Indeed the claim has been made that it does fail on this account [173], but for many researchers the case is unconvincing since the properties of quark matter are not sufficiently well known [146, 152]. We shall turn back to this issue in chapter 18, where sequences of compact strange stars and their continuous connection to a previously unexplored section of the sequences of collapsed stars – the strange white dwarfs – will be explored in great detail.

The strange white dwarfs constitute the strange counterparts of ordinary white dwarfs. They consist of a strange quark matter core in the star's center which is enveloped by a nuclear crust made up of ordinary atomic matter [28]. The crust is suspended out of contact with the quark core due to the existence of an electric dipole layer on the core's surface [28, 30], which insulates the crust from conversion to quark matter. Even so, the maximum density of the crust is strictly limited by the neutron drip density ($\epsilon_{\rm drip} = 4 \times 10^{11}$ g cm^{-3}), at which neutrons begin to drip out of the nuclei and would gravitate into the core where they would be dissolved into strange matter.

Strange white dwarfs comprise a hitherto unexplored consequence of the strange matter hypothesis. Depending on the amount of crust mass, their properties may differ considerably from those of ordinary white dwarfs. For instance, it is well known that the maximum density attained in the limiting mass white dwarf is about $\epsilon_{\rm wd} = 10^9$ g cm^{-3} [69, 70]. Above this density, the electron pressure is insufficient to support the star, and there are no stable equilibrium configurations until densities of the order of nuclear density ($\epsilon \gtrsim 10^{14}$ g cm^{-3}) are reached, which are neutron or strange stars. One class of strange dwarfs can be envisioned as consisting of a core of strange matter enveloped within what would otherwise be an ordinary white dwarf. They would be practically indistinguishable from ordinary white dwarfs. Of greater interest is the possible existence of a class of white dwarfs that contain nuclear material *up to the neutron drip density*, which would not exist without the stabilizing influence of the strange quark core [174]. These strange dwarfs could have densities of the nuclear envelope or crust, at its inner edge, that fall in the range of $\epsilon_{\rm wd} < \epsilon_{\rm crust} < \epsilon_{\rm drip}$. The maximum inner crust density therefore can be 400 times the central density of the limiting-mass white dwarf and 4×10^4 times that of the typical $0.6\, M_\odot$ white dwarf. An investigation of the stability of such very dense dwarf configurations to acoustical vibrations (radial oscillations), which will be performed in chapter 18, reveals stability over a large mass range from $\sim 10^{-3}\, M_\odot$ to slightly more than $1\, M_\odot$. This is the same range as ordinary white dwarfs except that the lower mass limit is smaller by a factor of $\sim 10^{-2}$. This is because of the influence of the strange core, to which the entire class owes its stability.

In summing up we note that beyond the specific properties of strange stars and strange dwarfs, several of which are exotic, their existence or nonexistence carries information of a fundamental nature as to the true ground state of the strong interaction that no other experiment or fact has yet revealed [20, 27, 142, 146, 152]. So such stars are extremely interesting objects to study in their many facets.

2.3.2 Peculiarities of strange quark matter

Of course, there is more to the absolute stability of strange quark matter than outlined in the previous section [142, 145, 146, 152]. An overview of numerous additional such aspects is given in table 2.1. Search experiments for strange quark matter are summarized in table 2.2. One class of such experiments searches for nuggets of strange matter which got stuck in terrestrial matter. Such nuggets will be less massive than $\sim 0.3 \times 10^{-9}$ g (carrying a baryon number of $A \sim 2 \times 10^{14}$, radius $R \sim 10^{-8}$ cm), otherwise

Table 2.1. Strange matter phenomenology (not complete).

Phenomenon	References
Centauro cosmic ray events	[27, 177, 178, 179]
High-energy gamma ray sources:	
Cyg X–3 and Her X–3	[180, 181]
Strange matter hitting the Earth:	
Strange meteors	[175]
Nuclearite-induced earthquakes	[175]
Strange nuggets in cosmic rays	[182, 183]
Strange matter in supernovae	[184, 185, 186]
Strange star (pulsar) phenomenology	[20, 28, 29, 30, 62, 144, 146, 148]
	[187, 188, 189, 190, 191, 192]
Strange dwarfs:	
Static properties and stability	[188, 193, 194, 195, 196]
Thermal evolution	[197]
Strange planets	[174, 188, 193, 195]
Strange MACHOS	[198]
Strangeness production in dense stars	[199]
Burning of neutron stars to strange stars	[170, 171, 172]
Gamma-ray bursts	[28, 200]
Cosmological aspects of strange matter	[27, 141, 146, 166, 167, 201]
Strange matter as compact energy source	[156]
Strangelets in nuclear collisions	[153, 154, 155]

they will not be slowed down and stopped in the crust of the Earth. Such encounters could take the form of unusual meteorite events,[3] earthquakes, and peculiar particle tracks in ancient mica, in meteorites and in cosmic-ray detectors, as argued by De Rújula and Glashow [175]. In particular, they were interested in the possibility of detecting a Galactic dark matter halo of strange nuggets, where typical velocities are a few hundred kilometers per second, given by the depth of the gravitational potential. For a dark matter halo of nuggets the expected flux at the Earth is on the order of $10^6 A^{-1} v_{250}\,\mathrm{cm}^{-2}\,\mathrm{s}^{-1}\,\mathrm{sr}^{-1}$, where v_{250} is the speed in units of $250\,\mathrm{km\,s}^{-1}$ [142]. Experiments sensitive at this flux level or better have been able to rule out quark nuggets as being the dark matter for baryon numbers $10^8 < A < 10^{25}$ [152, 176]. This however does not rule out a low flux level either left over from the Big Bang or arising from collisions of strange stars. If the strange matter hypothesis is valid, one should indeed expect a

[3] The apparent magnitude of a 20 g nugget at a distance of 10 km is -1.4, equal to that of the brightest star, Sirius. Another distinguishing feature would be the meteorite's velocity which is smaller than about $70\,\mathrm{km\,s}^{-1}$ for an ordinary meteorite bound to the solar system, but as high as about $250\,\mathrm{km\,s}^{-1}$ for a strange meteorite of galactic origin.

Table 2.2. Overview of search experiments for strange matter.

Experiment	References
Cosmic ray searches for strange nuggets:	
Balloon-borne experiment CRASH †	[205, 206, 207]
HADRON	[208]
MACRO	[209]
IMB	[210]
Tracks in ancient mica	[175, 211]
Rutherford backscattering of ^{238}U and ^{208}Pb	[212]
Heavy-ion experiments at BNL:	
Strangelet searches E858, E864, E878, E882-B, E896-A, E886	[213, 214]
H-dibaryon search E888	[215, 216]
Heavy-ion experiments at CERN: NA52	[213, 217, 218, 219]

† CRASH stands for Cosmic Ray And Strange Hadronic matter.

significant background flux from stellar collisions, since it is well known that a fraction of the observed pulsars are members of binary systems, where the two components are ultimately going to collide. If such stellar collisions spread as little as $0.1\,M_\odot$ of strangelets with baryon number A, a single collision will lead to a flux of $10^{-6}A^{-1}v_{250}\,\mathrm{cm}^{-2}\,\mathrm{s}^{-1}\,\mathrm{sr}^{-1}$ [142], assuming strangelets to be spread homogeneously in a halo of radius 10 kpc.

Glendenning estimated that nuggets with $A \sim 10^3$ emitted into space from strange star–strange star collisions taking place in our Galaxy would lead to a concentration of nuggets in our Galaxy of less than 10^{-8} nuggets cm^3, leading to a nugget concentration in terrestrial crust matter of at most 10^9 cm^{-3}, which corresponds to a concentration of nuggets per nucleon $\ll 10^{-14}$. The upper limit on the concentration of strange nuggets per nucleon in terrestrial matter established by experiment is 10^{-14}, which falls short of the conservative upper limit estimated by Glendenning. So we conclude that the concentration that may have developed over time in the Earth's crust is unlikely to be high enough to be detectable.

A limit on the total amount of strange matter in the Universe follows from the observed abundance of light isotopes. This is so because the strange nuggets formed in the Big Bang would have absorbed free neutrons which reduces the neutron to proton ratio, N_n/N_p. This effect in turn would lower the rate of production of the isotope ^{4}He, whose abundance is well known from observation. For a given mass of the strange nuggets, this constrains their total surface area. To be consistent with the missing dark matter, assumed to be strange quark matter, and the observed abundance of light isotopes, the primordial quark nuggets had to be made of more than $\sim 10^{23}$ quarks. According to what has been said above, quark nuggets that massive are not stopped by the Earth.

The masses of nuggets accumulated in terrestrial matter can range from those of atomic nuclei to the upper limit of 0.3×10^{-9} g. The passage of nuggets through detectors below the Earth's surface, like the Irvine–Michigan–Brookhaven (IMB) proton-decay detector, or the Gran Sasso detector designed to search for cosmic magnetic monopoles, would lead to observable signals for which these detectors have been designed to search. Nuggets of more than 10^{23} quarks would have too much momentum to be stopped by the encounter and thus would pass through the Earth.

The detection of strange drops in relativistic heavy-ion collisions is conceptually rather simple, because of the drop's low Z/A ratio. Technically, however, such an attempt is very difficult because of several reasons. First of all one has to defeat the finite-number destabilizing effect of strange nuggets, that is, the number of free quarks produced in a nuclear collision must be sufficiently large enough such that surface and shell effects do not dominate over the volume energy of strange matter, destroying its possible absolute stability. Moreover, since the dense and hot matter is produced for $\sim 10^{-22}$ s, there is no time to develop a net strangeness. However, it is believed that strange nuggets will result from two types of simultaneous fluctuations that separate strange and antistrange quarks, and that also cool the nuggets so that they do not evaporate. Finally, we mention the huge multiplicity of particles produced in such collisions, which makes the particle identification rather cumbersome. (For an overview, see [202].)

As shown in table 2.2, during the past few years several experiments have been using high-energy heavy-ion collisions to attempt to create cold strange quark matter in the laboratory. Thus far the only candidates that have been claimed are for the lightest strangelets, the H-dibaryon. High-sensitivity experiments have come on-line and should be able to provide stringent tests of plasma fragmentation, and coalescence calculations for the production of multiply-strange composite particles. The NA52 (Newmass) experiment is searching for long-lived massive strange quark matter particles as well as for antinuclei in Pb–Pb collisions at CERN. No evidence for the production of charged strange particles with a mass to charge ratio, m/Z, between 5 and 120 GeV/c^2 and lifetimes greater than 1.2 μs was found in the 1994 data sample of this experiment. It is planned for the future to increase the sensitivity by at least a factor of ten. During the 1995 data taken with lead, the accumulated statistics was already increased by about a factor of 20. A similar sensitivity is planned to be reached for positively charged strange particles in a run which took place in the fall of 1996. According to the present data analysis [203], one very serious candidate for a strange particle of mass $m = (7.4 \pm 0.3)$ GeV, electric charge $Z = -1$, and laboratory lifetime $\tau > 0.85 \times 10^{-6}$ s was found. After that time the candidate, which could be made up of $6u + 6d + 9s$ quarks (carrying a baryon number of $A = 7$) or $7u + 7d + 10s$ quarks ($A = 8$), either decayed or underwent an interaction. The latest Newmass strangelet experiment performed in the fall of 1988, which has improved statistics by about a factor of 20 in this mass range, is expected to shed light on this most unusual event [203].

Anomalous massive particles, which can be interpreted as strangelets, have been apparently observed in a number of independent cosmic ray experiments [204]. An overview of such experiments is given in table 2.2. Two of these anomalous events, which are consistent with electric charge values $Z \simeq 14$ and baryon numbers $A \simeq 350$, have been observed by a balloon-borne counter experiment devoted to the study of primary cosmic rays by Saito *et al* [205]. It remains to be seen whether or not the new balloon-borne experiment of the Italian/Japanese collaboration [206] (the CRASH collaboration) will confirm this most striking discovery. Evidence for the presence of strangelets in cosmic rays has recently been pointed out too by Shaulov [208]. This experiment, known as HADRON, was carried out at Tien-Shan Station between 1985 and 1991. It is based on a combination of extensive air shower arrays and large emulsion chambers. The strangelet component in this experiment was estimated to be about $1\ \mathrm{m}^{-2}\,\mathrm{yr}^{-1}$. Besides that, the so-called Price event [220] with $Z \simeq 46$ and $A > 1000$, regarded previously as a possible candidate for the magnetic monopole, turned out to be fully consistent with the Z/A ratio for strange quark matter [221]. Finally, an exotic track event with $Z \simeq 20$ and $A \simeq 460$, observed in an emulsion chamber exposed to cosmic rays on a balloon, has been reported by Miyamura [207]. This exotic track event motivated the balloon-borne emulsion chamber JACEE [222] and Concorde aircraft [223] experiments.

As known from the previous section, the possibility that strange matter is absolutely stable raises the possibility that strange stars are made of exclusively strange matter, and that the surface of the star is exposed strange matter. The thickness of the quark surface would be just ~1 fm, the length scale of the strong interaction. Electrons are held to quark matter electrostatically, and the thickness of the electron surface is several hundred fermis. Since neither component, electrons or quark matter, is held in place gravitationally, the Eddington limit[4] to the luminosity that a static surface may emit does not apply, and thus the object may have photon luminosities much greater than 10^{38} erg s^{-1}. Very recently it was shown

[4] Consider gas accreting steadily at a rate $\dot{M}$ onto a neutron star of mass M. The *Eddington limit*, defined as the critical luminosity at which photon radiation pressure from the surface of the star equals gravity, is then given by [224]

$$L_{\mathrm{Edd}} = 1.3 \times 10^{38} \big(M/M_\odot\big) \ \mathrm{erg\ s^{-1}} \,.$$

by Usov that this value may be exceeded by many orders of magnitude by the luminosity of e^+e^- pairs produced by the Coulomb barrier at the surface of a hot strange star [160]. For a surface temperature of $\sim 10^{11}$ K, the luminosity in the outflowing pair plasma was calculated to be as high as $\sim 3 \times 10^{51}$ erg s^{-1}. Such an effect may be a good observational signature of bare strange stars. If the strange star is enveloped by a nuclear crust however, which is gravitationally bound to the strange star, the surface made up of ordinary atomic matter would be subject to the Eddington limit and therefore the photon emissivity would be the same for an ordinary neutron star.

A very high-luminosity γ-ray burst took place in the Large Magellanic Cloud (LMC), some 55 kpc away, on 5 March 1979. The mysterious γ-ray bursts are cosmic events of a few seconds duration, coming from unidentified astrophysical sources which appear to be at extragalactic distances [225, 226, 227, 228]. γ-ray bursts at truly cosmological distances may be due to collisions of two neutron or two strange stars [229], depending on the true ground state of strongly interacting matter, in a binary system, each collision releasing $\sim 10^{52}$ ergs in the form of gamma rays over a time scale of about 0.2 s.

The inferred luminosity of the 5 March 1979 event was $\sim 10^6$ times the Eddington limit for a solar mass object, and the rise time is very much smaller than the time needed to drop $\sim 10^{25}$ g (about $10^{-8}\, M_\odot$) of normal material onto a neutron star. Alcock *et al* [28] suggested a detailed model for this burst which involves the particular properties of strange matter (see also Horvath *et al* [200]). The model assumes that a lump of strange matter of $\sim 10^{-8}\, M_\odot$ fell onto a rotating strange star. Since the lump is entirely made up of self-bound high-density matter, there would be only little tidal distortion of the lump, and so the duration of the impact can be very short, around $\sim 10^{-6}$ s, which would explain the observed rapid onset of the γ-ray flash. This and other observed features concerning energetics and time scales can be nicely explained under the assumption that the burster is located in a supernova remnant in the LMC, as position measurements seem to indicate.

Evidence for the possible existence of strange matter in cosmic rays may come from Centauro cosmic-ray events [27, 177, 178, 179, 230]. Such events are seen in mountain top emulsion chamber experiments. The typical energy of such an event is of order $\sim 10^3$ TeV, and the typical particle multiplicity is 50 to 100 particles. The typical transverse momentum (poorly determined) of about ~ 1 GeV seems to be larger than that for a typical event of the same energy. The striking feature is that there seem to be no photons produced in the primary interaction which makes the Centauro. This is unusual because in high-energy collisions, π^0 mesons are always produced, and they decay into photons. A Centauro event is much like a nuclear fragmentation. If a nucleus were to fragment, then there would be many nucleons, and if the interaction which produced the fragmentation were sufficiently peripheral, there would be few pions. This hypothesis is ruled out because the typical transverse momentum is so large, and more importantly because a nucleus would not survive to such a great depth in the atmosphere. Being much more tightly bound together than an ordinary nucleus, a strangelet with a baryon number around $A \sim 10^3$ explains many of these unusual features. So it is conceivable that strangelets incident upon the top of the atmosphere or produced at the top of the atmosphere could survive to mountain altitude. It may have lost a significant amount of baryon number before getting to this depth however. A peripheral interaction might be sufficient to unbind it, since it certainly will not be so tightly bound with reduced baryon number. The problem with this explanation is that it does not explain the high transverse momenta. At transverse momenta of ~1 GeV one would surely expect final-state interactions to generate some pions, and therefore an electromagnetic component, which, as mentioned above, is not observed [161].

Besides the peculiar Centauro events which possibly act as agents for strange matter, the high-energy γ-ray sources Cygnus X–3 and Hercules X–1 may give additional independent support for the existence of strange matter. The reason is that air shower experiments on Her X–1 and Cyg X–3 indicate that these air showers have a muon content typical of cosmic rays. This muon content is a surprising result. Typical cosmic rays at energies between 10 and 10^5 TeV are protons. To arrive from the direction of Cyg X–1 or Her X–1, the particle must be electrically neutral. To survive the trip, it must be long lived. The only known particle which can quantitatively satisfy this constraint is the photon. Photon air showers however have only a small muon component. Photons shower by producing electron pairs. When only underground data were known, it was proposed that the most likely candidate for the initiating particle was a hadron, and in order for interstellar magnetic fields not to alter its direction relative to that of the source, the hadron – known in the literature as the cygnet – must be neutral. A natural candidate for the cygnet appears to be the so-called H particle, the electrically neutral strangeness-2 di-baryons with the quantum numbers of two lambdas ($Z = 0$, $A = 2$) proposed by Jaffe [180].[5] To make the H sufficiently long lived, it is necessary to make the H have a mass below single weak decay stability. To generate a large enough flux of

[5] In the theory of hadrons composed of colored quarks and antiquarks, combinations other than the usual qqq and $q\bar{q}$ are allowed as long as they are color singlets. Jaffe found that a six-quark $uuddss$ color-singlet state (H) might have strong enough color-magnetic binding to be stable against strong decay. That is, $m_{\rm H}$ could be less than the strong-decay threshold, twice the Λ^0 (uds) mass, $m_{\rm H} < 2m_{\Lambda^0}$. Estimated lifetimes for H range from $\sim 10^{-10}$ s for $m_{\rm H}$ near the $\Lambda^0\Lambda^0$ threshold to $> 10^7$ for light H particles near the nn threshold. The potentially long lifetimes raise the possibility that H particles may be present as components of existing neutral particle beams (e.g. E888 experiment listed in table 2.2).

H particles, the source is assumed to be a strange star. Studies of Her X–1 however seem to rule out this hypothesis, since studies of the correlation in arrival time with the known period of Her X–1 give an upper limit of the particle mass of about 100 MeV. The source of radiation must be either due to anomalous interactions of photons or neutrinos, or from some exotic as yet undiscovered light-mass, almost stable particles. The problem with Cyg X–3 may be that it is accreting mass and thus has a crust, such that there is no exposed strange matter surface where small strangelets could be produced and subsequently accelerated electrodynamically to high energies into the atmosphere of the companion star where H particles were created via spallation reactions.

Finally we mention the possible creation of strange quark matter in the early Universe. If the Universe cooled through the first-order phase transition, then bulk thermodynamic equilibrium of the coexistence phase between the quark phase and the hadron phase (at a temperature of $\sim$100 MeV) required exchange of entropy and baryon number across the phase boundaries of unconfined quark matter and confined hadronic matter. Entropy would have been exchanged by neutrinos and photons, exchange of baryon number could only occur via the association of three quarks into confined hadrons at the phase boundary. The length scale for this process is $\sim$1 fm. If this process of association were inefficient, then the baryon number of the Universe would have been trapped in smaller regions of the quark phase. One possible outcome of this would have been the creation of lumps of strange matter, or quark nuggets. These would form cold dark matter, since their velocities with respect to the mean Hubble expansion would have been non-relativistic. The criticism of the above scenario is that the nuggets may have evaporated as the universe cooled down to a few MeV. Alcock and Farhi [164] argued that all lumps of strange matter with baryon number $A \lesssim 10^{52}$ would have evaporated. (Lumps with $A > 10^{52}$ could not have formed without violating causality at the epoch of formation.) Shortly afterwards, Madsen *et al* demonstrated that the critical baryon number at which evaporation of lumps of strange matter occurs is shifted down to 10^{46} if evaporation of baryons at the surface of lumps is taken into account [166, 167].

Chapter 3

Observed neutron star properties

The detection of thermal photons from a stellar surface serves as a principal window on the properties of a normal star: its radius, surface gravity, chemical composition, magnetic field strength, temperature, and other physical attributes are derivable from the measured flux and spectrum. In the case of neutron stars, X-ray astronomy has provided several possible approaches to measuring fundamental neutron star parameters in addition to the masses derived for those in binary star systems. In the following we give an overview of a number of important neutron star properties, each of which is known to be sensitive to the adopted microscopic model for the nucleon–nucleon interaction and therefore to the model constructed for the nuclear equation of state (EOS) [58, 62].

3.1 Masses

The gravitational mass is of special importance since it can be inferred directly from observations of X-ray binaries [231], binary pulsars, e.g. the Hulse–Taylor radio pulsar PSR 1913+16 [232], and possibly from supernova explosions [233, 234]. The determination of their masses, as well as those of other compact stars in X-ray binaries, is useful for at least four reasons [231]. Firstly, the observed values of the masses of neutron stars provide a unique test of the combined predictions of theories of nuclear matter and of general relativity (or of alternative theories of gravity). As mentioned in chapter 1, many important theoretical investigations have been devoted to sharpening these predictions. Secondly, the observed masses provide information about the final stages of stellar evolution. The determination of the masses of X-ray sources enables one to learn something about the cores of stars that have passed through the final stages of stellar evolution when neutron stars and black holes are formed. Thirdly, the values of the masses are important for answering questions about the nature of the X-ray source itself. Cyg X–1 is a good example of this kind of application

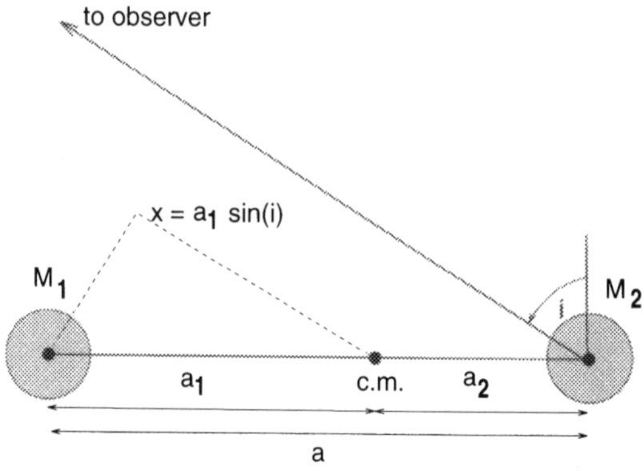

Figure 3.1. Parameters of a binary stellar system as viewed in the orbital plane. The projection of the orbital velocity of M_1 along the line of sight is $v_1 = (2\pi/P_b)x$, where P_b is the orbital period. (After Shapiro and Teukolsky [224].)

since the mass inferred for the X-ray star exceeds the standard theoretical limits for the masses of neutron stars and of white dwarfs (see below). Finally, the numerical values of the masses are required for understanding specific phenomena, such as the spin-up rate of the pulsar period, various other processes that depend on mass transfer, and the applicability of the Eddington limit [231].

At the time of writing this book, the radio pulsar PSR 1913+16 [232] is one out of five double neutron star binaries known. The others are J1518+4904, B1534+12, B2127+11C, and B2303+46. In each case, five Keplerian parameters can be very precisely measured by pulse timing techniques [12]. These are the binary period P_b, the projection of the orbital semimajor axis on the line of sight $x \equiv a_1 \sin i$ (see figure 3.1), the eccentricity e, and the time and longitude of periastron, T_0 and ω_0, respectively. These parameters are related to the pulsar and companion masses, M_1 and M_2, through the mass function f [32, 224, 231],

$$f(M_1, M_2, i) \equiv \frac{(M_2 \sin i)^3}{(M_1 + M_2)^2} = \frac{P_b\, v_1^3}{2\,\pi\, G} = \frac{4\,\pi^2\, x^3}{G\, P_b^2}\,. \tag{3.1}$$

In each case, the relativistic advance of the angle of periastron, $\dot{\omega}$, has also been measured, which yields an estimate of the system's total mass, $M_t = M_1 + M_2$. For three systems (PSRs B1534+12, B1913+16, and B2127+11C) the measurement of the combined effects of the transverse Doppler shift and the gravitational redshift allow the individual determination of the pulsar and companion masses (see figure 3.2).

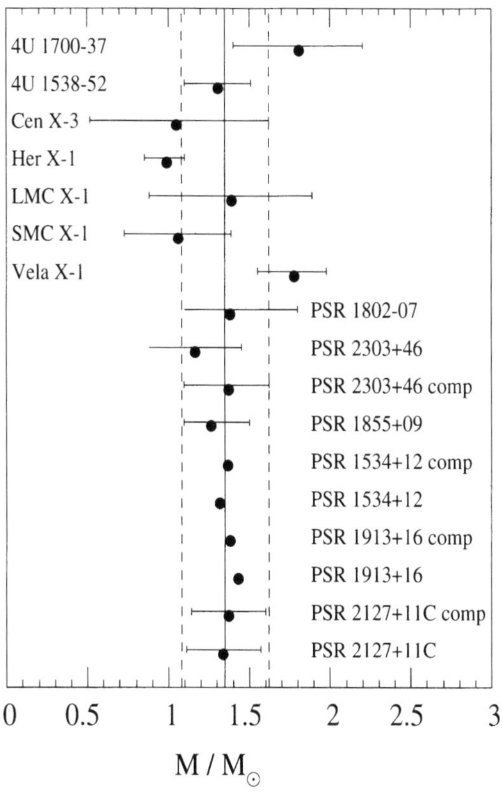

Figure 3.2. Observed masses of neutron stars [31]. Objects in massive X-ray binaries (binary X-ray pulsars) are at the top, radio pulsars and their companions (abbreviated to 'comp') are at the bottom.

Bahcall *et al* [231] and Rappaport and Joss [235, 236] were the first who deduced neutron star masses for several X-ray binaries. A reexamination of these masses became possible owing to the improved determinations of orbital parameters [15, 32]. The improved values are shown in the upper half of figure 3.2. Representatives of non-pulsating X-ray binaries are 3U1700–37 and Cyg X–1 whose masses are $0.6 \lesssim M/M_\odot$ and $9 \lesssim M/M_\odot \lesssim 15$, respectively. It is expected that at least one of these objects, likely Cyg X–1, is a black hole. Very remarkable is the extremely accurately determined mass of the Hulse–Taylor binary pulsar, PSR 1913+16, which is given by $(1.444 \pm 0.003)\, M_\odot$ [232]. Figure 3.2 suggests that the most probable value of neutron star masses, as derived from observations of binary X-ray pulsars, is close to about $1.4\, M_\odot$, and that the masses of individual neutron stars are likely to be in the range of $1.1 \lesssim M/M_\odot \lesssim 1.8$ [15, 32]. Indications for the existence of very heavy neutron stars, possibly as massive as about $2\,M_\odot$, may come from the observation of quasi-periodic oscillations (QPOs) in luminosity in low-mass X-ray binaries [237, 238]. This finding, however, is currently controversial since it is yet to be confirmed that the largest observed frequency is indeed the Keplerian frequency of the marginally stable orbit (see next paragraph). Aside from this current frailty, the QPOs may also yield information on other fundamental neutron star properties such as their rotational periods or radii, and may also serve as unique probes of strong field general relativity.

The quasi-periodic oscillation frequency is interpreted as a measure of the difference between the rotation frequency of the neutron star and the orbital frequencies of the plasma in the inner disk. Theoretically, the phenomenon is described by the magnetospheric beat-frequency model, which was originally developed to interpret the 5–50 Hz QPOs from high-luminosity Z-type low-mass X-ray binaries [239, 240, 241]. This model assumes that clumped plasma is accreting from an accretion disk onto a presumably weakly magnetic neutron star. Such clumping can be caused by magnetic, thermal, or shear instabilities. Once formed, clumps drift radially inward and are stripped of plasma by interaction with the magnetospheric field. Plasma stripped from the clump is quickly brought into corotation with the neutron star and falls to the stellar surface, where it produces X-rays. Because the clumps of plasma and the magnetosphere are rotating at different frequencies, the strength of the magnetic field seen by a given clump will vary at the so-called beat frequency or one of its harmonics, causing the X-ray emission to vary at the same frequency [241]. According to this concept, the beat frequency varies between the Kepler frequency, $\Omega_{\mathrm{K}}^{\mathrm{disk}}$, of the innermost region of the accretion disk (at distance r_{in} from the star's center) at the magnetospheric boundary, and the neutron star's spin frequency, Ω. As r_{in} decreases with increasing accretion rates, $\Omega_{\mathrm{K}}^{\mathrm{disk}}$ increases and so does the QPO beat frequency $\Omega_{\mathrm{beat}} = \Omega_{\mathrm{K}}^{\mathrm{disk}} - \Omega$.

Recently a team of astrophysicists at Columbia University used the QPO technique to determine the mass of neutron star 4U 1636–536, which too turns out to be significantly heavier, $M \sim 2\,M_\odot$ [238], than neutron stars of average mass (figure 3.2). The X-ray pulses from 4U 1636–536 were observed by the Rossi X-ray Timing Explorer satellite RXTE launched by NASA in late 1995. The quasi-periodic variability of the observed X-rays is believed to be produced by material falling from the orbiting companion of 4U 1636–536 into the neutron star's innermost stable orbit and then into 4U 1636–536 itself, as described just above.

Aside from having an additional handle on the determination of neutron star masses, which does not take recourse to the mass-function technique, the existence of an innermost stable orbit attains its particular interest because it is a unique prediction of Einstein's theory of general relativity. Hence, if the interpretation of the origin of QPOs is correct, another prediction of Einstein's theory would be confirmed, supplementing the excellent confirmations established for relativistic binary pulsars by Hulse and Taylor and the 'classical' tests of Einstein's theory [242, 243]

A general, theoretical estimate of the maximal possible mass of a stable neutron star was performed by Rhoades and Ruffini [244] on the basis that (1) Einstein's theory of general relativity is the correct theory of gravity, (2) the equation of state satisfies both the microscopic stability condition $\partial P/\partial\epsilon \geq 0$ (Le Chatelier's principle) and the causality condition $\partial P/\partial\epsilon \leq c^2$, and (3) that the equation of state below some matching density is known. They found that the maximum mass of the equilibrium configuration of a neutron star cannot be larger than $3.2\,M_\odot$. This value increases to about $5\,M_\odot$ if one abandons the causality constraint $\partial P/\partial\epsilon \leq c^2$ [245, 246], since it allows the equation of state to behave more stiffly at supernuclear densities. More recently, Woosley and Weaver predicted neutron star masses from type-II supernova explosion simulations in the range 1.15 to $2.0\,M_\odot$ with a preponderance between 1.3 and $1.6\,M_\odot$.

3.2 Pulsar distances and dispersion measure

The pulsed nature of the emission and the slightly dispersive nature of the interstellar medium combine to provide an index – the so-called dispersion measure, DM – which allows distance estimates to be made for individual pulsars. The index of refraction of a plasma is given by

$$n = \sqrt{1 - \frac{\omega_p^2}{\omega^2}} \approx 1 - \frac{\omega_p^2}{2\,\omega^2}\,, \tag{3.2}$$

where ω_p is the plasma frequency,

$$\omega_p^2 = \frac{e^2\,\rho^e}{\epsilon_0\, m_e} \approx 3.2 \times 10^9 \left(\frac{\rho^e}{1/\mathrm{cm}^3}\right)\ \mathrm{s}^{-2}\,, \tag{3.3}$$

and ρ^e is the electron density, m_e is the electron mass, and ϵ_0 ($= 8.85 \times 10^{-12}$ A s V^{-1} m^{-1}) is the vacuum permittivity. This form for the index of refraction neglects finite temperature and magnetic fields, but is, however, an excellent approximation for the low densities and weak field strengths appropriate for the interstellar medium. Typical values for ρ^e are now recognized to be about 0.03 cm^{-3}, giving a plasma frequency (3.3) of $\omega_p = 10^4\ \mathrm{s}^{-1}$. Since the index of refraction is less than unity, the phase velocity exceeds slightly the velocity of light and the group velocity is slightly less,

$$v_g = n\,c \simeq c\left(1 - \frac{\omega_p^2}{2\,\omega^2}\right). \tag{3.4}$$

Consequently a sharp pulse of radio emission is *dispersed* so that the high-frequency components reach a terrestrial radio telescope before the low-frequency components. By measuring the pulse arrival times at different frequencies, one measures the accumulated time difference caused by the difference in group velocity over the path length. Thus one directly determines the dispersion measure,

$$\mathrm{DM} = \int_0^L \mathrm{d}l\, \rho^e(l) \approx L\, \langle \rho^e \rangle\,, \tag{3.5}$$

which is conveniently expressed in units of pc cm^{-3}. The second, less rigorous expression for DM ignores possible variations of the electron number density ρ^e along the path of the ray. As mentioned just above, its average value, $\langle \rho^e \rangle$, over large distances within our Galaxy is believed to be between 0.02 and 0.03 cm^{-3}. Thus a DM of 30, combined with the above estimates for $\langle \rho^e \rangle$, implies a distance of about 1000 pc. The DM for the Crab, for instance, is 57. The good agreement with independently determined distances to the nebula, about 2 kpc, is only slightly tempered by the fact that this pulsar is one of about 30 used to estimate the average value for ρ^e in our Galaxy.

3.3 Rotational periods of pulsars

The rotational periods of fast pulsars provide a *double* constraint on the equation of state when combined with the observed masses of neutron stars [26, 108]. As already mentioned in chapter 1, the fastest pulsars observed so far have rotational periods of 1.6 ms (that is, 620 Hz). The successful model for the nuclear equation of state, therefore, must account for rotational neutron star periods of at least $P = 1.6$ ms as well as masses that lie in the mass range emphasized in figure 3.2. As we shall see later in this text, the constraints on the equation of state become the more stringent the smaller the star's rotational period and the larger its mass.

On the observational side, the discovery rate of new pulsars is pleasingly so that the number of known pulsars is growing at an impressive pace. Unfortunately, the searches for rapidly rotating radio pulsars are biased, being least sensitive to short periods [19, 20, 21, 22]. Therefore the world's data of radio pulsars, an impressive list of which is compiled in tables 1.1 and 1.2, do not represent the true underlying population of fast pulsars. Most of the large surveys have had very poor sensitivity to millisecond pulsars with rotational periods below about 4 ms, thus presumably distorting the statistics on pulsar periods [19, 21, 22, 23]. The present cutoff in short periods at about 1 ms, as evident in figures 3.3 and 3.4, is therefore possibly only an *artifact* of the search sensitivity. The growing number of pulsar discoveries with periods right down to

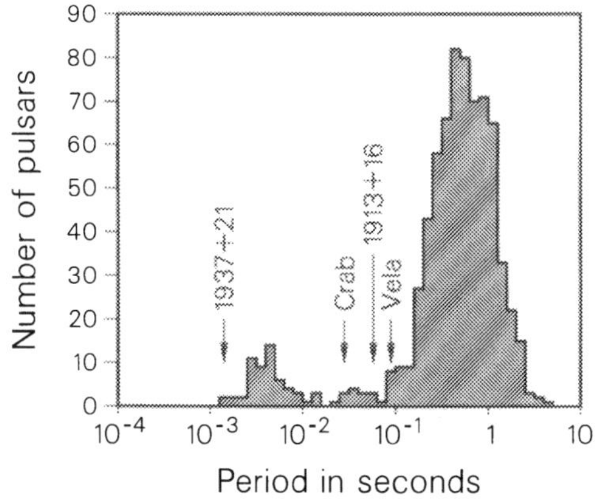

Figure 3.3. Distribution of pulsar periods. (Reprinted courtesy of Glendenning [247] and *Nucl. Phys.*)

this cutoff suggests that this is so and presents a special opportunity and challenge to the present generation of radio astronomical searches targeting the observation of (sub-) millisecond pulsars [22, 23].

Another interesting feature shown in figure 3.4 concerns the evolution of pulsars. Pulsars are born as rotating magnetic dipoles with magnetic field strengths of typically $B \sim 10^{12}$–10^{13} G (see also section 3.9). The tremendous store of rotational energy keeps them spinning for millions of years. Because of loss of angular momentum to radiation, P is gradually lengthening so that young pulsars evolve from the upper-left corner into the shaded region around $P \sim 1$ s and $\dot{P} \sim 10^{-15}$, where most of the observed pulsars lie. The continued growth of the pulsar periods finally brings the electromagnetic beaming effect to a halt, which renders any pulsar located to the right of the turn-off line radio silent.

Obviously, the millisecond pulsars cannot have been formed by simple aging of young pulsars. They have both much shorter periods and much weaker magnetic fields than the old, normal pulsars. It is generally accepted that the millisecond pulsars acquired their small rotational periods by mass accretion from low-density companions (e.g. white dwarfs), which spun up the pulsars because of transfer of angular momentum from the matter infalling onto their surfaces. A pulsar may have aquired a companion either right at birth, or may have captured one some time afterwards in the course of stellar evolution. Radio quiet pulsars, located to the right of the turn-off line, which are being spun up by accretion, evolve along the route of silent evolution. Eventually they cross the turn-off line, which ignites electromagnetic beaming again. Such pulsars, referred to as recycled ones [24], evolve right into the area populated by the millisecond pulsars.

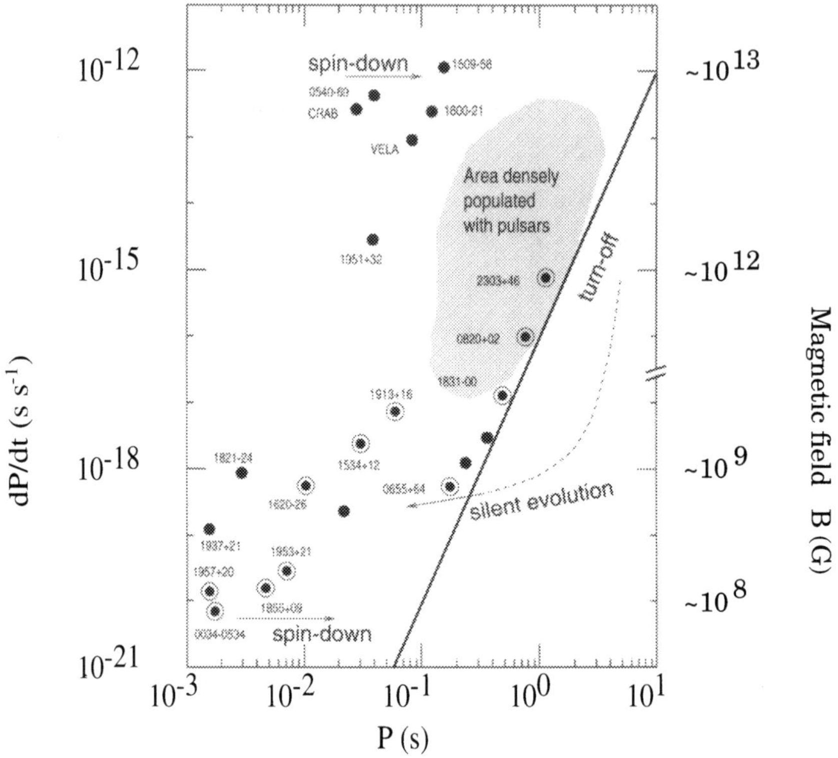

Figure 3.4. First time derivative of pulsar period versus pulsar period, P. Pulsars in the upper-left corner are associated with supernova remnants. Binary pulsars are indicated by ringed points. The magnetic fields of these pulsars are indicated on the right axis. (After Manchester [33] and Glendenning [247].)

This scenario provides a natural interpretation of millisecond pulsars as old neutron stars spun up by mass accretion. The magnetic field decay ($B \sim 10^8$ G) is possibly a result of the accretion process [248].

In the second part of this book (chapters 14, 17 and 18), it will be shown that the internal constitution of pulsars depends very sensitively on the star's central core density. Since the latter varies with P, we point out already at this stage that the internal constitution of the pulsars located in the different regions of figure 3.4 appears to be strikingly different!

As a last point on the distribution of pulsar periods, it needs to be noted that the pulsar searches at the Arecibo observatory, for instance, cover only 5% of the sky visible from Arecibo. A combination of all surveys to date covers at most 15%. Since this figure is expected to rise dramatically over the next few years, one may anticipate that a considerable number of new pulsars will be discovered in the near future. The history of astronomical discoveries cautions us that the laws of nature are explored by the Universe in many unanticipated ways. If such exotic systems such as sub-millisecond pulsars really exist, astronomers may find out within the next ten years or so.

As outlined by Lyne [21, 22] and Camilo [23], the main reason which made it impossible for radio astronomers to perform large-scale surveys of the sky in the 1980s was that the searches were severely limited by the data storage and computer technology available at that time. The problem lies in the dispersion of the pulsed signals by the ionized component of the interstellar medium, preventing the search process being simply a periodicity search of a time sequence. Because the pulses at low frequency are delayed relative to higher frequencies, recording a single, wide receiver band, which is required to provide high sensitivity, results in narrow pulses being broadened, possibly to such an extent that they cover more than the whole pulse period so that no periodicity would be observed at all. The most effective way of combating this is to split the receiver band into many narrow channels and to record the output of all of these in order to permit subsequent searches over many values of dispersion measure. Higher time resolution requires the use of narrower filters, so that, roughly speaking, in order to cope with dispersion, the size of the datasets and the magnitude of the processing tasks both increase as the square of the sampling rate. Thus, moving from the earlier searches for normal pulsars, which sampled at about 100 Hz, to the millisecond pulsar searches sampling at 3 kHz required *three orders of magnitude* increase in computing resources [21]. As an example, a 2.5 minute observation of the present Parkes survey [22] would result in a dataset of 128 million data points to be stored and processed. It would have taken in excess of one day to process such a dataset on a Vax 700 computer, the standard computer available to astronomers 10 years ago. The small number of millisecond pulsars that were found in these early years were mostly discovered with the aid of supercomputers. The second discovery burst of millisecond pulsars started as computer technology permitted widespread searches of the sky with high sensitivity and has run up to the present time.

We conclude this section by drawing attention to the recent optical observation of the binary millisecond pulsar J0034–0534 by Bell *et al* [249]. The present rotational period of this pulsar is 1.877 ms (cf. table 1.2). From the effective temperature $T \geq 3250$ K predicted for the companion of PSR J0034–0534, a cooling age of 4.0 Gyr was inferred which would imply an initial rotational period for J0034–0534 of 0.6 ms, which, if true, would have dramatic consequences for its interpretation as a rapidly spinning neutron star (cf. section 17.3). Accounting for the uncertainties in several of the parameters used to derive this value, the initial spin period of this pulsar appears to have been much less than 1.6 ms [249].

3.4 Radii

Measurement of a 10 km radius at a distance of at least 10^{15} km is a daunting experimental problem. So direct radius determinations for neutron stars do not exist. However, combinations of data of 10 well observed X-ray bursters with special theoretical assumptions led Van Paradijs [250] to the conclusion that the emitting surface has a radius of about 8.5 km. This value, as pointed out in [224], may be underestimated by a factor of two. Fujimoto and Taam [251] derived from the observational data of the X-ray burst source MXB 1636−536[1] under several uncertain theoretical assumptions a neutron star mass and radius of 1.45 $M_\odot$ and 10.3 km. Depending on the degree up to which their theoretical assumptions are satisfied, an error analysis led them to predict a mass and radius range from 1.28 to 1.65 $M_\odot$ and 9.1 to 11.3 km, respectively. A new technique for measuring a star's radius by detecting the effects of gravitational light bending near the stellar surface on pulsed emission from a hot magnetic polar cap on a nearby pulsar was demonstrated recently by Yancopoulos *et al* [252]. The result, $R = (9 \pm 3)$ km, is comforting but insufficiently precise to provide a serious constraint on the equation of state. Higher-sensitivity measurements with future instruments could well yield an answer to an accuracy of about $\pm 10\%$.

Finally we mention the rather promising development of the discovery of the nearby neutron star RXJ 185 635–3754 [42, 43, 44]. The observed X-ray flux and temperature, if originating from a uniform black body, appears to correspond to an effective stellar radius of only ~7 km. This is obtained from the relation

$$R_\infty = R \left(1 - \frac{2M}{R}\right)^{-1/2} \approx 7 \, \frac{D}{120 \text{ pc}} \text{ km}, \tag{3.6}$$

where R is the neutron star radius, M its mass, and D the distance to the star. One major uncertainty in (3.6) is buried in the numerical coefficient, which could increase by 50% or more if the star's surface temperature is not uniform. Moreover, the existence of an atmosphere on this neutron star could alter the coefficient probably even more strongly. The other major uncertainty in obtaining a radius is the distance, but the low extinction and the location of the star in the direction of the R CrA molecular cloud limits the distance to $D < 120$ pc. Upcoming parallax and proper motion observations [253] will pin down the radius of this star more accurately. If a radius value of ~7 km should be confirmed, then the underlying equation of state surely must be *extremely* soft up to several times nuclear matter density in order to get such a small neutron star (cf. chapter 14). The possible condensation of K^- mesons at high densities in neutron stars,

[1] MXB stands for MIT X-ray Burst Source.

which then go into a Bose condensate, could be the cause for such an extreme softening, as pointed out in [43, 44, 254, 255, 256]. We note that the possibly very soft behavior of the equation of state required by this radius measurement, if correct, is not in contradiction to the possible need for a rather stiff behavior at very high densities, as required by the possible existence of very heavy neutron stars of masses around $\sim 2\,M_\odot$ (see section 3.1).

In concluding this subsection, we mention the possibility of determining a semiempirical mass–radius relationship of neutron stars from the cyclotron line features observed in several X-ray pulsars with the Ginga satellite [45]. What is needed to perform this analysis are the rotational period P, the spin-up rate $\dot{P}$, the X-ray luminosity L_X, and the cyclotron line energy at infinity, ΔE_∞, of an X-ray emitting neutron star [257]. The semiempirical mass–radius relationship is derived under several assumptions, the least of which is that pulsars are assumed to be rotating neutron stars with dipolar magnetic fields accreting matter from an accretion disk. Another assumption is that cyclotron lines, discussed in some detail in section 3.9 below, arise from a stellar plasma at rest at the magnetic poles of a pulsar. With these assumptions, the magnetic field strength at the magnetic pole of a neutron star can be expressed in terms of the cyclotron line energy as

$$B = \frac{m_e\, c}{e\, \hbar} \, \frac{\Delta E_\infty}{\sqrt{1 - x^{-1}}} \,, \tag{3.7}$$

where x is the neutron star radius in units of the Schwarzschild radius, i.e. $x = R/(2GMc^{-2})$, and M and R denote the star's mass and radius. On the other hand, the field strength B at the magnetic pole can be expressed as [257]

$$B = \frac{3\,|\boldsymbol{m}|\,c^6}{4(G\,M)^3} \left(-\ln(1 - x^{-1}) - \frac{(1 + x^{-1}/2)}{x} \right), \tag{3.8}$$

with $|\boldsymbol{m}|$ the magnitude of the magnetic moment. If $|\boldsymbol{m}|$ is calculated for a given magnetospheric model combined with the observed data for P, $\dot{P}$ and L_X, the semiempirical mass–radius relationship can be determined [45] and its predictions compared with the mass–radius relationships computed for different competing models for the equation of state of superdense matter.

3.5 Pulsar braking

The rotational frequency Ω of isolated pulsars decreases slowly but measurably over time because of various energy loss mechanisms that can be at play such as the dipole radiation, part of which is detected on each revolution, as well as other losses such as ejection of charged particles [258]. Usually the first and occasionally also the second time derivative of Ω, that is, $\dot{\Omega}$ and $\ddot{\Omega}$, can also be measured, out of which a dimensionless quantity known as the braking index,

$$n(\Omega) \equiv \frac{\Omega\ddot{\Omega}}{\dot{\Omega}^2} = \frac{\nu\ddot{\nu}}{\dot{\nu}^2}, \tag{3.9}$$

is formed. Its value would be equal to 3 if the energy loss from a pulsar were due to pure magnetic dipole radiation and the star's moment of inertia is taken to be constant during slowdown, both of which are idealizations as we shall see in chapter 17. Lyne *et al* found a surprisingly low braking index of $n = 1.4 \pm 0.2$ for Vela [259], which suggests that the braking cannot be attributed entirely to radiation from a constant magnetic dipole but is probably due to a changing magnetic moment or an effective momentum of inertia. It may also indicate that the Vela pulsar is much older than previously thought, or that other yet unknown processes are at work (cf. section 17.3.3). The other three presently known braking indices are considerably closer to the canonical value of 3, as shown in table 3.1. It is interesting to note that if one accepts the current values of $\ddot{\nu}$ of pulsars B2127+11A and B1757–24, listed in table 3.2, their braking indices were

Table 3.1. Measured braking indices of pulsars.

Pulsar	n	Reference
B0531+21	2.51 ± 0.01	[260]
B1509–58	2.837 ± 0.001	[261]
B0540–69	2.24 ± 0.04	[262]
B0833–45	1.4 ± 0.2	[259]

Table 3.2. Rotational frequency ν, and its first and second time derivatives, $\dot{\nu}$ and $\ddot{\nu}$, of pulsars. In the features column, 'C' indicates globular cluster association, 'G' indicates glitches, and 'S' indicates supernova association.†

Pulsar	ν (s^{-1})	$\dot{\nu}$ (s^{-2})	$\ddot{\nu}$ (s^{-3})	Features
B0531+21	29.94	-3.77×10^{-10}	$(9.76 \pm 0.07) \times 10^{-21}$	GS
B1509–58	6.64	-6.77×10^{-11}	$(1.9587 \pm 0.0009) \times 10^{-21}$	S
B0540–69	19.85	-1.88×10^{-10}	$(3.66 \pm 0.04) \times 10^{-21}$	S
B0833–45	11.20	-1.56×10^{-11}	3.04×10^{-23}	GS
B2127+11A	9.04	1.72×10^{-15}	$(4.8 \pm 0.1) \times 10^{-25}$	C
B1757–24	8.01	-8.20×10^{-12}	$(3.2 \pm 0.4) \times 10^{-22}$	S

† Data taken from http://pulsar.princeton.edu/ftp/pub/catalog/.

$n = 1.5 \times 10^6$ and $n = 38$, respectively, the former of which deviating dramatically from $n = 3$.

3.6 Pulsar glitches

Based on 279 pulsars in a timing program at Jodrell Bank, Shemar and Lyne monitored 2500 years of pulsar rotation [263]. They detected a total of 25 glitches in 10 pulsars. Together with the 7 glitches that have been detected by other observing programs in 5 other pulsars, this increases the total number of detected glitches to 32 observed for 15 pulsars. Only a small fraction of the change of the rotation rate is recovered in the subsequent relaxation epoch. In several pulsars, the relaxations are found to consist of a single exponential-like decay, with time-constant of order 100 days, together with a long-term recovery. Particularly striking are so-called giant glitches which are characterized by relative changes in a pulsar's rotational frequency of $\Delta\Omega/\Omega \sim 10^{-6}$ [33]. One example of a pulsar that exhibits giant glitches of $\Delta\Omega/\Omega = 2 \times 10^{-6}$ is the young pulsar PSR B1757–24 (PSR J1801–2451) [264]. Glitches of that magnitude are seen only in about 7 pulsars (including Vela) of the 800 presently known.

A popular model [18, 265, 266, 267] that has been invented to explain the observed glitch behavior of pulsars is based on the formation of vortex lines in the superfluid neutron drip regime that is predicted [268, 269] to exist in the inner crust of neutron stars (see figure 2.1). In the inner crust, which extends from the neutron drip density of $\sim 10^{11}$ g cm^{-3} to about $\sim 10^{14}$ g cm^{-3}, that is, somewhat less than nuclear matter density, a gas of free neutrons coexists with a lattice of neutron-rich nuclei. For stellar densities $\sim 10^{12}$–10^{14} g cm^{-3}, corresponding to free neutron densities of $\sim 10^{-3}$–10^{-1} fm^{-1}, the free neutrons are believed to pair and form an isotropic S-wave fluid (more details will be given in section 19.4). In a rotating star the neutron superfluid rotates by forming vortex lines, singular regions in which the superfluid vanishes and around which the superfluid circulation is non-zero. Because the density of vortices per unit area is given by $n_{\rm vor} = 2m_n\Omega/(\pi\hbar)$, where m_n denotes the neutron mass, the superfluid velocity in the neutron star is completely determined by the spatial arrangement of the vortex lines. That is, a change in the velocity field requires a corresponding change in the vortex-line distribution, either a distortion of the configuration of the existing lines, or the creation of new lines. Were the vortex lines in a region of the inner crust immobilized, for example by pinning to the crystal lattice, the angular velocity of the superfluid in this region would be largely fixed. As the angular velocity of the solid crust of the neutron star changes, due to magnetic braking or mass accretion, the differences between the angular velocities of the superfluid and the solid crust grows. This differential motion may be the source of free energy that powers glitches [269]. A glitch may result

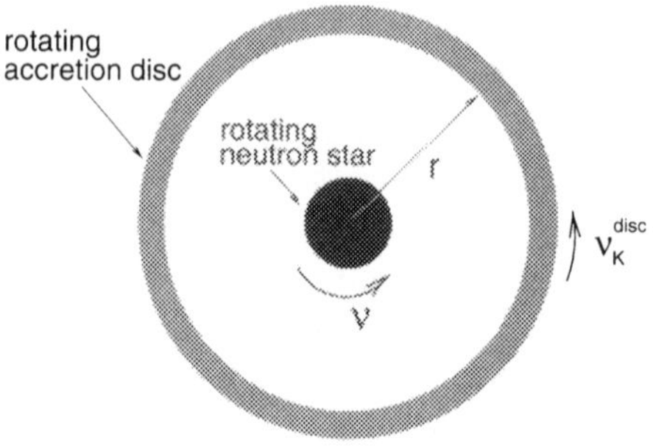

Figure 3.5. Geometrical illustration underlying the derivation of equation (3.10).

from the sudden transfer of angular momentum from the superfluid to the crust caused by the catastrophic unpinning of many vortex lines, or by cracking of the crust to which the vortex lines are pinned. The post glitch relaxation may be understood in terms of variations in the slow outward vortex creep rate [270, 271, 272, 273]. Surely, a qualitative investigation of these effects is complicated since, among other issues, the interaction of a single nucleus with a vortex line is to be determined by computing the change in the superfluid pairing energy and the penetration of the superfluid into a nucleus by minimizing a Ginzburg–Landau energy functional [269].

3.7 Moment of inertia

Another global neutron star property is the moment of inertia, I. Early estimates of the energy-loss rate from pulsars [55] spanned a wide range of I, from 7×10^{43} g cm^2 to about an order of magnitude higher. From the luminosity of the Crab nebula, which is $\sim$2–4 $\times 10^{38}$ erg s^{-1} (i.e. 2–4$\times 10^{31}$ W), several authors have found a lower bound on the moment of inertia of the Crab pulsar given by $I \gtrsim 4$–8×10^{44} g cm^2 [274, 275, 276]. As for neutron star radii, this range is insufficiently precise to provide a serious constraint on the equation of state.

Another handle on the determination of I may originate from the Lense–Thirring (frame dragging) precession caused by rotating neutron stars in low-mass X-ray binaries. As we shall see in section 15.4, the Lense–Thirring precession frequency, $\omega_{\rm LT}$, is connected with the moment of inertia of a neutron star rotating at frequency Ω ($= 2\pi\nu$) by the relation $\omega_{\rm LT}(r) = (2I/r^3)\Omega$ 9cf. equation (15.87)), where r denotes the radial distance of a local inertial frame located outside of the star. If r is chosen to be equal to the innermost region of the accretion disk at the magnetospheric boundary, i.e. $r = r_{\rm in}$ as shown in figure 3.5 (see also section 3.1), then the Kepler frequency of the innermost region is given by $\Omega_K^{\rm disk} = 2\pi\nu_K^{\rm disk} = \sqrt{M/r^3}$, leading to a Lense–Thirring frequency of $\nu_{\rm LT} \equiv \omega_{\rm LT}/2\pi = 8\pi^2 I(\nu_K^{\rm disk})^2\nu/M$ [277]. This expression can be rewritten as

$$\nu_{\rm LT} = 13\left(\frac{I}{10^{45}\ {\rm g\ cm}^2}\right)\left(\frac{M}{M_\odot}\right)^{-1}\left(\frac{\nu_K^{\rm disk}}{10^3\ {\rm Hz}}\right)^2\left(\frac{\nu}{300\ {\rm Hz}}\right)\ {\rm Hz}\,, \qquad (3.10)$$

where ν is the star's frequency and $\nu_K^{\rm disk}$ the Kepler frequency of the disk. Equation (3.10) may conversely be used to shed light on the magnitude of $\nu_{\rm LT}$, which is possibly extractable from the QPOs. This is readily accomplished by substituting representative values of I, M, $\nu_K^{\rm disk}$, and ν into (3.10), in which case each term on the right-hand side of this equation gives ~1. Hence one may expect that $\nu_{\rm LT}$ ~10 Hz.

3.8 Gravitational redshift

Another important source of information about the structure of neutron stars is provided by the neutron star redshifts, z. The expression of z, which will be derived for a more general situation in section 15.6, can be found heuristically as follows. Consider a photon created at the surface of a neutron star (emitter) and leaving its gravitational field toward a detector located at infinity. Then the photon's frequency at the emitter, ν_E, is just the inverse of the proper time (cf. equation (12.37)) between two wave crests, $d\tau_E$, as measured in the frame of the emitter [224], i.e. $\nu_E = 1/d\tau_E = (-g_{\mu\nu}dx^\mu dx^\nu)_E^{-1/2}$, where $dx^1 = dx^2 = dx^3 = 0$ because the emitter stays at a fixed position while emitting the photon. The same expression but written down for the infinitely far receiver reads $\nu_\infty = 1/d\tau_\infty = (-g_{\mu\nu}dx^\mu dx^\nu)_\infty^{-1/2}$. Hence, one obtains for ratio $\nu_\infty/\nu_E = [(-g_{00})^{1/2}dx^0]_E/[(-g_{00})^{1/2}dx^0]_\infty$. If we assume that the coordinate time dx^0 between two wave crests is the same at the star's surface and at the receiver, which is the case if the gravitational field is static so that whatever the worldline of one photon is from the star to the receiver, the next photon follows a congruent path, merely displaced by dx^0 at all points, this ratio simplifies to $\nu_\infty/\nu_E = [(-g_{00})^{1/2}]_E/[(-g_{00})^{1/2}]_\infty$. Making use of the definition of the gravitational redshift, $z = (\nu_E/\nu_\infty) - 1$, leads immediately to $z = [(-g_{00})^{1/2}]_\infty/[(-g_{00})^{1/2}]_E$ which can be written as

$$z = \left(1 - \frac{2M}{R}\right)^{-1/2} - 1\,, \qquad (3.11)$$

because the g_{00} component of the metric tensor at the surface of a non-rotating star is given by $g_{00}(R) = (1 - 2M/R)^{1/2}$ (cf. equations (14.15) and (14.94)).

Liang [46] has considered the neutron star redshift data base provided by measurements of γ-ray burst redshifted annihilation lines in the range of

300–511 keV. These bursts have widely been interpreted as gravitationally redshifted 511 keV $e^{\pm}$ pair annihilation lines from the surfaces of neutron stars. From this he showed that there is tentative evidence, if the interpretation is correct, to support a neutron star redshift range of $0.2 \leq z \leq 0.5$, with the highest concentration in the narrower range $0.25 \leq z \leq 0.35$. A particular role is played by the source of the 1979 March 5 γ-ray burst source, which has been identified with supernova remnant N49 by its position. From the interpretation of its emission, which has a peak at ~430 keV, as the 511 keV $e^{\pm}$ annihilation line [278, 279] the resulting gravitational redshift has a value of $z = 0.23 \pm 0.05$.

3.9 Magnetic fields and cyclotron lines

Pulsar theory leans heavily on the expectation that neutron stars are highly magnetized objects [55, 63, 137, 280]. There is a general plausibility argument to the effect that, if internal magnetic flux in a star is conserved owing to high conductivity, then the surface magnetic field increases as R^{-2} where R is the stellar radius. Thus, if the Sun were to collapse to neutron star dimensions (roughly a factor of 7×10^4 in radius), one would magnify the general 1 G solar field to about 5×10^9 G. This value, while being in good agreement with the magnetic field strength $B \sim 4\text{–}6 \times 10^{11}$ G estimated for the X-ray pulsar 1 E 2259+586 [36], the central source in the supernova remnant G 109.1–1.0, is a little less than 10^{12} G, a value inferred by Trümper *et al* [38, 39] and Wheaton *et al* [40] from line features in the pulsed hard X-ray spectrum of Her X–1 and 4U 0115–63, respectively. These line features are generally assumed to arise from electron cyclotron resonances, which occur when an electron in the magnetized stellar plasma makes a transition between different Landau levels. The nth Landau level energy, U_n, in a uniform magnetic field B is approximately given by [41]

$$U_n = \frac{1}{1+z} \left\{ m\,c^2 + \frac{p^2}{2\,m_e} + \left(n + \frac{1}{2}\right) E_0 \right\}, \tag{3.12}$$

where m_e is the electron mass, p the electron momentum along the magnetic field, and z the gravitational redshift. The quantity $E_0 = e\hbar B/m_e = 11.6 \times B/(10^{12}\text{ G})$ keV denotes the electron cyclotron resonance energy. Assuming $E_0 \ll mc^2$ and neglecting the p^2 term in (3.12) leads to equally spaced Landau levels, being separated by integer multiples of $E_0/(1+z)$. In particular, the fundamental resonance occurs at $E^* = E_0/(1+z) \sim E_0$, which predicts for the above given range of E_0 a cyclotron resonance scattering feature in the energy range $E^* \sim 1\text{–}10$ keV. On the basis of this analysis it was found that the line feature observed in the X-ray spectrum of Her X–1 and 4U 0115–63, interpreted as electron cyclotron resonance lines, implies magnetic field strengths of $B \sim 4\text{–}6 \times 10^{12}$ G for Her X–1 and $B \sim 2 \times 10^{12}$ G for 4U 0115–63. (Of course, the original analysis is more complicated than indicated above.) An overview of the magnetic field strengths at the surfaces of millisecond pulsars is given in tables 1.1 and 1.2.

Of course, one may argue that the Sun is not necessarily the best example of the pre-supernova object. The implication, then, is that the pre-supernova core, with a radius of say 10^4 km, will have a 'surface' field (actually deep within a massive giant star) of 10^6 G. Such empiricism, however, has not yet helped much to elucidate which physical mechanism might be important to create the extraordinarily strong magnetic fields of neutron stars. One possible mechanism may be differential rotation of the stellar core (as proposed for planetary magnetic fields). Besides that, ferromagnetism has been examined as a possible magnetic field source, while differential rotation between say superfluid protons and normal electrons has been another suggestion [281]. The influence of very strong magnetic fields, up to $\sim 10^{20}$ G, on the properties of nuclear matter as well as chemically equilibrated neutron star matter was studied recently by Chakrabarty *et al* [282]. In this text we shall not pursue a discussion of any of these interesting issues but refer to the literature for more details [137].

3.10 Cooling data

The detection of thermal photons from a stellar surface via X-ray observatories like Einstein, EXOSAT, ASCA, and ROSAT serves as our principal window on the properties of a star. Besides its radius, surface gravity, chemical composition, and magnetic field strength, the surface temperatures of stars are derivable from the measured flux and spectrum. The cooling rate of a hot, young neutron star is primarily dependent at early times on the neutrino emissivity of the core's composition. The possible existence of Bose–Einstein condensates or quark matter, for example, enhance neutrino emissivity, leading to more rapid early cooling. Superfluidity, on the other hand, has the opposite effect on cooling. Quantitative constraints have been hampered, however, by the small number of young pulsars known, the complication that several of them also display non-thermal, beamed X-ray emission from their magnetospheres [137], and uncertainties in distance and interstellar absorption.

At the time of writing this text, the X-ray spectra of at least 14 pulsars have been observed by Einstein, EXOSAT, ASCA, and ROSAT. Einstein has only little spatial resolution and so the spectral measurement may contain contributions from magnetosphere emission and synchrotron nebula emission, besides the thermal surface radiation. As a consequence, the calculated temperatures constitute only upper bounds. Moreover, when determining the star's surface temperature, first its absolute brightness is to be determined from the apparent brightness which involves knowledge of the distance of the star and the hydrogen column density in space. Additional uncertainties arise when the surface temperature is determined from the absolute brightness because assumptions about the neutron star's radius and the radiation itself (black-body type) are to be made. ROSAT avoids the determination of the absolute brightness. In this case the spectral fits are performed assuming a black-body spectrum, or the spectrum of a model atmosphere consisting of hydrogen or helium. Moreover, the thermal surface radiation emitted from a neutron star can be discriminated from the magnetospheric radiation and radiation coming from a synchrotron nebula, provided the pulsar's distance from ROSAT is not too large.

It is useful to classify the spectra into three distinct categories [37] as listed immediately below:

(1) The photon yields of the three pulsars PSR 1706–44, 1823–13, and 2334+61 contain too few photons in order to perform spectral fits. The luminosities are calculated by using the totally detected photon fluxes. When discussing the cooling behavior of neutron stars in chapter 19, these pulsars will be marked with triangles in the figures.
(2) The spectra of seven pulsars, including Crab (PSR 0531+21), can only be fitted by a power-law type spectrum or a black-body spectrum with a very high effective temperature, and an effective area much smaller than the neutron star surface. Their X-ray emission is predominated by magnetospheric emission. Therefore their temperatures, determined from the spectral fits, are probably too high. Theses pulsars will be marked with solid dots.
(3) Finally, there are the four pulsars PSR 0833–45 (Vela), 0656+14, 0630+18 (Geminga), and 1055–52 that allow two-component spectral fits, where the softer black-body components correspond to the actual surface emission of the neutron star and the harder to some magnetospheric emission. These pulsars will be marked with squares.

The inferred luminosities together with the pulsar spin-down ages are contained in table 3.3. The latter is obtained from the measured pulsar frequency and its time derivative as

$$\tau = -\frac{1}{n-1}\frac{\Omega}{\dot{\Omega}}, \tag{3.13}$$

with $n = 3$ the braking index for energy loss governed by magnetic dipole radiation. Equation (3.13) is obtained by integrating the general formula for the rate of change of the pulsar frequency (K denotes a constant),

$$\dot{\Omega} = -K\,\Omega^n, \tag{3.14}$$

from the star's birth to the present time, assuming that Ω/Ω^n at birth is negligibly small compared to its ratio at the time of observation.

When discussing spectra of neutron stars one should keep in mind that neutron stars are not black bodies, because the hydrogen and helium

Table 3.3. Survey of ages, τ, and luminosities, $L\dagger$, of observed pulsars.

Pulsar	Name	$\log_{10} \tau$ (yr)	$\log_{10} L$ (erg s^{-1})	References
Not enough data available for spectral analysis				
1706–44		4.25	32.8 ± 0.7	[283]
1823–13		4.50	33.2 ± 0.6	[284]
2334+61		4.61	33.1 ± 0.4	[285]
Power-law-type spectra or spectra with only a high-temperature component				
0531+21	Crab	3.09	35.5 ± 0.3	[286]
1509–58	SNR MSH 15-52	3.19	33.6 ± 0.4	[287, 288]
0540–69		3.22	36.2 ± 0.2	[289]
1951+32	SNR CTB 80	5.02	33.8 ± 0.5	[290]
1929+10		6.49	28.9 ± 0.5	[37, 291]
0950+08		7.24	29.6 ± 1.0	[292]
J0437–47		8.88	30.6 ± 0.4	[293]
Spectrum dominated by a soft component				
0833–45	Vela	4.05	32.9 ± 0.2	[294]
0656+14		5.04	32.6 ± 0.3	[295]
0630+18	Geminga	5.51	31.8 ± 0.4	[296]
1055–52		5.73	33.0 ± 0.6	[297]

† The luminosities are determined either by spectral fits or the totally detected photon flux, depending on the categories (1) to (3) explained in the text.

in their atmospheres modifies the black-body spectrum. Besides that, strong magnetic fields can also affect the surface emission [137]. The luminosities, L, determined by using hydrogen atmospheric or helium atmospheric models are generally lower by a factor of about three than those determined via black-body spectral fits. Similarly, the temperature, T, is reduced by about the same factor when realistic atmospheres are being used. An uncertainty in temperature by say a factor of two changes the luminosity by a factor of 16, according to the Stefan–Boltzmann law, $L \propto T^4$. It is therefore not very appropriate to perform a comparison with the surface temperatures of stars [37] but should be using the luminosities instead, as will be the case here. Aside from the fact that the atmospheric composition affects the interpretation of the observed luminosities and temperatures, the atmospheric composition also affects the cooling curves (stellar temperature as a function of time) since the atmospheric opacity depends upon composition.

Chapter 4

Physics of neutron star matter

4.1 Low-density regime

For neutron stars the density of matter spans an enormous range, from about ten times the density of normal nuclear matter in the star's core down to the density of iron, 7.9 g cm^{-3}, at the star's surface. Most of the mass of the star is contributed by highly compressed matter at nuclear and supernuclear densities, as can be seen in figure 4.1. It shows the energy density profiles of several neutron star models constructed for different equations of state, which will be introduced later in chapter 12 (cf. table 12.4 for the labeling and the characteristic features of these equations of state). Gravity compresses the matter in the cores of these neutron stars to densities between 9 to 13 times the density of ordinary nuclear matter. Each stellar model is constructed for the largest possible central density beyond which these stars become unstable against radial oscillations. The corresponding stellar masses, which therefore are maximal in each case, are listed in table 14.3. The pressure profiles of these neutron stars are shown in figure 4.2.

Depending on density, the structure of neutron stars, schematically illustrated in figure 2.1, is presently understood to be as follows [56, 58, 70, 71, 298]:

- *Surface:* Matter at mass densities in the range of 10^4 g cm^{-3} $< \epsilon <$ 10^6 g cm^{-3} is composed of normal nuclei and non-relativistic electrons. The outermost layer, with an optical depth of $\tau \sim 1$, is called the photosphere. The thermal radiation, observed in X-ray telescopes, is emitted from this region. This radiation dominates the cooling of neutron stars older than $\sim 10^6$ years. During the first $\sim 10^6$ years cooling is dominated by emission of neutrinos from the core region (see below). Depending on the neutrino reactions there, one distinguishes between standard and enhanced cooling (cf. chapter 19). Both the

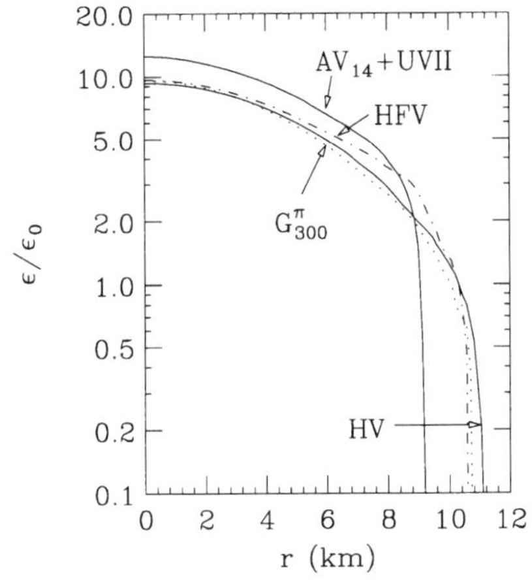

Figure 4.1. Energy density ϵ (in units of normal nuclear matter density, ϵ_0) versus radial distance in non-rotating maximum-mass neutron star models.

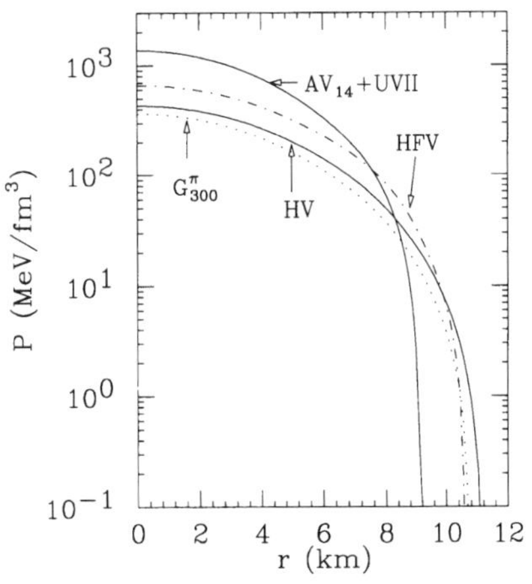

Figure 4.2. Pressure P versus radial distance in non-rotating maximum-mass neutron star models.

outer and inner crust act as thermal insulators between the cooling core and the surface.

- *Outer crust:* At densities of 7×10^6 g cm^{-3} $< \epsilon <$ 4.3×10^{11} g cm^{-3} the electrons become relativistic and form a relativistic electron gas. The atomic nuclei (lighter metals), while becoming more and more neutron rich, form a solid Coulomb lattice.
- *Inner crust:* It ranges from densities of about 4.3×10^{11} g cm^{-3} to 2×10^{14} g cm^{-3}. At 4.3×10^{11} g cm^{-3} neutrons begin to drip out of the neutron-saturated nuclei and populate free states outside of them. This density value is therefore referred to as neutron drip density. For increasing density matter clusters into extremely neutron rich nuclei (heavy metals) that are arranged on a lattice and immersed in a gas of neutrons and relativistic electrons.
- *Core region:* For densities beyond 2×10^{14} g cm^{-3} the clusters begin to dissolve and neutrons, protons, and electrons form a relativistic Fermi fluid. Model calculations show that hyperon production sets in at about twice nuclear matter density, 5×10^{14} g cm^{-3}. As we shall see later, the population of Δ's appears to be favored by many-body approximations that go beyond the relativistic mean-field approximation (relativistic Hartree–Fock, relativistic Brueckner–Hartree–Fock). Other unresolved issues concern the formation of meson condensates in the cores of neutron stars and the possible transition of confined hadronic matter into quark matter. Finally,

the possible absolute stability of 3-flavor strange quark matter, as described in section 2.3, would make 'conventional' neutron star matter, made up of either only hadrons or hadrons in equilibrium with quarks, metastable with respect to the lower energy state occupied by strange quark matter. In this case probably all 'neutron' stars would be made up entirely of pure 3-flavor strange quark matter, except a thin nuclear crust that may envelope the strange matter core. More than that, new and distinct classes of compact stars that are entirely different from neutron stars and white dwarfs should exist.

The surface and crust regions of neutron stars, whose equations of state are rather reliably known as opposed to the equation of state of the core region, are so thin that these contribute only little to the bulk properties like masses, radii, moments of inertia, and limiting rotational periods of the more massive members (that is, $M \gtrsim 1\, M_\odot$) of a neutron star sequence [61]. The equations of state of these density regimes therefore are only of minor importance for our studies. We shall use two models for the equation of state of the surface and crust region published in the literature. The first has been computed by Harrison–Wheeler [69] and Negele–Vautherin [72], the second by Baym–Pethick–Sutherland [70] and Baym–Bethe–Pethick [71]. Details of these models are listed in table 4.1. Hereafter we shall refer to these two models as HW–NV and BPS–BBP, respectively. In the stellar structure calculations that will be performed in

Table 4.1. Two models for the EOS of neutron star matter at subnuclear densities.

Equation of state	Density range † (g cm^{-3})	Composition
Harrison, Wheeler (HW) ‡	$7.9 \le \epsilon \le 10^{11}$	Crystalline; light metals, electron gas
Negele, Vautherin (NV)	$10^{11} \le \epsilon \le 10^{13}$	Crystalline; heavy metals, relativistic electron gas
Baym, Pethick, Sutherland (BPS) ‡	$7.9 \le \epsilon \le 4.3 \times 10^{11}$	Similar to HW but with Coulomb lattice correction
Baym, Bethe, Pethick (BBP)	$4.3 \times 10^{11} \le \epsilon \le 10^{14}$	Electrons, neutrons, and equilibrium nuclei

† For the conversion from g cm^{-3} to MeV fm^{-3}, see table 1.4.

‡ The low-density part, from 7.9 g cm^{-3} to 10^4 g cm^{-3}, has been computed by Feynman, Metropolis, and Teller [68].

the second part of this volume, these equations of state are joined with the models for the equation of state of the superdense core matter at densities between $10^{-2}\epsilon_0$ and $10^{-1}\epsilon_0$.[1]

4.2 High-density regime

At densities greater than the density of nuclear matter, ϵ_0, the Fermi momenta of the nucleons, N, in neutron star matter are so high that particle reactions such as

$$N + N \longrightarrow N + H + M \tag{4.1}$$

become possible, where H denotes a hyperon and M a meson. A sample reaction for (4.1) is, for instance,

$$N + N \longrightarrow N + \Lambda + K\,. \tag{4.2}$$

As described by Glendenning [61], strangeness is conserved by the strong interaction but not by the weak force, which enables the corresponding mesons, such as the K, for instance, to transform as

$$K^0 \longrightarrow 2\gamma\,, \qquad K^- \longrightarrow \mu^- + \nu\,, \tag{4.3}$$

$$\mu^- + K^+ \longrightarrow \mu^- + \mu^+ + \nu \longrightarrow 2\gamma + \nu\,. \tag{4.4}$$

The situation is different if the meson is driven by a phase transition, as will be discussed below in this section. Aside from the very early stages of a newly formed neutron star, the star's energy is lowered through the leakage of the photons and neutrinos, γ and ν respectively. Consequently, the hyperon, of which we choose without loss of generality the Λ, becomes Pauli blocked, and a net strangeness can evolve for sufficiently dense neutron stars. Other particle states, ranging from the more massive hyperons $\Sigma^{\pm}$, Σ^0, Ξ^0, Ξ^- and the Δ-resonance states (cf. table 5.1) to the up, down, and strange quarks, will become populated as the density of nucleons increases further. More than that, this may be accompanied by the formation of meson condensates, of which the K^- condensate has attracted a great deal of interest lately (see below). The governing principle that determines the complex particle population is known as *chemical equilibrium*, which is also referred to as β-equilibrium. Fortunately solving the problem of chemical equilibrium does not require that all kinds of these individual reactions be

[1] The intense magnetic fields ($B \sim 10^{12}$ G) that are believed to exist on the surfaces of neutron stars will plausibly modify the structure of bulk matter. Abrahams and Shapiro improved on previous statistical calculations of the equation of state of matter subject to such strong magnetic fields. Moreover, finite temperature corrections were taken into account too [299].

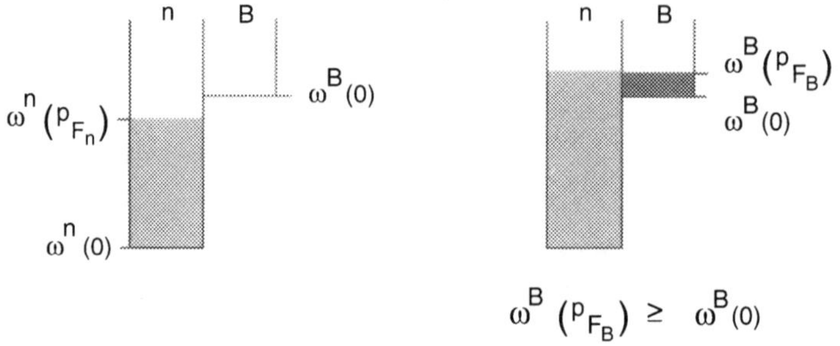

Figure 4.3. Condition for the onset of hyperon population in chemically equilibrated matter. Left: the single-particle energy is not high enough for the neutrons to transform to baryons of type B. Right: high-energy neutrons overcome the baryon threshold, that is, $\mu^B \equiv \omega^B(p_{F_B}) \geq \omega^B(0)$, and therefore transform to particle state B.

studied, as pointed out by Glendenning [61]. What is only required instead is the recognition of which charges are conserved by the system. For neutron star matter these charges are baryon number, q_B, and electric charge, $q_B^{\rm el}$. Associated with these two conserved charges are two independent chemical potentials, μ^n and μ^e respectively. All other particle chemical potentials can be expressed as a linear combination of these two. This considerably eases the problem of determining the composition of superdense neutron star matter in the ground state. For an arbitrary particle, chemical equilibrium in a star can then be expressed as

$$\mu^\chi = q_\chi \, \mu^n \; - \; q_\chi^{\rm el} \, \mu^e \,, \tag{4.5}$$

where χ stands for the various hadronic and quark fields, that is, $\chi = B, Q, L, M$ with baryons like $B = p, n, \Sigma^{\pm,0}, \Lambda, \Xi^{0,-}, \Delta^-$, quarks $Q = u, d, s$, leptons $L = e^-, \mu^-$, and mesons $M = \pi^-, K^-$. A particle state χ will be populated when its chemical potential μ^χ exceeds the particle's lowest-energy eigenstate in the medium, that is, only if

$$\mu^\chi \geq \omega^\chi(\boldsymbol{p} = \boldsymbol{0}) \,. \tag{4.6}$$

The situation is schematically illustrated in figure 4.3.

As long as the neutrinos and photons do not accumulate inside the star, as will be the case for the stars studied here, their respective chemical potentials μ^ν and μ^γ are equal to zero. This implies for the weak and electromagnetic decays (4.3) and (4.4)

$$\mu^\gamma = 0 \quad \Longrightarrow \quad \mu^{K^0} = 0 \tag{4.7}$$

for the first reaction of (4.3), and

$$\mu^{\nu} = 0 \quad \Longrightarrow \quad \mu^{K^-} = \mu^{\mu^-} \tag{4.8}$$

for the second reaction of (4.3). Reaction (4.4) tells us that

$$\mu^{K^+} = -\,\mu^{\mu^-}\,. \tag{4.9}$$

To illuminate the above principles in more detail, let us consider, as a first example, chemically equilibrated matter at such densities where it is made up of only protons and neutrons, which is the case around nuclear matter density. The protons and neutrons then obey $n \leftrightarrow p + e^- + \bar{\nu}_e$, which, for vanishing antineutrino population (that is, $\mu^{\bar{\nu}_e} = 0$) leads to $\mu^n = \mu^p + \mu^e$. As the neutron density increases, so does that of protons and electrons. Eventually μ^e reaches a value equal to the muon mass. If so the muon then too will be populated. Equilibrium with respect to $e^- \leftrightarrow \mu^- + \nu_e + \bar{\nu}_\mu$ is assured when

$$\mu^{\mu^-} = \mu^e\,. \tag{4.10}$$

Note that because of (4.8), this leads for the chemical potential of the K^- meson to

$$\mu^{K^-} = \mu^e\,. \tag{4.11}$$

If this condition is fulfilled at a certain density, then the K^- mesons begin to form a condensate, according to the reaction $e^- \rightarrow K^- + \nu$. Similarly, the condition for π^- condensation is obtained by replacing μ^{K^-} with μ^{π^-} in equation (4.11), that is,

$$\mu^{\pi^-} = \mu^e\,. \tag{4.12}$$

The particle reaction underlying π^- condensation is $n \rightarrow p + \pi^-$.

As a second example, let us proceed to densities where hyperon population is expected to set in. According to what has been said above in connection with equation (4.6), hyperons are energetically favored at densities for which the threshold condition $\omega^B(p_{F_B}) \equiv \mu^B \geq \omega^B(0)$ has a real solution. In case of the Λ hyperon, for instance, this condition reads $\omega^\Lambda(p_{F_\Lambda}) \equiv \mu^\Lambda \geq \omega^\Lambda(0)$. Since Λ is electrically neutral, equation (4.5) reduces to $\mu^\Lambda = \mu^n$. So the threshold condition for a gas of free, relativistic Λ's is given by $m_n^2 + p_{F_n}^2 \geq m_\Lambda^2$, from which it follows that for neutron Fermi momenta $p_{F_n} \geq \sqrt{m_\Lambda^2 - m_n^2}$ neutrons begin to leak into the Λ potential pot.

Equations (4.11) and (4.12) specifying the onset of meson condensation in neutron star matter are special cases of the general relation (4.5) applied to the possible formation of meson condensates in neutron stars. For mesons $q_B = 0$ and so one obtains from (4.5) as threshold condition for the onset of meson condensation

$$\mu^M = -\,q_M^{\rm el}\,\mu^e, \quad \text{where } M = \pi,\, K\,, \tag{4.13}$$

with $q_M^{\rm el} = -1$ for the negatively charged mesons π^- and K^-. The condensation of mesons other than π^- and K^- is strongly unfavored because of electric charge reasons. A brief description of the microphysical processes associated with the condensation of mesons can be found in section 7.9.

We recall that *electric charge neutrality* of neutron star matter is an absolute constraint on the composition of such matter, since it is imposed by a long-range force. If the net charge on a star is Ze and an additional charged particle is added, stability requires that the particle's gravitational attraction to the star dominates over the Coulomb repulsion [61], that is,

$$\frac{G\,(Am)\,m}{R} \geq \frac{Z\,e^2}{R}\,. \tag{4.14}$$

Here m denotes the mass of a nucleon, A is the star's baryon number, Am the star's mass, and R its radius. For net positive (proton) or negative (electron) charge, this means that

$$\frac{Z}{A} = \begin{cases} 10^{-36} & \text{(positive charge)}, \\ 10^{-39} & \text{(negative charge)}. \end{cases} \tag{4.15}$$

Therefore the electric charge density must be effectively zero. Otherwise the Coulomb repulsion would always win over gravity and the extra particle would not be bound to the star.

Accordingly, the particle populations must arrange themselves in such a way as to minimize the energy density in accord with electric charge neutrality and chemical equilibrium. As we shall see later, at normal nuclear matter density, neutron star matter consists primarily of neutrons and a small admixture of protons. The positive charge carried by the protons is neutralized at each density by a corresponding number of electrons. Since the number of protons increases with density, so does the number

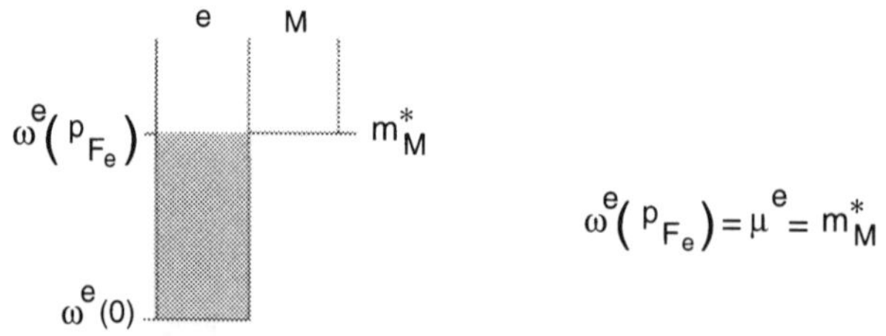

Figure 4.4. Condition for the onset of meson condensation in chemically equilibrated matter. For $\mu^e \equiv \omega^e(p_{F_e}) \geq m_M^*$ high-energy electrons can be replaced with negatively charged mesons ($M = \pi^-, K^-$), which, being bosons, can condense collectively into the ground state.

of electrons and thus μ^e. This trend is followed until μ^e is equal to the effective meson mass in matter, $\mu^e = m_M^*$. As illustrated in figure 4.4, at this density it may become energetically more favorable to fulfill the constraint of electric charge neutrality by means of populating negatively charged meson states rather than keep increasing the number of electrons. Being bosons, any number of mesons can go into the same quantum state, in contrast to the electrons whose Fermi energy would build up monotonically with density, which is energetically less favorable of course. Of course there may be other sources like the Σ^- or deconfined d and s quarks that deliver negative electric charge to the system before the meson threshold is reached. If so the chemical electron potential will saturate (rather than increase monotonically) before the threshold for meson condensation is reached and then drop with increasing density, possibly ruling out a condensate.

Whether or not mesons actually condense depends decisively on the density dependence of the effective meson mass m_M^* in neutron star matter, since, as outlined just above, to trigger meson condensation the meson energy must cross the electron chemical potential. In the case of K^- mesons, for instance, only then can highly degenerated electrons change through the reaction

$$e^- \longrightarrow K^- + \nu\,. \tag{4.16}$$

Once this reaction becomes possible in a neutron star, the star can lower its energy by replacing electrons with K^- mesons. How does the mass of the K^- in dense matter behave? It is known that the K^- has a mass of $m_{K^-} = 495$ MeV in the middle of the ^{56}Ni nucleus. On the other hand, the study of kaonic atoms indicates that the kaon appears to be bound by -200 ± 20 MeV [300]. That is, the attraction from nuclear matter at a density ρ_0 seems to be sufficient to greatly lower the mass of the kaon. The initial value of -200 MeV however turned out to be too large in magnitude, as has become clear from an analysis of the high quality K^- kinetic energy spectra extracted from Ni+Ni collisions at SIS (Schwerionen Synchrotron) energies, measured by the KaoS collaboration [301] at the Gesellschaft für Schwerionenforschung (GSI). An analysis of the Kaos data shows that the attraction at nuclear matter density is somewhat less, around -100 MeV [302, 255, 256], but nonetheless sufficient to bring the in-medium K^- mass down to $m_{K^-}^* \sim 200$ MeV at $\rho \sim 3\,\rho_0$, according to the relation [254]

$$m_{K^-}^* \sim m_{K^-}\left(1 - 0.2\,\frac{\rho}{\rho_0}\right) \tag{4.17}$$

for neutron rich matter. A value of $m_{K^-}^* \sim 200$ MeV lies in the vicinity of the value of the electron chemical potential in neutron star matter, for which competing theories predict values in the range from $\mu^e \sim 120$ to 220 MeV [61, 62, 79]. Whether or not the conditions for the transformation of electrons to K^- mesons are fulfilled in the dense pressure environment inside neutron stars remains to be seen. The extension of the K^-–nucleus interaction from ^{56}Ni to matter at densities $\rho \sim 3\,\rho_0$ surely is quite an extrapolation, which upcoming relativistic heavy-ion experiments may or may not confirm [254]. Concerning the present theoretical status of dense matter calculations, we have repeatedly mentioned that there exists a number of unresolved open issues that enter in such calculations. This makes it very hard to come up with stringent quantitative predictions [303].

Finally we mention the possibility of the transition of confined hadronic matter into quark matter in the high-pressure environment of neutron stars. Quarks have baryon number $q_Q = \frac{1}{3}$. Equation (4.5) thus leads for the quark chemical potentials to

$$\mu^Q = \frac{1}{3}\,\mu^n - \frac{2}{3}\,\mu^e\,, \quad \text{if } Q = u, c,$$
$$\mu^Q = \frac{1}{3}\,\mu^n + \frac{1}{3}\,\mu^e\,, \quad \text{if } Q = d, s. \tag{4.18}$$

Modelling the transition of confined hadronic matter to quark matter in a neutron star as a first-order one, then, according to Gibbs criteria, phase equilibrium will exist if the pressure of both phases is equal, that is,

$$P_H(\{\chi\}, \mu^n, \mu^e) = P_Q(\mu^n, \mu^e)\,. \tag{4.19}$$

Note that because of relation (4.18) no additional unknowns enter other than the two independent chemical potentials μ^e and μ^n and the unknown matter fields, $\{\chi\}$, when solving equation (4.19) for the region of phase equilibrium between hadronic matter and quark matter. To find this region, one has to solve (4.19) in the three-space spanned by these two chemical potentials and the pressure. This has been done for the first time only a few years ago by Glendenning [88]. We shall come back to this issue in greater detail in chapter 8.

Chapter 5

Relativistic field-theoretical description of neutron star matter

5.1 Choice of Lagrangian

According to what has been outlined in chapter 4, neutron star matter at supernuclear densities constitutes a very complex many-body system whose fundamental constituents will be protons, neutrons, hyperons, eventually even more massive baryons like the Δ, possibly an admixture of u, d, s quarks (other quark flavors are too massive to become populated in stable neutron stars), and eventually condensed mesons. The dynamics of the baryonic degrees of freedom, summarized in table 5.1, is described by a Lagrangian of the following type [79]:

$$\begin{aligned}\mathcal{L}(x) = & \sum_{B=p,n,\Sigma^{\pm,0},\Lambda,\Xi^{0,-},\Delta^{++,+,0,-}} \mathcal{L}_B^0(x) \\ & + \sum_{M=\sigma,\omega,\pi,\rho,\eta,\delta,\phi} \Big\{ \mathcal{L}_M^0(x) + \sum_{B=p,n,\ldots,\Delta^{++,+,0,-}} \mathcal{L}_{BM}(x) \Big\} \\ & + \mathcal{L}^{(\sigma^4)}(x) + \sum_{L=e^-,\mu^-} \mathcal{L}_L(x) . \end{aligned} \tag{5.1}$$

Summed are all baryon states B whose thresholds will be reached in the dense interiors of neutron stars. The summation also includes the Δ resonance whose appearance is favored by many-body theories that go beyond the relativistic mean-field approximation, as we shall see later in section 7. One may wonder to what extent hyperons and the Δ resonance may each be treated as a *separate* species. Such a treatment, however, seems to be well vindicated not only in finite nuclei, but also in nuclear matter at as high a density as encountered in neutron stars [304, 305]. Constraints, if any, due to anti-symmetrization between nucleons and the

Table 5.1. Masses (m_B) and quantum numbers (spin, J_B; isospin, I_B; strangeness, S_B; hypercharge, Y_B; third component of isospin, I_{3B}; electric charge, $q_B^{\rm el}$) of those baryons that have been found to become populated in the cores of neutron stars.

Baryon (B)	m_B (MeV)	J_B	I_B	S_B	Y_B	I_{3B}	$q_B^{\rm el}$
n	939.6	1/2	1/2	0	1	−1/2	0
p	938.3	1/2	1/2	0	1	1/2	1
Σ^+	1189	1/2	1	−1	0	1	1
Σ^0	1193	1/2	1	−1	0	0	0
Σ^-	1197	1/2	1	−1	0	−1	−1
Λ	1116	1/2	0	−1	0	0	0
Ξ^0	1315	1/2	1/2	−2	−1	1/2	0
Ξ^-	1321	1/2	1/2	−2	−1	−1/2	−1
Δ^{++}	1232	3/2	3/2	0	1	3/2	2
Δ^+	1232	3/2	3/2	0	1	1/2	1
Δ^0	1232	3/2	3/2	0	1	−1/2	0
Δ^-	1232	3/2	3/2	0	1	−3/2	−1

possible nucleon content of the resonances were found to be negligibly small. Besides that, the number density of a given nucleon resonance in a large assembly of nucleons and pions was found to obey the usual equation of thermal equilibrium $\mu^B = \mu^\pi + \mu^N$ in terms of chemical potentials. (The equilibrium concept and, thus, chemical potentials will be introduced in chapter 4.) Finally, we mention that the largeness of resonance widths would not affect their elementarity. Its effects may be interpreted as part of the interaction between the resonance species and the nucleon or meson species [305].

The interaction between the baryons is described by the exchange of mesons with masses up to about 1 GeV, depending on the many-body approximation. At the level of the simplest approximation – the relativistic mean-field (or Hartree) approximation – these are the σ, ω and ρ mesons only [78, 79, 84, 92]. Because of their spin and parity quantum numbers, which are listed in table 5.2, this approximation is also referred to as scalar–vector–isovector theory. The relativistic Hartree–Fock (RHF) approximation differs from the mean-field theory because of the exchange (Fock) term, which, by definition, is absent in the mean-field theory. For that reason the π meson, which contributes only to the exchange term of the self-energy, does not contribute to the self-energy computed at the relativistic mean-field level. The whole set of mesons summed in (5.1) is generally employed in the construction of relativistic meson-exchange models for the nucleon–nucleon interaction, of which the

Table 5.2. Mesons and their quantum numbers [306]. The entries are: spin J_M, parity π, isospin I_M, and mass m_M of meson M.

Meson (M)	J_M^π	I_M	Coupling	Mass (MeV)	Dominant decay mode
σ	0^+	0	scalar	550	–
ω	1^-	0	vector	783	3π
$\pi^\pm$	0^-	1	pseudovector	140	$\mu^\pm\nu$
π^0	0^-	1	pseudovector	135	$\gamma\gamma$
ρ	1^-	1	vector	769	2π
η	0^-	0	pseudovector	549	$\gamma\gamma$, $3\pi^0$
δ	0^+	1	scalar	983	$\eta\pi$, $K\bar{K}$
ϕ	1^-	0	vector	1020	K^+K^-
K^+	0^-	1/2	pseudovector	494	$\mu^+\nu$, $\pi^+\pi^0$
K^-	0^-	1/2	pseudovector	494	$\mu^-\nu$, $\pi^-\pi^0$

Bonn meson exchange model [119, 307] is a particularly sophisticated representative. Among other features, it not only accounts for single-meson exchange processes among the nucleons but also for explicit 2π-exchange contributions, involving the Δ isobar in intermediate states, and $\pi\rho$-exchange diagrams, which replace to a large extent the fictitious σ exchange used in former one-boson-exchange interactions OBEP [104, 119, 120, 308]. Such potentials can only be used in many-body methods that account for dynamical two-nucleon correlations calculated from the nucleon–nucleon scattering matrix in matter (**T**-matrix), as is the case for the relativistic Brueckner–Hartree–Fock (RBHF) approximation. In contrast to RBHF, the Hartree and Hartree–Fock approximations account only for what are called statistical correlations.

Models for the equation of state derived in the framework of the linear mean-field model are extremely stiff and cannot be reconciled with the empirical value for the incompressibility [92]. This can be cured by either introducing derivative couplings in the Lagrangian, which shall be done in section 7.3, or by means of adding nonlinear terms to it. Here we shall follow the suggestion of Boguta and Bodmer [309] and Boguta and Rafelski [310] and add cubic and quartic self-interactions of the σ field to the Lagrangian.

The equations of motion of the baryon and meson fields, which shall be derived below, are to be solved subject to the two constraints imposed on neutron star matter, outlined in section 4.2. These are electric charge neutrality and β-equilibrium. Both constraints imply the presence of leptons in neutron star matter. Mathematically we account for them by adding the Lagrangian of free relativistic leptons, $\mathcal{L}_L$, to the system's total Lagrangian, given in equation (5.1).

The individual terms in equation (5.1) will be given next. We begin with the Lagrangians of free baryon and meson fields which are given by

$$\mathcal{L}^0_B(x) = \bar{\psi}_B(x)\ (\mathrm{i}\gamma^\mu\partial_\mu\ -\ m_B)\ \psi_B\,, \tag{5.2}$$

$$\mathcal{L}^0_\sigma(x) = \frac{1}{2}\left(\partial^\mu\sigma(x)\,\partial_\mu\sigma(x) - m^2_\sigma\,\sigma^2(x)\right)\,, \tag{5.3}$$

$$\mathcal{L}^0_\omega(x) = -\ \frac{1}{4}F^{\mu\nu}(x)\,F_{\mu\nu}(x) + \frac{1}{2}\,m^2_\omega\,\omega^\nu(x)\,\omega_\nu(x)\,, \tag{5.4}$$

$$\mathcal{L}^0_\pi(x) = \frac{1}{2}\left(\partial^\mu\boldsymbol{\pi}(x)\cdot\partial_\mu\boldsymbol{\pi}(x)\ -\ m^2_\pi\,\boldsymbol{\pi}(x)\cdot\boldsymbol{\pi}(x)\right)\,, \tag{5.5}$$

$$\mathcal{L}^0_\rho(x) = -\ \frac{1}{4}\boldsymbol{G}^{\mu\nu}(x)\cdot\boldsymbol{G}_{\mu\nu}(x)\ +\ \frac{1}{2}\,m^2_\rho\,\boldsymbol{\rho}^\mu(x)\cdot\boldsymbol{\rho}_\mu(x)\,. \tag{5.6}$$

The interaction Lagrangians read as follows,

$$\mathcal{L}_{B\sigma}(x) = -\sum_B(\mathbf{1}\,g_{\sigma B})\,\bar{\psi}_B(x)\,\sigma(x)\,\psi_B(x)\,, \tag{5.7}$$

$$\begin{aligned}\mathcal{L}_{B\omega}(x) = &-\sum_B g_{\omega B}\,\bar{\psi}_B(x)\,\gamma^\mu\omega_\mu(x)\,\psi_B(x)\\ &-\sum_B \frac{f_{\omega B}}{4\,m_B}\,\bar{\psi}_B(x)\,\sigma^{\mu\nu}\,F_{\mu\nu}(x)\,\psi_B(x)\,,\end{aligned} \tag{5.8}$$

$$\mathcal{L}_{B\pi}(x) = -\sum_B \frac{f_{\pi B}}{m_\pi}\,\bar{\psi}_B(x)\,\gamma^5\gamma^\mu\,\left(\partial_\mu\boldsymbol{\tau}\cdot\boldsymbol{\pi}(x)\right)\,\psi_B(x)\,, \tag{5.9}$$

$$\begin{aligned}\mathcal{L}_{B\rho}(x) = &-\sum_B g_{\rho B}\,\bar{\psi}_B(x)\,\gamma^\mu\boldsymbol{\tau}\cdot\boldsymbol{\rho}_\mu(x)\,\psi_B(x)\\ &-\sum_B \frac{f_{\rho B}}{4\,m_B}\,\bar{\psi}_B(x)\,\sigma^{\mu\nu}\,\boldsymbol{\tau}\cdot\boldsymbol{G}_{\mu\nu}(x)\,\psi_B(x)\,.\end{aligned} \tag{5.10}$$

The quantity **1** in (5.7) denotes the unity matrix in Dirac space (cf. (5.132)). The second terms in (5.8) and (5.10) describe so-called tensor couplings, all other couplings are of standard Yukawa type. For the π meson we choose the pseudovector coupling scheme, since pseudoscalar coupling is known to lead to several inconsistencies when applied to nuclear matter calculations [92]. These originate from the circumstance that pseudoscalar coupling gives so much repulsion that the Hartree–Fock approximation becomes inadequate for the description of the properties of nuclear matter. As an example, the nucleon self-energy in nuclear matter calculated for the pseudoscalar coupling is about 40 times larger than for the pseudovector case. This leads to a ground state configuration at saturation density which is a Fermi-shell state rather than a Fermi sphere. The pseudovector coupling, on the other hand, is much weaker. The lowest energy configuration calculated for it is again a Fermi sphere. Finally we note that the pseudoscalar coupling is equivalent to pseudovector coupling for on-shell nucleons if one uses a pseudovector coupling constant, $f_{\pi N}$, which satisfies the so-called equivalence principle $g_{\pi N}/(2m_N) = f_{\pi N}/m_\pi$ in free space, that is, $f_{\pi N}^2/(4\pi) \approx 0.08$ [92]. It should be kept in mind, however, that the equivalence principle applied to dense nuclear matter can only be regarded as a guideline, since it may not be correct in dense matter.

As a final point on the coupling 'constants', we note that they may plausibly change with increasing density and/or temperature of the matter. The pion constant $f_{\pi N}$, for instance, is expected to decrease in the nuclear medium [311], according to the Brown–Rho scaling. The incorporation of density and/or temperature effects into fully self-consistent dense matter calculations constitutes an extremely cumbersome problem that has not been solved yet, though significant progress has been made in recent years toward accomplishing this problem [303, 312, 313]. This is different for the influence of such effects on the meson and baryon masses in dense matter [303, 312, 313], the latter of which will be discussed in great detail immediately below.

After these remarks, let us turn back to the field-theoretical description of dense matter. The still undefined field tensors $F^{\mu\nu}$ and $\boldsymbol{G}^{\mu\nu}$ are given by

$$F^{\mu\nu}(x) = \partial^\mu \, \omega^\nu(x) - \partial^\nu \, \omega^\mu(x) \,, \tag{5.11}$$

$$\boldsymbol{G}^{\mu\nu}(x) = \partial^\mu \, \boldsymbol{\rho}^\nu(x) - \partial^\nu \, \boldsymbol{\rho}^\mu(x) \,. \tag{5.12}$$

The latter tensor is of vectorial nature because the ρ meson is a vector in isospin space (cf. table 5.2). The quantity $\sigma^{\mu\nu}$ is an abbreviation for the commutator made up of a pair of γ matrices,

$$\sigma^{\mu\nu} = \frac{\mathrm{i}}{2} \, [\gamma^\mu, \gamma^\nu] \,, \tag{5.13}$$

from which one reads off that $\sigma^{\nu\mu} = -\sigma^{\mu\nu}$. The γ matrices are defined in appendix A, where an overview of some of their properties can be found too. The above mentioned cubic and quartic self-interactions of the σ field are described by a Lagrangian of the form

$$\mathcal{L}^{(\sigma^4)}(x) = -\frac{1}{3} \, m_N \, b_N \, \{g_{\sigma N} \, \sigma(x)\}^3 - \frac{1}{4} \, c_N \, \{g_{\sigma N} \, \sigma(x)\}^4 \,. \tag{5.14}$$

Finally the Lagrangian of free leptons reads

$$\mathcal{L}_L(x) = \bar{\psi}_L(x) \, (\mathrm{i} \, \gamma^\mu \partial_\mu \, - \, m_L) \, \psi_L(x) \,. \tag{5.15}$$

Above we have restricted ourselves to listing the Lagrangians of σ, ω, π, and ρ mesons only. The Lagrangians of δ, ϕ, and η mesons, which enter in one-boson-exchange interactions in addition, will not be given explicitly. Their form can be easily inferred however by looking at the quantum numbers of these mesons given in table 5.2. This reveals, for instance, that the δ meson has the same spin and parity as the σ meson, namely 0^+. Apart from isopin, which requires multiplication of the δ field with the Pauli matrix $\boldsymbol{\tau}$ $[= (\tau^1, \tau^2, \tau^3)]$ in the interaction Lagrangian, the Lagrangians of the δ are then obtained from equations (5.3) and (5.7) by replacing σ with δ in (5.3), and σ with $\boldsymbol{\tau}\delta$ in (5.7). Similarly, the Lagrangians of ϕ and η mesons are obtainable from those of ω and π mesons by replacing ω with ϕ, and π with η, respectively.

The baryon fields obey the anti-commutation relations

$$\begin{aligned} \{\bar{\psi}_\varsigma(x^0, \boldsymbol{x}), \psi_\varsigma(x^0, \boldsymbol{x}')\} &= \gamma^0_{\varsigma\varsigma'}\, \delta^3(\boldsymbol{x} - \boldsymbol{x}')\,, \\ \{\bar{\psi}_\varsigma(x^0, \boldsymbol{x}), \bar{\psi}_\varsigma(x^0, \boldsymbol{x}')\} &= \{\psi_\varsigma(x^0, \boldsymbol{x}), \psi_\varsigma(x^0, \boldsymbol{x}')\} = 0\,. \end{aligned} \tag{5.16}$$

The commutator relations of the scalar meson field, $\sigma(x)$, read

$$\begin{aligned} [\Pi_\sigma(x^0, \boldsymbol{x}), \sigma(x^0, \boldsymbol{x}')] &= -\,\mathrm{i}\,\delta^3(\boldsymbol{x} - \boldsymbol{x}')\,, \\ [\Pi_\sigma(x^0, \boldsymbol{x}), \Pi_\sigma(x^0, \boldsymbol{x}')] &= [\sigma(x^0, \boldsymbol{x}), \sigma(x^0, \boldsymbol{x}')] = 0\,, \end{aligned} \tag{5.17}$$

where $\Pi_\sigma(x)$ denotes the conjugate momentum of the σ field,

$$\Pi_\sigma(x) \equiv \frac{\partial \mathcal{L}}{\partial \dot{\sigma}(x)} = \dot{\sigma}(x)\,. \tag{5.18}$$

The fields Π_σ and σ commute with the baryon field operators,

$$\begin{aligned} [\sigma(x^0, \boldsymbol{x}), \psi(x^0, \boldsymbol{x}')] &= [\Pi_\sigma(x^0, \boldsymbol{x}), \psi(x^0, \boldsymbol{x}')] = 0\,, \\ [\sigma(x^0, \boldsymbol{x}), \bar{\psi}(x^0, \boldsymbol{x}')] &= [\Pi_\sigma(x^0, \boldsymbol{x}), \bar{\psi}(x^0, \boldsymbol{x}')] = 0\,. \end{aligned} \tag{5.19}$$

Since the interaction Lagrangians of the π, ω and ρ mesons contain derivatives of these fields, as can be seen in equations (5.8) to (5.10), the corresponding conjugate momenta possess a somewhat more complicated structure than (5.18), derived for the σ field. One obtains ($j = 1, 2, 3$):

$$\boldsymbol{\Pi}_\pi(x) = \dot{\boldsymbol{\pi}}(x) - \sum_B \frac{f_{\pi B}}{m_\pi}\, \bar{\psi}_B(x)\, \gamma^0\, \gamma^5\, \boldsymbol{\tau}\, \psi_B(x)\,, \tag{5.20}$$

$$\Pi_{\omega_j}(x) = F^{j0}(x) + \sum_B \frac{f_{\omega B}}{2\, m_B}\, \bar{\psi}_B(x)\, \sigma^{j0}\, \psi_B(x)\,, \tag{5.21}$$

$$\boldsymbol{\Pi}_{\rho_j}(x) = \boldsymbol{G}^{j0}(x) + \sum_B \frac{f_{\rho B}}{2\, m_B}\, \bar{\psi}_B(x)\, \boldsymbol{\tau}\, \sigma^{j0}\, \psi_B(x)\,, \tag{5.22}$$

with

$$\Pi_{\omega_j}(x) \equiv \frac{\partial \mathcal{L}}{\partial(\partial_0\, \omega_j)}\,. \tag{5.23}$$

Because of the γ matrices in equations (5.20) through (5.22), the quantities $\dot{\boldsymbol{\pi}}$, G^{j0} and $\boldsymbol{G}^{j0}$ do not commute with the nucleon field operators any longer, as was the case in (5.19) for $\dot{\sigma}$. One gets instead

$$[\dot{\boldsymbol{\pi}}(x^0, \boldsymbol{x}), \psi_{B'\zeta'}(x^0, \boldsymbol{x}')] = -\frac{f_{\pi B'}}{m_\pi}\, \delta^3(\boldsymbol{x} - \boldsymbol{x}')\,(\gamma_5 \otimes \boldsymbol{\tau})_{\zeta'\zeta}\, \psi_{B\zeta}(x)\,, \quad (5.24)$$

and for the ω meson

$$[F^{j0}(x^0, \boldsymbol{x}), \psi_{B'\zeta'}(x^0, \boldsymbol{x}')] = -\mathrm{i}\,\frac{f_{\omega B'}}{2\, m_{B'}}\, \delta^3(\boldsymbol{x} - \boldsymbol{x}')\,(\gamma^j)_{\zeta'\zeta}\, \psi_{B'\zeta}(x)\,, \quad (5.25)$$

$$[F^{j0}(x^0, \boldsymbol{x}), \bar{\psi}_{B'\zeta'}(x^0, \boldsymbol{x}')] = -\mathrm{i}\,\frac{f_{\omega B'}}{2\, m_{B'}}\, \delta^3(\boldsymbol{x} - \boldsymbol{x}')\, \bar{\psi}_{B'\zeta}(x)\,(\gamma^j)_{\zeta\zeta'}\,. \quad (5.26)$$

The corresponding expressions for the ρ meson follow from (5.25) and (5.26) by replacing F^{j0} with $\boldsymbol{G}^{j0}$, and σ^{j0} with $\sigma^{j0} \otimes \boldsymbol{\tau}$ etc.

5.2 Field equations

In this section we shall derive the equations of motion for the numerous particle fields from the Euler–Lagrange equation, which is a condition on the Lagrangian which guarantees that the action I, defined as

$$I \equiv \int \mathrm{d}^4x\; \mathcal{L}(\chi(x), \partial_\mu \chi(x)) \quad (5.27)$$

is an extremum, that is, $\delta I = 0$. We only consider the case where $\mathcal{L}$ depends explicitly on the matter fields and their derivatives, χ and $\partial_\mu \chi$ respectively, but not on the coordinates x^μ themselves. Writing out the variation of (5.27) explicitly gives

$$\int \mathrm{d}^4x\; \{\mathcal{L}(\chi + \delta\chi, \partial_\mu\chi + \delta\, \partial_\mu\chi) - \mathcal{L}(\chi, \partial_\mu\chi)\} = 0\,, \quad (5.28)$$

where the replacements

$$\begin{aligned} \chi(x) \rightarrow \chi'(x) &= \chi(x) + \delta\chi(x)\,, \\ \partial_\mu\chi(x) \rightarrow \partial_\mu\chi'(x) &= \partial_\mu\chi(x) + \partial_\mu\delta\chi(x)\,, \end{aligned} \quad (5.29)$$

denote variations of the fields. Taylor expansion of the first integrand in (5.28) leads to

$$\mathcal{L}(\chi + \delta\chi, \partial_\mu\chi + \delta\, \partial_\mu\chi) = \mathcal{L}(\chi, \partial_\mu\chi) + \frac{\partial\mathcal{L}}{\partial\chi}\, \delta\chi + \frac{\partial\mathcal{L}}{\partial(\partial_\mu\chi)}\, \delta(\partial_\mu\chi)\,. \quad (5.30)$$

Substituting (5.30) into (5.28) and making use of

$$\delta(\partial_\mu\chi) = \partial_\mu(\chi + \delta\chi) - \partial_\mu\chi = \partial_\mu\,(\delta\chi) \quad (5.31)$$

gives

$$\int \mathrm{d}^4x \left\{ \frac{\partial \mathcal{L}}{\partial \chi} \delta\chi + \frac{\partial \mathcal{L}}{\partial(\partial_\mu \chi)} \partial_\mu(\delta\chi) \right\} = 0\,. \tag{5.32}$$

Upon integrating the second term by parts, one obtains

$$\int \mathrm{d}^4x \left\{ \left[\frac{\partial \mathcal{L}}{\partial \chi} - \partial_\mu \left(\frac{\partial \mathcal{L}}{\partial(\partial_\mu \chi)} \right) \right] \delta\chi \right\} = 0\,, \tag{5.33}$$

provided the contribution from the surface of spacetime may be dropped. Thus, for arbitrary variations of the fields $\delta\chi$, the condition for the action to be stationary ($\delta I = 0$) reads

$$\partial_\mu \left(\frac{\partial \mathcal{L}}{\partial(\partial_\mu \chi)} \right) - \frac{\partial \mathcal{L}}{\partial \chi} = 0\,. \tag{5.34}$$

This is the Euler–Lagrange equation for given fields χ, which, in our case, are the fermion and boson fields ψ_B, ψ_L and $\sigma, \omega, \pi, \rho, \eta, \delta, \phi$.

We begin with deriving the equation of motion for the baryon fields ψ_B from (5.34). Since (5.34) does not contain derivatives of $\bar{\psi}_B$, the first term of the Euler–Lagrange equation gives no contribution. The second term leads to

$$\begin{aligned}
\frac{\partial \mathcal{L}}{\partial \bar{\psi}_B} = {} & (\mathrm{i}\gamma^\mu \partial_\mu - m_B)\, \psi_B(x) \; + \; g_{\sigma B}\, \sigma(x)\, \psi_B(x) \\
& - \left\{ g_{\omega B} \gamma^\mu \omega_\mu(x) + \frac{f_{\omega B}}{4 m_B} \sigma^{\mu\nu} F_{\mu\nu}(x) \right\} \psi_B(x) \\
& - \left\{ g_{\rho B}\, \gamma^\mu\, \boldsymbol{\tau} \cdot \boldsymbol{\rho}_\mu(x) + \frac{f_{\rho B}}{4\, m_B}\, \sigma^{\mu\nu}\, \boldsymbol{\tau} \cdot \boldsymbol{G}_{\mu\nu}(x) \right\} \psi_B(x) \\
& - \frac{f_{\pi B}}{m_\pi}\, \gamma^\mu \gamma^5\, \left(\partial_\mu\, \boldsymbol{\tau} \cdot \boldsymbol{\pi}(x) \right)\, \psi_B(x)\,.
\end{aligned} \tag{5.35}$$

From $\partial\mathcal{L}/\partial\bar{\psi}_B = 0$ one gets as the final result for the inhomogeneous Dirac equation

$$\begin{aligned}
(\mathrm{i}\gamma^\mu \partial_\mu - m_B)\, \psi_B(x) = {} & g_{\sigma B}\, \sigma(x)\, \psi_B(x) \\
& + \left\{ g_{\omega B} \gamma^\mu \omega_\mu(x) + \frac{f_{\omega B}}{4 m_B} \sigma^{\mu\nu} F_{\mu\nu}(x) \right\} \psi_B(x) \\
& + \left\{ g_{\rho B}\, \gamma^\mu\, \boldsymbol{\tau} \cdot \boldsymbol{\rho}_\mu(x) + \frac{f_{\rho B}}{4\, m_B}\, \sigma^{\mu\nu}\, \boldsymbol{\tau} \cdot \boldsymbol{G}_{\mu\nu}(x) \right\} \psi_B(x) \\
& + \frac{f_{\pi B}}{m_\pi}\, \gamma^\mu \gamma^5\, \left(\partial_\mu\, \boldsymbol{\tau} \cdot \boldsymbol{\pi}(x) \right)\, \psi_B(x)\,.
\end{aligned} \tag{5.36}$$

To find the equation of motion for the scalar σ field, we differentiate (5.1) with respect to σ, which leads to

$$\frac{\partial \mathcal{L}}{\partial \sigma} = - \, m_\sigma^2 \sigma - \sum_B g_{\sigma B}\, \bar{\psi}_B \psi_B - m_N\, b_N\, g_{\sigma N}\, (g_{\sigma N} \sigma)^2 - c_N\, g_{\sigma N}\, (g_{\sigma N} \sigma)^3\,. \tag{5.37}$$

The last two terms, which originate from $\mathcal{L}^{(\sigma^4)}$, shall be kept only when solving the equations of motion at the mean-field level. The differentiation of $\mathcal{L}$ with respect to $\partial_\mu\sigma$ is slightly more complicated since the partial derivative carries a covariant four-index. So when differentiating (5.1) with respect to $\partial_\mu\sigma$ we have to make sure that all the relevant partial derivatives are written in covariant form. This is accomplished via the metric tensor of flat spacetime (see appendix A) which allows us to write for a contravariant derivative $\partial^\kappa = g^{\kappa\nu}\partial_\nu$. Bearing this in mind, one readily verifies that

$$\frac{\partial \mathcal{L}}{\partial(\partial_\mu\sigma)} = \frac{1}{2}\,\frac{\partial}{\partial(\partial_\mu\sigma))}\,\{(g^{\kappa\nu}\partial_\nu\sigma)\,(\partial_\kappa\sigma)\} \tag{5.38}$$

$$= \frac{1}{2}\left\{g^{\kappa\nu}\frac{\partial(\partial_\nu\sigma)}{\partial(\partial_\mu\sigma)}\,(\partial_\kappa\sigma) + (g^{\kappa\nu}\partial_\nu\sigma)\,\frac{\partial(\partial_\kappa\sigma)}{\partial(\partial_\mu\sigma)}\right\} \tag{5.39}$$

$$= \frac{1}{2}\,\{g^{\kappa\nu}\,\delta_{\mu\nu}\,\partial_\kappa\sigma + g^{\kappa\nu}\,\delta_{\kappa\mu}\,\partial\sigma\} \tag{5.40}$$

$$= \frac{1}{2}\,\{g^{\kappa\mu}\,\partial_\kappa\sigma + g^{\kappa\nu}\,\partial_\nu\sigma\} = \partial^\mu\sigma\,. \tag{5.41}$$

From (5.39) to (5.40) we have used that $\partial(\partial_\nu\sigma)/\partial(\partial_\mu\sigma)$ only contributes if the subscripts obey $\nu = \mu$. Letting ∂_μ act on (5.41) leads to

$$\partial_\mu\frac{\partial\mathcal{L}}{\partial(\partial_\mu\sigma)} = \partial_\mu\partial^\mu\sigma\,. \tag{5.42}$$

Subtracting (5.37) from (5.42) leads to the equation of motion for the σ field,

$$\begin{aligned}(\partial^\mu\partial_\mu + m_\sigma^2)\,\sigma(x) = &-\sum_B g_{\sigma B}\,\bar\psi_B(x)\psi_B(x) - m_N\,b_N\,g_{\sigma N}\,(g_{\sigma N}\sigma(x))^2 \\ &- c_N\,g_{\sigma N}\,(g_{\sigma N}\sigma(x))^3\,,\end{aligned} \tag{5.43}$$

which constitutes an inhomogeneous Klein–Gordon equation.

To derive the equation of motion for the ω field, we proceed in a similar fashion as just above. The main difference with respect to the σ field arises from the vectorial nature of the ω field. Via the metric tensor, the fields and derivatives are transformed to their covariant or contravariant representations, as the case may be, and as before derivatives like $\partial\omega_{\mu'}/\partial\omega_\mu$ lead to factors of $\delta_{\mu\mu'}$. One then obtains

$$\frac{\partial\mathcal{L}}{\partial\omega_\mu} = -\sum_B g_{\omega B}\,\bar\psi_B\gamma^\mu\psi_B + m_\omega^2\,\omega^\mu\,, \tag{5.44}$$

where use of

$$\frac{\partial}{\partial\omega_\mu}\,(\omega^\nu\omega_\nu) = \left(\frac{\partial}{\partial\omega_\mu}g^{\nu\lambda}\omega_\lambda\right)\omega_\nu + \omega^\nu\frac{\partial\omega_\nu}{\partial\omega_\mu} = 2\,\omega^\mu \tag{5.45}$$

has been made. The other term of the Euler–Lagrange equation gives

$$\begin{aligned}\frac{\partial \mathcal{L}}{\partial(\partial_\lambda \omega_\mu)} &= -\frac{1}{4}\frac{\partial(F^{\kappa\nu}F_{\kappa\nu})}{\partial(\partial_\lambda\omega_\mu)} - \sum_B \frac{f_{\omega B}}{4m_B}\bar{\psi}_B\,\sigma^{\kappa\nu}\frac{\partial\,(F_{\kappa\nu}\,\psi_B)}{\partial(\partial_\lambda\omega_\mu)} \\ &= -\frac{1}{4}\frac{\partial}{\partial(\partial_\lambda\omega_\mu)}\{(\partial^\kappa\omega^\nu - \partial^\nu\omega^\kappa)\,(\partial_\kappa\omega_\nu - \partial_\nu\omega_\kappa)\} \\ &\quad -\sum_B \frac{f_\omega B}{4m_B}\bar{\psi}_B\sigma^{\kappa\nu}\left\{\frac{\partial}{\partial(\partial_\lambda\omega_\mu)}\,(\partial_\kappa\omega_\nu - \partial_\nu\omega_\kappa)\right\}\psi_B\,. \end{aligned} \tag{5.46}$$

Since the partial derivatives in (5.46) lead to

$$\frac{\partial(\partial^\kappa\omega^\nu)}{\partial(\partial_\lambda\omega_\mu)} = \frac{\partial}{\partial(\partial_\lambda\omega_\mu)}\,g^{\kappa\epsilon}\partial_\epsilon g^{\nu\tau}\omega_\tau = g^{\kappa\lambda}\,g^{\nu\mu}\,, \tag{5.47}$$

equation (5.46) can be rewritten as

$$\begin{aligned}\frac{\partial \mathcal{L}}{\partial(\partial_\lambda \omega_\mu)} &= -\frac{1}{4}\,\{(g^{\kappa\lambda}g^{\nu\lambda} - g^{\nu\lambda}g^{\kappa\mu})\,(\partial_\kappa\omega_\nu - \partial_\nu\omega_\kappa) \\ &\quad + (\partial^\kappa\omega^\nu - \partial^\nu\omega^\kappa)\,(\delta_{\kappa\lambda}\delta_{\nu\mu} - \delta_{\nu\lambda}\delta_{\kappa\mu})\} \\ &\quad - \sum_B \frac{f_{\omega B}}{4m_B}\,\bar{\psi}_B\,\sigma^{\kappa\nu}\,(\delta_{\kappa\lambda}\delta_{\nu\mu} - \delta_{\nu\lambda}\delta_{\kappa\mu})\,\psi_B \end{aligned} \tag{5.48}$$

$$= \partial^\mu\omega^\lambda - \partial^\lambda\omega_\mu - \sum_B \frac{f_{\omega B}}{4m_B}\,\bar{\psi}_B\left(\sigma^{\lambda\mu} - \sigma^{\mu\lambda}\right)\psi_B\,. \tag{5.49}$$

The quantity $\sigma^{\lambda\mu}$ in (5.49) is antisymmetric with respect to interchanging λ and μ, which follows readily from (5.13) as

$$\sigma^{\lambda\mu} = \frac{\mathrm{i}}{2}\left[\gamma^\lambda,\gamma^\mu\right] = \frac{\mathrm{i}}{2}\left(\gamma^\lambda\gamma^\mu - \gamma^\mu\gamma^\lambda\right) = -\frac{\mathrm{i}}{2}\left[\gamma^\mu,\gamma^\lambda\right] = -\sigma^{\mu\lambda}\,. \tag{5.50}$$

This enables us to write equation (5.49) as

$$\frac{\partial \mathcal{L}}{\partial(\partial_\lambda \omega_\mu)} = \partial^\mu\omega^\lambda - \partial^\lambda\omega_\mu - 2\sum_B \frac{f_{\omega B}}{4m_B}\bar{\psi}_B\sigma^{\lambda\mu}\psi_B\,. \tag{5.51}$$

Combining (5.44) and (5.51) gives for the equation of motion of the ω field

$$\begin{aligned}\partial^\mu F_{\mu\nu}(x) + m_\omega^2\,\omega_\nu(x) = \sum_B \Big\{ & g_{\omega B}\,\bar{\psi}_B(x)\gamma_\nu\psi_B(x) \\ & - \frac{f_{\omega B}}{2m_B}\,\partial^\mu\left(\bar{\psi}_B(x)\sigma_{\mu\nu}\psi_B(x)\right)\Big\}\,, \end{aligned} \tag{5.52}$$

which constitutes an inhomogeneous Proca equation.

The equation of motion of ρ mesons is similar to (5.52). The only differences originate from the isovectorial nature of the ρ, which has $I_\rho = 1$

(table 5.2), as opposed to the ω meson which is an isoscalar. This manifests itself in the occurrence of the Pauli isopin-matrix $\boldsymbol{\tau}$ in the equation of motion for the ρ meson,

$$\partial^\mu \boldsymbol{G}_{\mu\nu}(x) + m_\rho^2\, \boldsymbol{\rho}_\nu(x) = \sum_B \Big\{ g_{\rho B}\, \bar{\psi}_B(x) \boldsymbol{\tau} \gamma_\nu \psi_B(x) - \frac{f_{\rho B}}{2 m_B}\, \partial^\lambda \left(\bar{\psi}_B(x) \boldsymbol{\tau} \sigma_{\mu\nu} \psi_B(x) \right) \Big\} . \quad (5.53)$$

The still missing meson, whose equation of motion will be derived next, is the pion. Differentiating $\mathcal{L}$ with respect to π leads to

$$\frac{\partial \mathcal{L}}{\partial \boldsymbol{\pi}} = -\, m_\pi^2\, \boldsymbol{\pi} , \quad (5.54)$$

where use of the derivative of the scalar product $\boldsymbol{\pi} \cdot \boldsymbol{\pi} = \sum_i \pi^i \pi^i$ was made, from which one calculates

$$\frac{\partial}{\partial \pi^j} \left(\pi^i \pi^i \right) = 2\, \delta_{ij}\, \pi^i = 2\, \pi^j . \quad (5.55)$$

With the aid of the metric tensor, which, as before, is being used to shuffle indices up or down, one finds

$$\frac{\partial \mathcal{L}}{\partial(\partial_\mu \boldsymbol{\pi})} = \frac{1}{2} \frac{\partial}{\partial(\partial_\mu \boldsymbol{\pi})} \left(g^{\nu\lambda} \partial_\lambda \boldsymbol{\pi} \cdot \partial_\nu \boldsymbol{\pi} \right) - \sum_B \frac{f_{\pi B}}{m_\pi} \bar{\psi}_B \gamma^5 \gamma^\mu \boldsymbol{\tau} \psi_B$$
$$= \partial^\mu \boldsymbol{\pi} - \sum_B \frac{f_{\pi B}}{m_\pi} \bar{\psi}_B \gamma^5 \gamma^\mu \boldsymbol{\tau} \psi_B . \quad (5.56)$$

Letting ∂_μ act on (5.56) gives

$$\partial_\mu \frac{\partial \mathcal{L}}{\partial(\partial_\mu \boldsymbol{\pi})} = \partial_\mu \partial^\mu \boldsymbol{\pi} - \sum_B \frac{f_{\pi B}}{m_\pi} \partial_\mu \left(\bar{\psi}_B \gamma^5 \gamma^\mu \boldsymbol{\tau} \psi_B \right) , \quad (5.57)$$

which, combined with (5.54), leads to the equation of motion for the pion field. It is of the form

$$\left(\partial^\mu \partial_\mu + m_\pi^2 \right) \boldsymbol{\pi}(x) = \sum_B \frac{f_{\pi B}}{m_\pi}\, \partial^\mu \left(\bar{\psi}_B(x)\, \gamma_5\, \gamma_\mu\, \boldsymbol{\tau}\, \psi_B(x) \right) . \quad (5.58)$$

The equations of motion of all meson fields other than those already discussed above possess a mathematical structure that, depending on the meson's quantum nature which can be inferred from table 5.2, coincide with one of the above equations of motion. The equation of motion of the δ meson, for instance, coincides with the one of the σ meson except for the nonlinear self-interactions and the isopin. Subject to these modifications one obtains

$$(\partial^\mu \partial_\mu + m_\delta^2)\,\boldsymbol{\delta}(x) = \sum_B g_{\delta B}\,\bar{\psi}_B(x)\,\boldsymbol{\tau}\,\psi_B(x)\,. \tag{5.59}$$

The equations of motion of ϕ and η meson fields are given by

$$\partial^\mu F_{\mu\nu} + m_\phi^2\,\phi_\nu = \sum_B \Big\{ g_{\phi B}\,\bar{\psi}_B \gamma_\nu \psi_B - \frac{f_{\phi B}}{2m_B}\,\partial^\mu\left(\bar{\psi}_B \sigma_{\mu\nu} \psi_B\right) \Big\}, \tag{5.60}$$

and

$$\left(\partial^\mu \partial_\mu + m_\eta^2\right)\,\eta(x) = \sum_B \frac{f_{\eta B}}{m_\eta}\,\partial^\mu\left(\bar{\psi}_B(x)\,\gamma_5\,\gamma_\mu\,\psi_B(x)\right)\,. \tag{5.61}$$

Solving the coupled equations of motion derived above for the numerous matter fields $(\psi_B, \sigma, \omega, \pi, \rho, \ldots)$ constitutes an extremely complicated problem. An exact numerical solution is probably out of reach for the foreseeable future. So, to carry the problem beyond the formal equations for the fields, it is unavoidable at this stage to introduce suitable approximation schemes. This can be accomplished by means of introducing the so-called Green function technique [79, 84, 85, 125, 314]. Green functions are made up of time-ordered products of baryon or meson field operators. Instead of studying the equations of motion for the baryon fields themselves, one then deals with the equation of motion for the Green functions. On a first glance this may leave one with the impression that this renders the problem even more cumbersome than attempting to solve the field equations directly. This however is not true. As we shall see in the next section, the Green function technique will allow us to introduce physically motivated many-body approximations, which, combined with additional mathematical techniques (e.g. a spectral representation of the two-point Green function), will finally render the equations of motion numerically tractable. The many-body approximations are (1) relativistic Hartree, (2) relativistic Hartree–Fock, and (3) the relativistic ladder approximation to the **T**-matrix. The latter will be solved for the so-called $\boldsymbol{\Lambda}^{00}$ propagator as well as the more physical Brueckner–Hartree–Fock propagator. The level of sophistication and complexity of these three approximations increase considerably from (1) through (3).

5.3 Relativistic Green functions

The general definition of the $2n$-point Green function is given as the ground state expectation value of the time-ordered product of n baryon

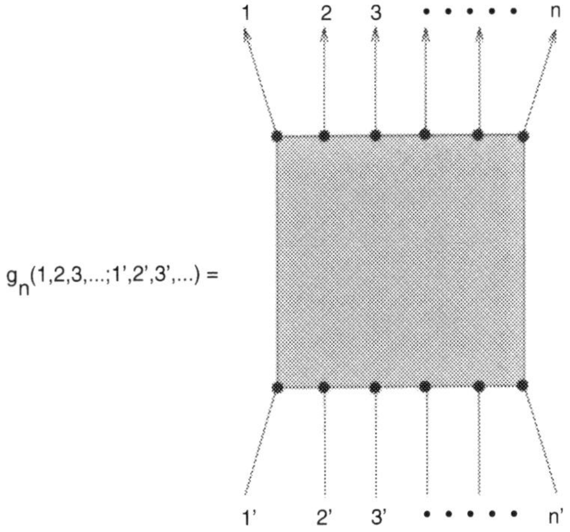

Figure 5.1. Graphical representation of the $2n$-point Green function defined in equation (5.62). The vertical lines denote the propagation of baryons in and out of the many-body vertex (shaded area).

field operators, ψ_B, and n' operators $\bar{\psi}_B$ ($\equiv \psi_B^\dagger \gamma^0$) in the form [117, 118, 125, 314, 315]

$$\begin{aligned} &\boldsymbol{g}_n^{B_1,\ldots,B_{n'}}(1,\ldots,n;1',\ldots,n') \\ &\quad = \mathrm{i}^n \,\langle \boldsymbol{\Phi}_0 | \hat{T}\left(\psi_{B_1}(1)\ldots\psi_{B_n}(n)\,\bar{\psi}_{B_{n'}}(n')\ldots\bar{\psi}_{B_{1'}}(1')\right) | \boldsymbol{\Phi}_0\rangle\,. \end{aligned} \tag{5.62}$$

The quantity $|\boldsymbol{\Phi}_0\rangle$ denotes the ground state of infinite nuclear matter, the integers $1 \equiv (x_1;\zeta_1)$ to $n \equiv (x_n;\zeta_n)$ stand for the spacetime coordinates $x_1 = (x_1^0, \boldsymbol{x}_1), \ldots, x_n = (x_n^0, \boldsymbol{x}_n)$ and spin and isospin quantum numbers $\zeta_1, \ldots, \zeta_n$. Physically, the $2n$-point Green function describes the propagation of n baryons relative to a many-particle background, which, in our case, is the nuclear matter ground state $|\boldsymbol{\Phi}_0\rangle$. Its graphical representation, shown in figure 5.1, is characterized by n' ingoing and n outgoing baryon lines. The quantity $\hat{T}$ is the time-ordering operator. It orders the field operators according to their value of x^0, with the smallest at the right. $\hat{T}$ also includes the signature factor $(-1)^P$, where P is the number of permutations of fermion field operators needed to restore the original ordering. Of particular interest is the two-point Green function obtained from (5.62) by setting $n=1$, i.e.

$$\boldsymbol{g}_1^{BB'}(x,\zeta;x',\zeta') \equiv \boldsymbol{g}_{\zeta\zeta'}^{BB'}(x,x') = \mathrm{i}\,\langle \boldsymbol{\Phi}_0 | \hat{T}\left(\psi_B(x,\zeta)\,\bar{\psi}_{B'}(x',\zeta')\right) | \boldsymbol{\Phi}_0\rangle\,. \tag{5.63}$$

The physical interpretation of $\boldsymbol{g}_1^{BB'}$ is illustrated in figure 5.2. It is this Green function that attains particular attention in the field-theoretical

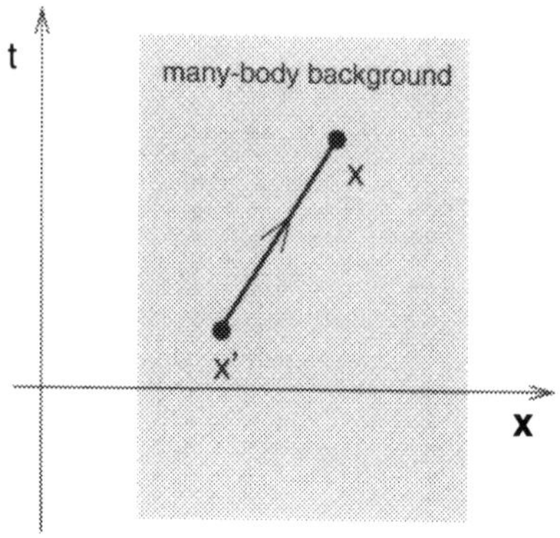

Figure 5.2. Physical interpretation of two-point Green function $\boldsymbol{g}_{>}(x, x')$ defined in equation (5.65). A baryon is created relative to the many-body background (shaded area) at spacetime point x', then propagates to x, where it is removed again.

treatment of the many-body system, for all the relevant observables of the system can be calculated from it. Writing out the time-ordering operator in (5.63) leads to

$$\boldsymbol{g}^{BB'}_{\zeta\zeta'}(x;x') = \Theta(x_0 - x'_0)\, \boldsymbol{g}^{BB'}_{>}(x,\zeta;x',\zeta') + \Theta(x'_0 - x_0)\, \boldsymbol{g}^{BB'}_{<}(x,\zeta;x',\zeta')\,, \tag{5.64}$$

with the definitions

$$\boldsymbol{g}^{BB'}_{>}(x,\zeta;x',\zeta') \equiv \mathrm{i}\,\langle \boldsymbol{\Phi}_0 | \psi_B(x,\zeta)\, \bar{\psi}_{B'}(x',\zeta') | \boldsymbol{\Phi}_0 \rangle\,, \tag{5.65}$$

$$\boldsymbol{g}^{BB'}_{<}(x,\zeta;x',\zeta') \equiv -\mathrm{i}\,\langle \boldsymbol{\Phi}_0 | \bar{\psi}_{B'}(x',\zeta')\, \psi_B(x,\zeta) | \boldsymbol{\Phi}_0 \rangle\,. \tag{5.66}$$

To find the equation of motion of $\boldsymbol{g}^{BB'}_1$ we apply the operator $(\mathrm{i}\gamma^\mu \partial_{\mu,1} - m_B)$ to the two-point Green function (5.64), which gives

$$(\mathrm{i}\gamma^\mu \partial_{\mu,1} - m_B)\, \boldsymbol{g}^{BB'}(x_1, x'_1) = (\mathrm{i}\gamma^\mu \partial_{\mu,1} - m_B) \times \mathrm{i}\left\{ \Theta(t_1 - t'_1) \langle \psi_B(x_1) \bar{\psi}_{B'}(x'_1) \rangle - \Theta(t'_1 - t_1) \langle \bar{\psi}_{B'}(x'_1) \psi_B(x_1) \rangle \right\}. \tag{5.67}$$

The subscript '1' attached to the partial derivative in (5.67) indicates that the derivative, explicitly given by

$$\gamma^\mu \partial_{\mu,1} \equiv \gamma^\mu \frac{\partial}{\partial x^\mu_1} = \gamma^0 \frac{\partial}{\partial t_1} + \boldsymbol{\gamma} \cdot \boldsymbol{\nabla}_1\,, \tag{5.68}$$

is to be performed with respect to the spacetime coordinate x_1. Equation (5.67) thus reads

$$\begin{aligned}(\mathrm{i}\,\gamma^\mu\partial_{\mu,1} - m_B)\,\boldsymbol{g}^{BB'}(x_1,x_1') = &-\gamma^0\frac{\partial}{\partial t_1}\left\{\Theta(t_1-t_1')\langle\psi_B(x_1)\bar{\psi}_{B'}(x_1')\rangle\right\}\\ &+\gamma^0\frac{\partial}{\partial t_1}\left\{\Theta(t_1'-t_1)\langle\bar{\psi}_{B'}(x_1')\psi_B(x_1)\rangle\right\}\\ &+\mathrm{i}\,(\mathrm{i}\,\boldsymbol{\gamma}\cdot\boldsymbol{\nabla}_1 - m_B)\,\Theta(t_1-t_1')\langle\psi_B(x_1)\bar{\psi}_{B'}(x_1')\rangle\\ &-\mathrm{i}\,(\mathrm{i}\,\boldsymbol{\gamma}\cdot\boldsymbol{\nabla}_1 - m_B)\,\Theta(t_1'-t_1)\langle\bar{\psi}_{B'}(x_1')\psi_B(x_1)\rangle\,. \end{aligned} \tag{5.69}$$

Performing the time derivatives in (5.69) gives

$$\begin{aligned}&\frac{\partial}{\partial t_1}\left\{\Theta(t_1-t_1')\langle\psi(x_1)\bar{\psi}(x_1')\rangle\right\}\\ &= \delta(t_1-t_1')\langle\psi(x_1)\bar{\psi}(x_1')\rangle + \Theta(t_1-t_1')\left\langle\frac{\partial\psi(x_1)}{\partial t_1}\bar{\psi}(x_1')\right\rangle\,, \end{aligned} \tag{5.70}$$

and

$$\begin{aligned}&\frac{\partial}{\partial t_1}\left\{\Theta(t_1'-t_1)\langle\bar{\psi}(x_1')\psi(x_1)\rangle\right\}\\ &= -\delta(t_1'-t_1)\langle\bar{\psi}(x_1')\psi(x_1)\rangle + \Theta(t_1'-t_1)\left\langle\bar{\psi}(x_1')\frac{\partial\psi(x_1)}{\partial t_1}\right\rangle\,. \end{aligned} \tag{5.71}$$

For the sake of brevity, we have dropped the subscripts and superscripts B, B' in the side-calculations (5.70) and (5.71). Hereafter, use of this simplification will be made occasionally without further notice. With the aid of (5.70) and (5.71), equation (5.69) can now be written as

$$\begin{aligned}&(\mathrm{i}\,\gamma^\mu\partial_{\mu,1} - m_B)\,\boldsymbol{g}^{BB'}(x_1,x_1') = -\gamma^0\delta(t_1-t_1')\langle\{\psi_B(x_1),\bar{\psi}_{B'}(x_1')\}\rangle\\ &-\gamma^0\Theta(t_1-t_1')\left\langle\frac{\partial\psi_B(x_1)}{\partial t_1}\bar{\psi}_{B'}(x_1')\right\rangle + \gamma^0\Theta(t_1'-t_1)\left\langle\bar{\psi}_{B'}(x_1')\frac{\partial\psi_B(x_1)}{\partial t_1}\right\rangle\\ &+\mathrm{i}\,\Theta(t_1-t_1')\langle(\mathrm{i}\,\boldsymbol{\gamma}\cdot\boldsymbol{\nabla}_1 - m_B)\,\psi_B(x_1)\bar{\psi}_{B'}(x_1')\rangle\\ &-\mathrm{i}\,\Theta(t_1'-t_1)\langle\bar{\psi}_{B'}(x_1')\,(\mathrm{i}\,\boldsymbol{\gamma}\cdot\boldsymbol{\nabla}_1 - m_B)\,\psi_B(x_1)\rangle\,. \end{aligned} \tag{5.72}$$

In the next step we employ the equation of motion for the baryon fields, derived in (5.36), to get rid of the two time derivatives in (5.72). For this purpose, we write (5.36) in the form

$$\begin{aligned}&\left(\mathrm{i}\,\gamma^0\frac{\partial}{\partial t_1} + \mathrm{i}\,\boldsymbol{\gamma}\cdot\boldsymbol{\nabla}_1 - m_B\right)\psi_B(x_1)\\ &\qquad = g_{\sigma B}\sigma(x_1)\psi_B(x_1) + g_{\omega B}\gamma^\mu\omega_\mu(x_1)\psi_B(x_1) \pm \dots\,, \end{aligned} \tag{5.73}$$

which, upon multiplying through with $-\mathrm{i}$ and rearranging terms, leads to

$$\begin{aligned}&\gamma^0\frac{\partial}{\partial t_1}\psi_B(x_1) - \mathrm{i}\,(\mathrm{i}\,\boldsymbol{\gamma}\cdot\boldsymbol{\nabla}_1 - m_B)\,\psi_B(x_1)\\ &\qquad = -\,\mathrm{i}\,g_{\sigma B}\,\sigma(x_1)\,\psi_B(x_1) - \mathrm{i}\,g_{\omega B}\,\gamma^\mu\omega_\mu(x_1)\,\psi_B(x_1) \pm \dots \end{aligned} \tag{5.74}$$

By means of substituting equation (5.74) into (5.72) and noticing that $\{\bar{\psi}(x), \psi(x')\} = \gamma^0 \delta^3(\boldsymbol{x} - \boldsymbol{x}')$, according to (5.16), we then obtain

$$
\begin{aligned}
(\mathrm{i}\,\gamma^\mu \partial_{\mu,1} - m_B)\, \boldsymbol{g}^{BB'}(x_1, x_1') = &- \delta^4(x_1 - x_1')\, \delta_{BB'} \\
&+ \mathrm{i}\, g_{\sigma B}\big\{\Theta(t_1 - t_1')\langle \psi_B(x_1)\sigma(x_1)\bar{\psi}_{B'}(x_1')\rangle \\
&- \Theta(t_1' - t_1)\langle \bar{\psi}_{B'}(x_1')\sigma(x_1)\psi_B(x_1)\rangle\big\} \\
&+ \mathrm{i}\, g_{\omega B}\big\{\Theta(t_1 - t_1')\langle \gamma^\mu \omega_\mu(x_1)\psi_B(x_1)\bar{\psi}_{B'}(x_1')\rangle \\
&- \Theta(t_1' - t_1)\langle \bar{\psi}_{B'}(x_1')\gamma^\mu \omega_\mu(x_1)\psi_B(x_1)\rangle\big\}\,, \qquad (5.75)
\end{aligned}
$$

which, upon introducing the time-ordering operator (cf. equations (5.63) and (5.64)) into this equation, leads to

$$
\begin{aligned}
&(\mathrm{i}\,\gamma^\mu \partial_{\mu,1} - m_B)\, \boldsymbol{g}^{BB'}(x_1, x_1') = - \delta^4(x_1 - x_1')\, \delta_{BB'} \\
&+ \mathrm{i}\, g_{\sigma B} \left\langle \hat{T}\left[\psi_B(x_1)\sigma(x_1)\bar{\psi}_{B'}(x_1')\right]\right\rangle \\
&+ \mathrm{i}\, g_{\omega B} \left\langle \hat{T}\left[\gamma^\mu \omega_\mu(x_1)\psi_B(x_1)\bar{\psi}_{B'}(x_1')\right]\right\rangle \\
&+ \mathrm{i}\, g_{\rho B} \Big\langle \hat{T}\Big[\Big(\gamma^\mu \psi_B(x_1)\boldsymbol{\tau}\cdot\boldsymbol{\rho}_\mu(x_1) \\
&+ \frac{f_{\rho B}}{4 m_B}\sigma^{\mu\nu}\psi_B(x_1)\boldsymbol{\tau}\cdot\boldsymbol{G}_{\mu\nu}(x_1)\Big)\bar{\psi}_{B'}(x_1')\Big]\Big\rangle \\
&+ \mathrm{i}\,\frac{f_{\pi B}}{m_\pi} \left\langle \hat{T}\left[\gamma^5 \gamma^\mu \psi_B(x_1)\,(\partial_\mu \boldsymbol{\tau}\cdot\boldsymbol{\pi}(x_1))\,\bar{\psi}_{B'}(x_1')\right]\right\rangle . \qquad (5.76)
\end{aligned}
$$

Equation (5.76) constitutes the Green-function analog to the inhomogeneous Dirac equation derived in (5.36). It still depends on the numerous unknown meson field operators, which we shall eliminate next. For this purpose we invert the meson field equations, derived in section 5.2, for the fields which are then substituted into (5.76). This will lead to the occurrence of higher-order Green functions in (5.76), which, however, can be approximated by lower-order ones.

To accomplish the inversion of the meson field equations, note that all meson field equations constitute partial differential equations for the fields,

$$
D^{(M)}(x)\, M(x) = R^{(M)}(x)\,, \qquad (5.77)
$$

where $D^{(M)}$ is a linear differential operator whose mathematical structure varies from meson to meson (M). The operator acts on a meson field $M(x)$. $R^{(M)}$ stands for the inhomogeneous part of each differential equation. Partial differential equations of this type can immediately be inverted if the free Green function Δ^{0M} associated with (5.77) is known. Δ^{0M} is defined as the solution of

$$
D^{(M)}(x)\, \Delta^{0M}(x, x') = \delta(x - x')\,, \qquad (5.78)
$$

from which it then follows that $M(x)$ of equation (5.77) is given by

$$M(x) = \int \mathrm{d}^4y\, \Delta^{0M}(x,y)\, R^{(M)}(y)\,. \tag{5.79}$$

To make this trick applicable to our problem the equations of motion for the meson Green functions need to be derived first.

We begin with defining the two-point Green function associated with the scalar σ mesons, which, in analogy to the two-point baryon Green function of equation (5.63), is defined as

$$\Delta^{\sigma}(x,x') = \mathrm{i}\,\langle \mathbf{\Phi}_0 | \hat{T}\,(\sigma(x)\sigma(x'))\,|\mathbf{\Phi}_0\rangle\,. \tag{5.80}$$

A comparison of (5.77) with the σ-meson field equation (5.43) shows that the differential operator $D^{(M)}$ is given by

$$D^{(\sigma)} \equiv \partial^{\mu}\partial_{\mu} + m_{\sigma}^2 = \frac{\partial^2}{\partial t^2} - \boldsymbol{\nabla}^2 + m_{\sigma}^2\,. \tag{5.81}$$

To find the result of $D^{(\sigma)}\Delta^{\sigma}$ let us consider first the action of the time derivative operator on the propagator Δ^{σ}, that is,

$$\frac{\partial^2}{\partial t^2}\,\{\Theta(t-t')\langle\sigma(x)\sigma(x')\rangle + \Theta(t'-t)\langle\sigma(x')\sigma(x)\rangle\}\,. \tag{5.82}$$

The chain rule then leads for (5.82) to

$$\begin{aligned}
&\frac{\partial}{\partial t}\,\{\Theta(t-t')\langle\dot\sigma(x)\sigma(x')\rangle + \Theta(t'-t)\langle\sigma(x')\dot\sigma(x)\rangle\}\\
&\quad = \delta(t-t')\langle\dot\sigma(x)\sigma(x')\rangle + \Theta(t-t')\langle\ddot\sigma(x)\sigma(x')\rangle\\
&\quad - \delta(t'-t)\langle\sigma(x')\dot\sigma(x)\rangle + \Theta(t'-t)\langle\sigma(x')\ddot\sigma(x)\rangle \qquad (5.83)\\
&\quad = \delta(t-t')\langle[\dot\sigma(x),\sigma(x')]\rangle + \langle\hat{T}\,(\ddot\sigma(x)\sigma(x'))\rangle\,. \qquad (5.84)
\end{aligned}$$

To get from equation (5.83) to (5.84), use of the commutator relation

$$[\dot\sigma(x),\sigma(x')] = \dot\sigma(x)\sigma(x') - \sigma(x')\dot\sigma(x) \tag{5.85}$$

and the definition of $\hat{T}$ has been made. To calculate the commutator in (5.84), let us replace $\dot\sigma$ with its associated conjugate field Π_{σ},

$$\Pi_{\sigma}(t,\boldsymbol{x}) = \frac{\partial\mathcal{L}}{\partial\partial_0\sigma(t,\boldsymbol{x})} = \dot\sigma(t,\boldsymbol{x})\,, \tag{5.86}$$

which leads for the commutator to (see equation (5.17))

$$[\dot\sigma(x),\sigma(x')] = [\Pi_{\sigma}(x),\sigma(x')] = -\,\mathrm{i}\,\delta^3(\boldsymbol{x}-\boldsymbol{x}')\,. \tag{5.87}$$

With the aid of (5.87), we arrive for (5.82) at the final result,

$$\frac{\partial^2}{\partial t^2} \langle \hat{T}\left(\sigma(x)\sigma(x')\right)\rangle = -\,\mathrm{i}\,\delta^4(x-x') + \langle \hat{T}\left(\ddot{\sigma}(x)\sigma(x')\right)\rangle \,. \tag{5.88}$$

Now we have all ingredients at hand that are required to calculate $D^{(\sigma)}\Delta^\sigma$. With the help of (5.88), one then gets for the Δ^σ propagator,

$$\begin{aligned}\left(\partial^\mu\partial_\mu + m_\sigma^2\right)\Delta^\sigma(x,x') = \delta^4(x-x') + \mathrm{i}\,\langle \hat{T}\left(\ddot{\sigma}(x)\sigma(x')\right)\rangle \\ + \left(-\Delta_x + m_\sigma^2\right)\Delta^\sigma(x,x')\,.\end{aligned} \tag{5.89}$$

Substituting equation (5.80) for Δ^σ then leads for the right-hand side of this equation to

$$\delta^4(x-x') + \mathrm{i}\,\langle \hat{T}\left\{\left[\left(\partial^\mu\partial_\mu + m_\sigma^2\right)\sigma(x)\right]\sigma(x')\right\}\rangle \,. \tag{5.90}$$

The expression in square brackets can be replaced with its source term, equation (5.43), which leads to the desired result for the equation of motion of the full σ-meson propagator, given by

$$\begin{aligned}\left(\partial^\mu\partial_\mu + m_\sigma^2\right)\Delta^\sigma(x,x') = \delta^4(x-x') \\ -\,\mathrm{i}\sum_B g_{\sigma B}\langle \hat{T}\left(\bar{\psi}_B(x^+)\psi_B(x)\sigma(x')\right)\rangle \,.\end{aligned} \tag{5.91}$$

By definition, the free meson Green function associated with (5.91), denoted by $\Delta^{0\sigma}$, is given as the solution of

$$\left(\partial^\mu\partial_\mu + m_\sigma^2\right)\Delta^{0\sigma}(x,x') = \delta^4(x-x')\,. \tag{5.92}$$

Four-dimensional Fourier transformation of (5.92) into energy–momentum space, as outlined in section B.2 of appendix B, leads for the meson propagator to

$$\Delta^{0\sigma}(p) = -\frac{1}{p_0^2 - \boldsymbol{p}^2 - m_\sigma^2 + \mathrm{i}\,\eta}\,. \tag{5.93}$$

Now we have all ingredients at hand to invert the equation of motion of the σ field. Proceeding as described in equations (5.77) through (5.79), we get for the σ-meson field

$$\sigma(x) = -\sum_{B'} g_{\sigma B'} \int \mathrm{d}^4x'\;\Delta^{0\sigma}(x,x')\;\bar{\psi}_{B'}(x')\psi_{B'}(x')\,. \tag{5.94}$$

In the next step we invert the field equations of the vector mesons ω^μ and ρ^μ. Their associated two-point Green functions are given by

$$\left(\mathcal{D}^\omega\right)^{\mu\nu}(x,x') = \mathrm{i}\,\langle \boldsymbol{\Phi}_0|\hat{T}\left(\omega^\mu(x)\omega^\nu(x')\right)|\boldsymbol{\Phi}_0\rangle \tag{5.95}$$

and

$$(\mathcal{D}^{\rho})^{\mu\nu}(x,x';r,r') = \mathrm{i}\,\langle\mathbf{\Phi}_0|\hat{T}\left(\rho^{r\mu}(x)\rho^{r'\nu}(x')\right)|\mathbf{\Phi}_0\rangle\,, \tag{5.96}$$

respectively. The equations of motion of these two propagators are obtained in complete analogy to the σ field. What complicates matters is the vectorial nature of these mesons. Moreover the ρ field additionally is a three-vector in isospin space. We begin with writing the left-hand side of (5.52) as

$$\partial^{\lambda} F_{\lambda\nu} + m_{\omega}^2 \omega_{\nu} = \left(\partial^{\lambda}\partial_{\lambda}\delta_{\nu}{}^{\mu} - \partial^{\mu}\partial_{\nu} + m_{\omega}^2\delta_{\nu}{}^{\mu}\right)\omega_{\mu}\,, \tag{5.97}$$

which leads for the field equation of the ω meson to

$$\left(\partial^{\lambda}\partial_{\lambda}\delta_{\nu}{}^{\mu} - \partial^{\mu}\partial_{\nu} + m_{\omega}^2\delta_{\nu}{}^{\mu}\right)\omega_{\mu} = \sum_{B}\left\{g_{\omega B}\bar{\psi}_B\gamma_{\nu}\psi_B - \frac{f_{\omega B}}{2m_B}\,\partial^{\lambda}\left(\bar{\psi}_B\sigma_{\lambda\nu}\psi_B\right)\right\}. \tag{5.98}$$

Next let us define

$$\mathcal{D}^{0\omega}_{\mu\kappa}(x,x') = \left(g_{\mu\kappa} + \frac{\partial_{\mu}\partial_{\kappa}}{m_{\omega}^2}\right)\Delta^{0\omega}(x,x')\,, \tag{5.99}$$

whose Fourier transform reads (cf. appendix B.2)

$$\mathcal{D}^{0\omega}_{\mu\kappa}(p) = \left(g_{\mu\kappa} - \frac{p_{\mu}p_{\kappa}}{m_{\omega}^2}\right)\Delta^{0\omega}(p) \tag{5.100}$$

with $\Delta^{0\omega}(p)$ as in (5.93). It is readily shown that propagator (5.99) obeys

$$\left(\partial^{\lambda}\partial_{\lambda}\delta_{\nu}{}^{\mu} - \partial^{\mu}\partial_{\nu} + m_{\omega}^2\delta_{\nu}{}^{\mu}\right)\mathcal{D}^{0\omega}_{\mu\kappa}(x,x') = g_{\nu\kappa}\,\delta^4(x-x')\,. \tag{5.101}$$

The field equation (5.52) can now be inverted following the procedure outlined just above. One obtains

$$\omega_{\mu}(x) = \int \mathrm{d}^4x'\;\mathcal{D}^{0\omega}_{\mu\kappa}(x,x') \times \sum_{B}\left\{g_{\omega B}\bar{\psi}_B(x')\gamma^{\kappa}\psi_B(x') - \frac{f_{\omega B}}{2m_B}\,\partial^{\lambda}\left(\bar{\psi}_B(x')\sigma_{\lambda}{}^{\kappa}\psi_B(x')\right)\right\}. \tag{5.102}$$

The corresponding expressions for the ρ-meson field are very similar to those of the ω-meson field derived in equations (5.97) to (5.102). The only differences arise from the isovectorial nature of the ρ-meson field. It therefore carries an extra index r (=1,2,3) which discriminates between the meson's three isospin components. Bearing this in mind, one can proceed in complete analogy to the above. The individual equations are then given by

$$\left(\partial^{\lambda}\partial_{\lambda}\delta_{\nu}{}^{\mu} - \partial^{\mu}\partial_{\nu} + m_{\rho}^2\delta_{\nu}{}^{\mu}\right)\rho^{r}_{\mu} = \sum_{B}\left\{g_{\rho B}\,\bar{\psi}_B\tau^{r}\gamma_{\nu}\psi_B - \frac{f_{\rho B}}{2m_B}\,\partial^{\lambda}\left(\bar{\psi}_B\tau^{r}\sigma_{\lambda\nu}\psi_B\right)\right\}, \tag{5.103}$$

which defines the free ρ-meson Green function via the equation

$$\left(\partial^\lambda \partial_\lambda \delta_\nu{}^\mu - \partial^\mu \partial_\nu + m_\rho^2 \delta_\nu{}^\mu\right) \mathcal{D}^{0\rho}_{\mu\kappa}(x, x'; r, r') = g_{\nu\kappa}\, \delta^4(x - x')\delta_{rr'}\,, \quad (5.104)$$

with

$$\mathcal{D}^{0\rho}_{\mu\kappa}(x, x'; r, r') = \left(g_{\mu\kappa} + \frac{\partial_\mu \partial_\kappa}{m^2_{\rho,r}}\right) \Delta^{0\rho}(x, x'; r, r')\,, \quad (5.105)$$

and, with $\Delta^{0\rho}(p)$ as in equation (5.93),

$$\mathcal{D}^{0\rho}_{\mu\kappa}(p) = \left(g_{\mu\kappa} - \frac{p_\mu p_\kappa}{m^2_{\rho,r}}\right) \Delta^{0\rho}(p)\,. \quad (5.106)$$

The ρ-meson field is therefore given by

$$\rho^r_\mu(x) = \sum_{r'} \int \mathrm{d}^4x'\, \mathcal{D}^{0\rho}_{\mu\kappa}(x, x'; r, r')$$
$$\times \sum_B \left\{ g_{\rho B} \bar\psi_B(x') \tau^{r'} \gamma^\kappa \psi_B(x') - \frac{f_{\rho B}}{2m_B}\, \partial^\lambda \left(\bar\psi_B(x') \tau^{r'} \sigma_\lambda{}^\kappa \psi_B(x')\right)\right\}. \quad (5.107)$$

The π meson, being an isovector particle too, also carries an index r. Its two-point function is given by

$$\Delta^{0\pi}(x, x'; r, r') = \mathrm{i} \left\langle \mathbf{\Phi}_0 \left| \hat T \left(\pi^r(x) \pi^{r'}(x')\right) \right| \mathbf{\Phi}_0 \right\rangle\,, \quad (5.108)$$

which obeys

$$\left(\partial^\mu \partial_\mu + m_\pi^2\right) \Delta^{0\pi}(x, x'; r, r') = \delta^4(x - x')\delta_{rr'}\,. \quad (5.109)$$

The momentum-space representation of $\Delta^{0\pi}(x, x'; r, r')$ is given by

$$\Delta^{0\pi}(p) = -\frac{1}{p_0^2 - \boldsymbol{p}^2 - m^2_{\pi,r} + \mathrm{i}\,\eta}\,. \quad (5.110)$$

The equation for the pion field then follows as

$$\pi^r(x) = \sum_{B,r'} \frac{f_{\pi B}}{m_\pi} \int \mathrm{d}^4x'\, \Delta^{0\pi}(x, x'; r, r')\, \partial_{\mu, x'} \left(\bar\psi_B(x') \gamma^5 \gamma^\mu \tau^{r'} \psi_B(x')\right)\,. \quad (5.111)$$

With the aid of the explicit expressions for the meson fields derived in equations (5.94), (5.102), (5.107) and (5.111), the meson fields in (5.76) can now be replaced with meson Green functions. Dropping the tensor part of the ρ meson (term $\propto f_{\rho B}$) for the moment, which can be easily restored again, as we shall see later, this yields for (5.76) to ($\partial\!\!\!/ \equiv \gamma^\mu \partial_\mu$)

$$\left(\mathrm{i}\,\partial\!\!\!/_{x_1} - m_B\right) \boldsymbol{g}_1^{BB'}(x_1, x_1') = -\,\delta^4(x_1 - x_1')\,\delta_{BB'} + F^{BB'}(x_1, x_1')\,, \quad (5.112)$$

where, with $\boldsymbol{\tau}\Delta^{0\pi}(x,x')\boldsymbol{\tau} \equiv \sum_{r,r'} \tau^r \Delta^{0\pi}(x,x';r,r')\tau^{r'}$, then $F^{BB'}$ is given by

$$
\begin{aligned}
F^{BB'}(x_1,x_1') = \mathrm{i}\sum_{B''}\int \mathrm{d}^4x' \Big\{ & - g_{\sigma B}\, g_{\sigma B''}\, \Delta^{0\sigma}(x_1,x') \\
& \times \langle \hat{T}\left(\psi_B(x_1)\bar{\psi}_{B''}(x'^+)\psi_{B''}(x')\bar{\psi}_{B'}(x_1')\right)\rangle \\
& + g_{\omega B}\, g_{\omega B''}\, \gamma^\mu \mathcal{D}^{0\omega}_{\mu\kappa}(x_1,x')\langle \hat{T}\left(\psi_B(x_1)\bar{\psi}_{B''}(x'^+)\gamma^\kappa \psi_{B''}(x')\bar{\psi}_{B'}(x_1')\right)\rangle \\
& + g_{\rho B}\, g_{\rho B''}\, \gamma^\mu \boldsymbol{\tau}\mathcal{D}^{0\rho}_{\mu\kappa}(x_1,x')\langle \hat{T}\left(\psi_B(x_1)\bar{\psi}_{B''}(x'^+)\boldsymbol{\tau}\,\gamma^\kappa \psi_{B''}(x')\bar{\psi}_{B'}(x_1')\right)\rangle \\
& + \frac{f_{\pi B}}{m_\pi}\,\frac{f_{\pi B''}}{m_\pi}\,\gamma^5\gamma^\mu\left(\partial_{\mu,x_1}\boldsymbol{\tau}\Delta^{0\pi}(x_1,x')\right) \\
& \quad \times \langle \hat{T}\left(\psi_B(x_1)\partial_{\kappa,x'}\left[\bar{\psi}_{B''}(x'^+)\gamma^5\gamma^\kappa\,\boldsymbol{\tau}\psi_{B''}(x')\right]\bar{\psi}_{B'}(x_1')\right)\rangle \Big\}. \qquad (5.113)
\end{aligned}
$$

The major mathematical advantage of (5.112) over (5.76) is that instead of the meson fields themselves, we are now dealing with the expectation values of time-ordered products of baryon-field operators which, upon closer inspection (cf. equation (5.62)), turn out to constitute noting else but four-point Green functions, $\boldsymbol{g}_2$. These have the advantage over the meson fields that physically motivated many-body approximations can be introduced that allow one to solve equation (5.112) in a physically transparent manner, as will be discussed subsequently.

Before, however, we shall introduce two more field-theoretical concepts, namely the Dyson equation and the self-energy $\boldsymbol{\Sigma}^B$ of a baryon, which is also known as mass operator, or effective single-particle potential. Both the Dyson equation as well as $\boldsymbol{\Sigma}^B$ play an equally important role in field theory as the baryon propagators, $\boldsymbol{g}_1$. Let us begin with decomposing $F^{BB'}$

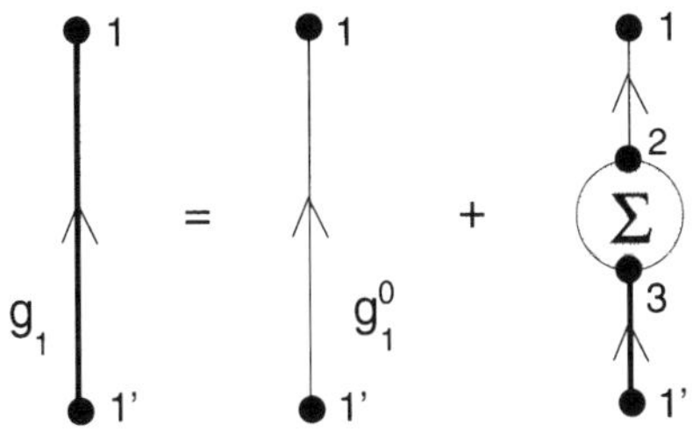

Figure 5.3. Graphical representation of Dyson's equation for the self-consistent two-point baryon Green function $\boldsymbol{g}_1$. The quantity $\boldsymbol{g}_1^0$ denotes the propagator of free baryons, which do not feel the nuclear medium (i.e. $\boldsymbol{\Sigma} \equiv \mathbf{0}$). The momentum-space representation of Dyson's equation is given in (5.126).

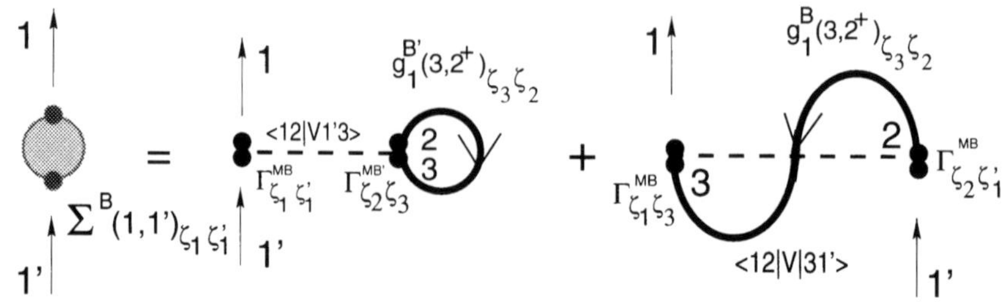

Figure 5.4. Diagrammatic equation of baryon self-energy, $\boldsymbol{\Sigma}^B$, in Hartree–Fock approximation. The matrix elements $\langle 12|\mathbf{V}^{BB'}|1'3\rangle$ and $\langle 12|\mathbf{V}^{BB'}|31'\rangle$, defined in equation (5.151), describe the meson-exchange interaction in the direct (Hartree) and exchange (Fock) term of $\boldsymbol{\Sigma}^B$, respectively. Γ^{MB} and $\Gamma^{MB'}$ denote baryon–meson vertices ($M = \sigma, \omega, \ldots$; $B = p, n, \Sigma^{\pm}, \ldots$). The analytic expression of $\boldsymbol{\Sigma}^B(1,1')$ can be inferred from (5.115) in reference to (5.120), or, alternatively, from equations (7.2) and (7.3).

of (5.113) in the following manner,

$$F^{BB'}(x_1, x_1') \equiv \sum_{B''} \int \mathrm{d}^4x' \, \boldsymbol{\Sigma}^{BB''}(x_1, x') \, \boldsymbol{g}_1^{B''B'}(x', x_1') \,. \qquad (5.114)$$

The equation of motion for $\boldsymbol{g}_1^{BB'}$ derived in (5.112) can then be written as

$$(\mathrm{i}\,\not\partial_{x_1} - m_B)\, \boldsymbol{g}_1^{BB'}(x_1, x_1') = -\,\delta^4(x_1 - x_1')\,\delta_{BB'} + \sum_{B''} \int \mathrm{d}^4x' \, \boldsymbol{\Sigma}^{BB''}(x_1, x') \, \boldsymbol{g}_1^{B''B'}(x', x_1') \,. \qquad (5.115)$$

Employing the method outlined in connection with (5.77), equation (5.115) can be readily transformed into the alternative form

$$\boldsymbol{g}_1^{BB'}(x_1, x_1') = \boldsymbol{g}_1^{0BB'}(x_1, x_1') - \sum_{B_2, B_3} \int \mathrm{d}^4x_2 \int \mathrm{d}^4x_3 \, \boldsymbol{g}_1^{0BB_2}(x_1, x_2) \times \boldsymbol{\Sigma}^{B_2B_3}(x_2, x_3)\, \boldsymbol{g}_1^{B_3B'}(x_3, x_1') \,. \qquad (5.116)$$

Since we shall be dealing with scenarios where a given baryon does not transform into another baryon along its path $x_1' \to x_1$ (see figure 5.2), we may write $\boldsymbol{g}^{BB'} = \delta_{BB'}\, \boldsymbol{g}^B$. Incorporating this feature into (5.116) leads for the Dyson equation to

$$\boldsymbol{g}_1^B(x_1, x_1') = \boldsymbol{g}_1^{0B}(x_1, x_1') - \int \mathrm{d}^4x_2 \int \mathrm{d}^4x_3 \, \boldsymbol{g}_1^{0B}(x_1, x_2)\, \boldsymbol{\Sigma}^B(x_2, x_3)\, \boldsymbol{g}_1^B(x_3, x_1') \,. \qquad (5.117)$$

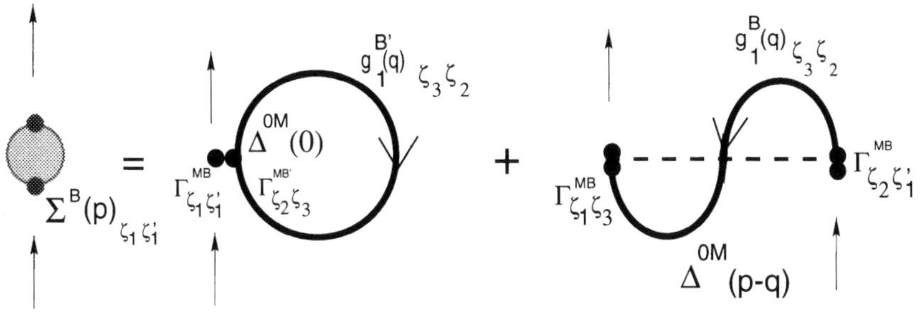

Figure 5.5. Diagrammatic representation of Hartree–Fock baryon self-energy $\mathbf{\Sigma}^B$ in momentum space. $\Delta^{0M}(0)$ and $\Delta^{0M}(p-q)$ denote meson propagators, derived, for instance, in equations (5.93) and (5.100). All other quantities are explained in figure 5.4. For the analytic form of $\mathbf{\Sigma}^B(p)$, see, for example, equations (5.134) through (5.142).

Its graphical representation is illustrated in figure 5.3, with the corresponding diagrammatic representation of the self-energy shown in figure 5.4. The representation of the latter diagrams in momentum space is given in figure 5.5.

5.4 Relativistic Hartree and Hartree–Fock approximation

The two simplest many-body approximations that will be introduced in this volume are the relativistic Hartree and the relativistic Hartree–Fock (HF) approximations. The mathematical structure of the latter is already considerably more complicated than the former whose self-energies, as we shall see later, depend only on density but neither on energy nor momentum. Besides that there are quantitative differences between both approximations which originate from the Fock terms contained in the HF approximation and the different coupling constants of both theories. This is specifically the case for the coupling constant of the ρ meson, which plays a crucial role for the composition of neutron star matter.

The relativistic HF approximation is obtained by factorizing the four-point baryon Green functions in (5.113), given by

$$\begin{aligned}
&\langle \hat{T}\left(\psi_B(x_1)\bar{\psi}_{B''}(x'^{+})\psi_{B''}(x')\bar{\psi}_{B'}(x'_1)\right)\rangle \\
&\qquad = -\langle \hat{T}\left(\psi_B(x_1)\psi_{B''}(x')\bar{\psi}_{B''}(x'^{+})\bar{\psi}_{B'}(x'_1)\right)\rangle \\
&\qquad = \boldsymbol{g}_2(x_1 B, x' B''; x'^{+} B'', x'_1 B'), \qquad (5.118)
\end{aligned}$$

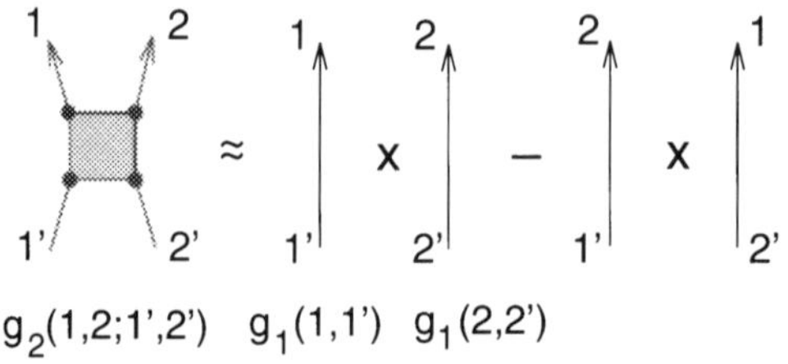

Figure 5.6. Factorization scheme of four-point baryon Green function, $\boldsymbol{g}_2$, into antisymmetrized products of two-point baryon Green functions, $\boldsymbol{g}_1 \times \boldsymbol{g}_1$ (cf. (5.119)). Direct (Hartree) and exchange (Fock) contribution are shown. This factorization scheme truncates the many-body equations at the Hartree–Fock level.

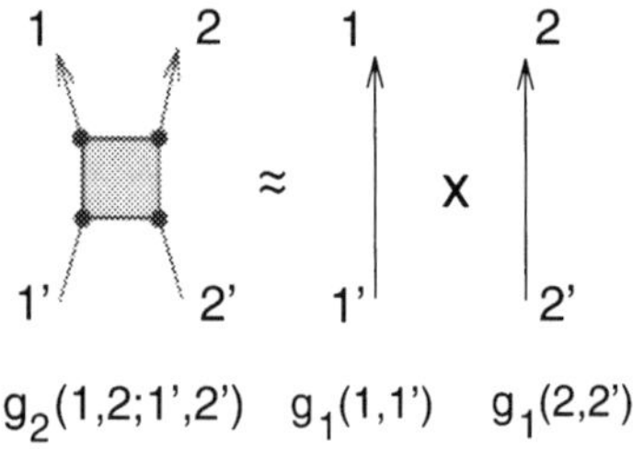

Figure 5.7. Factorization scheme of four-point baryon Green function, $\boldsymbol{g}_2$, into a product of two-point baryon Green functions, $\boldsymbol{g}_1 \times \boldsymbol{g}_1$. Keeping only the first term of (5.119), this truncates the many-body equations at the Hartree level.

into products of two-point baryon Green functions,

$$\begin{aligned} &\boldsymbol{g}_2(x_1 B, x' B''; x'^{+} B'', x_1' B) \\ &\approx \boldsymbol{g}_1(x_1 B, x_1' B')\, \boldsymbol{g}_1(x' B'', x'^{+} B'') - \boldsymbol{g}_1(x_1 B, x'^{+} B'')\, \boldsymbol{g}_1(x' B'', x_1' B')\, \delta_{BB''} \\ &\equiv \boldsymbol{g}_1^{BB'}(x_1, x_1')\, \boldsymbol{g}_1^{B''B''}(x', x'^{+}) - \boldsymbol{g}_1^{BB''}(x_1, x'^{+})\, \boldsymbol{g}_1^{B''B'}(x', x_1')\, \delta_{BB''}\,. \end{aligned} \tag{5.119}$$

A graphical illustration of the factorization scheme is displayed in figure 5.6. The first term on the right-hand side of (5.119), referred to as the Hartree (or direct) term, truncates the many-body equations at the relativistic Hartree level. The second term, referred to as the Fock (or exchange) contribution, whose final states are interchanged, leads to the HF approximation. Neglecting the Fock term in (5.119) leads to the frequently used Hartree approximation (figure 5.7). A characteristic feature of these approximations is that both baryons propagate independent from each other in the medium, aside from effects stemming from the Pauli exclusion principle. Any *dynamical* correlations between the baryons are completely lost for these approximations, in sharp contrast to the relativistic ladder (Brueckner–Hartree–Fock type) approximation where an effective **T**-matrix in matter is calculated from one-boson-exchange interactions. We shall follow up this approximation in section 5.5 and chapters 9 and 10.

Substituting the HF approximated $\boldsymbol{g}_2$ function of (5.119) into equation (5.113) leads to an equation of motion for the two-point baryon Green function given by

$$
\begin{aligned}
(\mathrm{i}\,\not{\partial}_{x_1} - m_B)\, \boldsymbol{g}_1^{BB'}(x_1, x_1') &= -\,\delta^4(x_1 - x_1')\,\delta_{BB'} \\
+\mathrm{i}\sum_{B''} \int \mathrm{d}^4x' \Big\{ &-g_{\sigma B}\, g_{\sigma B''}\, \Delta^{0\sigma}(x_1, x') + \mathrm{i}\, g_{\omega B}\, g_{\omega B''}\, \gamma^\mu \gamma^\kappa\, \mathcal{D}^{0\omega}_{\mu\kappa}(x_1, x') \\
&+ g_{\rho B}\, g_{\rho B''}\, \big(\gamma^\mu \boldsymbol{\tau}\big)\big(\gamma^\kappa \boldsymbol{\tau}\big)\, \mathcal{D}^{0\rho}_{\mu\kappa}(x_1, x') \\
&+ \frac{f_{\pi B}}{m_\pi}\frac{f_{\pi B''}}{m_\pi}\big(\gamma^5\gamma^\mu \boldsymbol{\tau}\partial_{\mu, x_1}\big)\big(\gamma^5\gamma^\kappa \boldsymbol{\tau}\partial_{\kappa, x'}\big)\Big)\, \Delta^{0\pi}(x_1, x')\Big\} \\
\times \big\{ \boldsymbol{g}_1^{BB'}(x_1, x_1')\, \boldsymbol{g}_1^{B''B''}(x', x'^+) &- \boldsymbol{g}_1^{BB''}(x_1, x'^+)\, \boldsymbol{g}_1^{B''B'}(x', x_1')\, \delta_{BB''} \big\}\,.
\end{aligned}
\tag{5.120}
$$

In the next step we transform (5.120) into four-momentum space. There the equations become much simpler, since we are dealing with a spatially uniform system that is invariant under translations. All functions in (5.120) therefore depend only on the coordinate differences, as already indicated for the argument of the δ-function in (5.120). The four-dimensional Fourier transforms in these coordinates are given for the Hartree term by expressions like (cf. appendix B.2)

$$
\begin{aligned}
&\int \mathrm{d}^4x'\, \Delta^{0\sigma}(x_1 - x')\, \boldsymbol{g}_1^{BB'}(x_1 - x_1')\, \boldsymbol{g}_1^{B''B''}(x' - x'^+) \\
&= \int \frac{\mathrm{d}^4q}{(2\pi)^4} \int \frac{\mathrm{d}^4p}{(2\pi)^4}\, \mathrm{e}^{\mathrm{i}\eta q^0}\, \mathrm{e}^{-\mathrm{i}p(x_1 - x_1')}\, \Delta^{0\sigma}(0)\, \boldsymbol{g}_1^{BB'}(p)\, \boldsymbol{g}_1^{B''B''}(q)\,,
\end{aligned}
\tag{5.121}
$$

depending on the meson propagator, and for the respective Fock terms by

$$
\begin{aligned}
&\int \mathrm{d}^4x'\, \Delta^{0\sigma}(x_1 - x')\, \boldsymbol{g}_1^{BB''}(x_1 - x'^+)\, \boldsymbol{g}_1^{B''B'}(x' - x_1') \\
&= \int \frac{\mathrm{d}^4q}{(2\pi)^4} \int \frac{\mathrm{d}^4p}{(2\pi)^4}\, \mathrm{e}^{\mathrm{i}\eta q^0}\, \mathrm{e}^{-\mathrm{i}p(x_1 - x_1')}\, \Delta^{0\sigma}(p - q)\, \boldsymbol{g}_1^{BB''}(p)\, \boldsymbol{g}_1^{B''B'}(q)\,.
\end{aligned}
\tag{5.122}
$$

Equation (5.120) can then be written very neatly as

$$
(\not{p} - m_B)\, \boldsymbol{g}_1^B(p) = -\,1 + \boldsymbol{\Sigma}^B(p)\, \boldsymbol{g}_1^B(p)\,, \tag{5.123}
$$

which constitutes Dyson's equation (5.115) in momentum space, with the baryon self-energy in the Hartree–Fock approximation given by (see also the results derived in equations (5.135) through (5.142) as well as in appendix D)

$$\begin{aligned}\boldsymbol{\Sigma}^B(p) \equiv & -\mathrm{i}\sum_{B'} g_{\sigma B}\, g_{\sigma B'} \int \frac{\mathrm{d}^4q}{(2\pi)^4}\, \mathrm{e}^{\mathrm{i}\eta q^0} \big\{\Delta^{0\sigma}(0)\, \boldsymbol{g}_1^{B'}(q) \\ & \qquad - \delta_{BB'}\Delta^{0\sigma}(p-q)\, \boldsymbol{g}_1^{B}(q)\big\} \\ & +\mathrm{i}\sum_{B'} g_{\omega B}\, g_{\omega B'} \int \frac{\mathrm{d}^4q}{(2\pi)^4}\, \mathrm{e}^{\mathrm{i}\eta q^0} \big\{\gamma^\mu\, \gamma^\nu\, D^{0\omega}_{\mu\nu}(0)\, \boldsymbol{g}_1^{B'}(q) \\ & \qquad - \delta_{BB'}\gamma^\mu\, \gamma^\nu\, D^{0\omega}_{\mu\nu}(p-q)\, \boldsymbol{g}_1^{B}(q)\big\} \\ & +\mathrm{i}\sum_{B'} g_{\rho B}\, g_{\rho B'} \int \frac{\mathrm{d}^4q}{(2\pi)^4}\, \mathrm{e}^{\mathrm{i}\eta q^0} \big\{(\gamma^\mu \boldsymbol{\tau})\, (\gamma^\nu \boldsymbol{\tau})\, D^{0\rho}_{\mu\nu}(0)\, \boldsymbol{g}_1^{B'}(q) \\ & \qquad - \delta_{BB'}(\gamma^\mu \boldsymbol{\tau})\, (\gamma^\nu \boldsymbol{\tau})\, D^{0\rho}_{\mu\nu}(p-q)\, \boldsymbol{g}_1^{B}(q)\big\} \\ & +\mathrm{i}\left(\frac{f_{\pi B}}{m_B}\right)^2 \int \frac{\mathrm{d}^4q}{(2\pi)^4}\, \mathrm{e}^{\mathrm{i}\eta q^0} \big\{(\gamma^5\gamma^\lambda \boldsymbol{\tau})(\gamma^5\gamma^\mu \boldsymbol{\tau})(p-q)_\lambda \\ & \qquad \times (p-q)_\mu\, \Delta^{0\pi}(p-q)\, \boldsymbol{g}_1^{B}(q)\big\}\,. \end{aligned} \tag{5.124}$$

The two-point baryon function associated with a non-interacting many-body system, characterized by $\boldsymbol{\Sigma}^B \equiv 0$, follows from (5.123) in the form

$$\boldsymbol{g}_1^{0B}(p) = -\,(\not p - m_B)^{-1}\,. \tag{5.125}$$

Substituting (5.125) into (5.123) and multiplying both sides with $\boldsymbol{g}_1^{0B}$ gives the momentum-space analog of Dyson's equation in coordinate space, derived in (5.117), in the form

$$\boldsymbol{g}^B_{1\,\zeta_1\zeta_2}(p) = \boldsymbol{g}^{0B}_{1\,\zeta_1\zeta_2}(p) - \boldsymbol{g}^{0B}_{1\,\zeta_1\zeta_1'}(p)\, \boldsymbol{\Sigma}^B_{\zeta_1'\zeta_2'}(p)\, \boldsymbol{g}^B_{1\,\zeta_2'\zeta_2}(p)\,. \tag{5.126}$$

Equations (5.123) and (5.126) constitute matrix equations in Dirac (spin) and isospin space. The corresponding indices are denoted by α and i, respectively, which we combine frequently to the single symbol ζ [$\equiv (\alpha, i)$]. Assigning the spin and isospin indices to $\boldsymbol{g}_1^B$ and $\boldsymbol{\Sigma}^B$ leads for (5.123) to

$$(\not p - m_B)_{\zeta_1\zeta_1'}\, \boldsymbol{g}^B_{\zeta_1'\zeta_2}(p) \;=\; -\,\delta_{\zeta_1\zeta_2} + \boldsymbol{\Sigma}^B_{\zeta_1\zeta_1'}(p)\, \boldsymbol{g}^B_{\zeta_1'\zeta_2}(p)\,, \tag{5.127}$$

with the σ-meson self-energy matrix $\boldsymbol{\Sigma}^B_{\zeta_1\zeta_1'}(p)$ given by

$$\begin{aligned}\boldsymbol{\Sigma}^B_{\zeta_1\zeta_1'}(p)\Big|_\sigma = & -\mathrm{i}\,(g_{\sigma B}\,\mathbf{1})_{\zeta_1\zeta_1'} \sum_{B'} (g_{\sigma B'}\,\mathbf{1})_{\zeta_3\zeta_4} \int \frac{\mathrm{d}^4q}{(2\pi)^4}\, \mathrm{e}^{\mathrm{i}\eta q^0}\, \Delta^{0\sigma}(0)\, \boldsymbol{g}^{B'}_{\zeta_4\zeta_3}(q) \\ & +\mathrm{i}\,(g_{\sigma B}\,\mathbf{1})_{\zeta_1\zeta_3} \int \frac{\mathrm{d}^4q}{(2\pi)^4}\, \mathrm{e}^{\mathrm{i}\eta q^0}\, \Delta^{0\sigma}(p-q)\, \boldsymbol{g}^{B'}_{\zeta_3\zeta_2}(q)\, (g_{\sigma B}\,\mathbf{1})_{\zeta_2\zeta_1'}\,. \end{aligned} \tag{5.128}$$

The other mesons of our collection lead to the same matrix structure for $\boldsymbol{\Sigma}$ as in (5.128), aside from deviations that originate from the different baryon–meson couplings. As known from (5.7) through (5.10), these are simplest for the scalar σ and most complicated for the vector mesons ω and ρ. Before we shall turn our interest to the latter mesons, however, let us point out a few notational simplifications concerning the summations over the spin and isospin indices in (5.128). For instance, the coupling constant $(g_{\sigma B'}\mathbf{1})_{\zeta_3\zeta_4}$ in the Hartree term of (5.128) can be combined with the two-point Green function $\boldsymbol{g}^{B'}_{\zeta_4\zeta_3}(q)$ there according to

$$(g_{\sigma B'}\,\mathbf{1})_{\zeta_3\zeta_4}\;\boldsymbol{g}^{B'}_{\zeta_4\zeta_3}(q) \equiv g_{\sigma B'}\;\mathrm{Tr}\left(\mathbf{1}\,\boldsymbol{g}^{B'}(q)\right) \equiv g_{\sigma B'}\;\mathrm{Tr}\;\boldsymbol{g}^{B'}(q)\,. \qquad (5.129)$$

The trace in (5.129), denoted Tr, sums the diagonal elements of the matrix $\mathbf{1}\,\boldsymbol{g}^B(q) = \boldsymbol{g}^B(q)$. One of its interesting properties, which we shall encounter in chapter 6, is that the trace of a product of two matrices A and B is independent of the order of multiplication (cyclic behavior of trace),

$$\mathrm{Tr}\;(\mathsf{A}\,\mathsf{B}) \equiv \sum_i (\mathsf{A}\,\mathsf{B})_{ii} = \sum_{ij} a_{ij}\,b_{ji} = \sum_{ji} b_{ji}\,a_{ij} = \sum_j (\mathsf{B}\,\mathsf{A})_{jj} = \mathrm{Tr}\;(\mathsf{B}\,\mathsf{A})\,. \qquad (5.130)$$

The symbol $\mathbf{1}$ in (5.128) stands either for the unity matrix in Dirac (spin) space or both Dirac-spin and isospin space combined. In the latter case it reads

$$\mathbf{1} \equiv \mathbf{1}^{\mathrm{Dirac}} \otimes \mathbf{1}^{\mathrm{iso}}\,, \qquad (5.131)$$

where $\otimes$ denotes the direct tensor (Kronecker) product of the 4×4 Dirac matrix $\mathbf{1}^{\mathrm{Dirac}}$ with the 2×2 isospin matrix $\mathbf{1}^{\mathrm{iso}}$. Thus $\mathbf{1}$ is a 8×8 matrix with matrix elements

$$(\mathbf{1})_{\zeta\zeta'} = \left(\mathbf{1}^{\mathrm{Dirac}}\right)_{\alpha\alpha'}\,\left(\mathbf{1}^{\mathrm{iso}}\right)_{ii'}\,, \qquad (5.132)$$

which is equivalent to

$$\delta_{\zeta\zeta'} = \delta_{\alpha\alpha'}\,\delta_{ii'}\,. \qquad (5.133)$$

The factor $(g_{\sigma B}\mathbf{1})_{\zeta_1\zeta_1'}$ in the Hartree term of (5.128) can therefore be replaced with $g_{\sigma B}\,\delta_{\zeta_1\zeta_1'}$. This is a particular feature that holds only for the scalar coupling case. For the more complicated couplings involving Dirac matrices one gets instead factors like $(\gamma^\mu)_{\zeta_1\zeta_1'}$, or combinations thereof. Finally we add that mesons like the ρ particle, which is a vector in isospin space, require the additional occurrence of Pauli matrices $\boldsymbol{\tau}$ in the respective coupling constants (cf. equation (5.124)).

Taking these considerations into account, equation (5.128) can then be brought into the alternative, somewhat more compact form

$$\boldsymbol{\Sigma}^B_{\zeta_1\zeta_1'}(p)\Big|_\sigma = -\mathrm{i}\,\delta_{\zeta_1\zeta_1'}\,\Delta^{0\sigma}(0)\,g_{\sigma B}\sum_{B'} g_{\sigma B'}\int \frac{\mathrm{d}^4 q}{(2\pi)^4}\,\mathrm{e}^{\mathrm{i}\eta q^0}\;\mathrm{Tr}\;\boldsymbol{g}^{B'}(q)$$
$$+\,\mathrm{i}\,g^2_{\sigma B}\int \frac{\mathrm{d}^4 q}{(2\pi)^4}\,\mathrm{e}^{\mathrm{i}\eta q^0}\,\Delta^{0\sigma}(p-q)\,\left(\mathbf{1}\otimes\boldsymbol{g}^{B'}(q)\otimes\mathbf{1}\right)_{\zeta_1\zeta_1'}\,. \qquad (5.134)$$

Hence, the contribution of the σ meson to the baryon self-energy can be written in the following manner:

$$\Sigma^{\mathrm{H},B}_{\zeta_1\zeta_1'}(p)\Big|_{\sigma} = -\mathrm{i}\,\delta_{\zeta_1\zeta_1'}\,\Delta^{0\sigma}(0)\,g_{\sigma B}\sum_{B'} g_{\sigma B'}\int\frac{\mathrm{d}^4q}{(2\pi)^4}\,\mathrm{e}^{\mathrm{i}\eta q^0}\,\mathrm{Tr}\,\boldsymbol{g}^{B'}(q)\,, \tag{5.135}$$

and

$$\Sigma^{\mathrm{F},B}_{\zeta_1\zeta_1'}(p)\Big|_{\sigma} = \mathrm{i}\,g^2_{\sigma B}\int\frac{\mathrm{d}^4q}{(2\pi)^4}\,\mathrm{e}^{\mathrm{i}\eta q^0}\,\Delta^{0\sigma}(p-q)\left(\mathbf{1}\otimes\boldsymbol{g}^B(q)\otimes\mathbf{1}\right)_{\zeta_1\zeta_1'}\,, \tag{5.136}$$

where $\Sigma^{\mathrm{H},B}$ and $\Sigma^{\mathrm{F},B}$ denote the Hartree and the Fock contributions to Σ^{B}, respectively. From equation (5.124) one finds that the other mesons, inclusive of the tensor coupling term of the ρ meson, contribute to Σ^{B} as follows:

$$\Sigma^{\mathrm{H},B}_{\zeta_1\zeta_1'}(p)\Big|_{\omega} = \mathrm{i}\,\gamma^{\mu}_{\zeta_1\zeta_1'}\,D^{0\omega}_{\mu\nu}(0)\,g_{\omega B}\sum_{B'} g_{\omega B'}\int\frac{\mathrm{d}^4q}{(2\pi)^4}\,\mathrm{e}^{\mathrm{i}\eta q^0}\,\mathrm{Tr}\left(\gamma^{\nu}\boldsymbol{g}^{B'}(q)\right)\,, \tag{5.137}$$

$$\Sigma^{\mathrm{F},B}_{\zeta_1\zeta_1'}(p)\Big|_{\omega} = -\,\mathrm{i}\,g^2_{\omega B}\int\frac{\mathrm{d}^4q}{(2\pi)^4}\,\mathrm{e}^{\mathrm{i}\eta q^0}\,\gamma^{\mu}_{\zeta_1\zeta_3}\,\gamma^{\nu}_{\zeta_2\zeta_1'}\,D^{0\omega}(p-q)_{\mu\nu}\,\boldsymbol{g}^B_{\zeta_3\zeta_2}(q)\,, \tag{5.138}$$

$$\Sigma^{\mathrm{H},B}_{\zeta_1\zeta_1'}(p)\Big|_{\pi} = 0\,, \tag{5.139}$$

$$\begin{aligned}\Sigma^{\mathrm{F},B}_{\zeta_1\zeta_1'}(p)\Big|_{\pi} = {}&\mathrm{i}\left(\frac{f_{\pi B}}{m_\pi}\right)^2\int\frac{\mathrm{d}^4q}{(2\pi)^4}\,\mathrm{e}^{\mathrm{i}\eta q^0}\left(\gamma_5\gamma_\mu\otimes\boldsymbol{\tau}\right)_{\zeta_1\zeta_3}\left(\gamma_5\gamma_\nu\otimes\boldsymbol{\tau}\right)_{\zeta_2\zeta_1'}\\ &\times(p-q)^{\mu}\,(p-q)^{\nu}\,\Delta^{0\pi}(p-q)\,\boldsymbol{g}^B_{\zeta_3\zeta_2}(q)\,,\end{aligned} \tag{5.140}$$

$$\begin{aligned}\Sigma^{\mathrm{H},B}_{\zeta_1\zeta_1'}(p)\Big|_{\rho} = {}&\mathrm{i}\sum_{B'}\int\frac{\mathrm{d}^4q}{(2\pi)^4}\,\mathrm{e}^{\mathrm{i}\eta q^0}\left[\left(g_{\rho B}\gamma_\mu - \mathrm{i}\,\frac{f_{\rho B}}{2m_B}\,(p-q)^{\lambda}\,\sigma_{\lambda\mu}\right)\otimes\boldsymbol{\tau}\right]_{\zeta_1\zeta_1'}\\ &\times D^{0\rho}(0)^{\mu\nu}\,\mathrm{Tr}\left\{\left[\left(g_{\rho B'}\gamma_\nu + \mathrm{i}\,\frac{f_{\rho B'}}{2m_{B'}}\,(p-q)^{\kappa}\,\sigma_{\kappa\nu}\right)\otimes\boldsymbol{\tau}\right]\boldsymbol{g}^{B'}(q)\right\}\,,\end{aligned} \tag{5.141}$$

and

$$\begin{aligned}\Sigma^{\mathrm{F},B}_{\zeta_1\zeta_1'}(p)\Big|_{\rho} = {}&-\mathrm{i}\int\frac{\mathrm{d}^4q}{(2\pi)^4}\,\mathrm{e}^{\mathrm{i}\eta q^0}\left[\left(g_{\rho B}\gamma_\mu - \mathrm{i}\,\frac{f_{\rho B}}{2m_B}\,(p-q)^{\lambda}\,\sigma_{\lambda\mu}\right)\otimes\boldsymbol{\tau}\right]_{\zeta_1\zeta_3}\\ &\times\left[\left(g_{\rho B}\gamma_\nu + \mathrm{i}\,\frac{f_{\rho B}}{2m_B}\,(p-q)^{\kappa}\,\sigma_{\kappa\nu}\right)\otimes\boldsymbol{\tau}\right]_{\zeta_2\zeta_1'} D^{0\rho}(p-q)^{\mu\nu}\,\boldsymbol{g}^B_{\zeta_3\zeta_2}(q)\,.\end{aligned} \tag{5.142}$$

5.5 Relativistic T-matrix approximation

5.5.1 T-matrix approximation versus Hartree–Fock

The parameters of the Hartree and the Hartree–Fock approximation, that is, the meson masses and coupling constants, enter as free parameters

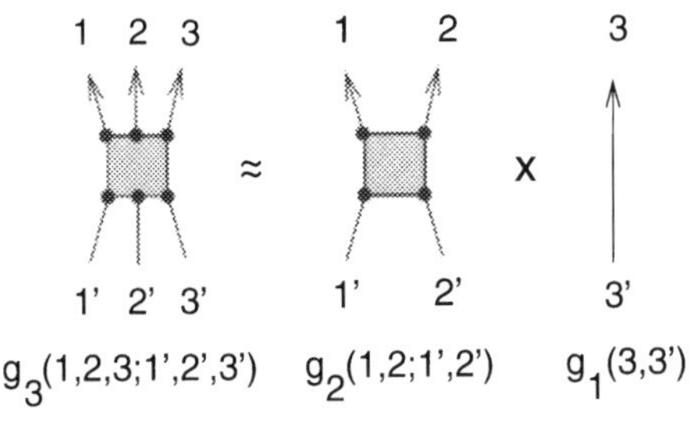

Figure 5.8. Factorization of six-point Green function, $\boldsymbol{g}_3$, into a product of four-point, $\boldsymbol{g}_2$, and two-point, $\boldsymbol{g}_1$, baryon Green functions, which truncates the many-body equations at the level of the **T**-matrix (ladder) approximation. Dynamical two-particle correlations contained in $\boldsymbol{g}_2$ are kept. Particle '3' contributes only as a spectator.

in these theories. Their values are therefore chosen such that the bulk properties of nuclear matter are reproduced by these approximations. For instance, in the case of the nonlinear scalar–vector–isovector Lagrangian solved for the Hartree approximation, the meson–nucleon coupling constants of the three fields σ, ω, ρ plus the cubic and quartic σ-meson self-interactions enter. These five parameters are then adjusted to the binding energy, effective nucleon mass, symmetry energy, and incompressibility of nuclear matter at saturation density [92, 100, 123, 124]. The application of the parameters of a given one-boson-exchange interaction, which are deduced from the analysis of two-nucleon scattering in free space and the properties of the deuteron, fails for these two approximations, since the resulting nuclear matter properties do not agree with the empirical values. Depending on the choice for the one-boson-exchange potential, for which there are different sets available in the literature, nuclear matter does not saturate or may be even unbound for some of the latter [117, 118, 125].

This feature is remedied by the relativistic **T**-matrix approximation, which goes beyond the relativistic Hartree and HF approximation insofar as it also accounts for so-called *dynamical* two-particle correlations, which are contained in the four-point Green function $\boldsymbol{g}_2$, and not only for statistical correlations as is the case for the Hartree or HF approximation. We recall that for both Hartree and HF, the dynamical correlations are lost because of the factorization of $\boldsymbol{g}_2$ into independent products of $\boldsymbol{g}_1$ functions (figures 5.7 and 5.6, respectively). The factorization scheme in the case of the **T**-matrix approximation, illustrated in figure 5.8, keeps the dynamical correlations contained in $\boldsymbol{g}_2$ [117, 118, 315, 316]. Naturally this approximation can only be justified as long as the densities are not considerably higher than the density of nuclear matter. Only in this case is the range of the strong interaction of the same size as the mean nucleon distance. Otherwise correlations other than those taken into account in the **T**-matrix approximation are likely to become important, which makes an independent pair approximation questionable [317].

As we shall see below, the **T**-matrix sums repeated two-particle scattering processes in matter up to infinite order for which reason it is also referred to as a ladder ($\boldsymbol{\Lambda}$) approximation. Mathematically **T** is given as the solution of a Bethe–Salpeter-type integral equation whose Born term is made up of a model for the relativistic one-boson-exchange interaction in free space, **V**. Since the parameters of **V** are adjusted to the nucleon–nucleon scattering problem in free space and the properties of the deuteron, the **T**-matrix approximation is referred to as a *parameter-free* approach. There exist a number of unresolved issues on which the applicability of the **T**-matrix approximation to dense matter calculations hinges upon. Some of the open issues, for instance, concern [109, 117, 118, 318]:

- Correlations other than those taken into account in the **T**-matrix approximation (i.e. three- and higher-body ones) may become important for the higher-density regime, that is, densities $\epsilon \gtrsim (2\text{–}3)\epsilon_0$. (Three-body correlations are less important in pure neutron matter than in symmetric nuclear matter, since these arise primarily from tensor forces acting in isospin-zero states [58].)
- The applicability of the meson-exchange model to the nucleon–nucleon force in superdense matter is not known.
- The lack of knowledge of the baryon–baryon and baryon–meson coupling strengths. A meson-exchange model for the hyperon–nucleon interaction has been presented, however, by Holzenkamp, Holinde, and Speth [307].
- The variation of the baryon–baryon and baryon–meson coupling strengths with density [313].
- The variation of meson masses with density [303, 312, 313].
- Because of derivative baryon–meson couplings, renormalizability of the theory is irrevocably lost, that is, corrections arising from the Dirac sea (see figure 9.3) cannot be calculated systematically. One is left with the so-called no-sea approximation.

These complications, some of which hold also for the Hartree and HF approximation, should not prevent one of course from applying the **T**-matrix approximation to dense matter calculations. On the contrary, it is exactly this application which enables one to test the applicability of the **T**-matrix formalism to dense matter calculations and explore possible quantitative differences with respect to the two less sophisticated methods, relativistic Hartree and relativistic HF. Naturally up to densities of say $(2\text{–}3)\epsilon_0$ one may certainly argue that the outcome of a relativistic **T**-matrix calculation is to be preferred over a relativistic Hartree or HF calculation [109], which totally ignores dynamical correlations. Up to such densities neutron star matter consists to a very good approximation of mainly chemically equilibrated protons and neutrons. Hence for neutron star matter up to such densities the **T**-matrix approximation may certainly qualify as a reliable many-body approximation. At densities beyond about $2\,\rho_0$ one is then left with the Hartree or HF approximation. In the matching density regime one has to make sure that the different many-body techniques lead to properties of many-baryon matter (e.g. chemical potentials, particle densities, effective masses) that are consistent with each other.

5.5.2 Martin–Schwinger hierarchy

Let us go back to the exact equation of motion for $\boldsymbol{g}_1^B$, as derived in equations (5.112) and (5.117). Considering only σ mesons for the moment, it is seen that the equation of motion is of the following type,

$$
\begin{aligned}
(\mathrm{i}\,\gamma^\mu \partial_{\mu,x_1} - m_B)\,\boldsymbol{g}^B(x_1, x_1') = &-\delta^4(x_1 - x_1') \\
&+ \mathrm{i} \sum_{B'} \int \mathrm{d}^4x' \; \Gamma^{\sigma B}\,\Gamma^{\sigma B'}\,\Delta^{0\sigma}(x_1, x')\,\boldsymbol{g}_2(x_1 B, x' B'; x'^+ B', x_1' B)\,, \quad (5.143)
\end{aligned}
$$

where we have introduced $\Gamma^{\sigma B} \equiv \mathrm{i}\mathbf{1}\,g_{\sigma B}$ and $\Gamma^{\sigma B'} \equiv \mathrm{i}\mathbf{1}\,g_{\sigma B'}$, which will turn out to be a useful notation. Equation (5.143) can readily be solved for $\boldsymbol{g}^B(x_x, x_1')$ using the method outlined in (5.77) to (5.79). Introducing the free baryon Green function associated with (5.143), which obeys

$$
(\mathrm{i}\,\gamma^\mu \partial_{\mu,x_1} - m_B)\,\boldsymbol{g}^{0B}(x_1, x_1') = -\,\delta^4(x_1 - x_1')\,, \qquad (5.144)
$$

the full Green function $\boldsymbol{g}^B$ is given by the expression

$$
\begin{aligned}
\boldsymbol{g}^B(x_1, x_1') = &\int \mathrm{d}^4y\; \boldsymbol{g}^{0B}(x_1, y)\{\delta(y - x_1') \\
&+ \sum_{B'} \int \mathrm{d}^4x' \; \Gamma^{\sigma B}\,\Gamma^{\sigma B'}\,\Delta^{0\sigma}(x_1, x')\,\boldsymbol{g}_2(x_1 B, x' B'; x'^+ B', y B)\}\,, \quad (5.145)
\end{aligned}
$$

which, upon performing the integral over the δ-function, leads to

$$
\begin{aligned}
\boldsymbol{g}^B(x_1, x_1') = \boldsymbol{g}^{0B}(x_1, x_1') + \mathrm{i} \sum_{B'} \int \mathrm{d}^4y \int \mathrm{d}^4x' \; &\{\boldsymbol{g}^{0B}(x_1, y) \\
\times\, \Gamma^{\sigma B}\,\Gamma^{\sigma B'}\,\Delta^{0\sigma}(x_1, x')\,&\boldsymbol{g}_2(x_1 B, x' B'; x'^+ B', y B)\}\,. \quad (5.146)
\end{aligned}
$$

The next step consists in deriving the equation of motion for $\boldsymbol{g}_2$, which, in analogy to the equation of motion for $\boldsymbol{g}_1$, couples to the baryon Green functions of the next higher as well as the next lower order, which are $\boldsymbol{g}_3$ and

$\boldsymbol{g}_1$, respectively. This feature repeats itself for all the higher-order baryon Green functions too. In general, the $2n$-point baryon Green function can be written as (cf. figure 5.1)

$$\begin{aligned} &\boldsymbol{g}_n^{B_1,\ldots,B_{n'}}(1,\ldots,n;1',\ldots,n') \\ &= \sum_{k=1}^{n} (-)^{k+1}\, \boldsymbol{g}_1^{0B_1,B_{k'}}(1;k')\; \boldsymbol{g}_{n-1}^{B_2,\ldots,B_{n'}}(2,\ldots,n;1',\ldots \text{omit}\, k' \ldots,n') \\ &+ \mathrm{i}\boldsymbol{g}_1^{0B_1,B_{m_1}}(1;m_1)\, \langle m_1,m_2|\mathbf{V}^{B_{m_1}B_{m_2}B_{m_3}B_{m_4}}|m_3,m_4\rangle \\ &\times \boldsymbol{g}_{n+1}^{B_2,\ldots,B_n,B_{m_3},B_{m_4},B_{m_2},B_{1'},\ldots,B_{n'}}(2,\ldots,n,m_3,m_4;m_2^+,1',\ldots,n'). \end{aligned} \tag{5.147}$$

Equation (5.147) is known as the Martin–Schwinger hierarchy, derived first by these authors for non-relativistic Green functions [319]. The first to study the relativistic case was Wilets [315]. The quantity $\mathbf{V}$ in (5.147) denotes the two-body interaction, which, for the σ-meson case discussed above, is given by $\Gamma^{\sigma B}\Delta^{0\sigma}\Gamma^{\sigma B'}$. According to the nature of the strong interaction between two nucleons, the mesons that contribute to $\mathbf{V}$ have masses less than about 1 GeV, the lightest of which being the π meson and the heaviest the ϕ, as can be seen in table 5.2. All these mesons can be accounted for in equation (5.146) by writing it in the form

$$\begin{aligned} \boldsymbol{g}_1^{B}(1,1') = \boldsymbol{g}_1^{0B}(1,1') + \mathrm{i}\sum\!\!\!\!\!\!\int \mathrm{d}^4x_1'' \sum\!\!\!\!\!\!\int \mathrm{d}^4x_2 \sum\!\!\!\!\!\!\int \mathrm{d}^4x_3 \sum\!\!\!\!\!\!\int \mathrm{d}^4x_4\; \boldsymbol{g}_1^{0B}(1,1'') \\ \times \langle 1''2|\mathbf{V}^{BB'}|34\rangle\, \boldsymbol{g}_2^{B'B}(34;2^+1'). \end{aligned} \tag{5.148}$$

The free baryon propagator $\boldsymbol{g}_1^{0B}(1,1')$ in (5.148) is given by [note that $k^\mu = \mathrm{i}\partial^\mu$ (cf. section B.2))

$$\boldsymbol{g}_1^{0B}(x_1 - x_1')^{-1}_{\zeta\zeta'} = \left(-\gamma^\mu\, k_{\mu,x_1} + m_B\right)_{\zeta\zeta'}\, \delta^4(x_1 - x_1'), \tag{5.149}$$

which follows from (5.144) in combination with

$$\int \mathrm{d}^4y\, g_1^{0B}(x_1,y)\, g_1^{0B}(y,x_2)^{-1} = \delta^4(x_1 - x_2). \tag{5.150}$$

The matrix elements $\langle 12|\mathbf{V}^{BB'}|34\rangle$ in (5.148) for the numerous types of mesons can be written in the compact form

$$\langle 12|\mathbf{V}^{BB'}|34\rangle \equiv \sum_{M=\sigma,\omega,\pi,\rho,\eta,\delta,\phi} \langle 12|\mathbf{V}_M^{BB'}|34\rangle, \tag{5.151}$$

where

$$\langle 12|\mathbf{V}_M^{BB'}|34\rangle \equiv \delta^4(x_1 - x_3)\, \delta^4(x_2 - x_4)\, \big\langle \zeta_1\zeta_2|\mathbf{V}_M^{BB'}(x_1,x_2)|\zeta_3\zeta_4\big\rangle. \tag{5.152}$$

The mathematical structure of $\mathbf{V}_M^{BB'}(x_1, x_2)$ depends on the nature of the baryon–meson couplings. These are, with increasing level of complexity, the scalar meson coupling ($M = \sigma, \delta$),

$$\langle \zeta_1 \zeta_2 | \mathbf{V}_M^{BB'}(x_1, x_2) | \zeta_3 \zeta_4 \rangle = \left(\Gamma^{MB}\right)_{\zeta_1\zeta_3} \left(\Gamma^{MB'}\right)_{\zeta_2\zeta_4} \Delta^{0M}(x_1, x_2)\,, \tag{5.153}$$

the pseudovector coupling ($M = \pi, \eta$),

$$\langle \zeta_1 \zeta_2 | \mathbf{V}_M^{BB'}(x_1, x_2) | \zeta_3 \zeta_4 \rangle = \left(\Gamma^{MB}_{x_1}\right)_{\zeta_1\zeta_3} \left(\Gamma^{MB'}_{x_1}\right)_{\zeta_2\zeta_4} \Delta^{0M}(x_1, x_2)\,, \tag{5.154}$$

and the vector coupling ($M = \omega, \rho, \phi$),

$$\langle \zeta_1 \zeta_2 | \mathbf{V}_M^{BB'}(x_1, x_2) | \zeta_3 \zeta_4 \rangle = \left(\Gamma^{MB}_{\mu,x_1}\right)_{\zeta_1\zeta_3} \left(\Gamma^{MB'}_{\mu,x_2}\right)_{\zeta_2\zeta_4} \left(\Delta^{0M}(x_1, x_2)\right)^{\mu\nu}\,. \tag{5.155}$$

The individual baryon–meson vertices Γ^{MB} in equations (5.153) through (5.155) are given by

$$\left(\Gamma^{MB}\right)_{\zeta_1\zeta_3} = \mathrm{i}\, g_{MB}\, (\mathbf{1})_{\zeta_1\zeta_3} \qquad \text{(scalar coupling)}\,, \tag{5.156}$$

$$\left(\Gamma^{MB}_{x_1}\right)_{\zeta_1\zeta_3} = \frac{f_{MB}}{m_M} \left(\gamma_5\, \gamma_\mu\, \partial^\mu_{x_1}\right)_{\zeta_1\zeta_3} \qquad \text{(pseudovector coupling)}\,, \tag{5.157}$$

$$\left(\Gamma^{MB}_{\mu,x_1}\right)_{\zeta_1\zeta_3} = \left(g_{MB}\, \gamma_\mu + \frac{\mathrm{i}}{2} \frac{f_{MB}}{2 m_B}\, \partial^\lambda_{x_1}\, [\gamma_\lambda, \gamma_\mu]\right)_{\zeta_1\zeta_3} \qquad \text{(vector coupling)}\,. \tag{5.158}$$

The vertices (5.156) through (5.158) refer to mesons which are isoscalars in isovector space. In the case of isovector mesons, like the ρ, π or δ meson (cf. table 5.2), the vertices are to be multiplied by a Pauli matrix $\boldsymbol{\tau}^r$ and summed over r, where $r = 1, 2, 3$ refer to the three isospin components of an isospin-1 (that is, $T = 1$) particle. Explicitly one has

$$\ldots\, (\mathbf{1})_{\zeta_1\zeta_3} \longrightarrow \sum_r \ldots\, (\mathbf{1} \otimes \boldsymbol{\tau}^r)_{\zeta_1\zeta_3}\,, \tag{5.159}$$

$$\ldots\, (\gamma_5)_{\zeta_1\zeta_3} \longrightarrow \sum_r \ldots\, (\gamma_5 \otimes \boldsymbol{\tau}^r)_{\zeta_1\zeta_3}\,. \tag{5.160}$$

As in sections 5.3 and 5.4, the functions Δ^{0M} and $\mathcal{D}^{0M}$ in (5.153) to (5.155) denote the free meson propagators. They are defined in equations (5.80), (5.95), (5.96), and (5.108).

5.5.3 Parameters of relativistic meson-exchange interactions

In this volume we shall study several different meson-exchange potentials, which have been constructed to model the nucleon–nucleon interaction

Table 5.3. Cutoff masses Λ_M and parameter κ_M of the Bonn and HEA boson-exchange interactions.

Quantity	Mesons (M) σ	ω	π	ρ	η	δ	ϕ
	Bonn potential						
Λ_M (GeV)	2.0	1.5	1.3	2.0	1.5	2.0	–
κ_M	1	1	1	2	1	1	–
	HEA potential						
Λ_M (GeV)	1.95	1.9	1.95	1.95	1.95	1.95	1.95
$\Lambda_M^{(v)}$ (GeV)	–	1.25	–	1.25	–	–	1.25

Table 5.4. Cutoff masses Λ_M and form factors F^{MN} of the BM boson-exchange interactions A through C.

Quantity	Mesons (M) σ	ω	π	ρ	η	δ
	Brockmann–Machleidt potential A					
Λ_M (GeV)	2.0	1.5	1.05	1.3	1.5	2.0
	Brockmann–Machleidt potential B					
Λ_M (GeV)	2.0	2.5	1.2	1.3	1.5	1.5
	Brockmann–Machleidt potential C					
Λ_M (GeV)	1.8	1.5	1.2	1.3	1.5	1.5

in terms of meson exchange. These interactions are the HEA and Bonn meson-exchange interactions, and the Brockmann–Machleidt (BM) potentials A, B, C. As already mentioned, these potentials allow for so-called parameter-free nuclear matter calculations, which, however, are to be performed at the level of the rather complicated ladder approximation to the $\mathbf{T}$-matrix. The incorporation of such forces into relativistic Hartree and HF calculations fails to reproduce the properties of infinite nuclear matter [118]. All the one-boson-exchange (OBE) amplitudes in (5.151), which enter in the determination of the scattering $\mathbf{T}$-matrix, are badly behaved asymptotically. As we shall see explicitly in chapter 10, they display polynomial divergences which make it impossible to obtain solutions of $\mathbf{T}$. The physical reason for the divergent behavior of the OBE amplitudes can be attributed to the fact that point-like nucleons are employed in the OBE model. To overcome this problem, so-called form factors, $F^{MN}(k', k)$ are

Table 5.5. Coupling constants and masses of relativistic one-boson-exchange interactions that enter in relativistic Brueckner–Hartree–Fock type calculations.

			Brockmann–Machleidt potentials		
Quantity	Bonn	HEA	*A*	*B*	*C*
m_N (MeV)	938.926	938.9	938.926	938.926	938.926
m_σ (MeV)	550 †	500	550	550	550
m_ω (MeV)	782.6	782.8	782.6	782.6	782.6
m_π (MeV)	138.03	138.5	138.03	138.03	138.03
m_ρ (MeV)	769	763	769	769	769
m_η (MeV)	548.8	548.5	548.8	548.8	548.8
m_δ (MeV)	983	960	983	983	983
m_ϕ (MeV)	–	1020	–	–	–
$g^2_{\sigma N}/4\pi$	8.2797	4.63	8.3141	8.0769	8.0279
$g^2_{\omega N}/4\pi$	20	14	20	20	20
$g^2_{\pi N}/4\pi$	14.6	13.00	14.9	14.6	14.6
$g^2_{\rho N}/4\pi$	0.81	1.5	0.99	0.95	0.95
$f_{\rho N}/g_{\rho N}$	6.1	3.5	6.1	6.1	6.1
$g^2_{\eta N}/4\pi$	5	6.0	7	5	3
$g^2_{\delta N}/4\pi$	1.1075	4.74	0.7709	3.1155	5.0742
$g^2_{\phi N}/4\pi$	–	7.0	–	–	–
Reference	[119]	[120]	[104]	[104]	[104]

† The parameters for the σ meson apply only to the isospin $T = 1$ potential. For $T = 0$ one has $m_\sigma = 720$ MeV and $g^2_\sigma/4\pi{=}16.9822$.

inserted at each vertex of the OBE interaction, as illustrated in figure 5.9. Mathematically, this results in replacing the baryon–meson vertices (5.156) through (5.158) according to the scheme [92, 119, 320, 321]

$$\Gamma^{MB} \longrightarrow F^{MN}(k - k')\,\Gamma^{MB}\,, \tag{5.161}$$

where M refers to the exchanged mesons listed in table 5.2, and $B = N \equiv p, n$. Alternatively to (5.161), one may also replace the meson propagators as

$$\Delta^{0M}(q) \longrightarrow \tilde{\Delta}^0 M(q) \equiv \left(F^{MB}(q)\right)^2 \Delta^{0M}(q)\,. \tag{5.162}$$

These replacements modify the OBE amplitudes in such a way that they converge sufficiently rapidly as $|\boldsymbol{k}| \to \infty$ to make the **T**-matrix equation of Fredholm type. Physically, the form factors account for the extended structure of the nucleon. In the absence of a fully fledged relativistic theory capable of describing the 'blob' in figure 5.9, the mathematical structure of the strong interaction form factor remains largely arbitrary. The form

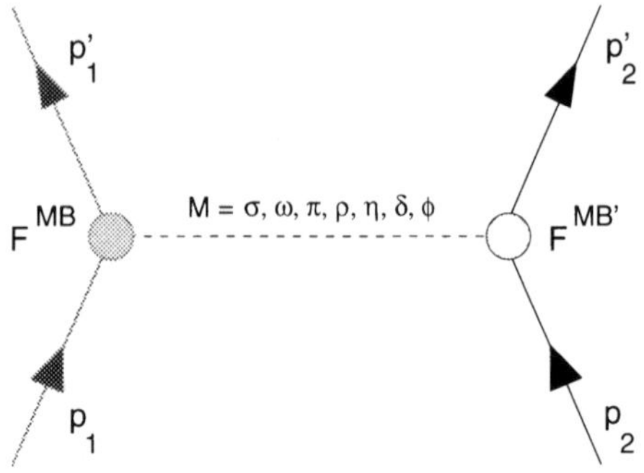

Figure 5.9. Born-term Feynman graph for the two-nucleon interaction with 'blobs' at the vertices representing the nucleon form factors, F^{MB}, which account for the extended structure of the baryons B, B'. The quantity M stands for the meson that is being exchanged between B and B'.

factors adopted for the OBE interactions are of the following type. In the case of the Bonn potential, they are given by

$$F^{MN}(\boldsymbol{k},\boldsymbol{k}') = \left(\frac{\Lambda_M^2 - m_M^2}{\Lambda_M^2 + (\boldsymbol{k}-\boldsymbol{k}')^2}\right)^{\kappa_M}, \tag{5.163}$$

while for the HEA potential,

$$F^{MN}(k,k') = \frac{\Lambda_M^2}{\Lambda_M^2 - (k-k')^2}, \tag{5.164}$$

if $M = \sigma, \pi, \delta, \eta$, and

$$F^{MN}(k,k') = \frac{\Lambda_M^2}{\Lambda_M^2 - (k-k')^2} \left(\frac{(\Lambda_M^{(v)})^2 - m_M^2}{(\Lambda_M^{(v)})^2 - (k-k')^2}\right)^{1/2}, \tag{5.165}$$

if $M = \omega, \rho, \phi$. For the Brockmann–Machleidt potentials A to C, monopole form factors of the type

$$F^{MN}(\boldsymbol{k},\boldsymbol{k}') = \frac{\Lambda_M^2 - m_M^2}{\Lambda_M^2 + (\boldsymbol{k}-\boldsymbol{k}')^2} \tag{5.166}$$

were chosen. The difference $\boldsymbol{k}-\boldsymbol{k}'$ denotes the three-momentum transfer. The values of the cutoff masses, Λ_M, and the exponent κ_M are given in tables 5.3 and 5.4. Table 5.5 reveals that the ratio f_{MN}/g_{MN} of the tensor/vector coupling of the vector bosons to the nucleon is either 3.5 or 6.1. In the strict interpretation of the vector dominance model [322], in which one assumes that the photon couples to the nucleon through a vector boson (i.e. ω or ρ meson), this ratio should be 3.7 for the ρ meson which describes the (anomalous) Pauli isovector form factor. The large value, $f_{\rho N}/g_{\rho N} \approx 6$, which is used in meson-exchange models for the nuclear force and which is also supported by empirical information from dispersion analyses [323, 324], seems to violate the vector dominance assumption. However, this can be understood in a simple way by assuming that there is also a direct vector coupling of the photon to the nucleon (i.e. no vector boson is involved in the interaction and hence no coupling goes through the ρ meson). Furthermore, the vector dominance model implies for the ω meson, which describes the Pauli isoscalar form factor, $f_{\omega N}/g_{\omega N} = -0.12$, which is a negligibly small value. Therefore most meson-exchange models for the nuclear force use $f_{\omega N} = 0$.

Without going into too much detail regarding the choice of the mathematical structure of F^{MN} [320, 321], we mention that the form factor (5.163) of the Bonn potential is of dipole structure. This is in close agreement with fits to the electromagnetic form factor of the nucleon, for which a dipole is preferred and from which a value for the cutoff mass on the order of 2 GeV is obtained [322]. The pionic form factor is the only one whose mathematical structure is empirically known to a certain extent [320]. Its form, given by

$$F^{\pi N}(\boldsymbol{k}, \boldsymbol{k}') = \left\{ \left(\Lambda_\pi^2 - m_\pi^2\right) / \left(\Lambda_\pi^2 - (\boldsymbol{k} - \boldsymbol{k}')^2\right) \right\}^n , \tag{5.167}$$

with the cutoff Λ_π and power n as free parameters, resembles the form factors of both the Bonn and BM interactions. These form factors would be of standard monopole form if $m_M = 0$ in (5.163) and (5.166).

It is illustrating to recall several of the advantages of relativistic boson-exchange interactions over phenomenological non-relativistic models [325] for the nucleon–nucleon interactions [320, 307]:

- The potential expression is uniquely determined for a given meson–nucleon interaction Lagrangian. However, form factors F^{MN}, up to now more or less phenomenological, which take into account the extended structure of the nucleon, need to be introduced.
- The interactions contain few parameters, which, moreover, are physical quantities and, in principle, are determined from other sources.
- The boson-exchange interactions correlate a great amount of data, providing a link between elementary particle physics and nuclear physics.
- Meson retardation (i.e. the energy dependence of the meson propagators) effects are significant.
- The saturation properties of nuclear matter predicted by relativistic **T**-matrix calculations performed for modern boson-exchange interactions are in agreement with the empirical values, as we shall see in

chapters 11 and 12. This is known to be a problem for many-body calculations performed for non-relativistic, phenomenological nucleon–nucleon interactions [326] (cf. Coester band in chapter 12).

5.5.4 Factorization of six-point Green function

Introducing the function L for the correlated part of g_2 according to

$$g_2(1,2;1',2') = g_1(1,1')\,g_1(2,2') - g_1(1,2')\,g_1(2,1') + \mathsf{L}(1,2;1',2') \quad (5.168)$$

and substituting this relation into the equation for the two-particle Green function, which is obtained from the Martin–Schwinger hierarchy (5.147), leads for L to

$$\begin{aligned}\mathsf{L}(1,2;1',2') = {} & (g_1^0(1,1') - g_1(1,1'))\,g_1(2,2') \\ & - (g_1^0(1,2') - g_1(1,2'))\,g_1(2,1') \\ & + \mathrm{i}\,g_1^0(1,3)\,\langle 34|\mathbf{V}|56\rangle\,g_3(2,5,6;4^+,1',2')\,. \end{aligned} \quad (5.169)$$

With the help of the equation for the two-point Green function (5.148), the first two terms of (5.169) can be expressed in terms of $g^0\mathbf{V}g_2$, which leads to

$$\begin{aligned}\mathsf{L}(1,2;1',2') = {} & \mathrm{i}\,g_1^0(1,3)\,\langle 34|\mathbf{V}|56\rangle\{g_3(2,5,6;4^+,1',2') \\ & - g_1(2,2')\,g_2(5,6;4^+,1') + g_1(2,1')\,g_2(5,6;4^+,2')\}\,. \end{aligned} \quad (5.170)$$

In the next step we decouple (5.170) from the Martin–Schwinger hierarchy by means of factorizing the three-point Green function as

$$\begin{aligned} g_3(2,5,6;4^+,1',2') \approx {} & g_1(2,4^+)\,g_2(5,6;1',2') \\ & - g_1(2,1')\,g_2(5,6;4^+,2') + g_1(2,2')\,g_2(5,6;4^+,1')\,, \end{aligned} \quad (5.171)$$

that is, only the dynamical two-particle correlations between particles '5' and '6' of g_2 are kept and particle '2' is treated as a spectator. For reasons that will become clear later, the factorization scheme (5.171) leads to the so-called $\mathbf{\Lambda}^{10}$ approximation. Inserting (5.171) into (5.170) and antisymmetrizing the resulting expression leads for L to

$$\begin{aligned}\mathsf{L}(1,2;1',2') = {} & \frac{\mathrm{i}}{2}\,\{g_1^0(1,3)\,g_1(2,4^+) - g_1^0(1,4)\,g_1(2,3^+)\} \\ & \times \langle 34|\mathbf{V}|56\rangle\,g_2(5,6;1',2')\,. \end{aligned} \quad (5.172)$$

The equation for the $\mathbf{T}$-matrix can now be derived from (5.172). For this purpose, we multiply (5.168) from the left with the boson-exchange matrix element $\langle 3'4'|\mathbf{V}|12\rangle$ and integrate over doubly occurring coordinates, which leads to

$$\begin{aligned}\langle 3'4'|\mathbf{V}|12\rangle\,g_2(12;1'2') = {} & \langle 3'4'|\mathbf{V}|12\rangle\{g_1(1,1')\,g_1(2,2') \\ & - g_1(1,2')\,g_1(2,1')\} + \langle 3'4'|\mathbf{V}|12\rangle\,\mathsf{L}(12;1'2')\,. \end{aligned} \quad (5.173)$$

From (5.172) one finds for the $\mathbf{V} \times \mathbf{L}$ term in (5.173)

$$\langle 3'4'|\mathbf{V}|12\rangle\, \mathbf{L}(1,2;1',2') = \langle 3'4'|\mathbf{V}|12\rangle \\ \times \left\{ \frac{\mathrm{i}}{2} \left[g_1^0(1,3)\, g_1(2,4^+) - g_1^0(1,4)\, g_1(2,3^+) \right] \right\} \langle 34|\mathbf{V}|56\rangle\, g_2(5,6;1',2') \,. \tag{5.174}$$

Making use of the defining relation for the **T**-matrix, given by

$$\langle 12|\mathbf{T}|34\rangle\, g_1(3,1')\, g_1(4,2') = \langle 12|\mathbf{V}|34\rangle\, g_2(3,4;1',2') \,, \tag{5.175}$$

the products $\mathbf{V}g_2$ in (5.173) and (5.174) can be replaced with

$$\langle 3'4'|\mathbf{V}|12\rangle\, g_2(12;1'2') = \langle 3'4'|\mathbf{T}|12\rangle\, g_1(1,1')\, g_1(2,2') \,, \tag{5.176}$$

$$\langle 3'4'|\mathbf{V}|56\rangle\, g_2(56;1'2') = \langle 3'4'|\mathbf{T}|56\rangle\, g_1(5,1')\, g_1(6,2') \,, \tag{5.177}$$

respectively. Substituting (5.176) and (5.177) into (5.173) and relabeling dummy variables then leads to

$$\langle 3'4'|\mathbf{T}|12\rangle\, g_1(1,1)\, g_1(2,2') = \langle 3'4'|\mathbf{V}|12\rangle \{ g_1(1,1')\, g_1(2,2') \\ - g_1(1,2')\, g_1(2,1') \} + \langle 3'4'|\mathbf{V}|34\rangle \left\{ \frac{\mathrm{i}}{2} \left[g_1^0(3,5)\, g_1(4,6^+) \right.\right. \\ \left.\left. - g_1^0(3,6)\, g_1(4,5^+) \right] \right\} \langle 56|\mathbf{T}|12\rangle\, g_1(1,1')\, g_1(2,2') \,. \tag{5.178}$$

Hence the equation for the (antisymmetrized) **T**-matrix itself reads

$$\langle 12|\mathbf{T}|1'2'\rangle = \langle 12|\mathbf{V}|1'2'\rangle - \langle 12|\mathbf{V}|2'1'\rangle \\ + \langle 12|\mathbf{V}|34\rangle\, \mathbf{\Lambda}^{10}(34;56)\, \langle 56|\mathbf{T}|1'2'\rangle \,, \tag{5.179}$$

where

$$\mathbf{\Lambda}^{10}(34;56) = \frac{\mathrm{i}}{2} \left\{ g_1^0(3,5)\, g(4,6) + g_1^0(3,6)\, g_1(4,5) \right\} \tag{5.180}$$

denotes the $\mathbf{\Lambda}^{10}$ propagator, which describes the propagation of baryons in the intermediate scattering states. The graphical illustration of (5.179) is given in figure 5.10. One sees from equation (5.180) that for the $\mathbf{\Lambda}^{10}$ approximation, one of the two baryons propagates as a free particle (described by g_1^0) while the other, g_1, feels the many-body background, which enters into g_1 via the self-energy $\mathbf{\Sigma}^B$. Thus the propagator for this approximation is to be determined self-consistently. This is also the case for the $\mathbf{\Lambda}^{11}$ approximation, for which the factorization of g_3 is given by

$$g_3(2,5,6;4^+,1',2') \approx g_1(2,4^+)\, g_2(5,6;1',2') \\ + g_2(5,6;4^+,7)\, g_1^{-1}(7,8)\, g_2(8,2;1',2') \,. \tag{5.181}$$

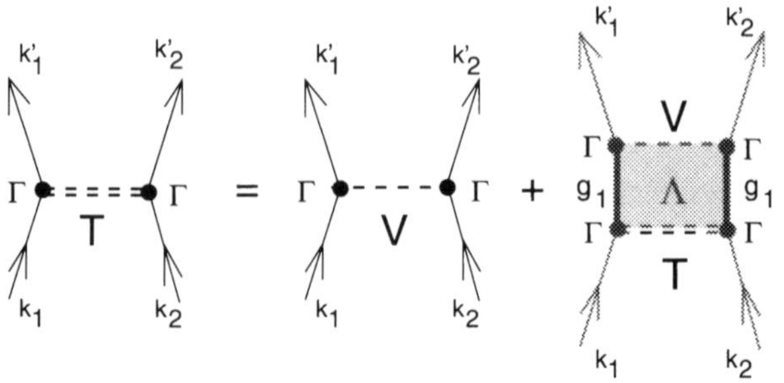

Figure 5.10. Diagrammatic representation of the self-consistent ladder summation that determines the **T**-matrix (dashed double lines). The first term on the right-hand side illustrates the Born-term Feynman graph for the scattering of two baryons via a given OBEP, **V** (dashed line). The second term sums repeated two-baryon scattering processes in the ladder approximation. $\mathbf{\Lambda}$ (shaded area), made up of a product of two-point functions $\mathbf{g}_1$, describes the propagation of two baryons in intermediate scattering states. k_i and k'_i are the four-momenta of incoming and outgoing baryons, respectively, and Γ the baryon–meson vertices.

The equation for the **T**-matrix has the same mathematical form as for the $\mathbf{\Lambda}^{10}$ approximation, except that the intermediate particle propagator now reads

$$\mathbf{\Lambda}^{11}(34;56) = \mathrm{i}\,\boldsymbol{g}_1(3,5)\,\boldsymbol{g}_1(4,6^+)\,. \tag{5.182}$$

It is seen from (5.182) that now the propagation of both particles in the intermediate scattering states is described by the full two-point Green function, $\boldsymbol{g}_1$, as opposed to $\mathbf{\Lambda}^{10}$ where only one of the two particles feels the medium. Obviously the simplest approximation results if the motion of none of the two baryons is coupled to the medium. In this case the propagation of both particles is simply described by a product of two free Green functions, that is,

$$\mathbf{\Lambda}^{00}(34;56) = \mathrm{i}\,\boldsymbol{g}_1^0(3,5)\,\boldsymbol{g}_1^0(4,6^+)\,. \tag{5.183}$$

This approximation is known as the $\mathbf{\Lambda}^{00}$ approximation. It constitutes the simplest of the three $\mathbf{\Lambda}$ approximations. All three approximations have been solved for the properties of nuclear matter in the non-relativistic case in [325].

We complete this section with deriving the expression for the self-energy in the **T**-matrix approximation. For this purpose we go back to equation (5.148) and replace $\mathbf{V}\boldsymbol{g}_2$ with $\mathbf{T}\boldsymbol{g}_1\boldsymbol{g}_1$ there, according to

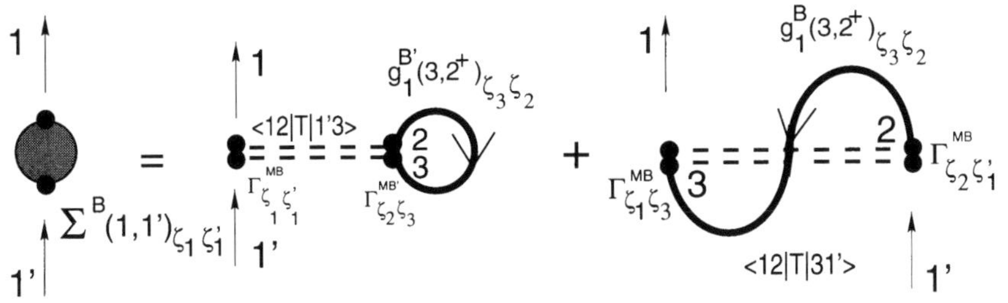

Figure 5.11. Diagrammatic representation of baryon self-energy $\boldsymbol{\Sigma}^B$ in the **T**-matrix (dashed double lines) approximation. The mathematical expression of $\boldsymbol{\Sigma}^B$ is given in (5.186). Its Hartree–Fock counterpart is shown in figure 5.4.

equation (5.175). We are thus left with

$$\boldsymbol{g}_1^B(1,1') = \boldsymbol{g}_1^{0B}(1,1') + \mathrm{i}\sum\!\!\!\!\!\!\int \mathrm{d}^4x_1'' \sum\!\!\!\!\!\!\int \mathrm{d}^4x_2 \sum\!\!\!\!\!\!\int \mathrm{d}^4x_3 \sum\!\!\!\!\!\!\int \mathrm{d}^4x_4\, \boldsymbol{g}_1^{0B}(1,1'') \times \langle 1''2|\mathbf{T}^{BB'}|34\rangle\, \boldsymbol{g}_1^{B'}(3,2^+)\, \boldsymbol{g}_1^B(4,1') . \tag{5.184}$$

A comparison of (5.184) with the Dyson equation derived in (5.117) then shows that

$$\mathrm{i}\,\boldsymbol{g}_1^{0B}(1,1'')\,\langle 1''2|\mathbf{T}^{BB'}|34\rangle\, \boldsymbol{g}_1^{B'}(3,2^+)\, \boldsymbol{g}_1^B(4,1') = -\boldsymbol{g}_1^{0B}(1,1'')\,\boldsymbol{\Sigma}(1'',4)\,\boldsymbol{g}_1^{B'}(4,1') . \tag{5.185}$$

Upon relabeling the variables, the expression for the self-energy reads

$$\boldsymbol{\Sigma}^B(1,2) = -\mathrm{i}\sum_{B'}\int \mathrm{d}^4x_3 \int \mathrm{d}^4x_4\, \langle 13|\mathbf{T}^{BB'}|42\rangle\, \boldsymbol{g}_1^{B'}(4,3^+) . \tag{5.186}$$

Its graphical representation is given in figure 5.11. Equation (5.186) is very similar in structure to its HF counterpart (5.124). Instead of the Born term of the **T**-matrix, however, which enters in (5.124), now the full scattering amplitude in the medium enters in the calculation of the self-energy.

In leaving this section, let us write down Dyson's equation (5.117) for *asymmetric* nuclear matter made up of protons and neutrons, $B = p, n$. In this case one finds from (5.117),

$$\left(-\gamma^\mu k_\mu + m^p + \boldsymbol{\Sigma}^p(k)\right)_{\alpha_1\alpha_3} \boldsymbol{g}^p_{\alpha_3\alpha_2}(k) = \delta_{\alpha_1\alpha_2} , \tag{5.187}$$

$$\left(-\gamma^\mu k_\mu + m^n + \boldsymbol{\Sigma}^n(k)\right)_{\alpha_1\alpha_3} \boldsymbol{g}^n_{\alpha_3\alpha_2}(k) = \delta_{\alpha_1\alpha_2} , \tag{5.188}$$

where $\boldsymbol{g}^p$ and $\boldsymbol{g}^p$ denote the proton and neutron two-point Green function, respectively. It is diagonal in isospin space,

$$\boldsymbol{G} \equiv \left(\frac{1+\tau_3}{2}\right)_{i_1i_2} \boldsymbol{g}^p_{\alpha_1\alpha_2} + \left(\frac{1-\tau_3}{2}\right)_{i_1i_2} \boldsymbol{g}^n_{\alpha_1\alpha_2} , \tag{5.189}$$

since the strong interaction is independent of isospin. The matrix notation of equations (5.187) and (5.188) reads

$$\begin{pmatrix} -\not{k} + m^p + \mathbf{\Sigma}^p & 0 \\ 0 & -\not{k} + m^n + \mathbf{\Sigma}^n \end{pmatrix} \begin{pmatrix} \boldsymbol{g}_1^p & 0 \\ 0 & \boldsymbol{g}_1^n \end{pmatrix} = \begin{pmatrix} 1 & 0 \\ 0 & 1 \end{pmatrix} . \quad (5.190)$$

5.5.5 T-matrix in momentum space

In what follows, we shall mostly be dealing with the non-antisymmetrized **T**-matrix,

$$\begin{aligned} \langle 12|\hat{\mathbf{T}}|1'2'\rangle &= \langle 12|\mathbf{V}|1'2'\rangle \\ &\quad + \langle 12|\mathbf{V}|34\rangle\, \mathbf{\Lambda}^{10}(3,4;5,6)\, \langle 56|\hat{\mathbf{T}}|1'2'\rangle\,, \end{aligned} \quad (5.191)$$

since the antisymmetrized solution is obtained from (5.191) as [327]

$$\langle 12|\mathbf{T}|1'2'\rangle = \langle 12|\hat{\mathbf{T}}|1'2'\rangle - \langle 12|\hat{\mathbf{T}}|2'1'\rangle\,. \quad (5.192)$$

The momentum-space representation of (5.191) is given by

$$\begin{aligned} \hat{\mathbf{T}}(K,K';k,k') &= \mathbf{V}(K,K';k,k') + \int \frac{d^4K''}{(2\pi)^4} \int \frac{d^4K'''}{(2\pi)^4} \int \frac{d^4k''}{(2\pi)^4} \int \frac{d^4k'''}{(2\pi)^4} \\ &\quad \times \{\mathbf{V}(K,K'';k,k'')\, \mathbf{\Lambda}(K'',K''';k'',k''')\, \hat{\mathbf{T}}(K''',K';k''',k')\}\,, \end{aligned} \quad (5.193)$$

where k, k', K, K' denote relative and center-of-mass four-momenta defined as

$$\begin{aligned} K &= k_1 + k_2\,, & k &= \frac{1}{2}(k_1 - k_2)\,, \\ K' &= k_1' + k_2'\,, & k' &= \frac{1}{2}(k_1' - k_2')\,, \end{aligned} \quad (5.194)$$

and for the Fourier transform

$$\begin{aligned} \hat{\mathbf{T}}(K,K';k,k') &= \int d^4x_1 \int d^4x_2 \int d^4x_1' \int d^4x_2' \langle x_1,x_2|\hat{\mathbf{T}}|x_1',x_2'\rangle \\ &\quad \times e^{i(K/2+k)x_1}\, e^{i(K/2-k)x_2}\, e^{-i(K'/2+k')x_1'}\, e^{-i(K'/2-k')x_2'}\,. \end{aligned} \quad (5.195)$$

Because of the homogeneity of space and time, which implies translational invariance of space and time, total three-momentum ($\boldsymbol{K}$) and total energy (K^0) of a pair of baryons that undergo a scattering process is conserved, i.e. $K' = K$. The **T**-matrix therefore obeys (note that $K \equiv K^\mu = (K^0, \boldsymbol{K})$, etc)

$$\hat{\mathbf{T}}(K,K';k,k') = (2\,\pi)^4\, \delta^4(K - K')\, \hat{\mathbf{T}}(K;k,k')\,. \quad (5.196)$$

As we have seen in (5.152), the potential matrix elements possess the structure

$$\langle x_1,x_2|\mathbf{V}|x_1',x_2'\rangle = \delta^4(x_1 - x_1')\, \delta^4(x_2 - x_2')\, \mathbf{V}(x_1 - x_2)\,. \quad (5.197)$$

Hence the potential is independent of the center-of-mass momentum and depends only on the difference of the relative momenta, that is, $\mathbf{V}(K, K'; k, k')$ can be written as

$$\mathbf{V}(K; k, k') = \mathbf{V}(k - k') = \int d^4x \, \mathbf{V}(x) \, e^{i(k-k')x} \,. \tag{5.198}$$

The dependence of the two-point Green function on its relative coordinates, that is, $\boldsymbol{g}_1(1, 1') = \boldsymbol{g}_1(x_1 - x_1')$, leads for the $\boldsymbol{\Lambda}$ propagators to

$$\boldsymbol{\Lambda}(k_3, k_4; k_5, k_6) = (2\,\pi)^8 \, \delta^4(k_3 - k_5) \, \delta^4(k_4 - k_6) \, \boldsymbol{\Lambda}(k_3, k_4) \,, \tag{5.199}$$

with

$$\boldsymbol{\Lambda}^{00}(k_3, k_4) = i \, e^{i\eta k_4^0} \, \boldsymbol{g}_1^0(k_3) \, \boldsymbol{g}_1^0(k_4) \,, \tag{5.200}$$

$$\boldsymbol{\Lambda}^{10}(k_3, k_4) = \frac{i}{2} \left\{ e^{i\eta k_4^0} \, \boldsymbol{g}_1^0(k_3) \, \boldsymbol{g}_1(k_4) + e^{i\eta k_3^0} \, \boldsymbol{g}_1^0(k_4) \, \boldsymbol{g}_1(k_3) \right\} , \tag{5.201}$$

$$\boldsymbol{\Lambda}^{11}(k_3, k_4) = i \, e^{i\eta k_4^0} \, \boldsymbol{g}_1(k_3) \, \boldsymbol{g}_1(k_4) \,. \tag{5.202}$$

Inserting (5.196), (5.198) and (5.199) into (5.193) results in the **T**-matrix equation in momentum space given by

$$\begin{aligned} \hat{\mathbf{T}}(K; k, k') &= \mathbf{V}(k - k') \\ &+ \int \frac{d^4k''}{(2\pi)^4} \, \mathbf{V}(k - k'') \, \boldsymbol{\Lambda}\Big(\frac{K}{2} + k'', \frac{K}{2} - k''\Big) \, \hat{\mathbf{T}}(K; k'', k') \,. \end{aligned} \tag{5.203}$$

This equation constitutes an operator equation in the spin and isospin space of the two nucleons. The momenta k', k'', k denote the relative four-momenta in the initial, intermediate and final state. The timelike components of k and k'', i.e. k_0 and k_0'', are in general unequal to zero, because this equation is an off-mass-shell equation. Equation (5.203) may be considered as the relativistic, quantum field theoretical generalization of the Lippmann–Schwinger equation, well known from the non-relativistic two-body scattering problem in free space [328]. We shall come back to an analysis of the non-relativistic limit of **T** briefly in section 12.5.

The most orthodox, manifestly covariant approach to the relativistic two-nucleon potential in quantum field theory constitutes the four-dimensional Bethe–Salpeter equation, given in operator notation by [329]

$$\mathbf{T} = \mathbf{K} + i \int \mathbf{K} \, \boldsymbol{g} \, \boldsymbol{g} \, \mathbf{T} \,. \tag{5.204}$$

Obviously, our scattering equation (5.203) bears a strong similarity with the full two-body scattering amplitude (5.204) in that the full two-body scattering kernel **K**, which in general is the sum of all connected two-particle-irreducible diagrams [330], has become replaced by the one-boson-exchange interaction: $\mathbf{K} \to \mathbf{V} \equiv \mathbf{V}_{\mathrm{OBE}}$. This constitutes a simpler parametrization and allows the important Lorentz structure of the interaction to be maintained [122]. More than that, this replacement is a necessary step, enforced by the fact that the full kernel-$\mathbf{K}$ sum cannot be expressed in closed form, to render the Bethe–Salpeter equation solvable. Without this replacement any attempt to solve the Bethe–Salpeter exactly would be evidently hopeless.

Despite the approximation $\mathbf{K} \to \mathbf{V} \equiv \mathbf{V}_{\mathrm{OBE}}$, the scattering equation (5.204) still constitutes a four-dimensional integral equation, which demands the introduction of additional simplifying approximations. Its three-dimensional reduction, for instance, is known to be accomplishable by defining a unitarized *two-particle* propagator $\boldsymbol{g}_2$ and a quasi-potential $\mathbf{U}$ (not to be confused with $\mathbf{U}$ in section 10.2 where this symbol denotes the mixed nucleon–antinucleon matrix elements) according to

$$\mathbf{T} = \mathbf{U} + \int \mathbf{U}\, g_2\, \mathbf{T}\,, \tag{5.205}$$

$$\mathbf{U} = \mathbf{V} + \int \mathbf{V}\,(\mathrm{i}\, g_1\, g_1 - g_2)\, \mathbf{U}\,. \tag{5.206}$$

These two equations are equivalent to (5.204) with $\mathbf{K}$ replaced with $\mathbf{V}$. If $\boldsymbol{g}_2$ is a good approximation to $\mathrm{i}\boldsymbol{g}_1\boldsymbol{g}_1$, which is a product made up of one-particle propagators, it is reasonable to take $\mathbf{U} \simeq \mathbf{V}$ [92, 93]. Corrections may be evaluated using equation (5.206) [331, 332]. Various choices for $\boldsymbol{g}_2$, which must maintain both Lorentz covariance and the unitarity of the scattering amplitude in the medium, were discussed in the literature [92, 93, 320, 331]. A popular choice for the covariant, unitarized $\boldsymbol{g}_2$ function is the Blankenbecler–Sugar propagator [333], which has the form [92]

$$\boldsymbol{g}_2^{\mathrm{BBS}}(k'', K_0) = \delta(k''_0)\, \bar{\boldsymbol{g}}_2(\boldsymbol{k}'', K_0)\,, \tag{5.207}$$

with k'' the relative four-momentum of the nucleon pair. The mathematical form of $\bar{\boldsymbol{g}}_2$ is of no importance right now. (We shall come back to this issue in chapter 9.) What should be noted, however, is the δ-function in (5.207) which reduces the four-dimensional integral equation (5.204) to a three-dimensional one, which is then tractable. As we shall see in chapter 9, the mathematical structure of $\boldsymbol{g}_2^{\mathrm{BBS}}$ displayed in (5.207) carries over to the two-particle propagator $\boldsymbol{\Lambda}$, rendering the $\mathbf{T}$-matrix equation (5.203) tractable too.

5.5.6 Spectral representation of T-matrix

To explore the analytic properties of the **T**-matrix, we go back to (5.191) and rewrite the scattering matrix formally as

$$\tilde{\mathbf{T}}(z) = \frac{\mathbf{V}}{1 - \mathbf{V}\,\tilde{\mathbf{\Lambda}}(z)} \,. \qquad (5.208)$$

The tildes denote analytically continued functions into the complex energy plane. The momentum dependence of **T** and $\mathbf{\Lambda}$ is unimportant for what follows. So we ignore these variables here. As we shall see later, the $\mathbf{\Lambda}^{00}$ propagator has the mathematical structure

$$\tilde{\mathbf{\Lambda}}^{00}(z) \propto \frac{1}{z - \omega^{0B}(k_3) - \omega^{0B}(k_4)} \,, \qquad (5.209)$$

where $\omega^{0B}(k) = \sqrt{m_B^2 + \boldsymbol{k}^2}$ is the energy–momentum relation of a free baryon of type B. Inspection of equation (5.208) in the case of the $\mathbf{\Lambda}^{00}$ propagator reveals that the **T**-matrix possesses a cut along the real energy axis for $\mathrm{Re}\, z \in [m_B, \infty)$ but behaves analytically otherwise, as shown in figure 5.12. In addition **T** will have isolated poles ω_k along the real axis too if there are bound particle states in the medium [334]. We shall assume that all bound states have energies greater than the chemical potential, that is, $\omega_k > \mu$. The other case $\omega_k < \mu$ corresponds to Cooper pairs in matter [334]. We shall not pursue this possibility here but assume that neutron star matter is normal fluid. In fact whether or not neutron star matter is superfluid has practically no consequences for the global properties of neutron stars, such as the range of possible masses, radii, moment of inertia, limiting rotational periods, since superfluidity changes the equation of state of neutron star matter only insignificantly.[1] The situation is different however for phenomena like the cooling of neutron stars and very likely the phenomenon of pulsar glitches.

The analytic properties of the **T**-matrix in the case of the $\mathbf{\Lambda}^{10}$ approximation are similar to those established just above for the $\mathbf{\Lambda}^{00}$ approximation. The analytically continued $\mathbf{\Lambda}^{10}$ propagator behaves analytically in the entire complex energy plane except for $\mathrm{Re}\, z \in [m_B + \mu^B, \infty)$. So the analytic properties of **T** are the same as above, except that the onset of the cut has been shifted from m_B to $m_B + \mu^B$.

To find the spectral representation of **T**, we note that asymptotically

$$\tilde{\mathbf{T}}(z) \stackrel{z\to\infty}{\longrightarrow} \mathbf{V} + \mathcal{O}\left(\frac{1}{z}\right) , \qquad (5.210)$$

[1] We recall that the pairing energy is typically less than about one percent of the total interaction energy which makes only very little difference to the pressure versus density relation.

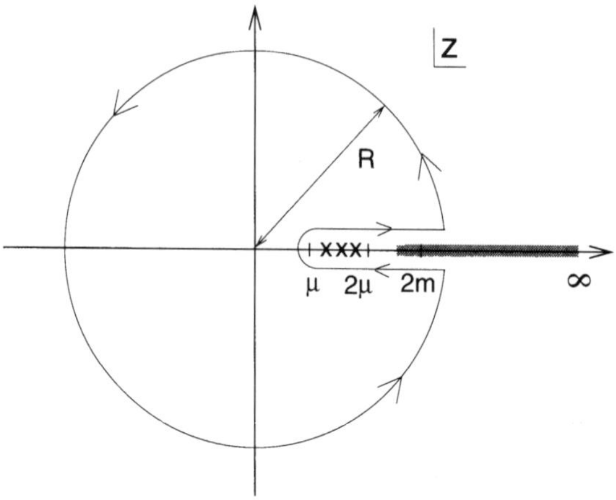

Figure 5.12. Contour of integration applied to the function $F(z)$ to obtain equation (5.211). Crosses ($\times$) correspond to bound states, the shaded area denotes the cutline along the real energy axis.

which follows from equations (5.208) and (5.209). Thus by means of applying Cauchy's integral formula to the function $F(z) \equiv \tilde{\mathbf{T}}(z) - \mathbf{V}$ and choosing a contour of integration as shown in figure 5.12 [335] leads to

$$\tilde{\mathbf{T}}(z) = \mathbf{V} + \int_{\mu}^{+\infty} \mathrm{d}\omega \, \frac{\mathcal{T}(\omega)}{z - \omega}, \tag{5.211}$$

with

$$\mathcal{T}(\omega) = \frac{\mathrm{i}}{2\pi}\{\tilde{\mathbf{T}}(\omega + \mathrm{i}\eta) - \tilde{\mathbf{T}}(\omega - \mathrm{i}\eta)\}. \tag{5.212}$$

Because of Schwarz's reflection principle [336], the spectral function $\mathcal{T}(\omega)$ is a real function. Moreover $\mathcal{T}(\omega) = 0$ for $\omega < \mu^B$ so that the integration in (5.211) can be extended to $[0, +\infty)$.

We conclude this section by turning back to the expression for the self-energy in the case of the $\mathbf{T}$-matrix approximation, derived in (5.186). Fourier transformation of this expression to four-momentum space leads to

$$\begin{aligned}\mathbf{\Sigma}^B_{\zeta_1\zeta_2}(k) = \mathrm{i}\sum_{B'}\sum_{\zeta_3\zeta_4}\int \frac{\mathrm{d}^4k'}{(2\pi)^4}\mathrm{e}^{\mathrm{i}\eta k_0'}\Big\{\Big\langle \zeta_1\zeta_3\frac{k-k'}{2}\Big|\hat{\mathbf{T}}^{BB'}(k+k')\Big|\frac{k-k'}{2}\zeta_2\zeta_4\Big\rangle \\ - \Big\langle \zeta_1\zeta_3\frac{k-k'}{2}\Big|\hat{\mathbf{T}}^{BB'}(k+k')\Big|\frac{k'-k}{2}\zeta_4\zeta_2\Big\rangle\Big\}\, \boldsymbol{g}^{B'}_{\zeta_4\zeta_3}(k').\end{aligned} \tag{5.213}$$

As in connection with (6.86), we perform an analytical continuation of equation (5.213) into the complex energy plane, and then replace the

T-matrix and two-point Green function with their respective spectral representations, given in (5.211) and (6.71). One gets

$$\tilde{\boldsymbol{\Sigma}}(z,\boldsymbol{k}) = \int \frac{\mathrm{d}^3\boldsymbol{k}'}{(2\pi)^3} \int \mathrm{d}\omega \,\{\hat{\mathbf{T}}(z+\omega+\mu) - \hat{\mathbf{T}}^{\mathrm{ex}}(z+\omega+\mu)\}\, \boldsymbol{\Xi}(\omega)\,\Theta(-\omega)$$
$$- \int \frac{\mathrm{d}^3\boldsymbol{k}'}{(2\pi)^3} \int \mathrm{d}\omega \int \mathrm{d}\omega' \,\frac{\mathcal{T}(\omega+\mu) - \mathcal{T}^{\mathrm{ex}}(\omega+\mu)}{z-\omega-\omega'-\mu}\, \boldsymbol{\Xi}(\omega')\,\Theta(-\omega)\,. \quad (5.214)$$

The second term in (5.214) does not contribute to $\tilde{\boldsymbol{\Sigma}}$ since $\mathcal{T}(\omega+\mu) = 0$ for $\omega < 0$, as pointed out just above.

Chapter 6

Spectral representation of the two-point Green function

6.1 Finite-temperature two-point function

For technical purposes, it is extremely useful to introduce a spectral representation for the two-point baryon Green function $\boldsymbol{g}_1^B$, for it will enables us to perform the energy integrations in the numerous baryon self-energy expressions, derived in equations (5.135) to (5.142), analytically. Instead of having to deal with $\boldsymbol{g}_1^B$, we are then left with the determination of the baryon spectral function associated with $\boldsymbol{g}_1^B$. This technique is particularly useful for systems whose self-energies are pure real functions, since then a *single-particle* description for the baryons in matter holds. Mathematically, this reflects itself in a spectral function which separates a δ-function which contributes only for energies equal to the single-particle energy of a baryon in matter. This feature renders the integrations over the energy variable mentioned just above nearly trivial. The single-particle behavior, which is exact for pure real self-energies, breaks down if the self-energy becomes complex. In this case the δ-function spreads out over a certain finite energy range, which is the broader the larger the imaginary part of the baryon self-energy. Nevertheless for not too large imaginary parts the single-particle picture appears to be well applicable [337].

In the following discussion we shall be somewhat more general than in section 5.3, where we have introduced the two-point baryon Green function at zero temperature only, by extending its definition to finite temperatures. It is given by

$$\boldsymbol{g}^{BB'}(x,\zeta;x',\zeta') = \mathrm{i}\,\frac{\mathrm{Tr}\left(\mathrm{e}^{-\beta\mathcal{H}}\,\hat{T}\big(\psi_B(x,\zeta)\,\bar{\psi}_{B'}(x',\zeta')\big)\right)}{\mathrm{Tr}\;\mathrm{e}^{-\beta\mathcal{H}}}\,, \tag{6.1}$$

which denotes the quantum mechanical average of time-ordered baryon field operators over a canonical ensemble [338, 339, 340]. The auxiliary functions

$g_<$ and $g_>$, defined in equations (5.65) and (5.66), are now given by

$$g_>^{BB'}(x,\zeta;x',\zeta') \equiv \mathrm{i}\,\langle psi_B(x,\zeta)\,\bar{\psi}_{B'}(x',\zeta')\rangle_\beta = \mathrm{i}\,\frac{\mathrm{Tr}\left(\mathrm{e}^{-\beta\mathcal{H}}\;\psi_B(x,\zeta)\;\bar{\psi}_{B'}(x',\zeta')\right)}{\mathrm{Tr}\;\mathrm{e}^{-\beta\mathcal{H}}}\,, \tag{6.2}$$

$$g_<^{BB'}(x,\zeta;x',\zeta') \equiv -\,\mathrm{i}\,\langle\bar{\psi}_{B'}(x',\zeta')\,\psi_B(x,\zeta)\rangle_\beta = -\,\mathrm{i}\,\frac{\mathrm{Tr}\left(\mathrm{e}^{-\beta\mathcal{H}}\;\bar{\psi}_{B'}(x',\zeta')\;\psi_B(x,\zeta)\right)}{\mathrm{Tr}\;\mathrm{e}^{-\beta\mathcal{H}}}\,, \tag{6.3}$$

where $\langle\ldots\rangle_\beta$ refers to the definition at finite temperatures. The quantity $\mathcal{H}$ denotes the system's Hamiltonian. The time-development operator $\mathrm{e}^{-\mathrm{i}\mathcal{H}x_0}$ contained in the fields ψ_B and $\bar{\psi}_B$,

$$\psi_B(x_0,\boldsymbol{x}) = \mathrm{e}^{\mathrm{i}\mathcal{H}x_0}\;\psi_B(0,\boldsymbol{x})\;\mathrm{e}^{-\mathrm{i}\mathcal{H}x_0}\,, \tag{6.4}$$

bears a strong formal similarity to the weighting factor $\mathrm{e}^{-\beta\mathcal{H}}$ that occurs in the canonical average of (6.1) to (6.3). By means of considering the time variables x_0, x_0' of $g^B(x,x')$ to being restricted to $0 \le \mathrm{i}x_0,\ \mathrm{i}x_0' \le \beta$ and, secondly, extending the definition of the time-ordering operator to mean $\mathrm{i}x_0$ and $\mathrm{i}x_0'$ ordering when times are imaginary, then the Green functions are again well defined in the interval $\mathrm{i}x_0,\ \mathrm{i}x_0' \in (0,\beta)$ [338, 339].

By making use of the cyclic property of the trace, shown in (5.130), one then readily verifies the following relations for imaginary times in the interval $(0,-\mathrm{i}\beta)$:

$$g^B(x,\zeta;x',\zeta')\big|_{x_0=0} = g_<^B(x,\zeta;x',\zeta')\big|_{x_0=0}\,, \tag{6.5}$$

$$g^B(x,\zeta;x',\zeta')\big|_{x_0=-\mathrm{i}\beta} = g_>^B(x,\zeta;x',\zeta')\big|_{x_0=-\mathrm{i}\beta}\,. \tag{6.6}$$

This remarkable periodicity of the finite-temperature baryon Green function in the limited imaginary-time domain will be fundamental to all of the subsequent work. For definiteness, let us consider the case that $\mathrm{i}x_0'$ is fixed ($0 < \mathrm{i}x_0' < \beta$). It is then verified that (as before, we drop the superscript B's carried by g and the baryon field operators in side-calculations):

$$\begin{aligned}
\left(\mathrm{Tr}\;\mathrm{e}^{-\beta\mathcal{H}}\right)\,g_<(x,x')|_{x_0=0} &= -\,\mathrm{i}\;\mathrm{Tr}\left[\mathrm{e}^{-\beta\mathcal{H}}\,\bar{\psi}(x')\,\psi(x)\right]_{x_0=0}\\
&= -\,\mathrm{i}\;\mathrm{Tr}\left[\mathrm{e}^{-\beta\mathcal{H}}\,\bar{\psi}(x')\,\mathrm{e}^{-\mathrm{i}\mathcal{H}x_0}\,\psi(0,\boldsymbol{x})\,\mathrm{e}^{\mathrm{i}\mathcal{H}x_0}\right]\\
&= -\,\mathrm{i}\;\mathrm{Tr}\left[\mathrm{e}^{\mathrm{i}\mathcal{H}x_0}\,\mathrm{e}^{-\beta\mathcal{H}}\,\bar{\psi}(x')\,\mathrm{e}^{-\mathrm{i}\mathcal{H}x_0}\,\psi(x)\right]_{x_0=-\mathrm{i}\beta}\\
&= -\,\mathrm{i}\;\mathrm{Tr}\left[\mathrm{e}^{-\beta\mathcal{H}}\,\psi(x)\,\bar{\psi}(x')\right]_{x_0=-\mathrm{i}\beta}\\
&= -\,\left(\mathrm{Tr}\;\mathrm{e}^{-\beta\mathcal{H}}\right)\,g_>(x,x')|_{x_0=-\mathrm{i}\beta}\,,
\end{aligned} \tag{6.7}$$

from which it follows that

$$g_<^B(x,\zeta;x',\zeta')\big|_{x_0=0} = -\,g_>^B(x,\zeta;x',\zeta')\big|_{x_0=-\mathrm{i}\beta}\,, \tag{6.8}$$

and

$$\boldsymbol{g}^B(x,\zeta;x',\zeta')\big|_{x_0=0} = -\,\boldsymbol{g}^B(x,\zeta;x',\zeta')\big|_{x_0=-\mathrm{i}\beta}\,. \tag{6.9}$$

The latter relation follows via substituting (6.8) into equations (6.5) and (6.6). The minus signs in (6.8) and (6.9) is a consequence of the Fermi statistic obeyed by the baryons.

In the next step we incorporate the anti-periodic behavior of (6.9) into $\boldsymbol{g}^B(x,x')$, which is accomplished by introducing Fourier series and integrals as follows. For discrete frequencies of $\omega_n \equiv \frac{2n+1}{-\mathrm{i}\beta}\,\pi$ with $n = 0, \pm 1, \ldots$, the Fourier representation of the two-point baryon function reads [338, 341, 342, 343]

$$\boldsymbol{g}^B(x,\zeta;x',\zeta') = \frac{1}{(-\mathrm{i}\beta)^2}\sum_{n,n'} \mathrm{e}^{-\mathrm{i}\omega_n x_0 + \mathrm{i}\omega_{n'} x'_0} \times \int\frac{\mathrm{d}^3\boldsymbol{k}}{(2\pi)^3}\int\frac{\mathrm{d}^3\boldsymbol{k}'}{(2\pi)^3}\,\mathrm{e}^{\mathrm{i}\boldsymbol{k}\cdot\boldsymbol{x} - \mathrm{i}\boldsymbol{k}'\cdot\boldsymbol{x}'}\,\boldsymbol{g}^B(\omega_n,\boldsymbol{k},\zeta;\omega_{n'},\boldsymbol{k}',\zeta')\,, \tag{6.10}$$

with its inverse given by

$$\boldsymbol{g}^B(\omega_n,\boldsymbol{k},\zeta;\omega_{n'},\boldsymbol{k}',\zeta') = \int_0^{-\mathrm{i}\beta}\mathrm{d}x_0\int_0^{-\mathrm{i}\beta}\mathrm{d}x'_0\,\mathrm{e}^{\mathrm{i}\omega_n x_0 - \mathrm{i}\omega_{n'} x'_0} \times \int\mathrm{d}^3\boldsymbol{x}\int\mathrm{d}^3\boldsymbol{x}'\,\mathrm{e}^{-\mathrm{i}\boldsymbol{k}\cdot\boldsymbol{x} + \mathrm{i}\boldsymbol{k}'\cdot\boldsymbol{x}'}\,\boldsymbol{g}^B_{\zeta\zeta'}(x-x')\,. \tag{6.11}$$

The δ-function for discrete energies has the form

$$\delta(x_0) = \frac{1}{-\mathrm{i}\beta}\sum_n \mathrm{e}^{-\mathrm{i}\omega_n x_0}\,, \qquad \delta_{nn'} = \frac{1}{-\mathrm{i}\beta}\int_0^{-\mathrm{i}\beta}\mathrm{d}x_0\,\mathrm{e}^{\mathrm{i}x_0(\omega_n-\omega_{n'})}\,. \tag{6.12}$$

Because of translational invariance of space and time, the arguments of the two-point function obey the relation $\boldsymbol{g}^B(x,x') = \boldsymbol{g}^B(x-x')$, which implies for $\boldsymbol{g}^B(\omega_n,\omega_{n'})$ of equation (6.10),

$$\boldsymbol{g}^B(\omega_n,\boldsymbol{k},\zeta;\omega_{n'},\boldsymbol{k}',\zeta') = (-\mathrm{i}\beta)\,(2\pi)^3\,\delta_{nn'}\,\delta^3(\boldsymbol{k}-\boldsymbol{k}')\,\boldsymbol{g}^B_{\zeta\zeta'}(\omega_n,\boldsymbol{k})\,. \tag{6.13}$$

The momentum-space representation of the free baryon propagator, $\boldsymbol{g}^{0B}$, is given by

$$(\gamma^\mu k_\mu - m_B)_{\zeta\zeta''}\,\boldsymbol{g}^{0B}_{\zeta''\zeta'}(k) = -\,\delta_{\zeta\zeta'}\,. \tag{6.14}$$

This expression is formally identical with the Fourier transformed of (5.144). Here, however, the four-momenta are given by $k = (k^0,\boldsymbol{k}) = (\omega_n,\boldsymbol{k})$. So there is no ambiguity in the division process when solving (6.14) for $\boldsymbol{g}^{0B}(k)$, that is $(\gamma^\mu k_\mu - m_B) \neq 0$ [338].

In the next step we combine the anti-periodicity properties derived for $\boldsymbol{g}^B$ above with the real-time formalism to represent $\boldsymbol{g}^B(x - x')$ by Fourier integrals. With the definitions for Fourier integrals given in appendix B.2, one obtains

$$\boldsymbol{g}_<(k_0, \boldsymbol{k}) = \int \mathrm{d}^4x \; \mathrm{e}^{\mathrm{i}kx} \; \boldsymbol{g}_<(x^0, \boldsymbol{x}) \tag{6.15}$$

$$= - \int \mathrm{d}x_0 \int \mathrm{d}^3\boldsymbol{x} \; \mathrm{e}^{\mathrm{i}(k_0 x_0 - \boldsymbol{k}\cdot\boldsymbol{x})} \; \boldsymbol{g}_>(x_0 - \mathrm{i}\beta, \boldsymbol{x}) \, . \tag{6.16}$$

To get from (6.15) to (6.16) use of the anti-periodicity condition (6.8) was made. The Fourier transform of the integrand $\boldsymbol{g}_>(x_0 - \mathrm{i}\beta, \boldsymbol{x})$ is given by

$$\boldsymbol{g}_>(x_0 - \mathrm{i}\beta, \boldsymbol{x}) = \int \frac{\mathrm{d}^4k'}{(2\pi)^4} \, \mathrm{e}^{-\mathrm{i}(k'_0(x_0 - \mathrm{i}\beta) - \boldsymbol{k}'\cdot\boldsymbol{x})} \; \boldsymbol{g}_>(k'_0, \boldsymbol{k}') \tag{6.17}$$

$$= \int \frac{\mathrm{d}^4k'}{(2\pi)^4} \, \mathrm{e}^{-\mathrm{i}(k'_0 x_0 - \boldsymbol{k}'\cdot\boldsymbol{x})} \; \mathrm{e}^{-\beta k'_0} \; \boldsymbol{g}_>(k'_0, \boldsymbol{k}') \, . \tag{6.18}$$

Substituting (6.18) into (6.16) then leads to

$$\boldsymbol{g}_<(k_0, \boldsymbol{k}) = - \int \mathrm{d}x_0 \int \mathrm{d}^3\boldsymbol{x} \int \frac{\mathrm{d}k'_0}{2\pi} \int \frac{\mathrm{d}^3\boldsymbol{k}'}{(2\pi)^3} \; \mathrm{e}^{\mathrm{i}(k_0 x_0 - \boldsymbol{k}\cdot\boldsymbol{x})}$$
$$\times \; \mathrm{e}^{-\mathrm{i}(k'_0 x_0 - \boldsymbol{k}'\cdot\boldsymbol{x})} \; \mathrm{e}^{-\beta k'_0} \; \boldsymbol{g}_>(k'_0, \boldsymbol{k}') \, , \tag{6.19}$$

which, upon integrating over the two δ-functions $\delta(k_0 - k'_0)$ and $\delta^3(\boldsymbol{k} - \boldsymbol{k}')$ inherently contained in (6.19), can be written in the following manner,

$$\boldsymbol{g}^B_<(k_0, \boldsymbol{k})_{\zeta\zeta'} = - \, \mathrm{e}^{-\beta k_0} \; \boldsymbol{g}^B_>(k_0, \boldsymbol{k})_{\zeta\zeta'} \, . \tag{6.20}$$

Denoting the difference between $\boldsymbol{g}^B_<$ and $\boldsymbol{g}^B_>$ as

$$\boldsymbol{\Xi}^B(k) \equiv \frac{1}{2\mathrm{i}\pi} \left(\boldsymbol{g}^B_>(k) \; - \; \boldsymbol{g}^B_<(k) \right) , \tag{6.21}$$

and replacing $\boldsymbol{g}^B_<$ by $\boldsymbol{g}^B_>$ with the aid of equation (6.20) leads for (6.21) to

$$2\mathrm{i}\pi \, \boldsymbol{\Xi}^B(k) = \left(1 \, + \, \mathrm{e}^{-\beta k_0} \right) \, \boldsymbol{g}^B_>(k) \, . \tag{6.22}$$

With the aid of the Fermi–Dirac function, given by

$$f(k_0) = \frac{1}{\mathrm{e}^{\beta k_0} + 1} \, , \tag{6.23}$$

equation (6.22) can be written as

$$\boldsymbol{g}^B_>(k)_{\zeta\zeta'} = 2\mathrm{i}\pi \; \boldsymbol{\Xi}^B_{\zeta\zeta'}(k) \; \left(1 - f(k_0) \right) \, . \tag{6.24}$$

Substituting (6.24) into (6.20) gives

$$\boldsymbol{g}^B_<(k)_{\zeta\zeta'} = -\,2\mathrm{i}\pi\ \Xi^B_{\zeta\zeta'}(k)\ f(k_0)\,. \tag{6.25}$$

The Fourier representation of $\boldsymbol{g}^B(\omega_n, \boldsymbol{k})$ reads

$$\boldsymbol{g}(\omega_n, \boldsymbol{k}) = \int_0^{-\mathrm{i}\beta} \mathrm{d}x_0\ \mathrm{e}^{-\frac{(2n+1)\pi}{\beta}x_0} \int \mathrm{d}^3\boldsymbol{x}\ \mathrm{e}^{-\mathrm{i}\boldsymbol{k}\cdot\boldsymbol{x}} \boldsymbol{g}(x_0, \boldsymbol{x})\,, \tag{6.26}$$

with

$$\boldsymbol{g}(x_0, \boldsymbol{x}) = \Theta(\mathrm{i}x_0)\,\boldsymbol{g}_>(x_0, \boldsymbol{x}) + \Theta(-\mathrm{i}x_0)\,\boldsymbol{g}_<(x_0, \boldsymbol{x})\,. \tag{6.27}$$

Because $x_0 \in (0, -\mathrm{i}\beta)$ in (6.26), equation (6.27) simplifies to $\boldsymbol{g}(x_0, \boldsymbol{x}) = \boldsymbol{g}_>(x_0, \boldsymbol{x})$, and therefore from (6.26),

$$\boldsymbol{g}(\omega_n, \boldsymbol{k}) = \int_0^{-\mathrm{i}\beta} \mathrm{d}x_0\ \mathrm{e}^{-\frac{(2n+1)\pi}{\beta}x_0} \int \mathrm{d}^3\boldsymbol{x}\ \mathrm{e}^{-\mathrm{i}\boldsymbol{k}\cdot\boldsymbol{x}}\ \boldsymbol{g}_>(x_0, \boldsymbol{x})\,. \tag{6.28}$$

Inspection of the second integral in (6.28) shows that this expression is nothing but the Fourier transform of $\boldsymbol{g}_>(x_0, \boldsymbol{k})$. Replacing $\boldsymbol{g}_>(x_0, \boldsymbol{k})$ with its Fourier transform $\boldsymbol{g}_>(k_0, \boldsymbol{k})$ in (6.28) leads to

$$\boldsymbol{g}(\omega_n, \boldsymbol{k}) = \int_0^{-\mathrm{i}\beta} \mathrm{d}x_0\ \mathrm{e}^{-\frac{(2n+1)\pi}{\beta}x_0} \int_{-\infty}^{+\infty} \frac{\mathrm{d}k_0}{2\pi}\ \mathrm{e}^{-\mathrm{i}k_0x_0}\ g > (k_0, \boldsymbol{k})\,. \tag{6.29}$$

Rearranging terms and substituting (6.24) for $\boldsymbol{g}^B_>(k_0, \boldsymbol{k})$ then gives

$$\boldsymbol{g}(\omega_n, \boldsymbol{k}) = \int_{-\infty}^{+\infty} \frac{\mathrm{d}k_0}{2\pi} \left\{ \int_0^{-\mathrm{i}\beta} \mathrm{d}x_0\ \mathrm{e}^{-\frac{(2n+1)\pi}{\beta}x_0 - \mathrm{i}k_0x_0} \right\} \{2\mathrm{i}\pi\ \Xi(k)\,[1 - f(k_0)]\}\,. \tag{6.30}$$

The first term in curly brackets can be integrated which results in

$$\int_0^{-\mathrm{i}\beta} \mathrm{d}x_0\ \mathrm{e}^{-(\frac{(2n+1)\pi}{\beta} + \mathrm{i}k_0)\,x_0} = \frac{\mathrm{i}}{\omega_n - k_0}\ \left(\mathrm{e}^{-\beta k_0} + 1\right)\,. \tag{6.31}$$

Equation (6.30) can thus be brought to the form

$$\boldsymbol{g}^B_{\zeta\zeta'}(\omega_n, \boldsymbol{k}) = -\int_{-\infty}^{+\infty} \mathrm{d}k_0\ \frac{\Xi^B_{\zeta\zeta'}(k)}{\omega_n - k_0}\,. \tag{6.32}$$

Replacing ω_n with the continuous, complex variable z leads to the analytically continued spectral representation of $\boldsymbol{g}^B$, given by[1]

$$\tilde{\boldsymbol{g}}^B_{\zeta\zeta'}(z, \boldsymbol{k}) = \int_{-\infty}^{+\infty} \mathrm{d}\omega\ \frac{\Xi^B_{\zeta\zeta'}(\omega, \boldsymbol{k})}{\omega - z}\,. \tag{6.33}$$

[1] Throughout this text, analytically continued functions carry a tilde.

The quantity $\boldsymbol{\Xi}^B$ is referred to as the spectral function. It will be calculated for dense nuclear matter in section 6.2. With help of the relation

$$\frac{1}{x \pm \mathrm{i}\eta} = \mathsf{P}\,\frac{1}{x} \mp \mathrm{i}\,\pi\,\delta(x)\,, \tag{6.34}$$

with P denoting the principal value, one readily verifies that the spectral function is given in terms of $\tilde{\boldsymbol{g}}^B$ as

$$\Xi^B_{\zeta\zeta'}(\omega,\boldsymbol{k}) = \frac{\tilde{g}^B_{\zeta\zeta'}(\omega+\mathrm{i}\eta,\boldsymbol{k}) \;-\; \tilde{g}^B_{\zeta\zeta'}(\omega-\mathrm{i}\eta,\boldsymbol{k})}{2\mathrm{i}\pi}\,. \tag{6.35}$$

Expressions (6.33) and (6.35) are formally identical with their non-relativistic counterparts [317, 319, 339, 344]. Here however we are dealing with spectral functions that possess a Dirac–Lorentz structure, that is, $\boldsymbol{\Xi}^B$ consists of a number of individual functions which altogether form $\boldsymbol{\Xi}^B$. This is in sharp contrast to the non-relativistic case where $\boldsymbol{\Xi}^B$ is a single scalar function.

The Fourier transform of $\boldsymbol{g}(x) = \Theta(x_0)\boldsymbol{g}_>(x) + \Theta(-x_0)\boldsymbol{g}_<(x)$ is given by

$$\boldsymbol{g}(k) = \int \mathrm{d}^4x\; \mathrm{e}^{\mathrm{i}kx}\,\left[\Theta(x_0)\,\boldsymbol{g}_>(x) + \Theta(-x_0)\,\boldsymbol{g}_<(x)\right]\,. \tag{6.36}$$

Expressing the Heaviside step function as

$$\Theta(\pm\, x_0) = \frac{\mathrm{i}}{2\pi}\int\limits_{-\infty}^{+\infty} \mathrm{d}\omega\;\frac{\mathrm{e}^{\mp\mathrm{i}\omega x_0}}{\omega+\mathrm{i}\eta}\,, \tag{6.37}$$

and replacing $\boldsymbol{g}_>(x)$ and $\boldsymbol{g}_<(x)$ with their Fourier transforms leads to

$$\begin{aligned}\boldsymbol{g}(k) = \frac{\mathrm{i}}{2\pi}\int \mathrm{d}x_0 \int \mathrm{d}^3\boldsymbol{x} \int \mathrm{d}\omega \int \frac{\mathrm{d}k_0'}{2\pi}\int\frac{\mathrm{d}^3\boldsymbol{k}'}{(2\pi)^3}\;\mathrm{e}^{-\mathrm{i}(\boldsymbol{k}-\boldsymbol{k}')\cdot\boldsymbol{x}} \\ \times\left\{\mathrm{e}^{\mathrm{i}(k_0-k_0'-\omega)x_0}\,\frac{\boldsymbol{g}_>(k_0',\boldsymbol{k}')}{\omega+\mathrm{i}\eta} + \mathrm{e}^{\mathrm{i}(k_0-k_0'+\omega)x_0}\,\frac{\boldsymbol{g}_<(k_0',\boldsymbol{k}')}{\omega+\mathrm{i}\eta}\right\}.\end{aligned} \tag{6.38}$$

The integrals over x_0 and $\boldsymbol{x}'$ constitute δ-functions of the form $\delta^3(\boldsymbol{k}-\boldsymbol{k}')$ and $\delta(k_0-k_0'-\omega)$, respectively. Introducing them in (6.38) leads to

$$\begin{aligned}\boldsymbol{g}(k) = \frac{\mathrm{i}}{2\pi}\int \mathrm{d}\omega \int \frac{\mathrm{d}k_0'}{2\pi}\int\frac{\mathrm{d}^3\boldsymbol{k}'}{(2\pi)^3}\;(2\pi)^4\,\delta^3(\boldsymbol{k}-\boldsymbol{k}') \\ \times\left\{\delta(k_0-k_0'-\omega)\frac{\boldsymbol{g}_>(k_0',\boldsymbol{k}')}{\omega+\mathrm{i}\eta} + \delta(k_0-k_0'+\omega)\frac{\boldsymbol{g}_<(k_0',\boldsymbol{k}')}{\omega+\mathrm{i}\eta}\right\},\end{aligned} \tag{6.39}$$

which, upon carrying out the integrals containing δ-functions, leads to

$$g^B_{\zeta\zeta'}(k) = \frac{\mathrm{i}}{2\pi}\int\limits_{-\infty}^{+\infty}\mathrm{d}k_0'\left\{\frac{g^B_>(k_0',\boldsymbol{k}')_{\zeta\zeta'}}{k_0-k_0'+\mathrm{i}\eta} \;-\; \frac{g^B_<(k_0',\boldsymbol{k}')_{\zeta\zeta'}}{k_0-k_0'-\mathrm{i}\eta}\right\}. \tag{6.40}$$

In the last step we replace the functions $\boldsymbol{g}^B_>$ and $\boldsymbol{g}^B_<$ in (6.40) by the expressions derived for them in equations (6.24) and (6.25), respectively. This results in

$$\boldsymbol{g}^B_{\zeta\zeta'}(k) = -\int\limits_{-\infty}^{+\infty} \mathrm{d}k_0' \left\{ \frac{1-f(k_0')}{k_0 - k_0' + \mathrm{i}\eta} + \frac{f(k_0')}{k_0 - k_0' - \mathrm{i}\eta} \right\} \Xi^B_{\zeta\zeta'}(k_0', \boldsymbol{k}) \,. \quad (6.41)$$

By means of the well-known mathematical relation

$$\frac{1}{k_0 - k_0' + \mathrm{i}\eta} - \frac{1}{k_0 - k_0' - \mathrm{i}\eta} = -2\mathrm{i}\pi\, \delta(k_0 - k_0') \,, \quad (6.42)$$

which is a consequence of equation (6.34), equation (6.41) can be brought into the form

$$\boldsymbol{g}^B_{\zeta\zeta'}(k_0, \boldsymbol{k}) = \int\limits_{-\infty}^{+\infty} \mathrm{d}\omega\, \frac{\Xi^B_{\zeta\zeta'}(\omega, \boldsymbol{k})}{\omega - k_0 - \mathrm{i}\eta} - 2\mathrm{i}\pi\, \Xi^B_{\zeta\zeta'}(k_0, \boldsymbol{k})\, f(k_0) \,. \quad (6.43)$$

The numerator in (6.43) can be written in a somewhat different fashion. For this purpose we formally add

$$0 \equiv -\int\limits_{-\infty}^{0} \mathrm{d}\omega\, \frac{\Xi(\omega, \boldsymbol{k})}{k_0 - \omega - \mathrm{i}\eta} + \int\limits_{-\infty}^{0} \mathrm{d}\omega\, \frac{\Xi(\omega, \boldsymbol{k})}{k_0 - \omega - \mathrm{i}\eta} \quad (6.44)$$

to (6.43), which leads to

$$\boldsymbol{g}(k) = -\left\{ \int\limits_{-\infty}^{0} \mathrm{d}\omega\, \frac{\Xi(\omega, \boldsymbol{k})}{k_0 - \omega + \mathrm{i}\eta} + \int\limits_{0}^{+\infty} \mathrm{d}\omega\, \frac{\Xi(\omega, \boldsymbol{k})}{k_0 - \omega + \mathrm{i}\eta} \right.$$
$$\left. - \int\limits_{-\infty}^{0} \mathrm{d}\omega\, \frac{\Xi(\omega, \boldsymbol{k})}{k_0 - \omega - \mathrm{i}\eta} + \int\limits_{-\infty}^{0} \mathrm{d}\omega\, \frac{\Xi(\omega, \boldsymbol{k})}{k_0 - \omega - \mathrm{i}\eta} \right\} - 2\mathrm{i}\pi\, f(k_0)\, \Xi(k_0, \boldsymbol{k}) \,. \quad (6.45)$$

Introducing the signum function $\Theta(-\omega)$ in the first and third integral of (6.45), the interval of integration can be extended from $(-\infty, 0)$ to $(-\infty, +\infty)$. The second and fourth integral can be combined to one integral. This is not immediately clear from equation (6.45). To see this, note that the integrand of the second integral has a pole only if $k_0 > 0$. Therefore, without loss of generality, we can multiply $\mathrm{i}\eta$ of the denominator of this integrand with k_0. Similarly, the pole of the integrand of the fourth integral occurs only if $k_0 < 0$, and correspondingly we may multiply $\mathrm{i}\eta$ of this integrand with $-k_0$. Hence we are left with

$$\begin{aligned} \boldsymbol{g}(k) = & - \Bigg\{ \int_{-\infty}^{+\infty} \mathrm{d}\omega \, \Xi(\omega, \boldsymbol{k}) \left[\frac{1}{k_0 - \omega + \mathrm{i}\eta} - \frac{1}{k_0 - \omega - \mathrm{i}\eta} \right] \Theta(-\omega) \\ & - \left[\int_{-\infty}^{0} \mathrm{d}\omega \, \frac{\Xi(\omega, \boldsymbol{k})}{\omega - k_0 + \mathrm{i}\eta \cdot (-k_0)} + \int_{0}^{+\infty} \mathrm{d}\omega \, \frac{\Xi(\omega, \boldsymbol{k})}{\omega - k_0 - \mathrm{i}\eta \cdot (k_0)} \right] \Bigg\} \\ & - 2\,\mathrm{i}\pi \, f(k_0) \, \Xi(k_0, \boldsymbol{k}) \, . \end{aligned} \tag{6.46}$$

The integrand of the first integral in (6.46) can be replaced with $-2\mathrm{i}\pi\delta(k_0 - \omega)$. So the integral over ω simply gives $2\mathrm{i}\pi\Theta(k_0)\Xi(k_0, \boldsymbol{k})$, which we combine with the last term in (6.46). One gets $-2\mathrm{i}\pi\mathrm{sign}(k_0) f(|k_0|)\Xi(k)$, where use of the relation

$$f(x) - \Theta(-x) = \mathrm{sign}(x) \; f(|x|) \tag{6.47}$$

was made. Equation (6.47) is readily verified by means of making use of $f(-|x|) - 1 = -f(|x|)$ and the definition of the signum function, $\mathrm{sign}(x) = \Theta(x) - \Theta(-x)$. The remaining two integrals in (6.46) can be combined, which leads to the final result for the spectral representation of $\boldsymbol{g}^B$ in the form

$$\boldsymbol{g}^B_{\zeta\zeta'}(k) = \int_{-\infty}^{+\infty} \mathrm{d}\omega \; \frac{\Xi^B_{\zeta\zeta'}(\omega, \boldsymbol{k})}{\omega - k_0\,(1 + \mathrm{i}\eta)} - 2\mathrm{i}\pi \; \mathrm{sign}(k_0) \; f(|k_0|) \; \Xi^B_{\zeta\zeta'}(k) \, . \tag{6.48}$$

As an easy illustration, let us apply the technique developed just above to the derivation of the finite-density, finite-temperature expression of the free baryon propagator. The free baryon propagator is known from equation (5.125). It reads

$$\boldsymbol{g}^{0B}(k) = - \frac{1}{\not{k} - m_B} \, , \tag{6.49}$$

where $\not{k}$ is given by $\not{k} \equiv \gamma^\mu k_\mu = \gamma^0 k_0 - \boldsymbol{\gamma} \cdot \boldsymbol{k}$. Multiplying both numerator and denominator of (6.49) with $\not{k} + m$ and making use of $\not{k}^2 = k^2$, which follows from the relation (see also appendix A.2)

$$\not{k}^2 = (\gamma^\mu k_\mu)^2 = \gamma^\mu k_\mu \, \gamma^\nu k_\nu = k_\mu k_\nu \{ 2 g^{\mu\nu} - \gamma^\nu \gamma^\mu \} = 2 k^\nu k_\nu - \not{k}^2 \, , \tag{6.50}$$

one gets

$$\boldsymbol{g}^{0B}(k_0, \boldsymbol{k}) = - \frac{\gamma^0 k_0 - \boldsymbol{\gamma} \cdot \boldsymbol{k} + m_B}{k_0^2 - \boldsymbol{k}^2 - m_B^2} \, , \tag{6.51}$$

and for the analytically continued propagator,

$$\tilde{\boldsymbol{g}}^{0B}(z, \boldsymbol{k}) = -\frac{\gamma^0 z - \boldsymbol{\gamma}\cdot\boldsymbol{k} + m_B}{z^2 - \boldsymbol{k}^2 - m_B^2}\,. \tag{6.52}$$

Evaluating equation (6.52) for energies $z = k_0 \pm \mathrm{i}\eta$ leads to

$$\tilde{\boldsymbol{g}}^{0B}(k_0, \boldsymbol{k}) = -\frac{\gamma^0(k_0 \pm \mathrm{i}\eta) - \boldsymbol{\gamma}\cdot\boldsymbol{k} + m_B}{(k_0 \pm \mathrm{i}\eta)^2 - \boldsymbol{k}^2 - m_B^2}\,, \tag{6.53}$$

which can be rewritten as

$$\tilde{\boldsymbol{g}}^{0B}(k_0, \boldsymbol{k}) = -\frac{\not{k} + m_B}{k^2 - m_B^2 \pm \mathrm{i}\,\eta\,\mathrm{sign}(k_0)}\,. \tag{6.54}$$

Here we have made use of the fact that the term whose denominator is proportional to $\pm\mathrm{i}\eta$ does not give a contribution, as is the case for the term in the numerator proportional to η^2. A straightforward evaluation of $\tilde{\boldsymbol{g}}^{0B}$ at the cut along the k_0 axis gives

$$\begin{aligned}\tilde{\boldsymbol{g}}^{0B}(k_0 + \mathrm{i}\eta, \boldsymbol{k}) - \tilde{\boldsymbol{g}}^{0B}(k_0 - \mathrm{i}\eta, \boldsymbol{k}) &= -(\not{k} + m_B)\\ &\times \left[\frac{1}{k^2 - m_B^2 + \mathrm{i}\,\eta\,\mathrm{sign}(k_0)} - \frac{1}{k^2 - m_B^2 - \mathrm{i}\,\eta\,\mathrm{sign}(k_0)}\right]\\ &= 2\,\mathrm{i}\,\pi\,(\not{k} + m_B)\,\delta(k^2 - m_B^2)\,\mathrm{sign}(k_0)\,,\end{aligned} \tag{6.55}$$

from which we find for the spectral function associated with the free baryon Green functions,

$$\begin{aligned}\boldsymbol{\Xi}^{0B}_{\zeta\zeta'}(k) &= \frac{\tilde{\boldsymbol{g}}^{0B}_{\zeta\zeta'}(k_0 + \mathrm{i}\eta, \boldsymbol{k}) - \tilde{\boldsymbol{g}}^{0B}_{\zeta\zeta'}(k_0 - \mathrm{i}\eta, \boldsymbol{k})}{2\,\mathrm{i}\,\pi}\\ &= (\not{k} + m_B)_{\zeta\zeta'}\,\delta(k^2 - m_B^2)\,\mathrm{sign}(k_0)\,.\end{aligned} \tag{6.56}$$

The representation of the free baryon Green functions is found by means of substituting $\boldsymbol{\Xi}^{0B}$ of (6.56) into equation (6.48). This leads to (as usual, to keep the notation at a minimum, the superscript B is dropped)

$$\begin{aligned}\boldsymbol{g}^0(k) = &\int_{-\infty}^{+\infty} \mathrm{d}\omega\, \frac{(\gamma^0\omega - \boldsymbol{\gamma}\cdot\boldsymbol{k} + m)\,[\delta(\omega - \omega^0(\boldsymbol{k})) + \delta(\omega + \omega^0(\boldsymbol{k}))]\,\mathrm{sign}(\omega)}{2\,\omega(\boldsymbol{k})\,[\omega - k_0(1 + \mathrm{i}\,\eta)]}\\ &- \frac{\mathrm{i}\pi}{\omega^0(\boldsymbol{k})}\, f(|k_0|)\,(\not{k} + m)\,[\delta(k_0 - \omega^0(\boldsymbol{k})) + \delta(k_0 + \omega^0(\boldsymbol{k}))]\,,\end{aligned} \tag{6.57}$$

with the free single-particle energy given by $\omega^0(\boldsymbol{k}) = \sqrt{m^2 + \boldsymbol{k}^2}$. To arrive at equation (6.57), use of

$$\begin{aligned}\delta(k^2 - m^2) &= \delta\left(k_0^2 - \left(\omega^0(\boldsymbol{k})\right)^2\right)\\ &= \frac{1}{2\,\omega^0(\boldsymbol{k})}\,[\delta(k_0 - \omega^0(\boldsymbol{k})) + \delta(k_0 + \omega^0(\boldsymbol{k}))]\end{aligned} \tag{6.58}$$

and $\delta(ax) = \frac{1}{|a|}\delta(x)$ has been made. The expression $[\delta(\omega - \omega(\boldsymbol{k})) + \delta(\omega + \omega(\boldsymbol{k}))]\,\mathrm{sign}(\omega)$ in the numerator of (6.57) can be rewritten as

$$[\delta(\omega - \omega^0(\boldsymbol{k})) + \delta(\omega + \omega^0(\boldsymbol{k}))]\,[\Theta(\omega) - \Theta(-\omega)] \\ = [\delta(\omega - \omega^0(\boldsymbol{k})) - \delta(\omega + \omega^0(\boldsymbol{k}))]\,. \quad (6.59)$$

Substituting (6.59) into (6.57) and integrating over ω gives for the integrand

$$\frac{\gamma^0\omega^0(\boldsymbol{k}) - \boldsymbol{\gamma}\cdot\boldsymbol{k} + m}{\omega^0(\boldsymbol{k}) - (k_0 + \mathrm{i}\,\eta\,k_0)} + \frac{-\gamma^0\omega^0(\boldsymbol{k}) - \boldsymbol{\gamma}\cdot\boldsymbol{k} + m}{\omega^0(\boldsymbol{k}) + (k_0 + \mathrm{i}\,\eta\,k_0)} = \frac{2\,\omega^0(\boldsymbol{k})\,(\not{k} + m)}{-k^2 + m^2 - \mathrm{i}\,\eta}\,, \quad (6.60)$$

and thus for $\boldsymbol{g}^{0B}$,

$$\boldsymbol{g}^{0B}_{\zeta\zeta'}(k) = \frac{(\gamma^\mu k_\mu + m_B)_{\zeta\zeta'}}{k^2 - m_B^2 + \mathrm{i}\eta} + \mathrm{i}\pi\,f(|k_0|)\,\frac{(\gamma^\mu k_\mu + m_B)_{\zeta\zeta'}}{\omega^{0B}(\boldsymbol{k})} \\ \times \left[\,\delta(k_0 - \omega^{0B}(\boldsymbol{k})) \;+\; \delta(k_0 + \bar{\omega}^{0B}(\boldsymbol{k}))\,\right]. \quad (6.61)$$

The poles of $\boldsymbol{g}^{0B}(k)$ in the absence of a medium are graphically illustrated in figure 6.1. The presence of a medium doubles the number of poles, from two to four. To see this it is illustrative to rewrite (6.61) as follows. First, expand the first term of (6.61) as

$$\frac{(\not{k} + m_B)_{\zeta\zeta'}}{k^2 - m_B^2 + \mathrm{i}\eta} = \frac{\not{k} + m_{B\,\zeta\zeta'}}{2\,\omega^{0B}(\boldsymbol{k})}\,\frac{2\,\omega^{0B}(\boldsymbol{k})}{k^2 - m_B^2 + \mathrm{i}\eta}\,, \quad (6.62)$$

and then write for the second term of this expansion

$$\frac{2\,\omega^0(\boldsymbol{k})}{k^2 - m^2 + \mathrm{i}\,\eta} = \frac{1}{k_0 - \omega^0(\boldsymbol{k}) + \mathrm{i}\,\eta} - \frac{1}{k_0 + \omega^0(\boldsymbol{k}) - \mathrm{i}\,\eta}\,. \quad (6.63)$$

Substituting (6.62) and (6.63) into equation (6.61) leads to

$$\boldsymbol{g}^{0B}_{\zeta\zeta'}(k) = \frac{(\gamma^\mu k_\mu + m_B)_{\zeta\zeta'}}{2\,\omega^{0B}(\boldsymbol{k})}\left\{\frac{1 - f^B(\boldsymbol{k})}{k_0 - \omega^{0B}(\boldsymbol{k}) + \mathrm{i}\,\eta} + \frac{f^B(\boldsymbol{k})}{k_0 - \omega^{0B}(\boldsymbol{k}) - \mathrm{i}\eta}\right. \\ \left. - \frac{1 - \bar{f}^B(\boldsymbol{k})}{k_0 + \omega^{0B}(\boldsymbol{k}) - \mathrm{i}\eta} - \frac{\bar{f}^B(\boldsymbol{k})}{k_0 + \omega^{0B}(\boldsymbol{k}) + \mathrm{i}\eta}\right\}. \quad (6.64)$$

In the above equations, the single-particle energy of free baryons is given by

$$\omega^{0B}(\boldsymbol{k}) = +\sqrt{m_B^2 + \boldsymbol{k}^2} = -\,\bar{\omega}^{0B}(\boldsymbol{k})\,, \quad (6.65)$$

and for the Fermi–Dirac functions,

$$f^B(\boldsymbol{k}) \equiv f(\omega^{0B}(\boldsymbol{k})) = \frac{1}{\mathrm{e}^{\beta\omega^{0B}(k)+1}}\,, \quad \bar{f}^B(\boldsymbol{k}) \equiv \bar{f}(\bar{\omega}^{0B}(\boldsymbol{k})) = \frac{1}{\mathrm{e}^{\beta|\bar{\omega}^{0B}(k)|+1}}\,. \quad (6.66)$$

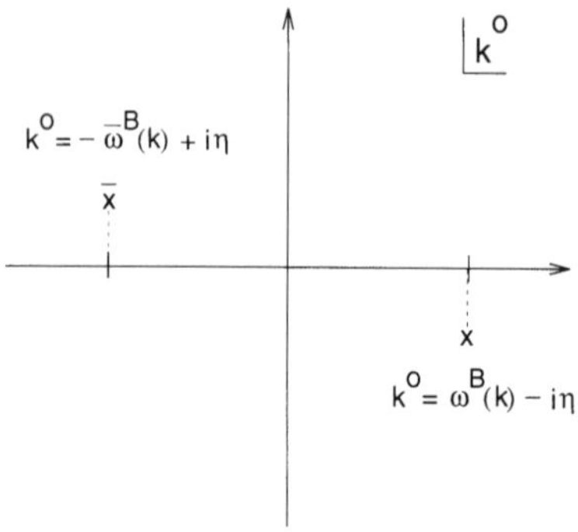

Figure 6.1. Poles of baryon propagator in free space. The symbols 'x' and 'x̄' refer to the locations of particle and antiparticle poles, respectively.

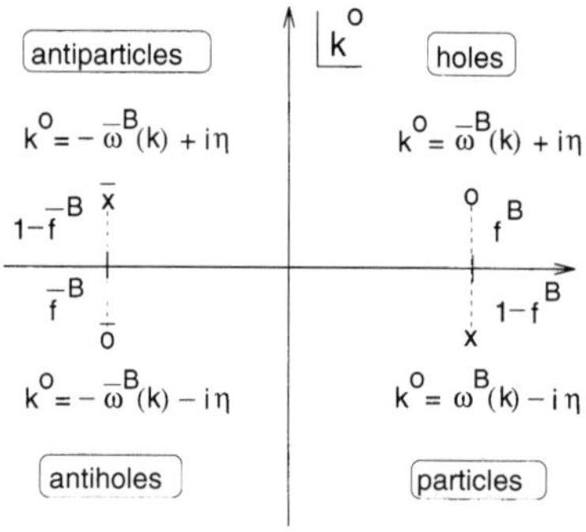

Figure 6.2. Poles of baryon propagator in case of a medium. The crosses 'x' and 'x̄' refer to the locations of particle and antiparticle poles, respectively. The circles 'o' and 'ō' denote hole and antihole poles, respectively.

Introducing the chemical potential of baryons, μ^B, and antibaryons, $\bar{\mu}^B$, into the the Fermi–Dirac functions of baryons and antibaryons leads to

$$f^B(\boldsymbol{k}) = \frac{1}{\mathrm{e}^{\beta(\omega^B(k)-\mu^B)}+1}, \quad \bar{f}^B(\boldsymbol{k}) = \frac{1}{\mathrm{e}^{\beta|\bar{\omega}^B(k)+\bar{\mu}^B|}+1}. \tag{6.67}$$

The physical interpretation of $\boldsymbol{g}^{0B}(k)$ is as follows [345]. Both particle and antiparticle states occur as in the usual (causal) Feynman propagator. But due to the nuclear (stellar) medium two new states corresponding to holes in the particle Fermi sea (unfilled states in the Fermi sea of particles) and antiholes in the antiparticle Fermi sea (unfilled states in the Fermi sea of antiparticles) result, as illustrated in figures 6.2 and 6.3. Thus, the principal

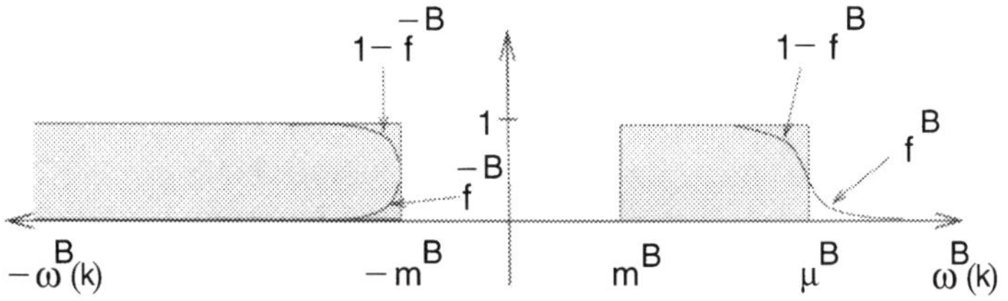

Figure 6.3. Depletion of single-particle states at finite temperature.

effect of finite temperatures on baryon propagation results in states with momenta $|\boldsymbol{k}| > k_F$ and $|\boldsymbol{k}| > \bar{k}_F$ (that is, states outside the Fermi seas of particles and antiparticles become populated), as do hole (antihole) states in the corresponding Fermi seas of particles (antiparticles).

The zero-temperature limit of (6.64) is obtained by noticing that $\bar{f} \to 0$ and

$$f(\omega(\boldsymbol{k}) - \mu) \longrightarrow \Theta(k_F - |\boldsymbol{k}|) \,. \tag{6.68}$$

Equation (6.64) then reduces to

$$\begin{aligned} g^{0B}_{\zeta\zeta'}(k) = & \frac{(\gamma^\mu k_\mu + m_B)_{\zeta\zeta'}}{2\,\omega^{0B}(\boldsymbol{k})} \\ & \times \left\{ \frac{1 - \Theta(k_F - |\boldsymbol{k}|)}{\omega^{0B}(\boldsymbol{k}) - k_0 - \mathrm{i}\,\eta} + \frac{\Theta(k_F - |\boldsymbol{k}|)}{\omega^{0B}(\boldsymbol{k}) - k_0 - \mathrm{i}\eta} \right\}. \end{aligned} \tag{6.69}$$

With the help of (6.42) it is seen that the two terms proportional to $\Theta(k_F - |\boldsymbol{k}|)$ can be combined to a δ-function. So in the zero-temperature limit

$$\begin{aligned} g^{0B}_{\zeta\zeta'}(k) = \frac{(\not{k} + m_B)_{\zeta\zeta'}}{2\,\omega^B(\boldsymbol{k})} & \left\{ \frac{1}{\omega^B(\boldsymbol{k}) - k_0 - \mathrm{i}\,\eta} \right. \\ & \left. - 2\,\mathrm{i}\,\pi\,\delta(\omega^B(\boldsymbol{k}) - k_0)\,\Theta(k_{F_B} - |\boldsymbol{k}|) \right\}. \end{aligned} \tag{6.70}$$

The first term is the usual Dirac propagator of free fermions, which is corrected for the medium by the second term. This follows from the zero density of (6.69), in which case the Fermi momentum becomes zero and therefore $\Theta(k_F - |\boldsymbol{k}|) \to \Theta(-|\boldsymbol{k}|) = 0$. Hence only the first term of (6.70) survives for particles in free space.

In the interacting particle case the spectral function Ξ^B has a more complicated structure than in (6.56). The corresponding baryon Green function however is similar in structure to (6.64). This is specifically the case for the relativistic Hartree (mean-field) approximation, as will be shown in section 6.2.2. The modifications of (6.64) can formally be taken into account by making the following replacements: $k_0 \to k_0 - \boldsymbol{\Sigma}_0^{\mathrm{H,B}}$, $m_B \to m_B + \boldsymbol{\Sigma}_S^{\mathrm{H,B}}$, and $\omega^{0B} \to \omega^{\mathrm{H,B}}$ (cf. equation (6.170)). That is, the coupling of the motion of a baryon to the nuclear background, which implies non-vanishing self-energy components, modifies the baryon masses and single-particle energy spectra. The single-particle description carries over as long as the many-particle system is treated for approximation schemes for which the self-energy does not become complex, as is the case for the relativistic Hartree, Hartree–Fock, and some versions of the **T**-matrix approximation [337].

6.2 Determination of baryon spectral function

In many-body treatments it is customary and useful to measure energies relative to the chemical potential, μ. The concept of chemical potentials has already been discussed in great detail in chapter 4 in connection with the particle composition of neutron star matter at equilibrium. In the previous section we have seen how the single-particle distribution changes with temperature relative to the zero-temperature distribution, whereby the chemical potential plays a most intuitive role. To introduce μ^B into the spectral representation of $\boldsymbol{g}^B(k)$ we rescale the energy argument in (6.48) according to the replacement $k_0 \to k_0 - \mu^B$, which gives

$$\begin{aligned} g^B_{\zeta\zeta'}(k_0, \boldsymbol{k}) &= \int_{-\infty}^{+\infty} \mathrm{d}\omega \, \frac{\Xi^B_{\zeta\zeta'}(\omega, \boldsymbol{k})}{\omega - (k_0 - \mu^B)\,(1 + \mathrm{i}\,\eta)} \\ &- 2\,\mathrm{i}\,\pi \, \mathrm{sign}(k_0 - \mu^B)\, f(|k_0 - \mu^B|)\, \Xi^B_{\zeta\zeta'}(k_0 - \mu^B, \boldsymbol{k})\,. \end{aligned} \tag{6.71}$$

To ensure compatibility between (6.71) and its analytically continued representation, derived in equation (6.33), we must ensure that

$$\boldsymbol{g}^B(k_0 + \mu^B, \boldsymbol{k}) = \int_{-\infty}^{+\infty} \mathrm{d}\omega \, \frac{\boldsymbol{\Xi}^B(\omega, \boldsymbol{k})}{\omega - k_0(1 + \mathrm{i}\,\eta)} = \tilde{\boldsymbol{g}}^B(k_0(1 + \mathrm{i}\eta), \boldsymbol{k})\,, \tag{6.72}$$

with the identification $z = k_0(1 + \mathrm{i}\eta)$. We know that $\boldsymbol{g}^B(k_0, \boldsymbol{k})$ is obtained from Dyson's equation, given in (5.123). So $\boldsymbol{g}^B(k_0 + \mu^B, \boldsymbol{k})$ of (6.72) must be made compatible with

$$\left(\gamma^0 k_0 - \boldsymbol{\gamma} \cdot \boldsymbol{k} - m_B \, \mathbf{1} - \boldsymbol{\Sigma}^B(k_0, \boldsymbol{k})\right) \boldsymbol{g}^B(k_0, \boldsymbol{k}) = -\,\mathbf{1}\,, \tag{6.73}$$

which is accomplished by the replacement $k_0 \to k_0 + \mu^B$ in (6.73),

$$\left\{\gamma^0(k_0 + \mu^B) - \boldsymbol{\gamma} \cdot \boldsymbol{k} - m_B \, \mathbf{1} - \boldsymbol{\Sigma}^B(k_0 + \mu^B, \boldsymbol{k})\right\} \boldsymbol{g}^B(k_0 + \mu^B, \boldsymbol{k}) = -\,\mathbf{1}\,. \tag{6.74}$$

Because of (6.72) we have

$$\boldsymbol{\Sigma}^B(k_0+\mu^B,\boldsymbol{k}) = \tilde{\boldsymbol{\Sigma}}^B(k_0(1+\mathrm{i}\eta),\boldsymbol{k})\,, \tag{6.75}$$

$$\boldsymbol{g}^B(k_0+\mu^B,\boldsymbol{k}) = \tilde{\boldsymbol{g}}^B(k_0(1+\mathrm{i}\eta),\boldsymbol{k})\,, \tag{6.76}$$

and thus for Dyson's equation

$$\{\gamma^0(k_0+\mu^B) - \boldsymbol{\gamma}\cdot\boldsymbol{k} - m_B\,\mathbf{1} - \tilde{\boldsymbol{\Sigma}}^B(k_0(1+\mathrm{i}\eta),\boldsymbol{k})\}\tilde{\boldsymbol{g}}^B(k_0(1+\mathrm{i}\eta),\boldsymbol{k}) = -\mathbf{1}\,,$$

which, upon replacing $k_0(1+\mathrm{i}\eta)$ with z, leads to the desired analytically continued representation of Dyson's equation,

$$\{\gamma^0(z+\mu^B) - \boldsymbol{\gamma}\cdot\boldsymbol{k} - m_B\,\mathbf{1} - \tilde{\boldsymbol{\Sigma}}^B(z,\boldsymbol{k})\}\,\tilde{\boldsymbol{g}}^B(z,\boldsymbol{k}) = -\,\mathbf{1}\,. \tag{6.77}$$

Finally we note that from equation (6.76), the *physical* two-point baryon Green function and self-energy, $\boldsymbol{g}^B$ and $\boldsymbol{\Sigma}^B$, are obtained from their analytically continued counterparts as

$$\boldsymbol{g}^B(k_0,\boldsymbol{k}) = \tilde{\boldsymbol{g}}^B((k_0-\mu^B)(1+\mathrm{i}\eta),\boldsymbol{k})\,, \tag{6.78}$$

and

$$\boldsymbol{\Sigma}^B(k_0,\boldsymbol{k}) = \tilde{\boldsymbol{\Sigma}}^B((k_0-\mu^B)(1+\mathrm{i}\eta),\boldsymbol{k})\,. \tag{6.79}$$

As outlined in reference [92], because of the translational and rotational invariance in the rest frame of infinite nuclear matter and the assumed invariance under parity and time reversal, the self-energy may be written quite generally as

$$\begin{aligned}\boldsymbol{\Sigma}^B(k) &= \Sigma^B_S(k) + \gamma^\mu\,\Sigma^B_\mu(k)\\ &\equiv \Sigma^B_S(k^0,|\boldsymbol{k}|) + \gamma^0\,\Sigma^B_0(k^0,|\boldsymbol{k}|) + \boldsymbol{\gamma}\cdot\hat{\boldsymbol{k}}\,\Sigma^B_V(k^0,|\boldsymbol{k}|)\,,\end{aligned} \tag{6.80}$$

The functions Σ^B_S, Σ^B_V, and Σ^B_0 are referred to as scalar, vector, and timelike components of the baryon self-energy. The proof of this decomposition is as follows [92]. At finite density, the self-energy may depend on two four-vectors, k^μ and B^μ, and three Lorentz scalars, k^2, B^2, and kB. In the rest frame of nuclear matter ($B^\mu = \delta^{\mu 0}\rho$), the latter may be replaced with $\boldsymbol{k}^2$, ρ, and k^0, which leads to the arguments of the Lorentz-scalar functions in (6.80). The matrix structure of $\boldsymbol{\Sigma}^B$ is determined by combining k^μ and B^μ with gamma matrices, which leads to the four independent, parity-conserving choices

$$1,\qquad \gamma_\mu\,k^\mu,\qquad \gamma_\mu\,B^\mu,\qquad \sigma_{\mu\nu}\,k^\mu\,B^\nu\,, \tag{6.81}$$

or

$$1,\qquad \gamma_\mu\,k^\mu,\qquad \gamma^0\,\rho,\qquad \sigma_{0i}\,k^i\,\rho^\nu\,. \tag{6.82}$$

The tensor piece proportional to σ_{0i} does not contribute if one assumes time-reversal invariance and the hermiticity of $\mathbf{\Sigma}^B$. Linear combination of the other three forms then results in the three terms in (6.80).

Attaching the spin and isospin indices to (6.80), i.e.

$$\mathbf{\Sigma}^B_{\zeta\zeta'} \equiv (\mathbf{1})_{\zeta\zeta'}\, \Sigma^B_S + (\boldsymbol{\gamma}\cdot\hat{\boldsymbol{k}})_{\zeta\zeta'}\, \Sigma^B_V + (\gamma^0)_{\zeta\zeta'}\, \Sigma^B_0\,, \tag{6.83}$$

and substituting this result into Dyson's equation (6.73) leads to

$$\begin{aligned}\{\mathbf{1}\, [m_B + \Sigma^B_S(k_0,\boldsymbol{k})] \;+\; \boldsymbol{\gamma}\cdot\hat{\boldsymbol{k}}\; [|\boldsymbol{k}| + \Sigma^B_V(k_0,\boldsymbol{k})] \\ +\; \gamma^0\, [\Sigma^B_0(k_0,\boldsymbol{k}) - k_0]\}\; \boldsymbol{g}^B(k_0,\boldsymbol{k}) \;=\; 1\,,\end{aligned} \tag{6.84}$$

and for the analytically continued Dyson equation

$$\begin{aligned}\{\mathbf{1}\, [m_B + \tilde{\Sigma}^B_S(z,\boldsymbol{k})] \;+\; \boldsymbol{\gamma}\cdot\hat{\boldsymbol{k}}\; [|\boldsymbol{k}| + \tilde{\Sigma}^B_V(z,\boldsymbol{k})] \\ +\; \gamma^0\, [\tilde{\Sigma}^B_0(z,\boldsymbol{k}) + (z+\mu^B)]\}\; \tilde{\boldsymbol{g}}^B(z,\boldsymbol{k}) \;=\; 1\,.\end{aligned} \tag{6.85}$$

To derive the explicit form of the baryon spectral function, which follows from equation (6.35), we need to know the analytic properties of Dyson's equation, that is, of the functions $\tilde{\mathbf{\Sigma}}^B$ and $\tilde{\boldsymbol{g}}^B$. For this purpose we express $\tilde{\mathbf{\Sigma}}^B$ via a spectral representation of the form

$$\tilde{\mathbf{\Sigma}}^B_{\zeta\zeta'}(z,\boldsymbol{k}) \;=\; \mathbf{\Sigma}^{(\infty)\,B}_{\zeta\zeta'} \;+\; \int\limits_{-\infty}^{+\infty} \mathrm{d}\omega\; \frac{\boldsymbol{S}^B_{\zeta\zeta'}(\omega,\boldsymbol{k})}{\omega - z}\,, \tag{6.86}$$

with

$$\mathbf{\Sigma}^{(\infty)\,B}_{\zeta\zeta'} \;\equiv\; (\gamma_\mu\, k^\mu \;-\; m_B)_{\zeta\zeta'}\,. \tag{6.87}$$

The associated spectral function, $\boldsymbol{S}^B$, is then obtained as

$$\tilde{\mathbf{\Sigma}}(\omega + \mathrm{i}\eta,\boldsymbol{k}) - \tilde{\mathbf{\Sigma}}(\omega - \mathrm{i}\eta,\boldsymbol{k}) \;=\; 2\,\mathrm{i}\,\pi\; \boldsymbol{S}(\omega,\boldsymbol{k})\,. \tag{6.88}$$

Since $\mathbf{\Sigma}^{(\infty)B}$ and $\boldsymbol{S}$ are real functions, one readily finds from (6.86) that the self-energy obeys the relation

$$\tilde{\mathbf{\Sigma}}(z^*,\boldsymbol{k}) = \tilde{\mathbf{\Sigma}}^*(z,\boldsymbol{k})\,, \tag{6.89}$$

and, similarly, for the analytically continued two-point baryon Green function,

$$\tilde{\boldsymbol{g}}(z^*,\boldsymbol{k}) = \tilde{\boldsymbol{g}}^*(z,\boldsymbol{k})\,. \tag{6.90}$$

Moreover one has

$$\mathrm{Re}\; \tilde{\mathbf{\Sigma}}(\omega + \mathrm{i}\eta,\boldsymbol{k}) = \mathrm{Re}\; \tilde{\mathbf{\Sigma}}(\omega - \mathrm{i}\eta,\boldsymbol{k})\,, \tag{6.91}$$

and

$$\text{Im}\ \mathbf{\Sigma}(\omega + \mathrm{i}\eta, \boldsymbol{k}) = -\text{Im}\ \mathbf{\Sigma}(\omega - \mathrm{i}\eta, \boldsymbol{k}) = -\pi\, \mathcal{S}(\omega, \boldsymbol{k})\,. \tag{6.92}$$

Because of equations (6.79) and (6.91), we can write for the real part of the physical self-energy

$$\mathbf{\Lambda}(\omega, \boldsymbol{k}) \equiv \text{Re}\ \tilde{\mathbf{\Sigma}}(\omega - \mu - \mathrm{i}\eta, \boldsymbol{k}) = \text{Re}\ \tilde{\mathbf{\Sigma}}(\omega - \mu + \mathrm{i}\eta, \boldsymbol{k})\,, \tag{6.93}$$

which, upon rescaling the energy argument according to $\omega \to \omega + \mu$, reads

$$\mathbf{\Lambda}(\omega + \mu, \boldsymbol{k}) = \text{Re}\ \tilde{\mathbf{\Sigma}}(\omega - \mathrm{i}\eta, \boldsymbol{k}) = \text{Re}\ \tilde{\mathbf{\Sigma}}(\omega + \mathrm{i}\eta, \boldsymbol{k})\,. \tag{6.94}$$

The mathematical structure of the imaginary part of the physical self-energy demands for somewhat more consideration. Turning back to (6.79), one finds

$$\begin{aligned}\text{Im}\ \mathbf{\Sigma}(\omega, \boldsymbol{k}) &= \text{Im}\ \tilde{\mathbf{\Sigma}}((\omega - \mu)(1 + \mathrm{i}\eta), \boldsymbol{k}) \\ &= \text{Im}\ \tilde{\mathbf{\Sigma}}(\omega - \mu + \mathrm{i}\eta\, \text{sign}(\omega - \mu), \boldsymbol{k})\,.\end{aligned} \tag{6.95}$$

Since $\tilde{\mathbf{\Sigma}}(\omega - \mu + \mathrm{i}\eta, \boldsymbol{k})$ and $\tilde{\mathbf{\Sigma}}(\omega - \mu - \mathrm{i}\eta, \boldsymbol{k})$ differ only by a sign, as can be seen from (6.92), equation (6.95) may be written as

$$\text{Im}\ \mathbf{\Sigma}(\omega, \boldsymbol{k}) = \text{sign}(\mu - \omega)\ \text{Im}\ \tilde{\mathbf{\Sigma}}(\omega - \mu - \mathrm{i}\eta, \boldsymbol{k}) \equiv \mathbf{\Gamma}(\omega, \boldsymbol{k})\,. \tag{6.96}$$

Multiplying both sides of (6.96) with $\text{sign}(\mu - \omega)$ leads for $\text{Im}\ \tilde{\mathbf{\Sigma}}$ to

$$\text{Im}\ \tilde{\mathbf{\Sigma}}(\omega - \mu \mp \mathrm{i}\eta, \boldsymbol{k}) \ = \ \pm\ \text{sign}(\mu - \omega)\ \mathbf{\Gamma}(\omega, \boldsymbol{k})\,, \tag{6.97}$$

which, upon rescaling the energy argument as just above, $\omega \to \omega + \mu$, reads

$$\text{Im}\ \tilde{\mathbf{\Sigma}}(\omega \mp \mathrm{i}\eta, \boldsymbol{k}) = \pm\ \text{sign}(-\omega)\ \mathbf{\Gamma}(\omega + \mu, \boldsymbol{k})\,. \tag{6.98}$$

To keep the notation in the subsequent analysis to a minimum, let us introduce an auxiliary functions F_S for the scalar part of the baryon self-energy, defined as

$$\begin{aligned}F_S^{\pm} &\equiv F_S(\omega \pm \mathrm{i}\eta, \boldsymbol{k}) \equiv m + \tilde{\Sigma}_S(\omega \pm \mathrm{i}\eta, \boldsymbol{k}) \\ &= m + \text{Re}\ \tilde{\Sigma}_S(\omega \pm \mathrm{i}\eta, \boldsymbol{k}) + \mathrm{i}\ \text{Im}\ \tilde{\Sigma}_S(\omega \pm \mathrm{i}\eta, \boldsymbol{k}) \\ &= m + \Lambda_S(\omega + \mu, \boldsymbol{k}) \pm \mathrm{i}\ \text{sign}(+\omega)\Gamma_S(\omega + \mu, \boldsymbol{k}) \\ &\equiv m + \Lambda_S^{+} \pm \mathrm{i}\,\sigma^{+}\Gamma_S^{+}\,,\end{aligned} \tag{6.99}$$

where use of (6.94) and (6.98), and of the relation $\text{sign}(\omega) = -\text{sign}(-\omega)$ was made. The plus signs attached as superscripts to Λ and Γ in (6.99) refer to the plus signs that occur in their arguments, i.e. $\omega + \mu$. Similarly we introduced the abbreviation $\sigma^{\pm} \equiv \mathrm{sign}(\pm\omega)$. In close analogy to $F_S^{\pm}$, we introduce for the vector component of the baryon self-energy

$$\begin{aligned} F_V^{\pm} &\equiv F_V(\omega \pm \mathrm{i}\eta, \boldsymbol{k}) \equiv |\boldsymbol{k}| + \tilde{\Sigma}_V(\omega \pm \mathrm{i}\eta, \boldsymbol{k}) \\ &= |\boldsymbol{k}| + \mathrm{Re}\, \tilde{\Sigma}_V(\omega \pm \mathrm{i}\eta, \boldsymbol{k}) + \mathrm{i}\, \mathrm{Im}\, \tilde{\Sigma}_V(\omega \pm \mathrm{i}\eta, \boldsymbol{k}) \\ &= |\boldsymbol{k}| + \Lambda_V(\omega + \mu, \boldsymbol{k}) \pm \mathrm{i}\, \mathrm{sign}(+\omega)\Gamma_V(\omega + \mu, \boldsymbol{k}) \\ &\equiv |\boldsymbol{k}| + \Lambda_V^{+} \pm \mathrm{i}\,\sigma^{+}\Gamma_V^{+}\,, \end{aligned} \tag{6.100}$$

and for the timelike component of the self-energy,

$$\begin{aligned} F_0^{\pm} &\equiv F_0(\omega \pm \mathrm{i}\eta, \boldsymbol{k}) \equiv \tilde{\Sigma}_0(\omega \pm \mathrm{i}\eta, \boldsymbol{k}) - \mu \\ &= \mathrm{Re}\, \tilde{\Sigma}_0(\omega \pm \mathrm{i}\eta, \boldsymbol{k}) + \mathrm{i}\, \mathrm{Im}\, \tilde{\Sigma}_0(\omega \pm \mathrm{i}\eta, \boldsymbol{k}) - \mu \\ &= \Lambda_0(\omega + \mu, \boldsymbol{k}) \pm \mathrm{i}\, \mathrm{sign}(+\omega)\Gamma_0(\omega + \mu, \boldsymbol{k}) - \mu \\ &\equiv \Lambda_0^{+} - \mu \pm \mathrm{i}\,\sigma^{+}\Gamma_0^{+}\,. \end{aligned} \tag{6.101}$$

From equations (6.99) through (6.101) one calculates that

$$\begin{aligned} &\left(F_S^{\pm}\right)^2 + \left(F_V^{\pm}\right)^2 - \left(F_0^{\pm} - (\omega \pm \mathrm{i}\eta)\right)^2 \\ &\qquad = \left[m + \Lambda_S^{+} \pm \mathrm{i}\,\sigma^{+}\Gamma_S^{+}\right]^2 + \left[|\boldsymbol{k}| + \Lambda_V^{+} \pm \mathrm{i}\,\sigma^{+}\Gamma_V^{+}\right]^2 \\ &\qquad\quad - \left[\Lambda_0^{+} - \mu - \omega \pm \mathrm{i}(\sigma^{+}\Gamma_0^{+} - \eta)\right]^2 , \end{aligned} \tag{6.102}$$

which can be written as

$$\begin{aligned} &\left(F_S^{\pm}\right)^2 + \left(F_V^{\pm}\right)^2 - \left(F_0^{\pm} - (\omega \pm \mathrm{i}\eta)\right)^2 \\ &\qquad = \Big[(m + \Lambda_S^{+})^2 - {\Gamma_S^{+}}^2 + (\boldsymbol{k} + \Lambda_V^{+})^2 - {\Gamma_V^{+}}^2 \\ &\qquad\quad - (\Lambda_0^{+} - \mu - \omega)^2 + {\Gamma_0^{+}}^2 - 2\,\sigma^{+}\Gamma_0^{+}\eta + \eta^2\Big] \\ &\qquad\quad \pm \mathrm{i}\,\sigma^{+}\Big[2(m + \Lambda_S^{+})\Gamma_S^{+} + 2(|\boldsymbol{k}| + \Lambda_V^{+})\Gamma_V^{+} \\ &\qquad\quad - 2(\Lambda_0^{+} - \mu - \omega)(\Gamma_0^{+} - \sigma^{+}\eta)\Big]\,. \end{aligned} \tag{6.103}$$

With the above definitions at our disposal, we now solve Dyson's equation (6.85) for the analytically continued two-point Green function, which gives

$$\tilde{g}(z) = \frac{1}{\mathbf{1}\, F_S(z) \;+\; \boldsymbol{\gamma}\cdot\hat{\boldsymbol{k}}\; F_V(z) \;+\; \gamma^0\, (F_0(z) - z)}\,. \tag{6.104}$$

By means of multiplying both numerator and denominator of (6.104) with $\mathbf{1}F_S(z) + \boldsymbol{\gamma}\cdot\hat{\boldsymbol{k}}F_V(z) + \gamma^0(F_0(z) - z)$ we can get rid of the Dirac matrices, as outlined in appendix A.2. One then obtains

$$\tilde{g}(z) = \frac{\mathbf{1}\, F_S(z) \;-\; \boldsymbol{\gamma}\cdot\hat{\boldsymbol{k}}\; F_V(z) \;-\; \gamma^0\, (F_0(z) - z)}{F_S^2(z) \;+\; F_V^2(z) \;-\; (F_0(z) - z)^2}\,. \tag{6.105}$$

In order to derive the expression for the spectral function Ξ, we need to calculate the discontinuity of $\tilde{\boldsymbol{g}}(z)$ across the real energy axis. With the aid of (6.105), one arrives for Ξ at

$$\begin{aligned}\Xi(\omega,\boldsymbol{k}) &= \frac{1}{2\,\mathrm{i}\,\pi}\left\{\tilde{\boldsymbol{g}}(\omega+\mathrm{i}\eta,\boldsymbol{k}) - \tilde{\boldsymbol{g}}(\omega-\mathrm{i}\eta,\boldsymbol{k})\right\} \\ &= \frac{1}{2\,\mathrm{i}\,\pi}\left\{\frac{\mathbf{1}\,F_S^+ - \boldsymbol{\gamma}\cdot\hat{\boldsymbol{k}}\,F_V^+ - \gamma^0\,(F_0^+ - \omega + \mathrm{i}\eta)}{\left(F_S^+\right)^2 + \left(F_V^+\right)^2 - \left(F_0^+ - \omega + \mathrm{i}\eta\right)^2}\right. \\ &\qquad \left. - \frac{\mathbf{1}\,F_S^- - \boldsymbol{\gamma}\cdot\hat{\boldsymbol{k}}\,F_V^- - \gamma^0\,(F_0^- - \omega - \mathrm{i}\eta)}{\left(F_S^-\right)^2 + \left(F_V^-\right)^2 - \left(F_0^- - \omega - \mathrm{i}\eta\right)^2}\right\} \qquad (6.106) \\ &\equiv \frac{1}{2\,\mathrm{i}\,\pi}\left\{a^+ - a^-\right\}. \qquad (6.107)\end{aligned}$$

The calculation of the difference between a^+ and a^- in (6.107) is somewhat lengthy and cumbersome. To keep it as easy to survey as possible, let us define the following additional auxiliary functions:

$$\begin{aligned}N &\equiv \left[(m+\Lambda_S)^2 + (|\boldsymbol{k}|+\Lambda_V)^2 - (\Lambda_0 - (\omega+\mu))^2\right. \\ &\quad \left. - \Gamma_S^2 - \Gamma_V^2 + \Gamma_0^2\right]^2 + 4\big[\Gamma_S(m+\Lambda_S) + \Gamma_V(|\boldsymbol{k}|+\Lambda_V) \\ &\quad - (\Gamma_0 - \sigma(\omega)\eta)(\Lambda_0 - (\omega+\mu))\big]^2 \\ &\equiv \left[F - \Gamma_S^2 - \Gamma_V^2 + \Gamma_0^2\right]^2 + \hat{\Gamma}^2, \qquad (6.108)\end{aligned}$$

where

$$F \equiv F(\omega+\mu) = (m+\Lambda_S)^2 + (|\boldsymbol{k}|+\Lambda_V)^2 - (\Lambda_0 - (\omega+\mu))^2, \qquad (6.109)$$

$$\begin{aligned}\hat{\Gamma} &\equiv \hat{\Gamma}(\omega+\mu) \\ &= 2\big[\Gamma_S(m+\Lambda_S) + \Gamma_V(|\boldsymbol{k}|+\Lambda_V) - (\Gamma_0 - \sigma(\omega)\eta)(\Lambda_0 - (\omega+\mu))\big], \qquad (6.110)\end{aligned}$$

$$F_1 \equiv \frac{1}{2}\,\sigma(-\omega)\,\hat{\Gamma}(\omega+\mu), \qquad (6.111)$$

and

$$F_2 \equiv F(\omega+\mu) - \Gamma_S^2 - \Gamma_V^2 + \Gamma_0^2 - 2\sigma(\omega)\Gamma_0^+\eta + \eta^2. \qquad (6.112)$$

With the definitions (6.108) through (6.112) the functions $a^\pm$ in (6.107) are given by

$$a^\pm = \frac{\left[\mathbf{1}\,F_S^\pm - \boldsymbol{\gamma}\cdot\hat{\boldsymbol{k}}\,F_V^\pm - \gamma^0\,(F_0^\pm - (\omega\pm\mathrm{i}\eta))\right]\left[F_2 \pm \mathrm{i}\,\sigma^-\,\hat{\Gamma}\right]}{N}. \qquad (6.113)$$

Substituting (6.99) to (6.101) for $F_S^\pm$, $F_V^\pm$ and $F_0^\pm$ into (6.113) leads to

$$N\,a^\pm = \left[\mathbf{1}\,\left((m+\Lambda_S)\mp \mathrm{i}\sigma^-\Gamma_S\right) - \boldsymbol{\gamma}\cdot\hat{\boldsymbol{k}}\,\left((|\boldsymbol{k}|+\Lambda_V)\mp\mathrm{i}\sigma^-\Gamma_V\right)\right.$$
$$\left. -\gamma^0\left((\Lambda_0-(\omega+\mu\pm\mathrm{i}\eta))\mp\mathrm{i}\sigma^-\Gamma_0\right)\right]\left[F_2\pm\mathrm{i}\sigma^-\hat{\Gamma}\right]. \quad (6.114)$$

Collecting terms according to their matrix structure then gives

$$N\,a^\pm = \mathbf{1}\left\{F_2\left[m+\Lambda_S\mp\mathrm{i}\sigma^-\Gamma_S\right]\pm\mathrm{i}\sigma^-\hat{\Gamma}\left[m+\Lambda_S\mp\mathrm{i}\sigma^-\Gamma_S\right]\right\}$$
$$-\boldsymbol{\gamma}\cdot\hat{\boldsymbol{k}}\left\{F_2\left[|\boldsymbol{k}|+\Lambda_V\mp\mathrm{i}\sigma^-\Gamma_V\right]\pm\mathrm{i}\sigma^-\hat{\Gamma}\left[|\boldsymbol{k}|+\Lambda_V\mp\mathrm{i}\sigma^-\Gamma_V\right]\right\}$$
$$-\gamma^0\left\{F_2\left[\Lambda_0-(\omega+\mu\pm\mathrm{i}\eta)\mp\mathrm{i}\sigma^-\Gamma_0\right]\right.$$
$$\left.\pm\mathrm{i}\sigma^-\hat{\Gamma}\left[\Lambda_0-(\omega+\mu\pm\mathrm{i}\eta)\mp\mathrm{i}\sigma^-\Gamma_0\right]\right\}. \quad (6.115)$$

Putting this expression back into (6.107) results in

$$\frac{a^+-a^-}{2\,\mathrm{i}\,\pi} = \mathbf{1}\,\frac{1}{\pi\,N}\left\{\sigma^-\hat{\Gamma}(m+\Lambda_S)-F_2\sigma^-\Gamma_S\right\}$$
$$-\boldsymbol{\gamma}\cdot\hat{\boldsymbol{k}}\,\frac{1}{\pi\,N}\left\{\sigma^-\hat{\Gamma}(|\boldsymbol{k}|+\Lambda_V)-F_2\sigma^-\Gamma_V\right\}$$
$$-\gamma^0\,\frac{1}{\pi\,N}\left\{\sigma^-\hat{\Gamma}(\Lambda_0-(\omega+\mu))-F_2\sigma^-\Gamma_0\right\}. \quad (6.116)$$

From equation (6.116) one sees that the spectral function splits up into the same three different Dirac–Lorentz components as the self-energy (6.83). We therefore introduce the decomposition

$$\Xi(\omega,\boldsymbol{k}) = \mathbf{1}\;\Xi_S(\omega,\boldsymbol{k})+\boldsymbol{\gamma}\cdot\hat{\boldsymbol{k}}\;\Xi_V(\omega,\boldsymbol{k})+\gamma^0\;\Xi_0(\omega,\boldsymbol{k}). \quad (6.117)$$

The expressions for the individual terms in (6.117) can then be identified by comparing (6.117) with (6.116). This leads to

$$\Xi_S(\omega,\boldsymbol{k}) = \frac{\sigma^-}{\pi}\,\frac{\hat{\Gamma}(m+\Lambda_S)-\left[F-\Gamma_S^2-\Gamma_V^2+\Gamma_0^2\right]\Gamma_S}{\left[F-\Gamma_S^2-\Gamma_V^2+\Gamma_0^2\right]+\hat{\Gamma}^2}, \quad (6.118)$$

$$\Xi_V(\omega,\boldsymbol{k}) = -\frac{\sigma^-}{\pi}\,\frac{\hat{\Gamma}(|\boldsymbol{k}|+\Lambda_V)-\left[F-\Gamma_S^2-\Gamma_V^2+\Gamma_0^2\right]\Gamma_V}{\left[F-\Gamma_S^2-\Gamma_V^2+\Gamma_0^2\right]+\hat{\Gamma}^2}, \quad (6.119)$$

$$\Xi_S(\omega,\boldsymbol{k}) = -\frac{\sigma^-}{\pi}\,\frac{\hat{\Gamma}(\Lambda_0-(\omega+\mu))-\left[F-\Gamma_S^2-\Gamma_V^2+\Gamma_0^2\right]\Gamma_0}{\left[F-\Gamma_S^2-\Gamma_V^2+\Gamma_0^2\right]+\hat{\Gamma}^2}. \quad (6.120)$$

The spectral functions derived in equations (6.118) to (6.120) look considerably simpler for systems whose baryon self-energies are real functions, as will be demonstrated next. The real-self-energy limit is obtained by taking $\Gamma_i \to 0$, with $i = S, V, 0$. So the second terms in (6.118) to (6.120) vanish trivially. Care, however, is to be taken with respect to the terms proportional to $\sigma^{-}\hat{\Gamma} \equiv \mathrm{sign}(-\omega)\hat{\Gamma}$. By making use of the general relation

$$|x|\,\mathrm{sign}(x\,y) = x\,\mathrm{sign}(y) \tag{6.121}$$

and setting $x = \hat{\Gamma}$ and $y = -\omega$, these products can be written as

$$|\hat{\Gamma}(\omega+\mu)|\;\mathrm{sign}\big(-\omega\,\hat{\Gamma}(\omega+\mu)\big) = \hat{\Gamma}(\omega+\mu)\;\mathrm{sign}(-\omega)\,. \tag{6.122}$$

Substituting (6.122) into (6.118) to (6.120) then gives for the spectral functions

$$\begin{aligned}\Xi_S(\omega,\boldsymbol{k}) = {} & \frac{1}{\pi}\frac{|\hat{\Gamma}(\omega+\mu)|}{\left[F-\Gamma_S^2-\Gamma_V^2+\Gamma_0^2\right]^2+\hat{\Gamma}^2}\,\mathrm{sign}(-\omega\hat{\Gamma})(m+\Lambda_S) \\ & -\frac{1}{\pi}\frac{|\hat{\Gamma}_S|}{\left[F-\Gamma_S^2-\Gamma_V^2+\Gamma_0^2\right]^2+\hat{\Gamma}^2} \\ & \times\,\mathrm{sign}(-\omega\hat{\Gamma}_S)\big[F-\Gamma_S^2-\Gamma_V^2+\Gamma_0^2\big] & (6.123)\\ \longrightarrow {} & \delta[F(\omega+\mu,\boldsymbol{k})]\;\mathrm{sign}[-\omega\,\hat{\Gamma}(\omega+\mu,\boldsymbol{k})] \\ & \times\,[m+\Lambda_S(\omega+\mu,\boldsymbol{k})] \qquad \text{for} \quad \hat{\Gamma}\to 0\,, & (6.124)\end{aligned}$$

$$\begin{aligned}\Xi_V(\omega,\boldsymbol{k}) = {} & -\frac{1}{\pi}\frac{|\hat{\Gamma}(\omega+\mu)|}{\left[F-\Gamma_S^2-\Gamma_V^2+\Gamma_0^2\right]^2+\hat{\Gamma}^2}\,\mathrm{sign}(-\omega\hat{\Gamma})(|\boldsymbol{k}|+\Lambda_V) \\ & +\frac{1}{\pi}\frac{|\hat{\Gamma}_V|}{\left[F-\Gamma_S^2-\Gamma_V^2+\Gamma_0^2\right]^2+\hat{\Gamma}^2} \\ & \times\,\mathrm{sign}(-\omega\hat{\Gamma}_V)\big[F-\Gamma_S^2-\Gamma_V^2+\Gamma_0^2\big] & (6.125)\\ \longrightarrow {} & -\,\delta[F(\omega+\mu,\boldsymbol{k})]\;\mathrm{sign}[-\omega\,\hat{\Gamma}(\omega+\mu,\boldsymbol{k})] \\ & \times\,[|\boldsymbol{k}|+\Lambda_V(\omega+\mu,\boldsymbol{k})] \qquad \text{for} \quad \hat{\Gamma}\to 0\,, & (6.126)\end{aligned}$$

and

$$\begin{aligned}\Xi_0(\omega,\boldsymbol{k}) = {} & -\frac{1}{\pi}\frac{|\hat{\Gamma}(\omega+\mu)|}{\left[F-\Gamma_S^2-\Gamma_V^2+\Gamma_0^2\right]^2+\hat{\Gamma}^2}\,\mathrm{sign}(-\omega\hat{\Gamma})(\Lambda_0-(\omega+\mu)) \\ & +\frac{1}{\pi}\frac{|\hat{\Gamma}_S|}{\left[F-\Gamma_S^2-\Gamma_V^2+\Gamma_0^2\right]^2+\hat{\Gamma}^2} \\ & \times\,\mathrm{sign}(-\omega\hat{\Gamma}_0)\big[F-\Gamma_S^2-\Gamma_V^2+\Gamma_0^2\big] & (6.127)\\ \longrightarrow {} & -\,\delta[F(\omega+\mu,\boldsymbol{k})]\;\mathrm{sign}[-\omega\,\hat{\Gamma}(\omega+\mu,\boldsymbol{k})] \\ & \times\,[\Lambda_0(\omega+\mu,\boldsymbol{k})-(\omega+\mu)] \qquad \text{for} \quad \hat{\Gamma}\to 0\,. & (6.128)\end{aligned}$$

The quantity $\hat{\Gamma}$ reduces for $\Gamma_i \to 0$ to

$$\hat{\Gamma}(\omega+\mu,\boldsymbol{k}) \longrightarrow 2\eta\,\mathrm{sign}(\omega)\;[\Lambda_0(\omega+\mu,\boldsymbol{k})-(\omega+\mu)]\;, \tag{6.129}$$

where ω is to be set equal to the respective single-particle energy, that is, $\omega_1 \equiv \omega(\boldsymbol{k}) - \mu$ for particles and $\omega_2 \equiv \bar{\omega}(\boldsymbol{k}) - \mu$ for antiparticles. Thus

$$\hat{\Gamma}(\omega + \mu, \boldsymbol{k})\big|_{\omega=\omega_1-\mu} \longrightarrow 2\eta\, \mathrm{sign}(\omega_1 - \mu)\ [\Lambda_0(\omega_1, \boldsymbol{k}) - \omega_1]\,, \quad (6.130)$$

and

$$\hat{\Gamma}(\omega + \mu, \boldsymbol{k})\big|_{\omega=\omega_2-\mu} \longrightarrow 2\eta\, \mathrm{sign}(\omega_2 - \mu)\ [\Lambda_0(\omega_2, \boldsymbol{k}) - \omega_2]\,. \quad (6.131)$$

Inspection of the signs in (6.130) and (6.131) leads to

$$\mathrm{sign}\big[\hat{\Gamma}(\omega + \mu, \boldsymbol{k})\big|_{\omega=\omega_{1/2}-\mu}\big]_{\Gamma_i=0} = \pm 1\,, \quad (6.132)$$

where the plus (minus) sign refers to particles (antiparticles). Via equations (6.130) and (6.131) we find for the sign of the expression $\mathrm{sign}(-\omega\hat{\Gamma})$ the result

$$\begin{aligned}\mathrm{sign}&\left[\Big(-\omega\,\hat{\Gamma}(\omega + \mu, \boldsymbol{k})\Big)\Big|_{\omega=\omega_{1/2}-\mu}\right]_{\Gamma_i=0}\\ &= \mathrm{sign}\,\big[\mp(\omega_{1/2} - \mu)\big]\,\Gamma(\omega_{1/2}, \boldsymbol{k})\big]_{\Gamma_i=0}\\ &= \mathrm{sign}(\pm\,|\omega_{1/2} - \mu|) = \pm\,1\,. \end{aligned} \quad (6.133)$$

Finally, we are left with evaluating the function $\delta[F(\omega + \mu, \boldsymbol{k})]$ in equations (6.124), (6.126), and (6.128). To find the zeroes of the argument of this δ-function,

$$\begin{aligned}F(\omega + \mu, \boldsymbol{k}) = {}& [m_B + \Lambda_S(\omega + \mu, \boldsymbol{k})]^2 + [|\boldsymbol{k}| + \Lambda_V(\omega + \mu, \boldsymbol{k})]^2\\ &- [\Lambda_0(\omega + \mu, \boldsymbol{k}) - (\omega + \mu)]^2\,,\end{aligned} \quad (6.134)$$

we make use of the general mathematical relation

$$\delta[F(\omega + \mu, \boldsymbol{k})] = \sum_{l=1}^{2} \left|\left(\frac{\partial F}{\partial \omega}\right)_{\omega_l(k)}\right| \delta(\omega + \mu - \omega_l(k))\,, \quad (6.135)$$

where ω_l denotes the two solutions for which $F(\omega_l) = 0$. This is the case for

$$\omega_{1/2} = \Lambda_0(\omega_{1/2}) \pm \sqrt{[m + \Lambda_S(\omega_{1/2})]^2 + [|\boldsymbol{k}| + \Lambda_V(\omega_{1/2})]^2}\,, \quad (6.136)$$

as can be inferred from (6.134). The partial derivative in (6.135) is readily found to read

$$\begin{aligned}\frac{\partial F(\omega, |\boldsymbol{k}|)}{\partial \omega} = 2\Big\{&[m + \Lambda_S(\omega, \boldsymbol{k})]\,\frac{\partial \Lambda_S}{\partial \omega} + [|\boldsymbol{k}| + \Lambda_V(\omega, \boldsymbol{k})]\,\frac{\partial \Lambda_V}{\partial \omega}\\ &+ [\Lambda_0(\omega, \boldsymbol{k}) - \omega]\Big[1 - \frac{\partial \Lambda_0}{\partial \omega}\Big]\Big\}\,.\end{aligned} \quad (6.137)$$

In the case of the relativistic Hartree approximation the self-energies are energy independent and so the partial derivatives $\partial\Lambda_i/\partial\omega$ vanish. Equation (6.137) therefore simplifies for this approximation to

$$\left|\left(\frac{\partial F(\omega,\boldsymbol{k})}{\partial\omega}\right)_{\omega_1}\right| = 2\sqrt{[m+\Lambda_S]^2+[|\boldsymbol{k}|+\Lambda_V]^2} = \left|\left(\frac{\partial F(\omega,\boldsymbol{k})}{\partial\omega}\right)_{\omega_2}\right| . \tag{6.138}$$

Substituting equations (6.135) and (6.138) into equations (6.124), (6.126) and (6.128) leads for the components of the spectral function to

$$\begin{aligned}\Xi_S(\omega,\boldsymbol{k}) = {} & [m+\Lambda_S(\omega_1(\boldsymbol{k}),\boldsymbol{k})] \left.\frac{\partial F(\omega,\boldsymbol{k})}{\partial\omega}\right|_{\omega_1(k)} \delta[\omega+\mu-\omega_1(\boldsymbol{k})] \\ & - [m+\Lambda_S(\omega_2(\boldsymbol{k}),\boldsymbol{k})] \left.\frac{\partial F(\omega,\boldsymbol{k})}{\partial\omega}\right|_{\omega_2(k)} \delta[\omega+\mu-\omega_2(\boldsymbol{k})] ,\end{aligned} \tag{6.139}$$

$$\begin{aligned}\Xi_V(\omega,\boldsymbol{k}) = {} & -[|\boldsymbol{k}|+\Lambda_V(\omega_1(\boldsymbol{k}),\boldsymbol{k})] \left.\frac{\partial F(\omega,\boldsymbol{k})}{\partial\omega}\right|_{\omega_1(k)} \delta[\omega+\mu-\omega_1(\boldsymbol{k})] \\ & + [|\boldsymbol{k}|+\Lambda_V(\omega_2(\boldsymbol{k}),\boldsymbol{k})] \left.\frac{\partial F(\omega,\boldsymbol{k})}{\partial\omega}\right|_{\omega_2(k)} \delta[\omega+\mu-\omega_2(\boldsymbol{k})] ,\end{aligned} \tag{6.140}$$

and

$$\begin{aligned}\Xi_0(\omega,\boldsymbol{k}) = {} & -[\Lambda_0(\omega_1(\boldsymbol{k}),\boldsymbol{k})-\omega_1] \left.\frac{\partial F(\omega,\boldsymbol{k})}{\partial\omega}\right|_{\omega_1(k)} \delta[\omega+\mu-\omega_1(\boldsymbol{k})] \\ & + [\Lambda_0(\omega_2(\boldsymbol{k}),\boldsymbol{k})-\omega_2] \left.\frac{\partial F(\omega,\boldsymbol{k})}{\partial\omega}\right|_{\omega_2(k)} \delta[\omega+\mu-\omega_2(\boldsymbol{k})] .\end{aligned} \tag{6.141}$$

The mathematical structure of equations (6.139) to (6.141) suggests a decomposition of the spectral function according to

$$\begin{aligned}\left(\boldsymbol{\Xi}_i^B\right)_{\zeta\zeta'}(\omega,\boldsymbol{k}) \equiv {} & \delta\big(\omega+\mu^B-\omega^B(\boldsymbol{k})\big)\left(\boldsymbol{\Xi}_i^B\right)_{\zeta\zeta'}(\boldsymbol{k}) \\ & + \delta\big(\omega+\bar{\mu}^B-\bar{\omega}^B(\boldsymbol{k})\big)\left(\bar{\boldsymbol{\Xi}}_i^B\right)_{\zeta\zeta'}(\boldsymbol{k}) ,\end{aligned} \tag{6.142}$$

with the individual, energy independent spectral functions $\Xi_i^B(\boldsymbol{k})$ ($i = S,V,0$) given by

$$\Xi_S^B(\boldsymbol{k}) \equiv \frac{m_B+\Sigma_S^B(\omega^B(\boldsymbol{k}),\boldsymbol{k})}{2\sqrt{(m_B+\Sigma_S^B(\omega^B(\boldsymbol{k}),\boldsymbol{k}))^2+(|\boldsymbol{k}+\Sigma_V^B(\omega^B(\boldsymbol{k}),\boldsymbol{k}))^2}} , \tag{6.143}$$

$$\Xi_V^B(\boldsymbol{k}) \equiv -\frac{|\boldsymbol{k}|+\Sigma_V^B(\omega^B(\boldsymbol{k}),\boldsymbol{k})}{2\sqrt{(m_B+\Sigma_S^B(\omega^B(\boldsymbol{k}),\boldsymbol{k}))^2+(|\boldsymbol{k}+\Sigma_V^B(\omega^B(\boldsymbol{k}),\boldsymbol{k}))^2}} , \tag{6.144}$$

$$\Xi_0^B(\boldsymbol{k}) \equiv \frac{1}{2} . \tag{6.145}$$

The corresponding expressions for the antibaryons are given by

$$\bar{\Xi}_S^B(\boldsymbol{k}) \equiv -\frac{m_B + \Sigma_S^B(\bar{\omega}^B(\boldsymbol{k}), \boldsymbol{k})}{2\sqrt{(m_B + \Sigma_S^B(\bar{\omega}^B(\boldsymbol{k}), \boldsymbol{k}))^2 + (|\boldsymbol{k} + \Sigma_V^B(\bar{\omega}^B(\boldsymbol{k}), \boldsymbol{k}))^2}}\,, \tag{6.146}$$

$$\bar{\Xi}_V^B(\boldsymbol{k}) \equiv \frac{|\boldsymbol{k}| + \Sigma_V^B(\bar{\omega}^B(\boldsymbol{k}), \boldsymbol{k})}{2\sqrt{(m_B + \Sigma_S^B(\bar{\omega}^B(\boldsymbol{k}), \boldsymbol{k}))^2 + (|\boldsymbol{k} + \Sigma_V^B(\bar{\omega}^B(\boldsymbol{k}), \boldsymbol{k}))^2}}\,, \tag{6.147}$$

$$\bar{\Xi}_0^B(\boldsymbol{k}) \equiv \frac{1}{2}\,. \tag{6.148}$$

The single-particle energies in (6.143) through (6.145) read

$$\omega^B(\boldsymbol{k}) = \Sigma_0^B(\omega^B(\boldsymbol{k}), \boldsymbol{k}) + \left\{\left(m_B + \Sigma_S^B(\omega^B(\boldsymbol{k}), \boldsymbol{k})\right)^2 + (|\boldsymbol{k}| + \Sigma_V^B(\omega^B(\boldsymbol{k}), \boldsymbol{k}))^2\right\}^{1/2} \tag{6.149}$$

$$\equiv \Sigma_0^B(\boldsymbol{k})\left\{(m_B^*)^2 + (\boldsymbol{k}^*)^2\right\}^{1/2}, \tag{6.150}$$

and for the antibaryons

$$\bar{\omega}^B(\boldsymbol{k}) = \Sigma_0^B(\bar{\omega}^B(\boldsymbol{k}), \boldsymbol{k}) - \left\{\left(m_B + \Sigma_S^B(\bar{\omega}^B(\boldsymbol{k}), \boldsymbol{k})\right)^2 + (|\boldsymbol{k}| + \Sigma_V^B(\bar{\omega}^B(\boldsymbol{k}), \boldsymbol{k}))^2\right\}^{1/2}. \tag{6.151}$$

Finally the chemical potentials of baryons and antibaryons propagating in the medium with Fermi momenta k_{F_B} and $\bar{k}_{F_B}$ are given by

$$\mu^B = \omega^B(k_{F_B})\,, \qquad \text{and} \qquad \bar{\mu}^B = \bar{\omega}^B(\bar{k}_{F_B})\,, \tag{6.152}$$

respectively.

6.2.1 Application to free lepton propagator

The mathematical content of the spectral formalism developed in the preceding part of section 6.2 can be nicely demonstrated for the simple case of a gas of free, relativistic fermions, for which we chose the leptons. We recall that leptons, whose masses and quantum numbers are listed in table 6.1, are present in neutron star matter because of chemical equilibrium, the guiding principle by means of which neutron star matter settles down into the lowest possible energy state. The stating point is the spectral representation for the fermion propagator derived in equation (6.40). Replacing the label B with L, where $L = e^-, \mu^-$ gives for equation (6.40)

$$\boldsymbol{g}^L_{\zeta\zeta'}(k) = \frac{\mathrm{i}}{2\pi}\int_{-\infty}^{+\infty} \mathrm{d}\omega \left\{\frac{\boldsymbol{g}^L_>(\omega, \boldsymbol{k})_{\zeta\zeta'}}{k_0 - \omega - \mu^L + \mathrm{i}\eta} - \frac{\boldsymbol{g}^L_<(\omega, \boldsymbol{k})_{\zeta\zeta'}}{k_0 - \omega - \mu^L - \mathrm{i}\eta}\right\}. \tag{6.153}$$

Table 6.1. Masses m_L, spin quantum numbers J_L, and electric charges $q_L^{\rm el}$ of those leptons which contribute to the EOS of neutron star matter.

Lepton (L)	m_L (MeV)	J_L	$q_L^{\rm el}$
e^-	0.511	1/2	−1
μ^-	106	1/2	−1

For simplicity, we shall restrict ourselves to zero temperature, in which case the Fermi–Dirac functions reduce to

$$f^L(\omega - \mu^L) \longrightarrow \Theta(\mu^L - \omega) \, . \tag{6.154}$$

Equation (6.41) then reads

$$\boldsymbol{g}^L_{\zeta\zeta'}(k) = - \int\limits_{-\infty}^{+\infty} \mathrm{d}\omega \left\{ \frac{1 - \Theta(\mu^L - k_0)}{k_0 - \omega - \mu^L + \mathrm{i}\eta} + \frac{\Theta(\mu^L - k_0)}{k_0 - \omega - \mu^L - \mathrm{i}\eta} \right\} \boldsymbol{\Xi}^L_{\zeta\zeta'}(\omega, \boldsymbol{k}) \, , \tag{6.155}$$

which can be brought to the shorter form

$$\boldsymbol{g}^L_{\zeta\zeta'}(k) = \int\limits_{-\infty}^{+\infty} \mathrm{d}\omega \frac{\boldsymbol{\Xi}^L_{\zeta\zeta'}(\omega, \boldsymbol{k})}{\omega - (k_0 - \mu^L)\,(1 + \mathrm{i}\eta)} \, . \tag{6.156}$$

Equation (6.156) is the analog to (6.48), except that the energy argument has been shifted by the chemical potential of the leptons, μ^L, in analogy to the rescaling procedure for the baryon chemical potential outlined at the beginning of section 6.2.

The mathematical form of the lepton spectral function, $\boldsymbol{\Xi}^L$, is obtained from the cutline of the analytically continued two-point lepton function,

$$\tilde{\boldsymbol{g}}^L_{\zeta\zeta'}(z, \boldsymbol{k}) = \int\limits_{-\infty}^{+\infty} \mathrm{d}\omega \, \frac{\boldsymbol{\Xi}^L_{\zeta\zeta'}(\omega, \boldsymbol{k})}{\omega - z} \, , \tag{6.157}$$

along the real energy axis, that is,

$$\boldsymbol{\Xi}^L_{\zeta\zeta'}(\omega, \boldsymbol{k}) = \frac{\tilde{\boldsymbol{g}}^L_{\zeta\zeta'}(\omega + \mathrm{i}\eta, \boldsymbol{k}) - \tilde{\boldsymbol{g}}^L_{\zeta\zeta'}(\omega - \mathrm{i}\eta, \boldsymbol{k})}{2\,\mathrm{i}\,\pi} \, . \tag{6.158}$$

Leptons do not carry isospin. So for them the unity matrix introduced in (5.131) consists only of the Dirac part, and therefore the labels ζ, ζ' reduce to α, α'. Hence one has for the matrix elements of the unity matrix

$$\left(\boldsymbol{1}\right)_{\alpha\alpha'} \equiv \left(\boldsymbol{1}^{\rm Dirac}\right)_{\alpha\alpha'} = \delta_{\alpha\alpha'} \, . \tag{6.159}$$

Inversion of the lepton Dyson equation,

$$\{\mathbf{1} m_B + \boldsymbol{\gamma} \cdot \hat{\boldsymbol{k}}\, |\boldsymbol{k}| - \gamma^0 k_0\} \, \boldsymbol{g}^L(k_0, \boldsymbol{k}) = 1 , \tag{6.160}$$

gives for the lepton two-point function (cf. equation (6.105))

$$\tilde{\boldsymbol{g}}^L(z, \boldsymbol{k}) = \frac{m_L \mathbf{1} - |\boldsymbol{k}| \gamma \cdot \hat{\boldsymbol{k}} + (z + \mu^L) \gamma^0}{m_L^2 + \boldsymbol{k}^2 - (z + \mu^L)^2} . \tag{6.161}$$

The *physical* lepton propagator, $\boldsymbol{g}^L$, is obtained from the analytically continued expression (6.157) as

$$\boldsymbol{g}^L_{\alpha\alpha'}(k) = \tilde{\boldsymbol{g}}^L_{\alpha\alpha'}((k_0 - \mu^L)(1 + \mathrm{i}\eta), \boldsymbol{k}) . \tag{6.162}$$

Substituting (6.161) into (6.158) then leads for $\boldsymbol{\Xi}^L$ to

$$\boldsymbol{\Xi}^L_{\alpha\alpha'}(\omega, \boldsymbol{k}) = \delta\big(\omega + \mu^L - \omega^L(\boldsymbol{k})\big) \, \boldsymbol{\Xi}^L_{\alpha\alpha'}(\boldsymbol{k}) + \delta\big(\omega + \bar{\mu}^L - \bar{\omega}^L(\boldsymbol{k})\big) \, \bar{\boldsymbol{\Xi}}^L_{\alpha\alpha'}(\boldsymbol{k}) , \tag{6.163}$$

where the first term on the right-hand side corresponds to particles and the second to antiparticles. The energy-independent spectral functions in (6.163) are given by

$$\boldsymbol{\Xi}^L_{\alpha\alpha'}(\boldsymbol{k}) = + \frac{m_L \, (\mathbf{1})_{\alpha\alpha'} - |\boldsymbol{k}| \, (\gamma \cdot \hat{\boldsymbol{k}})_{\alpha\alpha'} + \omega^L(\boldsymbol{k}) \, (\gamma^0)_{\alpha\alpha'}}{2\sqrt{m_L^2 + \boldsymbol{k}^2}} \tag{6.164}$$

$$\equiv \Xi^L_S \, (\mathbf{1})_{\alpha\alpha'} + \Xi^L_V \, (\gamma \cdot \hat{\boldsymbol{k}})_{\alpha\alpha'} + \Xi^L_0 \, (\gamma)^0_{\alpha\alpha'} \tag{6.165}$$

for particles, and by

$$\bar{\boldsymbol{\Xi}}^L_{\alpha\alpha'}(\boldsymbol{k}) = - \frac{m_L \, (\mathbf{1})_{\alpha\alpha'} - |\boldsymbol{k}| \, (\gamma \cdot \hat{\boldsymbol{k}})_{\alpha\alpha'} + \bar{\omega}^L(\boldsymbol{k}) \, (\gamma^0)_{\alpha\alpha'}}{2\sqrt{m_L^2 + \boldsymbol{k}^2}} \tag{6.166}$$

$$\equiv \bar{\Xi}^L_S \, (\mathbf{1})_{\alpha\alpha'} + \bar{\Xi}^L_V \, (\gamma \cdot \hat{\boldsymbol{k}})_{\alpha\alpha'} + \bar{\Xi}^L_0 \, (\gamma)^0_{\alpha\alpha'} \tag{6.167}$$

for the antiparticles. Finally, the energy–momentum relation of free leptons and antileptons read

$$\omega^L(\boldsymbol{k}) = + \sqrt{m_L^2 + \boldsymbol{k}^2} = - \bar{\omega}^L(\boldsymbol{k}) , \tag{6.168}$$

so that the lepton (antilepton) chemical potentials are given by

$$\mu^L = \omega^L(k_{F_L}) \qquad \text{and} \qquad \bar{\mu}^L = \bar{\omega}^L(\bar{k}_{F_L}) , \tag{6.169}$$

where k_{F_L} and $\bar{k}_{F_L}$ denote the respective lepton Fermi momenta.

6.2.2 Baryon propagator in relativistic Hartree approximation

As a second example, we consider the explicit mathematical structure of the baryon two-point function in the interacting particle case. The relativistic Hartree approximation is chosen as the underlying many-body approach [314, 346]. The spectral function for this approximation has already been derived in equations (6.142) through (6.145). Repeating the steps outlined at the end of section 6.1, which have led us to the two-point baryon function in the non-interacting particle case (equation (6.64)), one arrives at

$$
\begin{aligned}
-\, \boldsymbol{g}^{\mathrm{H},B}_{\zeta\zeta'}(k^\mu) &= \frac{\gamma^0_{\zeta\zeta'}\,(k_0 - \Sigma^{\mathrm{H},B}_0(\boldsymbol{k})) - (\boldsymbol{\gamma}\cdot\boldsymbol{k})_{\zeta\zeta'} + (\mathbf{1})_{\zeta\zeta'}\,(m_B + \Sigma^{\mathrm{H},B}_S(\boldsymbol{k}))}{2\,\epsilon^{\mathrm{H},B}(\boldsymbol{k})} \\
&\times \left\{ \frac{1 - f^B(\boldsymbol{k})}{k_0 - \Sigma^{\mathrm{H},B}_0(\boldsymbol{k}) - \epsilon^{\mathrm{H},B}(\boldsymbol{k}) + \mathrm{i}\eta} + \frac{f^B(\boldsymbol{k})}{k_0 - \boldsymbol{\Sigma}^{\mathrm{H},B}_0(\boldsymbol{k}) - \epsilon^{\mathrm{H},B}(\boldsymbol{k}) - \mathrm{i}\eta} \right. \\
&\left. - \frac{1 - \bar{f}^B(\boldsymbol{k})}{k_0 - \Sigma^{\mathrm{H},B}_0(\boldsymbol{k}) + \epsilon^{\mathrm{H},B}(\boldsymbol{k}) - \mathrm{i}\eta} - \frac{\bar{f}^B(\boldsymbol{k})}{k_0 - \Sigma^{\mathrm{H},B}_0(\boldsymbol{k}) + \epsilon^{\mathrm{H},B}(\boldsymbol{k}) + \mathrm{i}\eta} \right\},
\end{aligned}
\tag{6.170}
$$

with the single-particle energy $\epsilon^{\mathrm{H},B}$ given by

$$
\epsilon^{\mathrm{H},B}(\boldsymbol{k}) = \sqrt{(m + \Sigma^{\mathrm{H},B}_S)^2 + \boldsymbol{k}^2} = \frac{1}{2}\left|\frac{\partial F^B}{\partial \omega}\right|_{\omega^B(|\boldsymbol{k}|)} . \tag{6.171}
$$

The physical interpretation of the individual terms in (6.170) is given too at the end of section 6.1. The free-particle limit is obtained from equations (6.170) and (6.171) if the interactions among the baryons are switched off, which implies that $\boldsymbol{\Sigma}^B \to 0$.

6.3 Baryon number density

In the next step we outline how the number density of baryons is obtained from the baryon two-point Green function [79, 118, 125, 314, 347]. Let us begin with defining the total baryon number operator,

$$
A \equiv \int \mathrm{d}^3\boldsymbol{x}\ [\psi^\dagger_B(x,\zeta), \psi_B(x,\zeta)] , \tag{6.172}
$$

from which the definition of the density of baryons follows as

$$
\rho^B \equiv \frac{1}{\Omega}\, \langle \boldsymbol{\Phi}_0 | A | \boldsymbol{\Phi}_0 \rangle , \tag{6.173}
$$

with Ω a volume element. Substituting (6.172) into (6.173) gives

$$
\rho^B = \frac{1}{\Omega} \int_\Omega \mathrm{d}^3\boldsymbol{x}\, \left\{ \langle \boldsymbol{\Phi}_0 | \psi^\dagger_B(x,\zeta)\psi_B(x,\zeta) - \psi_B(x,\zeta)\psi^\dagger_B(x,\zeta) | \boldsymbol{\Phi}_0 \rangle \right\} . \tag{6.174}
$$

Inspection of the defining relation for the two-point function, equation (5.63), shows that the field-operator products in (6.174) can be expressed as

$$\boldsymbol{g}^B(x,\zeta;x'=x^+,\zeta') = -\,\mathrm{i}\,\langle\bar{\psi}_B(x^+,\zeta')\psi_B(x,\zeta)\rangle$$
$$= -\,\mathrm{i}\,\gamma^0\langle\psi_B^\dagger(x^+,\zeta')\psi_B(x,\zeta)\rangle\,, \qquad (6.175)$$

and

$$\boldsymbol{g}^B(x,\zeta;x'=x^-,\zeta') = \mathrm{i}\,\langle\bar{\psi}_B(x,\zeta)\psi_B(x^-,\zeta')\rangle$$
$$= \mathrm{i}\,\gamma^0\langle\psi_B(x,\zeta)\psi_B^\dagger(x^-,\zeta')\rangle\,. \qquad (6.176)$$

With the aid of these relations, equation (6.174) can be written as

$$\rho^B = \mathrm{i}\,\gamma^0\frac{1}{\Omega}\int_\Omega \mathrm{d}^3\boldsymbol{x}\,\{\boldsymbol{g}^B(x,\zeta;x^+,\zeta') + \boldsymbol{g}^B(x,\zeta;x^-,\zeta')\}\,, \qquad (6.177)$$

for which we introduce the more compact notation

$$\rho^B \equiv \mathrm{i}\,(\gamma^0)_{\zeta\zeta'}\,\frac{1}{\Omega}\sum_{y=x^+,x^-}\lim_{x'\to y}\int_\Omega \mathrm{d}^3\boldsymbol{x}\;\boldsymbol{g}^B_{\zeta'\zeta}(x;y)\,. \qquad (6.178)$$

The Fourier transform of (6.178) reads

$$\lim_{x'\to x^\pm}\frac{1}{\Omega}\int_\Omega \mathrm{d}^3\boldsymbol{x}\;\boldsymbol{g}(x;x') = \int\frac{\mathrm{d}^4q}{(2\pi)^4}\,\mathrm{e}^{\pm\,\mathrm{i}\eta q^0}\;\boldsymbol{g}(q)\,, \qquad (6.179)$$

and therefore

$$\rho^B = \mathrm{i}\,(\gamma^0)_{\zeta\zeta'}\sum_{s=+,-}\int\frac{\mathrm{d}^4q}{(2\pi)^4}\,\mathrm{e}^{\mathrm{i}s\eta q^0}\;\boldsymbol{g}^B_{\zeta'\zeta}(q)\,. \qquad (6.180)$$

After contour integration and rearranging terms one gets (cf. appendix B.1)

$$\rho^B = (\gamma^0)_{\zeta\zeta'}\int\frac{\mathrm{d}^3\boldsymbol{q}}{(2\pi)^3}\,\boldsymbol{\Xi}^B_{\zeta\zeta'}(\boldsymbol{q})\;f(\omega^B(\boldsymbol{q})-\mu^B)$$
$$-\,(\gamma^0)_{\zeta\zeta'}\int\frac{\mathrm{d}^3\boldsymbol{q}}{(2\pi)^3}\,\bar{\boldsymbol{\Xi}}^B_{\zeta\zeta'}(\boldsymbol{q})\;f(-(\bar{\omega}^B(\boldsymbol{q})-\mu^B))\,. \qquad (6.181)$$

The traces in (6.181) are evaluated as follows:

$$(\gamma^0)_{\zeta\zeta'}\;\boldsymbol{\Xi}^B_{\zeta'\zeta} = \mathrm{Tr}\,(\gamma^0\,\boldsymbol{\Xi}^B)$$
$$= \mathrm{Tr}\,\{\gamma^0\,(\mathbf{1}\,\Xi^B_S + \boldsymbol{\gamma}\cdot\hat{\boldsymbol{q}}\,\Xi^B_V + \gamma^0\,\Xi^B_0)\otimes\mathbf{1}^{\mathrm{iso}}\}$$
$$= \mathrm{Tr}\,\{(\gamma^0)^2\;\Xi^B_0\}\;\mathrm{Tr}\,\{\mathbf{1}^{\mathrm{iso}}\}$$
$$= 2\,(2I_B+1)\,(2J_B+1)\;\Xi^B_0 \equiv 2\,\nu_B\,\Xi^B_0\,, \qquad (6.182)$$

where the quantity ν_B, defined as

$$\nu_B \equiv 2\,(2I_B + 1)\,(2J_B + 1)\,, \tag{6.183}$$

accounts for the spin and isospin degeneracy of the baryon in question (cf. table 5.1). Expression (6.183) applies to the nuclear matter case. In the case of neutron star matter one has $I_B = 0$ and therefore (6.183) reduces to

$$\nu_B \equiv 2\,(2J_{B'} + 1)\,. \tag{6.184}$$

Substituting (6.182) into (6.181) gives for the baryon number density

$$\rho^B = 2\,\nu_B \int \frac{d^3\boldsymbol{q}}{(2\pi)^3}\,\{\Xi_0^B(\boldsymbol{q})\,f^B(\boldsymbol{q}) - \bar{\Xi}_0^B(\boldsymbol{q})\,\bar{f}^B(\boldsymbol{q})\}\,, \tag{6.185}$$

with f^B and $\bar{f}^B$ defined in (6.67). Equation (6.185) simplifies at zero temperature to the expression

$$\rho^B = 2\,(2J_B + 1)(2I_B + 1) \int \frac{d^3\boldsymbol{q}}{(2\pi)^3}\,\Xi_0^B(\boldsymbol{q})\,\Theta^B(\boldsymbol{q})\,. \tag{6.186}$$

For theories with $\Xi_0^B = \frac{1}{2}$, one readily verifies from (6.186) the expression

$$\rho^B = \frac{\nu_B}{2}\,\frac{k_{F_B}^3}{3\pi^2}\,, \tag{6.187}$$

which leads to the well-known relations

$$\rho = \frac{2k_F^3}{3\pi^2}\,, \qquad \text{nuclear matter case}\,, \tag{6.188}$$

$$\rho = \frac{k_F^3}{3\pi^2}\,, \qquad \text{neutron matter case}\,. \tag{6.189}$$

Finally, the total baryon number density, ρ, is obtained by summing the partial number densities,

$$\rho \equiv \sum_B \rho^B\,. \tag{6.190}$$

Another frequently encountered quantity, besides the baryon number density, is the scalar baryon density, $\bar{\rho}^B$. It is defined, somewhat similarly to the baryon density, by

$$\bar{A} \equiv \int d^3\boldsymbol{x}\,[\bar{\psi}_B(x,\zeta), \psi_B(x,\zeta)]\,, \tag{6.191}$$

with the decisive difference, however, that $\psi_B^\dagger$ is replaced with $\bar{\psi}_B$. $\bar{\rho}^B$ is then obtained from (6.191) as

$$\bar{\rho}^B \equiv \frac{1}{\Omega}\,\langle\boldsymbol{\Phi}_0|\bar{A}|\boldsymbol{\Phi}_0\rangle\,. \tag{6.192}$$

Repeating the steps as for ρ^B just above, one gets

$$\bar{\rho}^B = \frac{1}{\Omega}\int_{\Omega} \mathrm{d}^3\boldsymbol{x}\ \{\langle \boldsymbol{\Phi}_0 | \bar{\psi}_B(x,\zeta)\psi_B(x,\zeta) - \psi_B(x,\zeta)\bar{\psi}_B(x,\zeta)\boldsymbol{\Phi}_0\rangle\}\,. \quad (6.193)$$

Substituting the baryon fields by the associated baryon two-point function gives for (6.193)

$$\bar{\rho}^B = \mathrm{i}\frac{1}{\Omega}\int_{\Omega} \mathrm{d}^3\boldsymbol{x}\ \{\boldsymbol{g}^B(x,\zeta;x^+,\zeta) + \boldsymbol{g}^B(x,\zeta;x^-,\zeta)\} \quad (6.194)$$

$$\equiv \mathrm{i}\frac{1}{\Omega}\sum_{y=x^+,x^-}\lim_{x'\to y}\int_{\Omega}\mathrm{d}^3\boldsymbol{x}\ \boldsymbol{g}^B_{\zeta\zeta}(x;y)\,. \quad (6.195)$$

Fourier transformation of (6.195) gives

$$\bar{\rho}^B = \mathrm{i}\sum_{s=+,-}\int \frac{\mathrm{d}^4 q}{(2\pi)^4}\,\mathrm{e}^{\mathrm{i}s\eta q^0}\ \boldsymbol{g}^B_{\zeta\zeta}(q)\,. \quad (6.196)$$

In accordance with the standard procedure, the next step consists in replacing the two-point function with its spectral representation. Contour integration then leaves us with (cf. equations (B.8) and (B.10))

$$\bar{\rho}^B = \int \frac{\mathrm{d}^3\boldsymbol{q}}{(2\pi)^3}\,\{\boldsymbol{\Xi}^B_{\zeta\zeta}(\boldsymbol{q})\ f^B(\boldsymbol{q}) - \bar{\boldsymbol{\Xi}}^B_{\zeta\zeta}(\boldsymbol{q})\ \bar{f}^B(\boldsymbol{q})\}\,, \quad (6.197)$$

with the trace given by

$$\boldsymbol{\Xi}^B_{\zeta\zeta} = \mathrm{Tr}\ (\boldsymbol{\Xi}^B) \equiv \mathrm{Tr}\left\{\mathbf{1}\ \Xi^B_S + \boldsymbol{\gamma}\cdot\hat{\boldsymbol{q}}\ \Xi^B_V + \gamma^0\ \Xi^B_0] \otimes \mathbf{1}^{\mathrm{iso}}\right\}$$

$$= \mathrm{Tr}\left\{\mathbf{1}\ \Xi^B_S\right\}\ \mathrm{Tr}\left\{\mathbf{1}^{\mathrm{iso}}\right\} = 2\,\nu_B\ \Xi^B_S\,. \quad (6.198)$$

Substituting (6.198) into (6.197) leads to

$$\bar{\rho}^B = 2\,\nu_B\int \frac{\mathrm{d}^3\boldsymbol{q}}{(2\pi)^3}\,\{\Xi^B_S(\boldsymbol{q})\ f^B(\boldsymbol{q}) - \bar{\Xi}^B_S(\boldsymbol{q})\ \bar{f}^B(\boldsymbol{q})\}\,. \quad (6.199)$$

Replacing the spectral functions $\boldsymbol{\Xi}^B$ and $\bar{\boldsymbol{\Xi}}^B$ in (6.199) with their explicit representations (6.143) and (6.146) finally gives for the scalar density

$$\bar{\rho}^B = 2\,\nu_B\int \frac{\mathrm{d}^3\boldsymbol{q}}{(2\pi)^3}\,\frac{m^*_B}{2\sqrt{(m^*_B)^2 + (\boldsymbol{q}^*)^2}}$$

$$\times\ \{f(\omega^B(\boldsymbol{q}) - \mu^B) + f(-(\bar{\omega}^B(\boldsymbol{q}) - \mu^B))\}\,. \quad (6.200)$$

The zero-temperature limit of (6.200) can be handled analytically for the relativistic Hartree approximation, since the masses and self-energies are independent of momentum for this approximation. With the aid of the momentum integrals given in appendix B.3, one readily calculates from

$$\bar{\rho}^B = 2\,(2J_B+1)(2I_B+1)\int\frac{\mathrm{d}^3\boldsymbol{q}}{(2\pi)^3}\;\Xi^B_S(\boldsymbol{q})\;\Theta^B(\boldsymbol{q}) \tag{6.201}$$

the relation

$$\bar{\rho}^B \;\overset{T\to 0}{\longrightarrow}\; \frac{\nu_B}{6\pi^2}\,m^*_B\left\{\frac{k_{F_B}}{2}\,\sqrt{(m^*_B)^2+k^2_{F_B}} - \frac{(m^*_B)^2}{2}\,\ln\left|\frac{k_{F_B}+\sqrt{(m^*_B)^2+k^2_{F_B}}}{m^*_B}\right|\right\}. \tag{6.202}$$

We close this section by noting that the total scalar density, $\bar{\rho}$, is obtained from (6.200) as

$$\bar{\rho} \equiv \sum_B \bar{\rho}^B\,. \tag{6.203}$$

Chapter 7

Dense matter in the relativistic Hartree and Hartree–Fock approximations

7.1 Self-energies in the Hartree–Fock approximation

We recall that the HF approximation to the many-body system is obtained by keeping only the Born term of the T-matrix equation (5.179), that is by replacing $\mathsf{T} \rightarrow \mathsf{T}^{\mathrm{HF}} \equiv \mathsf{V} - \mathsf{V}^{\mathrm{ex}}$, with the matrix elements of V explicitly given in equations (5.151) through (5.155). The antisymmetrized HF T-matrix has the form

$$\langle 1\,2|\mathsf{T}^{\mathrm{HF},BB'}|3\,4\rangle \equiv \langle 1\,2|\mathsf{V}^{BB'}|3\,4\rangle - \langle 1\,2|\mathsf{V}^{BB'}|4\,3\rangle\,, \tag{7.1}$$

where the first respectively second term on the right-hand side constitute the direct (Hartree) and exchange (Fock) contribution to T^{HF}. The graphical illustration of T^{HF} is displayed in figure 7.1. A comparison with the structure of the full scattering matrix, displayed in figure 5.10, shows that the repeated two-baryon scattering processes in matter summed in the full T-matrix approximation are absent for the HF approach. Substituting (7.1) into (5.186), which defines $\boldsymbol{\Sigma}$, leads to

$$\boldsymbol{\Sigma}^{\mathrm{H},B}(1,1') = \mathrm{i}\sum_{B'} \langle 1\,2|\mathsf{V}^{BB'}|1'\,3\rangle\, \boldsymbol{g}_1^{B'}(3,2^+)\,, \tag{7.2}$$

$$\boldsymbol{\Sigma}^{\mathrm{F},B}(1,1') = -\,\mathrm{i}\sum_{B'} \langle 1\,2|\mathsf{V}^{BB'}|3\,1'\rangle\, \boldsymbol{g}_1^{B'}(3,2^+)\,, \tag{7.3}$$

where (7.2) is the Hartree self-energy and (7.3) the Fock contribution to the self-energy. Both expressions added together give the total HF self-energy in the form

$$\boldsymbol{\Sigma}^{\mathrm{HF},B}(1,1') \equiv \boldsymbol{\Sigma}^{\mathrm{H},B}(1,1') + \boldsymbol{\Sigma}^{\mathrm{F},B}(1,1')\,. \tag{7.4}$$

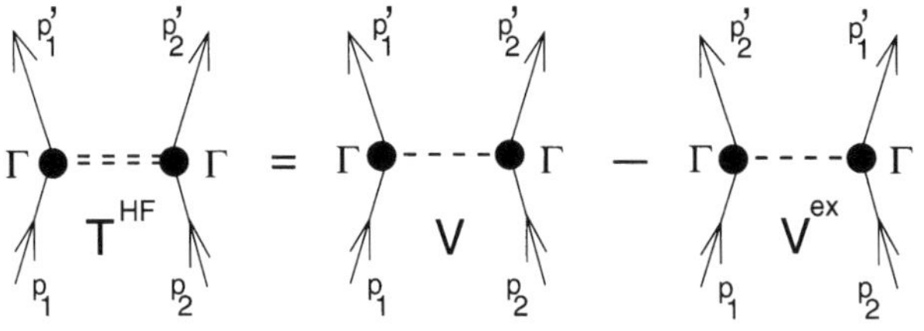

Figure 7.1. Graphical representation of the Hartree–Fock **T**-matrix, $\mathbf{T}^{\text{HF}}$, obtained by restriction to the Born term of the **T**-matrix equation (5.203). Γ denote baryon–meson vertices listed in (5.156) through (5.158). The analytic form of $\mathbf{T}^{\text{HF}}$ is given in equation (7.1).

The explicit expressions of $\boldsymbol{\Sigma}^{\text{H},B}$ and $\boldsymbol{\Sigma}^{\text{F},B}$ in momentum space [79, 125, 314] are derived in equations (5.135) through (5.142). For the σ meson, for instance, we obtained for the Hartree term

$$\boldsymbol{\Sigma}^{\text{H},B}_{\zeta_2\zeta_1}\Big|_{\sigma} = -\,\mathrm{i}\,\delta_{\zeta_2\zeta_1}\,\Delta^{0\sigma}(0)\,g_{\sigma B}\sum_{B'} g_{\sigma B'}\int\frac{\mathrm{d}^4 q}{(2\pi)^4}\,\mathrm{e}^{\mathrm{i}\eta q^0}\ \mathrm{Tr}\ \boldsymbol{g}^{B'}(q)\,. \tag{7.5}$$

Upon replacing $\boldsymbol{g}^{B'}(q)$ with its spectral representation, derived in equation (6.71), and performing the contour integrations (cf. figure D.1) over the energy variable q^0, as described in appendix B, we find for (7.5)

$$\boldsymbol{\Sigma}^{\text{H},B}_{\zeta_2\zeta_1}\Big|_{\sigma} = -\ \delta_{\zeta_2\zeta_1}\left(\frac{g_{\sigma B}}{m_\sigma}\right)^2\sum_{B'}\left(\frac{g_{\sigma B'}}{g_{\sigma B}}\right)\int\frac{\mathrm{d}^3\boldsymbol{q}}{(2\pi)^3}\ \mathrm{Tr}\ \boldsymbol{\Xi}^{B'}(\boldsymbol{q})\,\Theta(q_{F_{B'}} - |\boldsymbol{q}|)\,. \tag{7.6}$$

The σ-meson propagator in (7.5) at zero energy and momentum has been replaced with $\Delta^{0\sigma}(0) = 1/m_\sigma^2$, which follows from (5.93). Moreover, to get from (7.5) to (7.6) we have restricted ourselves to the zero-temperature limit, in which case the thermal distribution function can be replaced with the step function, $f(\omega^B(\boldsymbol{k}) - \mu^B) \longrightarrow \Theta(k_{F_B} - |\boldsymbol{k}|)$, and no thermally excited antibaryons contribute (that is, $\bar{f}^B \rightarrow 0$), as discussed in connection with the physical interpretation of the zero-temperature two-point Green function (6.69). The baryon self-energies at finite temperatures are summarized in appendix D. The trace in (7.6) is to be calculated with respect to the Dirac (spin) and isospin indices carried by $\boldsymbol{\Xi}^B$. Since the Dirac matrices γ^0 and γ^i are traceless (cf. appendix A.2), that is,

$$\mathrm{Tr}\ \gamma^0 = \mathrm{Tr}\ \gamma^i = 0\,, \tag{7.7}$$

we arrive for the trace of $\boldsymbol{\Xi}^B$ of (6.117) at

$$\mathrm{Tr}\ \boldsymbol{\Xi}^{B'}(\boldsymbol{q}) \equiv \mathrm{Tr}\left\{\left(\mathbf{1}\,\Xi_S^{B'}(\boldsymbol{q}) + \gamma\cdot\hat{\boldsymbol{q}}\,\Xi_V^{B'}(\boldsymbol{q}) + \gamma^0\,\Xi_0^{B'}(\boldsymbol{q})\right)\otimes\mathbf{1}^{\text{iso}}\right\}$$

$$= \mathrm{Tr}\,\left(\mathbf{1}\otimes\mathbf{1}^{\mathrm{iso}}\right)\,\Xi_S^{B'}(\boldsymbol{q})\,. \tag{7.8}$$

As already known from the discussion of the baryon self-energy in section 5.4, the expression $\mathrm{Tr}(\mathbf{1}\otimes\mathbf{1}^{\mathrm{iso}})$ denotes the direct product of the two matrices $\mathbf{1}$ and $\mathbf{1}^{\mathrm{iso}}$, which can be written as (cf. appendix A.3) $\mathrm{Tr}(\mathbf{1}\otimes\mathbf{1}^{\mathrm{iso}}) = \mathrm{Tr}(\mathbf{1})\ \mathrm{Tr}(\mathbf{1}^{\mathrm{iso}})$. Equation (7.8) therefore can be brought into the form

$$\mathrm{Tr}\ \boldsymbol{\Xi}^{B'}(\boldsymbol{q}) = \mathrm{Tr}\,(\mathbf{1})\ \mathrm{Tr}\,(\mathbf{1}^{\mathrm{iso}})\ \Xi_S^{B'}(\boldsymbol{q})\,. \tag{7.9}$$

Upon evaluating the traces of the two individual matrices in the Dirac (spin) and isospin space in the form

$$\mathrm{Tr}\ (\mathbf{1})\ \mathrm{Tr}\ \left(\mathbf{1}^{\mathrm{iso}}\right) = 2\,(2I_{B'}+1)\,(2J_{B'}+1)\,, \tag{7.10}$$

equation (7.9) can be written as

$$\mathrm{Tr}\ \boldsymbol{\Xi}^{B'}(\boldsymbol{q}) = 2\,(2I_{B'}+1)\,(2J_{B'}+1)\ \Xi_S^{B'}(\boldsymbol{q}) \equiv 2\,\nu_{B'}\ \Xi_S^{B'}(\boldsymbol{q})\,. \tag{7.11}$$

Substituting the result (7.11) into (7.6) gives for the baryon self-energy

$$\boldsymbol{\Sigma}^{\mathrm{H},B}_{\zeta_2\zeta_1}(\{p_{F_B}\})\Big|_\sigma = -\,2\,\delta_{\zeta_2\zeta_1}\left(\frac{g_{\sigma B}}{m_\sigma}\right)^2\sum_{B'}\frac{g_{\sigma B'}}{g_{\sigma B}}\,\nu_{B'} \times \int\frac{\mathrm{d}^3\boldsymbol{q}}{(2\pi)^3}\,\Xi_S^{B'}(\boldsymbol{q})\,\Theta(p_{F_{B'}}-|\boldsymbol{q}|)\,. \tag{7.12}$$

One sees from (7.12) that in order to determine $\boldsymbol{\Sigma}^{\mathrm{H},B}|_\sigma$, knowledge of the Fermi momenta $p_{F_{B'}}$ (and thus the densities) of all the other baryons, predicted to be present in the system at a given total baryon density, is necessary. This functional dependence is indicated by $\{p_{F_{B'}}\}$ as the argument of $\boldsymbol{\Sigma}^{\mathrm{H},B}|_\sigma$. Alternatively to equation (7.12), the self-energy contribution can be expressed as

$$\Sigma^{\mathrm{H},B}_S(\{p_{F_B}\})\Big|_\sigma = -\,2\left(\frac{g_{\sigma B}}{m_\sigma}\right)^2\sum_{B'}\frac{g_{\sigma B'}}{g_{\sigma B}}\,\nu_{B'} \times \int\frac{\mathrm{d}^3\boldsymbol{q}}{(2\pi)^3}\,\Xi_S^{B'}(\boldsymbol{q})\,\Theta(p_{F_{B'}}-|\boldsymbol{q}|)\,, \tag{7.13}$$

which follows immediately by comparing (7.12) with the general decomposition for $\boldsymbol{\Sigma}^B$ given in (6.83). This leaves us with a simple scalar function for $\Sigma^{\mathrm{H},B}_S|_\sigma$, in contrast to $\boldsymbol{\Sigma}^{\mathrm{H},B}_{\zeta_2\zeta_1}|_\sigma$ of (7.12) which is a matrix equation in spin–isospin space. Besides that we note the general feature that after contour integration an explicit determination of the baryon two-point function is no longer necessary, either here or for the determination of the equation of state of the many-body system, as we shall see in chapter 12.

Having calculated the Hartree expression of the baryon self-energy which originates from σ-meson exchange among the baryons, we proceed now to the calculation of the self-energy that originates from ω-meson exchange. (Contributions arising from the exchange of π and ρ mesons are listed in appendices C and D for matter at zero and finite temperatures, respectively.) In close analogy to the above, we go back to the momentum-space representation of $\boldsymbol{\Sigma}^{\mathrm{H},B}(1,1')$, now, however, derived for the case of ω-meson exchange. This expression is given in (5.137). Replacing $\boldsymbol{g}^{B'}(q)$ with its associated spectral representation (6.71) and performing the contour integrations over q^0, exactly as in (7.5), leaves us with

$$\boldsymbol{\Sigma}^{\mathrm{H},B}_{\zeta_2\zeta_1}(\{p_{F_B}\})\Big|_{\omega} = \gamma^{\mu}_{\zeta_2\zeta_1}\,\frac{g_{\mu\nu}}{m_\omega^2}\,g_{\omega B}\sum_{B'} g_{\omega B'} \times \int \frac{\mathrm{d}^3\boldsymbol{q}}{(2\pi)^3}\,\mathrm{Tr}\left(\gamma^\nu\,\boldsymbol{\Xi}^{B'}(\boldsymbol{q})\right)\Theta(p_{F_{B'}}-|\boldsymbol{q}|)\,. \tag{7.14}$$

The calculation of the trace in (7.14),

$$\mathrm{Tr}\left(\gamma^\nu\boldsymbol{\Xi}^{B'}\right) = \mathrm{Tr}\left\{\left[\gamma^\nu\left(\mathbf{1}\,\Xi_S^{B'} + \boldsymbol{\gamma}\cdot\hat{\boldsymbol{q}}\,\Xi_V^{B'} + \gamma^0\,\Xi_0^{B'}\right)\right]\otimes\mathbf{1}^{\mathrm{iso}}\right\}, \tag{7.15}$$

is somewhat more complicated than for the scalar σ meson in (7.6). We begin with the trace of the first term in (7.15), which is trivially found to given

$$\mathrm{Tr}\left(\mathbf{1}\,\gamma^\nu\right) = \mathrm{Tr}\;\gamma^\nu = 0\,, \tag{7.16}$$

since the traces of the γ matrices vanish. When calculating the contribution of the trace of the second term in (7.15), we note that (see appendices A.2 and A.3):

$$\begin{aligned}\mathrm{Tr}\left(\gamma^\nu\,\boldsymbol{\gamma}\cdot\hat{\boldsymbol{q}}\right) &= \mathrm{Tr}\left(\gamma^\nu\,\gamma^i\,\hat{q}^i\right)\\ &= \hat{q}^i\;\mathrm{Tr}\left(\gamma^\nu\,\gamma^i\right)\\ &= \hat{q}^i\;\mathrm{Tr}\left(2\,g^{\nu i}\,\mathbf{1} - \gamma^i\,\gamma^\nu\right)\\ &= \hat{q}^i\;\mathrm{Tr}\left(8\,g^{\nu i} - \mathrm{Tr}\left(\gamma^i\,\gamma^\nu\right)\right)\\ &= 8\,\hat{q}^i\,g^{\nu i} - \mathrm{Tr}\left(\gamma^\nu\,\boldsymbol{\gamma}\cdot\hat{\boldsymbol{q}}\right)\,,\end{aligned} \tag{7.17}$$

and therefore, by moving the second term on the right-hand side of (7.17) to the left-hand side of this equation,

$$\mathrm{Tr}\left(\gamma^\nu\,\boldsymbol{\gamma}\cdot\hat{\boldsymbol{q}}\right) = 4\,\hat{q}^i\,g^{\nu i}\,. \tag{7.18}$$

Multiplying both sides of (7.18) with $g_{\mu\nu}$ and summing over ν leads to

$$\sum_\nu g_{\mu\nu}\;\mathrm{Tr}\left(\gamma^\nu\,\boldsymbol{\gamma}\cdot\hat{\boldsymbol{q}}\right) = 4\sum_\nu \hat{q}^i\,g_{\mu\nu}\,g^{\nu i} = 4\sum_{i=1}^{3}\hat{q}^i\,\delta_\mu^{\;i}\,. \tag{7.19}$$

To pick up the last Dirac matrix, γ^μ, of equation (7.14), we multiply (7.19) with γ^μ and sum over the doubly occurring index μ. This leads to the final result

$$\sum_\mu \gamma^\mu \Big\{ \sum_\nu g_{\mu\nu} \, \mathrm{Tr}\,(\gamma^\nu \boldsymbol{\gamma}\cdot\hat{\boldsymbol{q}}) \Big\} = 4 \sum_\mu \sum_{i=1}^{3} \gamma^\mu \, \hat{q}^i \, \delta_\mu^{\ i} = 4\, \boldsymbol{\gamma}\cdot\hat{\boldsymbol{q}} \,. \tag{7.20}$$

The Dirac algebra thus gives a non-vanishing contribution for the $\boldsymbol{\gamma}\cdot\hat{\boldsymbol{q}}$ term in (7.14). Nevertheless, we can forget about this term because of the vanishing integrals over the solid angle,

$$\begin{aligned} &\int_{(4\pi)} \mathrm{d}\Omega_{\boldsymbol{q}} \; \boldsymbol{\gamma}\cdot\hat{\boldsymbol{q}} \\ &= \boldsymbol{\gamma} \cdot \int_0^{2\pi} \mathrm{d}\phi \int_0^{\pi} \mathrm{d}\theta \, \sin\theta \times (\sin\theta\,\cos\phi, \sin\theta\,\sin\phi, \cos\theta) = 0 \,, \end{aligned} \tag{7.21}$$

with $\hat{\boldsymbol{q}} = \boldsymbol{q}/|\boldsymbol{q}| = (\sin\theta\,\cos\phi, \sin\theta\,\sin\phi, \cos\theta)$. Each integral of (7.21) vanishes because of symmetry reasons. Finally, the contribution of the trace of the third term of (7.15) is readily found by noticing that

$$\begin{aligned} \mathrm{Tr}\,(\gamma^\nu\,\gamma_0) &= \mathrm{Tr}\,(g^{\nu\mu}\,\gamma_\mu\,\gamma_0) \\ &= \mathrm{Tr}\,(g^{\nu\mu}\,(2\,g_{\mu\nu}\,\mathbf{1} - \gamma_0\,\gamma_\mu)) \\ &= 2\,g^{\nu\mu}\,g_{\mu 0}\ \mathrm{Tr}\,\mathbf{1} - \mathrm{Tr}\,(\gamma_0\,\gamma^\nu)\,. \end{aligned} \tag{7.22}$$

Since $\mathrm{Tr}\,\mathbf{1} = 4$ and $g^{\nu\mu}\,g_{\mu 0} = \delta^\nu_{\ 0}$, we find from (7.22) for the desired trace

$$\mathrm{Tr}\,(\gamma^\nu\,\gamma_0) = 4\,\delta^\nu_{\ 0}\,. \tag{7.23}$$

Having the results of all relevant traces at our disposal, we now proceed with the calculation of $\boldsymbol{\Sigma}^{\mathrm{H},B}_{\zeta_2\zeta_1}|_\omega$. Upon substituting the results derived in equations (7.16), (7.21) and (7.23) into (7.14), we arrive for $\boldsymbol{\Sigma}^{\mathrm{H},B}_{\zeta_2\zeta_1}|_\omega$ at the expression

$$\begin{aligned} \boldsymbol{\Sigma}^{\mathrm{H},B}_{\zeta_2\zeta_1}(\{p_{F_B}\})\Big|_\omega &= 2\,\gamma^0_{\zeta_2\zeta_1} \left(\frac{g_{\omega B}}{m_\omega^2}\right)^2 \sum_{B'} \nu_{B'}\,\frac{g_{\omega B'}}{g_{\omega B}} \\ &\quad \times \int \frac{\mathrm{d}^3\boldsymbol{q}}{(2\pi)^3}\,\Xi_0^{B'}(\boldsymbol{q})\,\Theta(p_{F_{B'}} - |\boldsymbol{q}|)\,. \end{aligned} \tag{7.24}$$

A comparison with the general decomposition of $\boldsymbol{\Sigma}^B$ of (6.83) shows that

$$\begin{aligned} \Sigma_0^{\mathrm{H},B}(\{p_{F_B}\})\Big|_\omega &= 2 \left(\frac{g_{\omega B}}{m_\omega^2}\right)^2 \sum_{B'} \nu_{B'}\,\frac{g_{\omega B'}}{g_{\omega B}} \\ &\quad \times \int \frac{\mathrm{d}^3\boldsymbol{q}}{(2\pi)^3}\,\Xi_0^{B'}(\boldsymbol{q})\,\Theta(p_{F_{B'}} - |\boldsymbol{q}|)\,. \end{aligned} \tag{7.25}$$

Note that at the level of the relativistic Hartree approximation, there is neither a scalar nor a vector-component contribution to $\mathbf{\Sigma}^B$.

We conclude this section with giving the expressions for the Fock contributions $\mathbf{\Sigma}^{\mathrm{F},B}(1,1')$ of (7.3) which arise from the exchange of σ and ω mesons. In analogy to the Hartree case, we turn back to the momentum-space representations of $\mathbf{\Sigma}^{\mathrm{F},B}(1,1')|_\sigma$ and $\mathbf{\Sigma}^{\mathrm{F},B}(1,1')|_\omega$ given in (5.136) and (5.138), respectively. Replacing $\boldsymbol{g}^{B'}(q)$ in these equations with its associated spectral representation, derived in (6.71), and subsequently performing the contour integrations leads to

$$\mathbf{\Sigma}^{\mathrm{F},B}_{\zeta_1\zeta_1'}(k)\Big|_\sigma = g^2_{\sigma B}\int\frac{\mathrm{d}^3\boldsymbol{q}}{(2\pi)^3}\{\delta_{\zeta_1\zeta_1'}\;\Xi^B_S(\omega^B(\boldsymbol{q}),\boldsymbol{q}) + (\gamma\cdot\hat{\boldsymbol{k}})_{\zeta_1\zeta_1'}\;\hat{\boldsymbol{k}}\cdot\hat{\boldsymbol{q}}\;\times$$
$$\Xi^B_V(\omega^B(\boldsymbol{q}),\boldsymbol{q}) + \gamma^0_{\zeta_1\zeta_1'}\;\Xi^B_0(\omega^B(\boldsymbol{q}),\boldsymbol{q})\}\;\Delta^{0\sigma}(k_0-\omega^B(\boldsymbol{q}),\boldsymbol{k}-\boldsymbol{q})\;\Theta^B(\boldsymbol{q})\,. \tag{7.26}$$

In the case of ω mesons one arrives at

$$\mathbf{\Sigma}^{\mathrm{F},B}_{\zeta_1\zeta_1'}(k)\Big|_\omega = g^2_{\omega B}\int\frac{\mathrm{d}^3\boldsymbol{q}}{(2\pi)^3}\Big\{-\delta_{\zeta_1\zeta_1'}\Big[4-\frac{(k^0-\omega^B(\boldsymbol{q}))^2-(\boldsymbol{k}-\boldsymbol{q})^2}{m^2_\omega}\Big]$$
$$\times\,\Xi^B_S(\omega^B(\boldsymbol{q}),\boldsymbol{q}) + (\gamma\cdot\hat{\boldsymbol{k}})_{\zeta_1\zeta_1'}\Big[\Big(\hat{\boldsymbol{k}}\cdot\hat{\boldsymbol{q}}\Big(2-\frac{\boldsymbol{k}^2+\boldsymbol{q}^2+(k^0-\omega^B(\boldsymbol{q}))^2}{m^2_\omega}\Big)$$
$$+\frac{2\,|\boldsymbol{k}|\,|\boldsymbol{q}|}{m^2_\omega}\Big)\,\Xi^B_V(\omega^B(\boldsymbol{q}),\boldsymbol{q}) \;-\; \frac{2}{m^2_\omega}\,\hat{\boldsymbol{k}}\cdot(\boldsymbol{k}-\boldsymbol{q})\,(k^0-\omega^B(\boldsymbol{q}))\,\Xi^B_0(\omega^B(\boldsymbol{q}),\boldsymbol{q})\Big]$$
$$+\,\gamma^0_{\zeta_1\zeta_1'}\Big[\frac{2}{m^2_\omega}\,\hat{\boldsymbol{q}}\cdot(\boldsymbol{k}-\boldsymbol{q})\,(k^0-\omega^B(\boldsymbol{q}))\,\Xi^B_V(\omega^B(\boldsymbol{q}),\boldsymbol{q})$$
$$+\Big(2+\frac{(k^0-\omega^B(\boldsymbol{q}))^2+(\boldsymbol{k}-\boldsymbol{q})^2}{m^2_\omega}\Big)\,\Xi^B_0(\omega^B(\boldsymbol{q}),\boldsymbol{q})\Big]\Big\}$$
$$\times\;\Delta^{0\omega}(k^0-\omega^B(\boldsymbol{q}),\boldsymbol{k}-\boldsymbol{q})\,\Theta(q_{F_B}-|\boldsymbol{q}|)\,. \tag{7.27}$$

The Hartree and Fock contributions to $\mathbf{\Sigma}^B$ which originate from π and ρ meson exchange among the baryons are listed in appendix C. The extension of $\mathbf{\Sigma}^B$ to finite temperatures is performed in appendix D. In closing this section, we note that the effective baryon mass, m^*_B, is defined as

$$m^*_B \equiv m^*_B(\omega^B(\boldsymbol{k}),\boldsymbol{k}) \equiv m^*_B + \Sigma^{\mathrm{HF},B}_S(\omega^B(\boldsymbol{k}),\boldsymbol{k})\,. \tag{7.28}$$

7.2 Self-energies in the Hartree approximation (Walecka model)

At the level of the relativistic Hartree approximation, the mathematical structure of the baryon self-energies becomes extremely simple. This originates primarily from the very simple form of the baryon spectral functions $\mathbf{\Xi}^B$ for this approximation, which, as demonstrated at the end of section (6.2), simplify to

$$\Xi_S^{\mathrm{H},B}(\boldsymbol{q}) = \frac{m_B^*}{2\,\epsilon^{\mathrm{H},B}(\boldsymbol{q})}\,, \quad \Xi_V^{\mathrm{H},B}(\boldsymbol{q}) = \frac{-\,|\boldsymbol{q}|}{2\,\epsilon^{\mathrm{H},B}(\boldsymbol{q})}\,, \quad \Xi_0^{\mathrm{H},B}(\boldsymbol{q}) = \frac{1}{2}\,. \tag{7.29}$$

The quantity m_B^* denotes the effective, medium-modified mass of a baryon in dense matter, defined, in accordance with (7.28), as

$$m_B^* \equiv m_B + \Sigma_S^{\mathrm{H},B}\,. \tag{7.30}$$

Moreover we have introduced the auxiliary quantity $\epsilon^{\mathrm{H},B}$ given by (cf. equation (6.171))

$$\epsilon^{\mathrm{H},B}(\boldsymbol{q}) = \sqrt{(m_B^*)^2 + \boldsymbol{q}^2}\,, \tag{7.31}$$

by means of which the single-baryon energy (6.150) can be expressed as

$$\omega^{\mathrm{H},B}(\boldsymbol{q}) = \Sigma_0^{\mathrm{H},B}(\boldsymbol{q}) + \epsilon^{\mathrm{H},B}(\boldsymbol{q})\,. \tag{7.32}$$

The above equations follow immediately from (6.143) through (6.145) by noticing that the baryon self-energies (and thus m^*) at the Hartree level are independent of both energy and momentum, as we know from the expressions for $\boldsymbol{\Sigma}^B$ derived in equations (7.13) and (7.25). Only the density dependence survives because of the proportionality $\boldsymbol{\Sigma}^B \propto k_{F_B}^3$. This simplification is lost for the HF approximation, where the exchange term depends on both energy and momentum.

Upon substituting (7.29) into equation (7.13), one obtains for the scalar component of the *nucleon* self-energy in dense nuclear matter

$$\begin{aligned}\Sigma_S^{\mathrm{H},N} = {} & \frac{1}{4\pi^2}\left(\frac{g_{\sigma N}}{m_\sigma}\right)^2 \sum_B \frac{g_{\sigma B}}{g_{\sigma N}}\,\nu_B \left\{ k_{F_B}\sqrt{(m_B^*)^2 + k_{F_B}^2} \right. \\ & \left. - (m_B^*)^2 \ln\left|\frac{k_{F_B} + \sqrt{(m_B^*)^2 + k_{F_B}^2}}{m_B^*}\right| \right\} \\ & + \left(\frac{g_{\sigma N}}{m_\sigma}\right)^2 \left\{ b_N\, m_N\, (\boldsymbol{\Sigma}_S^N)^2 - c_N\, (\boldsymbol{\Sigma}_S^N)^3 \right\}. \end{aligned} \tag{7.33}$$

The timelike component, $\Sigma_0^{\mathrm{H},N}$, follows from (7.25) as

$$\Sigma_0^{\mathrm{H},N} = \frac{1}{6\,\pi^2}\left(\frac{g_{\omega N}}{m_\omega}\right)^2 \sum_B \left(\frac{g_{\omega B}}{g_{\omega N}}\right) \nu_B\, k_{F_B}^3\,. \tag{7.34}$$

Equations (7.33) and (7.34) are special cases of the more general expressions

$$\Sigma_0^{\mathrm{H},B} = \left(\frac{g_{\omega B}}{m_\omega}\right)^2 \sum_{B'} \frac{g_{\omega B'}}{g_{\omega B}}\,\rho^{H,B'}\,, \tag{7.35}$$

and

$$\Sigma_S^{\mathrm{H},B} = -\left(\frac{g_{\sigma B}}{m_\sigma}\right)^2 \left\{ \sum_{B'} \frac{g_{\sigma B'}}{g_{\sigma B}}\, \bar{\rho}^{\mathrm{H},B'} - b_B\, m_N \left(\Sigma_S^{\mathrm{H},B}\right)^2 + c_B \left(\mathbf{\Sigma}_S^{\mathrm{H},B}\right)^3 \right\}, \tag{7.36}$$

with the definitions

$$b_B \equiv \left(\frac{g_{\sigma N}}{g_{\sigma B}}\right)^4 b_N\,, \qquad c_B = \left(\frac{g_{\sigma N}}{g_{\sigma B}}\right)^5 c_N\,. \tag{7.37}$$

These follow from equations (7.13) and (7.25) by making use of the baryon number densities ρ^B and $\bar{\rho}^B$, derived in equations (6.185) and (6.199), for the relativistic Hartree approximation.

Note that there is no contribution to the vector component Σ_V^B at the level of the relativistic Hartree approximation. The results (7.33) and (7.34) are nothing but Walecka's σ–ω mean-field equations in their non-linear form [92, 348, 349, 350] for uniform static matter, in which space and time derivatives of the fields can be dropped. This can be readily verified by setting [79, 125, 314]

$$\Sigma_S^B = -g_{\sigma B}\,\langle\sigma_0\rangle \qquad \text{and} \qquad \Sigma_0^B = g_{\omega B}\,\langle\omega_0\rangle\,, \tag{7.38}$$

where $\langle\sigma_0\rangle$ and $\langle\omega_0\rangle$ denote the static amplitudes of the meson-field equations (5.43) and (5.44), with space and time derivatives ignored. The baryon source currents in these meson-field equations are replaced with their ground-state expectation values, with the ground state defined as having the single-particle momentum eigenstates of the Dirac equations filled to the top of the Fermi sea of each baryon species, in accord with the condition of chemical equilibrium and electric charge neutrality.

The last term in equation (7.33) contributes only if cubic and quartic self-interactions of the σ field are included in the Lagrangian (5.1). This leads to a self-energy contribution, denoted by $\mathbf{\Sigma}^{(\sigma^4)}$, which has the form [79, 125, 314]

$$\mathbf{\Sigma}^{(\sigma^4)}_{\zeta_2\zeta_1} = \delta_{\zeta_2\zeta_1} \left(\frac{g_{\sigma N}}{m_\sigma}\right)^2 \left\{ m_N\, b_N \left(\Sigma_S^{\mathrm{H},N}\right)^2 - c_N \left(\Sigma_S^{\mathrm{H},N}\right)^3 \right\}. \tag{7.39}$$

Such cubic and quartic terms are known to be important since the linear σ–ω theory fails to account for an effective nucleon mass in matter, m_N^*, and a incompressibility, K, which are compatible with experimental values [100, 124, 309, 310]. Alternatively to supplementing the Lagrangian with non-linear terms, it has recently been pointed out by Zimanyi and Moszkowski [351] and Glendenning, Weber, and Moszkowsi [86] that if the scalar field is coupled to the derivative of the nucleon field, these two nuclear properties are automatically in fairly reasonable accord with present knowledge of their values. We introduce this model in the following section.

7.3 Derivative coupling model

The linear σ–ω nuclear field theory has been broadly studied in both spherical and deformed nuclei. However, in the linear version [348, 349, 350] it has too small a nucleon effective mass ($\sim 0.55\, m_N$) at saturation density of nuclear matter and too large an incompressibility ($\sim$ 560 MeV). As discussed in section 5.1, these properties can be brought under control at the cost of two additional parameters by the addition of scalar cubic and quartic self-interactions in the so-called non-linear model [309]. Alternatively it has been recently noticed by Zimanyi and Moszkowski [351] that, if the scalar field is coupled to the derivative of the nucleon field, these two nuclear properties are automatically in reasonable accord with our present knowledge of their values, the two coupling constants of the theory being fixed by the empirical saturation density and binding as in the linear σ–ω theory. The agreement with bulk nuclear properties can be further improved by a slight modification of the model of Zimanyi and Moszkowski, which we shall call the hybrid derivative coupling model, and which we shall discuss below. Renormalization is irrevocably lost in derivative coupling models, but since (strong interacting) nuclear field theory is usually regarded as an effective one, this does not seem to be a weighty objection.

In place of the purely derivative coupling of the scalar field to the baryons and vector meson of the Zimanyi-Moszkowski model, we couple it here by both Yukawa point and derivative coupling to baryons and both vector fields. This improves the agreement with the incompressibility and effective nucleon mass at saturation density. The nuclear matter properties obtained for the hybrid derivative coupling model will be listed below. To account for the symmetry force, we include the coupling of the ρ meson to the isospin current. The ρ-meson contribution to this current vanishes in the mean-field approximation and so we do not write its formal contribution in the Lagrangian [352]:

$$\begin{aligned}\mathcal{L} = \sum_B \Big\{ \Big(1 + \frac{g_{\sigma B}\sigma}{2m_B}\Big) \Big[\bar{\psi}_B (\mathrm{i}\gamma_\mu \partial^\mu - g_{\omega B}\gamma_\mu \omega^\mu - \frac{1}{2} g_{\rho B}\, \gamma_\mu \boldsymbol{\tau} \cdot \boldsymbol{\rho}^\mu) \psi_B \Big] \\ - \Big(1 - \frac{g_{\sigma B}\sigma}{2m_B}\Big) m_B \bar{\psi}_B \psi_B \Big\} + \frac{1}{2}\left(\partial_\mu \sigma \partial^\mu \sigma - m_\sigma^2 \sigma^2\right) - \frac{1}{4} F_{\mu\nu} F^{\mu\nu} \\ + \frac{1}{2} m_\omega^2\, \omega_\mu \omega^\mu - \frac{1}{4} \boldsymbol{G}_{\mu\nu} \boldsymbol{G}^{\mu\nu} + \frac{1}{2} m_\rho^2\, \boldsymbol{\rho}_\mu \cdot \boldsymbol{\rho}^\mu + \sum_{L = e^-, \mu^-} \mathcal{L}_L \,. \qquad (7.40)\end{aligned}$$

In the first term one sees the coupling of the scalar field to the derivatives of the baryon fields and to the vector mesons. The Yukawa point coupling to the baryon fields is contained in the second term. In the last line one recognizes the free scalar, vector, and vector-isovector mesons, and the lepton Lagrangian of (5.15). As we know from section 4.2, leptons must be present because of electric charge neutrality and chemical equilibrium of neutron star matter. (The notation in (7.40) is the same as at the beginning of section 5.1 where we introduced the standard Lagrangian of neutron star matter.) The baryon Lagrangian is in the first line together with the interaction terms with the above-mentioned mesons. The sum over B in (7.40) is extended over all higher-mass baryons listed in table 5.1 for which the baryon chemical potential exceeds their rest mass in dense matter, i.e. corrected for interactions and electric charge. The solution is most easily obtained by means of transforming all baryon fields as

$$\psi_B = \left(1 + \frac{g_{\sigma B}\sigma}{2m_B}\right)^{-1/2} \Psi_B \,. \tag{7.41}$$

The equivalent Lagrangian is then given by

$$\begin{aligned}\mathcal{L} = \sum_B \bar{\Psi}_B \Big(\mathrm{i}\gamma_\mu \partial^\mu - m_B^* - g_{\omega B}\,, \gamma_\mu \omega^\mu - \frac{1}{2} g_{\rho B}\, \gamma_\mu \boldsymbol{\tau} \cdot \boldsymbol{\rho}^\mu\Big) \Psi_B \\ + \frac{1}{2}\left(\partial_\mu \sigma \partial^\mu \sigma - m_\sigma^2 \sigma^2\right) - \frac{1}{4} F_{\mu\nu} F^{\mu\nu} + \frac{1}{2} m_\omega^2\, \omega_\mu \omega^\mu \\ - \frac{1}{4} \boldsymbol{G}_{\mu\nu} \cdot \boldsymbol{G}^{\mu\nu} + \frac{1}{2} m_\rho^2\, \boldsymbol{\rho}_\mu \cdot \boldsymbol{\rho}^\mu + \sum_L \bar{\psi}_L \left(i\gamma_\mu \partial^\mu - m_L\right) \psi_L \,. \end{aligned} \tag{7.42}$$

It is evident that the baryons now have effective masses

$$m_B^* = \left(1 - \frac{g_{\sigma B}\sigma}{2m_B}\right)\left(1 + \frac{g_{\sigma B}\sigma}{2m_B}\right)^{-1} m_B \,. \tag{7.43}$$

In the next step we solve the field equations in the mean-field (Hartree) approximation, introduced in section 7.2. The meson-field equations in uniform static matter, in which space and time derivatives can be dropped, are then given by

$$\langle \omega_0 \rangle = \sum_B \frac{g_{\omega B}}{m_\omega^2}\, \rho^B \,, \tag{7.44}$$

$$\langle \boldsymbol{\rho}_{03} \rangle = \sum_B \frac{g_{\rho B}}{m_\rho^2}\, I_{3B}\, \rho^B \,, \tag{7.45}$$

$$\begin{aligned} m_\sigma^2\, \sigma &= \sum_B g_{\sigma B} \left(1 + \frac{g_{\sigma B}\sigma}{2m_B}\right)^{-2} \langle \boldsymbol{\Phi}_0 | \bar{\Psi}_B \Psi_B | \boldsymbol{\Phi}_0 \rangle \\ &= \sum_B g_{\sigma B} \left(1 + \frac{g_{\sigma B}\sigma}{2m_B}\right)^{-2} \frac{2J_B + 1}{2\pi^2} \int_0^{k_{F_B}} \mathrm{d}k\, k^2\, \frac{m_B^*}{\sqrt{k^2 + (m_B^*)^2}} \,. \end{aligned} \tag{7.46}$$

As we know from section 7.2, the spacelike components of both vector fields vanish, for the physical reason that the ground state is isotropic and has definite charge [61]. The baryon density ρ^B is given by

$$\rho^B \equiv \langle \mathbf{\Phi}_0 | \Psi_B^\dagger \Psi_B | \mathbf{\Phi}_0 \rangle = \frac{1}{6\pi^2} \, (2J_B + 1) \, k_{F_B}^3 \, . \tag{7.47}$$

The condition of electric charge neutrality is expressed by

$$\frac{1}{6\pi^2} \sum_B \, (2J_B + 1) \, q_B^{\mathrm{el}} \, k_{F_B}^3 - \frac{1}{3\pi^2} \sum_L \, k_{F_L}^3 = 0 \, , \tag{7.48}$$

where the first sum is over the baryons whose electric charges are listed in table 5.1, and the second sum is over the leptons e^- and μ^-. Chemical equilibrium is imposed through the two independent chemical potentials μ^n and μ^e, which lead for the baryon chemical potential to $\mu^B = \mu^n - q_B^{\mathrm{el}} \mu^e$.

7.4 Coupling constants and masses

At the level of the relativistic Hartree and relativistic HF approximation, the parameters (i.e. coupling constants and particle masses) of the Lagrangian (5.1) are not determined by the nucleon–nucleon interaction in free space combined with the data of the deuteron [92, 118], as for the **T**-matrix approximation, but are to be adjusted to the bulk properties of infinite nuclear matter at saturation density, ρ_0 [92, 100, 123, 124]. These properties are the binding energy E/A, effective nucleon mass m_N^*/m_N, incompressibility K, and the symmetry energy a_{sym} whose respective values are given by,

$$\rho_0 = 0.16 \ \mathrm{fm}^{-3} \, , \quad E/A = -16.0 \ \mathrm{MeV} \, , \quad a_{\mathrm{sym}} = 32.5 \ \mathrm{MeV} \, ,$$
$$K = 265 \ \mathrm{MeV} \, , \quad m_N^*/m_N = 0.796 \, . \tag{7.49}$$

Of the five, the value for the incompressibility of nuclear matter carries some uncertainty. Its value is currently believed to lie in the range between about 200 and 300 MeV.

At first sight it seems as if $\mathcal{L}$ of (5.1) would contain an enormous number of unknowns, which is in fact not the case. If one imposes the principal of universal coupling, which consists in setting the baryon couplings to the meson fields, g_{MB}, equal to the nucleon couplings to the respective meson field, g_{MN}, then there remain only a few unknown parameters. These are the four meson masses

$$m_\sigma \, , \quad m_\omega \, , \quad m_\pi \, , \quad m_\rho \, , \tag{7.50}$$

and the seven baryon–meson coupling constants

$$g_{\sigma N} \, , \quad g_{\omega N} \, , \quad f_{\pi N} \, , \quad g_{\rho N} \, , \quad f_{\rho N} \, , \quad b_N \, , \quad c_N \, . \tag{7.51}$$

As for the baryons, the meson masses usually are taken to be equal to their physical values [353], except for the hypothetical σ meson, which is introduced to simulate the correlated 2π exchange. For it one generally takes a tentative value of about 550 MeV. The ρ-meson vector coupling constant, $g_{\rho N}$, can be deduced from the description of the nucleon–nucleon interaction, and the ratio of the tensor to the vector coupling strength, that is $f_{\rho N}/g_{\rho N}$, can be obtained from the vector dominance model [354] which leads to $f_{\rho N}/g_{\rho N} \approx 3.7$. Hence, there remain four undetermined coupling strengths in the theory,

$$g_{\sigma N}\,, \quad g_{\omega N}\,, \quad b_N\,, \quad c_N\,. \tag{7.52}$$

This set reduces to only the first two if σ^4 self-interactions are taken into account, in which case $b_N = c_N = 0$. It is these four, respectively two, coupling constants that are to be adjusted to the ground-state properties of nuclear matter quoted in equation (7.49), in so far as they are left undetermined by the nucleon–nucleon interaction data, of course. Recall that the latter can be used to determine the ρ-meson vector coupling constant which, in turn, fixes $a_{\rm sym}$. A parameter set adjusted along these lines, which allows for HF calculations based on the scalar–vector–isovector Lagrangian, but without the σ^4 terms, has been given by Bouyssy *et al* [98]. We shall adopt this parameter set, which is denoted by HFV [84]. Its parameter values are given in table 7.1.

In the framework of the non-linear Hartree approximation the nuclear forces are described via the exchange of σ, ω, π mesons among the baryons. There are no π-meson contributions because of parity reasons. This leaves one with a one-to-one correspondence between the number of coupling constants, $g_{\sigma N}$, $g_{\omega N}$, $g_{\rho N}$, b_N, c_N, and the nuclear matter properties of (7.49). To determine these couplings for nuclear matter near saturation, one simply needs to fix the Fermi momenta $k_{F_n} = k_{F_p} \equiv k_F$. The scalar and vector coupling constants are then fixed by the known saturation density, ρ_0, and the binding energy per nucleon, $E/A = (\epsilon/\rho)_0 - m_N$. The ρ-meson vector-coupling constant is adjusted to give the empirical symmetry coefficient which is given by the expression [86]

$$a_{\rm sym} = \frac{1}{2}\left(\frac{\partial^2(\epsilon/\rho)}{\partial\delta^2}\right)_{\delta=0} = \left(\frac{g_\rho}{m_\rho}\right)^2 \frac{k_{F_0}^3}{12\,\pi^2} + \frac{k_{F_0}^2}{6\sqrt{k_{F_0}^2 + (m_N^*)^2}}\,, \tag{7.53}$$

where $\delta \equiv (\rho^n - \rho^p)/\rho$, and k_{F_0} the Fermi momentum of symmetric nuclear matter at saturation density, ρ_0. Finally, the non-linear σ-meson self-interactions, proportional to b_N and c_N, are chosen such that a consistent value for the incompressibility of nuclear matter is obtained. Parameter sets adjusted in this way by Glendenning are listed in table 7.1. (The nuclear matter data associated with these parametrizations, which individually

Table 7.1. Coupling constants and masses of several different parameter sets applicable to relativistic Hartree and HF calculations (see table 12.4).† The corresponding nuclear matter properties are listed in table 12.7.

Quantity	HV	HFV	G^{K240}_{B180}	G_{300}	G^{K300}_{B180}	G^{DCM2}_{265}
m_N (MeV)	939	939	938	939	938	939
m_σ (MeV)	550	550	550	600	550	550
m_ω (MeV)	783	783	783	783	783	783
m_π (MeV)	–	138	–	–	–	–
m_ρ (MeV)	770	770	770	770	770	770
$g^2_{\sigma N}/4\pi$	6.16	7.10	6.14	6.644	7.29	5.34
$g^2_{\omega N}/4\pi$	6.71	6.80	6.04	5.930	8.96	5.15
$f^2_{\pi N}/4\pi$	–	0.08	–	–	–	–
$g^2_{\rho N}/4\pi$	7.51	0.55	5.81	5.846	5.34	5.50
$f_{\rho N}/g_{\rho N}$	–	6.6	–	–	–	–
$10^3\ b_N$	4.14	–	8.65	3.305	2.95	–
$10^3\ c_N$	7.16	–	−2.42	15.29	−1.07	–
References	[61]	[98]	[66]	[355]	[66]	[86]

† HFV is a relativistic Hartree–Fock parametrization, all others are relativistic Hartree parameter sets.

vary about the properties quoted in (7.49), will be surveyed in table 12.7.) The parameter sets of table 7.1, which will be applied to stellar structure calculations in the second part of the book, are complemented by several extra Hartree and HF parameter sets, listed in table 7.2, which have been widely used in the literature for the calculation of the properties of finite nuclei as well as nuclear and neutron matter. Listed are the nucleon mass m_N, σ-meson mass m_σ, ω-meson mass m_ω, the respective baryon–meson coupling constants $g_{\sigma N}$, $g_{\omega N}$, and the parameters (if any) of cubic and quartic σ-meson self-interactions. In should be noted that none of these parameter sets accounts for ρ-meson exchange, for which reason the value of the symmetry energy coefficient remain practically uncontrolled, aside from the contribution to $a_{\rm sym}$ that originates from the Fermi momentum [second term in equation (7.53)]. This becomes very obvious from table 7.3. These parameter sets should therefore not be applied to neutron star matter calculations, whose properties depend rather crucially on $a_{\rm sym}$.

The ratio of hyperon to nucleon couplings to the meson fields,

$$x_\sigma = g_{\sigma H}/g_{\sigma N}\,, \quad x_\omega = g_{\omega H}/g_{\omega N}\,, \quad x_\rho = g_{\rho H}/g_{\rho N}\,, \tag{7.54}$$

are not defined by the ground-state properties of normal nuclear matter and so must be chosen according to other considerations [86, 362]. In studies of hypernuclear levels [363, 364, 365], these ratios are typically

Table 7.2. Model parameter sets applicable to relativistic Hartree and Hartree–Fock calculations based on the *standard* (i.e. restriction to only σ and ω meson exchange) scalar–vector Lagrangian. The coupling constants are obtained by fitting the binding energy and density of equilibrium nuclear matter in the relativistic Hartree (HI, HII, HIII, HIV) and relativistic Hartree–Fock (HFI, HFII) approximation (cf. table 7.3).

	HI	HII	HIII	HIV	HFI	HFII
m_N (MeV)	939	939	938	939	939	939
m_σ (MeV)	570	550	492.36	550	550	550
m_ω (MeV)	782.8	783	795.36	783	783	783
$g^2_{\sigma N}/4\pi$	7.826	6.718	8.180	5.958	6.614	8.658
$g^2_{\omega N}/4\pi$	10.824	8.650	14.049	5.678	8.598	11.889
$10^4\left(\frac{g_{\sigma N}}{m_\sigma}\right)^2$	3.0267	2.7906	4.2424	2.4749	2.7474	3.5967
$10^4\left(\frac{g_{\omega N}}{m_\omega}\right)^2$	2.2197	1.7730	2.7908	1.1638	1.7624	2.4368
$10^3\ b_N$	–	1.8	2.46	8.95	–	–
$10^4\ c_N$	–	2.87	−34.3	36.89	–	–
References	[92, 356]	[357, 358, 359]	[360]	[361]	[92]	[92]

Table 7.3. Energy per nucleon E/A, Fermi momentum k_{F_0}, incompressibility K, effective nucleon mass m^*_N/m_N, and symmetry energy $a_{\rm sym}$ (MeV) of equilibrium nuclear matter obtained for the different Hartree (labels 'H') and Hartree–Fock (labels 'HF') parameter sets listed in table 7.2.

	E/A (MeV)	k_{F_0} (fm^{-1})	K (MeV)	m^*_N/m_N	$a_{\rm sym}$ (MeV)
HI	−15.74	1.42	540	0.56	22.1
HII	−15.75	1.34	360	0.693	16.6
HIII	−16.34	1.31	195	0.582	18.4
HIV	−15.95	1.29	237	0.798	13.6
HFI	−15.75	1.42	540	0.529	36.5
HFII	−15.75	1.30	580	0.515	33.6

taken to be equal. In that case, small values between 0.33 and 0.4 are required. These are too small as regards neutron star masses, as shown in table 7.4. We recall that the most accurately determined mass, which is not necessarily the maximal possible mass, is that of PSR 1913+16 with $M = (1.442\pm0.003)M_\odot$ [232] (see also figure 3.2). There is another relevant measurement, that of Vela X-1 (4U 0900–40) with $M = 1.79^{+0.19}_{-0.24}\,M_\odot$ [31, 236]. However, the error is so large that many authors take the

Table 7.4. Values of the hyperon to nucleon scalar and vector coupling that are compatible with the binding of −28 MeV for the lambda hyperon in nuclear matter and the corresponding maximum neutron star mass, as determined by Glendenning *et al* [86, 362]. Agreement with the lower bound on the observed maximum neutron star masses is achieved for hyperon-to-nucleon scalar couplings $x_\sigma > 0.65$, which implies for the vector coupling $x_\omega > 0.75$.

x_σ	x_ω	$M/M_\odot$
0.3	0.262	1.08
0.4	0.415	1.13
0.5	0.566	1.23
0.6	0.714	1.36
0.7	0.859	1.51
0.8	1.00	1.66
0.9	1.14	1.79
1.0	1.27	1.88

other measurement as the limit. The actual number of known masses at the present time is about 20 and we cannot exclude that a more massive neutron star will be found, as indicated by the observation of QPOs for neutron star 4U 1636–536 (cf. section 3.1). However, to the imperfect extent to which the type-II supernova mechanism is understood, it appears that neutron stars are created in a fairly narrow range of masses, around $M \sim 1.4\,M_\odot$, so that independent of whether or not the true equation of state would support more massive neutron stars, none may be made in type-II supernovae explosions.

As noted just above, when hypernuclear levels are analyzed with the constraint $x_\sigma = x_\omega$, the result is a small hyperon coupling leading to a neutron star family with much too small a limiting mass. However, one is not compelled to take the ratios in equation (7.54) to be equal, but there are large correlation errors in $x_\sigma = 0.464 \pm 0.255$ and $x_\omega = 0.481 \pm 0.315$, in the published analysis of hypernuclear levels that leave them uncorrelated [365]. These correlation errors are probably due to the degeneracy with respect to the Λ binding in nuclear matter which we derive next. As noted elsewhere [362], this binding energy serves to strictly correlate the values of x_σ and x_ω but leaves a continuous pairwise ambiguity which hypernuclear levels may be able to resolve. The published analysis so far does not take account of this [365]. Millener, Dover, and Gall inferred in [366] the binding of the Λ hyperon in nuclear matter to be −28 MeV. To impose this constraint on the values of x_σ and x_ω, we need to derive the expression for this binding in the derivative coupling model. From the Weisskopf relation [367] between the Fermi energy and the energy per nucleon of a self-bound system at saturation density, $\omega(k_F) = (\epsilon/\rho)_0$, which is a special case of the Hugenholtz–Van Hove theorem [368], we obtain for the binding energy of the lowest Λ level in nuclear matter, for which $k_{F_\Lambda} = 0$, the relation [362]

$$\begin{aligned} \left.\frac{E}{A}\right|_\Lambda &= x_\omega \, \Sigma_0^N + m_\Lambda^* - m_\Lambda \\ &= x_\omega \, \Sigma_0^N - \frac{x_\sigma \, \Sigma_S^N}{1 + x_\sigma \, \Sigma_S^N/(2\,m_\Lambda)}, \end{aligned} \tag{7.55}$$

where we have made use of

$$\Sigma_S^N \equiv -g_{\sigma N} \, \langle \sigma_0 \rangle \,, \qquad \Sigma_0^N \equiv g_{\omega N} \, \langle \omega_0 \rangle \,, \tag{7.56}$$

which are special cases of equation (7.38), and of equations (6.149) and (7.43). The first line in (7.55) holds for both the linear and non-linear σ–ω theory as well as for this one. The second line specializes to this theory. Thus, as far as the Λ binding in nuclear matter is concerned, the scalar and vector ratios x_σ and x_ω need not be equal, but when so, they must be small, about 0.37. We show a few typical values in table 7.4. Since the neutron star mass limit must exceed a value of about 1.44 to $1.5\,M_\odot$, and as it depends on the hyperon coupling, we infer that $x_\sigma > 0.65$ and a corresponding value of x_ω, as given by equation (7.55).

There are additional constraints on the values of the hyperon constants that can be invoked. There is good reason to believe [82] that these ratios are less than unity. Moreover, according to the analysis of hypernuclear levels in finite nuclei, it is found that when the ratios are taken unequal, the maximum likely value is $x_\sigma < 0.719$ [365]. It is not clear how strong this last constraint is because it applies to the non-linear field theory [309] whose results would carry over only approximately to the present one. For such relatively simple theories of matter, perhaps one should not insist that when the interest is focused on bulk matter, the level spacings of finite nuclei are compelling constraints. In any case, for x_σ and x_ω chosen to be compatible with the Λ binding in nuclear matter, neutron star masses place a *lower* bound on the coupling, and hypernuclear levels appear to place an *upper* bound, but so far less well determined. Within this range hyperons have a large population in neutron stars and neutrons have a bare majority.[1]

We have assumed that other hyperons in the lowest SU(6) octet have the same coupling as the Λ, and also we have arbitrarily taken $x_\rho = x_\sigma$. This choice produces results that are very close to another possible choice, namely $x_\rho = x_\omega$ [362].

We add here a parenthetical note on the analysis of hypernuclei, involving both the Λ hyperon and any other hyperon. We quoted above the

[1] This in the case for universal coupling of the hyperons too.

$\sim$ 50% correlation error found in the least-square fit of x_σ and x_ω to the hypernuclear levels when these parameters are treated independently [365]. But these are not independent parameters as derived just above. They are correlated in a specific way to the binding of the Λ in saturated nuclear matter, a binding that can be inferred quite accurately by an extrapolation from hypernuclear levels in finite-A nuclei [366]. The correlation found in the least-square fit is simply a reflection of the fact that, as a function of A, the finite nuclei are 'pointing' to this binding in $A \to \infty$ matter. It is clear, therefore, that it would be advantageous in the analysis of hypernuclei to take into account the relation that x_σ and x_ω must obey, if the Λ binding in nuclear matter is to come out right [362]. In the linear [348, 349, 350] and non-linear scalar version [309] of nuclear field theory, the difference in masses entering the first line of equation (7.55) is $m^*_H - m_H = x_\sigma \Sigma^N_S$, whereas in the present hybrid derivative coupling model it is given by the second line of equation (7.55).

7.5 Summary of the many-body equations

In the following we summarize those many-body equations that determine the properties of dense nuclear matter as well as dense neutron star matter treated in the framework of either the relativistic Hartree approximation or the relativistic HF approximation [61, 79]. The sets of equations are to be solved self-consistently for a given density until numerical convergency is achieved. The individual equations are:

1) Equations (6.143) through (6.145) which determine the baryon spectral functions Ξ^B, and (6.149) which expresses the medium-modified energy–momentum relation ω^B of a baryon B propagating in dense matter. B stands for

$$B = p\,,\ n\,,\ \Sigma^{\pm,0}\,,\ \Lambda\,,\ \Xi^{0,-}\,,\ \Delta^{++,+,0,-}\,. \tag{7.57}$$

Calculating the energy–momentum relation at the Fermi momentum of each respective baryon listed in (7.57), that is, at

$$k_{F_p}\,,\ k_{F_n}\,,\ k_{F_{\Sigma^{\pm,0}}}\,,\ k_{F_\Lambda}\,,\ k_{F_{\Xi^{0,-}}}\,,\ k_{F_{\Delta^{++,+,0,-}}}\,, \tag{7.58}$$

determines the baryon chemical potentials via the relation $\mu^B = \omega^B(k_{F_B})$. This constitutes a maximum number of $b = 13$ unknowns (see table 5.1). Whether or not a given baryon state becomes actually populated depends, among other attributes, on the total baryon density, ρ, of the system. The expression for ρ is derived in equation (6.190).

2) Chemical equilibrium is imposed through the chemical potentials. Only two independent chemical potentials, μ^n and μ^e, corresponding

to baryon and electric charge conservation [61], are involved. For a baryon of type B, the baryon chemical potential can be inferred from (4.5) to be given by

$$\mu^B = \mu^n - q_B^{\rm el}\, \mu^e \,. \tag{7.59}$$

Hence, only the knowledge of μ^n and μ^e is necessary for the determination of the baryon chemical potentials μ^B. The chemical potentials of the leptons (listed in table 6.1) obey

$$\mu^\mu = \mu^e \,. \tag{7.60}$$

The lepton energy–momentum relation (6.168) determines the lepton Fermi momenta,

$$k_{F_e}\,, \qquad k_{F_\mu}\,. \tag{7.61}$$

3) Equations (7.13), (7.25), (7.26), and (7.27) determine the baryon self-energies, $\mathbf{\Sigma}^B$, in case of the linear σ–ω field theory. The self-energies which arise from the exchange of π and ρ mesons among the baryons are listed in appendices C and D. The individual, non-vanishing self-energy contributions at the level of the HF approximation are

$$\mathbf{\Sigma}^{{\rm H},B}\Big|_{\sigma,\omega,\rho}\,, \qquad \mathbf{\Sigma}^{{\rm F},B}\Big|_{\sigma,\omega,\pi,\rho}\,, \tag{7.62}$$

which constitutes seven unknown functions.

4) The constraint of electric charge neutrality on neutron star matter, that is, $\rho^{\rm el}_{\rm tot} = \rho^{\rm el}_{\rm Bary} + \rho^{\rm el}_{\rm Lep} \equiv 0 + \rho^{\rm el}_{\rm Mes}$, leads to additional constraints on the Fermi momenta of the form

$$\sum_B q_B^{\rm el}(2J_B+1)\frac{k^3_{F_B}}{6\pi^2} - \sum_{L=e,\mu}\frac{k^3_{F_L}}{3\pi^2} - \rho_M \Theta(\mu^M - m_M) = 0\,, \tag{7.63}$$

where the last term accounts for the negative electric charges carried by condensed mesons of type M. As discussed in section 4.2, the only mesons that could plausibly condense in neutron star matter are the π^- [58, 61, 352, 369, 370, 371] or, alternatively, the currently more favored K^- [372, 373, 374]. Relation (7.63) follows readily from the total particle number densities of baryons and leptons, ρ and $\rho^{\rm Lep}$ respectively, given in equations (6.190) and (12.99).

In summary, the total number of unknowns encountered in either the relativistic Hartree or the relativistic HF treatment equals $(7+b)$ and $(11+b)$, respectively. Once these unknowns have been computed self-consistently from the matter equations compiled in items 1) through 4), the equation of state of the system can be computed via simple numerical integration techniques.

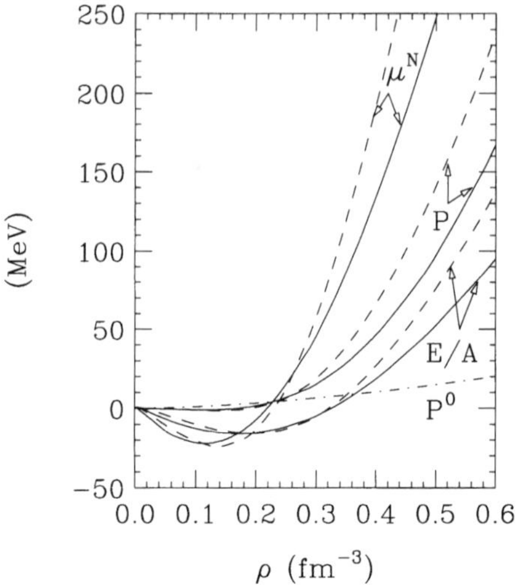

Figure 7.2. Energy per nucleon E/A, chemical nucleon potential μ^N, and pressure P of cold nuclear matter versus density, for Hartree parameter sets HI (dashed curves) and HII (solid curves) of table 7.2 [125]. The dot–dashed curve labeled P^0 shows the pressure of a free relativistic nucleon gas. (Reprinted courtesy of *Z. Phys.*)

5) The total energy density, ϵ, at zero temperature follows from equation (12.44), or one of the relations derived from it, such as (12.53) and (12.54). The total pressure, P, follows from equation (12.63), or one of the relations derived from this expression, such as (12.65). Antiparticles make a contribution only at non-zero temperatures (section 12.2). Combining different ϵ values with their associated P values leads to the equation of state in the parametric form $P(\epsilon)$.
6) Equations (12.93) and (12.95) determine the total lepton energy density and lepton pressure, $\epsilon_{\rm Lep}$ and $P_{\rm Lep}$, respectively, which can be combined to the functional dependence $P_{\rm Lep}(\epsilon_{\rm Lep})$.

7.6 Properties of nuclear and neutron matter at zero and finite temperatures

In figure 7.2 we show the energy per nucleon, pressure, and chemical potential computed for the relativistic Hartree approximation HI, Walecka's original parametrization of $\mathcal{L}$ in the scalar–vector approximation [92, 350]. Each quantity increases rather rapidly at higher density, that is, shows a rather stiff behavior, which is known to be a generic feature of the scalar–vector Lagrangian. Even the inclusion of π and ρ mesons does not change this behavior significantly [123, 124, 356, 361]. Preliminary experimental determinations of the nuclear equation of state at supernuclear densities [375, 376, 377, 378, 379, 380, 381] combined with astrophysical data [382], however, indicate that the true equation of state may behave considerably more softly than predicted by HI. The stiffness of a given equation of state is generally expressed in terms of its associated incompressibility K, or, more properly, by its effective nucleon mass, m_N^*. The latter appears to be a unique measure for the stiffness of an equation of state, in contrast to K [84]. From table 7.3 it is seen that for HI the incompressibility is very high, $K = 540$ MeV, and so the effective mass rather small, $m_N^* = 0.56 m_N$. Such an inverse interplay between K and m_N^* at saturation can be demonstrated analytically in the framework of the relativistic Hartree theory [84, 383], as we shall see later in equation (12.134).

A simple way to soften the equation of state consists in adding to the linear Lagrangian cubic and quartic self-interactions of the σ-meson field, as contained in $\mathcal{L}^{(\sigma^4)}$ of (5.14). The inclusion of such terms requires a readjustment of the parameters of the theory, whose number increases from two ($g_{\sigma N}$, $g_{\omega N}$) to four (b_N, c_N in addition to $g_{\sigma N}$, $g_{\omega N}$). Several possible parametrizations, each of which is compatible with the properties of nuclear matter at saturation density, were listed in tables 7.2 and 7.1. The equation of state computed for one such parametrization, denoted by HII, is graphically illustrated in figure 7.2. The softer behavior of the nuclear matter properties computed for this parametrization is obvious. The softness of HII manifests itself in an incompressibility of 360 MeV, in comparison with 540 MeV obtained for HI. Depending on the parametrization of the non-linear terms (see table 7.3), incompressibilities as small as ~ 200 MeV can be obtained. Such rather soft models for the equation of state are also obtained in the framework of non-relativistic many-body calculations [384].

Inclusion of the exchange term of the nucleon self-energy is known to stiffen the equation of state [92, 356]. This is graphically illustrated in figure 7.3, where we compare the outcome obtained for the non-linear Hartree model HII with the Hartree–Fock HFI results. The nuclear matter properties at lower densities do not differ significantly for the Hartree (that is, mean-field) and HF approximation, as can be seen in figure 7.3, provided of course the parameters of both theories are properly constrained by the nuclear matter data. (Despite this agreement, however, there may be striking differences in the single-particle spectrum [356].) This agreement between Hartree and HF carries over to the behavior of the equation of state of dense matter in the low-temperature portion of its phase diagram as shown in figures 7.4 through 7.6, which display the internal energy, $\mathcal{E}^{\rm int}(\rho, T)/\rho$ [to be derived in equation (12.107)], at zero and finite temperatures. Only at higher temperatures and greater nuclear densities does one obtain deviations between the HF and the mean-field treatments.

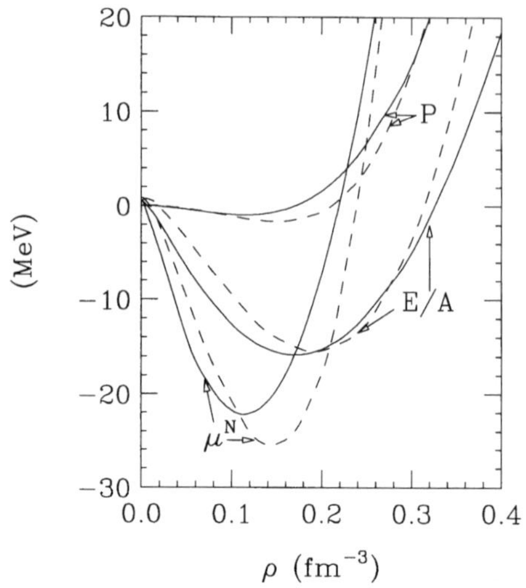

Figure 7.3. Energy per nucleon E/A, chemical nucleon potential μ^N, and pressure P as a function of density of cold nuclear matter, computed for Hartree parameter set HII (solid curves) and HF parameter set HFI (dashed curves) listed in table 7.2 [125]. (Reprinted courtesy of *Z. Phys.*)

This is specifically the case for the critical temperature of nuclear matter, as we shall see below.

In figures 7.7 and 7.8 we compare the energy per nucleon computed for different Hartree and HF parametrizations at zero and finite temperature. The sequence of small crosses in figure 7.7 indicates the behavior of a soft equation of state with an incompressibility of $K \simeq 240$ MeV, while the sequence made up of the bigger crosses refers to a stiff one with $K \simeq 380$ MeV, as indicated by the analysis of sideways flow in the Vlasov–Uhling–Uhlenbeck formalism [375, 376, 377, 378] and by the π-data analysis [381, 385], respectively. In accordance with what has been said above, the HF equations of state behave more stiffly than the linear Hartree model HI which, itself is stiffer than the non-linear Hartree models HII and HIV. This trend remains unchanged at finite temperatures, as can be seen from figure 7.8. The pressure isotherms, which are of particular interest when discussing phase transitions of nuclear matter, associated with figures 7.7 and 7.8 are shown in figures 7.9 and 7.10.

In the case of neutron matter we find greater differences between the Hartree and the HF approximation than for nuclear matter. First, we show in figure 7.11 the energy per nucleon of neutron matter at zero temperature, computed for several different parametrizations. It is most striking that the local minimum of the energy per baryon obtained for

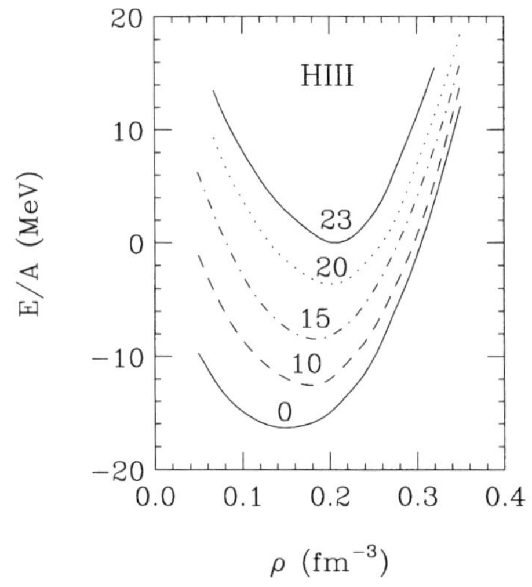

Figure 7.4. Energy per nucleon of nuclear matter versus density, for Hartree parameter set HIII listed in table 7.2. The labels refer to temperature in MeV [125]. (Reprinted courtesy of *Z. Phys.*)

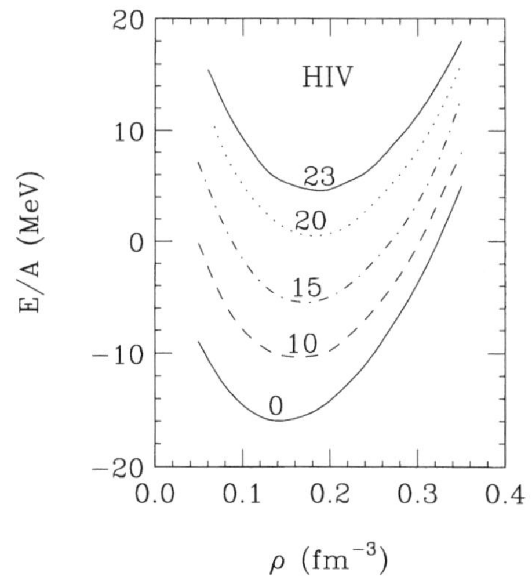

Figure 7.5. Same as figure 7.4, but for Hartree parameter set HIV [125]. (Reprinted courtesy of *Z. Phys.*)

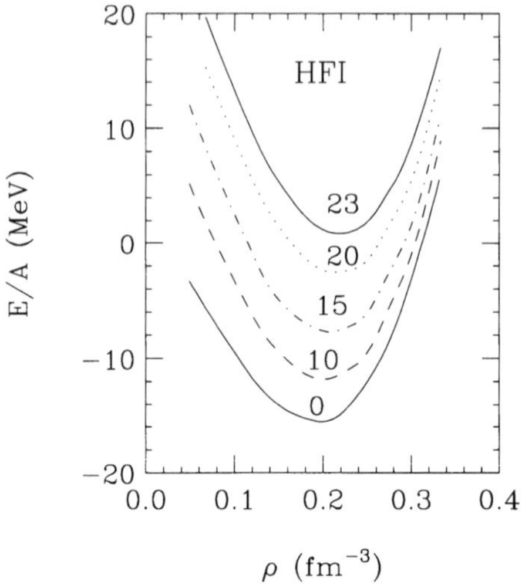

Figure 7.6. Same as figure 7.4, but for Hartree–Fock parameter set HFI [125]. (Reprinted courtesy of *Z. Phys.*)

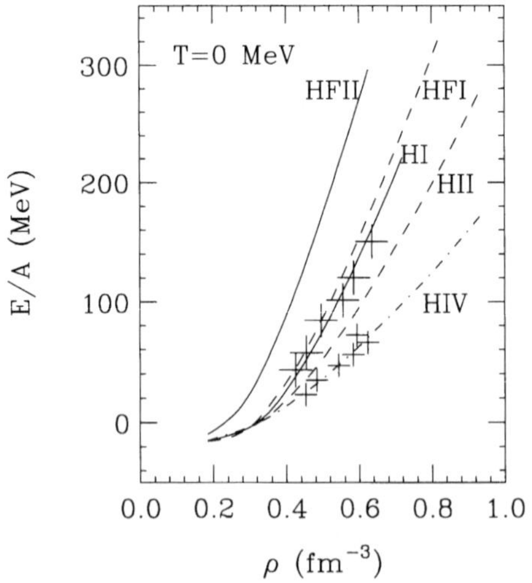

Figure 7.7. Energy per nucleon in cold nuclear matter versus density, computed for different Hartree and Hartree–Fock parameter sets listed in table 7.2 [125]. (Reprinted courtesy of *Z. Phys.*)

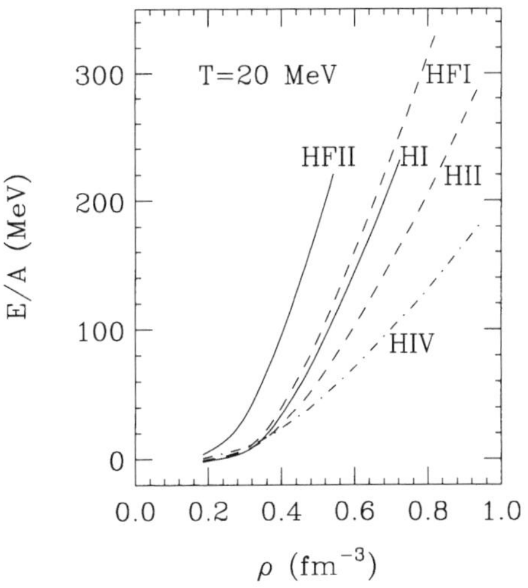

Figure 7.8. Same as figure 7.7, but for a temperature of $T = 20$ MeV [125]. (Reprinted courtesy of *Z. Phys.*)

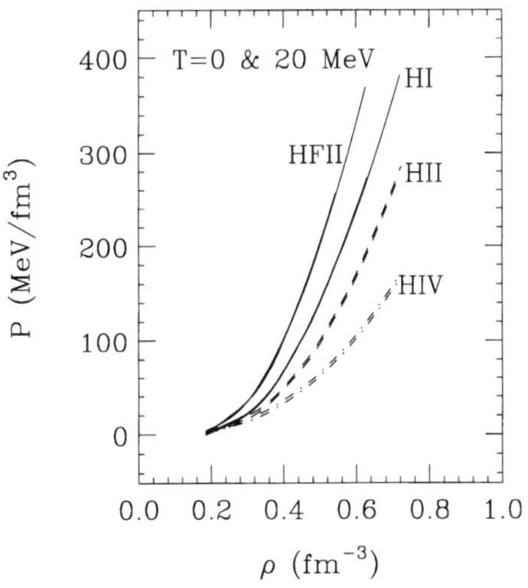

Figure 7.9. Pressure isotherms of nuclear matter versus density, for different Hartree and Hartree–Fock parameter sets of table 7.2 [125]. The temperatures are $T = 0$ (lower-lying curves) and 20 MeV (upper-lying curves). (Reprinted courtesy of *Z. Phys.*)

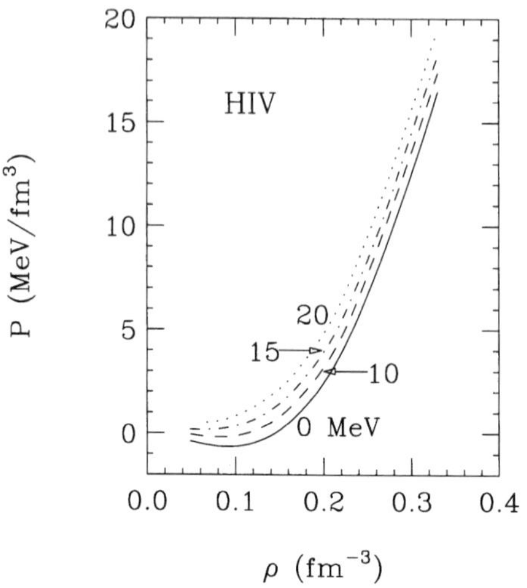

Figure 7.10. Pressure isotherms of nuclear matter versus density, computed for Hartree parameter set HIV of table 7.2 [125]. The labels refer to temperature in MeV. (Reprinted courtesy of *Z. Phys.*)

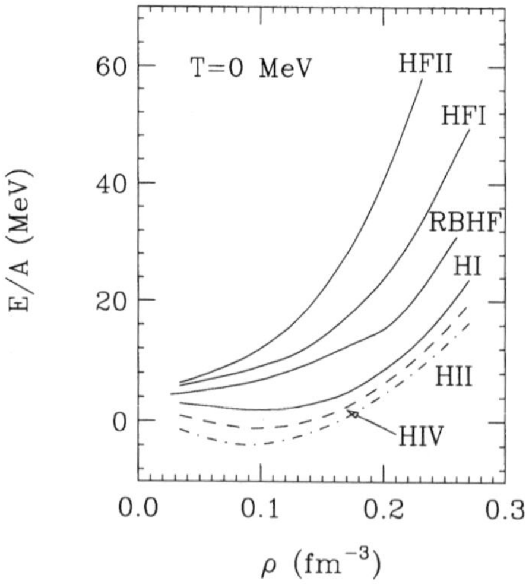

Figure 7.11. Energy per nucleon of cold neutron matter versus density, for different Hartree and HF parameter sets of table 7.2 [125]. The curve labeled RBHF refers to a relativistic Brueckner–Hartree–Fock calculation performed for BM potential B (cf. section 5.5 and chapters 9 and 10). (Reprinted courtesy of *Z. Phys.*)

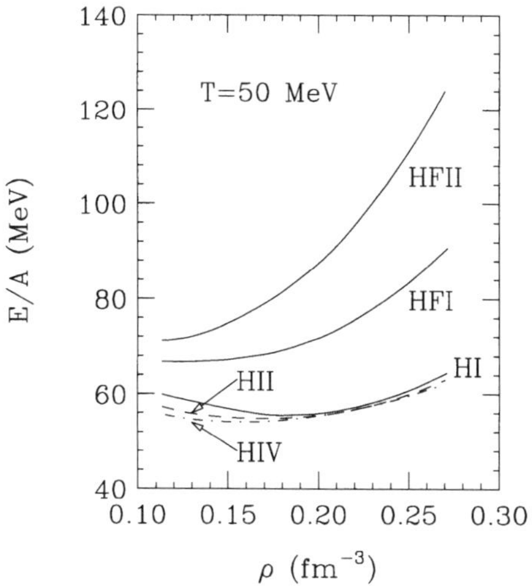

Figure 7.12. Same as figure 7.11, but for $T = 50$ MeV [125]. (Reprinted courtesy of *Z. Phys.*)

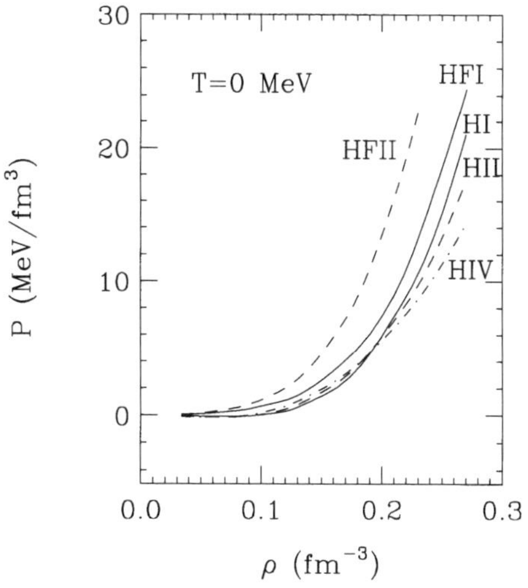

Figure 7.13. Pressure of cold neutron matter as a function of density, computed for different Hartree and HF parameter sets of table 7.2 [125]. (Reprinted courtesy of *Z. Phys.*)

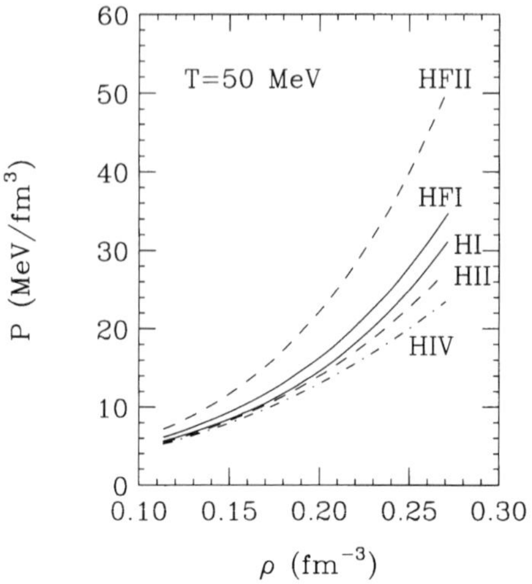

Figure 7.14. Same as figure 7.13, but for $T = 50$ MeV [125]. (Reprinted courtesy of *Z. Phys.*)

Table 7.5. Critical temperatures, T_c, of nuclear and neutron matter calculated for the different Hartree and HF parameter sets of table 7.2 [125].

	T_c (MeV) nuclear matter	T_c (MeV) neutron matter
HI	19	9
HII	16	10
HIII	14	8.5
HIV	15	10
HFI	21	absent
HFII	17.5	absent

the Hartree approximations disappears for both the HF and the BHF approximation, which is a consequence of the self-consistent exchange terms contained in the latter two approximations. This feature carries over to finite temperatures, as can be seen by inspection of figure 7.12. The pressure associated with figures 7.11 and 7.12 is shown in figures 7.13 and 7.14, respectively. Figure 7.15 shows the reduction of pressure over a certain density range for neutron matter treated in the Hartree approximation. Because of the exchange term, this feature is absent in case of the HF approximation.

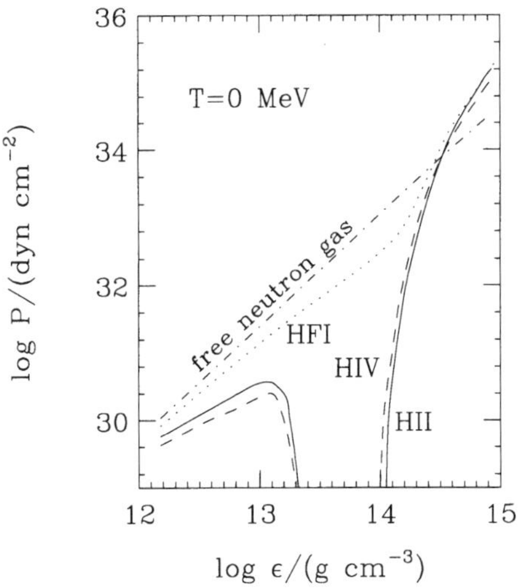

Figure 7.15. Pressure versus energy density of cold neutron matter computed for the Hartree parameter sets HII (solid lines) and HIV (dashed lines), and the HF parameter set HFI (dotted line) listed in table 7.2 [125]. For comparison, the EOS of a free relativistic neutron gas is shown too. (Reprinted courtesy of *Z. Phys.*)

The self-consistent effective nucleon mass in matter, defined as $m_N^* = m_N + \Sigma_S^N$, is displayed in figures 7.16 and 7.17. It is increased by the occurrence of non-linear σ-meson self-interactions, which were introduced in order to soften the equation of state. A softening, i.e. a reduction of the incompressibility K, is generally accompanied by an increase of m_N^* as can be demonstrated analytically in the framework of the relativistic Hartree theory [84, 383] (see equation (12.134)). Because of this inverse interplay between K and m_N^*, the K values associated with the curves in figure 7.16 *decrease* monotonically from 580 MeV for HFII to 237 MeV for HIV (cf. table 7.3). These figures also reveal that m^* increases only very weakly with temperature, even weaker than for the relativistic BHF theory [93, 386].

Finally, figure 7.18 serves to discuss the liquid–gas phase transition of nuclear matter. Plotting the nucleon chemical potential as a function of pressure results in the well-known zig-zag shape, which becomes the less pronounced the higher the temperature. Phase equilibrium between the nuclear liquid and the nuclear gas phase exists right at the point where the two curves intersect each other. The system's *critical* temperature, beyond which only one phase of matter (gas in our case) exists, is found from that isotherm whose zig-zag structure has turned over into a kink. In the present calculation performed for parameter set HII, this is the case

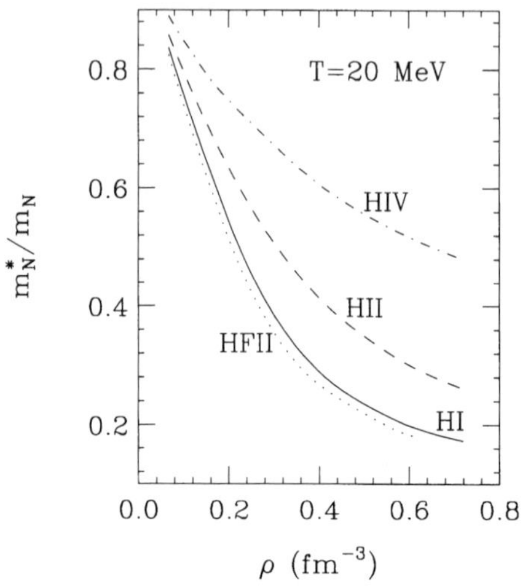

Figure 7.16. Effective nucleon mass m_N^*/m_N in nuclear matter at $T = 20$ MeV versus density, for different Hartree and HF parameter sets of table 7.2 [125]. (Reprinted courtesy of *Z. Phys.*)

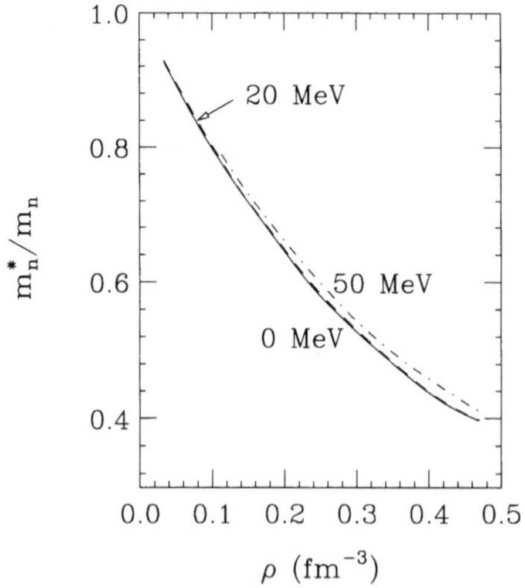

Figure 7.17. Effective neutron mass m_n^*/m_n in neutron matter versus density, for Hartree parameter set HII of table 7.2 [125]. The labels refer to temperature. (Reprinted courtesy of *Z. Phys.*)

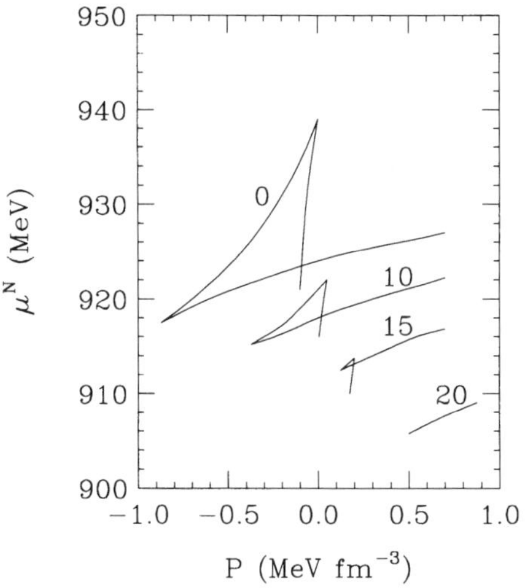

Figure 7.18. Chemical nucleon potential versus pressure for nuclear matter, for Hartree parameter set HII of table 7.2 [125]. The labels refer to temperature in MeV. (Reprinted courtesy of *Z. Phys.*)

for $T_c \simeq 16$ MeV. The critical temperatures computed for the other Hartree and Hartree–Fock parameter sets are summarized in table 7.5. It is seen that T_c varies between 14 and 21 MeV for nuclear matter but falls into a considerably smaller range for neutron matter. The corresponding densities are about 45% to 75% of normal nuclear matter density, which is in agreement with the outcome of other microscopic calculations reported in the literature [350, 384, 387, 388, 389, 390]. Somewhat lower critical temperatures of around $T_c \simeq 12$ MeV are obtained in the framework of RBHF calculations [386, 391]. In any event, however, a phase transition is always found for nuclear matter, independent of the many-body approximation and the forces. This is not so for neutron matter, which only exhibits a phase transition when treated in the relativistic Hartree approximation. There is no phase transition if the underlying many-body approximation is relativistic HF. That a critical temperature exists for neutron matter for some approximations and forces may indicate that they are unphysical. However, it needs to be recalled that neither Hartree (mean-field) nor HF may be sufficiently adequate to properly model the nuclear matter phase transition, since it may be accompanied by large density fluctuations, preventing us from discriminating between physical and unphysical approximations.

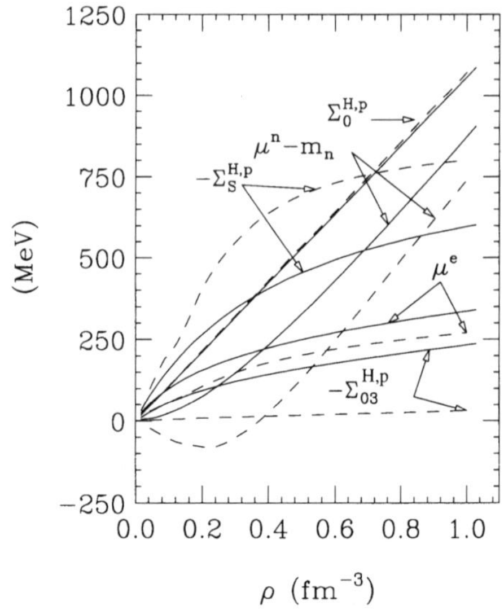

Figure 7.19. Hartree self-energies $\Sigma_S^{H,p}$, $\Sigma_0^{H,p}$, $\Sigma_{03}^{H,p}$, calculated from (7.13), (7.25), (C.1) for the parameter sets HV (solid lines) and HFV (dashed lines) [79]. The chemical potentials of electrons μ^e and neutrons μ^n are defined in (7.60) and (6.152). Only $B = p, n$ are taken into account. (Reprinted courtesy of *Nucl. Phys.*)

7.7 Baryon self-energies

In figures 7.19 and 7.20 we show the Hartree self-energy components of protons, along with the chemical potentials of neutrons and electrons, self-consistently calculated for the relativistic Hartree and HF parametrizations HV and HFV listed in table 7.1 [79]. The first of these figures, 7.19, refers to stellar matter made up of chemically equilibrated protons and neutrons only, which however is not the ground state of neutron star matter. Complete chemical equilibrium is taken into account in figure 7.20, where all higher-mass baryons of table 5.1 are included for which the baryon chemical potential exceeds their rest mass (corrected for interactions and electric charge, as will be discussed in section 7.8). Qualitatively similar to the nuclear matter case, one sees that the scalar component of the self-energy at saturation density, $\rho_0 = 0.16\ \mathrm{fm}^{-3}$, is large and negative, typically $\Sigma_S \sim -250$ MeV, while the timelike component is large and positive, $\Sigma_0 \sim 150$ MeV. Observe that at higher densities, the Hartree part of Σ_S computed for HFV becomes considerably more negative than for HV. This manifests itself in a rather strong reduction of the effective nucleon mass m_N^* from its vacuum value, $m = 939$ MeV, as is shown in

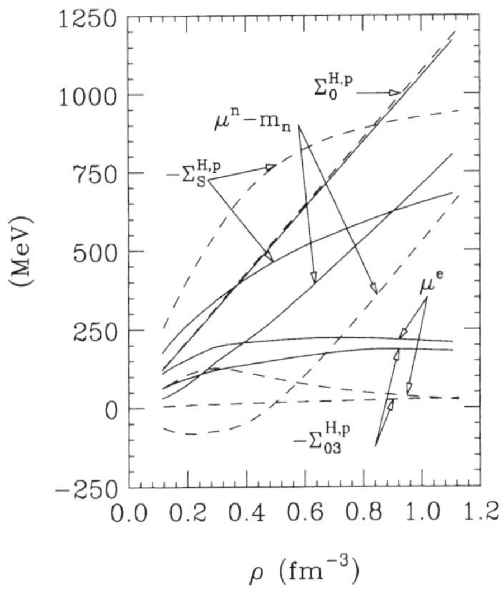

Figure 7.20. Same as figure 7.19, but for all baryon states of table 5.1 whose threshold densities are reached (universal coupling) in β-stable, electrically charge neutral neutron star matter [79]. (Reprinted courtesy of *Nucl. Phys.*)

figure 7.21. In fact, the reduction of m_N^* in fully chemically equilibrated stellar matter turns out to be so strong that m_N^* even turns negative at $\rho \sim 1.5$ fm^{-3}. This value, however, lies considerably beyond the central density encountered in the most massive, stable neutron star constructed for this model, which is 1.16 fm^{-3}, or 1350 MeV fm^{-3} (cf. table 14.3). Negative effective nucleon masses are obtained neither for the relativistic Hartree approximation, because of the smaller (in magnitude) scalar self-energies Σ_S, nor, trivially, when the more massive baryon states are excluded from the calculation. This leaves us with the conjecture that the population of the Δ^- resonance state, obtained for HFV but not for HV for reasons that will be discussed in greater detail in section 7.8, may be driving the effective nucleon mass toward negative values. Indeed, the phenomenon that the effective masses of the lower vacuum-mass baryons in non-strange baryonic matter will pass zero when the Δ resonance state becomes populated for increasing density has already been pointed out years ago in the literature [84, 392]. A negative effective baryon mass (and thus negative contribution to the scalar density) however appears not to be in contradiction with theory, since the scalar density has no probabilistic meaning.

A characteristic difference between the results shown in figures 7.19 and 7.20 originates from the number of electrons that are present in the matter. Obviously, if the presence of negatively charged higher-mass baryons is

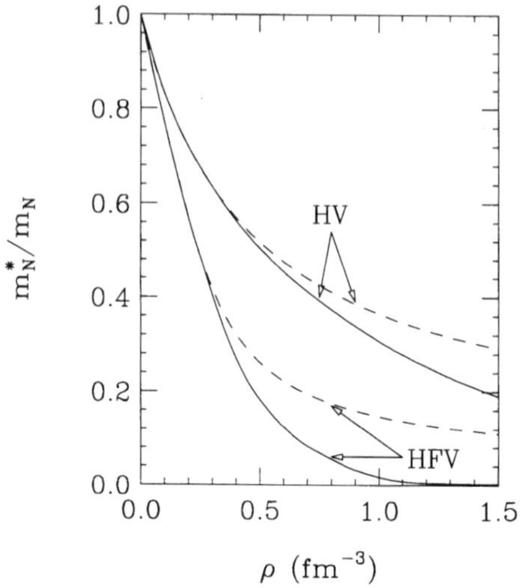

Figure 7.21. Effective nucleon mass m_N^*/m_N as a function of baryon density. The curves indicate results in the relativistic Hartree approximation HV, and the relativistic HF approximation HFV. The HF masses, defined in (7.28), are computed at $k = k_F$. The solid curves are for matter made up of chemically equilibrated protons, neutrons plus all more massive baryon states that become populated at the densities shown. In contrast, the dashed curves account for matter made up of only protons and neutrons.

forbidden, the positive charge carried by the protons must be entirely balanced by electrons. Being fermions, the number of protons increases monotonically with density, and so does the number of electrons in order to render the system at each density electrically charge neutral. The electron chemical potential, μ^e, which is a measure for the number of electrons present in a system, must therefore be monotonically increasing at *all* densities, as can be seen from figure 7.19. This is strikingly different if there are other negatively charged degrees of freedom accessible to the system – such as hyperons, quarks, or meson condensates. The impact of hyperon population on μ^e is shown in figure 7.20. One observes that in both cases, with and without hyperons, μ^e is monotonically increasing at low densities, in such a way that the corresponding number of electrons (plus some muons) equals the number of the protons at each density (compare with figures 7.23, 7.24, and 7.25). As soon as the threshold of a negatively charged hyperon is reached however, which is typically the case around twice nuclear matter density (figures 7.23 and 7.24), negative charge is being delivered by these particles and so fewer electrons are needed to keep the matter electrically charge neutral. Fewer electrons means however that μ^e will saturate at a density a few times ρ_0 and finally fall off toward smaller values, depending on how much negative charge is being delivered by hyperons, higher-mass baryons like the Δ^-, possibly d and s quarks, or negatively charged meson condensates.

The fact that the lepton population will be strongly suppressed by negatively charged hyperons, the Δ^-, quarks, or a meson condensate may have a strong effect on the electrical conductivity of the stellar matter [393, 394, 395]. This is be so because of the smaller number of electrons that will be present in the stellar matter in case that (some of) these degrees of freedom play a role. The electrical conductivity of the stellar matter will then be lower than if these degrees of freedom were absent. The electrical conductivity, in turn, determines the decay rate of the strong magnetic field [396] needed to produce the pulsar beam effect [137]. Consequently the presence of hyperons, or of any other of the above mentioned degrees of freedom, in a pulsar may register itself in the active lifetime of the pulsar [61]. This is estimated from astrophysical data to be less than several hundred million years [12]. Of course, from the observation of an anomaly in the decay time alone one may not be able to draw any stringent conclusions about the phenomenon that is actually causing the anomaly. Only the simultaneous observation of several physical properties, which complement one another, may put us into this position.

As a last point on these issues, we mention the important rôle played by leptons in the viability of hot, newly formed, lepton-rich neutron stars. According to the numerical simulations of the neutrino-driven deleptonization and cooling of such objects performed by Keil and Janka [397], neutron stars are stabilized by the confined leptons for an interim period lasting from a few seconds up to about ten, which is followed, after the loss of a significant fraction of the leptons, by gravitational collapse of the stellar core.

Our calculation shows that the presence of negatively charged hyperons saturates the electron chemical potential at $\mu^e \sim 220$ MeV in the case of the HV parameter set, and at about ~ 150 MeV for the HFV parameter set. Knowledge of the saturation value as well as the density at which the maximum is attained is of decisive importance to the question of whether or not electrons in neutron star matter may be replaced with negatively charged mesons, such as possibly the π^- [58, 140, 352, 369] or the K^- [254, 372, 373, 374], by means of which Bose condensates will be formed. Glendenning found in an earlier investigation [140] that in the framework of the relativistic mean-field (Hartree) approximation the π^- meson goes into a Bose condensate in neutron star matter slightly above nuclear matter density. Mesons such as π^0 and π^+ are prevented from condensation because of the electric charge they are carrying, which, according to equation (4.13), implies for their chemical potentials that $\mu^{\pi^0} = 0$ and $\mu^{\pi^+} = -\mu^e$. Hence condensation of such mesons can occur only if their respective effective masses in neutron star matter either vanish or would become negative (see figure 4.4). Both is not the case.

As shall be discussed in more detail in section 7.9, in the case of pion condensation, the source current in the pion-field equation (5.58) acquires a finite value. Consequently the π meson ceases to be free and is driven to have a finite amplitude [58, 61]. To indicate this here mathematically, let use write the pion-field equation (5.58) as

$$\left[-k^\mu k_\mu + m_\pi^2 + \Pi(k_0, \boldsymbol{k})\right] \Delta^\pi(k_0, \boldsymbol{k}) = 1\,, \tag{7.64}$$

where m_π and Π denote the pion mass and the pion polarization operator, respectively. At the threshold density of a possible phase transition, at which there exists a k_c, the critical wave vector, one has for the pion propagator $(\Delta^\pi(k_0 = \mu^e, k_c))^{-1} = 0$. Hence we read off from (7.64) that

$$-\,k_0^2 + \boldsymbol{k}^2 + m_\pi^2 + \Pi(k_0, \boldsymbol{k}) = 0\,. \tag{7.65}$$

The threshold density for the onset of a π^- condensate is that density for which equation (7.65) first has a solution for real $\boldsymbol{k}$ and $k_0 = \mu^e$. If μ^e attains a value on the order of the pion mass, negative pions will condense, and, being bosons, arbitrarily many of them can settle into the same energy state. The condensation point would be precisely m_π if the pion did not interact with other hadrons. In the presence of interactions the discussion is more complicated [58, 303]. Earlier calculations of pion condensation in neutron star matter indicated that a condensate could be expected at $k_F \approx 1.5\ \mathrm{fm}^{-1}$ and that μ^e saturates at 177 MeV [61]. The saturation value would be precisely the pion mass, $\mu^e = m_\pi$, with $m_\pi = 138$ MeV, if interaction effects of the pions with the nuclear medium are completely neglected.

As an important consequence of the possible saturation of μ^e near the pion mass, the pion may foreclose other types of phase transitions, like the K^- condensation, provided the corresponding polarization operator is not very large and attractive so as to overcome the K^-'s large vacuum mass. Recent experimental evidence [398, 399] on kaon–nucleon interactions indeed seem to indicate that the effective K^- mass may be greatly lowered in dense nuclear matter, eventually bringing it down to $m^*_{K^-} \sim 200$ MeV. If so the condensation of the K^- meson may not be ruled out because of too large a mass, depending on the value of μ^e. A maximum value of $\mu^e =$ 155 MeV as obtained for the relativistic HF calculation presented above (cf. figure 7.20) would probably be too small to initiate K^- condensation. The situation looks better for the Hartree calculation HV in which case the maximum value is shifted up to $\mu^e = 220$ MeV, for universally coupled hyperons. Otherwise, in the case of non-universal hyperon-to-nucleon coupling based on quark counting arguments as $x_\sigma = x_\omega = x_\rho = \sqrt{2/3}$, this value drops down to 177 MeV [61]. To put all this together, we show

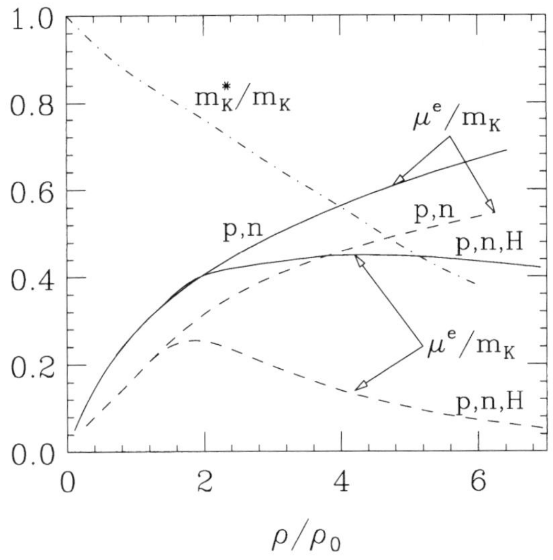

Figure 7.22. Density dependence of K^- mass, m_K^*, in neutron star matter. The electron chemical potentials μ^e (see figure 7.20) are calculated in relativistic Hartree (solid lines) and relativistic HF (dashed lines). The curves labeled p, n refer to β-equilibrated neutron star matter made up of protons and neutrons only. Hyperons are taken into account in the curves labeled p, n, H. (After Waas, Rho, and Weise [400].)

in figure 7.22 the behavior of the K^- mass in neutron star matter as recently calculated by Waas, Rho, and Weise [400]. One sees that m_K^* drops significantly with density, crossing the electron chemical potential curves μ^e at densities between four and five times the density of nuclear matter, depending on the underlying many-body approximation and whether or not hyperon population in neutron star matter is taken into account. Most strikingly, the rapid drop of μ^e beyond about twice nuclear matter density obtained for the relativistic HF calculation HFV prevents K^- condensation in neutron stars. Such a rapid drop is also obtained for the relativistic **T**-matrix calculations to be discussed later, or by the inclusion of hyperon–hyperon interactions in the framework of the Hartree approximation [401].

If one accepts the chemical potentials plotted in figure 7.22 as reliable bounds on μ^e in neutron star matter, then the conclusion to be drawn is that the relativistic HF approximation HFV (as well as relativistic **T**-matrix calculations in general) tend to disfavor the formation of meson condensates in neutron star matter. In any case, what is required to achieve condensation within these models is a strong reduction of the meson mass with density, to such an amount that the effective meson mass is considerably reduced from its vacuum value. The final answer to this problem clearly hinges on the interaction strength of the meson field with the neutron matter medium whereby the meson mass is changed from its vacuum value.

7.8 Baryon–lepton composition

Having explored the important role of electrons in neutron star matter in the previous section, we turn now to the discussion of the baryon–lepton composition of such matter. The underlying microscopic many-body theories will be again the relativistic Hartree and HF approximations [79].

As repeatedly stressed before, neutron star matter at densities around nuclear matter density consists primarily of chemically equilibrated protons and neutrons ($B = p, n$) and leptons ($L = e^-, \mu^-$). At densities greater than the density of nuclear matter more massive baryon states B (with $B = \Sigma^{\pm,0}, \Lambda, \Xi^{0,-}, \ldots$) become successively populated as soon as their chemical potentials μ^B exceed the lowest-lying energy-eigenstates, that is, $\mu^B \geq \omega^B(|\boldsymbol{k}| = 0)$, as schematically illustrated in figure 4.3. In the trivial case of non-interacting particles, $\omega^B(|\boldsymbol{k}| = 0)$ is equal to the particle's vacuum mass, m^B. Interactions modify the energy-eigenstate of a baryon, as we know from our considerations performed in chapter 5, according to

$$\omega^B(|\boldsymbol{k}|) = \Sigma_0^B(\omega^B(|\boldsymbol{k}|), |\boldsymbol{k}|) + I_{3B}\, \Sigma_{03}^B(\omega^B(|\boldsymbol{k}|), |\boldsymbol{k}|) \\ + \sqrt{[m_B + \Sigma_S^B(\omega^B(|\boldsymbol{k}|), |\boldsymbol{k}|)]^2 + [\boldsymbol{k} + \Sigma_V^B(\omega^B(|\boldsymbol{k}|), |\boldsymbol{k}|)]^2}\,. \quad (7.66)$$

This relation tells us that the baryon thresholds are determined by attributes like the medium-modified masses of these particles, their electric charges and their isospin quantum numbers. Making use of $\mu^B \geq \omega^B(|\boldsymbol{k}| = 0)$ and the fact that μ^B can be expressed in terms of the two independent chemical potentials μ^n and μ^e, which is a direct manifestation of the conservation of baryon number and electric charge in neutron star matter (cf. equation (4.5)), we get the following threshold condition for baryon B,

$$\mu^B \equiv \mu^n - q_B^{\rm el}\, \mu^e \geq \omega^B(|\boldsymbol{k}| = 0)\,. \quad (7.67)$$

Substituting (7.66) into (7.67) then leads to

$$\mu^B = \mu^n - q_B^{\rm el}\, \mu^e \geq \Sigma_0^B(\omega^B(0), 0) + I_{3B}\, \Sigma_{03}^B(\omega^B(0), 0) \\ + \sqrt{[m_B + \Sigma_S^B(\omega^B(0), 0)]^2 + [\Sigma_V^B(\omega^B(0), 0)]^2}\,. \quad (7.68)$$

For the sake of completeness, we also give the non-relativistic reduction of the single-baryon energy (7.66) (more details can be found in section 12.5), given by

$$\omega^B(|\boldsymbol{k}|) = \frac{1}{2\, m_B}\, \boldsymbol{k}^2 + \Sigma^B(|\boldsymbol{k}|)\,, \quad (7.69)$$

where $\Sigma^B(|\boldsymbol{k}|)$ denotes the effective, non-relativistic one-particle potential felt by a baryon inside the stellar medium.

Since we are dealing with particles, the chemical potentials μ^n and μ^e are positive quantities. The self-energies Σ_{03}^B, accounting for the difference of the density of protons minus the one of neutrons, carry negative signs. It thus follows from equation (7.68) that,

(i) negatively charged baryons are *charge-favored*,

(ii) baryons having the same isospin projection as the neutron, that is, $I_{3B} < 0$, are *isospin-unfavored* [61].

The physical reason behind item (i) is that negatively charged baryons can replace electrons at the top of the Fermi sea by means of which a lower-energy state is reached. Item (ii) is a consequence of the extreme isospin asymmetry of neutron star matter, that is, the extreme difference in the number of protons and neutrons, which pushes the energy per baryon of neutron star matter way above the one of isospin-symmetric nuclear matter. Neutron star matter therefore constitutes a highly excited state of matter relative to symmetric nuclear matter. Hence, as soon as there are new degrees of freedom available to the system which allow neutron star matter to become more isospin symmetric – and thus get away from this energetically unfavorable high energy state – it will make use of them. Besides these two guiding principles, which play a crucial role for the composition of neutron star matter, the nuclear many-body background created by the presence of about 10^{57} baryons which form a neutron star, influences the baryon thresholds – as well as those of meson and quarks degrees of freedom, of course – too. The nuclear background enters the threshold relation via the negative baryon self-energies Σ_S^B, the positive self-energies Σ_0^B (cf. figures 7.19 and 7.20), and Σ_V^B which is negligibly small compared to the other two [118]. The effective, medium-modified baryon masses in matter, defined as $m_B^* = m_B + \Sigma_S^B$, are therefore smaller than the vacuum masses, and decrease for increasing density. According to (7.68) this behavior tends to depress the threshold of each baryon species. The precise way, however, in which the numerous thresholds are reached can only be found by solving the many-body equations, summarized for the relativistic Hartree and HF cases in section 7.5, self-consistently.

As representative examples, we show in figures 7.23 and 7.24 the relative baryon and lepton populations of neutron star matter calculated for the relativistic Hartree and HF parameter sets HV and HFV, respectively, which are listed in table 7.1. The populations are shown up to the highest densities encountered in the cores of the heaviest neutron star models constructed from the equations of state that are associated with these populations (details will be discussed in chapters 12 and 14). These populations bear several similarities as well as striking differences. With respect to their common features, one sees that both approximations

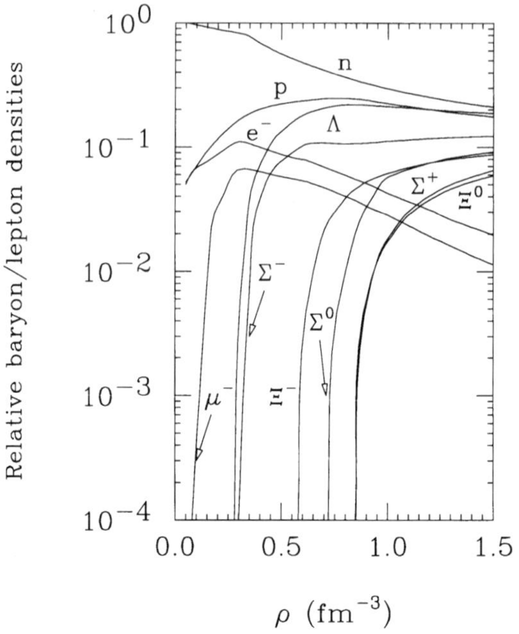

Figure 7.23. Baryon–lepton densities ρ^B/ρ and ρ^L/ρ of neutron star matter normalized to the total baryon density ρ [79]. The densities ρ^B and ρ^L are defined in (6.187) and (12.98). The calculation is performed for the relativistic Hartree parameter set HV of table 7.1. (Reprinted courtesy of *Nucl. Phys.*)

predict that neutron star matter at densities around nuclear matter density, $\rho_0 = 0.16$ fm^{-3}, consists of nearly 100% neutrons, a small admixture of protons plus – because of the requirement of electric charge neutrality – a certain number of leptons. The concentration of the latter therefore is exactly such that the total electric charge, carried by the protons plus the leptons, adds up to zero at each respective density. For increasing values of density, high-momentum neutrons undergo β-decay into protons and electrons or muons. The μ^- threshold varies a little bit with the many-body approximation. It occurs just below $\rho_0 = 0.15$ fm^{-3} for HV and increases to about 0.20 fm^{-3} for HFV, as can be seen in figures 7.23 and 7.24 respectively.

Severe differences between both many-body approximations occur at still higher densities, as we shall discuss next. In figure 7.23, for instance, it is seen that the Λ hyperon, which remains unaffected by the effects (i) and (ii) since its electric charge $q^{\rm el}_\Lambda = 0$ and the third component of isospin $I_{3\Lambda} = 0$ (cf. table 5.1) has the lowest threshold of all baryons that are predicted to become populated in dense neutron star matter. The

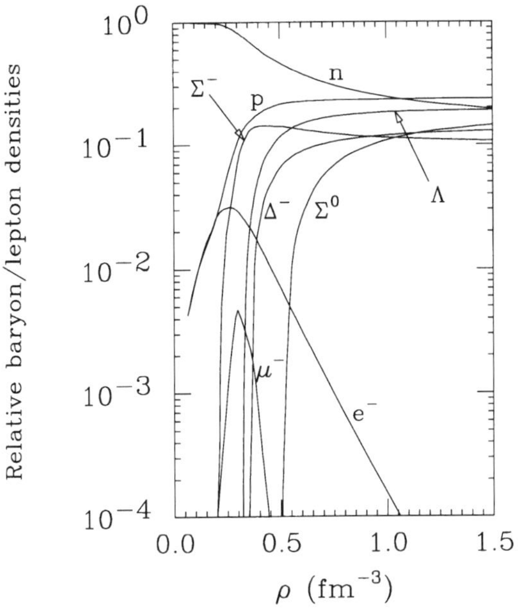

Figure 7.24. Same as figure 7.23, but calculated for the relativistic Hartree–Fock parameter set HFV of table 7.1 [79]. (Reprinted courtesy of *Nucl. Phys.*)

threshold density of the Λ is $\rho_\Lambda^{\rm HV} = 0.28$ fm^{-3}, which corresponds to a total energy density of about $\epsilon_\Lambda^{\rm HV} = 280$ MeV fm^{-3}. Hence the Λ comes in at a density somewhat less than twice nuclear matter density. The next two hyperon species that become populated are the negatively charged, but charge-favored Σ^- and Ξ^-. The threshold of the charge neutral Σ^0 hyperon, which is neutral to the above two effects (i) and (ii), is lower than the one for the charge-unfavored, but isospin-favored Σ^+ hyperon. Finally, the hyperon which comes in last is the Ξ^0. Being neutral to (i) and (ii) too, it has to overcome its relatively large mass which is only possible at high densities, since only then has $\Sigma_S^{\Xi^0}$ grown sufficiently large and negative. The concentration of the Ξ^0 is rather close to that of the Σ^+. It should be noted that, for the density range in question, none of the electrically charged states of the Δ particle becomes populated in the framework of the relativistic Hartree approximation, in sharp contrast to RHF and RBHF calculations as we shall see below.

The important net effect of each additional nuclear degree of freedom – each of which contributes to the total energy density, baryon number density and pressure of the stellar matter – is a *softening* of the associated equation of state, which has immediate consequences for observable properties of neutron stars. We shall turn back to this issue repeatedly in the second part of this volume.

As an interesting side remark, the significant fraction of Λ hyperons in neutron star matter makes one wonder whether the formation of H-dibaryons (or other exotic hyperonic configurations) may become possible in such an environment. Tamagaki pointed out that such systems may condense to form so-called H-matter [402]. Recently it was shown by Sakai *et al* [403] that improvements in the interaction between H-dibaryons lower the critical density at which formation of H-matter becomes energetically favorable below $6\,\epsilon_0$. The latter value was established in Tamagaki's original work [402]. Such densities are reached in stable neutron stars (cf. chapter 14). H-matter could thus exist in the cores of sufficiently massive neutron stars. If indeed formed, however, it may not remain dormant there forever because of its instability against compression. For that reason, H-matter possibly triggers the conversion of a neutron star into a strange star [403, 404, 405]. Another recent study on the formation of a possible H-dibaryon condensate in neutron stars was performed by Glendenning and Schaffner-Bielich [406]. They find that if the limiting neutron star mass is about that of the Hylse–Taylor pulsar PSR 1913+16, i.e. $M = 1.444\,M_\odot$, a condensate of H-dibaryons of vacuum mass $m_{\rm H} \sim 2.2$ GeV and a moderately attractive H-dibaryon potential in the medium of about $U_{\rm H} = -30$ MeV could not be ruled out. Conversely, if the medium potential were moderately repulsive, say around $U_{\rm H} \sim +30$ MeV, the H-dibaryon would not likely exist in neutron stars, since then the threshold density for the H-condensate, found to be around $\sim 5\,\epsilon_0$, may not be reached in neutron stars of mass $M = 1.444\,M_\odot$.

In the next step we compare the above population, calculated for relativistic Hartree HV, with the outcome of a relativistic HF calculation performed for parameter set HFV. The population is graphically illustrated in figure 7.24. In this case the charge-favored, but isospin-unfavored Σ^- hyperon becomes populated before the Λ. Its threshold density is $\rho_{\Sigma^-}^{\rm HFV} = 0.22$ fm^{-3}, which corresponds to a total energy density of about $\epsilon_{\Sigma^-}^{\rm HFV} = 210$ MeV fm^{-3}. The Λ, which is somewhat lighter than the Σ^- (cf. table 5.1), follows next at $\rho_\Lambda^{\rm HFV} = 0.32$ fm^{-3}. This is another quantitative example which underlines the qualitative arguments given above in connection with items (i) and (ii) that it is by no means the baryon's mass which determines the threshold but attributes like electric charge and orientation in isospin space.

At still higher densities, the HFV population looks radically different from the HV population, as can be seen by comparing figures 7.23 and 7.24 with each other. The striking difference arises from the presence of the Δ resonance state, which becomes populated at a density of $\rho_{\Delta^-}^{\rm HFV} = 0.35$ fm^{-3}. The presence of the Δ^- has its origin in the different modeling of the nuclear forces in both treatments. Firstly, the HFV parameter set comprises a somewhat larger class of mesons that are being exchanged among the baryons (see table 7.1). This leads to different density dependences of the self-energies $\mathbf{\Sigma}^B$ in both approximations. To be more precise, the HV parameter set describes the nuclear forces via the exchange of σ, ω, and ρ mesons only. In contrast to this, HFV accounts for π-meson exchange too. With the exception of the π, all mesons contribute to the direct (Hartree) terms of $\mathbf{\Sigma}^B$, while all four types of meson contribute to the exchange (Fock) terms. Thereby the large attractive and repulsive self-energies that are typical for the Hartree approximation[2] are somewhat reduced in the case of the HF treatment. This has an important impact on the threshold relation (7.68), which is such that the charge-favored Σ^- is no longer prevented from becoming populated by reasons of isospin orientation arguments. Similarly, the isospin of the Δ resonance is no longer a liability, and the most favored charge state of the Δ particle, that is, the Δ^-, becomes populated first. The only problem for the Δ^- in becoming populated is to overcome its large mass, which, however, is reduced by the surrounding nuclear environment, that is, mathematically by the large magnitude of $\Sigma_S^{\Delta^-}$. The Hartree part of $\Sigma_S^{\Delta^-}$ (which coincides with Σ_S^p in the case of universal coupling, which is assumed here) is shown in figure 7.20 (dashed line). One immediately sees that the effective mass $m^*_{\Delta^-}$, and thus the expression on the right-hand side of equation (7.68), is more reduced for the HFV Hartree–Fock calculation (dashed line in figure 7.20) than it is the case for the HV Hartree calculation (solid line in figure 7.20). As a consequence, the population of the Δ^- state is more favored for the HFV treatment [79]. This trend resembles the non-relativistic Schrödinger-based neutron star matter calculations performed by Pandharipande [407] and Bethe and Johnson [408].

Secondly, aside from the larger class of mesons that enters the HF description, there is another generic difference between Hartree and HF that concerns the coupling strength of the ρ meson, as becomes evident from table 7.1. It is well known that Fock terms are essential in obtaining the correct value for the symmetry energy coefficient, $a_{\rm sym}$, for which a value around 32 MeV is generally accepted [409]. The most recent value, calculated for the new Thomas–Fermi model of Myers and Swiatecki, is 32.65 MeV [409]. (Details of this model will be given in section 12.6.) There is therefore no need in HF calculations to use large values of $g_{\rho N}$, in contrast to Hartree calculations [99]. In the latter case, where the Fock term is absent by definition, large values of $g_{\rho N}$ are needed to obtain the right $a_{\rm sym}$ coefficient. For this reason the HV ρ-meson coupling constant is therefore about 13 times larger than for the HFV parametrization,

[2] This is specifically the case for Walecka's σ–ω model discussed in section 7.2, which accounts only for σ- and ω-meson exchange among the baryons.

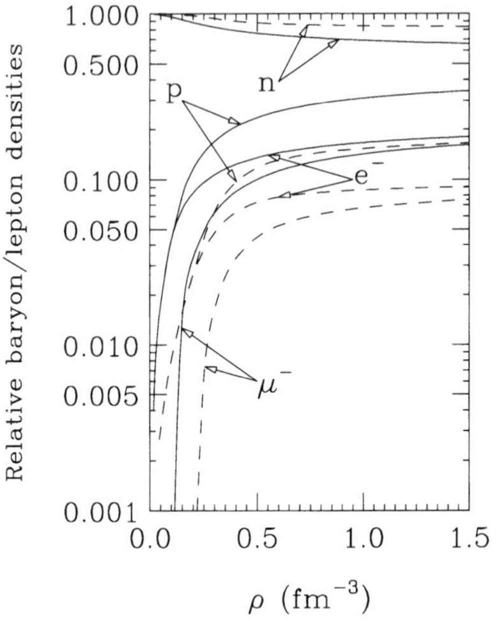

Figure 7.25. Same as figure 7.23, but for chemically equilibrated protons and neutrons only ($B = p, n$) [79]. Hyperons are artificially suppressed. The underlying parameter sets are HV (solid line) and HFV (dashed line) listed in table 7.1. (Reprinted courtesy of *Nucl. Phys.*)

that is, $(g_{\rho N}^{\rm HV})^2/(4\pi) = 7.51$ compared with $(g_{\rho N}^{\rm HFV})^2/(4\pi) = 0.55$ in the case of HFV. This trend is not a peculiarity of these two parameter sets. Hartree–Fock parametrizations adjusted to the many-body outcome of RHBF calculations, for instance, all have $g_{\rho N}^2/(4\pi) = 0.25$ to 0.35 (table 11.2). Somewhat larger values of $g_{\rho N}^2/(4\pi) \sim 1$ enter the relativistic OBE interactions, table 5.5, which are however still small compared to the HV case. The important implication of all this for astrophysical dense matter calculations is that the smaller ρ-meson coupling diminishes the Σ_{03}^B term in (7.68) favoring the population of the Δ^-, as established for HFV. One may speculate that RBHF calculations, being based on OBE interactions, should make a rather strong case for an early onset of Δ^- population too. In chapter 11 we shall see that Hartree–Fock calculations based on RBHF parametrizations indeed confirm this conjecture.

For the purpose of comparison, we display in figure 7.25 the particle population of chemically equilibrated matter where hyperon population has been arbitrarily *suppressed*. That is, the only types of baryon that are taken into account are protons and neutrons. The calculations are performed for the relativistic Hartree and HF parameter sets HV and HFV. As already pointed out above, such matter does *not* constitute the lowest-energy state of neutron star matter [61, 410]. In fact it constitutes a state of matter that is highly excited relative to fully equilibrated neutron star matter and would rapidly weakly transform into the latter. Another consequence is that, because of the suppression of degrees of freedom that would be accessible to the system, the equation of state associated with proton–neutron matter behaves more stiffly (that is, one obtains more pressure for a given density) than the equation of state of fully equilibrated neutron star matter.

Another interesting point, with far-reaching consequences for the possible formation of meson condensates, concerns the behavior of the lepton fraction. In sharp contrast to figures 7.23 and 7.24, where the full set of baryons of table 5.1 was allowed to contribute, the relative lepton densities ρ^e/ρ and ρ^μ/ρ are now monotonically increasing functions of density, *without* possessing an upper bound. This means that for increasing values of density the number of electrons and muons present in the system increases without bound, and consequently the number of protons of the system increases as well to maintain electric charge neutrality. The number of neutrons on the other hand decreases with density because they undergo β^--decay into protons, electrons, and muons. The steady increase of the number of leptons implies that their chemical potentials, μ^e and μ^μ, too are *monotonically increasing* functions. This can be seen graphically in figures 7.19 and 7.20.

7.9 Microphysics behind π^- and K^- condensation

As we have seen in section 7.8, more massive particles such as the hyperons Λ, Σ, etc as well as some of the Δ states are predicted by modern many-body calculations to be present in neutron star matter at supernuclear densities. How these particles interact with nucleons and among themselves is not too well understood yet, and so the consequences of their presence for the dense-matter equation of state can only be estimated from particular models of their interactions [411]. Aside from the numerous baryon states quoted just above, the meson fields that mediate the interactions between baryons may also become dynamical degrees of freedom with increasing density. If mesons appear in the matter in its ground state, they will, because they are bosons, macroscopically occupy the lowest available energy state (cf. figure 4.4), that is, form a meson condensate. A meson condensate is characterized by a macroscopic excitation of the meson field rather than a sea of mesons, in sharp contrast to the fermions (cf. figure 9.3). As described in section 4.2, the two types of condensate that may exist in neutron stars are those associated with the lowest-mass mesons of table 5.2, namely the π^- and the strangeness carrying K^- meson [303, 412].

Ignoring the interactions of pions with the medium, one would conclude

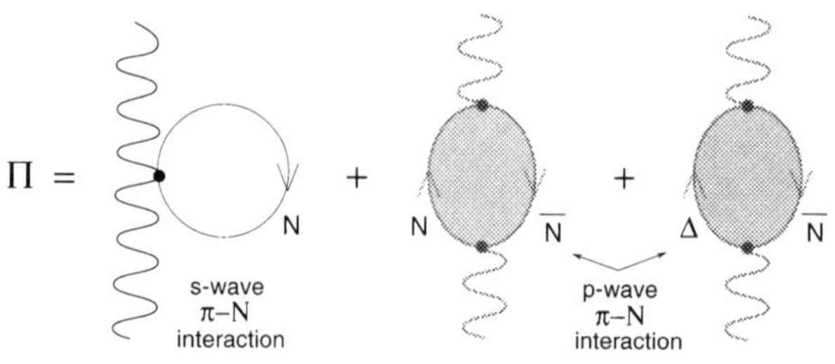

Figure 7.26. Processes contributing to pion self-energy, Π (cf. (7.64)). The coupling of the pion to intermediate nucleon particle-hole ($N\bar{N}$) states and Δ-particle nucleon hole ($\Delta\bar{N}$) states leads to p-wave ($\boldsymbol{k} \neq \boldsymbol{0}$) contributions. These are absent at the level of the s-wave approximation, in which case $\boldsymbol{k} = \boldsymbol{0}$.

that charged pions make their appearance once the electron chemical potential, given by the difference between neutron and proton chemical potentials, $\mu^e = \mu^n - \mu^p$, exceeds $m_\pi = 140$ MeV, for it is then energetically favorable (cf. figure 4.4) for a neutron at the top of the neutron Fermi sea to turn into a proton and a π^-, according to the reaction $n \rightarrow p + \pi^-$. As we know from section 7.7, in chemically equilibrated neutron star matter, $\mu^n - \mu^p$ attains a value in the range between about 100 MeV and 140 MeV at nuclear matter density, ϵ_0, with the individual value depending rather strongly on the underlying microscopic many-body method. The relativistic HF approximation, for instance, appears to favor smaller μ^e values than the relativistic Hartree approximation. For the bare pion mass of 140 MeV one might expect the π^- to condense at densities slightly higher than ϵ_0 [61]. However, as pointed out later in the literature, the coupling of the nucleon particle-hole states and Δ-particle nucleon hole states to the pionic degrees of freedom (cf. figure 7.26), which contribute to Π only at the level of the p-wave π–N ineraction, produce an important mixing of these states which leads to a somewhat different physical picture of the onset of condensation [303, 412].

In this text we shall not go into the complicated details concerning the condensation of mesons, but refer the interested reader to the literature instead. An overview of the theory of condensation and numerical calculations can be found, for instance, in [58, 303, 413, 414, 415, 416]. In particular we draw attention to the review of Baym [412] which we closely follow next. As outlined there, nuclear and neutron star matter has a *collective mode* with the quantum number of the charged pion. This mode constitutes an oscillation of the matter with spatially-varying nucleon spin ($S = 1$) and isospin ($T = 1$) which, at a certain critical density ρ_π, goes to zero frequency at a critical wave vector, k_c. This 'softening' of the collective mode causes the nucleon eigenstates to become rotated in isospin space. Instead of the states being pure neutron and proton, $|n\rangle$ and $|p\rangle$ respectively, the eigenstates spontaneously undergo a chiral SU(2)⊗SU(2) rotation to become linear superpositions of neutron and proton states. This superposition is of the form

$$|N'\rangle = \cos\theta\,|n\rangle + \sin\theta\,|p\rangle\,, \qquad |P'\rangle = \cos\theta\,|p\rangle - \sin\theta\,|n\rangle\,, \tag{7.70}$$

where the condensation angle θ grows from zero as the density increases above ρ_π. To conserve charge as the nucleon eigenstates are rotated, the system develops a macroscopic spatially varying p-wave pion field, i.e. the condensate, of net negative charge. Mathematically, the condensate is written as $\langle\pi\rangle \sim \exp(\mathrm{i}\boldsymbol{k}\cdot\boldsymbol{x})$, with $\boldsymbol{k}$ the condensation wave vector. Its magnitude, $|\boldsymbol{k}|$, begins as k_c at density ρ_π. An analogous neutral-pion condensed state can be formed through softening of the neutron particle-hole collective mode. The neutral condensed state is characterized by a spatially varying finite expectation value of the neutral-pion field [417]. Neutral and charged-pion condensates can in principle coexist.

Early estimates predicted the onset of charged-pion condensation at $\rho^\pi \sim 2\rho_0$. However, these estimates are very sensitive to the strength of the effective nucleon particle-hole repulsion in the $T = 1$, $S = 1$ channel, described by the Landau Fermi-liquid parameter g' [418], which tends to suppress the condensation mechanism. Measurements in nuclei tend to indicate that the repulsion is too strong to permit condensation [419, 420] in nuclear matter. Nevertheless, some authors like Tatsumi and Muto *et al* [421], and Umeda *et al* [422] argue to the contrary in the case of neutron star matter, so that the question of whether or not pions condense in neutron star matter may still be considered as an open issue. As we shall see in section 12.6 and chapters 14, 15, and 19, pion condensation would have two important effects on neutron stars. Firstly, it slightly softens the equation of state above the critical density ρ_π for onset of condensation, which, with respect to the properties of neutron stars constructed for such an equation of state, reduces the maximum stellar mass. At the same time, however, the central stellar density increases, because of the softening. Secondly, a pion condensate would lead to a neutrino luminosity of $\epsilon_\nu \sim T_9^6$ which is enhanced over that of the normal modified Urca process, for which $\epsilon_\nu \sim T_9^8$ (see table 19.4).

A further and very actively debated form of condensation involves the spontaneous formation of K mesons, as already mentioned in section 4.2. The underlying chiral SU(3)⊗SU(3) symmetry of the strong interaction, which is exact in the limit that the u, d, s-quark masses (table 18.1) vanish, implies that K mesons have an effective attractive interaction with nucleons of the form $\mathcal{H}_\mathrm{eff} \sim -\rho\bar{K}K$, where ρ is the baryon number density, and K the K-meson field. This interaction acts as a density-dependent term in the kaon effective mass which lowers the kaon energy in the matter

[303, 412]. As first noted by Kaplan and Nelson [372], the energy of a K^+ falls below $-\mu^e = \mu^p - \mu^n$, the chemical potential for a K^+, at a critical density of $\rho_K \sim 2.5 - 3\rho_0$. Above this density the system should form a kaon or strangeness condensate, with a macroscopic expectation value of the charged K field. In the case of kaon condensation the nucleons undergo a chiral SU(3)⊗SU(3) rotation, analogous to the rotation of the nucleon eigenstates in the case of pion condensation, equation (7.70), in which a neutron state becomes a linear superposition of a neutron and a Σ^-, mathematically expressed as

$$|N'\rangle = \cos\theta_K\,|n\rangle + \sin\theta_K\,|\Sigma^-\rangle\,, \tag{7.71}$$

while a proton state is rotated into a linear superposition of a proton, Σ^0, and a Λ. From the point of view of the underlying quark structure of the baryons ($n = udd$, $p = uud$, $\Sigma^- = dds$, $\Sigma^0 = uds$, $\Lambda = uds$), the u and s components are rotated into each other by the angle θ_K. The rotation leads to a non-zero field expectation value in matter, $\langle\bar{s}u\rangle$, with the quantum numbers of the K^+. The condensed state spontaneously breaks the chiral SU(3)⊗SU(3) symmetry. In addition, as pointed out in [412], there is also the possibility that the matter forms a η-meson condensate, a state with a non-vanishing expectation value, $\langle\bar{s}s\rangle$, with the quantum numbers of the η meson [412].

Similar to the π^- condensate, a K^- condensate would also soften the equation of state and enhance the neutrino luminosity [423, 424]. The softening of the equation of state can be quite substantial, as pointed out by Brown and collaborators, reducing the maximum neutron star mass from $\sim 2\,M_\odot$ to about $1.5\,M_\odot$ [255, 256]. Based on this, Brown and Bethe proposed the interesting possibility of the existence of a large number of low-mass black holes in the Galaxy [134]. The extent to which this is the case depends on the parameter values of the effective interaction of kaons and the effective strange-baryon nucleon-hole couplings in neutron star matter. Presently these parameters carry significant uncertainties which lead to uncertainties in the condensation angle expected at a given density and hence the magnitude of the effects of kaon condensation on neutron star properties, cautioning one from drawing final conclusions about the existence of meson condensates in compact stars at the present time.

In this text we shall restrict ourselves to mainly studying the impact of a π^- condensate on the global and thermal properties of neutron stars. Since the impact of a K^- condensate on the equation of state and neutrino luminosity is rather similar to a π^- condensate, the findings established for the latter carry over to the K^- case. Possible exceptions were pointed out by Thorsson *et al* [425], Brown [139, 254], and Prakash *et al* [43, 67].

Chapter 8

Quark–hadron phase transition

It is generally believed that the very early Universe was filled with fast-moving particles, including quarks. But about one-millionth of a second after the big bang, the temperature of the cosmos cooled sufficiently to allow the quarks to cross the so-called quark–hadron phase transition and combine into the familiar nuclear particles. Very recently, it has been demonstrated that this transition may be taking place in reverse in today's Universe – but this time inside pulsars [426]. The physical reason behind this is easy to understand. After their violent birth, pulsars gradually slow down, and the loss of their outward centrifugal force compresses their interiors. As we shall see in chapter 17, theory predicts that the interior pressure could become so great that the neutrons 'burst', releasing the quarks within them.

The possibility of quark deconfinement in the cores of neutron stars had already been suggested in the 1970s by several authors (see, for instance, [73, 74, 75, 76, 77, 427]). However until recently no observational signal has ever been proposed. This is so because whether or not the quark–hadron phase transition occurs in such stars makes only a small difference to their static properties, like the range of possible masses, radii, or even their limiting rotational periods. This, however, turns out to be strikingly different for the timing structure of pulsars that develop quark matter cores in the course of spin-down (spin-up). As we shall see in section 17.3.3, such pulsars can have braking indices in the range $-\infty < n < +\infty$, generically different from $n \simeq 3$ accessible to stars made up of confined hadronic matter.

Parenthetically, we point out that the above scenario is not to be confused with the possible QCD phase transition that may already happen when (proto) neutron stars are formed in supernova core collapse. In this event, a first-order phase transition actually may give rise to two shocks which quickly coalesce [428]. More importantly, it was shown that there may be significant differences in the evolution of cores with or without first- or second-order quark–hadron phase transitions which may eventually lead to observational signatures in the neutrino signal [428]. Still another issue concerns the formation of black holes from collapsing neutron stars, which is likely to be delayed if quarks are present [429].

Of course, at present one does not know from experiment at what density the expected phase transition to quark matter occurs, and one has no guide yet from lattice QCD simulations [430]. From simple geometrical considerations it follows that nuclei begin to touch each other at densities of $\sim (4\pi r_N^3/3)^{-1}$, which, for a characteristic nucleon radius of $r_N \sim 1$ fm, indicates this to happen at just a few times nuclear matter density [190]. Above this density one might speculate that the nuclear boundaries of particles like p, n, Σ^-, Λ, K^- will dissolve and quarks populate free states outside of baryons. The high-pressure environment that exists in the cores of neutron stars, with densities up to an order of magnitude higher than that of nuclear matter, constitute ideal sites where hadrons may transform into quark matter, forming a *permanent* component of matter inside such stars.

8.1 Conserved charges and internal forces

The arguments provided just above in favor of the existence of quark matter in neutron stars, plausibly coined hybrid stars in the literature [88], is in sharp contrast to what had been thought for the past 25 years. As pointed out by Glendenning [88], the reason being that in all earlier work on the quark–hadron phase transition in neutron star matter, assumed to be a first-order transition, a degree of freedom was frozen out which yielded a description of the transition as a constant pressure one. This had the explicit consequence of excluding the coexistence phase hadrons and quarks from neutron stars. The degree of freedom frozen out is the possibility of reaching the lowest energy state by rearranging electric charge between the regions of hadronic matter and quark matter in phase equilibrium. The microphysics behind this preference for charge rearrangement is the *charge-symmetric nuclear force* which acts to relieve the high isospin asymmetry of neutron star matter as soon as it is in equilibrium with quark matter. This introduces a net positive charge on the hadronic regions and a compensatory net negative charge on the quark matter. Because of this preference for charge rearrangement exploited by neutron star matter, the pressure in the mixed phase varies as the proportions of the phases and is *not a constant* in the coexistence phase! The situation is schematically illustrated in figures 8.1 and 8.2 which show qualitatively the inconsistent respectively consistent modelling of the quark–hadron phase transition in neutron star matter. Varying pressure in the mixed phase is of crucial importance for neutron stars, inside of which radial pressure can *never* be constant

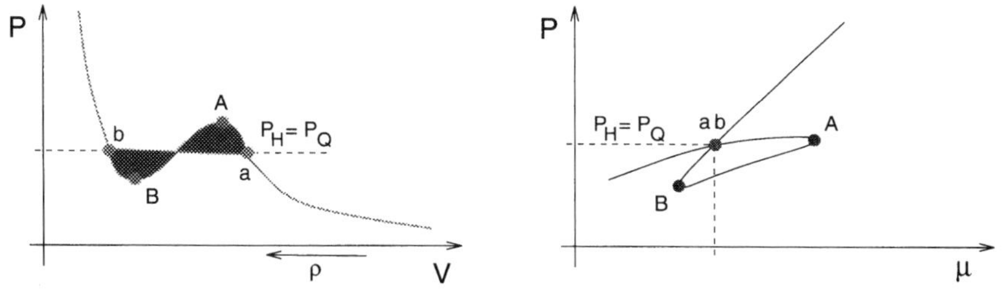

Figure 8.1. Schematic illustration of a phase transition in a *simple* substance (only one conserved entity) for which pressure stays constant in the transition region *a–b*. Left: volume dependence of pressure for a given temperature. Points *A* and *B* denote metastable states. Right: same as left-hand side, but for the chemical potential as independent variable. The labels refer to the same points as in the illustration to the left. Phase transition is mapped onto the point *ab* where the curves intersect.

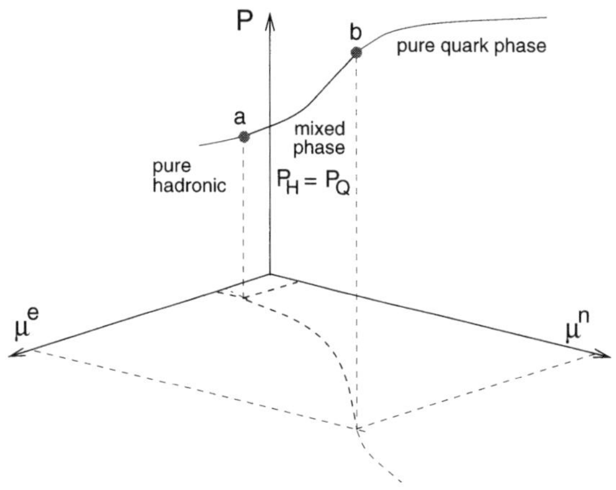

Figure 8.2. Schematic illustration of a phase transition in a body with two conserved entities, like neutron star matter. The entities are electric charge and baryon charge. In contrast to the phase transition in a body with only one conserved charge, shown in figure 8.1, now pressure in the mixed phase (region *a–b*) *varies* with density.

(cf. figure 18.21), no matter how small the radial regime, as dictated by the principle of hydrostatic equilibrium. For that reason, the mixed phase is strictly (and incorrectly) excluded from neutron stars constructed for equations of state of the mathematical form shown in figure 8.1, in sharp contrast to the correct modelling of the first-order phase transition, figure 8.2, in which case a significant fraction of the star's mass can be made up of deconfined quark matter.

If the dense interiors of neutron stars are converted to quark matter [149, 191, 431], it must be 3-flavor quark matter since it has lower energy than 2-flavor quark matter, and just as in the discussion of the hyperon content of neutron stars, strangeness is not conserved on macroscopic time scales which allows evolving neutron stars to convert confined hadronic matter to 3-flavor quark matter until equilibrium brings this process to an end. Many of the earlier investigations have treated the neutron star as pure in neutrons, and the quark phase as consisting of the equivalent number of u and d quarks [74, 75, 76, 77, 149, 191, 427, 431, 432, 433]. We have repeatedly stressed that pure neutron matter is not the ground state of a neutron star, nor is a mixture of $\rho^d = 2\,\rho^u$ (as required by electric charge neutrality) the ground state of quark matter! In fact the latter constitutes a highly excited state of quark matter, which will quickly weakly decay to an approximate equal mixture of u, d and s quarks. Several others have approximated the mixed phase as two components which are *separately* charge neutral. Both of these approximations are hiding the possible quark–hadron phase transition in neutron star matter [88].

8.2 Chemical thermodynamics

In the trivial case of a star made from pure neutron matter, electric charge neutrality is automatically satisfied and so its matter is characterized by a *single* chemical potential, namely that for the neutrons, μ^n. According to chemical thermodynamics, the Gibbs condition for phase equilibrium between neutron matter and quark matter at a given temperature would then read (cf. figure 8.1)

$$P_{\rm H}(\mu^n, \{\phi\}, T) = P_{\rm Q}(\mu^n, T)\,, \tag{8.1}$$

where the subscripts H and Q attached to the pressure symbol denote the confined hadronic and quark phase, respectively, and $\{\phi\}$ stands collectively for the field variables and Fermi momenta that characterize a solution to the equations of confined hadronic matter (cf. section 7.5) exclusive of the chemical potential [88]. As pointed out by Glendenning [66, 88], such a treatment, however, incorrectly describes neutron star matter, the reason being that the composition of such matter is governed by *two* conserved laws, one which conserves baryon charge and the other which conserves electric charge.[1] The Gibbs condition for phase equilibrium then is that the

[1] This principle is also inherent in the long-established equilibrium that exists between nuclei and nucleons (the so-called nuclear coexistence regime) below nuclear matter density [71, 434, 435, 436].

two chemical potentials μ^n and μ^e, which reflect baryon and electric charge conservation, and the pressure in the two phases be equal (cf. figure 8.2),

$$P_{\rm H}(\mu^n, \mu^e, \{\phi\}, T) = P_{\rm Q}(\mu^n, \mu^e, T) . \qquad (8.2)$$

As known from (4.18), the quark chemical potentials are related to the baryon and charge chemical potentials in (8.2) via the relations

$$\mu^u = \mu^c = \frac{1}{3}\,\mu^n - \frac{2}{3}\,\mu^e , \qquad \mu^d = \mu^s = \frac{1}{3}\,\mu^n + \frac{1}{3}\,\mu^e . \qquad (8.3)$$

Equation (8.2) is to be supplemented with the conditions of baryon charge conservation and electric charge conservation. It was again Glendenning [66, 88] who first realized the astrophysical importance of the fact that the underlying conservation laws are in general *global* laws rather than local ones. The reason being that the global nature constitutes a weaker condition on the arrangement of the microscopic degrees of freedom of a given system than the assumption of local conservation. Mathematically, the global conservation of baryon charge within an unknown volume, V, containing A baryons is expressed as

$$\rho \equiv \frac{A}{V} = (1-\chi)\,\rho_{\rm H}(\mu^n, \mu^e, T) + \chi\,\rho_{\rm Q}(\mu^n, \mu^e, T) , \qquad (8.4)$$

where $\chi \equiv V_{\rm Q}/V$ denotes the volume proportion of quark matter, $V_{\rm Q}$, in the unknown volume V [88]. By definition, χ varies between 0 and 1, depending on how much confined hadronic matter has been converted to quark matter. The global neutrality of electric charge within the volume V is mathematically expressed as [88]

$$0 = \frac{Q}{V} = (1-\chi)\,q_{\rm H}(\mu^n, \mu^e, T) + \chi\,q_{\rm Q}(\mu^n, \mu^e, T) + q_{\rm L} , \qquad (8.5)$$

where $q_{\rm H} = \sum_B q^{\rm el}_B \rho^B$ and $q_{\rm Q} = \sum_i q^{\rm el}_i \rho^i$ denote the net electric charges carried by hadronic (table 5.1) and quark (table 18.1) matter, and $q_{\rm L} = \sum_L q^{\rm el}_L \rho^L$ stands for the electric charge density of the leptons (table 6.1). Equation (8.5) follows from the integral over the charge density $q(r)$, given by

$$Q \equiv 4\pi \int_V {\rm d}r\, r^2\, q(r) , \qquad (8.6)$$

which, as required just above, must vanish rather than $q(r)$ itself [88]. If the nuclear substances are assumed to be uniform, which is well justified in any small locally inertial region V inside a star that develops quark deconfinement, the integral in (8.6) can readily be calculated. It then takes the simple form [66, 89, 90]

$$Q = (V - V_{\rm Q})\,q_{\rm H}(\mu^n, \mu^e) + V_{\rm Q}\,q_{\rm Q}(\mu^n, \mu^e) . \qquad (8.7)$$

Dividing this expression by V and introducing the definition for χ leads to the result given in (8.5).

For a given temperature T, equations (8.2) through (8.5) serve to determine the two independent chemical potentials μ^n and μ^e, and the volume V for a specified volume fraction χ of the quark phase in equilibrium with the hadronic phase. After completion, $V_{\rm Q}$ is obtained as $V_{\rm Q} = \chi V$. Through equations (8.2) to (8.5) the chemical potentials μ^n and μ^e obviously depend on the volume fraction χ and thus on density, which renders all properties that depend on μ^n and μ^e – from the energy densities to the baryon and charge densities of each phase to the common pressure – density dependent too.

Parenthetically we note that the conditions of global conservation expressed by (8.4) and (8.5) are compatible, together with (8.2), with the number of unknowns to be determined. It would not be possible to satisfy the Gibbs condition if local conservation were demanded, for that would replace (8.5) by two equations, such as $Q_{\rm H}/V_{\rm H} = q_{\rm H}(\mu^n, \mu^e, T)$ and $Q_{\rm Q}/V_{\rm Q} = q_{\rm Q}(\mu^n, \mu^e, T)$, and so the problem would be overdetermined [66, 89, 90].

Solving the models of confined and deconfined phases, in both pure phases and in the mixed phase, we can compute the baryon, lepton and quark populations in charge-neutral β-stable neutron star matter from

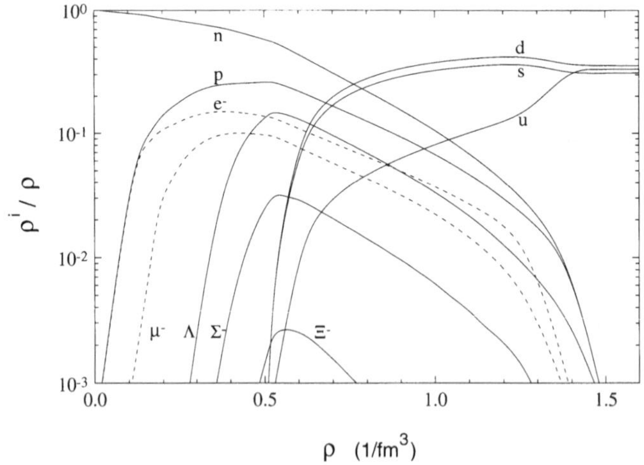

Figure 8.3. Composition of chemically equilibrated, stellar quark–hadron (hybrid star) matter as a function of baryon density. Hadronic matter is described by the relativistic Hartree theory HV; the bag constant is $B = 250$ MeV fm^{-3}. (Adapted from Hermann [437].)

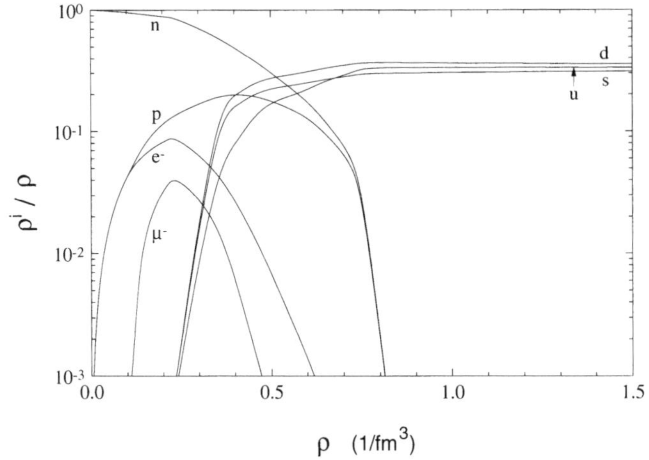

Figure 8.4. Composition of chemically equilibrated, stellar quark–hadron matter as a function of baryon density. Hyperons are artificially suppressed. Hadronic matter is described by the relativistic HF theory HFV; the bag constant is $B = 150$ MeV fm^{-3}. (Adapted from Hermann [437].)

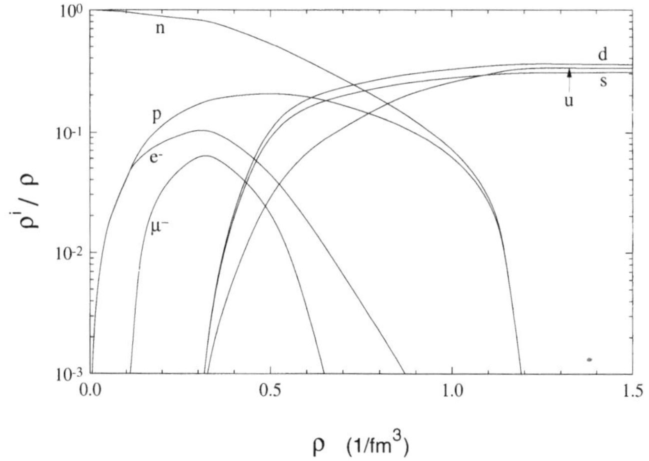

Figure 8.5. Same as figure 8.4, but for a bag constant of $B = 250$ MeV fm^{-3}. (Adapted from Hermann [437].)

equations (8.2) through (8.5). The outcome is shown in figures 8.3 to 8.5 for different representative bag constants as well as different many-body approximations employed to model confined hadronic matter (cf. table 12.4). The baryon density in the mixed phase region is defined in equation (8.4). Outside of the mixed phase region the baryon and quark densities have their standard definitions. Three features emerge immediately from these populations. Firstly, one sees that the transition from pure hadronic matter to the mixed phase occurs at rather a low density of about $3\,\rho_0$ or less [66, 88, 437]. Depending on parameters like the bag constant as well as the underlying nuclear many-body approximation, threshold values even as small as about $2\,\rho_0$ were found in the pioneering work of Glendenning [88, 89]. Secondly, we emphasize the saturation of the leptons as soon as quark matter appears. At this stage, charge neutrality is achieved more economically among the baryon-charge carrying particles themselves. Thirdly, the presence of quark matter enables the hadronic regions of the mixed phase to arrange to be more isospin symmetric (i.e. closer equality in proton and neutron number) than in the pure phase by transferring charge to the quark phase in equilibrium with it. Symmetry energy will be lowered thereby at only a small cost in rearranging the quark Fermi surfaces. Electrons play only a minor role when neutrality can be realized among the baryon-charge carrying particles. Thus the mixed phase region of the star will have positively charged regions of nuclear matter and negatively charged regions of quark matter.

8.3 Note on the structure of the quark–hadron phase

Because of the competition between the Coulomb and the surface energies associated with the positively charged regions of nuclear matter and negatively charged regions of quark matter, the mixed phase will develop geometrical structures, just as expected of the subnuclear liquid–gas phase transition [434, 435, 436]. This competition establishes the shapes, sizes and spacings of the rarer phase in the background of the other in order to minimize the lattice energy. A Coulomb lattice structure of varying geometry, a schematic illustration of which is shown in figure 8.6, may thus be introduced to the interior of neutron stars, as demonstrated quantitatively by Glendenning [66, 88, 438]. According to what has been outlined in the previous section, the formation of quark drops (q drops) will set in at around $3\,\epsilon_0$. At a somewhat greater density and therefore larger χ values the drops are more closely spaced and slightly larger in size. Still deeper in the star, the drops are no longer the energetically favored configuration but merge to form rods of varying diameter and spacing. At still greater depth, the rods grow together to slabs. Beyond this density the forms are repeated in reverse order until at the inner edge of the mixed phase hadronic drops (h drops) of finite size but separated from each other are immersed in quark matter. At densities around $10\,\epsilon_0$, the hadronic drops have completely dissolved into pure quark matter [66, 88]. In all

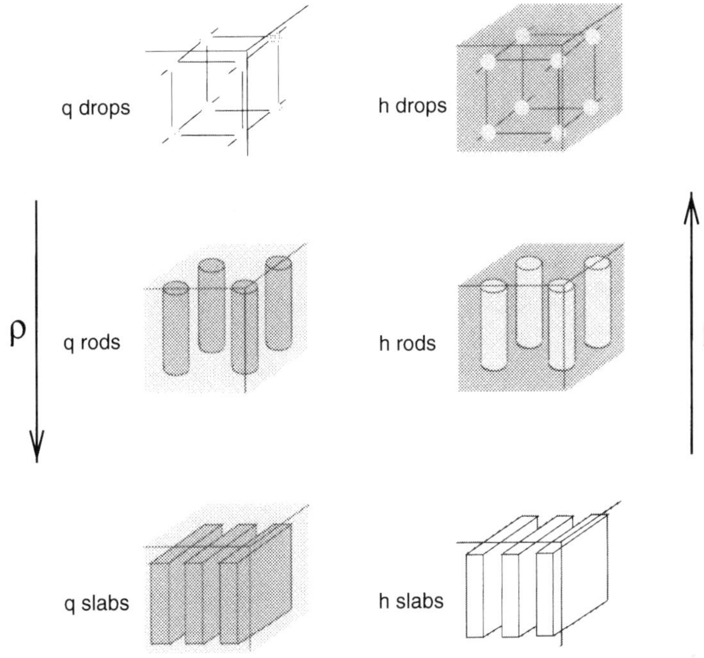

Figure 8.6. Coulomb lattice structure introduced in the interior of 'neutron' stars because of the competition of the Coulomb and surface energies, as demonstrated by Glendenning *et al* [88, 438]. The arrows point in the direction of increasing density.

cases the geometric forms lie between about 10 and 25 fm [66, 438]. The change in energy accompanied by developing such geometrical structures is likely to be very small in comparison with the volume energy [88, 163, 439] and, thus, cannot much affect the global properties of a neutron (i.e. hybrid) star. Nevertheless, the geometrical structure of the mixed phase may be very important for transport phenomena [66, 88].

8.4 Model for the equation of state

We conclude this chapter by presenting a representative model for the equation of state of hybrid-star matter. It is shown in figure 8.7 [437]. The hadronic phase is modelled, as in figure 8.3, by the relativistic Hartree approximation HV, the quark phase by the bag model with a bag constant of $B = 150$ MeV fm^{-3} and a strange quark mass of 150 MeV. The only difference with respect to figure 8.3 is the lower B value which lowers the onset of quark deconfinement from $3\,\rho_0$ to $2\,\rho_0$ [437]. Up to neutron

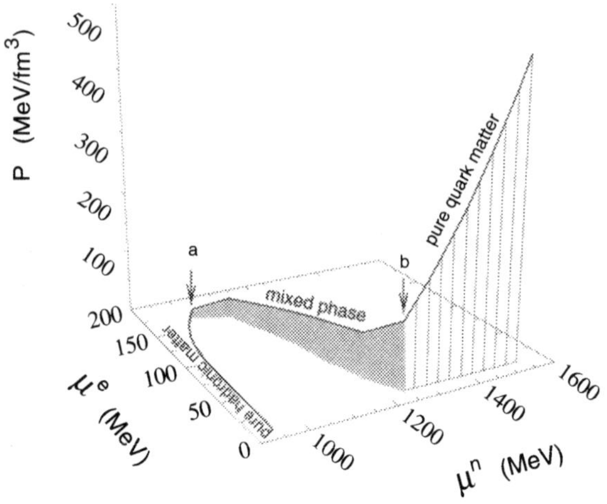

Figure 8.7. EOS of dense neutron star matter which accounts for the quark–hadron phase transition. (Adapted from Hermann [437].)

chemical potentials $\mu^n \sim 10^3$ MeV, the matter stays in the pure hadronic phase. The onset of quark deconfinement, which saturates the electron density (figure 8.3), occurs at point a in the diagram where μ^e attains its maximum, $\mu^e \simeq 180$ MeV. As remarked just above, this value corresponds to a baryon number density of about $2\,\rho_0$. Beyond this density, μ^e falls off toward rather small values because fewer and fewer electrons are present in dense hybrid star matter. The mixed phase region (a–b) exists, in the direction of increasing density, for electron chemical potential in the range 180 MeV$\gtrsim \mu^e \gtrsim$25 MeV, which corresponds to neutron chemical potential of 10^3 MeV$\lesssim \mu^n \lesssim 1.2 \times 10^3$ MeV. For this range, the volume proportion of quark matter varies in the range $0 \le \chi \le 1$. The energy density in the mixed phase is the same linear combination of the two phases as the charge and baryon number [66, 88], namely

$$\epsilon = (1-\chi)\,\epsilon_{\rm H}(\mu^n, \mu^e, \{\phi\}, T) + \chi\,\epsilon_{\rm Q}(\mu^n, \mu^e, T)\,. \tag{8.8}$$

Most importantly, as pointed out above qualitatively, the pressure in the mixed phase region *varies* with density rather than being constant, which would be the case if the conservation of electron charge were ignored (in which case the μ^e-axis is absent). The pure quark matter phase ($\chi = 1$) sets in at point b where the density has grown to about $6\,\rho_0$. It is characterized by a relatively steep increase of pressure with density. Whether or not this phase exists in neutron stars constructed for this equation of state depends on the star's central density and thus on its mass.

Chapter 9

Ladder approximation in the self-consistent baryon–antibaryon basis

9.1 Self-consistent baryon–antibaryon basis

In principle, the integral equation that determines the nucleon–nucleon **T**-matrix has 256 elements with respect to the spin indices which renders its numerical calculation extremely complicated. One is therefore interested in introducing a basis which *decouples* the integral equations for these elements, and makes the two-particle propagators $\mathbf{\Lambda}$ diagonal. This can be achieved by introducing a complete particle basis made up of baryon–antibaryon spinors $\Phi_l^B(k)$ and $\Theta_l^B(k)$ respectively [440], where the quantity $l = \pm 1/2$ stands for the helicity eigenvalues. We shall denote this helicity basis as $\{\Phi_l^B, \Theta_l^B\}$. For the baryon spinors we use their Dirac-to-Pauli reduced representation which is of the form [320]

$$\Phi_l^B(k) = \sqrt{\frac{\epsilon^B(k)}{2\, m_B^*(k)}} \begin{pmatrix} 1 \\ \frac{2\, l\, k_B^*(k)}{\epsilon^B(k)} \end{pmatrix} \otimes |\, l\, \rangle \,, \tag{9.1}$$

with $k \equiv (k_0, \boldsymbol{k}) = (\omega^B(\boldsymbol{k}), \boldsymbol{k})$. In accordance with the relativistic HF approximation, m_B^* and k_B^* are defined as

$$m_B^*(k) = m_B + \Sigma_S^B(k) \,, \qquad k_B^*(k) = |\boldsymbol{k}\,| + \Sigma_V^B(k) \,, \tag{9.2}$$

and

$$\epsilon^B(k) = W^B(k) + m_B^*(k) \,, \qquad W^B(k) = \sqrt{\left(m_B^*(k)\right)^2 + \left(k_B^*(k)\right)^2} \,. \tag{9.3}$$

This basis is generally referred to as self-consistent, since it is determined by the self-consistent quantities m_B^* and k_B^*. The quantity $|l\rangle$ in (9.1) denotes the helicity states, which are eigenstates of the helicity operator

$$\boldsymbol{s} \cdot \frac{\boldsymbol{k}}{|\boldsymbol{k}|} \equiv \frac{1}{2} \boldsymbol{\sigma} \cdot \hat{\boldsymbol{k}} , \tag{9.4}$$

that is,

$$\frac{1}{2} \boldsymbol{\sigma} \cdot \hat{\boldsymbol{k}} \, | \, l \, \rangle \, = l \, | \, l \, \rangle \, , \tag{9.5}$$

with eigenvalues $l = \pm\frac{1}{2}$. As usual, $\hat{\boldsymbol{k}}$ is the unit vector, i.e. $\hat{\boldsymbol{k}} = \boldsymbol{k}/|\boldsymbol{k}$. Because of (9.5), the baryon spinors obey the relation

$$\frac{1}{2} \boldsymbol{\sigma} \cdot \hat{\boldsymbol{k}} \, \Phi_l^B(k) = l \, \Phi_l^B(k) \, . \tag{9.6}$$

Finally we note that the helicity states $|l\rangle$ are connected with the Pauli spinors χ_l, given by

$$\chi_{+\frac{1}{2}} = \begin{pmatrix} 1 \\ 0 \end{pmatrix} , \qquad \chi_{-\frac{1}{2}} = \begin{pmatrix} 0 \\ 1 \end{pmatrix} , \tag{9.7}$$

via the rotation operator $\hat{D}(\hat{\boldsymbol{k}})$ [441] according to the relation

$$| \, l \, \rangle \; = \; \hat{D}(\hat{\boldsymbol{k}}) \cdot \chi_l \, . \tag{9.8}$$

The baryon spinors are given as solutions of

$$\left\{ \left(k_0 - \Sigma_0^B(k) \right) \gamma^0 \; - \; \boldsymbol{\gamma} \cdot \hat{\boldsymbol{k}} \, k^*(k) \; - \; m_B^*(k) \right\} \Phi_l^B(k) \; = \; 0 \, . \tag{9.9}$$

The antibaryon spinors are given by

$$\Theta_l^B(k) = \sqrt{\frac{\epsilon^B(k)}{2 \, m_B^*(k)}} \begin{pmatrix} \frac{2 \, l \, k_B^*(k)}{\epsilon^B(k)} \\ \\ 1 \end{pmatrix} \otimes | \, l \, \rangle \, , \tag{9.10}$$

which can be written as (see appendix A.2)

$$\Theta_l^B(k) = \gamma_5 \, \Phi_l^B(k) \, . \tag{9.11}$$

The spinors and antispinors obey the orthogonality relations [347]

$$\bar{\Phi}_l^B(k) \, \Phi_{l'}^{B'}(k) \; = \; \delta_{ll'} \, \delta_{BB'} \, , \qquad \bar{\Theta}_l^B(k) \, \Theta_{l'}^{B'}(k) \; = \; - \, \delta_{ll'} \, \delta_{BB'} \, , \tag{9.12}$$

$$\bar{\Theta}_l^B(k) \, \Phi_{l'}^{B'}(k) \; = \; \bar{\Phi}_l^B(k) \, \Theta_{l'}^{B'}(k) \; = \; 0 \, , \tag{9.13}$$

and form a complete basis,

$$\sum_{l} \Big\{ \bar{\Phi}_l^B(k;\zeta)\, \Phi_l^{B'}(k;\zeta') \;-\; \bar{\Theta}_l^B(k;\zeta)\, \Theta_l^{B'}(k;\zeta') \Big\} \;=\; \delta_{\zeta\zeta'}\, \delta_{BB'} \,. \tag{9.14}$$

To calculate the matrix elements of a given boson-exchange interaction, the spectral functions $\boldsymbol{\Xi}^B$ of (6.117), and the self-energies $\boldsymbol{\Sigma}^B$ of (6.83) in the self-consistent basis, knowledge of the following matrix elements is necessary:

$$\langle \Phi_l^B(k) | \gamma^0 | \Phi_{l'}^{B'}(k) \rangle = \delta_{ll'}\, \delta_{BB'}\, \frac{W^B(k)}{m_B^*(k)} \,, \tag{9.15}$$

$$\langle \Theta_l^B(k) | \gamma^0 | \Theta_{l'}^{B'}(k) \rangle = \delta_{ll'}\, \delta_{BB'}\, \frac{W^B(k)}{m_B^*(k)} \,, \tag{9.16}$$

$$\langle \Theta_l^B(k) | \gamma^0 | \Phi_{l'}^{B'}(k) \rangle \;=\; \langle \Phi_l^B(k) | \gamma^0 | \Theta_{l'}^{B'}(k) \rangle \;=\; \delta_{ll'}\, \delta_{BB'}\, \frac{2\, l\, k_B^*(k)}{m_B^*(k)} \,, \tag{9.17}$$

$$\langle \Phi_l^B(k) | \boldsymbol{\gamma} \cdot \hat{\boldsymbol{k}} | \Phi_{l'}^{B'}(k) \rangle = \delta_{ll'}\, \delta_{BB'}\, \frac{k_B^*(k)}{m_B^*(k)} \,, \tag{9.18}$$

$$\langle \Theta_l^B(k) | \boldsymbol{\gamma} \cdot \hat{\boldsymbol{k}} | \Theta_{l'}^{B'}(k) \rangle = \delta_{ll'}\, \delta_{BB'}\, \frac{k_B^*(k)}{m_B^*(k)} \,, \tag{9.19}$$

$$\langle \Theta_l^B(k) | \boldsymbol{\gamma} \cdot \hat{\boldsymbol{k}} | \Phi_{l'}^{B'}(k) \rangle \;=\; \langle \Phi_l^B(k) | \boldsymbol{\gamma} \cdot \hat{\boldsymbol{k}} | \Theta_{l'}^{B'}(k) \rangle \;=\; \delta_{ll'}\, \delta_{BB'}\, \frac{2\, l\, W^B(k)}{m_B^*(k)} \,. \tag{9.20}$$

9.2 Matrix elements of boson-exchange interactions

Using the baryon spinors of equations (9.1) and (9.10), it is a straightforward (but cumbersome) task to calculate the plane-wave helicity-state matrix elements of a given one-boson-exchange interaction. In the following we outline how this is accomplished for the case of *scalar* meson coupling (table 5.2). The results obtained for the other two coupling types, *pseudovector* and *vector*, are compiled in appendix E.

We begin with transforming the boson-exchange interactions (5.153) to (5.155) into momentum space. This leads to:

$$\big\langle \zeta_1\zeta_2 | \mathbf{V}_{\rm s}^{BB'}(q) | \zeta_3\zeta_4 \big\rangle = \big(\Gamma_{\rm s}^B\big)_{\zeta_1\zeta_3}\, \big(\Gamma_{\rm s}^{B'}\big)_{\zeta_2\zeta_4}\, \Delta_{\rm s}^0(q) \tag{9.21}$$

for the case of scalar meson coupling,

$$\big\langle \zeta_1\zeta_2 | \mathbf{V}_{\rm pv}^{BB'}(q) | \zeta_3\zeta_4 \big\rangle = \big(\Gamma_{\rm pv}^B\big)_{\zeta_1\zeta_3}\, \big(\Gamma_{\rm pv}^{B'}\big)_{\zeta_2\zeta_4}\, \Delta_{\rm pv}^0(q) \tag{9.22}$$

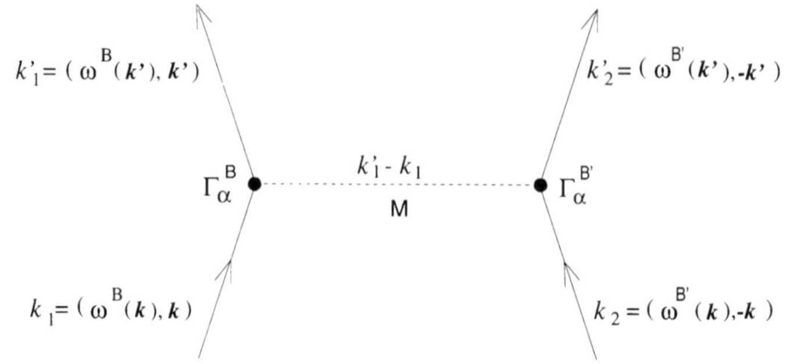

Figure 9.1. Two-baryon scattering in the center-of-mass system. k_1, k_2 and k_1', k_2' are the four-momenta of incoming and outgoing baryons, respectively. The dashed line represents the exchanged meson of type M carrying four-momentum $q \equiv k_1' - k_1$. Γ_α^B (where $\alpha = \mathrm{s, pv, v}$) denotes the baryon–meson vertices.

for pseudovector meson coupling, and to

$$\langle \zeta_1 \zeta_2 | \mathbf{V}_\mathrm{v}^{BB'}(q) | \zeta_3 \zeta_4 \rangle = \left(\Gamma_{\mathrm{v},\mu}^B\right)_{\zeta_1\zeta_3} \left(\Gamma_{\mathrm{v},\nu}^{B'}\right)_{\zeta_2\zeta_4} \left(\mathcal{D}_\mathrm{v}^0(q)\right)^{\mu\nu} \tag{9.23}$$

for the case of vector meson coupling. The momentum-space baryon–meson vertices Γ_α^B (where $\alpha = \mathrm{s, pv, v}$) in equations (9.21) to (9.23) are given by

$$\left(\Gamma_\mathrm{s}^B\right)_{\zeta_1\zeta_3} = \mathrm{i}\, g_{\mathrm{s}B} \, (\mathbf{1})_{\zeta_1\zeta_3} , \tag{9.24}$$

$$\left(\Gamma_\mathrm{pv}^B\right)_{\zeta_1\zeta_3} = -\,\mathrm{i}\, \frac{f_{\mathrm{pv}B}}{m_\mathrm{pv}} \left(\gamma_5\, \gamma_\mu\, q^\mu\right)_{\zeta_1\zeta_3} , \tag{9.25}$$

$$\left(\Gamma_{\mathrm{v},\mu}^B\right)_{\zeta_1\zeta_3} = \left(g_{\mathrm{v}B}\, \gamma_\mu + \frac{1}{2}\, \frac{f_{\mathrm{v}B}}{2m_B}\, q^\lambda \, [\gamma_\lambda, \gamma_\mu]\right)_{\zeta_1\zeta_3} , \tag{9.26}$$

respectively. Note that according to (5.13), $\sigma_{\lambda\mu} = \frac{\mathrm{i}}{2}[\gamma_\lambda, \gamma_\mu]$ which may be used to get rid of the commutator in (9.26). The quantity q stands for the four-momentum carried by the respective meson that is being exchanged between two incoming baryons. A geometrical illustration of such a scattering process is given in figure 9.1. The amplitudes' four-momenta k_1, k_2, k_1', k_2' are connected to their associated relative and center-of-mass four-momenta k, k', K and K' as (cf. (5.193))

$$k_1 = \frac{K}{2} + k\,, \quad k_2 = \frac{K}{2} - k\,, \quad k_1' = \frac{K'}{2} + k'\,, \quad k_2' = \frac{K'}{2} - k'\,. \tag{9.27}$$

Since the total four-momentum is a conserved quantity (see section 5.5.5), it follows from equation (9.27) that $k_1' - k_1 = k' - k \equiv q$, as indicated in figure 9.1. In the center-of-velocity frame, which is adopted here, one has

$K = K' = 0$ and thus from (9.27),

$$k_1 = k\,, \quad k_2 = -k\,, \quad k_1' = k'\,, \quad k_2' = -k'\,. \tag{9.28}$$

For the sake of completeness, we note that inversion of (9.27) leads to

$$k = \frac{k_1 - k_2}{2}\,, \quad k' = \frac{k_1' - k_2'}{2}\,, \quad K = k_1 + k_2\,, \quad K' = k_1' + k_2'\,. \tag{9.29}$$

A typical plane-wave boson-exchange matrix element, being of the form $\langle \Phi^{B_1'}_{l_1'}(k_1'), \Phi^{B_2'}_{l_2'}(k_2') | \mathbf{V}_\alpha | \Phi^{B_1}_{l_1}(k_1), \Phi^{B_2}_{l_2}(k_2)\rangle$ (cf. figure 9.1), may thus be written in terms of the center-of-velocity momenta (9.28) as

$$\begin{aligned} &\langle \Phi^{B_1'}_{l_1'}(k_1'), \Phi^{B_2'}_{l_2'}(k_2') | \mathbf{V}_\alpha | \Phi^{B_1}_{l_1}(k_1), \Phi^{B_2}_{l_2}(k_2)\rangle \\ &\qquad = \langle \Phi^{B_1'}_{l_1'}(k'), \Phi^{B_2'}_{l_2'}(k') | \mathbf{V}_\alpha | \Phi^{B_1}_{l_1}(k), \Phi^{B_2}_{l_2}(k)\rangle\,, \end{aligned} \tag{9.30}$$

which is conveniently abbreviated to

$$\langle l_1' l_2' k' | \mathbf{V}_\alpha | l_1 l_2 k \rangle \equiv \langle \Phi^{B_1'}_{l_1'}(k'), \Phi^{B_2'}_{l_2'}(k') | \mathbf{V}_\alpha | \Phi^{B_1}_{l_1}(k), \Phi^{B_2}_{l_2}(k)\rangle\,. \tag{9.31}$$

The quantity $\mathbf{V}_\alpha(k', k)$ in the relations above is given in equations (9.21) through (9.23). Aside from the boson-exchange matrix elements made up of only particle spinors, *mixed* baryon–antibaryon matrix elements enter into the treatment too. It is convenient to abbreviate the latter as follows,

$$\langle l_1' l_2' k' | \mathbf{U}_\alpha | l_1 l_2 k \rangle \equiv \langle \Theta^{B_1'}_{l_1'}(k'), \Phi^{B_2'}_{l_2'}(k') | \mathbf{V}_\alpha | \Phi^{B_1}_{l_1}(k), \Phi^{B_2}_{l_2}(k)\rangle\,, \tag{9.32}$$

$$\langle l_1' l_2' k' | \mathbf{W}_\alpha | l_1 l_2 k \rangle \equiv \langle \Theta^{B_1'}_{l_1'}(k'), \Phi^{B_2'}_{l_2'}(k') | \mathbf{V}_\alpha | \Phi^{B_1}_{l_1}(k), \Theta^{B_2}_{l_2}(k)\rangle\,, \tag{9.33}$$

$$\langle l_1' l_2' k' | \mathbf{Z}_\alpha | l_1 l_2 k \rangle \equiv \langle \Theta^{B_1'}_{l_1'}(k'), \Phi^{B_2'}_{l_2'}(k') | \mathbf{V}_\alpha | \Theta^{B_1}_{l_1}(k), \Phi^{B_2}_{l_2}(k)\rangle\,. \tag{9.34}$$

With these definitions at our disposal, let us calculate next several boson-exchange matrix elements explicitly. For this purpose, it is convenient to introduce the auxiliary function

$$S(k', k) \equiv \frac{1}{4}\sqrt{\frac{\epsilon'^{B_1'}}{m'^*_{B_1'}}\,\frac{\epsilon^{B_1}}{m^*_{B_1}}\,\frac{\epsilon'^{B_2'}}{m'^*_{B_2'}}\,\frac{\epsilon^{B_2}}{m^*_{B_2}}}\,, \tag{9.35}$$

where $(j = 1, 2)$

$$\begin{aligned} \epsilon^{B_j} &= m^*_{B_j}(k_j) + W^{B_j}(k_j)\,, \qquad & \epsilon'^{B_j'} &= m^*_{B_j'}(k_j') + W^{B_j'}(k_j')\,, \\ m^*_{B_j} &= m^*_{B_j}(k_j)\,, & m'^*_{B_j'} &= m^*_{B_j'}(k_j')\,. \end{aligned} \tag{9.36}$$

In the case of either symmetric nuclear matter or pure neutron matter, equation (9.35) reduces to

$$S(k',k) = \begin{cases} \dfrac{1}{4}\dfrac{\epsilon'^{N}\,\epsilon^{N}}{m_N'^{*}\,m_N^{*}} & \text{for symmetric nuclear matter } (B_j = B_j' = N)\,, \\[2ex] \dfrac{1}{4}\dfrac{\epsilon'^{n}\,\epsilon^{n}}{m_n'^{*}\,m_n^{*}} & \text{for pure neutron matter } (B_j' = B_j' = n)\,. \end{cases} \tag{9.37}$$

With these definitions, one arrives for the scalar meson coupling interaction, $\mathbf{V}_{\rm s}$, sandwiched between baryon (antibaryon) spinors, the graphical illustration of which is shown in figure 9.1, at the following results ($g_{\rm s} \equiv g_{{\rm s}B} = g_{{\rm s}B'}$),

$$\begin{aligned} \langle l_1' l_2' k' | \mathbf{V}_{\rm s} | l_1 l_2 k \rangle &= -\, g_{\rm s}^2\, \tilde{\Delta}_{\rm s}^0(k',k)\; S(k',k) \\ &\times \left(1 - \frac{4 l_1' l_1 k_{B_1'}'^{*} k_{B_1}^{*}}{\epsilon'^{B_1'} \epsilon^{B_1}} \right) \left(1 - \frac{4 l_2' l_2 k_{B_2'}'^{*} k_{B_2}^{*}}{\epsilon'^{B_2'} \epsilon^{B_2}} \right) \langle l_1'\, l_2' \,|\, 1 \,|\, l_1\, l_2 \rangle\,, \end{aligned} \tag{9.38}$$

$$\begin{aligned} \langle l_1' l_2' k' | \mathbf{U}_{\rm s} | l_1 l_2 k \rangle &= -\, g_{\rm s}^2\, \tilde{\Delta}_{\rm s}^0(k',k)\; S(k',k) \\ &\times \left(\frac{2 l_1' k_{B_1'}'^{*}}{\epsilon'^{B_1'}} - \frac{2 l_1 k_{B_1}^{*}}{\epsilon^{B_1}} \right) \left(1 - \frac{4 l_2' l_2 k_{B_2'}'^{*} k_{B_2}^{*}}{\epsilon'^{B_2'} \epsilon^{B_2}} \right) \langle l_1'\, l_2' \,|\, 1 \,|\, l_1\, l_2 \rangle\,, \end{aligned} \tag{9.39}$$

$$\begin{aligned} \langle l_1' l_2' k' | \mathbf{W}_{\rm s} | l_1 l_2 k \rangle &= g_{\rm s}^2\, \tilde{\Delta}_{\rm s}^0(k',k)\; S(k',k) \\ &\times \left(\frac{2 l_1' k_{B_1'}'^{*}}{\epsilon'^{B_1'}} - \frac{2 l_1 k_{B_1}^{*}}{\epsilon^{B_1}} \right) \left(\frac{2 l_2' k_{B_2'}'^{*}}{\epsilon'^{B_2'}} - \frac{2 l_2 k_{B_2}^{*}}{\epsilon^{B_2}} \right) \langle l_1'\, l_2' \,|\, 1 \,|\, l_1\, l_2 \rangle\,, \end{aligned} \tag{9.40}$$

and

$$\langle l_1' l_2' k' | \mathbf{Z}_{\rm s} | l_1 l_2 k \rangle = -\, \langle l_1' l_2' k' | \mathbf{V}_{\rm s} | l_1 l_2 k \rangle\,. \tag{9.41}$$

In accordance with (9.2) and (9.3), the following abbreviations were introduced in equations (9.38) through (9.41),

$$\begin{aligned} k_{B_j}^{*} &= k_{B_j}^{*}(k_j)\,, & k_{B_j'}'^{*} &= k_{B_j'}^{*}(k_j')\,, \\ W^{B_j} &= W^{B_j}(k_j)\,, & W'^{B_j'} &= W^{B_j'}(k_j')\,, \end{aligned} \tag{9.42}$$

with m_B^* and k_B^* defined in equation (9.2). As already mentioned, the helicity-state OBEP matrix elements that result for *pseudovector* and *vector* meson coupling are listed in appendix E. The helicity-state matrix elements form the basic quantities for the partial-wave decomposition of the OBEP, which will be the subject of chapter 10. Before we tackle this problem, however, we note that, upon carrying out a little spin algebra [320, 442], the spin-operator matrix elements $\langle l_1' \, l_2' \, |\mathbf{1}| \, l_1 \, l_2 \rangle$ and $\langle l_1' \, l_2' \, |\boldsymbol{\sigma}_1 \cdot \boldsymbol{\sigma}_2| \, l_1 \, l_2 \rangle$ in equations (9.38) to (9.41) and as listed in chapter 10 are given by

$$\begin{aligned} \langle l_1' \, l_2' \, | \, \mathbf{1} \, | \, l_1 \, l_2 \rangle = & \left[|l_1' + l_1| \cos\left(\frac{\vartheta}{2}\right) + (l_1' - l_1) \sin\left(\frac{\vartheta}{2}\right) \right] \\ \times & \left[|l_2' + l_2| \cos\left(\frac{\vartheta}{2}\right) - (l_2' - l_2) \sin\left(\frac{\vartheta}{2}\right) \right] \end{aligned} \tag{9.43}$$

and

$$\begin{aligned} \langle l_1' \, l_2' \, | \, \boldsymbol{\sigma}_1 \cdot \boldsymbol{\sigma}_2 \, | \, l_1 \, l_2 \rangle = & \left[(l_1' + l_1) \sin\left(\frac{\vartheta}{2}\right) + |l_1' - l_1| \cos\left(\frac{\vartheta}{2}\right) \right] \\ \times & \left[(l_2' + l_2) \sin\left(\frac{\vartheta}{2}\right) - |l_2' - l_2| \cos\left(\frac{\vartheta}{2}\right) \right] \\ - & \left[|l_1' + l_1| \sin\left(\frac{\vartheta}{2}\right) - (l_1' - l_1) \cos\left(\frac{\vartheta}{2}\right) \right] \\ \times & \left[|l_2' + l_2| \sin\left(\frac{\vartheta}{2}\right) + (l_2' - l_2) \cos\left(\frac{\vartheta}{2}\right) \right] \\ - & \left[(l_1' + l_1) \cos\left(\frac{\vartheta}{2}\right) - |l_1' - l_1| \sin\left(\frac{\vartheta}{2}\right) \right] \\ \times & \left[(l_2' + l_2) \cos\left(\frac{\vartheta}{2}\right) + |l_2' - l_2| \sin\left(\frac{\vartheta}{2}\right) \right] . \end{aligned} \tag{9.44}$$

9.3 Matrix elements of spectral functions

Having calculated the matrix elements of the γ matrices in the self-consistent basis $\{\Phi, \Theta\}$ in equations (9.15) through (9.20), it is straightforward to calculate the matrix elements of the particle spectral functions, Ξ^B, in this basis. One arrives at [117, 118, 440]

$$\langle \Phi_l^B(\omega^B(|\boldsymbol{k}|), \boldsymbol{k}) | \boldsymbol{\Xi}^B(\boldsymbol{k}) | \Phi_{l'}^B(\omega^B(|\boldsymbol{k}|), \boldsymbol{k}) \rangle = \delta_{ll'} \, n^B(|\boldsymbol{k}|) \, , \tag{9.45}$$

with

$$n^B(|\boldsymbol{k}|) = \frac{m_B^*}{\left| W^B - \left(m_B^* \, \frac{\partial \Sigma_S^B}{\partial \omega} + k_B^* \, \frac{\partial \Sigma_V^B}{\partial \omega} + W^B \, \frac{\partial \Sigma_0^B}{\partial \omega} \right) \right|} \Bigg|_{k_0 = \omega^B(|\boldsymbol{k}|)} \, , \tag{9.46}$$

$$\begin{aligned} & \langle \Phi_l^B(\omega^B(|\boldsymbol{k}|), \boldsymbol{k}) | \boldsymbol{\Xi}^B(\boldsymbol{k}) | \Theta_{l'}^B(\omega^B(|\boldsymbol{k}|), \boldsymbol{k}) \rangle \\ & \quad = \langle \Theta_l^B(\omega^B(|\boldsymbol{k}|), \boldsymbol{k}) | \boldsymbol{\Xi}^B(\boldsymbol{k}) | \Phi_{l'}^B(\omega^B(|\boldsymbol{k}|), \boldsymbol{k}) \rangle = \delta_{ll'} \, \mathcal{O}\Big(\frac{\partial \boldsymbol{\Sigma}}{\partial \omega}\Big) \, , \end{aligned} \tag{9.47}$$

$$\langle \Theta_l^B(\omega^B(|\boldsymbol{k}|), \boldsymbol{k}) | \boldsymbol{\Xi}^B(\boldsymbol{k}) | \Theta_{l'}^B(\omega^B(|\boldsymbol{k}|), \boldsymbol{k}) \rangle = \delta_{ll'} \, \mathcal{O}\Big(\frac{\partial \boldsymbol{\Sigma}}{\partial \omega}\Big) \, . \tag{9.48}$$

The matrix elements of the antiparticle spectral functions read

$$\langle \Phi_l^B(\bar{\omega}^B(|\boldsymbol{k}|), \boldsymbol{k}) | \bar{\Xi}^B(\boldsymbol{k}) | \Phi_{l'}^B(\bar{\omega}^B(|\boldsymbol{k}|), \boldsymbol{k}) \rangle = \delta_{ll'} \, \mathcal{O}\Big(\frac{\partial \boldsymbol{\Sigma}}{\partial \omega}\Big) , \tag{9.49}$$

$$\begin{aligned} &\langle \Phi_l^B(\bar{\omega}^B(|\boldsymbol{k}|), \boldsymbol{k}) | \bar{\Xi}^B(\boldsymbol{k}) | \Theta_{l'}^B(\bar{\omega}^B(|\boldsymbol{k}|), \boldsymbol{k}) \rangle \\ &\quad = \langle \Theta_l^B(\bar{\omega}^B(|\boldsymbol{k}|), \boldsymbol{k}) | \bar{\Xi}^B(\boldsymbol{k}) | \Phi_{l'}^B(\bar{\omega}^B(|\boldsymbol{k}|), \boldsymbol{k}) \rangle = \delta_{ll'} \, \mathcal{O}\Big(\frac{\partial \boldsymbol{\Sigma}}{\partial \omega}\Big) , \end{aligned} \tag{9.50}$$

and

$$\langle \Theta_l^B(\bar{\omega}^B(|\boldsymbol{k}|), \boldsymbol{k}) | \bar{\Xi}^B(\boldsymbol{k}) | \Theta_{l'}^B(\bar{\omega}^B(|\boldsymbol{k}|), \boldsymbol{k}) \rangle = \delta_{ll'} \, \bar{n}^B(|\boldsymbol{k}|) , \tag{9.51}$$

with

$$\bar{n}^B(|\boldsymbol{k}|) = \frac{m_B^*}{\Big| W^B - \big(m_B^* \frac{\partial \Sigma_S^B}{\partial \omega} + k_B^* \frac{\partial \Sigma_V^B}{\partial \omega} + W^B \frac{\partial \Sigma_0^B}{\partial \omega} \big) \Big|} \Bigg|_{k_0 = \bar{\omega}^B(|k|)} . \tag{9.52}$$

The above expressions are correct up to order $\mathcal{O}(\partial \boldsymbol{\Sigma} / \partial \omega)$ which, as can be inferred from figure 9.2, holds to a very good approximation [118, 123]. The functions n^B and $\bar{n}^B$ play the role of momentum distribution functions (occupation probabilities) of the single-particle (antiparticle) states, which follows from exploring their non-relativistic limits. According to the rules given in section 12.5, this limit results for $\Sigma_S^B \to 0$, $\Sigma_V^B \to 0$, and $\Sigma_0^B \to \Sigma^B$. This gives for n^B the well-known non-relativistic relation for the occupation probability of single-particle states [317]:

$$n^B(|\boldsymbol{k}|) = \frac{1}{\Big| 1 - \frac{\partial \boldsymbol{\Sigma}^B}{\partial \omega} \Big|} . \tag{9.53}$$

Note that because of $\partial \boldsymbol{\Sigma}^B / \partial \omega < 0$ [325], the momentum distribution function of (9.53) obeys $n^B < 1$. The same is the case for its relativistic counterpart n^B and the antiparticle contribution $\bar{n}$, derived in equations (9.46) and (9.52) [117, 118]. Parenthetically, we note further that all spectral matrix elements associated with antiparticle contributions vanish entirely in this approximation if retardation effects, that is, the energy dependence of $\boldsymbol{\Sigma}^B$, is neglected. This follows immediately from equations (9.45) through (9.52). With the exception of equations (9.47) to (9.50), the energy dependence of $\boldsymbol{\Sigma}^B$ shall be kept everywhere else in the $\mathbf{T}$-matrix approximation, however.

9.4 Propagators in the self-consistent basis

In this text we shall be dealing only with the simplest propagator, $\boldsymbol{\Lambda}^{00}$, of the ladder approximation as well as the physically more realistic BHF

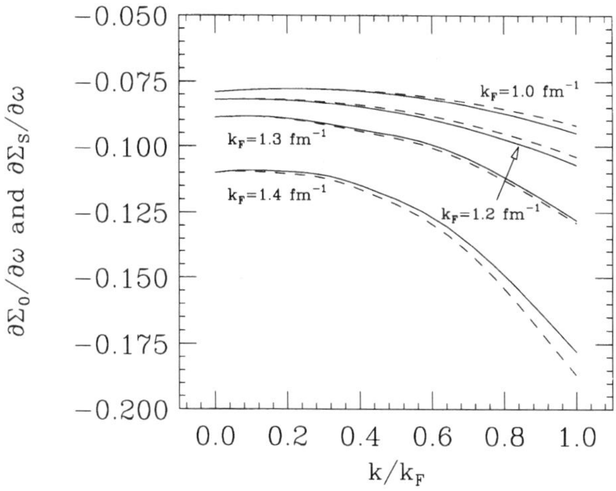

Figure 9.2. Partial derivatives of Σ_S^N (solid curves) and Σ_0^N (dashed curves) for symmetric nuclear matter treated in the relativistic $\mathbf{\Lambda}^{00}$ approximation. The underlying OBE interaction is HEA. (Data courtesy of Poschenrieder [440].)

propagator, $\mathbf{\Lambda}^{\mathrm{RBHF}}$. The former propagator was derived, in terms of the two-point Green function, in (5.200). These are now replaced with their spectral representations (6.71) and expressed in the self-consistent spinor–antispinor basis. Making use of the matrix elements (9.45) to (9.52), one gets

$$\mathbf{\Lambda}^{00}_{BB'}\Big(\frac{K}{2}+q,\frac{K}{2}-q\Big) = 2\pi\,\delta\Big(q^0-\omega^B\Big(\Big|\frac{\boldsymbol{K}}{2}+\boldsymbol{q}\Big|\Big)+\frac{K^0}{2}\Big) \\ \times\,\boldsymbol{\lambda}^{00}_{BB'}\Big(\Big|\frac{\boldsymbol{K}}{2}+\boldsymbol{q}\Big|,\Big|\frac{\boldsymbol{K}}{2}-\boldsymbol{q}\Big|,K^0\Big)\,, \tag{9.54}$$

with

$$\boldsymbol{\lambda}^{00}_{BB'}(\boldsymbol{k}_+,\boldsymbol{k}_-,K^0) \equiv \frac{n^0_B(\omega^{0B}(\boldsymbol{k}_+),\boldsymbol{k}_+)\;n^0_{B'}(K^0-\omega^{0B'}(\boldsymbol{k}_+),\boldsymbol{k}_-)}{K^0-\omega^{0B}(\boldsymbol{k}_+)-\omega^{0B'}(\boldsymbol{k}_-)+\mathrm{i}\eta} \tag{9.55}$$

and the definitions

$$\boldsymbol{k}_+ = \frac{1}{2}\boldsymbol{K}+\boldsymbol{q}\,,\qquad \boldsymbol{k}_- = \frac{1}{2}\boldsymbol{K}-\boldsymbol{q}\,. \tag{9.56}$$

The occupation probability, n^0, was defined in (9.46). Computing it for the spectral function of free baryons, $\mathbf{\Xi}^{0B}$, results in

$$n^0_B(k^0,\boldsymbol{k}) = \langle\Phi^B_l(k)|\mathbf{\Xi}^{0B}(\boldsymbol{k})|\Phi^B_l(k)\rangle\big|_{k^0=\omega^{0B}(k)}\,, \tag{9.57}$$

which can be written as

$$n_B^0(k^0, \boldsymbol{k}) = \Xi_S^B(\boldsymbol{k}) + \Xi_V^B(\boldsymbol{k})\, \frac{k_B^*(k^0, \boldsymbol{k})}{m_B^*(k^0, \boldsymbol{k})} + \Xi_0^B(\boldsymbol{k})\, \frac{W^B(k^0, \boldsymbol{k})}{m_B^*(k^0, \boldsymbol{k})} \,. \tag{9.58}$$

The $\boldsymbol{\Lambda}^{00}$ propagator is independent of the helicity eigenvalues and therefore can be evaluated for fixed values of l and l'. Moreover the regularization scheme, which will be introduced in section 5.5.3, restricts the propagator to intermediate positive-energy spinors only.

The Brueckner propagator is given by

$$\boldsymbol{\Lambda}_{BB'}^{\mathrm{RBHF}}(\boldsymbol{k}_+, \boldsymbol{k}_-, K^0) = 2\pi\, \frac{\Theta(|\boldsymbol{k}_-| - k_{F_B})\, \Theta(|\boldsymbol{k}_+| - k_{F_{B'}})}{K^0 - \omega^B(\boldsymbol{k}_+) - \omega^{B'}(\boldsymbol{k}_-)} \,, \tag{9.59}$$

with the single-particle energy $[\Sigma_i^B(\boldsymbol{k}_\pm) \equiv \Sigma_i^B(\omega^B(\boldsymbol{k}_\pm), \boldsymbol{k}_\pm)]$

$$\omega^B(\boldsymbol{k}_\pm) = \Sigma_0^B(\boldsymbol{k}_\pm) + \sqrt{(m_B + \Sigma_S^B(\boldsymbol{k}_\pm)^2 + (|\boldsymbol{k}_\pm| + \Sigma_V^B(\boldsymbol{k}_\pm))^2} \,. \tag{9.60}$$

The Pauli exclusion operator in (9.59), i.e.

$$Q(\boldsymbol{k}_+, \boldsymbol{k}_-) \equiv \Theta(|\boldsymbol{k}_+| - k_{F_B})\, \Theta(|\boldsymbol{k}_-| - k_{F_{B'}}) \,, \tag{9.61}$$

prevents scattering into occupied baryon states.

9.5 Matrix elements of baryon self-energies

The matrix elements of the baryon self-energies, Σ^B, expressed in the baryon basis $\{\Phi_l^B, \Theta_l^B\}$ are defined as

$$\Sigma_{\Phi\Phi}^B(k) \equiv \langle \Phi_l^B(k) | \boldsymbol{\Sigma}^B(k) | \Phi_l^B(k) \rangle \,, \tag{9.62}$$
$$\Sigma_{\Theta\Theta}^B(k) \equiv \langle \Theta_l^B(k) | \boldsymbol{\Sigma}^B(k) | \Theta_l^B(k) \rangle \,, \tag{9.63}$$
$$\Sigma_{\Theta\Phi}^B(k) \equiv \langle \Theta_{\frac{1}{2}}^B(k) | \boldsymbol{\Sigma}^B(k) | \Phi_{\frac{1}{2}}^B(k) \rangle \,. \tag{9.64}$$

Substituting equation (6.83) for $\boldsymbol{\Sigma}^B$ into (9.62) through (9.64) leads for the matrix elements of the self-energy in the self-consistent basis to

$$\Sigma_{\Phi\Phi}^B(k) = \Sigma_S^B(k) + \frac{k_B^*(k)}{m_B^*(k)}\, \Sigma_V^B(k) + \frac{\Omega^B(k)}{m_B^*(k)}\, \Sigma_0^B(k) \,, \tag{9.65}$$

$$\Sigma_{\Theta\Theta}^B(k) = -\, \Sigma_S^B(k) + \frac{k_B^*(k)}{m_B^*(k)}\, \Sigma_V^B(k) + \frac{\Omega^B(k)}{m_B^*(k)}\, \Sigma_0^B(k) \,, \tag{9.66}$$

$$\Sigma_{\Theta\Phi}^B(k) = \frac{\Omega^B(k)}{m_B^*(k)}\, \Sigma_V^B(k) + \frac{k_B^*(k)}{m_B^*(k)}\, \Sigma_0^B(k) \,. \tag{9.67}$$

Conversely, equations (9.65) through (9.67) can be solved for Σ_S^B, Σ_V^B, and Σ_0^B. One obtains immediately

$$\Sigma_S^B(k) = \frac{1}{2}\, \{\Sigma_{\Phi\Phi}^B(k) - \Sigma_{\Theta\Theta}^B(k)\} \,. \tag{9.68}$$

Adding m_B to both sides of (9.68) and making use of the definition of the effective baryon mass in matter, given in (9.2), leads to

$$m_B^* = m_B + \frac{1}{2}\left\{\Sigma^B_{\Phi\Phi}(k) - \Sigma^B_{\Theta\Theta}(k)\right\}. \tag{9.69}$$

To solve the above equations for Σ^B_V andΣ^B_0, it is convenient to introduce

$$\mathcal{W}^B(k) = \frac{\mathcal{Q}^B(k)\,k^*_B(k) - |\boldsymbol{k}|}{\mathcal{Z}^B(k)}, \tag{9.70}$$

with the definitions:

$$\mathcal{Q}^B(k) \equiv 1 + \frac{\Sigma^B_{\Phi\Phi}(k) + \Sigma^B_{\Theta\Theta}(k)}{2\,m^*_B(k)}, \qquad \mathcal{Z}^B(k) \equiv \frac{\Sigma^B_{\Theta\Phi}(k)}{m^*_B(k)}, \tag{9.71}$$

and

$$k^*_B(k) = \frac{\mathcal{Q}^B(k)\,|\boldsymbol{k}| \pm \mathcal{Z}^B(k)\sqrt{\boldsymbol{k}^2 + \left\{\left(\mathcal{Q}^B(k)\right)^2 - \left(\mathcal{Z}^B(k)\right)^2\right\}\left(m^*_B(k)\right)^2}}{\left(\mathcal{Q}^B(k)\right)^2 - \left(\mathcal{Z}^B(k)\right)^2}. \tag{9.72}$$

The vector and timelike self-energies are then given by

$$\Sigma^B_V(k) = k^*_B(k) - |\boldsymbol{k}|, \tag{9.73}$$

and

$$\Sigma^B_0(k) = \frac{1}{2}\frac{\mathcal{W}^B(k)}{m^*_B(k)}\left\{\boldsymbol{\Sigma}^B_{\Phi\Phi}(k) + \boldsymbol{\Sigma}^B_{\Theta\Theta}(k)\right\} - \frac{k^*_B(k)}{m^*_B(k)}\,\boldsymbol{\Sigma}^B_{\Theta\Phi}(k). \tag{9.74}$$

Let us proceed next to the calculation of the matrix elements of $\boldsymbol{\Sigma}$. The general expression of this quantity was derived in (5.214). Writing it out for the case of a baryon of type B that propagates in the nuclear background medium created by the baryons B' leads for $\boldsymbol{\Sigma}$ to

$$\tilde{\boldsymbol{\Sigma}}^B_{\zeta_1\zeta_2}(z,\boldsymbol{k}) = \sum_{B'}\sum_{\zeta_3\zeta_4}\int\frac{\mathrm{d}^3\boldsymbol{k}'}{(2\pi)^3}\int \mathrm{d}\omega\Bigg\{\left\langle \zeta_1\,\zeta_3\,\frac{k-k'}{2}\right|\tilde{\mathbf{T}}^{BB'}_{k+k'}\left|\frac{k-k'}{2}\,\zeta_2\,\zeta_4\right\rangle$$
$$- \left\langle \zeta_1\,\zeta_3\,\frac{k-k'}{2}\right|\tilde{\mathbf{T}}^{BB'}_{k+k'}\left|\frac{k'-k}{2}\,\zeta_4\,\zeta_2\right\rangle\Bigg\}\,\boldsymbol{\Xi}^{B'}_{\zeta_4\zeta_3}(\omega,\boldsymbol{k}')\,\Theta(-\omega), \tag{9.75}$$

where k and k' denote four-momenta whose energy components are $k_0 = z$ and $k'_0 = \omega + \mu$. Because of the δ-functions contained in $\boldsymbol{\Xi}^B(\omega,\boldsymbol{k})$ (cf. equation (6.142)), the integral over ω in (9.75) can be readily performed, which will be done next. Abbreviating the zero-temperature Fermi–Dirac distribution function as $\Theta^B(\boldsymbol{k}) \equiv \Theta(k_{F_B} - |\boldsymbol{k}|)$, one obtains from

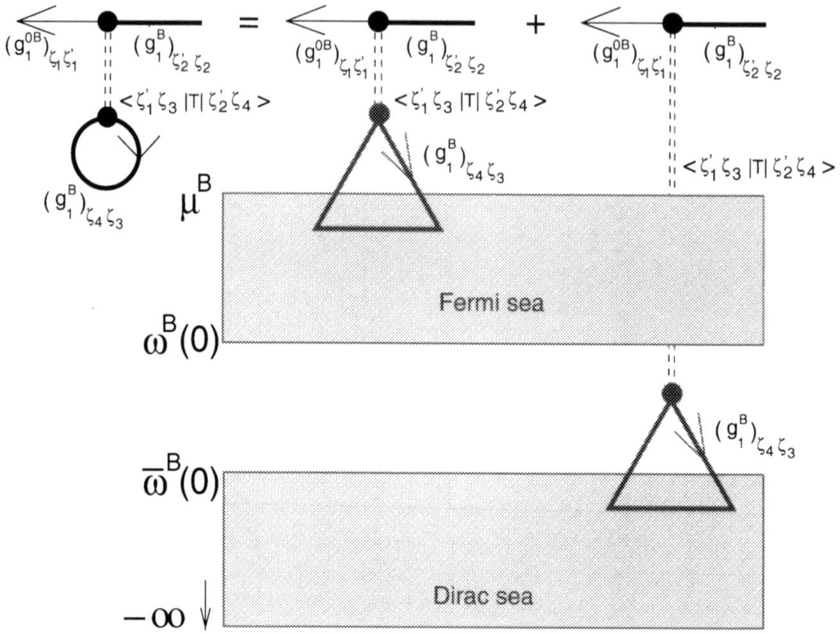

Figure 9.3. Graphical representation of $\boldsymbol{g}^{0B}_{1\,\zeta_1\zeta_1'}\boldsymbol{\Sigma}^B_{\zeta_1'\zeta_2'}\boldsymbol{g}^B_{1\,\zeta_2'\zeta_2}$ term of Dyson's equation (5.126). The Fermi sea of each baryon is filled up to the highest single-particle energy state, μ^B. The infinite Dirac sea is filled with antibaryons, which too modify the motion of the baryons. The Fermi-sea and Dirac-sea graphs are referred to as medium and vacuum polarization contributions, respectively. The no-sea approximation, which is adopted here, ignores the latter.

equation (9.75) the expression

$$\boldsymbol{\Sigma}^B_{\zeta_1\zeta_2}(k) = \sum_{B'}\sum_{\zeta_3\zeta_4}\int \frac{d^3\boldsymbol{k}'}{(2\pi)^3}\Bigg\{\left\langle \zeta_1\,\zeta_3\,\frac{k-k'}{2}\right|\mathsf{T}^{BB'}_{k+k'}\left|\frac{k-k'}{2}\,\zeta_2\,\zeta_4\right\rangle - \left\langle \zeta_1\,\zeta_3\,\frac{k-k'}{2}\right|\mathsf{T}^{BB'}_{k+k'}\left|\frac{k'-k}{2}\,\zeta_4\,\zeta_2\right\rangle\Bigg\}_{k^0=\omega^{B'}(k')} \boldsymbol{\Xi}^{B'}_{\zeta_4\zeta_3}(\boldsymbol{k}')\,\Theta^{B'}(\boldsymbol{k}')\,. \quad (9.76)$$

Both the Fermi-sea and the Dirac-sea contributions to the direct (Hartree) term of $\boldsymbol{\Sigma}^B$ are graphically depicted in figure 9.3. There the solid double lines represent the full two-point propagators $\boldsymbol{g}^B$, solid single lines the free ones $\boldsymbol{g}^{0,B}$. Nuclear matter can be viewed as a finite-density background system in which a particular baryon of type B propagates (upper graph of figure 9.3). The knowledge of $\boldsymbol{\Sigma}^B$ is necessary to determine $\boldsymbol{g}^B$. This self-consistency condition is mathematically imposed by Dyson's equation (5.126).

The Fermi-sea contributions to $\mathbf{\Sigma}^B$ are finite, while the terms that arise from the Dirac sea of antiparticles are not [92, 345, 346]. It is well known that in a relativistic treatment, interactions polarize the filled Dirac sea of negative-energy states, and this introduces virtual baryon–antibaryon intermediate states into the computation (e.g. self-energies, energy density, pressure). The no-sea approximation, which is used here, neglects these contributions by omitting them from the calculation of the self-energy and energy density. An investigation of negative-energy contributions in the present formalism has been performed in [122]. Though the computation of the effects of negative-energy states on the RBHF results is well defined in a renormalizable quantum hadron dynamic model, it is technically quite complicated to compute them. It requires the development of new techniques and introduces new uncertainties that are unlike any that have been studied in the non-relativistic problem. In view of this the authors found it preferable to estimate this new contribution before undertaking the considerable effort required by a detailed calculation. The way followed in [122] utilizes a simple approximation that should provide an orientation to the size of these effects and indicate whether a detailed calculation is necessary. To be more specific, (a) the relativistic Bethe–Brueckner–Goldstone (RBBG) formalism has been used to evaluate the interactions between positive-energy baryons; (b) it was assumed that the self-energy arising from the modified Dirac sea at finite density can be calculated in the mean-field approximation, which is accompanied by replacing the ladder-approximated $\mathbf{T}$-matrix with the direct one-boson-exchange term (Born-approximated $\mathbf{T}$-matrix), that is, $\mathbf{T} \to \mathbf{V}$.

A critical discussion of the influence of vacuum renormalization of relativistic nuclear field theory for both nuclear as well as neutron matter has been performed in reference [355]. It was found that these have negligible influence on the equation of state up to densities of at least ten times normal nuclear matter density, provided the baryon–meson coupling constants are tightly constrained by the saturation properties of nuclear matter.

After these remarks, let us turn back to (9.76) and extract from this relation the self-energy expression for the case of asymmetric nuclear matter. For this purpose we recall that ζ_j denotes the spin and isospin quantum numbers, i.e. $\zeta_j = \alpha_j, i_j$, of particle 'j'. Hence, by means of assigning $i = +\frac{1}{2}$ to the proton and $i = -\frac{1}{2}$ to the neutron, we obtain from (9.76) the expressions

$$\mathbf{\Sigma}^p_{\alpha_1\alpha_2}(k_{F_p}, k_{F_n}, k) = \int \frac{\mathrm{d}^3\boldsymbol{k}'}{(2\pi)^3} \Bigg\{ \Bigg[\Bigg\langle \alpha_1\frac{1}{2}, \alpha_3\frac{1}{2}, \frac{k-k'}{2} \Bigg| \mathbf{T}_K \Bigg| \frac{k-k'}{2}, \alpha_2\frac{1}{2}, \alpha_4\frac{1}{2} \Bigg\rangle - \Bigg\langle \alpha_1\frac{1}{2}, \alpha_3\frac{1}{2}, \frac{k-k'}{2} \Bigg| \mathbf{T}_K \Bigg| \frac{k'-k}{2}, \alpha_4\frac{1}{2}, \alpha_2\frac{1}{2} \Bigg\rangle \Bigg]_{k^0=\omega^p(k')} \mathbf{\Xi}^p_{\alpha_4\alpha_3}(\boldsymbol{k}')\, \Theta^p(\boldsymbol{k}')$$

$$+ \left[\left\langle \alpha_1 \frac{1}{2}, \alpha_3 - \frac{1}{2}, \frac{k-k'}{2} \middle| \mathbf{T}_K \middle| \frac{k-k'}{2}, \alpha_2 \frac{1}{2}, \alpha_4 - \frac{1}{2} \right\rangle \right.$$
$$\left. - \left\langle \alpha_1 \frac{1}{2}, \alpha_3 - \frac{1}{2}, \frac{k-k'}{2} \middle| \mathbf{T}_K \middle| \frac{k'-k}{2}, \alpha_4 - \frac{1}{2}, \alpha_2 \frac{1}{2} \right\rangle \right]_{k^0=\omega^n(k')}$$
$$\times \, \Xi^n_{\alpha_4\alpha_3}(\boldsymbol{k}') \, \Theta^n(\boldsymbol{k}') \Bigg\}, \tag{9.77}$$

$$\Sigma^n_{\alpha_1\alpha_2}(k_{F_p}, k_{F_n}, k)$$
$$= \int \frac{\mathrm{d}^3\boldsymbol{k}'}{(2\pi)^3} \Bigg\{ \left[\left\langle \alpha_1 - \frac{1}{2}, \alpha_3 - \frac{1}{2}, \frac{k-k'}{2} \middle| \mathbf{T}_K \middle| \frac{k-k'}{2}, \alpha_2 - \frac{1}{2}, \alpha_4 - \frac{1}{2} \right\rangle \right.$$
$$\left. - \left\langle \alpha_1 - \frac{1}{2}, \alpha_3 - \frac{1}{2}, \frac{k-k'}{2} \middle| \mathbf{T}_K \middle| \frac{k'-k}{2}, \alpha_4 - \frac{1}{2}, \alpha_2 - \frac{1}{2} \right\rangle \right]_{k^0=\omega^n(k')}$$
$$\times \, \Xi^n_{\alpha_4\alpha_3}(\boldsymbol{k}') \, \Theta^n(\boldsymbol{k}')$$
$$+ \left[\left\langle \alpha_1 - \frac{1}{2}, \alpha_3 \frac{1}{2}, \frac{k-k'}{2} \middle| \mathbf{T}_K \middle| \frac{k-k'}{2}, \alpha_2 - \frac{1}{2}, \alpha_4 \frac{1}{2} \right\rangle \right.$$
$$\left. - \left\langle \alpha_1 - \frac{1}{2}, \alpha_3 \frac{1}{2}, \frac{k-k'}{2} \middle| \mathbf{T}_K \middle| \frac{k'-k}{2}, \alpha_4 \frac{1}{2}, \alpha_2 - \frac{1}{2} \right\rangle \right]_{k^0=\omega^p(k')}$$
$$\times \, \Xi^p_{\alpha_4\alpha_3}(\boldsymbol{k}') \, \Theta^p(\boldsymbol{k}') \Bigg\}. \tag{9.78}$$

One sees explicitly that $\boldsymbol{\Sigma}^p$ and $\boldsymbol{\Sigma}^p$ are coupled with each other via the Fermi momenta of the protons and neutrons, k_{F_p} and k_{F_n}. Because of the assumed asymmetry of the matter one has $k_{F_p} \neq k_{F_n}$.

The matrix elements of $\boldsymbol{\Sigma}^B$, defined in equations (9.62) through (9.64), follow from (9.76) to (9.78) in the form

$$\Sigma^B_{\Phi\Phi}(k) = \frac{1}{2} \sum_{B'} \int \frac{\mathrm{d}^3\boldsymbol{k}'}{(2\pi)^3} \Big\{ \langle \Phi^B_l(k), \Phi^{B'}_{l'}(k') | \mathbf{T} | \Phi^B_l(k), \Phi^{B'}_{l'}(k') \rangle$$
$$- \langle \Phi^B_l(k), \Phi^{B'}_{l'}(k') | \mathbf{T} | \Phi^{B'}_{l'}(k'), \Phi^B_l(k) \rangle \Big\} \, n^B(|\boldsymbol{k}'|) \, \Theta^{B'}(\boldsymbol{k}'), \tag{9.79}$$

$$\Sigma^B_{\Theta\Theta}(k) = \frac{1}{2} \sum_{B'} \int \frac{\mathrm{d}^3\boldsymbol{k}'}{(2\pi)^3} \Big\{ \langle \Theta^B_l(k), \Phi^{B'}_{l'}(k') | \mathbf{T} | \Theta^B_l(k), \Phi^{B'}_{l'}(k') \rangle$$
$$- \langle \Theta^B_l(k), \Phi^{B'}_{l'}(k') | \mathbf{T} | \Phi^{B'}_{l'}(k'), \Theta^B_l(k) \rangle \Big\} \, n^{B'}(|\boldsymbol{k}'|) \, \Theta^{B'}(\boldsymbol{k}'), \tag{9.80}$$

$$\Sigma^B_{\Theta\Phi}(k) = \frac{1}{2} \sum_{B'} \int \frac{\mathrm{d}^3\boldsymbol{k}'}{(2\pi)^3} \Big\{ \langle \Theta^B_{\frac{1}{2}}(k), \Phi^{B'}_{l'}(k') | \mathbf{T} | \Phi^B_{\frac{1}{2}}(k), \Phi^{B'}_{l'}(k') \rangle$$
$$- \langle \Theta^B_{\frac{1}{2}}(k), \Phi^{B'}_{l'}(k') | \mathbf{T} | \Phi^{B'}_{l'}(k'), \Phi^B_{\frac{1}{2}}(k) \rangle \Big\} \, n^{B'}(|\boldsymbol{k}'|) \, \Theta^{B'}(\boldsymbol{k}'). \tag{9.81}$$

Here we have made use of the completeness relation (9.14) and the quantity n^B defined in equation (9.46).

$$\langle \Phi_l(k), \Phi_{l'}(k') \, | \, \mathsf{T} \, | \, \Phi_l(k), \Phi_{l'}(k') \rangle \qquad\qquad \langle \Theta_l(k), \Phi_{l'}(k') \, | \, \mathsf{T} \, | \, \Theta_l(k), \Phi_{l'}(k') \rangle$$

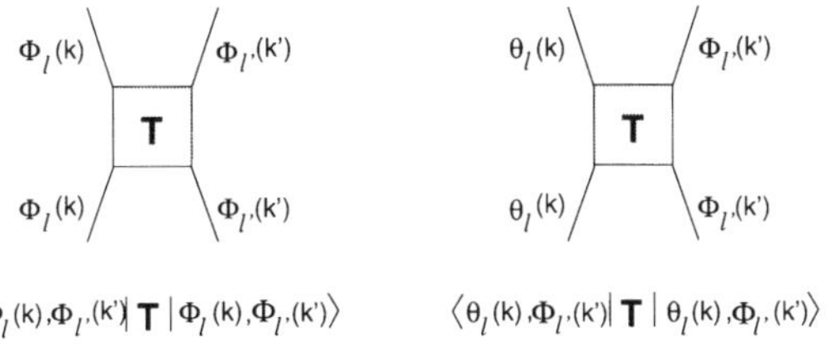

Figure 9.4. Graphical illustration of nucleon–nucleon and nucleon–antinucleon **T**-matrix amplitudes within the direct terms of $\Sigma^B_{\Phi\Phi}$ and $\Sigma^B_{\Theta\Theta}$ (equations (9.79) and (9.80)) in the case of *symmetric* nuclear matter. The quantities Φ and Θ denote nucleon and antinucleon spinors, respectively.

$$\langle \Theta_l(k), \Phi_{l'}(k') \, | \, \mathsf{T} \, | \, \Phi_l(k), \Phi_{l'}(k') \rangle \qquad\qquad \langle \Theta_l(k), \Phi_{l'}(k') \, | \, \mathsf{T} \, | \, \Phi_{l'}(k'), \Theta_l(k) \rangle$$

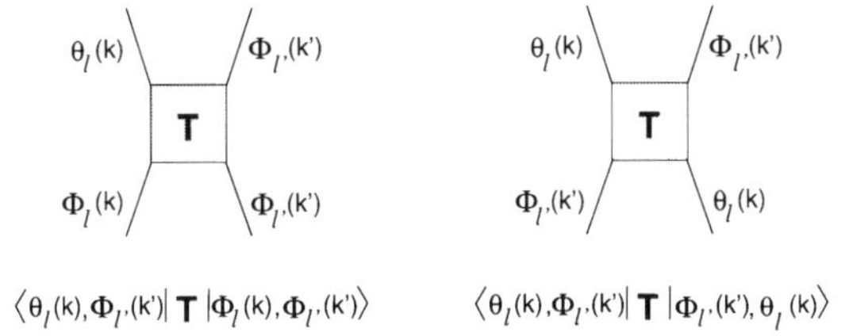

Figure 9.5. Left: Same as figure 9.4, but for the direct term of $\Sigma^B_{\Theta\Phi}$ (equation (9.81)). Right: Amplitude contributing to the exchange term of $\mathbf{\Sigma}^B$.

The **T**-matrix elements that enter (9.79) through (9.81) are graphically illustrated in figures 9.4 and 9.5. Note that knowledge of *both* the nucleon–nucleon scattering amplitude as well as the mixed nucleon–antinucleon amplitude is necessary to uniquely determine the matrix elements of $\mathbf{\Sigma}^B$ in the self-consistent particle basis.

In chapters 10 and 12, we shall adopt the relativistic BHF propagator, $\mathbf{\Lambda}^{\mathrm{RBHF}}$, together with the $\mathbf{\Lambda}^{00}$ propagator to the determination of the properties of neutron star matter in the framework of the **T**-matrix approximation. In the framework of this self-consistent many-body scheme, the equation of state, the asymmetry, and the baryon–lepton composition of such matter will be calculated. Among other boson-exchange interactions, the BM potentials A, B and C will be employed. Furthermore we shall appraise the possibility of parametrizing the BHF outcome in the framework of the relativistic Hartree and HF approximation. Such parametrizations will be very helpful in connection with investigations dealing, for instance, with the equations of state of proto-neutron stars or supernova matter, since much less numerical effort is needed when computing the equation of state for the latter two approximations.

The $\mathbf{\Lambda}^{10}$ and $\mathbf{\Lambda}^{11}$ propagators shall not be studied here. The former

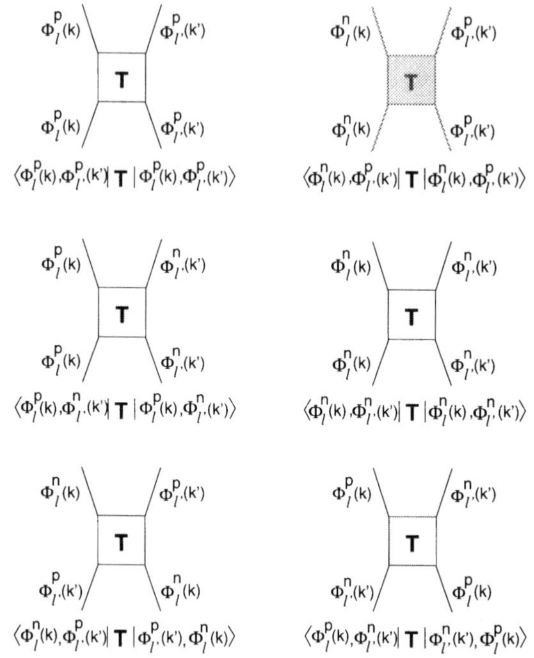

Figure 9.6. Graphical illustration of nucleon–nucleon $\mathbf{T}$-matrix amplitudes $\overset{1}{\mathbf{T}}$ through $\overset{6}{\mathbf{T}}$ (cf. equations (9.82) to (9.87)) which enter the calculation of $\Sigma^B_{\Phi\Phi}$ (equation (9.79)) in the case of *asymmetric* nuclear matter.

has been applied to nuclear matter calculations by Poschenrieder *et al* [117, 118]. The full $\mathbf{\Lambda}^{11}$ propagator has been studied so far, because of its complexity, in the non-relativistic limit only [325, 443, 444, 445].

9.6 Scattering matrix in the self-consistent basis

Some of the $\mathbf{T}$-matrix elements that enter the calculation of Σ^B in the case of asymmetric nuclear matter are as follows (cf. figure 9.6):

$$\langle l_1' l_2' k' | \overset{1}{\mathbf{T}} | l_1 l_2 k \rangle \equiv \langle \Phi^p_{l_1'}(k_1'), \Phi^n_{l_2'}(k_2') | \mathbf{T} | \Phi^p_{l_1}(k_1), \Phi^n_{l_2}(k_2) \rangle , \tag{9.82}$$

$$\langle l_1' l_2' k' | \overset{2}{\mathbf{T}} | l_1 l_2 k \rangle \equiv \langle \Phi^n_{l_1'}(k_1'), \Phi^p_{l_2'}(k_2') | \mathbf{T} | \Phi^n_{l_1}(k_1), \Phi^p_{l_2}(k_2) \rangle , \tag{9.83}$$

$$\langle l_1' l_2' k' | \overset{3}{\mathbf{T}} | l_1 l_2 k \rangle \equiv \langle \Phi^p_{l_1'}(k_1'), \Phi^n_{l_2'}(k_2') | \mathbf{T} | \Phi^n_{l_1}(k_1), \Phi^p_{l_2}(k_2) \rangle , \tag{9.84}$$

$$\langle l_1' l_2' k' | \overset{4}{\mathbf{T}} | l_1 l_2 k \rangle \equiv \langle \Phi^n_{l_1'}(k_1'), \Phi^p_{l_2'}(k_2') | \mathbf{T} | \Phi^p_{l_1}(k_1), \Phi^n_{l_2}(k_2) \rangle , \tag{9.85}$$

$$\langle l_1' l_2' k' | \overset{5}{\mathbf{T}} | l_1 l_2 k \rangle \equiv \langle \Phi^p_{l_1'}(k_1'), \Phi^p_{l_2'}(k_2') | \mathbf{T} | \Phi^p_{l_1}(k_1), \Phi^p_{l_2}(k_2) \rangle , \tag{9.86}$$

$$\langle l_1' l_2' k' | \overset{6}{\mathbf{T}} | l_1 l_2 k \rangle \equiv \langle \Phi^n_{l_1'}(k_1'), \Phi^n_{l_2'}(k_2') | \mathbf{T} | \Phi^n_{l_1}(k_1), \Phi^n_{l_2}(k_2) \rangle \,. \tag{9.87}$$

In analogy to equations (9.82) through (9.87), knowledge of the following nucleon–nucleon matrix elements is necessary,

$$\langle l_1' l_2' k' | \overset{1}{\mathbf{V}} | l_1 l_2 k \rangle \equiv \langle \Phi^p_{l_1'}(k_1'), \Phi^n_{l_2'}(k_2') | \mathbf{V} | \Phi^p_{l_1}(k_1), \Phi^n_{l_2}(k_2) \rangle \,, \tag{9.88}$$

$$\langle l_1' l_2' k' | \overset{2}{\mathbf{V}} | l_1 l_2 k \rangle \equiv \langle \Phi^n_{l_1'}(k_1'), \Phi^p_{l_2'}(k_2') | \mathbf{V} | \Phi^n_{l_1}(k_1), \Phi^p_{l_2}(k_2) \rangle \,, \tag{9.89}$$

$$\langle l_1' l_2' k' | \overset{3}{\mathbf{V}} | l_1 l_2 k \rangle \equiv \langle \Phi^p_{l_1'}(k_1'), \Phi^n_{l_2'}(k_2') | \mathbf{V} | \Phi^n_{l_1}(k_1), \Phi^p_{l_2}(k_2) \rangle \,, \tag{9.90}$$

$$\langle l_1' l_2' k' | \overset{4}{\mathbf{V}} | l_1 l_2 k \rangle \equiv \langle \Phi^n_{l_1'}(k_1'), \Phi^p_{l_2'}(k_2') | \mathbf{V} | \Phi^p_{l_1}(k_1), \Phi^n_{l_2}(k_2) \rangle \,, \tag{9.91}$$

$$\langle l_1' l_2' k' | \overset{5}{\mathbf{V}} | l_1 l_2 k \rangle \equiv \langle \Phi^p_{l_1'}(k_1'), \Phi^p_{l_2'}(k_2') | \mathbf{V} | \Phi^p_{l_1}(k_1), \Phi^p_{l_2}(k_2) \rangle \,, \tag{9.92}$$

$$\langle l_1' l_2' k' | \overset{6}{\mathbf{V}} | l_1 l_2 k \rangle \equiv \langle \Phi^n_{l_1'}(k_1'), \Phi^n_{l_2'}(k_2') | \mathbf{V} | \Phi^n_{l_1}(k_1), \Phi^n_{l_2}(k_2) \rangle \,, \tag{9.93}$$

in terms of which $\overset{1}{\mathbf{T}}$ is given by

$$\begin{aligned}
\langle l_1' l_2' k' | \overset{1}{\mathbf{T}}_K | l_1 l_2 k \rangle &= \langle l_1' l_2' k' | \overset{1}{\mathbf{V}} | l_1 l_2 k \rangle \\
&+ \int \frac{d^4 q}{(2\pi)^4} \langle l_1' l_2' k' | \overset{1}{\mathbf{V}} | l_3 l_4 q \rangle \boldsymbol{\Lambda}_{pn}\Big(\frac{K}{2} + q, \frac{K}{2} - q\Big) \langle l_3 l_4 q | \overset{1}{\mathbf{T}}_K | l_1 l_2 k \rangle \\
&+ \int \frac{d^4 q}{(2\pi)^4} \langle l_1' l_2' k' | \overset{3}{\mathbf{V}} | l_3 l_4 q \rangle \boldsymbol{\Lambda}_{np}\Big(\frac{K}{2} + q, \frac{K}{2} - q\Big) \langle l_3 l_4 q | \overset{4}{\mathbf{T}}_K | l_1 l_2 k \rangle \,.
\end{aligned} \tag{9.94}$$

We recall that *intermediate* antiparticle states do not occur in the $\mathbf{T}$-matrix equation (9.94). Such states are excluded from the construction of OBE interactions, such as Bonn, HEA, or the BM potentials. For that reason all intermediate antinucleon contributions must be dropped and the baryon–meson vertices, Γ_α, are to be regularized, as described in section 5.5.3. In contrast to the intermediate states, the initial and final scattering states in the $\mathbf{T}$-matrix equation contain both particle as well as antiparticle states. It is this circumstance that requires the determination of the complete set of spinors and antispinors.

Equation (9.94) represents a coupled integral equation for the amplitudes $\overset{1}{\mathbf{T}}$ and $\overset{4}{\mathbf{T}}$. All other amplitudes are obtained by simple integrations. According to the possible helicity degrees of freedom, each of the above amplitudes has 16 components. This number can be reduced significantly by exploiting the invariance properties of the $\mathbf{T}$-matrix, as we shall see in chapter 10. Despite this simplification, the calculation of the $\mathbf{T}$-matrix still is numerically very demanding. A widely used technique which makes the determination of the $\mathbf{T}$-matrix numerically tractable consists in its partial-wave expansion [440], which is the subject of the next chapter.

Chapter 10

Partial-wave expansions

10.1 Nucleon–nucleon matrix elements

In this chapter we describe the partial-wave decomposition of the q-space one-boson-exchange interaction as well as of the scattering matrix. Since the nucleon matrix elements (5.151) are invariant under Lorentz transformations, we perform the partial-wave expansions in the center-of-mass system. The Pauli helicity spinors of a baryon B in its initial ($|l_1\rangle$ and $|l_2\rangle$) and final ($|l_1'\rangle$ and $|l_2'\rangle$) states can then be represented as

$$|l_1\rangle = \chi_{l_1}\,, \qquad |l_2\rangle = (-1)^{\frac{1}{2}-l_2}\,\chi_{-l_2}\,, \tag{10.1}$$

and

$$|l_1'\rangle = \mathrm{e}^{-\mathrm{i}\frac{\vartheta}{2}\sigma_y}\,\chi_{l_1'} \qquad |l_2'\rangle = (-1)^{\frac{1}{2}-l_2'}\,\mathrm{e}^{-\mathrm{i}\frac{\vartheta}{2}\sigma_y}\,\chi_{-l_2'}\,, \tag{10.2}$$

respectively. The quantity ϑ denotes the scattering angle and, in the reference frame used (see figure 10.1), χ_l the Pauli spinors obeying

$$\frac{1}{2}\,\sigma_z\,\hat{D}\,\chi_l = l\,\hat{D}\,\chi_l\,. \tag{10.3}$$

In accordance with the phase convention of Jacob and Wick [446], the helicity states $|l_2\rangle$ and $|l_2'\rangle$ are multiplied with additional phase factors, that is, we perform the following replacements,

$$|l_2\rangle \longrightarrow (-1)^{\frac{1}{2}-l_2}\,|l_2\rangle\,, \qquad |l_2'\rangle \longrightarrow (-1)^{\frac{1}{2}-l_2'}\,|l_2'\rangle\,. \tag{10.4}$$

This convention ensures that for a particle at rest that

$$|l_2\rangle = \chi_{-l_2}\,. \tag{10.5}$$

Therefore the spinors of particles '1' and '2' occur in the matrix elements symmetrically.

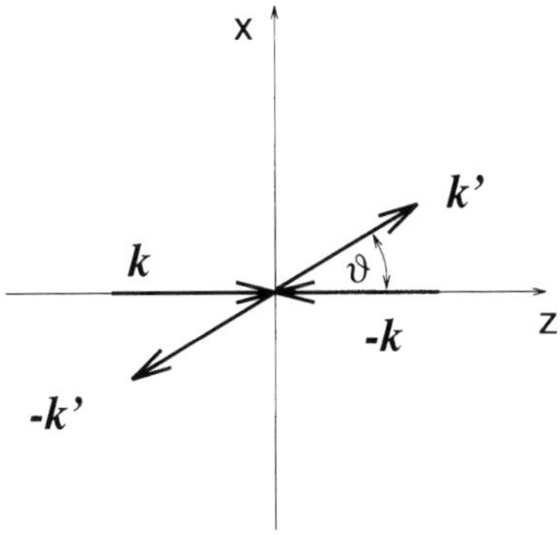

Figure 10.1. Two-particle scattering reference scheme in the center-of-mass frame.

Assuming rotational invariance for the OBEP, the expansion of a matrix element like $\langle \Phi_{l_1'}(k'), \Phi_{l_2'}(k')|\mathbf{V}|\Phi_{l_1}(k), \Phi_{l_2}(k)\rangle$ into angular momentum states $|jm, l_1 l_2\rangle$, given by [446]

$$|\hat{\boldsymbol{k}}, l_1 l_2\rangle = \frac{1}{2\pi}\sum_j \sum_m \sqrt{2j+1}\, D^j_{m,l}(\varphi, \vartheta, -\varphi)|jm, l_1 l_2\rangle\,, \qquad (10.6)$$

with $l = l_1 - l_2$, proceeds as follows. If we choose (without loss of generality) the incident direction along the z axis and outgoing momentum $\boldsymbol{k}'$ to be in the x–z plane, as shown in figure 10.1, the expansion may be written as

$$\begin{aligned}&\langle \Phi_{l_1'}(k'), \Phi_{l_2'}(k')|\mathbf{V}|\Phi_{l_1}(k), \Phi_{l_2}(k)\rangle \\ &\qquad = \frac{1}{4\pi}\sum_j (2j+1)\, d^j_{ll'}(\vartheta)\langle l_1' l_2'|\mathsf{V}^j(k', k)|l_1 l_2\rangle\,, \end{aligned}\qquad (10.7)$$

where ϑ is the scattering angle between $\boldsymbol{k}$ and $\boldsymbol{k}'$, and $l = l_1 - l_2$, $l' = l_1' - l_2'$. Besides that, use of

$$\langle l_1' l_2', jm|\mathbf{V}(k', k)|l_1 l_2, jm\rangle = \delta_{jj'}\,\delta_{mm'}\langle l_1' l_2'|\mathsf{V}^j(k', k)|l_1 l_2\rangle \qquad (10.8)$$

was made in arriving at (10.7), which is a consequence of the system's rotational invariance. With the aid of the orthogonality relation [441],

$$\int_{-1}^{+1} \mathrm{d}\cos\vartheta\; d^j_{ll'}(\vartheta)\, d^{j'}_{ll'}(\vartheta) = \frac{2}{2j+1}\delta_{jj'}\,, \qquad (10.9)$$

equation (10.7) can be inverted. One obtains

$$\begin{aligned}&\langle l_1' l_2'|\mathsf{V}^j(k', k)|l_1 l_2\rangle \\ &\quad = 2\pi \int_{-1}^{+1} \mathrm{d}\cos\vartheta\, \langle \Phi_{l_1'}(k'), \Phi_{l_2'}(k')|\mathbf{V}|\Phi_{l_1}(k), \Phi_{l_2}(k)\rangle d^j_{ll'}(\vartheta)\,. \end{aligned}\qquad (10.10)$$

The quantities $d^j_{ll'}(\vartheta)$ are the reduced rotation matrices, which can be expressed in terms of the usual Legendre polynomials $P_j(\cos\vartheta)$. The relations needed to carry out the angle integration in (10.10) read

$$\begin{aligned} d^j_{00}(\vartheta) &= P_j(\cos\vartheta)\,, \\ \sin\vartheta\; d^j_{10}(\vartheta) &= \frac{2\sqrt{j(j+1)}}{2j+1}\left(P_{j+1}(\cos\vartheta) - P_{j-1}(\cos\vartheta)\right)\,, \\ d^j_{01}(\vartheta) &= -\,d^j_{10}(\vartheta)\,, \end{aligned} \tag{10.11}$$

and

$$\begin{aligned} (1+\cos\vartheta)d^j_{11}(\vartheta) &= \frac{j+1}{2j+1}P_{j-1}(\cos\vartheta) + P_j(\cos\vartheta) + \frac{j}{2j+1}P_{j+1}(\cos\vartheta)\,, \\ (1-\cos\vartheta)d^j_{1-1}(\vartheta) &= \frac{j+1}{2j+1}P_{j-1}(\cos\vartheta) - P_j(\cos\vartheta) + \frac{j}{2j+1}P_{j+1}(\cos\vartheta)\,, \\ d^j_{-11}(\vartheta) &= d^j_{1-1}(\vartheta)\,. \end{aligned} \tag{10.12}$$

The last expressions in (10.11) and (10.12) are special cases of the general relations

$$d^j_{ll'}(\vartheta) = d^j_{-l',-l}(\vartheta) = (-)^{l-l'}\, d^j_{l'l}(\vartheta)\,. \tag{10.13}$$

In the next step we turn our interest to the invariance properties of the nucleon–nucleon scattering problem which imply the following important relations for the V^j matrix elements:

(i) Parity conservation,

$$\langle l_1' l_2' | \mathsf{V}^j(k',k) | l_1 l_2 \rangle = \langle -l_1' \; -l_2' | \mathsf{V}^j(k',k) | -l_1 \; -l_2 \rangle\,. \tag{10.14}$$

(ii) Conservation of total spin,

$$\langle l_1' l_2' | \mathsf{V}^j(k',k) | l_1 l_2 \rangle = \langle l_2' l_1' | \mathsf{V}^j(k',k) | l_2 l_1 \rangle\,. \tag{10.15}$$

(iii) Time-reversal invariance,

$$\langle l_1' l_2' | \mathsf{V}^j(k',k) | l_1 l_2 \rangle = \langle l_1 l_2 | \mathsf{V}^j(k',k) | l_1' l_2' \rangle\,. \tag{10.16}$$

Thus, parity and spin conservation alone reduces the sixteen V^j amplitudes to *six* independent ones in the case of symmetric nuclear matter, and to *eight* for asymmetric matter. Along the lines of Erkelenz [320], we choose the following six linear combinations (to keep notation to a minimum, $l = \pm\frac{1}{2}$ is abbreviated to $\pm$):

$${}^{0}\mathsf{V}^j_\alpha(k',k) = \langle ++|\mathsf{V}^j_\alpha(k',k)|+\rangle + - \langle ++\,|\mathsf{V}^j_\alpha(k',k)|--\rangle\,, \tag{10.17}$$

$${}^{12}\mathsf{V}^j_\alpha(k',k) = \langle ++|\mathsf{V}^j_\alpha(k',k)|++\rangle + \langle ++|\mathsf{V}^j_\alpha(k',k)|--\rangle\,, \tag{10.18}$$

$${}^{1}\mathsf{V}^j_\alpha(k',k) = \langle +-|\mathsf{V}^j_\alpha(k',k)|+-\rangle - \langle +-|\mathsf{V}^j_\alpha(k',k)|-+\rangle\,, \tag{10.19}$$

$${}^{34}\mathsf{V}^j_\alpha(k',k) = \langle +-|\mathsf{V}^j_\alpha(k',k)|+-\rangle + \langle +-|\mathsf{V}^j_\alpha(k',k)|-+\rangle\,, \tag{10.20}$$

$${}^{5}\mathsf{V}^j_\alpha(k',k) = \langle ++|\mathsf{V}^j_\alpha(k',k)|+-\rangle\,, \tag{10.21}$$

$${}^{6}\mathsf{V}^j_\alpha(k',k) = \langle +-|\mathsf{V}^j_\alpha(k',k)|++\rangle\,. \tag{10.22}$$

Hence, the off-shell scattering of two baryons propagating in symmetric nuclear matter is determined by six amplitudes, which, in the on-shell case, reduce to five. The asymmetric nuclear matter case is somewhat more involved insofar as eight amplitudes are to be determined. Instead of equation (10.21) we choose the amplitude

$$ {}^{5}\mathsf{V}^j_\alpha(k',k) = \langle ++|\mathsf{V}^j_\alpha(k',k)|+-\rangle - \langle ++|\mathsf{V}^j_\alpha(k',k)|-+\rangle\,, \qquad (10.23) $$

which vanishes for symmetric matter. The amplitude (10.22) is replaced by

$$ {}^{56}\mathsf{V}^j_\alpha(k',k) = \langle +-|\mathsf{V}^j_\alpha(k',k)|++\rangle + \langle +-|\mathsf{V}^j_\alpha(k',k)|--\rangle\,, \qquad (10.24) $$

which reduces to ${}^{56}\mathsf{V}^j_\alpha \rightarrow 2\times{}^{6}\mathsf{V}^j_\alpha$ for symmetric matter. Finally the following two additional amplitudes are introduced,

$$ {}^{3}\mathsf{V}^j_\alpha(k',k) = \langle +-|\mathsf{V}^j_\alpha(k',k)|++\rangle - \langle +-|\mathsf{V}^j_\alpha(k',k)|--\rangle\,, \qquad (10.25) $$

$$ {}^{78}\mathsf{V}^j_\alpha(k',k) = \langle ++|\mathsf{V}^j_\alpha(k',k)|+-\rangle + \langle ++|\mathsf{V}^j_\alpha(k',k)|-+\rangle\,, \qquad (10.26) $$

whose symmetric matter limits are zero and ${}^{78}\mathsf{V}^j_\alpha \rightarrow 2\times{}^{6}\mathsf{V}^j_\alpha = 2\times{}^{5}\mathsf{V}^j_\alpha$, respectively, the reasons being parity conservation (10.14) and conservation of total spin (10.15).

Collecting all results and applying them to equation (10.10) leads to the off-shell angular momentum state amplitudes ${}^{i}\mathsf{V}^j_\alpha(k',k)$ with $i = 0, 12, 1, 34, 5, 6$ and $\alpha = \mathrm{s}, \mathrm{pv}, \mathrm{v}$. When displaying the results for these amplitudes, it is convenient to decompose each of the amplitudes $\langle l_1'l_2'|\mathsf{V}^j(k',k)|l_1l_2\rangle$ according to the scheme

$$ \langle l_1'l_2'|\mathsf{V}^j_\alpha(k',k)|l_1l_2\rangle = \frac{1}{m'^*\,m^*}\{F_\alpha(l_1'l_2',l_1l_2;k',k)\langle l_1'l_2'|f^j_\alpha|l_1l_2\rangle $$
$$ \times\, G_\alpha(l_1'l_2',l_1l_2;k',k)\langle l_1'l_2'|g^j_\alpha|l_1l_2\rangle H_\alpha(l_1'l_2',l_1l_2;k',k)\langle l_1'l_2'|h^j_\alpha|l_1l_2\rangle\}\,, $$
$$ (10.27) $$

where F_α, G_α and H_α are ϑ-independent coefficients, and the ϑ-dependent functions f^j_α, g^j_α and h^j_α are defined as

$$ \langle l_1'l_2'|f^j_\alpha|l_1l_2\rangle \equiv 2\pi\int_{-1}^{+1}\mathrm{d}\cos\vartheta\,\langle l_1'l_2'|\mathbf{1}|l_1l_2\rangle F^2_\alpha(k',k)\,\mathcal{G}(k',k)d^j_{ll'}(\vartheta)\,, $$
$$ (10.28) $$

$$ \langle l_1'l_2'|g^j_\alpha|l_1l_2\rangle \equiv 2\pi\int_{-1}^{+1}\mathrm{d}\cos\vartheta\,\langle l_1'l_2'|\boldsymbol{\sigma}_1\cdot\boldsymbol{\sigma}_2|l_1l_2\rangle F^2_\alpha(k',k)\,\mathcal{G}(k',k)d^j_{ll'}(\vartheta)\,, $$
$$ (10.29) $$

$$ \langle l_1'l_2'|h^j_\alpha|l_1l_2\rangle \equiv 2\pi\int_{-1}^{+1}\mathrm{d}\cos\vartheta\;\cos\vartheta\,\langle l_1'l_2'|\mathbf{1}|l_1l_2\rangle F^2_\alpha(k',k)\,\mathcal{G}(k',k)d^j_{ll'}(\vartheta)\,. $$
$$ (10.30) $$

By making use of the recurrence formula for the Legendre polynomials, the integrals over ϑ in the last three equations can be finally expressed in terms of elementary integrals. For this purpose we introduce several additional definitions:

$$\tilde{Q}_j(z_\alpha; k', k) \equiv \frac{1}{2}\int_{-1}^{+1} \mathrm{d}\cos\vartheta \, \frac{P_j(\cos\vartheta)\, F_\alpha^2(k', k)}{z_\alpha - \cos\vartheta}\,, \tag{10.31}$$

$$R_j(k', k) \equiv \frac{1}{2}\int_{-1}^{+1} \mathrm{d}\cos\vartheta \, P_j(\cos\vartheta)\, F_\alpha^2(k', k)\,, \tag{10.32}$$

$$\tilde{Q}_j^{(1)}(z_\alpha; k', k) = z_\alpha\, \tilde{Q}_j(z_\alpha; k', k) - R_j(k', k)\,, \tag{10.33}$$

$$\tilde{Q}_j^{(2)}(z_\alpha; k', k) = \frac{1}{j+1}\{j\, z_\alpha\, \tilde{Q}_j(z_\alpha, k', k) - j\, R_j(k', k) + \tilde{Q}_{j-1}(z_\alpha, k', k)\} \quad \text{if } j \geq 1\,, \tag{10.34}$$

$$\tilde{Q}_j^{(3)}(z_\alpha; k', k) \equiv \frac{1}{2}\sqrt{\frac{j}{j+1}}\{j\, z_\alpha\, \tilde{Q}_j^{(1)}(z_\alpha, k', k) - \tilde{Q}_{j-1}(z_\alpha, k', k)\} \quad \text{if } j \geq 1\,, \tag{10.35}$$

$$S_j(k', k) \equiv \frac{1}{2}\int_{-1}^{+1} \mathrm{d}\cos\vartheta \, P_j(\cos\vartheta)\, F_\alpha^2(k', k)\,, \tag{10.36}$$

$$\tilde{Q}_j^{(4)}(z_\alpha; k', k) \equiv z_\alpha\, \tilde{Q}_j^{(1)}(z_\alpha; k', k) - S_j(k', k)\,, \tag{10.37}$$

$$\tilde{Q}_j^{(5)}(z_\alpha; k', k) \equiv \frac{j}{j+1}\{j\, \tilde{Q}_j^{(4)}(z_\alpha, k', k) + \tilde{Q}_{j-1}^{(1)}(z_\alpha, k', k)\} \quad \text{if } j \geq 1\,, \tag{10.38}$$

$$\tilde{Q}_j^{(6)}(z_\alpha; k', k) \equiv \frac{1}{2}\sqrt{\frac{j}{j+1}}\{\tilde{Q}_j^{(4)}(z_\alpha, k', k) - \tilde{Q}_{j-1}^{(1)}(z_\alpha, k', k)\} \quad \text{if } j \geq 1\,. \tag{10.39}$$

The argument z_α in the above expressions is defined as

$$z_\alpha \equiv \frac{k'^2 + k^2 - (k'^0 - k^0)^2 + m_\alpha^2}{2\, k\, k'}\,, \tag{10.40}$$

with m_α the meson mass. We note that the notation here is chosen such that $\tilde{Q}_j$ and $\tilde{Q}_j^{(i)}$ reduce for $F_\alpha \to 1$ to the corresponding functions (Q_j respectively $Q_j^{(i)}$) used by Erkelenz [320]. In this limiting case R_j and S_j obey the simple relations $R_j(k'^0\boldsymbol{k}', k^0\boldsymbol{k}) = \delta_{j0}$ and $S_j(k'^0\boldsymbol{k}', k^0\boldsymbol{k}) = \delta_{j1}$.

The final results for the nucleon–nucleon angular momentum state amplitudes then follow by inserting the matrix elements (9.38) to (9.41), plus those listed in appendix E, into equations (10.27) respectively (10.17) to (10.22). For the scalar coupling (i.e. $\alpha = \mathrm{s}$) one then arrives at,

$$ {}^{0}\mathsf{V}^{j}_{\mathrm{s}}(k',k) = g_{\mathrm{s}}^{2}\,\frac{\pi}{m'^{*}\,m^{*}}\left\{k'^{*}\,k^{*}\,\frac{\tilde{Q}^{(1)}_{j}}{|\boldsymbol{k}'||\boldsymbol{k}|} - (m'^{*}\,m^{*} + W'\,W)\frac{\tilde{Q}_{j}}{|\boldsymbol{k}'||\boldsymbol{k}|}\right\}, \tag{10.41} $$

$$ {}^{12}\mathsf{V}^{j}_{\mathrm{s}}(k',k) = g_{\mathrm{s}}^{2}\,\frac{\pi}{m'^{*}\,m^{*}}\left\{k'^{*}\,k^{*}\,\frac{\tilde{Q}_{j}}{|\boldsymbol{k}'||\boldsymbol{k}|} - (m'^{*}\,m^{*} + W'\,W)\frac{\tilde{Q}^{(1)}_{j}}{|\boldsymbol{k}'||\boldsymbol{k}|}\right\}, \tag{10.42} $$

$$ {}^{1}\mathsf{V}^{j}_{\mathrm{s}}(k',k) = g_{\mathrm{s}}^{2}\,\frac{\pi}{m'^{*}\,m^{*}}\left\{k'^{*}\,k^{*}\,\frac{\tilde{Q}^{(2)}_{j}}{|\boldsymbol{k}'||\boldsymbol{k}|} - (m'^{*}\,m^{*} + W'\,W)\frac{\tilde{Q}_{j}}{|\boldsymbol{k}'||\boldsymbol{k}|}\right\}, \tag{10.43} $$

$$ {}^{34}\mathsf{V}^{j}_{\mathrm{s}}(k',k) = g_{\mathrm{s}}^{2}\,\frac{\pi}{m'^{*}\,m^{*}}\left\{k'^{*}\,k^{*}\,\frac{\tilde{Q}_{j}}{|\boldsymbol{k}'||\boldsymbol{k}|} - (m'^{*}\,m^{*} + W'\,W)\frac{\tilde{Q}^{(2)}_{j}}{|\boldsymbol{k}'||\boldsymbol{k}|}\right\}, \tag{10.44} $$

$$ {}^{5}\mathsf{V}^{j}_{\mathrm{s}}(k',k) = g_{\mathrm{s}}^{2}\,\frac{\pi}{m'^{*}\,m^{*}}\left\{(m^{*}\,W' + m'^{*}\,W)\frac{\tilde{Q}^{(3)}_{j}}{|\boldsymbol{k}'||\boldsymbol{k}|}\right\}, \tag{10.45} $$

$$ {}^{6}\mathsf{V}^{j}_{\mathrm{s}}(k',k) = g_{\mathrm{s}}^{2}\,\frac{\pi}{m'^{*}\,m^{*}}\left\{(m^{*}\,W' + m'^{*}\,W)\frac{\tilde{Q}^{(3)}_{j}}{|\boldsymbol{k}'||\boldsymbol{k}|}\right\}. \tag{10.46} $$

Note that ${}^{5}\mathsf{V}^{j}_{\mathrm{s}} = {}^{6}\mathsf{V}^{j}_{\mathrm{s}}$ which is a consequence of the time-reversal invariance (10.16). The partial-wave expansions for the pseudovector and vector couplings are given in appendix F.

10.2 Nucleon–antinucleon matrix elements

In the previous section we performed the partial-wave decomposition of the one-boson-exchange interaction for the nucleon–nucleon sector. Now we turn to the partial-wave decomposition of the mixed *nucleon–antinucleon* matrix elements,

$$ \langle l'_1 l'_2, k'|\mathbf{U}|l_1 l_2, k\rangle \equiv \langle \Theta_{l'_1}(k'), \Phi_{l'_2}(k')|\mathbf{V}|\Phi_{l_1}(k), \Phi_{l_2}(k)\rangle\,, \tag{10.47} $$

$$ \langle l'_1 l'_2, k'|\mathbf{W}|l_1 l_2, k\rangle \equiv \langle \Theta_{l'_1}(k'), \Phi_{l'_2}(k')|\mathbf{V}|\Phi_{l_1}(k), \Theta_{l_2}(k)\rangle\,, \tag{10.48} $$

$$ \langle l'_1 l'_2, k'|\mathbf{Z}|l_1 l_2, k\rangle \equiv \langle \Theta_{l'_1}(k'), \Phi_{l'_2}(k')|\mathbf{V}|\Theta_{l_1}(k), \Phi_{l_2}(k)\rangle\,, \tag{10.49} $$

which is accomplished in close analogy to our analysis performed in section 10.1. Since the angular dependence of the spinors is in both cases the same, the expansions (10.7) and its inverse (10.10) remain valid unchanged. Differences arise however from the different invariance properties of the matrix elements, which are less constrained by symmetry relations than the particle–particle matrix elements. Consequently, the number of independent particle–antiparticle matrix elements is not as large as in the particle–particle case. To be more precise, the only symmetry relations obeyed by the matrix elements listed above are

$$\langle l_1' l_2' | \mathsf{U}^j(k',k) | l_1 l_2 \rangle = -\langle -l_1' \; -l_2' | \mathsf{U}^j(k',k) | -l_1 \; -l_2 \rangle \,, \tag{10.50}$$

and

$$\langle l_1' l_2' | W^j(k',k) | l_1 l_2 \rangle = -\langle -l_1' \; -l_2' | W^j(k',k) | -l_1 \; -l_2 \rangle \,. \tag{10.51}$$

The symmetry relations for the matrix elements $\langle l_1' l_2', k' | \mathbf{Z} | l_1 l_2, k \rangle$ are the same as for the particle–particle matrix elements (10.14) to (10.14). The net result is that there are *eight* independent matrix elements U^j and W^j, and *six* independent Z^j. In analogy to (10.17) to (10.22) we choose the following linear combinations for U^j (as before, the $\pm$ signs refer to helicity, $l = \pm\frac{1}{2}$):

$$^{0}\mathsf{U}^j_\alpha(k',k) = \langle ++|\mathsf{U}^j_\alpha(k',k)|++\rangle - \langle ++|\mathsf{U}^j_\alpha(k',k)|--\rangle \,, \tag{10.52}$$

$$^{12}\mathsf{U}^j_\alpha(k',k) = \langle ++|\mathsf{U}^j_\alpha(k',k)|++\rangle + \langle ++|\mathsf{U}^j_\alpha(k',k)|--\rangle \,, \tag{10.53}$$

$$^{1}\mathsf{U}^j_\alpha(k',k) = \langle +-|\mathsf{U}^j_\alpha(k',k)|+-\rangle - \langle +-|\mathsf{U}^j_\alpha(k',k)|-+\rangle \,, \tag{10.54}$$

$$^{34}\mathsf{U}^j_\alpha(k',k) = \langle +-|\mathsf{U}^j_\alpha(k',k)|+-\rangle + \langle +-|\mathsf{U}^j_\alpha(k',k)|-+\rangle \,, \tag{10.55}$$

$$^{3}\mathsf{U}^j_\alpha(k',k) = \langle +-|\mathsf{U}^j_\alpha(k',k)|++\rangle - \langle +-|\mathsf{U}^j_\alpha(k',k)|--\rangle \,, \tag{10.56}$$

$$^{56}\mathsf{U}^j_\alpha(k',k) = \langle +-|\mathsf{U}^j_\alpha(k',k)|++\rangle + \langle +-|\mathsf{U}^j_\alpha(k',k)|--\rangle \,, \tag{10.57}$$

$$^{5}\mathsf{U}^j_\alpha(k',k) = \langle ++|\mathsf{U}^j_\alpha(k',k)|+-\rangle - \langle ++|\mathsf{U}^j_\alpha(k',k)|-+\rangle \,, \tag{10.58}$$

$$^{78}\mathsf{U}^j_\alpha(k',k) = \langle ++|\mathsf{U}^j_\alpha(k',k)|+-\rangle + \langle ++|\mathsf{U}^j_\alpha(k',k)|-+\rangle . \tag{10.59}$$

For the eight independent components of W^j and Z^j we choose the same notation as for U^j in equations (10.52) to (10.59) just above. In the latter case, Z^j, we have ${}^3\mathsf{Z}^j_\alpha = {}^5\mathsf{Z}^j_\alpha = 0$. The angular momentum state amplitudes can then be calculated exactly as in the nucleon–nucleon case, treated in section 10.1. We begin with the results for mesons with *scalar* coupling, in which case one gets

$$^{0}\mathsf{U}^j_{\rm s}(k',k) = g_{\rm s}^2 \, \frac{\pi}{m'^* \, m^*} \left\{ k^* \, W' \, \frac{\tilde{Q}^{(1)}_j}{|\boldsymbol{k}'||\boldsymbol{k}|} - k'^* \, W \, \frac{\tilde{Q}_j}{|\boldsymbol{k}'||\boldsymbol{k}|} \right\} , \tag{10.60}$$

$$^{12}\mathsf{U}^j_{\rm s}(k',k) = g_{\rm s}^2 \, \frac{\pi}{m'^* \, m^*} \left\{ k^* \, W' \, \frac{\tilde{Q}_j}{|\boldsymbol{k}'||\boldsymbol{k}|} - k'^* \, W \, \frac{\tilde{Q}^{(1)}_j}{|\boldsymbol{k}'||\boldsymbol{k}|} \right\} , \tag{10.61}$$

$$^{1}\mathsf{U}^j_{\rm s}(k',k) = g_{\rm s}^2 \, \frac{\pi}{m'^* \, m^*} \left\{ k^* \, W' \, \frac{\tilde{Q}^{(2)}_j}{|\boldsymbol{k}'||\boldsymbol{k}|} - k'^* \, W \, \frac{\tilde{Q}_j}{|\boldsymbol{k}'||\boldsymbol{k}|} \right\} , \tag{10.62}$$

$$ {}^{34}\mathsf{U}^j_{\rm s}(k',k) = g_{\rm s}^2\,\frac{\pi}{m'^*\,m^*}\left\{k^*\,W'\,\frac{\tilde{Q}_j}{|\boldsymbol{k}'||\boldsymbol{k}|} - k'^*\,W\,\frac{\tilde{Q}_j^{(2)}}{|\boldsymbol{k}'||\boldsymbol{k}|}\right\}, \tag{10.63} $$

$$ {}^{3}\mathsf{U}^j_{\rm s}(k',k) = g_{\rm s}^2\,\frac{\pi}{m'^*\,m^*}\left\{-2\,m'^*\,k^*\,\frac{\tilde{Q}_j^{(3)}}{|\boldsymbol{k}'||\boldsymbol{k}|}\right\}, \tag{10.64} $$

$$ {}^{56}\mathsf{U}^j_{\rm s}(k',k) = g_{\rm s}^2\,\frac{\pi}{m'^*\,m^*}\left\{2\,m^*\,k'^*\,\frac{\tilde{Q}_j^{(3)}}{|\boldsymbol{k}'||\boldsymbol{k}|}\right\}, \tag{10.65} $$

$$ {}^{5}\mathsf{U}^j_{\rm s}(k',k) = g_{\rm s}^2\,\frac{\pi}{m'^*\,m^*}\left\{-2\,m'^*\,k^*\,\frac{\tilde{Q}_j^{(3)}}{|\boldsymbol{k}'||\boldsymbol{k}|}\right\}, \tag{10.66} $$

$$ {}^{78}\mathsf{U}^j_{\rm s}(k',k) = g_{\rm s}^2\,\frac{\pi}{m'^*\,m^*}\left\{2\,m^*\,k'^*\,\frac{\tilde{Q}_j^{(3)}}{|\boldsymbol{k}'||\boldsymbol{k}|}\right\}. \tag{10.67} $$

The angular momentum state amplitudes associated with (10.48) follow as:

$$ {}^{0}\mathsf{W}^j_{\rm s}(k',k) = g_{\rm s}^2\,\frac{\pi}{m'^*\,m^*}\left\{(W\,W' - m'^*\,m^*)\frac{\tilde{Q}_j}{|\boldsymbol{k}'||\boldsymbol{k}|} - k'^*\,k^*\,\frac{\tilde{Q}_j^{(1)}}{|\boldsymbol{k}'||\boldsymbol{k}|}\right\}, \tag{10.68} $$

$$ {}^{12}\mathsf{W}^j_{\rm s}(k',k) = g_{\rm s}^2\,\frac{\pi}{m'^*\,m^*}\left\{(W\,W' - m'^*\,m^*)\frac{\tilde{Q}_j^{(1)}}{|\boldsymbol{k}'||\boldsymbol{k}|} - k'^*\,k^*\,\frac{\tilde{Q}_j}{|\boldsymbol{k}'||\boldsymbol{k}|}\right\}, \tag{10.69} $$

$$ {}^{1}\mathsf{W}^j_{\rm s}(k',k) = g_{\rm s}^2\,\frac{\pi}{m'^*\,m^*}\left\{-(W\,W' - m'^*\,m^*)\frac{\tilde{Q}_j}{|\boldsymbol{k}'||\boldsymbol{k}|} + k'^*\,k^*\,\frac{\tilde{Q}_j^{(2)}}{|\boldsymbol{k}'||\boldsymbol{k}|}\right\}, \tag{10.70} $$

$$ {}^{34}\mathsf{W}^j_{\rm s}(k',k) = g_{\rm s}^2\,\frac{\pi}{m'^*\,m^*}\left\{-(W\,W' - m'^*\,m^*)\frac{\tilde{Q}_j^{(2)}}{|\boldsymbol{k}'||\boldsymbol{k}|} + k'^*\,k^*\,\frac{\tilde{Q}_j}{|\boldsymbol{k}'||\boldsymbol{k}|}\right\}, \tag{10.71} $$

$$ {}^{3}\mathsf{W}^j_{\rm s}(k',k) = 0\,, \qquad {}^{5}\mathsf{W}^j_{\rm s}(k',k) = 0\,, \tag{10.72} $$

$$ {}^{56}\mathsf{W}^j_{\rm s}(k',k) = g_{\rm s}^2\,\frac{\pi}{m'^*\,m^*}\left\{2(m^*\,W' - m'^*\,W)\frac{\tilde{Q}_j^{(3)}}{|\boldsymbol{k}'||\boldsymbol{k}|}\right\}, \tag{10.73} $$

$$ {}^{78}\mathsf{W}^j_{\rm s}(k',k) = g_{\rm s}^2\,\frac{\pi}{m'^*\,m^*}\left\{-2(m^*\,W' - m'^*\,W)\frac{\tilde{Q}_j^{(3)}}{|\boldsymbol{k}'||\boldsymbol{k}|}\right\}. \tag{10.74} $$

The angular momentum state amplitudes associated with *pseudovector* and *vector* meson exchange (cf. table 5.2) among baryons are listed in appendix F.

10.3 Isospin contribution of boson-exchange matrix elements

The boson-exchange matrix elements attain additional factors if the interaction between baryons is mediated by the isovector mesons listed in table 5.2. These factors will be calculated in the following. We begin with writing the baryon and antibaryon spinors as tensor products,

$$\Phi_l^i(k) \equiv \Phi_l(k) \otimes |i\rangle \,, \qquad \Theta_l^i(k) \equiv \Theta_l(k) \otimes |i\rangle \,, \tag{10.75}$$

where Φ_l and Θ_l denote the Dirac baryon spinors and antibaryon spinors defined in (9.1) and (9.10), and $|i\rangle$ stands for the baryon's isospin part. Guided by the isospin independence of the strong interaction, next we introduce two-baryon states that are eigenstates of total isospin. The latter is given by $\boldsymbol{T} \equiv \boldsymbol{T}^{(1)} + \boldsymbol{T}^{(2)}$, where $\boldsymbol{T}^{(1)}$ and $\boldsymbol{T}^{(2)}$ are the isospin operators of the two individual baryons, labeled '1' and '2', which undergo an interaction with each other. The eigenstates of $\boldsymbol{T}$ are then constructed from their spinors $\Phi_{l_1}^{i_1}(k_1)$ and $\Phi_{l_2}^{i_2}(k_2)$ as

$$|\Phi_{l_1}(k_1), \Phi_{l_2}(k_2), T\, i\rangle \equiv \sum_{i_1, i_2} \left(\begin{array}{cc|c} t_1 & t_2 & T \\ i_1 & i_2 & i \end{array} \right) |\Phi_{l_1}^{i_1}(k_1), \Phi_{l_2}^{i_2}(k_2)\rangle \,, \tag{10.76}$$

where the expression in large round brackets denotes a Clebsch–Gordon coefficient. Adopting the shorthand notation $|l_1\, l_2, T\, i, k\rangle \equiv |\Phi_{l_1}^{i_1}(k_1), \Phi_{l_2}^{i_2}(k_2)\rangle$, the diagonality of the particle–particle matrix elements in isospin space reads

$$\langle l_1'\, l_2', T'\, i', k'|\mathbf{V}|l_1\, l_2, T\, i, k\rangle = \delta_{T'\, T}\, \delta_{i'\, i}\, \langle l_1'\, l_2', k'|\mathbf{V}^T|l_1\, l_2, k\rangle \,. \tag{10.77}$$

One therefore gets for the boson-exchange matrix elements of isoscalar mesons

$$\langle l_1'\, l_2', k'|\mathbf{V}_\alpha^T|l_1\, l_2, k\rangle = \langle l_1'\, l_2', k'|\mathbf{V}_\alpha|l_1\, l_2, k\rangle \,, \tag{10.78}$$

and for the matrix elements of isovector mesons

$$\langle l_1'\, l_2', k'|\mathbf{V}_\alpha^T|l_1\, l_2, k\rangle = (4T - 3)\, \langle l_1'\, l_2', k'|\mathbf{V}_\alpha|l_1\, l_2, k\rangle \,, \tag{10.79}$$

with eigenvalues $T = 1$ or 0, depending on the isospin quantum numbers of the coupled two-baryon states. Finally, summing the matrix elements of the individual mesons, as indicated in (5.151), leads to the total baryon–baryon potential. The isospin part of particle–antiparticle states coincides with the one of the particle–particle states and hence will not be discussed separately.

The scheme outlined just above is not applicable to asymmetric nuclear matter, made up of different numbers of protons and neutrons, the reason being that, in this case, the boson-exchange matrix elements calculated for protons differ from those of neutrons which is a consequence of the self-consistent baryon–antibaryon basis. It is therefore of no advantage to proceed as described immediately above. Instead, we calculate the isospin factors of the individual matrix elements directly. We start again from the tensor products of equation (10.75), which remains unchanged of course. Now, however, $i = \frac{1}{2}$ denotes a proton or antineutron, whereas $i = -\frac{1}{2}$ stands for a neutron or an antiproton. The baryon–meson vertices of (9.24) through (9.26) are modified according to

$$\left(\Gamma^B_\alpha\right)_{\zeta_1\zeta_3}\left(\Gamma^B_\alpha\right)_{\zeta_2\zeta_4} \longrightarrow \left(\Gamma^B_\alpha \otimes \tau^r\right)_{\zeta_1\zeta_3}\left(\Gamma^B_\alpha \otimes \tau^r\right)_{\zeta_2\zeta_4}, \tag{10.80}$$

where τ^r $(r = 1, 2, 3)$ denote the Pauli matrices listed in appendix A. The individual matrix elements of the isospin parts of (10.80) are given by

$$\left\langle\frac{1}{2}\middle|\tau^1\middle|-\frac{1}{2}\right\rangle = 1\,,\quad \left\langle\frac{1}{2}\middle|\tau^2\middle|-\frac{1}{2}\right\rangle = -1\,,\quad \left\langle\frac{1}{2}\middle|\tau^3\middle|\frac{1}{2}\right\rangle = 1\,, \tag{10.81}$$

$$\left\langle-\frac{1}{2}\middle|\tau^1\middle|\frac{1}{2}\right\rangle = 1\,,\quad \left\langle-\frac{1}{2}\middle|\tau^2\middle|\frac{1}{2}\right\rangle = 1\,,\quad \left\langle-\frac{1}{2}\middle|\tau^3\middle|-\frac{1}{2}\right\rangle = -1\,. \tag{10.82}$$

This implies multiplying the corresponding boson-exchange matrix elements with one of the following factors:

$$\left\langle\frac{1}{2},\frac{1}{2}\middle|\mathbf{V}\middle|\frac{1}{2},\frac{1}{2}\right\rangle = \left\langle-\frac{1}{2},-\frac{1}{2}\middle|\mathbf{V}\middle|-\frac{1}{2},-\frac{1}{2}\right\rangle = 1\,, \tag{10.83}$$

$$\left\langle-\frac{1}{2},\frac{1}{2}\middle|\mathbf{V}\middle|-\frac{1}{2},\frac{1}{2}\right\rangle = \left\langle\frac{1}{2},-\frac{1}{2}\middle|\mathbf{V}\middle|\frac{1}{2},-\frac{1}{2}\right\rangle = -1\,, \tag{10.84}$$

$$\left\langle-\frac{1}{2},\frac{1}{2}\middle|\mathbf{V}\middle|\frac{1}{2},-\frac{1}{2}\right\rangle = \left\langle\frac{1}{2},-\frac{1}{2}\middle|\mathbf{V}\middle|-\frac{1}{2},\frac{1}{2}\right\rangle = 2\,. \tag{10.85}$$

10.4 Partial-wave expansion of scattering matrix

In this section we perform an expansion of the self-consistent relativistic **T**-matrix elements (9.82) to (9.87) into angular momentum states $|jm, l_1 l_2\rangle$. This expansion is achieved in complete analogy to the partial-wave expansion of the OBEP matrix elements performed in sections 10.1 and 10.2.

We begin with the ansatz for $\langle l_1' l_2' k'|\mathbf{T}|l_1 l_2 k\rangle$ which may be expanded, in complete analogy to (10.7), as

$$\langle l_1' l_2' k'|\overset{\mathrm{k}}{\mathbf{T}}_K|l_1 l_2 k\rangle = \sum_j \frac{2j+1}{4\,\pi}\, d^j_{ll'}(\theta)\langle l_1' l_2'|\overset{\mathrm{k}}{\mathbf{T}}{}^j(k,k';K)|l_1' l_2'\rangle\,. \tag{10.86}$$

The angular momentum states $|jm, l_1 l_2\rangle$ obey the completeness and orthonormality relations

$$\sum_{j,m} \langle\hat{\boldsymbol{k}} l_1 l_2|jm l_1 l_2\rangle\;\langle jm l_1 l_2|\hat{\boldsymbol{k}}' l_1 l_2\rangle = \delta(\Omega - \Omega') \tag{10.87}$$

and

$$\int \mathrm{d}\Omega \, \langle jml_1l_2|\hat{\boldsymbol{k}}l_1l_2\rangle \, \langle \hat{\boldsymbol{k}}l_1l_2|j'm'l_1l_2\rangle = \delta_{jj'} \, \delta_{mm'} \, , \tag{10.88}$$

respectively. With the aid of these two relations, the integral equation (9.94) for the self-consistent T-matrix amplitudes can be written as

$$\begin{aligned} &\langle l_1'l_2'k'|\overset{1}{\mathsf{T}}{}^j_K(k',k)|l_1l_2k\rangle = \langle l_1'l_2'k'|\overset{1}{\mathsf{V}}{}^j(k',k)|l_1l_2k\rangle \\ &+\sum_{j,m}\int \frac{\mathrm{d}^3\boldsymbol{q}}{(2\pi)^3}\,\langle l_1'l_2'|\overset{1}{\mathsf{V}}{}^j(k',q)|l_3l_4\rangle\,\langle jml_3l_4|l_3l_4\hat{\boldsymbol{q}}\rangle \\ &\times \boldsymbol{\Lambda}_{pn}\Big(\frac{\boldsymbol{K}}{2}+\boldsymbol{q},\frac{\boldsymbol{K}}{2}-\boldsymbol{q};K^0\Big)\langle l_3l_4\hat{\boldsymbol{q}}|j'm'l_3l_4\rangle\,\langle l_3l_4|\overset{1}{\mathsf{T}}{}^j_K(q,k')|l_1l_2\rangle \\ &+\sum_{j,m}\int \frac{\mathrm{d}^3\boldsymbol{q}}{(2\pi)^3}\,\langle l_1'l_2'|\overset{3}{\mathsf{V}}{}^j(k',q)|l_3l_4\rangle\,\langle jml_3l_4|l_3l_4\hat{\boldsymbol{q}}\rangle \\ &\times \boldsymbol{\Lambda}_{np}\Big(\frac{\boldsymbol{K}}{2}+\boldsymbol{q},\frac{\boldsymbol{K}}{2}-\boldsymbol{q};K^0\Big)\langle l_3l_4\hat{\boldsymbol{q}}|j'm'l_3l_4\rangle\,\langle l_3l_4|\overset{4}{\mathsf{T}}{}^j_K(q,k')|l_1l_2\rangle \, . \end{aligned} \tag{10.89}$$

Since the propagator $\boldsymbol{\Lambda}$ depends on the angle between the momenta $\boldsymbol{K}$ and $\boldsymbol{q}$, equation (10.89) constitutes a system of coupled equations for the $\overset{1}{\mathsf{T}}{}^j$'s. Replacing $\boldsymbol{\Lambda}$ with its momentum-averaged counterpart,

$$\bar{\boldsymbol{\Lambda}}(|\boldsymbol{q}|;K^0,|\boldsymbol{K}|) = \frac{1}{4\pi}\int \mathrm{d}\Omega \; \boldsymbol{\Lambda}\Big(\frac{\boldsymbol{K}}{2}+\boldsymbol{q},\frac{\boldsymbol{K}}{2}-\boldsymbol{q};K^0\Big)\, , \tag{10.90}$$

enables us to get rid of this additional complication, that is, we can immediately perform the integration over the solid angle in (10.89). Thus, replacing $\boldsymbol{\Lambda}$ with $\bar{\boldsymbol{\Lambda}}$ in (10.89) leads for the integral equation of the T-matrix to

$$\begin{aligned} &\langle l_1'l_2'k'|\overset{1}{\mathsf{T}}{}^j(k',k;K^0,|\boldsymbol{K}|)|l_1l_2k\rangle = \langle l_1'l_2'k'|\overset{1}{\mathsf{V}}{}^j(k',k)|l_1l_2k\rangle \\ &+\frac{1}{(2\pi)^2}\int_0^\infty \mathrm{d}q\,\boldsymbol{q}^2\langle l_1'l_2'|\overset{1}{\mathsf{V}}{}^j(k'^0|\boldsymbol{k}'|,q^0|\boldsymbol{q}|)|l_3l_4\rangle \\ &\times \bar{\boldsymbol{\Lambda}}_{pn}(|\boldsymbol{q}|;K^0,|\boldsymbol{K}|)\langle l_3l_4|\overset{1}{\mathsf{T}}{}^j(q^0,|\boldsymbol{q}|,k'^0,|\boldsymbol{k}'|;K^0,|\boldsymbol{K}|)|l_1l_2\rangle \\ &+\frac{1}{(2\pi)^2}\int_0^\infty \mathrm{d}q\,\boldsymbol{q}^2\langle l_1'l_2'|\overset{3}{\mathsf{V}}{}^j(k'^0,|\boldsymbol{k}'|,q^0,|\boldsymbol{q}|)|l_3l_4\rangle \\ &\times \bar{\boldsymbol{\Lambda}}_{np}(|\boldsymbol{q}|;K^0,|\boldsymbol{K}|)\langle l_3l_4|\overset{4}{\mathsf{T}}{}^j(q^0,|\boldsymbol{q}|,k'^0,|\boldsymbol{k}'|;K^0,|\boldsymbol{K}|)|l_1l_2\rangle \, . \end{aligned} \tag{10.91}$$

The replacement $\mathbf{\Lambda} \rightarrow \bar{\mathbf{\Lambda}}$ in (10.89) is known to be a very good approximation [317, 320] which implies a negligibly small error in the determination of $\overset{1}{\mathsf{T}}{}^{j}$. The integral equations for all other $\mathbf{T}$-matrix amplitudes are to be derived in complete analogy to the above. Resorting to the notation introduced in (10.17) through (10.26) for the boson-exchange matrix elements, the $\mathbf{T}$-matrix amplitudes $\overset{1}{\mathsf{T}}{}^{j}$, for instance, are given by

$$
{}^{0}\overset{1}{\mathsf{T}}{}^{j}(k',k) = \langle ++|\overset{1}{\mathsf{T}}{}^{j}(k',k)|++\rangle - \langle ++|\overset{1}{\mathsf{T}}{}^{j}(k',k)|--\rangle \,, \tag{10.92}
$$

in complete analogy to (10.17). Upon substituting the expressions for both $\langle ++|\overset{1}{\mathsf{T}}{}^{j}(k',k)|++\rangle$ and $\langle ++|\overset{1}{\mathsf{T}}{}^{j}(k',k)|--\rangle$, which are to be calculated from equation (10.91), into (10.92), one arrives at the following integral equation for ${}^{0}\overset{1}{\mathsf{T}}{}^{j}$,

$$
{}^{0}\overset{1}{\mathsf{T}}{}^{j} = {}^{0}\overset{1}{\mathsf{V}}{}^{j} + {}^{0}\overset{1}{\mathsf{V}}{}^{j}\,\bar{\mathbf{\Lambda}}\,{}^{0}\overset{1}{\mathsf{T}}{}^{j} + {}^{5}\overset{1}{\mathsf{V}}{}^{j}\,\bar{\mathbf{\Lambda}}\,{}^{3}\overset{1}{\mathsf{T}}{}^{j} + {}^{0}\overset{3}{\mathsf{V}}{}^{j}\,\bar{\mathbf{\Lambda}}\,{}^{0}\overset{4}{\mathsf{T}}{}^{j} + {}^{5}\overset{3}{\mathsf{V}}{}^{j}\,\bar{\mathbf{\Lambda}}\,{}^{3}\overset{4}{\mathsf{T}}{}^{j} \,. \tag{10.93}
$$

In a similar manner, one arrives for the other $\mathbf{T}^{j}$ amplitudes at the following coupled hierarchy:

$$
{}^{12}\overset{1}{\mathsf{T}}{}^{j} = {}^{12}\overset{1}{\mathsf{V}}{}^{j} + {}^{12}\overset{1}{\mathsf{V}}{}^{j}\,\bar{\mathbf{\Lambda}}\,{}^{12}\overset{1}{\mathsf{T}}{}^{j} + {}^{78}\overset{1}{\mathsf{V}}{}^{j}\,\bar{\mathbf{\Lambda}}\,{}^{56}\overset{1}{\mathsf{T}}{}^{j} + {}^{12}\overset{3}{\mathsf{V}}{}^{j}\,\bar{\mathbf{\Lambda}}\,{}^{12}\overset{4}{\mathsf{T}}{}^{j} + {}^{78}\overset{3}{\mathsf{V}}{}^{j}\,\bar{\mathbf{\Lambda}}\,{}^{56}\overset{4}{\mathsf{T}}{}^{j} \,, \tag{10.94}
$$

$$
{}^{1}\overset{1}{\mathsf{T}}{}^{j} = {}^{1}\overset{1}{\mathsf{V}}{}^{j} + {}^{1}\overset{1}{\mathsf{V}}{}^{j}\,\bar{\mathbf{\Lambda}}\,{}^{1}\overset{1}{\mathsf{T}}{}^{j} + {}^{3}\overset{1}{\mathsf{V}}{}^{j}\,\bar{\mathbf{\Lambda}}\,{}^{5}\overset{1}{\mathsf{T}}{}^{j} + {}^{1}\overset{3}{\mathsf{V}}{}^{j}\,\bar{\mathbf{\Lambda}}\,{}^{4}\overset{4}{\mathsf{T}}{}^{j} + {}^{3}\overset{3}{\mathsf{V}}{}^{j}\,\bar{\mathbf{\Lambda}}\,{}^{5}\overset{4}{\mathsf{T}}{}^{j} \,, \tag{10.95}
$$

$$
{}^{34}\overset{1}{\mathsf{T}}{}^{j} = {}^{34}\overset{1}{\mathsf{V}}{}^{j} + {}^{34}\overset{1}{\mathsf{V}}{}^{j}\,\bar{\mathbf{\Lambda}}\,{}^{34}\overset{1}{\mathsf{T}}{}^{j} + {}^{56}\overset{1}{\mathsf{V}}{}^{j}\,\bar{\mathbf{\Lambda}}\,{}^{78}\overset{1}{\mathsf{T}}{}^{j} + {}^{34}\overset{3}{\mathsf{V}}{}^{j}\,\bar{\mathbf{\Lambda}}\,{}^{34}\overset{4}{\mathsf{T}}{}^{j} + {}^{56}\overset{3}{\mathsf{V}}{}^{j}\,\bar{\mathbf{\Lambda}}\,{}^{78}\overset{4}{\mathsf{T}}{}^{j} \,, \tag{10.96}
$$

$$
{}^{3}\overset{1}{\mathsf{T}}{}^{j} = {}^{3}\overset{1}{\mathsf{V}}{}^{j} + {}^{3}\overset{1}{\mathsf{V}}{}^{j}\,\bar{\mathbf{\Lambda}}\,{}^{0}\overset{1}{\mathsf{T}}{}^{j} + {}^{1}\overset{1}{\mathsf{V}}{}^{j}\,\bar{\mathbf{\Lambda}}\,{}^{3}\overset{1}{\mathsf{T}}{}^{j} + {}^{3}\overset{3}{\mathsf{V}}{}^{j}\,\bar{\mathbf{\Lambda}}\,{}^{0}\overset{4}{\mathsf{T}}{}^{j} + {}^{1}\overset{3}{\mathsf{V}}{}^{j}\,\bar{\mathbf{\Lambda}}\,{}^{3}\overset{4}{\mathsf{T}}{}^{j} \,, \tag{10.97}
$$

$$
{}^{56}\overset{1}{\mathsf{T}}{}^{j} = {}^{56}\overset{1}{\mathsf{V}}{}^{j} + {}^{56}\overset{1}{\mathsf{V}}{}^{j}\,\bar{\mathbf{\Lambda}}\,{}^{12}\overset{1}{\mathsf{T}}{}^{j} + {}^{34}\overset{1}{\mathsf{V}}{}^{j}\,\bar{\mathbf{\Lambda}}\,{}^{56}\overset{1}{\mathsf{T}}{}^{j} + {}^{56}\overset{3}{\mathsf{V}}{}^{j}\,\bar{\mathbf{\Lambda}}\,{}^{12}\overset{4}{\mathsf{T}}{}^{j} + {}^{34}\overset{3}{\mathsf{V}}{}^{j}\,\bar{\mathbf{\Lambda}}\,{}^{56}\overset{4}{\mathsf{T}}{}^{j} \,, \tag{10.98}
$$

$$
{}^{5}\overset{1}{\mathsf{T}}{}^{j} = {}^{5}\overset{1}{\mathsf{V}}{}^{j} + {}^{0}\overset{1}{\mathsf{V}}{}^{j}\,\bar{\mathbf{\Lambda}}\,{}^{5}\overset{1}{\mathsf{T}}{}^{j} + {}^{5}\overset{1}{\mathsf{V}}{}^{j}\,\bar{\mathbf{\Lambda}}\,{}^{1}\overset{1}{\mathsf{T}}{}^{j} + {}^{0}\overset{3}{\mathsf{V}}{}^{j}\,\bar{\mathbf{\Lambda}}\,{}^{5}\overset{4}{\mathsf{T}}{}^{j} + {}^{5}\overset{3}{\mathsf{V}}{}^{j}\,\bar{\mathbf{\Lambda}}\,{}^{1}\overset{4}{\mathsf{T}}{}^{j} \,, \tag{10.99}
$$

$$
{}^{78}\overset{1}{\mathsf{T}}{}^{j} = {}^{78}\overset{1}{\mathsf{V}}{}^{j} + {}^{12}\overset{1}{\mathsf{V}}{}^{j}\,\bar{\mathbf{\Lambda}}\,{}^{78}\overset{1}{\mathsf{T}}{}^{j} + {}^{78}\overset{1}{\mathsf{V}}{}^{j}\,\bar{\mathbf{\Lambda}}\,{}^{34}\overset{1}{\mathsf{T}}{}^{j} + {}^{12}\overset{3}{\mathsf{V}}{}^{j}\,\bar{\mathbf{\Lambda}}\,{}^{78}\overset{4}{\mathsf{T}}{}^{j} + {}^{78}\overset{3}{\mathsf{V}}{}^{j}\,\bar{\mathbf{\Lambda}}\,{}^{34}\overset{4}{\mathsf{T}}{}^{j} \,. \tag{10.100}
$$

It is seen that the solutions of the first four of these equations, ${}^{0}\overset{1}{\mathsf{T}}{}^{j}$, ${}^{12}\overset{1}{\mathsf{T}}{}^{j}$, ${}^{1}\overset{1}{\mathsf{T}}{}^{j}$ and ${}^{34}\overset{1}{\mathsf{T}}{}^{j}$, enter the determination of the four latter amplitudes ${}^{3}\overset{1}{\mathsf{T}}{}^{j}$, ${}^{56}\overset{1}{\mathsf{T}}{}^{j}$, ${}^{5}\overset{1}{\mathsf{T}}{}^{j}$ and ${}^{78}\overset{1}{\mathsf{T}}{}^{j}$, and vice versa. In other words, the helicity quantum number couples the respective equations, that is (10.93) with (10.97),

(10.94) with (10.98) etc, pairwise with one another. Moreover, the isospin dependence of the amplitudes (10.93) to (10.100) involves a coupling of these amplitudes to the $\overset{4}{\mathsf{T}}{}^{j}$ amplitudes, which obey the following set of integral equations,

$$ {}^{0}\overset{4}{\mathsf{T}}{}^{j} = {}^{0}\overset{4}{\mathsf{V}}{}^{j} + {}^{0}\overset{4}{\mathsf{V}}{}^{j}\,\bar{\mathbf{\Lambda}}\,{}^{0}\overset{1}{\mathsf{T}}{}^{j} + {}^{5}\overset{4}{\mathsf{V}}{}^{j}\,\bar{\mathbf{\Lambda}}\,{}^{3}\overset{1}{\mathsf{T}}{}^{j} + {}^{0}\overset{2}{\mathsf{V}}{}^{j}\,\bar{\mathbf{\Lambda}}\,{}^{0}\overset{4}{\mathsf{T}}{}^{j} + {}^{5}\overset{2}{\mathsf{V}}{}^{j}\,\bar{\mathbf{\Lambda}}\,{}^{3}\overset{4}{\mathsf{T}}{}^{j}\,, \tag{10.101} $$

$$ {}^{12}\overset{4}{\mathsf{T}}{}^{j} = {}^{12}\overset{4}{\mathsf{V}}{}^{j} + {}^{12}\overset{4}{\mathsf{V}}{}^{j}\,\bar{\mathbf{\Lambda}}\,{}^{12}\overset{1}{\mathsf{T}}{}^{j} + {}^{78}\overset{4}{\mathsf{V}}{}^{j}\,\bar{\mathbf{\Lambda}}\,{}^{56}\overset{1}{\mathsf{T}}{}^{j} + {}^{12}\overset{2}{\mathsf{V}}{}^{j}\,\bar{\mathbf{\Lambda}}\,{}^{12}\overset{4}{\mathsf{T}}{}^{j} + {}^{78}\overset{2}{\mathsf{V}}{}^{j}\,\bar{\mathbf{\Lambda}}\,{}^{56}\overset{4}{\mathsf{T}}{}^{j}\,, \tag{10.102} $$

$$ {}^{1}\overset{4}{\mathsf{T}}{}^{j} = {}^{1}\overset{4}{\mathsf{V}}{}^{j} + {}^{1}\overset{4}{\mathsf{V}}{}^{j}\,\bar{\mathbf{\Lambda}}\,{}^{1}\overset{1}{\mathsf{T}}{}^{j} + {}^{3}\overset{4}{\mathsf{V}}{}^{j}\,\bar{\mathbf{\Lambda}}\,{}^{5}\overset{1}{\mathsf{T}}{}^{j} + {}^{1}\overset{2}{\mathsf{V}}{}^{j}\,\bar{\mathbf{\Lambda}}\,{}^{1}\overset{4}{\mathsf{T}}{}^{j} + {}^{3}\overset{2}{\mathsf{V}}{}^{j}\,\bar{\mathbf{\Lambda}}\,{}^{5}\overset{4}{\mathsf{T}}{}^{j}\,, \tag{10.103} $$

$$ {}^{34}\overset{4}{\mathsf{T}}{}^{j} = {}^{34}\overset{4}{\mathsf{V}}{}^{j} + {}^{34}\overset{4}{\mathsf{V}}{}^{j}\,\bar{\mathbf{\Lambda}}\,{}^{34}\overset{1}{\mathsf{T}}{}^{j} + {}^{56}\overset{4}{\mathsf{V}}{}^{j}\,\bar{\mathbf{\Lambda}}\,{}^{78}\overset{1}{\mathsf{T}}{}^{j} + {}^{34}\overset{2}{\mathsf{V}}{}^{j}\,\bar{\mathbf{\Lambda}}\,{}^{34}\overset{4}{\mathsf{T}}{}^{j} + {}^{56}\overset{2}{\mathsf{V}}{}^{j}\,\bar{\mathbf{\Lambda}}\,{}^{78}\overset{4}{\mathsf{T}}{}^{j}\,, \tag{10.104} $$

$$ {}^{3}\overset{4}{\mathsf{T}}{}^{j} = {}^{3}\overset{4}{\mathsf{V}}{}^{j} + {}^{3}\overset{4}{\mathsf{V}}{}^{j}\,\bar{\mathbf{\Lambda}}\,{}^{0}\overset{1}{\mathsf{T}}{}^{j} + {}^{1}\overset{4}{\mathsf{V}}{}^{j}\,\bar{\mathbf{\Lambda}}\,{}^{3}\overset{1}{\mathsf{T}}{}^{j} + {}^{3}\overset{2}{\mathsf{V}}{}^{j}\,\bar{\mathbf{\Lambda}}\,{}^{0}\overset{4}{\mathsf{T}}{}^{j} + {}^{1}\overset{2}{\mathsf{V}}{}^{j}\,\bar{\mathbf{\Lambda}}\,{}^{3}\overset{4}{\mathsf{T}}{}^{j}\,, \tag{10.105} $$

$$ {}^{56}\overset{4}{\mathsf{T}}{}^{j} = {}^{56}\overset{4}{\mathsf{V}}{}^{j} + {}^{56}\overset{4}{\mathsf{V}}{}^{j}\,\bar{\mathbf{\Lambda}}\,{}^{12}\overset{1}{\mathsf{T}}{}^{j} + {}^{34}\overset{4}{\mathsf{V}}{}^{j}\,\bar{\mathbf{\Lambda}}\,{}^{56}\overset{1}{\mathsf{T}}{}^{j} + {}^{56}\overset{2}{\mathsf{V}}{}^{j}\,\bar{\mathbf{\Lambda}}\,{}^{12}\overset{4}{\mathsf{T}}{}^{j} + {}^{34}\overset{2}{\mathsf{V}}{}^{j}\,\bar{\mathbf{\Lambda}}\,{}^{56}\overset{4}{\mathsf{T}}{}^{j}\,, \tag{10.106} $$

$$ {}^{5}\overset{4}{\mathsf{T}}{}^{j} = {}^{5}\overset{4}{\mathsf{V}}{}^{j} + {}^{0}\overset{4}{\mathsf{V}}{}^{j}\,\bar{\mathbf{\Lambda}}\,{}^{5}\overset{1}{\mathsf{T}}{}^{j} + {}^{5}\overset{4}{\mathsf{V}}{}^{j}\,\bar{\mathbf{\Lambda}}\,{}^{1}\overset{1}{\mathsf{T}}{}^{j} + {}^{0}\overset{2}{\mathsf{V}}{}^{j}\,\bar{\mathbf{\Lambda}}\,{}^{5}\overset{4}{\mathsf{T}}{}^{j} + {}^{5}\overset{2}{\mathsf{V}}{}^{j}\,\bar{\mathbf{\Lambda}}\,{}^{1}\overset{4}{\mathsf{T}}{}^{j}\,, \tag{10.107} $$

$$ {}^{78}\overset{4}{\mathsf{T}}{}^{j} = {}^{78}\overset{4}{\mathsf{V}}{}^{j} + {}^{12}\overset{4}{\mathsf{V}}{}^{j}\,\bar{\mathbf{\Lambda}}\,{}^{78}\overset{1}{\mathsf{T}}{}^{j} + {}^{78}\overset{4}{\mathsf{V}}{}^{j}\,\bar{\mathbf{\Lambda}}\,{}^{34}\overset{1}{\mathsf{T}}{}^{j} + {}^{12}\overset{2}{\mathsf{V}}{}^{j}\,\bar{\mathbf{\Lambda}}\,{}^{78}\overset{4}{\mathsf{T}}{}^{j} + {}^{78}\overset{2}{\mathsf{V}}{}^{j}\,\bar{\mathbf{\Lambda}}\,{}^{34}\overset{4}{\mathsf{T}}{}^{j}\,. \tag{10.108} $$

Similarly to the $\overset{1}{\mathsf{T}}{}^{j}$'s from above, the ${}^{0}\overset{4}{\mathsf{T}}{}^{j}$ equation is coupled with the one for ${}^{3}\overset{4}{\mathsf{T}}{}^{j}$, the ${}^{12}\overset{4}{\mathsf{T}}{}^{j}$ equation is coupled with the one for ${}^{56}\overset{4}{\mathsf{T}}{}^{j}$ etc, and each pair of equations is solved simultaneously in combination with the corresponding pair from (10.93) to (10.100) which determines the respective $\overset{1}{\mathsf{T}}{}^{j}$ amplitudes that enter in the $\overset{4}{\mathsf{T}}{}^{j}$'s. Hence the total number of integral equations to be solved simultaneously for a given angular momentum quantum number is four. To bring this out more clearly, the first set of these equations is given by

$$ {}^{0}\overset{1}{\mathsf{T}}{}^{j} = {}^{0}\overset{1}{\mathsf{V}}{}^{j} + {}^{0}\overset{1}{\mathsf{V}}{}^{j}\,\bar{\mathbf{\Lambda}}\,{}^{0}\overset{1}{\mathsf{T}}{}^{j} + {}^{5}\overset{1}{\mathsf{V}}{}^{j}\,\bar{\mathbf{\Lambda}}\,{}^{3}\overset{1}{\mathsf{T}}{}^{j} + {}^{0}\overset{3}{\mathsf{V}}{}^{j}\,\bar{\mathbf{\Lambda}}\,{}^{0}\overset{4}{\mathsf{T}}{}^{j} + {}^{5}\overset{3}{\mathsf{V}}{}^{j}\,\bar{\mathbf{\Lambda}}\,{}^{3}\overset{4}{\mathsf{T}}{}^{j}\,, \tag{10.109} $$

$$ {}^{3}\overset{1}{\mathsf{T}}{}^{j} = {}^{3}\overset{1}{\mathsf{V}}{}^{j} + {}^{3}\overset{1}{\mathsf{V}}{}^{j}\,\bar{\mathbf{\Lambda}}\,{}^{0}\overset{1}{\mathsf{T}}{}^{j} + {}^{1}\overset{1}{\mathsf{V}}{}^{j}\,\bar{\mathbf{\Lambda}}\,{}^{3}\overset{1}{\mathsf{T}}{}^{j} + {}^{3}\overset{3}{\mathsf{V}}{}^{j}\,\bar{\mathbf{\Lambda}}\,{}^{0}\overset{4}{\mathsf{T}}{}^{j} + {}^{1}\overset{3}{\mathsf{V}}{}^{j}\,\bar{\mathbf{\Lambda}}\,{}^{3}\overset{4}{\mathsf{T}}{}^{j}\,, \tag{10.110} $$

$$ {}^{0}\overset{4}{\mathsf{T}}{}^{j} = {}^{0}\overset{4}{\mathsf{V}}{}^{j} + {}^{0}\overset{4}{\mathsf{V}}{}^{j}\,\bar{\mathbf{\Lambda}}\,{}^{0}\overset{1}{\mathsf{T}}{}^{j} + {}^{5}\overset{4}{\mathsf{V}}{}^{j}\,\bar{\mathbf{\Lambda}}\,{}^{3}\overset{1}{\mathsf{T}}{}^{j} + {}^{0}\overset{2}{\mathsf{V}}{}^{j}\,\bar{\mathbf{\Lambda}}\,{}^{0}\overset{4}{\mathsf{T}}{}^{j} + {}^{5}\overset{2}{\mathsf{V}}{}^{j}\,\bar{\mathbf{\Lambda}}\,{}^{3}\overset{4}{\mathsf{T}}{}^{j}\,, \tag{10.111} $$

Table 10.1. Symmetric-matter limits of amplitudes and self-energies.

Asymmetric matter		Symmetric matter
${}^{78}\overset{1}{\mathsf{V}}{}^{j}$	$\longrightarrow$	$2\times{}^{5}\mathsf{V}^{j}$
${}^{78}\overset{1}{\mathsf{T}}{}^{j}$	$\longrightarrow$	$2\times{}^{5}\mathsf{T}^{j}$
${}^{56}\overset{1}{\mathsf{V}}{}^{j}$	$\longrightarrow$	$2\times{}^{6}\mathsf{V}^{j}$
${}^{56}\overset{1}{\mathsf{T}}{}^{j}$	$\longrightarrow$	$2\times{}^{6}\mathsf{T}^{j}$
${}^{3}\overset{1}{\mathsf{T}}{}^{j}$	$\longrightarrow$	0
${}^{5}\overset{1}{\mathsf{T}}{}^{j}$	$\longrightarrow$	0
$\overset{4}{\mathsf{T}}{}^{j}$	$\longrightarrow$	0
$\overset{4}{\mathsf{V}}{}^{j}$	$\longrightarrow$	0
$\boldsymbol{\Sigma}^{p}$, $\boldsymbol{\Sigma}^{n}$	$\longrightarrow$	$\boldsymbol{\Sigma}^{p}=\boldsymbol{\Sigma}^{n}=\boldsymbol{\Sigma}^{N}$

and

$$ {}^{3}\overset{4}{\mathsf{T}}{}^{j} = {}^{3}\overset{4}{\mathsf{V}}{}^{j} + {}^{3}\overset{4}{\mathsf{V}}{}^{j}\,\bar{\boldsymbol{\Lambda}}\,{}^{0}\overset{1}{\mathsf{T}}{}^{j} + {}^{1}\overset{4}{\mathsf{V}}{}^{j}\,\bar{\boldsymbol{\Lambda}}\,{}^{3}\overset{1}{\mathsf{T}}{}^{j} + {}^{3}\overset{2}{\mathsf{V}}{}^{j}\,\bar{\boldsymbol{\Lambda}}\,{}^{0}\overset{4}{\mathsf{T}}{}^{j} + {}^{1}\overset{2}{\mathsf{V}}{}^{j}\,\bar{\boldsymbol{\Lambda}}\,{}^{3}\overset{4}{\mathsf{T}}{}^{j}\,. \tag{10.112} $$

Before we proceed with introducing the $\mathbf{T}$-matrix elements which contain antibaryon spinors, let us consider equations (10.93) through (10.100) for the simpler case of symmetric nuclear matter. This special case is recovered from (10.93) through (10.100) by making use of the identifications listed in table 10.1. One then gets

$$ {}^{0}\mathsf{T}^{j} = {}^{0}\mathsf{V}^{j} + {}^{0}\mathsf{V}^{j}\,\bar{\boldsymbol{\Lambda}}\,{}^{0}\mathsf{T}^{j}\,, \tag{10.113} $$

$$ {}^{12}\mathsf{T}^{j} = {}^{12}\mathsf{V}^{j} + {}^{12}\mathsf{V}^{j}\,\bar{\boldsymbol{\Lambda}}\,{}^{12}\mathsf{T}^{j} + 4\,{}^{5}\mathsf{V}^{j}\,\bar{\boldsymbol{\Lambda}}\,{}^{6}\mathsf{T}^{j}\,, \tag{10.114} $$

$$ {}^{1}\mathsf{T}^{j} = {}^{1}\mathsf{V}^{j} + {}^{1}\mathsf{V}^{j}\,\bar{\boldsymbol{\Lambda}}\,{}^{1}\mathsf{T}^{j}\,, \tag{10.115} $$

$$ {}^{34}\mathsf{T}^{j} = {}^{34}\mathsf{V}^{j} + {}^{34}\mathsf{V}^{j}\,\bar{\boldsymbol{\Lambda}}\,{}^{34}\mathsf{T}^{j} + 4\,{}^{6}\mathsf{V}^{j}\,\bar{\boldsymbol{\Lambda}}\,{}^{5}\mathsf{T}^{j}\,, \tag{10.116} $$

$$ {}^{6}\mathsf{T}^{j} = {}^{6}\mathsf{V}^{j} + {}^{6}\mathsf{V}^{j}\,\bar{\boldsymbol{\Lambda}}\,{}^{12}\mathsf{T}^{j} + {}^{34}\mathsf{V}^{j}\,\bar{\boldsymbol{\Lambda}}\,{}^{6}\mathsf{T}^{j}\,, \tag{10.117} $$

$$ {}^{5}\mathsf{T}^{j} = {}^{5}\mathsf{V}^{j} + {}^{12}\mathsf{V}^{j}\,\bar{\boldsymbol{\Lambda}}\,{}^{5}\mathsf{T}^{j} + {}^{5}\mathsf{V}^{j}\,\bar{\boldsymbol{\Lambda}}\,{}^{34}\mathsf{T}^{j}\,. \tag{10.118} $$

There are no equations for the $\overset{4}{\mathsf{T}}{}^{j}$'s.

After these remarks let us turn back again to the asymmetric nuclear matter case and introduce the mixed baryon–antibaryon $\mathbf{T}$-matrix amplitudes as follows,

$$ \langle l_1' l_2' k' | \overset{1}{\mathbf{R}} | l_1 l_2 k \rangle \equiv \langle \Theta^{p}_{l_1'}(k_1'), \Phi^{n}_{l_2'}(k_2') | \mathbf{T} | \Phi^{n}_{l_1}(k_1), \Phi^{n}_{l_2}(k_2) \rangle\,, \tag{10.119} $$

$$ \langle l_1' l_2' k' | \overset{2}{\mathbf{R}} | l_1 l_2 k \rangle \equiv \langle \Theta^{p}_{l_1'}(k_1'), \Phi^{p}_{l_2'}(k_2') | \mathbf{T} | \Phi^{n}_{l_1}(k_1), \Phi^{p}_{l_2}(k_2) \rangle\,, \tag{10.120} $$

$$\langle l_1' l_2' k' | \overset{3}{\mathbf{R}} | l_1 l_2 k \rangle \equiv \langle \Theta^p_{l_1'}(k_1'), \Phi^p_{l_2'}(k_2') | \mathbf{T} | \Phi^p_{l_1}(k_1), \Phi^n_{l_2}(k_2) \rangle \,, \tag{10.121}$$

$$\langle l_1' l_2' k' | \overset{4}{\mathbf{R}} | l_1 l_2 k \rangle \equiv \langle \Theta^n_{l_1'}(k_1'), \Phi^p_{l_2'}(k_2') | \mathbf{T} | \Phi^p_{l_1}(k_1), \Phi^p_{l_2}(k_2) \rangle \,, \tag{10.122}$$

$$\langle l_1' l_2' k' | \overset{5}{\mathbf{R}} | l_1 l_2 k \rangle \equiv \langle \Theta^n_{l_1'}(k_1'), \Phi^n_{l_2'}(k_2') | \mathbf{T} | \Phi^n_{l_1}(k_1), \Phi^p_{l_2}(k_2) \rangle \,, \tag{10.123}$$

$$\langle l_1' l_2' k' | \overset{6}{\mathbf{R}} | l_1 l_2 k \rangle \equiv \langle \Theta^n_{l_1'}(k_1'), \Phi^n_{l_2'}(k_2') | \mathbf{T} | \Phi^p_{l_1}(k_1), \Phi^n_{l_2}(k_2) \rangle \,. \tag{10.124}$$

The $\mathbf{T}$-matrix amplitudes containing two antibaryon spinors are defined as,

$$\langle l_1' l_2' k' | \overset{1}{\mathbf{S}} | l_1 l_2 k \rangle \equiv \langle \Theta^p_{l_1'}(k_1'), \Phi^n_{l_2'}(k_2') | \mathbf{T} | \Phi^n_{l_1}(k_1), \Theta^p_{l_2}(k_2) \rangle \,, \tag{10.125}$$

$$\langle l_1' l_2' k' | \overset{2}{\mathbf{S}} | l_1 l_2 k \rangle \equiv \langle \Theta^p_{l_1'}(k_1'), \Phi^p_{l_2'}(k_2') | \mathbf{T} | \Phi^p_{l_1}(k_1), \Theta^n_{l_2}(k_2) \rangle \,, \tag{10.126}$$

$$\langle l_1' l_2' k' | \overset{3}{\mathbf{S}} | l_1 l_2 k \rangle \equiv \langle \Theta^p_{l_1'}(k_1'), \Phi^p_{l_2'}(k_2') | \mathbf{T} | \Phi^p_{l_1}(k_1), \Theta^p_{l_2}(k_2) \rangle \,, \tag{10.127}$$

$$\langle l_1' l_2' k' | \overset{4}{\mathbf{S}} | l_1 l_2 k \rangle \equiv \langle \Theta^n_{l_1'}(k_1'), \Phi^n_{l_2'}(k_2') | \mathbf{T} | \Phi^n_{l_1}(k_1), \Theta^n_{l_2}(k_2) \rangle \,, \tag{10.128}$$

and

$$\langle l_1' l_2' k' | \overset{1}{\mathbf{P}} | l_1 l_2 k \rangle \equiv \langle \Theta^p_{l_1'}(k_1'), \Phi^p_{l_2'}(k_2') | \mathbf{T} | \Theta^p_{l_1}(k_1), \Phi^p_{l_2}(k_2) \rangle \,, \tag{10.129}$$

$$\langle l_1' l_2' k' | \overset{2}{\mathbf{P}} | l_1 l_2 k \rangle \equiv \langle \Theta^n_{l_1'}(k_1'), \Phi^n_{l_2'}(k_2') | \mathbf{T} | \Theta^n_{l_1}(k_1), \Phi^n_{l_2}(k_2) \rangle \,, \tag{10.130}$$

$$\langle l_1' l_2' k' | \overset{3}{\mathbf{P}} | l_1 l_2 k \rangle \equiv \langle \Theta^p_{l_1'}(k_1'), \Phi^n_{l_2'}(k_2') | \mathbf{T} | \Theta^p_{l_1}(k_1), \Phi^n_{l_2}(k_2) \rangle \,, \tag{10.131}$$

$$\langle l_1' l_2' k' | \overset{4}{\mathbf{P}} | l_1 l_2 k \rangle \equiv \langle \Theta^n_{l_1'}(k_1'), \Phi^p_{l_2'}(k_2') | \mathbf{T} | \Theta^n_{l_1}(k_1), \Phi^p_{l_2}(k_2) \rangle \,. \tag{10.132}$$

Expanding each of these matrix elements into angular momentum states, in complete analogy to (10.89), and solving the corresponding equations for the numerous amplitudes $\overset{k}{\mathsf{R}}{}^j$, $\overset{k}{\mathsf{S}}{}^j$ and $\overset{k}{\mathsf{P}}{}^j$ finally enables one to compute the baryon self-energies of protons and neutrons. This is outlined in the following section.

10.5 Partial-wave expansion of baryon self-energies

Knowing the $\mathbf{T}$-matrix elements for the self-consistent baryon–antibaryon basis $\{\Phi, \Theta\}$, the expressions for the three self-energy components $\Sigma^B_{\Phi\Phi}$, $\Sigma^B_{\Theta\Theta}$ and $\Sigma^B_{\Theta\Phi}$, defined in equations (9.62) to (9.64) or, more explicitly, in (9.79) to (9.81), can be calculated. Substituting (10.86) into (9.79) gives for the $\Phi\Phi$-component of the proton self-energy, $\Sigma^p_{\Phi\Phi}$, the following result:

$$\Sigma^p_{\Phi\Phi}(k) = \frac{1}{4} \frac{1}{(2\,\pi)^2} \Bigg\{ \sum_j (2j+1) \int_0^{k_{F_p}} \frac{\mathrm{d}^3 \boldsymbol{k}'}{(2\pi)^3} \Big[{}^0\overset{5}{\mathsf{T}}{}^j \Big(\frac{k-k'}{2}, \frac{k-k'}{2}; k+k' \Big)$$

$$
\begin{aligned}
&\quad + {}^{12}\overset{5}{\mathsf{T}}{}^{j}\Big(\frac{k-k'}{2}, \frac{k-k'}{2}; k+k'\Big)\Big]_{k'^0=\omega^p(|k'|)} n^p(|\boldsymbol{k}'|) \\
&+ \sum_{j=1}(2j+1) \int_0^{k_{F_p}} \frac{\mathrm{d}^3\boldsymbol{k}'}{(2\pi)^3} \Big[{}^{1}\overset{5}{\mathsf{T}}{}^{j}\Big(\frac{k-k'}{2}, \frac{k-k'}{2}; k+k'\Big) \\
&\quad + {}^{34}\overset{5}{\mathsf{T}}{}^{j}\Big(\frac{k-k'}{2}, \frac{k-k'}{2}; k+k'\Big)\Big]_{k'^0=\omega^p(|k'|)} n^p(|\boldsymbol{k}'|) \\
&- \sum_{j}(2j+1)(-1)^j \int_0^{k_{F_p}} \frac{\mathrm{d}^3\boldsymbol{k}'}{(2\pi)^3} \Big[{}^{0}\overset{5}{\mathsf{T}}{}^{j}\Big(\frac{k-k'}{2}, \frac{k'-k}{2}; k+k'\Big) \\
&\quad + {}^{12}\overset{5}{\mathsf{T}}{}^{j}\Big(\frac{k-k'}{2}, \frac{k'-k}{2}; k+k'\Big)\Big]_{k'^0=\omega^p(|k'|)} n^p(|\boldsymbol{k}'|) \\
&- \sum_{j=1}(2j+1)(-1)^{j-1} \int_0^{k_{F_p}} \frac{\mathrm{d}^3\boldsymbol{k}'}{(2\pi)^3} \Big[{}^{1}\overset{5}{\mathsf{T}}{}^{j}\Big(\frac{k-k'}{2}, \frac{k'-k}{2}; k+k'\Big) \\
&\quad - {}^{34}\overset{5}{\mathsf{T}}{}^{j}\Big(\frac{k-k'}{2}, \frac{k'-k}{2}; k+k'\Big)\Big]_{k'^0=\omega^p(|k'|)} n^p(|\boldsymbol{k}'|) \\
&+ \sum_{j}(2j+1) \int_0^{k_{F_n}} \frac{\mathrm{d}^3\boldsymbol{k}'}{(2\pi)^3} \Big[{}^{0}\overset{1}{\mathsf{T}}{}^{j}\Big(\frac{k-k'}{2}, \frac{k-k'}{2}; k+k'\Big) \\
&\quad + {}^{12}\overset{1}{\mathsf{T}}{}^{j}\Big(\frac{k-k'}{2}, \frac{k-k'}{2}; k+k'\Big)\Big]_{k'^0=\omega^n(|k'|)} n^n(|\boldsymbol{k}'|) \\
&+ \sum_{j=1}(2j+1) \int_0^{k_{F_n}} \frac{\mathrm{d}^3\boldsymbol{k}'}{(2\pi)^3} \Big[{}^{1}\overset{1}{\mathsf{T}}{}^{j}\Big(\frac{k-k'}{2}, \frac{k-k'}{2}; k+k'\Big) \\
&\quad + {}^{34}\overset{1}{\mathsf{T}}{}^{j}\Big(\frac{k-k'}{2}, \frac{k-k'}{2}; k+k'\Big)\Big]_{k'^0=\omega^n(|k'|)} n^n(|\boldsymbol{k}'|) \\
&- \sum_{j}(2j+1)(-1)^j \int_0^{k_{F_n}} \frac{\mathrm{d}^3\boldsymbol{k}'}{(2\pi)^3} \Big[{}^{0}\overset{3}{\mathsf{T}}{}^{j}\Big(\frac{k-k'}{2}, \frac{k'-k}{2}; k+k'\Big) \\
&\quad + {}^{12}\overset{3}{\mathsf{T}}{}^{j}\Big(\frac{k-k'}{2}, \frac{k'-k}{2}; k+k'\Big)\Big]_{k'^0=\omega^n(|k'|)} n^n(|\boldsymbol{k}'|) \\
&- \sum_{j=1}(2j+1)(-1)^{j-1} \int_0^{k_{F_n}} \frac{\mathrm{d}^3\boldsymbol{k}'}{(2\pi)^3} \Big[{}^{1}\overset{3}{\mathsf{T}}{}^{j}\Big(\frac{k-k'}{2}, \frac{k'-k}{2}; k+k'\Big) \\
&\quad - {}^{34}\overset{3}{\mathsf{T}}{}^{j}\Big(\frac{k-k'}{2}, \frac{k'-k}{2}; k+k'\Big)\Big]_{k'^0=\omega^n(|k'|)} n^n(|\boldsymbol{k}'|)\Big\}. \qquad (10.133)
\end{aligned}
$$

The ΦΦ-component of the neutron self-energy is given by

$$
\begin{aligned}
\Sigma^n_{\Phi\Phi}(k) = \frac{1}{4}\,\frac{1}{(2\,\pi)^2}\Bigg\{ & \sum_j (2j+1) \int_0^{k_{F_n}} \frac{\mathrm{d}^3\boldsymbol{k}'}{(2\pi)^3} \Bigg[{}^{0}\overset{6}{\mathsf{T}}{}^{j}\left(\frac{k-k'}{2},\frac{k-k'}{2};k+k'\right) \\
& \qquad + {}^{12}\overset{6}{\mathsf{T}}{}^{j}\left(\frac{k-k'}{2},\frac{k-k'}{2};k+k'\right)\Bigg]_{k'^0=\omega^n(|k'|)} n^n(|\boldsymbol{k}'|) \\
& + \sum_{j=1} (2j+1) \int_0^{k_{F_n}} \frac{\mathrm{d}^3\boldsymbol{k}'}{(2\pi)^3} \Bigg[{}^{1}\overset{6}{\mathsf{T}}{}^{j}\left(\frac{k-k'}{2},\frac{k-k'}{2};k+k'\right) \\
& \qquad + {}^{34}\overset{6}{\mathsf{T}}{}^{j}\left(\frac{k-k'}{2},\frac{k-k'}{2};k+k'\right)\Bigg]_{k'^0=\omega^n(|k'|)} n^n(|\boldsymbol{k}'|) \\
& - \sum_j (2j+1)\,(-1)^j \int_0^{k_{F_n}} \frac{\mathrm{d}^3\boldsymbol{k}'}{(2\pi)^3} \Bigg[{}^{0}\overset{6}{\mathsf{T}}{}^{j}\left(\frac{k-k'}{2},\frac{k'-k}{2};k+k'\right) \\
& \qquad + {}^{12}\overset{6}{\mathsf{T}}{}^{j}\left(\frac{k-k'}{2},\frac{k'-k}{2};k+k'\right)\Bigg]_{k'^0=\omega^n(|k'|)} n^n(|\boldsymbol{k}'|) \\
& - \sum_{j=1} (2j+1)\,(-1)^{j-1} \int_0^{k_{F_n}} \frac{\mathrm{d}^3\boldsymbol{k}'}{(2\pi)^3} \Bigg[{}^{1}\overset{6}{\mathsf{T}}{}^{j}\left(\frac{k-k'}{2},\frac{k'-k}{2};k+k'\right) \\
& \qquad - {}^{34}\overset{6}{\mathsf{T}}{}^{j}\left(\frac{k-k'}{2},\frac{k'-k}{2};k+k'\right)\Bigg]_{k'^0=\omega^n(|k'|)} n^n(|\boldsymbol{k}'|) \\
& + \sum_j (2j+1) \int_0^{k_{F_p}} \frac{\mathrm{d}^3\boldsymbol{k}'}{(2\pi)^3} \Bigg[{}^{0}\overset{2}{\mathsf{T}}{}^{j}\left(\frac{k-k'}{2},\frac{k-k'}{2};k+k'\right) \\
& \qquad + {}^{12}\overset{2}{\mathsf{T}}{}^{j}\left(\frac{k-k'}{2},\frac{k-k'}{2};k+k'\right)\Bigg]_{k'^0=\omega^p(|k'|)} n^p(|\boldsymbol{k}'|) \\
& + \sum_{j=1} (2j+1) \int_0^{k_{F_p}} \frac{\mathrm{d}^3\boldsymbol{k}'}{(2\pi)^3} \Bigg[{}^{1}\overset{2}{\mathsf{T}}{}^{j}\left(\frac{k-k'}{2},\frac{k-k'}{2};k+k'\right) \\
& \qquad + {}^{34}\overset{2}{\mathsf{T}}{}^{j}\left(\frac{k-k'}{2},\frac{k-k'}{2};k+k'\right)\Bigg]_{k'^0=\omega^p(|k'|)} n^p(|\boldsymbol{k}'|) \\
& - \sum_j (2j+1)\,(-1)^j \int_0^{k_{F_p}} \frac{\mathrm{d}^3\boldsymbol{k}'}{(2\pi)^3} \Bigg[{}^{0}\overset{4}{\mathsf{T}}{}^{j}\left(\frac{k-k'}{2},\frac{k'-k}{2};k+k'\right) \\
& \qquad + {}^{12}\overset{4}{\mathsf{T}}{}^{j}\left(\frac{k-k'}{2},\frac{k'-k}{2};k+k'\right)\Bigg]_{k'^0=\omega^p(|k'|)} n^p(|\boldsymbol{k}'|)
\end{aligned}
$$

$$- \sum_{j=1} (2j+1)\,(-1)^{j-1} \int_0^{k_{F_p}} \frac{\mathrm{d}^3\boldsymbol{k}'}{(2\pi)^3} \Big[{}^{1}\overset{4}{\mathsf{T}}{}^{j}\Big(\frac{k-k'}{2}, \frac{k'-k}{2}; k+k'\Big)$$

$$- {}^{34}\overset{4}{\mathsf{T}}{}^{j}\Big(\frac{k-k'}{2}, \frac{k'-k}{2}; k+k'\Big)\Big]_{k'^0=\omega^p(|k'|)} n^p(|\boldsymbol{k}'|) \Big\}. \qquad (10.134)$$

Substituting (10.86) into (9.80) leads for the $\Theta\Theta$-component of the proton self-energy, $\Sigma^p_{\Theta\Theta}$, to the expression

$$\Sigma^p_{\Theta\Theta}(k) = \frac{1}{4}\,\frac{1}{(2\,\pi)^2} \Big\{ \sum_j (2j+1) \int_0^{k_{F_p}} \frac{\mathrm{d}^3\boldsymbol{k}'}{(2\pi)^3} \Big[{}^{0}\overset{4}{\mathsf{P}}{}^{j}\Big(\frac{k-k'}{2}, \frac{k-k'}{2}; k+k'\Big)$$

$$+ {}^{12}\overset{4}{\mathsf{P}}{}^{j}\Big(\frac{k-k'}{2}, \frac{k-k'}{2}; k+k'\Big)\Big]_{k'^0=\omega^p(|k'|)} n^p(|\boldsymbol{k}'|)$$

$$+ \sum_{j=1} (2j+1) \int_0^{k_{F_p}} \frac{\mathrm{d}^3\boldsymbol{k}'}{(2\pi)^3} \Big[{}^{1}\overset{4}{\mathsf{P}}{}^{j}\Big(\frac{k-k'}{2}, \frac{k-k'}{2}; k+k'\Big)$$

$$+ {}^{34}\overset{4}{\mathsf{P}}{}^{j}\Big(\frac{k-k'}{2}, \frac{k-k'}{2}; k+k'\Big)\Big]_{k'^0=\omega^p(|k'|)} n^p(|\boldsymbol{k}'|)$$

$$- \sum_j (2j+1)\,(-1)^{j} \int_0^{k_{F_p}} \frac{\mathrm{d}^3\boldsymbol{k}'}{(2\pi)^3} \Big[{}^{0}\overset{2}{\mathsf{S}}{}^{j}\Big(\frac{k-k'}{2}, \frac{k'-k}{2}; k+k'\Big)$$

$$+ {}^{12}\overset{2}{\mathsf{S}}{}^{j}\Big(\frac{k-k'}{2}, \frac{k'-k}{2}; k+k'\Big)\Big]_{k'^0=\omega^p(|k'|)} n^p(|\boldsymbol{k}'|)$$

$$- \sum_{j=1} (2j+1)\,(-1)^{j-1} \int_0^{k_{F_p}} \frac{\mathrm{d}^3\boldsymbol{k}'}{(2\pi)^3} \Big[{}^{1}\overset{2}{\mathsf{S}}{}^{j}\Big(\frac{k-k'}{2}, \frac{k'-k}{2}; k+k'\Big)$$

$$- {}^{34}\overset{2}{\mathsf{S}}{}^{j}\Big(\frac{k-k'}{2}, \frac{k'-k}{2}; k+k'\Big)\Big]_{k'^0=\omega^p(|k'|)} n^p(|\boldsymbol{k}'|)$$

$$+ \sum_j (2j+1) \int_0^{k_{F_n}} \frac{\mathrm{d}^3\boldsymbol{k}'}{(2\pi)^3} \Big[{}^{0}\overset{2}{\mathsf{P}}{}^{j}\Big(\frac{k-k'}{2}, \frac{k-k'}{2}; k+k'\Big)$$

$$+ {}^{12}\overset{2}{\mathsf{T}}{}^{j}\Big(\frac{k-k'}{2}, \frac{k-k'}{2}; k+k'\Big)\Big]_{k'^0=\omega^n(|k'|)} n^n(|\boldsymbol{k}'|)$$

$$+ \sum_{j=1} (2j+1) \int_0^{k_{F_n}} \frac{\mathrm{d}^3\boldsymbol{k}'}{(2\pi)^3} \Big[{}^{1}\overset{2}{\mathsf{P}}{}^{j}\Big(\frac{k-k'}{2}, \frac{k-k'}{2}; k+k'\Big)$$

$$+ {}^{34}\overset{2}{\mathsf{T}}{}^{j}\Big(\frac{k-k'}{2}, \frac{k-k'}{2}; k+k'\Big)\Big]_{k'^0=\omega^n(|k'|)} n^n(|\boldsymbol{k}'|)$$

$$
\begin{aligned}
&-\sum_{j}(2j+1)\,(-1)^{j}\int_{0}^{k_{F_n}}\frac{\mathrm{d}^3\boldsymbol{k}'}{(2\pi)^3}\Big[{}^{0}\overset{4}{\mathsf{S}}{}^{j}\Big(\frac{k-k'}{2},\frac{k'-k}{2};k+k'\Big)\\
&\qquad +{}^{12}\overset{4}{\mathsf{S}}{}^{j}\Big(\frac{k-k'}{2},\frac{k'-k}{2};k+k'\Big)\Big]_{k'^0=\omega^n(|k'|)}n^n(|\boldsymbol{k}'|)\\
&-\sum_{j=1}(2j+1)\,(-1)^{j-1}\int_{0}^{k_{F_n}}\frac{\mathrm{d}^3\boldsymbol{k}'}{(2\pi)^3}\Big[{}^{1}\overset{4}{\mathsf{S}}{}^{j}\Big(\frac{k-k'}{2},\frac{k'-k}{2};k+k'\Big)\\
&\qquad -{}^{34}\overset{4}{\mathsf{S}}{}^{j}\Big(\frac{k-k'}{2},\frac{k'-k}{2};k+k'\Big)\Big]_{k'^0=\omega^n(|k'|)}n^n(|\boldsymbol{k}'|)\Big\}. \qquad (10.135)
\end{aligned}
$$

The ΘΘ-component of the neutron self-energy is given by

$$
\begin{aligned}
\Sigma^{n}_{\Theta\Theta}(k) &= \frac{1}{4}\,\frac{1}{(2\,\pi)^2}\Big\{\sum_{j}(2j+1)\int_{0}^{k_{F_n}}\frac{\mathrm{d}^3\boldsymbol{k}'}{(2\pi)^3}\Big[{}^{0}\overset{3}{\mathsf{P}}{}^{j}\Big(\frac{k-k'}{2},\frac{k-k'}{2};k+k'\Big)\\
&\qquad +{}^{12}\overset{3}{\mathsf{P}}{}^{j}\Big(\frac{k-k'}{2},\frac{k-k'}{2};k+k'\Big)\Big]_{k'^0=\omega^n(|k'|)}n^n(|\boldsymbol{k}'|)\\
&+\sum_{j=1}(2j+1)\int_{0}^{k_{F_n}}\frac{\mathrm{d}^3\boldsymbol{k}'}{(2\pi)^3}\Big[{}^{1}\overset{3}{\mathsf{P}}{}^{j}\Big(\frac{k-k'}{2},\frac{k-k'}{2};k+k'\Big)\\
&\qquad +{}^{34}\overset{3}{\mathsf{P}}{}^{j}\Big(\frac{k-k'}{2},\frac{k-k'}{2};k+k'\Big)\Big]_{k'^0=\omega^n(|k'|)}n^n(|\boldsymbol{k}'|)\\
&-\sum_{j}(2j+1)\,(-1)^{j}\int_{0}^{k_{F_n}}\frac{\mathrm{d}^3\boldsymbol{k}'}{(2\pi)^3}\Big[{}^{0}\overset{1}{\mathsf{S}}{}^{j}\Big(\frac{k-k'}{2},\frac{k'-k}{2};k+k'\Big)\\
&\qquad +{}^{12}\overset{1}{\mathsf{S}}{}^{j}\Big(\frac{k-k'}{2},\frac{k'-k}{2};k+k'\Big)\Big]_{k'^0=\omega^n(|k'|)}n^n(|\boldsymbol{k}'|)\\
&-\sum_{j=1}(2j+1)\,(-1)^{j-1}\int_{0}^{k_{F_n}}\frac{\mathrm{d}^3\boldsymbol{k}'}{(2\pi)^3}\Big[{}^{1}\overset{1}{\mathsf{S}}{}^{j}\Big(\frac{k-k'}{2},\frac{k'-k}{2};k+k'\Big)\\
&\qquad -{}^{34}\overset{1}{\mathsf{S}}{}^{j}\Big(\frac{k-k'}{2},\frac{k'-k}{2};k+k'\Big)\Big]_{k'^0=\omega^n(|k'|)}n^n(|\boldsymbol{k}'|)\\
&+\sum_{j}(2j+1)\int_{0}^{k_{F_p}}\frac{\mathrm{d}^3\boldsymbol{k}'}{(2\pi)^3}\Big[{}^{0}\overset{1}{\mathsf{P}}{}^{j}\Big(\frac{k-k'}{2},\frac{k-k'}{2};k+k'\Big)\\
&\qquad +{}^{12}\overset{1}{\mathsf{P}}{}^{j}\Big(\frac{k-k'}{2},\frac{k-k'}{2};k+k'\Big)\Big]_{k'^0=\omega^p(|k'|)}n^p(|\boldsymbol{k}'|)
\end{aligned}
$$

$$+\sum_{j=1}(2j+1)\int_0^{k_{F_p}}\frac{d^3\boldsymbol{k}'}{(2\pi)^3}\Big[{}^1\overset{1}{\mathsf{P}}{}^j\Big(\frac{k-k'}{2},\frac{k-k'}{2};k+k'\Big)$$

$$+{}^{34}\overset{1}{\mathsf{P}}{}^j\Big(\frac{k-k'}{2},\frac{k-k'}{2};k+k'\Big)\Big]_{k'^0=\omega^p(|k'|)}n^p(|\boldsymbol{k}'|)$$

$$-\sum_{j}(2j+1)(-1)^j\int_0^{k_{F_p}}\frac{d^3\boldsymbol{k}'}{(2\pi)^3}\Big[{}^0\overset{3}{\mathsf{S}}{}^j\Big(\frac{k-k'}{2},\frac{k'-k}{2};k+k'\Big)$$

$$+{}^{12}\overset{3}{\mathsf{S}}{}^j\Big(\frac{k-k'}{2},\frac{k'-k}{2};k+k'\Big)\Big]_{k'^0=\omega^p(|k'|)}n^p(|\boldsymbol{k}'|)$$

$$-\sum_{j=1}(2j+1)(-1)^{j-1}\int_0^{k_{F_p}}\frac{d^3\boldsymbol{k}'}{(2\pi)^3}\Big[{}^1\overset{3}{\mathsf{S}}{}^j\Big(\frac{k-k'}{2},\frac{k'-k}{2};k+k'\Big)$$

$$-{}^{34}\overset{3}{\mathsf{S}}{}^j\Big(\frac{k-k'}{2},\frac{k'-k}{2};k+k'\Big)\Big]_{k'^0=\omega^p(|k'|)}n^p(|\boldsymbol{k}'|)\Big\}. \quad (10.136)$$

Finally, substituting (10.86) into (9.81) leads for the mixed $\Theta\Phi$-component of proton self-energy, $\Sigma^p_{\Theta\Phi}$, to

$$\Sigma^p_{\Theta\Phi}(k)=\frac{1}{4}\frac{1}{(2\pi)^2}\Big\{\sum_{j}(2j+1)\int_0^{k_{F_p}}\frac{d^3\boldsymbol{k}'}{(2\pi)^3}\Big[{}^0\overset{4}{\mathsf{R}}{}^j\Big(\frac{k-k'}{2},\frac{k-k'}{2};k+k'\Big)$$

$$+{}^{12}\overset{4}{\mathsf{R}}{}^j\Big(\frac{k-k'}{2},\frac{k-k'}{2};k+k'\Big)\Big]_{k'^0=\omega^p(|k'|)}n^p(|\boldsymbol{k}'|)$$

$$+\sum_{j=1}(2j+1)\int_0^{k_{F_p}}\frac{d^3\boldsymbol{k}'}{(2\pi)^3}\Big[{}^1\overset{4}{\mathsf{R}}{}^j\Big(\frac{k-k'}{2},\frac{k-k'}{2};k+k'\Big)$$

$$+{}^{34}\overset{4}{\mathsf{R}}{}^j\Big(\frac{k-k'}{2},\frac{k-k'}{2};k+k'\Big)\Big]_{k'^0=\omega^p(|k'|)}n^p(|\boldsymbol{k}'|)$$

$$-\sum_{j}(2j+1)(-1)^j\int_0^{k_{F_p}}\frac{d^3\boldsymbol{k}'}{(2\pi)^3}\Big[{}^0\overset{4}{\mathsf{R}}{}^j\Big(\frac{k-k'}{2},\frac{k'-k}{2};k+k'\Big)$$

$$+{}^{12}\overset{4}{\mathsf{R}}{}^j\Big(\frac{k-k'}{2},\frac{k'-k}{2};k+k'\Big)\Big]_{k'^0=\omega^p(|k'|)}n^p(|\boldsymbol{k}'|)$$

$$-\sum_{j=1}(2j+1)(-1)^{j-1}\int_0^{k_{F_p}}\frac{d^3\boldsymbol{k}'}{(2\pi)^3}\Big[{}^1\overset{4}{\mathsf{R}}{}^j\Big(\frac{k-k'}{2},\frac{k'-k}{2};k+k'\Big)$$

$$-{}^{34}\overset{4}{\mathsf{R}}{}^j\Big(\frac{k-k'}{2},\frac{k'-k}{2};k+k'\Big)\Big]_{k'^0=\omega^p(|k'|)}n^p(|\boldsymbol{k}'|)$$

$$
\begin{aligned}
&+\sum_{j}(2j+1)\int_{0}^{k_{F_n}}\frac{\mathrm{d}^3\boldsymbol{k}'}{(2\pi)^3}\Big[{}^{0}\overset{6}{\mathsf{R}}{}^{j}\Big(\frac{k-k'}{2},\frac{k-k'}{2};k+k'\Big)\\
&\qquad+{}^{12}\overset{6}{\mathsf{R}}{}^{j}\Big(\frac{k-k'}{2},\frac{k-k'}{2};k+k'\Big)\Big]_{k'^0=\omega^n(|\boldsymbol{k}'|)}n^n(|\boldsymbol{k}'|)\\
&+\sum_{j=1}(2j+1)\int_{0}^{k_{F_n}}\frac{\mathrm{d}^3\boldsymbol{k}'}{(2\pi)^3}\Big[{}^{1}\overset{6}{\mathsf{R}}{}^{j}\Big(\frac{k-k'}{2},\frac{k-k'}{2};k+k'\Big)\\
&\qquad+{}^{34}\overset{6}{\mathsf{R}}{}^{j}\Big(\frac{k-k'}{2},\frac{k-k'}{2};k+k'\Big)\Big]_{k'^0=\omega^n(|\boldsymbol{k}'|)}n^n(|\boldsymbol{k}'|)\\
&-\sum_{j}(2j+1)(-1)^j\int_{0}^{k_{F_n}}\frac{\mathrm{d}^3\boldsymbol{k}'}{(2\pi)^3}\Big[{}^{0}\overset{5}{\mathsf{R}}{}^{j}\Big(\frac{k-k'}{2},\frac{k'-k}{2};k+k'\Big)\\
&\qquad+{}^{12}\overset{5}{\mathsf{R}}{}^{j}\Big(\frac{k-k'}{2},\frac{k'-k}{2};k+k'\Big)\Big]_{k'^0=\omega^n(|\boldsymbol{k}'|)}n^n(|\boldsymbol{k}'|)\\
&-\sum_{j=1}(2j+1)(-1)^{j-1}\int_{0}^{k_{F_n}}\frac{\mathrm{d}^3\boldsymbol{k}'}{(2\pi)^3}\Big[{}^{1}\overset{5}{\mathsf{R}}{}^{j}\Big(\frac{k-k'}{2},\frac{k'-k}{2};k+k'\Big)\\
&\qquad-{}^{34}\overset{5}{\mathsf{R}}{}^{j}\Big(\frac{k-k'}{2},\frac{k'-k}{2};k+k'\Big)\Big]_{k'^0=\omega^n(|\boldsymbol{k}'|)}n^n(|\boldsymbol{k}'|)\Big\}. \qquad (10.137)
\end{aligned}
$$

The mixed $\Theta\Phi$-component of the neutron self-energy is given by

$$
\begin{aligned}
\Sigma^n_{\Theta\Phi}(k)=\frac{1}{4}\frac{1}{(2\,\pi)^2}\Big\{&\sum_{j}(2j+1)\int_{0}^{k_{F_n}}\frac{\mathrm{d}^3\boldsymbol{k}'}{(2\pi)^3}\Big[{}^{0}\overset{1}{\mathsf{R}}{}^{j}\Big(\frac{k-k'}{2},\frac{k-k'}{2};k+k'\Big)\\
&\qquad+{}^{12}\overset{1}{\mathsf{R}}{}^{j}\Big(\frac{k-k'}{2},\frac{k-k'}{2};k+k'\Big)\Big]_{k'^0=\omega^n(|\boldsymbol{k}'|)}n^n(|\boldsymbol{k}'|)\\
&+\sum_{j=1}(2j+1)\int_{0}^{k_{F_n}}\frac{\mathrm{d}^3\boldsymbol{k}'}{(2\pi)^3}\Big[{}^{1}\overset{1}{\mathsf{R}}{}^{j}\Big(\frac{k-k'}{2},\frac{k-k'}{2};k+k'\Big)\\
&\qquad+{}^{34}\overset{1}{\mathsf{R}}{}^{j}\Big(\frac{k-k'}{2},\frac{k-k'}{2};k+k'\Big)\Big]_{k'^0=\omega^n(|\boldsymbol{k}'|)}n^n(|\boldsymbol{k}'|)\\
&-\sum_{j}(2j+1)(-1)^j\int_{0}^{k_{F_n}}\frac{\mathrm{d}^3\boldsymbol{k}'}{(2\pi)^3}\Big[{}^{0}\overset{1}{\mathsf{R}}{}^{j}\Big(\frac{k-k'}{2},\frac{k'-k}{2};k+k'\Big)\\
&\qquad+{}^{12}\overset{1}{\mathsf{R}}{}^{j}\Big(\frac{k-k'}{2},\frac{k'-k}{2};k+k'\Big)\Big]_{k'^0=\omega^n(|\boldsymbol{k}'|)}n^n(|\boldsymbol{k}'|)
\end{aligned}
$$

$$-\sum_{j=1}(2j+1)\,(-1)^{j-1}\int_0^{k_{F_n}}\frac{\mathrm{d}^3\boldsymbol{k}'}{(2\pi)^3}\Big[{}^{1}\overset{1}{\mathsf{R}}{}^{j}\Big(\frac{k-k'}{2},\frac{k'-k}{2};k+k'\Big)$$

$$-\,{}^{34}\overset{1}{\mathsf{R}}{}^{j}\Big(\frac{k-k'}{2},\frac{k'-k}{2};k+k'\Big)\Big]_{k'^0=\omega^n(|k'|)}n^n(|\boldsymbol{k}'|)$$

$$+\sum_{j}(2j+1)\int_0^{k_{F_p}}\frac{\mathrm{d}^3\boldsymbol{k}'}{(2\pi)^3}\Big[{}^{0}\overset{2}{\mathsf{R}}{}^{j}\Big(\frac{k-k'}{2},\frac{k-k'}{2};k+k'\Big)$$

$$+\,{}^{12}\overset{2}{\mathsf{R}}{}^{j}\Big(\frac{k-k'}{2},\frac{k-k'}{2};k+k'\Big)\Big]_{k'^0=\omega^p(|k'|)}n^p(|\boldsymbol{k}'|)$$

$$+\sum_{j=1}(2j+1)\int_0^{k_{F_p}}\frac{\mathrm{d}^3\boldsymbol{k}'}{(2\pi)^3}\Big[{}^{1}\overset{2}{\mathsf{R}}{}^{j}\Big(\frac{k-k'}{2},\frac{k-k'}{2};k+k'\Big)$$

$$+\,{}^{34}\overset{2}{\mathsf{R}}{}^{j}\Big(\frac{k-k'}{2},\frac{k-k'}{2};k+k'\Big)\Big]_{k'^0=\omega^p(|k'|)}n^p(|\boldsymbol{k}'|)$$

$$-\sum_{j}(2j+1)\,(-1)^{j}\int_0^{k_{F_p}}\frac{\mathrm{d}^3\boldsymbol{k}'}{(2\pi)^3}\Big[{}^{0}\overset{3}{\mathsf{R}}{}^{j}\Big(\frac{k-k'}{2},\frac{k'-k}{2};k+k'\Big)$$

$$+\,{}^{12}\overset{3}{\mathsf{R}}{}^{j}\Big(\frac{k-k'}{2},\frac{k'-k}{2};k+k'\Big)\Big]_{k'^0=\omega^p(|k'|)}n^p(|\boldsymbol{k}'|)$$

$$-\sum_{j=1}(2j+1)\,(-1)^{j-1}\int_0^{k_{F_p}}\frac{\mathrm{d}^3\boldsymbol{k}'}{(2\pi)^3}\Big[{}^{1}\overset{3}{\mathsf{R}}{}^{j}\Big(\frac{k-k'}{2},\frac{k'-k}{2};k+k'\Big)$$

$$-\,{}^{34}\overset{3}{\mathsf{R}}{}^{j}\Big(\frac{k-k'}{2},\frac{k'-k}{2};k+k'\Big)\Big]_{k'^0=\omega^p(|k'|)}n^p(|\boldsymbol{k}'|)\Big\}. \qquad (10.138)$$

Chapter 11

Dense matter in the relativistic ladder approximation

11.1 Summary of the many-body equations

Before we begin with a discussion of the properties of dense matter treated in the relativistic ladder approximation to the **T**-matrix, we briefly summarize those many-body equations that determine the properties of nuclear matter and neutron star matter for this approximation. These matter equations will be solved up to intermediate nuclear densities that are about a factor of three higher than the density of nuclear matter, ϵ_0. Up to such densities neutron star matter consists to a large percentage of chemically equilibrated neutrons and protons only [78, 79]. Restriction to neutrons and protons implies that in our notation $B = n, p$. Both particles are in chemical equilibrium with the leptons, $L = e^-, \mu^-$. The case of symmetric nuclear matter is readily recovered by setting $B = N$ and ignoring the leptons, in which case the mathematical formalism as well as the computational demands simplify considerably.

The following set of equations is to be solved self-consistently for the properties of dense matter described in the relativistic **T**-matrix approximation [62, 447]:

1) Equation (5.203) which determines the in-medium scattering **T**-matrix in momentum space. For technical reasons, it is convenient to introduce a self-consistent spinor–antispinor (helicity) basis $\{\Phi_l^B, \Theta_l^B\}$, calculate the matrix elements of all the relevant quantities (i.e. **V**, **Ξ**, **Λ**, **Σ**) of the theory in this basis (chapter 9), and lastly perform a partial-wave expansion of all these quantities in the angular momentum state basis $|jm, l_1 l_2\rangle$ (chapter 10). The **T**-matrix in the helicity basis is given in equation (9.94), its partial-wave expansion in equation (10.91).

2) Several different boson-exchange interactions (e.g., Bonn, HEA, and the modern BM potentials A, B, C), which serve as an input when solving the $\mathbf{T}$-matrix equation, will be employed in our calculations. The matrix elements of a given boson-exchange interaction in the basis $\{\Phi_l^B, \Theta_l^B\}$ are calculated in chapter 9, and its partial-wave expansion in the angular-momentum state representation is performed in chapter 10.
3) The matrix elements of the spectral functions Ξ^B in the self-consistent basis are to be calculated from equations (9.45) through (9.52).
4) The baryon propagators, $\mathbf{\Lambda}$, in the self-consistent basis are given in equations (9.54) and (9.59).
5) The baryon self-energies Σ_i^B follow from equations (9.68), (9.73) and (9.74) once $\Sigma_{\Phi\Phi}^B$, $\Sigma_{\Theta\Theta}^B$ and $\Sigma_{\Theta\Phi}^B$ have been computed from the partial-wave expanded equations (9.79) through (9.81). These are given explicitly in (10.133) through (10.138) for the case of asymmetric nucleonic matter, with Σ^n and Σ^p the self-energies of neutrons and protons, respectively,
6) The equation of state, $P(\epsilon)$, follows from equations (12.103) and (12.105).

The conditions of chemical equilibrium and electric charge neutrality, which are to be imposed for neutron star matter, were already listed in items 2) and 4) of section 7.5, and are therefore not repeated here again.

11.2 Properties of symmetric and asymmetric nuclear matter

Examples of these three self-energy matrix elements $\Sigma_{\Phi\Phi}^B$, $\Sigma_{\Theta\Theta}^B$ and $\Sigma_{\Theta\Phi}^B$ are shown for a number of selected nuclear matter densities in figures 11.1 through 11.3. The nucleon–nucleon matrix element $\Sigma_{\Phi\Phi}^N$ exhibits the standard parabolic form with a depth of ~ -80 MeV around nuclear matter density ($k_F \sim 1.45$ fm^{-1}), which becomes considerably shallower with decreasing density. The extremely weak momentum dependence of $\Sigma_{\Theta\Theta}^N$, which is independent of density, is quite striking. Another striking feature is the practically linear variation of the mixed antinucleon–nucleon matrix element $\Sigma_{\Theta\Phi}^N$ with momentum, which, however, changes with density. The scalar and timelike self-energies, Σ_S and Σ_0 respectively, are obtained from $\Sigma_{\Phi\Phi}^B$, $\Sigma_{\Theta\Theta}^B$ and $\Sigma_{\Theta\Phi}^B$ by means of inverting equations (9.65) to (9.67). We have repeatedly stressed that relativistic nuclear matter calculations with inclusion of dynamical two-body correlations require knowledge of the effective scattering matrix $\mathbf{T}$ in the nuclear-matter frame. To determine $\mathbf{T}$ in a non-ambiguous manner, use of a complete representation of the nucleon–nucleon amplitude in the *full* Dirac space, spanned by the self-consistent spinors Φ and Θ, is made [117, 118, 448]. This avoids the ambiguities which arise from restriction to the positive-energy sector [93, 104, 449], which will be discussed in greater detail in section 12.4. As

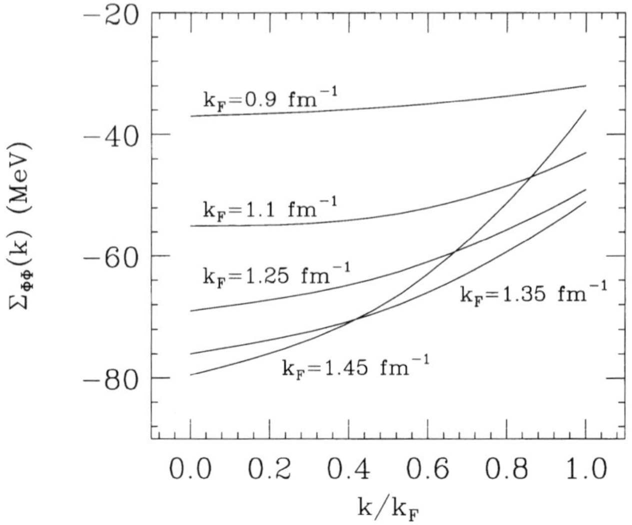

Figure 11.1. Momentum dependence of $\Sigma^N_{\Phi\Phi}$ for symmetric nuclear matter treated in the relativistic $\mathbf{\Lambda}^{00}$ approximation. k_F denotes the Fermi momentum. The underlying OBE interaction is HEA. (Data courtesy of Poschenrieder [440].)

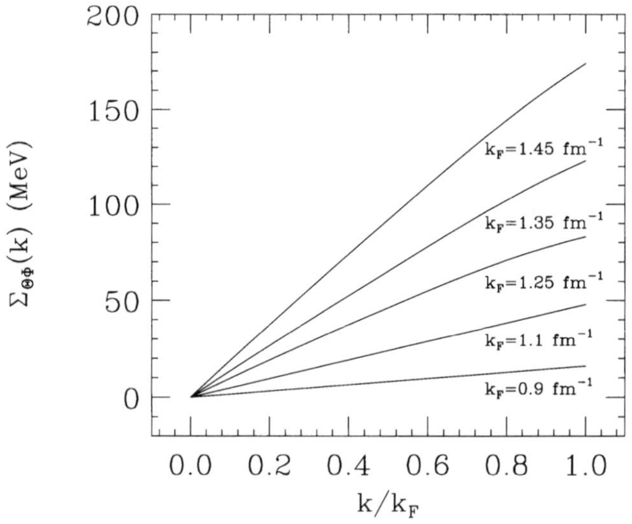

Figure 11.2. Momentum dependence of $\Sigma^N_{\Theta\Phi}$ for symmetric nuclear matter treated in the relativistic $\mathbf{\Lambda}^{00}$ approximation. k_F denotes the Fermi momentum. The underlying OBE interaction is HEA. (Data courtesy of Poschenrieder [440].)

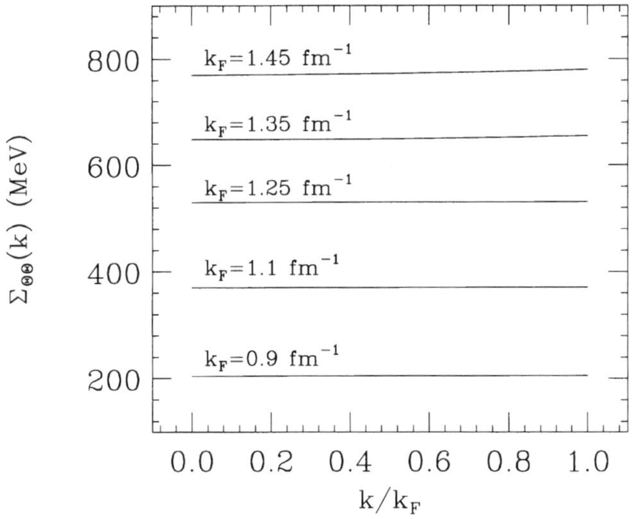

Figure 11.3. Momentum dependence of $\Sigma^N_{\ominus\ominus}$ for symmetric nuclear matter treated in the relativistic $\mathbf{\Lambda}^{00}$ approximation. k_F denotes the Fermi momentum. The underlying OBE interaction is HEA. (Data courtesy of Poschenrieder [440].)

input interactions we employ three modern one-boson-exchange interactions constructed by Brockmann and Machleidt. These potentials differ mainly with respect to the tensor force. For that reason one expects differences of the properties of symmetric matter, while those of pure neutron matter should remain almost identical, which is confirmed by our calculations. The bulk properties of symmetric nuclear matter, calculated for these three potentials, are in good agreement with the semi-empirical values, as we shall see below.

Figure 11.4 shows the momentum dependence of the scalar and timelike self-energy components, computed for infinite symmetric nuclear matter. The extremely weak dependence of these quantities on momentum, which is largely independent of density, is quite striking. Naturally such a behavior motivates approximations which treat $\mathbf{\Sigma}^B$ as being either as entirely momentum independent, or account for the momentum dependence in a momentum-averaged fashion only. A comparison of such momentum-averaged self-energies with their non-averaged counterparts, shown in figure 11.5, reveals that momentum-averaging increases the magnitude of both the scalar and timelike self-energy component, Σ_S and Σ_0 respectively. This trend increases with density, and originates from the fact that Σ_S increases near the Fermi momentum, in contrast to Σ_0 which decreases in the vicinity of the Fermi surface. Momentum-averaging suppresses this

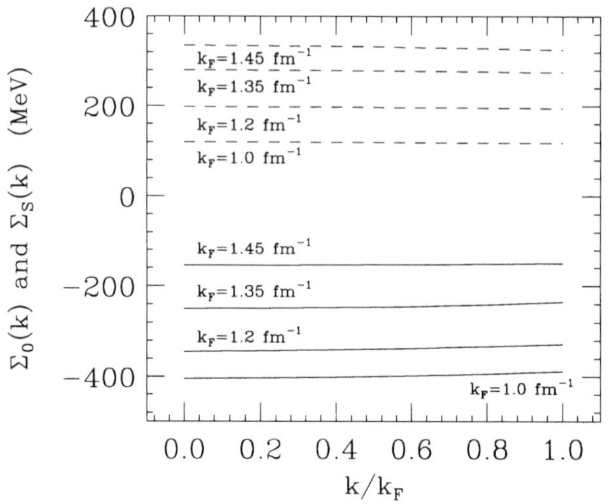

Figure 11.4. Momentum dependence of Σ_S^N (solid curves) and Σ_0^N (dashed curves) for symmetric nuclear matter treated in the relativistic $\mathbf{\Lambda}^{00}$ approximation. k_F denotes the Fermi momentum. The underlying OBE interaction is HEA. (Data courtesy of Poschenrieder [440].)

effect which leads to the differences outlined just above. This has non-negligible implications when calculating the baryon composition, and thus the equation of state, of dense stellar matter in β-equilibrium. The reason is that, in this case, one needs to know the single-particle energy at the top of the Fermi sea of each respective baryon, which, as we know from (6.136), is given for protons (p) and neutrons (n) by

$$\mu^{p,n} = \omega^{p,n}(k_F^{p,n}) = \Sigma_0^{p,n}(\omega^{p,n}(k_F^{p,n}), k_F^{p,n}) + \sqrt{[m + \Sigma_S^{p,n}(\omega^{p,n}(k_F^{p,n}), k_F^{p,n})]^2 + [k_F^{p,n} + \Sigma_V^{p,n}(\omega^{p,n}(k_F^{p,n}), k_F^{p,n})]^2}\,. \tag{11.1}$$

The Dirac mass of the nucleon, defined as $m_N^* = m_N + \Sigma_S^N(k_{F_N})$, is about $0.66\, m_N$ for all three BM potentials A, B, C, which is somewhat larger than the Dirac mass obtained for the momentum-averaged approximation, in which case $m_N^* \approx 0.62\, m_N$. Parenthetically we note that density dependent parametrizations of $\mathbf{\Sigma}^B$ have been applied with some success to nuclear structure calculations of finite nuclei [451, 452]. A comparison within the extended Thomas–Fermi scheme, which gives *a priori* smaller radii for nuclei, does not allow one to draw clear conclusions with respect to the density dependence of $\mathbf{\Sigma}^B$, since either the energies or the radii are better reproduced by the parametrizations of the different nuclear matter schemes.

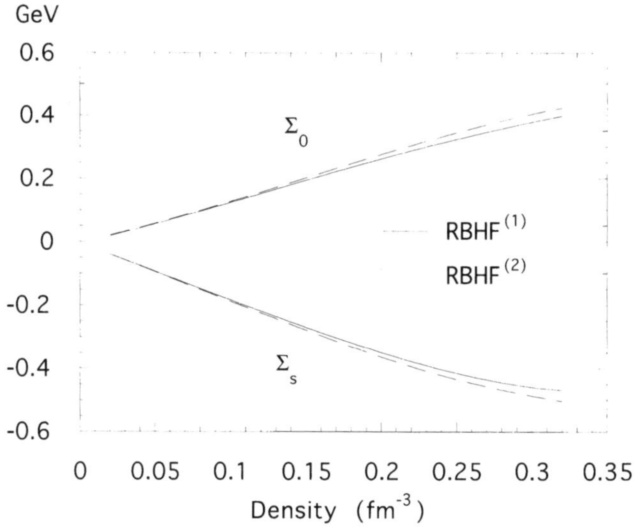

Figure 11.5. Density dependence of scalar and timelike self-energy components in symmetric nuclear matter for full ($\text{RBHF}^{(1)}$) and momentum-averaged ($\text{RBHF}^{(2)}$) treatments [450]. The adopted boson-exchange interaction is BM B. (Reprinted courtesy of *Phys. Rev.*)

It appears that, for finite nuclei, the choice of the σ-meson mass is of greater importance. This choice, however, is less restrictive since the properties of nuclear matter are mainly influenced by the ratio g_σ/m_σ [451, 453]. The poor reproducibility of the surface properties of semi-infinite nuclear matter utilizing Brockmann's parametrization seems to confirm this [453, 454].

The density dependence of the self-energies is shown for several selected asymmetries in figures 11.6 and 11.7. As one already expects from the behavior of the direct terms of the self-energies, Σ_0^n increases in asymmetric matter, while Σ_0^p is reduced in such matter by approximately the same amount. The differences in the scalar parts of $\mathbf{\Sigma}^B$ arise primarily from the exchange contributions to $\mathbf{\Sigma}^B$. (We recall that for the Hartree theory $\Sigma_S^p = \Sigma_S^n$.) A closer inspection of the exchange terms of the relativistic HF treatment reveals that the contributions of ρ, δ, and π mesons to the proton self-energy is larger than for the neutron self-energy [448]. For pure neutron matter the Fock contribution to $\mathbf{\Sigma}^p$ is about twice as large as the contribution to $\mathbf{\Sigma}^n$. Therefore one expects, as demonstrated in figure 11.6, more pronounced change of Σ_0 for growing values of δ than for Σ_S, with Σ_S^p more influenced by deviations from symmetric nuclear matter than Σ_S^n.

As has become clear in the previous chapters, the physical content of the relativistic BHF approximation goes significantly beyond what is

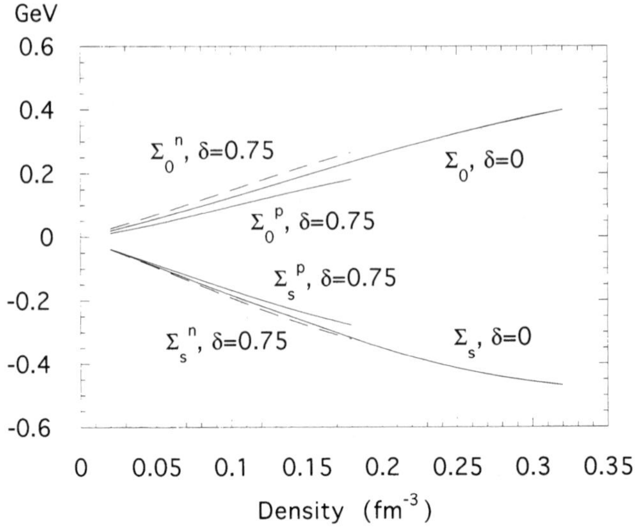

Figure 11.6. Density dependence of proton and neutron self-energies for different proton-to-neutron asymmetries δ, in the RBHF$^{(1)}$ approximation [450]. The adopted boson-exchange interaction is BM B. (Reprinted courtesy of *Phys. Rev.*)

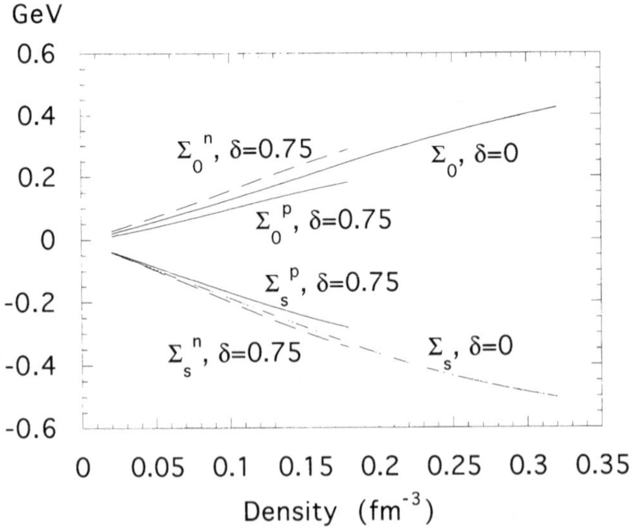

Figure 11.7. Same as figure 11.6, but for the RBHF$^{(2)}$ approximation [450]. (Reprinted courtesy of *Phys. Rev.*)

contained in the relativistic Hartree or HF approximation. So, naturally, it is tempting to implement this method into the mathematical description of dense neutron star matter, over that density range for which the RBHF constitutes a justifiable many-body approximation [450, 455, 456, 457, 458, 459, 460]. The upper bound of this density range appears to be around two to three times nuclear matter density. At still higher densities, correlations other than the simultaneous two-nucleon correlations taken into account in the ladder-approximated **T**-matrix may plausibly become important [315, 317], not to mention the applicability of one-boson-exchange interactions themselves. Another open issue concerns the inclusion of hyperonic degrees of freedom at higher densities in the ladder approximation which requires knowledge of the hyperon–hyperon and hyperon–nucleon potentials. A further problem concerns the non-renormalizability of the ladder approximation [117, 118, 122]. In a first step, one can render the theory finite of course by neglecting contributions arising from the negative-energy Dirac sea by restriction to the so-called no-sea approximation explained in figure 9.3 [92, 122]. This approximation has been shown to be very well justifiable at the level of the relativistic Hartree approximation [355], in which case vacuum renormalization has negligible influence on the equation of state of both nuclear as well as neutron matter up to about 10 times the density of nuclear matter, provided the coupling constants are tightly constrained by the five saturation properties (7.49) of nuclear matter.

In contrast to our analysis of the properties of asymmetric nuclear matter performed immediately above, the treatment of neutron star matter requires imposing the additional conditions of electric charge neutrality and β-equilibrium on the relativistic BHF method. The fundamental constituents of neutron star matter up to about two to three times nuclear matter density have been shown to be primarily protons and neutrons. Being in chemical equilibrium with one another requires the presence of a certain admixture of electrons and muons too. Such matter, although complicated, can be successfully treated in the framework in the RBHF method. The extrapolation to still neutron star densities is then plausibly performed in the framework of the simpler relativistic Hartree and HF approximations. To perform these extrapolations on a solid ground, it is necessary to investigate, first of all, whether these two approximations will be capable of reproducing the results of asymmetric relativistic BHF calculations, such that the corresponding neutron star matter outcomes can be joined neatly with each other at a certain density. Whether this is possible at all and, if so, to which extent such parametrizations will work was an open problem until recently. One basic problem is, for instance, whether the parameters of the standard HF Lagrangian, adjusted to the properties of symmetric nuclear matter [84, 85, 100, 461] or finite nuclei [98, 99, 462, 463], are not too restrictive to simultaneously reproduce the

Table 11.1. Properties of nuclear matter in relativistic Brueckner–Hartree–Fock RBHF. Bro A, Bro B, and Bro C denote the outcome of RBHF$^{(2)}$ (i.e. full basis, momentum averaged self-energies) calculations performed for the BM potentials A, B, C. BrM A, BrM B, and BrM C are RBHF calculations performed for the same potentials but restricted to positive-energy spinors only and momentum independent self-energies [462].

Method	E/A (MeV)	ρ_0 (fm^{-3})	K (MeV)	a_{sym} (MeV)	L (MeV)	K^* (MeV)
Bro A	−16.49	0.174	280	34.4	81.9	−66.4
Bro B	−15.73	0.172	249	32.8	90.2	9.97
Bro C	−14.38	0.170	258	31.5	76.1	−35.1
BrM A	−15.59	0.185	290	–	–	–
BrM B	−13.60	0.174	249	–	–	–
BrM C	−12.26	0.155	185	–	–	–
RHF$^{(1)}$	−15.75	0.148	610	28.9	132	466
RHF$^{(2)}$	−15.75	0.148	360	43.3	135	105
RHF$^{(3)}$	−15.75	0.148	460	38.6	138	276

properties of highly asymmetric neutron star matter too. So far, the only attempt made in this direction, which investigates the properties of asymmetric matter in the framework of the relativistic HF theory employing a parametrization adjusted to the properties of symmetric nuclear matter, was not very encouraging [461], as can be seen from table 11.1 which compares several of the properties of symmetric and asymmetric nuclear matter computed for the RBHF approximation with those obtained in relativistic HF [461, 463]. The notation is as follows: RHF$^{(1)}$: σ and ω mesons exchange only; RHF$^{(2)}$: σ, ω, π and ρ mesons exchange, $f_{\rho N}/g_{\rho N} = 6.6$; RHF$^{(3)}$: σ, ω, π and ρ mesons exchange, $f_{\rho N}/g_{\rho N} = 3.7$. While the HF results show that symmetric nuclear matter parametrizations are not sufficient to reproduce the properties properly, the outcome obtained for the RBHF approximation (rows labeled Br) agrees rather well with the semiempirical values. This comprises even the values of L which follow the expected pattern [100]. One problem of the HF approximation, for instance, may originate from the occurrence of the δ meson (which has isospin $I = 1$ and $J^{\pi} = 0^{+}$) in the OBE interaction. The δ meson is absent in the relativistic HF Lagrangian which may cause a discrepancy in the description of the scalar self-energies at large proton-to-neutron asymmetries.

In what follows we shall restrict ourselves to the treatment of asymmetric, β-equilibrated nuclear matter in the framework of the relativistic BHF scheme, and the possibility of reproducing the properties of such matter in the framework of the considerably simpler relativistic

Hartree and HF approximations. As it will turn out, the relativistic Hartree approximation has not enough flexibility to achieve this goal. Fortunately, however, the situation is different for relativistic Hartree–Fock.

11.3 Properties of chemically equilibrated stellar matter

Here we study the properties of dense chemically equilibrated matter treated in different versions of the relativistic ladder approximation. The BM potentials A, B, and C shall serve as input quantities. The latter were frequently used in recent relativistic many-body calculations of nuclear matter. These potentials are adjusted to the two-nucleon scattering data and the properties of the deuteron, and they differ mainly by the strength of the tensor force, which increases from A to C [104]. The many-body equations, which we surveyed in section 11.1, will be solved in two versions. In the *full* treatment, labeled as RBHF$^{(1)}$ [101, 102], the many-body equations are iterated until self-consistency for the momentum dependent self-energy $\boldsymbol{\Sigma}(p^0, \boldsymbol{p})$ is achieved. In contrast to this, the simpler version, denoted by RBHF$^{(2)}$, ignores the momentum dependence of the self-energy [101, 102]. In this case the momentum dependent self-energies are replaced with momentum averaged ones in the iteration procedure. This has the advantage that their self-consistent numerical determination is far less time consuming than in the case when the full momentum dependence is kept. The agreement of the saturation properties of symmetric matter with the results of Brockmann and Machleidt is in both cases rather good. We obtain a little more binding, and the saturation density varies less [101, 102]. A comparison with the outcome of the Brockmann–Machleidt treatment as well as other methods is given in table 11.1, where we also give the results for the volume asymmetry properties. The parameters are defined by the nuclear mass formula [101, 102],

$$e(\rho, \delta) = \Big(\frac{E}{A} + \frac{1}{18} K \Delta^2 \pm \ldots\Big) + \delta^2 \Big(a_{\rm sym} + \frac{1}{3} L \Delta + \frac{1}{18} K^* \, \Delta^2 + \ldots\Big) \,, \tag{11.2}$$

with the definition $\Delta = (\rho - \rho_0)/\rho_0$, and ρ_0 the saturation density of infinite symmetric nuclear matter. The quantity δ denotes the neutron-to-proton asymmetry,

$$\delta = \frac{\rho^n - \rho^p}{\rho} \,, \quad \text{with} \quad \rho = \rho^p + \rho^n \,. \tag{11.3}$$

K stands for the incompressibility of infinite, symmetric nuclear matter,

$$K = 9 \, \rho^2 \left(\frac{\partial^2 e}{\partial \rho^2} \right)_{\rho = \rho_0, \delta = 0} \,. \tag{11.4}$$

The quantity $a_{\rm sym}$ denotes the bulk symmetry-energy coefficient of nuclear matter at saturation density, which follows from the semi-empirical mass formula (11.2) as

$$a_{\rm sym} \equiv e^{\rm sym}(\rho = \rho_0), \quad \text{where} \quad e^{\rm sym}(\rho) \equiv \frac{1}{2} \frac{\partial^2 e(\rho, \delta)}{\partial \delta^2}\Big|_{\delta=0} . \tag{11.5}$$

$e^{\rm sym}(\rho)$ is known in the literature as the symmetry energy. The quantities L and K^* are related to the slope and the curvature of the symmetry energy at ρ_0,

$$L = 3\,\rho\, \frac{\partial e^{\rm sym}}{\partial \delta}\Big|_{\rho=\rho_0}, \qquad K^* \equiv 9\,\delta^2\, \frac{\partial^2 e^{\rm sym}}{\partial \rho^2}\Big|_{\rho=\rho_0} . \tag{11.6}$$

For small values of δ one obtains the approximate formulae for the equilibrium values of density and volume energy (i.e. energy per baryon),

$$\rho_{\rm eq}^{(\delta)} \simeq \rho_0 \left(1 - \frac{3\,L}{K}\,\delta^2\right), \qquad e(\rho_{\rm eq}, \delta) \simeq \frac{E}{A} + a_{\rm sym}\,\delta^2 . \tag{11.7}$$

Since the agreement is best for potentials A and B, we will restrict ourselves mainly to these two nucleon–nucleon interactions.

To achieve our aim, namely to reproduce and extend the relativistic BHF results within the framework of the simpler relativistic Hartree and HF approximations, one is tempted to try, because of its simplicity, the relativistic Hartree approximation first. Within this approximation the symmetry-energy coefficient $a_{\rm sym}$ is given by an analytical formula [86, 92], as known from equation (7.53). However, as it turns out, this approximation is only poorly capable of reproducing the outcome of asymmetric RBHF calculations for β-stable matter. As an example, we mention the slope of δ as a function of density which turns out to be too steep. For that reason we are left with the relativistic HF scheme parametrization as a possible alternative capable of reproducing the asymmetric RBHF outcome. Unfortunately all parametrizations of the relativistic HF Lagrangian published in the literature turned out to be adjusted only to the properties of symmetric nuclear matter (see, for instance, references [100, 461]), or to finite nuclei [98, 99, 462, 463]. (Some of these investigations leave out the tensor force [463].) For that reason we have to readjust the coupling constants of the relativistic HF scheme to the outcome of asymmetric relativistic BHF calculations, for which an additional numerical procedure for the determination of the symmetry-energy coefficient $a_{\rm sym}$ is necessary.

A first test of whether a certain parametrization of the relativistic HF Lagrangian is suitable or not concerns the reproducibility of the volume properties of asymmetric nuclear matter. The relevant parameters of the relativistic HF Lagrangian are (cf. section 7.4)

$$g_{\sigma N}/m_\sigma, \quad g_{\omega N}/m_\omega, \quad g_{\rho N}, \quad f_{\rho N}/g_{\rho N}, \quad b_N, \quad c_N ,$$

which are used to fix the volume parameters of symmetric matter, i.e. E/A, ρ, K, and the nucleon Dirac mass m_N^*. Besides these quantities, it is of crucial importance for the calculation of the properties of asymmetric stellar matter to control the value of $a_{\rm sym}$ too, which is intimately linked to the coupling strength of the ρ meson. Typical values for its tensor coupling strength, i.e. $f_{\rho N}/g_{\rho N}$, range from 3.7 to 6.6 [98, 99]. The value 3.7 emerges from the vector dominance model. Such a value, however, does not lead to agreement with the π–nucleus scattering data, which require a larger value of about 6.6. However since the HF approximation overestimates the short-range contributions – due to the use of uncorrelated wave functions by means of which only statistical correlations are included – one expects that smaller values of $f_{\rho N}/g_{\rho N}$ will somehow simulate the outcome of a calculation which accounts for dynamical correlations [98, 99]. This amounts to a certain freedom for the choice of $f_{\rho N}/g_{\rho N}$. Modern HF parametrizations use $f_{\rho N} = 0$ [463], which, as a beneficial side issue, implies a simpler numerical evaluation than would be the case for a finite $f_{\rho N} = 0$ value. For the reasons given above we shall restrict ourselves to the small values for $f_{\rho N}/g_{\rho N}$, and to $f_{\rho N} = 0$. Our calculations performed elsewhere [102] confirm the presumption that the smaller values give better fits to the RBHF results, and that the latter choice, $f_{\rho N} = 0$, gives the best agreement. Using $f_{\rho N}/g_{\rho N} = 3.7$ requires relatively large values for $g_{\rho N}$, as can be seen in table 11.2, which is necessary to reproduce the correct value for the symmetry coefficient, since the Fock contribution of the ρ meson is strongly reduced by the tensor coupling term [121].

To demonstrate the reproducibility of the outcome of asymmetric RBHF calculations in the framework of relativistic HF calculations, we illustrate, as a first test, in figure 11.8 the energy per baryon versus density

Table 11.2. Parameters for relativistic Hartree–Fock Lagrangian which reproduce the outcome of RBHF$^{(2)}$ calculations performed for BM potentials A and B. RHFA 1 and RHFA 2 reproduces the outcome of a RBHF$^{(2)}$ calculation performed for potential A, RHFB 1 and RHFB 2 are for potential B. The Dirac masses at the Fermi surface are $m_N^* = 617.8$ MeV and 621.8 MeV for potentials A and B, respectively. The particle masses have values of $m_N = 939$ MeV, $m_\sigma = 550$ MeV, $m_\omega = 738$ MeV, $m_\pi = 138$ MeV, and $m_\rho = 770$ MeV ($g_\pi = 1.00265$, $f_\pi^2/4\pi = 0.08$).

Parametrization	$g_\sigma^2/4\pi$	$g_\omega^2/4\pi$	$g_\rho^2/4\pi$	$10^3\, b_N$	$10^3\, c_N$	f_ρ/g_ρ
RHFA 1	6.858 30	5.579 99	0.351 21	3.333 67	−2.152 39	0.0
RHFA 2	6.798 07	5.423 44	0.384 49	2.965 14	−2.686 14	3.7
RHFB 1	6.984 26	5.621 22	0.250 23	3.743 54	−3.184 56	0.0
RHFB 2	6.931 08	5.510 58	0.275 54	3.443 06	−3.462 61	3.7

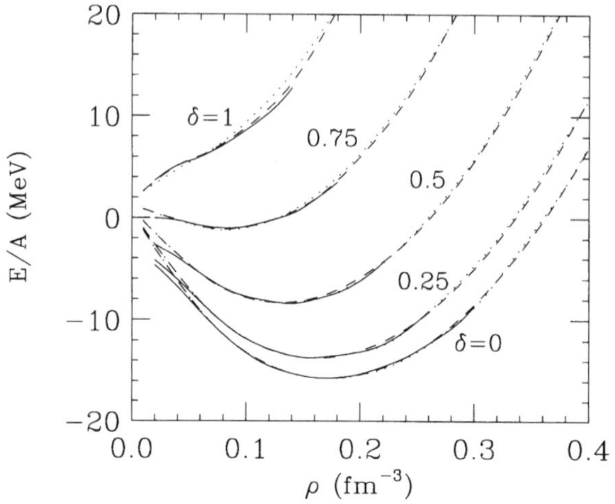

Figure 11.8. Comparison of the RBHF$^{(2)}$ EOS of asymmetric nuclear matter (solid line), computed for BM potential B, with its relativistic Hartree–Fock parametrized counterparts RHFB 1 (dashed curves) and RHFB 2 (dotted curves) [102]. The asymmetry parameter δ ranges from 0 to 1. RHFB 1 and RHFB 2 differ with respect to the tensor coupling strength chosen for the ρ meson, which is $f_\rho/g_\rho = 0$ and $f_\rho/g_\rho = 3.7$ respectively (table 11.2). Both HF parametrizations give very good agreement over the whole asymmetry range. (Reprinted courtesy of *Nucl. Phys.*)

for several different selected asymmetries δ. One sees that parametrizations of the relativistic HF Lagrangian over the whole theoretical range of $f_{\rho N}/g_{\rho N}$ have the capability to reproduce the outcome of asymmetric relativistic BHF calculations rather well. Lower values of $f_{\rho N}/g_{\rho N}$ give a slightly better fit. The coupling constants and meson masses are given in table 11.2. Similarly good agreement is obtained for the Brockmann–Machleidt potentials A and C [102].

In figure 11.9 we show the agreement of the energy per particle at equilibrium density as a function of asymmetry. The agreement is again very good, as is the case for the density dependence of the symmetry energy, which, therefore, is not shown separately.

More delicate to reproduce are the properties of electrically charge neutral asymmetric stellar matter in β-equilibrium. In this case the asymmetry is not specified from the outset by assigning a certain value to the asymmetry parameter δ but results self-consistently from the proton-to-neutron ratio, which varies with density. Models for the equation of state of such matter are shown, for different tensor coupling strengths $f_{\rho N}/g_{\rho N}$, in figures 11.10 and 11.11. The upper three curves correspond to matter made

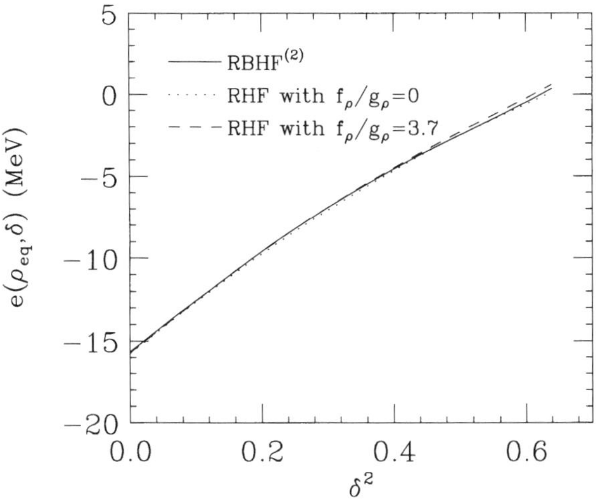

Figure 11.9. Comparison of the energy per particle at equilibrium nuclear matter density as a function of asymmetry, δ [102]. The solid curve shows the outcome obtained for an RBHF$^{(2)}$ calculation performed for BM potential B. The dotted and dashed curves refer to the outcome obtained for the relativistic HF parametrizations RHFB 1 (dotted curve) and RHFB 2 (dashed curve), which differ with respect to the f_ρ/g_ρ strength (table 11.2). (Reprinted courtesy of *Nucl. Phys.*)

up of protons and neutrons in chemical equilibrium with only electrons. The presence of muons has been arbitrarily suppressed in these calculations, which increases the energy per baryon relative to fully equilibrated matter (lower three curves) in which the muon is present too. In the latter case, the number of degrees of freedom accessible to the system is larger which enables fully equilibrated matter to settle in a lower energy state.

The density dependence of the asymmetry parameter δ is shown in figure 11.12. Figure 11.13 displays the baryon–lepton composition of fully equilibrated matter, whose equations of state were shown in figure 11.10 and 11.11 just above. It is seen from figure 11.12 that at lower densities the agreement between RBHF and its RHF parametrized outcome is less satisfactory. The high-density behavior however is quite well reproduced by RHF. In general we conclude that an RHF parametrization with vanishing tensor force, that is $f_{\rho N} = 0$, gives the best agreement with the RBHF outcome. At higher densities we do not obtain a decrease of the proton, electron, and muon fractions as found in a non-relativistic treatment [464]. This effect appears to be caused by the increased short-range repulsion in isospin-triplet pairs at higher densities. Surprisingly it is foreign to

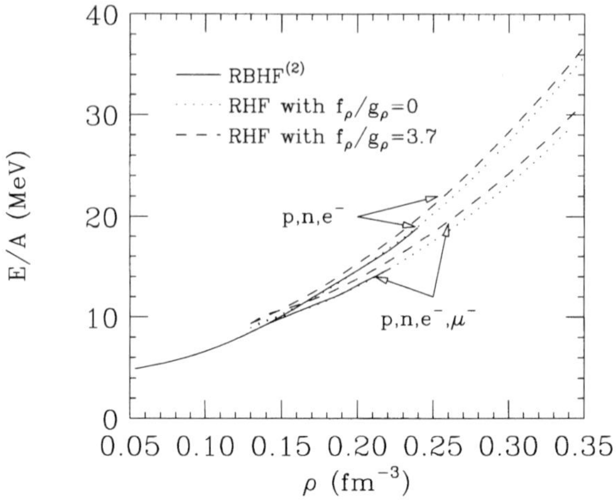

Figure 11.10. Energy per particle of β-equilibrated matter computed for RBHF$^{(2)}$ (BM potential B) and its relativistic HF parametrizations (cf. table 11.2) [102]. The lower three curves are for fully catalyzed matter, made up of protons and neutrons in equilibrium with electrons and muons. Ignoring the muon shifts each curve up by several MeV, depending on density, which leads to a somewhat stiffer EOS. A value of $f_\rho/g_\rho = 0$ leads to better agreement with RBHF$^{(2)}$ than $f_\rho/g_\rho = 3.7$. (Reprinted courtesy of *Nucl. Phys.*)

relativistic theories of dense matter [78, 79, 101, 461]. One may speculate that this effect is also responsible for the bending of the symmetry energy at higher densities found in non-relativistic many-body theories [461, 464, 465].

Finally, in figure 11.14 we show the density dependence of the self-energies computed for the relativistic HF approximation RHFB 1 for several different neutron-to-proton asymmetries δ, defined in (11.3). As one may expect from relativistic Hartree calculations, the timelike self-energies of protons and neutrons, Σ_0^p and Σ_0^n respectively, are almost symmetric about $\delta = 0$. The asymmetric behavior of the scalar self-energies Σ_S^p and Σ_S^p is caused by the exchange terms, to which the proton part – due to the occurrence of π and ρ mesons – contributes more strongly than the neutron part [101, 448, 450]. The quantitatively very similar behavior of the self-energies is of key importance for the equivalence of the different many-body approximations. Since the structure of both the relativistic HF and relativistic BHF approximation is formally similar, one may expect a quantitatively similar behavior of the self-energies for these two approximations, which our studies confirm for symmetric matter as well as asymmetric nuclear matter. In the latter case the deviations are less than about 5%, and are practically perfect for the former case.

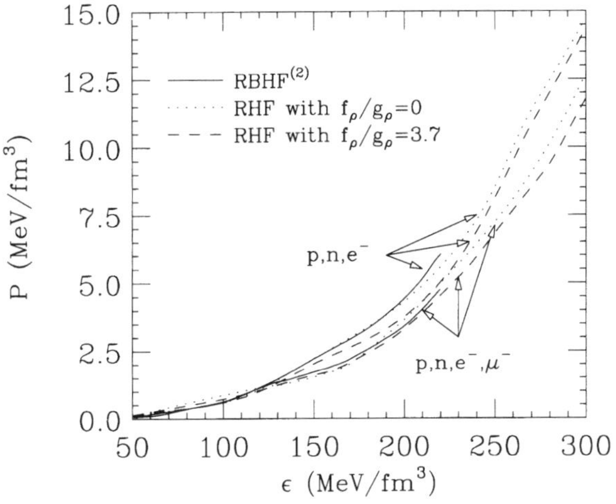

Figure 11.11. EOS of β-equilibrated matter computed for RBHF$^{(2)}$ (BM potential B) and its relativistic HF parametrizations, listed in table 11.2 [102]. The saturation density of normal nuclear matter is about 140 MeV fm^{-3}. (Reprinted courtesy of *Nucl. Phys.*)

Somewhat surprising is the good agreement of the scalar self-energies, which is better than 3%, though the isovector–scalar meson in the relativistic HF Lagrangian was neglected. Besides that, we recall that it is, roughly speaking, the sum of Σ_S and Σ_0 which decisively determines the equation of state. Hence, possible small deviations of these self-energies from the RBHF outcome tend to cancel each other in the equation of state.

In summary, above we have presented calculations of asymmetric nuclear matter as well as chemically equilibrated neutron star matter performed in the relativistic BHF approximation. As mentioned before, this treatment is *parameter-free* in the sense that the OBE interactions are adjusted to the two-nucleon data. For the OBE interactions we use the modern BM potentials. Furthermore we tested the possibility whether the outcome of these microscopic calculations, for both sorts of matter, can be reproduced in the framework of relativistic HF calculations. The latter are considerably less time consuming than RBHF calculations.

From our results it follows that the fits for asymmetric nuclear matter with fixed asymmetry are very good. The agreement with the outcome for chemically equilibrated neutron star matter is not as perfect as for asymmetric matter with fixed asymmetry but still acceptably accurate. The origin of the remaining deviations is difficult to locate. They may be caused by the structure of the OBE Lagrangian of the relativistic BHF treatment which is more complex than the relativistic HF Lagrangian, and/or the

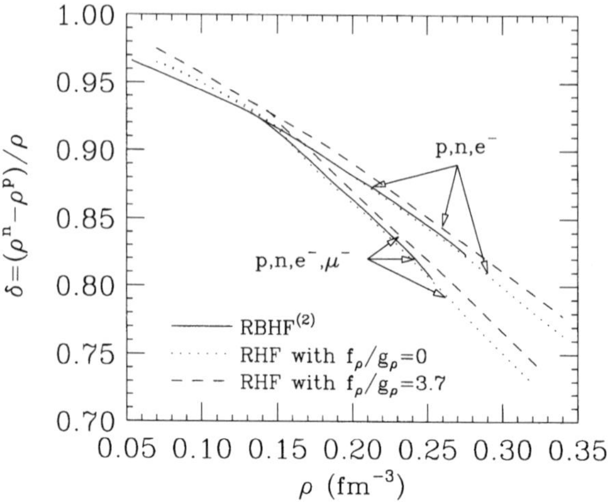

Figure 11.12. Asymmetry δ computed for RBHF$^{(2)}$ (BM potential B) and its relativistic HF parametrization, listed in table 11.2 [102]. The smaller value of f_ρ/g_ρ leads to a better agreement with RBHF$^{(2)}$. (Reprinted courtesy of *Nucl. Phys.*)

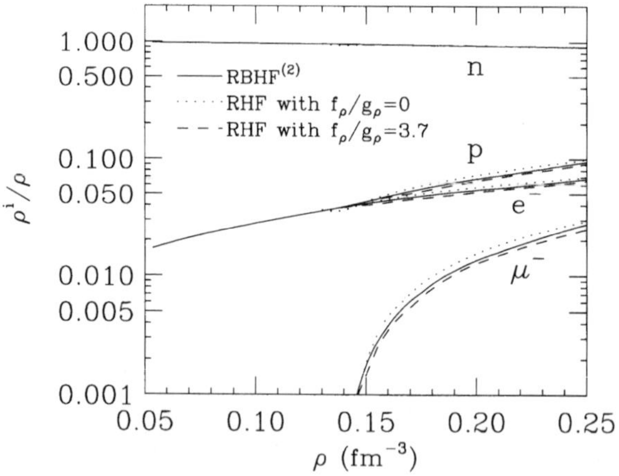

Figure 11.13. Nucleon–lepton compositions of β-equilibrated matter computed for RBHF$^{(2)}$ (BM potential B) and its relativistic HF parametrizations, listed in table 11.2 [102]. (Reprinted courtesy of *Nucl. Phys.*)

different energy dependence of the self-energies in both approximations. With respect to optionally including an additional isovector–scalar meson,

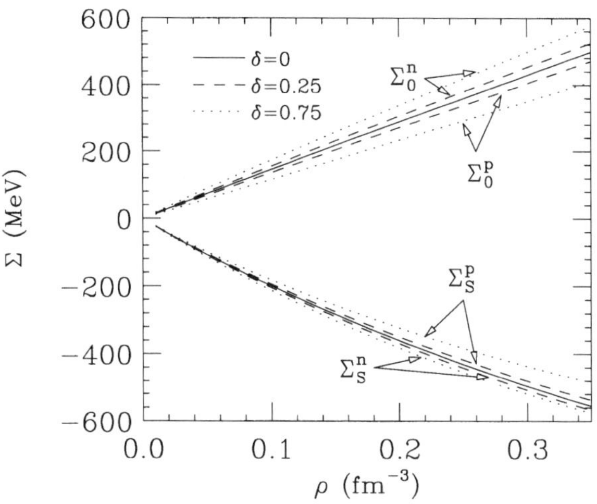

Figure 11.14. Density dependence of scalar and timelike self-energies, Σ_S^B and Σ_0^B respectively, for different asymmetries δ in relativistic HF based on parameter set RHFB 1 (cf. table 11.2) [102]. The other HF parametrizations of table 11.2, which have different tensor strengths, lead to almost identical results (below the resolution of this figure). The deviations from the self-energies computed for the full relativistic BHF calculation RBHF$^{(2)}$, shown in figure 11.7, are less than 5%. (Reprinted courtesy of *Nucl. Phys.*)

like the δ (table 5.2), in the standard HF Lagrangian, it turns out that the scalar part of the self-energy, Σ_S, is already reproducible with the same level of agreement as the vector part, Σ_V. For that reason it appears unnecessary to incorporate such an extra meson into the relativistic HF treatment, which, contrary to one's intention, would impose new problems (concerning its coupling constant and mass, for instance) if one extends the model to multi-baryon systems, like neutron star matter. Finally we mention that for potential A we obtain an agreement of similar quality [102].

As representative examples, we showed in figures 7.23 and 7.24 the relative baryon–lepton populations calculated at the level of the relativistic Hartree and HF approximations, respectively. The hyperon-to-nucleon coupling was chosen to be universal, i.e. $x_\sigma = x_\omega = x_\rho = 1$. This calculation is repeated here for the HF approximation, but for the case of *non-universal* coupling. Aside from this, the parameters of the present calculations, RHFB 1 of table 11.2, are adjusted to the outcome of asymmetric relativistic BHF calculations, whose physical content, as we have seen repeatedly, stretches considerably beyond the content of relativistic HF.

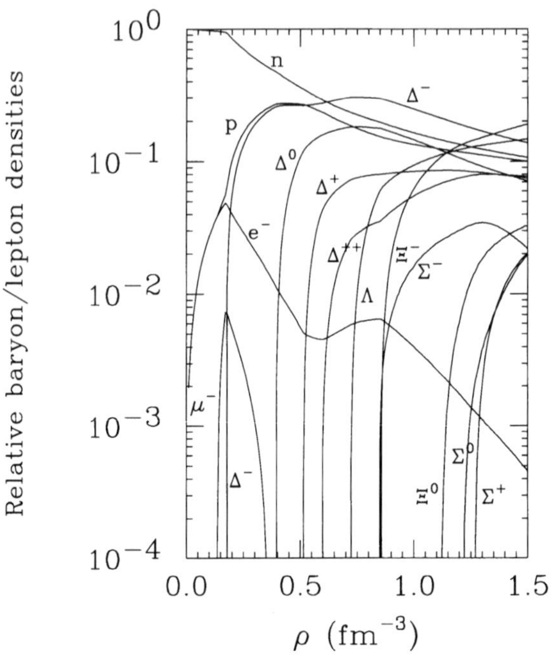

Figure 11.15. Baryon–lepton densities ρ^B/ρ and ρ^L/ρ, normalized to the total baryon density ρ, of neutron star matter, computed in relativistic HF based on parameter set RHFB 1 [391]. The hyperons are non-universally coupled, the Δ's universally. (Reprinted courtesy of *Int. J. Mod. Phys.*)

Figure 11.15 shows the baryon–lepton populations computed for σ couplings adjusted to the hyperon potential depths [391],

$$x_{\sigma\Sigma} = x_{\sigma\Lambda} = 0.44\,, \quad x_{\sigma\Xi} = 0.26\,,$$
$$x_{\omega\Sigma} = x_{\omega\Lambda} = \frac{2}{3}\,, \quad x_{\omega\Xi} = \frac{1}{3}\,, \quad x_{\sigma\Delta} = x_{\omega\Delta} = 1\,. \tag{11.8}$$

Figure 11.16 is the analog to figure 11.15 but computed for non-universal Δ coupling constants, i.e. $x_{\sigma\Delta} = x_{\omega\Delta} = 0.625$ instead of $x_{\sigma\Delta} = x_{\omega\Delta} = 1$, too. The latter non-universal choice, suggested by Rapp *et al* [466], shifts the onset of Δ population to higher densities while hyperon population sets in at lower densities. The principal physical reasons that determine the baryon–lepton population of neutron star matter have already been discussed in section 7.8 and so will not be repeated here again. We only recall that because of energetic factors, electrons at the top of the Fermi sea will be replaced by negatively charged baryons as soon as the corresponding baryon thresholds are reached. This process increases the baryon number density

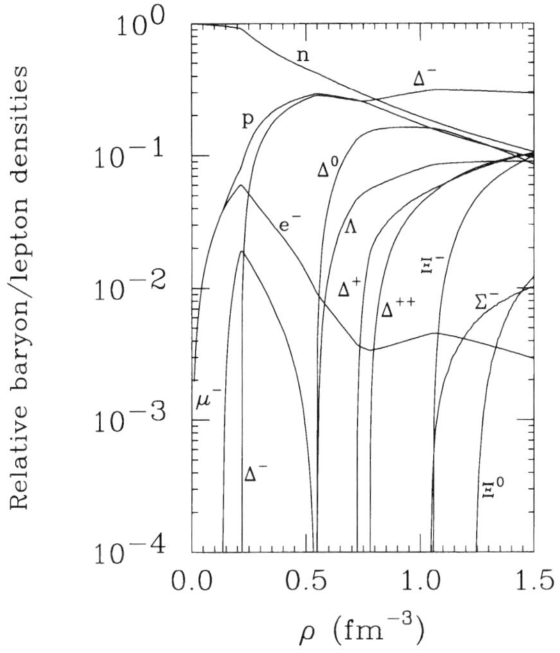

Figure 11.16. Same as figure 11.15, but for non-universally coupled hyperons and Δ's [391]. (Reprinted courtesy of *Int. J. Mod. Phys.*).

while at the same time the pressure of neutron star matter drops. Secondly, the baryon population will be such that the extreme isospin asymmetry of neutron star matter at lower nuclear densities ($\sim$ 100% neutrons and only very few protons around nuclear matter density, $\rho_0 = 0.16$ fm^{-3}) becomes as small as possible at higher densities, which too reduces the energy per baryon of such matter. It is these two principles together with the very small ρ-meson coupling constant characteristic of relativistic HF and BHF calculations that determine the populations shown in figures 11.15 and 11.16. These populations clearly bear some resemblance to the HF population shown in figure 7.24 where all particles were universally coupled, the reason being the small ρ-meson coupling strength in all three cases, which lowers the Δ^- threshold considerably in comparison to relativistic Hartree calculations (cf. figure 7.23).

As a final point to be made in connection with the populations shown in figures 11.15 and 11.16, we mention that the effective Dirac masses of the nucleons are driven toward negative values, because of the relatively early occurrence of negatively charged Δ states. This feature can be avoided, for instance, by introducing in the underlying non-linear HF description a modified σ-meson mass in combination with new self-interaction coupling constants [391],

$$m_\sigma^2 \to m_\sigma^2 + \frac{g_\sigma^2}{4\,c_N}\, m_N^2\, b_N^2 \Big(\frac{\alpha}{c_N} - 1\Big)\,,$$

$$b_N \to \alpha\,\frac{b_N}{c_N}\,, \qquad c_N \to \alpha\,, \tag{11.9}$$

which lead to a moderate drop of the nucleon Dirac masses beyond four times nuclear matter saturation density. In the case of figures 11.15 and 11.16 the constant $\alpha = 0.02$ was chosen. Obviously, the choice $\alpha = c_N$ leads to the original HF dynamics. The linear Walecka model is recovered for $\alpha = 0$.

Chapter 12

Models for the equation of state

We recall that the Lagrangian given in equation (5.1) depends on the spacetime coordinates x only through the fields and their gradients. Under the transformation $x'_\mu = x_\mu + \epsilon_\mu$ we have $\mathcal{L}' \equiv \mathcal{L}(x') \equiv \mathcal{L}[\chi(x'), \partial_\mu \chi(x')]$, and therefore

$$\delta \mathcal{L} = \mathcal{L}' - \mathcal{L} = \epsilon_\mu \, \partial^\mu \mathcal{L} \,. \tag{12.1}$$

Taylor expansion of $\delta \mathcal{L}$ gives

$$\delta \mathcal{L}(\chi, \partial^\mu \chi) = \frac{\partial \mathcal{L}}{\partial \chi} \, \delta \chi + \frac{\partial \mathcal{L}}{\partial(\partial^\mu \chi)} \, \delta(\partial^\mu \chi) \,, \tag{12.2}$$

with $\delta\chi = \chi(x+\epsilon) - \chi(x) = \epsilon_\mu \partial^\mu \chi$. Equations (12.1) and (12.2) can be combined to give

$$\epsilon_\mu \partial^\mu \mathcal{L} = \frac{\partial \mathcal{L}}{\partial \chi} \, \delta \chi + \frac{\partial \mathcal{L}}{\partial(\partial^\mu \chi)} \, \delta(\partial^\mu \chi) \,. \tag{12.3}$$

From (5.31) it is know that the variation of $\partial^\mu \chi$ obeys $\delta(\partial^\mu \chi) = \partial^\mu(\delta\chi)$. Hence, upon replacing $\partial \mathcal{L}/\partial \chi$ with the Euler–Lagrange equation $\partial \mathcal{L}/\partial \chi = \partial^\mu [\partial \mathcal{L}/\partial(\partial^\mu \chi)]$, we obtain from (12.3) the relation

$$\epsilon_\mu \partial^\mu \mathcal{L} = \partial^\mu \left(\frac{\partial \mathcal{L}}{\partial(\partial^\mu \chi)} \, \delta \chi \right) = \partial^\mu \left(\frac{\partial \mathcal{L}}{\partial(\partial^\mu \chi)} \, \epsilon_\nu \, \partial^\nu \chi \right) . \tag{12.4}$$

For arbitrary ϵ_μ it follows from (12.4) that

$$-\,\partial_\nu \mathcal{L} + \partial^\mu \Big(\frac{\partial \mathcal{L}}{\partial(\partial^\mu \chi)} \, \partial^\nu \chi \Big) = 0 \,, \tag{12.5}$$

which we write as

$$\partial^\mu \, T_{\mu\nu}(x) \, = \, 0 \,, \tag{12.6}$$

with the energy–momentum tensor defined as

$$T_{\mu\nu}(x) \equiv - g_{\mu\nu}\, \mathcal{L}(x) + \frac{\partial \mathcal{L}}{\partial(\partial^\mu \chi(x))}\, \partial_\nu \chi(x) \,. \tag{12.7}$$

In the case of neutron star matter the relevant matter fields are baryon, lepton, and quark fields. For the baryon fields, for instance, equation (12.7) takes on the form

$$T_{\mu\nu}(x) \equiv - g_{\mu\nu}\, \mathcal{L}(x) + \sum_B \frac{\partial \mathcal{L}(x)}{\partial[\partial^\mu \psi_B(x)]}\, \partial_\nu \psi_B(x) \,. \tag{12.8}$$

Equation (12.6) constitutes a conservation law for $T_{\mu\nu}$, which follows from the invariance under spacetime transformations. The four quantities

$$P^\nu \equiv \int \mathrm{d}^3\boldsymbol{x}\, T^{0\nu}(\boldsymbol{x}, t) \,, \tag{12.9}$$

which correspond to total energy ($\nu = 0$) and three-momentum ($\nu = 1, 2, 3$), are time independent since

$$\dot{P}^\nu = \int \mathrm{d}^3\boldsymbol{x}\, \partial_0\, T^{0\nu}(\boldsymbol{x}, t) = - \int \mathrm{d}^3\boldsymbol{x} \sum_{i=1}^{3} \partial_i\, T^{i\nu}(\boldsymbol{x}, t) = 0 \,, \tag{12.10}$$

provided that the fields vanish sufficiently rapidly for large arguments (that is, no energy or momentum escape at infinity). Finally we note that from (12.7),

$$T^{00} = -\mathcal{L} + \frac{\partial \mathcal{L}(x)}{\partial(\partial_0 \chi)}\, \partial^0 \chi \,. \tag{12.11}$$

Replacing $\partial\mathcal{L}/\partial(\partial_0\chi)$ with the associated conjugate field $\Pi(\boldsymbol{x}, t)$ gives

$$T^{00} = -\mathcal{L} + \Pi\, \dot{\chi} = \mathcal{H}(\boldsymbol{\pi}, \chi) \,, \tag{12.12}$$

where $\mathcal{H}$ denotes the Hamiltonian density. For the total energy density one thus obtains

$$\epsilon \equiv \langle \boldsymbol{\Phi}_0 | T^{00} | \boldsymbol{\Phi}_0 \rangle = \langle \boldsymbol{\Phi}_0 | \mathcal{H} | \boldsymbol{\Phi}_0 \rangle \,, \tag{12.13}$$

and for total pressure

$$\boldsymbol{P} = \int \mathrm{d}^3\boldsymbol{x}\, \psi_B^\dagger(x)\, (-\mathrm{i}\, \boldsymbol{\nabla})\, \psi_B(x) \,. \tag{12.14}$$

After these introductory remarks we turn to the main topic of this chapter, namely the calculation of the equation of state of neutron star matter described by the Lagrangian of (5.1). Because of $\partial\mathcal{L}/\partial(\partial^\mu\psi_B) = \mathrm{i}\bar{\psi}_B\gamma_\mu$ we obtain from (12.8)

$$
\begin{aligned}
T_{\mu\nu}(x) = \sum_B \bar{\psi}_B(x)\, \{\mathrm{i}\,\gamma_\mu\partial_\nu - g_{\mu\nu}\,[\mathrm{i}\,\gamma^\lambda\partial_\lambda \, - \, m_B - g_{\sigma B}\,\sigma(x) \\
- g_{\omega B}\,\gamma^\lambda\,\omega_\lambda(x)]\}\,\psi_B(x) \\
- g_{\mu\nu}\Big\{-\frac{1}{2}\,\sigma(x)\,[\partial_\lambda\partial^\lambda \, + \, m_\sigma^2]\,\sigma(x) \, + \, \frac{1}{2}\,\partial_\lambda[\sigma(x)\partial^\lambda\sigma(x)] + \ldots \\
- \frac{1}{2}\,\partial_\lambda[\omega_\kappa(x)\,F^{\lambda\kappa}(x)] + \frac{1}{2}\,\omega_\lambda(x)\,\big[\partial_\kappa F^{\kappa\lambda}(x) \, + \, m_\omega^2\,\omega^\lambda(x)\big]\Big\} \\
+ g_{\mu\nu}\,\Big\{\frac{1}{3}\,b_N\,m_N\,[g_{\sigma N}\sigma(x)]^3 \, + \, \frac{1}{4}\,c_N\,[g_{\sigma N}\sigma(x)]^4\Big\},
\end{aligned}
\tag{12.15}
$$

where use of

$$
\partial_\mu(\sigma\partial^\mu\sigma) = (\partial_\mu\sigma)\,(\partial^\mu\sigma) + \sigma\partial_\mu\partial^\mu\sigma \tag{12.16}
$$

and

$$
\partial_\mu(\omega_\nu\,F^{\mu\nu}) = \frac{1}{2}\,F_{\mu\nu}F^{\mu\nu} + \omega_\nu\partial_\mu F^{\mu\nu} \tag{12.17}
$$

was made. For the sake of brevity, we have dropped in (12.15) the contributions that arise from π- and ρ-meson exchange among the baryons. Below we shall see that it is rather straightforward to incorporate their contributions again. The divergences $\partial_\lambda[\sigma\partial^\lambda\sigma]$ and $\partial_\lambda[\omega_\kappa F^{\lambda\kappa}]$ in (12.15) can be discarded, since the diagonal matrix elements of a total divergence are zero. The remaining expressions are simplified by making use of the field equations for ψ_B, σ and ω_κ derived in (5.36), (5.43), and (5.52) respectively. One obtains

$$
\begin{aligned}
T_{\mu\nu}(x) = \sum_B \Big\{\mathrm{i}\,\bar{\psi}_B(x)\gamma_\mu\partial_\nu\psi_B(x) - \frac{1}{2}\,g_{\mu\nu}\,g_{\sigma B}\,\bar{\psi}_B(x)\sigma(x)\psi_B(x) \\
- \frac{1}{2}\,g_{\mu\nu}\,g_{\omega B}\,\bar{\psi}_B(x)\gamma^\kappa\omega_\kappa(x)\psi_B(x)\Big\} + \ldots .
\end{aligned}
\tag{12.18}
$$

In the next step we shall replace the baryon field products with their associated two-point baryon Green functions, $\boldsymbol{g}_1^B$. Before however we need to make sure that the ordering of the field operators in (12.15) remains unchanged under the action of the time-ordering operator $\hat{T}$. This is readily accomplished by adding infinitesimal increments to the time arguments of the baryon field operators. We are then left with ($\partial_\nu = \partial/\partial x^\nu$)

$$
\begin{aligned}
T_{\mu\nu}(x) = \sum_B \Big\{\mathrm{i}\,\lim_{x'\to x^+}\,\partial_\nu\,\hat{T}\big(\bar{\psi}_B(x')\gamma_\mu\psi_B(x)\big) \\
- \frac{1}{2}\,g_{\mu\nu}\,g_{\sigma B}\,\hat{T}\big(\bar{\psi}_B(x^{++})\sigma(x)\psi_B(x^+)\big) \\
- \frac{1}{2}\,g_{\mu\nu}\,g_{\omega B}\,\hat{T}\big(\bar{\psi}_B(x^{++})\gamma^\mu\omega_\mu(x)\psi_B(x^+)\big)\Big\}.
\end{aligned}
\tag{12.19}
$$

The expectation value of $T_{\mu\nu}$ is then given by

$$\langle\mathbf{\Phi}_0|T_{\mu\nu}(x)|\mathbf{\Phi}_0\rangle = \sum_B \Big\{ \mathrm{i} \lim_{x'\to x^+} \partial_\nu \langle\mathbf{\Phi}_0|\hat{T}\big(\bar{\psi}_B(x')\gamma_\mu\psi_B(x)\big)|\mathbf{\Phi}_0\rangle$$
$$- \frac{1}{2} g_{\mu\nu}\, g_{\sigma B}\, \langle\mathbf{\Phi}_0|\hat{T}\big(\bar{\psi}_B(x^{++})\sigma(x)\psi_B(x^+)\big)|\mathbf{\Phi}_0\rangle$$
$$- \frac{1}{2} g_{\mu\nu}\, g_{\omega B}\, \langle\mathbf{\Phi}_0|\hat{T}\big(\bar{\psi}_B(x^{++})\gamma^\lambda\omega_\lambda(x)\psi_B(x^+)\big)|\mathbf{\Phi}_0\rangle \Big\}. \qquad (12.20)$$

Explicit expressions for the meson fields $\sigma(x)$ and $\omega_\mu(x)$ were derived in equations (5.94) and (5.102). Substituting them into (12.20) gives

$$\langle\mathbf{\Phi}_0|T_{\mu\nu}(x)|\mathbf{\Phi}_0\rangle = \sum_B \mathrm{i} \lim_{x'\to x^+} \partial_\nu \langle\mathbf{\Phi}_0|\hat{T}\big(\bar{\psi}_B(x')\gamma_\mu\psi_B(x)\big)|\mathbf{\Phi}_0\rangle$$
$$+ \frac{1}{2} g_{\mu\nu} \sum_{B,B'} g_{\sigma B}\, g_{\sigma B'} \int \mathrm{d}^4x'\, \Delta^{0\sigma}(x,x')$$
$$\times \langle\mathbf{\Phi}_0|\hat{T}\big(\bar{\psi}_B(x^{++})\bar{\psi}_{B'}(x'^+)\psi_{B'}(x')\psi_B(x^+)\big)|\mathbf{\Phi}_0\rangle \qquad (12.21)$$
$$- \frac{1}{2} g_{\mu\nu} \sum_{B,B'} g_{\omega B}\, g_{\omega B'} \int \mathrm{d}^4x'\, \mathcal{D}^{0\omega}_{\lambda\kappa}(x,x')$$
$$\times \langle\mathbf{\Phi}_0|\hat{T}\big(\bar{\psi}_B(x^{++})\gamma^\lambda\bar{\psi}_{B'}(x'^+)\gamma^\kappa\psi_{B'}(x')\psi_B(x^+)\big)|\mathbf{\Phi}_0\rangle\,.$$

12.1 Equation of state in relativistic Hartree and Hartree–Fock approximations

With the technique developed in section 5.3 (cf. discussion of figure 5.1) the latter two expectation values in (12.21) can be replaced with four-point baryon Green functions, $\boldsymbol{g}_2$. From (5.62) one reads off that

$$\langle\mathbf{\Phi}_0|\hat{T}\big(\bar{\psi}_B(x^{++})\bar{\psi}_{B'}(x'^+)\psi_{B'}(x')\psi_B(x^+)\big)|\mathbf{\Phi}_0\rangle$$
$$= -\,\boldsymbol{g}_2(x^+B, x'B'; x'^+B', x^{++}B)\,, \qquad (12.22)$$

while the first expectation value in (12.21) can be written in terms of the two-point baryon Green function,

$$\lim_{x'\to x^+} \boldsymbol{g}^B(x,x') = -\,\mathrm{i}\,\langle\mathbf{\Phi}_0|\hat{T}\big(\bar{\psi}_B(x')\psi_B(x)\big)|\mathbf{\Phi}_0\rangle\,. \qquad (12.23)$$

Upon substituting (12.22) and (12.23) into (12.21) and recalling the matrix structure $\bar{\psi}(x')\gamma_\mu\psi(x') \equiv \bar{\psi}_{\varsigma'}(x')(\gamma_\mu)_{\varsigma'\varsigma}\psi_\varsigma(x') = (\gamma_\mu)_{\varsigma'\varsigma}\bar{\psi}_{\varsigma'}(x')\psi_\varsigma(x')$, one gets

$$\langle\mathbf{\Phi}_0|T_{\mu\nu}(x)|\mathbf{\Phi}_0\rangle = -\lim_{x'\to x^+} \sum_B \partial_\nu\, (\gamma_\mu)_{\varsigma'\varsigma}\, \boldsymbol{g}^B_{\varsigma\varsigma'}(x,x') - \frac{1}{2} g_{\mu\nu}$$
$$\times \sum_{BB'} \int \mathrm{d}^4x'\, \big\{ g_{\sigma B}\, g_{\sigma B'}\, \Delta^{0\sigma}(x,x') - g_{\omega B}\, g_{\omega B'}\, (\gamma^\lambda)_{\varsigma''\bar{\varsigma}}\, \mathcal{D}^{0\omega}_{\lambda\kappa}(x,x')\, (\gamma^\kappa)_{\varsigma'\bar{\varsigma}'} \big\}$$
$$\times\, \boldsymbol{g}_2(x^+B\bar{\varsigma}, x'B'\bar{\varsigma}'; x'^+B'\varsigma', x^{++}B\varsigma'')\,. \qquad (12.24)$$

In the next step we replace $\boldsymbol{g}_2$ with its HF approximated counterpart, which amounts to replacing $\boldsymbol{g}_2$ with an antisymmetrized product of $\boldsymbol{g}_1$ functions, as given in equations (5.118) and (5.119). We then arrive for (12.24) at the expression

$$\begin{aligned}
\langle \boldsymbol{\Phi}_0 | T_{\mu\nu}(x) | \boldsymbol{\Phi}_0 \rangle = & - \lim_{x' \to x^+} \sum_B \partial_\nu \left(\gamma_\mu\right)_{\varsigma'\varsigma} \boldsymbol{g}^B_{\varsigma\varsigma'}(x, x')) - \frac{1}{2}\, g_{\mu\nu} \\
& \times \sum_{BB'} \int \mathrm{d}^4 x' \, \{ g_{\sigma B}\, g_{\sigma B'}\, \Delta^{0\sigma}(x, x') - g_{\omega B}\, g_{\omega B'} \left(\gamma^\lambda\right)_{\varsigma''\bar{\varsigma}} \mathcal{D}^{0\omega}_{\lambda\kappa}(x, x') \left(\gamma^\kappa\right)_{\varsigma'\bar{\varsigma}'} \} \\
& \times \{ \boldsymbol{g}^B_{\bar{\varsigma}\varsigma''}(x^+, x^{++})\, \boldsymbol{g}^{B'}_{\bar{\varsigma}'\varsigma'}(x', x'^+) - \delta_{BB'}\, \boldsymbol{g}^B_{\bar{\varsigma}\varsigma'}(x^+, x'^+)\, \boldsymbol{g}^B_{\bar{\varsigma}'\varsigma''}(x', x^{++}) \} ,
\end{aligned} \tag{12.25}$$

which reads in four-momentum space (cf. equations (B.23) and (B.24))

$$\begin{aligned}
\langle \boldsymbol{\Phi}_0 | T_{\mu\nu}(x) | \boldsymbol{\Phi}_0 \rangle = & - \lim_{x' \to x^+} \sum_B \partial_\nu \int \frac{\mathrm{d}^4 p}{(2\pi)^4}\, \mathrm{e}^{-\mathrm{i}p(x-x')} \left(\gamma_\mu\right)_{\varsigma'\varsigma} \boldsymbol{g}^B_{\varsigma\varsigma'}(p) \\
& - \frac{1}{2}\, g_{\mu\nu} \sum_{BB'} \int \frac{\mathrm{d}^4 p}{(2\pi)^4} \int \frac{\mathrm{d}^4 q}{(2\pi)^4}\, \mathrm{e}^{\mathrm{i}\eta(p^0+q^0)} \{ [g_{\sigma B}\, g_{\sigma B'}\, \Delta^{0\sigma}(0) - g_{\omega B}\, g_{\omega B'} \\
& \times \left(\gamma^\lambda\right)_{\varsigma''\bar{\varsigma}} \mathcal{D}^{0\omega}_{\lambda\kappa}(0) \left(\gamma^\kappa\right)_{\varsigma'\bar{\varsigma}'}]\, \boldsymbol{g}^{B'}_{\bar{\varsigma}'\varsigma'}(q)\, \boldsymbol{g}^B_{\bar{\varsigma}\varsigma''}(p) - \delta_{BB'} [g^2_{\sigma B}\, \Delta^{0\sigma}(p-q) - g^2_{\omega B} \\
& \times \left(\gamma^\lambda\right)_{\varsigma''\bar{\varsigma}} \mathcal{D}^{0\omega}_{\lambda\kappa}(0) \left(\gamma^\kappa\right)_{\varsigma'\bar{\varsigma}'}]\, \boldsymbol{g}^{B'}_{\bar{\varsigma}\varsigma'}(q)\, \boldsymbol{g}^B_{\bar{\varsigma}'\varsigma''}(p) \} .
\end{aligned} \tag{12.26}$$

A comparison of (12.26) with the expression of the HF self-energy,

$$\begin{aligned}
\boldsymbol{\Sigma}^B_{\varsigma''\bar{\varsigma}}(p) = & - \mathrm{i} \sum_{B'} \int \frac{\mathrm{d}^4 q}{(2\pi)^4}\, \mathrm{e}^{\mathrm{i}\eta q^0} \{ g_{\sigma B}\, g_{\sigma B'}\, \Delta^{0\sigma}(0) \\
& \qquad - g_{\omega B}\, g_{\omega B'} \left(\gamma^\lambda\right)_{\varsigma''\bar{\varsigma}} \mathcal{D}^{0\omega}_{\lambda\kappa}(0) \left(\gamma^\kappa\right)_{\varsigma'\bar{\varsigma}'} \}\, \boldsymbol{g}^{B'}_{\bar{\varsigma}'\varsigma'}(q) \\
& + \mathrm{i} \int \frac{\mathrm{d}^4 q}{(2\pi)^4}\, \mathrm{e}^{\mathrm{i}\eta q^0} \{ g^2_{\sigma B}\, \Delta^{0\sigma}(p-q) \\
& \qquad - g^2_{\omega B} \left(\gamma^\lambda\right)_{\varsigma''\bar{\varsigma}} \mathcal{D}^{0\omega}_{\lambda\kappa}(p-q) \left(\gamma^\kappa\right)_{\varsigma'\bar{\varsigma}'} \}\, \boldsymbol{g}^B_{\bar{\varsigma}\varsigma'}(q) \\
& + \ \ldots
\end{aligned} \tag{12.27}$$

derived in equation (5.124), reveals that (12.26) can be expressed in terms of the baryon self-energy in the following manner,

$$\begin{aligned}
\langle \boldsymbol{\Phi}_0 | T_{\mu\nu}(x) | \boldsymbol{\Phi}_0 \rangle = & - \lim_{x' \to x^+} \sum_B \partial_\nu \int \frac{\mathrm{d}^4 p}{(2\pi)^4}\, \mathrm{e}^{-\mathrm{i}p(x-x')}\, \mathrm{Tr} \left(\gamma_\mu\, \boldsymbol{g}^B(p)\right) \\
& - \frac{\mathrm{i}}{2}\, g_{\mu\nu} \sum_B \int \frac{\mathrm{d}^4 p}{(2\pi)^4}\, \mathrm{e}^{\mathrm{i}\eta p^0}\, \mathrm{Tr} \left(\boldsymbol{\Sigma}^B(p)\, \boldsymbol{g}^B(p)\right) .
\end{aligned} \tag{12.28}$$

The traces sum spin and isospin matrix indices, as illustrated in equation (5.129). Since the system's total energy density is given by $\epsilon = \langle \boldsymbol{\Phi}_0 | T_{00} | \boldsymbol{\Phi}_0 \rangle$, according to equation (12.13), we read off from (12.28) that

$$\epsilon = -\sum_B \lim_{x' \to x^+} \partial_0 \int \frac{\mathrm{d}^4 p}{(2\pi)^4} \, \mathrm{e}^{-\mathrm{i}p(x-x')} \, (\gamma_0)_{\zeta\zeta'} \, \boldsymbol{g}^B_{\zeta'\zeta}(p) - \frac{\mathrm{i}}{2} \sum_B \int \frac{\mathrm{d}^4 p}{(2\pi)^4} \, \mathrm{e}^{\mathrm{i}\eta p^0} \, \boldsymbol{\Sigma}^B_{\zeta\zeta'}(p) \, \boldsymbol{g}^B_{\zeta'\zeta}(p) \,. \tag{12.29}$$

Performing the differentiation with respect to the time coordinate x_0 in the first term simply gives a factor of $-\mathrm{i}p^0$. Moreover, the summations over the spin–isospin indices in both terms of (12.29) can be written as traces, as described in section 5.4, which leaves us with

$$\epsilon = \mathrm{i} \sum_B \int \frac{\mathrm{d}^4 p}{(2\pi)^4} \, \mathrm{Tr} \left\{ \mathrm{e}^{-\mathrm{i}p(x-x')} \, p^0 \, \gamma_0 \, \boldsymbol{g}^B(p) - \frac{1}{2} \, \mathrm{e}^{\mathrm{i}\eta p^0} \, \boldsymbol{\Sigma}^B(p) \, \boldsymbol{g}^B(p) \right\} . \tag{12.30}$$

Lastly, we replace $\boldsymbol{g}^B(p)$ with its spectral representation given in (6.71) and perform the integration over the energy variable p^0 analytically. Details are outlined in appendix B. Restricting ourselves to the zero-temperature case for the moment (the extension to finite temperatures will be discussed in section 12.2), which implies closing the integration path in the upper half of the complex energy plane (cf. figure D.1), the result for the energy density then reads

$$\epsilon = \sum_B \int \frac{\mathrm{d}^3 \boldsymbol{p}}{(2\pi)^3} \, \mathrm{Tr} \left\{ \left(\gamma_0 \, \boldsymbol{\Xi}^B(p) \right) \omega^B(\boldsymbol{p}) - \frac{1}{2} \, \boldsymbol{\Sigma}^B(p) \, \boldsymbol{\Xi}^B(p) \right\} \Theta(p_{F_B} - |\boldsymbol{p}|) \,. \tag{12.31}$$

Note from (12.31) that the two-point baryon Green function $\boldsymbol{g}^B_1$ need not be determined explicitly when ϵ is being calculated. The quantity that needs to be determined instead is the baryon spectral function $\boldsymbol{\Xi}^B$ associated with $\boldsymbol{g}^B_1$, which is considerably easier to accomplish than calculating $\boldsymbol{g}_1$ itself. We recall that this feature holds not only for ϵ but all the other properties of the many-body system too. After this parenthetical remark, let us turn to the traces in (12.31). These are to be evaluated with respect to the spin–isospin indices carried by the γ matrices and $\boldsymbol{\Xi}^B$. Proceeding as in section 7.1, one obtains

$$\begin{aligned} \mathrm{Tr}\,(\gamma^0 \, \boldsymbol{\Xi}^B) &= \mathrm{Tr}\,\{\gamma^0 \otimes (\mathbf{1}\, \Xi^B_S + \boldsymbol{\gamma} \cdot \hat{\boldsymbol{p}} \, \Xi^B_V + \gamma^0 \, \Xi^B_0)\} \\ &= \mathrm{Tr}\,(\gamma^0 \otimes \mathbf{1}^{\mathrm{iso}}) \, \Xi^B_S + \mathrm{Tr}\,(\gamma^0 \, \boldsymbol{\gamma} \cdot \hat{\boldsymbol{p}} \otimes \mathbf{1}^{\mathrm{iso}}) \, \Xi^B_V \\ &\quad + \mathrm{Tr}\,(\mathbf{1}^{\mathrm{Dirac}} \otimes \mathbf{1}^{\mathrm{iso}}) \, \Xi^B_0 = 2\,\nu_B \, \Xi^B_0 \,, \end{aligned} \tag{12.32}$$

since for the traces (cf. appendix A.3)

$$\mathrm{Tr}\,(\gamma^0 \otimes \mathbf{1}^{\mathrm{iso}}) = \mathrm{Tr}\,(\gamma^0)\ \mathrm{Tr}\,(\mathbf{1}^{\mathrm{iso}}) = 0\,, \tag{12.33}$$

$$\mathrm{Tr}\,(\gamma^0\,\boldsymbol{\gamma}\cdot\hat{\boldsymbol{p}} \otimes \mathbf{1}^{\mathrm{iso}}) = \mathrm{Tr}\,(\gamma^0\,\boldsymbol{\gamma}\cdot\hat{\boldsymbol{p}})\ \mathrm{Tr}\,(\mathbf{1}^{\mathrm{iso}}) = 0\,, \tag{12.34}$$

and

$$\mathrm{Tr}\,(\mathbf{1} \otimes \mathbf{1}^{\mathrm{iso}}) = \mathrm{Tr}\,(\mathbf{1})\ \mathrm{Tr}\,(\mathbf{1}^{\mathrm{iso}}) = 2\,(2J_B+1)(2I_B+1) \equiv 2\,\nu_B\,. \tag{12.35}$$

Similarly, one calculates for the trace of the second term in (12.31)

$$\begin{aligned}\mathrm{Tr}\,(\boldsymbol{\Xi}^B\,\boldsymbol{\Sigma}^B) &= \mathrm{Tr}\left\{\left(\mathbf{1}\,\Xi_S^B + \boldsymbol{\gamma}\cdot\hat{\boldsymbol{p}}\,\Xi_V^B + \gamma^0\,\Xi_0^B\right)\left(\mathbf{1}\,\Sigma_S^B + \boldsymbol{\gamma}\cdot\hat{\boldsymbol{p}}\,\Sigma_V^B + \gamma^0\,\Sigma_0^B\right)\right\} \\ &= 2\,\nu_B\left\{\Sigma_S^B\,\Xi_S^B - \Sigma_V^B\,\Xi_V^B + \Sigma_0^B\,\Xi_0^B\right\},\end{aligned} \tag{12.36}$$

where we have made use of

$$\begin{aligned}\mathrm{Tr}\left((\boldsymbol{\gamma}\cdot\hat{\boldsymbol{p}})\,(\boldsymbol{\gamma}\cdot\hat{\boldsymbol{p}})\right) &= \mathrm{Tr}\ \hat{p}^i\,\hat{p}^j\ \gamma^i\gamma^j \\ &= 2\ \mathrm{Tr}\ \hat{p}^i\,\hat{p}^j\ g^{ij}\,\mathbf{1} - \mathrm{Tr}\ \hat{p}^i\,\hat{p}^j\ \gamma^j\gamma^i\,, \end{aligned} \tag{12.37}$$

$$2\ \mathrm{Tr}\left((\boldsymbol{\gamma}\cdot\hat{\boldsymbol{p}})\,(\boldsymbol{\gamma}\cdot\hat{\boldsymbol{p}})\right) = -\,2\,\hat{\boldsymbol{p}}\,\hat{\boldsymbol{p}}\ \mathrm{Tr}\,\mathbf{1}\,, \tag{12.38}$$

$$\mathrm{Tr}\,(\boldsymbol{\gamma}\cdot\hat{\boldsymbol{p}})^2 = -\ 4\,\hat{\boldsymbol{p}}^2 = -\,4\,, \tag{12.39}$$

and

$$\mathrm{Tr}\left((\boldsymbol{\gamma}\cdot\hat{\boldsymbol{p}})^2 \otimes \mathbf{1}^{\mathrm{iso}}\right) = -\,2\,\nu_B\,. \tag{12.40}$$

Substituting (12.32) and (12.36) in (12.31) gives for the energy density

$$\begin{aligned}\epsilon = {} & 2\sum_B \nu_B \int \frac{\mathrm{d}^3\boldsymbol{p}}{(2\pi)^3}\,\omega^B(\boldsymbol{p})\,\Xi_0^B(\boldsymbol{p})\,\Theta(p_{F_B} - |\boldsymbol{p}|) \\ & - \sum_B \nu_B \int \frac{\mathrm{d}^3\boldsymbol{p}}{(2\pi)^3}\,\left\{\Sigma_S^B(\omega^B(\boldsymbol{p}),\boldsymbol{p})\,\Xi_S^B(\boldsymbol{p}) - \Sigma_V^B(\omega^B(\boldsymbol{p}),\boldsymbol{p})\,\Xi_V^B(\boldsymbol{p})\right. \\ & \left. + \,\Sigma_0^B(\omega^B(\boldsymbol{p}),\boldsymbol{p})\,\Xi_0^B(\boldsymbol{p})\right\}\,\Theta(p_{F_B} - |\boldsymbol{p}|)\,.\end{aligned} \tag{12.41}$$

Alternatively, we may substitute the expression for the single-particle energy ω^B, derived in (6.149), into the first term of (12.41). Adopting $\Xi_0^B = \frac{1}{2}$ for the HF approximation (cf. (6.145)), one finds

$$\begin{aligned}\omega^B\,\Xi_0^B &= \Sigma_0^B\,\Xi_0^B + \frac{1}{2}\sqrt{(m_B^*)^2 + (p_B^*)^2} \\ &= \Sigma_0^B\,\Xi_0^B + \frac{(m_B^*)^2}{2\sqrt{(m_B^*)^2 + (p_B^*)^2}} + \frac{(p_B^*)^2}{2\sqrt{(m_B^*)^2 + (p_B^*)^2}}\,.\end{aligned} \tag{12.42}$$

A comparison of (12.42) with the spectral functions in HF approximation, given in equations (6.143) through (6.145), reveals that

$$\omega^B\,\Xi_0^B = m_B^*\,\Xi_S^B - p_B^*\,\Xi_V^B + \Sigma_0^B\,\Xi_0^B\,. \tag{12.43}$$

Substituting (12.43) into (12.41) and rearranging terms then gives

$$\epsilon = 2 \sum_B \nu_B \int \frac{d^3\boldsymbol{p}}{(2\pi)^3} \left\{ m_B \, \Xi^B_S(\boldsymbol{p}) - |\boldsymbol{p}| \, \Xi^B_V(\boldsymbol{p}) \right\} \Theta(p_{F_B} - |\boldsymbol{p}|)$$
$$+ \sum_B \nu_B \int \frac{d^3\boldsymbol{p}}{(2\pi)^3} \left\{ \Sigma^B_S(\omega^B(\boldsymbol{p}), \boldsymbol{p}) \, \Xi^B_S(\boldsymbol{p}) - \Sigma^B_V(\omega^B(\boldsymbol{p}), \boldsymbol{p}) \, \Xi^B_V(\boldsymbol{p}) \right.$$
$$\left. + \, \Sigma^B_0(\omega^B(\boldsymbol{p}), \boldsymbol{p}) \, \Xi^B_0(\boldsymbol{p}) \right\} \Theta(p_{F_B} - |\boldsymbol{p}|) \,. \tag{12.44}$$

The contributions of the cubic and quartic terms of the σ-meson field in (12.15) to the total energy density follow as

$$\epsilon^{(\sigma^4)} = \frac{1}{2} \left\{ \frac{1}{3} \, b_N \, m_N \left(\Sigma^{\mathrm{H}.N}_S \right)^3 - \frac{1}{2} \, c_N \left(\Sigma^{\mathrm{H},N}_S \right)^4 \right\} . \tag{12.45}$$

As we shall see in the second part of this book (see, for instance, chapter 13), knowledge of the total energy density is necessary when solving Einstein's field equations of general relativity, since it is the total energy density (besides other quantities like pressure) which enters the source term of Einstein's field equations. The energy per baryon, measured relative to the particle masses, is obtained from the total energy density ϵ as

$$\frac{E}{A} = \frac{1}{\rho} \left\{ \epsilon - \sum_B m_B \, \rho^B \right\} , \tag{12.46}$$

where ρ^B are the partial particle densities that were calculated in equations (6.187) and (6.186).

In the non-relativistic limit we have $\Sigma^B_S \to 0$ and $\Sigma^B_V \to 0$. Hence the single-particle energy takes on the familiar form

$$\sqrt{m_B^2 + \boldsymbol{p}^2} \approx m_B \left(1 + \frac{\boldsymbol{p}^2}{2 \, m_B} \right) . \tag{12.47}$$

Substituting (12.47) into (12.41) gives for the non-relativistic limit of the energy density (for more details, see section 12.5)

$$\epsilon = \sum_B \nu_B \int \frac{d^3\boldsymbol{p}}{(2\pi)^3} \left\{ m_B + \frac{\boldsymbol{p}^2}{2 \, m_B} + \frac{1}{2} \, \Sigma^B_0(\omega^B(\boldsymbol{p}) - \mu^B, \boldsymbol{p}) \right\} \Theta^B(\boldsymbol{p}) \,, \tag{12.48}$$

where $\Theta(\boldsymbol{p}) \equiv \Theta(p_{F_B} - |\boldsymbol{p}|)$. The non-relativistic expression for the energy per baryon thus reads

$$\frac{E}{A} = \frac{1}{\rho} \sum_B \nu_B \int \frac{d^3\boldsymbol{p}}{(2\pi)^3} \left\{ \frac{\boldsymbol{p}^2}{2 \, m_B} + \frac{1}{2} \, \Sigma^B_0(\omega^B(\boldsymbol{p}) - \mu^B, \boldsymbol{p}) \right\} \Theta^B(\boldsymbol{p}) \,. \tag{12.49}$$

The expression of the total energy density (12.41) can be brought into a more transparent form if the many-body system is studied in the relativistic

Hartree approximation, as will be done next. For this purpose we replace both ω^B and the spectral functions Ξ_S^B, $\Xi_V^B = 0$ and Ξ_0^B by their Hartree approximated relations, given in (7.32) and (7.30) respectively. One readily finds (note that $\Sigma_V^B = 0$ for this approximation)

$$\epsilon^{\mathrm{H}} = 2\sum_B \nu_B \int \frac{\mathrm{d}^3\boldsymbol{p}}{(2\pi)^3} \left\{ \frac{1}{2}\left(\Sigma_0^{\mathrm{H},B} + \sqrt{(m_B^*)^2 + \boldsymbol{p}^2} \right) \right.$$
$$\left. - \left[\frac{\Sigma_S^{\mathrm{H},B}\, m_B^*}{2\sqrt{(m_B^*)^2 + \boldsymbol{p}^2}} + \frac{1}{2}\Sigma_0^B \right] \right\} \Theta(p_{F_B} - |\boldsymbol{p}|) \,. \qquad (12.50)$$

Making use of the circumstance that the self-energies, and thus the effective baryon masses, are momentum independent for the relativistic Hartree approximation, we arrive for (12.50) at

$$\epsilon^{\mathrm{H}} = \frac{1}{2}\sum_B \left(\Sigma_0^{\mathrm{H},B}\, \rho^B\right) - \frac{1}{2}\sum_B \left(\Sigma_S^{\mathrm{H},B}\, \bar{\rho}^B\right)$$
$$+ \sum_B \nu_B \int \frac{\mathrm{d}^3\boldsymbol{p}}{(2\pi)^3} \sqrt{(m_B^*)^2 + \boldsymbol{p}^2}\; \Theta(p_{F_B} - |\boldsymbol{p}|) \,. \qquad (12.51)$$

The quantities ρ and $\bar{\rho}$ denote baryon number density and scalar density, respectively, defined in equations (6.186) and (6.201). Finally after some straightforward algebraic manipulations (see appendix B.3), equation (12.51) can be brought into the alternative form

$$\epsilon^{\mathrm{H}} = \sum_B \left\{ \frac{\nu_B}{12\,\pi^2}\, p_{F_B}^3\, \Sigma_0^{\mathrm{H},B} + \frac{\nu_B}{8\,\pi^2}\left((m_B^*)^2 + m_B\, m_B^*\right) \right.$$
$$\times \left(p_{F_B}\, \epsilon_F^{\mathrm{H},B} - (m_B^*)^2 \ln\left| \frac{p_{F_B} + \epsilon_F^{\mathrm{H},B}}{m_B^*} \right| \right) + \frac{\nu_B}{8\,\pi^2}$$
$$\left. \times \left(p_{F_B}\, (\epsilon_F^{\mathrm{H},B})^3 - \frac{5}{2}\,(m_B^*)^2\, p_{F_B}\, \epsilon_F^{\mathrm{H},B} + \frac{3}{2}\,(m_B^*)^4 \ln\left| \frac{p_{F_B} + \epsilon_F^{\mathrm{H},B}}{m_B^*} \right| \right) \right\} , \qquad (12.52)$$

where $\epsilon_F^{\mathrm{H},B} \equiv \epsilon^{\mathrm{H},B}(p_{F_B})$, with $\epsilon^{\mathrm{H},B}$ defined in (7.31).

As a final point, we derive from (12.41) the energy density of asymmetric *neutron star matter* at zero temperature. It is illustrative to split up the result into the Hartree, ϵ^{H}, and Fock, ϵ^{F}, contribution [79]. One then finds for the Hartree density

$$\epsilon^{\mathrm{H}} = \frac{1}{2}\sum_B \left(\Sigma_0^{\mathrm{H},B} + I_{3B}\, \Sigma_{03}^{\mathrm{H},B}\right) \rho^{\mathrm{H},B} - \frac{1}{2}\sum_B \left(\Sigma_S^{\mathrm{H},B}\, \bar{\rho}^{\mathrm{H},B}\right)$$
$$+ \frac{1}{2\pi^2}\sum_B (2J_B + 1) \int_0^{p_{F_B}} \mathrm{d}p\, \boldsymbol{p}^2 \sqrt{\left(m_B + \Sigma_S^{\mathrm{H},B}\right)^2 + \boldsymbol{p}^2}$$

$$
-\frac{1}{2}\Big\{-\frac{1}{3}\,b_N\,m_N\,(\Sigma_S^{\mathrm{H},N})^3+\frac{1}{2}\,c_N\,(\Sigma_S^{\mathrm{H},N})^4\Big\}
$$

$$
+\frac{1}{2\pi^2}\sum_{L=e,\mu}(2J_L+1)\int_0^{p_{F_B}}\mathrm{d}p\,\boldsymbol{p}^2\,\sqrt{m_L^2+\boldsymbol{p}^2}\,. \tag{12.53}
$$

This result agrees, as must be the case, with the energy density of Walecka's mean-field theory [61]. The Fock density is given by

$$
\epsilon^{\mathrm{F}}=\sum_B(2J_B+1)\int\frac{\mathrm{d}^3\boldsymbol{p}}{(2\pi)^3}\,\{\boldsymbol{\Sigma}_S^{\mathrm{F},B}(\boldsymbol{p})\,\Xi_S^{\mathrm{H},B}(\boldsymbol{p})-\boldsymbol{\Sigma}_V^{\mathrm{F},B}(\boldsymbol{p})\,\Xi_V^{\mathrm{H},B}(\boldsymbol{p})
$$

$$
+\boldsymbol{\Sigma}_0^{\mathrm{F},B}(\boldsymbol{p})\,\Xi_0^{\mathrm{H},B}(\boldsymbol{p})\}\,\Theta^B(\boldsymbol{p})\,. \tag{12.54}
$$

To calculate the system's pressure, we proceed in analogy to the calculation of the total energy density. The starting point is again the expression for the energy–momentum tensor derived in (12.28), whose diagonal elements T_{jj} specify the pressure P according to the relation (cf. (14.13))

$$
P=\frac{1}{3}\sum_{j=1}^{3}\langle\boldsymbol{\Phi}_0|T_{jj}|\boldsymbol{\Phi}_0\rangle\,. \tag{12.55}
$$

Substituting T_{jj} of (12.28) into (12.55) leads to [$g_{ii}=-1$ according to (A.2)]

$$
P=-\frac{1}{3}\sum_{j=1}^{3}\Big\{\lim_{x'\to x^+}\partial_j\sum_B\int\frac{\mathrm{d}^4p}{(2\pi)^4}\,\mathrm{e}^{-\mathrm{i}p(x-x')}\,(\gamma_j)_{\zeta\zeta'}\,\boldsymbol{g}^B_{\zeta'\zeta}(p)
$$

$$
+\frac{\mathrm{i}}{2}\sum_B\int\frac{\mathrm{d}^4p}{(2\pi)^4}\,\mathrm{e}^{\mathrm{i}\eta p^0}\,\boldsymbol{\Sigma}^B_{\zeta\zeta'}(p)\,\boldsymbol{g}^B_{\zeta'\zeta}(p)\Big\}\,. \tag{12.56}
$$

Since for the derivative operator $\partial_j=-\mathrm{i}p_j$, and $\lim_{x'\to x^+}\mathrm{e}^{-\mathrm{i}p(x-x')}=\mathrm{e}^{\mathrm{i}\eta p^0}$, equation (12.56) transforms into

$$
P=\mathrm{i}\sum_B\int\frac{\mathrm{d}^4p}{(2\pi)^4}\,\mathrm{e}^{\mathrm{i}\eta p^0}\,\Big\{\frac{1}{3}\,(\gamma\cdot\boldsymbol{p})_{\zeta\zeta'}+\frac{1}{2}\,\boldsymbol{\Sigma}^B_{\zeta\zeta'}(p)\Big\}\,\boldsymbol{g}^B_{\zeta'\zeta}(p)\,. \tag{12.57}
$$

The integrals over p^0 can be evaluated via contour integration, as outlined in appendix B.1, which results essentially in Θ-functions. Making use of the expressions (B.9) and (B.19), equation (12.57) can be written as

$$
P=\sum_B\int\frac{\mathrm{d}^3\boldsymbol{p}}{(2\pi)^3}\,\mathrm{Tr}\Big\{\frac{1}{3}\gamma\cdot\boldsymbol{p}\,\boldsymbol{\Xi}^B(\boldsymbol{p})+\frac{1}{2}\,\boldsymbol{\Sigma}^B(\omega^B(\boldsymbol{p}),\boldsymbol{p})\,\boldsymbol{\Xi}^B(\boldsymbol{p})\Big\}\,\Theta^B(\boldsymbol{p})\,. \tag{12.58}
$$

The traces of the first term in (12.58) lead to

$$\mathrm{Tr}\left\{\boldsymbol{\gamma}\cdot\boldsymbol{p}\left(\mathbf{1}\,\Xi_S+\boldsymbol{\gamma}\cdot\hat{\boldsymbol{p}}\,\Xi_V+\gamma^0\,\Xi_0\right)\otimes\mathbf{1}^{\mathrm{iso}}\right\}$$
$$=\mathrm{Tr}\left(\boldsymbol{\gamma}\cdot\boldsymbol{p}\,\boldsymbol{\gamma}\cdot\hat{\boldsymbol{p}}\,\Xi_V\right)\,\mathrm{Tr}\left(\mathbf{1}^{\mathrm{iso}}\right), \qquad (12.59)$$

where

$$\mathrm{Tr}\left(\boldsymbol{\gamma}\cdot\boldsymbol{p}\,\boldsymbol{\gamma}\cdot\hat{\boldsymbol{p}}\,\Xi_V\right)= -\,4p^j\hat{p}^j\,\Xi_V=-4\,|\boldsymbol{p}|\,\Xi_V$$
$$= -\,2\,(2J_B+1)\,|\boldsymbol{p}|\,. \qquad (12.60)$$

Hence one gets for (12.59) the relation

$$\mathrm{Tr}\left\{\boldsymbol{\gamma}\cdot\boldsymbol{p}\left(\mathbf{1}\,\Xi_S+\boldsymbol{\gamma}\cdot\hat{\boldsymbol{p}}\,\Xi_V+\gamma^0\,\Xi_0\right)\otimes\mathbf{1}^{\mathrm{iso}}\right\}=-2\,\nu_B\,|\boldsymbol{p}|\,\Xi_V\,. \qquad (12.61)$$

The trace of the second term in (12.58) has already been calculated in (12.36). Substituting both results in (12.58) then leads to the expression

$$P=-\,\frac{2}{3}\sum_B\nu_B\int\frac{\mathrm{d}^3\boldsymbol{p}}{(2\pi)^3}\,|\boldsymbol{p}|\,\Xi_V^B(\boldsymbol{p})\,\Theta(p_{F_B}-|\boldsymbol{p}|)$$
$$+\sum_B\nu_B\int\frac{\mathrm{d}^3\boldsymbol{p}}{(2\pi)^3}\,\{\Sigma_S^B(\omega^B(\boldsymbol{p}),\boldsymbol{p})\,\Xi_S^B(\boldsymbol{p})-\Sigma_V^B(\omega^B(\boldsymbol{p}),\boldsymbol{p})\,\Xi_V^B(\boldsymbol{p})$$
$$+\,\Sigma_0^B(\omega^B(\boldsymbol{p}),\boldsymbol{p})\,\Xi_0^B(\boldsymbol{p})\}\,\Theta(p_{F_B}-|\boldsymbol{p}|)\,. \qquad (12.62)$$

Alternatively, (12.62) may be written in the following manner,

$$P=\sum_B\nu_B\int\frac{\mathrm{d}^3\boldsymbol{p}}{(2\pi)^3}\left\{\Sigma_S^B(\omega^B(\boldsymbol{p}),\boldsymbol{p})\,\Xi_S^B(\boldsymbol{p})-\left[\Sigma_V^B(\omega^B(\boldsymbol{p}),\boldsymbol{p})+\frac{2}{3}\,|\boldsymbol{p}|\right]\right.$$
$$\left.\times\,\Xi_V^B(\boldsymbol{p})+\Sigma_0^B(\omega^B(\boldsymbol{p}),\boldsymbol{p})\,\Xi_0^B(\boldsymbol{p})\right\}\Theta(p_{F_B}-|\boldsymbol{p}|)\,. \qquad (12.63)$$

The non-linear σ-meson interactions contribute to pressure as follows:

$$P^{(\sigma^4)}=\frac{1}{2}\left\{-\,\frac{1}{3}\,b_N\,m_N\left(\Sigma^{\mathrm{H},N}\right)^3+\frac{1}{2}\,c_N\left(\Sigma^{\mathrm{H},N}\right)^4\right\}. \qquad (12.64)$$

Above we have seen that for the relativistic Hartree approximation, the expression for the energy density could be brought into a more transparent form (cf. equations (12.50) and (12.51)). Repeating these steps here again leads for the pressure given in (12.63) to

$$P^{\mathrm{H}}=\frac{1}{2}\sum_B\left(\Sigma_0^{\mathrm{H},B}+I_{3B}\,\Sigma_{03}^{\mathrm{H},B}\right)\rho^{\mathrm{H},B}+\frac{1}{2}\sum_B\left(\Sigma_S^{\mathrm{H},B}\,\bar{\rho}^{\mathrm{H},B}\right)$$
$$+\,\frac{1}{6\pi^2}\sum_B(2J_B+1)\int_0^{p_{F_B}}\mathrm{d}p\,\frac{\boldsymbol{p}^4}{\sqrt{(m_B+\Sigma_S^{\mathrm{H},B})^2+\boldsymbol{p}^2}}$$

$$+ \frac{1}{2}\Big\{-\frac{1}{3}\,b_N\,m_N\big(\Sigma_S^{\mathrm{H},N}\big)^3 + \frac{1}{2}\,c_N\,\big(\Sigma_S^{\mathrm{H},N}\big)^4\Big\}$$

$$+ \frac{1}{6\pi^2}\sum_{L=e,\mu}(2J_L+1)\int\limits_0^{p_{F_L}} \mathrm{d}p\,\frac{\boldsymbol{p}^4}{\sqrt{m_L^2+\boldsymbol{p}^2}}\,, \tag{12.65}$$

which, upon performing the momentum integrations (appendix B.3) and rearranging terms, takes on the final form

$$P^{\mathrm{H}} = \frac{1}{2\,\pi^2}\sum_B \nu_B\Big\{\frac{1}{12}\,p_{F_B}^3\,\epsilon_F^{\mathrm{H},B} - \frac{1}{8}\,(m_B^*)^2\,p_{F_B}\,\epsilon_F^{\mathrm{H},B} + \frac{1}{8}\,(m_B^*)^4\,\ln\frac{p_{F_B}+\epsilon_F^{\mathrm{H},B}}{m_B^*}\Big\}. \tag{12.66}$$

The Fock contribution to pressure turns out to be given by $P^{\mathrm{F}} = \epsilon^{\mathrm{F}}$ [79].

12.2 Thermal bosons and antibaryons

In section 6.1 we have seen that at finite temperatures not only particle and antiparticle states occur in the baryon propagator $\boldsymbol{g}^B(k)$, but because of the nuclear medium two new states, corresponding to holes in the particle Fermi sea and antiholes in the antiparticle Fermi sea, result, as illustrated in figures 6.2 and 6.3. The contribution of the antiholes, to which we refer as thermally excited antibaryons, are easily included in the self-energies and the equation of state by simply extending the path of contour integration in such a way that the antihole pole of figure 6.2 is enclosed too. The resulting paths are shown in figure D.1. This is ensured by simply adding $\lim_{x'\to x^-}$ to the expression $\lim_{x'\to x^+}$, the latter accounting for the medium-corrected baryon propagation only, everywhere in the text where it applies. As an example, the generalized expectation value of the energy–momentum tensor (12.28) then reads

$$\langle\boldsymbol{\Phi}_0|T_{\mu\nu}(x)|\boldsymbol{\Phi}_0\rangle = \Big(\lim_{x'\to x^+} + \lim_{x'\to x^-}\Big)\Big\{-\partial_\nu\,\mathrm{Tr}\left(\gamma_\mu\,\boldsymbol{g}(x,x')\right) - \frac{g_{\mu\nu}}{2}\int \mathrm{d}^4y\;\mathrm{Tr}\left(\boldsymbol{\Sigma}(x,y)\,\boldsymbol{g}(y,x')\right)\Big\}. \tag{12.67}$$

Aside from the thermally excited antibaryons, there will be thermal contributions of bosons too at finite temperature [325, 467]. Their contribution is derivable, for instance, from the energy–momentum tensor associated with these bosons,

$$T_{\mu\nu} \equiv \sum_{M=\sigma,\omega,\dots} T_{\mu\nu}^{(M)}\,, \tag{12.68}$$

with

$$T^{(\sigma)}_{\mu\nu}(x) = (\partial_\mu \sigma(x))\,(\partial_\nu \sigma(x))\,, \quad T^{(\omega)}_{\mu\nu}(x) = (\partial_\nu \omega^\lambda(x))\,F_{\lambda\mu}(x)\,. \quad (12.69)$$

The expectation value of (12.68) can be written as [125]

$$\begin{aligned}\langle \mathbf{\Phi}_0 | \mathcal{T}_{\mu\nu}(x) | \mathbf{\Phi}_0 \rangle &= \mathrm{i} \lim_{x' \to x^+} \{ \nu_\sigma \partial_\mu \partial_\nu \Delta(x, x') + \nu_\omega (\partial_\nu \partial_\lambda \mathcal{D}^\lambda{}_\mu(x, x') \\ &\quad - \partial_\nu \partial_\mu \mathcal{D}^\lambda{}_\lambda(x, x')) \}\,, \qquad (12.70)\end{aligned}$$

with ν_σ and ν_ω defined in equation (12.80). To transform (12.70) into a quantitatively tractable form let us introduce the expression for the boson two-point function, $\Delta(x, x')$. This is accomplished by deriving from the field equations (5.36) and (5.43), treated at the level of the HF approximation, the following relation for the propagator of scalar mesons [125],

$$\begin{aligned}\Delta(x, x') &= \Delta^0(x, x') - \mathrm{i}\, g^2_{\sigma N} \int \mathrm{d}^4 y \int \mathrm{d}^4 y'\ \Delta^0(x, y) \\ &\times \{ \boldsymbol{g}(y, y^+) \boldsymbol{g}(y', y'^+) - \boldsymbol{g}(y, y'^+) \boldsymbol{g}(y', y^+) \}\, \Delta^0(y', x')\,. \ (12.71)\end{aligned}$$

With the aid of this relation, one can replace the full meson propagator Δ (similarly for $\mathcal{D}$) in (12.70) with their free counterparts, Δ^0 and $\mathcal{D}^0$, since the contribution of the second term in equation (12.71) is zero in the Hartree approximation, and the exchange correction is known to be very small [468]. The momentum-space expression of $\Delta^0(x, x')$, which is recognized as the two-point Green function of free, scalar mesons at finite density and temperature, is found in close analogy to the derivation of the two-point Green function of baryons at finite temperature and density (cf. section 6.2.2). The only differences that arise originate from the different particle statistics. One arrives at [62]

$$\begin{aligned}\Delta^{0\sigma}(k) = \frac{-1}{2\,\omega^\sigma(\boldsymbol{k})} \Big\{ & \frac{1 + f^\sigma(\boldsymbol{k})}{k^0 - \omega^\sigma(\boldsymbol{k}) + \mathrm{i}\,\eta} - \frac{f^\sigma(\boldsymbol{k})}{k^0 - \omega^\sigma(\boldsymbol{k}) - \mathrm{i}\,\eta} \\ & - \frac{1 + f^\sigma(\boldsymbol{k})}{k^0 + \omega^\sigma(\boldsymbol{k}) - \mathrm{i}\,\eta} + \frac{f^\sigma(\boldsymbol{k})}{k^0 + \omega^\sigma(\boldsymbol{k}) + \mathrm{i}\,\eta} \Big\}\,. \qquad (12.72)\end{aligned}$$

The structure of the free vector propagator $\mathcal{D}^0_{\mu\nu}(k)$ is determined by (5.100). The Bose–Einstein distribution function, f^M, of thermal mesons of type M propagating with energy

$$\omega^M(\boldsymbol{k}) = \sqrt{\boldsymbol{k}^2 + m^2_M}\,, \qquad (12.73)$$

is given by

$$f^M(\boldsymbol{k}) = \frac{1}{\mathrm{e}^{\beta\,\omega^M(\boldsymbol{k})} - 1}\,. \qquad (12.74)$$

The total energy density of the system is given by

$$\epsilon = \langle \mathbf{\Phi}_0 | T_{00}(x) + \mathcal{T}_{00}(x) | \mathbf{\Phi}_0 \rangle \,, \tag{12.75}$$

from which one derives (cf. equation (12.29))

$$\begin{aligned}
\epsilon = & -\sum_B \Big(\lim_{x' \to x^+} + \lim_{x' \to x^-} \Big) \, \partial_0 \int \frac{\mathrm{d}^4 q}{(2\pi)^4} \mathrm{e}^{-\mathrm{i}q(x-x')} \, (\gamma^0)_{\zeta\zeta'} \, \boldsymbol{g}^B_{\zeta'\zeta}(q) \\
& - \frac{\mathrm{i}}{2} \sum_B \int \frac{\mathrm{d}^4 q}{(2\pi)^4} \, (\mathrm{e}^{\mathrm{i}\eta q^0} + \mathrm{e}^{-\mathrm{i}\eta q^0}) \, \boldsymbol{\Sigma}^B_{\zeta\zeta'}(q) \, \boldsymbol{g}^B_{\zeta'\zeta}(q) \\
& - \frac{1}{2} \Big(\frac{1}{3} \, b_N \, m_N \, (-\Sigma^{\mathrm{H},N}_S)^3 + \frac{1}{2} \, c_N \, (\Sigma^{\mathrm{H},N}_S)^4 \Big) \\
& - \mathrm{i} \sum_M \nu_M \int \frac{\mathrm{d}^4 q}{(2\pi)^4} \, \mathrm{e}^{\mathrm{i}\eta k^0} \, q_0 \, q_0 \, \Delta^{0M}(q) \,,
\end{aligned} \tag{12.76}$$

and for the system's pressure (cf. equation (12.55)),

$$\begin{aligned}
P = & \frac{\mathrm{i}}{3} \sum_B \int \frac{\mathrm{d}^4 q}{(2\pi)^4} (\mathrm{e}^{\mathrm{i}q^0} + \mathrm{e}^{-\mathrm{i}\eta q^0}) \, (\boldsymbol{\gamma} \cdot \hat{\boldsymbol{p}})_{\zeta\zeta'} \, \boldsymbol{g}^B_{\zeta'\zeta}(q) \\
& + \frac{\mathrm{i}}{2} \sum_B \int \frac{\mathrm{d}^4 q}{(2\pi)^4} \, (\mathrm{e}^{\mathrm{i}\eta q^0} + \mathrm{e}^{-\mathrm{i}\eta q^0}) \, \boldsymbol{\Sigma}^B_{\zeta\zeta'}(q) \, \boldsymbol{g}^B_{\zeta'\zeta}(q) \\
& + \frac{1}{2} \Big(\frac{1}{3} \, b_N \, m_N \, (-\,\Sigma^{\mathrm{H},N}_S)^3 + \frac{1}{2} \, c_N \, (\Sigma^{\mathrm{H},N}_S)^4 \Big) \\
& - \frac{\mathrm{i}}{3} \sum_M \sum_{j=1}^{3} \nu_M \int \frac{\mathrm{d}^4 q}{(2\pi)^4} \, \mathrm{e}^{\mathrm{i}\eta q^0} \, q_j \, q_j \, \Delta^{0M}(q) \,.
\end{aligned} \tag{12.77}$$

Equations (12.76) and (12.77) ignore terms of the order $\mathcal{O}(\bar{\rho} - \bar{\rho}^{\mathrm{H}})$, where $\bar{\rho}^{\mathrm{H}}$ denotes the scalar density (6.200) calculated for the relativistic Hartree approximation. The last terms in (12.76) and (12.77) are the contributions to energy density and pressure that arise from thermal bosons. Performing the integrations over q^0 leads for the latter contributions to

$$\mathcal{E} = \sum_M \nu_M \int \frac{\mathrm{d}^3 \boldsymbol{q}}{(2\pi)^3} \, \sqrt{\boldsymbol{q}^2 + m_M^2} \, f^M(\boldsymbol{q}) \,, \tag{12.78}$$

and

$$\mathcal{P} = \frac{1}{3} \sum_M \nu_M \int \frac{\mathrm{d}^3 \boldsymbol{q}}{(2\pi)^3} \, \frac{\boldsymbol{q}^2}{\sqrt{\boldsymbol{q}^2 + m_M^2}} \, f^M(\boldsymbol{q}) \,. \tag{12.79}$$

The spin–isospin degeneracy factor of bosons is defined as

$$\nu_M = (2I_M + 1)\,(2J_M + 1) \,, \tag{12.80}$$

with spin and isospin quantum numbers, J_M and I_M, listed in table 5.2.

Finally, we insert the spectral decomposition derived for $\boldsymbol{g}^B$ in (6.71) into equations (12.76) and (12.77), leading for energy density and pressure to

$$\begin{aligned}\epsilon = \sum_B \nu_B \int \frac{d^3\boldsymbol{q}}{(2\pi)^3} \Big\{ \big[2\big(m_B\, \Xi_S^B(\boldsymbol{q}) - |\boldsymbol{q}|\, \Xi_V^B(\boldsymbol{q})\big) + \big(\Sigma_S^B(\omega^B(\boldsymbol{q}),\boldsymbol{q})\, \Xi_S^B(\boldsymbol{q}) \\ - \Sigma_V^B(\omega^B(\boldsymbol{q}),\boldsymbol{q})\, \Xi_V^B(\boldsymbol{q}) + \Sigma_0^B(\omega^B(\boldsymbol{q}),\boldsymbol{q})\, \Xi_0^B(\boldsymbol{q})\big)\big]\, f^B(\boldsymbol{q}) \\ - \big[2\big(m_B\, \bar{\Xi}_S^B(\boldsymbol{q}) - |\boldsymbol{q}|\, \bar{\Xi}_V^B(\boldsymbol{q})\big) + \big(\Sigma_S^B(\bar{\omega}^B(\boldsymbol{q}),\boldsymbol{q})\, \bar{\Xi}_S^B(\boldsymbol{q}) \\ - \Sigma_V^B(\bar{\omega}^B(\boldsymbol{q}),\boldsymbol{q})\, \bar{\Xi}_V^B(\boldsymbol{q}) + \Sigma_0^B(\bar{\omega}^B(\boldsymbol{q}),\boldsymbol{q})\, \bar{\Xi}_0^B(\boldsymbol{q})\big)\big]\, \bar{f}^B(\boldsymbol{q}) \Big\} \\ - \frac{1}{2}\Big(-\frac{1}{3}\, b_N\, m_N\, \big(\Sigma_S^{\mathrm{H},N}\big)^3 + \frac{1}{2}\, c_N\, \big(\Sigma_S^{\mathrm{H},N}\big)^4 \Big) + \mathcal{E}\,, \qquad (12.81)\end{aligned}$$

and

$$\begin{aligned}P = \sum_B \nu_B \int \frac{d^3\boldsymbol{q}}{(2\pi)^3} \Big\{ -\frac{2}{3} |\boldsymbol{q}| \big(\Xi_V^B(\boldsymbol{q})\, f^B(\boldsymbol{q}) - \bar{\Xi}_V^B(\boldsymbol{q})\, \bar{f}^B(\boldsymbol{q}) \big) \\ + \big[\Sigma_S^B(\omega^B(\boldsymbol{q}),\boldsymbol{q})\, \Xi_S^B(\boldsymbol{q}) - \Sigma_V^B(\omega^B(\boldsymbol{q}),\boldsymbol{q})\, \Xi_V^B(\boldsymbol{q}) \\ + \Sigma_0^B(\omega^B(\boldsymbol{q}),\boldsymbol{q})\, \Xi_0^B(\boldsymbol{q}) \big]\, f^B(\boldsymbol{q}) - \big[\Sigma_S^B(\bar{\omega}^B(\boldsymbol{q}),\boldsymbol{q})\, \bar{\Xi}_S^B(\boldsymbol{q}) \\ - \Sigma_V^B(\bar{\omega}^B(\boldsymbol{q}),\boldsymbol{q})\, \bar{\Xi}_V^B(\boldsymbol{q}) + \Sigma_0^B(\bar{\omega}^B(\boldsymbol{q}),\boldsymbol{q})\, \bar{\Xi}_0^B(\boldsymbol{q}) \big]\, \bar{f}^B(\boldsymbol{q}) \Big\} \\ + \frac{1}{2}\Big(-\frac{1}{3}\, b_N\, m_N \big(\Sigma_S^{\mathrm{H},N}\big)^3 + \frac{1}{2}\, c_N \big(\Sigma_S^{\mathrm{H},N}\big)^4 \Big) + \mathcal{P}\,. \qquad (12.82)\end{aligned}$$

With the aid of the individual expressions derived for Ξ_i^B and $\bar{\Xi}_i^B$ ($i = S, V, 0$) in (7.29), expressions (12.81) and (12.82) can be written in the relativistic *Hartree* approximation as

$$\begin{aligned}\epsilon^{\mathrm{H}} = \frac{1}{2} \sum_B \big(\Sigma_0^{\mathrm{H},B}\, \rho^B \big) - \frac{1}{2} \sum_B \big(\Sigma_S^{\mathrm{H},B}\, \bar{\rho}^B \big) \\ + \frac{1}{2\,\pi^2} \sum_B \nu_B \int_0^\infty dq\, q^2 \sqrt{(m_B + \Sigma_S^{\mathrm{H},B})^2 + \boldsymbol{q}^2}\, \big(f^B(\boldsymbol{q}) + \bar{f}^B(\boldsymbol{q}) \big) \\ - \frac{1}{2}\Big(-\frac{1}{3}\, b_N\, m_N\, \big(\Sigma_S^{\mathrm{H},N}\big)^3 + \frac{1}{2}\, c_N\, \big(\Sigma_S^{\mathrm{H},N}\big)^4 \Big) + \mathcal{E}\,, \qquad (12.83)\end{aligned}$$

and

$$\begin{aligned}P^{\mathrm{H}} = \frac{1}{2} \sum_B \big(\Sigma_0^{\mathrm{H},B}\, \rho^B \big) + \frac{1}{2} \sum_B \big(\Sigma_S^{\mathrm{H},B}\, \bar{\rho}^B \big) \\ + \frac{1}{6\pi^2} \sum_B \nu_B \int_0^\infty dq\, \frac{q^4}{\sqrt{(m_B + \Sigma_S^{\mathrm{H},B})^2 + \boldsymbol{q}^2}} \big(f^B(\boldsymbol{q}) + \bar{f}^B(\boldsymbol{q}) \big) \\ + \frac{1}{2}\Big(-\frac{1}{3}\, b_N\, m_N\, \big(\Sigma_S^{\mathrm{H},N}\big)^3 + \frac{1}{2}\, c_N\, \big(\Sigma_S^{\mathrm{H},N}\big)^4 \Big) + \mathcal{P}\,. \qquad (12.84)\end{aligned}$$

Another attractive simplification of expressions (12.81) and (12.82), which works very reasonably because of the rather weak energy and momentum dependence of the self-energies Σ_S and Σ_0 and the smallness of Σ_V [84, 125, 314, 469], consists in replacing the HF spectral functions (6.139) through (6.141) with their Hartree approximated counterparts derived in equations (6.143) through (6.148). Substituting the latter into (12.81) and (12.82), we find for ϵ and P in relativistic *Hartree–Fock*,

$$\begin{aligned}
\epsilon^{\mathrm{H}} = 2 \sum_B \nu_B \int \frac{\mathrm{d}^3\boldsymbol{q}}{(2\pi)^3} \left(m_B\, \Xi_S^{\mathrm{H},B}(\boldsymbol{q}) - |\boldsymbol{q}|\, \Xi_V^{\mathrm{H},B}(\boldsymbol{q})\right) \left(f^B(\boldsymbol{q}) + \bar{f}^B(\boldsymbol{q})\right) \\
+ \sum_B \nu_B \int \frac{\mathrm{d}^3\boldsymbol{q}}{(2\pi)^3} \left\{\left(\Sigma_S^{\mathrm{H},B}(\boldsymbol{q})\, \Xi_S^{\mathrm{H},B}(\boldsymbol{q}) - \Sigma_V^{\mathrm{H},B}(\boldsymbol{q})\, \Xi_V^{\mathrm{H},B}(\boldsymbol{q})\right)\right. \\
\left.\times \left(f^B(\boldsymbol{q}) + \bar{f}^B(\boldsymbol{q})\right) + \Sigma_0^{\mathrm{H},B}(\boldsymbol{q})\, \Xi_0^{\mathrm{H},B}(\boldsymbol{q}) \left(f^B(\boldsymbol{q}) - \bar{f}^B(\boldsymbol{q})\right)\right\} \\
- \frac{1}{2}\Big(-\frac{1}{3} b_N\, m_N \left(\Sigma_S^{\mathrm{H},N}\right)^3 + \frac{1}{2} c_N \left(\boldsymbol{\Sigma}_S^{\mathrm{H},N}\right)^4\Big) + \mathcal{E}\,,
\end{aligned} \tag{12.85}$$

$$\begin{aligned}
\epsilon^{\mathrm{F}} = \sum_B \nu_B \int \frac{\mathrm{d}^3\boldsymbol{q}}{(2\pi)^3} \left\{\left(\Sigma_S^{\mathrm{F},B}(\boldsymbol{q})\, \Xi_S^{\mathrm{H},B}(\boldsymbol{q}) - \Sigma_V^{\mathrm{F},B}\, \Xi_V^{\mathrm{H},B}(\boldsymbol{q})\right)\right. \\
\left.\times \left(f^B(\boldsymbol{q}) + \bar{f}^B(\boldsymbol{q})\right) + \Sigma_0^{\mathrm{F},B}(\boldsymbol{q})\, \Xi_0^{\mathrm{H},B}(\boldsymbol{q})\left(f^B(\boldsymbol{q}) - \bar{f}^B(\boldsymbol{q})\right)\right\},
\end{aligned} \tag{12.86}$$

and

$$\begin{aligned}
P^{\mathrm{H}} = -\frac{2}{3} \sum_B \nu_B \int \frac{\mathrm{d}^3\boldsymbol{q}}{(2\pi)^3} |\boldsymbol{q}|\, \Xi_V^{\mathrm{H},B}(\boldsymbol{q}) \left(f^B(\boldsymbol{q}) + \bar{f}^B(\boldsymbol{q})\right) \\
+ \sum_B \nu_B \int \frac{\mathrm{d}^3\boldsymbol{q}}{(2\pi)^3} \left\{\left(\Sigma_S^{\mathrm{H},B}(\boldsymbol{q})\, \Xi_S^{\mathrm{H},B}(\boldsymbol{q}) - \Sigma_V^{\mathrm{H},B}(\boldsymbol{q})\, \Xi_V^{\mathrm{H},B}(\boldsymbol{q})\right)\right. \\
\left.\times \left(f^B(\boldsymbol{q}) + \bar{f}^B(\boldsymbol{q})\right) + \Sigma_0^{\mathrm{H},B}(\boldsymbol{q})\, \Xi_0^{\mathrm{H},B}(\boldsymbol{q})\left(f^B(\boldsymbol{q}) - \bar{f}^B(\boldsymbol{q})\right)\right\} \\
+ \frac{1}{2}\Big(-\frac{1}{3} b_N\, m_N \left(\Sigma_S^{\mathrm{H},N}\right)^3 + \frac{1}{2} c_N \left(\Sigma_S^{\mathrm{H},N}\right)^4\Big) + \mathcal{P}\,,
\end{aligned} \tag{12.87}$$

$$P^{\mathrm{F}} = \epsilon^{\mathrm{H}}\,. \tag{12.88}$$

The total contributions to energy density and pressure in relativistic Hartree–Fock then follow from (12.85) to (12.88) as

$$\epsilon = \epsilon^{\mathrm{H}} + \epsilon^{\mathrm{F}}\,, \quad \text{and} \quad P = P^{\mathrm{H}} + \epsilon^{\mathrm{F}}\,. \tag{12.89}$$

The simpler mathematical structure of these expressions arises to a great extent from the symmetry relations between the particle and antiparticle baryon spectral functions (6.143) to (6.148). As they should, these relations bear a strong resemblance with the corresponding ones derived in a somewhat different mathematical framework by Serot and Walecka [92].

12.3 Equation of state of a relativistic lepton gas

The calculation of energy density and pressure of a free, relativistic lepton gas may serve as a further simple case to demonstrate the principal ideas behind the Green function technique. We start from the energy–momentum tensor of such a system, which follows from equation (12.7) for $\mathcal{L} = \mathcal{L}_L$ (cf. (5.15)) and by replacing the summation over B with the summation over $L = e^-, \mu^-$. One then obtains for the momentum tensor's expectation value [79]

$$\langle \mathbf{\Phi}_0 | T^{\mathrm{Lep}}_{\mu\nu}(x) | \mathbf{\Phi}_0 \rangle = - \lim_{x' \to x^+} \sum_{L=e,\mu} \partial_\nu \ \mathrm{Tr}\left(\gamma_\mu \, \boldsymbol{g}^L(x, x')\right), \tag{12.90}$$

whose Fourier transform is given by

$$\langle \mathbf{\Phi}_0 | T^{\mathrm{Lep}}_{\mu\nu}(x) | \mathbf{\Phi}_0 \rangle = - \lim_{x' \to x^+} \partial_\nu \sum_{L=e,\mu} \int \frac{\mathrm{d}^4 q}{(2\pi)^4} \mathrm{e}^{-\mathrm{i}q(x-x')} \ \mathrm{Tr}\left(\gamma_\mu \, \boldsymbol{g}^L(q)\right). \tag{12.91}$$

Substituting $\boldsymbol{g}^L$ by its spectral representation, derived in (6.156), performing the contour integration, and lastly inserting the lepton spectral functions (6.165) into the resulting expression gives for the energy density $\epsilon_{\mathrm{Lep}} \equiv \langle \mathbf{\Phi}_0 | T^{\mathrm{Lep}}_{00} | \mathbf{\Phi}_0 \rangle$ of the lepton gas

$$\epsilon_{\mathrm{Lep}} = \frac{1}{2\pi^2} \sum_{L=e,\mu} (2J_L + 1) \int_0^{k_{F_L}} \mathrm{d}q \, \boldsymbol{q}^2 \sqrt{m_L^2 + \boldsymbol{q}^2}\,. \tag{12.92}$$

The factor $2J_L+1$ accounts for the spin degeneracy of leptons (see table 6.1), and k_{F_L} denote their Fermi momenta. It is a simple exercise to calculate the momentum integral in (12.92) analytically. The result is given in equation (B.30). Substituting this expression into (12.92) leads to the final result for ϵ_{Lep},

$$\epsilon_{\mathrm{Lep}} = \frac{1}{16\,\pi^2} \sum_{L=e,\mu} (2J_L + 1) \left\{ k_{F_L} \left(2\,k_{F_L}^2 + m_L^2\right) \sqrt{m_L^2 + k_{F_L}^2} \right.$$
$$\left. - \, m_L^4 \, \ln \left(\frac{k_{F_L} + \sqrt{m_L^2 + k_{F_L}^2}}{m_L} \right) \right\}. \tag{12.93}$$

The pressure, obtained as $P_{\mathrm{Lep}} \equiv \frac{1}{3} \sum_{i=1}^{3} \langle \mathbf{\Phi}_0 | T^{\mathrm{Lep}}_{ii} | \mathbf{\Phi}_0 \rangle$, is given by

$$P_{\mathrm{Lep}} = \frac{1}{6\pi^2} \sum_{L=e,\mu} (2J_L + 1) \int_0^{k_{F_L}} \mathrm{d}q \, \frac{\boldsymbol{q}^4}{\sqrt{m_L^2 + \boldsymbol{q}^2}}\,. \tag{12.94}$$

Making use of the analytical result for the momentum integral, given in (B.29), leads to

$$P_{\rm Lep} = \frac{1}{6\,\pi^2} \sum_{L=e,\mu} (2J_L+1) \left\{ \frac{k_{F_L}}{4} \left\{ k_{F_L}^2 - \frac{3}{2} m_L^2 \right\} \sqrt{m_L^2 + k_{F_L}^2} \right.$$
$$\left. + \frac{3}{8} m_L^4 \ln \left(\frac{k_{F_L} + \sqrt{m_L^2 + k_{F_L}^2}}{m_L} \right) \right\} . \qquad (12.95)$$

The calculation of the lepton density, denoted by ρ^L, can be performed in close analogy to the calculation of the baryon number density in section 12.4. One gets

$$\rho^L = {\rm i} \lim_{x' \to x^+} \gamma^0_{\zeta\zeta'} \, \boldsymbol{g}^L_{\zeta'\zeta}(x, x') \qquad (12.96)$$
$$= {\rm i} \int \frac{{\rm d}^4 q}{(2\pi)^4} \, {\rm e}^{{\rm i}\eta q^0} \, \gamma^0_{\zeta\zeta'} \, \boldsymbol{g}^L_{\zeta'\zeta}(q) \qquad (12.97)$$
$$= 2\,(2J_L+1) \int \frac{{\rm d}^3 \boldsymbol{q}}{(2\pi)^3} \, \Xi^L_0(\boldsymbol{q}) \, \Theta^L(\boldsymbol{q}) = \frac{2J_L+1}{2} \, \frac{k_{F_L}^3}{3\,\pi^2} . \qquad (12.98)$$

The lepton distribution function has been abbreviated to $\Theta^L(\boldsymbol{q}) \equiv \Theta(k_{F_L} - |\boldsymbol{q}|)$. Finally the total number density of leptons, $\rho^{\rm Lep}$, is obtained as

$$\rho^{\rm Lep} \equiv \sum_L \rho^L . \qquad (12.99)$$

12.4 Equation of state in the relativistic ladder approximation

To derive the equation of state for this many-body approximation, we choose, for the sake of illustration, a mathematical route different from the one adopted for the HF case. We begin with going back to equation (12.21). Since no approximations have been introduced in deriving this relation, it serves as the starting point for deriving the equation of state for both the relativistic HF and relativistic ladder approximation. Here, however, we shall put the derivation of the equation of state on more intuitive grounds by starting from the system's Hamiltonian density, complementary to sketching the derivation of the equation of state from the energy–momentum tensor too.

The energy density of the system described by the Lagrangian (5.1) can be split up into the following three contributions [117, 118, 122]:

$$\epsilon \equiv \frac{1}{\Omega} \int_\Omega {\rm d}^3 \boldsymbol{x} \, \langle \boldsymbol{\Phi}_0 | \mathcal{H}^0_{\rm B}(x) + \mathcal{H}^0_{\rm M}(x) + \mathcal{H}^{\rm I}(x) | \boldsymbol{\Phi}_0 \rangle \, , \qquad (12.100)$$

where Ω denotes the volume. The individual terms in (12.100) originate from free baryons, free mesons, and from the interactions between the baryons mediated by mesons. For what follows, it is convenient to write (12.100) in the form $\epsilon \equiv \epsilon^0_{\rm B} + \epsilon^0_{\rm M} + \epsilon^{\rm I}$. The first term of this decomposition, $\epsilon^0_{\rm B}$, constitutes the ground-state expectation value, that is $\langle \mathbf{\Phi}_0 | \mathcal{H}^0_{\rm B} | \mathbf{\Phi}_0 \rangle$, of the Hamilton density of free baryons, which follows from (12.12) in the form

$$\mathcal{H}^0_{\rm B}(x) = \sum_B \bar{\psi}_B(x) \left(\gamma^\mu p_\mu + m_B \right) \psi_B(x) . \tag{12.101}$$

When calculating $\langle \mathbf{\Phi}_0 | \mathcal{H}^0_{\rm B} | \mathbf{\Phi}_0 \rangle$ one encounters the ground-state expectation value $\langle \mathbf{\Phi}_0 | \bar{\psi}_B \psi_B | \mathbf{\Phi}_0 \rangle$ which can be replaced with its associated two-point Green function, derived in (5.63). Subsequent Fourier transformation then leads to the analog of (12.23) given by [125, 314]

$$\epsilon^0_{\rm B} = - \lim_{x' \to x^+} \sum_B \partial_0 \int \frac{{\rm d}^4 q}{(2\pi)^4} \, {\rm e}^{-{\rm i}q(x-x')} \, \gamma^0_{\zeta\zeta'} \, \boldsymbol{g}^B_{\zeta'\zeta}(q) . \tag{12.102}$$

A comparison of this expression with equation (12.29) reveals that the first term there is the energy density that originates from free baryons. Since this term transforms, after contour integration and calculation of the trace, into the first term in (12.44), we can write for (12.102) the relation

$$\epsilon^0_{\rm B} = 2 \sum_B \nu_B \int \frac{{\rm d}^3 \boldsymbol{q}}{(2\pi)^3} \, \{ m_B \, \Xi^B_S(\boldsymbol{q}) - |\boldsymbol{q}| \, \Xi^B_V(\boldsymbol{q}) \} \Theta^B(\boldsymbol{q}) . \tag{12.103}$$

The energy densities $\epsilon^0_{\rm M}$ and $\epsilon^{\rm I}$ that originate from the other two terms in (12.100) remain to be calculated. These contributions are found most readily by noticing that all the terms in (12.21) that arise from baryon–meson interactions (that is, all terms except the first expression, which, as seen just above, corresponds to free baryon propagation) can be written in accordance with equations (5.153) through (5.155) in the form $\Gamma^{MB} \Gamma^{MB'} \Delta^M \boldsymbol{g}_2^{B'B} = \mathbf{V}^{BB'} \boldsymbol{g}_2^{B'B}$. Hence we arrive for $\epsilon^0_{\rm M} + \epsilon^{\rm I}$ at

$$\begin{aligned} \epsilon^0_{\rm M} + \epsilon^{\rm I} = -\frac{1}{2\Omega} \sum_{BB'} \int_\Omega {\rm d}^3 \boldsymbol{x} \, \langle x_1 \zeta_1, x_2 \zeta_2 | \mathbf{V}^{BB'} | x_3 \zeta_3, x_4 \zeta_4 \rangle \\ \times \, \boldsymbol{g}_2^{B'B}(x_3 \zeta_3, x_4 \zeta_4; x_1^+ \zeta_1, x_2^+ \zeta_2) . \end{aligned} \tag{12.104}$$

Because equations (5.175) and (5.186) can be combined to the operator equation $\mathbf{V} \boldsymbol{g}_2 = {\rm i} \, \mathbf{\Sigma} \, \boldsymbol{g}_1$, we may bring (12.104) into the form

$$\begin{aligned} \epsilon^0_{\rm M} + \epsilon^{\rm I} = \sum_B \nu_B \int \frac{{\rm d}^3 \boldsymbol{q}}{(2\pi)^3} \{ \Sigma^B_S(\omega^B(\boldsymbol{q}), \boldsymbol{q}) \, \Xi^B_S(\boldsymbol{q}) - \Sigma^B_V(\omega^B(\boldsymbol{q}), \boldsymbol{q}) \, \Xi^B_V(\boldsymbol{q}) \\ + \Sigma^B_0(\omega^B(\boldsymbol{q}), \boldsymbol{q}) \, \Xi^B_0(\boldsymbol{q}) \} \, \Theta^B(\boldsymbol{q}) . \end{aligned} \tag{12.105}$$

The integrals in equations (12.103) and (12.105) can be calculated via straightforward numerical methods once the baryon spectral functions and self-energies have been computed self-consistently from the equations summarized in section 11.1. Parenthetically we note that the result obtained above for the total energy density, $\epsilon = \epsilon_{\rm B}^0 + \epsilon_{\rm M}^0 + \epsilon^{\rm I}$, coincides of course with (12.44) derived, however, from the energy–momentum tensor.

The energy per baryon, E/A, internal energy density, $\mathcal{E}^{\rm int}$, and pressure, P, are obtainable from the total energy density, ϵ. For the former two, one has

$$\frac{E(\rho)}{A} = \frac{\mathcal{E}^{\rm int}(\rho)}{\rho}, \tag{12.106}$$

where

$$\frac{\mathcal{E}^{\rm int}(\rho)}{\rho} = \frac{\epsilon(\rho)}{\rho} - m_N ,$$

and m_N is the nucleon mass. One sees here that the internal energy is defined as the volume density of the energy per baryon. Generally $\mathcal{E}^{\rm int}$ includes all forms of energy except the rest mass of the baryons. We shall encounter this quantity again when calculating the total baryon mass of stars (cf. equations (15.90) and (15.91)). Finally, at zero temperature the pressure of the system follows from the thermodynamic relation

$$P = \rho^2 \frac{\partial}{\partial \rho} \frac{E(\rho)}{A} = \rho^2 \frac{\partial}{\partial \rho} \frac{\epsilon(\rho)}{\rho} . \tag{12.107}$$

Having derived the expression for the equation of state, we turn now to the numerical outcome for the energy per baryon as a function of total baryon density ρ and asymmetry δ, computed for the modern boson-exchange interactions A, B and C of Brockmann and Machleidt (BM), and the Groningen B interaction. The BM potentials differ mainly with respect to the strength of the tensor force, which increases from A to C. Since this force is the main agent that determines the location of the saturation point of nuclear matter, it is interesting to see whether the saturation energies predicted by these potentials, plotted as a function of baryon density (or Fermi momentum), fall in a narrow band – known as the Coester band, as in the case for non-relativistic theories of matter. The answer is given in figure 12.1. The correct binding energy of nuclear matter at the empirical saturation density, $E/A(\rho) \simeq 15$ MeV, is only obtained for BM B, while the other interactions overbind or underbind nuclear matter.

As described in chapter 10, we perform the calculation of the **T**-matrix in the full Dirac space spanned by the components of the nucleon spinors, which is therefore a matrix in the 16×16 direct product space of the two particles. The **T**-matrix is calculated from the partial wave expanded integral equation (5.203). McNeil, Shephard, and Wallace [470] have suggested a decomposition of **T** into scalar, vector, tensor, pseudovector, and axial Fermi invariants, ${}^S\mathbf{T}$, ${}^V\mathbf{T}$, ${}^T\mathbf{T}$, ${}^P\mathbf{T}$, and ${}^A\mathbf{T}$ respectively, according

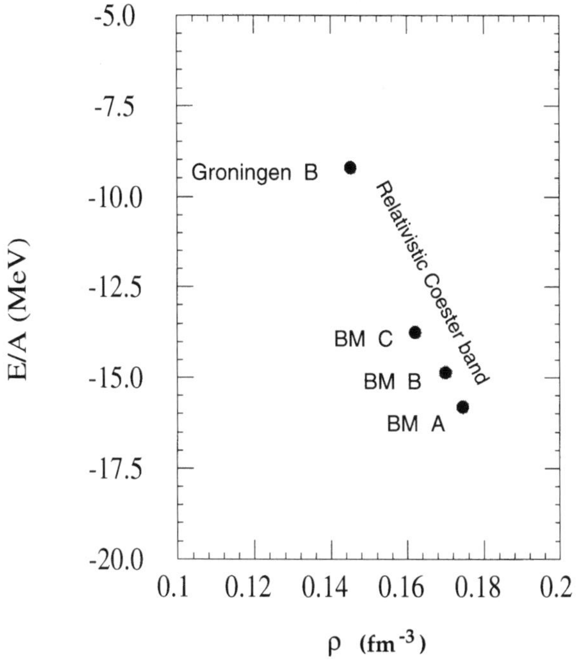

Figure 12.1. Relativistic Coester band associated with $RBHF^{(1)}$ calculations performed for potentials BM A through BM C, and Groningen B. The data are given in tables 12.1 and 12.2.

to the scheme

$$\mathbf{T} = \sum_{S,V,T,P,A} {}^{S}\mathbf{T}\, \mathbf{1}^{(1)}\mathbf{1}^{(2)} + {}^{V}\mathbf{T}\, \gamma_{\mu}^{(1)}\gamma^{(2)\mu} + {}^{T}\mathbf{T}\, \sigma_{\mu\nu}^{(1)}\sigma^{(2)\mu\nu}$$
$$+ {}^{P}\mathbf{T}\, \gamma_5^{(1)}\gamma_5^{(2)} + {}^{A}\mathbf{T}\, \gamma_5^{(1)}\gamma_{\mu}^{(1)}\gamma_5^{(2)}\gamma^{(2)\mu}\,. \tag{12.108}$$

This, however, constitutes a non-unique ansatz for $\mathbf{T}$ that may lead, depending on the decomposition, to different results for the baryon self-energy [103, 119]. The equation of state, on the other hand, appears to be rather insensitive against the chosen decomposition [469]. The superscripts in (12.108) denote the particle (1 or 2) acted on by the matrix, and all spacetime variables are contracted to create Lorentz scalars. Our determination of the $\mathbf{T}$-matrix also avoids another popular approximation, namely the fitting procedure of Machleidt, Holinde, and Elster [119] which ignores the momentum dependence of the baryon self-energies Σ_S and Σ_0 completely, together with the well-justified approximation $\Sigma_V = 0$ [469]. The – then constant – self-energies Σ_S and Σ_0 are extracted from the positive-energy spinor matrix elements $\Sigma_{\Phi\Phi}$ of (9.65) via a fitting procedure.

Table 12.1. Saturation properties of symmetric nuclear matter computed for different many-body approximations. The nucleon–nucleon interactions are the BM potentials A, B, C [104]. The notation is as follows, $\mathrm{RBHF}^{(1)}$: RBHF calculation in full basis and for momentum dependent self-energies; $\mathrm{RBHF}^{(2)}$: RBHF calculation in full basis but for momentum-averaged self-energies; $\mathrm{RBHF}^{(3)}$: RBHF calculation for positive-energy spinors only and momentum independent self-energies; $\mathbf{\Lambda}^{00(2)}$: same as $\mathrm{RBHF}^{(2)}$ but for $\mathbf{\Lambda}^{00}$ propagator instead of $\mathbf{\Lambda}^{\mathrm{RBHF}}$.

Method	Potential	E/A (MeV)	ρ_0 (fm^{-3})	k_{F_0} (fm^{-1})	K (MeV)
$\mathrm{RBHF}^{(1)}$	A	−15.72	0.174	1.37	336
$\mathrm{RBHF}^{(2)}$	A	−16.49	0.174	1.37	280
$\mathrm{RBHF}^{(3)}$	A	−15.59	0.185	1.40	290
$\mathbf{\Lambda}^{00(2)}$	A	−23.51	0.215	1.47	297
$\mathrm{RBHF}^{(1)}$	B	−14.81	0.170	1.36	264
$\mathrm{RBHF}^{(2)}$	B	−15.73	0.172	1.37	249
$\mathrm{RBHF}^{(3)}$	B	−13.60	0.174	1.37	249
$\mathbf{\Lambda}^{00(2)}$	B	−21.90	0.210	1.46	260
$\mathrm{RBHF}^{(1)}$	C	−13.73	0.162	1.34	268
$\mathrm{RBHF}^{(2)}$	C	−14.38	0.170	1.36	258
$\mathrm{RBHF}^{(3)}$	C	−12.26	0.155	1.32	185
$\mathbf{\Lambda}^{00(2)}$	C	−20.57	0.206	1.45	293

These different approximation techniques imposed onto one and the same many-body approximation, RBHF, render a comparison of the corresponding numerical outcome non-trivial, even when the underlying nucleon–nucleon interaction is the same. A close similarity between our treatment and Brockmann's is accomplished by replacing in each iteration step the momentum dependent self-energies by their momentum-averaged counterparts. This version is denoted by $\mathrm{RBHF}^{(2)}$. It is numerically far less time consuming than $\mathrm{RBHF}^{(1)}$, the iteration procedure which keeps the full momentum dependence. Calculations performed for positive-energy spinors only and momentum independent self-energies are denoted by $\mathrm{RBHF}^{(3)}$.

The impact of the different approximation techniques $\mathrm{RBHF}^{(1)}$ to $\mathrm{RBHF}^{(3)}$ on the saturation properties of symmetric nuclear matter is shown in tables 12.1 and 12.2, and figure 12.2 [101, 459]. One sees that agreement between the full treatment, $\mathrm{RBHF}^{(1)}$, and the results of Brockmann and Machleidt, $\mathrm{RBHF}^{(2)}$, is relatively good. The full treatment leads to somewhat less binding, and the saturation density appears to vary only little too. The less physical $\mathbf{\Lambda}^{00}$ approximation, which is the simplest of the ladder approximations, is known to give more binding than the RHBF

Table 12.2. Saturation properties of symmetric nuclear matter obtained for the Groningen B potential [471]. The different methods are explained in the text.

Method	E/A (MeV)	ρ_0 (fm^{-3})	k_{F_0} (fm^{-1})	K (MeV)
$\mathrm{RBHF}^{(1)}$	-9.21	0.145	1.29	191
$\mathrm{RBHF}^{(2)}$	-9.68	0.152	1.31	183
$\mathbf{\Lambda}^{00(1)}$	-13.92	0.181	1.39	264
$\mathbf{\Lambda}^{00(2)}$	-14.53	0.189	1.41	178

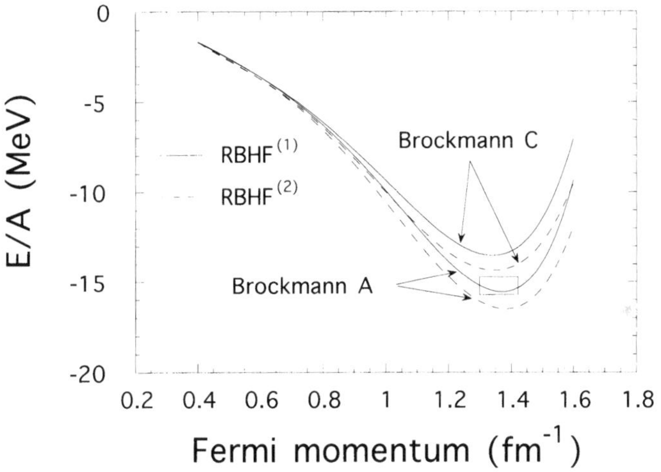

Figure 12.2. Energy per nucleon versus Fermi momentum of symmetric nuclear matter in RBHF for BM potentials A and C [450]. The solid (dashed) curves correspond to the treatment with (averaged) momentum dependency of the self-energies. The square indicates empirical saturation values. (Reprinted courtesy of *Phys. Rev.*)

approximation [117, 118]. This feature is confirmed by our calculations too, as can be seen from tables 12.1 and 12.2. The other two $\mathbf{\Lambda}$ approximations, $\mathbf{\Lambda}^{10}$ and $\mathbf{\Lambda}^{11}$, are numerically so demanding that they have not been applied to asymmetric nuclear matter calculations yet. But from non-relativistic calculations one would expect them to give too little ($\mathbf{\Lambda}^{10}$) or about the correct ($\mathbf{\Lambda}^{10}$) binding energy [325]. The nuclear matter properties computed for the Groningen potential B are compiled in table 12.2, and graphically illustrated in figure 12.3. There, we also perform a comparison with the results of ter Haar and Malfliet [93], which are based on the decomposition of the scattering $\mathbf{T}$-matrix into the five Fermi invariants

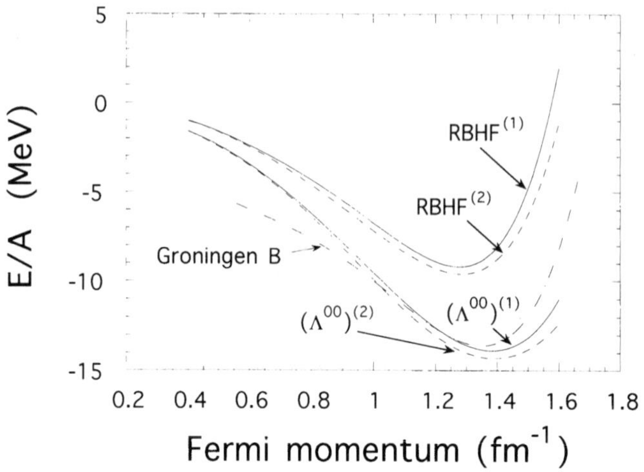

Figure 12.3. Energy per nucleon of nuclear matter computed for different many-body approximations [450]. The underlying potential in each case is Groningen B. The individual approximations are, RBHF$^{(1)}$ and $\mathbf{\Lambda}^{00(1)}$: relativistic **T**-matrix calculation in RBHF and $\mathbf{\Lambda}^{00}$ approximation where full momentum dependence of self-energy is kept; RBHF$^{(2)}$ and $\mathbf{\Lambda}^{00(2)}$: same as RBHF$^{(1)}$ and $\mathbf{\Lambda}^{00(1)}$ but for momentum-averaged self-energy; Groningen B: RBHF calculation based on the decomposition of the **T**-matrix into the Fermi invariants listed in equation (12.108). (Reprinted courtesy of *Phys. Rev.*)

of (12.108). This leads to deviations from the non-composed treatment, RBHF$^{(1)}$, of about 5 MeV at saturation density.

Next, let us turn to the interesting case of asymmetric nuclear matter. The energy per nucleon of such matter as a function of baryon density is displayed in figures 12.4 and 12.5. The asymmetric parameter $\delta = (\rho^n - \rho^p)/\rho$ varies from 0 to 1, where the limiting values correspond to symmetric nuclear matter and pure neutron matter, respectively [101, 459]. One recognizes that the results are almost identical for pure neutron matter, since the different tensor forces of the potentials are not relevant for such matter. Figure 12.6 shows the energy per baryon computed for the $\mathbf{\Lambda}^{00}$ approximation. Engvik *et al* have calculated the properties of asymmetric matter for the BM potential A, adopting the Brockmann–Machleidt approximation but a different Pauli exclusion operator [455]. This makes an immediate comparison difficult. Subject to this caveat, a comparison with the E/A curves shows an agreement of similar quality as for the Brockmann–Machleidt calculations of symmetric matter (figure 12.2). To give an example, the deviation at nuclear matter saturation is approximately 2 MeV for pure neutron matter.

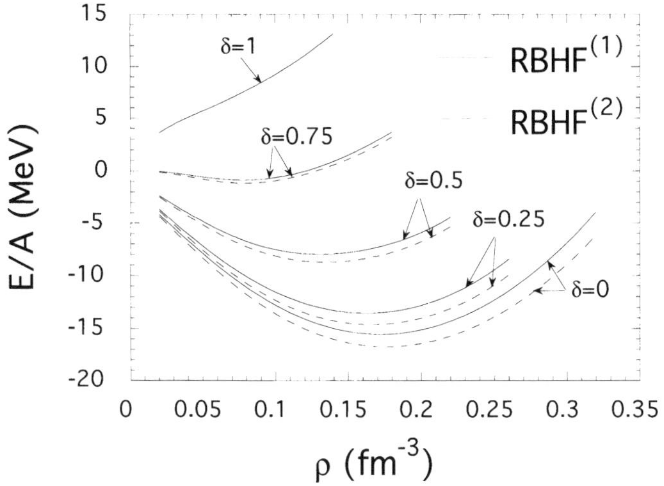

Figure 12.4. Binding energy per nucleon versus density for different asymmetries in the RBHF approximation (BM potential A) [450]. (Reprinted courtesy of *Phys. Rev.*)

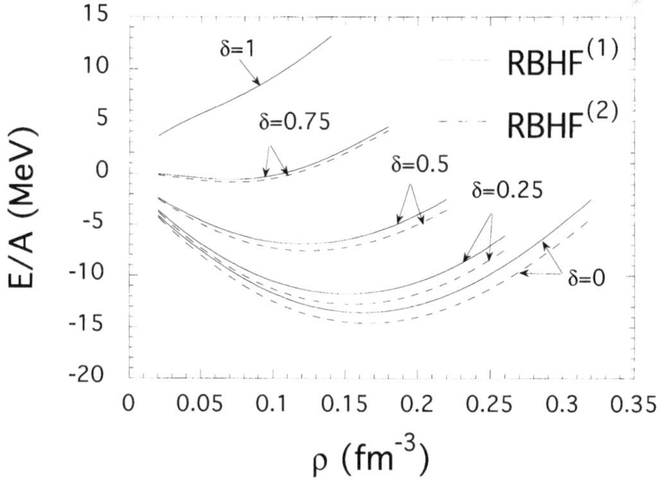

Figure 12.5. Binding energy per nucleon versus density for different asymmetries in the RBHF approximation (BM potential C) [450]. (Reprinted courtesy of *Phys. Rev.*)

Of great interest for neutron star calculations is the behavior of the energy per particle at equilibrium density as a function of asymmetry,

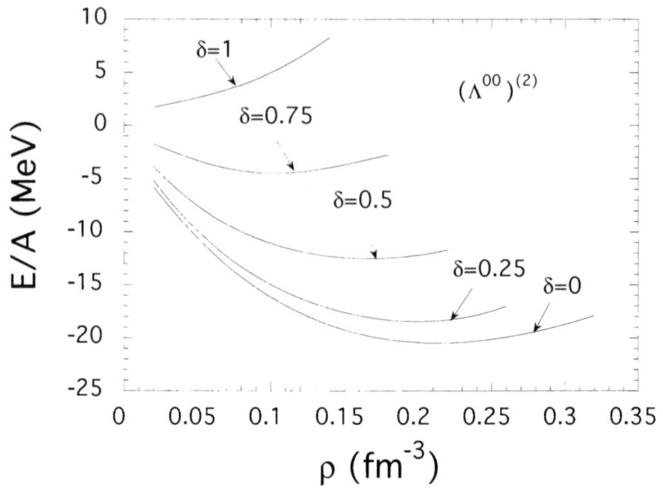

Figure 12.6. Binding energy per nucleon versus density for different asymmetries in the $\mathbf{\Lambda}^{00}$ approximation (BM potential C) [450]. (Reprinted courtesy of *Phys. Rev.*)

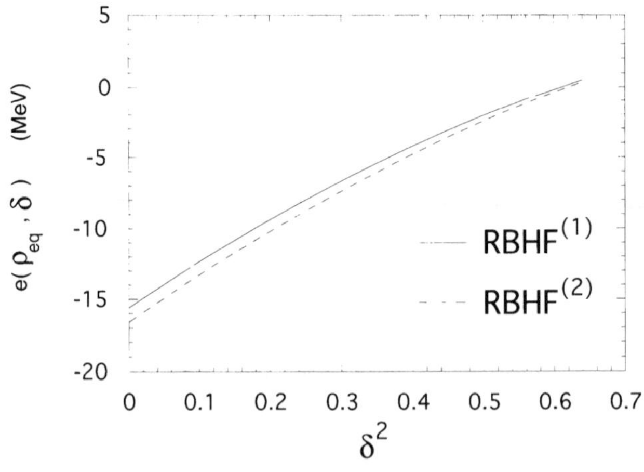

Figure 12.7. Energy per particle (11.4) at equilibrium as a function of asymmetry δ, computed for BM potential A [450]. (Reprinted courtesy of *Phys. Rev.*)

displayed in figure 12.7. Of even greater interest may be the density dependence of the symmetry energy, e^{sym}, because of its importance for astrophysics (see below). The latter is shown in figure 12.8. The monotonic

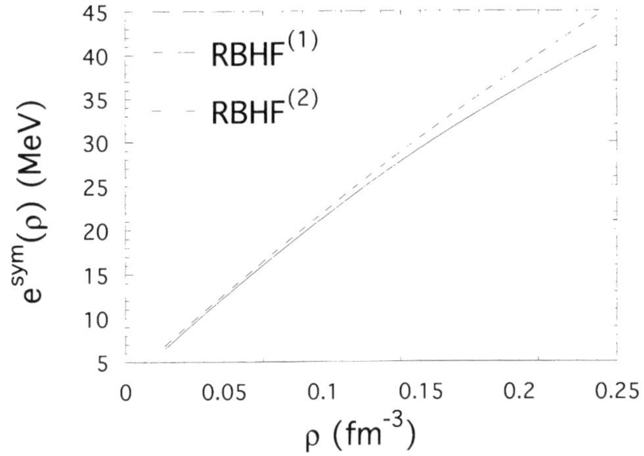

Figure 12.8. Symmetry energy (11.5) as a function of density computed for BM potential A [450]. (Reprinted courtesy of *Phys. Rev.*)

growth of these curves with density, which is also obtained for relativistic Hartree and HF calculations, seems to be a generic feature of relativistic theories of dense matter [472]. Relativistic HF and BHF calculations, for instance, predict an almost linear increase of the symmetry energy with density up to typically $5\,\rho_0$ followed by a slight flattening at still higher densities [474]. This appears to be quite different for several non-relativistic many-body calculations, like the ones based on Skyrme forces (e.g. force S III) or the modern Argonne–Urbana nucleon–nucleon interaction AV_{14}+UVII, where the symmetry energy bends over at higher nuclear densities ($\rho_0 = 0.15$ fm^{-3} [461] and about $6\,\rho_0$ [464], respectively). What is really causing this behavior appears to be not completely clear yet. It could be caused by the employed nucleon–nucleon interaction rather than the many-body method itself [473]. The accurate knowledge of the behavior of $e^{\rm sym}(\rho)$ is of great astrophysical importance for two reasons. The first is that it is this quantity which effectively determines the proton fraction Y_p of neutron star matter. Theories predicting a monotonic increase of $e^{\rm sym}(\rho)$, for instance, lead preferably to critical proton fractions $Y_p \gtrsim 11\%$, beyond which the protons have large enough Fermi momenta so that the reactions $n \to p + e^- + \bar{\nu}_e$ and $p + e^- \to n + \nu_e$, known as direct Urca processes (section 19.5.2), can occur without a bystander particle. By means of these reactions neutron stars can cool very efficiently, as we shall see in chapter 19. The second reason concerns the radii and the crustal extent of neutron stars, which is determined by the density dependence of $e^{\rm sym}(\rho)$ too (section 14.4).

Table 12.3. Properties of symmetric and asymmetric nuclear matter computed for different approximation techniques. Bro A, Bro B and Bro C refer to RBHF calculations (RBHF$^{(2)}$) performed for BM potentials A, B and C. All other entries are explained in the text.

	E/A (MeV)	ρ_0 (fm^{-3})	K (MeV)	a_{sym} (MeV)	L (MeV)	K^* (MeV)
Bro A	−16.49	0.174	280	34.4	81.9	−66.4
Bro B	−15.73	0.172	249	32.8	90.2	9.97
Bro C	−14.38	0.170	258	31.5	76.1	−35.1
RHF$^{(1)}$	−15.75	0.148	610	28.9	132	466
RHF$^{(2)}$	−15.75	0.148	360	43.3	135	105
RHF$^{(3)}$	−15.75	0.148	460	38.6	138	276
NL 1	−16.42	0.152	212	43.5	140	143.0
NL-SH	−16.35	0.146	356	36.1	114	079.82
SkM*	−15.78	0.160	217	30.0	45.8	−155.9
SIII	−15.86	0.145	355	28.2	9.9	−393.7
FRDM	−16.25	0.152	240	32.7	0	–
ETFSI-1	−15.87	0.161	235	27.0	−9.29	−336.8

The properties of symmetric as well as asymmetric nuclear matter computed for RBHF are compared with those computed for a broad variety of competing dense matter calculations in table 12.3. These range from relativistic Hartree and HF calculations to several non-relativistic calculations [461, 475]. The Hartree calculations are based on two frequently used parameter sets, LN1 and NL-SH. The HF results are taken from [461]. SkM* and SIII denote two well-known Skyrme forces, FRDM is the latest and most sophisticated version of the droplet-model mass formulae, while ETFSI-1 denotes the first mass formula based entirely on microscopic forces [475]. As mentioned before the agreement of the bulk properties with the accepted values of the mass formula is quite satisfactory. Also the value of the symmetry parameter a_{sym} is located within the accepted boundaries. The values of L and K^*, which are much more uncertain than the value of a_{sym}, lie between the values computed for the relativistic Hartree parametrization NL-SH and the Skyrme parametrization SkM*, and comply nicely with the systematics already established by non-relativistic fits to the nuclear data. For instance, increasing values of a_{sym} are accompanied by increasing values of L [476]. Although these parametrizations have been used very often in the past, they nevertheless have some deficiencies [454, 475, 477].

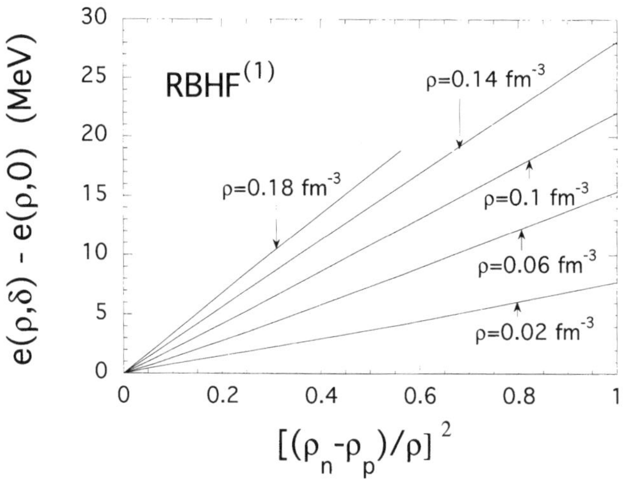

Figure 12.9. Test of the quadratic approximation of the symmetry energy, for five selected densities ρ [450]. The calculations are performed for the RBHF$^{(1)}$ approximation (BM potential A). In accordance with (12.109), the slope of each curve is to a good approximation equal to $e^{\text{sym}}(\rho)$. (Reprinted courtesy of *Phys. Rev.*)

We close this section by testing the validity of the quadratic approximation for the symmetry energy, which is of the form (cf. equation (11.2))

$$e(\rho,\delta) = e(\rho,0) + e^{\text{sym}}(\rho)\,\delta^2\,. \tag{12.109}$$

This approximation is known to hold for the non-relativistic Brueckner approximation [465]. It is evident from figure 12.9 that this approximation is also obeyed to a very high degree by the RBHF approximation.

12.5 Non-relativistic limit

In the non-relativistic treatment the self-energies of equation (6.83) satisfy

$$\Sigma_S^B(k) \to 0\,, \qquad \Sigma_V^B(k) \to 0\,, \qquad \Sigma_0^B(k) \to \Sigma^B(k)\,, \tag{12.110}$$

which leads to the replacements

$$m_B^* \to m_B\,, \quad k_B^* \to k\,, \quad W^B \to \sqrt{m_B^2 + \boldsymbol{k}^2} \approx m_B + \frac{\boldsymbol{k}^2}{2\,m_B}\,. \tag{12.111}$$

The relativistic energy–momentum relation of equation (6.149) is to be replaced according to

$$\omega^B(\boldsymbol{k}) \to \omega^B(\boldsymbol{k}) - m_B \equiv \epsilon^B(\boldsymbol{k}) = \frac{\boldsymbol{k}^2}{2\,m_B} + \Sigma^B(\epsilon^B(\boldsymbol{k}) - \mu^B, \boldsymbol{k})\,. \quad (12.112)$$

The spectral function $\Xi^B(\boldsymbol{k})$ of equation (6.163) plays the role of a momentum–density function, that is,

$$\Xi^B(k^0, \boldsymbol{k}) = n^B(|\boldsymbol{k}|)\,\delta(k^0 - \epsilon^B(\boldsymbol{k}) + \mu^B)\,, \quad (12.113)$$

with (cf. equation (9.46))

$$n^B(|\boldsymbol{k}|) \equiv \left| 1 - \frac{\partial \Sigma^B}{\partial \omega} \right|^{-1}_{\omega^B = \epsilon^B(k)}\,. \quad (12.114)$$

Equations (12.113) and (12.114) are well-known results of the non-relativistic Green function theory [317, 340]. The expressions of energy density and baryon number density follow from equations (12.103) and (12.105), and (6.186) respectively, as

$$\epsilon = 2 \sum_B \nu_B \int \frac{\mathrm{d}^3\boldsymbol{k}}{(2\pi)^3} \left\{ \frac{\boldsymbol{k}^2}{2\,m_B} + \frac{1}{2}\,\Sigma^B(\epsilon^B(\boldsymbol{k}) - \mu^B, \boldsymbol{k}) \right\} n^B(|\boldsymbol{k}|)\,\Theta^B(\boldsymbol{k})\,, \quad (12.115)$$

and

$$\rho = 2 \sum_B \nu_B \int \frac{\mathrm{d}^3\boldsymbol{k}}{(2\pi)^3}\, n^B(|\boldsymbol{k}|)\,\Theta(k_{F_B} - |\boldsymbol{k}|)\,. \quad (12.116)$$

We recall that in the non-relativistic $\boldsymbol{\Lambda}$ theory the single-particle bases are *a priori* given plane-wave functions. In the relativistic approach the basis consists of self-consistent, effective Dirac spinors. Their density dependence has an important influence on the saturation mechanism of nuclear matter [117] and hence on the nuclear equation of state itself. Introducing the definitions

$$\boldsymbol{k}_+ = \frac{1}{2}\boldsymbol{K} + \boldsymbol{q}\,, \qquad \boldsymbol{k}_- = \frac{1}{2}\boldsymbol{K} - \boldsymbol{q}\,, \quad (12.117)$$

the integral equation of the $\mathbf{T}$-matrix in the plane-wave basis reads [317]

$$\begin{aligned} \langle \boldsymbol{k}|\mathbf{T}_{\boldsymbol{K}}(E)|\boldsymbol{k}'\rangle &= \langle \boldsymbol{k}|2\,\mathbf{V}^{\mathrm{a}}|\boldsymbol{k}'\rangle \\ &+ \int \frac{\mathrm{d}^3\boldsymbol{q}}{(2\pi)^3}\,\langle \boldsymbol{k}|\mathbf{V}^{\mathrm{a}}|\boldsymbol{q}\rangle\,\boldsymbol{\Lambda}(\boldsymbol{k}_+, \boldsymbol{k}_-; E)\,\langle \boldsymbol{q}|\mathbf{T}_{\boldsymbol{K}}(E)|\boldsymbol{k}'\rangle\,, \end{aligned} \quad (12.118)$$

where $2\,\mathbf{V}^{\mathrm{a}} = \mathbf{V} - \mathbf{V}^{\mathrm{ex}}$ denotes the antisymmetrized nucleon–nucleon interaction in free space [317]. The non-relativistic $\boldsymbol{\Lambda}^{00}$ nuclear matter propagator has the form [compare with equation (9.54)]

$$\boldsymbol{\Lambda}^{00}(\boldsymbol{k}_+, \boldsymbol{k}_-; E) = \left[E - \frac{\boldsymbol{k}_+^2}{2\,m_B} - \frac{\boldsymbol{k}_-^2}{2\,m_B} + 2\,\mu^B \right]^{-1}\,, \quad (12.119)$$

and the non-relativistic Brueckner propagator reads

$$\boldsymbol{\Lambda}^{\rm BHF}(\boldsymbol{k}_+, \boldsymbol{k}_-; E) = 2\,\pi\, \frac{\Theta(|\boldsymbol{k}_+| - k_{F_B})\,\Theta(|\boldsymbol{k}_-| - k_{F_B})}{E - \epsilon^B(|\boldsymbol{k}_+|) - \epsilon^B(|\boldsymbol{k}_-|) + {\rm i}\,\eta}\,, \qquad (12.120)$$

whose relativistic counterpart was given in (9.60). The on-shell mass operator in the $\boldsymbol{\Lambda}^{00}$ approximation is given by

$$\Sigma^B(\hat\epsilon_1, \boldsymbol{k}_1) = \frac{1}{2}\sum_{B'} \int \frac{{\rm d}^3\boldsymbol{k}_2}{(2\pi)^3} \Big\{ \Big\langle \frac{\boldsymbol{k}_1 - \boldsymbol{k}_2}{2} \Big| \mathbf{T}^{00}_{\boldsymbol{k}_1+\boldsymbol{k}_2}(\hat\epsilon^B_1 + \hat\epsilon^{B'}_2) \Big| \frac{\boldsymbol{k}_1 - \boldsymbol{k}_2}{2} \Big\rangle$$
$$- \Big\langle \frac{\boldsymbol{k}_1 - \boldsymbol{k}_2}{2} \Big| \mathbf{T}^{00}_{\boldsymbol{k}_1+\boldsymbol{k}_2}(\hat\epsilon^B_1 + \hat\epsilon^{B'}_2) \Big| \frac{\boldsymbol{k}_2 - \boldsymbol{k}_1}{2} \Big\rangle \Big\}\,, \qquad (12.121)$$

where $\hat\epsilon^B_i \equiv \epsilon^B(\boldsymbol{k}_i) - \mu^B$ and $i = 1, 2$.

12.6 Collection of selected neutron star matter equations of state

In this section we introduce a broad collection of different competing models for the equation of state of superdense neutron (star) matter. Some of these models describe conventional neutron star matter in phase equilibrium with quarks. Others account for meson condensates or variations in the hyperon population depending on the underlying many-body method. This broad sample of equations of state will be applied in the second part of the book, starting with chapter 13, to the analysis of the structure and stability of models of rapidly spinning neutron and quark matter stars. Non-rotating stellar models will be treated as a byproduct. In doing so, we shall be particularly concerned with testing the compatibility of these equations of state, which predict quite different neutron star properties, with the body of observed data of pulsars – like rotational periods, masses, radii, redshifts, or cooling data. Not all equations of state may accommodate the observed data, specifically the just mentioned rotational periods and masses. This attains its particular interest in view of the rapid pace of discovery of millisecond pulsars [478], which imposes the double constraint of fast rotation and a large enough neutron star mass. Moreover, the determination of the smallest possible rotational pulsar period sheds light on the true ground state of strongly interacting matter.

The features of the total collection of equations of state are summarized in table 12.4, and their properties are compiled in tables 12.5 to 12.7. The graphical illustrations are given in figures 12.10 to 12.14. This collection is divided into two categories, that is, non-relativistic potential model equations of state, and relativistic equations of state determined in the framework of the relativistic nuclear field theories discussed in the previous chapters of this book. For the non-relativistic models 12 to 17 of table 12.4, the starting point is a phenomenological nucleon–nucleon interaction, $\boldsymbol{V}_{ij}$.

Table 12.4. Overview of the broad collection of neutron star matter EOSs. For their tabulated representations, see appendix J. Further details of these EOSs are compiled in tables 12.5, 12.6, and 12.7.

Label	EOS	Many-body approximation †	References
	Relativistic field-theoretical equations of state		
1	G_{300}	RH	[355]
	$G_{300}^{K^-}$	RH	[355]
	G_{B180}^{K300}	RH + MIT	[66]
2	HV	RH	[61, 79]
3	G_{B180}^{DCM2}	RH + MIT	[88]
	G_{B180}^{K240}	RH + MIT	[66]
4	G_{265}^{DCM2}	RH	[86]
	G_{M78}^{K240}	RH	[66]
5	G_{300}^{π}	RH	[355]
6	G_{200}^{π}	RH	[91]
7	Λ_{Bonn}^{00}+HV	RBHF + RH	[109]
8	G_{225}^{DCM1}	RH	[86]
9	G_{B180}^{DCM1}	RH + MIT	[88]
10	HFV	RHF	[79]
11	Λ_{HEA}^{00}+HFV	RBHF + RHF	[109]
	Λ_{BroB}^{RBHF}+HFV	RBHF + RHF	[460, 450]
	Non-relativistic potential-model equations of state		
12	BJ(I)	Var	[408]
13	UV_{14}+TNI	Var	[464]
14	V_{14}+TNI	Var	[384]
15	UV_{14}+UVII	Var	[464]
16	AV_{14}+UVII	Var	[464]
17	Pan(C)	Var	[407]
18	TF96	TF	[472]

† The following abbreviations are used:
RH: Relativistic Hartree, RHF: Relativistic Hartree–Fock,
RBHF: Relativistic Brueckner–Hartree–Fock, MIT: MIT bag model,
Var: Variational method, TF: Thomas–Fermi method.

In the case of the equations of state reported here, different two-nucleon potentials, denoted by $\boldsymbol{V}_{ij}$, which fit the nucleon–nucleon scattering data and deuteron properties, have been employed. Most of these two-nucleon potentials are supplemented with three-nucleon interactions, $\boldsymbol{V}_{ijk}$. The Hamiltonian $\mathcal{H}$ is therefore of the form

$$\mathcal{H} = \sum_i \frac{-\hbar^2}{2m} \boldsymbol{\nabla}_i^2 + \sum_{i<j} \boldsymbol{V}_{ij} + \sum_{i<j<k} \boldsymbol{V}_{ijk} \,. \tag{12.122}$$

Table 12.5. Properties of the relativistic, field-theoretical EOSs listed in table 12.4.

Label	EOS	Composition	Interaction (meson exchange)
1	G_{300}	$p, n, \Lambda, \Sigma^{\pm,0}, \Xi^{0,-}, e^-, \mu^-$	σ, ω, ρ
	$G_{300}^{K^-}$	$p, n, \Lambda, \Sigma^{\pm,0}, \Xi^{0,-}, e^-, \mu^-$ + K^- condensate	σ, ω, ρ
	G_{B180}^{K300}	$p, n, \Lambda, \Sigma^{\pm,0}, \Xi^{0,-}, e^-, \mu^-$ + u, d, s quark matter	σ, ω, ρ
2	HV	$p, n, \Lambda, \Sigma^{\pm,0}, \Xi^{0,-}, e^-, \mu^-$	σ, ω, ρ
3	G_{B180}^{DCM2}	$p, n, \Lambda, \Sigma^{\pm,0}, \Xi^{0,-}, e^-, \mu^-$ + u, d, s quark matter	σ, ω, ρ
	G_{B180}^{K240}	$p, n, \Lambda, \Sigma^{\pm,0}, \Xi^{0,-}, e^-, \mu^-$ + u, d, s quark matter	σ, ω, ρ
4	G_{265}^{DCM2}	$p, n, \Lambda, \Sigma^{\pm,0}, \Xi^{0,-}, e^-, \mu^-$	σ, ω, ρ
	G_{M78}^{K240}	$p, n, \Lambda, \Sigma^{\pm,0}, \Xi^{0,-}, e^-, \mu^-$	σ, ω, ρ
5	G_{300}^{π}	$p, n, \Lambda, \Sigma^{\pm,0}, \Xi^{0,-}, e^-, \mu^-$ + π^- condensate	σ, ω, ρ
6	G_{200}^{π}	$p, n, \Lambda, \Sigma^{\pm,0}, \Xi^{0,-}, e^-, \mu^-$ + π^- condensate	σ, ω, ρ
7	Λ_{Bonn}^{00}+HV	$p, n, \Lambda, \Sigma^{\pm,0}, \Xi^{0,-}, e^-, \mu^-$	$\sigma, \omega, \pi, \rho, \eta, \delta$
8	G_{225}^{DCM1}	$p, n, \Lambda, \Sigma^{\pm,0}, \Xi^{0,-}, e^-, \mu^-$	σ, ω, ρ
9	G_{B180}^{DCM1}	$p, n, \Lambda, \Sigma^{\pm,0}, \Xi^{0,-}, e^-, \mu^-$ + u, d, s quark matter	σ, ω, ρ
10	HFV	$p, n, \Lambda, \Sigma^{0,-}, \Delta^-, e^-, \mu^-$	$\sigma, \omega, \pi, \rho$
11	Λ_{HEA}^{00}+HFV	$p, n, \Lambda, \Sigma^{0,-}, \Delta^-, e^-, \mu^-$	$\sigma, \omega, \pi, \rho, \eta, \delta, \phi$
	Λ_{BroB}^{RBHF}+HFV	$p, n, \Lambda, \Sigma^{0,-}, \Delta^-, e^-, \mu^-$	$\sigma, \omega, \pi, \rho, \eta, \delta$

The many-body method adopted to solve the Schrödinger equation is based on the variational approach [479, 480, 481] where a variational trial function $|\Psi_v\rangle$ is constructed from a symmetrized product of two-body correlation operators (F_{ij}) acting on an unperturbed ground state, that is,

$$|\Psi_v\rangle = \left[\hat{S} \prod_{i<j} F_{ij}\right] |\Phi\rangle \,, \tag{12.123}$$

where $|\Phi\rangle$ denotes the antisymmetrized Fermi-gas wave function,

$$|\Phi\rangle = \hat{A} \prod_{j} \exp(\mathrm{i}\,\boldsymbol{p}_j \cdot \boldsymbol{x}_j) \,. \tag{12.124}$$

The correlation operator contains variational parameters which are varied

Table 12.6. Properties of the non-relativistic EOSs listed in table 12.4.

Label	EOS	Composition	Interaction
12	BJ(I)	$p, n, \Lambda, \Sigma^{\pm,0}, \Xi^{0,-}, \Delta, e^-, \mu^-$	2-nucleon potential
13	UV_{14}+TNI	p, n, e^-, μ^-	2-nucleon potential V_{14} +3-nucleon interaction
14	V_{14}+TNI	n	2-nucleon potential V_{14} +3-nucleon interaction
15	UV_{14}+UVII	p, n, e^-, μ^-	2-nucleon potential V_{14} +3-nucleon potential VII
16	AV_{14}+UVII	p, n, e^-, μ^-	2-nucleon potential AV_{14} +3-nucleon potential VII
17	Pan(C)	$p, n, \Lambda, \Sigma^{\pm,0}, \Xi^{0,-}, \Delta, e^-, \mu^-$	2-nucleon potential
18	TF96	p, n, e^-, μ^-	Seyler–Blanchard

to minimize the energy per baryon for a given density ρ [464, 479, 480, 481],

$$E_{\text{var}}(\rho) = \min \left\{ \frac{\langle \Psi_v \,|\, \mathcal{H} \,|\, \Psi_v \rangle}{\langle \Psi_v \,|\, \Psi_v \rangle} \right\} \geq E_0 . \tag{12.125}$$

As indicated in (12.125), E_{var} constitutes an upper bound to the ground-state energy E_0. The energy density $\epsilon(\rho)$ and pressure $P(\rho)$ are obtained from equations (12.107) and (12.107).

The TF96 equation of state of table 12.4, recently calculated by Strobel *et al* [472], is based on the new Thomas–Fermi approach of Myers and Swiatecki [409, 482, 483, 484, 485]. The effective interaction v_{12} of this new approach consists of the Seyler–Blanchard potential of [486], generalized by the addition of one momentum dependent and one density dependent term [409],

$$v_{12} = -\frac{2\,T_0}{\rho_0}\, Y(r_{12}) \left\{ \frac{1}{2}(1 \mp \xi)\,\alpha - \frac{1}{2}(1 \mp \zeta)\left(\beta \left(\frac{p_{12}}{k_{F_0}}\right)^2 - \gamma\,\frac{k_{F_0}}{p_{12}} + \sigma \left(\frac{2\bar{\rho}}{\rho_0}\right)^{2/3}\right) \right\} . \tag{12.126}$$

The upper (lower) sign in (12.126) corresponds to nucleons with equal (unequal) isospin. The quantities k_{F_0}, T_0 $(= k_{F_0}^2/2m)$, and ρ_0 are the Fermi momentum, the Fermi energy, and the particle density of symmetric nuclear matter. The potential's radial dependence is describe by the normalized Yukawa interaction

$$Y(r_{12}) = \frac{1}{4\,\pi\,a^3}\,\frac{\mathrm{e}^{-\,r_{12}/a}}{r_{12}/a} . \tag{12.127}$$

Its strength depends both on the magnitude of the particles' relative momentum, p_{12}, and on an average of the densities at the locations of

Table 12.7. Nuclear matter properties at saturation density of the EOSs compiled in table 12.4. The listed quantities are: energy per baryon E/A, incompressibility K, effective nucleon mass M^* ($\equiv m_N^*/m_N$), symmetry energy $a_{\rm sym}$.

Label	EOS	E/A (MeV)	ρ_0 (fm^{-3})	K (MeV)	M^* (MeV)	$a_{\rm sym}$ (MeV)	ϵ/ϵ_0 †
		Relativistic field-theoretical EOSs					
1	G_{300}	−16.3	0.153	300	0.78	32.5	–
	G_{B180}^{K300}	−16.3	0.153	300	0.70	32.5	–
2	HV	−15.98	0.145	285	0.77	36.8	–
3	G_{B180}^{DCM2}	−16.0	0.16	265	0.796	32.5	–
	G_{B180}^{K240}	−16.3	0.153	240	0.78	32.5	–
4	G_{265}^{DCM2}	−16.0	0.16	265	0.796	32.5	–
	G_{M78}^{K240}	−16.3	0.153	240	0.78	32.5	–
5	G_{300}^{π}	−16.3	0.153	300	0.78	32.5	–
6	G_{200}^{π}	−15.95	0.145	200	0.8	36.8	–
7	$\Lambda_{\rm Bonn}^{00}$+HV	−11.9	0.134	186	0.79	21.3	–
8	G_{225}^{DCM1}	−16.0	0.16	225	0.796	32.5	–
9	G_{B180}^{DCM1}	−16.0	0.16	225	0.796	32.5	–
10	HFV	−15.54	0.159	376	0.62	30	–
11	$\Lambda_{\rm HEA}^{00}$+HFV	−8.7	0.132	115	0.82	29	–
	$\Lambda_{\rm BroB}^{\rm RBHF}$+HFV	−15.73	0.172	249	0.73	34.3	–
		Non-relativistic potential-model EOSs					
12	BJ(I)	~ -10	~ 0.18	–	–	–	23.1
13	UV_{14}+TNI	−16.6	0.157	261	0.65	30.8	14
14	V_{14}+TNI	−16.00	0.159	240	0.64	–	5.6
15	UV_{14}+UVII	−11.5	0.175	202	0.79	29.3	6.5
16	AV_{14}+UVII	−12.4	0.194	209	0.66	27.6	7.2
17	Pan(C)	~ -10	~ 0.18	60	–	35	23.6
18	TF96	−16.24	0.161	234	–	32.7	13

† Energy density in units of normal nuclear matter density beyond which the velocity of sound in neutron matter becomes superluminal, that is, $v_s/c > 1$. No entry means that causality is not violated.

the particles. The parameters ξ and ζ, generally taken to be different from one another, were introduced in order to achieve better agreement with asymmetric nuclear systems, and the behavior of the optical potential is improved by the term $\sigma(2\bar{\rho}/\rho_0)^{2/3}$. Here the average density is defined by $\bar{\rho}^{2/3} = (\rho_1^{2/3} + \rho_2^{2/3})/2$, where ρ_1 and ρ_2 are the relevant densities of the interacting particle (neutron or protons) at points 1 and 2. For the seven free parameters – adjusted to the properties of finite nuclei, the parameters of the mass formula, and the behavior of the optical potential – the following

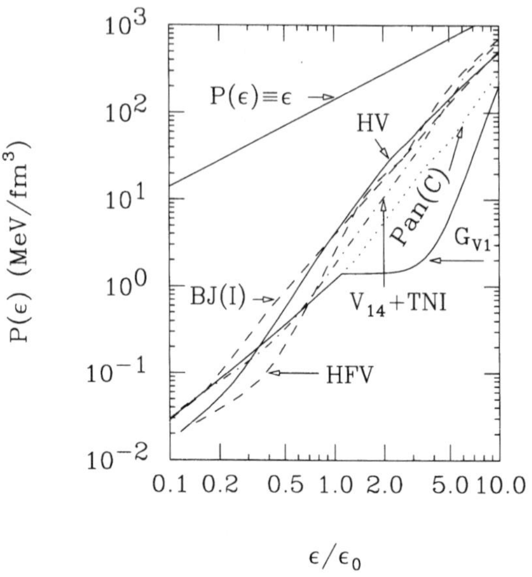

Figure 12.10. Graphical illustration of EOSs BJ(I), Pan(C), V_{14}+TNI, HV, and HFV [447]. Details of these EOSs are listed in table 12.4.

values were deduced [409]:

$$\alpha = 1.946\,84\,, \quad \beta = 0.153\,11\,, \quad \gamma = 1.136\,72\,, \quad \sigma = 1.05,$$
$$\xi = 0.279\,76\,, \quad \zeta = 0.556\,65\,, \quad a = 0.592\,94 \text{ fm}\,. \tag{12.128}$$

This set of parameters leads to the nuclear matter properties at saturation listed in the last line of table 12.7.

The new force has the advantage over the standard Seyler–Blanchard interaction [486, 487] to not only reproduce the ground-state properties of finite nuclei and infinite symmetric nuclear matter, but also the optical potential and, as revealed by a comparison with the theoretical investigations of Friedman and Pandharipande [384], the properties of pure neutron matter (model V_{14}+TNI in figure 12.10) too. These features make the new Thomas–Fermi model a very attractive method for the investigation of the properties of dense nuclear matter treated in the non-relativistic framework.[1] Aside from V_{14}+TNI, we shown in figure 12.10 also the older, and by now outdated, Pan(C) model for the equation of state of neutron star matter proposed by Pandharipande as early as 1971

[1] For an early application of the Thomas–Fermi method to the stellar matter problem, see Hartmann *et al* [488]. More recent Thomas–Fermi calculations of the properties of infinite nuclear matter as well as finite nuclei based on the Seyler–Blanchard force can be found in [489, 490].

[407]. Nevertheless this model attains its particular interest because of its extreme softness – i.e. relatively little pressure for a given density – for it constitutes a lower bound on the pressure that must be provided by a given model for the nuclear equation of state that successfully accommodates both pulsars with rotational periods down to the smallest rotational periods yet known, 1.6 ms, and masses larger than typically $1.5\,M_{\odot}$. While the former constraint is easily fulfilled by Pan(C) [106, 107], the mass constraint is not. One sees that Pan(C) is considerably softer than the other two non-relativistic equations of state shown in this figure, BJ(I) and V_{14}+TNI. The BJ(I) and Pan(C) models account for baryon population in neutron star matter which leads to a weak flattening of the pressure curves at densities greater than two and four times normal nuclear matter density, respectively. The pressure associated with the relativistic equations of state HV and HFV is shown too for the purpose of comparison. The Hartree–Fock equation of state HFV becomes stiffer than Hartree HV at $\epsilon \gtrsim 3\epsilon_0$ which has its origin in the exchange (Fock) contribution $\Sigma^{\mathrm{F},B}$ to the baryon self-energy. By definition, the exchange term is absent in the Hartree treatment. The structures in the HFV and HV equations of state in the form of a slight softening at densities of around $1.6\,\epsilon_0$ and $2\,\epsilon_0$, respectively, corresponds to the onset of hyperon populations, which set in somewhat earlier than for the two non-relativistic models described just above. As known from figures 7.23 and 7.24, those hyperons that become populated first are the Σ^- and Λ, respectively. A further constraining model on the pressure as a function of density is the equation of state denoted G_{V1} in figure 12.10. It is based on an investigation of the limiting rotational Kepler period of neutron stars that is performed *without* taking recourse to any particular models of dense matter but derives the limit only on the general principles that (a) Einstein's equations describe stellar structure, (b) matter is microscopically stable, and (c) causality is not violated [491]. On this basis, a lower bound for the smallest possible Kepler period for a $M = 1.442\,M_{\odot}$ neutron star of $P_{\mathrm{K}} = 0.33$ ms was established. Hence, the G_{V1} curve sets an *absolute* limit on rapid rotation on any star bound by gravity. Of course the equation of state that nature has chosen need not be the one that allows stars to rotate most rapidly, so the above is a strict model independent limit.

Figure 12.11 compares the non-relativistic equations of state of Wiringa, Fiks, and Fabrocini (WFF) with two relativistic ones. We recall that only the latter two describe neutron star matter in terms of baryons in generalized β-equilibrium with leptons. The eventual condensation of pions in dense neutron star matter is taken into account in equation of state G^{π}_{300}, additionally to baryon population. According to this equation of state, condensation is predicted to set in at about $1.5\,\epsilon_0$. This can be seen by comparing the dash–dotted and dotted curves in figure 12.11 with one another. At densities $\epsilon \gtrsim 4\,\epsilon_0$ the non-relativistic WFF models behave more stiffly than the relativistic ones, violating causality – an issue that will be

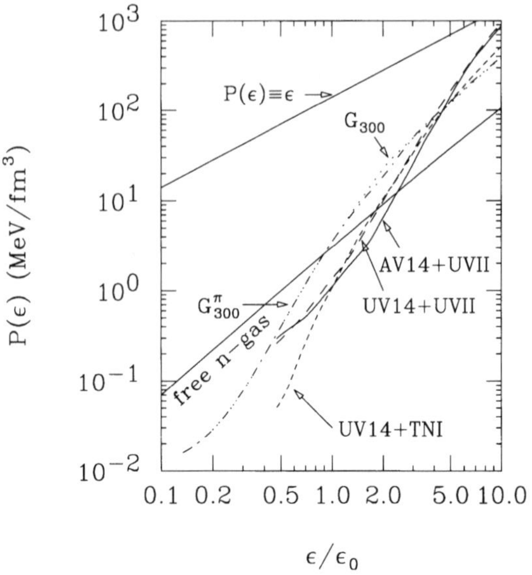

Figure 12.11. Graphical illustration of EOSs $AV_{14}+UVII$, $UV_{14}+UVII$, UV_{14}+TNI, G_{300}, and G^{π}_{300} [447]. Details of these EOSs are listed in table 12.4.

discussed immediately below – at densities around $8\,\epsilon_0$. The AV_{14}+UVII and UV_{14} + UVII equations of state are rather similar to each other at subnuclear densities which is not the case for the third WFF potential model, UV_{14}+TNI, which accounts for three-nucleon interactions in matter.

A very comforting feature of the relativistic equations of state is that they do not violate causality, that is, the velocity of a signal, given by

$$v_s = c\sqrt{\mathrm{d}P/\mathrm{d}\epsilon}\ , \tag{12.129}$$

is smaller than the velocity of light, c, at all densities. This becomes very obvious by looking at figure 12.12 which shows the behavior of v_s in dense neutron star matter for a few selected equations of state, both relativistic as well as non-relativistic ones. It is, however, not entirely clear how serious a constraint on the equation of state this constitutes, the reasons being that (12.129) holds exactly only if neutron star matter is neither dispersive nor absorptive [246]. Neither case is rigorous. Hence, more accurately stated, what is being expressed in (12.129) holds only if the hydrodynamic phase velocity of sound waves, given by $v_\varphi \equiv c\sqrt{\mathrm{d}P/\mathrm{d}\epsilon}$, is equal to the velocity of light, which, as just stated, is only the case if effects arising from dispersion and absorption are either absent or insignificantly small. Otherwise one has for the signal velocities $v_{\rm signal} = v_\varphi < c$. If one does accept the

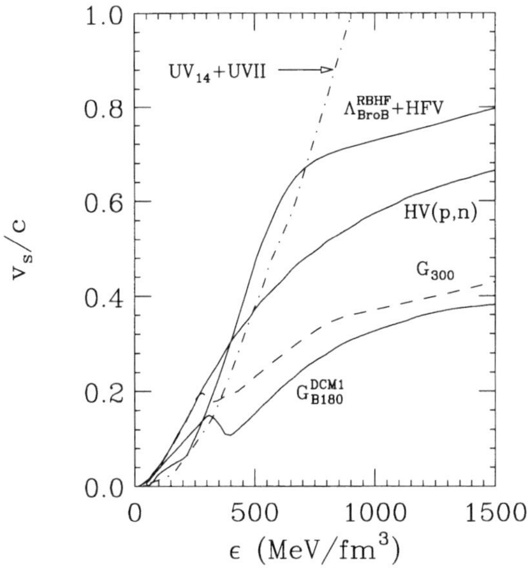

Figure 12.12. Velocity of sound, v_s, in units of the speed of light, c, as a function of energy density calculated for several selected equations of state of table 12.4 .

relation (12.129) as a criteria for causality violation, then it follows that all the non-relativistic models of our collection of equations of state violate causality, some at smaller densities than others as shown in the last column of table 12.7. However, two of those models, UV$_{14}$+TNI and TF96, do not violate causality up to the highest densities relevant for the construction of models of neutron stars from them. The equations of state BJ(I) and Pan(C) violate causality at about 23 times the density of normal nuclear matter, not much above the central densities of the maximum-mass neutron stars constructed for these equations of state. The equations of state V$_{14}$+TNI, AV$_{14}$+UVII, and UV$_{14}$+UVII become superluminal at considerably smaller densities, between six to seven times normal nuclear matter density, which is less than the central densities encountered in the maximum neutron star mass models constructed from these equations of state. Again, the extent to which these conclusions apply rests entirely on (12.129), whose validity as yet seems not to be too compelling.

Asymptotically, the relativistic equations of state approach $P \to \epsilon$ because the repulsion arises from the exchange of vector mesons. Such a behavior of vector meson interactions has been remarked on by Zel'dovich [492, 493]. It can be seen explicitly by examining equations (12.53) and (12.65) for ϵ and P in the limit of large density. As $k_{F_B} \to \infty$, the mass

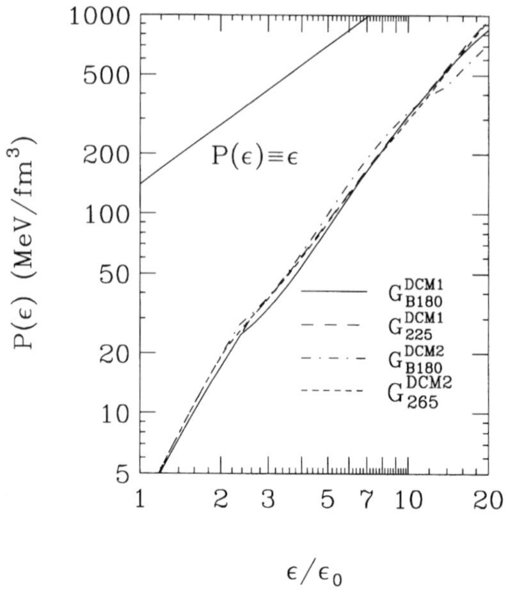

Figure 12.13. Graphical illustration of EOSs $G^{\rm DCM1}_{\rm B180}$, $G^{\rm DCM1}_{225}$, $G^{\rm DCM2}_{\rm B180}$, and $G^{\rm DCM2}_{265}$ [447]. Models $G^{\rm DCM1}_{\rm B180}$ and $G^{\rm DCM2}_{\rm B180}$ account for quark deconfinement. Details of these EOSs are listed in table 12.4.

terms in the integrals can be ignored. The σ field is bounded by the order of the baryon mass. Then it follows that [61]

$$\epsilon \longrightarrow \frac{1}{2}\sum_B \Big(\frac{g_{\omega B}}{m_\omega}\,\rho^B\Big)^2 + \frac{1}{2}\sum_B \Big(\frac{g_{\rho B}}{m_\rho}\,I_{3B}\,\rho^B\Big)^2 + \sum_B \frac{2J_B+1}{8\,\pi^2}\,k^4_{F_B}\,, \tag{12.130}$$

and

$$P \longrightarrow \frac{1}{2}\sum_B \Big(\frac{g_{\omega B}}{m_\omega}\,\rho^B\Big)^2 + \frac{1}{2}\sum_B \Big(\frac{g_{\rho B}}{m_\rho}\,I_{3B}\,\rho^B\Big)^2 + \frac{1}{3}\sum_B \frac{2J_B+1}{8\,\pi^2}\,k^4_{F_B}\,. \tag{12.131}$$

Since $\rho^B \propto k^3_{F_B}$, we find from these two relations that P approaches ϵ from *below*, and the speed of sound, $\sqrt{{\rm d}P/{\rm d}\epsilon}$, approaches but will stay below the speed of light.

The relativistic derivative coupling Lagrangian of Zimanyi and Moszkowski [351] was adopted in the determination of the equations of state denoted by DCM1 in figure 12.13, while those labeled DCM2 correspond to the *hybrid coupling* model. Both types of coupling were

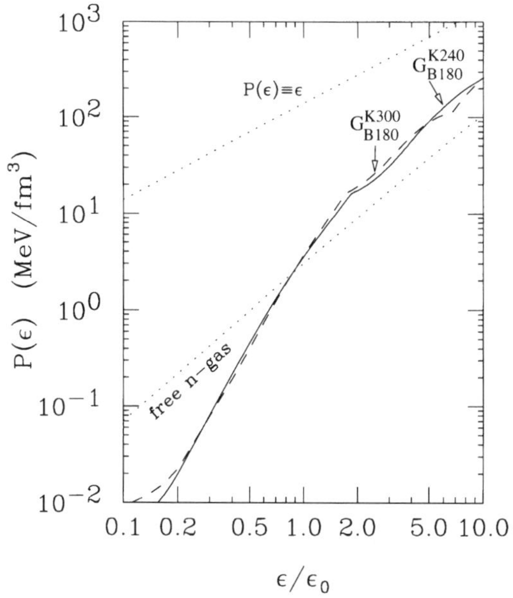

Figure 12.14. Graphical illustration of EOSs G_{B180}^{K240} and G_{B180}^{K300} accounting for quark deconfinement [66]. Details of these EOSs are listed in table 12.4.

described in detail in section 7.3. We recall that in the framework of the latter coupling, the scalar field (σ meson) is coupled to both the Yukawa point *and* derivative coupling to baryons and the ω- and ρ-meson vector fields. This improves the agreement with the incompressibility of nuclear matter and effective nucleon mass at saturation density. Zimanyi and Moszkowski originally have introduced a purely derivative coupling of the scalar field to the baryons and mesons. The possibility of a phase transition of the dense neutron star core to 3-flavor quark matter, outlined in chapter 8, is taken into account in models G_{B180}^{DCM1} and G_{B180}^{DCM2} of figure 12.13, and G_{B180}^{K240} and G_{B180}^{K300} of figure 12.14. The transition sets in typically somewhat about $2\,\epsilon_0$ and, because it introduces additional degrees of freedom, lowers the pressure for a given density relative to those equations of state which stay in the confined hadronic matter phase. The mixed quark–hadron phase ends, that is, the pure quark phase begins, between ~ 7 and $15\,\epsilon_0$, depending on the input parameters (cf. table 7.1). We stress again that these density values are rather different from those determined by others in earlier investigations on the quark–hadron phase transition in neutron star matter for reasons that have been outlined in chapter 8. As we shall see in section 17.3.3, neutron star models constructed for these equations of state contain – besides nucleons and hyperons – a large percentage of u, d and s quarks in their interiors. For that reason such objects are called *hybrid stars* [88]. We recall that the notion *hybrid* is used for both the chosen type of nuclear coupling, i.e. hybrid derivative coupling or standard Yukawa coupling as discussed in section 7.3, as well as the quark–hadron composition of neutron star matter. If not stated otherwise, henceforth the notion hybrid will take reference to the star's composition. A bag constant of $B^{1/4} = 180$ MeV has been used in each case for the determination of the above hybrid-star matter equations of state. This choice places the energy per baryon of u, d, s quark matter at about 1100 MeV, way above the energy per baryon in infinite nuclear matter. Hence, u, d, s quark matter here is far from being absolutely stable. The latter option, absolute stability, was discussed in section 2.3.1, where we found that strange quark matter, made up entirely of chemically equilibrated u, d, s quarks, can be absolutely stable in the framework of the bag model for some range of parameters, like bag constants that lie in the range $145 \leq B^{1/4} \leq 160$ MeV for a free, relativistic massless quark gas. Of course, the bag model computations are no doubt unreliable. So whether or not strange matter is indeed absolutely stable is probably outside the predictive ability of the bag model. Nevertheless one may argue that some of the qualitative features extracted from the perturbative analysis may surely have some correspondence in the full non-perturbative theory. In any event, we can conclude for sure that *if* strange quark matter should indeed be absolutely stable, then a mixed quark–hadron phase can not exist inside neutron stars (cf. section 2.3).

A further difference between the two categories of equations of state of our collection originates not only from the quark degrees of freedom taken into account in a few of these models, but from the different baryonic degrees of freedom too. So not all the non-relativistic equations of state describe neutron star matter in full β-equilibrium between nucleons and hyperons, as can be seen from entries 13 through 16 and 18 in table 12.4. What these models describe is neutron star matter composed of only neutrons, or made up of neutrons and protons in β-equilibrium with electrons and muons, neither of which however is the true ground state of neutron star matter predicted by theory [61, 62, 407, 408]. Indeed already the very early discussion of Ambartsumyan and Saakyan [494] based on a Fermi gas made a very plausible case for the existence of a hyperon charge on neutron stars. This was confirmed by later calculations which included effects of nuclear forces in the Schrödinger theory [82, 407, 408, 495, 496, 497].

Table 12.7 summarizes the nuclear matter properties at saturation density associated with the equations of state of table 12.4. As we have seen in section 7.4, the coupling constants underlying the relativistic Hartree and Hartree–Fock equations of state are to be chosen such that these are in accordance with the empirical nuclear matter data of energy per baryon, incompressibility, effective nucleon mass, and symmetry energy, each at the correct saturation density. This is in sharp contrast to the relativistic **T**-matrix approximation, which, as we know from section 5.5, accounts for dynamical two-particle correlations between nucleons that are calculated for a given model for the nucleon–nucleon interaction in free space (that is, a relativistic one-boson-exchange interaction). In this case the parameters of the theory are entirely fixed by the two-nucleon scattering data and the properties of the deuteron (section 5.5.3). Hence, all parameters are already fixed from the first, and so we are left with a parameter-free treatment. Model equations of state belonging to the latter category are $\Lambda^{00}_{\rm Bonn}$+HV, $\Lambda^{00}_{\rm HEA}$+HFV, and $\Lambda^{\rm RBHF}_{\rm BroB}$+HFV. The chosen one-boson-exchange interactions in these cases are the relativistic Bonn, the HEA, and the Brockmann–Machleidt interaction, respectively. At densities greater than about three times the density of nuclear matter, these model equations of state were joined smoothly to HV and HFV (cf. table 12.4), a procedure very well justified by our analysis performed in section 11.3 where we demonstrated that the properties of neutron star matter – from its various thermodynamic quantities to the particle composition – computed for the relativistic **T**-matrix approximation can very well be reproduced in the simpler HF approach, provided the parameters of the underlying HF Lagrangian are properly chosen.

The inclusion of dynamical correlations leads to equations of state that are markedly soft in the vicinity of ρ_0, which is indicated by the small incompressibilities K (or, conversely, large effective nucleon masses). A similar trend is observed for the potential model equations of state too, in which case the impact of dynamical two-particle correlations originate from nucleon–nucleon potentials like Urbana (denoted by U) or Argonne (denoted by A). Like the relativistic **T**-matrix equations of state mentioned just above, the potential model equations of state, constructed for a given non-relativistic nucleon–nucleon potential, belong to the parameter-free category too. From table 12.7 it is seen that nuclear matter is underbound by about 4 MeV for two potential model equations of state. The corresponding saturation densities, ρ_0, are in the range between 0.17 to 0.19 fm^{-3}, of which the latter value is significantly larger than the empirical saturation density of about 0.15 fm^{-3} [487]. The equations of state $\rm UV_{14}$+TNI and $\rm V_{14}$+TNI, labeled 13 and 14 in table 12.7, lead to binding energies and saturation densities that are in good agreement with the empirical values, which have their original dynamical *three-nucleon interactions* (denoted by TNI) included in these models. In this connection, we recall the well-known circumstance that two-particle correlations alone fail in reproducing the empirical binding energy at the correct saturation density. More than that, the saturation points calculated from both the (non-relativistic) standard BHF [479, 498, 499] and the ladder ($\boldsymbol{\Lambda}$) approximation [325, 500, 501] with realistic nucleon–

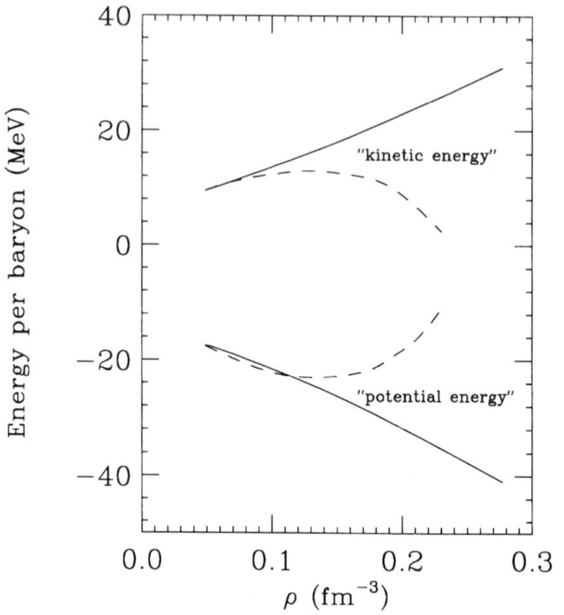

Figure 12.15. Dirac ('kinetic') and 'potential' energy of symmetric nuclear matter as a function of density, computed for RBHF. The underlying OBE interaction is HEA. The solid curves result from using vacuum spinors, the dashed curves from making use of the fully self-consistent spinor–antispinor basis introduced in chapter 9. (Data courtesy of Poschenrieder [440].)

nucleon interactions plotted as a function of the Fermi momentum at saturation fall in a narrow band, often called the *Coester band* [326]. It appears likely that this band would contain the calculated saturation point for *any* realistic nucleon–nucleon interaction [502, 503]. A comparison with the relativistic, parameter-free equations of state shows that these saturate nuclear matter at somewhat smaller densities, in the range between 0.13 and 0.16 fm^{-3}. This shift is caused by the relativistic saturation mechanism [92, 93, 118, 121, 504], which originates from the large self-energy components Σ_S that enter the lower components of the (self-consistent) Dirac spinors $\Phi(k)$ of the nucleons (cf. equation (9.1)). This dramatically alters the density dependence of the 'kinetic' and 'potential' energy parts, as shown in figure 12.15, which together constitute the energy per baryon of a nucleon in the matter. Note that, even though translational invariance in the infinite medium implies that the single-particle spatial wave functions are plane waves in both the relativistic and non-relativistic treatments, the self-consistent determination of the

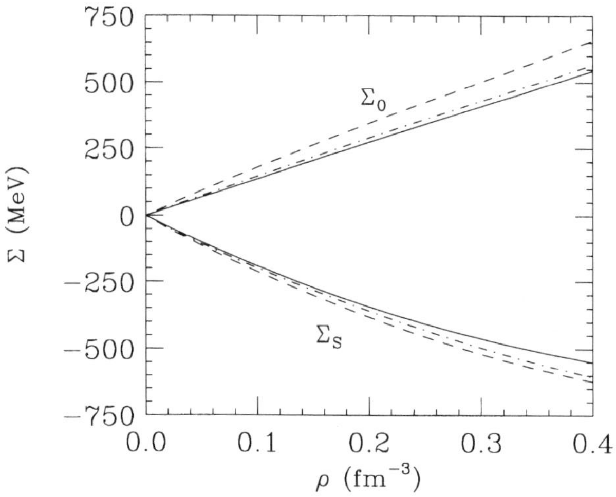

Figure 12.16. Scalar and timelike self-energy components, Σ_S and Σ_0 respectively, computed in relativistic Hartree (parameter set HII) and relativistic BHF (BM potential B). Solid and dash–dotted curves refer to symmetric nuclear matter treated in HII and RBHF, respectively; the dashed curves are for pure neutron matter treated in RBHF.

Dirac spinors (or equivalently, the baryon propagator $\boldsymbol{g}$) goes significantly beyond the usual self-consistency in the particle spectrum required in non-relativistic Brueckner-type calculations. Therefore, because Σ_S is large and negative, as shown in figures 12.16 and 11.4 for instance, the attraction between nucleons diminishes with increasing density, while the repulsion between them becomes more pronounced. To see this in more quantitative terms, let us look at the density dependence of the self-consistent matrix elements of the nucleon–nucleon interaction. The density dependence of the vector-nucleon–nucleon vertex is proportional to $\Phi(\boldsymbol{k})\Phi^\dagger(\boldsymbol{k})$, while the scalar-nucleon–nucleon vertex depends on

$$\bar{\Phi}^B(\boldsymbol{k})\,\Phi^B(\boldsymbol{k}) = \frac{1-\left(k_B^*/\epsilon^B\right)^2}{1+\left(k_B^*/\epsilon^B\right)^2}\,\Phi^{\dagger B}(\boldsymbol{k})\Phi^B(\boldsymbol{k})\,. \tag{12.132}$$

It is straightforward to show that (12.132) reduce to leading order in m_B^*/k_B^* to

$$\bar{\Phi}^B(\boldsymbol{k})\,\Phi^B(\boldsymbol{k}) \simeq \frac{m_B^*}{k_B^*}\,\Phi^{\dagger B}(\boldsymbol{k})\Phi^B(\boldsymbol{k})\,, \tag{12.133}$$

with m_B^* and k_B^* defined in (9.2). Thus, as the density increases and therefore m_B^* $(= m_B + \Sigma_S^B)$ decreases (figure 12.16) while $k_B^* \simeq |\boldsymbol{k}|$, there is a decrease in the relative strength of the one-boson-exchange scalar contribution, leading to a more repulsive interaction between two nucleons [122]. (Again, in non-relativistic treatments this relative strength is the same at finite density and in free space.) This is the reason why the relativistic **T**-matrix calculation performed for the HEA potential, i.e. our $\Lambda^{00}_{\rm HEA}$+HFV model of table 12.7, saturates nuclear matter at a density of $\rho_0 = 0.13\ {\rm fm}^{-3}$ considerably smaller than non-relativistic (lowest-order helicity state) **T**-matrix calculations performed for the same potential, for which $\rho_0 = 0.25\ {\rm fm}^{-3}$ [120, 320]. The saturation energy in the latter case drops by several MeV down to $E/A = 12.4$ MeV. The relativistic **T**-matrix calculation performed for the Bonn potential leads to a better agreement with the empirical nuclear matter data, as calculated for TF96, than the HEA potential, and is nearly perfect for the modern BM potential B entering $\Lambda^{\rm RBHF}_{\rm BroB}$+HFV (cf. table 12.7).

As can be seen from table 12.7, the incompressibilities of nuclear matter at saturation obtained for the parameter-free **T**-matrix calculations based on the Bonn, HEA, and Brockmann–Machleidt B OBE interactions are relatively small. Besides that $K_{\rm HEA} < K_{\rm Bonn}$, and the effective nucleon mass in matter obeys $m^*_{\rm HEA} > m^*_{\rm Bonn}$. Such an interplay between incompressibility and effective nucleon mass at saturation can be demonstrated analytically for the relativistic Hartree theory [84, 383]. In this case the energy per nucleon in symmetric nuclear matter at saturation density is given by (cf. section 7.2)

$$\frac{E}{A} = \omega^N(k_{F_0}) = g_{\omega N}\,\langle\omega_0(k_{F_0})\rangle + \sqrt{\left(m^*_N(k_{F_0})\right)^2 + k^2_{F_0}}\,, \qquad (12.134)$$

from which it follows that an increase of m^*_N must be accompanied by a simultaneous *decrease* of the vector coupling constant $g_{\omega N}$ in order to keep the values of E/A and k_{F_0} unchanged at saturation. A decrease of the vector coupling constant however leads to less repulsion and, thus, to a softer behavior of the equation of state, characterized by a lower value for K. Obviously the situation is more complex for the relativistic **T**-matrix theory than for relativistic Hartree. Nevertheless the quantitative features established between m^*_B and K for the relativistic Hartree approximation must plausibly carry over (at least to some extent) to the **T**-matrix theory, since the nature of the underlying repulsive forces in both treatments is the same. To be more quantitative, we find that $m^*_{\rm HEA} = 0.82$ and $m^*_{\rm Bonn} = 0.79$ in the case of nuclear matter, which increase to 0.91 and 0.88 respectively for neutron matter, indicating that the HEA equation of state behaves more softly at saturation density than the Bonn equation of state. This behavior is reflected in figure 12.17. It exhibits the somewhat weaker density dependence of both energy per particle and pressure obtained for HEA relative to Bonn. The slope of the E/A curve

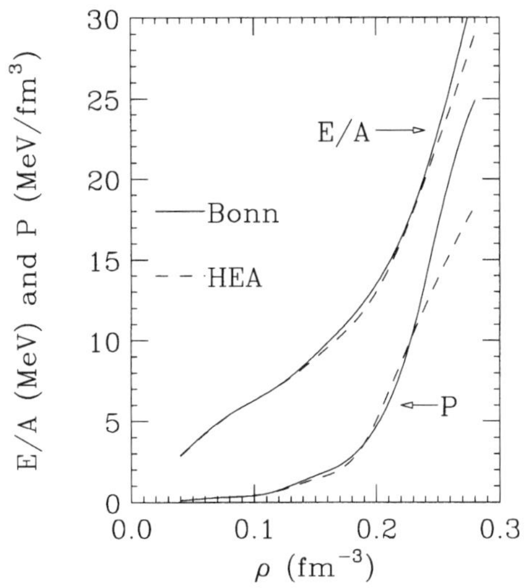

Figure 12.17. Energy per nucleon, E/A, and pressure, P, as a function of baryon number density, ρ, for pure neutron matter (neutrons only) in $\mathbf{\Lambda}^{00}$ approximation [62]. The underlying boson-exchange potentials are Bonn and HEA. (Reprinted courtesy of World Scientific.)

enters the thermodynamic relation (12.107) which determines the pressure of matter at zero temperature.

As already mentioned, the self-consistent nuclear matter incompressibilities at saturation obtained for the **T**-matrix calculations as well as those based on non-relativistic nucleon–nucleon potentials are relatively small in comparison to those associated with relativistic Hartree or Hartree–Fock. (Recall that in the latter two cases K enters as a parameter.) Such an impact on K is well known for theories that account for dynamical two-particle correlations [93, 117, 118, 121, 122], as is the case for the relativistic **T**-matrix and all the non-relativistic models of our collection. On the other hand, it is known that such equations of state develop a rather stiff behavior at higher nuclear densities, typically beyond about $2\,\epsilon_0$ [93, 109, 122, 318]. These features are plausibly linked to the influence of the nuclear medium on the dynamical interaction between two nucleons [505]. For one thing, the Pauli principle forbids scattering processes in which one of the intermediate nucleon states is already occupied (see the Brueckner propagator given in (9.59)), which eliminates some of the attraction between two nucleons arising from pion exchange. Secondly, because the particles are not in free space, their motion in the intermediate particle states acquires 'dispersion' corrections, leading to a dependence on m_B^* as well as on energy of these states, as can be inferred from inspecting (9.59) together with (9.60). The net result of both features is a decrease in the intermediate range attraction in the medium, which becomes more important with increasing density [58].

The softness of the nuclear equation of state is of great importance in astrophysics, where a controversy is concerned with the softness of the equation of state as constrained by supernova calculations [382]. According to the present understanding of the mechanism of supernova explosions, only relatively soft equations of state allow in supernova calculations for a sufficiently large enough energy release necessary to catalyze the prompt-bounce mechanisms (see, however, references [8, 47]). From what has been outlined just above, one sees that nuclear matter theories which account for dynamical two-particle correlations predict equations of state that exhibit a rather soft behavior at not too large nuclear matter densities. We also note that a soft behavior of the equation of state at lower nuclear densities accompanied by a rather stiff behavior at the high-density regime ($\epsilon \gtrsim 4 - 5\,\epsilon_0$) is important in order to obtain large enough neutron star masses and to withstand fast rotation [108, 410]. More than that, the behavior of the equation of state above nuclear matter density intimately manifests itself in the interplay between radius, mass, and central density of a neutron star, as we shall see in section 14.4 and chapter 15. This interplay also concerns the thickness and volume of the stellar crust. Finally, with a stiffer equation of state the maximum neutron star mass increases, as mentioned above.

Chapter 13

General relativity in a nutshell

With central mass densities of up to several 10^{15} g cm^{-3}, neutron stars and their hypothetical strange counterparts constitute objects of highly compressed matter. As already remarked at the beginning of this book (see, for instance, figure 3.2), such objects possess masses of $M \sim 1.5\, M_{\odot}$ and radii of $R \sim 10$ km making the ratio $2M/R$, which may be considered as a critical measure for the 'strength' of gravity, as large as about 40% or more.[1] For such large values of $2M/R$ the geometry of spacetime is changed considerably from flat space. It is therefore imperative to construct models of neutron and strange quark matter stars in the framework of Einstein's general theory of relativity, in which the gravitational force is replaced by the concept of curved spacetime, as caused by a compact star. Besides the crucial classical tests of Einstein's theory, which allow one to discriminate between the predictions of the general theory of relativity and that of Newtonian theory, each one deciding unequivocally in favor of the general theory of relativity, recently Hulse and Taylor were able to test the validity of Einstein's theory from studying the motion of the binary pulsars system PSR 1913+16 [133]. Their study confirmed the theory with unprecedented accuracy.

In general relativity the curvature in the geometry of four-dimensional spacetime, which manifests itself as gravitation (and vice versa), is the result of mass. In the case of a compact stellar configuration, it is the star's mass that acts as the source which curves the geometry of spacetime inside and outside of the star. If the star is rotating, then the so-called Lense–Thirring [506] or frame dragging effect comes into play, which describes the onset of rotation of the local inertial frames, induced by the mass of the rotating star, in the direction of the star's rotation (see, for example, references [507, 508, 509, 510]). This effect has no analogy in classical

[1] For the purpose of comparison, this ratio is about 10^{-6} for the Sun, 10^{-9} for the Earth, and 10^{-25} for a human body.

Newtonian mechanics and therefore may be hard to imagine intuitively. At the bottom of the heart of this effect lies the question of whether or not non-uniform motion is, like uniform motion, relative too. It was studied by Newton in 1686 using a vessel filled with water. He argued that a bucket's rotation was absolute because water in it will be forced up the sides of the vessel due to the centrifugal force. If the bucket is considered fixed and the cosmos rotating about it, then what causes the water to rise up the sides of the vessel? In 1883, E. Mach reexamined Newton's discussion of inertial forces on a fluid contained in a rotating vessel, in an attempt to understand better how inertial forces arise. He suggested that the shape of a water surface may depend on the rotation of the vessel if the sides of the vessel increase in thickness and mass till they were ultimately several leagues thick. The calculation of such effects became possible when Einstein formulated his general theory of relativity in 1916, and was carried out in 1918 by Thirring [506]. Using the weak-field approximation to Einstein's equations, he found that a slowly rotating mass shell drags along the inertial frames within it. More recently, Brill and Cohen clarified the connection of the frame dragging effect with Mach's principle [509]. Thirring's result is valid only when the induced rotation rate is small compared to the rotation rate of the shell. For decades the frame dragging problem remained dormant, for it appeared to have only little physical significance for actual stellar phenomena. The observation of rapidly spinning neutron stars whose enormous mass concentrations in their centers cause the local inertial frames there to corotate at about half the star's rotational frequency, as we shall see in chapters 15 and 17, has renewed considerable interest in this phenomenon.

Before we proceed to the construction of models of neutron and strange matter stars, the general theory of relativity, which culminates in Einstein's field equations, will be reviewed next. We begin with putting together some basic elements of tensor analysis.

13.1 Some formulae of tensor analysis

As in flat spacetime, generalized coordinates in curved four-dimensional space are denoted as $x^\mu = x^0, x^1, x^2, x^3$, $x'^\mu = x'^0, x'^1, x'^2, x'^3$, etc, where $\mu, \nu, \ldots$ assume values 0,1,2,3. In our case, the zero-components of x^μ generally refer to time, t, while x^1, x^2, x^3 stand for r, θ, ϕ. The new set of coordinates x'^μ are functions of the old coordinates by the functions $x'^\mu = x'^\mu(x^0, x^1, x^2, x^3)$. Generally, superscripts refer to *contravariant* quantities, while subscripts indicate *covariant* quantities. Multiplication of a contravariant vector a^μ with a covariant vector b_μ gives [511]

$$a^\mu\, b_\mu \equiv \sum_{\mu=0}^{4} a^\mu\, b_\mu = a^0\, b_0 + a^1\, b_1 + a^2\, b_2 + a^3\, b_3 \,. \tag{13.1}$$

As indicated in (13.1) the double occurrence of dummy suffixes in a given term of a tensor expression will always be taken to denote summation over the four values 1,2,3,4. Multiplication of contravariant and covariant tensors of rank two gives

$$a^{\mu\nu}\, b_{\mu\nu} \equiv \sum_{\mu=0}^{4} \sum_{\nu=0}^{4} a^{\mu\nu}\, b_{\mu\nu} = a^{00}\, b_{00} + a^{01}\, b_{01} + \ldots + a^{33}\, b_{33}\,, \quad (13.2)$$

where generally the rank r of a tensor is equal to the total number of indices $\mu, \nu, \ldots$ carried by a quantity. The values of a contravariant tensor of rank two, $T^{\alpha\beta}$, are transformed in accordance with

$$T'^{\mu\nu} = \frac{\partial x'^{\mu}}{\partial x^{\alpha}} \frac{\partial x'^{\nu}}{\partial x^{\beta}}\, T^{\alpha\beta}\,, \quad (13.3)$$

while for a covariant tensor of the same rank

$$T'_{\mu\nu} = \frac{\partial x^{\alpha}}{\partial x'^{\mu}} \frac{\partial x^{\beta}}{\partial x'^{\nu}}\, T_{\alpha\beta}\,. \quad (13.4)$$

Tensors of mixed contravariant and covariant nature or of higher rank can be similarly defined in accordance with the general expression

$$T'^{\mu\nu\ldots}_{\rho\sigma\ldots} = \frac{\partial x'^{\mu}}{\partial x^{\alpha}} \frac{\partial x'^{\nu}}{\partial x^{\beta}} \frac{\partial x^{\gamma}}{\partial x'^{\rho}} \frac{\partial x^{\delta}}{\partial x'^{\sigma}} \ldots T^{\alpha\beta\ldots}_{\gamma\delta\ldots}\,. \quad (13.5)$$

We recall that the requirement that any relativistic equation must remain invariant under coordinate transformation (covariance), that is, its mathematical structure must be the same in all coordinate systems is one of the key elements of general relativity. This is not to be confused, however, with the fact that the content of the equations of physics may change when we change to new coordinate systems as due to a change in gravitational field rather than to a change in the absolute motion of the spatial framework, as the principle of equivalence permits us. The requirement of covariance is automatically fulfilled by tensor equations. Examples of tensors of different rank are listed in the following. Let us begin with the simplest case, namely a tensor of rank zero (scalar invariant). It transforms under (13.5) as:

$$s' = s\,. \quad (13.6)$$

Contravariant tensor of rank one (vector):

$$a'^{\mu} = \frac{\partial x'^{\mu}}{\partial x^{\alpha}}\, a^{\alpha}\,. \quad (13.7)$$

Covariant tensor of rank one:

$$a'_{\mu} = \frac{\partial x^{\alpha}}{\partial x'^{\mu}}\, a_{\alpha}\,. \quad (13.8)$$

Mixed tensor of rank two:

$$T'^{\nu}_{\mu} = \frac{\partial x'^{\nu}}{\partial x^{\alpha}} \frac{\partial x^{\beta}}{\partial x'^{\mu}} T^{\alpha}_{\ \beta} . \tag{13.9}$$

Symmetric tensor:

$$T^{\mu\nu} = T^{\nu\mu} . \tag{13.10}$$

Metric tensor:

$$g^{\mu\nu} = g^{\nu\mu} . \tag{13.11}$$

An infinitesimal difference in coordinate position is given by $\mathrm{d}x^{\mu} \equiv \mathrm{d}x^0, \mathrm{d}x^1, \mathrm{d}x^2, \mathrm{d}x^3$ from which one obtains for the scalar interval (line element) $\mathrm{d}s$ associated with $\mathrm{d}x^{\mu}$ the expression

$$\mathrm{d}s^2 = g_{\mu\nu}\, \mathrm{d}x^{\mu}\, \mathrm{d}x^{\nu} . \tag{13.12}$$

The determinant formed from the components of $g_{\mu\nu}$ is abbreviated to

$$g \equiv \det(g_{\mu\nu}) \equiv |g_{\mu\nu}| . \tag{13.13}$$

Finally we note that the components of the mixed tensor g^{ν}_{μ} are given by

$$g^{\nu}_{\mu} = \delta^{\nu}_{\mu} = \begin{cases} 1 & \text{if } \ \mu = \nu \ , \\ 0 & \text{if } \ \mu \neq \nu \ . \end{cases} \tag{13.14}$$

The Galilean values of $g_{\mu\nu}$ in flat space are $\delta_{\mu\nu} = \pm 1, 0$, where only the diagonal components are different from zero.

13.2 Tensor manipulations

Raising, lowering, and change of indices is accomplished via the metric tensor $g^{\mu\nu}$ as follows:

$$a^{\nu} = g^{\nu\mu}\, a_{\mu} , \quad a_{\mu} = g^{\mu\nu}\, a^{\nu} , \quad a^{\nu} = g^{\nu}_{\mu}\, a^{\mu} . \tag{13.15}$$

Other frequently encountered manipulations are:
Contraction,

$$T^{\ \nu}_{\nu} = g_{\nu\mu}\, T^{\nu\mu} = T^{\ 0}_{0} + T^{\ 1}_{1} + T^{\ 2}_{2} + T^{\ 3}_{3} . \tag{13.16}$$

Addition,

$$a_{\mu} = b_{\mu} + c_{\mu} = (b_0 + c_0), (b_1 + c_1), (b_2 + c_2), (b_3 + c_3) . \tag{13.17}$$

Outer product,

$$\begin{aligned} a^{\nu}_{\mu} = b_{\mu}\, c^{\nu} = {} & b_0\, c^0 \quad b_0\, c^1 \quad b_0\, c^2 \quad b_0\, c^3 \\ & b_1\, c^0 \quad b_1\, c^1 \quad b_1\, c^2 \quad b_1\, c^3 \\ & b_2\, c^0 \quad b_2\, c^1 \quad b_2\, c^2 \quad b_2\, c^3 \\ & b_3\, c^0 \quad b_3\, c^1 \quad b_3\, c^2 \quad b_3\, c^3 . \end{aligned} \tag{13.18}$$

Inner product,

$$a = a^{\nu}_{\nu} = b_{\nu}\, c^{\nu} = b_0\, c^0 + b_1\, c^1 + b_2\, c^2 + b_3\, c^3\,. \tag{13.19}$$

13.3 Einstein's field equations

Equipped with the mathematical formalism summarized in sections 13.1 and 13.2, we now proceed to the formulation of Einstein's general theory of relativity, which, as mentioned before, culminates in his field equations. The basis elements of this theory can be formulated as follows [512]:

(i) The geometry of spacetime is described by the line element, $\mathrm{d}s$, given by $\mathrm{d}s^2 = g_{\mu\nu}\,\mathrm{d}x^{\mu}\,\mathrm{d}x^{\nu}$. The metric tensor is itself a function of spacetime coordinates.

(ii) The amount of mass energy in a unit volume is determined by the stress–energy (energy–momentum) tensor $T^{\mu\nu}$. For a gas or perfect fluid,

$$T^{\mu\nu} = (\epsilon + P)\, u^{\mu}\, u^{\nu} + P\, g^{\mu\nu}\,, \tag{13.20}$$

where ϵ is the energy density of the matter as measured in its restframe, P is the pressure, and u^{μ} is the four-velocity of the gas. A perfect fluid is a fluid or gas that moves through spacetime with a four-velocity which may vary from event to event, and exhibits a density of mass energy ϵ and an isotropic pressure P in the restframe of each fluid element. Shear stresses, anisotropic pressure, and viscosity must be absent, or the fluid is not perfect.

(iii) Conservation of energy–momentum,

$$\frac{1}{\sqrt{-g}}\, \frac{\partial}{\partial x^{\nu}} \left(\sqrt{-g}\, T^{\nu}_{\ \mu}\right) - \frac{1}{2}\, T^{\nu\lambda}\, \frac{\partial}{\partial x^{\mu}}\, g_{\nu\lambda} = 0\,, \tag{13.21}$$

where $g = \det(g_{\mu\nu})$. For a perfect fluid in flat space one immediately derives from (13.20)

$$\frac{\partial}{\partial x^{\nu}} T^{\mu\nu} = \frac{\partial P}{\partial x_{\mu}} + \frac{\partial}{\partial x^{\nu}} \left[(\epsilon + P)\, u^{\mu}\, u^{\nu}\right] = 0\,, \tag{13.22}$$

and by substituting $u^i = v^i\, u^0$ it becomes Euler's equation,

$$\frac{\partial \boldsymbol{v}}{\partial t} + (\boldsymbol{v} \cdot \boldsymbol{\nabla})\, \boldsymbol{v} = - \frac{1 - \boldsymbol{v}^2}{\epsilon + P} \left(\boldsymbol{\nabla} P + \boldsymbol{v}\, \frac{\partial P}{\partial t} \right). \tag{13.23}$$

(iv) Riemann–Christoffel curvature tensor, $R^{\tau}_{\ \mu\nu\sigma}$, is the agent by which curves in spacetime produce the relative acceleration of geodesics,

$$R^{\tau}_{\ \mu\nu\sigma} = \frac{\partial}{\partial x^{\nu}}\, \Gamma^{\tau}_{\mu\sigma} - \frac{\partial}{\partial x^{\sigma}}\, \Gamma^{\tau}_{\mu\nu} + \Gamma^{\kappa}_{\mu\sigma}\, \Gamma^{\tau}_{\kappa\nu} - \Gamma^{\kappa}_{\mu\nu}\, \Gamma^{\tau}_{\kappa\sigma}\,. \tag{13.24}$$

(v) Christoffel symbols (of the second kind), which are not third-rank tensors, are defined by the relations

$$\Gamma^{\sigma}_{\mu\nu} \equiv \frac{1}{2}\, g^{\sigma\lambda} \left(\frac{\partial}{\partial x^{\nu}}\, g_{\mu\lambda} + \frac{\partial}{\partial x^{\mu}}\, g_{\nu\lambda} - \frac{\partial}{\partial x^{\lambda}}\, g_{\mu\nu} \right) , \tag{13.25}$$

where

$$\Gamma^{\sigma}_{\lambda\sigma} = \frac{\partial}{\partial x^{\lambda}} \log_{\mathrm{e}} \sqrt{-g} = \frac{1}{\sqrt{-g}}\, \frac{\partial}{\partial x^{\lambda}} \sqrt{-g}\, , \tag{13.26}$$

which alternatively can be written as

$$\Gamma^{\sigma}_{\lambda\sigma} = \Gamma^{\sigma}_{\sigma\lambda} = \frac{1}{2}\, g^{\mu\nu}\, \frac{\partial}{\partial x^{\lambda}}\, g_{\mu\nu}\, . \tag{13.27}$$

Notice that the Christoffel symbols obey $\Gamma^{\sigma}_{\mu\nu} = \Gamma^{\sigma}_{\nu\mu}$. The Christoffel symbols of the first kind are defined as

$$\begin{aligned}
[\lambda\nu, \mu] &\equiv g_{\mu\sigma}\, \Gamma^{\sigma}_{\lambda\nu} \\
&= \frac{1}{2}\, g_{\mu\sigma}\, g^{\sigma\tau} \left(\frac{\partial}{\partial x^{\nu}}\, g_{\lambda\tau} + \frac{\partial}{\partial x^{\lambda}}\, g_{\nu\tau} - \frac{\partial}{\partial x^{\tau}}\, g_{\lambda\nu} \right) \\
&= \frac{1}{2} \left(\frac{\partial}{\partial x^{\nu}}\, g_{\lambda\mu} + \frac{\partial}{\partial x^{\lambda}}\, g_{\nu\mu} - \frac{\partial}{\partial x^{\mu}}\, g_{\lambda\nu} \right) = [\lambda\nu, \mu]\, .
\end{aligned} \tag{13.28}$$

(vi) Using the Christoffel symbols, the equation expressing the conservation of energy–momentum becomes

$$T^{\mu\nu}{}_{;\sigma} = \frac{\partial}{\partial x^{\sigma}}\, T^{\mu\nu} + \Gamma^{\mu}_{\lambda\sigma}\, T^{\lambda\nu} + \Gamma^{\nu}_{\lambda\sigma}\, T^{\mu\lambda}\, , \tag{13.29}$$

or

$$T^{\mu\nu}{}_{;\mu} = \frac{\partial}{\partial x^{\mu}}\, T^{\mu\nu} + \Gamma^{\mu}_{\lambda\mu}\, T^{\lambda\nu} + \Gamma^{\nu}_{\lambda\mu}\, T^{\mu\lambda}\, , \tag{13.30}$$

$$= \frac{1}{\sqrt{-g}}\, \frac{\partial}{\partial x^{\mu}} \left(\sqrt{-g}\, T^{\mu\nu} \right) + \Gamma^{\nu}_{\lambda\mu}\, T^{\mu\lambda}\, . \tag{13.31}$$

(vii) Above, the semicolon denotes the covariant divergence. For a contravariant vector, a^{μ}, and a covariant vector, a_{μ}, the covariant derivatives are

$$a^{\mu}{}_{;\nu} = \frac{\partial}{\partial x^{\nu}}\, a^{\mu} + \Gamma^{\mu}_{\lambda\nu}\, a^{\lambda} \tag{13.32}$$

and

$$a_{\mu;\,\nu} = \frac{\partial}{\partial x^{\nu}}\, a_{\mu} - \Gamma^{\lambda}_{\mu\nu}\, a_{\lambda}\, . \tag{13.33}$$

For any vector

$$a^{\mu}{}_{;\mu} = \frac{\partial}{\partial x^{\mu}}\, a^{\mu} + \Gamma^{\mu}_{\lambda\mu}\, a^{\lambda} = \frac{1}{\sqrt{-g}}\, \frac{\partial}{\partial x^{\lambda}} \left(\sqrt{-g}\, a^{\lambda} \right) . \tag{13.34}$$

The following properties hold:

$$(A\,a^{\mu}{}_{\nu} + B\,b^{\mu}{}_{\nu})_{;\,\lambda} = A\,a^{\mu}{}_{\nu;\,\lambda} + B\,b^{\mu}{}_{\nu;\,\lambda}\,, \tag{13.35}$$

$$(a^{\mu}{}_{\nu}\,b^{\lambda})_{;\,\kappa} = a^{\mu}{}_{\nu;\,\kappa}\,b^{\lambda} + a^{\mu}{}_{\nu}\,b^{\lambda}{}_{;\,\kappa}\,. \tag{13.36}$$

For tensors we note that

$$\left(T^{\,\dots\nu\dots}_{\,\dots\mu\dots}\right)_{;\,\sigma} = \frac{\partial T^{\,\dots\nu\dots}_{\,\dots\mu\dots}}{\partial x^{\sigma}} + \Gamma^{\nu}_{\sigma\lambda}\,T^{\,\dots\lambda\dots}_{\,\dots\mu\dots} + \dots - \Gamma^{\lambda}_{\mu\sigma}\,T^{\,\dots\nu\dots}_{\,\dots\lambda\dots} - \dots\,, \tag{13.37}$$

where '$+\dots$' ('$-\dots$') indicates that a similar term is to be added (subtracted) for each additional contravariant (covariant) index. Examples of (13.37) are

$$T^{\mu\lambda}{}_{\lambda;\,\sigma} = \frac{\partial}{\partial x^{\sigma}}\,T^{\mu\lambda}{}_{\lambda} + \Gamma^{\mu}_{\sigma\kappa}\,T^{\kappa\lambda}{}_{\lambda}\,, \tag{13.38}$$

$$g_{\mu\nu;\,\lambda} = \frac{\partial}{\partial x^{\lambda}}\,g_{\mu\nu} - \Gamma^{\kappa}_{\mu\lambda}\,g_{\kappa\nu} - \Gamma^{\kappa}_{\nu\lambda}\,g_{\mu\kappa} = 0\,. \tag{13.39}$$

Moreover we note for tensors that

$$g_{\mu\nu,\lambda} \equiv \frac{\partial}{\partial x^{\lambda}}\,g_{\mu\nu} = \Gamma^{\kappa}_{\lambda\mu}\,g_{\kappa\nu} + \Gamma^{\kappa}_{\lambda\nu}\,g_{\kappa\mu}\,, \tag{13.40}$$

$$g^{\mu\nu}{}_{,\lambda} \equiv \frac{\partial}{\partial x^{\lambda}}\,g^{\mu\nu} = -\,\Gamma^{\mu}_{\kappa\lambda}\,g^{\kappa\nu} - \Gamma^{\nu}_{\kappa\lambda}\,g^{\kappa\mu}\,, \tag{13.41}$$

$$g^{\nu\mu}{}_{;\lambda} = g^{\mu\nu}{}_{;\lambda} = 0\,. \tag{13.42}$$

(viii) The production of curvature by mass energy is specified by Einstein's field equations,

$$G_{\mu\nu} = 8\,\pi\,T_{\mu\nu}\,, \quad \text{where} \quad G_{\mu\nu} \equiv R_{\mu\nu} - \frac{1}{2}\,g_{\mu\nu}\,R \tag{13.43}$$

is the Einstein tensor, which for empty space reduces to

$$R_{\mu\nu} = 0\,. \tag{13.44}$$

The Ricci tensor is obtained from the Riemann tensor by contraction, that is, $R_{\mu\nu} = R^{\tau}{}_{\mu\sigma\nu}\,g^{\sigma}{}_{\tau} = R^{\tau}{}_{\mu\tau\nu}$, which leads to

$$R_{\mu\nu} = \frac{\partial}{\partial x^{\nu}}\,\Gamma^{\sigma}_{\mu\sigma} - \frac{\partial}{\partial x^{\sigma}}\,\Gamma^{\sigma}_{\mu\nu} + \Gamma^{\kappa}_{\mu\sigma}\,\Gamma^{\sigma}_{\kappa\nu} - \Gamma^{\kappa}_{\mu\nu}\,\Gamma^{\sigma}_{\kappa\sigma} \tag{13.45}$$

$$= \frac{\partial}{\partial x^{\sigma}}\,\Gamma^{\sigma}_{\mu\nu} + \Gamma^{\sigma}_{\mu\lambda}\,\Gamma^{\lambda}_{\sigma\nu} + \frac{\partial^2}{\partial x^{\mu}\partial x^{\nu}}\,\log_e\sqrt{-g} - \Gamma^{\lambda}_{\mu\nu}\,\frac{\partial}{\partial x^{\lambda}}\,\log_e\sqrt{-g}\,. \tag{13.46}$$

The scalar curvature of spacetime, R, also known as the Ricci scalar follows from Ricci's tensor as

$$R = R_{\mu\nu}\,g^{\mu\nu} = R^{\mu}{}_{\mu}. \tag{13.47}$$

Einstein's field equations follow from the assumptions that the ratio of gravitational and inertial mass is a universal constant, that the laws of nature are expressed in the simplest possible set of equations that are covariant for all systems of spacetime coordinates, and that the laws of special relativity hold locally in a coordinate system with a vanishing gravitational field.

(ix) In a gravitational field, a particle moves along a geodesic line which is specified by the geodesic differential equation

$$\frac{\mathrm{d}^2 x^\mu}{\mathrm{d}s^2} + \Gamma^\mu_{\kappa\sigma} \frac{\mathrm{d}x^\kappa}{\mathrm{d}s} \frac{\mathrm{d}x^\sigma}{\mathrm{d}s} = 0 . \tag{13.48}$$

Light rays are represented by null geodesics for which $\mathrm{d}s = 0$.

(x) An observer's proper reference frame is formed by an orthonormal tetrad which keeps its time-leg tangent to the observer's world-line. Expressed in less mathematical terms, the measurements performed by an observer, who moves along a world-line through spacetime, in his own neighborhood (with distances small compared to the radii of curvature of spacetime) are called the values of the measured quantities relative to the observer's proper reference frame. The laws of physics expressed in the proper reference frame are those of flat spacetime (i.e. special relativity) as augmented by an inertial (or, according to the equivalence principle, gravitational) acceleration.

(xi) An observer's proper time, τ, is governed by the metric along his world-line,

$$\mathrm{d}\tau = \sqrt{-\,\mathrm{d}s^2} = \sqrt{-\,g_{\mu\nu}\,\mathrm{d}x^\mu\,\mathrm{d}x^\nu}\,, \tag{13.49}$$

where the world-line is described by any time parameter, t, $x^\nu = x^\nu(t)$.

(xii) In any infinitesimal neighborhood of any point in spacetime, the proper time intervals must satisfy the laws of special relativity. That is, locally the line element, $\mathrm{d}s$, becomes the Minkowski metric given in rectangular coordinates by

$$\mathrm{d}s^2 = -\mathrm{d}t^2 + \mathrm{d}x^2 + \mathrm{d}y^2 + \mathrm{d}z^2\,, \tag{13.50}$$

and by

$$\mathrm{d}s^2 = -\mathrm{d}t^2 + \mathrm{d}r^2 + r^2\,\mathrm{d}\theta^2 + r^2\,\sin^2\theta\,\mathrm{d}\phi^2 \tag{13.51}$$

in the case of spherical coordinates. In each case the coordinates are measured in an inertial frame of reference. In Minkowski spacetime, $\mathrm{d}s$ is invariant under the Lorentz transformation, geometry is Euclidean, space is flat, and the interval of proper time, $\mathrm{d}\tau$, is given by

$$\mathrm{d}\tau = \sqrt{1 - \boldsymbol{V}^2}\,\mathrm{d}t\,, \tag{13.52}$$

which is the interval read by a clock moving at the velocity $\boldsymbol{V}$,

$$|\boldsymbol{V}| = \sqrt{\left(\frac{\mathrm{d}x}{\mathrm{d}t}\right)^2 + \left(\frac{\mathrm{d}y}{\mathrm{d}t}\right)^2 + \left(\frac{\mathrm{d}z}{\mathrm{d}t}\right)^2} . \tag{13.53}$$

(xiii) For a spherically symmetric gravitational field outside a massive non-rotating body in vacuum (where $R_{\mu\nu} = 0$), the line element, $\mathrm{d}s$, becomes the Schwarzschild metric given by

$$\mathrm{d}s^2 = -\left(1 - \frac{2\,M}{r}\right) \mathrm{d}t^2 + \left(1 - \frac{2\,M}{r}\right)^{-1} \mathrm{d}r^2 + r^2\,\mathrm{d}\theta^2 + r^2\,\sin^2\theta\,\mathrm{d}\phi^2 . \tag{13.54}$$

Here, r, θ, and ϕ are spherical coordinates whose origin is at the center of the massive object, and M is the mass which determines the Newtonian gravitational field M/r.

The exact solution of Einstein's field equations (13.43) in empty space, $R_{\mu\nu} = 0$, obtained by Schwarzschild in 1916 [513], describes the geometry of spacetime outside of a spherically symmetric, non-pulsating distribution of matter. In this case the metric functions, which enter the underlying line element, are particularly simple. Proceeding one step further and solving the field equations (13.43) for a a spherically symmetric stellar object made up of a perfect fluid, whose energy–stress tensor has a relatively simple form, one arrives at the so-called Tolman–Oppenheimer–Volkoff (TOV) equations. These can be solved, by means of applying straightforward numerical techniques (such as Runge–Kutta integration), for a given model for the equation of state of the stellar matter, as will be discussed in great detail in chapter 14. Rotation complicates the construction of stellar models considerably, as we shall see in chapter 15. The reason for this is threefold. For one, rotating stars are rotationally deformed, that is, they are flattened at the pole but grow in size in the equatorial direction, which leads to a dependence of the metric on the polar angle, in addition to the radial dependence. Secondly, rotation stabilizes a star against gravitational collapse. It can therefore carry more mass than would be the case if the star would be non-rotating. Being more massive, however, means that the geometry of spacetime will changed too. This makes the line element of a rotating star depend on the star's rotational frequency. Thirdly, there is the above mentioned frame dragging effect which imposes an additional self-consistency condition on the stellar structure, since the strength of the dragging of the local inertial frames depends on the star's properties, like mass (profile) and rotational frequency. So in order to construct models of rotating compact stars, one has to put the Einstein equations on a two-dimensional numerical grid, spanned by θ and r, and solve them self-consistently until convergence is achieved for a given model for the equation of state. These points render stellar structure calculations of rotationally deformed bodies considerably more complicated than those of non-rotating, spherically symmetric bodies. In the latter case, as shall be seen immediately below, one can carry out the analytical analysis all the way to the condition of hydrostatic equilibrium in general relativity theory, known as the TOV equations.

Chapter 14

Structure equations of non-rotating stars

In this chapter we shall derive the stellar structure equations of non-rotating, spherically symmetric objects in the framework of Einstein's general theory of relativity. For such stars the metric functions are particularly simple because they depend on the time and radial coordinate only. Another simplification arises from treating the stellar matter as a perfect fluid.

As already mentioned in chapter 13, at the heart of Einstein's theory lies, besides the equivalence principle, the requirement that any relativistic equation must remain covariant under coordinate transformation, that is, its mathematical structure must be the same in all coordinate systems. This requirement is automatically fulfilled by tensor equations. The chief objective of the next section will be to demonstrate this explicitly for the energy–momentum tensor of a perfect fluid.

14.1 Energy–momentum tensor in covariant form

The energy–momentum tensor is a tensor of rank two. To derive its covariant representation, we perform a coordinate transformation of the energy–momentum tensor $\overset{0}{T}{}^{\alpha\beta}$ given in a proper reference frame $\overset{0}{x}{}^{\mu} = (\overset{0}{x}{}^{0}, \overset{0}{x}{}^{1}, \overset{0}{x}{}^{2}, \overset{0}{x}{}^{3})$ to the coordinate system $x^{\mu} = (x^0, x^1, x^2, x^3)$ of actual interest, which moves relative to $\overset{0}{x}{}^{\mu}$. Approximating the stellar matter by an *ideal* (or perfect) fluid, which is an excellent approximation as long as one is interested in the global structural properties of compact stars, as is the case here, then the components of $\overset{0}{T}{}^{\alpha\beta}$ in a momentarily comoving reference frame are given by (for the sake of brevity, henceforth we shall drop the notion momentarily),

$$\overset{0}{T}{}^{\alpha\beta} = \begin{pmatrix} \overset{0}{\epsilon} & 0 & 0 & 0 \\ 0 & \overset{0}{P}_{xx} & \overset{0}{P}_{xy} & \overset{0}{P}_{xz} \\ 0 & \overset{0}{P}_{yx} & \overset{0}{P}_{yy} & \overset{0}{P}_{yz} \\ 0 & \overset{0}{P}_{zx} & \overset{0}{P}_{zy} & \overset{0}{P}_{zz} \end{pmatrix} = \begin{pmatrix} \overset{0}{\epsilon} & 0 & 0 & 0 \\ 0 & \overset{0}{P} & 0 & 0 \\ 0 & 0 & \overset{0}{P} & 0 \\ 0 & 0 & 0 & \overset{0}{P} \end{pmatrix} . \tag{14.1}$$

Equation (14.1) is a direct manifestation of the definition of a perfect fluid in relativity, defined as a fluid that has *no viscosity* and *no heat conduction* in the comoving reference frame. The quantities $\overset{0}{\epsilon}$ and $\overset{0}{P}$ in (14.1) denote energy density and pressure measured by a local observer comoving with $\overset{0}{x}{}^{\mu}$. No heat conduction implies that $\overset{0}{T}{}^{0i} = \overset{0}{T}{}^{i0} = 0$ in (14.1). Energy can flow only if particles flow. Viscosity is a force parallel to the interface between particles. Its absence means that the forces should always be perpendicular to the interface, i.e. $\overset{0}{T}{}^{ij}$ should be zero unless $i = j$. This means that $\overset{0}{T}{}^{ij}$ should be a diagonal matrix. Moreover, it must be diagonal in *all* comoving reference frames, since no viscosity is a statement independent of the spatial axes. The only matrix diagonal in all frames is a multiple of the identity: all its diagonal terms are equal. Thus, an x surface will have across it only a force in the x direction, and similarly for y and z. These forces-per-unit-area are all equal, and are called the pressure, P. So one has $\overset{0}{T}{}^{ij} = \overset{0}{P}\,\delta^{ij}$. From six possible quantities (the number of independent elements in the 3×3 symmetric matrix $\overset{0}{T}{}^{ij}$) the zero-viscosity assumption therefore reduces the number of functions to just one in (14.1), the pressure [514].

The two restrictions – no heat conduction and no viscosity – made in the definition of (14.1) enormously simplify the energy–momentum tensor, as we now see. Let us consider the fluid to be at rest in the coordinate system $\overset{0}{x}{}^{\mu}$, that is,

$$\frac{d\overset{0}{x}{}^{1}}{d\tau} = \frac{d\overset{0}{x}{}^{2}}{d\tau} = \frac{d\overset{0}{x}{}^{3}}{d\tau} = 0 . \tag{14.2}$$

The transformation of $\overset{0}{T}{}^{\alpha\beta}$ to another comoving reference frame is accomplished by the following manipulations,

$$T^{\mu\nu} = \frac{\partial x^{\mu}}{\partial \overset{0}{x}{}^{\alpha}} \frac{\partial x^{\nu}}{\partial \overset{0}{x}{}^{\beta}} \overset{0}{T}{}^{\alpha\beta} \tag{14.3}$$

$$= \frac{\partial x^{\mu}}{\partial \overset{0}{x}{}^{\alpha}} \frac{\partial x^{\nu}}{\partial \overset{0}{x}{}^{\alpha}} \overset{0}{T}{}^{\alpha\alpha} \tag{14.4}$$

$$= \frac{\partial x^{\mu}}{\partial \overset{0}{x}{}^{0}} \frac{\partial x^{\nu}}{\partial \overset{0}{x}{}^{0}} \overset{0}{\epsilon} + \frac{\partial x^{\mu}}{\partial \overset{0}{x}{}^{i}} \frac{\partial x^{\nu}}{\partial \overset{0}{x}{}^{i}} \overset{0}{P} , \tag{14.5}$$

where (14.3) is nothing but the transformation law for a rank-two tensor. Equation (14.4) reflects the diagonal structure of the energy–momentum tensor, which, via (14.1), leads immediately to (14.5). To simplify (14.5) we write in the first place for the contravariant components of the metric tensor in the new coordinates in terms of their values in the old coordinates

$$g^{\mu\nu} = \frac{\partial x^\mu}{\partial \overset{0}{x}{}^\alpha} \frac{\partial x^\nu}{\partial \overset{0}{x}{}^\beta} \overset{0}{g}{}^{\alpha\beta} , \tag{14.6}$$

which on substituting the simple values of $\overset{0}{g}{}^{\alpha\beta}$ gives

$$g^{\mu\nu} = -\frac{\partial x^\mu}{\partial \overset{0}{x}{}^0} \frac{\partial x^\nu}{\partial \overset{0}{x}{}^0} + \frac{\partial x^\mu}{\partial \overset{0}{x}{}^i} \frac{\partial x^\nu}{\partial \overset{0}{x}{}^i} . \tag{14.7}$$

In the second place, we write for the macroscopic velocity of the fluid with respect to the new coordinate system

$$\frac{\mathrm{d}x^\mu}{\mathrm{d}\tau} = \frac{\partial x^\mu}{\partial \overset{0}{x}{}^\nu} \frac{\mathrm{d}\overset{0}{x}{}^\nu}{\mathrm{d}\tau} , \tag{14.8}$$

which reduces to

$$\frac{\mathrm{d}x^\mu}{\mathrm{d}\tau} = \frac{\partial x^\mu}{\partial \overset{0}{x}{}^\nu} \delta^{\nu 0} = \frac{\partial x^\mu}{\partial \overset{0}{x}{}^0} , \tag{14.9}$$

owing to the value of zero for the spatial components of velocity ($\mathrm{d}\overset{0}{x}{}^i/\mathrm{d}\tau = 0$) and the value of unity ($\mathrm{d}\overset{0}{x}{}^0/\mathrm{d}\tau = 1$) for its temporal component in the old coordinates. Substituting (14.7) and (14.9) into equation (14.5) leads to

$$\begin{aligned} T^{\mu\nu} &= \frac{\mathrm{d}x^\mu}{\mathrm{d}\tau} \frac{\mathrm{d}x^\nu}{\mathrm{d}\tau} \overset{0}{\epsilon} + \frac{\partial x^\mu}{\partial \overset{0}{x}{}^i} \frac{\partial x^\nu}{\partial \overset{0}{x}{}^i} \overset{0}{P} \\ &= \frac{\mathrm{d}x^\mu}{\mathrm{d}\tau} \frac{\mathrm{d}x^\nu}{\mathrm{d}\tau} \overset{0}{\epsilon} + \left\{ \frac{\partial x^\mu}{\partial \overset{0}{x}{}^0} \frac{\partial x^\nu}{\partial \overset{0}{x}{}^0} + g^{\mu\nu} \right\} \overset{0}{P} \\ &= \frac{\mathrm{d}x^\mu}{\mathrm{d}\tau} \frac{\mathrm{d}x^\nu}{\mathrm{d}\tau} \overset{0}{\epsilon} + \left\{ \frac{\mathrm{d}x^\mu}{\mathrm{d}\tau} \frac{\mathrm{d}x^\nu}{\mathrm{d}\tau} + g^{\mu\nu} \right\} \overset{0}{P} , \end{aligned} \tag{14.10}$$

which allows us to express the energy–momentum tensor for a perfect fluid in the very useful and general form

$$T^{\mu\nu} = u^\mu u^\nu \left(\overset{0}{\epsilon} + \overset{0}{P} \right) + g^{\mu\nu} \overset{0}{P} . \tag{14.11}$$

The quantities u^μ and u^ν are four-velocities, defined as

$$u^\mu \equiv \frac{\mathrm{d}x^\mu}{\mathrm{d}\tau} , \qquad u^\nu \equiv \frac{\mathrm{d}x^\nu}{\mathrm{d}\tau} . \tag{14.12}$$

They are the components of the macroscopic velocity of the fluid with respect to the actual coordinate system that is being used. For the coordinate system in which the fluid is at rest one has $\mathrm{d}x^\mu/\mathrm{d}\tau = (\mathrm{d}x^0/\mathrm{d}\tau, \mathrm{d}\boldsymbol{x}/\mathrm{d}\tau) = (1,0,0,0) = \delta^{\mu 0}$, and thus from equation (14.11)

$$\overset{0}{\epsilon} = T^{00} \quad \text{and} \quad \overset{0}{P} = \frac{1}{3}\sum_{i=1}^{3} T^{ii} \,. \tag{14.13}$$

Note that equation (14.11) is a tensor equation and thus is valid in any comoving (locally inertial) coordinate system. The equivalence principle guarantees that equation (14.11) holds in flat spacetime as well as in curved spacetime. Having a manifestly covariant expression at hand for the source term of Einstein's field equations (13.43), we now proceed to solve them for spherically symmetric stellar objects.

14.2 Tolman–Oppenheimer–Volkoff equation

After these considerations we proceed to derive the stellar structure equations of a non-rotating, static (that is, non-pulsating), spherically symmetric compact star. The metric of such an object has the form [510, 512, 513]

$$\mathrm{d}s^2 = -\mathrm{e}^{2\,\Phi(r)}\,\mathrm{d}t^2 + \mathrm{e}^{2\,\Lambda(r)}\,\mathrm{d}r^2 + r^2\,\mathrm{d}\theta^2 + r^2\sin^2\!\theta\,\mathrm{d}\phi^2 \,, \tag{14.14}$$

where $\Phi(r)$ and $\Lambda(r)$ are the radially varying metric functions. Hereafter we shall drop the arguments of these functions quite often. Introducing the covariant components of the metric tensor as

$$g_{tt} = -\,\mathrm{e}^{2\,\Phi}\,, \quad g_{rr} = \mathrm{e}^{2\,\Lambda}\,, \quad g_{\theta\theta} = r^2\,, \quad g_{\phi\phi} = r^2\sin^2\!\theta\,, \tag{14.15}$$

the line element (14.14) can be written in the generally covariant expression for interval:

$$\mathrm{d}s^2 = g_{\mu\nu}\,\mathrm{d}x^\mu\,\mathrm{d}x^\nu \,. \tag{14.16}$$

The contravariant components of the metric tensor are obtained via the relation

$$g^{\mu\nu}\,g_{\nu\lambda} = \delta^\mu{}_\nu \,, \tag{14.17}$$

where $\delta^\mu{}_\nu$ is the four-dimensional Kronecker delta (cf. equation (13.14)). One then finds

$$g^{tt} = -\,\mathrm{e}^{-\,2\,\Phi}\,, \quad g^{rr} = \mathrm{e}^{-\,2\,\Lambda}\,, \quad g^{\theta\theta} = r^{-2}\,, \quad g^{\phi\phi} = \frac{1}{r^2\,\sin^2\!\theta} \,. \tag{14.18}$$

Note that because of the underlying symmetries, the only functional dependence that enters the metric is the dependence on radial distance

r, measured from the star's origin. The components of the metric tensor can be grouped together into a 4×4 matrix as

$$(g^{\mu\nu}) = \begin{pmatrix} -\,\mathrm{e}^{-2\Phi} & 0 & 0 & 0 \\ 0 & \mathrm{e}^{-2\Lambda} & 0 & 0 \\ 0 & 0 & r^{-2} & 0 \\ 0 & 0 & 0 & (r\sin\theta)^{-2} \end{pmatrix}, \qquad (14.19)$$

with column and row labels μ and ν running from $0, 1, 2, 3$, or alternatively from t, r, θ, ϕ. Finally, noticing that $g^{\mu}{}_{\lambda} = g^{\mu\nu} g_{\nu\lambda}$, according to the transformation law (13.15), we find from (14.17) that the mixed components of the metric tensor obey $g^{\mu}{}_{\lambda} = \delta^{\mu}{}_{\nu}$, and therefore

$$g^{t}{}_{t} = g^{r}{}_{r} = g^{\theta}{}_{\theta} = g^{\phi}{}_{\phi} = 1\,. \qquad (14.20)$$

The determinant of $g_{\mu\nu}$ is readily found from (14.15),

$$g \equiv \det(g_{\mu\nu}) = -\,\mathrm{e}^{2\Phi}\,\mathrm{e}^{2\Lambda}\,r^4\,\sin^2\!\theta\,. \qquad (14.21)$$

Having specified the metric, we can now proceed to calculate the Einstein tensor associated with our problem. Note that all quantities whose knowledge is necessary to calculate the Einstein tensor – from the Riemann tensor $R^{\tau}{}_{\mu\sigma\nu}$ to the curvature scalar R – are given in terms of the components of the metric tensor and derivatives thereof. Our chief task therefore will be to perform step by step the numerous (in general partial) differentiations of the metric tensor with respect to the coordinate variables t, r, θ, ϕ, which ultimately will lead us to the expression for the Einstein tensor associated with the metric (14.15).

Those quantities to be determined first in this step-by-step analysis are the Christoffel symbols $\Gamma^{\sigma}_{\mu\nu}$ introduced in (13.25). The non-vanishing symbols to the form of line element (14.15) of a spherically symmetric body are

$$\begin{aligned} &\Gamma^{r}_{tt} = \mathrm{e}^{2\Phi-2\Lambda}\,\Phi'\,, \quad \Gamma^{t}_{tr} = \Phi'\,, \quad \Gamma^{r}_{rr} = \Lambda'\,, \\ &\Gamma^{\theta}_{r\theta} = r^{-1}\,, \quad \Gamma^{\phi}_{r\phi} = r^{-1}\,, \quad \Gamma^{r}_{\theta\theta} = -\,r\,\mathrm{e}^{-2\Lambda}\,, \quad \Gamma^{\phi}_{\theta\phi} = \frac{\cos\theta}{\sin\theta}\,, \\ &\Gamma^{r}_{\phi\phi} = -\,r\,\sin^2\!\theta\,\mathrm{e}^{-2\Lambda}\,, \quad \Gamma^{\theta}_{\phi\phi} = -\sin\theta\,\cos\theta\,. \end{aligned} \qquad (14.22)$$

Here, and in the following, primes denote differentiation with respect to the radial coordinate r, as, for instance, $\Phi' \equiv \mathrm{d}\Phi/\mathrm{d}r$ and $\Phi'' \equiv \mathrm{d}^2\Phi/\mathrm{d}r^2$. Because of the relatively simple mathematical form of the metric (14.14), the number of non-vanishing Christoffel symbols turns out to be of manageable size. As already mentioned at the beginning of this chapter, this is no longer the case for a rotationally deformed star, whose metric functions, because of rotational deformation, depend also on the polar angle

θ. Moreover the dragging of local inertial frames manifests itself in the occurrence of an additional non-diagonal metric term, $g^{t\phi}$.

With the Christoffel symbols at our disposal, we now proceed with the calculation of the Riemann–Christoffel tensor $R^{\tau}{}_{\mu\nu\sigma}$, the Ricci tensor $R_{\mu\nu}$ and the Ricci scalar R, given in equations (13.24), (13.45), and (13.47) respectively, of which the Einstein tensor is composed. The following combinations of Christoffel symbols enter the calculation of $R^{\tau}{}_{\mu\nu\sigma}$:

$$\Gamma^{\lambda}_{t\kappa}\,\Gamma^{\kappa}_{\lambda t} = (\Gamma^{t}_{tt})^2 + 2\,\Gamma^{t}_{ti}\,\Gamma^{i}_{tt} + \Gamma^{i}_{tj}\,\Gamma^{j}_{it} = 2\,(\Phi')^2\,\mathrm{e}^{2\,\nu - 2\,\Lambda}\,,$$
$$\Gamma^{\lambda}_{r\kappa}\,\Gamma^{\kappa}_{\lambda r} = (\Gamma^{t}_{rt})^2 + 2\,\Gamma^{t}_{ri}\,\Gamma^{i}_{tr} + \Gamma^{i}_{rj}\,\Gamma^{j}_{ir} = (\Phi')^2 + (\Lambda')^2 + \frac{2}{r^2}\,,$$
$$\Gamma^{\lambda}_{\theta\kappa}\,\Gamma^{\kappa}_{\lambda\theta} = (\Gamma^{t}_{\theta t})^2 + 2\,\Gamma^{t}_{\theta i}\,\Gamma^{i}_{t\theta} + \Gamma^{i}_{\theta j}\,\Gamma^{j}_{i\theta} = -\,2\,\mathrm{e}^{-2\,\Lambda} + \cot^2\theta\,,$$
$$\Gamma^{\lambda}_{\phi\kappa}\,\Gamma^{\kappa}_{\lambda\phi} = (\Gamma^{t}_{\phi t})^2 + 2\,\Gamma^{t}_{\phi i}\,\Gamma^{i}_{t\phi} + \Gamma^{i}_{\phi j}\,\Gamma^{j}_{i\phi} = -\,2\left(\sin^2\theta\,\mathrm{e}^{-2\,\Lambda} + \cos^2\theta\right).$$

Substituting these results into (13.24) gives for the non-vanishing components of the Riemann–Christoffel tensor:

$$\begin{aligned}
&R^{t}{}_{rtr} = -\,\Phi'' - \left(\Phi'\right)^2 + \Phi'\,\Lambda'\,,\\
&R^{t}{}_{\theta t\theta} = -\,r\,\Phi'\,\mathrm{e}^{-2\,\Lambda}\,, \qquad R^{t}{}_{\phi t\phi} = -\,r\,\Phi'\,\sin^2\theta\,\mathrm{e}^{-2\,\Lambda}\,,\\
&R^{r}{}_{ttr} = \left\{-\,\Phi'' - \left(\Phi'\right)^2 + \Phi'\,\Lambda'\right\}\mathrm{e}^{2\,\Phi - 2\,\Lambda}\,,\\
&R^{r}{}_{\theta r\theta} = r\,\Lambda'\,\mathrm{e}^{-2\,\Lambda}\,, \qquad R^{r}{}_{\phi r\phi} = \Lambda'\,r\,\sin^2\theta\,\mathrm{e}^{-2\,\Lambda}\,,\\
&R^{\theta}{}_{tt\theta} = -\,\Phi'\,r\,\mathrm{e}^{2\,\Phi - 2\,\Lambda}\,, \qquad R^{\theta}{}_{rr\theta} = -\frac{1}{r}\,\Lambda'\,,\\
&R^{\theta}{}_{\phi\theta\phi} = \sin^2\theta\left(1 - \mathrm{e}^{-2\,\Lambda}\right), \qquad R^{\phi}{}_{tt\phi} = -\,\Phi'\,r\,\mathrm{e}^{2\,\Phi - 2\,\Lambda}\,,\\
&R^{\phi}{}_{rr\phi} = -\frac{1}{r}\,\Lambda'\,, \qquad R^{\phi}{}_{\theta\theta\phi} = -\,1 + \mathrm{e}^{-2\,\Lambda}\,.
\end{aligned} \tag{14.23}$$

For the sake of completeness, we also give the components of the pure covariant Riemann–Christoffel tensor. These are given by

$$\begin{aligned}
&R_{trtr} = \Phi''\,\mathrm{e}^{2\,\Phi} + \left(\Phi'\right)^2\mathrm{e}^{2\,\Phi} - \Phi'\,\mathrm{e}^{2\,\Phi}\,\Lambda'\,,\\
&R_{t\theta t\theta} = \Phi'\,r\,\mathrm{e}^{2\,\Phi - 2\,\Lambda}\,, \qquad R_{t\phi t\phi} = \Phi'\,r\,\sin^2\theta\,\mathrm{e}^{2\,\Phi - 2\,\Lambda}\,,\\
&R_{r\theta r\theta} = r\,\Lambda'\,, \qquad R_{r\phi r\phi} = \Lambda'\,r\,\sin^2\theta\,,\\
&R_{\theta\phi\theta\phi} = r^2\sin^2\theta\left(1 - \mathrm{e}^{-2\,\Lambda}\right).
\end{aligned} \tag{14.24}$$

Using the values of the Christoffel symbols listed in (14.22), the components of the Ricci tensor read

$$\begin{aligned}
&R_{tt} = \left\{-\Phi'\,\Lambda' + \Phi'' + \left(\Phi'\right)^2 + 2\,r^{-1}\,\Phi'\right\}\mathrm{e}^{2\,\Phi - 2\,\Lambda}\,,\\
&R_{rr} = -\,\Phi'' - \left(\Phi'\right)^2 + \Phi'\,\Lambda' + \frac{2}{r}\,\Lambda'\,,\\
&R_{\theta\theta} = \left\{-\,r\,\Phi' + r\,\Lambda' + \mathrm{e}^{2\,\Lambda} - 1\right\}\mathrm{e}^{-2\,\Lambda}\,,\\
&R_{\phi\phi} = -\,\sin^2\theta\left\{r\Phi' - r\,\Lambda' - \mathrm{e}^{2\,\Lambda} + 1\right\}\mathrm{e}^{-2\,\Lambda}\,.
\end{aligned} \tag{14.25}$$

The components of the mixed Ricci tensor are obtained from

$$R^{\sigma}{}_{\lambda} = g^{\sigma\tau}\, R_{\tau\lambda}\,, \tag{14.26}$$

with $R_{\tau\lambda}$ given by equation (13.45). One arrives for the individual components at,

$$\begin{aligned}
R^{t}{}_{t} &= \Big\{-\frac{1}{4}\,(\Phi')^2 + \frac{1}{4}\,\Phi'\,\Lambda' - \frac{1}{2}\,\Phi'' - \frac{1}{r}\,\Phi'\Big\}\,\mathrm{e}^{-\Lambda}\,,\\
R^{r}{}_{r} &= -\Big\{\frac{1}{4}\,(\Phi')^2 - \frac{1}{4}\,\Phi'\,\Lambda' + \frac{1}{2}\,\Phi'' - \frac{1}{r}\,\Lambda'\Big\}\,\mathrm{e}^{-\Lambda}\,,\\
R^{\theta}{}_{\theta} &= -\frac{1}{r^2}\,\mathrm{e}^{-\Lambda}\Big\{1 - \frac{1}{2}\,r\,\Lambda' + \frac{1}{2}\,r\,\Phi'\Big\} + \frac{1}{r^2}\\
R^{\phi}{}_{\phi} &= R^{\theta}{}_{\theta}\,.
\end{aligned} \tag{14.27}$$

Finally, the Ricci scalar, which follows from (13.47), has the form

$$\begin{aligned}
R = \big\{ &+ 2\,\Phi'\,\Lambda'\,r^2 - 2\,\Phi''\,r^2 - 2\,(\Phi')^2\,r^2 - 4\,r\,\Phi'\\
&+ 4\,r\,\Lambda' + 2\,\mathrm{e}^{2\,\Lambda} - 2\big\}\, r^{-2}\,\mathrm{e}^{-2\,\Lambda}\,.
\end{aligned} \tag{14.28}$$

The Einstein tensor, $G_{\mu\nu}$, could now be calculated from equations (14.25) and (14.28). However, it is more convenient to transform $G_{\mu\nu}$ to its mixed representation $G^{\mu}{}_{\nu}$, which is readily accomplished, according to the rules outlined in section 13.1, by multiplying the Einstein tensor $G_{\kappa\nu}$ with the metric tensor $g^{\mu\kappa}$ and summing over κ. This leads for the metric tensor in (13.43) to $g^{\mu\kappa} g_{\mu\nu} = \delta^{\kappa}{}_{\nu}$, where, as known from equation (13.14), the elements of $\delta^{\kappa}{}_{\nu}$ possess a particularly simple form. The Einstein tensor (13.43) in the mixed representation thus reads

$$G^{\mu}{}_{\nu} \equiv R^{\mu}{}_{\nu} - \frac{1}{2}\,\delta^{\mu}{}_{\nu}\,R = 8\,\pi\,T^{\mu}{}_{\nu}\,, \tag{14.29}$$

with

$$T^{\mu}{}_{\nu} = (\epsilon + P)\,\frac{\mathrm{d}x^{\mu}}{\mathrm{d}\tau}\,\frac{\mathrm{d}x_{\nu}}{\mathrm{d}\tau} + \delta^{\mu}{}_{\nu}\,P\,. \tag{14.30}$$

As already noted, the connection between the distribution of matter and energy, contained in the energy–momentum tensor, with the geometry of spacetime is the physical content of Einstein's field equations (14.29). The left-hand side of this equation gives a quantity whose tensor divergence is known to be identically equal to zero. In accordance with the rules of covariant differentiation introduced in (13.37), we may write as an immediate consequence of (14.29),

$$T^{\mu}{}_{\nu\,;\,\mu} \equiv \frac{\partial}{\partial x^{\mu}} T^{\mu}{}_{\nu} + \Gamma^{\mu}_{\kappa\mu}\,T^{\kappa}{}_{\nu} - \Gamma^{\kappa}_{\nu\mu}\,T^{\mu}{}_{\kappa} = 0\,. \tag{14.31}$$

This expression reduces in *flat* spacetime, that is, in a local inertial frame, to its special relativistic form given by

$$T^{\mu}{}_{\nu;\,\mu} = \frac{\partial}{\partial x^{\mu}} T^{\mu}{}_{\nu} = 0\,. \tag{14.32}$$

For the metric of (14.14) the covariant derivative of $T^{\mu}{}_{\nu}$ is given by

$$T^{\mu}{}_{\nu;\,\mu} = (\epsilon + P)\,\Phi' + P'\,. \tag{14.33}$$

Below we shall demonstrate that already from (14.31) alone one can draw many important conclusions as to the behavior of matter and energy.

Before, however, we give the components of the Einstein tensor in the mixed representation. These are obtained by substituting the expressions listed in (14.27) into (14.29), leading to

$$G^{t}{}_{t} = R^{t}{}_{t} - \frac{1}{2}\,R = \mathrm{e}^{-2\,\Lambda}\Big(\frac{1}{r^2} - 2\,\frac{\Lambda'}{r}\Big) - \frac{1}{r^2}\,, \tag{14.34}$$

$$G^{r}{}_{r} = R^{r}{}_{r} - \frac{1}{2}\,R = \mathrm{e}^{-2\,\Lambda}\Big(2\,\frac{\Phi'}{r} + \frac{1}{r^2}\Big) - \frac{1}{r^2}\,, \tag{14.35}$$

$$G^{\theta}{}_{\theta} = R^{\theta}{}_{\theta} - \frac{1}{2}\,R = \mathrm{e}^{-2\,\Lambda}\Big(\Phi'' - \Phi'\,\Lambda' + (\Phi')^2 + \frac{\Phi' - \Lambda'}{r}\Big)\,, \tag{14.36}$$

and

$$G^{\phi}{}_{\phi} = G^{\theta}{}_{\theta}\,. \tag{14.37}$$

For the sake of completeness, we also list the purely covariant components of the Einstein tensor, which are given by

$$\begin{aligned}
G_{tt} &= \big(2\,r\,\Lambda' + \mathrm{e}^{2\,\Lambda} - 1\big)\,r^{-2}\,\mathrm{e}^{2\,\Phi - 2\,\Lambda}\,,\\
G_{rr} &= \big(2\,r\,\Phi' - \mathrm{e}^{2\,\Lambda} + 1\big)\,r^{-2}\,,\\
G_{\theta\theta} &= \big\{r\big(\Phi' - \Lambda' - \Phi'\,\Lambda'\,r + r\,\Phi'' + \big(\Phi'\big)^2\,r\big)\big\}\,\mathrm{e}^{-2\,\Lambda}\,,\\
G_{\phi\phi} &= \big\{r\sin^2\!\theta\big(\Phi' - \Lambda' - \Phi'\,\Lambda'\,r + r\,\Phi'' + \big(\Phi'\big)^2\,r\big)\big\}\,\mathrm{e}^{-2\,\Lambda}\,.
\end{aligned} \tag{14.38}$$

The purely contravariant components are given by

$$\begin{aligned}
G^{tt} &= \frac{2\,\Lambda'}{\mathrm{e}^{2\,\Phi}\,r\,\mathrm{e}^{2\,\Lambda}} + \frac{1}{\mathrm{e}^{2\,\Phi}\,r^2} - \frac{1}{\mathrm{e}^{2\,\Phi}\,r^2\,\mathrm{e}^{2\,\Lambda}}\,,\\
G^{rr} &= \frac{2\,\Phi'}{\mathrm{e}^{4\,\Lambda}\,r} - \frac{1}{r^2\,\mathrm{e}^{2\,\Lambda}} + \frac{1}{\mathrm{e}^{4\,\Lambda}\,r^2}\,,\\
G^{\theta\theta} &= \frac{\Phi'}{r^3\,\mathrm{e}^{2\,\Lambda}} - \frac{\Lambda'}{r^3\,\mathrm{e}^{2\,\Lambda}} - \frac{\Phi'\,\Lambda'}{r^2\,\mathrm{e}^{2\,\Lambda}} + \frac{\Phi''}{r^2\,\mathrm{e}^{2\,\Lambda}} + \frac{\big(\Phi'\big)^2}{r^2\,\mathrm{e}^{2\,\Lambda}}\,,\\
G^{\phi\phi} &= \frac{\Phi'}{r^3\,\sin^2\!\theta\,\mathrm{e}^{2\,\Lambda}} - \frac{\Lambda'}{r^3\,\sin^2\!\theta\,\mathrm{e}^{2\,\Lambda}} - \frac{\Phi'\,\Lambda'}{r^2\,\sin^2\!\theta\,\mathrm{e}^{2\,\Lambda}}\\
&\quad + \frac{\Phi''}{r^2\,\sin^2\!\theta\,\mathrm{e}^{2\,\Lambda}} + \frac{\big(\Phi'\big)^2}{r^2\,\sin^2\!\theta\,\mathrm{e}^{2\,\Lambda}}\,.
\end{aligned} \tag{14.39}$$

Those components of the Einstein tensor that are not listed above are understood to be equal to zero. Combining the expressions derived in equations (14.34) to (14.37) with (14.29) and (14.30) leads to the following field equations:

$\mu = \nu = t$:

$$e^{-2\Lambda}\Big(2\,\frac{\Lambda'}{r} - \frac{1}{r^2}\Big) + \frac{1}{r^2} = 8\,\pi\,\epsilon\,, \tag{14.40}$$

$\mu = \nu = r$:

$$e^{-2\Lambda}\Big(2\,\frac{\Phi'}{r} + \frac{1}{r^2}\Big) - \frac{1}{r^2} = 8\,\pi\,P\,, \tag{14.41}$$

$\mu = \nu = \theta$:

$$e^{-2\Lambda}\Big(\Phi'' - \Phi'\,\Lambda' + (\Phi')^2 + \frac{\Phi' - \Lambda'}{r}\Big) = 8\,\pi\,P\,, \tag{14.42}$$

$\mu = \nu = \phi$:

$$G^{\phi}{}_{\phi} = G^{\theta}{}_{\theta}, \quad \text{and} \quad T^{\phi}{}_{\phi} = T^{\theta}{}_{\theta}\,. \tag{14.43}$$

In deriving equations (14.40) through (14.43), we made use of the fact that we are dealing with a *static* stellar configuration, in which case one derives from the line element (14.16)

$$\frac{dr}{d\tau} = \frac{d\theta}{d\tau} = \frac{d\phi}{d\tau} = 0 \quad \Rightarrow \quad \frac{dt}{d\tau} = e^{-\Phi}\,. \tag{14.44}$$

This implies for the mixed components of the energy–momentum tensor of (14.30) that

$$T^{t}{}_{t} = -\,\epsilon\,, \qquad T^{i}{}_{i} = P\,. \tag{14.45}$$

The first relation in (14.45) follows because

$$\frac{dx^t}{d\tau} \equiv \frac{dt}{d\tau} = e^{-\Phi}\,, \tag{14.46}$$

and therefore

$$\frac{dx_t}{d\tau} = g_{t\sigma}\,\frac{dx^\sigma}{d\tau} = e^{-2\Phi}\,\frac{dt}{d\tau} = -\,e^{\Phi}\,. \tag{14.47}$$

The second relation follows trivially from (14.30) since $dx^i/d\tau = 0$ ($i = 1, 2, 3$) for the static stellar configuration.

In the next step let us turn our interest for a moment toward the general properties of the energy–momentum tensor, for some of its properties will be very useful to bring the stellar structure equations (14.40) through (14.43) into a more illuminating form. We begin with abbreviating the fluid's four-velocity as $u^\mu = \mathrm{d}x^\mu/\mathrm{d}\tau$ and $u_\nu = \mathrm{d}x_\nu/\mathrm{d}\tau$. Covariant differentiation of (14.30) then leads to

$$0 = T^\mu{}_{\nu;\,\mu} = (\epsilon + P)_\mu \, u^\mu \, u_\nu + (\epsilon + P)(u_{\nu;\,\mu} \, u^\mu + u_\nu \, u^\mu{}_{;\,\mu}) + \delta^\mu{}_\nu \, P_{;\,\mu} \,, \tag{14.48}$$

where we have used that $\delta^\mu{}_{\nu;\,\mu} = 0$, which follows from $\delta^\mu{}_{\nu;\,\mu} = (g^{\mu\kappa} g^{\kappa\nu})_{;\,\mu} = 0$. Noticing that the four-velocity is given by

$$u^\mu = (u^0, u^1, u^2, u^3) = \frac{\mathrm{d}x^\mu}{\mathrm{d}\tau} = \left(\frac{\mathrm{d}x^0}{\mathrm{d}\tau}, \frac{\mathrm{d}x^1}{\mathrm{d}\tau}, \frac{\mathrm{d}x^2}{\mathrm{d}\tau}, \frac{\mathrm{d}x^3}{\mathrm{d}\tau} \right), \tag{14.49}$$

and upon calculating the square of it, one arrives at

$$u^\mu \, u_\mu = u^\mu \, g_{\mu\nu} \, u^\nu = g_{00} \, (u^0)^2 + g_{ii} \, (u^i)^2 = -\,1 \,. \tag{14.50}$$

In flat space, equation (14.50) reduces to the familiar relation

$$u^\mu \, u_\mu = -\,(u^0)^2 + \boldsymbol{u}^2 = -1 \,. \tag{14.51}$$

To extract the Euler equation of relativistic hydrodynamics from $T^\mu{}_{\nu;\,\mu} = 0$, we project equation (14.48) perpendicular to $\boldsymbol{u}$ by applying the projection tensor [510]

$$\mathsf{P}^{\lambda\nu} \equiv u^\lambda \, u^\nu + g^{\lambda\nu} \,, \tag{14.52}$$

to $T^\mu{}_{\nu;\,\mu}$, that is $\mathsf{P}^{\lambda\nu} T^\mu{}_{\nu;\,\mu} = 0$. This leaves us with the expression

$$\mathsf{P}^{\lambda\nu} \{ (\epsilon + P)_\mu \, u_\nu \, u^\mu + (\epsilon + P)(u_{\nu;\,\mu} \, u^\mu + u_\nu \, u^\mu{}_{;\,\mu}) + \delta_\nu{}^\mu P_{,\,\mu} \} = 0 \,. \tag{14.53}$$

The first term on the left-hand side of (14.53) vanishes, which follows from the normalization condition of the four-velocity, $u^\nu \, u_\nu = -1$, and the relation $g^{\lambda\nu} \, u_\nu = u^\lambda$ applied to

$$(u^\lambda \, u^\nu + g^{\lambda\nu}) \, u_\nu \, u^\mu = 0 \,. \tag{14.54}$$

We are thus left with

$$(\epsilon + P)(u^\lambda \, u^\nu \, u_{\nu;\,\mu} \, u^\mu + g^{\lambda\nu} u_{\nu;\,\mu} \, u^\mu) + (u^\lambda \, u^\mu P_{,\,\mu} + g^{\lambda\mu} P_{,\,\mu}) = 0 \,, \tag{14.55}$$

where $P_{,\,\mu}$ denotes the usual subscripted notation to indicate the differentiation of P with respect to x^μ, i.e. $\partial P / \partial x^\mu \equiv P_{,\,\mu}$.

To carry the evaluation of (14.55) further, we note that

$$g^{\lambda\nu} \, u_{\nu;\,\mu} = (g^{\lambda\nu} \, u_\nu)_{;\mu} = u^\lambda{}_{;\,\mu} \,, \tag{14.56}$$

and from the normalization condition (14.50) of u^μ,

$$0 = (u^\nu\, u_\nu)_{;\,\mu} = u^\nu{}_{;\,\mu}\, u_\nu + u^\nu\, u_{\nu;\,\mu} = 2\, u^\nu{}_{;\,\mu}\, u_\nu\,, \tag{14.57}$$

from which it follows that

$$u^\nu{}_{;\,\mu}\, u_\nu = u^\nu\, u_{\nu;\,\nu} = 0\,. \tag{14.58}$$

To arrive at the last equality in (14.57) use of

$$u^\nu\, u_{\nu;\,\mu} = g^{\nu\tau}\, u_\tau\, (g_{\nu\sigma}\, u^\sigma)_{;\,\mu} = g^{\nu\tau}\, g_{\nu\sigma}\, u_\tau\, u^\sigma{}_{;\,\mu} = \delta^\tau{}_\sigma\, u_\tau\, u^\sigma{}_{\;\;\mu} = u_\nu\, u^\nu{}_{;\,\mu} \tag{14.59}$$

was made. With the aid of (14.56) and (14.58), equation (14.55) can be written as

$$(\epsilon + P)\, u^\lambda{}_{;\mu}\, u^\mu + (u^\lambda\, u^\mu\, P_{,\,\mu} + g^{\lambda\mu}\, P_{,\,\mu}) = 0\,, \tag{14.60}$$

which, upon multiplication with $g_{\lambda\sigma}$, transforms after some algebraic manipulations to

$$(\epsilon + P)\, u_{\sigma;\,\mu}\, u^\mu + P_{,\sigma} + P_{,\mu}\, u^\mu\, u_\sigma = 0\,. \tag{14.61}$$

This relation is known as Euler's equation, which determines the flow lines to which u is tangent. It has precisely the same form as the corresponding flat-spacetime Euler equation. Note that the pressure gradient, not gravity, is responsible for all deviations of flow lines from geodesics. Let us now choose the fluid's rest frame and compute the zero-component of the equation of motion from $T^\mu{}_\nu$. In this case $u^0 = 1$, $\boldsymbol{u} = 0$, and $u^0{}_{;\,\mu} = 0$. Hence the relation $0 = u^\nu\, T^\mu{}_{\nu;\,\mu}$ reduces to

$$0 = u^\nu\, T^\mu{}_{\nu;\,\mu} = T^\mu{}_{0;\,\mu}\,, \tag{14.62}$$

which, on substituting the energy–momentum tensor (14.30), can be written in the manner

$$\begin{aligned} 0 &= \left((\epsilon + P)\, u^\mu\, u_0 + \delta^\mu{}_0\, P\right)_{;\mu} \\ &= -\,(\epsilon + P)_{,\,\mu}\, u^\mu - (\epsilon + P)\, u^\mu{}_{;\,\mu} + \delta^\mu{}_0\, P_{,\,\mu}\,. \end{aligned} \tag{14.63}$$

Performing the summation over μ leaves us with

$$0 = \frac{\mathrm{d}\epsilon}{\mathrm{d}t} + (\epsilon + P)\, u^\mu{}_{;\mu}\,. \tag{14.64}$$

With the baryon number density ρ in the fluid's rest frame, defined as $\rho = A/V$, the number flux vector of baryons, $\rho\, u^\mu$, is conserved. Hence, in the rest frame we have

$$0 = (\rho\, u^\mu)_{;\,\mu} = \rho_{,\,\mu}\, u^\mu + \rho\, u^\mu{}_{;\,\mu} = \frac{\mathrm{d}\rho}{\mathrm{d}t} + \rho\, u^\mu{}_{;\,\mu}\,. \tag{14.65}$$

Equation (14.65) enables us to eliminate the term $u^\mu{}_{;\,\mu}$ in (14.64), leading to

$$\frac{\mathrm{d}\epsilon}{\mathrm{d}t} = \frac{\epsilon + P}{\rho}\,\frac{\mathrm{d}\rho}{\mathrm{d}t}\,. \tag{14.66}$$

Equation (14.66) is recognized as the first law of thermodynamics for a relativistic fluid,

$$\mathrm{d}\epsilon = \frac{\epsilon + P}{\rho}\,\mathrm{d}\rho + T\,\rho\,\mathrm{d}s\,, \tag{14.67}$$

for which the entropy per baryon, s, is conserved along a flow line, $\mathrm{d}s/\mathrm{d}\tau = 0$. The values of ρ, ϵ, P, T and s in (14.67) measure, as usual, the physical quantities in the rest frame of a fluid element. That is, the baryon number density ρ is understood to give the number of baryons per unit three-dimensional volume of rest frame, with antibaryons (if any) counted negatively (cf. section 6.3). In accordance with the field-theoretical part of this book, ϵ denotes the total mass energy – including rest mass, thermal energy, compressional energy, etc – contained in a unit three-dimensional volume of the rest frame. The quantity s is the entropy per baryon, $s = S/A$, in the rest frame. Thus, the entropy per unit volume is given by $\rho\,s$. Finally, P and T are the isotropic pressure and temperature in the rest frame. With these conventions the law of energy conservation in flat spacetime dictates that

$$\mathrm{d}(\epsilon\,A/\rho) = -\,P\,\mathrm{d}(A/\rho) + T\,\mathrm{d}(A\,s)\,, \tag{14.68}$$

which, for a fixed number, A, of baryons leads immediately to (14.67). There is no reason for surprise at this circumstance, for, by virtue of the principle of equivalence, the first law of thermodynamics, expressed in the proper reference frame of a fluid element, is identical to the first law in flat spacetime. We shall encounter equation (14.67) in chapter 19 again when discussing the cooling behavior of compact stars.

Let us turn back to the Euler equation (14.61) for a moment. For the fluid of a static star we have $\mathrm{d}\epsilon/\mathrm{d}t = \mathrm{d}\rho/\mathrm{d}t = 0$, and a fluid four-velocity of $u^\nu = (u^t, 0, 0, 0)$. The expression of u^t is readily obtained from the normalization condition (14.50),

$$-\,1 = u^\nu\,u_\nu = u^\nu\,g_{\nu\mu}\,u^\mu = (u^t)^2\,g_{tt} = -\,(u^t)^2\,\mathrm{e}^{2\,\Phi}\,, \tag{14.69}$$

with g_{tt} known from equation (14.15), as

$$u^t = \mathrm{e}^{-\Phi}\,. \tag{14.70}$$

Because only $P_{,r}$ is non-zero, the only non-trivial component of the Euler equation (14.61) is

$$(\epsilon + P)\,u_{r;\,\mu}\,u^\mu + P_{,r} = 0\,. \tag{14.71}$$

Writing the covariant derivative $u_{r;\,\mu}$ in the form

$$u_{r;\,\mu} = \frac{\partial u_r}{\partial x^\mu} - \Gamma^\lambda_{r\mu}\, u_\lambda = -\,\Gamma^\lambda_{r\mu}\, u_\lambda \tag{14.72}$$

enables us to express (14.71) as

$$\frac{\mathrm{d}P}{\mathrm{d}r} = (\epsilon + P)\,\Gamma^\lambda_{r\mu}\, u_\lambda\, u^\mu = -\,(\epsilon + P)\,\frac{\mathrm{d}\Phi}{\mathrm{d}r}\,. \tag{14.73}$$

To arrive at the latter equality, we have made use of (only u^t is non-zero)

$$\Gamma^\lambda_{r\mu}\, u_\lambda\, u^\mu = \Gamma^t_{rt}\, u_t\, u^t = \Phi'\, u_t\, u^t = -\,\Phi'\,, \tag{14.74}$$

with Γ^t_{rt} given in equation (14.22).

The stellar structure equations in their final form are now readily found as follows. Let us introduce the quantity $m(r)$ as

$$m(r) \equiv 4\,\pi \int_0^r \mathrm{d}r\, r^2\, \epsilon(r) \quad \Rightarrow \quad \frac{\mathrm{d}m}{\mathrm{d}r} = 4\,\pi\, r^2\, \epsilon(r)\,, \tag{14.75}$$

which can be interpreted as the amount of mass energy contained in a sphere of radius r. At the star's origin we impose the condition $m(0) = 0$. With this definition Einstein's field equation (14.40) can then be integrated. This is accomplished by multiplying both sides of (14.40) with r^2 and noticing that

$$\mathrm{e}^{-2\,\Lambda}\, 2\,\Lambda'\, r - \mathrm{e}^{-2\,\Lambda} + 1 = -\,\frac{\mathrm{d}}{\mathrm{d}r}\left(\mathrm{e}^{-2\,\Lambda}\, r - r\right)\,.$$

The integration then yields

$$\mathrm{e}^{-2\,\Lambda} = 1 - \frac{2\,m}{r}\,. \tag{14.76}$$

In the next step we add the field equations (14.40) and (14.41) which gives

$$8\,\pi\,(\epsilon + P) = \frac{2}{r}\,\mathrm{e}^{-2\,\Lambda}\,(\Lambda' + \Phi')\,. \tag{14.77}$$

The metric function Λ in (14.77) can be eliminated with the help of (14.76). For this purpose we differentiate (14.76) with respect to r, which gives

$$-2\,\Lambda'\,\mathrm{e}^{-2\,\Lambda} = \frac{2}{r}\left(\frac{m}{r} - m'\right)\,, \tag{14.78}$$

and substitute this result into (14.77). After some straightforward algebraic manipulations one arrives at

$$8\,\pi\, P = -\,\frac{2\,m}{r^3} + 2\left(1 - \frac{2\,m}{r}\right)\frac{\Phi'}{r}\,. \tag{14.79}$$

Solving this expression for Φ gives

$$\Phi' = \frac{4\,\pi\,r^3\,P + m}{r^2\,(1 - 2\,m/r)}\,, \tag{14.80}$$

with the boundary condition at the star's surface

$$\Phi(r = R) = \frac{1}{2}\ln\Big(1 - \frac{2\,M}{R}\Big)\,, \tag{14.81}$$

where M and R denote the star's mass and radius, respectively (details will be discussed immediately below). Finally substituting from (14.73) for Φ', we arrive for the pressure gradient inside a spherically symmetric configuration at the final result

$$\frac{\mathrm{d}P(r)}{\mathrm{d}r} = -\frac{\{\epsilon(r) + P(r)\}\,\{4\pi r^3 P(r) + m(r)\}}{r^2\,(1 - 2\,m(r)/r)}\,, \tag{14.82}$$

with the central pressure $P(r = 0) \equiv P(\epsilon_c)$ and ϵ_c as the star's central mass energy density. Equation (14.82) is know as the Tolman–Oppenheimer–Volkoff (TOV) equation [515]. This equation is fundamental to the description of the structure of a hydrostatically stable stellar configuration treated in the framework of Einstein's general theory of relativity. As in classical Newtonian mechanics, so also in Einstein's theory, the force that acts on a mass shell inside the star is the pressure force of the stellar matter enclosed in that shell, which acts in the radial outward direction. Gravity pulls on that mass shell in the radial inward direction such that both forces, because of hydrostatic equilibrium, counterbalance each other. In the classical limit one has $P \ll \epsilon$, $P \ll m$, and $m \ll r$ in (14.82), which leads to a pressure gradient given by $\mathrm{d}P/\mathrm{d}r = \epsilon\, m/r^2$. This relation reveals that Einstein's theory increases (the magnitude of) the pressure gradient over what one obtains from the Newtonian treatment, which is quite crucial for stellar bodies, for there are *no* mass or radius limits in Newtonian theory. Neutron stars could therefore be made as massive as one pleases, in sharp contrast to Einstein's theory where neutron stars with too high a central density are subject to gravitational collapse. Let us see how this comes out when integrating (14.82). First one has to specify a model for the equation of state in the form $P(\epsilon)$ and a value of the central density, ϵ_c. This determines P_c while P' and m vanish at $r = 0$. Equation (14.75) then determines m for an infinitesimal increase in r. Plugging in these values for m, ϵ_c and P_c into (14.82) determines the value of P', allowing the determination of P at the next step. For the P thus found the equation of state determines ϵ, and we go over the whole process once again to determine the values of the variables at the next step. In this way, the computation of P, ϵ, and m for successively increasing values of

r goes on until we arrive at $P = 0$,[1] which is identified as the radius, R, of the star, and the value of m there is the star's total mass energy, M. If desired, the metric function Φ can be obtained by simultaneously integrating (14.80) too, for an arbitrarily specified value of Φ at the star's center, $\Phi_0 \equiv \Phi(r = 0)$. After having reached the surface, Φ is to be renormalized by adding a constant to it everywhere, so that it obeys the boundary condition (14.81). The iteration procedure of equations (14.75) and (14.82) is repeated for different values of ϵ_c, leading each time to a particular relativistic stellar model, whose structure functions Φ, m, ϵ, P, ρ satisfy the equations of stellar structure. Notice that for any fixed choice of the equation of state of the form $P(\epsilon)$, or $P = P(\rho)$, $\epsilon = \epsilon(\rho)$, the stellar models form a *one-parameter* sequence (parameter ϵ_c). Once the central density has been specified, the model is determined uniquely.

We finish this section with pointing out that the mass $M \equiv m(r = R)$, contained inside a sphere of Schwarzschild radius $r = R$ is to be identified with the star's *gravitational mass*. From (14.75) one sees that M is given by

$$M = 4\,\pi \int_0^R \mathrm{d}r\, r^2\, \epsilon(r)\,. \tag{14.83}$$

It is this quantity which has to be identified with the total mass energy of the system because it governs the geometry exterior to the matter and therefore fixes such observables as the period of a planetary orbit, precession of perihelion, and gravitational bending of a light ray [69].

For later use we also define the *proper* star mass given by

$$M_{\mathrm{P}} = 4\pi \int_0^R \mathrm{d}r\; r^2\, \frac{\epsilon(r)}{\sqrt{1 - 2\,m(r)/r}}\,. \tag{14.84}$$

In contrast to (14.75), the volume element in the integral of (14.84) is the *proper* volume. Hence the name proper mass for the mass defined in (14.84). The gravitational mass differs from the proper mass on two accounts [69]. Firstly, the quantity $\epsilon(r)$ in (14.83) is not simply the baryon number density, $\rho(r)$, multiplied by the mass per baryon in cold, catalyzed matter at zero pressure ($\frac{1}{56}$ of the mass of one ^{56}Fe atom),

$$\epsilon \neq \rho \times \frac{1}{56} M(^{56}\mathrm{Fe})\,, \tag{14.85}$$

but the mass energy density *exceeds* that product by an amount equivalent to the mass energy of compression. Secondly, what is being integrated in

[1] The Tolman–Oppenheimer–Volkoff equation (14.82) guarantees that the pressure will decrease monotonically so long as the chosen model for the equation of state obeys the reasonable restriction $\epsilon \geq 0$ for all $P \geq 0$ (cf. section 3.1).

(14.75) to get M is not the energy density $\epsilon(r)$ itself, but energy density multiplied by a factor which is less than 1. This can be seen by writing (14.75) formally as

$$M = \int \mathrm{d}^3\boldsymbol{r} \, \frac{1}{\sqrt{1 - 2m(r)/r}} \, \epsilon(r) \, \sqrt{1 - 2m(r)/r} \, \Theta(R - r) \quad (14.86)$$

and introducing the proper volume element,

$$\mathrm{d}V_{\mathrm{prop}} = 4\,\pi\,r^2 \, \frac{1}{\sqrt{1 - 2m(r)/r}} \, \mathrm{d}r \, . \quad (14.87)$$

Equation (14.86) then takes on the form

$$M = \int \mathrm{d}V_{\mathrm{prop}} \, \epsilon(r) \, \sqrt{1 - 2m(r)/r} \, \Theta(R - r) \, . \quad (14.88)$$

The square root in (14.88) in effect corrects for the negative gravitational potential energy of the interaction of the mass with itself. Hence, the factor which multiplies the proper volume – and which in this sense constitutes the integrand – is $\epsilon(r)\sqrt{1 - 2m(r)/r}$, a quantity evidently *smaller* than $\epsilon(r)$. Equation (14.75), superficially identical with the non-relativistic integral for the mass, is evidently quite subtle. It allows both for the work of compression (positive) and the potential energy of gravitation (negative), as pointed out by Harrison and Wheeler [69].

We next introduce the total baryon number, A, of a neutron star. The expression for A as an integral over the number density of baryons, ρ, follows directly from the differential form of the baryon conservation law $[(-g)^{1/2}\rho u^\mu]_{,\mu} = 0$ [516]. The expression is given by

$$A = \int \mathrm{d}^3\boldsymbol{x} \, \sqrt{-g} \, \rho \, u^t \, . \quad (14.89)$$

Substituting the expressions for $\sqrt{-g}$ and the four-velocity u^t, given in (14.21) and (14.70) respectively, into equation (14.89) and performing straightforward algebraic manipulations, with $\mathrm{e}^{-2\Lambda}$ taken from (14.76), then gives

$$A = 4\pi \int_0^R \mathrm{d}r \, r^2 \, \frac{\rho(r)}{\sqrt{1 - 2m(r)/r}} \, . \quad (14.90)$$

Multiplication of A with the rest mass-energy associated with a single neutron, m_n, leads to the so-called star's *baryon mass*,

$$M_{\mathrm{A}} = m_n \, A \, . \quad (14.91)$$

The binding energy, E_B, of a relativistic star is defined to be the difference between its gravitational mass and the mass of all its matter when cold and dispersed, $m_n A$. Thus, with the aid of (14.91), one has

$$E_B = M - M_{\mathrm{A}} \,. \tag{14.92}$$

A calculation of the binding energy is, therefore, equivalent to a calculation of the total baryon number. Finally, we note that combining (14.80) with the Tolman–Oppenheimer–Volkoff equation (14.82) leads to the following, alternative differential equation for the metric function Φ,

$$\frac{\mathrm{d}\Phi(r)}{\mathrm{d}r} = -\frac{1}{\epsilon(r) + P(r)} \frac{\mathrm{d}P(r)}{\mathrm{d}r} \,. \tag{14.93}$$

Since $m(r) = M$ and $P(r) = 0$ for $r \geq R$, one obtains from (14.93) for $\Phi(r)$ outside of the star

$$\mathrm{e}^{2\,\Phi(r)} = 1 - \frac{2\,M}{r} \,, \qquad r \geq R \,. \tag{14.94}$$

The other metric function, $\Lambda(r)$, in the line element (14.14) is known from (14.76) to read

$$\mathrm{e}^{-2\,\Lambda(r)} = 1 - \frac{2\,m(r)}{r} \,, \qquad r \geq 0 \,, \tag{14.95}$$

inside and outside of the star.

14.3 Stability against radial oscillations

Stellar bodies that are in hydrostatic equilibrium are not automatically stable against oscillations about their equilibrium configurations or other – more complex – types of vibration, such as torsional or octupole eigenmodes. The so-called *radial* vibrations [517, 518, 519] have been studied most in the literature, while, in contrast to this, the quantitative stability analysis of stars against *non-radial* pulsations appears to be less developed [520, 521], the reason being that radial oscillations are associated with vibrations about a given stellar equilibrium configuration that preserve the star's spherical symmetry, while non-radial oscillations deform a star away from spherical symmetry (cf. figures 14.1 and 14.2) which is accompanied by the emission of gravitational waves.

Very recently, a simple but efficient method to adequately analyze the vibrational and seismological properties of compact stars performing non-radial oscillations was developed by Bastrukov *et al* [522, 523, 524, 525]. It is basically an elastodynamic continuum approach that is suitable for the

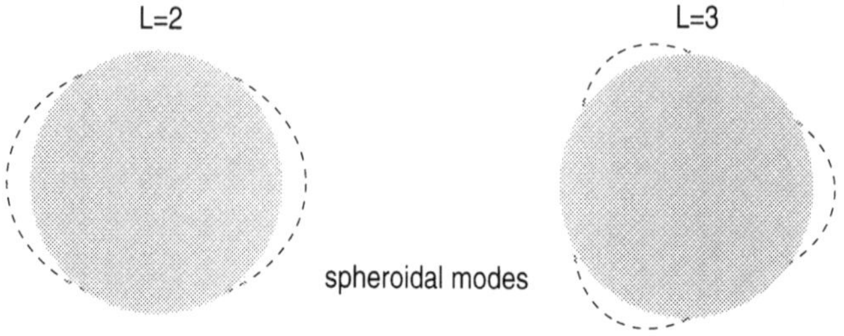

Figure 14.1. Schematic illustration of the quadrupole ($L = 2$) and octupole ($L = 3$) spheroidal, non-radial eigenmodes of a neutron star [525]. (Reprinted courtesy of *J. Phys. G: Nucl. Part. Phys.*)

description of the large-scale motion of highly compressed stellar matter, as it exists in white dwarfs, neutron stars, and stars made up of strange matter.[2] Figures 14.1 and 14.2 illustrate two different non-radial vibrational modes established for neutron stars, which can be classified as spheroidal and torsional eigenmodes. These modes may register themselves in the spectra of stars. Evidence for such motions appears to be provided by the electromagnetic activity of fairly well-studied pulsating white dwarfs, whose electromagnetic variability is plausibly associated with non-radial vibrations [530]. In particular, the hundred-second variability of some white dwarfs could be assigned to possible torsional vibrations, as suggested in [531, 532]. Moreover, non-radial pulsations are thought to be responsible for the drifting sub-pulses and micropulses detected in some radio sources, and of the quasi-periodic variability seen in some X-ray burst sources and in several bright X-ray sources [533]. Another association of non-radial oscillations and the elastodynamic behavior of superdense stellar matter with observed phenomena may be pulsar glitches.

A quantitative analysis of the periods of non-radial modes shows that these are $P_L \sim 10^{-4}$–10^{-3} s for $L = 2, 3, 4$ for both spheroidal as well as torsional motions [525]. This range is between one and two orders of magnitude smaller than the rotational periods of radio pulsars,

[2] The discovery of elastic properties of macroscopic nuclear matter [526] constitutes one of the major landmarks in the current development of laboratory nuclear physics. Modeling a heavy nucleus as a piece of nuclear matter, it was found that the nuclear response bears a strong resemblance to the behavior of an elastic sphere, but not to a liquid drop. The modern macroscopic theory of nuclear collective motions interprets the giant magnetic resonances as non-radial torsional elastic eigenmodes of heavy nuclei, and the giant electric resonances as spheroidal non-radial elastic eigenmodes. The most successful feature of the elastodynamic nuclear model consists in explaining the empirical regular dependence of energy, total probability, and width on mass number [527]. The fingerprints of an elastic-like behavior of macroscopic nuclear matter have also been disclosed in the dynamics of nuclear fusion [528] and fission [529].

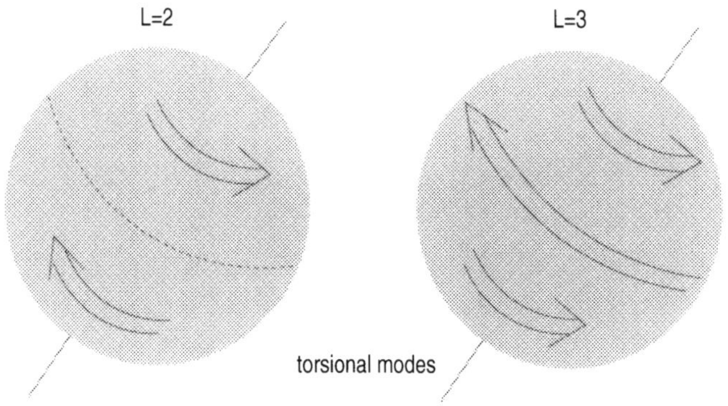

Figure 14.2. Schematic illustration of the quadrupole ($L = 2$) and octupole ($L = 3$) torsional, non-radial eigenmodes of a neutron star. [525]. (Reprinted courtesy of *J. Phys. G: Nucl. Part. Phys.*)

which lie in the range between 1.6 ms and about 5 s. Hence non-radial gravitational oscillations, as well as the radial ones which have been fairly well studied over the years, have nothing to do with the pulsar phenomenon. This may be quite different for the radio emission spectrum of a wide class of pulsars known as complex or c-pulsars [12]. Such pulsars exhibit characteristic substructures within the pulse window, the physical nature of which however is not well understood. In particular, along with the well-recognized interpulse, located approximately at the center between two main pulses, there are clearly distinguishable sub-pulses with the length of period of the order of 10^{-4} to 10^{-3} seconds. Further inspection of the sub-pulse window reveals weak peaks of micropulses with periods less than 10^{-4} seconds [12, 224, 534]. The range of pulsation periods of non-radial oscillations, $P_L \sim 10^{-4}$ to 10^{-3} seconds, as cited above, falls right into this latter window. This supports the early hypothesis that the weak variability of neutron stars, observable as fine details in the sub-pulse region of their spectra, can be assigned to their gravitational vibrations [519].

After these remarks on non-radial oscillations, we now turn to the quantitative description of radial stellar oscillations. The equations which govern these modes were derived, in the framework of Einstein's theory, by Chandrasekhar [535] and Bardeen *et al* [536]. The latter authors provided a very useful catalogue of methods for studying the normal modes of radial oscillations. Therein the adiabatic motion of a star in its nth normal eigenmode ($n = 0$ is the fundamental mode) is expressed in terms of an amplitude $\xi_n(r)$ given by

$$\delta r(r,t) = \mathrm{e}^{2\,\Phi(r)}\,\xi_n(r)\,\mathrm{e}^{\mathrm{i}\,\omega_n\,t}\,r^{-2}\,, \qquad (14.96)$$

where $\delta r(r,t)$ denotes small Lagrangian perturbations in r.[3] The quantity ω_n is the star's oscillation frequency, which we want to compute. The eigenequation for $\xi_n(r)$, which governs the normal modes, is of Sturm–Liouville type,

$$\frac{\mathrm{d}}{\mathrm{d}r}\left(\Pi(r)\,\frac{\mathrm{d}\xi_n(r)}{\mathrm{d}r}\right) + \left(Q(r) + \omega_n^2\, W(r)\right)\, \xi_n(r) = 0\,, \qquad (14.97)$$

which implies that the eigenvalues ω_n^2 are all real and form an infinite, discrete sequence $\omega_0^2 < \omega_1^2 < \omega_2^2 < \ldots$. Another consequence is that the eigenfunction ξ_n corresponding to ω_n^2 has n nodes in the radial interval $0 < r < R$ (see figure 14.5). Hence, the eigenfunction ξ_0 is free of nodes in this interval. The functions $\Pi(r)$, $Q(r)$, and $W(r)$ are expressed in terms of the equilibrium configurations of the star by

$$\Pi = \mathrm{e}^{2\,\Lambda + 6\,\Phi}\, r^{-2}\, \Gamma\, P\,, \qquad (14.98)$$

$$Q = -\,4\,\mathrm{e}^{2\,\Lambda + 6\,\Phi}\, r^{-3}\, \frac{\mathrm{d}P}{\mathrm{d}r} - 8\,\pi\, \mathrm{e}^{6\,(\Lambda + 2\,\Phi)}\, r^{-2}\, P\,(\epsilon + P)$$
$$+\, \mathrm{e}^{2\,\Lambda + 6\,\Phi}\, r^{-2}\, (\epsilon + P)^{-1} \left(\frac{\mathrm{d}P}{\mathrm{d}r}\right)^2 , \qquad (14.99)$$

$$W = \mathrm{e}^{6\,\Lambda + 2\,\Phi}\, r^{-2}\, (\epsilon + P)\,. \qquad (14.100)$$

The quantities ϵ and P in the above equations denote, as usual, the energy density (total mass energy density) and the pressure of the stellar equilibrium configuration as measured by a local observer. The pressure gradient, $\mathrm{d}P/\mathrm{d}r$, is obtained from the Tolman–Oppenheimer–Volkoff equation (14.82). The symbol Γ denotes the adiabatic index at constant entropy, s, given by

$$\Gamma = \frac{\partial \ln P(\rho, s)}{\partial \ln \rho} = \frac{(\epsilon + P)}{P}\, \frac{\partial P(\epsilon, s)}{\partial \epsilon}\,, \qquad (14.101)$$

which varies throughout the star's interior. The boundary conditions for equation (14.97) are

$$\xi_n \sim r^3 \qquad \text{at star's origin,} \quad r = 0\,, \qquad (14.102)$$

$$\frac{\mathrm{d}\xi_n}{\mathrm{d}r} = 0 \qquad \text{at star's surface,} \quad r = R\,. \qquad (14.103)$$

[3] The description of fluid perturbations is divided into two categories [224]. The first is a 'macroscopic' point of view, where one considers changes in fluid variables at a particular point in space, i.e. $\delta Q \equiv Q(x^\mu) - Q_0(x^\mu)$ where Q is any attribute of the perturbed fluid and Q_0 is the attribute for the unperturbed flow. Such perturbations, which compare values of Q at the same point of spacetime, are referred to as Eulerian changes. In the 'microscopic' approach, one defines a Lagrangian displacement $\boldsymbol{\xi}(x^\mu)$ which connects fluid elements in the unperturbed state to corresponding elements in the perturbed state. The Lagrangian change δQ in an attribute Q is defined as $\Delta Q \equiv Q[t, \boldsymbol{x} + \boldsymbol{\xi}(t, \boldsymbol{x})] - Q_0(t, \boldsymbol{x})$, that is, the fluid element at $\boldsymbol{x}$ is displaced to $\boldsymbol{x} + \boldsymbol{\xi}$, and one compares Q at the same fluid element.

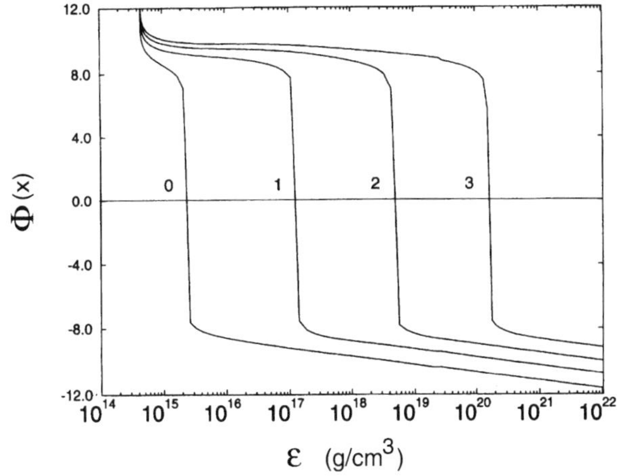

Figure 14.3. Pulsation frequencies of the lowest four ($n = 0, 1, 2, 3$) normal radial modes of quark matter stars as a function of central star density [30]. For convenience, on the vertical axis the quantity $\Phi(x) \equiv \text{sign}(x) \log_{10}(1 + |x|)$ with $x \equiv (\omega_n/\text{s}^{-1})^2$ is plotted instead of ω_n^2 itself. (Reprinted courtesy of *Phys. Rev.*)

Solving equation (14.97) subject to the boundary conditions (14.102) and (14.103) leads to the ordered frequency spectrum $\omega_n^2 < \omega_{n+1}^2$ ($n = 0, 1, 2, \ldots$) of the normal radial modes of a given stellar model. If any of these is negative for a particular star, the frequency is imaginary to which there corresponds an exponentially growing amplitude of oscillation. Such a star would be unstable.

As an example of such a calculation, we show in figure 14.3 the four lowest-lying eigenfrequencies of quark stars, whose structure and theoretical description will be discussed in great detail in chapter 18. A comparison with the mass-central density relationship of such objects, displayed in figure 14.4, shows that these equilibrium configurations possess a characteristic mode of vibration of *zero* frequency, that is, $\omega_n^2 = 0$, when and only when the star's mass attains an extremum (turning point associated with an extremum of the mass), in agreement with the theorems of Harrison and Wheeler [69]. The presupposition inherent in the turning-point method is that the stellar model is constructed for a one-parameter equation of state, where one makes the approximation that the adiabatic index Γ, defined in (14.101), governing the pulsation is the polytropic index of the equilibrium star [537],

$$\Gamma = \frac{\mathrm{d}P/\mathrm{d}r}{P} \frac{\epsilon + P}{\mathrm{d}\epsilon/\mathrm{d}r} . \tag{14.104}$$

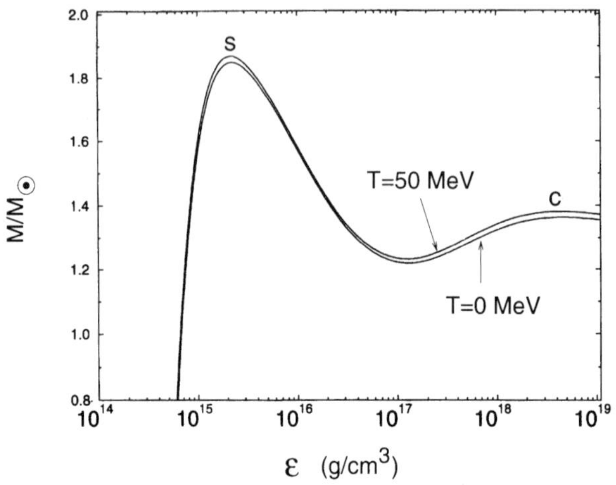

Figure 14.4. Gravitational mass of quark stars (in units of solar mass) at zero and finite ($T = 50$ MeV) temperature versus central star density [30]. The two mass peaks labeled S and C denote the heaviest strange- and charm-quark stars, respectively. (Reprinted courtesy of *Phys. Rev.*)

For dynamical oscillations of neutron stars the adiabatic index does not coincide with the polytropic index of the star, and the turning point method locates a *secular* instability, one whose growth time is long compared with the typical dynamical time of stellar oscillation. The turning-point instability proceeds on a time scale slow enough to accommodate the nuclear reactions and energy transfer that accompany the change to a nearby equilibrium. Finally, we note that when the star's mass has a maximum, the total baryon number A turns over as well, because of the relation $\mathrm{d}M = \mu \mathrm{d}A$ [537, 538].

What is not automatically known from the theorems of Harrison and Wheeler, however, is *which* mode is changing stability. Of course it must be the lowest-lying one which was previously stable, that is, for which $0 < \omega_n^2$. We find that it is the $n = 0$ mode which becomes zero first at a density which corresponds to the maximum-mass quark star, labeled S in figure 14.4. Since ω_0^2 remains negative at all densities larger than this one, it follows that no quark matter stars can exist in nature that are more compact than the hypothetical strange stars, whose central densities are at most several 10^{15} g cm^{-3}, two orders of magnitude smaller than the density at which charm quarks become populated in β-stable quark matter. This rules out the possible existence of *charm* stars. (We shall come back to this issue in chapter 18.) In fact, as one sees from figure 14.3, going to higher and higher central stellar densities leads to the successive excitation of more and more unstable modes, that is, modes for which $\omega_n^2 < 0$ with $n = 1, 2, 3, \ldots$. Thus no stars more compact than strange stars fulfill the condition $\omega_n^2 > 0$ for *all* $n \geq 0$, which is necessary for stability.

This situation is analogous to that of hydrostatic equilibrium configurations in the neutron star sequence with central densities above that of the maximum-mass neutron star. The diagram of M versus R of such stellar objects – from planets to white dwarfs to neutron stars – was shown in figure 2.2. At large radii (low density), all eigenmodes are stable. The first critical point that is reached is at point c, the maximum white dwarf mass. At this point, the lowest-lying eigenmode, $n = 0$, is changing stability, i.e. ω_0^2 becomes negative, while all other eigenfrequencies remain positive. ω_0^2 stays negative until the radius has shrunk to point b, where ω_0^2 becomes positive again. Hence all stars between points c and b are unstable against oscillation, while those between b and the next inflection point of mass at a are stable again. The latter range constitutes the sequence of stable neutron stars. Point b corresponds to the minimum neutron star mass, and a to the maximum neutron star mass. The critical point c is analogous to a and ω_0^2 becomes negative again, remaining negative no matter how big the central stellar density gets. In fact, increasing ϵ_c beyond its value at a leads to the successive excitation of higher and higher eigenmodes. There is never a de-excitation again as occurs at point b, where unstable 'white dwarfs' become stable neutron stars. This renders *all* stars more compact than neutron stars (or their hypothetical strange counterparts discussed above) unstable against radial oscillations, which therefore collapse to black holes.

The instability of the charm stars is already indicated by the small values of the adiabatic index Γ, defined in (14.101), of an ultra-relativistic quark gas [29]. By means of (18.17) and figures 18.6 and 14.4 it is seen that $\Gamma \approx 4/3$ at charm-star densities. Stability of a stellar model with respect to small radial perturbations in the *post-Newtonian approximation* requires that $\Gamma > 4/3 + 2M\kappa/R$, where $\kappa \sim 1$ [224]. For typical masses and radii of charm stars of $\sim 1.3\,M_\odot$ and ~ 8 km (see figures 14.4 and 18.11), one finds $2M\kappa/R \sim 0.5$, leading to $\Gamma \gtrsim 5/3$ for such stars. Therefore, the less deep analysis of the adiabatic index, performed in the post-Newtonian approximation, indicates that the family of charm stars may be unstable against radial oscillations. Of course, from this simplified analysis alone one cannot definitely conclude that charm stars are unstable.

Figure 14.5 exhibits the oscillation amplitudes of the first few vibrational modes of a strange-quark star with mass $M \sim 0.5\,M_\odot$. One sees that the number of nodes associated with an oscillation is equal to its order, n, as determined by the mathematical structure of the eigenvalue equation. Specifically, the $n = 0$ mode of oscillation is free of nodes. The corresponding periods of radial oscillation, τ_n $(\equiv 2\pi/\omega_n)$, whose values are listed in the figure caption, lie in the millisecond range, which is consistent

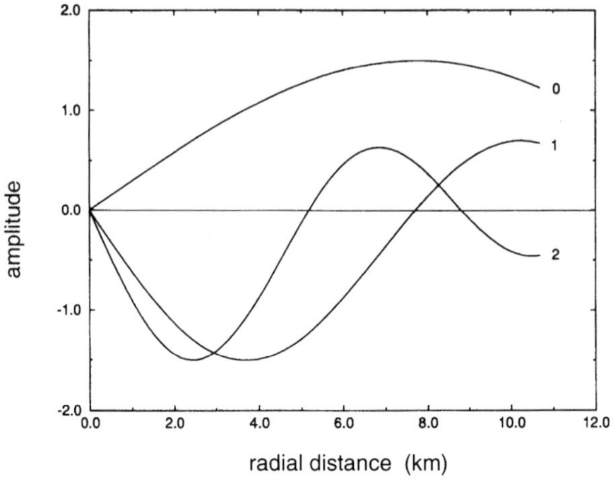

Figure 14.5. Amplitudes of the three lowest-lying eigenmodes of oscillation ($n = 0, 1, 2$) of a strange star possessing a representative mass of $M \sim 1.5\, M_{\odot}$ [30]. The associated periods of radial oscillation are $\tau_0 = 0.334$ ms, $\tau_1 = 0.117$ ms, and $\tau_2 = 0.075$ ms. (Reprinted courtesy of *Phys. Rev.*)

Table 14.1. Periods of radial oscillation, τ_n, of a few selected strange-star models.

$M/M_{\odot}$	τ_0 (ms)	τ_1 (ms)	τ_2 (ms)
10^{-3}	0.0123	0.006165	0.00411
10^{-2}	0.0270	0.0134	0.000895
0.1	0.0614	0.0301	0.0200
0.5	0.125	0.0581	0.0382
1.0	0.197	0.084	0.055
1.4	0.279	0.106	0.068
1.6	0.353	0.121	0.0769
1.85	∞	0.160	0.0974

with the findings in [539]. Table 14.1 shows that the lighter the strange star, the smaller the oscillation periods. Indeed, in the limit of vanishing star masses, which are obtained for $\epsilon \to 4B$, the periods of all modes of strange stars go to zero, as shown in [540]. This behavior arises because $\Gamma \to \infty$ when $\epsilon \to 4B$, that is, $p \to 0$, when the central density of the strange star tends toward its smallest possible value [29]. As known from figure 14.3, the frequency of the fundamental mode, ω_0, of the most massive

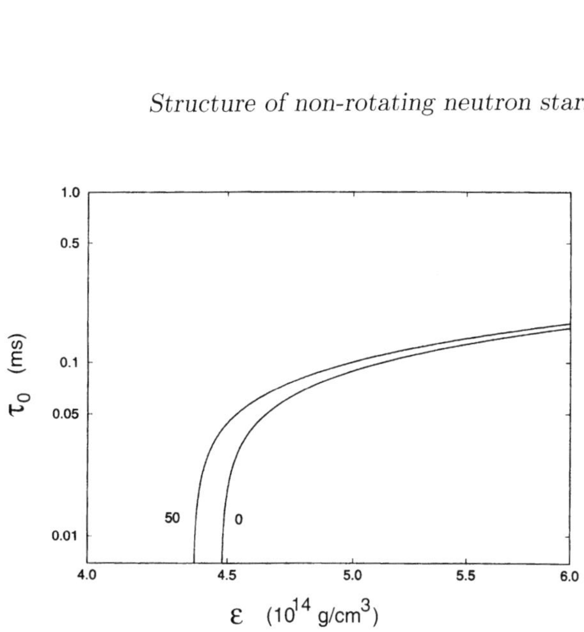

Figure 14.6. Lowest-lying periods of radial oscillation, τ_0, of strange stars versus central star density, for star temperatures $T = 0$ and 50 MeV [30]. (Reprinted courtesy of *Phys. Rev.*)

strange star (labeled S in figure 14.4) is zero. Therefore $\tau_0 = \infty$ for that star model. The higher acoustical modes of vibration of the maximum-mass star are non-zero. They pass through zero at higher densities (figure 14.3).

The impact of finite temperatures on the periods of the fundamental radial oscillation, τ_0, of light strange stars is illustrated in figure 14.6. It is significant only for stars with masses $M \lesssim 1\, M_\odot$ (cf. figure 14.4), whose properties are most sensitive against variations of temperature. It is interesting to compare these periods with those of neutron stars, which reveals that the oscillation periods of neutron stars, constructed for a few selected nuclear equations of state [540], attain a minimum value of $\tau_0 \sim (0.3\text{–}0.4)$ ms at intermediate star masses. This is different for strange stars, due to their different generic mass–radius relationship, for which τ_0 is the smaller the lighter the star (cf. table 14.1). Both types of star with masses $M \gtrsim 1\, M_\odot$ seem to possess periods of oscillation of the same magnitude.

14.4 Structure of non-rotating neutron stars

Having derived in the first part of this chapter the structure equations of non-rotating compact stars, we can now proceed to comparing the properties of actual neutron star models – such as their gravitational masses, radii, redshifts, and moments of inertia, as summarized in chapter 3 – that result when different equations of state are employed in the TOV

Table 14.2. Properties of non-rotating neutron stars of mass $M = 1.4\,M_{\odot}$, calculated for several relativistic equations of state. The entries are: central energy density ϵ_c, gravitational mass M, radius R, moment of inertia I, star's baryon number A, injection energy β, gravitational redshift z. The influence of rotation on these properties is shown for rotational periods of $P = 1.6$ ms and $P = P_{\rm K}$ (mass shedding) in tables 17.5 and 17.6, respectively.

Quantity	HV	HFV	$\Lambda^{\rm RBHF}_{\rm BroB}$ +HFV	$\rm G^{K240}_{M78}$	$\rm G^{K240}_{M78}$(NP)
ϵ_c (MeV fm^{-3})	363.78	478.09	497.61	596.65	428.56
$M/M_{\odot}$	1.40	1.40	1.40	1.40	1.40
R (km)	14.059	12.384	11.557	12.613	13.025
$\log_{10} I$ (g cm^2)	45.2624	45.1849	45.1451	45.1832	45.2187
$\log_{10} A$	57.2493	57.2743	57.2796	57.2639	57.2637
β	0.706	0.666	0.642	0.672	0.683
z	0.1903	0.2253	0.2479	0.2197	0.2104

equation (14.82). We restrict ourselves to the properties of isotropic neutron star models. Properties of non-isotropic models were studied by Hillebrandt and Steinmetz [541]. As input equations of state we shall be using the broad collection of modern equations of state of neutron star matter introduced in table 12.4.

We begin with figure 14.7 which exhibit the gravitational mass of non-rotating neutron stars, M, in units of the solar mass, as a function of central energy density, ϵ_c. Each stellar sequence is computed for a particular model for the equation of state. While it is evident that each model fulfills the minimal constraint of accommodating a neutron star as massive as pulsar PSR 1913+16, whose mass is given by $(1.444 \pm 0.003)M_{\odot}$ [232], the properties predicted for individual neutron stars of a given mass depend sensitively on the equation of state. This is shown in table 14.2 for neutron stars with a fixed mass $M = 1.40\,M_{\odot}$ but constructed for several different models for the equation of state. The sequences in figure 14.7 are shown up to central densities that are slightly larger than those of the heaviest neutron star (indicated by a tick mark) of each sequence. As we know from section 14.3, it is this star which terminates the sequence of stable neutron stars. The maximum neutron star mass depends quite sensitively on the behavior of the equation of state in the high-density domain. An asymptotic stiffening of the equation of state, as for HFV for instance, generally increases the maximum mass. At the same time, such stars are not necessarily much denser than their less massive counterparts obtained for softer equations of state (e.g. $\rm G^{\pi}_{300}$). The mass increase originates from the fact that a stiffer equation of state provides more pressure at a given density

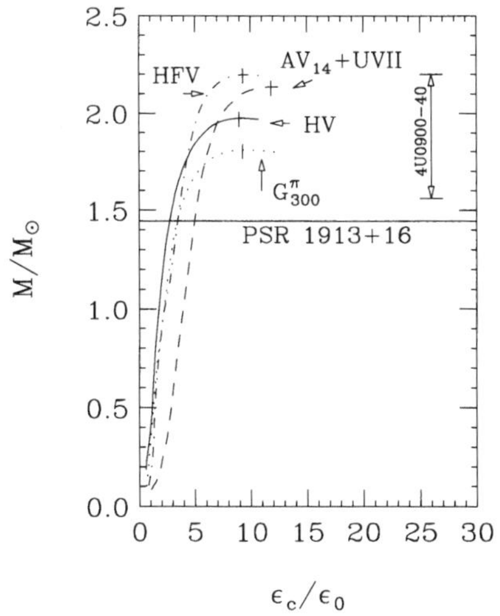

Figure 14.7. Gravitational neutron star mass, $M/M_{\odot}$, as a function of central density, ϵ_c, for a sample of selected EOSs listed in table 12.4 [187]. (Reprinted courtesy of Institute of Physics Publishing.)

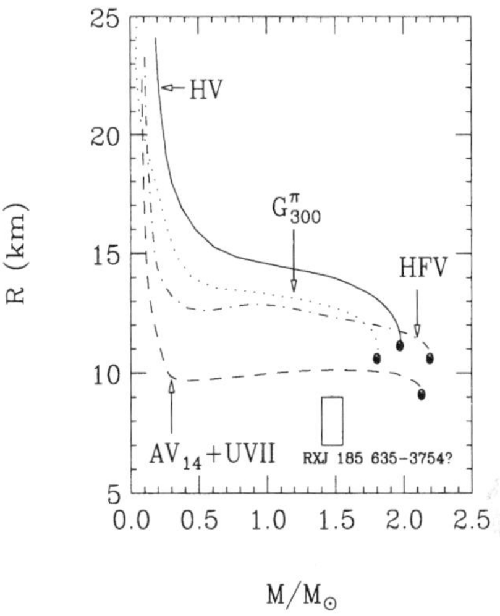

Figure 14.8. Neutron star radius, R, as a function of gravitational mass, calculated for a sample of selected EOSs listed in table 12.4.

than a soft version. The matter inside of a neutron star constructed for such an equation of state is thus more repelling, so that the star can carry more mass before instability against gravitational collapse occurs. The second feature, concerning the behavior of the central density ϵ_c in limiting-mass stars, can easily be understand qualitatively too. Just consider two heavy stellar models, one constructed for a soft equation of state, the other for a stiff one. Independent of whether the equation of state is soft or stiff, the pressure P_c at the centers of these stars must be sufficiently large in order to counterbalance the gravitational pull of the star. For the stiffer equation of state, this particular P_c value is reached by definition at a somewhat *lower* density ϵ_c than is the case for the softer equation of state. This can be verified quantitatively from figures 12.10 through 12.13. Hence the limiting-mass star constructed for the stiffer equation of state has the lower central density.

The interplay between mass, central density, and radius is more delicate at intermediate nuclear densities, where it is sometimes argued that the stiffening of the equation of state above nuclear matter density makes neutron stars of a given mass larger in size and lower in density, which also increases the crust of the star in mass and volume. This however appears to be a misconception, which can be easily refuted by reference to figure 14.8.

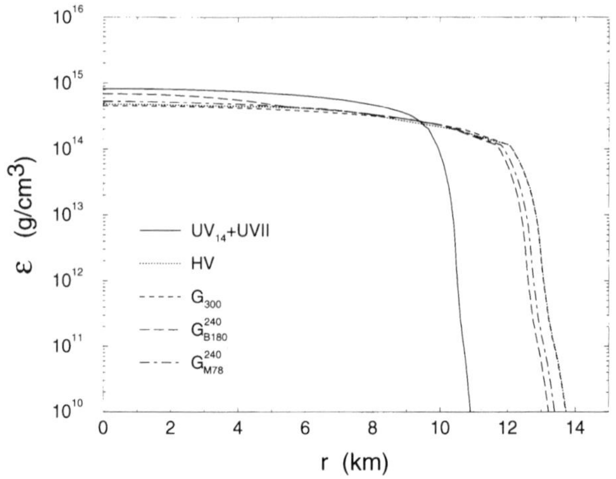

Figure 14.9. Mass density profiles of neutron stars of mass $1.0\,M_{\odot}$, for a sample of selected EOSs [543]. (Reprinted courtesy of *Nucl. Phys.*)

The quantity that in fact determines the radius, and the crustal extent, of 'normal' neutron stars in the mass range of $\sim$1–1.5 $M_{\odot}$ is the density dependence of the nuclear symmetry energy, $a_{\rm sym}(\rho)$, not how stiff the equation of state is at supernuclear density. The arguments for this are made in a forthcoming paper by An, Prakash, Walter, and Lattimer [542]. The notion 'normal' refers here to stars that are not inordinately soft, such as those with kaon condensates or strange-quark-matter stars.

Figures 14.9 through 14.11 are examples of how changes in the central mass density of neutron stars, originating from using different models for the equation of state of neutron star matter, modify their radii. It is evident that variations in ϵ_c, originating from the uncertainties inherent in the models for the equation of state of superdense matter, can change the radii of a stellar model of one and the same mass by several kilometres. More than that, it is seen that the stars' outer regions, at densities less than about $\sim 10^{14}$ g cm^{-3}, are influenced rather strongly too by these uncertainties. Possible phenomena that cause a distinct softening of the equation of state of neutron star matter at supernuclear densities are the condensation of mesons, such as the π^-, or the very actively debated K^- condensate as outlined in section 7.9. The K^-, as pointed out by Li, Lee, and Brown [255, 256], can soften the equation of state to such an extent that the limiting neutron star mass drops by about 20%. The effect of a π^- condensate is taken into account in model G^{π}_{300} of figures 14.7 and 14.8, where, because of the loss of pressure associated with condensation, this model sets the lower bound on the maximal possible neutron star

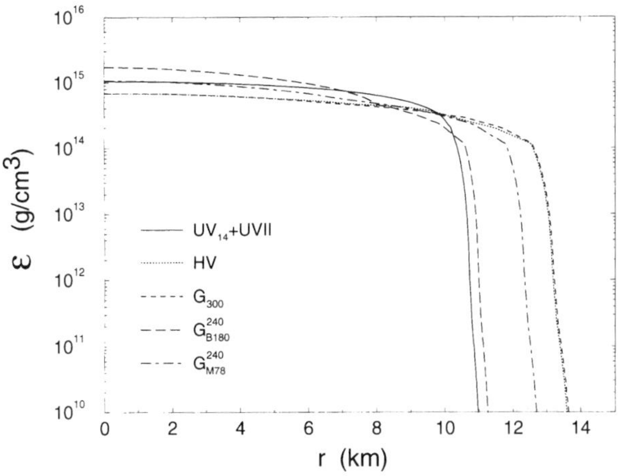

Figure 14.10. Same as figure 14.9, but for a neutron star of mass $1.4\,M_\odot$ [543]. (Reprinted courtesy of *Nucl. Phys.*)

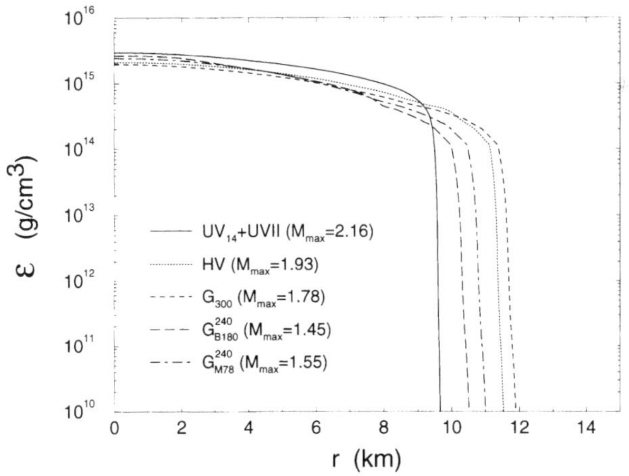

Figure 14.11. Same as figure 14.9, but for the heaviest neutron star obtainable for each EOS [543]. (Reprinted courtesy of *Nucl. Phys.*)

mass. Its radius is somewhat larger than about 10 km. Considerably smaller radii, possibly as small as ~7 km previously derived for the nearby neutron star RXJ 185 635–3754 (see, however, our discussion of the caveat concerning this radius determination from equation (3.6)) can only be obtained if the equation of state softens dramatically over an extended density range a few times beyond the saturation density of nuclear matter.

Brown and collaborators pointed out that the possible condensation of K^- mesons at high densities in neutron stars, which then go into a Bose condensate, could be the cause for such an extreme softening [43, 44, 254]. The final conclusion surely must await the self-consistent determination of the effective K^- mass in dense, chemically equilibrated neutron star matter, which has not been performed yet (see figure 7.22). Another physical cause that was shown in section 12.6 to considerably soften the equation of state, in the range between $\sim \epsilon_0$ up to several times this value, is dynamical two-nucleon correlations. The mass–radius relationship in figure 14.8 computed for AV_{14}+UVII accounts for such correlations. Two generic features are observed. Firstly, the softening of the equation of state caused by the dynamical correlations makes neutron stars constructed from such an equation of state relatively small, with radii ~10 km. This holds over a significant stellar mass range. More than that, light neutron stars, $M \sim 0.3\,M_\odot$ in the present case, can have about the same radii as neutron stars at the mass peaks (indicated by dots). The reason being the relatively weak increase of pressure already below ϵ_0. Secondly, the maximum neutron star masses obtained for such models are typically on the large side, as such equations of state become rather stiff at high densities. While it is true that non-universal coupling schemes, the inclusion of additional microscopic degrees of freedom other than those already contained in AV_{14}+UVII (cf. table 12.7), and causality corrections reduce the maximum neutron star mass [78, 491], the exchange-term contributions *always* stiffen the equation of state which are pushing up the stellar mass toward the gravitational neutron star mass limit. The overall behavior of equations of state which account for dynamical two-nucleon correlations is therefore such that these resolve what may appear as a discrepancy at a first glance. Firstly, they tend to lead to relatively small neutron star models consistent with what may be required by the radius measurements of X-ray bursters, provided their neutron star interpretation is correct of course, since about 20 years ago [250, 251] (see, however, reference [224]). Secondly, their rather stiff behavior at high supernuclear densities guarantees the stability of very heavy neutron stars, of masses around $\sim 2\,M_\odot$, against gravitational collapse, as possibly required from the observation of quasi-periodic oscillations in luminosity in low-mass X-ray binaries [237, 238]. In leaving this issue, we mention that small radii are also of great importance for neutron stars to withstand rapid rotation, in the millisecond range or below. This topic will be investigated in great detail in chapters 16 and 17.

The upper bound on the theoretically predicted neutron star mass is set by the HFV equation of state, because of its extreme stiffness which originates from the exchange contributions of this model (cf. section 12.6). Depending on the stiffness of the equation of state in the supernuclear density regime, as shown in figures 12.10 to 12.13, all other equations of state lead to limiting masses that lie between these two extreme values.

Table 14.3. Properties of non-rotating maximum-mass neutron stars. The quantities listed are: gravitational mass, M; radius, R; central density (in units of density of normal nuclear matter), ϵ_c; moment of inertia, I; gravitational redshift, z. The labeling of the EOSs is explained in table 12.4.

Label	EOS	$M/M_\odot$	R (km)	ϵ_c/ϵ_0	I (10^{45} g cm^2)	z
Models based on relativistic field-theoretical equations of state						
1	G_{300}	1.792	11.12	9.37	1.7586	0.2147
2	HV	1.976	11.34	9.29	2.0398	0.2389
3	G_{B180}^{DCM2}	1.605	10.48	11.36	1.2924	0.2006
4	G_{265}^{DCM2}	1.510	10.14	11.99	1.0991	0.1934
5	G_{300}^{π}	1.808	10.97	9.57	1.7282	0.2214
6	G_{200}^{π}	1.458	10.12	12.10	1.0515	0.1852
7	Λ_{Bonn}^{00}+HV	1.969	10.97	9.29	2.0165	0.2490
8	G_{225}^{DCM1}	1.525	9.84	12.79	1.0807	0.2038
9	G_{B180}^{DCM1}	1.494	9.48	13.93	0.9884	0.2084
10	HFV	2.198	10.70	9.64	2.3963	0.3028
11	Λ_{HEA}^{00}+HFV	2.195	10.63	9.64	2.3670	0.3052
Models based on non-relativistic potential-model equations of state						
12	BJ(I)	1.843	9.77	12.50	1.5521	0.2672
13	UV_{14}+TNI	1.848	9.24	12.79	1.4962	0.2910
14	V_{14}+TNI	1.971	9.29	12.16	1.6819	0.3180
15	UV_{14}+UVII	2.197	9.56	11.54	2.1696	0.3604
16	AV_{14}+UVII	2.135	9.17	12.50	1.9823	0.3680

A compilation of all limiting neutron star masses computed for the collection of equations of state of table 12.4 is contained in table 14.3. With the exception of the relatively soft equations of state G_{200}^{π} and G_{B180}^{DCM1}, all equations of state are able to support non-rotating neutron star models of gravitational masses up to $M \approx 1.5\, M_\odot$. On the other hand, rather massive non-rotating stars of say $M \gtrsim 2\, M_\odot$ can only be obtained for a few equations of state. As already mentioned in section 3.1, indications for the possible existence of very heavy neutron stars, with masses around $2\, M_\odot$, may come from the observation of quasi-periodic oscillations in luminosity in low-mass X-ray binaries [237, 238]. If neutron stars that massive should indeed exist, a significant fraction of equations of state of our collection could be ruled out, and the underlying theories should be seriously revised. Finally, the observation of neutron stars with masses significantly larger than $\sim 2\, M_\odot$ would be in clear contradiction to our compilation of maximum neutron star masses.

It is readily verified that the ratio $2M/R < 8/9$ for all the stellar models listed in table 14.3. This limit, originally proven for stars of uniform density, applies in fact to spheres of arbitrary density profiles, as long as the density decreases monotonically in the radial outward direction [243, 510]. The principle of hydrostatic equilibrium guarantees that this is the case for all the stellar models compiled in table 14.3. Graphical illustrations of their density profiles are shown in figures 14.9 through 14.11. Being of purely relativistic origin, no such M/R limit occurs in Newtonian theory.

The limiting neutron star mass of a theory is of importance for two reasons. The first is that some neutron star masses are known, as discussed in chapter 3, and the largest of these imposes a lower bound on the limiting mass of theoretical models. The current lower bound is about $(1.55 \pm 0.1)\, M_{\odot}$, which, as we have just seen, does not set too stringent a constraint on the nuclear equation of state. The situation could easily change if an accurate future determination of the mass of neutron star 4U 0900–40 should result in a mass value that is close to its present upper bound of $2.2\, M_{\odot}$. In this case most of the equations of state of our collection would be ruled out. The second reason is that the limiting mass can be useful in identifying black hole candidates [544]. That is, if the mass of a compact companion of an optical star is determined to exceed the limiting mass of a neutron star, it must be a black hole. The limiting mass of stable neutron stars in our theory is $\sim 2.2\, M_{\odot}$! Hence, the two non-pulsating X-ray binaries 3U 1700–37 and Cyg X–1 are predicted to be black holes.

The neutron star mass as a function of gravitational redshift, z, whose expression was derived in equation (3.11) (see also equation (15.108)), is displayed in figure 14.12. One sees that maximum-mass stars can have redshifts in the broad range of $0.4 \lesssim z \lesssim 0.8$. Neutron stars with typical masses of $M \sim 1.5\, M_{\odot}$ are predicted to have redshifts in the considerably narrower range $0.2 \leq z \leq 0.32$. The solid rectangle covers masses and redshifts in the range of $1.30 \leq M_s/M_{\odot} \leq 1.65$ and $0.25 \leq z \leq 0.35$. This mass range has been determined from observational data of X-ray burst source MXB 1636$-$536 [251], while the latter is based on the neutron star redshift data base provided by measurements of gamma-ray burst pair annihilation lines in the energy range 300 to 511 keV [46]. These bursts have widely been interpreted as gravitationally redshifted 511 keV $e^{\pm}$ pair annihilation lines from the surfaces of neutron stars. So the just mentioned tentative evidence for a neutron star redshift range of $0.2 \leq z \leq 0.5$ is true only if this interpretation is correct. A particular role is played by the source of the 1979 March 5 γ-ray burst source, which has been identified with SNR N49 by its position. From the interpretation of its emission, which has a peak at $\sim$ 430 keV, as the 511 keV $e^{\pm}$ annihilation line [278, 279] the resulting gravitational redshift has a value of $z = 0.23 \pm 0.05$. The small square in figure 14.12 corresponds to $M = 1.45\, M_{\odot}$ and $z = 0.31$ [251]. This extraordinarily precise mass value for MXB 1636$-$536 hinges

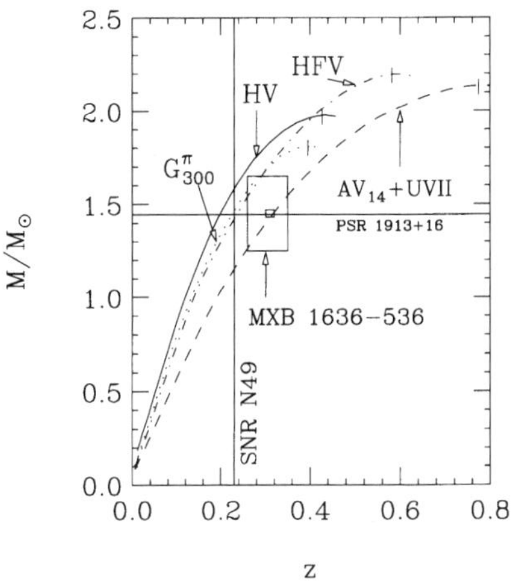

Figure 14.12. Neutron star mass, $M/M_{\odot}$, versus gravitational redshift, z, for a sample of selected EOSs listed in table 12.4 [187]. (Reprinted courtesy of Institute of Physics Publishing.)

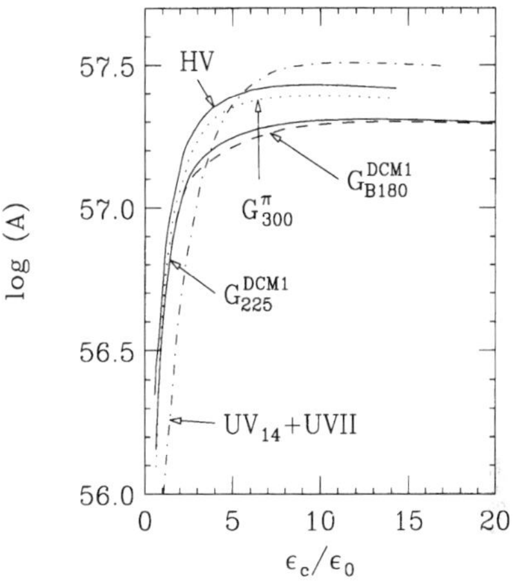

Figure 14.13. Baryon number, A, of neutron stars as a function of central energy density, for a sample of selected EOSs listed in table 12.4.

on certain model assumptions whose validities are however not known. For that reason the latter two values may not yet constrain the behavior of the nuclear equation of state from observed data too seriously, and the fact that the AV_{14}+UVII equation of state is rather consistent with these two values supposedly is purely accidental. With respect to the calculations of the properties of rotating neutron stars, to be presented later in chapter 17, we note that z then not only depends on the star's mass, but also accounts for the rotational Doppler shift effect caused by the star's rotation, as can be seen in (15.103). The rotational Doppler shift can lead to both redshifted as well as blueshifted photon emission lines.

Figure 14.13 shows the total baryon number A, defined in equation (14.90), for several different neutron star models as a function of central density. It is seen that stable neutron stars consist of typically $A \sim 10^{56}$ to several 10^{57} baryons. The baryon number cannot be significantly higher. Otherwise the gravitational attraction among the baryons would dominate over the nuclear repulsion, rendering neutron stars unstable against gravitational collapse. From the baryon number of a given stellar model and its gravitational mass, one readily calculates the star's binding energy E_B defined in (14.92). Figure 14.14 shows the behavior of E_B as a function of gravitational mass. One sees that E_B changes sign for stellar masses in the range 0.1–0.2 $M_{\odot}$, leading to a positive energy per

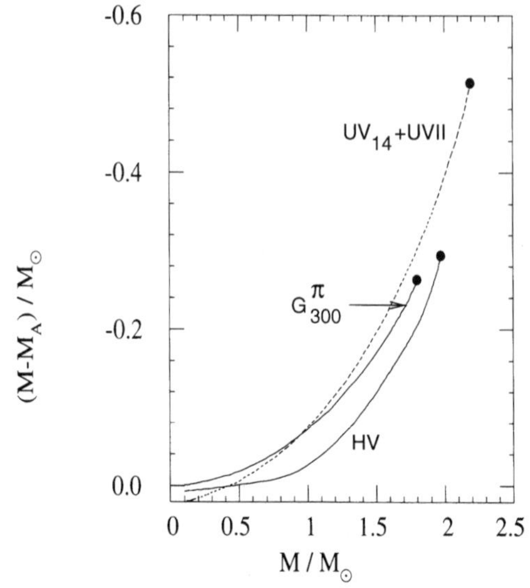

Figure 14.14. Binding energy, $E_B = M - M_{\rm A}$, in sequences of neutron stars constructed for three different equations of state listed in table 12.4. The dots denote the limiting-mass model of each sequence.

baryon of $E_B/A \sim +50$ MeV or greater for very light neutron stars. This value decreases to $E_B/A \sim -100$ MeV for neutron stars of canonical mass, $M \sim 1.4\,M_\odot$ [66]. Observationally, the binding energy is an interesting quantity because it is linked to the neutrino signal from a supernova which derives its energy from the gravitational binding of the neutron star formed in the explosion [66]. The neutrino signal therefore carries information about the binding energy of a neutron star, which could be measurable [67, 545].

In figures 14.15 and 14.16 we show the moment of inertia, I, whose expression will be derived in equation (15.41), as a function of neutron star mass and central density, respectively. From classical Newtonian mechanics one would expect that $I \propto R^2\,M$. Inspection of figure 14.16 in combination with figure 14.8 reveals that this functional dependence carries over to the general relativistic case [57]. Nevertheless, the general relativistic frame dragging effect and the curvature of spacetime modify the quantitative behavior of I significantly [148]. Since both R and M are the smaller the softer the nuclear equation of state, the moment of inertia decreases with increasing softness of the equation of state. This is specifically pronounced for I computed for the equation of state $\mathrm{G}^{\mathrm{DCM1}}_{\mathrm{B180}}$, whose softening originates from the possible transition of confined hadronic matter into quark matter.

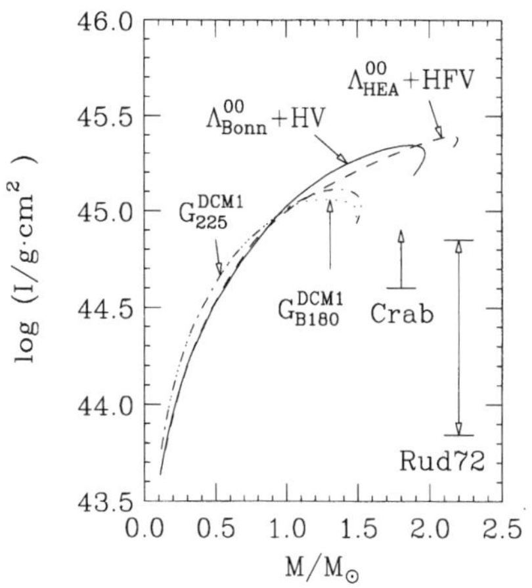

Figure 14.15. Moment of inertia, I, of neutron stars versus gravitational mass, for a sample of selected EOSs listed in table 12.4 [187]. (Reprinted courtesy of Institute of Physics Publishing.)

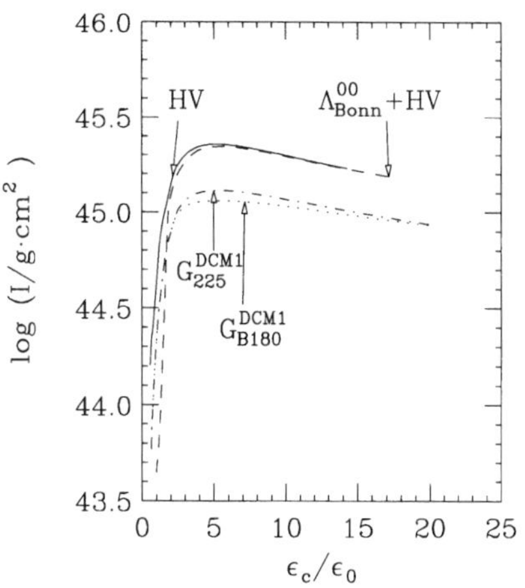

Figure 14.16. Moment of inertia, I, of neutron stars as a function of central energy density, for a sample of EOSs listed in table 12.4.

This model therefore sets the lower bound on I in figures 14.15 and 14.16. The arrow labeled 'Crab' in figure 14.15 places the lower bound on the moment of inertia for the pulsar in the Crab nebula, as calculated by Trimble and Rees [275]. This estimate is based on the nebula's luminosity. Consideration of additional observational data led Börner and Cohen to the suggestion that the lower bound on I for the Crab pulsar may approach $I_{\rm Crab} \simeq 10^{45}$ g cm^2 [276]. Estimates for the upper and lower bounds on the moment of inertia of the Crab pulsar derived from the pulsar's energy loss rate, labeled Rud72 in figure 14.15, were performed by Ruderman [55]. (The arrows refer only to the value of $I_{\rm Crab}$ and not to its mass, which is not known.) Since the Crab pulsar is rotating at a period of 33 Hz, one readily estimates for a moment of inertia of that magnitude that the pulsar's rotational energy is on the order of 10^{55} MeV. This value increases by two more orders of magnitude for the two fastest pulsars known to date, PSRs 1937+21 and 1957+20. Pulsars consume part of this immense store of energy for the emission of magnetic dipole radiation and a wind of electron–positron pairs [546].

A value for the moment of inertia of $I \simeq 10^{45}$ g cm^2 appears to be a problem for stellar models based on very soft equations of state, since they clearly tend to yield $I < 10^{45}$ g cm^2. Figure 14.15 also shows that, independent of whether the equation of state is stiff or soft, the lighter neutron stars of a stable stellar sequence, particularly those with masses less than $M \lesssim 1\, M_\odot$, cannot have moments of inertia as large as $I \simeq 10^{45}$ g cm^2.

As we know from section 3.6, glitches are sudden relatively small changes in the period of pulsars, which otherwise increase very slowly with time due to the loss of rotational energy through radiation and the emission of an electron–positron wind . They occur in various pulsars at intervals of days to months or years, and in some pulsars are small (Crab), and in others large (Vela) and infrequent ($\Delta\Omega/\Omega \sim 10^{-8}$ and 10^{-6} respectively). In the framework of the standard star quake model for glitches proposed already in the late 1960s [168, 169], the characteristic time between two glitches (or quakes), the so-called interglitch time, is given by

$$t_q = 3.553 \times 10^6 \, \frac{\big((M_\mathrm{P} - M)/M_\odot\big)^2}{R^3 \, I_{45}} \text{ yr} . \tag{14.105}$$

Here R is the radius in kilometres, I_{45} the moment of inertia in units of 10^{45} g cm^2, and M_P and M denote proper and gravitational mass, defined in equations (14.84) and (14.83), respectively [547]. As indicated in (14.105), the interglitch time is given in years. Furthermore we note that in order to arrive at (14.105), an iron crust ($Z = 26$) has been assumed, and for Ω and $\mathrm{d}\Omega/\mathrm{d}t$ the observational data from the Crab pulsar ($T \approx 2260$ yr) and an oblateness change of $\Delta\epsilon \approx 0.9 \times 10^{-9}$ has been taken.

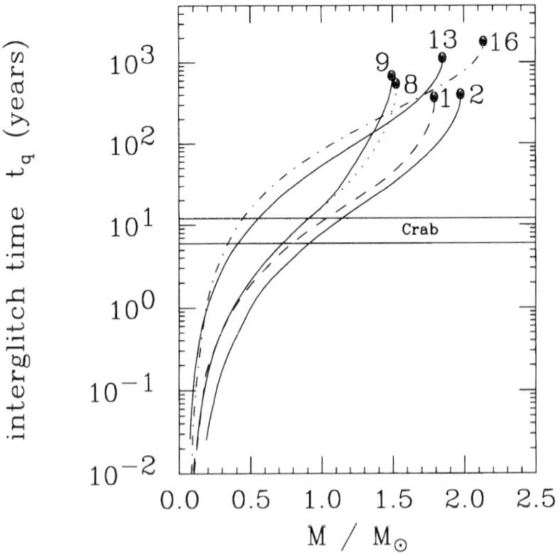

Figure 14.17. Interglitch time, t_q, between two successive pulsar glitches as a function of star mass, M [187]. The labeling of the EOSs is as in table 12.4. (Reprinted courtesy of Institute of Physics Publishing.)

The interglitch time calculated from (14.105) is shown in figure 14.17 for a few selected equations of state. It is striking that t_q depends rather sensitively on M. For example, considering $t_q \sim 10$ yr, as compatible with observational evidence for glitches, we obtain masses in the range between about $0.3\,M_\odot$ and $1.1\,M_\odot$. Conversely, if we assume values of $M \approx 1.4$–$1.5\,M_\odot$ as the most probable Crab pulsar mass, the interglitch time ranges from 30 to 700 years! If the Crab pulsar were a medium massive neutron star of say about $1\,M_\odot$, the interglitch times obtained for the relativistic equations of state are compatible with the observed times. To obtain compatibility with star models constructed for the non-relativistic equations of state, the pulsar's mass must be $M \approx 0.4\,M_\odot$! Of course, the above analysis applies only if the star quake model is correct. Depending to which extent modern developments of the theoretical description of the star quake phenomenon (such as vertex pinning) will modify some of the model's ingredients, our predictions drawn will be subject to modifications too. In any case, one may expect that some of the qualitative features established just above will carry over to a more updated treatment.

Chapter 15

Structure equations of rotating stars

The structure equations of rotating compact stars are considerably more complicated that those of non-rotating compact stars derived in chapter 14. These complications have their cause in the rotational deformation, that is, a flattening at the pole accompanied with a radial blowup in the equatorial direction, which leads to a dependence of the star's metric on the polar coordinate, θ, in addition to the dependence on the radial coordinate, r. Secondly, rotation stabilizes a star against gravitational collapse. It can therefore carry more mass than would be the case if the star were non-rotating. Being more massive, however, means that the geometry of spacetime will change too. This makes the line element of a rotating star depend on the star's rotational frequency. Finally, the general relativistic effect of the dragging of local inertial frames implies the occurrence of an additional non-diagonal term, $g^{t\phi}$, in the metric tensor. This term imposes a self-consistency condition on the stellar structure equations, since the extent to which the local inertial frames are dragged along in the direction of the star's rotation (see figure 15.1) is determined by that star's properties, like its mass and rotational frequency. These three attributes render the construction of rotating stellar models rather cumbersome. So it is not surprising that until the beginning of this decade most of the neutron star calculations reported in the literature were performed for non-rotating, spherical bodies, whose properties are determined by the rather simple TOV equation (14.82), though the discovery of the first millisecond pulsar, PSR 1937+215, by Backer *et al* in 1982, rotating at about 620 Hz, initiated great interest in the properties of extremely rapidly spinning neutron stars.

Table 15.1 gives an overview of representative theoretical studies of the properties of rotating neutron stars reported in the literature. These studies can be divided into three categories: (1) restriction to weak gravitational fields, which is not justified for neutron stars however, (2) restriction to rotational frequencies small in comparison with the Kepler frequency, i.e.

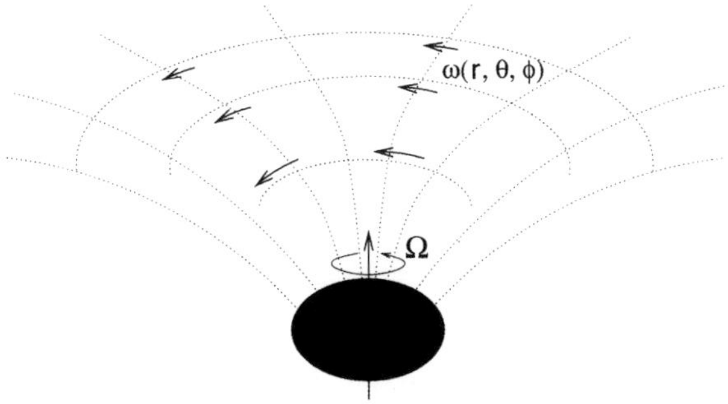

Figure 15.1. Rotating, compact star in general relativity theory. The star, rotating at frequency Ω, becomes rotationally deformed, deforms the geometry of spacetime, and sets the local inertial frames into rotation by a position dependent angular velocity $\omega(r, \theta, \phi)$ calculated in equation (15.87).

$\Omega \ll \Omega_K$ (cf. section 16.1), at which mass shedding from the star sets in, and (3) limited predictive power because of assuming an oversimplified density dependence of pressure. Only a few studies listed in table 15.1 are unrestricted with respect to these three items. All others are restricted, some more severely than others. The slow-rotation investigations, for instance, are based on Hartle's perturbative method, which will be outlined in detail below. As shown elsewhere [62, 66, 109, 112], the notion of 'slow' here is quite misleading, since Hartle's method is capable of predicting not only a variety of important stellar properties of slowly rotating stars (i.e. $\Omega \ll \Omega_K$) very well, but also those of extremely rapidly spinning stars with frequencies close to the mass shedding frequency Ω_K too. Depending on the behavior of superdense nuclear matter, Ω_K may be as large as 1000 to 2000 Hz, which corresponds to rotational Kepler periods of $P_K = 1$ ms to 0.5 ms, respectively [143]. A detailed investigation concerning the applicability of Hartle's method to the construction of rapidly spinning neutron star models will be performed in section 17.2, where a comparison is performed between neutron star properties computed by solving Einstein's equations numerically exactly and computed within Hartle's perturbative method.

After these introductory remarks, we turn now to the derivation of the stellar structure equations of compact, rotationally deformed stars. To be consistent with the notation in the previous chapter, R is reserved for the radius of a non-rotating spherical star. The equatorial and polar radii of rotationally deformed stars are denoted by $R_{\rm eq}$ and $R_{\rm p}$. Most of the mathematical expressions that follow from Einstein's equations for a rotating stellar configuration are too lengthy to be listed in the main body of this text. They can be found in appendix G.

Table 15.1. Survey of theoretical studies of rotating neutron stars.

Gravitational field	Rotational frequency	Density distribution	References
		Restricted applications	
strong	'slow'	arbitrary	[116]
strong	rapid	ad hoc §	[548]
strong	slow	arbitrary	[549]
weak †	rapid	$n = \frac{3}{2}$ polytrope	[550]
strong	rapid	homogeneous ‡	[551, 552, 553]
strong	rapid	polytropic ‖	[554]
strong	'slow'	arbitrary	[555]
strong	'slow'	arbitrary	[62, 109, 111, 112, 148]
		Unrestricted applications	
strong	rapid	arbitrary	[106, 107, 143, 537]
strong	rapid	arbitrary	[108]
strong	rapid	arbitrary	[556, 557]
strong	rapid	arbitrary	[558]
strong	rapid	arbitrary	[559]
strong	rapid	arbitrary	[560, 561]

§ The pressure is taken to be equal to one-third the thermal energy density.
† Parametrized post-Newtonian framework.
‡ Constant energy density, $\epsilon =$ const, inside the star.
‖ The pressure P is related to the energy density in the form $P = K\epsilon^{1+1/n}$, where K is a constant, and n the polytropic index.

Let us begin with the metric in spherical coordinates of a stationary rotating, axially symmetric equilibrium configuration, which can be written in the form [106, 115, 116, 562]

$$ds^2 = -\,e^{2\,\nu}\,(dt)^2 + e^{2\,\psi}\,\big(d\phi - \omega\,dt\big)^2 + e^{2\,\mu}\,(d\theta)^2 + e^{2\,\lambda}\,(dr)^2\,, \quad (15.1)$$

where each metric function, i.e. ν, ψ, μ, and λ, depends on the radial coordinate r, polar angle θ, and implicitly on the star's angular velocity Ω. The quantity ω denotes the angular velocity of the local inertial frames, dragged along in the direction of the star's rotation. This frequency too depends on the radial coordinate as well as on the polar angle. It is also an implicit function of the star's frequency, Ω, since the rotation rates of the local inertial frames intimately depend on the star's mass concentration inside and outside of it, which varies with Ω. Of particular interest is the *relative* angular velocity $\bar{\omega}$ defined as

$$\bar{\omega}(r, \theta, \Omega) \equiv \Omega - \omega(r, \theta, \Omega), \tag{15.2}$$

which is the angular velocity of the fluid, Ω, relative to the angular velocity of the local inertial frame, $\omega(r)$. It is this frequency $\bar{\omega}(r)$ that is of relevance when discussing the rotational flow of the fluid inside the star, since the magnitude of the centrifugal force acting on a fluid element is governed – in general relativity as well as in Newtonian gravitational theory – by the rate of rotation of the fluid element relative to a local inertial frame [115]. In contrast to Newtonian theory, however, the inertial frames inside a general relativistic fluid are not at rest with respect to the distant stars. Rather, the local inertial frames are dragged along by the rotating fluid. We shall come across the frame dragging effect in chapter 17 when discussing the properties of rotating stars.

The metric functions in (15.1) are independent of both time and azimuthal angle ϕ, reflecting *stationary rotation* and *axial symmetry* about the axis of rotation, respectively. Restriction to axially symmetric objects is required if the star is assumed not to radiate away rotational energy in the form of gravitational radiation, as is the case here. Otherwise, that is, if there are time-dependent moments of the mass distribution, the star could not remain in equilibrium over time. We also assume *reflection symmetry* of the configuration about a plane perpendicular to the axis of rotation. From one's experience with the Newtonian theory of figures of equilibrium it is plausible that both of these assumptions – axial and reflection symmetry – are really consequences of slowly rotating bodies and not restrictions at all. The situation is different for rapidly spinning bodies, which are subject to non-radial oscillations superimposed on their rotational motion. We shall come back to this issue in chapter 16.

The star's frequency Ω is assumed to be constant throughout the stellar fluid, that is, the stellar configurations are assumed to be *uniformly rotating* (rigid body rotation). This assumption appears to be well justified in hot as well as cold neutron stars for the following reasons. In hot, newly formed neutron stars, for instance, it has been shown that Ekman pumping (turbulent mixing) [563] damps out any amount of differential rotation initially present within at most a few days. This is supported by the viscosity of neutron star matter, however on a longer time scale of about $\tau \approx 10^7\, T_9^2$ s [564]. Since the interior temperature is expected to drop to $T_9 \approx 1$ within a few years after its birth [543, 565], a neutron star is expected to be rigidly rotating after this time. In cold neutron stars, superfluid eddy viscosity, viscosity arising from electron–vortex scattering, or degenerate particle viscosity will mix vortices by means of which rigid-body rotation is guaranteed for such objects too. Besides these arguments, it is a comforting feature that every configuration which minimizes the total mass energy, e.g. the stable configurations, must rotate uniformly

[562]. This will become important below, since u^ϕ and u^t are related for uniformly rotating bodies by the simple relation $u^\phi = \Omega u^t$, which simplifies the mathematical analysis.

From the point of view of numerical calculations, the form of the line element (15.1) has the advantage of having only one coordinate with an infinite range. In the absence of rotation, it also reduces directly to the Schwarzschild line element (14.14). Equation (15.1) thus appears as a natural starting point when studying rotation as a perturbation on a spherically symmetric problem. The mathematical structure of the metric (15.1) is determined by the so-called Killing vectors associated with the rotating system [106], but can be derived on more physical grounds too [115, 562]. Without going into too much detail, we point out that for the description of an axially symmetric equilibrium configuration of a rotating fluid, two coordinates are conveniently taken to be the time $t = x^0$ and an angle $\phi = x^3$. The stationary nature of the configuration and its axial symmetry are then expressed by requiring that the metric coefficients be independent of t and ϕ, i.e. $g_{\mu\nu}{=}g_{\mu\nu}(x^1, x^2)$. The metric remains a function of x^1 and x^2 *alone* under coordinate transformations of the form $t = t' + f_t(x^1, x^2)$ and $\phi = \phi' + f_\phi(x^1, x^2)$. By an appropriate choice of f_t and f_ϕ, one can always find a coordinate system in which $g_{12} = g_{13} = 0$, as pointed out by Hartle and Sharp [562]. In addition to the independence of the metric on the angle ϕ, the assumption of axial symmetry also entails the velocity distribution has only an angular component, so that the fluid four-velocity may be written as $u^\mu = (u^0, 0, 0, u^\phi)$. This means that the source terms $T^{\mu\nu}$ of the gravitational field is left unchanged by the simultaneous inversion of time and azimuthal angle, i.e. by the transformation $t \to -t$ and $\phi \to -\phi$. This symmetry must be reflected in the solution of the field equations, and consequently we must require the line element to be left unchanged under this inversion. This requires one to set $g_{01} = g_{02} = 0$. Combining the arguments that have led us to this result and to $g_{12} = g_{13} = 0$ from above, one sees that the line element must have the general form [562],

$$\begin{aligned} \mathrm{d}s^2 = {} & g_{00}\,(\mathrm{d}t)^2 + 2\,g_{03}\,\mathrm{d}\phi\,\mathrm{d}t + g_{33}\,(\mathrm{d}\phi)^2 + g_{11}\,(\mathrm{d}x^1)^2 \\ & + 2\,g_{23}\,\mathrm{d}x^2\,\mathrm{d}x^3 + g_{33}\,(\mathrm{d}x^3)^2\,. \end{aligned} \qquad (15.3)$$

This form of the line element remains unchanged under coordinate transformations of the form

$$x^1 = x^1(x^{1'}, x^{2'})\,, \qquad x^2 = x^2(x^{1'}, x^{2'})\,. \qquad (15.4)$$

By a coordinate transformation of the form (15.4), one can always put the quadratic form $g_{AB}(x^1, x^2)\mathrm{d}x^A\mathrm{d}x^B$ $(A, B = 1, 2)$ into the form $f(x^1, x^2)[(\mathrm{d}x^1)^2 + (\mathrm{d}x^2)^2]$ [566]. Writing $x^1 = r$ and $x^2 = \theta$, one can thus put the line element in the form given in (15.1).

From the line element (15.1) we can infer the covariant components of the associated metric tensor. These read

$$g_{tt} = -\, e^{2\,\nu(r,\theta)} + e^{2\,\psi(r,\theta)}\,\omega^2(r,\theta)\,, \quad g_{t\phi} = -\, e^{2\,\psi(r,\theta)}\,\omega(r,\theta)\,,$$

$$g_{rr} = e^{2\,\lambda(r,\theta)}\,, \quad g_{\theta\theta} = e^{2\,\mu(r,\theta)}\,, \quad g_{\phi\phi} = e^{2\,\psi(r,\theta)}\,. \tag{15.5}$$

The non-rotating counterparts of these expressions, listed in (14.15), are obtained by taking $\omega(r,\theta) \rightarrow 0$. The contravariant components of the metric tensor are given by

$$g^{tt} = -\, e^{-2\,\nu(r,\theta)}\,, \quad g^{t\phi} = -\, e^{-2\,\nu(r,\theta)}\,\omega(r,\theta)\,,$$

$$g^{rr} = e^{-2\,\lambda(r,\theta)}\,, \quad g^{\theta\theta} = e^{-2\,\mu(r,\theta)}\,,$$

$$g^{\phi\phi} = e^{-2\,\psi(r,\theta)} - \omega^2(r,\theta)\, e^{-2\,\nu(r,\theta)}\,. \tag{15.6}$$

The determinant of $g_{\mu\nu}$ follows as

$$g \equiv \det(g_{\mu\nu}) = -\, e^{2\,\lambda(r,\theta)+2\,\mu(r,\theta)+2\,\psi(r,\theta)+2\,\nu(r,\theta)}\,. \tag{15.7}$$

The non-vanishing contravariant components of the fluid's four-velocity, $u^\mu = dx^\mu/d\tau$, read for the metric of (15.1)

$$u^t = \frac{1}{\sqrt{e^{2\,\nu(r,\theta)} - (\Omega - \omega(r,\theta))^2\, e^{2\,\psi(r,\theta)}}}\,,$$

$$u^\phi = \frac{\Omega}{\sqrt{e^{2\,\nu(r,\theta)} - (\Omega - \omega(r,\theta))^2\, e^{2\,\psi(r,\theta)}}}\,,$$

$$u^r = u^\theta = 0\,, \tag{15.8}$$

where we have defined $\Omega \equiv d\phi/dt$. Note that the first two relations of (15.8) are related by $u^\phi = \Omega\, u^t$. The non-vanishing covariant components of the four-velocity are given by

$$u_t = -\,\frac{e^{2\,\nu(r,\theta)} - e^{2\,\psi(r,\theta)}\,\omega^2(r,\theta) + e^{2\,\psi(r,\theta)}\,\omega(r,\theta)\,\Omega}{\sqrt{e^{2\,\nu(r,\theta)} - (\Omega - \omega(r,\theta))^2\, e^{2\,\psi(r,\theta)}}}\,,$$

$$u_\phi = \frac{e^{2\,\psi(r,\theta)}\,(\Omega - \omega(r,\theta))}{\sqrt{e^{2\,\nu(r,\theta)} - (\Omega - \omega(r,\theta))^2\, e^{2\,\psi(r,\theta)}}}\,. \tag{15.9}$$

From equations (15.8) and (15.9) one readily verifies the normalization condition $u^\mu\, u_\mu = -1$. The calculation of the Christoffel symbols (13.25) for the above metric is straightforward but more cumbersome than for the metric of a spherically symmetric star. The individual components of $\Gamma^{\sigma}_{\mu\nu}$ are as follows:

$$\Gamma^{r}_{tt} = -\,e^{-2\lambda(r,\theta)}\Big\{-\Big(\frac{\partial}{\partial r}\nu(r,\theta)\Big)e^{2\nu(r,\theta)} + \Big(\frac{\partial}{\partial r}\psi(r,\theta)\Big)e^{2\psi(r,\theta)}\omega^2(r,\theta) + e^{2\psi(r,\theta)}\omega(r,\theta)\frac{\partial}{\partial r}\omega(r,\theta)\Big\},$$

$$\Gamma^{\theta}_{tt} = -\,e^{-2\mu(r,\theta)}\Big\{-\Big(\frac{\partial}{\partial\theta}\nu(r,\theta)\Big)e^{2\nu(r,\theta)} + \Big(\frac{\partial}{\partial\theta}\psi(r,\theta)\Big)e^{2\psi(r,\theta)}\omega^2(r,\theta) + e^{2\psi(r,\theta)}\omega(r,\theta)\frac{\partial}{\partial\theta}\omega(r,\theta)\Big\},$$

$$\Gamma^{t}_{tr} = -\frac{1}{2}\Big\{-2\Big(\frac{\partial}{\partial r}\nu(r,\theta)\Big)e^{2\nu(r,\theta)} + e^{2\psi(r,\theta)}\omega(r,\theta)\frac{\partial}{\partial r}\omega(r,\theta)\Big\}e^{-2\nu(r,\theta)},$$

$$\Gamma^{\phi}_{tr} = -\frac{1}{2}\Big\{-2\,\omega(r,\theta)\Big(\frac{\partial}{\partial r}\nu(r,\theta)\Big)e^{2\nu(r,\theta)} + e^{2\psi(r,\theta)}\omega^2(r,\theta)\frac{\partial}{\partial r}\omega(r,\theta) + 2\,e^{2\nu(r,\theta)}\Big(\frac{\partial}{\partial r}\psi(r,\theta)\Big)\omega(r,\theta) + e^{2\nu(r,\theta)}\frac{\partial}{\partial r}\omega(r,\theta)\Big\}e^{-2\nu(r,\theta)},$$

$$\Gamma^{t}_{t\theta} = -\frac{1}{2}\Big\{-2\Big(\frac{\partial}{\partial\theta}\nu(r,\theta)\Big)e^{2\nu(r,\theta)} + e^{2\psi(r,\theta)}\omega(r,\theta)\frac{\partial}{\partial\theta}\omega(r,\theta)\Big\}e^{-2\nu(r,\theta)},$$

$$\Gamma^{\phi}_{t\theta} = -\frac{1}{2}\Big\{-2\,\omega(r,\theta)\Big(\frac{\partial}{\partial\theta}\nu(r,\theta)\Big)e^{2\nu(r,\theta)} + e^{2\psi(r,\theta)}\omega^2(r,\theta)\frac{\partial}{\partial\theta}\omega(r,\theta) + 2\,e^{2\nu(r,\theta)}\Big(\frac{\partial}{\partial\theta}\psi(r,\theta)\Big)\omega(r,\theta) + e^{2\nu(r,\theta)}\frac{\partial}{\partial\theta}\omega(r,\theta)\Big\}e^{-2\nu(r,\theta)},$$

$$\Gamma^{r}_{t\phi} = \frac{1}{2}\Big\{2\Big(\frac{\partial}{\partial r}\psi(r,\theta)\Big)\omega(r,\theta) + \frac{\partial}{\partial r}\omega(r,\theta)\Big\}e^{2\psi(r,\theta)-2\lambda(r,\theta)},$$

$$\Gamma^{\theta}_{t\phi} = \frac{1}{2}\Big\{2\Big(\frac{\partial}{\partial\theta}\psi(r,\theta)\Big)\omega(r,\theta) + \frac{\partial}{\partial\theta}\omega(r,\theta)\Big\}e^{2\psi(r,\theta)-2\mu(r,\theta)},$$

$$\Gamma^{r}_{rr} = \frac{\partial}{\partial r}\lambda(r,\theta), \quad \Gamma^{\theta}_{rr} = -\Big(\frac{\partial}{\partial\theta}\lambda(r,\theta)\Big)e^{-2\mu(r,\theta)+2\lambda(r,\theta)},$$

$$\Gamma^{r}_{r\theta} = \frac{\partial}{\partial\theta}\lambda(r,\theta), \quad \Gamma^{\theta}_{r\theta} = \frac{\partial}{\partial r}\mu(r,\theta),$$

$$\Gamma^{t}_{r\phi} = \frac{1}{2}\Big(\frac{\partial}{\partial r}\omega(r,\theta)\Big)e^{2\psi(r,\theta)-2\nu(r,\theta)},$$

$$\Gamma^{\phi}_{r\phi} = \frac{1}{2}\Big\{e^{2\psi(r,\theta)}\omega(r,\theta)\frac{\partial}{\partial r}\omega(r,\theta) + 2\Big(\frac{\partial}{\partial r}\psi(r,\theta)\Big)e^{2\nu(r,\theta)}\Big\}e^{-2\nu(r,\theta)},$$

$$\Gamma^{r}_{\theta\theta} = -\Big(\frac{\partial}{\partial r}\mu(r,\theta)\Big)e^{-2\lambda(r,\theta)+2\mu(r,\theta)}, \quad \Gamma^{\theta}_{\theta\theta} = \frac{\partial}{\partial\theta}\mu(r,\theta),$$

$$\Gamma^{t}_{\theta\phi} = \frac{1}{2}\Big(\frac{\partial}{\partial\theta}\omega(r,\theta)\Big)e^{2\psi(r,\theta)-2\nu(r,\theta)},$$

$$\Gamma^{\phi}_{\theta\phi} = \frac{1}{2}\Big\{e^{2\psi(r,\theta)}\omega(r,\theta)\frac{\partial}{\partial\theta}\omega(r,\theta) + 2\Big(\frac{\partial}{\partial\theta}\psi(r,\theta)\Big)e^{2\nu(r,\theta)}\Big\}e^{-2\nu(r,\theta)},$$

$$\Gamma^{r}_{\phi\phi} = -\Big(\frac{\partial}{\partial r}\psi(r,\theta)\Big)e^{2\psi(r,\theta)-2\lambda(r,\theta)},$$

$$\Gamma^{\theta}_{\phi\phi} = -\Big(\frac{\partial}{\partial\theta}\psi(r,\theta)\Big)\mathrm{e}^{2\,\psi(r,\theta)-2\,\mu(r,\theta)}\,. \tag{15.10}$$

Having the Christoffel symbols at our disposal, the components of the Riemann tensor (13.24) can be computed next. As an example, we give the expression for ${R^t}_{tt\phi}$, which reads

$$\begin{aligned}{R^t}_{tt\phi} = &\frac{1}{4}\Big\{\omega(r,\theta)\Big[\Big(\frac{\partial}{\partial r}\omega(r,\theta)\Big)^2\mathrm{e}^{2\,\mu(r,\theta)+2\,\psi(r,\theta)} + 4\Big(\frac{\partial}{\partial r}\psi(r,\theta)\Big)\\ &\times\Big(\frac{\partial}{\partial r}\nu(r,\theta)\Big)\mathrm{e}^{2\,\mu(r,\theta)+2\,\nu(r,\theta)} + \Big(\frac{\partial}{\partial\theta}\omega(r,\theta)\Big)^2\mathrm{e}^{2\,\lambda(r,\theta)+2\,\psi(r,\theta)}\\ &+ 4\Big(\frac{\partial}{\partial\theta}\psi(r,\theta)\Big)\Big(\frac{\partial}{\partial\theta}\nu(r,\theta)\Big)\mathrm{e}^{2\,\lambda(r,\theta)+2\,\nu(r,\theta)}\Big]\\ &\times\ \mathrm{e}^{-2\,\nu(r,\theta)-2\,\lambda(r,\theta)+2\,\psi(r,\theta)-2\,\mu(r,\theta)}\Big\}\,.\end{aligned} \tag{15.11}$$

All other non-vanishing components are listed in appendix G. The individual components of the Ricci tensor, Einstein tensor, energy–momentum tensor and the Ricci scalar (scalar curvature) are listed there too. A comparison of (15.11) with the components of the Riemann tensor derived in equation (14.23) for the non-rotating case shows that (15.11) has no non-rotating counterpart, since the frame-dragging frequency $\omega(r,\theta) = 0$ in that case. Moreover we emphasize the explicit dependence of all stellar quantities on the polar angle θ, because of the star's rotational deformation. Rotational symmetry about the star's symmetry axis however still exists so that there is no explicit dependence on the azimuthal angle ϕ. Finally, since the star is considered to be statically rotating, that is, no simultaneous pulsations superimposed on the fluid's rotational motion are allowed, all components are independent of time too.

A very elegant approximation that enables us to simplify the above equations considerably was introduced by Hartle [115]. This approximation, which accounts for the basic features of rotating relativistic stars, that is, rotational deformation, frame dragging, and mass increase, each of which is to be found self-consistently, has the valuable feature of leading to a considerably more transparent treatment than numerically solving Einstein's equations for a rotating body exactly. We shall outline this approximation in detail immediately below.

15.1 Rotational perturbations

The basic idea in Hartle's treatment is the development of a perturbation solution that is based on the Schwarzschild metric (14.14). Let us assume that under the influence of rotation the star distorts and so the pressure, energy density, and baryon number density change by amounts of ΔP,

$\Delta\epsilon$, and $\Delta\rho$. These changes will modify the energy–momentum density tensor by $\Delta T_{\mu\nu}$, and therefore the system's total energy–momentum tensor becomes [115, 567]

$$T_{\mu\nu} \equiv T^0_{\mu\nu} + \Delta T_{\mu\nu} \,, \tag{15.12}$$

where

$$\begin{aligned} T^0_{\mu\nu} &\equiv (\epsilon + P)\, u_\mu\, u_\nu + P\, g_{\mu\nu} \,, \\ \Delta T_{\mu\nu} &\equiv (\Delta\epsilon + \Delta P)\, u_\mu\, u_\nu + \Delta P\, g_{\mu\nu} \,. \end{aligned} \tag{15.13}$$

The quantity $T^0_{\mu\nu}$ denotes the perfect-fluid energy–momentum tensor (14.11) of the non-rotating stellar configuration, where P, ϵ, and ρ are understood to be measured by an observer in a local inertial frame comoving with the fluid at the instant of measurement. Henceforth, we shall drop the zeros originally attached to ϵ and P in (14.11). As repeatedly mentioned, the fluid's four-velocity, u^μ, is normalized as $u^\nu\, u_\nu = -1$. For the distortion functions in equations (15.12) and (15.13) a multipole expansion is performed. Assuming axial symmetry, one can write for the changes $\Delta\epsilon$, ΔP, and $\Delta\rho$ in lowest order,

$$\Delta P = (\epsilon + P)\,\big(p_0 + p_2\, P_2(\cos\theta)\big) \,, \tag{15.14}$$

and

$$\Delta\epsilon = \Delta P\, \frac{\partial \epsilon}{\partial P} \,, \quad \Delta\rho = \Delta P\, \frac{\partial \rho}{\partial P} \,. \tag{15.15}$$

These contributions brought about by rotation are second order in the star's angular velocity [115] and are to be determined by solving Einstein's equations to this order. The quantity $P_2(\cos\theta)$ in equation (15.14) is the second order Legendre polynomial, defined as

$$P_2(x) = \frac{1}{2}\,(3\,x^2 - 1) \,. \tag{15.16}$$

The perturbed line element of the rotationally deformed fluid,

$$\begin{aligned} ds^2 = \; & - e^{2\,\nu(r,\theta,\Omega)}\, (dt)^2 + e^{2\,\psi(r,\theta,\Omega)}\, \big(d\phi - \omega(r,\theta,\Omega)\, dt\big)^2 + e^{2\,\mu(r,\theta,\Omega)}\, (d\theta)^2 \\ & + e^{2\,\lambda(r,\theta,\Omega)}\, (dr)^2 + \mathcal{O}(\Omega^3) \,, \end{aligned} \tag{15.17}$$

has formally the same mathematical structure as the line element (15.1) used in the exact numerical methods of table 15.1. In the perturbation approach, however, the metric functions ν, ψ, μ, and λ are expanded through second order in the star's rotational frequency Ω, while an expansion of the frame dragging term is only of first order. This is a consequence of the fact that the metric of a stationary, axially symmetric system will behave in the same way under a reversal in the direction of rotation as under a reversal in the direction of time [115]. The perturbed metric functions in (15.17) have the form [115, 116, 568],

$$e^{2\nu(r,\theta,\Omega)} = e^{2\Phi(r)} \{1 + 2(h_0(r,\Omega) + h_2(r,\Omega)\, P_2(\cos\theta))\} \,, \tag{15.18}$$

$$e^{2\psi(r,\theta,\Omega)} = r^2 \sin^2\theta \,\{1 + 2(v_2(r,\Omega) - h_2(r,\Omega))\; P_2(\cos\theta)\} \,, \tag{15.19}$$

$$e^{2\mu(r,\theta,\Omega)} = r^2 \,\{1 + 2(v_2(r,\Omega) - h_2(r,\Omega))\; P_2(\cos\theta)\} \,, \tag{15.20}$$

$$e^{2\lambda(r,\theta,\Omega)} = e^{2\Lambda(r)} \left\{1 + \frac{2}{r}\, \frac{m_0(r,\Omega) + m_2(r,\Omega)\, P_2(\cos\theta)}{1 - \Upsilon(r)}\right\} , \tag{15.21}$$

where we have introduced the second-order terms

$$h(r,\theta,\Omega) = h_0(r,\Omega) + h_2(r,\Omega)\, P_2(\cos\,\theta) + \ldots \,, \tag{15.22}$$

$$v(r,\theta,\Omega) = v_0(r,\Omega) + v_2(r,\Omega)\, P_2(\cos\,\theta) + \ldots \,, \tag{15.23}$$

$$m(r,\theta,\Omega) = m_0(r,\Omega) + m_2(r,\Omega)\, P_2(\cos\,\theta,\Omega) + \ldots \,, \tag{15.24}$$

to be computed from the Einstein equations. A simplification of the metric follows from transformations of the type $r \to f(r)$ which do not change the form of the metric (15.1). This guarantees that $v_0 = 0$. To keep the notation at a minimum, we have introduced in (15.21) the abbreviation

$$\Upsilon(r) \equiv \frac{2\,m(r)}{r} \,, \tag{15.25}$$

which will be used henceforth quite frequently. Note that from (15.25) $\Upsilon_R \equiv \Upsilon(r = R) = 2M/R$. As mentioned just above, the metric functions (15.18) through (15.21) are independent of both time t and azimuthal angle ϕ, which reflects stationary rotation and axial symmetry about the axis of rotation, respectively.

We need to say a few clarifying words about the star's critical angular velocity up to which the perturbation treatment – based on fractional changes in pressure, energy density, and gravitational field – may be applicable. Naturally, one may expect that the classical expression for the onset of mass shedding from the star's equator, determined by the equality between centrifugal force and gravity as

$$\Omega_c = \sqrt{\left(\frac{c}{R}\right)^2 \frac{M\,G}{R\,c^2}} \simeq 36 \sqrt{\frac{M/M_\odot}{(R/\mathrm{km})^3}} \times 10^4 \ \mathrm{s}^{-1} \,, \tag{15.26}$$

constitutes a plausible upper bound, that is, only such stellar frequencies are allowed when $\Omega \ll \Omega_c$. More than that, the expression on the right of (15.26) is the only multiplicative combination of M, R, G, and c which goes over into the Newtonian expression for the critical angular velocity as the velocity of light c becomes large. For the unperturbed configuration the factor GM/Rc^2 is less than unity [569, 570]. Consequently the condition in (15.26) also implies $R\Omega \ll c$. That is, every particle is to move at non-relativistic velocities if the perturbation of the geometry is to be small in terms of percentage. Later we shall see that the velocity of a particle at the equator of a star rotating at its mass-shedding frequency is between about 50 and 60% of the velocity of light, depending on the model chosen for the equation of state. Although such velocities already come close to c, Hartle's method nevertheless predicts a number of stellar properties surprisingly reliably (cf. section 17.2).

In general, the specification of instability criteria of rapidly spinning bodies is a non-trivial point complicated issue in the framework of Einstein's general theory of relativity. Later we shall be discussing the onset of so-called gravitational-radiation reaction driven instabilities in rapidly rotating neutron stars, which are accompanied by the emission of gravity waves from the neutron star. Since these carry away angular momentum from the star, the neutron star rotating at this instability frequency slows down. A second criterion is given by the onset of mass shedding from the equator of a rapidly spinning neutron star, the above mentioned Kepler frequency, Ω_K. If a star rotates that fast, mass loss drives the star out of hydrostatic equilibrium and, thus, stability is irrevocably lost. Since no star can spin more rapidly than at the mass-shedding frequency, this limit sets an *absolute* bound on rapid rotation. As shall be seen in chapter 16, the determination of Ω_K for a particular stellar configuration imposes a self-consistency condition on Einstein's field equations formulated for a rotating stellar body. Hartle and Thorne [115, 116], as well as all other authors who subsequently applied Hartle's method to the construction of models of rotating stars (cf. table 15.1), have avoided this complication by choosing (15.26) for the maximal possible critical frequency, which expresses the balance of centrifugal force and gravity in classical Newtonian mechanics! It is therefore not too surprising that this choice leads to neutron star properties that are in rather poor agreement with the properties obtained from an exact numerical method solved for the general relativistic expression for Ω_K [106, 107, 143, 537]. Because of this failure, Hartle's method was dismissed in the literature for more than two decades. The situation changed with the investigations performed in [62, 109, 111, 112], where it was demonstrated that the reason for the discrepancy between Hartle's method and the exact numerical method lies in applying the classical Newtonian expression (15.26) as a stability criterion rather than its general relativistic counterpart, which will be derived in section 16.1. There we shall see that (15.26) is an extremely bad approximation to its general relativistic expression, which becomes strongly modified by the star's rotational deformation and the dragging of the local inertial frames (see section 16.2), effects on the stellar body that are to be determined *self-consistently* and which are irrevocably lost when making use of the oversimplified expression (15.26), where M and R denote the star's non-rotating mass and radius.

After these remarks, we now turn to the derivation of the stellar structure equations in the framework of Hartle's method. The equation that determines the frame dragging function $\omega(r,\theta)$ is found from the Einstein equation (note that $\delta_\phi{}^t = 0$)

$$G_\phi{}^t = R_\phi{}^t = 8\,\pi\,T_\phi{}^t\,, \tag{15.27}$$

expanded to first order in the angular velocity Ω [115]. The expression of $R_\phi{}^t$ is found from $R_\phi{}^t = R_{\phi\nu}\,g^{\nu t}$ with $R_{\phi\nu}$ given in section G.2, and $g^{\nu t}$ in (15.6). The result is

$$\begin{aligned}
R_\phi{}^t = &-\frac{1}{2}\Big\{ e^{2\,\mu(r,\theta)}\frac{\partial^2}{\partial r^2}\omega(r,\theta) + e^{2\,\lambda(r,\theta)}\frac{\partial^2}{\partial \theta^2}\omega(r,\theta) + \Big(\frac{\partial}{\partial\theta}\lambda(r,\theta)\Big) \\
&\times e^{2\,\lambda(r,\theta)}\frac{\partial}{\partial\theta}\omega(r,\theta) + \Big(\frac{\partial}{\partial r}\mu(r,\theta)\Big)e^{2\,\mu(r,\theta)}\frac{\partial}{\partial r}\omega(r,\theta) \\
&+ 3\Big(\frac{\partial}{\partial\theta}\psi(r,\theta)\Big)e^{2\,\lambda(r,\theta)}\frac{\partial}{\partial\theta}\omega(r,\theta) - \Big(\frac{\partial}{\partial r}\lambda(r,\theta)\Big)e^{2\,\mu(r,\theta)} \\
&\times \frac{\partial}{\partial r}\omega(r,\theta) - \Big(\frac{\partial}{\partial r}\omega(r,\theta)\Big)e^{2\,\mu(r,\theta)}\frac{\partial}{\partial r}\nu(r,\theta) - \Big(\frac{\partial}{\partial\theta}\omega(r,\theta)\Big) \\
&\times e^{2\,\lambda(r,\theta)}\frac{\partial}{\partial\theta}\nu(r,\theta) + 3\Big(\frac{\partial}{\partial r}\psi(r,\theta)\Big)e^{2\,\mu(r,\theta)}\frac{\partial}{\partial r}\omega(r,\theta) \\
&- \Big(\frac{\partial}{\partial\theta}\mu(r,\theta)\Big)e^{2\,\lambda(r,\theta)}\frac{\partial}{\partial\theta}\omega(r,\theta)\Big\} \\
&\times\ e^{2\,\psi(r,\theta)-2\,\nu(r,\theta)-2\,\lambda(r,\theta)-2\,\mu(r,\theta)}\,.
\end{aligned} \tag{15.28}$$

The component $T_\phi{}^t$ is given in (G.73) which, up to order $\mathcal{O}(\Omega^3)$, simplifies to

$$T_\phi{}^t = (\epsilon + P)\,\bar\omega\,r^2\,\sin^2\theta\,e^{-2\,\nu} + \mathcal{O}(\Omega^3)\,. \tag{15.29}$$

Substituting (15.28) and (15.29) into equation (15.27) gives

$$\begin{aligned}
\frac{1}{r^4}\frac{\partial}{\partial r}\Big(r^4\,e^{-(\nu+\lambda)}\frac{\partial\bar\omega}{\partial r}\Big) + \frac{e^{\lambda-\nu}}{r^2\,\sin^3\theta}\frac{\partial}{\partial\theta}\Big(\sin^3\theta\,\frac{\partial\bar\omega}{\partial\theta}\Big)& \\
-\,16\,\pi\,(\epsilon+P)\,\bar\omega\,e^{\lambda-\nu} = 0\,,&
\end{aligned} \tag{15.30}$$

where we have made use of $\partial\lambda/\partial\theta = \partial\nu/\partial\theta = \partial\mu/\partial\theta = 0$ which holds up to $\mathcal{O}(\Omega^3)$. Use may be made of the zero-order field equations (14.75), (14.82), and (14.93) and the definition

$$j(r) = e^{-(\Phi(r)+\Lambda(r))} = e^{-\Phi(r)}\,\sqrt{1-\Upsilon(r)}\,, \tag{15.31}$$

to express the coefficients of $\bar\omega$ in equation (15.30) completely in terms of the unperturbed metric. Noticing that the radial derivative of (15.31) is given by

$$\frac{\mathrm{d}j(r)}{\mathrm{d}r} = -\,4\,\pi\,r\,\frac{\epsilon(r)+P(r)}{\sqrt{1-\Upsilon(r)}}\,\mathrm{e}^{-\Phi(r)}\,, \tag{15.32}$$

which follows by differentiating (15.31) with respect to r and replacing the gradients m' and Φ' with the expressions derived in (14.75) and (14.80), respectively, we then arrive at

$$\frac{1}{r^4}\frac{\partial}{\partial r}\Big(r^4\,j\,\frac{\partial\bar{\omega}}{\partial r}\Big) + \frac{4}{r}\frac{\mathrm{d}j}{\mathrm{d}r}\,\bar{\omega} + \frac{\mathrm{e}^{\lambda-\nu}}{r^2\,\sin^3\theta}\,\frac{\partial}{\partial\theta}\Big(\sin^3\theta\,\frac{\partial\bar{\omega}}{\partial\theta}\Big) = 0\,. \tag{15.33}$$

Since ω transforms like a vector under rotation, equation (15.33) may be separated by expanding ω in vector spherical harmonics,

$$\bar{\omega}(r,\theta) = \sum_{l=1}^{\infty}\bar{\omega}_l(r)\,\Big(-\frac{1}{\sin\theta}\,\frac{\mathrm{d}P_l}{\mathrm{d}\theta}\Big)\,. \tag{15.34}$$

The functions $\bar{\omega}_l(r)$ then satisfy the equation

$$\frac{1}{r^4}\frac{\mathrm{d}}{\mathrm{d}r}\Big(r^4\,j(r)\,\frac{\mathrm{d}\bar{\omega}_l(r)}{\mathrm{d}r}\Big) + \Big(\frac{4}{r}\,\frac{\mathrm{d}j(r)}{\mathrm{d}r} - \mathrm{e}^{\lambda-\nu}\,\frac{l(l+1)-2}{r^2}\Big)\bar{\omega}_l(r) = 0\,. \tag{15.35}$$

From rather general arguments concerning the asymptotic behavior of the solutions to this equation, it follows that all coefficients in the expansion of $\bar{\omega}$ vanish except $l=1$ [115]. Consequently $\bar{\omega}$ is a function of r alone. It obeys the differential equation

$$\frac{\mathrm{d}}{\mathrm{d}r}\left(r^4\,j(r)\,\frac{\mathrm{d}\bar{\omega}(r)}{\mathrm{d}r}\right) + 4\,r^3\,\frac{\mathrm{d}j(r)}{\mathrm{d}r}\,\bar{\omega}(r) = 0\,,\qquad r<R\,. \tag{15.36}$$

Equation (15.36) is to be solved subject to the boundary conditions that (1) $\bar{\omega}$ is regular at the star's origin $r=0$, and (2) $\mathrm{d}\bar{\omega}/\mathrm{d}r=0$ there. In practice, one integrates equation (15.36) outward from the star's origin, where the boundary condition (1) is imposed by choosing an arbitrary value for $\bar{\omega}$ at $r=0$, which we abbreviate as $\bar{\omega}_c \equiv \bar{\omega}(r=0)$. Arnett and Bowers [57], for instance, use a value of $\bar{\omega}_c = 1.823\,42\ \mathrm{s}^{-1}$. Outside the star one has

$$\bar{\omega}(r,\Omega) = \Omega - \frac{2}{r^3}\,J(\Omega)\,,\qquad r>R\,, \tag{15.37}$$

where $J(\Omega)$ is the total angular momentum of the star, defined by

$$J(\Omega) = I(\Omega)\,\Omega = \frac{1}{6}\,R^4\,\Big(\frac{\mathrm{d}\bar{\omega}}{\mathrm{d}r}\Big)_R\,. \tag{15.38}$$

From equations (15.37) and (15.38) it follows that the angular velocity Ω, which corresponds to a given boundary value of $\bar{\omega}_c$, is then determined by

$$\Omega(\bar{\omega}_c) = \bar{\omega}_c(R) + \frac{2}{R^3}\,J(\Omega)\,. \tag{15.39}$$

The linearity of the differential equation (15.36) implies that $\bar{\omega}$ corresponding to an Ω value different from that for which $\bar{\omega}$ was computed from (15.36) can be obtained by simply rescaling $\bar{\omega}$. That is, say $\bar{\omega}$ was computed for an angular velocity $\Omega^{\rm old}$, then $\bar{\omega}$ associated with an angular velocity $\Omega^{\rm new}$ is obtained according to

$$\bar{\omega}^{\rm new}(r) = \bar{\omega}^{\rm old}(r) \, \frac{\Omega^{\rm new}}{\Omega^{\rm old}} \, . \tag{15.40}$$

The star's moment of inertia, I, is given as the ratio J/Ω. Alternatively, an explicit expression for I, which, at first glance, does not bear a strong similarity with the expression known from classical Newtonian mechanics, can be derived by integrating equation (15.36) from the star's origin radially outward to its surface and replacing $\mathrm{d}\bar{\omega}/\mathrm{d}r$ with $J = I\Omega$ by making use of equation (15.38). The result is given by

$$I = \frac{J(\Omega)}{\Omega} = \frac{8\,\pi}{3} \int_0^R \mathrm{d}r \; r^4 \, \frac{\epsilon(r) + P(r)}{\sqrt{1 - \Upsilon(r)}} \, \frac{\Omega - \omega(r)}{\Omega} \; \mathrm{e}^{-\Phi(r)} \, , \tag{15.41}$$

where use of $j(R) = 1$ was made, which follows immediately from (15.31). Relativistic corrections to the Newtonian expression for I, which, for a sphere of uniform mass density, $\epsilon(r) = \text{const}$, is given by $I = \frac{2}{5} M R^2$, come from the dragging of local inertial frames ($\bar{\omega}/\Omega < 1$), and the redshift ($\mathrm{e}^{-\Phi}$) and space curvature factors ($(1 - \Upsilon)^{-1/2}$), with Υ defined in (15.25). While the dragging effect can be quite large in the center of compact stars, as we shall see in chapter 17, its effect becomes suppressed in the radial outward direction by a factor of r^{-3} (equation (15.37)). For that reason, the moment of inertia is sometimes evaluated for $\bar{\omega}(r)/\Omega \approx 1$ throughout the star. This approximation can only be well justified, however, if the stellar model is slowly rotating as well as of low mass. Only in this case are frame dragging and rotational deformation negligible to a very good approximation. Neutron stars with masses that scatter about the canonical, observed mass value of $1.45\, M_\odot$, figure 3.2, are already too dense for this approximation.

A severe disadvantage of the expression for the moment of inertia derived in (15.41) is that it does not account for the rotational deformation of a rapidly spinning star. The stellar properties that enter in (15.41) are those of a non-rotating, spherically symmetric body, independent of the star's frequency Ω. Naturally, this assumption is only justified as long as the star is slowly rotating relative to its Kepler frequency, $\Omega \ll \Omega_{\rm K}$. Stars rotating at frequencies that constitute an appreciable fraction of the Kepler frequency, or even at the Kepler limit itself, are known to be significantly rotationally deformed [106]. The above expression thus appears not to be adequate to our purpose, since we wish to study various phenomena that are connected with rapid neutron star rotation, and the temporal slowing-down of such stars from $\Omega_{\rm K}$ to zero rotation. The moment of inertia of such stars will not be constant in time but will respond to changes in rotational frequency, in accord with the star's internal constitution, that is, the density dependence of the nuclear symmetry energy, the softness or stiffness of the equation of state that is being adopted to construct the stellar model, and according to whether the stellar mass is small or large.

To derive the expression for the moment of inertia of a rotationally deformed, axisymmetric star in hydrostatic equilibrium, we start from the following expression [571],

$$I(\mathcal{A},\Omega) \equiv \frac{1}{\Omega} \int_{\mathcal{A}} \mathrm{d}r\, \mathrm{d}\theta\, \mathrm{d}\phi\, T_{\phi}{}^{t}(r,\theta,\phi,\Omega)\, \sqrt{-g(r,\theta,\phi,\Omega)}\,. \quad (15.42)$$

As before we assume *stationary* rotation, which is well justified for our investigations. The quantity $\mathcal{A}$ denotes an axially symmetric region in the interior of a stellar body where all matter is rotating with the same angular velocity Ω, and $\sqrt{-g} = \mathrm{e}^{\lambda+\mu+\nu+\psi}$ (cf. equation (15.7)). The component $T_{\phi}{}^{t}$ of the energy–momentum tensor is found from (14.30) to be given by

$$T_{\phi}{}^{t} = (\epsilon + P)\, u_{\phi}\, u^{t}\,. \quad (15.43)$$

Let us focus next on the determination of the fluid's four-velocity. From the general normalization relation $u^{\mu} u_{\mu} = -1$ one readily derives

$$-1 = (u^{t})^{2}\, g_{tt} + 2\, u^{t}\, u^{\phi}\, g_{tt} + (u^{\phi})^{2}\, g_{\phi\phi}\,. \quad (15.44)$$

It can be further rewritten by noticing that

$$u^{\phi} = \Omega u^{t}\,. \quad (15.45)$$

This relation extremizes the total mass energy of the stationary stellar fluid subject to the constraint that the angular momentum about the star's symmetry axis, J_z, and its baryon number, A, remain fixed [562]. Equation (15.44) then transforms to

$$-1 = (u^{t})^{2}\, (g_{tt} + 2\, g_{t\phi}\, \Omega + g_{\phi\phi}\, \Omega^{2})\,, \quad (15.46)$$

which can be solved for u^{t}. This gives

$$u^{t} = \frac{1}{\sqrt{-(g_{tt} + 2\, g_{t\phi}\, \Omega + g_{\phi\phi}\, \Omega^{2})}}\,. \quad (15.47)$$

Replacing $g_{\mu\nu}$ with its components given in (15.5) and rearranging terms leads to

$$u^{t} = \frac{\mathrm{e}^{-\nu}}{\sqrt{1 - (\omega - \Omega)^{2}\, \mathrm{e}^{2\psi - 2\nu}}}\,. \quad (15.48)$$

An expression for u_ϕ, which enters (15.42), is obtained from $u_\phi = g_{\phi\kappa}\, u^\kappa$, which, for a metric of the form (15.5), leads to $u_\phi = g_{\phi t}\, u^t + g_{\phi\phi}\, u^\phi$. Upon substituting the expressions for the metric components $g_{\phi t}$ and $g_{\phi\phi}$ into this relation, we get

$$u_\phi = (\Omega - \omega)\, \mathrm{e}^{2\,\psi}\, u^t\,. \tag{15.49}$$

Substituting the four-velocities (15.48) and (15.49) into (15.43) gives the desired expression for the energy–momentum tensor,

$$T_\phi{}^t = \frac{(\epsilon + P)\,(\Omega - \omega)\,\mathrm{e}^{2\,\psi}}{\mathrm{e}^{2\,\nu} - (\omega - \Omega)^2\,\mathrm{e}^{2\,\psi}}\,. \tag{15.50}$$

Inserting this expression into (15.42) leaves us with the expression for the moment of inertia of a rotationally deformed star given by

$$I(\Omega) = 2\,\pi \int_0^\pi \mathrm{d}\theta \int_0^{R(\theta)} \mathrm{d}r\; \mathrm{e}^{\lambda+\mu+\nu+\psi}\, \frac{\epsilon + P(\epsilon)}{\mathrm{e}^{2\nu-2\psi} - (\omega - \Omega)^2}\, \frac{\Omega - \omega}{\Omega}\,. \tag{15.51}$$

The spherical limit of (15.51) is recovered upon performing the following replacements,

$$\mathrm{e}^{2\,\nu} \longrightarrow \mathrm{e}^{2\,\Phi}\,, \quad \mathrm{e}^{2\,\psi} \longrightarrow r^2\, \sin^2\!\theta\,,$$

$$\mathrm{e}^{2\,\mu} \longrightarrow r^2\,, \quad \mathrm{e}^{2\,\lambda} \longrightarrow \mathrm{e}^{2\,\Lambda} = \left(1 - \frac{2\,m(r)}{r}\right)^{-1}\,, \tag{15.52}$$

which can be read off from the metric tensor given in (15.18) through (15.21). These replacements imply that for $\Omega \rightarrow 0$,

$$\mathrm{e}^{\lambda+\mu+\nu+\psi}\, \frac{\epsilon + P}{\mathrm{e}^{2\nu-2\psi} - (\omega - \Omega)^2} \longrightarrow \frac{r^4\, \sin^3\!\theta\, \mathrm{e}^{-\Phi}}{\sqrt{1 - \frac{2\,m(r)}{r}}}\, (\epsilon + P)\,. \tag{15.53}$$

With the aid of (15.53), a straightforward calculation then shows that (15.51) reduces to (15.41) in the spherical limit.

15.2 Monopole equations

The monopole ($l = 0$) equations which determine p_0 of (15.14) and the functions h_0 and m_0 of expansions (15.22) through (15.24) follow from $G_t{}^t = 8\,\pi\, T_t{}^t$ and $G_r{}^r = 8\,\pi\, T_r{}^r$, where $G_\mu{}^\nu = R_\mu{}^\nu - \frac{1}{2}\,\delta_\mu{}^\nu\, R$ [115, 116]. The resulting set of equations can be integrated once $\bar{\omega}(r)$ has been calculated from (15.36). The differential equation for the monopole mass perturbation function, m_0, given by [115, 116, 567, 572]

$$\frac{\mathrm{d}m_0}{\mathrm{d}r} = 4\pi\, r^2\, \frac{\partial \epsilon}{\partial P}\, (\epsilon + P)\, p_0 + \frac{1}{12}\, j^2\, r^4\, \left(\frac{\mathrm{d}\bar{\omega}}{\mathrm{d}r}\right)^2 + \frac{8\pi}{3}\, r^4\, j^2\, \frac{\epsilon + P}{1 - \Upsilon}\, \bar{\omega}^2\,, \tag{15.54}$$

is coupled to the monopole pressure perturbation function, p_0, through

$$\begin{aligned} \frac{\mathrm{d}p_0}{\mathrm{d}r} = & - \frac{1 + 8\pi r^2 P}{r^2\,(1-\Upsilon)^2}\, m_0 - 4\pi \frac{(\epsilon + P)\, r}{1 - \Upsilon}\, p_0 + \frac{1}{12}\, \frac{r^3 j^2}{1 - \Upsilon} \left(\frac{\mathrm{d}\bar{\omega}}{\mathrm{d}r}\right)^2 \\ & + \frac{1}{3}\, \frac{\mathrm{d}}{\mathrm{d}r} \left(\frac{r^2 j^2 \bar{\omega}^2}{1 - \Upsilon}\right) . \end{aligned} \tag{15.55}$$

The boundary conditions are that $m_0 \rightarrow 0$ and $p_0 \rightarrow 0$ for $r \rightarrow 0$. In the exterior star region, $\bar{\omega}$ in equations (15.54) and (15.55) can be expressed in terms of $J(\Omega)$ by making use of (15.37). One then obtains for (15.54) the relation

$$m_0(\Omega) = \Delta M(\Omega) - \frac{1}{r^3}\, J(\Omega)^2\,, \qquad r > R\,, \tag{15.56}$$

where ΔM is the change in gravitational mass due to rotation. Evaluation of equation (15.56) at the star's surface gives for ΔM

$$\Delta M(\Omega) = m_0(R) + \frac{1}{R^3}\, J(\Omega)^2\,. \tag{15.57}$$

Hence, ΔM is known once the mass perturbation function m_0 at the equator and the star's moment of inertia have been calculated. Numerical solutions to equations (15.54) and (15.55) are found by integrating them radially outward from the origin with the boundary conditions (as usual, the subscript c refers to the star's center)

$$p_0(r) \longrightarrow \frac{1}{3}\, (j_c\, \bar{\omega}_c)^2\, r^2\,, \qquad \text{for } r \rightarrow 0\,, \tag{15.58}$$

and

$$m_0(r) \longrightarrow \frac{4\,\pi}{15}\, (\epsilon_c + P_c) \left(2 + \left.\frac{\partial \epsilon}{\partial P}\right|_c\right) (j_c\, \bar{\omega}_c)^2\, r^5\,, \qquad \text{for } r \rightarrow 0\,. \tag{15.59}$$

The function h_0 can be calculated from the algebraic relations

$$h_0(r) = -\, p_0 + \frac{r^2}{3}\, \bar{\omega}^2\, \mathrm{e}^{-2\,\Phi} + h_{0c}\,, \qquad r < R\,, \tag{15.60}$$

and

$$h_0(r) = -\frac{\Delta M}{r\,(1-\Upsilon)} + \frac{J^2}{r^4\,(1-\Upsilon)}\,, \qquad r > R\,, \tag{15.61}$$

once $\bar{\omega}$, ΔM, J, and p_0 are known. The quantity h_{0c} in (15.60) is a constant of integration which is to be specified by demanding that $h_0(r)$ is continuous across the star's surface.

15.3 Quadrupole equations

The quadrupole ($l = 2$) perturbation functions h_2, m_2, and v_2, introduced in (15.22) through (15.24), determine the shape of the rotating star. They are determined in this order by the three field equations, chosen with a view to yielding the simplest non-trivial expressions, $R_\theta{}^\theta - R_\phi{}^\phi = 8\,\pi(T_\theta{}^\theta - T_\phi{}^\phi)$, $G_r{}^r = 8\,\pi\,P$, and $R_r{}^\theta = 0$. From them one derives the following coupled set of differential equations [115, 116, 567]:

$$\frac{dv_2}{dr} = -\,2\,\frac{d\Phi}{dr}\,h_2 + \left\{\frac{1}{r} + \frac{d\Phi}{dr}\right\}\left\{-\frac{r^3}{3}\,\frac{dj^2}{dr}\,\bar{\omega}^2 + \frac{j^2}{6}\,r^4\left(\frac{d\bar{\omega}}{dr}\right)^2\right\}\,, \tag{15.62}$$

and

$$\begin{aligned}\frac{dh_2}{dr} = &\left\{-\,2\,\frac{d\Phi}{dr} + \frac{2}{1-\Upsilon}\left(\frac{d\Phi}{dr}\right)^{-1}\left(2\pi\kappa\,(\epsilon+P) - \frac{m}{r^3}\right)\right\} h_2 \\ &- \frac{2}{r^2\,(1-\Upsilon)}\left(\frac{d\Phi}{dr}\right)^{-1} v_2 \\ &+ \frac{1}{6}\left\{r\,\frac{d\Phi}{dr} - \frac{1}{2\,r\,(1-\Upsilon)}\left(\frac{d\Phi}{dr}\right)^{-1}\right\} r^3\,j^2\left(\frac{d\bar{\omega}}{dr}\right)^2 \\ &- \frac{1}{3}\left\{r\,\frac{d\Phi}{dr} + \frac{1}{2\,r\,(1-\Upsilon)}\left(\frac{d\Phi}{dr}\right)^{-1}\right\}(r\,\bar{\omega})^2\,\frac{dj^2}{dr}\,.\end{aligned} \tag{15.63}$$

Their boundary conditions are $h_2(0) = v_2(0) = 0$ and $h_2(\infty) = v_2(\infty) = 0$. Numerical solutions of (15.62) and (15.63) are computed by simultaneously integrating them radially outward from the star's origin. At the origin the solutions must be regular. An examination of (15.62) and (15.63) for $r \to 0$ shows that

$$h_2(r) \longrightarrow A_2\,r^2\,, \qquad \text{for } r \to 0\,, \tag{15.64}$$

and

$$v_2(r) \longrightarrow B_2\,r^4\,, \qquad \text{for } r \to 0\,, \tag{15.65}$$

where A_2 and B_2 are any dimensionless constants related with each other by

$$B_2 + 2\pi\,A_2\left(P_c + \frac{1}{3}\,\epsilon_c\right) = -\,\frac{4\pi}{3}\,(\epsilon_c + P_c)\,(j_c\,\bar{\omega}_c)^2\,. \tag{15.66}$$

Thus, if an arbitrary value of B_2 is chosen, A_2 is given by

$$A_2 = \left(B_2 - \frac{4\pi}{3}(\epsilon_c + P_c)(j_c\bar{\omega}_c)^2\right)\left(2\pi\left(P_c + \frac{\epsilon_c}{3}\right)\right)^{-1} . \quad (15.67)$$

The solutions of the homogeneous differential equations associated with equations (15.62) and (15.63), denoted $h_2^{(\mathrm{h})}$ and $v_2^{(\mathrm{h})}$, are given for $r \to 0$ by

$$h_2^{(\mathrm{h})}(r) \longrightarrow A_2\, r^2 , \qquad \text{for } r \to 0 , \quad (15.68)$$

and

$$v_2^{(\mathrm{h})}(r) \longrightarrow B_2\, r^4 , \qquad \text{for } r \to 0 , \quad (15.69)$$

with A_2 and B_2 arbitrarily chosen, but now related by

$$B_2 + 2\pi\, A_2 \left(P_c + \frac{1}{3}\,\epsilon_c\right) = 0 . \quad (15.70)$$

The general solutions to equations (15.62) and (15.63) are then given as superpositions of the homogeneous and particular solutions,

$$v_2 = C_2\, v_2^{(\mathrm{h})} + v_2^{(\mathrm{p})} , \qquad h_2 = C_2\, h_2^{(\mathrm{h})} + h_2^{(\mathrm{h})} . \quad (15.71)$$

The solutions v_2 and h_2 of (15.71) are to be matched to the exterior solutions (15.62) and (15.63). The exterior solutions are given by

$$v_2^{>} = -\left(\frac{J}{r^2}\right)^2 + A_2\, \frac{\Upsilon}{\sqrt{1-\Upsilon}}\, Q_2^1\left(\frac{2}{\Upsilon} - 1\right) , \quad (15.72)$$

and

$$h_2^{>} = \left(\frac{J}{r^2}\right)^2 \left(1 + \frac{2}{\Upsilon}\right) + A_2\, Q_2^2\left(\frac{2}{\Upsilon} - 1\right) . \quad (15.73)$$

The quantities Q_2^1 and Q_2^2 in (15.72) and (15.73) are associated Legendre polynomials of the second kind [573], defined as

$$Q_2^1(x) = \sqrt{x^2-1}\left\{\frac{3x^2-2}{x^2-1} - \frac{3}{2}\,x\,\ln\frac{x+1}{x-1}\right\} , \qquad x > 1 ,$$
$$Q_2^2(x) = \left\{\frac{3}{2}\,(x^2-1)\,\ln\frac{x+1}{x-1} - \frac{3x^3-5x}{x^2-1}\right\} , \qquad x > 1 . \quad (15.74)$$

By matching the general exterior solution $h_2^{>}(r)$ of (15.73) and its first derivative to its interior solution (15.71) at $r = R$, i.e. requiring that

$$h_2^{<}(R) = h_2^{>}(R) , \qquad \frac{\mathrm{d}}{\mathrm{d}r}\, h_2^{<}(R) = \frac{\mathrm{d}}{\mathrm{d}r}\, h_2^{>}(R) , \quad (15.75)$$

the integration constants A_2 and C_2 can be determined. One finds

$$A_2 = \Big\{ \Big[\Big(\frac{J}{R^2}\Big)^2 \Big(1 + \frac{\Upsilon}{2}\Big)^{-1} - h_2^{(P)} \Big] \frac{1}{h_2^{(H)}(R)} \frac{dh_2^{(H)}(R)}{dr} + \frac{dh_2^{(P)}(R)}{dr} - \Big(\frac{J}{R^2}\Big)^2 \frac{1}{R} \Big(1 + \frac{\Upsilon}{2}\Big)^{-1} \Big[\frac{\Upsilon/2}{1 + \Upsilon/2} - 4 \Big] \Big\} \times \Big\{ \frac{dQ_2^2(2/\Upsilon - 1)}{dr} - \frac{Q_2^2(2/\Upsilon - 1)}{h_2^{(H)}(R)} \frac{dh_2^{(H)}(R)}{dr} \Big\}^{-1}, \tag{15.76}$$

and

$$C_2 = \Big\{ \Big(\frac{J}{R^2}\Big)^2 \Big(1 + \frac{\Upsilon}{2}\Big)^{-1} + A_2\, Q_2^2(2/\Upsilon - 1) - h_2^{(P)}(R) \Big\} \Big\{ h_2^{(H)}(R) \Big\}^{-1}. \tag{15.77}$$

The quadrupole mass and pressure perturbation functions, m_2 and p_2, can be calculated once h_2 is known. The expression for m_2 is given by

$$m_2 = r\,(1 - \Upsilon) \Big\{ -h_2 - \frac{r^3}{3} \Big(\frac{dj^2}{dr}\Big) \bar{\omega}^2 + \frac{r^4\, j^2}{6} \Big(\frac{d\bar{\omega}}{dr}\Big)^2 \Big\}. \tag{15.78}$$

Finally, the pressure perturbation function p_2 satisfies

$$p_2 = -h_2 - \frac{1}{3}\,(r\,\bar{\omega})^2\, e^{-2\Phi}. \tag{15.79}$$

The properties of rotationally deformed stars can be calculated from the functions p_0, p_2, v_2, h_2 and the integration constant A_2. We begin with the spherical stretching, ξ_0, due to rotation. This quantity follows from the monopole function p_0 in the form

$$\xi_0 = -p_0\,(\epsilon + P) \Big(\frac{\partial P}{\partial r}\Big)^{-1}. \tag{15.80}$$

Knowledge of the quadrupole stretching function, ξ_2, given by

$$\xi_2 = -p_2\,(\epsilon + P) \Big(\frac{\partial P}{\partial r}\Big)^{-1}, \tag{15.81}$$

is necessary to determine the displacement of the star's surface of constant given density that lies on radius r in the non-rotating configuration. An invariant parametrization of a surface of constant density is given by [116]

$$r^*(\theta^*) = r + \xi_0(r) + \big\{ \xi_2(r) + r\,\big(v_2(r) - h_2(r)\big) \big\}\, P_2(\cos\theta^*), \tag{15.82}$$

where θ^* denotes the polar angle and P_2 defined in (15.16). The star's eccentricity as defined by Hartle and Thorne is given by

$$e^{\rm HT} = \sqrt{\Big(\frac{R_{\rm eq}}{R_{\rm p}}\Big)^2 - 1} \approx \sqrt{-3 \Big\{ v_2(R) - h_2(R) + \frac{\xi_2(R)}{R} \Big\}}. \tag{15.83}$$

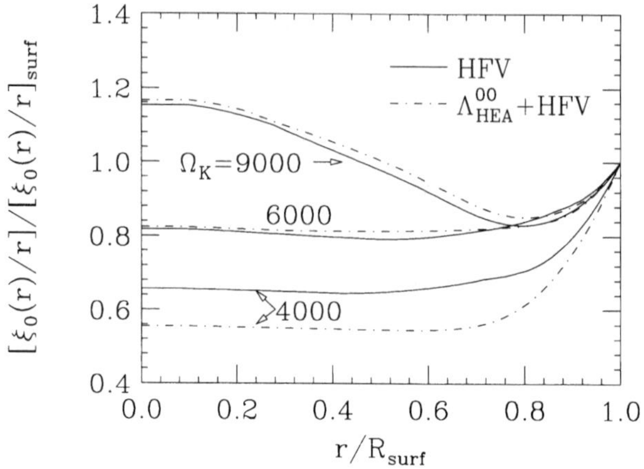

Figure 15.2. Spherical stretching, ξ_0 [defined in (15.80)], relative to its surface value, plotted as a function of radial distance from the star's center to the surface. The stellar models, each one rotating at its respective Kepler frequency Ω_K (equation (16.7)), are constructed for the EOSs HFV and $\Lambda^{00}_{\rm HEA}$+HFV of table 12.4. The corresponding Kepler periods are $P_K = 1.6$ ms, 1 ms, and 0.7 ms. [62]. (Reprinted courtesy of World Scientific.)

We shall mainly be making use of the eccentricity's definition given by Friedman, Ipser, and Parker [106],

$$e = \sqrt{1 - \left(\frac{R_{\rm p}}{R_{\rm eq}}\right)^2} \,. \tag{15.84}$$

The mass quadrupole moment of a rotating star follows from

$$\Pi = \frac{8}{5} A_2 \left(\frac{\Upsilon}{2}\right)^2 + \left(\frac{J}{R}\right)^2 . \tag{15.85}$$

The spherical stretching of a star due to rotation, $\xi_0(r)$, defined in equation (15.80), as a function of radius is shown in figure 15.2 for three different neutron star models. The masses of these models are, in the direction of increasing Ω_K values, $\sim$0.4 to $\sim$1.3 to $\sim 2.2\, M_\odot$, and their central densities are $\epsilon_c \sim 180$ MeV fm^{-3} $\epsilon_c \sim 350$ MeV fm^{-3} and $\epsilon_c \sim 600$ MeV fm^{-3}, respectively. The accurate values, together with numerous other stellar properties, can be found in tables 17.2 to 17.4 (see also appendix I). Note the rather sensitive dependence of ξ_0 on the star's rotational frequency (and thus on ϵ_c) in the star's central region.

Figure 15.3 illustrates the eccentricity, $e^{\rm HT}(r)$, of surfaces of constant density plotted as a function of radial distance, for the same stellar models as in figure 15.2. The intrinsic geometries of the surfaces in the Hartle

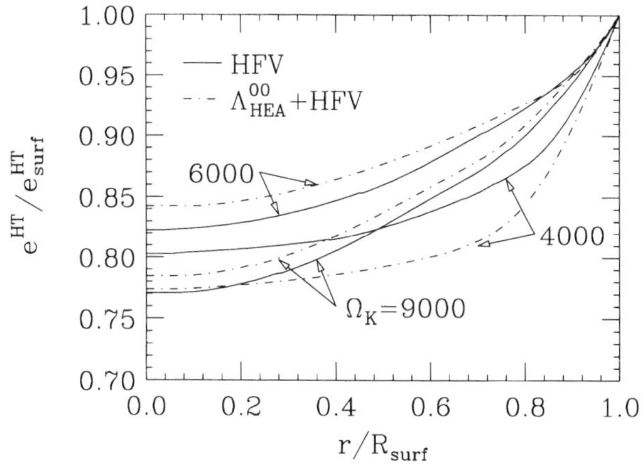

Figure 15.3. Eccentricity of surfaces of constant density at radius r. Plotted is the eccentricity $e^{\rm HT}$, as defined by Hartle and Thorne (equation (15.83)), relative to its surface value $e^{\rm HT}_{\rm surf} = e^{\rm HT}(R)$, for the EOSs HFV and $\Lambda^{00}_{\rm HEA}$+HFV [62]. (Reprinted courtesy of World Scientific.)

method are spheroids, accurate to order Ω^2 (the 3-surface in flat space is given by equation (15.82)). The eccentricity of these spheroids is related to the amount of quadrupole deformation of the star. It is evident from figure 15.3 that the deformation is largest at the surface of a star, as one would expect. However, it may appear striking that the $e^{\rm HT}(r)$ curves cross each other for sufficiently large values of $\Omega_{\rm K}$. The reason for this is that $e^{\rm HT}(r)$ inside very rapidly rotating neutron stars ($\Omega_{\rm K} = 9000$ s^{-1} in the present case) is significantly reduced in the star's core, which is caused by the large central energy density necessary in this case to keep the star together. With increasing distance from the core, $e^{\rm HT}(r)$ increases rather rapidly since the core's strong gravitational attraction is the more counterbalanced by centrifugal forces the larger the radial distance from the star's center.

To summarize, the following set of equations is to be solved when determining the properties of rotating, compact stars in general relativity theory:

(1) Equations (14.82) through (14.83), which determine radius, R, and mass, M, of the spherical, non-rotating, stellar configuration on which the perturbation solution is built upon. One obtains the radial dependence of pressure, $P(r)$, and of energy density, $\epsilon(r)$, and the metric function $\Phi(r)$ inside the spherical star ($0 \le r \le R$). The equation of state, needed in the one-parameter form $P(\epsilon)$, serves as an input.

(2) Equation (15.36), which determines the effective angular velocity $\bar{\omega}(r)$ of the fluid relative to the angular velocity of the local inertial frames (Lense–Thirring effect).
(3) Equations (15.54), (15.55), (15.60) and (15.61) which determine the monopole perturbation functions $m_0(r)$, $p_0(r)$ and $h_0(r)$. These functions specify the star's mass increase due to rotation.
(4) Equations (15.62), (15.63), (15.78) and (15.79) which determine the quadrupole perturbation functions $v_2(r)$, $h_2(r)$, $m_2(r)$ and $p_2(r)$. These functions determine the shape of the rotationally deformed star.
(5) These equations are supplemented by an additional transcendental relation for the Kepler frequency, Ω_K, if the stellar model is computed for $\Omega = \Omega_K$. The expression for Ω_K, which will be derived in section 16.1, imposes a self-consistency condition on the stellar equations, which renders the numerical method of solution of Hartle's equations somewhat more complicated, as will be described in section 17.1.1.
(6) If Ω_K is computed for a star with a *fixed* baryon number, $A = A_0$, a sequence of stellar models with $A = A_0 =$ const can be computed for rotational frequencies that cover the entire, maximal possible frequency range $0 \le \Omega \le \Omega_K$. In this case the star's central density is to be determined self-consistently for $A = A_0$ and a chosen value of $\Omega \in [0, \Omega_K]$ (further details will be given in section 17.1.2).

15.4 Dragging of local inertial frames inside rotating neutron stars

The general relativistic phenomenon that the local inertial frames inside as well as outside of a compact rotating star are dragged along in the direction of the star's rotation has been described qualitatively at the beginning of chapter 13. This phenomenon has an manifest influence on the properties of rotating stars, as discussed for instance in section 16.2, since it is the amount of rotation relative to the local inertial frames that matters to the star's global properties [115]. This relative angular frequency is given by $\bar{\omega}(r) \equiv \Omega - \omega(r)$ where the individual frequencies have the same meaning as in chapter 15, and as summarized below [116],

- Ω: angular velocity of the star's fluid relative to the distant stars.
- $\omega(r)/\Omega$: angular velocity of the local inertial frames at radius r, as measured by a distant observer, divided by the angular velocity Ω of the fluid with respect to the distant stars.
- $\bar{\omega}(r)/\Omega \equiv (\Omega - \omega(r))/\Omega$: fluid angular velocity at radius r relative to the local inertial frames there, as measured by a distant observer, divided by the angular velocity Ω of the fluid with respect to the distant stars (equation (15.36)).

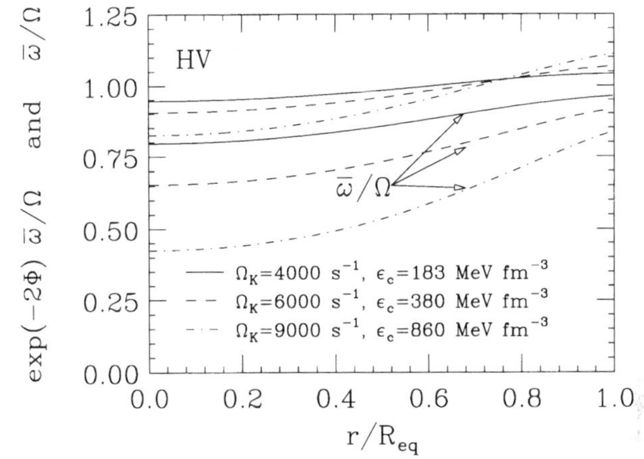

Figure 15.4. Dragging of the local inertial frames inside rotating neutron stars, from origin to the equator, calculated for the HV EOS whose properties are explained in table 12.4 [62]. Ω_{K} and ϵ_c denote Kepler frequency (defined in (16.7)) and central star density, respectively. (Reprinted courtesy of World Scientific.)

In the limit of negligible dragging, which is the case for rotating white dwarfs [574, 575] but not for neutron stars, one has $\omega(r) \ll \Omega$ and hence $\bar{\omega}(r) \simeq \Omega$. Only in this case the effective rotational frequency is equal to the frequency of the star's fluid.

The dragging frequencies as a function of radial distance from the star's center to its surface are exhibited for several sample stars in figures 15.4 to 15.7. In each figure the masses of these stars range, in the direction of increasing central density ϵ_c, from $\sim$0.4 to $\sim$1.3 to $\sim 2.2\,M_\odot$. The precise values, which are of no interest here, can be found together with numerous other stellar quantities in tables 17.2 through 17.4. As described just above, the angular velocities Ω, $\omega(r)$ and $\bar{\omega}(r)$ are measured by the *distant* observer. The associated angular velocities measured by a *local* observer are obtained by multiplication with the time dilation factor, $\mathrm{e}^{-2\Phi(r)}$ (see equation (15.18)). This gives $\mathrm{e}^{-2\Phi(r)}\Omega$, $\mathrm{e}^{-2\Phi(r)}\omega(r)$, and $\mathrm{e}^{-2\Phi(r)}\bar{\omega}(r)$, which too are shown in figures 15.4 through 15.7. The rotational frequency of each stellar model is the Kepler frequency, Ω_{K}, at which mass shedding from the equator sets in which renders the stellar rotation unstable. (Details will be discussed in section 16.1.) The Ω_{K} values correspond to rotational Kepler periods of $P_{\mathrm{K}} = 1.6$ ms, 1 ms, and 0.7 ms. As pointed out by Hartle and Thorne [116], at any given radius, the frequencies Ω, ω, and $\bar{\omega}$ are greater when measured by a local observer than when measured by an observer far away, who looks down into the star. Moreover, according to a theorem of Hartle, the dragging of inertial frames with respect to a distant observer

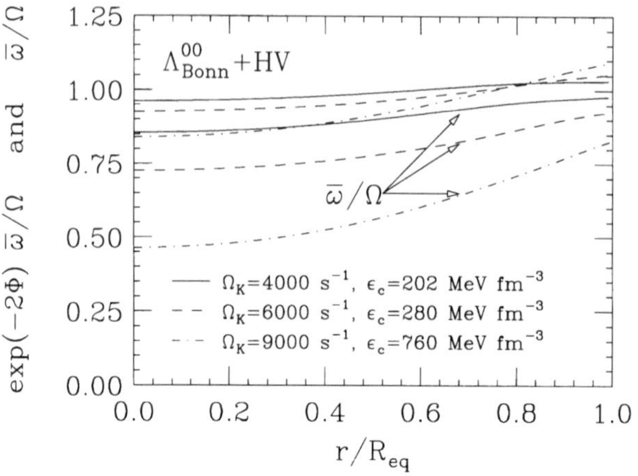

Figure 15.5. Same as figure 15.4, but calculated for the $\Lambda^{00}_{\rm Bonn}$+HV EOS [62]. (Reprinted courtesy of World Scientific.)

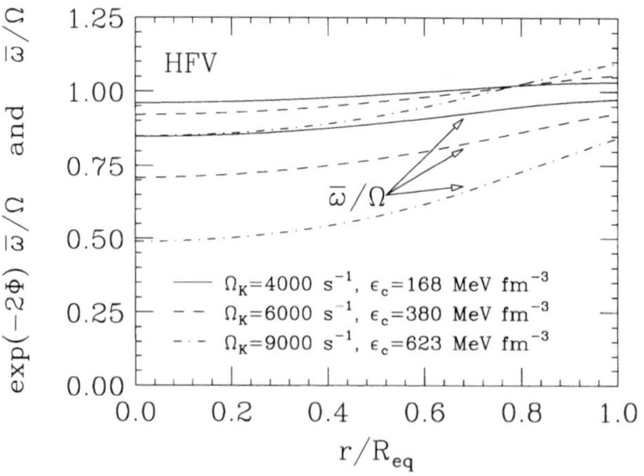

Figure 15.6. Same as figure 15.4, but calculated for the HFV EOS [62]. (Reprinted courtesy of World Scientific.)

is always greater at the star's center, where the mass–energy density is largest, and is decreasing in the radial outward direction. Both features are confirmed in figures 15.4 through 15.7.

The smooth behavior of $\bar\omega(r)/\Omega$ and $\mathrm{e}^{-2\Phi(r)}\bar\omega(r)/\Omega$ as a function of r is typical for a stable star configuration. Otherwise the dragging of the local inertial frames would be sharply peaked at the core [116]. The actual

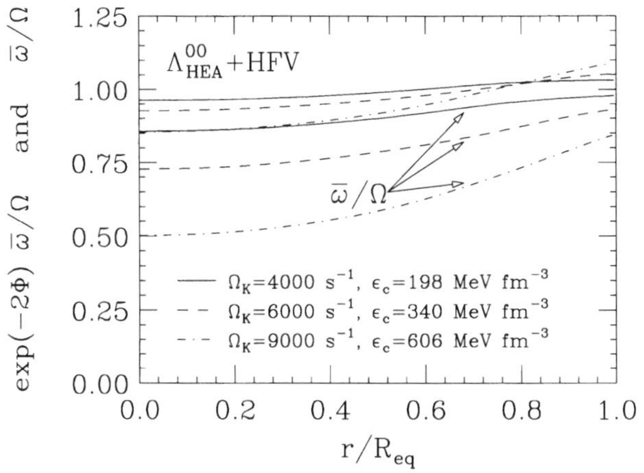

Figure 15.7. Same as figure 15.4, but calculated for the $\Lambda^{00}_{\rm HEA}$+HFV EOS [62]. (Reprinted courtesy of World Scientific.)

Table 15.2. Relative angular velocities of the local inertial frames at the center, $\tilde{\omega}_c$ ($\equiv \omega_c/\Omega_{\rm K}$), and surface, $\tilde{\omega}_s$ ($\equiv \omega_s/\Omega_{\rm K}$), of neutron stars rotating at their respective Kepler periods, $P_{\rm K}$, measured by a distant observer ($P_{\rm K} \equiv 2\pi/\Omega_{\rm K}$).

	$P_{\rm K}$ (10^{-3} s)	HV	$\Lambda^{00}_{\rm Bonn}$+HV	HFV	$\Lambda^{00}_{\rm HEA}$+HFV	$\Lambda^{\rm RBHF}_{\rm BroB}$+HFV
	1.6	0.191	0.127	0.130	0.127	0.125
$\tilde{\omega}_c$	1.0	0.393	0.315	0.326	0.299	0.266
	0.7	0.615	0.578	0.559	0.545	0.515
	1.6	0.035	0.017	0.022	0.018	0.018
$\tilde{\omega}_s$	1.0	0.102	0.091	0.087	0.080	0.063
	0.7	0.177	0.184	0.179	0.177	0.169

strength of frame dragging depends on the star's central density, as is very evident from the figures: the higher ϵ_c, the stronger the dragging of the inertial frames. Relative dragging values, calculated at the center ($\tilde{\omega}_c$) and at the surface ($\tilde{\omega}_s$) of the star models shown in figures 15.4 through 15.7 are listed in table 15.2. These are $\tilde{\omega}_c \approx$ 54–61% at the star's center and fall off to $\tilde{\omega}_s \approx$ 18% at the surface with the highest central density, i.e. the smallest Kepler period, $P_{\rm K}$ = 0.7 ms. Considerably smaller values of $\tilde{\omega}_c \approx$ 14–20% and $\tilde{\omega}_s \approx$ 2–4% are encountered for the less dense stars rotating at $P_{\rm K}$ = 1.6 ms. Bounds on the effective rotational frequencies,

$\bar{\omega}_c \equiv 1-\omega_c/\Omega_{\rm K}$ and $\bar{\omega}_s \equiv 1-\omega_s/\Omega_{\rm K}$, acting on a mass element at the star's center and the surface, respectively, can be read off from table 16.2, and are given by 18–42% and 64–85%. This confirms that frame dragging varies as the star's mass, but inversely to its radius. This can be easily understood since in the case of two stars of comparable mass but different radii (e.g. entries 10 and 11, 12 and 13 of table 16.2) the one with the smaller radius possesses the higher concentration of mass energy, leading to a stronger dragging of the local inertial frames [509].

For comparison, we quote the outcome of two sample stars (V_γ configurations) computed by Hartle and Thorne in their early investigation [116]. There values up to $\tilde{\omega}_c \approx 75\%$ and $\tilde{\omega}_s \approx 25\%$ were found for a heavy neutron star of mass $1.95\,M_\odot$ and central density $\epsilon_c = 1685$ MeV fm^{-3} rotating nearly at its classical mass shedding frequency $\Omega_{\rm K}$, defined by the Newtonian expression (15.26). As we shall see later, adopting the classical expression, however, ignores important self-consistent effects brought along with the general relativistic expression for $\Omega_{\rm K}$, such as the dragging of the local inertial frames, the star's mass increase and its non-sphericity due to rotation. All three effects influence the value of $\Omega_{\rm K}$ considerably, as we shall see in section 16.2 (see also figure 16.1). For that reason the mass shedding frequency of the above model star, $\Omega = 16\,400$ s^{-1}, was considerably overestimated in this early application. The corresponding dragging values for a light neutron star of $0.1\,M_\odot$ and $\Omega = 113$ s^{-1} were found by Hartle and Thorne to be $\tilde{\omega}_c \approx 3\%$ and $\tilde{\omega}_s \approx 0.002\%$. Large values of ω_c and ω_s clearly underline the importance of relativity when constructing models of neutron stars. This is contrary to the case of white dwarfs for which the dragging of inertial frames, leading to $\tilde{\omega}_c \leq 0.3\%$ and $\tilde{\omega}_s \leq 0.01\%$, is negligibly small [116].

From the differential equation (15.36) for $\bar{\omega}$ one readily derives an explicit relation for the frame dragging frequency ω of a local inertial frame at the surface of a compact rotating star. For this purpose one merely needs to substitute $\mathrm{d}j/\mathrm{d}r$ from (15.32) into (15.36) and make use of (15.41). One gets

$$r^4\, j\, \frac{\mathrm{d}\bar{\omega}}{\mathrm{d}r} = 6\,J\,, \qquad r = R\,. \tag{15.86}$$

For $r \geq R$ one has $j \equiv 1$, as known from (15.31), and one therefore obtains from (15.36)

$$\bar{\omega} = -\frac{A}{r^3} + B\,, \tag{15.87}$$

where A and B are integration constants. Since frame dragging vanishes infinitely far from the rotating star, that is, $\omega(r) \to 0$ for $r \to \infty$, it follows that $\bar{\omega}(r) \equiv \Omega - \omega(r) \to \Omega$ in that limit. Hence, one gets from (15.87) for the integration constant $B = \Omega$. To determine the other constant, A, we compute $\mathrm{d}\bar{\omega}/\mathrm{d}r$ from (15.87) and make use of (15.86), evaluated at $r = R$.

This leads to $A = 2J$. The angular velocity of the dragged inertial frames at the star's equator is therefore given by

$$\omega(R) = \frac{2\,I}{R^3}\,\Omega\,. \tag{15.88}$$

15.5 Baryon number of a rotating star

The total baryon number of a spherical neutron star is known from equation (14.90). The change in total baryon number due to rotation, denoted by ΔA, follows from (m_n is the mass of a neutron given in table 5.1)

$$m_n\,\Delta A = m_0(R) + 4\,\pi \int_0^R \mathrm{d}r\, r^2\,\mathcal{B}(r)\,, \tag{15.89}$$

with the quantity $\mathcal{B}$ defined as

$$\begin{aligned}\mathcal{B}(r) = (\epsilon + P)\,p_0 &\left\{ \frac{\partial\epsilon}{\partial P}\Big(\frac{1}{\sqrt{1-\Upsilon}} - 1\Big) - \frac{\partial\mathcal{E}^{\mathrm{int}}}{\partial P}\frac{1}{\sqrt{1-\Upsilon}} \right\} \\ &+ \frac{(\epsilon - \mathcal{E}^{\mathrm{int}})}{\sqrt{1-\Upsilon}^{\,3}}\Big(\frac{m_n}{r} + \frac{1}{3}\,(j\,r\,\bar{\omega})^2\Big) \\ &- \frac{1}{4\,\pi\,r^2}\Big(\frac{1}{12}\Big(j\,r^2\,\frac{\mathrm{d}\bar{\omega}}{\mathrm{d}r}\Big)^2 - \frac{1}{3}\,\frac{\mathrm{d}j^2}{\mathrm{d}r}\,r^3\,\bar{\omega}^2\Big)\,. \end{aligned} \tag{15.90}$$

The quantities m_0 and p_0 are the monopole perturbation functions of mass and pressure defined in (15.54) and (15.55). The functions $\bar{\omega}$ and j were introduced in equations (15.36) and (15.31), and $P = P(\epsilon)$ is the equation of state. Finally, the internal energy $\mathcal{E}^{\mathrm{int}}$ in equation (15.90) is given in terms of the energy density ϵ and baryon number density ρ according to

$$\mathcal{E}^{\mathrm{int}} \equiv \epsilon\, -\, m_n\,\rho\,. \tag{15.91}$$

15.6 Redshift, blueshift, injection energy, stability parameter

Further important stellar quantities, accessible to observations, are the redshift, blueshift, injection energy, and the stability parameter. Their expressions will be derived next. As discussed at the beginning of this chapter (cf. equation (15.8)), the four-velocity of an axisymmetric, rotationally deformed star is given by $u^\mu = (u^t, 0, 0, u^\phi)$, where $u^t = \mathrm{e}^{-\nu}\sqrt{1-V^2}$ with the three-velocity $V = (\Omega - \omega)\,\mathrm{e}^{\psi-\nu}$. Moreover, u^ϕ is connected with u^t as $u^\phi = \Omega\,u^t = \Omega\,\mathrm{e}^{-\nu}/\sqrt{1-V^2}$. Substituting the latter expressions into the first expression for the four-velocity u^μ gives

$$u^\mu = \frac{\mathrm{e}^{-\nu}}{\sqrt{1-V^2}}\,(1, 0, 0, \Omega)\,. \tag{15.92}$$

The Killing vectors underlying the metric of (15.5) have components

$$t^\mu = (1,0,0,0) = \delta^{\mu t}, \quad \phi^\mu = (0,0,0,1) = \delta^{\mu\phi}, \tag{15.93}$$

which gives for $t^\mu + \Omega\phi^\mu$

$$t^\mu + \Omega\,\phi^\mu = (1,0,0,\Omega). \tag{15.94}$$

Substituting equation (15.94) into (15.92) gives for the four-velocity

$$u^\mu = \frac{e^{-\nu}}{\sqrt{1-V^2}}\left(t^\mu + \Omega\,\phi^\mu\right). \tag{15.95}$$

Equation (15.95) gives the fluid's four-velocity at the equator.

The four-momentum of light emitted forward (backward) at the equator of a rotating star is given by [106]

$$p^\mu = \text{const} \times \left[t^\mu + (\omega \pm e^{\nu-\psi})\,\phi^\mu\right], \tag{15.96}$$

where the plus (minus) sign refers to forward (backward) emission. The emitted frequency is given by

$$\omega_E = p_\mu\, u^\mu, \tag{15.97}$$

and for the frequency observed at infinity

$$\omega_\infty = p_\mu\, t^\mu. \tag{15.98}$$

One then obtains for the ratio ω_∞/ω_E the relation

$$\frac{\omega_\infty}{\omega_E} = \frac{[t_\mu + (\omega \pm e^{\nu-\psi})\,\phi_\mu]t^\mu}{[t_\mu + (\omega \pm e^{\nu-\psi})\,\phi_\mu]t^\mu}. \tag{15.99}$$

Writing all vectors in a contravariant fashion and rearranging terms by making use of (15.93) gives

$$\begin{aligned}\frac{\omega_\infty}{\omega_E} &= \frac{g_{\mu\kappa}[t^\kappa t^\mu + (\omega \pm e^{\nu-\psi})\,\phi^\kappa t^\mu]}{g_{\mu\kappa}[t^\kappa u^\mu + (\omega \pm e^{\nu-\psi})\,\phi^\kappa u^\mu]} \\ &= \frac{g_{tt} + (\omega \pm e^{\nu-\psi})\,g_{t\phi}\,\phi^\phi\,t^t}{g_{tt}\,t^t\,u^t + g_{\phi t}\,t^t\,u^\phi + (\omega \pm e^{\nu-\psi})(g_{t\phi}\,\phi^\phi\,u^t + g_{\phi\phi}\,\phi^\phi\,u^\phi)}.\end{aligned} \tag{15.100}$$

After some algebra this relation can be brought into the form

$$\frac{\omega_\infty}{\omega_E} = e^\nu\sqrt{\frac{1\mp V}{1\pm V}}\left(1 \pm \omega\, e^{\psi-\nu}\right), \tag{15.101}$$

which, by defining

$$z \equiv \frac{\omega_E}{\omega_\infty} - 1, \tag{15.102}$$

can be written in the form

$$z_{\mathrm{B,F}}(\Omega) = \mathrm{e}^{-\nu(\Omega)} \sqrt{\frac{1 \pm V(\Omega)}{1 \mp V(\Omega)}} \left(1 \pm \omega(\Omega)\, \mathrm{e}^{\psi(\Omega)-\nu(\Omega)}\right)^{-1} - 1\,. \quad (15.103)$$

This expression accounts for the shifts in spectral line frequency of photons which are caused by the presence of a strong gravitational field as well as the Doppler shift. The metric functions that enter (15.103) are given in equations (15.18) and (15.19), and are to be determined from Einstein's field equations of a rotating star.

To find the redshift of photons emitted at the star's pole, note that an observer there has a four-velocity given by

$$u^\mu = \frac{t^\mu}{\sqrt{t^\kappa\, t_\kappa}} = \frac{t^\mu}{\sqrt{|g_{tt}|}}\,, \quad (15.104)$$

where the second equality follows because of $t^\kappa\, t_\kappa = g_{00}$. The fact that $t^\mu \propto u^\mu$ (cf. (15.95) with $\Omega = 0$ at the pole) implies that the polar redshift is independent of the photon's direction:

$$\frac{\omega_\infty}{\omega_E} = \frac{p_\mu\, t^\mu}{p_\mu\, u^\mu}\,, \quad (15.105)$$

where $u^\mu = \mathrm{e}^{-\nu}\, t^\mu$ at the pole. This leads for (15.105) to

$$\frac{\omega_\infty}{\omega_E} = \frac{\mathrm{e}^\nu}{\sqrt{1-V^2}} = \mathrm{e}^\nu\,, \quad \text{at the star's pole}\,, \quad (15.106)$$

and with the notation of (15.102)

$$z_{\mathrm{p}}(\Omega) = \mathrm{e}^{-\nu(\Omega)} - 1\,. \quad (15.107)$$

The frequency dependence of z_{p} is supposed to indicate the change of the star's global properties with frequency, that is, a flattening at the equator and a radial growth at the equator. The non-rotating limit equation (15.107) reads

$$z = \mathrm{e}^{-\,\Phi(R)} - 1 = \left(1 - \frac{2\,M}{R}\right)^{-1/2} - 1\,. \quad (15.108)$$

In the comoving frame, the product of the chemical potential, $\mathrm{d}\epsilon/\mathrm{d}\rho$, measuring the increase in energy of a unit volume as $\mathrm{d}\rho$ particles are added to it [69], and $(u^t)^{-1}$ of equation (15.46) becomes [562]

$$\beta \equiv \frac{\mathrm{d}\epsilon}{\mathrm{d}\rho} \sqrt{-\,g_{00}}\,. \quad (15.109)$$

This product denotes the so-called injection energy, β, which, physically, is the energy needed in a comoving frame to move a baryon from some standard point outside the fluid to any point inside. β can be expressed in terms of $z_{\rm P}$, given in equation (15.107), as

$$\beta(\Omega) \equiv e^{2\nu(\Omega)}\Big|_{\rm pole} = \frac{1}{(z_{\rm P}(\Omega) + 1)^2} \,. \qquad (15.110)$$

Finally, we conclude this section by defining the stability parameter t as

$$t(\Omega) \equiv \frac{T(\Omega)}{|W(\Omega)|} \,, \qquad (15.111)$$

where

$$T(\Omega) \equiv \frac{1}{2} J(\Omega)\, \Omega \,, \qquad W(\Omega) \equiv M_{\rm P}(\Omega) + T(\Omega) - M(\Omega) \,, \qquad (15.112)$$

which is the ratio of the star's rotational energy, T, to gravitational energy, W. The stability parameter is of importance for investigating the instability of rotating neutron stars against non-radial toroidal modes [537, 576, 577]. From a normal-mode analysis of Maclaurin spheroids – uniformly rotating, self-gravitating homogeneous ellipsoids – it is known that such toroidal modes go dynamically unstable (i.e. the eigenfrequencies become complex because $\omega^2 < 0$) in such objects for $t > 0.2738$ [224]. An instability that requires the presence of dissipation is called a *secular instability*, to distinguish it from a *dynamical instability*. Such a secular instability sets in earlier along the Maclaurin sequence, at $t = 0.1375$. This value, in fact, is a point of bifurcation, where a whole new sequence of equilibrium configurations branches off the Maclaurin sequence. This new sequence consists of the Jacobi ellipsoids, which are similar in structure to the Maclaurin spheroids but of triaxial shape (rotating footballs) [578]. Beyond the point of bifurcation, Maclaurin spheroids should be unstable to becoming Jacobi ellipsoids. In reality, however, the situation is considerably more complicated, since both viscosity and gravitational radiation are operative [579]. As we shall see in section 16.3, both effects are in fact counter-operative. Gravitational radiation alone would renders all rotating bodies unstable, that is, the only stable stellar configurations would be non-rotating stars. Viscosity, on the other hand, tends to damp out gravity-wave emission and, thus, stabilizes stellar rotation. The extent to which this is the case depends on the strength of viscosity relative to gravitational radiation [537], which links this problem to the temperature of a star.

As reference values for our analysis to be performed in section 16.3.3, we shall adopt the criteria $t \gtrsim 0.14$ for secular instability and $t \gtrsim 0.26$ for dynamical instability, as these values are close to the exact Maclaurin spheroid values, and seem to hold for a wide range of angular momentum distributions and equations of state [224].

Chapter 16

Criteria for maximum rotation

16.1 Kepler frequency

No simple stability criteria are known for rapidly rotating stellar configurations in general relativity. An *absolute* upper limit on stable neutron star rotation is set by the Kepler frequency Ω_K [106], which is the maximum frequency a star can have before mass loss (mass shedding) at the equator sets in. The Kepler frequency, determined by the equality between centrifugal force and gravity, is readily obtained in classical mechanics (cf. equation 15.26). In order to derive the expression for Ω_K in the framework of general relativity theory, we apply the extremal principle to the circular orbit of a point mass rotating at the star's equator. Since $r = \theta = \text{const}$ for a point mass there one thus has $\mathrm{d}r = \mathrm{d}\theta = 0$. The line element (15.1) therefore reduces to

$$\mathrm{d}s^2 = \left(\mathrm{e}^{2\,\nu} - \mathrm{e}^{2\,\psi}\,(\Omega - \omega)^2\right)\,(\mathrm{d}t)^2\,. \tag{16.1}$$

Substituting this expression into $J \equiv \int_{s_1}^{s_2} \mathrm{d}s$, where s_1 and s_2 refer to points located at that particular orbit for which J becomes extremal, gives

$$J = \int_{s_1}^{s_2} \mathrm{d}t\,\sqrt{\mathrm{e}^{2\,\nu} - \mathrm{e}^{2\,\psi}\,(\Omega - \omega)^2}\,. \tag{16.2}$$

Applying the extremal condition $\delta J = 0$ to (16.2) leads to

$$\int_{s_1}^{s_2} \mathrm{d}t\;\delta\sqrt{\mathrm{e}^{2\,\nu} - \mathrm{e}^{2\,\psi}(\Omega - \omega)^2} = 0\,,$$

which, upon performing the variation of the integrand, takes on the form

$$\int_{s_1}^{s_2} \mathrm{d}t\,\delta r\;\frac{\frac{\partial \nu}{\partial r}\,\mathrm{e}^{2\,\nu} - \left(\frac{\partial \psi}{\partial r}(\Omega - \omega) - \frac{\partial \omega}{\partial r}\right)(\Omega - \omega)\mathrm{e}^{2\,\psi}}{\sqrt{\mathrm{e}^{2\,\nu} - \mathrm{e}^{2\,\psi}(\Omega - \omega)^2}} = 0\,. \tag{16.3}$$

Because the variation δr with respect to radial coordinate is arbitrary, it follows from (16.3) that the integrand's numerator must vanish identically,

$$\frac{\partial \psi}{\partial r}\, \mathrm{e}^{2\,\nu}\, V^2 - \frac{\partial \omega}{\partial r}\, \mathrm{e}^{\nu+\psi}\, V - \frac{\partial \nu}{\partial r}\, \mathrm{e}^{2\,\nu} = 0\,. \tag{16.4}$$

Equation (16.4) constitutes a simple quadratic equation for the orbital velocity V of a comoving observer at the star's equator relative to a locally non-rotating observer with zero angular momentum in the ϕ-direction, henceforth simply referred to as equatorial three-velocity. It is a straightforward task to solve equation (16.4) for V, which gives

$$V_{\pm} = \frac{\partial \omega/\partial r}{2\,\partial \psi/\partial r}\, \mathrm{e}^{\psi-\nu} \pm \sqrt{\left(\frac{\partial \omega/\partial r}{2\,\partial \psi/\partial r}\right)^2 \mathrm{e}^{2(\psi-\nu)} + \frac{\partial \nu/\partial r}{\partial \psi/\partial r}}\,. \tag{16.5}$$

On the other hand, the three-velocity, written in terms of Ω, follows immediately from (16.1) as

$$V = \mathrm{e}^{\psi-\nu}\,(\Omega - \omega)\,. \tag{16.6}$$

The time component of the mass point's four-velocity, u^t ($\equiv \mathrm{d}t/\mathrm{d}s$), is then given by the familiar expression $\mathrm{d}t/\mathrm{d}s = \mathrm{e}^{-\nu}\sqrt{1-V^2}$. Recall that the other components of the fluid's four-velocity vanish, i.e. $u^r = u^\theta = 0$. Solving (16.6) for $\Omega = \Omega_{\mathrm{K}}$ gives the fluid's general relativistic Kepler frequency in terms of V, the metric functions ν and ψ, and the frame dragging frequency ω, each of which is a complicated function of all the other quantities. In this manner Ω_{K} is given by

$$\Omega_{\mathrm{K}} = \left\{ V(\Omega_{\mathrm{K}})\, \mathrm{e}^{\nu(\Omega_{\mathrm{K}})-\psi(\Omega_{\mathrm{K}})} + \omega(\Omega_{\mathrm{K}}) \right\}_{\mathrm{eq}}, \tag{16.7}$$

with $V = V_+$ given in (16.5). The subscript 'eq' indicates evaluation of (16.7) at the star's equator. The numerical outcome of the properties of neutron stars rotating at their respective general relativistic Kepler periods, defined as $P_{\mathrm{K}} = 2\pi/\Omega_{\mathrm{K}}$, is compiled in tables 16.1 and 16.2. The calculations are performed for the heaviest neutron star of each sequence, each one based on a different model for the equation of state. These tables reveal that equatorial velocity V_{eq} reached by these stars is between ~50% and 70% of the velocity of light, depending on the stiffness of the equation of state. A stiffer equation of state generally leads to larger values of V_{eq}, since the corresponding neutron stars are less compressed by gravity and, therefore, have bigger radii than neutron stars constructed for soft equations of state (cf. section 14.4). It is also apparent that none of these stellar models can rotate at periods significantly below 0.5 ms. If the observation of a 0.5 ms (1968 Hz, to be precise) pulsar in SN 1987A had not been spurious, the non-relativistic models of our collection of equations

Table 16.1. Some properties of rotating maximum-mass neutron star models. The entries are: EOS (the labeling is explained in table 12.4), Kepler period P_K, gravitational mass M, equatorial radius R_{eq}, central density (in units of nuclear matter density) ϵ_c/ϵ_0, moment of inertia I. Further properties are listed in table 16.2.

Label	EOS	P_K (10^{-3} s)	$M/M_\odot$	R_{eq} (km)	ϵ_c/ϵ_0	I (10^{45} g cm^2)
1	G_{300}	0.676	2.04	13.60	7.16	1.8922
2	HV	0.635	2.25	13.43	7.83	2.1459
3	G_{B180}^{DCM2}	0.622	1.87	12.42	9.43	1.3338
4	G_{265}^{DCM2}	0.622	1.72	12.17	10.29	1.1644
5	G_{300}^{π}	0.616	2.06	12.87	8.67	1.8087
6	G_{200}^{π}	0.610	1.67	11.91	10.84	1.0687
7	Λ_{Bonn}^{00}+HV	0.604	2.24	12.97	7.96	2.1153
8	G_{225}^{DCM1}	0.593	1.73	11.91	10.61	1.1555
9	G_{B180}^{DCM1}	0.556	1.69	11.31	12.07	1.0165
10	HFV	0.542	2.47	12.34	7.12	2.4526
11	Λ_{HEA}^{00}+HFV	0.532	2.47	12.24	7.66	2.4418
12	BJ(I)	0.561	2.09	11.59	10.81	1.6081
13	UV_{14}+TNI	0.487	2.09	10.97	10.02	1.5320
14	V_{14}+TNI	0.459	2.21	10.82	10.62	1.7307
15	UV_{14}+UVII	0.433	2.44	10.70	10.07	2.2114
16	AV_{14}+UVII	0.416	2.38	10.31	10.51	2.0188

of state would have appeared to have less trouble in accommodating such a rapidly spinning pulsar, provided the pulsar's mass is assumed to be close to the mass peak. This appears implausible because the observed masses clearly cluster about $1.4\,M_\odot$, as was shown in figure 3.2. So, in the wake of the possible future discovery of sub-millisecond pulsars by radio or X-ray astronomers, we conclude that, based on our collection of equations of state, pulsar periods in the half-millisecond regime or even below appear to be hardly accessible to such objects, provided their mass is around $1.4\,M_\odot$ and they are made up of confined baryonic matter, which may or may not be in phase equilibrium with deconfined quark matter. The situation is different for stars made up of *self-bound* matter, of which absolutely stable 3-flavor strange quark matter (cf. section 2.2) fits most comfortably with our present theoretical concept of the behavior of superdense matter. The phenomenology of such stars will be discussed in chapter 18.

In order to avoid confusion, the statement made just above about the smallest possible rotational periods accessible to gravitationally bound stars is based on particular models for the equation of state and so must not be

Table 16.2. Further properties of rotating maximum-mass neutron star models (continued from table 16.1). The entries are: EOS, stability parameter t, equatorial velocity $V_{\rm eq}$, eccentricity e, relative angular velocities of the local inertial frames at center, $\bar{\omega}_c \equiv \Omega_{\rm K} - \omega_c$, and surface $\bar{\omega}_s \equiv \Omega_{\rm K} - \omega_s$, of the star in units of Keper frequency $\Omega_{\rm K}$ $(= 2\pi/P_{\rm K})$.

Label	EOS	t	$V_{\rm eq}/c$	e	$\bar{\omega}_c/\Omega_{\rm K}$	$\bar{\omega}_s/\Omega_{\rm K}$
1	G_{300}	0.10	0.50	0.70	0.42	0.83
2	HV	0.10	0.54	0.69	0.33	0.80
3	$G^{\rm DCM2}_{\rm B180}$	0.09	0.51	0.69	0.39	0.84
4	$G^{\rm DCM2}_{265}$	0.09	0.49	0.69	0.40	0.85
5	G^{π}_{300}	0.10	0.54	0.69	0.36	0.81
6	G^{π}_{200}	0.09	0.48	0.69	0.40	0.85
7	$\Lambda^{00}_{\rm Bonn}$+HV	0.11	0.55	0.69	0.33	0.78
8	$G^{\rm DCM1}_{225}$	0.09	0.50	0.69	0.39	0.84
9	$G^{\rm DCM1}_{\rm B180}$	0.09	0.51	0.68	0.36	0.83
10	HFV	0.11	0.60	0.69	0.24	0.72
11	$\Lambda^{00}_{\rm HEA}$+HFV	0.12	0.61	0.68	0.24	0.71
12	BJ(I)	0.10	0.57	0.68	0.28	0.78
13	UV_{14}+TNI	0.11	0.59	0.69	0.26	0.73
14	V_{14}+TNI	0.12	0.62	0.68	0.23	0.71
15	UV_{14}+UVII	0.13	0.66	0.64	0.18	0.65
16	AV_{14}+UVII	0.13	0.67	0.65	0.18	0.64

understood to be final [580]. In this respect the model-independent limit of $P_{\rm K} \sim 0.3$ ms [491], derived for neutron stars on the basis of rather general physical principles rather than a particular model for the equation of state (cf. section 12.6), attains its particular interest. Of course, while being strictly model independent, the equation of state that nature has chosen need not be the one that allows neutron stars to rotate most rapidly, as expressed in the just quoted $P_{\rm K}$ value.

To summarize, equation (16.7) in combination with (16.5) constitutes the general relativistic expression for the rotational frequency of a massive particle rotating in a last stable orbit of constant radial distance (i.e. $r = R_{\rm eq}$ and $\theta = \pi/2$). Below we shall verify that these relations reduce to the classical expression for the Kepler frequency given in (15.26). In sharp contrast to the classical expression, however, its general relativistic counterpart is to be solved *self-consistently* in combination with the Einstein equations of the rotating stellar body, because it is these equations which determine the input quantities ν, ψ and ω whose knowledge is necessary to compute $\Omega_{\rm K}$ from (16.7).

If one neglects the distortion functions in the line element (15.17)

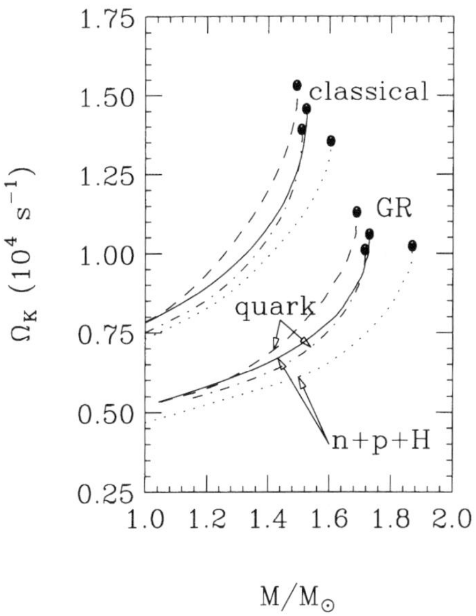

Figure 16.1. Kepler frequency, Ω_K, computed in classical Newtonian mechanics and general relativity theory for two sequences of compact stars [447]. One is populated by 'conventional' neutron stars (made up of chemically equilibrated neutrons, protons, hyperons), the other by quark-hybrid stars. The underlying equations of state are G_{225}^{DCM1} (solid line), G_{B180}^{DCM1} (dashed), G_{265}^{DCM2} (dash–dotted), and G_{B180}^{DCM2} (dotted). Details are given in section 12.6.

and approximates $\bar{\omega} \approx \Omega$, that is, frame dragging of the local inertial frames is neglected, then the orbital velocity at the equator V_{eq} $(= V_+)$ of equation (16.5) takes on the form

$$V_{eq} = \sqrt{\frac{\Upsilon_{eq}}{2}} \, \frac{1}{\sqrt{1 - \Upsilon_{eq}}} \longrightarrow R_{eq}\, \Omega_c \quad \text{for} \quad \Upsilon_{eq} \ll 1 \,, \tag{16.8}$$

with Ω_c given in equation (15.26) and $\Upsilon_{eq} \equiv \Upsilon(R_{eq})$. From equation (16.8) it is seen that the classical Newtonian expression for V is recovered by neglecting the curvature of spacetime (that is, $\Upsilon_{eq} \ll 1$), the rotational deformation of the star, and the frame dragging of the local inertial frames. The classical limit of (16.7), given by $\Omega_K \to \Omega_c$, is readily found too by noticing that $v_2 \ll 1$ and $h_2 \ll 1$, and $e^{\psi} \to r$ for a particle at the star's equator. The importance of the curvature of spacetime, the star's rotational deformation, and the dragging of the local inertial frames in the direction of the star's rotation, each of which is to be determined self-consistently, on Ω_K and V becomes very obvious from figure 16.1.

An approximate solution of (16.7) for the *heaviest* rotating neutron star of a stable sequence can be obtained from the empirically established expression of Haensel and Zdunik [189], and Friedman, Ipser, and Parker [107, 143] (see also Lattimer *et al* [108]), which is given by

$$\Omega_{\rm K} = 24 \sqrt{\frac{M/M_\odot}{(R/{\rm km})^3}} \times 10^4 \ {\rm s}^{-1} = \frac{2}{3}\, \Omega_{\rm c} \, . \tag{16.9}$$

This formula has been established from the exact numerical solution of Einstein's field equations for rapidly rotating neutron stars. It attains its particular importance for two reasons. Firstly, only knowledge of the mass and radius of the *non-rotating* maximum-mass star is necessary, which are relatively easy to calculate in contrast to solving Einstein's equations for a rotating body. Secondly, (16.9) apparently predicts the Kepler frequency not only for a particular model for the equation of state of superdense matter, but, surprisingly, for a broad range of entirely different ones, from the equation of state of pure neutron matter to hybrid-star matter to strange quark matter. In the next section 16.2 immediately below, we shall carefully analyze the formula (16.9), motivating the dependence of $\Omega_{\rm K}$ on R and M quantitatively and shedding light on its robustness against microphysical variations [111, 112]. As it turns out, the proportionality factor in (16.9) has its origin in equal parts in the dragging effect of the local inertial frames and the star's rotational deformation.

For the purpose of completeness we also give the expression for the limiting rotational frequency chosen by Ray and Datta in their analysis of the properties of rotating neutron stars constructed for Hartle's method [572, 581],

$$\Omega = 19 \sqrt{\frac{M/M_\odot}{(R/{\rm km})^3}} \times 10^4 \ {\rm s}^{-1} \, . \tag{16.10}$$

This expression has its origin in the onset of a secular instability in uniformly rotating, self-gravitating configurations of uniform density, that is, the Maclaurin spheroids described briefly in section 15.6. This instability was found, in the framework of classical Newtonian analysis, by Chandrasekhar [578]. Of course, its extrapolation to rotating, general relativistic systems of non-uniform density appears questionable. Besides that, applying an expression of the form as (16.10) ignores very important self-consistency aspects that are crucial for obtaining stellar properties that are in accord with the exact numerical method, as stressed repeatedly.

16.2 Empirical formula for Kepler frequency

As pointed out just above, the general relativistic expression for the Kepler frequency of a star, derived in (16.7), can be found only through the application of this expression as a self-consistency condition on the solution of the relativistic equations for rotating stars. Haensel and Zdunik [189], Friedman, Ipser, and Parker [106, 107], and Lattimer *et al* [108] have shown that the *relativistic* Kepler frequency of a neutron star rotating at its mass limit can be estimated to better than 10% from the mass and radius of the corresponding (i.e. same equation of state) *non-rotating* maximum-mass star. This 'empirical' relation is given by

$$\Omega_{\rm K} = C\,\sqrt{(M/M_\odot)/(R/10\,{\rm km})^3} = (0.63\text{–}0.67)\,\Omega_{\rm c}\,, \tag{16.11}$$

with $\Omega_{\rm c} = \sqrt{M/R^3}$ given in (15.26). The quantities M and R denote the gravitational mass and radius, respectively, of the spherical star of limiting mass. The quantity C is a constant for which values of $C_{\rm HZ} = 7700\ {\rm s}^{-1}$ (Haensel and Zdunik [189]) and $C_{\rm FIP} = 7200\ {\rm s}^{-1}$ (Friedman, Ipser, and Parker [107]) have been extracted from a comparison with exact numerical integration of Einstein's equations for a sample of equations of state [106, 107]. As indicated in equation (16.11), these values reduce $\Omega_{\rm K}$ from its classical value $\Omega_{\rm c}$ by factors of 0.67 and 0.63, respectively. Equation (16.11) approximates the exact value of $\Omega_{\rm K}$ to better than 10% in the case of $C = C_{\rm FIP}$ and 5% for $C = C_{\rm HZ}$.

A striking feature is the independence of C from the particular stellar model (and hence of the equation of state itself). The numerical value of C will be derived from theory immediately below. To the best of our knowledge, the only motivation for an expression such as equation (16.11) has been given by Shapiro, Teukolsky, and Wasserman [582] from the study of homogeneous stars (Maclaurin spheroids) in Newtonian mechanics and centrally condensed stars (Roche models) in general relativity, for which, however, an approximation is made which omits the important frame dragging effect. It is extremely astonishing that the simple, empirically established expression for $\Omega_{\rm K}$ so successfully approximates the exact solution of equation (16.7). Obviously the rotating and non-rotating star models are two rather different physical objects which are not related by any quantity with each other, aside from being determined from the same equation of state. One is therefore particularly tempted to understand the relationship (16.11) from theory, which is our next goal. We shall proceed in two steps. In the first we will neglect the purely general relativistic effect of the dragging of local inertial frames. By this a rather straightforward investigation of $\Omega_{\rm K}$ can be performed in which the effect of the equatorial flattening due to rotation can be incorporated. In the second step the dragging effect is taken into account.

16.2.1 Dragging of local inertial frames neglected

The dragging effect of the local inertial frames, expressed by the frequency $\omega(r)$, describes the onset of rotation of these inertial frames induced by a rotating stellar mass. Values of fractional dragging at the center and surface of rotating neutron stars, $\bar{\omega}_c/\Omega$ and $\bar{\omega}_s/\Omega$ respectively, are listed in table 16.2 for rotation at $\Omega = \Omega_{\rm K}$, where $\bar{\omega} = \Omega - \omega$. The dragging effect was shown to be largest (hence the effective rotational frequency $\bar{\omega}$ is smallest) at the center of rotating neutron stars with decreasing magnitude toward its surface [62, 106, 109, 116] (see also chapter 17). Table 16.2 reveals that $\bar{\omega}_s \approx (2\text{–}3)\,\bar{\omega}_c$, depending on the equation of state. In the limit when the rotational frequencies of the local inertial frames $\omega(r)$ are small in comparison with the star's rotational frequency Ω, the dragging effect can be ignored. Only in this limit is the angular velocity, that determines the magnitude of the centrifugal force acting on the star's matter, equal to the star's angular frequency Ω [115]. We restrict ourselves in the first part of our analysis $\omega/\Omega \ll 1$! By substituting equations (15.18) and (15.19) into the expression for the Kepler frequency, equation (16.7), one obtains for the orbital velocity of a mass element rotating at the star's equator [112],

$$V_{\text{no dragging}} = \Omega_{\rm c} \frac{R}{\sqrt{1 - 2MG/R}} \tag{16.12}$$

$$\longrightarrow R\,\Omega_{\rm c} \qquad \text{(Newtonian limit)}\,. \tag{16.13}$$

Again, the critical frequency $\Omega_{\rm c}$ in equations (16.12) and (16.13) is the one defined in (15.26). From (16.12) one arrives for the general relativistic Kepler frequency, with the neglect of frame dragging, that is, $\Omega \approx {\rm e}^{\nu-\psi}\, V_{\text{no dragging}}$ where ${\rm e}^{\nu-\psi} = \sqrt{1 - 2MG/R}/R$,

$$\Omega_{\rm K}|_{\text{no dragging}} = \sqrt{MG/R^3}\,. \tag{16.14}$$

One sees that the relativistic expression for $\Omega_{\rm K}$, with the neglect of frame dragging, coincides with the one of classical mechanics [510], expressing the circular movement of a massive particle in that stable orbit for which balance between centrifugal force and gravity occurs.

In the following we estimate the impact of *non-sphericity* on $\Omega_{\rm K}$. For this purpose we apply equation (16.14) to a mass element rotating at the star's equator. The quantities M and R in (16.14) are understood as the mass and equatorial radius of the rotating star. From equation (15.84) and the approximate relation $R_s \approx (R_{\rm p} + R_{\rm eq})/2$, one obtains $R_s \approx (1 - e^2/4) R_{\rm eq}$, with e the star's eccentricity. Likewise we express the mass of the rotating star as $M \approx (1 + \delta)\, M_s$. This leads for $\Omega_{\rm K}$ of equation (16.14) to

$$\Omega_{\rm K}|_{\text{no dragging}} \approx \left(1 + \frac{1}{2}\delta\right)\left(1 - \frac{3}{8}e^2\right)\Omega_{\rm c} \quad \text{(deformed star)}\,. \tag{16.15}$$

From table 16.1 we see that δ and e lie in fairly narrow ranges and that the most typical values are $\delta \approx 0.14$ and $e \approx 0.69$, yielding

$$\Omega_{\rm K}|_{\rm no\ dragging} \approx 0.88 \times \Omega_{\rm c} \,. \tag{16.16}$$

16.2.2 Inclusion of the frame dragging effect

16.2.2.1 Restriction to Schwarzschild metric

To obtain an analytic solution to the problem, we shall, in a first step, take the metric which corresponds to that of a static spherically symmetric star, that is, the Schwarzschild metric (14.14). This will serve to provide a first orientation. Corrections to this metric will be considered in the next sections. Thus at the equator we take

$$e^{2\nu} = 1 - \frac{2M}{R}\,, \qquad e^{2\psi} = R^2\,, \tag{16.17}$$

where for our approximate solution to equation (16.7) we take M to be the mass of the rotating star and R its equatorial radius. (The second of these equations looks somewhat strange, but we follow an old precedent so as not to introduce confusion [106, 562, 583].) Combined with the condition (15.88) that outside the star $\omega(r) = 2I\Omega/r^3$ $(r > R)$, where I is the moment of inertia, we are able to write an approximate solution to the transcendental equation for $\Omega_{\rm K}$, namely

$$\begin{aligned}
{\Omega_{\rm K}}^2 &= \left(1 + \frac{\omega(R)}{\Omega_{\rm K}} - 2\Big(\frac{\omega(R)}{\Omega_{\rm K}}\Big)^2\right)^{-1} \frac{M}{R^3} \\
&= \left(1 + \frac{2I}{R^3} - 2\Big(\frac{2I}{R^3}\Big)^2\right)^{-1} \frac{M}{R^3}\,.
\end{aligned} \tag{16.18}$$

This approximate result has a very interesting structure, for it shows the classical result modified by a prefactor. The prefactor leads to a *reduction* of the relativistic Kepler frequency when $\omega(R)/\Omega_{\rm K} < 1/2$ or, equivalently, $4I/R^3 < 1$. There is no apparent reason why this limit *must* be obeyed, even if in practice it is [62, 106, 112]. Therefore we proceed to an improved metric.

16.2.2.2 Monopole corrected metric

Here, we carry the analytic investigation one step further by taking monopole corrections to the Schwarzschild metric into account [112, 115, 116]. Since the monopole function h_0 is given by

$$e^{2\Phi(r)}\, h_0(r) = -\,\frac{\Delta M}{r} + \frac{J^2}{r^4}\,, \qquad r \geq R\,, \tag{16.19}$$

one obtains for the metric functions at the star's equator ($J \equiv I\Omega$)

$$e^{2\nu(R)} = 1 - \frac{2M}{R} + \frac{2J^2}{R^4}\,, \qquad e^{2\psi(R)} = R^2\,. \tag{16.20}$$

In this case the first relation in (16.17) reads

$$e^{2\nu} = 1 - \frac{2M}{R} + \frac{2J^2}{R^4}\,, \tag{16.21}$$

while the second, $e^{2\psi} = R^2$, remains unchanged. From (16.7) one finds for the Kepler frequency

$$\begin{aligned} {\Omega_K}^2 &= \left(1 + \frac{\omega(R)}{\Omega_K} - \left(\frac{\omega(R)}{\Omega_K}\right)^2\right)^{-1} \frac{M}{R^3} \\ &= \left(1 + \frac{2I}{R^3} - \left(\frac{2I}{R^3}\right)^2\right)^{-1} \frac{M}{R^3}\,. \end{aligned} \tag{16.22}$$

The prefactor in (16.22) *always* leads to a *reduction* of the Kepler frequency below its classical value because $\omega(R)/\Omega_K < 1$. The dragging frequency cannot exceed the star's frequency [115]. This universal limit is what the improved metric has bought. Finally, it may be of interest that equation (15.88) places a limit involving the moment of inertia and radius of a star, $2I/R^3 < 1$.

16.2.2.3 Quadrupole corrected metric

At the level of quadrupole corrections there are certain terms in the metric that we can investigate only numerically. For a broad sample of equations of state the terms not susceptible to an analytic analysis, which are those proportional to the associated Legendre polynomials, are shown to alter the Kepler frequency, equation (16.7), generally by less than 2 to 3%, depending on the equation of state. So we ignore them. Then the relevant metric functions through to quadrupole corrections due to rotation are given by

$$\begin{aligned} e^{2\nu} &= 1 - \frac{2M}{R} + \frac{J^2}{R^4}\left(3 + \frac{2M_s}{R} - \frac{R}{M_s}\right), \\ e^{2\psi} &= R^2\left(1 + \frac{J^2}{R^4}\left(2 + \frac{R}{M_s}\right)\right), \end{aligned} \tag{16.23}$$

where M_s is the mass of the star at the mass limit of the *non-rotating* and therefore spherical sequence (solution to the TOV equation). After considerable but straightforward algebra, an equation similar to those derived above is obtained,

$$\Omega_K^2 = \left(1 + (1+\eta_1)\left(\frac{\omega(R)}{\Omega_K}\right) - (2+\eta_2)\left(\frac{\omega(R)}{\Omega_K}\right)^2\right)^{-1} \frac{M}{R^3}\,, \tag{16.24}$$

where the auxiliary functions η_1 and η_2 are given by

$$\eta_1 = \frac{5}{2}\left(1 + \frac{2}{5}\frac{R}{M_s}\right) R^2\,\omega^2(R)\,, \tag{16.25}$$

and

$$\eta_2 = \eta_1 + \frac{1}{2}\left(1 + \frac{1}{4}\frac{R}{M_s}\right) R^2\,\omega^2(R)\left(\frac{{\Omega_{\rm K}}^2}{\omega^2(R)} - 1\right) - \frac{3}{2}\left(1 + \frac{5}{6}\frac{M_s}{R} - \frac{1}{4}\frac{R}{M_s}\right). \tag{16.26}$$

By means of the empirical formula (16.11) in the form $\Omega_{\rm K} = \alpha\sqrt{M_s/R_s^3}$, with $\alpha = 0.625 \simeq \sqrt{2/5}$, one obtains for equations (16.25) and (16.26),

$$\eta_1 = \frac{2}{5}\left(1 + \frac{\Delta R}{R_s}\right)^3 \left(1 + \frac{10}{4}\frac{M_s}{R}\right)\left(\frac{\omega(R)}{\Omega_{\rm K}}\right)^2, \tag{16.27}$$

and

$$\eta_2 = \frac{1}{20}\left(1 + \frac{\Delta R}{R_s}\right)^3 \left\{1 + 4\frac{M_s}{R} + 7\left(1 + \frac{16}{7}\frac{M_s}{R}\right)\left(\frac{\omega(R)}{\Omega_{\rm K}}\right)^2\right\} - \frac{3}{2}\left(1 + \frac{5}{6}\frac{M_s}{R} - \frac{1}{4}\frac{R}{M_s}\right), \tag{16.28}$$

with the definition $\Delta R \equiv R - R_s$, where R_s denotes the radius of the non-rotating maximum-mass star. For a wide selection of models [106, 107, 112], we have computed these parameters which we record in tables 16.3 and 16.4, together with the ratio of frame dragging to Kepler frequency of the limiting mass star as computed in general relativity. The phenomenon of frame dragging causes a *reduction* in the Kepler frequency if $\omega/\Omega_{\rm K} < (1+\eta_1)/(2+\eta_2)$ (obtained from equation (16.24), which we see is indeed satisfied by a comfortable margin in all cases). The results derived for $\Omega_{\rm K}$ in (16.24) and in the previous two subsections reduce to the classical expression (15.26) for a particle in stable orbit around a static relativistic star, since in that case $\omega(r) \equiv 0$.

To summarize, we have shown above how the effect of frame dragging on the Kepler frequency can be expressed as a factor, slightly model-dependent, times the classical expression for the balance between gravity and centrifuge at the equator R of a rotating star of mass M at the termination of the stable sequence. The empirical expression involves only the radius and mass of the corresponding spherically symmetric, non-rotating star [107, 108, 189],

$$\Omega_{\rm K} = \alpha\sqrt{\frac{M_s}{R_s^3}}\,, \qquad \alpha \simeq 0.63\text{–}0.67\,. \tag{16.29}$$

Table 16.3. The model-dependent parameters η_1, η_2 and the ratio $\omega(R)/\Omega$, all computed in general relativity. The last column is the limiting value of the ratio that leads to a reduction in Kepler frequency due to frame dragging.

Label †	EOS	η_1	η_2	ω/Ω_K	$(1+\eta_1)/(2+\eta_2)$
1	G_{300}	0.031	0.356	0.17	0.44
2	HV	0.040	0.110	0.20	0.49
3	G^{DCM2}_{B180}	0.025	0.358	0.16	0.43
4	G^{DCM2}_{265}	0.022	0.448	0.15	0.42
5	G^{π}_{300}	0.035	0.191	0.19	0.47
6	G^{π}_{200}	0.021	0.468	0.15	0.41
7	Λ^{00}_{Bonn}+HV	0.049	0.058	0.22	0.51
8	G^{DCM1}_{225}	0.026	0.388	0.16	0.43
9	G^{DCM1}_{B180}	0.029	0.318	0.17	0.44
10	HFV	0.078	−0.209	0.28	0.60
11	Λ^{00}_{HEA}+HFV	0.084	−0.217	0.29	0.61
12	BJ(I)	0.050	−0.024	0.22	0.53
13	UV_{14}+TNI	0.078	−0.102	0.27	0.57
14	V_{14}+TNI	0.087	−0.234	0.29	0.62
15	UV_{14}+UVII	0.119	−0.415	0.35	0.71
16	AV_{14}+UVII	0.128	−0.418	0.36	0.71
17	Pan(C)	0.073	−0.201	0.27	0.60

† These labels refer to the equations of state listed in table 12.4.

The M/R^3 term in (16.24) is reduced to the final form in (16.29) by accounting for the radius and mass augmentation caused by rotation [112]. Taking into account these augmentations in terms of the two prefactors introduced in equation (16.15), the total reduction of Ω_K can be written as

$$\Omega_K\big|_{\text{dragging}} \approx \left(1+\frac{1}{2}\delta\right)\left(1-\frac{3}{8}e^2\right) D\,\Omega_c \tag{16.30}$$

$$\approx (0.61\text{–}0.71)\,\Omega_c \qquad \text{(deformed star)}\,. \tag{16.31}$$

The first factor in (16.30) accounts for the mass increase due to rotation, the second for the non-sphericity, and $D \approx [1+(3M_s/M_\odot)/(2R_s/\text{km})]^{-1}$ $(0.69 \lesssim D \lesssim 0.81)$ as an estimate of the frame dragging factors in (16.24) [112]. The rotational flattening of the star and the general relativistic frame dragging effect reduce Ω_K from the classical value Ω_c by roughly the same amount. Most importantly, for the maximum-mass star, all three factors in (16.30) vary within a *narrow* range for the broad collection of equations of state studied which accounts for the validity and usefulness of the empirical formula. The frame dragging effect naturally occurs because centrifugal effects depend on the difference of the star's rotational

Table 16.4. See caption of table 16.3.

EOS †	η_1	η_2	$\omega/\Omega_{\rm K}$	$(1+\eta_1)/(2+\eta_2)$
L	0.060	0.074	0.23	0.51
PAL1	0.015	0.660	0.10	0.38
D	0.028	0.390	0.14	0.43
C	0.041	0.232	0.17	0.47
PAL3	0.021	0.445	0.13	0.42
FP	0.061	−0.013	0.21	0.53
F	0.023	0.348	0.13	0.44
A	0.052	0.082	0.20	0.51
π	0.063	−0.040	0.22	0.54
B	0.049	0.100	0.19	0.50
G	0.046	−0.003	0.20	0.52

† The notation for the equations of state is the same as in reference [107].

frequency Ω and the rotational frequencies of the local inertial frames ω, rather than on Ω alone. The result (16.31) is in excellent agreement with equation (16.11). The C-values associated with (16.31) lie in the range of $7000\ {\rm s}^{-1} \lesssim C \lesssim 8200\ {\rm s}^{-1}$, which compares rather well with the values $7200\ {\rm s}^{-1}$ and $7700\ {\rm s}^{-1}$ established by Friedman, Ipser, and Parker, and Haensel and Zdunik respectively. Elsewhere we established a value of $C \approx 8500\,{\rm s}^{-1}$ in the framework of Hartle's method [111]. This value appears to be somewhat on the large side.

In summary, the above analysis has shown that the dragging of local inertial frames caused by the rotation of any massive star *reduces* its Kepler (mass shedding) frequency relative to the Kepler period of a satellite in circular orbit around a non-rotating star, contrary to the intuitive expectation that naturally follows from the fact that the centrifugal force on fluid elements of the star is determined by the frequency of the star relative to the local inertial frames which are dragged along in the direction of the star's rotation. This counter-intuitive behavior can be understood mathematically as following from the fact that equation (16.7) is not a formula for $\Omega_{\rm K}$, but a transcendental equation, in which all quantities on the right depend also on $\Omega_{\rm K}$ and on $\omega(r)$. Thus to say that the centrifugal effect on a fluid element of the star at r depends on $\Omega_K - \omega(r)$, while true, does not inform one that there is a reduction in the centrifugal effect with corresponding increase in the Kepler frequency. We mention that this counter-intuitive behavior of the role of frame dragging, though a peculiar effect of rotation, has nothing to do with the still more bizarre 'change in sign of the centrifugal force' in the vicinity of black holes which, as the discoverers of this latter effect emphasize, has nothing to do with frame dragging since it holds for a satellite in orbit around a Schwarzschild black hole [584, 585].

16.3 Gravitational-radiation reaction driven instability

Besides the absolute upper limit on rotation set by the Kepler frequency, there is another (non-axisymmetric) instability that sets in at a somewhat *lower* rotational frequency, and which therefore appears to set a more stringent limit on stable rotation than mass shedding [586]. The instability is driven by gravitational waves. It was discovered by Chandrasekhar [587] and shown to be generic – i.e. it tends to make *all* rotating stars unstable if no other physical mechanism, like viscous damping, were operating – by Friedman and Schutz [576, 588]. The instability, known as the gravitational-radiation reaction (GRR) driven instability, or simply gravity-wave instability, will be described in the following sections.

16.3.1 Qualitative picture

The GRR driven instability originates from counter-rotating surface vibrational modes on rotating neutron stars, which at sufficiently high rotational star frequencies are dragged forward along in the star's rotational direction. Following Lindblom [589], the action of the GRR driven instability can be understood as follows. Consider a neutron star that rotates with angular velocity Ω, and a small perturbation of this star having time dependence $e^{-i\omega_m(\Omega)t}$ and angular dependence $e^{im\phi}$, with m an integer. This perturbation propagates with angular velocity ω/m in the direction opposite to the star's rotation (assuming $m > 0$) in sufficiently slowly rotating stars. Figure 16.2 illustrates this situation graphically. These

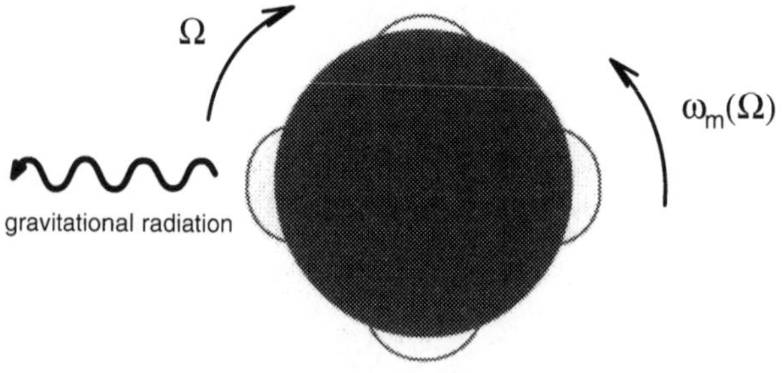

Figure 16.2. Representation of an $m = 4$ perturbation of a rotating neutron star. Ω denotes the star's rotational frequency, ω_m is the frequency of the counter-rotating perturbation (after Lindblom [589]).

perturbations create time-dependent mass-multipoles, which cause the star to emit gravitational radiation having negative angular momentum. The perturbation itself has negative angular momentum – since it propagates in the direction opposite to the star's rotation – and so the gravitational radiation reduces the perturbation's amplitude in order to conserve angular momentum. Thus gravitational radiation damps out the perturbation. In sufficiently rapidly rotating stars these perturbations are forced to move in the opposite direction. The waves are in effect dragged along by the fluid in the star. In this case the star emits gravitational radiation having positive angular momentum. Since the angular momentum in the perturbation is negative – it still propagates against the rotational flow of the star – the perturbation's amplitude must *grow* in order to conserve angular momentum. Thus, any counter-rotating perturbation will become unstable when the star rotates rapidly enough to force it to corotate with the star.

This GRR driven instability is quite generic. These perturbations are rather superficial and propagate much like waves on the surface of the ocean. Thus, the speed of the wave relative to the matter is rather independent of the rotation of the star. So, to a fairly good approximation, the angular velocity dependence of the frequency of a perturbation is given by $\omega_m(\Omega) \approx \omega_m(0) + m\Omega$ [589], where $\omega_m(0)$ is the frequency when the star is not rotating. These perturbations will reverse direction at that star's angular velocity where $\omega_m(\Omega)$ passes zero, that is, when $\Omega \approx -\omega_m(0)/m$. From this formula, it is easy to see why the GRR driven instability is generic. The frequencies of the modes on non-rotating stars increase with m roughly as $\omega_m(0) \propto \sqrt{m}$. Thus the angular velocity where a perturbation becomes unstable varies with m approximately as $\Omega \propto 1/\sqrt{m}$. Hence, an unstable perturbation can be found in *any* rotating star, simply by choosing m sufficiently large. It is known of course that all rotating stars are not unstable. The argument outlined above merely shows that some other physical mechanism must be operative to prevent gravitational radiation from driving these perturbations unstable. One such mechanism is internal dissipation in the stellar matter [590]. Viscosity and thermal conductivity quickly damp out any large gradients in the velocity or thermal perturbations. Those perturbations with angular dependence $e^{im\phi}$ have gradients that increase as m increases. Thus, the time scales for the internal dissipation mechanisms to damp out a perturbation tend to decrease as m increases. In contrast, the time scale for gravitational radiation to drive a perturbation unstable becomes very long as m gets large. This is because the radiation couples more weakly to the higher-mass multipoles. Thus, for sufficiently large m, viscosity will suppress the GRR driven instability [591]. As a consequence, sufficiently slowly rotating stars are *stable*. This behavior manifests itself in the time dependence of a perturbation as

$$\propto \exp(-i\omega_m(\Omega)t + im\phi - t/\tau_m(\Omega)) , \tag{16.32}$$

where $\omega_m(\Omega)$ is the frequency of the surface mode which depends on the star's angular velocity Ω, ϕ denotes the azimuthal angle, and τ_m is the time scale for the mode which determines its *growth* or *damping*. Both $\omega_m(\Omega)$ and $\tau(\Omega)$ are real. The rotation frequency Ω at which $\tau(\Omega)$ changes sign, causing an exponential growth of the perturbation (16.32), is the critical frequency for the particular mode m. The mathematical expression which determines this critical frequency is derived in the next section.

16.3.2 Oscillations of rotating stars

In order to determine which rotating neutron stars are actually stable, a detailed analysis of their perturbations must be carried out which includes the influence of gravitational radiation and viscosity. Following closely the excellent reviews of Friedman and Ipser [537], and Lindblom [589], we next outline the techniques that have been developed during the past several years by Ipser and Managan [592], Managan [593], Ipser and Lindblom [594, 595, 596, 597], and Cutler and Lindblom [598] for treating the normal-mode equations of rapidly rotating stars. The formal development is complete only in the Newtonian limit, where it has been successfully used to obtain, for the first time, the non-axisymmetric modes of rotating stars. (Up to now the analysis has not been carried out in general relativity theory.)

This method involves a reformulation of the stellar-pulsation equations in terms of two independent potential functions. These are $\delta\Phi$, the Eulerian perturbation of the gravitational potential, and δU, the difference between the Eulerian perturbation of the enthalpy and $\delta\Phi$. When the eigenequations are rewritten in terms of δU and $\delta\Phi$, a relatively simple fourth-order system of equations is obtained, which can be solved successfully for both the eigenfrequencies and eigenfunctions of normal modes. The basic equations governing the evolution of a Newtonian fluid configuration are the continuity equation, Euler's equation, and Poisson's equation (cf. (14.61)),

$$\frac{\partial \epsilon}{\partial t} + \nabla_i(\epsilon\, v^i) = 0\,, \tag{16.33}$$

$$\epsilon\left(\frac{\partial v^i}{\partial t} + v^j \nabla_j v^i\right) = -\nabla^i P + \epsilon \nabla^i \Phi\,, \tag{16.34}$$

$$\nabla^i \nabla_i \Phi = -4\pi\,\epsilon\,. \tag{16.35}$$

The variables ϵ and P denote, as usual, mass (energy) density and pressure. v^i and Φ are the three-velocity vector and the gravitational potential, respectively. $\nabla_i = \partial/\partial x^i$ is the standard Euclidean covariant derivative (compare with table 1.7). The three-dimensional Euclidean metric g_{ij} (i.e. the identity matrix in Cartesian coordinates) and its inverse g^{ij} are used to raise and lower tensor indices. We are interested in the pulsations of an equilibrium stellar model that is axisymmetric and rotating, perhaps differentially, about its z-axis. Hence the unperturbed velocity field is of the form $v^i = \Omega\phi^i$, where Ω is the equilibrium angular velocity and ϕ^i is a rotational Killing vector field that satisfies

$$\nabla_i \phi_j + \nabla_j \phi_i = 0\,. \tag{16.36}$$

Our demand that meridional circulation be absent from the equilibrium state implies that (16.34) has the first integral [537]

$$\frac{1}{2}\Omega^2 \nabla_i \varpi^2 = \frac{(\nabla_i P)}{\epsilon} - \nabla_i \Phi\,, \tag{16.37}$$

where ϖ is the standard cylindrical radial coordinate.

Given an equilibrium configuration, we focus on small perturbations of its structure and motions away from equilibrium. The evolution equations governing these perturbations are obtained by linearizing equations (16.33) through (16.35) in the perturbations. This yields the following perturbed equations [537],

$$\frac{\partial \delta\epsilon}{\partial t} + v^i \nabla_i \delta\epsilon + \nabla_i(\epsilon\, \delta v^i) = 0\,, \tag{16.38}$$

$$\left(\frac{\partial \delta v^i}{\partial t} + v^j \nabla_j \delta v^i + \delta v^j\, \nabla_j \delta v^i\right) = -\frac{\nabla^i \delta P}{\epsilon} + \delta\epsilon\, \frac{\nabla^i P}{\epsilon^2} + \nabla^i\, \delta\Phi\,, \tag{16.39}$$

$$\nabla^i \nabla_i\, \delta\Phi = -4\pi\, \delta\epsilon\,. \tag{16.40}$$

The symbol δ denotes the Eulerian perturbation of a quantity. Quantities not preceded by a δ are understood to be equilibrium values. The system of perturbed equations is completed by assuming that the Lagrangian change in pressure is proportional to the Lagrangian change in the density,

$$\Delta P = \delta P + \xi^i \nabla_i P = \frac{\Gamma P}{\epsilon} \Delta\epsilon = \frac{\Gamma P}{\epsilon}(\delta\epsilon + \xi^i \nabla_i \epsilon)\,, \tag{16.41}$$

where ξ^i is the Lagrangian displacement [537]. It is related to the velocity perturbation by the relation

$$\delta v^i = \frac{\partial \xi^i}{\partial t} + v^j \nabla_j\, \xi^i - \xi^j \nabla_j\, v^i\,. \tag{16.42}$$

The traditional analysis of the perturbation equations involves eliminating δv^i in favor of ξ^i. The alternative method that has been developed recently proceeds the other way round, by eliminating ξ^i in favor of δv^i, and by then eliminating δv^i itself in a way that is described next.

Let us focus on the normal-mode solutions to the above perturbation equations, i.e. those solutions with time dependence $e^{i\omega t}$ and azimuthal-angle dependence $e^{im\phi}$, where ω is the mode frequency and m is an integer, as in the previous section. For these modes, (16.42) yields an algebraic expression for ξ^i in terms of δv^j (or vice versa) [537],

$$\xi_i = -\,\mathrm{i}\left(g_{ij}\,\hat{\omega}^{-1} - (\mathrm{i}\,\phi_i \nabla_j \Omega)\,\hat{\omega}^{-2}\right)\delta v^j\,, \tag{16.43}$$

where $\hat{\omega} \equiv \omega + m\Omega$. Equation (16.41) now takes the form

$$\delta\epsilon = \frac{\epsilon}{\Gamma P}\,\delta P + \mathrm{i}\,\frac{\epsilon^2}{\hat{\omega}}\,A_i \delta v^i\,, \tag{16.44}$$

where

$$A_i \equiv \frac{\nabla_i \epsilon}{\epsilon^2} - \frac{\nabla_i P}{\Gamma \epsilon P}\,. \tag{16.45}$$

Note that $A_i = 0$ for barotropic configurations, which have an adiabatic index $\Gamma = (\epsilon/P)(\mathrm{d}P/\mathrm{d}\epsilon)$. For general configurations, such as those in which the pressure is not a unique function of density, A_i does not vanish. With $\delta\epsilon$ eliminated via (16.44), equation (16.41) has the representation [537],

$$\begin{aligned} \mathrm{i}\,\hat{\omega}\,Q^{-1}_{ij}\,\delta v^j &\equiv \left(\mathrm{i}\,\hat{\omega} g_{ij} + 2\nabla_j v_i - \phi_i \nabla_j \Omega - \mathrm{i}\,\frac{\nabla_i P A_j}{\hat{\omega}}\right)\delta v^j \\ &= -\;\nabla_i \delta U - \epsilon\,(\delta U + \delta\Phi)\,A_i\,, \end{aligned} \tag{16.46}$$

where

$$\delta U \equiv \frac{\delta P}{\epsilon} - \delta\Phi\,. \tag{16.47}$$

Equation (16.46) is algebraic in δv^i and can be solved for δv^i as long as the tensor Q^{-1}_{ij} is invertible. In this case the solution to (16.46) for δv^i is [537]

$$\delta v^i = \mathrm{i}\,Q^{ij}\,\nabla_j \delta U + \mathrm{i}\,\epsilon\,(\delta U + \delta\Phi)\,Q^{ij}\,A_j\,, \tag{16.48}$$

with

$$\begin{aligned} Q^{ij} = \frac{\lambda}{\hat{\omega}^3}\{&(\hat{\omega}^2 - A^k \nabla_k P) g^{ij} - 2\omega^i \Omega^j + \mathrm{i}\hat{\omega}\phi^i \nabla^j \Omega - 2\mathrm{i}\hat{\omega}\Omega \nabla^i \phi^j \\ &+ \nabla^i P A^j + (\mathrm{i}/\hat{\omega})\phi^i \phi^{jk} \nabla_k P A^l \omega_l - (2\mathrm{i}/\hat{\omega})\phi^j \phi^{ik} A_k \Omega^l \nabla_l P\}\,, \end{aligned} \tag{16.49}$$

and the definition

$$\det\, Q^{-1\,i}{}_j \equiv \frac{\hat{\omega}^3}{\lambda} = \frac{1}{\hat{\omega}}\left(\hat{\omega}^4 - \hat{\omega}^2 A^i \nabla_i P - 2\hat{\omega}^2 \Omega^i \omega_i + 2A^i \omega_i \Omega^j \nabla_j P\right). \tag{16.50}$$

Here $\omega^i = \epsilon^{ijk}\nabla_j v_k$ is the fluid's velocity, ϵ_{ijk} is the antisymmetric tensor (whose components are ± 1 in Cartesian coordinates), the fluid angular velocity vector $\Omega^i = \Omega z^i$ with z^i a unit vector parallel to the rotation axis, and $\phi^{ij} = \epsilon^{ijk}\phi_k/\phi^l\phi_l$. We use (16.48) to eliminate δv^i in favor of δU and $\delta\Phi$, and the perturbation $\delta\epsilon$ can be eliminated by combining equations (16.38), (16.47), and (16.44), which yields

$$\delta\epsilon = \Psi_1\,(\delta U + \delta\Phi) - \frac{\epsilon^2}{\hat{\omega}}\,A_i\,Q^{ij}\,\nabla_j\delta U\,, \tag{16.51}$$

where $\Psi_1 = \epsilon^2/\Gamma P - (\epsilon^3/\hat{\omega})A_iQ^{ij}A_j$. We have not yet made use of equations (16.38) and (16.40). It is these that now provide the fundamental set of coupled eigenequations for the two potentials δU and $\delta\Phi$. Using (16.47), (16.48), and (16.51) to eliminate δP, $\delta\epsilon$, and δv^i, we are able to re-express equations (16.38) and (16.40) as the fourth-order system [537],

$$\nabla_i(\epsilon\,Q^{ij}\nabla_j\delta U) + \Psi_3\,\delta U = -\,\epsilon^2\,Q^{ij}A_j\nabla_i\delta\Phi - \Psi_2\,\delta\Phi\,, \tag{16.52}$$

$$\nabla^i\nabla_i\,\delta\Phi + 4\,\pi\,\Psi_1\,\delta\Phi = \frac{4\,\pi\,\epsilon^2}{\hat{\omega}}A_iQ^{ij}\nabla_j\delta U - 4\,\pi\,\Psi_1\,\delta U\,, \tag{16.53}$$

where

$$\Psi_2 = \hat{\omega}\,\Psi_1 + \nabla_i(\epsilon^2\,Q^{ij}\,A_j)\,,$$
$$\Psi_3 = \Psi_2 - \frac{m\lambda}{\hat{\omega}^3}\epsilon^2\Big(\hat{\omega}A^i\nabla_i\Omega + \frac{4\hat{\omega}\Omega}{\varpi}A^i\nabla_i\varpi + \frac{1}{\hat{\omega}}A_i\phi^{ij}\nabla_jPA^k\omega_k\Big)\,.$$

It is easy to verify that (16.52) and (16.53) are real equations for $\delta U(z,\varpi)$ and $\delta\Phi(z,\varpi)$, where $\delta U = \delta U(z,\varpi)\mathrm{e}^{\mathrm{i}\omega t + \mathrm{i}m\phi}$ and $\delta\Phi = \delta\Phi(z,\varpi)\mathrm{e}^{\mathrm{i}\omega t + \mathrm{i}m\phi}$. We complete the specification of the eigenvalue problem by imposing appropriate boundary conditions at the stellar surface and at $r = \infty$, where r is the spherical radial coordinate. At the stellar surface the appropriate boundary condition is that the Lagrangian change in the pressure vanishes, i.e. that

$$\nabla P = \delta P + \xi^i\nabla_i P = 0 \tag{16.54}$$

at the stellar surface. It follows from (16.43), (16.47), and (16.48) that this condition can be rewritten as

$$\delta U + \delta\Phi + \frac{\nabla_i P}{\epsilon}Q^{ij}\{\nabla_j\delta U + \epsilon A_j(\delta U + \delta\Phi)\} = 0\,. \tag{16.55}$$

The remaining boundary condition is that $\delta\Phi \to 0$ sufficiently rapidly as $r \to \infty$. For a perturbation with ϕ-coordinate dependence $\mathrm{e}^{\mathrm{i}m\phi}$, examination of the expansion of $\delta\Phi$ as a power series in negative powers of r reveals that this condition can be expressed as [537]

$$\frac{\partial}{\partial r}\,\delta\Phi + \sum_{l\geq|m|}\frac{(l+1)(2l+1)(l-m)!}{2r(l+m)!}\,P_l^m(\mu)\int_{-1}^{+1}\mathrm{d}\mu'\;\delta\Phi(r,\mu')\,P_l^m(\mu') = 0\,, \tag{16.56}$$

where $\mu = \cos\theta$, where θ is the polar angular coordinate, and P_l^m the associated Legendre function.

The numerical method that has been developed and used to solve the eigenequations directly for the frequency ω and the eigenfunctions δU and $\delta\Phi$ is described in detail by Ipser and Lindblom [594, 595]. Once these quantities have been determined, it is relatively easy to evaluate the effects of (weak) dissipation on the pulsation of a star. To this end, it is useful to introduce the following 'energy' associated with the pulsations [537, 589],

$$E(t) = \frac{1}{2}\int \mathrm{d}^3\boldsymbol{x}\,\left(\epsilon\,\delta v^i\delta v_i^* + \frac{1}{2}\left(\delta\epsilon\delta U^* + \delta\epsilon^*\delta U\right)\right), \tag{16.57}$$

where δv_i^* etc represents the complex conjugate quantity. This energy is conserved, $\mathrm{d}E/\mathrm{d}t = 0$, in the absence of dissipation. In general its time derivative is determined by the equations for the evolution of a viscous fluid coupled to gravitational radiation,

$$\frac{\mathrm{d}E}{\mathrm{d}t} = -\int \mathrm{d}^3\boldsymbol{x}\,\left(2\,\eta\,\delta\sigma^{ij}\delta\sigma_{ij}^* + \zeta\,\delta\sigma\delta\sigma^*\right) - \hat{\omega}\sum_{l=l_{\rm min}}^{\infty} N_l\,\omega^{2l+1}\,\delta D_l^m\delta D_l^{*m}\,. \tag{16.58}$$

In this expression $l_{\rm min}$ is the larger of 2 or $|m|$. The functions ζ and η are the bulk viscosity coefficient and shear viscosity coefficient, while $\delta\sigma^{ij}$ and $\delta\sigma$ are the shear and expansion of the perturbed fluid motion,

$$\begin{aligned}\delta\sigma^{ij} &= \frac{1}{2}\left(\nabla^i\delta v^j + \nabla^j\delta v^i - \frac{2}{3}\,g^{ij}\,\nabla_k\delta v^k\right),\\ \delta\sigma &= \nabla_i\delta v^i\,.\end{aligned} \tag{16.59}$$

The gravitational-radiation energy loss is determined by the mass multipole moment

$$\delta D_l^m = \int \mathrm{d}^3\boldsymbol{x}\;\delta\epsilon\, r^l\left(Y_l^m\right)^* \tag{16.60}$$

and the coupling constant N_l,

$$N_l = \frac{4\,\pi}{c^{2l+1}}\,\frac{(l+1)(l+2)}{l(l-1)[(2l+1)!!]^2}\,. \tag{16.61}$$

When dissipation is present in the star, it is convenient to represent the time dependence of a perturbation in the form $\mathrm{e}^{-\mathrm{i}\omega(\Omega)t - t/\tau(\Omega)}$, as described in equation (16.32), where ω and τ are real. The energy $E(t)$ defined in (16.57) is a real function that is quadratic in the perturbation variables. Thus, its time derivative is readily found to be given by

$$\frac{\mathrm{d}E}{\mathrm{d}t} = -\,\frac{2\,E}{\tau}\,. \tag{16.62}$$

This formula can be used to evaluate $1/\tau$. The integrals in (16.57) and (16.58) which determine E and $\mathrm{d}E/\mathrm{d}t$ may be evaluated to lowest order in the strength of the dissipative forces by using non-dissipative values of the frequency ω and the potentials δU and $\delta\Phi$. Once evaluated, these integrals determine $1/\tau$ via equation (16.62). It is convenient to decompose the imaginary part of the frequency into contributions from each of the dissipative forces as $1/\tau(\Omega) \equiv 1/\tau_\zeta(\Omega) + 1/\tau_\eta(\Omega) + 1/\tau_{\mathrm{GR}}(\Omega)$, where the individual damping times are defined, using equations (16.58) and (16.62), by the following integrals,

$$\frac{1}{\tau_\zeta} = \frac{1}{2\,E} \int \mathrm{d}^3\boldsymbol{x}\; \zeta\, \delta\sigma\, \delta\sigma^* \,, \tag{16.63}$$

$$\frac{1}{\tau_\eta} = \frac{1}{E} \int \mathrm{d}^3\boldsymbol{x}\; \eta\, \delta\sigma^{ij}\, \delta\sigma^*_{ij} \,, \tag{16.64}$$

$$\frac{1}{\tau_{\mathrm{GR}}} = \frac{\hat{\omega}}{2\,E} \sum_{l=l_{\mathrm{min}}}^{\infty} N_l\, \omega^{2l+1}\, \delta D_l^m\, \delta D_l^{*m} \,. \tag{16.65}$$

A perturbation on a star is stable whenever the imaginary part of the frequency of that perturbation $1/\tau$ is positive. Stars whose angular velocities do not exceed the smallest root of the equation $1/\tau(\Omega_{\mathrm{GRR}}) = 0$ are *stable.* The problem of determining the maximum angular velocity of a neutron star has been reduced, therefore, to finding the values of the critical angular velocities, Ω_{GRR}, at which the gravitational-radiation reaction instability sets in, which are given as the roots of the equation

$$0 = \frac{1}{\tau(\Omega_{\mathrm{GRR}})} = \frac{1}{\tau_\zeta(\Omega_{\mathrm{GRR}})} + \frac{1}{\tau_\eta(\Omega_{\mathrm{GRR}})} + \frac{1}{\tau_{\mathrm{GR}}(\Omega_{\mathrm{GRR}})} \,. \tag{16.66}$$

The integrals in equations (16.57) and (16.63) to (16.65) are easily evaluated once the non-dissipative pulsation problem has been solved. Then Ω_{GRR} is determined from (16.66) for each solution to the perturbation equations. The *smallest* of these critical angular velocities is, therefore, the maximum angular velocity of a stable neutron star.

Before the dissipation time scales $\tau_\zeta(\Omega)$ and $\tau_\eta(\Omega)$ can be determined, expressions for the bulk and shear viscosity coefficients ζ and η must be given [564]. *Bulk viscosity* arises in neutron star matter because the pressure and density perturbations become slightly out of phase due to the long time scale needed for the weak interaction to reestablish local thermodynamic equilibrium. Sawyer calculates the bulk viscosity of neutron star matter to be $\zeta = 6.0 \times 10^{-59} \epsilon^2 \omega^{-2} T^6$ in cgs units [599]. *Shear viscosity* in neutron star matter is primarily the result of neutron–neutron scattering, provided the temperature exceeds the superfluid transition temperature. Flowers and Itoh calculate this form of shear viscosity to be approximately given by $\eta = 347 \epsilon^{9/4} T^{-2}$ [600, 601]. If the temperature does not exceed the superfluid transition temperature, the neutron star matter is in the superfluid phase and the shear viscosity arises primarily from electron scattering in the superfluid, in which case $\eta = 6.0 \times 10^{18}(\epsilon/10^{15}\,\mathrm{g\ cm^{-3}})^2\ T^{-2}$ [600]. Lindblom and Mendell, however, claim that at such temperatures friction between electrons and neutron vortices (electron–vortex scattering) is the dominant dissipative mechanism and that it is always large enough to damp the non-axisymmetric instability [602, 603]. We shall come back to this in the following section.

16.3.3 Application to $l = m$ f-modes

As the discussion in the previous section 16.3.2 indicates, the modes of primary interest here are those which propagate in the direction opposite to the star's rotation. These are the modes which may become unstable via the emission of gravitational radiation in sufficiently rapidly rotating stars. The modes that are the most susceptible to this instability are the $l = m$ f-modes [604], those having no radial nodes in the non-rotating limit of uniformly rotating polytropes [595]. These configurations have an equation of state and adiabatic index given by $P = K\epsilon^{1+1/n}$ and $\Gamma = 1 + 1/n$, respectively, which implies that $A_i = 0$. Realistic equations of state for neutron star matter have $n \simeq 1$. The following discussion will focus on this value. Most of the results computed for the $l = m$ f-mode instability are independent of the parameter K in the polytropic equation of state. For numerical purposes its value is chosen to make the physical size of these models comparable to those based on more realistic equations of state.

The dependence of the eigenfrequency ω on the star's angular velocity Ω is conveniently expressed in terms of the function

$$\alpha_m(\Omega) = \frac{\omega(\Omega) - m\,\Omega}{\omega(0)}, \tag{16.67}$$

with $m \geq 2$. The $\alpha_m(\Omega)$ are very slowly varying functions of Ω with $\alpha_m \approx 1$ over the entire range of angular velocities [595]. This fact justifies the argument given in subsection 16.3.2, that the frequency of these modes is given approximately by $\omega_m(\Omega) \approx \omega(0) + m\Omega$. A reasonable first step is to replace the $\alpha_m(\Omega)$ by their corresponding Maclaurin spheroid functions α_m [586, 605]. We therefore take $\alpha_m(\Omega)$ as calculated in [594, 595] for the oscillations of rapidly rotating inhomogeneous Newtonian stellar models (polytropic index $n = 1$). Post-Newtonian corrections have been shown to modify α_m only very little [598].

The angular velocity dependences of the damping times is most conveniently expressed as dimensionless functions,

$$\gamma_m(\Omega) = \frac{\omega_m(\Omega)}{\omega_m(0)} \left(\frac{\tau_\eta(0)}{\tau_{\mathrm{GR}}(0)} \, \frac{\tau_{\mathrm{GR}}(\Omega)}{\tau_\eta(\Omega)} \right)^{1/(2m+1)},$$

$$\epsilon_m(\Omega) = \frac{\tau_\zeta(0)}{\tau_\eta(0)} \frac{\tau_\eta(\Omega)}{\tau_\zeta(\Omega)} . \tag{16.68}$$

These functions are independent of the temperature of the neutron star matter and the parameter K that appears in the polytropic equation of state. These functions too are very slowly varying except for the very highest angular velocities Ω. As for α_m, in a reasonable first step we replace $\gamma_m(\Omega)$ by their Maclaurin spheroid counterparts [586, 605], that is, calculated for uniform-density Maclaurin spheroids (polytropic index $n = 0$). Managan has shown that the critical angular frequencies $\Omega_{\rm GRR}$ depend much more strongly on the equation of state and the mass of the neutron star model – through $\omega_m(0)$ and $\tau_{{\rm GR},m}$ – than on the polytropic index assumed in calculating α_m [606].

The effects of bulk viscosity are suppressed in spherical stars because the non-radial pulsations have very little expansion $\delta\sigma$ associated with them. In very rapidly rotating stars, however, spherical symmetry is broken and the pulsations are no longer constrained to have small $\delta\sigma$. As a consequence the τ_ζ are much shorter in rapidly rotating stars [589].

With the angular velocity dependence of the damping times $\tau_\zeta(\Omega)$, $\tau_\eta(\Omega)$, and $\tau_{\rm GR}(\Omega)$, equation (16.66) can be solved for the critical angular velocities $\Omega_{\rm GRR}$ where the perturbation becomes unstable. The numerical determination of $\Omega_{\rm GRR}$ is made easier by transforming equation (16.66) into the form [586, 589],

$$\Omega_{\rm GRR} \equiv \Omega_m^\nu = \frac{\omega_m(0)}{m} \times \left\{ \alpha_m(\Omega_m^\nu) + \gamma_m(\Omega_m^\nu) \Big(\frac{\tau_{{\rm GR},m}(0)}{\tau_{\nu,m}(0)} + \frac{\tau_{{\rm GR},m}(0)}{\tau_{\zeta,m}(0)} \, \epsilon_m(\Omega_m^\nu) \Big)^{1/(2m+1)} \right\} , \tag{16.69}$$

where

$$\omega_m(0) \equiv \sqrt{\frac{2m(m-1)}{2m+1} \frac{M}{R^3}} \tag{16.70}$$

is the frequency of the vibrational mode in a non-rotating star. The time scales for gravitational-radiation reaction [607], $\tau_{{\rm GR},m}$, and for viscous damping time [608], $\tau_{\nu,m}$, are given by

$$\tau_{{\rm GR},m} = \frac{2}{3} \frac{(m-1)\,[(2m+1)!!]^2}{(m+1)\,(m+2)} \Big(\frac{2m+1}{2m(m-1)}\Big)^m \Big(\frac{R}{M}\Big)^{m+1} R , \tag{16.71}$$

$$\tau_{\nu,m} = \frac{R^2}{(2m+1)\,(m-1)} \frac{1}{\nu} . \tag{16.72}$$

The shear viscosity is denoted by ν. It depends on the star's temperature as $\nu(T) \propto T^{-2}$. It is small in very hot ($T \approx 10^{10}$ K) and therefore young

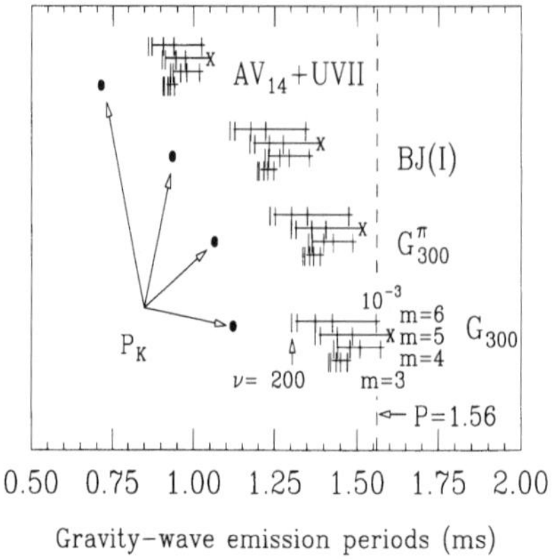

Figure 16.3. Critical rotational periods P_m^ν and P_K of a $M = 1.3\,M_\odot$ star at which instability against gravitational-radiation reaction (GRR) and mass shedding sets in, respectively, computed for different EOSs, table 12.4 [447]. The dashed vertical line indicates the rotational period of the fast pulsar PSR 1937+21 which is $P = 1.56$ ms.

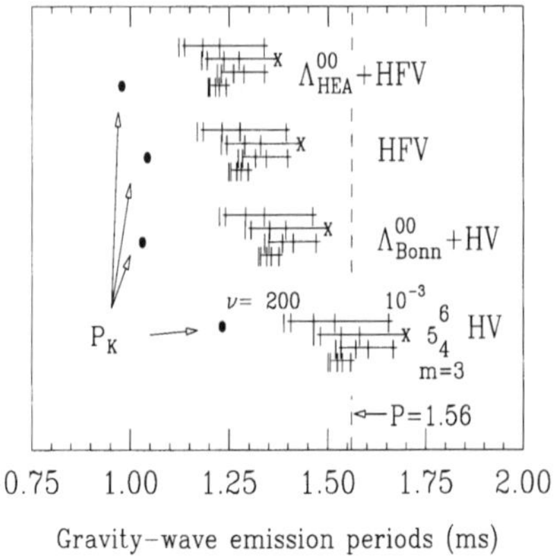

Figure 16.4. Critical rotational periods, as for figure 16.3, also for $M = 1.3\,M_\odot$, but for a different sample of EOSs [447].

stars and larger in cold ones. A characteristic feature of (16.69) to (16.72) is that Ω_m^ν merely depends on radius and mass (R and M) of the spherical star model.

The numerical outcome for the critical rotational star periods P_m^ν, defined in terms of Ω_m^ν as $P_m^\nu = 2\pi/\Omega_m^\nu$, at which the GRR driven instability is excited, is surveyed in figures 16.3 to 16.8 for neutron star masses ranging from $M = 1.30\,M_\odot$ up to the limiting mass [110]. In this connection, we recall the mass of the binary pulsar PSR 1913+16 which has been accurately determined to be $(1.442 \pm 0.003)\,M_\odot$ [232], and that the observed neutron star masses roughly cover the range $1.1 \lesssim M/M_\odot \lesssim 1.8$ (see chapter 3). Furthermore, typical rotational frequencies of fast pulsars lie in the millisecond range, as shown in tables 1.1 and 1.2. As an example, we pick out the frequency of the fast pulsar PSR 1937+21, given by $\Omega_{1937+21} = 4033\ \mathrm{s}^{-1}$, which corresponds to a rotational period of $P_{1937+21} = 1.56$ ms [48].

The stellar properties needed to solve (16.69) for Ω_m^ν are computed for the broad collection of equations of state listed in table 12.4. Each figure displays the instability periods, indicated by tick marks, associated with the instability modes $m = 3, 4, 5, 6$ (in this order from bottom to

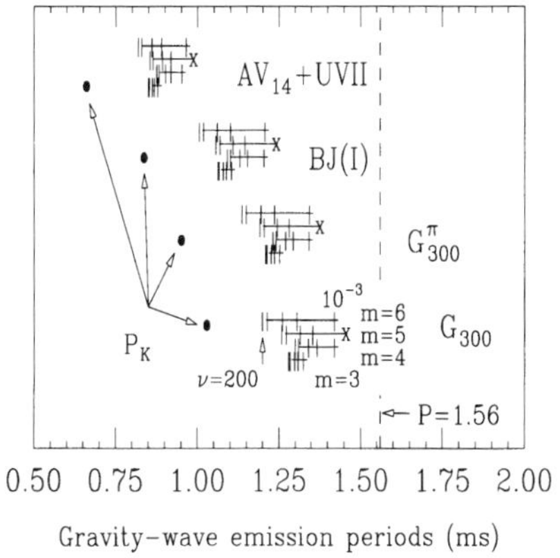

Figure 16.5. Same as figure 16.3, but for a stellar mass of $M = 1.5\,M_{\odot}$ and the same sample of EOSs [447].

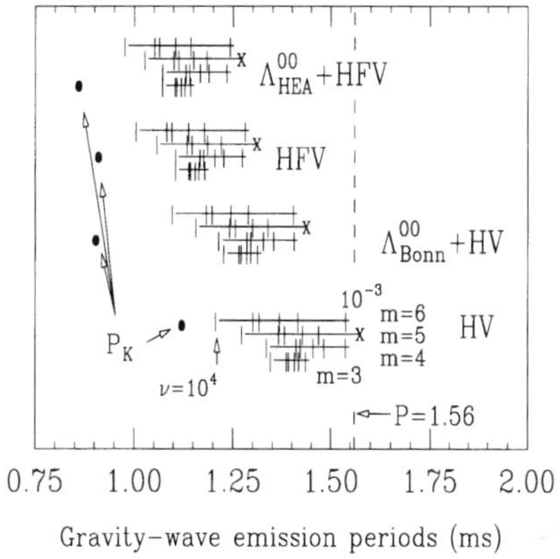

Figure 16.6. Same as figure 16.4, but for a stellar mass of $M = 1.5\,M_{\odot}$ and the same sample of EOSs [110]. (Reprinted courtesy of *Z. Phys.*)

top), which are the relevant ones that set the limit for rapidly spinning neutron stars. This is known from earlier investigations [106, 577, 586] where it was found that the instability modes with $m \gtrsim 6$ are damped by viscosity, which is confirmed by our results. The dependence of the instability periods on viscosity, and thus on temperature, is shown by the variation of the tick marks along a given horizontal line, along which m is fixed, from left to right. In this order the shear viscosity decreases for each m from $\nu = 200$ cm^2 s^{-1} to 100, 10, 1, to 10^{-3} cm^2 s^{-1}. The latter value was adopted to describe zero-shear-viscosity matter. For the purpose of comparison, we included in figure 16.6 the case of $\nu = 10^4$ cm^2 s^{-1} too. Shear viscosity values as large as $\nu = 10^7$ cm^2 s^{-1} will be discussed later in figures 16.10 through 16.17. This range of ν values, from $\nu = 10^7$ cm^2 s^{-1} to $\nu \sim 0$, corresponds to stellar temperatures between $T \sim 10^6$ K and $T \sim 10^{10}$ K, respectively.

Temperatures in the range of $T \sim 10^6$–10^7 K refer to old neutron stars that have already lost a significant amount of thermal energy acquired at birth due to the emission of neutrinos and, at the later stages of their thermal evolution, photons [609]. (Details will be discussed in chapter 19.) In contrast to this, the temperature range addressed by $\nu \sim 0$ corresponds to the very early stages of a hot neutron star [610] newly born, as the final stage of stellar collapse, in supernova explosions, or, alternatively, formed from the collapse of accreting white dwarfs. In both cases the stellar temperature is on the order of $T \sim 10^{10}$ K. The distinction between newly formed neutron stars and those which are old, possibly reheated ones is of great astrophysical importance since it is now believed that globular clusters

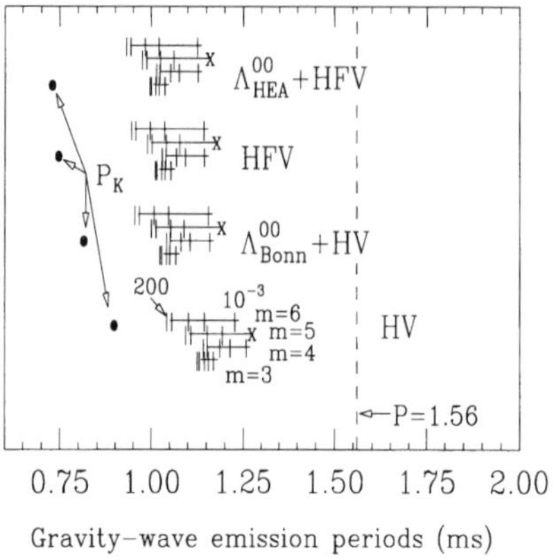

Figure 16.7. Same as figure 16.3, but for a stellar mass of $M = 1.9\,M_\odot$ and a different sample of EOSs [447].

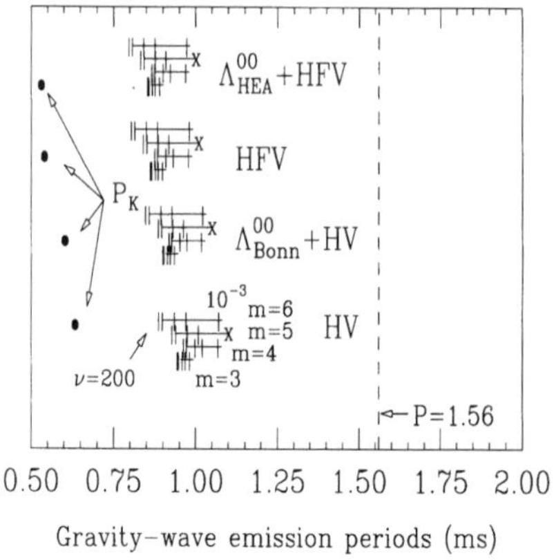

Figure 16.8. Same as figure 16.3, but for a stellar mass of $M = 1.9\,M_\odot$ and a different sample of EOSs [447].

are rich in old neutron stars that have been spun up via mass accretion from a companion captured in the dense environment of the cluster [25]. In that case, as we shall see immediately below, shear viscosity is very efficient in damping the GRR instability modes. Therefore old and cold stars can rotate at smaller rotational periods than young and hot ones. This conclusion, however, hinges on the role of bulk viscosity of neutron star matter, which, as mentioned in subsection 16.3.2, may become very large at high temperatures. Sawyer has pointed out that it goes as the sixth power of the temperature as compared with a T^{-2} dependence for the shear viscosity [599]. This means that at temperatures $T \gtrsim 10^9$ K the bulk viscosity may dominate over the shear viscosity and regulate the GRR driven instability in rapidly rotating neutron stars, pushing their critical rotational periods toward smaller values, possibly even as small as the Kepler period (we shall come back to this at the end of this subsection). On the other hand, one needs to admit that calculations of the viscosity of neutron star matter are rather complex and to date may not have reached a state that allows one to draw final conclusions on this issue.

Figures 16.3 to 16.8 reveal that independently of the equations of state and the star's mass, P_m^ν is largest (and hence Ω_m^ν is smallest) for the $m = 5$ eigenmodes in the case of negligible viscosity, indicated by the crossed endpoints to the right of the $m = 5$ lines in figures 16.3 through 16.8 [110, 113]. Hence, only stars rotating at periods $P > P_{m=5}^{\nu=0}$ (that is, rotational frequencies $\Omega < \Omega_{m=5}^{\nu=0}$) do *not* emit gravitational radiation and therefore perform stable rotations. Those with smaller rotational periods excite the GRR driven instability which makes their rotational motion unstable. None of these stellar models can rotate stably at the

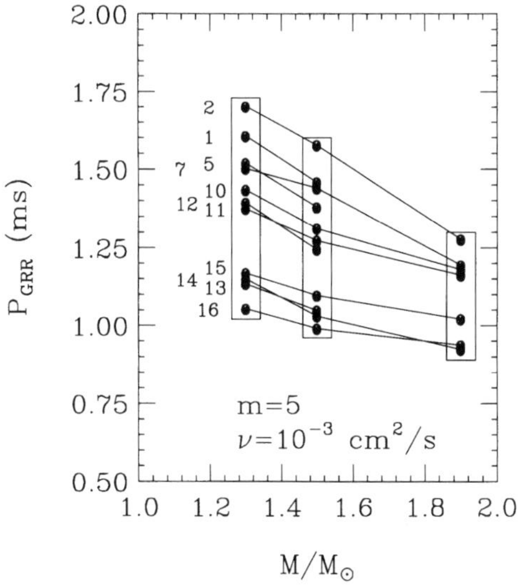

Figure 16.9. Critical rotational periods $P_{\rm GRR} = P^{\nu}_{m=5}$ at which the GRR driven instability in *hot*, rotating neutron stars of gravitational masses $M = 1.30$, 1.50, and $1.90\,M_\odot$ sets in [447]. The results are computed for the total set of EOSs listed in table 12.4. $\nu = 10^{-3}$ cm^2 s^{-1} denotes the shear viscosity.

corresponding Kepler period, whose location is shown by solid dots in these figures. This conclusion would only be altered if the viscosity of the matter becomes very large such that the GRR driven instability can be damped very efficiently. Shear viscosity may become that large in cold neutron stars with temperatures $T \sim 10^6$ K, as indicated in figures 16.6 ($\nu = 10^4$ cm^2 s^{-1}) and 16.11 ($\nu = 10^7$ cm^2 s^{-1}). Friedman *et al* [106, 577] and Lindblom [586] found in earlier studies that the $m = 3$ or 4, or $m = 4$ or 5 modes are likely to set the limit on stable rotation. The results presented here are in accord with those of Lindblom, because of a comparable choice for the frequency dependence of α_m and equations of state which are not that different from those employed by Lindblom. This is not the case for the equations of state adopted in the study by Friedman *et al*, all of which are older non-relativistic potential model equations of state.

Figure 16.9 collects the calculated $m = 5$ instability periods in young and hot stars, $P^{\nu=0}_{m=5}$, for a representative sample of neutron star masses of 1.3, 1.5, and $1.9\,M_\odot$. (The shear viscosity has an insufficiently small value of $\nu = 10^{-3}$ cm^2 s^{-1}.) Solid dots referring to rotational periods that are calculated for one and the same equation of state are joined by straight lines. Again, it is these rotational periods that define the smallest stable rotational periods of young and hot neutron stars, provided bulk viscosity

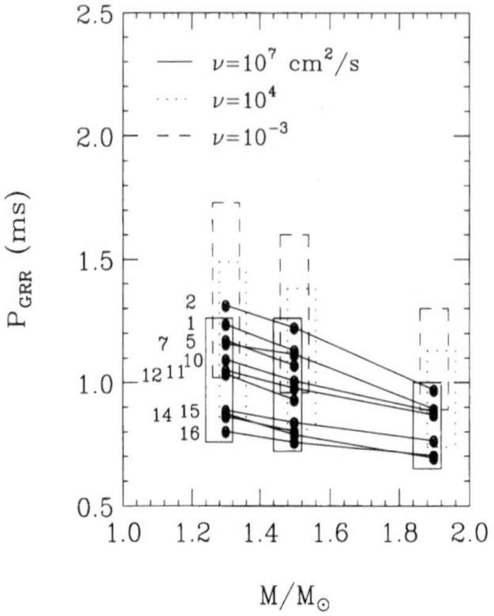

Figure 16.10. Same as figure 16.9, but for neutron stars at three different temperatures: $T = 10^{10}$ K ($\nu \approx 10^{-3}$ cm^2 s^{-1}), $T = 10^7$ K ($\nu \approx 10^4$ cm^2 s^{-1}), and $T = 10^6$K ($\nu \approx 10^7$ cm^2 s^{-1}) [447].

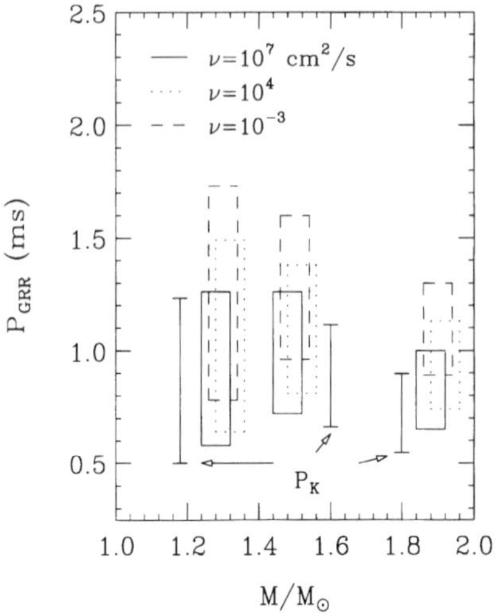

Figure 16.11. Comparison of the GRR driven instability periods $P_{\rm GRR}$ of figure 16.10 with the Kepler periods $P_{\rm K}$, calculated for the total set of EOSs listed in table 12.4 [447].

does not play a significant role in such stars (see below). Modes other than $m = 5$ (± 1) appear to be set in at higher frequencies. This is specifically so for the Dedekind bar instability found by Chandrasekhar [587], which is the $m = 2$ case of the GRR driven instability. Several higher modes are excited first before the $m = 2$ mode becomes unstable [537].

After the discussion of the GRR instability in hot neutron stars, we turn now to the inverse situation of large viscosity values as encountered in old and therefore cold ($T \sim 10^6$–10^7 K) neutron stars in binary systems that are being spun up – and thereby reheated to $T \lesssim 10^7$ K – by mass accretion from a companion. For our studies we take temperatures lying in the range of $\sim 10^6$–10^7 K, which correspond to viscosity values on the order of $\sim 10^5$–10^7 cm^2 s^{-1} [600, 601]. The GRR driven instability periods in old rotating neutron stars are exhibited in figures 16.10 and 16.11 for the two representative viscosity values of $\nu = 10^4$ and 10^7 cm^2 s^{-1}. The stellar masses are 1.3, 1.5, and 1.9 $M_\odot$. A comparison with the critical periods in young and hot stars ($\nu = 10^{-3}$ cm^2 s^{-1}) is given too. The latter periods, calculated for numerous equations of state of our collection listed in table 12.4, are contained in the dashed rectangles. The solid lines join the instability periods that are calculated for one and the same equation of state. One immediately sees the impact of the damping effect of viscosity on the instability modes, which is the more pronounced the colder (that is, the larger the shear viscosity, ν) a neutron star. Therefore, the colder a neutron star, the smaller its critical instability period and the faster it can rotate. The influence of the star's mass on $P_{\rm GRR}$ can be inferred from these figures too. One sees that the more massive a neutron star, the smaller its instability period, which originates primarily from the tighter shape (smaller radii) of the heavier neutron stars. Figure 16.11 compares the periods $P_{\rm GRR}$, again computed for shear viscosities of 10^{-3}, 10^4 and 10^7 cm^2 s^{-1}, with the general relativistic Kepler periods $P_{\rm K}$, discussed in section 16.1, for all equations of state compiled in table 12.4. One sees that only in old and cold neutron stars, $P_{\rm GRR}$ takes on values that come close to the mass shedding limit set by $P_{\rm K}$. In hotter neutron stars, $P_{\rm GRR}$ is distinctively larger than $P_{\rm K}$, and therefore the instability sets a much more stringent limit on stable neutron star rotation than the onset of mass shedding.

From the above analysis one readily establishes that the critical rotational frequencies $\Omega^{\nu=0}_{m=5}$ of hot ($T \sim 10^{10}$ K) rotating neutron stars with canonical masses around $1.45\,M_\odot$ set by the GRR instability are between 63 and 75% of the corresponding Kepler frequencies, depending on the equation of state [62, 110, 113]. Somewhat larger values for $\Omega^{\nu=0}_{m=5}$ than those derived here have been found by Lindblom [586], on the basis of older non-relativistic models for the nuclear equation of state. The situation is different in old and cold ($T \sim 10^6$–10^7 K) rotating neutron stars in which case Ω^{ν}_{m} is $\sim$ 77–92% of $\Omega_{\rm K}$ for stellar masses around $1.5\,M_\odot$. Because of the mass dependence, Ω^{ν}_{m} increases to 80–91% in neutron stars with masses around $1.9\,M_\odot$. As a consequence, an old neutron star, located in a globular cluster for example, can be spun up by mass accretion from a companion star to rotational frequencies which are closer to $\Omega_{\rm K}$ than it is the case for a young and hot rotating neutron star formed in a supernova explosion, or being born from a collapsing white dwarf. The actual value depends on the star's temperature to which it is reheated during the accretion process.

We next show in figures 16.12 through 16.17, in a somewhat more compact form than above, the dependence of the critical instability period $P_{\rm GRR}$ on both the star's mass as well as on temperature T. Figure 16.12 additionally exhibits which mode m it is that actually sets the limit on stable rotation. The temperature is kept constant along each stellar sequence. The solid dots refer to the heaviest neutron star, which terminates the gravitationally stable sequence. From figure 16.12 it follows that in cold neutron stars of temperature $T = 10^6$ K, the $m = 2$ mode is excited first, which is consistent with earlier work done by Lindblom [611], and Ipser and Lindblom [594]. In somewhat hotter neutron stars of $T = 10^8$ K, the instability period $P_{\rm GRR}$ is related to m values that change with gravitational mass M, such that for $M \lesssim 0.5\,M_\odot$ ($\gtrsim 0.5\,M_\odot$) the $m = 2$ ($m = 3$) mode sets the limit on stable rotation. This is the only

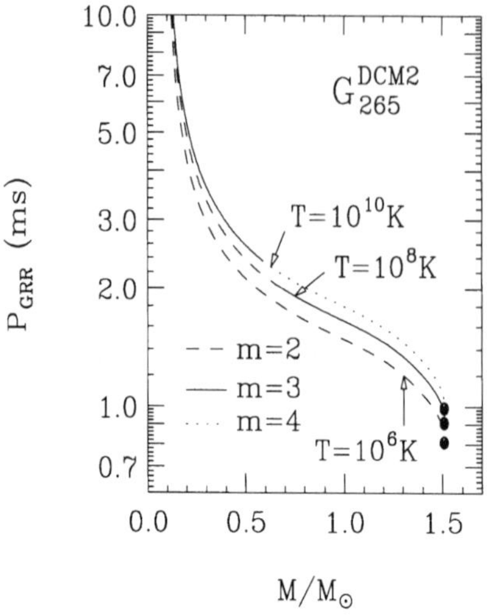

Figure 16.12. GRR driven instability periods $P_{\rm GRR}$ as a function of neutron star mass. Also shown is the dependence of $P_{\rm GRR}$ on the mode index, m, and the star's temperature, T, for EOS $\rm G_{265}^{DCM2}$ [447].

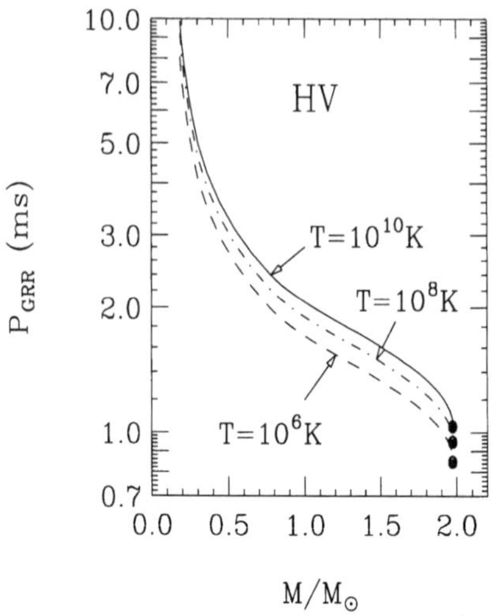

Figure 16.13. GRR driven instability periods $P_{\rm GRR}$ as a function of neutron star mass for stellar temperatures of $T = 10^6$, 10^8, and 10^{10} K. The EOS is HV [447].

case where the Dedekind bar instability [587], i.e. the $m = 2$ case of the GRR driven instability, appears to be of physical significance for rotating neutron stars, since higher modes always become unstable first. The $m = 4$ instability mode is excited first only in hot neutron stars of temperatures $T \sim 10^{10}$ K and masses $M \gtrsim 0.5\,M_\odot$. Recall from figures 16.3 through 16.8 that the $m = 5$ mode only plays a role in extremely hot rotating neutron stars, like pulsars in the very early stages of evolution after their formation in supernovae, or as formed from collapsing white dwarfs that have become too massive because of mass accretion from a companion.

In summing up, figures 16.12 through 16.17 make a strong case that the GRR driven instability limits stable neutron star rotation to periods, depending on the equation of state, larger than about $P_{\rm GRR} \sim$ (1–1.4×10^{-3} s if the stars are hot, and assuming a canonical neutron star mass around $1.5\,M_\odot$. This range moves down to $P_{\rm GRR} \sim 0.8$–1.2×10^{-3} s for cold neutron stars. A variety of stellar properties of such neutron stars are collected in tables 16.5 through 16.8. One also observes that non-relativistic models for the equation of state generally set the lower bounds on $P_{\rm GRR}$, while the relativistic equations of state set the upper ones. This is caused by the somewhat different functional dependence of radius on mass in both cases, as was shown in figure 14.8. (Neutron stars constructed for non-

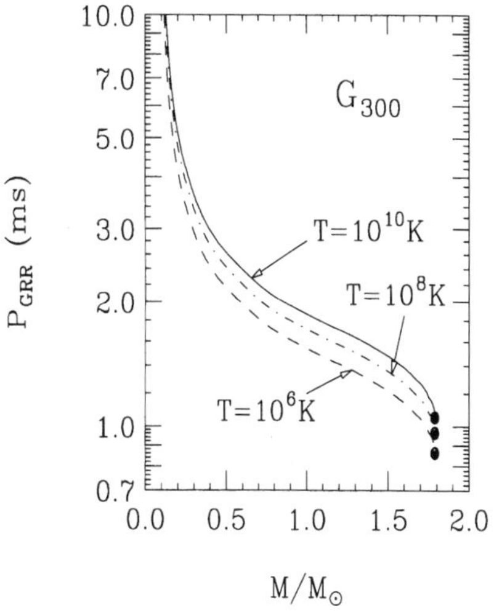

Figure 16.14. Same as figure 16.13, but for EOS G_{300} [447].

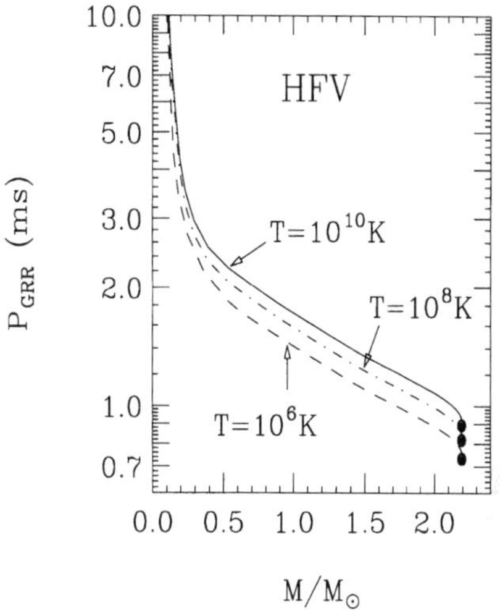

Figure 16.15. Same as figure 16.13, but for EOS HFV [447].

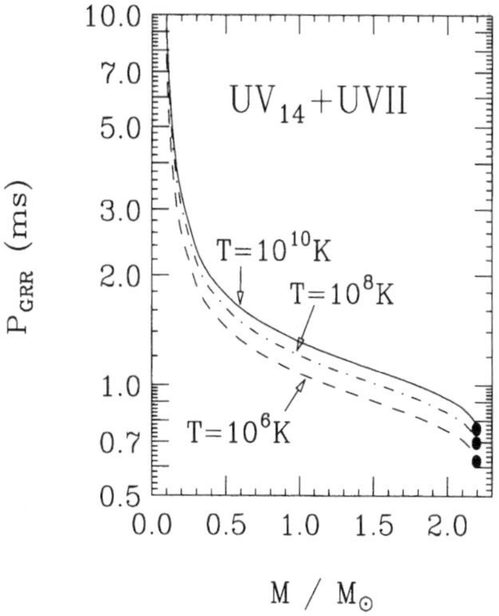

Figure 16.16. Same as figure 16.13, but for UV_{14}+UVII [447].

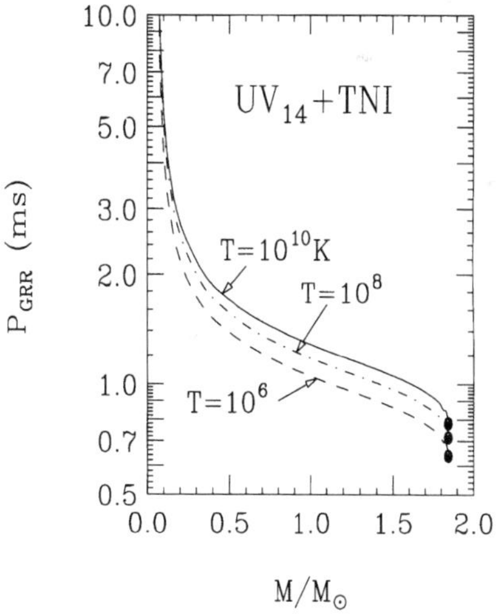

Figure 16.17. Same as figure 16.13, but for UV_{14}+TNI [447].

relativistic equations of state tend to be more compact for reasons discussed in section 14.4.) From these numbers it appears that the two fastest yet observed pulsars PSRs 1937+21 and 1957+20, each one rotating at 1.6 ms, come already very close to the limit set by the GRR instability. Of course,

Table 16.5. Properties of neutron stars of mass $M = 1.4\,M_\odot$ rotating at those rotational periods at which instability against emission of gravity waves in cold ($T = 10^6$ K) stars sets in. The entries are: central energy density ϵ_c (in MeV fm^{-3}), rotational period P (in ms), equatorial and polar radius $R_{\rm eq}$ and $R_{\rm p}$ (in km), moment of inertia I (in g cm^2), number of baryons A, stability parameter t, injection energy β, redshift at the pole $z_{\rm p}$, eccentricity e, quadrupole moment Π (in km^3).

Quantity	HV	$\Lambda^{00}_{\rm Bonn}$+HV	HFV	$\Lambda^{00}_{\rm HEA}$+HFV	$\Lambda^{\rm RBHF}_{\rm BroB}$+HFV
ϵ_c	302.85	313.79	412.34	427.45	454.08
P	1.402	1.222	1.152	1.107	1.028
$R_{\rm eq}$	15.99	14.46	14.24	13.74	12.93
$R_{\rm p}$	12.69	11.36	11.13	10.81	10.26
$\log_{10} I$	45.1893	45.1498	45.1219	45.1055	45.0647
$\log_{10} A$	57.2431	–	57.2645	–	57.2723
t	0.060	0.069	0.068	0.067	0.066
β	0.674	0.636	0.628	0.618	0.597
$z_{\rm p}$	0.2180	0.2541	0.2615	0.2725	0.2943
e	0.61	0.62	0.62	0.62	0.61
Π	14.63	14.56	13.23	12.14	10.34

Table 16.6. Properties of neutron stars of mass $M = 1.4\,M_\odot$ rotating at those rotational periods at which instability against emission of gravity waves in hot ($T = 10^{10}$ K) stars sets in. The entries are explained in table 16.5.

Quantity	HV	$\Lambda^{00}_{\rm Bonn}$+HV	HFV	$\Lambda^{00}_{\rm HEA}$+HFV	$\Lambda^{\rm RBHF}_{\rm BroB}$+HFV
ϵ_c	320.36	338.79	434.92	447.08	470.84
P	1.695	1.482	1.399	1.346	1.252
$R_{\rm eq}$	15.30	13.92	13.52	13.12	12.44
$R_{\rm p}$	13.19	11.89	11.60	11.28	10.72
$\log_{10} I$	45.2151	45.1809	45.1442	45.1288	45.0924
$\log_{10} A$	57.2456	–	57.2680	–	57.2747
t	0.040	0.046	0.044	0.044	0.044
β	0.686	0.652	0.643	0.633	0.614
$z_{\rm p}$	0.2070	0.2382	0.2466	0.2567	0.2761
e	0.51	0.52	0.51	0.51	0.51
Π	9.76	9.97	8.35	7.86	6.91

very heavy neutron stars can spin more rapidly. Periods that lie clearly in the sub-millisecond regime, however, are only found for neutron stars whose

Table 16.7. Properties of neutron stars of mass $M = 1.5\,M_\odot$ rotating at those rotational periods at which instability against emission of gravity waves in cold ($T = 10^6$ K) stars sets in. The entries are explained in table 16.5.

Quantity	HV	$\Lambda^{00}_{\rm Bonn}$+HV	HFV	$\Lambda^{00}_{\rm HEA}$+HFV	$\Lambda^{\rm RBHF}_{\rm BroB}$+HFV
ϵ_c	328.96	342.97	446.10	459.23	481.32
P	1.334	1.176	1.099	1.063	0.992
$R_{\rm eq}$	15.91	14.53	14.03	13.61	12.93
$R_{\rm p}$	12.60	11.38	11.05	10.77	10.26
$\log_{10} I$	45.2262	45.1955	45.1549	45.1426	45.1081
$\log_{10} A$	57.2773	–	57.3002	–	57.3067
t	0.062	0.070	0.070	0.070	0.068
β	0.648	0.610	0.599	0.589	0.568
$z_{\rm p}$	0.2420	0.2808	0.2922	0.3033	0.3266
e	0.61	0.62	0.62	0.61	0.61
Π	15.54	15.78	13.11	12.27	10.95

Table 16.8. Properties of neutron stars of mass $M = 1.5\,M_\odot$ rotating at those rotational periods at which instability against emission of gravity waves in hot ($T = 10^{10}$ K) stars sets in. The entries are explained in table 16.5.

Quantity	HV	$\Lambda^{00}_{\rm Bonn}$+HV	HFV	$\Lambda^{00}_{\rm HEA}$+HFV	$\Lambda^{\rm RBHF}_{\rm BroB}$+HFV
ϵ_c	351.88	369.96	470.09	480.57	492.42
P	1.614	1.427	1.335	1.293	1.208
$R_{\rm eq}$	15.20	13.94	13.36	13.02	12.43
$R_{\rm p}$	13.10	11.91	11.51	11.23	10.72
$\log_{10} I$	45.2516	45.2227	45.1775	45.1660	45.1358
$\log_{10} A$	57.2793	–	57.3037	–	57.3094
t	0.041	0.047	0.044	0.044	0.045
β	0.662	0.628	0.615	0.605	0.587
$z_{\rm p}$	0.2293	0.2619	0.2752	0.2852	0.3054
e	0.51	0.52	0.51	0.51	0.51
Π	10.23	10.53	8.33	7.96	7.28

masses are very close to the limiting mass (solid dots). This is so because it is the maximum-mass stars that have the smallest radii and therefore are the most compact members of each stellar sequence. Both features – small radii and a compact structure – are the necessary ingredients in order to achieve rapid rotation. On the other hand, the lighter a neutron star the weaker gravity and therefore the larger its size. Such stars have relatively

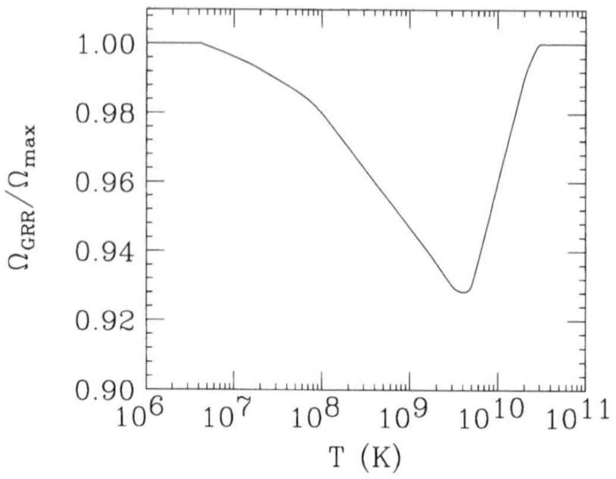

Figure 16.18. Critical angular velocities of rotating neutron stars, modeled as $n = 1$ polytropes (after Lindblom [589]).

larger radii, as can be seen, for instance, in figure 14.8. This increases the critical instability periods $P_{\rm GRR}$ significantly, specifically for stars lighter than $M \sim 0.4\,M_\odot$. The influence of temperature on $P_{\rm GRR}$ becomes less significant too.

Finally we say a few words about the possibly very significant role of *bulk* viscosity of neutron star matter, as pointed out by Sawyer [599]. Throughout the above discussion we have restricted ourselves to the shear viscosity as the only source that damps the GRR driven instability in rotating neutron stars. As Sawyer has shown (see previous subsection), the bulk viscosity, being proportional to T^6, may become very large in hot neutron stars, damping the GRR instability in such neutron stars very efficiently. Figure 16.18 illustrates this quantitatively. Shown are the smallest $\Omega_{\rm GRR}$ frequencies for a range of neutron star temperatures. These $\Omega_{\rm GRR}$ are displayed as ratios, with $\Omega_{\rm max}$ the angular velocity above which mass shedding occurs from neutron stars modeled as $n = 1$ polytropes. For these polytropes one has, in accordance with (16.11), $\Omega_{\rm max} = 0.639\sqrt{\pi\bar{\epsilon}}$ (where $\bar{\epsilon}$ is the average density). This figure shows that the GRR instability is completely suppressed in neutron stars except for those with temperature in the temperature window of 10^7 K $\lesssim T \sim 10^{10}$ K. Shear viscosity suppresses the instability for temperatures $T \lesssim 10^7$ K, while bulk viscosity suppresses it for higher temperatures!

The analysis shown in figure 16.18 does not taken into account the plausible *superfluid* nature of neutron star matter at $T \lesssim 10^9$ K. The investigation of Lindblom and Mendell indicates that dissipation in the superfluid state due to electron–vortex scattering completely suppresses the GRR in all neutron stars cooler than the superfluid transition temperature,

$T \approx 10^9$ K [602, 603]. Neutron stars formed from the collapse of accreting white dwarfs are therefore likely to have their rotation limited by the GRR driven instability. This appears not to be the case – if Lindblom and Mendell are correct – for neutron stars spun up by mass accretion from binary stars which will never be hot enough to become GRR unstable. The class of accretion-driven neutron stars may therefore have slightly higher limiting rotational frequency than the class of neutron stars with dwarf progenitors, as pointed out by Friedman and Ipser [537].

Chapter 17

Models of rotating neutron stars

17.1 Method of construction

17.1.1 Stars rotating at the mass shedding frequency

In the first step, let us outline how stellar models are constructed that rotate at their respective *self-consistent* Kepler frequencies [62]. Self-consistency is encountered because the general relativistic expression for $\Omega_{\rm K}$ is given in terms of stellar quantities – like the star's velocity at the equator and the strength of frame dragging there – which themselves depend on $\Omega_{\rm K}$. This becomes immediately obvious from equation (16.7).

As usual, at the outset a model for the equation of state of neutron star matter is to be chosen. This is followed by a suitable choice for the value of the star's central density, ϵ_c, where the outcome of non-rotating stellar structure calculations serves as a useful guideline. For this value of ϵ_c the frame dragging function $\bar{\omega}(r)$ is computed from equation (15.36) by

(1) choosing a boundary value for $\bar{\omega}_c \equiv \bar{\omega}(r=0)$. Equation (15.39) then determines the stellar frequency $\Omega(\epsilon_c, \bar{\omega}_c(\Omega))$ associated with these values of ϵ_c and $\bar{\omega}_c(\Omega)$. Of course, Ω will in general will be quite different from $\Omega_{\rm K}$ which we compute from (16.7), and hence is not the Kepler frequency of that particular stellar model.

(2) In order to find it, we choose a starting value for $\Omega_{\rm K}$, denoted by ${\Omega_{\rm K}}^{\rm start}$, and determine $\bar{\omega}(r)$ again, such that it corresponds to ${\Omega_{\rm K}}^{\rm start}$, that is, $\bar{\omega}(r) = \bar{\omega}(r, {\Omega_{\rm K}}^{\rm start})$. This is easily accomplished by simply rescaling $\bar{\omega}(r, \Omega)$ with respect to ${\Omega_{\rm K}}^{\rm start}$, as described in equation (15.40). Once this is done, the monopole and quadrupole perturbation functions corresponding to this choice for ${\Omega_{\rm K}}^{\rm start}$ follow in the form $m_0(\epsilon_c, {\Omega_{\rm K}}^{\rm start})$, $p_0(\epsilon_c, {\Omega_{\rm K}}^{\rm start})$, $h_0(\epsilon_c, {\Omega_{\rm K}}^{\rm start})$, and $h_2(\epsilon_c, {\Omega_{\rm K}}^{\rm start})$ and $v_2(\epsilon_c, {\Omega_{\rm K}}^{\rm start})$, respectively.

(3) The newly calculated functions m_0, p_0, h_0, h_2, and v_2 are then used to calculate a new value for Ω_K from the transcendental equation (16.7), i.e. ${\Omega_K}^{new} = \Omega_K(\epsilon_c, \bar{\omega}_c({\Omega_K}^{start}))$. It is this equation which leads to a new, generally improved value for the Kepler frequency.

(4) The next step in the iteration scheme consists in rescaling $\bar{\omega}$ with respect to ${\Omega_K}^{new}$ and determining the monopole and quadrupole perturbation functions associated with the rescaled frequency $\bar{\omega}$. Moreover care needs to be taken that Ω_K is constrained by (15.39).

The last three steps are to be repeated iteratively until sufficiently precise solutions for the above listed functions are obtained. The obtained solution $\Omega_K(\epsilon_c, \bar{\omega}_c(\Omega_K))$ then is the wanted Kepler frequency of a self-consistent neutron star model of rotational mass $M(\epsilon_c, \Omega_K(\epsilon_c))$ having a central density ϵ_c. For the sake of brevity, the Kepler frequency and rotational star mass are simply denoted by Ω_{GRR} and M; so are all the other quantities of stars rotating at their respective Kepler frequencies.

That particular ϵ_c value for which $M(\epsilon_c, \Omega_K(\epsilon_c))$ reaches its maximum value, i.e. $\partial M/\partial\epsilon_c = 0$, defines the maximum-mass model of the stellar

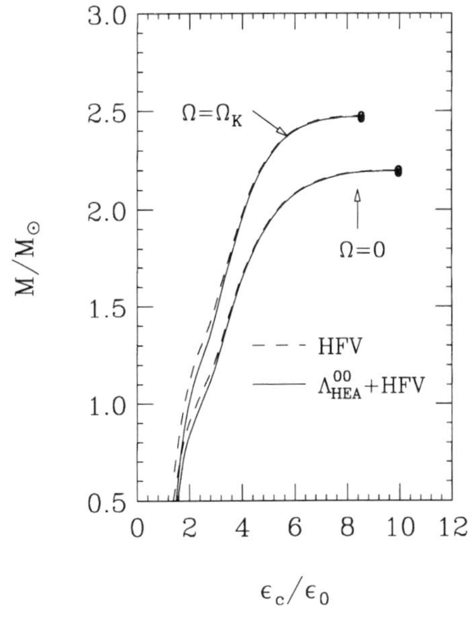

Figure 17.1. Gravitational mass as a function of central energy density of neutron star sequences constructed for Λ^{00}_{HEA}+HFV and HFV [62]. $\Omega = 0$ indicates non-rotating stars, Ω_K refers to rotation at the Kepler frequency which sets an absolute limit on rapid rotation. The solid dots denote the heaviest stellar model of each sequence. (Reprinted courtesy of World Scientific.)

sequence (turning-point method) [106, 109, 612, 613]. An example is shown in figure 17.1 which displays $M(\epsilon_c, \Omega_K(\epsilon_c))$ as a function of ϵ_c for a few different equations of state of table 12.4. The curves labeled with Ω show the increase of mass due to rotation at the Kepler frequency, where the stars become unstable because of mass shedding from the equator. Hence no star can rotate at frequencies $\Omega > \Omega_K$. Being gravitationally most compressed, neutron stars at the mass peak possess the largest Kepler frequencies (smallest periods) of all stars that populate a certain stellar family constructed for a given equation of state. The rotationally supported mass increase is therefore largest for stars at the mass peak, typically between 15% and 20%, depending on the equation of state, and correspondingly smaller for the lighter stars of the sequence, because Ω_K decreases with M.

The actual Ω_K value of each particular star along the rotating branch in figure 17.1 is a unique quantity to be determined self-consistently. Figure 17.2 shows the dependence of Ω_K on M for a large number of different equations of state [111, 112]. The rapid change of Ω_K in the vicinity of the limiting mass is apparent from this figure. Besides that, it is remarkable that the potential models of our collection listed in table 12.4 lead to Kepler frequencies that are larger than those obtained for the field-theoretical equations of state (see also table 16.1). The softness at low and intermediate nuclear densities but rather stiff behavior at high densities of the non-relativistic equations of state is responsible for this

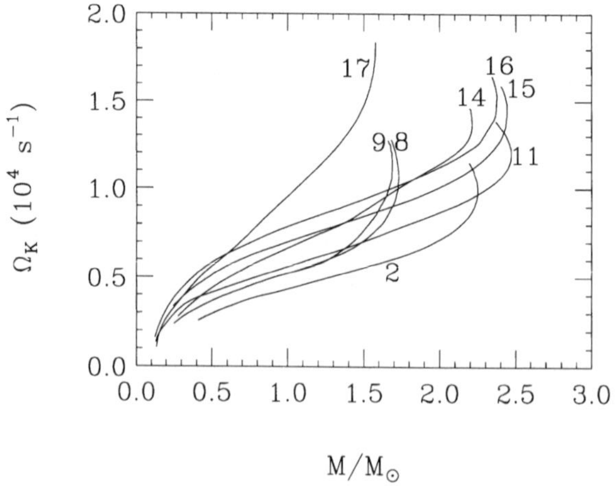

Figure 17.2. Kepler frequency, Ω_K, computed self-consistently from equation (16.7) as a function of rotational neutron star mass, shown for a sample of EOSs listed in table 12.4 [112]. (Reprinted courtesy of *Astrophys. J.*)

behavior [108, 109, 410]. Among the relativistic models, we find that the $\Lambda^{00}_{\rm HEA}$+HFV equation of state predicts the largest Kepler frequency, that is, $\Omega_{\rm K} = 1.18 \times 10^4$ s^{-1}. The softening of this equation of state , caused by the dynamical two-nucleon correlations at low and intermediate nuclear densities in combination with the (Fock) exchange contribution contained in HFV which stiffens the equation of state at high densities, is responsible for this behavior [62, 112]. Since these features – dynamical two-nucleon correlations and exchange contributions – are inherent in the non-relativistic equations of state too, is it therefore not surprising that it is the $\Lambda^{00}_{\rm HEA}$+HFV equation of state whose outcome comes closest to the potential models.

17.1.2 Stellar sequences of constant baryon number

Naturally, sequences of stars rotating at their respective Kepler frequencies attain their particular interest because it is these sequences that define which region in the three-parameter space spanned by ϵ_c, $\Omega_{\rm K}$, and M can actually be populated by physically existing stars (the physical region) and which not (unphysical region). One class of stars that are interesting in connection with the properties of isolated pulsars are those with *constant* baryon number, A_0.

The rotational frequencies of such pulsars, like those of any other pulsar, are constrained to the interval $\Omega \in [0, \Omega_{\rm K}]$, part of which they may be passing through in the course of their (active) lifetime because of the emission of electromagnetic dipole radiation and a wind of electron–positron pairs, both of which carry off rotational energy from the star. Models of such stars are constructed in close analogy to what has been outlined in section 17.1.1. Now, however, the star's central mass density is varied, for a given value of Ω which remains fixed during the self-consistent iteration scheme, until the stellar model possessing the desired baryon number A_0 is obtained. Changing ϵ_c implies however that now the relative frame dragging frequency $\bar{\omega}(r)$ is to be computed for each iteration step again. This step was not necessary in the computation of $\Omega_{\rm K}$, in which case ϵ_c was kept fixed. The whole set of stellar structure equations of rotating stars is therefore to be iterated now. The iteration procedure is finished if the outcome of two successive iteration steps agrees sufficiently well with each other. Repeating this procedure for other values of Ω, with $0 \le \Omega \le \Omega_{\rm K}$, leads to the stellar sequence of constant baryon number.

17.1.3 Stars of a given rotational mass

What has been said just above in section 17.1.2 carries over to the case of constructing models of rotating neutron stars of given gravitational masses. What is being varied is again ϵ_c, while Ω has a given, fixed value. The self-consistent iteration scheme is completed if $M(\epsilon_c, \Omega)$ agrees sufficiently well with the value wanted for it.

17.2 Exact versus approximate solution of Einstein's equations

We have not yet analyzed the applicability of Hartle's perturbative 'slow-rotation' formalism to the construction of stars spinning so rapidly that mass shedding from the equator sets in. This will be done in this section. For this purpose we compare a number of different properties of neutron stars, spinning at their respective Kepler frequencies, constructed from Hartle's improved method (i.e. supplemented by the self-consistency condition imposed by Ω_K) with the corresponding properties obtained from the numerical solution of Einstein's unapproximated equations. The latter is quite often referred to as the *exact* method. It is found that both methods, perturbative and exact, lead to compatible results down to Kepler periods of $P_K \approx 0.5$ ms, a value which is by far smaller than the smallest yet observed pulsar period. Because of its relatively simple structure, Hartle's method appears to be a very practical tool for testing models for the nuclear equation of state with data on pulsar periods. While leading to compatible results for numerous stellar quantities, some of which are in better agreement with the exact method than others, this should not deceive one that the improved perturbative method is not designed to be competitive with the exact high-precision method studied recently by Salgado, Bonazzola, Gourgoulhon, and Haensel [559].

We begin with comparing the properties of rotating neutron stars calculated for the equations of state of BJ(I), V_{14}+TNI, and Pan(C) of table 12.4. Figure 17.3 exhibits the rotational neutron star mass as a function of central energy density, ϵ_c, calculated for the BJ(I) equation of state (solid line). The crosses denote the mass values obtained from the exact calculation performed by Friedman, Ipser, and Parker [106]. For a sample of ϵ_c values the Kepler frequencies Ω_K (in units of 10^4 s^{-1}) are displayed too. The numbers in round brackets refer to the exact Ω_K values. It is seen that the rotational star masses obtained from Hartle's method are in very good agreement up to $M \approx 1.85\, M_\odot$ (which corresponds to $\Omega_K \approx 9000$ s^{-1}). Rotating star models of larger masses are characterized, in the exact treatment, by being slightly more massive. For the maximum-mass model a mass difference of $\approx 3\%$ between the two treatments is obtained. The location of the mass limit occurs, according to figure 17.3, in both treatments at more or less the same central energy density. The limiting Kepler frequency for BJ(I) has in our treatment a value of $\Omega_K = 1.11 \times 10^4$ s^{-1}, in close agreement with the exact value given by 1.12×10^4 s^{-1} (cf. table 17.1).

Figure 17.4 is the analog of figure 17.3, but calculated for the equation of state V_{14}+TNI. The rotating neutron star masses in this case are in

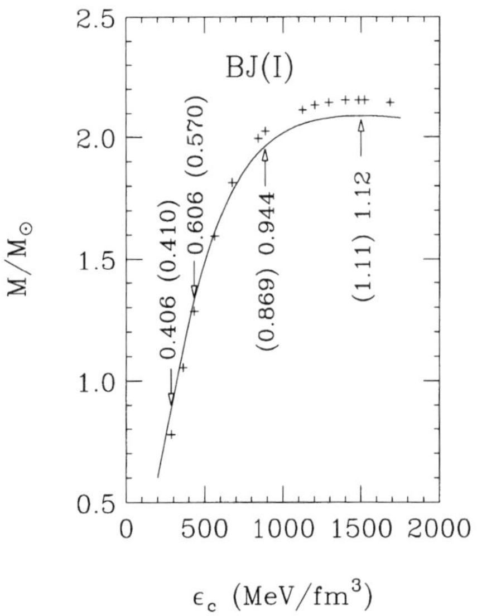

Figure 17.3. Comparison of rotational neutron star mass as a function of central energy density, ϵ_c, obtained from Hartle's method (solid line) [112] with the exact numerical results (crosses) of [106]. The underlying EOS is BJ(I) of table 12.4. The numbers refer to the Kepler frequency Ω_K in units of 10^4 s^{-1}. The values obtained for the exact method are quoted in round brackets. (Reprinted courtesy of *Astrophys. J.*)

good agreement up to $M \approx 2.1\, M_\odot$. The Kepler frequency associated with this star is 1.198×10^4 s^{-1} (Hartle) and 1.038×10^4 s^{-1} (exact). The limiting Hartle mass is $\approx 4\%$ smaller than the exact value. As for the BJ(I) equation of state, the mass limit occurs in both treatments practically at the same ϵ_c value.

In table 17.1 we summarize rotating neutron star properties derived from Hartle's method, where the equations of state V_{14}+TNI, BJ(I), and Pan(C) served as an input. The properties listed are: rotational star mass in units of the solar mass, $M/M_\odot$; central energy density, ϵ_c; Kepler period, P_K; equatorial radius, R_{eq}; percentage of central frame dragging, ω_c/Ω_K; ratio of rotational energy to gravitational energy, T/W; equatorial velocity, V_{eq}/c, and eccentricity, e. The exact results, taken from Friedman, Ipser, and Parker [106] are listed in the upper rows within each pair of rows. All Hartle models, shown in the lower rows, have been determined self-consistently such that these possess the same rotational mass as the exact models. From the above discussion it is known that Hartle's method leads to maximum-mass stars that are between 3 and 4% less massive than their exact counterparts. For that reason the maximum-mass models in

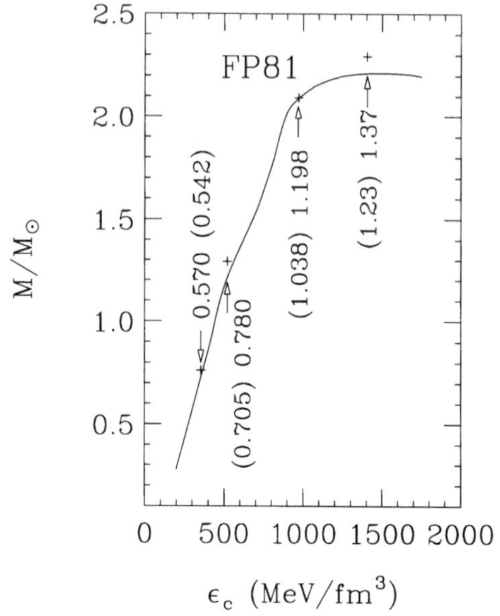

Figure 17.4. Same as figure 17.3, but calculated for the EOS of Friedman–Pandharipande, V_{14}+TNI [112]. (Reprinted courtesy of *Astrophys. J.*)

table 17.1 (indicated by a dagger) have slightly different masses. This table also reveals that the values of ϵ_c, Ω_K, ω_c/Ω_K, T/W, V_{eq}/c, and e of both methods are in good agreement. The equatorial radii of the Hartle models coincide to a lesser extent with the exact outcome, which is not too surprising because Hartle's method, being based on a perturbation expansion of the metric, neglects corrections higher than quadrupole. As concerns the maximum-mass model related to the Pan(C) equation of state, given in the last row of table 17.1, we find a value for Ω_K which is roughly 15% larger than the one of the exact numerical treatment ($\Omega_K = 1.57 \times 10^4$ s^{-1}). However, the determination of Ω_K for the Pan(C) equation of state is particularly complicated by the fact that Ω_K depends very sensitively – much more than is the case for all other equations of state of our collection – on M (cf. figure 17.2). Our determination of the mass limit gives $\Omega_K = 1.81 \times 10^4$ s^{-1}. This is compatible with the exact method if one assumes that the mass is determined only within 2% in the latter case [111].

In summary, figures 17.3 and 17.4, and the promising agreement of the perturbative rotating neutron star models with their exact counterparts, demonstrated in table 17.1, provide strong evidence for the applicability of Hartle's method up to Kepler frequencies of (at least) $\Omega_K \approx 1.2 \times 10^4$ s^{-1}, corresponding to a rotational Kepler period of $P_K \approx 0.5$ ms, which is a fraction of the smallest yet observed pulsar period of 1.6 ms. We emphasize once again that the self-consistency condition imposed by the general relativistic expression for Ω_K, as outlined in section 16.1, has not been previously imposed by other authors who have employed Hartle's method [70, 116, 555, 572, 581]. Our results clearly demonstrate the crucial importance of self-consistency, aside from employing the correct expression for Ω_K, of course. Only then is a rotation-induced mass increase between 15% and 25% obtained, depending on the equation of state [62, 109, 112]. Such a mass increase is in excellent agreement with the one established from the exact solution of Einstein's equations. The magnitude of this mass increase cannot be obtained without solving the full general relativistic stellar structure equations self-consistently. For that reason too small a mass increase – typically less than about 8% [555] – was obtained in earlier non-self-consistent applications of Hartle's method and was viewed as an inherent weakness of the method [106]. The findings presented above show that this objection loses its validity when self-consistency is imposed on Hartle's equations [112].

Table 17.1. Comparison of the properties of rotating neutron star models calculated from Hartle's method (lower rows within each pair of rows) with those of the exact numerical method (upper rows), for a sample of different EOSs listed in table 12.4.

EOS	$M/M_\odot$	$\epsilon_c/10^{15}$ (g cm^{-3})	P_K (ms)	R_{eq} (km)	ω_c/Ω	T/W	V_{eq}/c	e
V_{14}+TNI	0.77	0.63	1.159	15.1	0.27	0.102	0.27	0.72
	0.77	0.66	1.102	13.7	0.23	0.080	0.26	0.71
	1.30	0.92	0.891	14.9	0.43	0.120	0.35	0.77
	1.30	0.98	0.805	13.10	0.40	0.094	0.39	0.71
	2.30 †	2.5	0.511	12	0.83	0.133	0.49	0.67
	2.21 †	2.7	0.459	10.7	0.77	0.117	0.62	0.68
BJ(I)	0.78	0.51	1.532	18.1	0.25	0.071	0.25	0.71
	0.78	0.47	1.548	17.5	0.21	0.066	0.25	0.70
	1.29	0.77	1.102	16.9	0.40	0.093	0.38	0.74
	1.29	0.75	1.037	15.6	0.34	0.081	0.35	0.70
	2.16 †	2.7	0.566	12.9	0.79	0.110	0.47	0.68
	2.09 †	2.7	0.525	11.6	0.72	0.099	0.57	0.68
Pan(C)	1.65 †	5.16	0.400	9.2	–	0.107	–	0.66
	1.58 †	5.16	0.347	8.0	0.76	0.108	0.61	0.67

† Star rotating at the mass limit, $\partial M(\epsilon_c, \Omega_K(\epsilon_c))/\partial \epsilon_c = 0$.

17.3 Properties of rotating neutron stars

17.3.1 Absolute bounds on neutron star properties set by mass shedding

We analyze in the following numerous properties of sequences of rotating neutron stars.[1] The sequences are constructed for rotational frequencies up to the Kepler frequency Ω_K, defined in (16.7), where mass shedding from the equator sets an absolute limit on rapid rotation. We shall also come across the GRR driven instability in rotating neutron stars, which, as shown in section 16.3, sets a more stringent limit on stable rapid rotation than mass shedding, depending on the star's history. The analysis here, like in the previous chapters, is based on the total collection of nuclear equations of state introduced in table 12.4.

It has been shown in section 12.6 that dynamical two-particle correlations, irrespective of whether these are calculated from the relativistic ladder-approximated **T**-matrix or non-relativistic variational calculations, modify the equation of state most strongly at intermediate nuclear densities, in the range between $\epsilon \sim 20$ to about 300 MeV fm^{-3}. As a consequence of this, the properties of stellar models with central densities in that range are most sensitive to the dynamical correlations among baryons, since it is the behavior of the equation of state at a neutron star's core region which essentially determines its global properties, like mass and radius [78]. The rotating low-mass neutron star models listed in table 17.2, each of which having a central density less than 220 MeV fm^{-3}, bear witness to this feature. Specifically the stellar properties computed for HV (no dynamical correlations) and $\Lambda^{00}_{\rm Bonn}$ + HV (with dynamical correlations) are quite different, which has its origin in the significant softening of the equation of state around ϵ_0 accompanied by a pronounced stiffening at somewhat higher densities obtained for the Bonn OBE potential [109]. This behavior is less pronounced for the HEA OBE potential. The properties of neutron stars whose central densities lie beyond ~ 300 MeV fm^{-3} are influenced by the dynamical correlations too, as becomes clear from table 17.3 where the stellar masses are $\sim 1\,M_\odot$ and more. This is not the case anymore for very massive neutron stars, table 17.4, where the effect is practically integrated out, because of the huge density range encountered in this case. Another feature, which we have stressed repeatedly (see section 14.4), concerns the fact that the softening of the equation of state caused by the dynamical correlation makes neutron stars less massive, as can be inferred from the tables listed in appendix I, since less pressure is provided by such equations of state at a given density. This, in turn, makes neutron stars of comparable mass constructed for a soft equation of state more dense, as shown in the

[1] A comparison of the properties of rotating fermion stars with those of rotating boson stars can be found in [614].

Table 17.2. Properties of neutron stars rotating at a Kepler period of $P_K = 1.6$ ms, computed for several different EOSs. The entries are: central energy density ϵ_c (in MeV fm^{-3}), gravitational mass M, equatorial and polar radius R_{eq} and R_p (in km), moment of inertia I (in g cm^2), number of baryons A, stability parameter t, injection energy β, equatorial velocity V_{eq}, redshift in backward and forward direction z_B and z_F, redshift at the pole z_p, eccentricity e, quadrupole moment Π (in km^3).

Quantity	HV	$\Lambda^{00}_{\rm Bonn}$+HV	HFV	$\Lambda^{00}_{\rm HEA}$+HFV	$\Lambda^{\rm RBHF}_{\rm BroB}$+HFV
ϵ_c	183.2	202.4	168.1	198.3	220.3
$M/M_\odot$	0.839	0.396	0.462	0.394	0.390
R_{eq}	17.49	14.28	14.75	14.123	14.21
R_p	12.46	10.25	10.48	10.28	9.99
$\log_{10} I$	44.8751	44.2895	44.4270	44.2828	44.2903
$\log_{10} A$	57.0017	–	56.7518	–	56.6783
t	0.073	0.059	0.069	0.058	0.061
β	0.801	0.886	0.870	0.887	0.885
V_{eq}/c	0.25	0.19	0.20	0.19	0.19
z_B	0.3594	0.2350	0.2452	0.2341	0.2251
z_F	−0.1430	−0.1181	−0.1151	−0.1170	−0.1117
z_p	0.1172	0.0624	0.0722	0.0618	0.0632
e	0.70	0.70	0.70	0.69	0.71
Π	11.63	2.79	4.36	2.62	3.04

tables of appendix I. The central star density also increases for increasing values of Ω_K. This is because of the centrifugal force acting on the star's matter that must be counterbalanced by the attractive gravitational forces, which increase with mass density.

The stability parameter, t, defined in equation (15.111), turns to be rather insensitive, with the exception of very light neutron stars, to different equations of state, as follows from tables 17.3 and 17.4. More than that, table 16.2 reveals that $t \lesssim 0.13$ even for the most rapidly rotating neutron stars at their respective mass limits! This means that all neutron star sequences end before $t = 0.14$ beyond which Newtonian instability against a bar mode (related to $m = 2$) in a rotating Maclaurin spheroid is expected to set in [105, 106, 578]. A graphical illustration of t, shown as a function of Ω_K for sequences of neutron stars up to their mass limits (solid dots), is given in figure 17.5. The value of the stability parameter obtained for all maximum-mass models takes on values of $0.088 \le t_K < 0.13$, where the subscript K refers to rotation at Ω_K.

We recall that Maclaurin spheroids are systems of uniform density with a polytropic index of $n = 0$. Work on uniformly rotating polytropes

Table 17.3. Properties of neutrons rotating at a Kepler period of 1 ms. The entries are explained in table 17.2.

Quantity	HV	$\Lambda^{00}_{\rm Bonn}$+HV	HFV	$\Lambda^{00}_{\rm HEA}$+HFV	$\Lambda^{\rm RBHF}_{\rm BroB}$+HFV
ϵ_c	380.3	279.9	380.1	339.9	349.8
$M/M_\odot$	1.761	1.397	1.374	1.229	0.969
$R_{\rm eq}$	16.66	15.33	15.28	14.85	13.73
$R_{\rm p}$	11.59	10.47	10.48	10.14	9.51
$\log_{10} I$	45.2747	45.0965	45.0813	44.9960	44.7953
$\log_{10} A$	57.3528	–	57.2514	–	57.0922
t	0.095	0.102	0.098	0.099	0.090
β	0.551	0.606	0.613	0.642	0.699
$V_{\rm eq}/c$	0.39	0.36	0.36	0.34	0.31
$z_{\rm B}$	0.8206	0.6695	0.6644	0.6032	0.5166
$z_{\rm F}$	−0.2009	−0.1799	−0.1815	−0.1741	−0.1669
$z_{\rm p}$	0.3467	0.2851	0.2777	0.2481	0.1958
e	0.72	0.73	0.73	0.73	0.72
Π	25.43	20.95	19.47	16.81	10.33

Table 17.4. Properties of neutron stars rotating at a Kepler period of 0.7 ms. The entries are explained in table 17.2.

Quantity	HV	$\Lambda^{00}_{\rm Bonn}$+HV	HFV	$\Lambda^{00}_{\rm HEA}$+HFV	$\Lambda^{\rm RBHF}_{\rm BroB}$+HFV
ϵ_c	860.1	760.1	622.9	605.9	575.2
$M/M_\odot$	2.235	2.187	2.150	2.095	1.974
$R_{\rm eq}$	14.25	14.09	13.95	13.85	13.55
$R_{\rm p}$	10.27	9.97	9.87	9.75	9.47
$\log_{10} I$	45.3525	45.3457	45.3192	45.2999	45.2434
$\log_{10} A$	57.4792	–	57.4854	–	57.4430
t	0.097	0.106	0.106	0.107	0.108
β	0.358	0.352	0.357	0.365	0.384
$V_{\rm eq}/c$	0.52	0.51	0.50	0.50	0.48
$z_{\rm B}$	1.4531	1.4055	1.4043	1.3557	1.2773
$z_{\rm F}$	−0.2362	−0.2279	−0.2328	−0.2289	−0.2261
$z_{\rm p}$	0.6722	0.6863	0.6742	0.6547	0.6132
e	0.69	0.71	0.71	0.71	0.71
Π	18.38	21.36	20.36	20.38	19.35

[615, 593, 606], having $n > 0$, shows that the critical values t_3 and t_4 that mark the Newtonian instability points of the $m = 3$ and $m = 4$ modes (m takes reference to the order of the non-axisymmetric instability

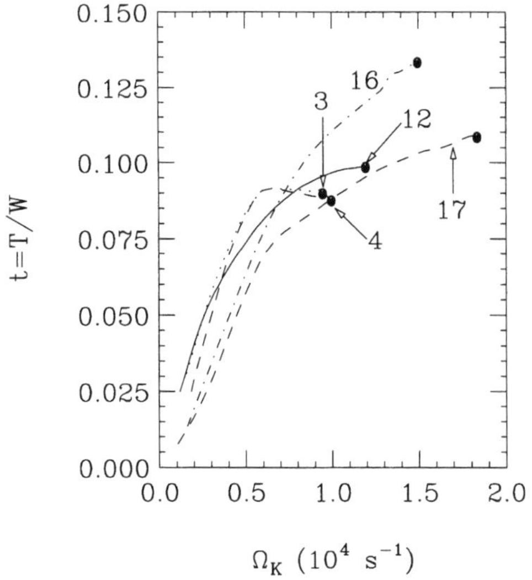

Figure 17.5. Stability parameter, t, as a function of Kepler frequency, for stellar sequences constructed for a sample of EOSs of table 12.4 [447].

modes discussed in section 16.3) decrease from $t_3 \approx 0.10$ and $t_4 \approx 0.08$ for homogeneous configurations to $t_3 \approx 0.08$ and $t_4 \approx 0.06$ for $n = 1$ polytropes [615, 593]. More than that, it was found that the critical values of the stability parameter as a function of polytropic index, $t(n)$, decrease with increasing n, in such a way that the $m = 3$ and $m = 4$ instabilities always appear before uniformly rotating polytropes reach the mass shedding frequency. Hence, as relativity probably decreases the critical t value slightly, and as n associated with the proposed, realistic equations of state corresponds to the range of $0.5 \lesssim n \lesssim 1$, one may expect that t_3 and t_4 lie somewhere between the boundaries given just above, i.e. $0.08 \lesssim t_3 \lesssim 0.10$ and $0.06 \lesssim t_4 \lesssim 0.08$ for relativistic neutron star models [105]. This range appears to be quite consistent with what we find for rotating neutron stars with canonical masses around 1.4 to $1.5\,M_\odot$, whose properties are listed in tables 17.5 and 17.6. From the latter table one reads off that $0.09 \lesssim t_K \lesssim 1.0$. While, in general, the values of t_K cannot be identified with those of t_3 and t_4, luckily for the HV equation of state the value of the Kepler period, $P_K \simeq 1.2$ ms, is close to $P_{GRR} \simeq 1.35$ ms (figure 16.13), the critical rotational period at which the $m = 2$ mode is excited. So, subject to this agreement, we establish for fully relativistic stellar neutron stars constructed for a realistic equation of state an upper bound on the t_2 range of $0.09 \lesssim t_2 \lesssim 1.0$. The higher eigenmodes $m = 3$ and $m = 4$ are excited first only at higher temperatures (see figure 16.12),

Table 17.5. Properties of neutron stars with the same baryon numbers A as those in table 14.2, but for a rotational period of $P = 1.6$ ms. The entries are: central energy density ϵ_c, gravitational mass M, mass increase relative to non-rotating stellar model $\Delta M/M$, equatorial and polar radius $R_{\rm eq}$ and $R_{\rm p}$, moment of inertia I, stability parameter t, injection energy β, gravitational redshift at pole $z_{\rm p}$, eccentricity e, quadrupole moment Π.

Quantity	HV	HFV	$\Lambda_{\rm BroB}^{\rm RBHF}$+HFV	$\rm G_{M78}^{K240}$	$\rm G_{M78}^{K240}$(NP)
ϵ_c (MeV fm^{-3})	319.55	450.40	482.88	489.55	390.93
$M/M_\odot$	1.413	1.412	1.408	1.410	1.407
$\Delta M/M$	0.93	0.86	0.57	0.71	0.50
$R_{\rm eq}$ (km)	15.449	13.183	12.070	13.709	14.023
$R_{\rm p}$ (km)	13.067	11.798	11.062	12.143	12.291
$\log_{10} I$ (g cm^2)	45.2141	45.1589	45.1175	45.1700	45.1829
t	0.045	0.032	0.027	0.034	0.037
β	0.681	0.646	0.624	0.657	0.662
$z_{\rm p}$	0.2121	0.244	0.2660	0.2338	0.2292
e	0.53	0.45	0.40	0.46	0.48
Π (km^3)	7.43	6.11	4.19	6.85	7.89

Table 17.6. Properties of neutron stars with the same baryon number as those in table 14.2, but for rotation at their respective self-consistent Kepler periods, $P_{\rm K}$. The entries not already explained in table 17.5 are: equatorial velocity $V_{\rm eq}$, redshift in backward and forward direction, $z_{\rm B}$ and $z_{\rm F}$.

Quantity	HV	HFV	$\Lambda_{\rm BroB}^{\rm RBHF}$+HFV	$\rm G_{M78}^{K240}$	$\rm G_{M78}^{K240}$(NP)
ϵ_c (MeV fm^{-3})	286.07	396.48	438.20	368.52	339.63
$P_{\rm K}$ (10^{-3} s)	1.157	0.965	0.835	0.984	1.032
$M/M_\odot$	1.427	1.440	1.431	1.420	1.419
$\Delta M/M$	1.93	2.86	2.21	1.43	1.36
$R_{\rm eq}$ (km)	17.082	15.183	13.771	15.823	15.869
$R_{\rm p}$ (km)	11.865	10.426	9.516	10.915	10.913
$\log_{10} I$ (g cm^2)	45.1596	45.1043	45.0349	45.1192	45.1186
t	0.090	0.099	0.100	0.095	0.096
β	0.645	0.592	0.556	0.616	0.616
$V_{\rm eq}/c$	0.34	0.37	0.39	0.36	0.355
$z_{\rm B}$	0.6208	0.7082	0.7968	0.6672	0.6637
$z_{\rm F}$	−0.1808	−0.1865	−0.1963	−0.1836	−0.1828
$z_{\rm p}$	0.2453	0.2996	0.3414	0.2744	0.2742
e	0.72	0.73	0.72	0.72	0.73
Π (km^3)	14.85	19.87	15.73	20.40	20.67

Table 17.7. Properties of rotating neutron stars of gravitational mass $M = 1.4\,M_\odot$, computed for several different EOSs. Each star rotates at its respective Kepler period, P_K. The entries are: central energy density ϵ_c (in MeV fm^{-3}), Kepler period P_K (in ms), equatorial and polar radius R_{eq} and R_p (in km), moment of inertia I (in g cm^2), number of baryons A, stability parameter t, injection energy β, equatorial velocity V_{eq}, redshift in backward and forward direction z_B and z_F, redshift at the pole z_p, eccentricity e, quadrupole moment Π (in km^3).

Quantity	HV	$\Lambda^{00}_{\rm Bonn}$+HV	HFV	$\Lambda^{00}_{\rm HEA}$+HFV	$\Lambda^{\rm RBHF}_{\rm BroB}$+HFV
ϵ_c	280.1	280.1	385.9	401.0	431.2
P_K	1.173	1.000	0.986	0.933	0.845
R_{eq}	17.11	15.33	15.25	14.71	13.77
R_p	11.88	10.47	10.46	10.11	9.51
$\log_{10} I$	45.1469	45.0965	45.0905	45.0675	45.0200
$\log_{10} A$	57.2387	–	57.2603	–	57.2682
t	0.089	0.102	0.099	0.099	0.099
β	0.653	0.606	0.605	0.594	0.566
V_{eq}/c	0.34	0.36	0.36	0.37	0.38
z_B	0.6042	0.6695	0.6803	0.7056	0.7740
z_F	−0.1788	−0.1799	−0.1833	−0.1860	−0.1943
z_p	0.2372	0.2850	0.2859	0.2979	0.3293
e	0.72	0.73	0.73	0.73	0.72
Π	21.46	20.95	19.67	18.23	15.42

for the present equation of state at rotational periods of $P_{\rm GRR} \simeq 1.5$ ms and $P_{\rm GRR} \simeq 1.6$ ms respectively, which implies $t_2 > t_3 > t_4$. We recall our analysis of section 16.3 which indicates that in realistic neutron star models the instability for large m (≥ 6) is damped out by viscosity, and that the $m = 3$ or 4, or $m = 4$ or 5 modes, depending on the star's temperature, likely set the upper limit on stable rotation, and thus, indirectly, on the mass as well [106, 110, 577, 586].

From the survey of rotating neutron star properties presented in table 16.1, it follows that the equations of state of our collection predict Kepler periods for the most massive stars as small as $P_K \gtrsim 0.4$ ms. Neutron stars of canonical mass, between 1.4 and $1.5\,M_\odot$, are predicted to not rotate considerably faster than $\sim$ 1 ms, as can be seen from tables 17.7 and 17.8. Rotational pulsar periods as small as say half a millisecond, as spuriously attributed to a pulsar in SN 1987A, the recent Magellanic Cloud supernova, or as theoretically inferred for PSR J0034–0534 (section 3.3), are completely excluded by our study. This finding is even strengthened by the GRR driven instability in rapidly spinning neutron stars, which, as

Table 17.8. Same as table 17.7 but for rotating neutron stars of gravitational mass $M = 1.5\,M_\odot$.

Quantity	HV	$\Lambda^{00}_{\rm Bonn}$+HV	HFV	$\Lambda^{00}_{\rm HEA}$+HFV	$\Lambda^{\rm RBHF}_{\rm BroB}$+HFV
ϵ_c	301.2	311.0	414.3	429.2	454.3
$P_{\rm K}$	1.125	0.975	0.935	0.887	0.816
$R_{\rm eq}$	17.00	15.43	15.07	14.58	13.76
$R_{\rm p}$	11.83	10.53	10.37	10.05	9.51
$\log_{10} I$	45.1863	45.1456	45.1236	45.1074	45.0645
$\log_{10} A$	57.2727	–	57.2952	–	57.3028
t	0.091	0.104	0.099	0.100	0.101
β	0.626	0.579	0.573	0.559	0.534
$V_{\rm eq}/c$	0.35	0.37	0.38	0.39	0.40
$z_{\rm B}$	0.6590	0.7221	0.7520	0.7824	0.8458
$z_{\rm F}$	−0.1852	−0.1849	−0.1912	−0.1936	−0.2002
$z_{\rm p}$	0.2636	0.3141	0.3208	0.3370	0.3682
e	0.72	0.73	0.73	0.72	0.72
Π	22.75	22.83	20.00	18.85	16.43

repeatedly stressed, sets a more stringent limit on rotation than rotation at the mass shedding limit. In the wake of the possible future observation of a sub-millisecond pulsar, it appears that it would be hard to explain such an object other than in terms of absolutely stable strange quark matter. (The properties of such stars will be discussed in chapter 18.) Independent of any particular model, what is required to withstand very rapid rotation, say in the half-millisecond regime or below, are extremely high central mass densities, which is a necessary ingredient to obtain a small stellar model. Only then will mass shedding (as well as probably the GRR driven instability) occur at correspondingly small rotational periods. Figure 17.6 is an impressive illustration of how large the central densities must become in order to enable neutron stars to rotate at half-millisecond Kepler periods or less ($\Omega_{\rm K} \gtrsim 1.3 \times 10^4$ s^{-1}). In addition, it is evident from this figure that the stars must be rotating very close to their respective mass limits (indicated by solid dots) too. Only very extremely soft equations of state, like Pan(C) (marked by label 17 in figure 17.6), could easily accommodate half-millisecond pulsars. The extreme softness of this equation of state, as well as of any other such soft model too, is inevitably accompanied by two generic features. Firstly, the derived neutron star masses may not be as high as required by observation – the conservative lower bound on mass is presently $1.444\,M_\odot$, as pointed out in section 3.1. This observation outdates the Pan(C) equation of state because it predicts a limiting mass of only $1.42\,M_\odot$. Secondly, matter compressed

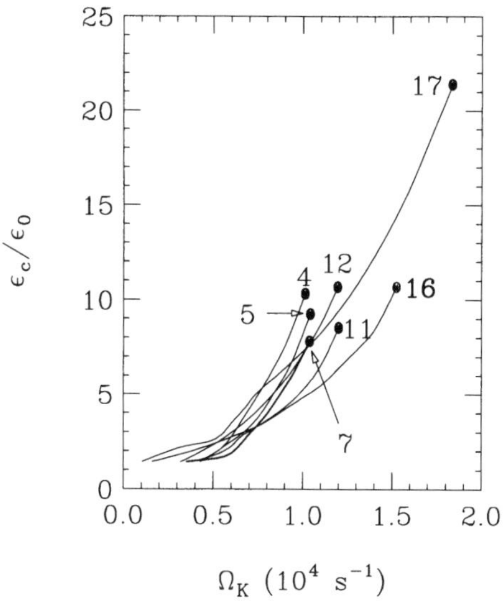

Figure 17.6. Central energy density ϵ_c (in units of normal nuclear matter density, ϵ_0) of neutron star sequences as a function of Kepler frequency, for a broad sample of EOSs listed in table 12.4 [447].

to such enormous densities as required in hypothetical half-millisecond pulsars is very unlikely to stay in the pure hadronic phase, made up of individual nucleons, hyperons and possibly Δ's, but will plausibly undergo a phase transition into deconfined quark matter, as discussed in chapter 8, and/or may have a K^- boson condensate. Models for the equation of state which account for this possibility are G^{K240}_{B180}, G^{DCM1}_{B180}, and G^{DCM2}_{B180} of table 12.4. They predict Kepler periods of $P_K \sim 0.6$ ms for stars at the mass peak (table 16.1), and periods around 0.9 ms for neutron stars of masses close to the canonical masses, making stable rotation at half-millisecond periods for this class of neutron stars unlikely too. So, subject to the uncertainties inherent in many-body calculations of superdense matter, which prevent us from drawing final conclusions yet, we carefully claim that half-millisecond pulsars, if existing, appear to be made up of some sort of *self-bound* matter, of which absolutely stable 3-flavor strange quark matter complies most comfortably with our present understanding of the behavior of superdense matter. The self-bound character of such matter implies that the corresponding stellar model is bound by the strong force, in sharp contrast to neutron stars which are bound by gravity only. The only role played by gravity is to make strange quark matter stars even more dense. Both features are of key importance to withstand rapid rotation (see chapter 18).

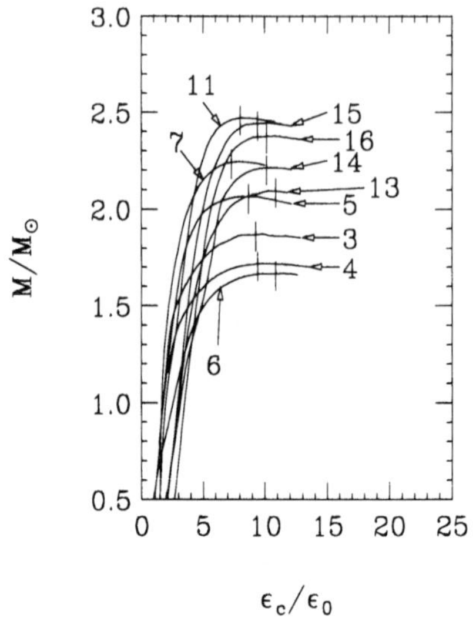

Figure 17.7. Rotational neutron star mass as a function of central energy density, ϵ_c, for a broad sample of EOSs of table 12.4 [114]. (Reprinted courtesy of World Scientific.)

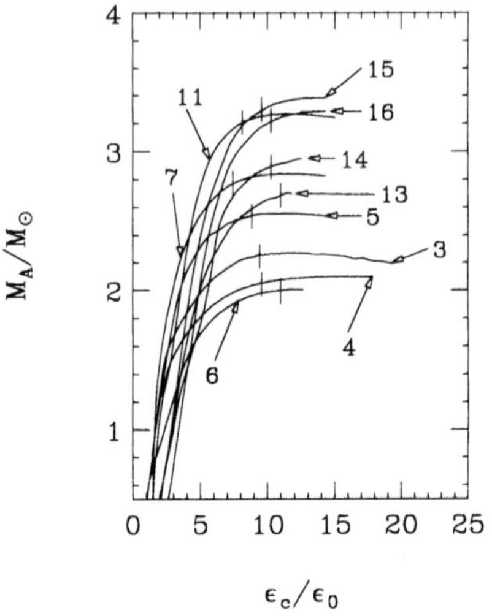

Figure 17.8. Rotational baryon mass of neutron stars as a function of central energy density, ϵ_c, for a sample of EOSs of table 12.4 [447].

As known from section 17.1.1, the gravitational mass $M \equiv M(\epsilon_c, \Omega_K(\epsilon_c))$ of neutron stars rotating at their respective Kepler frequencies can only be obtained by solving the stellar structure equations self-consistently. The numerical outcome of such calculations, performed for a broad collection of representative equations of state, is shown, as a function of central energy density ϵ_c, in figure 17.7. Figure 17.8 is the analog to figure 17.7, but for the self-consistent stellar baryon mass, $M_A \equiv M_A(\epsilon_c, \Omega_K(\epsilon_c))$. In general, rotation stabilizes a stellar model against gravitational collapse. For neutron stars, one sees that these can be at most ~ 14 to 20% more massive than at zero rotation, depending on stellar mass (the tick marks in figures 17.7 and 17.8 refer to the limiting-mass stars of each sequence) as well as on the underlying equation of state [106, 109]. If the gravitational mass of a neutron star increases beyond this limit, say because of mass accretion from a companion (e.g. a white dwarf or a blue giant) while at the same time being considerably spun up, stability against gravitational collapse is irrevocably lost, and the neutron star is condemned to collapse to a black hole. Another point of interest concerns the limiting masses obtainable for equations of state G^{π}_{200} and G^{DCM1}_{B180}. These are for non-rotating stars 1.46 and 1.50 $M_\odot$ (table 14.3), respectively. Specifically the first of these two values lies only slightly above the presently established lower bound on neutron star masses, $1.444\,M_\odot$. So, if the true critical neutron star mass were just slightly above this value, say around $1.45\,M_\odot$, then the limiting-mass model predicted computed for G^{π}_{200} would be very close to collapsing to a black hole. The mass increase induced by rotation can push up its mass to $1.67\,M_\odot$ (table 16.1), significantly above the non-rotating limit toward gravitational collapse.

Also, note from these figures that the central stellar densities are never significantly larger than about $10\,\epsilon_0$. The largest gravitational (as well as baryon) mass is obtained for equation of state $\Lambda^{00}_{\rm HEA}+{\rm HFV}$ (label 11), which is caused by its rather stiff behavior at supernuclear densities, stemming from the (Fock) exchange contribution [109]. The importance of the stiffness of the nuclear equation of state in connection with obtaining sufficiently large enough neutron star masses [108, 109, 410] was outlined qualitatively just above. Figures 17.7 and 17.8 exhibit this feature quantitatively. The equation of state associated with sequence 6, having the smallest limiting mass of all stellar sequences, is the softest equation of state of our collection listed in table 12.4. In contrast to this, the equation of state underlying sequence 11, accommodating the most massive neutron star, is the stiffest one. For the sequences in between, the stiffness increases from bottom to top.

Finally, as a last characteristic feature to be drawn from these figures, note that rotation reduces ϵ_c of a particular stellar model below its non-rotating value. Tables 14.3 and 16.1 show that for the non-rotating stars $9 < \epsilon_c/\epsilon_0 < 14$, which decreases to $7 < \epsilon_c/\epsilon_0 < 12$ in those stars rotating at their individual Kepler frequencies. The decrease of ϵ_c turns out to be more pronounced for the less dense stars (as obtained, for instance, for equations of state HFV and G_{300}) than for stars compressed to relatively high densities (e.g. G^{π}_{200} and $G^{\rm DCM1}_{\rm B180}$), because of the weaker gravitational binding. Typically, the reduction of ϵ_c, relative to the non-rotating value, drops from 24% to 10%, respectively. Recall that stars with relatively large (small) ϵ_c values are only obtained if the underlying equation of state is soft (stiff) at high nuclear densities, as is the case for G^{π}_{200} and $G^{\rm DCM1}_{\rm B180}$. The softness of the former is determined from the outset by adjusting the parameters of the many-body theory (relativistic Hartree) to a small incompressibility value. The condensation of pions, taken into account by this model too, softens the equation of state even further. The softness of the latter model, $G^{\rm DCM1}_{\rm B180}$, too originates from a low incompressibility value, and is further enhanced by quark deconfinement (table 12.7).

A reduction of ϵ_c may appear surprising at a first glance. It is not, however, but simply follows from the occurrence of the centrifugal force in rotating stars, which adds to the pressure force provided by the equation of state. Both of these forces must be counterbalanced by gravity which (roughly speaking) is of the same magnitude in rotating and non-rotating stars of comparable masses. So in the rotating case less pressure must

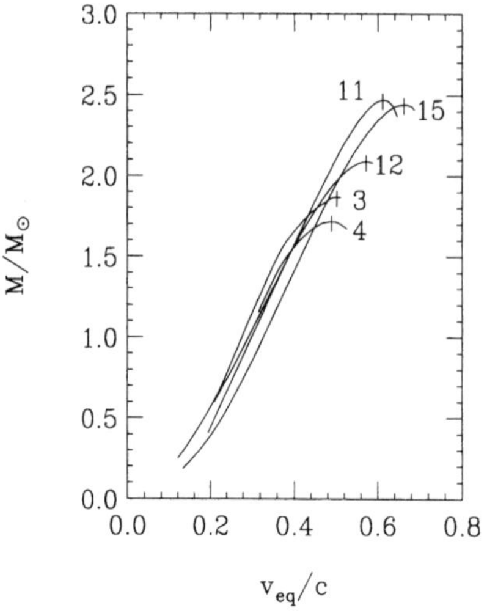

Figure 17.9. Gravitational mass of rotating neutron star sequences as a function of equatorial velocity, for a representative sample of EOSs listed in table 12.4 [447].

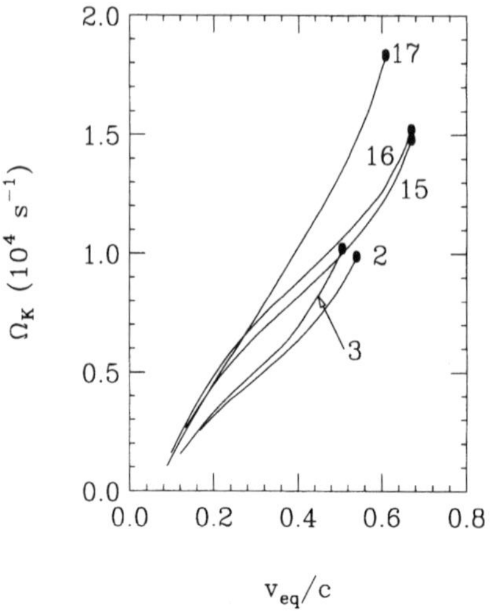

Figure 17.10. Kepler frequency as a function of equatorial velocity, for stellar sequences constructed for a sample of EOSs of table 12.4 [447].

be provided by the equation of state to counterbalance gravity. Less pressure, however, implies smaller energy densities, as $P(\epsilon)$ is monotonically decreasing for decreasing ϵ (figures 12.10 through 12.13), and thus less dense rotating star.

The results obtained for the equatorial velocities $V_{\rm eq}$, defined in equation (16.6), of stars rotating at $\Omega_{\rm K}$ are surveyed in figures 17.9 and 17.10. One sees that $V_{\rm eq}$ is up to about 70% of the velocity of light for the most massive star (indicated by tick marks) of the stellar sequence. Such enormous velocities stress again the importance of treating rotating neutron stars in the framework of Einstein's theory. The general trend of $V_{\rm eq}$ on star mass as well as on $\Omega_{\rm K}$ can be understood qualitatively in terms of the classical Newtonian expression $V_{\rm eq} = \sqrt{M/R_{\rm eq}} = R_{\rm eq}\Omega_{\rm K}$, whose mathematical form carries over, with restricted applicability of course, to the theory of general relativity. Note the dependence of $V_{\rm eq}$ on the ratio M/R which, as pointed out in section 16.2, is not too sensitive against variations of the equation of state, indicating a relatively narrow range for $V_{\rm eq}$ as computed for greatly different models for the equation of state.

The eccentricities e computed for rotationally deformed neutron stars are listed in table 16.2 too. From the definition of e, given in equation (15.84), one sees that e is given in terms of the ratio of polar to equatorial radius as $1-(R_{\rm p}/R_{\rm eq})^2$. This explains why e is decreasing

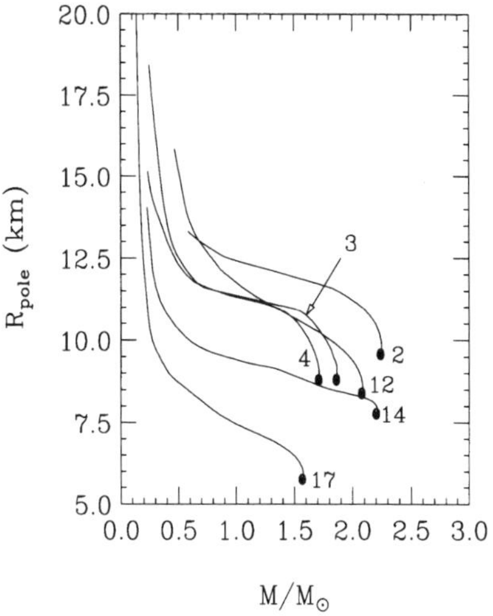

Figure 17.11. Polar radius versus mass for sequences of rotating neutron stars, for a sample of EOSs of table 12.4. [447]

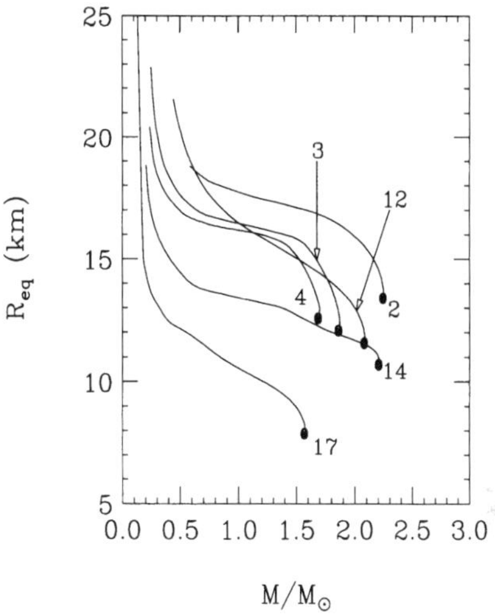

Figure 17.12. Same as figure 17.11, but for equatorial radius [447].

from top to bottom in table 16.2, being largest for the relativistic equations of state and smallest for the non-relativistic ones. Just recall that because of the smaller central mass densities, i.e. the weaker gravitational binding, of neutron stars constructed for the relativistic equations of state, rotation generally blows them up more significantly than is the case for neutron stars constructed for the non-relativistic equations of state. Despite this feature, one sees that the eccentricities are not that different for both classes of neutron stars, though the bulk properties of the maximum-mass models of our collection are rather different from each other. We find typically $R_{\rm p}/R_{\rm eq} \approx 3/4$.

The polar and equatorial radii as a function of gravitational mass are exhibited in figures 17.11 and 17.12, respectively. Each stellar sequence terminates at the location of the most massive neutron star, indicated by the solid dots. In accordance with what has been said just above about the dependence of e on the underlying equation of state (relativistic versus non-relativistic equation of state), the biggest stars and thus largest polar and equatorial radii are obtained for stellar models constructed for relatively stiff equations of state. In contrast to this, small radii result only from soft models for the equations of state. Neutron star radii as small as the one previously derived for neutron star RXJ 185 635–3754 (see, however, the caveat of this radius determination discussed in connection with equation (3.6)) can only be obtained if the equation of

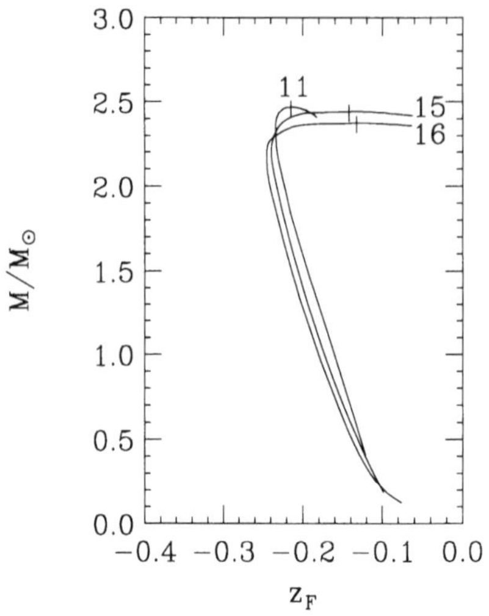

Figure 17.13. Gravitational mass of rotating neutron stars as a function of z_F. The labeling of the EOSs is explained in table 12.4.

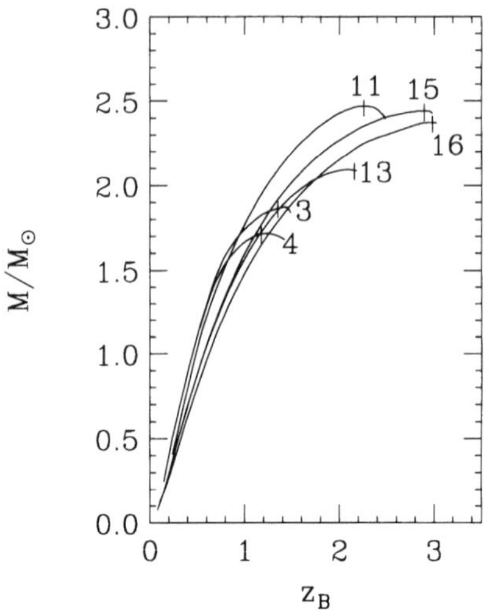

Figure 17.14. Same as figure 17.13, but calculated for the redshift z_B of photons emitted in the backward direction [447].

state softens dramatically over an extended density range a few times beyond the saturation density of nuclear matter, as already mentioned in section 14.4. For the sake of illustration, we illustrate that the extreme softness of the outdated Pan(C) model appears to be more than what is required by this neutron star. Again, in this respect the Pan(C) equation of state still attains its particular interest because it may serve to roughly indicate the lower bound on pressure at a given density necessary to attain neutron star data that are consistent with the body of observed data.

Figures 17.13 and 17.14 exhibit the change of z_F and z_B , derived in equation (15.103), with the gravitational mass of neutron stars rotating at their individual Kepler frequencies. Recall that z_F and z_B account for the shifts in spectral line frequency of photons, emitted in the forward and backward direction of the rotating star, which are caused by the presence of a strong gravitational field as well as the Doppler shift. In general, the red- and blueshifts of photons emitted from the surface of a neutron star are of interest because from this the observer gets information about the surface region of the object. For the most rapidly spinning neutron stars, we calculate for the outcome shown in figures 17.13 and 17.14 $|z_B/z_F| \approx 5$. Such a large magnitude for the backward to forward redshifts reflects large increases in radius for stars rotating near Ω_K [106].

The backward shift increases monotonically with the strength of the gravitational field (i.e. with neutron star mass), becoming maximal for the

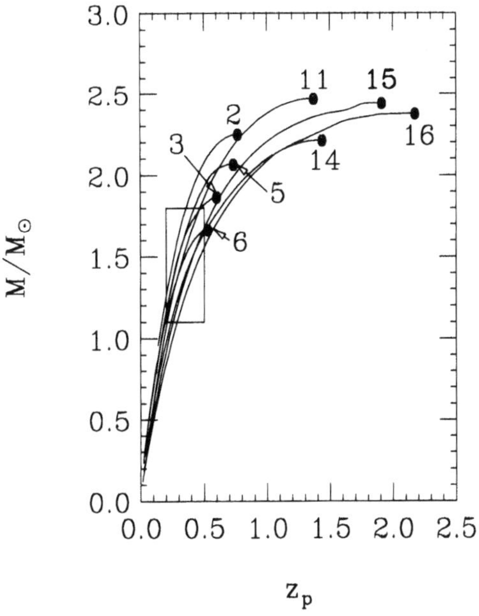

Figure 17.15. Rotational neutron star mass versus polar redshift. The solid dots refer to the maximum-mass stars [114]. (Reprinted courtesy of World Scientific.)

heaviest star (tick marks) of each sequence. These curves are qualitatively similar to those exhibiting the non-rotating neutron star mass as a function of gravitational redshift, which was shown in figure 14.12. The new feature which enters in the case of rotating neutron stars is that $z_{\rm F}$ and $z_{\rm B}$ contain both the redshift caused by gravity as well as the Doppler shift due to the star's rotation, as mentioned just above,. For that reason the $z_{\rm B}$ values are considerably larger than the redshifts of non-rotating stars, and the forward shifts $z_{\rm F}$ even have negative sign. The latter are the more negative the larger $\Omega_{\rm K}$ (i.e. the faster the star rotates), which increases with the star's mass. The decrease of $z_{\rm F}$ is however somewhat compensated by the gravitationally caused redshift which increases with mass. For that reason the decrease of $z_{\rm F}$ is reversed for rotating stars having masses that are close to the limiting ones. That is to say, gravity wins over the Doppler shift for stars near the mass peak.

Figure 17.15 shows the rotational neutron star mass plotted as a function of polar redshift, defined in (15.107). The neutron stars along each sequence rotate at their individual Kepler periods. Each sequence terminates at the most massive star (indicated by solid dots). Gamma-ray bursters are suspected, but not known, to be neutron stars. If interpreted as such, the measured gamma-ray burst pair annihilation lines give tentative evidence to support a neutron star redshift range of 0.2–0.5, with the highest concentration in the narrower range of 0.25–0.35 [46], as discussed in section 3.8. The former range combined with observed neutron star masses is exhibited by the rectangle in figure 17.15. Obviously, these data do not yet set too stringent a constraint on the equation of state.

17.3.2 Properties of rotating neutron star sequences with constant baryon number

Having already explored numerous attributes of rapidly rotating neutron stars, we next turn to the very interesting case of evolving neutron stars whose stellar baryon numbers over time vary only insignificantly, or not at all. Such scenarios apply to *isolated* neutron stars whose rotational periods are gradually lengthening in the course of time, because of the loss of rotational energy due to the emission of an electron–positron wind from the star accompanied by the emission of electromagnetic dipole radiation. The second scenario one may think of is neutron stars in *binary stellar systems* whose baryon numbers, while accreting mass from a companion star, do not change that much during not too large accretion time scales.

Figure 17.16 shows the change in gravitational mass of such neutron stars, spinning down from the Kepler period to zero rotation, or vice versa. As usual, the solid dots refer to the maximum-mass neutron star of each sequence. One reads off from this figure that during this process, for A = const, the star's gravitational mass changes by at most $\sim 0.05\,M_\odot$. More than that, an appreciable number of rapidly spinning, rotationally stabilized neutron stars do not have non-rotating (in fact, not even slowly rotating) counterparts and therefore will be collapsing to black holes when spinning down. This figure also reveals that light rotating neutron stars rotating at their individual Kepler periods possess about 2% fewer baryons than their non-rotating counterparts possessing the same gravitational mass. This number increases slightly with mass, but never amounting to more than 3% (G^{π}_{300}) to 5% (HV).

Note also that the heaviest non-rotating neutron star models of each sequence has hardly any correspondence with its rotating counterpart. For instance, the increase in baryon number caused by rotation is about 12% (for both equations of state). The rather different central densities of the two neutron star models becomes obvious from figure 17.17 (tick marks). Hence, the only physical quantity that is common to both the non-rotating and rotating maximum-mass neutron star models is the nuclear equation of state; neither their masses, nor baryon numbers, nor central densities are at least approximately the same. For that reason it really appears rather surprising from the outset that the general relativistic Kepler frequency of a rotating maximum-mass star can be obtained to such a good approximation

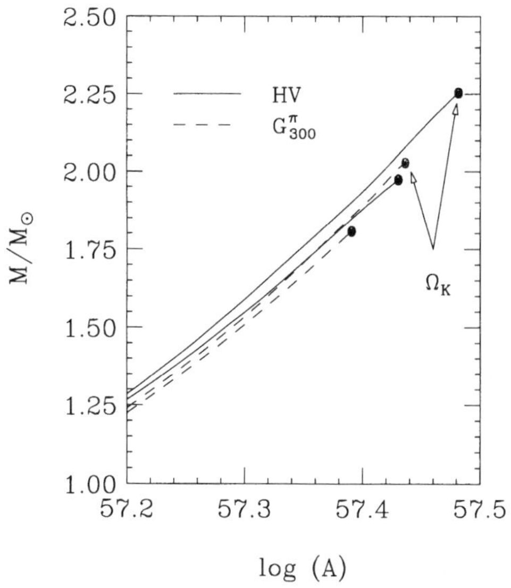

Figure 17.16. Gravitational mass of rotating ($\Omega = \Omega_K$) and non-rotating neutron star models as a function of star's baryon number, for EOSs HV and G^{π}_{300}. The solid dots denote the heaviest star of each sequence [447].

from the empirical expression for Ω_K, as explained in section 16.2.

A survey of the dependence of A on ϵ_c for a sample of selected equations of state is shown in figure 17.18, where the curves labeled 9 and 15 set the lower and upper bound on the number of baryons of which rapidly rotating neutron stars may be made of. The gravitational masses that correspond to these stellar sequences may be inferred from figure 17.7. The tick marks denote the maximum-mass star model of each stellar sequence.

17.3.3 Frequency dependence of hyperon thresholds and quark deconfinement

Generally, the weakening of centrifugal force accompanied by the slowing-down of a rotating body causes its central density to increase. This effect can be quite dramatic for neutron (hybrid) stars as shown in figures 17.19 and 17.20. There, along each curve the star's frequency varies from Kepler, Ω_K, to zero rotation, while the baryon number is kept constant as it should be for an isolated rotating neutron star that spins down in the course of time because of energy loss. The numbers associated with each curve denote the star's non-rotating gravitational mass. Hence, one reads off that the central density of the star with $M = 1.42\, M_\odot$ (figure 17.20), for instance, increases

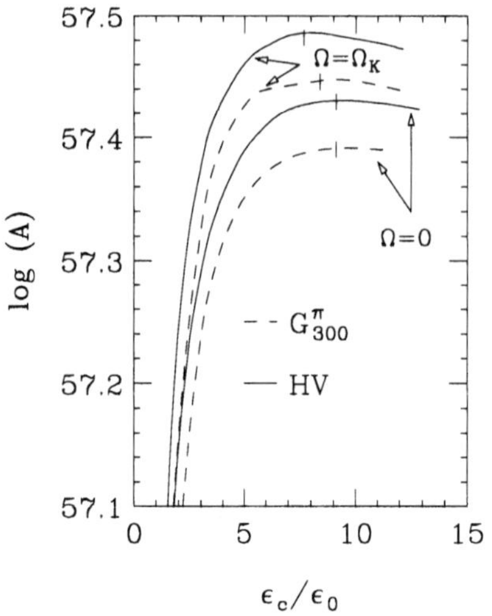

Figure 17.17. Baryon number of rotating (Ω_K) and non-rotating neutron stars as a function of central energy density, for EOSs HV and G^{π}_{300} [447].

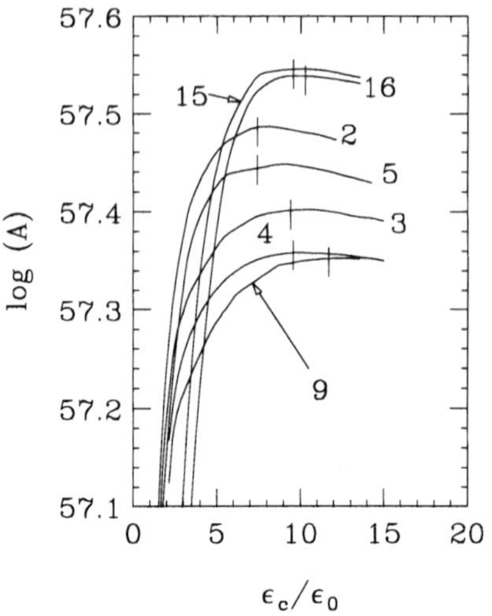

Figure 17.18. Baryon number of neutron stars rotating at Ω_K as a function of central energy density, for EOSs of table 12.4 [447].

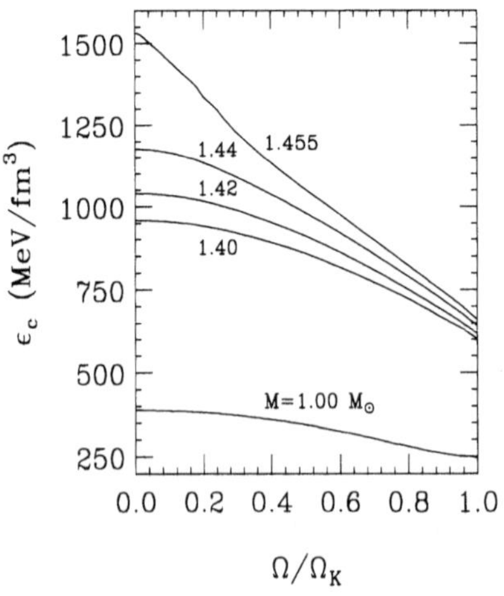

Figure 17.19. Central star density versus rotational frequency (in units of Kepler) for different sample stars. The EOS is G^{K240}_{B180}.

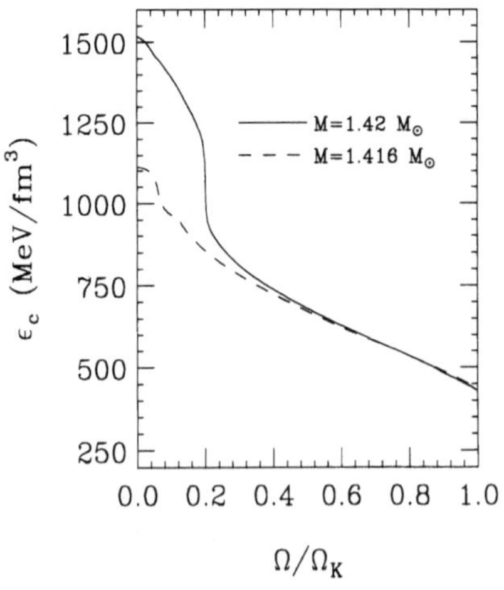

Figure 17.20. Same as figure 17.19, but for EOS G^{K300}_{B180}.

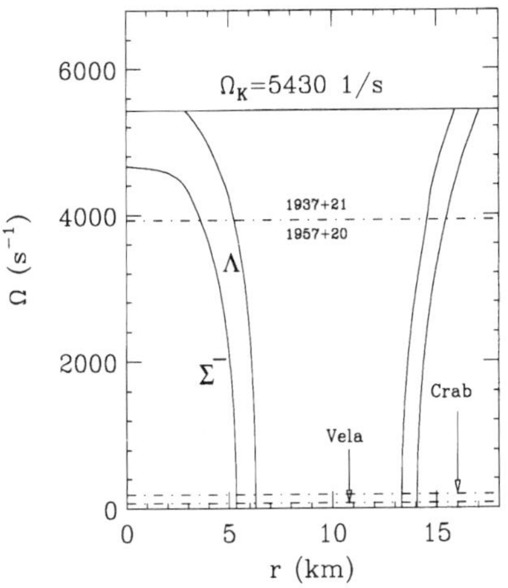

Figure 17.21. Frequency dependence of hyperon thresholds in equatorial direction, computed for HV. The star's non-rotating mass is $1.40\,M_\odot$ [616]. (Reprinted courtesy of Narosa Publishing House.)

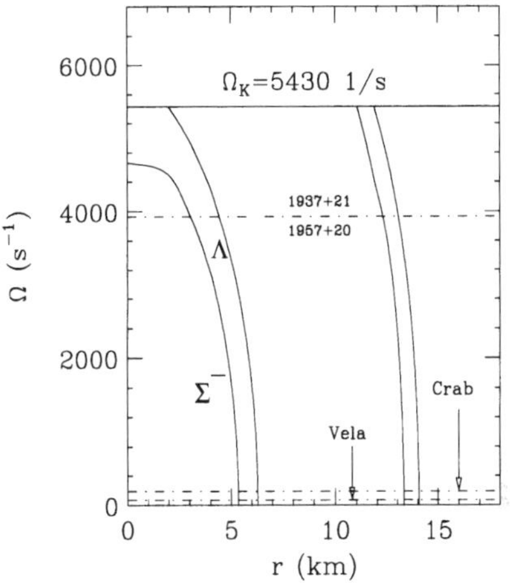

Figure 17.22. Same as figure 17.21, but in polar direction.

from about 450 MeV fm^{-3} for rotation at Ω_K to more than 1500 MeV fm^{-3} for $\omega = 0$, which is a $\sim$ 66% effect. Of course, this effect is smaller for the lighter stars, since for them the gravitational pull is correspondingly weaker. The Kepler frequencies of the two models of figure 17.20, which will be of particular interest in what follows, are $\Omega_K(1.42\ M_\odot) = 6168\ \mathrm{s}^{-1}$ and $\Omega_K(1.416\ M_\odot) = 6146\ \mathrm{s}^{-1}$.

Such dramatic changes in the interior density of slowing-down neutron stars imply profound changes of their interior composition, because of the sensitive variation of the particle composition with density as we have seen in figures 7.23, 7.24, 11.15, and 11.16 for 'conventional' neutron star matter made up exclusively of baryons, and in figures 8.3 to 8.5 for quark-hybrid star matter, which, by definition, accounts for quark deconfinement too. Figures 17.21 through 17.28 illustrate how these changes carry over to the internal structure of conventional neutron stars. Two different models for the equation of state, HV and HFV, were chosen. In each case the star's rotational frequency covers the maximal possible range $0 \leq \Omega \leq \Omega_K$. One sees that as the rotating stars spin down they become significantly less deformed, that is, one obtains closer equality between R_p and R_eq, and the central density rises from below to above the threshold density of individual hyperon species (such as Σ^-, Ξ^-, Σ^0) as well as possibly several of the Δ

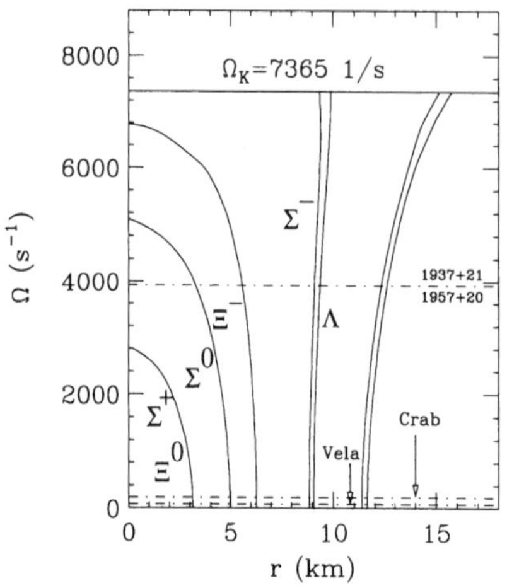

Figure 17.23. Frequency dependence of hyperon thresholds in equatorial direction, computed for HV. The star's non-rotating mass is 1.978 $M_\odot$.

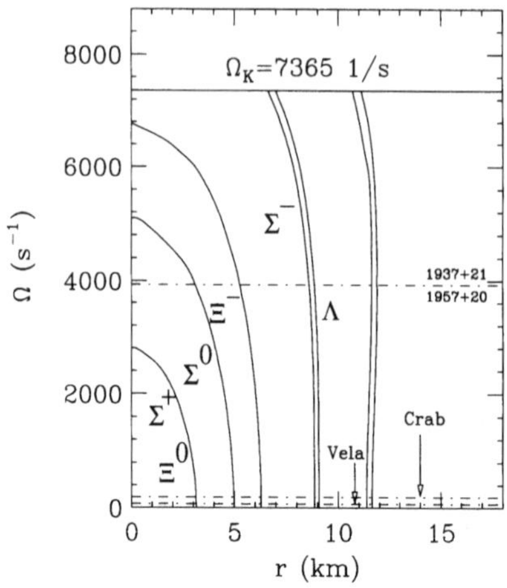

Figure 17.24. Same as figure 17.23, but in polar direction.

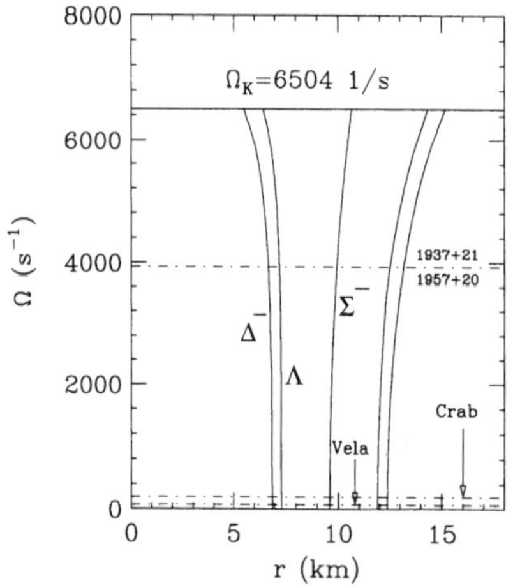

Figure 17.25. Frequency dependence of hyperon thresholds in equatorial direction, computed for HFV. The star's non-rotating mass is 1.40 $M_\odot$.

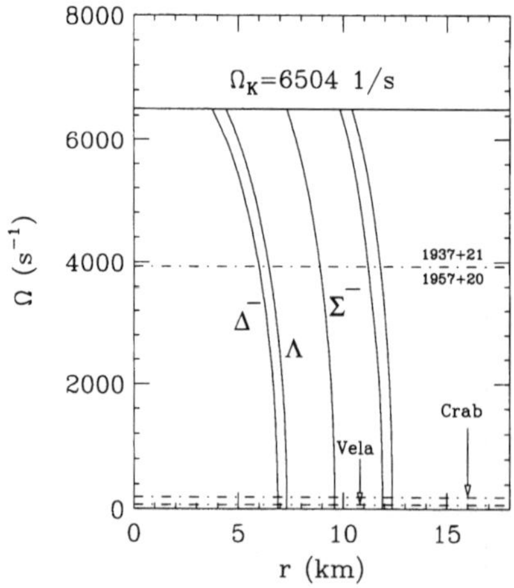

Figure 17.26. Same as figure 17.25, but in polar direction.

states too, depending on Ω, the star's mass, and the microscopic model adopted to derive the equation of state.

For some pulsars the mass and initial rotational frequency Ω may

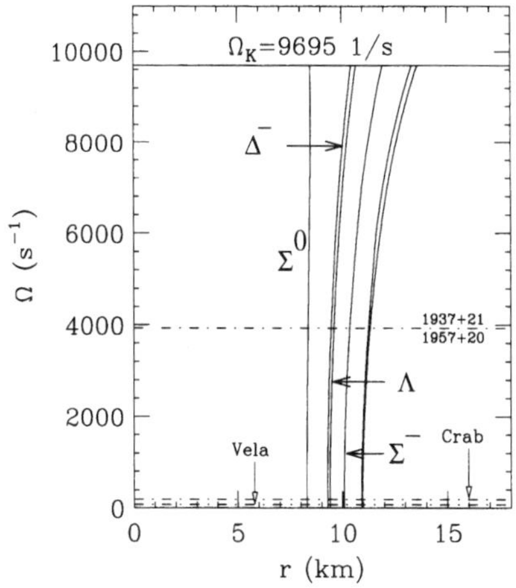

Figure 17.27. Frequency dependence of hyperon thresholds in equatorial direction, computed for HFV. The star's non-rotating mass is 2.20 $M_\odot$.

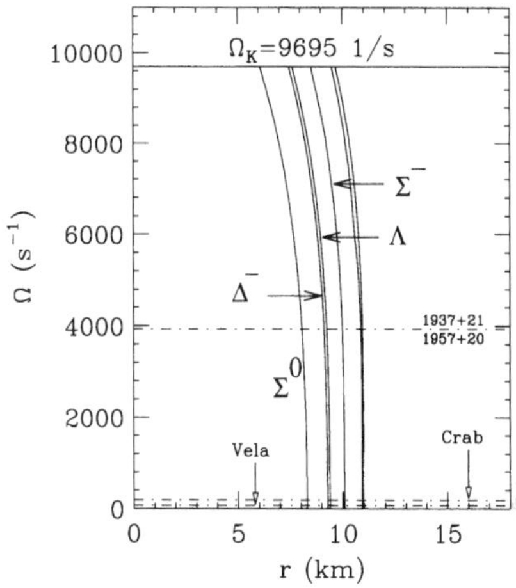

Figure 17.28. Same as figure 17.27, but in polar direction.

be such that the central density rises from below the critical density for dissolution of baryons into their quark constituents. Examples are shown in figures 17.29 through 17.32, which constitutes the hybrid-star counterparts to figure 17.21 to 17.28 computed for 'conventional' neutron stars [616]. As before, in all cases the stars' baryon number is kept constant during spin-down from the Kepler frequency to zero rotation. The evolution of the central mass densities of the stars of figures 17.29 through 17.32 can be inferred by reference to figure 17.20. For rotational frequencies $\Omega \lesssim 1250$ s^{-1} the hybrid star models consist of an inner region of purely quark matter ('q'), surrounded by a few-kilometer-thick shell of mixed phase of hadronic and quark matter – containing structures like hadronic drops ('hd'), hadronic rods ('hr'), hadronic slabs ('hs'), quark slabs ('qs'), quark rods ('qr'), quark drops ('qd'), in each case arranged in a lattice structure [66] – and this surrounded by a thin shell of hadronic liquid ('hl'), itself with a thin crust of heavy ions [426]. A schematic illustration of the different geometrical Coulomb structures has already been shown in figure 8.6. As described there, these structures introduced to the interior of neutron stars are a consequence of the competition of the Coulomb and surface energies of the hadronic and quark matter phase [66, 88]. This competition establishes the shapes, sizes, and spacings of the rarer phase in the background of the other, ranging, for decreasing density, from hadronic drops, hadronic rods, hadronic plates immersed in quark matter followed by quark plates, quark rods, to quark drops immersed in hadronic matter,

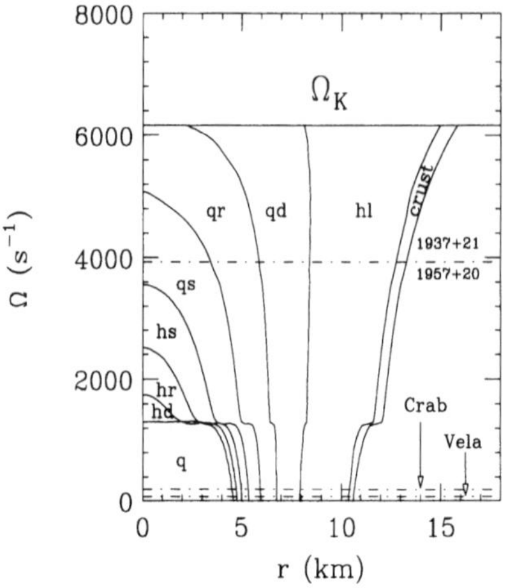

Figure 17.29. Frequency dependence of quark structure in equatorial direction, for G^{K300}_{B180} and non-rotating star mass 1.42 $M_\odot$ [616]. (Reprinted courtesy of Narosa Publishing House.)

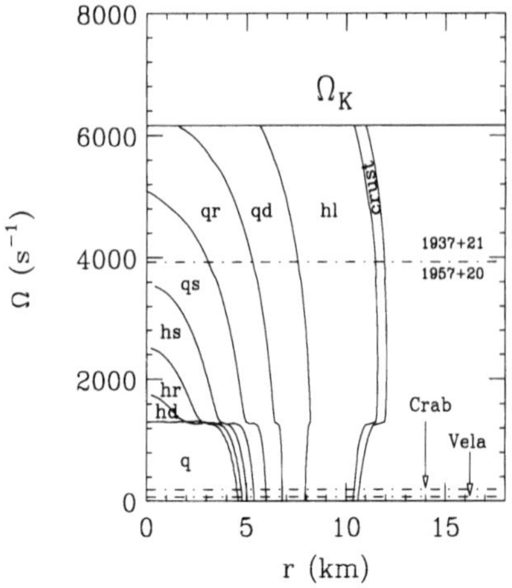

Figure 17.30. Same as figure 17.29, but in polar direction [616]. (Reprinted courtesy of Narosa Publishing House.)

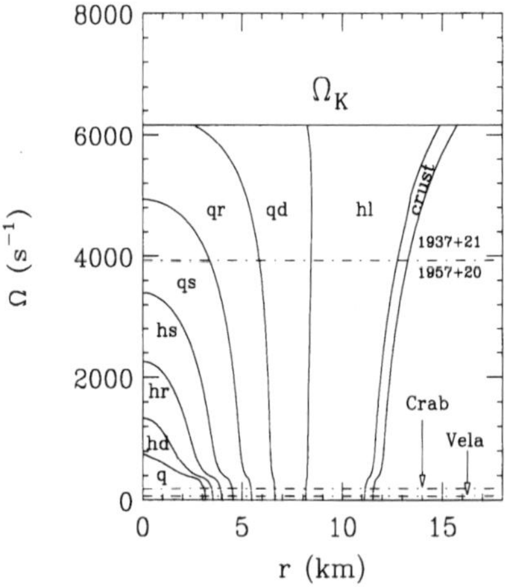

Figure 17.31. Frequency dependence of quark structure in equatorial direction, for EOS G^{K300}_{B180} and non-rotating star mass 1.416 $M_\odot$.

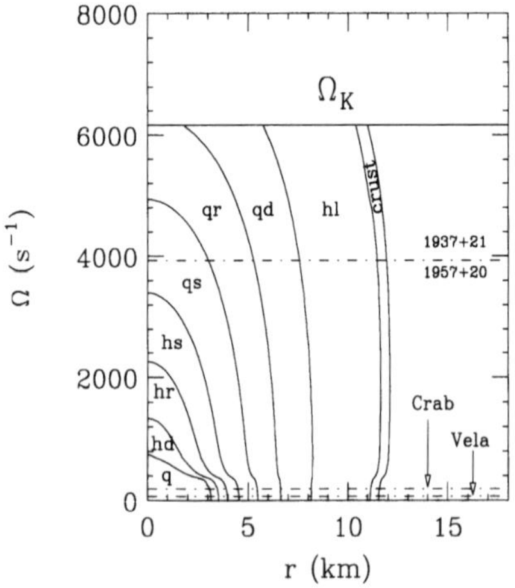

Figure 17.32. Same as figure 17.31, but in polar direction.

so as to minimize the lattice energy [66, 88, 438]. Glendenning has pointed out that these structures may have dramatic effects on pulsar observables

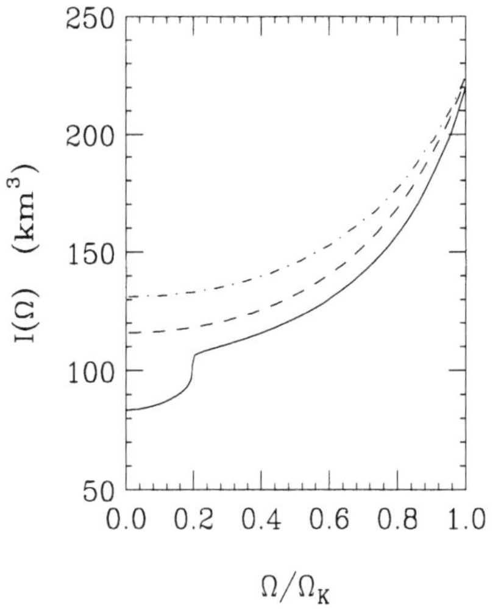

Figure 17.33. Moment of inertia versus rotational frequency for quark-hybrid star (solid curve), hyperon star (dashed), and 'neutron' star (dash–dotted) [616]. (Reprinted courtesy of Narosa Publishing House.)

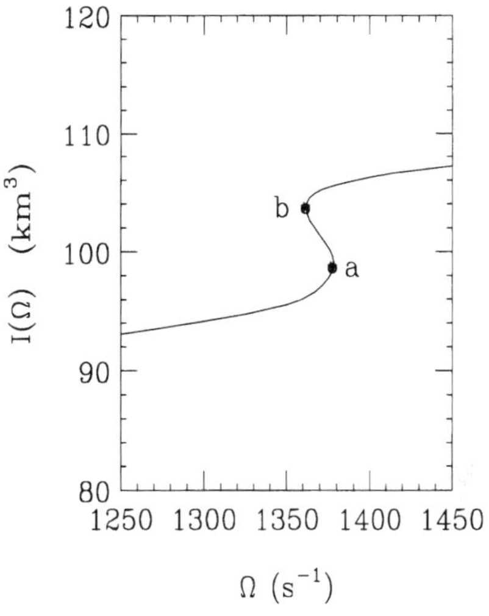

Figure 17.34. Same as figure 17.33, but for quark-hybrid star whose deconfined quark matter phase is sufficiently large to cause a 'backbending' of I in the region between points a and b.

including transport properties and the theory of glitches [66, 88].

If the mass and initial rotational frequency of a pulsar is such that during its slowing-down phase the interior density rises from below to above the critical density for the quark–hadron phase transition, first at the center where the density is highest (cf. figures 17.19 and 17.20) and then in an region expanding in the radial outward direction away from the star's center, matter will be converted from the relatively incompressible nuclear matter phase to the highly compressible quark matter phase. The tremendous weight of the overlaying layers of nuclear matter will compress the quark matter core, causing the entire star to *shrink* on a macroscopic length scale, as shown in figures 17.29 through 17.32. The mass concentration in the core of such a star will be further enhanced by the increasing gravitational attraction of the quark matter core on the overlaying nuclear matter. The moment of inertia thus decreases anomalously with decreasing rotational frequency as the new phase slowly engulfs a growing fraction of the star [426], as can be seen from figures 17.33 and 17.34. Figure 17.33 shows the moment of inertia, computed self-consistently from equation (15.51), of three sample stars having the same baryon number but different internal constitutions [616]. The quark-hybrid star, constructed for EOS G_{B180}^{K300}, is the same as in figures 17.29 and 17.30, the hyperon star is constructed for EOS G_{M78}^{K240}, and the 'neutron' star also for EOS G_{M78}^{K240} but for chemically equilibrated protons and neutrons only (hyperons ignored). The baryon number, $\log_{10} A = 57.27$, is again constant along each curve. This fixes the non-rotating stellar masses at $1.42\,M_\odot$ in case of the quark-hybrid star, and at $1.45\,M_\odot$ for both hyperon and neutron stars. The shrinkage of the hybrid star due to the development of a quark matter core at low frequencies is the more pronounced the bigger the quark matter core, i.e. the smaller Ω, in the center of the star. Model calculations indicate that the observed shrinkage and, thus, the dropping of I can not be obtained for physical scenarios other than the hypothetical quark–hadron phase transition in pulsars [617]. Parenthetically, we note that hyperon population alone obviously modifies the EOS much too little to cause any changes in I, no matter how small, as follows from model G_{M78}^{K240}.

The decrease of the moment of inertia, I, caused by the quark–hadron phase transition, shown in figure 17.33, is superimposed on the response of the stellar shape to a decreasing centrifugal force as the star spins down due to the loss of rotational energy. In order to conserve angular momentum not carried off by particle radiation from the star, the deceleration rate $\dot{\Omega}$ (< 0) must respond correspondingly by decreasing in absolute magnitude. More than that, $\dot{\Omega}$ may even change sign, as demonstrated by Glendenning, Pei, and Weber [426], carrying the important astrophysical information that an *isolated* pulsar may spin up during a certain period of its stellar evolution. The situation may be compared with an ice skater who spins up upon contraction of the arms before air resistance and friction of the skate on the ice reestablishes spin-down again. Such an anomalous decrease of I is analogous to the 'backbending' phenomenon known from nuclear physics, in which case the moment of inertia of an atomic nucleus changes anomalously because of a change in phase from a nucleon spin-aligned state at high angular momentum to a pair-correlated superfluid phase at low angular momentum. In the nuclear physics case, the backbending in the rotational bands of nuclei was predicted by Mottelson and Valatin [618] and then observed years later by [619, 620].

For neutron stars, the stellar backbending of I is shown in figure 17.34. Stars evolving from a towards b are rotationally decelerated ($\dot{\Omega} < 0$), while stars evolving from b towards a, say because of mass accretion from a companion, are rotationally accelerated ($\dot{\Omega} > 0$). As we shall see next, the backbending phenomenon dramatically modifies the timing structure of pulsar spin-down, linking quark deconfinement in neutron stars to radio astronomical observations.

17.3.4 Evolution of a pulsar's braking index

As discussed in chapter 3, pulsars are identified by their periodic signal believed to be due to a strong magnetic field fixed in the star and oriented

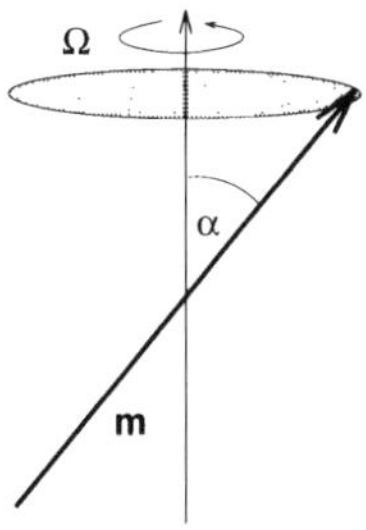

Figure 17.35. Schematic illustration of a magnetic dipole rotating with frequency Ω. $\boldsymbol{m}$ is the magnetic dipole moment, α the angle of inclination between magnetic moment and the rotation axis.

at an angle from the rotation axis. The period of the signal is therefore that of the rotation of the star. The angular velocity of rotation decreases slowly but measurably over time, and usually the first and occasionally second time derivative can also be measured. Various energy loss mechanisms could be at play such as the dipole radiation, part of which is detected on each revolution, as well as other losses such as ejection of charged particles [258]. The measured frequency and its time derivative have been used to estimate the spin-down time or age of pulsars (cf. section 3.5). The age is very useful for classifying and understanding pulsar phenomena such as glitch activity.

Without loss of generality, let us assume that the pulsar slow-down is governed by a single mechanism, or several mechanisms having the same power law. The energy balance equation can then be written in the form

$$\frac{\mathrm{d}E}{dt} = \frac{\mathrm{d}}{dt}\left\{\frac{1}{2}I\,\Omega^2\right\} = -\,C\,\Omega^{n+1}\,, \tag{17.1}$$

where, for magnetic dipole radiation, the constant C is equal to $C = \frac{2}{3}\boldsymbol{m}^2\sin^2\alpha$, $n = 3$, $\boldsymbol{m}$ the magnetic dipole moment, and α the angle of inclination between magnetic moment and the rotation axis. The geometry of a rotating, magnetic dipole is illustrated in figure 17.35. If, as is customary, the star's angular velocity Ω is regarded as the only time-dependent quantity, one obtains the usual formula for the rate of change of pulsar frequency, $\dot{\Omega} = -K\,\Omega^n$ (equation (3.14)) with K a constant and n the braking index. From the braking formula one usually defines from its solution the spin-down age of the pulsar given in (3.13).

However, the moment of inertia is not constant in time but responds to changes in rotational frequency, as can be seen in figure 17.33, more or less in accord with the softness or stiffness of the equation of state (that is, the star's internal constitution) and according to whether the stellar mass is small or large. This response changes the value of the braking index in a frequency dependent manner, even if the sole energy-loss mechanism were pure magnetic dipole, as expressed in equation (17.1). Thus during any epoch of observation, the braking index will be measured to be different from $n = 3$ by a certain amount. How much less depends, for any given pulsar, on its rotational frequency and for different pulsars of the same frequency, on their mass and on their internal constitution. When the frequency response of the moment of inertia is taken into account, equation (3.14) is replaced with

$$\dot{\Omega} = -2\,I\,K\,\frac{\Omega^n}{2\,I + I'\,\Omega} = -\,K\,\Omega^n\left\{1 - \frac{I'}{2\,I}\,\Omega + \left(\frac{I'}{2\,I}\,\Omega\right)^2 - \cdots\right\}, \tag{17.2}$$

where $I' \equiv \mathrm{d}I/\mathrm{d}\Omega$ and $K = C/I$. This explicitly shows that the frequency dependence of $\dot{\Omega}$ corresponding to any mechanism that absorbs (or deposits) rotational energy such as equation (17.1) cannot be a power law, as in (3.14) with K a constant. It must depend on the mass and internal constitution of the star through the response of the moment of inertia to rotation as in (17.2).

Equation (17.2) can be represented in the form of (3.14) – but now with a frequency dependent prefactor – by evaluating

$$n(\Omega) \equiv \frac{\Omega\,\ddot{\Omega}}{\dot{\Omega}^2} = n - \frac{3\,I'\,\Omega + I''\,\Omega^2}{2\,I + I'\,\Omega}\,. \tag{17.3}$$

Therefore the effective braking index depends explicitly and implicitly on Ω. The right side reduces to a constant n only if I is independent of frequency. But this cannot be, not even for slow pulsars if they contain a quark matter core. The centrifugal force ensures the response of I to Ω. As an example, we show in figure 17.36 the variation of the braking index with frequency for the rotating hybrid star of figures 17.29 and 17.30. For illustration we assume dipole radiation. As before, the baryon number of the star is kept constant. Because of the structure in the moment of inertia, driven by the phase transition of baryonic matter into deconfined quark matter, the braking index deviates dramatically from 3 for rotational frequencies $\Omega \sim 1370\ \mathrm{s}^{-1}$, where quark deconfinement leads to the formation of a pure quark matter core in the center of the star! Such an anomaly in $n(\Omega)$ (i.e. $-\infty < n < +\infty$) is not obtained for conventional neutron or hyperon stars because their moments of inertia increase smoothly with Ω, as shown in figure 17.33. The observation of such an anomaly in the timing structure of pulsar spin-down may thus be interpreted as a *signal of quark deconfinement* in the centers of pulsars. Of course, because of the extremely small temporal change of a pulsar's rotational period, one cannot measure the shape of the curve. This is indeed not necessary. Just a single anomalous value of n that differed significantly from the canonical value of 3 would suffice [247, 426].

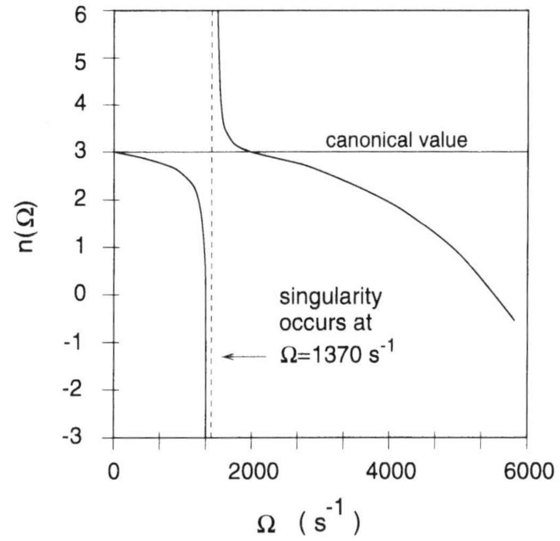

Figure 17.36. Braking index as a function of rotational frequency for a hybrid star. The singularity at $\Omega \sim 1370\ \mathrm{s}^{-1}$ is caused by *quark deconfinement.* The overall reduction of n below 3 is due to the frequency dependence of I and, thus, is independent of whether or not quark deconfinement takes place in pulsars.

Carried over to the observed pulsar data of Ω and $\dot{\Omega}$, it appears that the change in centrifugal force over the life of a canonical, slowly rotating pulsar is probably too meagre to span a significant change. The braking anomaly therefore is more likely to be restricted to millisecond pulsars. And even then, the phase change may occur only in such millisecond pulsars that rotate near the maximum mass peak. Otherwise the fraction of pure quark matter in their center may not be sufficient to cause the required shrinkage. The phase change itself may be first (as in our example) or second order. Both orders will cause a signal as long as quark deconfinement causes a sufficient softening of the EOS. A serious drawback may be that the braking indices of millisecond pulsars are very hard to measure, because of timing noise which renders the determination of $\ddot{\Omega}$ very complicated.

As a final but very important point on this subject, we estimate the duration over which the braking index is anomalous. It can be estimated from

$$\Delta T \simeq -\frac{\Delta\Omega}{\dot{\Omega}} = \frac{\Delta P}{\dot{P}}, \tag{17.4}$$

where $\Delta\Omega$ is the frequency interval of the anomaly. The range over which n is smaller than zero and larger than six (figure 17.36) is $\Delta\Omega \approx -100\ \mathrm{s}^{-1}$, or $\Delta P \approx -2\pi\Delta\Omega/\Omega^2 \approx 3 \times 10^{-4}$ s at $\Omega = 1370\ \mathrm{s}^{-1}$. So, for a millisecond pulsar whose period derivative is typically $\dot{P} \simeq 10^{-19}$, as can be seen from

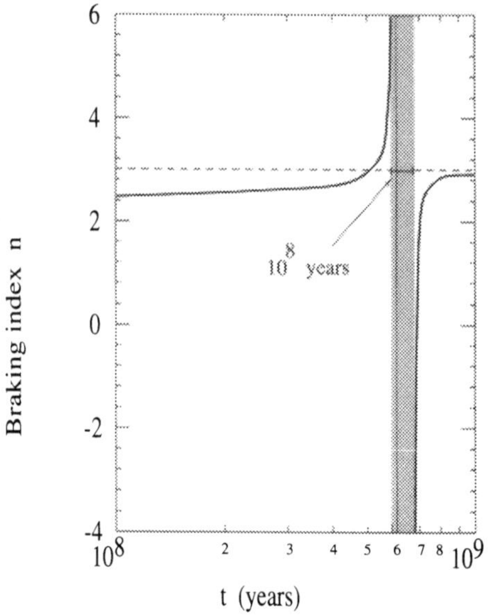

Figure 17.37. Braking index as a function of time. The epoch over which n is anomalous (i.e. $-\infty < n < +\infty$) because of quark deconfinement, $\sim 10^8$ years, is indicated by the shaded area (after Glendenning [247]).

tables 1.1 and 1.2, we find $\Delta T \simeq 10^8$ years (see figure 17.37). The dipole age of such pulsars is about 10^9 years. (Being conservative, we ignore the extension of both transition epoch and dipole age by the spin-up in making the estimate.) So as a rough estimate we may expect about 10% of the millisecond pulsars[2] to be in the transition epoch and so could be signaling the ongoing process of quark deconfinement in their cores! Note that the spin-up (region a–b in figure 17.34) is about 1/5 the epoch or about 1/50 of the spin-down time. To avoid confusion, we point out that the spin-up has nothing to do with the minuscule spin-up known as a pulsar glitch. In the latter case the relative change of the moment of inertia is very small, $\Delta I/I \simeq -\Delta\Omega/\Omega \simeq 10^{-6}$ or smaller, and approximates closely a continuous response of the star to changing frequency on any time scale that is large compared to the glitch and recovery interval. Excursion of such a magnitude as quoted would fall within the thickness of the line in figure 17.34.

It is probably needless to say that the observation of a pulsar with an anomalously large braking index would be a momentous discovery with most far-reaching consequences for nuclear and particle physics. It would

[2] Currently about 25 solitary millisecond pulsars are known, several of which are compiled in tables 1.1 and 1.2.

help to clarify how quark matter behaves, and give a boost to theories about the early Universe as well as laboratory searches for the production of quark matter in heavy-ion collisions. On an even more fundamental level, it would prove that the essentially free quark state predicted for matter at very high energy densities actually exists, and give us a picture of an early phase of the Universe that is based on observation, which may be coined *quark astronomy.*

17.3.5 Implications for searches for rapidly rotating pulsars

When discussing the limiting rotational periods of neutron stars it is very important to distinguish between their histories. Firstly, there are the hot neutron stars newly born in supernova explosions. The temperature of such objects is around $T = 10^{10}$ K. These are to be distinguished from the neutron stars formed from the collapse of accreting white dwarfs, whose temperatures, after birth, are around 10^{10} K too. The third category of neutron star evolution concerns old and therefore cold neutron stars in binary systems that are accreting mass from a companion, by which they are reheated. The temperature (and hence viscosity) in neutron stars differs in either case by orders of magnitude, which dramatically alters the time scales for the viscous dissipation mechanism to damp out stellar perturbations that are coupled to gravity wave emission (section 16.3).

Figures 17.38 and 17.39 survey the rotational instability periods P_{GRR} at which the gravitational-radiation reaction (GRR) driven instability is excited in rapidly rotating hot and cold neutron stars, respectively. Any observed, newly born neutron star having a canonical mass of $\sim 1.5\,M_{\odot}$ that rotates at periods appreciably below $\sim$ 1.6 ms, depending on the star's temperature and the equation of state used to compute the stellar model, appears to be subject to the GRR instability. For instance, newly born pulsars observed as the collapsed remnants of supernovae, or formed from collapsing white dwarfs are predicted to have stable rotational periods between about 1 and 1.6 ms, depending on equation of state, as long as their masses are close to the canonical value, $\sim 1.5\,M_{\odot}$ supported by supernova calculations [8].

The upper and lower bounds on the stable rotational period are somewhat smaller for cold neutron stars, as exhibited in figure 17.39, because of the significantly larger shear viscosity of cold neutron star matter, which damps out the GRR driven instability very efficiently. Hence, old and cold neutron stars of canonical mass, $M \sim 1.5\,M_{\odot}$, cannot be spun up to stable rotational periods smaller than about 0.8 ms. Half-millisecond periods of less are completely excluded by our collection of equations of state. We note that the 1.6 ms period of the two fastest yet observed pulsars are compatible with figure 17.39 as long as gravitational masses larger than $\sim 1\,M_{\odot}$ are assumed. Hot neutron stars rotating at 1.6 ms

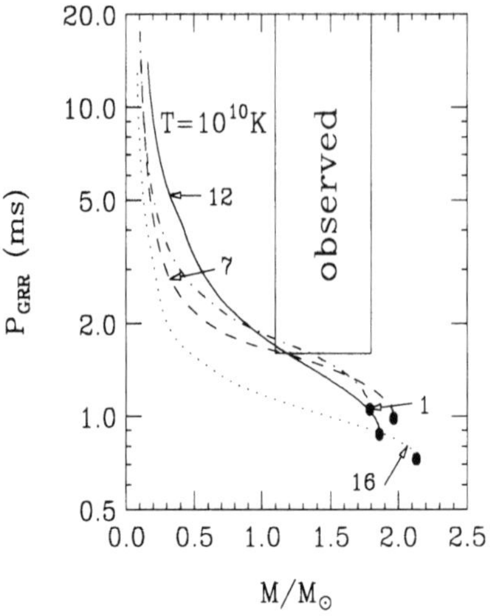

Figure 17.38. Bounds on rapid rotation of *hot*, $T = 10^{10}$ K, neutron stars set by the GRR driven instability period, P_{GRR}, for the EOSs of table 12.4 [187]. (Reprinted courtesy of Institute of Physics Publishing.)

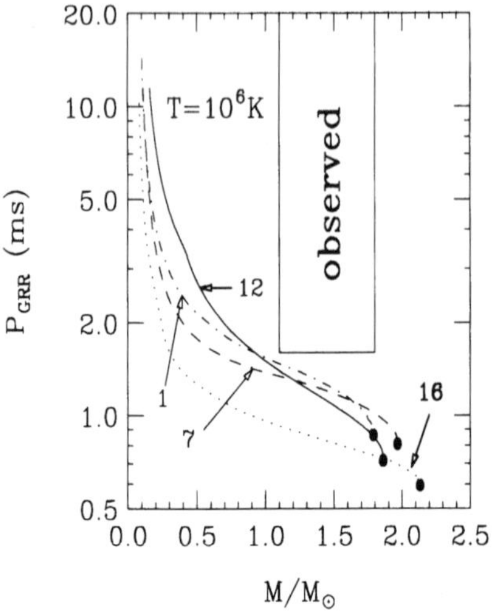

Figure 17.39. Bounds on rapid rotation of *cold*, $T = 10^6$ K, rotating neutron stars set by the GRR driven instability period, P_{GRR}, for the EOSs of table 12.4 [187]. (Reprinted courtesy of Institute of Physics Publishing.)

are predicted to possess gravitational masses larger than about $1.5\,M_\odot$. Table 17.9 summarizes theoretically established bounds on properties of rapidly spinning pulsars of masses in the range of 1.4 to $1.5\,M_\odot$, which are set by rotation at the GRR driven period as well as the Kepler period discussed next.

The Kepler period sets an absolute bound on rapid rotation, which cannot be overcome by any rotating star. Our predictions for this limit are graphically surveyed in figure 17.40. The rectangle indicates both the approximate range of observed neutron star masses (cf. section 3.1), as well as the observed rotational periods ($P \geq 1.6$ ms, see table 1.2). One clearly sees that all pulsar periods so far observed are larger than the absolute limiting Kepler values and, thus, can be understood as originating from rotating neutron stars made up of confined hadronic matter, independent of whether or not the densest fraction of their core matter has undergone a phase transition into deconfined quark matter. Half-millisecond periods are completely excluded for neutron stars of mass $1.4\,M_\odot$ [106, 107, 621]. The situation would be very different for 'neutron' stars made up of self-bound strange quark matter, the so-called strange stars, which will be discussed in the next chapter. Such stars appear to withstand stable rotation against

Table 17.9. Theoretically established lower and upper bounds on the properties of rotating neutron stars with masses in the range between 1.4 and 1.5 $M_\odot$, based on the total collection of EOSs compiled in table 12.4. The listed properties are: period at which the GRR driven instability sets in, $P_{\rm GRR}$; Kepler period, $P_{\rm K}$; central star density, ϵ_c; moment of inertia, I; redshift of photons emitted at the star's equator in backward ($z_{\rm B}$) and forward ($z_{\rm F}$) direction.

	$P_{\rm GRR}$ (ms)	$P_{\rm GRR}$ (ms)	$P_{\rm K}$ (ms)	ϵ_c/ϵ_0	$\log_{10} I$ (g cm^2)	$z_{\rm B}$	$z_{\rm F}$	$z_{\rm p}$
	10^6 K	10^{10} K						
upper bound	1.1	1.5	1	10	45.19	1.05	−0.18	0.45
lower bound	0.8	1.1	0.7	3	44.95	0.59	−0.21	0.23

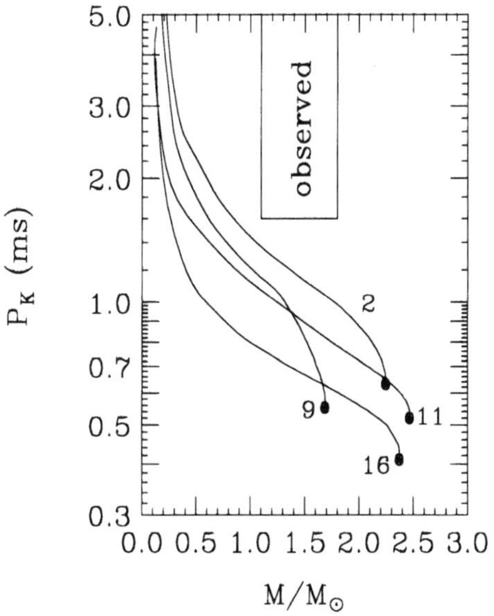

Figure 17.40. Kepler period as a function of rotational neutron star mass. The labeling of the curves is explained in table 12.4 [114]. (Reprinted courtesy of World Scientific.)

mass shedding down to rotational periods in the half-millisecond regime or even less [20]. As a consequence, the possible future discovery of a single sub-millisecond pulsar, say 0.5 ms, would give a strong hint that, firstly, strange stars actually exist and, secondly, the deconfined, self-bound phase of 3-flavor strange quark matter is the true ground state of the strong interaction [20], rather than nuclear matter, which constitutes one of the most startling possibilities revealed by modern particle physics. This conclusion is strengthened by the finding of Madsen [622] who pointed out that, in contrast to neutron stars, young strange stars appear not to be subject to the emission of gravitational radiation, which slows down rapidly rotating, hot neutron stars to periods of several milliseconds via gravitational-wave emission (figure 17.38).

In this connection it is to be recalled that most of the large searches for radio pulsars had no sensitivity below rotational periods of about 4 ms [21, 22, 23], for reasons outlined in section 3.3. However, there is a growing number of pulsar discoveries with periods right down to ~ 1 ms. Great efforts are being made presently by radio astronomers to detect pulsars with even shorter periods, this search having been motivated several years ago [20]. In view of the fundamental information carried by the limiting rotational period of rapidly spinning pulsars, these searches present a special opportunity and challenge to radio astronomy.

Chapter 18

Strange quark matter stars

18.1 Description of strange quark matter

In the following we present the description of electrically charge neutral quark-star matter in equilibrium with respect to the weak interactions at zero as well as finite external pressure and non-zero temperature. By strange quark-star matter we mean a Fermi gas of $3A$ quarks which together constitute a single color-singlet baryon with baryon number A. The dynamics of quark confinement is approximated by the bag model [623], which is graphically illustrated in figure 18.1. For this model, the pressure P^i of the individual quarks and leptons contained in the bag is counterbalanced by the total external bag pressure $P + B$ according to

$$P + B = \sum_{i=u,d,c,s;e^-,\mu^-} P^i \,, \tag{18.1}$$

while the total energy density of the particles is given by

$$\epsilon = \sum_{i=u,d,c,s;e^-,\mu^-} \epsilon^i + B \,. \tag{18.2}$$

The quantity B denotes the bag constant, and ϵ^i are the contributions of the individual quarks and leptons to the total energy density. The condition of electric charge neutrality among the particles reads

$$\sum_{i=u,d,c,s} q_i^{\mathrm{el}} \, \rho^i + \sum_{i=e^-,\mu^-} q_i^{\mathrm{el}} \, \rho^i = 0 \,. \tag{18.3}$$

The individual quark and lepton contributions to pressure, energy, and number density are determined by the thermodynamic potentials $\mathrm{d}\Omega^i =$

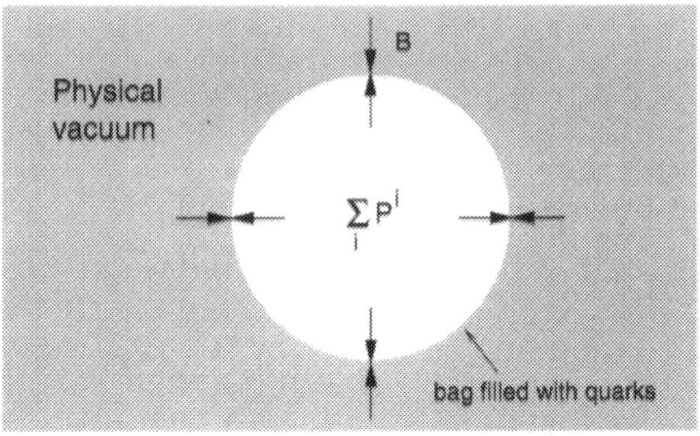

Figure 18.1. On the schematic illustration of the 'bag' model. The total pressure of the quarks confined in the bag, $\sum_i P^i$, is counterbalanced by the pressure of the physical vacuum, B, acting on the bag. The total energy density of the quarks, $\sum_i \epsilon^i$, relative to the physical vacuum [342, 624] is $\epsilon - B$ (cf. equations (18.1) and (18.2)).

$-S^i \mathrm{d}T - P^i \mathrm{d}V - A^i \mathrm{d}\mu^i$, from which one obtains

$$\omega^i = \frac{\partial \Omega^i}{\partial V} = -\frac{\nu_i T}{2\pi^2} \int_0^\infty \mathrm{d}k\, k^2 \,\ln\left(1 + \mathrm{e}^{-(E^i(k)-\mu^i)/T}\right), \tag{18.4}$$

$$P^i = -\,\omega^i = \frac{\nu_i}{6\pi^2} \int_{m_i}^\infty \mathrm{d}E\, (E^2 - m_i^2)^{3/2} f^i(E), \tag{18.5}$$

$$\rho^i = -\frac{\partial \omega^i}{\partial \mu^i} = \frac{\nu_i}{2\pi^2} \int_{m_i}^\infty \mathrm{d}E\, E\,\sqrt{E^2 - m_i^2} f^i(E), \tag{18.6}$$

where $E^i(k) \equiv \sqrt{k^2 + m_i^2}$, and m_i denotes the quark masses. The phase space factor ν_i is equal to 2 (spin) for leptons, and equal to 2 (spin) $\times$ 3 (color) $= 6$ for quarks. Antiparticle contributions are neglected for the moment. While this is well justified for the quarks, for their chemical potentials are much larger than the considered temperatures [625], the situation is more delicate for the positrons. These too, however, were found to contribute only very little to the equation of state. We shall turn back to this issue later in this chapter. The evaluation of the thermodynamic potential of a quark gas of A_c colors and A_f flavors to fourth order in the quark–gluon coupling was carried out in [626, 627]. For our purposes it is sufficient to perform the evaluation of Ω^i at a less sophisticated level. The expression for the energy density of the system reads

$$\epsilon^i = \frac{\nu_i}{2\pi^2} \int_{m_i}^\infty \mathrm{d}E\, E^2\, \sqrt{E^2 - m_i^2}\; f^i(E). \tag{18.7}$$

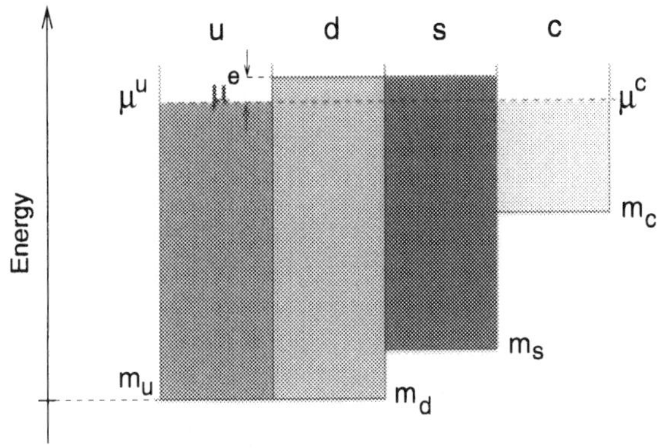

Figure 18.2. Chemical equilibrium among u, d, s, and c quarks.

The quantity $f^i(E) = 1/[1 + \exp((E - \mu^i)/T)]$ denotes the Fermi–Dirac distribution function introduced in (6.23). The baryon number is given by

$$A \equiv \sum_{i=u,d,s,c} A^i = \frac{\Omega}{3} \sum_{i=u,d,s,c} \rho^i \, , \tag{18.8}$$

where Ω denotes a volume element, in accordance with the notation introduced in section 6.3.

The principle of chemical equilibrium among the building blocks of confined hadronic matter has been discussed in great detail in section 4.2. Instead of hadrons, now we are dealing with quarks and leptons as the fundamental building blocks. Chemical equilibrium among the quark flavors and the leptons, illustrated in figure 18.2, is maintained by the following weak reactions (and their inverse),

$$d \to u + e^- + \bar{\nu}^e \, , \quad s \to u + e^- + \bar{\nu}^e \, , \quad s \to c + e^- + \bar{\nu}^e \, . \tag{18.9}$$

The reactions

$$s + u \leftrightarrow d + u \, , \qquad c + d \leftrightarrow u + d \, , \tag{18.10}$$

contribute to the equilibration of flavors. The loss of neutrinos by the star implies that their chemical potential is equal to zero, $\mu^\nu = \mu^{\bar{\nu}} = 0$. Hence, one obtains from the weak reactions (18.9) and (18.10)

$$\mu^d = \mu^u + \mu^e \, , \qquad \mu^c = \mu^u \, , \qquad \mu^d = \mu^s \, . \tag{18.11}$$

Finally, the conservation of electric charge implies that $\mu^e = \mu^{\mu^-}$ (cf. equation (4.10)). The third relation in (18.11) motivates defining

$$\mu \equiv \mu^d = \mu^s \, . \tag{18.12}$$

For notational purposes, it is convenient to introduce the additional definitions

$$\eta_i \equiv \frac{\mu^i}{\mu} = \begin{cases} 1-x & \text{if } i = u, c\,, \\ 1 & \text{if } i = d, s\,, \\ x & \text{if } i = e^-, \mu^-\,, \end{cases} \tag{18.13}$$

where x stands for

$$x \equiv \frac{\mu^e}{\mu}\,. \tag{18.14}$$

18.1.1 Cold matter consisting of massless quarks

It is illustrative to apply, in a first step, the mathematical framework introduced in the previous section to the derivation of the equation of state of relativistic quark matter at zero temperature made up of massless, non-interacting particles. Zero temperature implies that the Fermi–Dirac function $f^i(E) \stackrel{T\to 0}{\longrightarrow} \Theta(\mu^i - E)$. One thus calculates readily from equations (18.4) through (18.7):

$$P^i = \frac{\nu_i}{24\,\pi^2}\,\mu^4\,\eta_i^4 = \frac{1}{3}\,\epsilon^i\,, \tag{18.15}$$

$$\rho^i = \frac{\nu_i}{6\,\pi^2}\,\mu^3\,\eta_i^3\,, \tag{18.16}$$

with $\nu_i = 6$. The equation of state of such matter is obtained from equations (18.1) and (18.2). It is given by the very simple relationship

$$P = \frac{\epsilon - 4\,B}{3}\,. \tag{18.17}$$

The condition of electric charge neutrality, equation (18.3), reads

$$\frac{2}{3}\,\rho^u - \frac{1}{3}\,(\rho^d + \rho^s) = 0\,. \tag{18.18}$$

Note that, because of $m_s = 0$, no leptons are required to make the quark matter electrically charge neutral. Finally, for zero external pressure, $P = 0$, one derives from equation (18.1) $B = 3\mu^4/4\pi^2$, and for the energy per baryon number of strange matter (cf. section 2.3.1)

$$E_A \equiv \frac{\epsilon}{\rho} = \frac{4\,B}{(\rho^u + \rho^d + \rho^s)/3} = \frac{4\,B}{\rho^u} = \frac{4\,B\,\pi^2}{\mu^3}\,. \tag{18.19}$$

From this relation it follows that bag constants of $B = 57.5$ MeV fm^{-3} (i.e. $B^{1/4} = 145$ MeV) and $B = 85.3$ MeV fm^{-3} ($B^{1/4} = 160$ MeV) place the energy per baryon number of strange matter at $E_A = 829$ MeV

Table 18.1. Masses, m_i, and electric charges, q_i^{el}, of the quarks.

Quark flavor (i)	u	d	c	s	t	b
m_i (GeV)	~ 0.05	~ 0.01	~ 1.5	~ 0.2	~ 180	~ 4.7
q_i^{el}	$+\frac{2}{3}$	$-\frac{1}{3}$	$+\frac{2}{3}$	$-\frac{1}{3}$	$+\frac{2}{3}$	$-\frac{1}{3}$

and 915 MeV, respectively. These values represent strongly ($\sim$ 100 MeV) and weakly ($\sim$ 15 MeV) bound strange matter at zero external pressure, and in all cases correspond to strange matter being *absolutely* bound, i.e. $E_A < 930$ MeV, with respect to ^{56}Fe. We shall come back to this issue when discussing figures 18.3 and 18.4.

18.1.2 Cold matter consisting of massive quarks

The expressions (18.5) through (18.7) can be evaluated analytically too if the quarks are given their finite masses, which are listed in table 18.1. After some straightforward algebraic manipulations, one arrives for pressure, baryon density and mass density at the following expressions:

$$P^i = \frac{\nu_i \mu^4 \eta_i^4}{24\pi^2} \left\{ \sqrt{1 - z_i^2} \left(1 - \frac{5}{2} z_i^2\right) + \frac{3}{2} z_i^4 \, \ln \frac{1 + \sqrt{1 - z_i^2}}{z_i} \right\}, \quad (18.20)$$

$$\rho^i = \frac{\nu_i \mu^3 \eta_i^3}{6\pi^2} (1 - z_i^2)^{3/2} , \quad (18.21)$$

$$\epsilon^i = \frac{\nu_i \mu^4 \eta_i^4}{8\pi^2} \left\{ \sqrt{1 - z_i^2} \left(1 - \frac{1}{2} z_i^2\right) - \frac{z_i^4}{2} \, \ln \frac{1 + \sqrt{1 - z_i^2}}{z_i} \right\}, \quad (18.22)$$

where the suffix i stands again for the $u, d, c, \ldots$ quarks as well as for the leptons e^-, μ^-. The quantity z_i in (18.22) is defined as

$$z_i \equiv \frac{m_i}{\eta_i \, \mu} = \frac{m_i}{\mu^i} . \quad (18.23)$$

If we ignore the two most massive quark states, t and b, which are clearly excluded from becoming populated in compact stars because of their tremendously high masses, the condition of electric charge neutrality, equation (18.3), then reads

$$\frac{2}{3}(\rho^u + \rho^c) - \frac{1}{3}(\rho^d + \rho^s) - (\rho^e + \rho^\mu) = 0 . \quad (18.24)$$

Upon substituting (18.21) into (18.24) and making use of (18.14), this relation can be written as

$$2(1 - x^3)\big(1 + (1 - z_c^2)^{3/2}\big) - \big(1 + (1 - z_s^2)^{3/2}\big) - x^3\big(1 + (1 - z_\mu^2)^{3/2}\big) = 0 . \quad (18.25)$$

Substituting the result of (18.20) into equation (18.1) leads to

$$\frac{(P+B)\,4\pi^2}{\mu^4} = (1-x^4)\Big\{1+\sqrt{1-z_c^2}\Big(1-\frac{5}{2}z_c^2\Big)+\frac{3}{2}z_c^2\,\ln\frac{1+\sqrt{1-z_c^2}}{z_c}\Big\}$$
$$+\Big\{1+\sqrt{1-z_s^2}\Big(1-\frac{5}{2}z_s^2\Big)+\frac{3}{2}z_s^2\,\ln\frac{1+\sqrt{1-z_s^2}}{z_s}\Big\}$$
$$+\frac{x^4}{3}\Big\{1+\sqrt{1-z_\mu^2}\Big(1-\frac{5}{2}z_\mu^2\Big)+\frac{3}{2}z_\mu^2\,\ln\frac{1+\sqrt{1-z_\mu^2}}{z_\mu}\Big\}. \qquad (18.26)$$

The total energy density follows from (18.2) in the form

$$\epsilon = 3P+4B+\sum_{i=s,c,\mu^-}\frac{\nu_i\mu^4\eta_i^4}{4\pi^2}\,z_i^2\Big\{\sqrt{1-z_i^2}-z_i^2\,\ln\frac{1+\sqrt{1-z_i^2}}{z_i}\Big\}. \qquad (18.27)$$

The first two terms on the right-hand side of equation (18.27) represent the equation of state of massless quarks, given by equation (18.17). The third term accounts for the finite masses of muons and strange and charm quarks. We conclude the set of equations with the total baryon number density, which follows from (18.3) as

$$\rho = \frac{1}{3}\,(\rho^u+\rho^d+\rho^s+\rho^c)\,, \qquad (18.28)$$

which upon substituting (18.21) and making use of (18.14) transforms to

$$\rho = \frac{\mu^3}{3\pi^2}\{(1-x^3)(1+(1-z_c^2)^{3/2})+(1+(1-z_s^2)^{3/2})\}\,. \qquad (18.29)$$

Figure 18.3 shows the contours of the energy per baryon number, $E_A = \epsilon/\rho$, of strange matter at zero external pressure [625, 628], computed from equation (18.27). The influence of temperature is demonstrated for $T = 30$ MeV, which is typical for a newly formed neutron star in a supernova explosion [8, 53, 629]. The energy per baryon number of cold matter ranges from 830 to 950 MeV. We recall that the energy per baryon in ^{56}Fe is $M(^{56}\mathrm{Fe})c^2/56 = 930.4$ MeV, where $M(^{56}\mathrm{Fe})$ is the mass of the ^{56}Fe atom. Thus, with the exception of the 950 MeV contour, all the curves in figure 18.3 correspond to strange matter that is *absolutely* stable with respect to ^{56}Fe at zero external pressure. For a representative strange quark mass of $m_s = 150$ MeV this constrains the bag constant to values smaller than 75 MeV fm^{-3} ($B^{1/4} = 155$ MeV). The vertical line in figure 18.3 at $B = 57$ MeV fm^{-3} ($B^{1/4} = 145$ MeV) constitutes the lower bound on B, since for smaller bag constants the energy per baryon number of 2-flavor quark matter would drop below the energy per baryon of ^{56}Fe [628]. In this

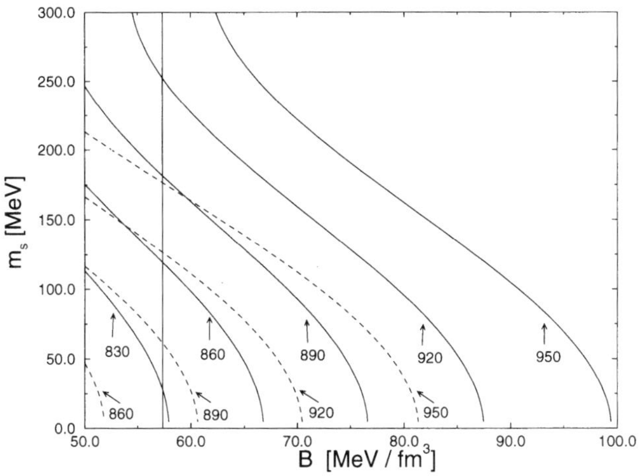

Figure 18.3. Contours of fixed energy per baryon number E_A (figures attached to the curves) of strange quark matter at zero external pressure, $P = 0$ [30]. The solid and dashed curves refer to $T = 0$ and $T = 30$ MeV, respectively. The range of physically allowed bag constants is constrained to $B \geq 57$ MeV fm^{-3}. (Reprinted courtesy of *Phys. Rev.*)

case ^{56}Fe would be made up of u and d quarks rather than nucleons, which is not observed of course. Finite temperatures – like finite quark masses, or external pressure (equation (18.27)) – increase both the energy density ϵ of the bag as well as the baryon number density, ρ. The impact is such that the energy contours are shifted toward smaller bag constants. This shift in B, as can be seen from figures 18.3 and 18.4, is quite large and is $\sim$20%, depending on the mass of the strange quark. The impact of finite external bag pressures, P, on the energy contours is illustrated in figure 18.4. A comparison with figure 18.3 shows that the energy contours are shifted toward smaller B values, too, which can be understood mathematically by combining equations (18.17) and (18.19) to $B = (\rho E_A - 3P)/4$. From the physical point of view, this feature becomes clear by remembering that finite P values increase the pressure which acts on the bag, equation (18.1). So B can be reduced on the account of P.

The relative quark–lepton composition of quark-star matter at zero temperature is shown in figure 18.5. All quark flavor states that become populated in such matter up to densities of 10^{19} g cm^{-3} are taken into account self-consistently. Since the Coulomb interaction is so much stronger than the gravitational, quark-star matter in the lowest energy state must be charge neutral to very high precision [146]. Therefore, any net positive quark charge must be balanced by a sufficiently large number of negatively

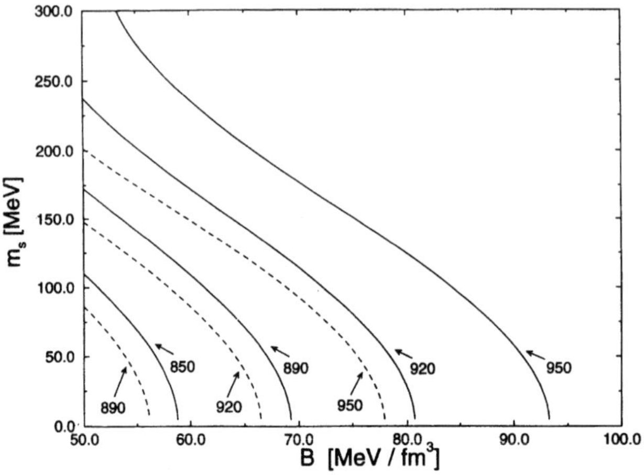

Figure 18.4. Same as figure 18.3, but for a finite external bag pressure of 50 MeV fm^{-3} [30]. (Reprinted courtesy of *Phys. Rev.*)

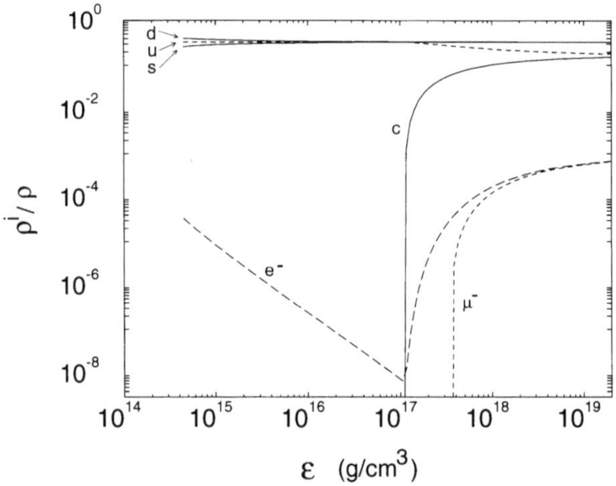

Figure 18.5. Relative quark and lepton densities in cold quark-star matter as a function of energy density [30]. Here, as in all subsequent figures of this chapter, a bag constant of $B^{1/4} = 145$ MeV was chosen. (Reprinted courtesy of *Phys. Rev.*)

charged quarks and leptons, which determines the lepton concentration shown in figure 18.5. One observes that at lower densities the number of d quarks is somewhat larger than the number of s quarks which is due to the finite mass of the latter. The behavior of ρ^d/ρ and ρ^s/ρ can be understood qualitatively from equation (18.21), which reveals that $\rho^d/\rho^s = (1 - m_s/\mu^s)^{-3/2}$. Since $m_s/\mu^s < 1$ it follows that $\rho^d > \rho^s$ at all densities, and, secondly, $\rho^d \stackrel{\epsilon\to\infty}{\longrightarrow} \rho^s$ from above since $m_s/\mu^s \stackrel{\epsilon\to\infty}{\longrightarrow} 0$ (cf. figure 18.10). (Strange and charm quark masses of respectively 0.15 GeV and 1.2 GeV are assumed.) In contrast to the sensitive density dependence of lepton number, the abundances of u, d, and s quarks in strange matter vary only rather weakly with density. The situation is different for the c quarks whose concentration increases at threshold density extremely rapidly. At still higher densities it tends against the concentration of u quarks, and charge neutrality is nearly achieved by appropriate concentrations of quarks of both charge states only. The slight deficit of negative quark charge is delivered to the system by electrons and muons, whose concentrations increase monotonically for all densities larger than the threshold density of the positively charged c quarks.

18.1.3 Quark matter at finite temperature

To derive the equation of state of quark-star matter at temperatures up to about $T \sim 50$ MeV, we perform a perturbation expansion of pressure $P^i = P^i(\mu, x, T)$ and baryon density $\rho^i = \rho^i(\mu, x, T)$ about their zero-temperature values, $P_0^i \equiv P^i(\mu_0, x_0, T_0)$ and $\rho_0^i \equiv \rho^i(\mu_0, x_0, T_0)$, where $T_0 = 0$ [625]. By means of writing these functions in the form $\chi^i(\mu, x, T) \equiv \chi^i(\mu_0 - \Delta\mu, x_0 + \Delta x, T_0 + \Delta T)$, where χ^i stands either for P^i or ρ^i, expanding them in a Taylor series and keeping only the lowest order terms, one obtains

$$\chi^i(\mu, x, T) \simeq \chi_0^i + \left.\frac{\partial\chi^i}{\partial\frac{\Delta\mu}{\mu_0}}\right|_{\mu_0, x_0, T_0} \frac{\Delta\mu}{\mu_0} + \left.\frac{\partial\chi^i}{\partial\Delta x}\right|_{\mu_0, x_0, T_0} \Delta x + \left.\frac{\partial\chi^i}{\partial\frac{T^2}{\mu_0^2}}\right|_{\mu_0, x_0, T_0} \frac{T^2}{\mu_0^2} \tag{18.30}$$

with $\chi_0^i \equiv \chi^i(\mu_0, x_0, T_0)$. Above, the definitions $\Delta\mu \equiv \mu_0 - \mu$ and $\Delta x \equiv x - x_0$ have been introduced, where $\mu_0 \equiv \mu(T_0)$ and $x_0 \equiv x(T_0)$. The major problem encountered now consists in calculating the expansion coefficients occurring in equation (18.30), $\partial\chi^i/\partial(\Delta\mu/\mu_0)$, $\partial\chi^i/\partial\Delta x$, and $\partial\chi^i/\partial(T/\mu_0)^2$. Their determination is outlined in more detail in appendix H. It should be noticed that because of $\partial\chi^i/\partial(T/\mu_0) = 0$, which is shown in reference [630], both pressure and particle density depend in lowest order only quadratically on temperature. After considerable algebra, one arrives for pressure, particle density, and total energy density at the relations

$$P^i = P_0^i + \frac{1}{6}\,\nu_i\, T^2\mu_0^2 \Big\{ -a\,\eta_i^4\,(1 - z_i^2)^{3/2} + b\,c_i\eta_i^3\,(1 - z_i^2)^{3/2} + \frac{1}{2}\,\eta_i^2\left(1 - \frac{1}{2} z_i^2\right) \Big\}, \tag{18.31}$$

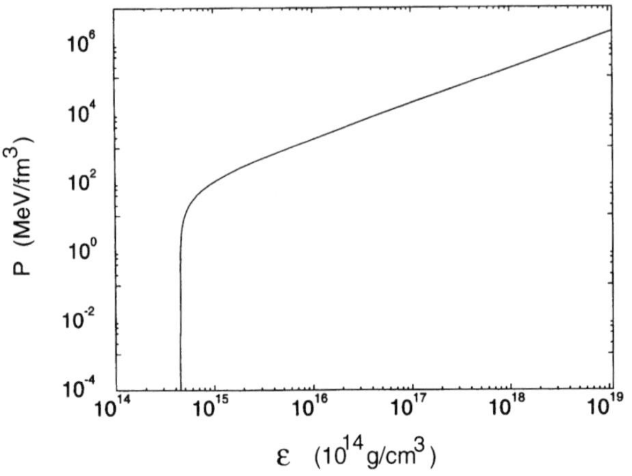

Figure 18.6. Pressure isotherm versus mass density of quark-star matter [30]. Temperatures $T \le 50$ MeV are significant only at densities $\epsilon \lesssim 5 \times 10^{14}$ g cm^{-3} as becomes clear from figure 18.7. (Reprinted courtesy of *Phys. Rev.*)

$$\rho^i = \rho_0^i + \frac{1}{2}\, \nu_i\, T^2 \mu_0 \left\{ -a\, \eta_i^3\, \sqrt{1 - z_i^2} + b\, c_i \eta_i^2\, \sqrt{1 - z_i^2} + \frac{1}{3}\eta_i \right\}, \quad (18.32)$$

$$\epsilon = 3P + 4B + \sum_{i=s,c,\mu^-} \frac{\nu_i \mu_0^2 \eta_i^2}{2} \left\{ \frac{\mu_0^2 \eta_i^2}{2\pi^2} z_i^2 \left(\sqrt{1 - z_i^2} - z_i^2 \ln \frac{1 + \sqrt{1 - z_i^2}}{z_i} \right) \right.$$
$$\left. + T^2 \left(z_i^2 \sqrt{1 - z_i^2}\, (b\, c_i \eta_i - a\, \eta_i^2) + \frac{1}{6} z_i^3 \right) \right\}, \quad (18.33)$$

where the quantities a and b are given as the solutions of two coupled, linear equations derived in appendix H. Moreover, we have introduced the definition

$$c_i \equiv \begin{cases} +1 & \text{if } i = e^-, \mu^-, \\ 0 & \text{if } i = d, s, \\ -1 & \text{if } i = u, c. \end{cases} \quad (18.34)$$

The zero-temperature limit of (18.33), derived in (18.27), is readily obtained by taking the limit $T \to 0$. The graphical illustration of the equation of state of strange matter at finite temperatures is shown in figure 18.6. There is a noticeable influence of temperature on $P(\epsilon)$ only at low nuclear densities, $\epsilon \simeq 4B$, because only this limit gives the finite-temperature term in equation (18.33), as well as the finite-mass terms too, an appreciable contribution comparable in magnitude to $\epsilon \sim 4B$. Figure 18.7 exhibits this feature for a few selected temperatures.

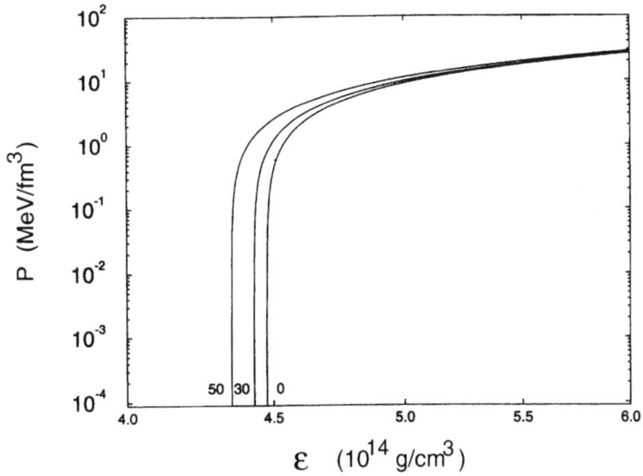

Figure 18.7. Enlargement of the left portion of figure 18.6 [30]. The pressure isotherms are computed for three different temperatures, which are quoted in units of MeV. (Reprinted courtesy of *Phys. Rev.*)

The expressions for electron density and electron pressure are given by

$$\rho^e = \frac{\mu_0^3 x_0^3}{3\pi^2} + T^2 \ \mu_0 \left(-a x_0^3 + b x_0^2 + \frac{1}{3} x_0\right), \tag{18.35}$$

$$P^e = \frac{\mu_0^4 x_0^4}{12\pi^2} + \frac{1}{3} \ T^2 \mu_0^2 \left(-a x_0^4 + b x_0^3 + \frac{1}{2} x_0^2\right), \tag{18.36}$$

which follow from equations (18.31) and (18.32) applied to massless electrons, for which $\eta_e = x_0$, $z_e = 0$, and $c_e = 1$, in accordance with equations (18.13), (18.23), and (18.34). Also, recall that $x(T) \equiv \mu^e(T)/\mu(T)$ reduces at zero temperature to $x_0 \equiv \mu_0^e/\mu_0$. The temperature dependence of ρ^e for zero and finite external bag pressure is exhibited in figure 18.8. Because finite P values increase the system's total energy density, in which case fewer electrons are known to be necessary in order to achieve electric charge neutrality, equation (18.33), the ρ^e isobars are shifted toward smaller densities. Temperatures typical for newly formed massive stars increase ρ^e by roughly two orders of magnitude, depending on external pressure. The quadratic dependence of ρ^e on T, equation (18.35), is significant at lower temperatures. For larger T, the term proportional to $T^2 x_0^3$ weakens the increase of ρ^e with temperature. The variation of electron chemical potential, μ^e, along the ρ^e isotherms is shown in figure 18.9. It is seen that μ^e deviates for temperatures $T \leq 50$ MeV from its zero-temperature value by at most 1 MeV. The decrease of μ^e with density reflects the fact that fewer electrons are needed in strange quark matter at higher densities (cf. figure 18.5). Furthermore we notice the downward shift

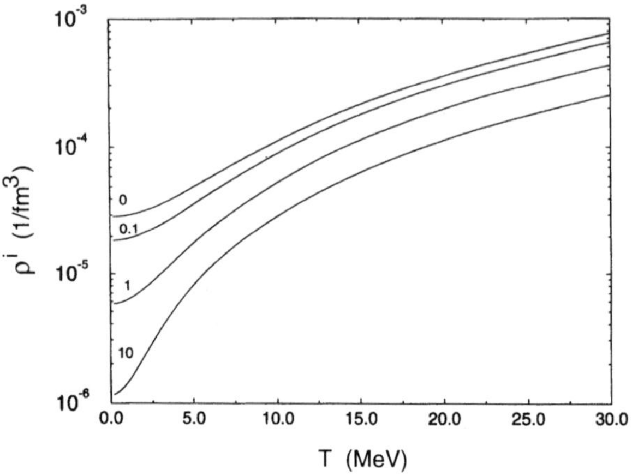

Figure 18.8. Density isobars of electrons versus temperature for different external pressure values of $P_{15} = 0, 0.1, 1.0, 10$ g cm^{-3} [30]. For the conversion of pressure from units of g cm^{-3} to MeV fm^{-3}, see table 1.4. (Reprinted courtesy of *Phys. Rev.*)

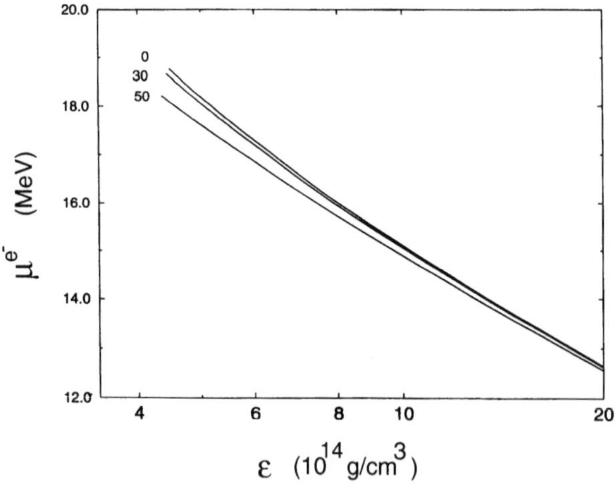

Figure 18.9. Chemical electron potential versus energy density for quark-star matter at temperatures of $T = 0$, 30, and 50 MeV [30]. T is constant along each curve. (Reprinted courtesy of *Phys. Rev.*)

of the μ^e isotherms, for a fixed density, with increasing temperature, which is due to the momentum tail of the Fermi–Dirac distribution function for $T > 0$.

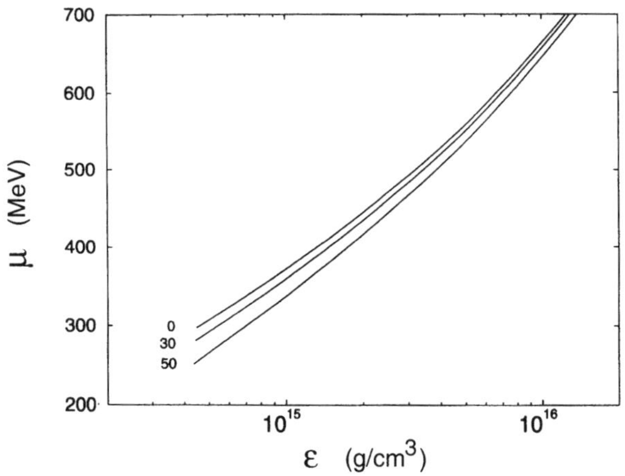

Figure 18.10. Same as figure 18.9, but for the chemical potential μ of d and s quarks (cf. equation (18.12)) [30]. (Reprinted courtesy of *Phys. Rev.*)

The density and temperature dependence of the chemical potential of d and s quarks, μ, is graphically depicted in figure 18.10. The density dependence can be inferred qualitatively from equation (18.21), from which one gets $\mu = m_s(1 - \rho^s/\rho^d)^{-2/3}$. Immediately below the threshold density of s quarks one has $\rho^s = 0$, and therefore $\mu^s = m_s$ there. The other extreme, high s quark densities, is characterized by $\rho^s \to \rho^d$, as is known from figure 18.5. This implies that μ becomes very large in the high-density regime. In section 18.2 it will be shown that stable strange stars possess central densities of at most $\sim 2 \times 10^{15}$ g cm^{-3}. Hence, we conclude from figure 18.10 that μ never exceeds $\sim$ 500 MeV in quark matter stars. This value is considerably smaller than the mass of the charm quark, $m_c \sim 1.2$ GeV. The impact of temperature on μ is most significant at densities $\epsilon \sim 4B$, for the same reasons as outlined in connection with figures 18.6 and 18.7. Finally finite temperatures reduce μ for a given density below its zero-temperature value. The reason lies again in the occurrence of the Fermi–Dirac function in equation (18.6) rather than the step function leading to smaller chemical potentials. For the selected temperatures, this reduction amounts at most to $\sim$100 MeV.

18.2 Hydrostatic equilibrium sequences of quark matter stars

The masses of strange and charm stars as a function of central density have already been shown in figure 14.4. We briefly recall that the strange-star sequence ends, and the charm stars begins, at a density of about 10^{17} g cm^{-3} where the charm-quark threshold is reached. Such additional families of stars at ultrahigh nuclear densities are obtained not only for quark matter but were shown to exist for the neutron star sequence extended to ultrahigh densities too [69, 631]. One of the most significant differences between both species of stars concerns the existence of a minimum-mass configuration of about $\sim 0.1\,M_\odot$ [70] for the neutron star. In sharp contrast to this, the sequence of (bare) strange stars, being bound by the strong interaction and not the gravitational force (the latter makes strange stars only somewhat more dense), does *not* possess a minimum-mass star. In fact, strange-matter objects can exist with baryon numbers in the enormous range of $10^2 \lesssim A \lesssim 10^{57}$ [628, 632]. The lower bound on A is determined by finite-size effects of strange nuggets, the upper bound is set by the gravitational attraction among the quarks, which increases with A, making strange stars with too large a central density unstable against gravitational collapse. The situation is the same as for the heaviest member of the stable neutron star sequence. Stars located beyond the mass peak, while in hydrostatic equilibrium, are unstable against radial vibrations, as investigated in section 14.3, and therefore cannot exist stably in the universe. We are thus left with the possible existence of strange quark stars only, which may coexist together with their non-strange counterparts, the neutron stars. As pointed out by Madsen, there exist however a number of arguments that if strange matter indeed constitutes the true ground state of the strong interaction, then it appears most likely that all neutron stars would in fact be strange stars [142].

Temperatures typical for newly formed pulsars influence the mass and radius of quark stars only rather weakly, as can be seen from figures 14.4 and 18.11. For masses larger than $\sim 0.5\,M_\odot$ the mass–radius relationship bears a strong similarity with the one of neutron stars, which was shown in figure 14.8. Temperatures $T \lesssim 50$ MeV modify the properties of the more massive stars of the sequence only slightly. All stars possessing central densities greater than the heaviest strange star, S, are unstable against radial oscillations. The same inwardly directed spiraling behavior is also obtained when extending the neutron star sequence to ultrahigh densities [69]. Hence this behavior is not generic for self-bound stars, but rather manifests the dominant role of gravity at ultrahigh densities.

18.3 Electrostatic surface properties of strange stars

As we have seen in section 18.1, equilibrium strange matter contains an approximately equal mixture of u, d, and s quarks, with a slight deficit of the latter. For that reason a certain number of electrons is necessary to make the matter electrically charge neutral. The electrons, being bound to quark matter by the electromagnetic interaction rather than the strong force, extend therefore several hundred fermis beyond the extremely sharp

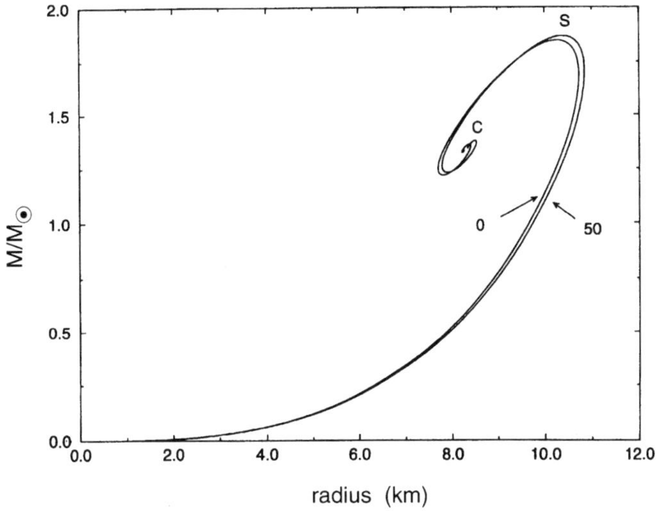

Figure 18.11. Gravitational mass (in units of solar mass) versus radius of the strange-star and charm-quark-star sequences of figure 14.4 [30]. The symbols S and C denote the maximum-mass model of each sequence. (Reprinted courtesy of *Phys. Rev.*)

surface of strange matter [28], which itself has a surface thickness of the order of the strong interaction range ($\sim$ 1 fm). A schematic illustration of this striking feature is shown in figure 18.12. Associated with this electron displacement at the surface region of strange matter is an extremely strong electric dipole field, on the order $|\boldsymbol{E}| \sim 10^{17}$ V cm^{-1}, which is radially outwardly directed.[1] Most importantly for the glitch behavior and the cooling behavior of strange stars, this layer can suspend a solid nuclear crust enveloping strange matter out of contact with it [20, 28, 148]. This suspension is of decisive importance for the stable existence of the nuclear crust on the strange star, for it prevents reactions between the atomic nucleons at the base of the crust and quark matter. Otherwise the atomic nuclei would be converted by hypothesis into the true ground state, strange matter. In the following, the behavior of the electrostatic potential of the electrons inside and in the close vicinity outside of strange stars is determined and its temperature dependence studied. Of particular importance will be the temperature dependence of the Coulomb barrier associated with the difference of the electrostatic potential at the surface

[1] Such strong fields would cause the spontaneous creation of e^+e^- pairs in free space. In the case of strange matter, however, the displaced electrons form a medium, and the Pauli principle obeyed by the displaced electrons ensures the stability of the vacuum.

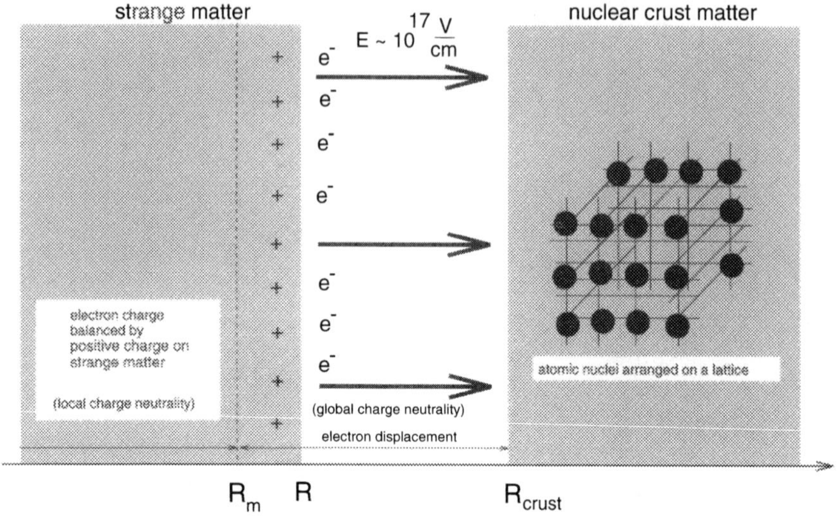

Figure 18.12. Surface region of strange quark matter. The region $R \leq r \leq R_{\rm crust}$ is filled with electrons that are electromagnetically bound to strange matter but extend beyond its surface, R, leading to a deficit of electrons in the range $R_m \leq r \leq R$ and therefore a net positive charge in this region. The associated electric field, $|\boldsymbol{E}| \sim 10^{17}$ V cm^{-1}, is sufficiently strong to avoid contact between atomic matter and strange matter, enabling strange matter to be enveloped by ordinary atomic matter.

of the strange core and the base of the inner nuclear crust. Depending on temperature, the Coulomb barrier may be a powerful source of e^+e^- pairs that are created in the extremely strong electric field associated with the barrier. The luminosity in the outflowing pair plasma may be as high as several 10^{51} erg s^{-1}, depending on the surface temperature [160].

18.3.1 Impact of finite temperatures on electron distribution

18.3.1.1 Inside strange quark matter

We begin with the determination of the electrostatic potential, $V(r)$, of the electrons inside strange quark matter, or, equivalently, inside a bare strange quark-matter star. A geometrical illustration of the region in question, i.e. $r \leq R$, is shown in figure 18.12. To determine the potential in this region, we recall that locally the energy of an electron sitting at the Fermi surface is given by $\mathcal{E}(r) \equiv \mu^e(r) - eV(r)$ [28, 625], where $\mu^e(r)$ denotes the electron's radially dependent chemical potential. In equilibrium one has $\mathrm{d}\mathcal{E}(r)/\mathrm{d}r = 0$. From the boundary conditions $V(r) \stackrel{r\to\infty}{\longrightarrow} 0$ and

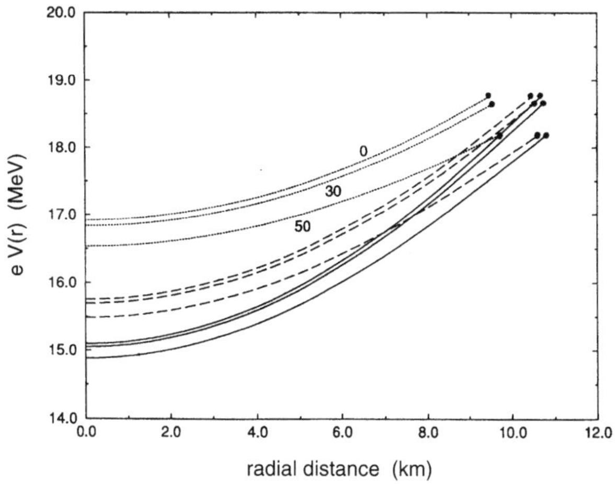

Figure 18.13. Electrostatic potential, $eV(r)$, of electrons inside ($r < R$) strange stars of masses $M = 1\,M_\odot$ (dotted curves), $1.4\,M_\odot$ (dashed curves), and $1.6\,M_\odot$ (solid curves) [30]. The temperatures in each case are $T = 0$, 30, and 50 MeV. (Reprinted courtesy of *Phys. Rev.*)

$\mu^e(r) \overset{r\to\infty}{\longrightarrow} 0$ [28] it follows that $eV(r) = \mu^e(r)$. This function is shown in figure 18.13 for a sample of strange stars. Since μ^e decreases with density, as known from figure 18.9, the electrostatic potential of electrons increases monotonically from the center to the surface of strange stars. Finite temperatures influence $eV(r)$ more significantly in the vicinity of the surface, because of the smaller density there, than at the stars' centers. For the heavier stars, possessing larger central densities, the isotherms are shifted downward, which is a consequence of the decreasing behavior of μ^e with density (figure 18.9). Another noteworthy feature is that independent of star mass (and thus, central star density), isotherms referring to the same temperature terminate at the *same* value of $eV(R)$. This is indicated by the solid dots which possess the same $eV(R)$ value for the same temperature. This independence of stellar mass (central star density) becomes clear from figure 18.9, which shows that the value of μ^e at the star's surface is determined only by the values of bag constant and the temperature. It also explains the shifts of the termination points for increasing temperatures toward larger radii.

18.3.1.2 Surface region of strange quark matter

We next turn to the discussion of $V(r)$ several hundred fermi inside and outside of the surface of strange matter, that is, the region $R_m \le r \le R_{\rm crust}$

shown in figure 18.12. Recall that due to the rearrangement of electrons there, the net positive charge of the quarks will be balanced *locally* by electrons only to radial distances $r \leq R_m$, the star's bulk-matter part, where R_m is slightly smaller than the star's radius, R. For distances beyond R_m, in the region $R_m \leq r < \infty$, the requirement of electric charge neutrality is a global constraint rather than a local one. To determine the value of R_m at which this happens, note that from Poisson's equation for radii in the range of $R_m < r < \infty$,

$$\frac{\mathrm{d}^2 eV}{\mathrm{d}r^2} = 4\pi e^2 [\rho^e(r) - \rho^q(r)]\, \Theta(r - R_m)\,.$$

The $\mathrm{d}V/\mathrm{d}r$ term can be neglected in here. The quantity ρ^q is defined as $3\rho^q \equiv 2\rho^u - \rho^d - \rho^s$ for $r < R_m$. In this region one also has $\rho^q - \rho^e = 0$. From (18.37) it follows that

$$\int_{R_m}^{R} \mathrm{d}r\ \rho^q(r) = \int_{R_m}^{\infty} \mathrm{d}r\ \rho^e(r)\,, \tag{18.37}$$

because $\mathrm{d}V(R_m)/\mathrm{d}r = \mathrm{d}V(\infty)/\mathrm{d}r = 0$. The first relation follows from the fact that $V(r)$ attains a maximum at R_m. The upper boundary in the second integral reflects the circumstance that the electrons extend beyond the surface, denoted as R in figure 18.12, of strange matter. Equation (18.37) can be transformed to

$$\int_{V(R_m)}^{V(R)} \mathrm{d}V\ \rho^q = \int_{V(R_m)}^{\infty} \mathrm{d}V\ \rho^e\,. \tag{18.38}$$

Making use of the mathematical relations $e\mathrm{d}V = \mathrm{d}\mu^e$ and $\rho^e = \partial P^e/\partial\mu^e$ (cf. equation (18.6)), (18.38) can be written as

$$\int_{V(R_m)}^{V(R)} \mathrm{d}V\ \rho^q = \frac{1}{e}\int_{\mu(R_m)}^{\mu(\infty)} \mathrm{d}\mu^e\ \frac{\partial P^e}{\partial\mu^e} = \frac{1}{e}\left[P^e(\infty) - P^e(R_m)\right]\,. \tag{18.39}$$

Because R and R_m differ only by a few hundred fermi [28], the density $\rho^q(r)$ in that range can be treated as being independent of r. Its value is therefore given, to a very good approximation, by $\rho^q(r) \simeq \rho^e(R_m)$. One thus obtains from equation (18.39),

$$e\,V(R,T) = \mu^e(R_m,T) - \frac{P^e(R_m,T)}{\rho^e(R_m,T)}\,. \tag{18.40}$$

By means of approximating $\mu^e(R_m,T) \simeq \mu^e(R,T)$ and substituting P^e/ρ^e with equations (18.35) and (18.36), one obtains for equation (18.40),

$$e\,V(R,T) \simeq \mu^e(R,T) - \frac{\left(\mu_0^e\right)^4(R) + 4\pi^2T^2\mu_0^2(R)[-ax_0^4(R) + bx_0^3(R) + \frac{1}{2}x_0^2(R)]}{4\left(\mu_0^e\right)^3(R) + 12\pi^2T^2\mu_0^2(R)[-ax_0^3(R) + bx_0^2(R) + \frac{1}{3}x_0^2(R)]}\,, \tag{18.41}$$

where the zero-temperature chemical potential of electrons, $\mu^e(T=0)$, is abbreviated to μ_0^e, and in accordance with equations (18.14) and (H.21), the electron chemical potential at finite temperature is given by

$$\mu^e(R,T) \equiv \mu(R,T)\; x(R,T) = \left(\mu_0(R) - a\pi^2 \frac{T^2}{\mu_0}\right) \left(x_0(R) + b\pi^2 \frac{T^2}{\mu_0^2}\right). \tag{18.42}$$

In the zero-temperature limit, equation (18.41) reduces immediately to the simple relation [28]

$$e\,V(R,T) \stackrel{T\to 0}{\longrightarrow} \frac{3}{4}\;\mu_0^e(R). \tag{18.43}$$

Hence, the electrostatic potential of electrons at the surface of the strange-matter core is reduced relative to its value obtained by imposing local (instead of global) charge neutrality [28], which, as explained above, holds only for radial distances out to R_m but not beyond, $R_m < r < R$. As we shall see below (cf. figure 18.16), finite temperatures lead to an even stronger reduction of the electrostatic electron potential, up to about 50% for the temperature range of newly formed pulsars. Equation (18.41) reveals that this decrease has its origin in the reduction of μ^e with temperature, shown in figure 18.10, which is additionally strengthened by the second term on the right-hand side of (18.41). As an example, the values $\mu^e(R,T=0) =$ 18.8 MeV and $eV(R,T=0) = 14.1$ MeV drop to 18.7 MeV and 9.5 MeV at $T = 30$ MeV.

Lastly, we determine $V(r)$ in the regions $R_m \le r \le R$ and $R \le r \le R_{\rm crust}$. The latter range lies beyond the surface of the star's strange-matter core (figure 18.12). There two regions are to be distinguished from each other. The first extends from the core's surface R to the crust's base located at $R_{\rm crust}$. The distance $R_{\rm gap} \equiv R_{\rm crust} - R$ is referred to henceforth as gap width. The second region begins at $R_{\rm crust}$ and extends in the radial outward direction toward infinity. The behavior of the electrostatic potential in the surface region is determined by Poisson's equation,

$$\frac{{\rm d}^2 eV}{{\rm d}r^2} + \frac{2}{r}\frac{{\rm d}eV}{{\rm d}r} = \frac{4\pi e^2}{3}\left\{\left[\frac{1}{\pi^2}[(eV)^3 - (eV(R_m))^3] + T^2[eV - eV(R_m)]\right] \times \Theta(r-R_m)\Theta(R-r) + \left[\frac{1}{\pi^2}(eV)^3 + T^2\, eV\right]\Theta(r-R)\Theta(R_{\rm crust}-r)\right\}. \tag{18.44}$$

Notice that in the first term on the right-hand side ($R_m < r < R$) the net charge density of electrons and quarks, $\rho^e(r) - \rho^q(r)$, enters, where $\rho^q(r) \simeq \rho^e(R_m) = (eV(R_m))^3/3\pi^2$. The quark density is zero in the second region, $R \le r \le R_{\rm crust}$. The expression of $\rho^e(r)$,

$$\rho^e(r) = \frac{1}{3\pi^2}\left(\mu^e\right)^3(r) + \frac{1}{3}\,\mu^e(r)\,T^2 = \frac{1}{3\pi^2}\,[eV(r)]^3 + \frac{1}{3}\,eV(r)\,T^2\,, \tag{18.45}$$

is computed exactly from equation (18.6), treating the electrons (and positrons) as massless particles.[2] An analytic representation of $V(r)$ in the gap region can be obtained at zero temperature if the $\mathrm{d}V/\mathrm{d}r$ term in equation (18.44) is ignored (a good approximation), leading to

$$\frac{\mathrm{d}^2 eV}{\mathrm{d}r^2} = \frac{4\,e^2}{3\,\pi}\,[e\,V(r)]^3\,\Theta(r-R)\,\Theta(R_{\mathrm{crust}}-r)\,. \tag{18.46}$$

Its solution is given by

$$e\,V(r) = \frac{C}{r - R + C/[e\,V(R)]}\,, \qquad R \le r \le R_{\mathrm{crust}}\,, \tag{18.47}$$

with $C \equiv \sqrt{3\pi/2}/e = 5.013 \times 10^3$ MeV fm. Equation 18.47 evaluated at $r = R_{\mathrm{crust}}$ leads to

$$R_{\mathrm{gap}} \equiv R_{\mathrm{crust}} - R = C\left(\frac{1}{e\,V_{\mathrm{crust}}} - \frac{1}{e\,V(R)}\right)\,, \tag{18.48}$$

where $V_{\mathrm{crust}} \equiv V(R_{\mathrm{crust}})$. Notice that a given value of V_{crust} determines R_{crust} and, thus, R_{gap}. It is obvious that the gap width shrinks to zero if the crust potential becomes equal to $V(R)$, the potential's value at the surface of strange matter. In this case atomic nuclei would come in contact with strange matter and be converted into the true ground state by hypothesis.

18.3.1.3 Outside of strange quark matter: crust region

The electrostatic potential in the nuclear crust, that is, the region $r \ge R_{\mathrm{crust}}$ in figure 18.12, is constant. This follows from the fact that the forces acting on the ions in the crust, gravitational and electric, must counterbalance each other at equilibrium. Since the former is tiny compared to the electric force in the gap, one obtains $\mathrm{d}V/\mathrm{d}r = 0$ which implies that the electrostatic potential is constant there, that is, $V(r) \equiv V_{\mathrm{crust}}$. For what follows, representative values of $V_{\mathrm{crust}} = 5$ and 10 MeV were chosen, together with zero external potential [28].

[2] The pressure of massless electrons and positrons can be calculated exactly in this limit, too. One obtains from equation (18.5) the expression

$$P^e(r) = \frac{1}{12\,\pi^2}\left(\mu^e(r)\right)^4 + \frac{1}{6}\,\mu^e(r)\,T^2 + \frac{7\,\pi^2}{180}\,T^4\,,$$

which leads for equation (18.40) to

$$e\,V(R,T) = \mu^e(R,T)\,\frac{3 + 2\,\pi^2\,T^2/\left(\mu^e(R,T)\right)^2}{4 + 4\,\pi^2\,T^2/\left(\mu^e(R,T)\right)^2}\,.$$

The electron pressure at the surface of strange matter at zero temperature thus follows as $P^e(\mu^e \sim 10\ \mathrm{MeV}) \sim 10^{-5}$ MeV fm^{-3}.

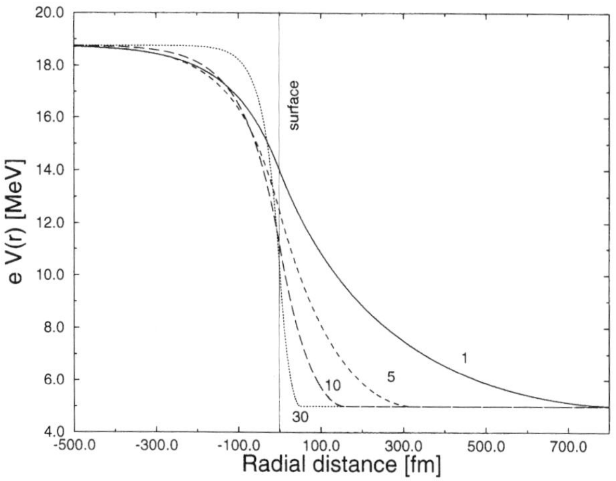

Figure 18.14. Electrostatic potential of electrons in the close vicinity inside and outside of the surface (vertical line) of strange matter [30]. The numbers assigned to the curves refer to temperature (in MeV). A representative value for the electrostatic crust potential, $eV_{\rm crust} = 5$ MeV (horizontal line), was chosen. The gap width extends from ~ 40 fm to about 800 fm, depending on temperature. (Reprinted courtesy of *Phys. Rev.*)

18.3.2 Gap width at finite temperature

Figures 18.14 to 18.16 exhibit the behavior of $eV(r)$ in the close vicinity inside and outside of R. The curves differ with respect to the temperature of strange matter and the value chosen for $V_{\rm crust}$. For zero temperature and a electrostatic crust potential of 5 MeV, as chosen in figure 18.14, a large gap of $R_{\rm gap} = 810$ fm is obtained. Larger values of $V_{\rm crust}$ reduce the potential difference between the surface of strange matter and the inner crust which leads to smaller gaps. For example, a value of $eV_{\rm crust} = 10$ MeV reduces $R_{\rm gap}$ to 280 fm [28], as can be seen in figure 18.15. Most interesting is the impact of *finite* temperatures on the gap. From equation (18.41) it is already known that the potential's value at the star's surface, $V(R)$, is reduced in this case. Figures 18.14 to 18.17 exhibit that this reduction is up to $\sim$50% for the temperatures under consideration. The corresponding reduction of $R_{\rm gap}$ with temperature is rather strong. In fact, we find that the gap even *shrinks to zero* for plausible values of $V_{\rm crust}$ and temperature that were typical for newly formed strange pulsars in supernovae.

Alcock *et al* established a minimum value of $R_{\rm gap} \sim 200$ fm as the lower bound on $R_{\rm gap}$ necessary to guarantee the crust's security against strong interactions with the star's strange-matter core [28]. From figure 18.17

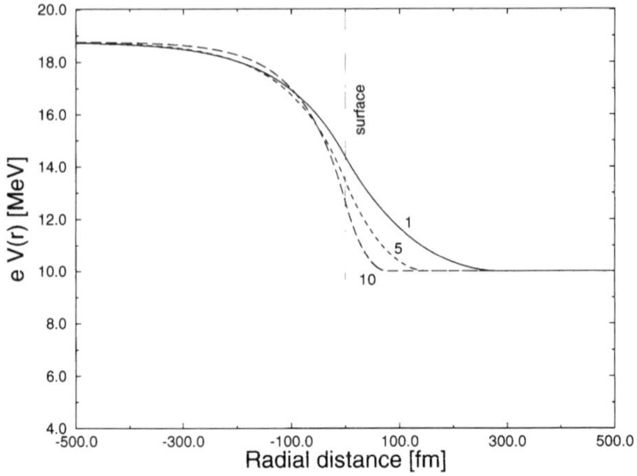

Figure 18.15. Same as figure 18.14, but for an electrostatic crust potential of $V_{\text{crust}} = 10$ MeV [30]. The labels refer to temperature in MeV. (Reprinted courtesy of *Phys. Rev.*)

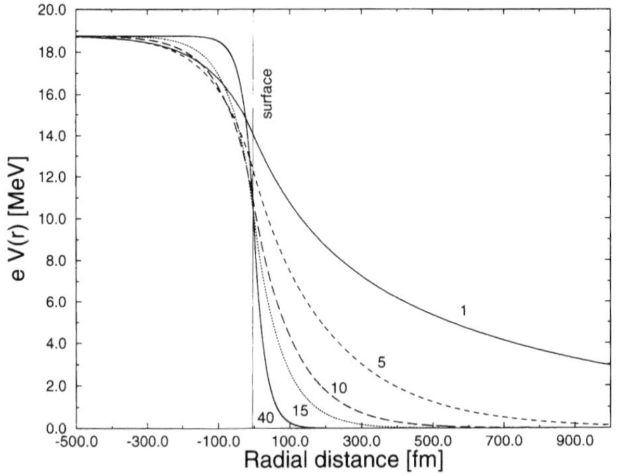

Figure 18.16. Same as figure 18.14, but for zero external electrostatic potential, $V_{\text{crust}} = 0$ MeV [30]. The labels refer to temperature in MeV. (Reprinted courtesy of *Phys. Rev.*)

one sees that a hot strange pulsar with say $T \sim 30$ MeV can only carry a nuclear crust whose electrostatic potential at the base is smaller than about $eV_{\text{crust}} \sim 0.1$ MeV. This value increases to about ~ 4 MeV for somewhat

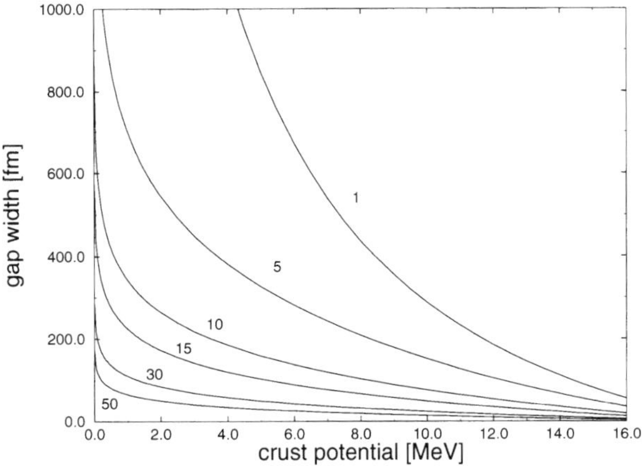

Figure 18.17. Gap width, $R_{\rm gap}$, versus electrostatic crust potential, $eV_{\rm crust}$ [30]. The labels refer to temperature in MeV. (Reprinted courtesy of *Phys. Rev.*)

cooler stars of $T \sim 10$ MeV. Finally, crust potentials between 8 and 12 MeV are possible only for stars of temperatures less than $T \sim 5$ MeV. In this connection it is interesting to recall that the upper limit on the density of the inner crust is determined by *neutron drip*, which occurs at about 4.3×10^{11} g cm^{-3}, where free neutrons begin to drip out of the most stable atomic nucleus, ^{118}Kr ($Z = 36$).[3] The electrostatic crust potential in atomic matter at drip density lies right in the above given density range. Hence, we conclude that the constraint $V_{\rm crust} \lesssim 10$ MeV established here provides another independent limit (besides the one set by neutron drip) on the maximal possible density of the nuclear crust that can be carried by strange matter. Accidentally, both issues – neutron drip and finite temperatures – lead to the same density limit. As to the formation of a crust on the surface of a bare strange pulsar formed in a supernova explosion, we are left with the important conclusion that the star's crust must be at rather low density in order to ensure a sufficiently wide enough gap. In terms of mass, the crust will be much lighter than $\sim 10^{-5}\, M_\odot$ [188] established for a strange pulsar possessing a nuclear crust whose density at the base is equal to neutron drip [28, 188].

As a further interesting aspect, the findings presented above serve to perform a few simple considerations concerning the accretion of matter onto the surface of a bare strange-matter star being bound in a binary system, whose companion star is made up of ordinary matter. Furthermore, since globular clusters are rather dirty environments, it seems plausible to assume

[3] The importance of neutron drip for the phenomenology of strange stars will be discussed in section 18.4.

Table 18.2. Kinetic energy, E_{p}, acquired by a proton falling toward the surface of a bare strange star, for a few selected strange star masses.

$M/M_{\odot}$	0.1	0.5	1.0	1.4	1.85
E_{p} (MeV)	36	95	187	252	350

that there might be strange stars that accrete some ambient (interstellar) material [28]. The idealized case of spherical accretion of a plasma, which consists of only protons and electrons, onto the surface of a bare, non-magnetized strange star has been considered in reference [633], assuming no dissipation in the radiation flow of the infalling matter which is not an issue for our studies either. There it was estimated that under these circumstances, the kinetic energy of protons hitting the surface of a bare strange star is given by

$$E_{\mathrm{p}} = \frac{138\ M/M_{\odot}}{R_6\ \sqrt{1 - 0.295\, M/(M_{\odot}\, R_6)}}\ \mathrm{MeV}\,, \tag{18.49}$$

where $R_6 = R/10^6$ cm and M the star's mass. By virtue of equation (18.49) we estimate that $E_{\mathrm{p}} \lesssim 250$ MeV for the strange star of figure 18.11 with a canonical pulsar mass of $M \sim 1.4\, M_{\odot}$. Further E_{p} values associated with a sample of other selected stellar masses are listed in table 18.2. Such kinetic energies enable a proton to penetrate the maximal possible Coulomb barrier easily, which has a height of $e\Delta V(R,T) \sim 15$ MeV, as can be read off from figures 18.14 through 18.16, and to undergo a reaction with strange matter. Accretion of less energetic protons or atomic nuclei onto the surfaces of initially bare strange stars, perhaps in proportion to the cosmic abundances in the interstellar medium, will plausibly lead to the development of (possibly thick) nuclear crusts in the course of stellar evolution. The height of the Coulomb barrier required in order to shield the strange matter from incoming atomic nuclei of electric charge Z_N is $Z_N \times eV(R,T)$, with $eV(R,T)$ to be read off again from figures 18.14 through 18.16. While most of this material swept up by a strange star will consist of the lightest elements, in the presence of the gravitational field of the core such material will undergo thermonuclear burning. For a core of a solar mass and radius of a few kilometers, the gravitational field is enormous compared to that found in normal stars that may be more massive but are also much larger. The burning processes may therefore resemble those considered in the evolution of ordinary stars as they burn from hydrogen through to iron cores except for at least two major differences. As already remarked, the highest possible inner density of the crust is the neutron drip density, much higher than the density ever found in ordinary stars and

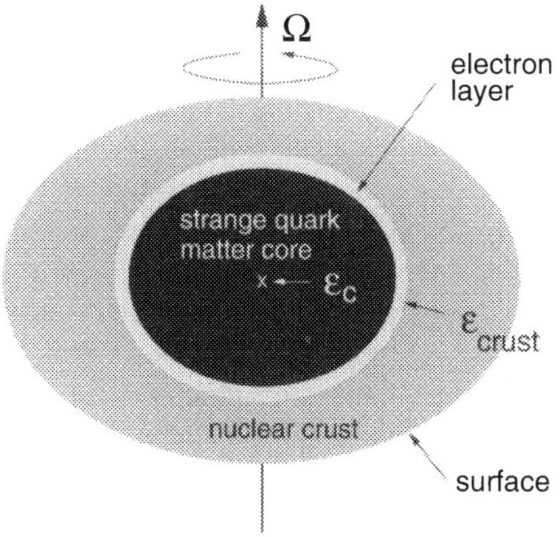

Figure 18.18. Cross section of a rotating strange star carrying a nuclear crust. The maximal possible range of the star's rotational frequency is $0 \le \Omega \le \Omega_K$. The quantities ϵ_c and ϵ_{crust} denote the star's central density and the density at the base of the crust, respectively.

even several orders of magnitude higher than that of white dwarfs. So the burning proceeds even further to exceedingly neutron rich nuclei. Secondly, the time scale for burning may be very fast by comparison, except it is likely to be limited by the accretion rate. There are obviously many interesting aspects of such a scenario to be investigated.

18.4 Rotating strange stars enveloped in nuclear crusts

In this section we study the thickness, mass, and moment of inertia of the nuclear crust carried by strange-quark-matter stars (cf. figure 18.18) as a function of mass and rotational frequency of the star. This topic is of particular interest for several reasons, the first one being that the crust is a possible site of the buildup of stress whose occasional release could be responsible for glitching or at least micro-glitching of pulsars made up of hypothetical strange matter (strange pulsars). Another motivation concerns the cooling rate of strange stars, which is altered by the presence of a crust [543, 634]. This topic will be explored in detail in chapter 19.

18.4.1 Equation of state of strange stars carrying nuclear crusts

In accordance with the analysis performed in section 18.3, a value of $R_{gap} = R_{crust} - R \sim 10^3$ fm as the lower conservative bound on the gap between the surface of strange matter and nuclear crust matter (cf. figure 18.12) appears to guarantee the crust's security against strong interactions with the strange-matter core of a hypothetical strange star. Combining this gap width with $e\Delta V \sim 10$ MeV for the potential difference of the electric field at the surface of strange matter, one obtains an extremely strong electric field strength there, on the order of $|\boldsymbol{E}| \sim 10^{17}$ V cm^{-1}. Pointing in the radial outward direction, this field can therefore suspend the crust out of contact with the core, as schematically illustrated in figures 18.12 and 18.18. The crust, which is made up of atomic matter, which itself is overall neutral but in which the charges of opposite sign are displaced by the field, is purely gravitationally bound to the core.

Because of the electric field, only crust particles that are electrically neutral, of which the neutron is the only candidate, can penetrate the core. For all other particles, the Coulomb barrier associated with the electric field prevents interactions between core and crust on astronomical time scales, as demonstrated by Alcock *et al* [28]. By hypothesis, strange matter is assumed to be absolutely stable (section 2.3). So neutrons, which come in contact with the strange-matter core, will be dissolved into quark matter as they gravitationally drip into the core. This peculiar feature strictly limits the maximum density of the crust to $\epsilon_{\rm drip} = 4.3{\times}10^{11}$ g cm^{-3} known as the *neutron drip density*. If the density at the base of the crust has grown to this value, the atomic nuclei (e.g. ^{118}Kr) there have become so neutron rich by electron capture reactions, ${}^{A}_{Z}X + e^- \rightarrow {}_{Z-1}{}^{A}Y$, that no additionally created neutrons can be accommodated inside the atomic nucleus Y, and thus begin to populate quantum states outside of them, leading to the formation of a relativistic neutron gas in which the nuclei are embedded. In what follows, we shall be interested in the nuclear crust of a strange star that has reached the final state of stellar evolution, namely cold catalyzed matter, no matter how the strange star acquired the crust, whether by accretion from the interstellar medium onto an initially, and possibly primordial bare strange star, or whether during its creation in a supernova, the crust in this case being debris from the progenitor star.

Because we are not interested in the stellar structure on the scale of the thickness of the gap between core and crust, $R_{\rm gap}$, of the order of several hundred fermis, the somewhat complicated situation described just above can be very simply modelled by a proper choice of the equation of state [148]. At densities below neutron drip it can be represented by the low-density equation of state of electrically charge-neutral nuclear matter. The most significant aspect of this density domain, as discussed in chapter 4, is that it consists of a Coulomb lattice of heavy ions immersed in an electron gas. The heavy ions become ever more neutron rich as the neutron drip density, $\epsilon_{\rm drip}$, is approached from below. A model for the equation of state of such matter has been calculated by Baym, Pethick, and Sutherland (cf. table 4.1), which will be adopted in our calculations

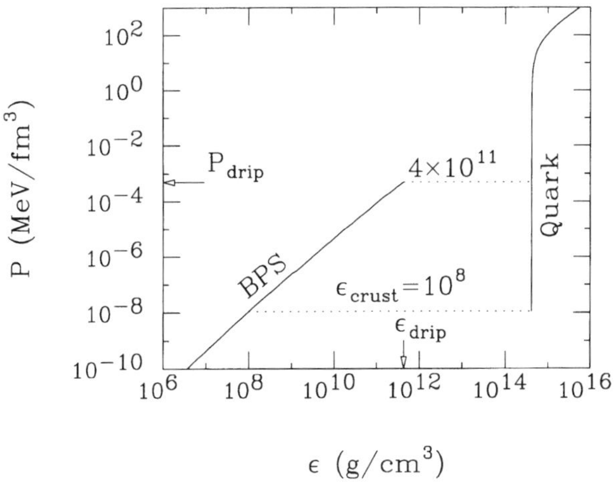

Figure 18.19. EOS of a strange star with nuclear crust [188]. The density at the base of the crust (inner crust density), $\epsilon_{\rm crust}$, is limited by neutron drip, $\epsilon_{\rm drip} = 4\times10^{11}$ g cm^{-3}. Only inner crust values lower than $\epsilon_{\rm drip}$ are possible. As an example, the EOS associated with $\epsilon_{\rm crust} = 10^8$ g cm^{-3} is shown. (Reprinted courtesy of *Astrophys. J.*)

below. To describe strange stars with maximal possible crust densities at their bases, i.e. the crust's inner edge, one encounters the low-density equation of state for pressures $P \leq P_{\rm drip}$. In the second distinct density regime, encountered for pressures above neutron drip pressure, $P_{\rm drip}$, the equation of state corresponds to the one of strange quark matter. In the calculations, we shall use the simplest form of the bag model equation of state for the description of strange matter, namely $P = (\epsilon - 4B)/3$ derived in equation (18.17), which is accurate to within 4% of the more complicated form (18.27) involving quark masses [28]. Medium effects in strange-quark matter possibly modify the equation of state more strongly [635]. The edge of the strange star, if bare, occurs at $P = 0$, or equivalently, $\epsilon = 4B$. In the presence of a crust, the quark core will be slightly squeezed, and the pressure at the edge of the core will be small but positive and equal to $P_{\rm drip}$. Correspondingly, the energy density at the edge of the core will be slightly larger than $4B$.

The equation of state of a strange star enveloped in a nuclear crust is illustrated in figure 18.19. Mathematically it is represented by the following expression [148, 188],

$$P(\epsilon) = \begin{cases} P_{\rm BPS}(\epsilon) & \text{if } P(\epsilon) < P_{\rm drip}\,, \\ \frac{1}{3}\,(\epsilon - 4\,B) & \text{if } P(\epsilon) \geq P_{\rm drip}\,. \end{cases} \tag{18.50}$$

Two different representative values for the bag constant, for which 3-flavor strange matter is absolutely stable, will be chosen. These are $B^{1/4}$=145 MeV (57 MeV fm^{-3}) and 160 MeV (85 MeV fm^{-3}). As can be seen from figure 18.3, for massless strange quarks these bag constants correspond to an equilibrium energy per baryon number of strange quark matter of about 830 MeV and 915 MeV, respectively. For strange quarks of m_s = 100 MeV, they correspond to about 855 MeV and 930 MeV, respectively. Hence, these choices represent strongly ($\sim$ 100 MeV) and weakly bound strange matter, and in all cases correspond to strange quark matter being absolutely bound with respect to ^{56}Fe, as required by the strange-matter hypothesis.

18.4.2 Generic properties of strange stars

The equation of state (18.50) of strange stars carrying nuclear crusts, shown in figure 18.19, serves to point out a generic feature between such stellar objects and their non-strange counterparts – the neutron stars, the reason being that neutron stars, as known from chapter 14, are constructed for one-parameter equations of state and, thus, constitute stellar one-parameter sequences, whose properties are uniquely determined by a proper choice of the central stellar density, ϵ_c. This is in sharp contrast to the strange stars with nuclear crusts which constitute *two-parameter* sequences, since their properties are determined by a properly chosen value for the inner crust density, $\epsilon_{\rm crust}$, and for a fixed value of that, the central density of the quark core, ϵ_c. As demonstrated by Glendenning *et al* [148, 174, 188], this feature leads to a whole variety of stellar strange-matter sequences whose properties are considerably more complex than those of the stellar one-parameter sequence, consisting of white dwarfs and neutron stars [148, 174, 188].

Keeping this important feature in mind, next we turn to an investigation of the properties of strange stars enveloped in nuclear crusts. A schematic illustration of the global structure of such objects is given in figure 18.18. In accordance with what has been noted just above, the amount of crust mass that can be carried by the core is uniquely determined for given values of ϵ_c and $\epsilon_{\rm crust}$ by Einstein's field equations of hydrostatic equilibrium, which reduce to the TOV equation (14.82) in the spherical limit, $\Omega \to 0$. The pressure profiles inside such objects will be a continuous and monotonically decreasing function of the Schwarzschild radial coordinate, r, as known from chapter 14. But for the strange stars with crust there will be a discontinuity in the energy density between strange quark matter and hadronic matter at the neutron drip pressure, $P_{\rm drip}$, of hadronic matter. The energy discontinuity between hadronic matter at the drip pressure and strange-quark matter at the same pressure can be read off from figure 18.19 to be given by $\Delta\epsilon \equiv (3P_{\rm drip} + 4B) - \epsilon_{\rm drip}$, with $\epsilon_{\rm drip}$ the energy density of nuclear matter at the neutron drip point.

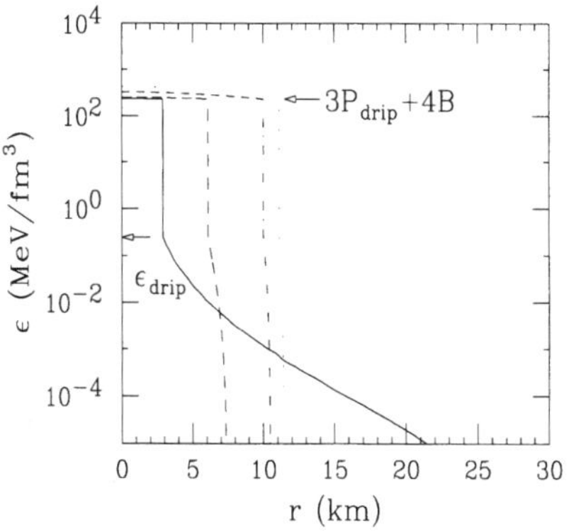

Figure 18.20. Energy density as a function of radial distance from the star's center for gravitational masses $M/M_\odot$=0.020 (solid line), 0.20 (dashed), 1.00 (dash–dotted), and 1.50 (dotted) [148]. The bag constant is $B^{1/4} = 145$ MeV. (Reprinted courtesy of *Astrophys. J.*)

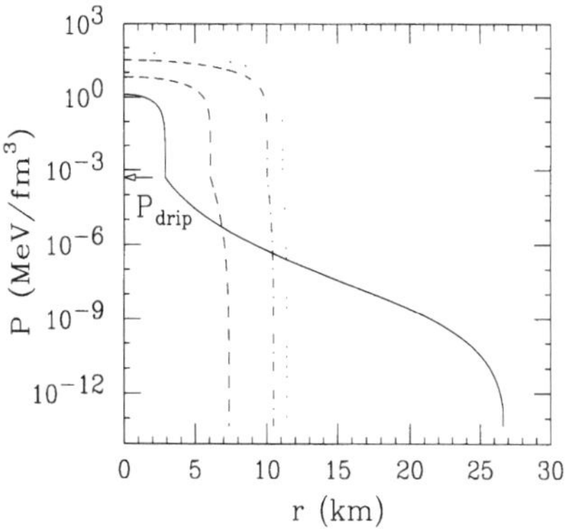

Figure 18.21. Pressure versus radial distance from the star's center for different gravitational masses [148]. The labeling is the same as in figure 18.20, $B^{1/4} = 145$ MeV. (Reprinted courtesy of *Astrophys. J.*)

The energy density in the star will suffer this discontinuity, across the several hundred fermi gap $R_{\rm gap}$ described in section 18.3.1, at that radial distance away from the star's center where the pressure falls from its central value to $P_{\rm drip}$. The energy density and pressure profiles of several such non-rotating sample stars are shown in figures 18.20 and 18.21 for gravitational masses in the range of $0.020\,M_\odot \leq M \leq 1.50\,M_\odot$. The boundaries of this range correspond to a very light strange star close to the lower mass limit and to one closer to the upper mass limit, respectively. The sharp drop in the energy density profile marks the boundary between the quark matter core and the nuclear crust. These profiles are evidently strikingly different from those of conventional neutron stars shown in figures 14.9 through 14.11. Another point to be remarked on concerns the dependence of the thickness of the crust on the core's size. One sees from figures 18.20 and 18.21 that for a given inner density of the crust, the crust will be the thinner the more massive the strange quark core, and the thicker the less massive it is. This behavior can be understood as a consequence of the way the core radius, $R_{\rm c}$, scales with the core mass, $M_{\rm c}$, as can be inferred as follows. For masses too small for gravity to play an important role, the energy–density relationship of strange matter can be approximated as $\epsilon(r) = \epsilon_*\,\Theta(R_{\rm c} - r)$, where $\epsilon_* = 4\,B$ is the equilibrium density of strange matter at zero pressure known from equation (18.17). Using this relation, the core's mass can be calculated analytically from equation (14.83) in the form $M_c = (4\pi/3)\, R_c^3\, \epsilon_*$. This relation is only somewhat modified near the most massive star in the stellar sequence, which gravity terminates. So, since $R_c \propto M_c^{1/3}$, the Newtonian gravitational force acting on unit mass at the surface of the core is $M_c/R_c^2 \propto M_c^{1/3}$, leading to thinner crusts for the heavier quark stars as noted above.[4] When ϵ_{crust} equals the neutron drip density, we find a minimum-mass star of mass $\sim 0.015\, M_\odot$, which is to be compared with $0.1\, M_\odot$ for the minimum mass of neutron stars [70]. So the thickening of the crust illustrated in figures 18.20 and 18.21 for $M = 0.02\, M_\odot$ corresponds to stars in the vicinity of the lower mass limit, provided the inner crust density is equal to the neutron drip density.

Parenthetically, we mention that a comparison of these figures for different bag constants B can be facilitated by noting the scaling laws that apply to strange stars [27, 636]. For a central energy density that is some fixed multiple of B, the mass and radius of bare strange stars which correspond to different values assumed for the bag constant scale as $M_c \propto 1/\sqrt{B}$ and $R_c \propto 1/\sqrt{B}$. The maximum-mass strange star for a given B corresponds to a central density of $19.2 \times B$.

The mass–radius relationship of strange stars carrying nuclear crusts is shown in figure 18.22. The crust thickness decreases very rapidly with mass, from more than about 20 km for very light strange stars of masses around $0.015\, M_\odot$ to several hundred meters for the star at the maximum mass. The crust mass varies between $10^{-5}\, M_\odot$ and $5\times 10^{-5}\, M_\odot$, depending on core mass and the star's rotational frequency. Moreover, the crust mass is about a factor of two larger for the smaller bag constant, $B^{1/4} = 145$ MeV [148]. Such small crust masses have only a little effect on the total gravitational mass of all but very light strange stars. Nevertheless, the light crusts are quite interesting because the strange quark core permits stable configurations of strange stars with crusts that can be significantly lighter than the least massive neutron star, $\sim 0.1\, M_\odot$ for the BPS equation of state. Since the crust is bound by the gravitational interaction, the mass–radius relationship is qualitatively similar to the one of neutron stars, shown in figure 14.8: the radius being largest for the lightest and smallest for the heaviest star of the sequence. More than that, the relationship is not necessarily monotonic at intermediate masses, just as for the neutron stars. The radius of the strange quark core is shown too in figure 18.22. Its dependence on the core's mass follows the inevitable behavior of objects that are self-bound, namely $R_c \propto M_c^{1/3}$ mentioned just above. This dependence is only somewhat modified near the limiting-mass star, where gravity terminates the stable sequence.

[4] Note that from $M_c = (4\pi/3)\, R_c^3\, \epsilon_*$, the strength of gravity, expressed as M/R, of the strange-matter core is given by $M_c/R_c = (16\pi/3)\, R_c^2\, B$. From this one finds $M_c/R_c \lesssim 10^{-3}$ for small strange-matter cores of radii $R_c \lesssim 1$ km. Hence, the curvature of spacetime does not play a significant role for objects of that size.

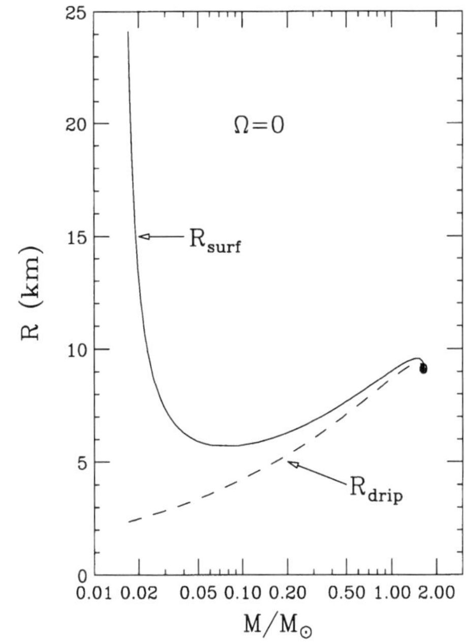

Figure 18.22. Radius as a function of mass of strange stars with crust, and radius of the strange star core for inner crust density equal to the neutron drip, for non-rotating stars [148]. Bag constant is $B^{1/4}$=160 MeV. The solid dots refer to the limiting-mass model of the sequence. (Reprinted courtesy of *Astrophys. J.*)

18.4.3 Properties of rotating strange stars

Having discussed some of the generic properties of static strange stars, let us turn now to the interesting question of how these and other properties are modified by rapid stellar rotation. In doing so, we shall restrict ourselves to the assumption that the star's crust and core (cf. figure 18.18) rotate with the same angular velocity, Ω. Since both crust and core are composed of charged particles, protons, electrons, and heavy ions in the case of the crust, quarks and electrons in the case of the core, the rigid coupling between both stellar regimes may plausibly be accomplished by the magnetic field of the strange pulsar. The properties of differentially rotating strange stars with crusts, that is, $\Omega_{\rm core} \neq \Omega_{\rm crust}$, have been explored only recently by Schaab [637], to which we refer for details.

In figure 18.23 we show the mass–central density sequences corresponding to static stars and to stars rotating at their respective general relativistic Kepler frequencies $\Omega_{\rm K}$, derived in equation (16.7), for more massive strange stars and for both bag constants, $B^{1/4} = 145$ MeV

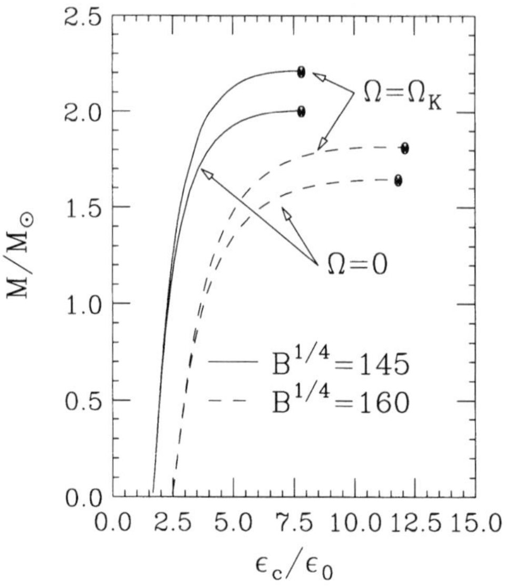

Figure 18.23. Gravitational masses of strange stars with hadronic crust as a function of central density of non-rotating sequences ($\Omega = 0$) and sequences rotating at their Kepler frequencies ($\Omega = \Omega_K$) [148]. The solid dots refer to the limiting-mass model of each sequence. (Reprinted courtesy of *Astrophys. J.*)

and $B^{1/4} = 160$ MeV. For the more massive stars, for which the Kepler frequency is greater, the mass that can be supported by a star of the same central density as a non-rotating one is larger for the rotating star by about 15%.

Figure 18.24 exhibits the crust mass as a function of total mass for three different, fixed frequencies. The choices are $\Omega = \Omega_K$, $\Omega_K/2$ and $\Omega = 0\ \mathrm{s}^{-1}$. The crust has the maximal possible density equal to neutron drip density, $\epsilon_{\rm drip}$. The solid dots refer to the limiting-mass model of each sequence. As already noted in section 18.4.2, the gravitational force per unit mass at the surface of the core is $\propto M_c^{1/3}$ so that the light cores have small attraction and therefore have thick and massive crusts. At the opposite extreme, near the upper termination of the sequence, the crust is thin and therefore light. In between the behavior is not monotonic because of the dependence of the star and core radius on mass, as shown in figure 18.22. It should be noted that the crust mass is smaller for larger bag constants which, again, can be understood in terms of the gravitational force at the surface of the core, $\propto M_c^{1/3}$, while the mass itself scales as $M_c \propto 1/\sqrt{B}$ [148].

Figure 18.25 displays the dependence of the nuclear crust mass on the Kepler period P_K $(= 2\pi/\Omega_K)$ of stars in two sequences with different bag

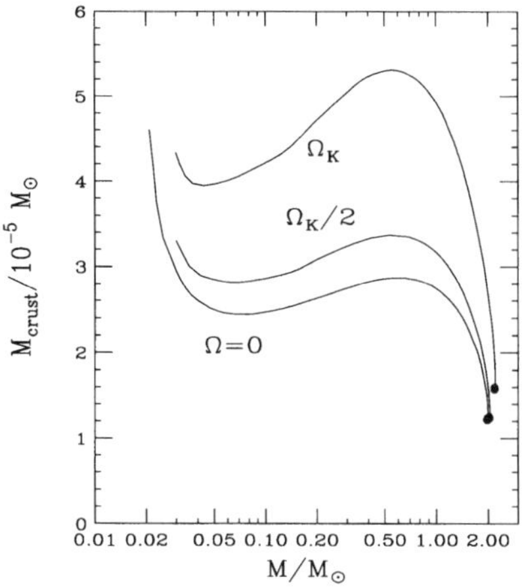

Figure 18.24. Mass of hadronic crust on a strange star as a function of total star mass for several rotational frequencies [148]. The bag constant is $B^{1/4} = 145$ MeV. (Reprinted courtesy of *Astrophys. J.*)

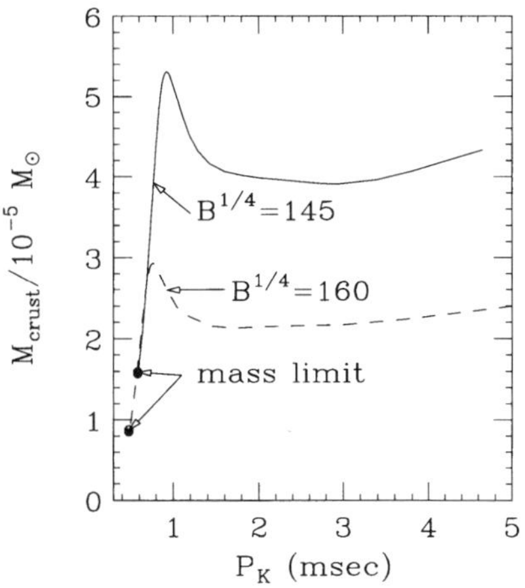

Figure 18.25. Mass of hadronic crust carried by rotating strange stars as a function of Kepler period, P_K [148]. (Reprinted courtesy of *Astrophys. J.*)

constants. As for neutron stars, the shortest rotational periods are attained by the limiting-mass stars, which is a consequence of their relatively small radii and the core's strong gravitational attraction. One reads off that these stars possess periods in the range of $0.5\ \text{ms} \lesssim P_K \lesssim 0.6$ ms, depending on the value of the bag constant. Larger bag constants generally enable quark stars to spin more rapidly, the reason being the somewhat higher mass densities in such stars (cf. figure 18.23), which makes them more compact than quark stars constructed for smaller bag constants. This feature follows readily from the bag-model equation of state derived in (18.17). To see this, let ϵ_1 and ϵ_2 denote the energy density inside of two sample stars constructed for equations of state with bag constants B_1 and B_2 ($B_1 < B_2$), respectively. Comparing locations where pressure is the same in both stars, i.e. $P_1 = P_2$, equation (18.17) then tells us that $\epsilon_2 - \epsilon_1 = 4(B_2 - B_1) > 0$, and thus $\epsilon_2 > \epsilon_1$. This inequality holds for any location inside the star, from the center to the surface. We conclude this paragraph by noting that the largest crust masses, which never exceed $5 \times 10^{-5}\ M_\odot$, are carried by stars whose masses are $\sim 0.7\ M_\odot$. It is striking that the crust mass depends only rather weakly on the Kepler period for most of the lighter stars, for which $P_K \gtrsim 2$ ms.

The impact of rotation on M_{crust} for star frequencies $0 \leq \Omega \leq \Omega_K$

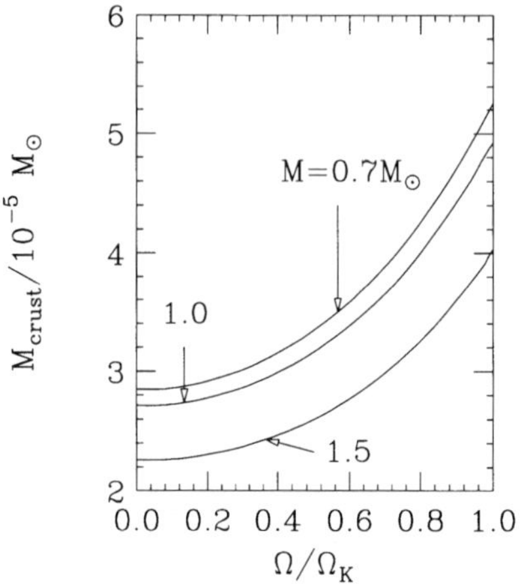

Figure 18.26. Hadronic crust mass as a function of rotational frequency Ω for different star masses [148]. The bag constant is $B^{1/4} = 145$ MeV. (Reprinted courtesy of *Astrophys. J.*)

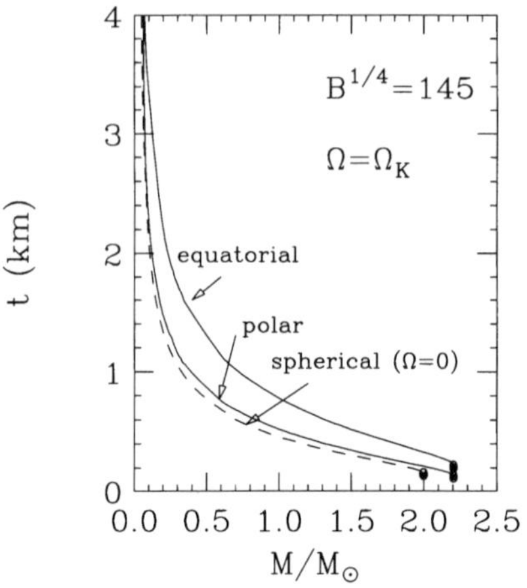

Figure 18.27. Crust thickness, t, at equator and pole versus star mass for rotation at Ω_K [148]. Results for non-rotating strange stars are labeled $\Omega = 0$. The bag constant is $B^{1/4} = 145$ MeV. (Reprinted courtesy of *Astrophys. J.*)

is shown in figure 18.26 for a representative sample of stellar masses, M. Again, as in figure 18.25, the larger B the smaller M_{crust}. The stellar mass is kept constant along the curves of this figure and therefore the crust mass is now a monotonically increasing function of the star's rotational frequency Ω. The constant-mass curves also emphasize the dependence of $M_{crust}(\Omega)$ on M.

The crust thickness as a function of strange-star mass is shown in figure 18.27. In accordance with one's expectation, the crust thickness at the star's equator is larger than at its pole due to the centrifugal force. The crust is thickest for the lighter stars because of the smaller gravitational force. For the purpose of comparison, the crust thickness of a non-rotating strange star ($\Omega = 0$) is exhibited too. It is interesting that even the polar thickness increases with rotational frequency when comparing stars of the same mass. As described by Glendenning and Weber [148], this feature originates from four physical processes that accompany stellar rotation. First, the centrifugal force opposes gravity, so that a rotating star of the same mass as its non-rotating counterpart is relatively decompressed. Second, the rotating lower-density star also has its mass redistributed, becoming oblate, that is $R_{eq} > R_p$. Third, the gravitational force of the mass in the sphere of polar radius, $R_{\rm p}$, acting on unit mass at $R_{\rm p}$ is less than at the surface of a non-rotating star [148],

$$F \sim \frac{M}{R_{\rm p}^2} \sim \frac{\left(\frac{4}{3}\pi R_{\rm p}^3\right)\bar{\epsilon}_{\rm rot}}{R_{\rm p}^2} \propto R_{\rm p}\,\bar{\epsilon}_{\rm rot} < R\,\bar{\epsilon}\,, \tag{18.51}$$

where $\bar{\epsilon}_{\rm rot}$ and $\bar{\epsilon}$ are the average densities for the decompressed, rotating and decompressed, non-rotating star respectively, which obey $\bar{\epsilon}_{\rm rot} < \bar{\epsilon}$. Moreover we have made use of the fact that $R_{\rm p} < R$. Fourth, the remainder of the mass contained in the oblate star outside the sphere of polar radius is at a greater distance to the pole compared to the same mass distributed over the spherical non-rotating star and therefore experiences a smaller gravitational attraction. So, in brief, the nuclear crust on a rotating strange star experiences a smaller gravitational attraction to the core, even at the pole and so is thicker everywhere when compared to a non-rotating star of the same core mass, which is essentially equal to the total stellar mass.

Figure 18.28 displays the crust thickness, t, at the equator and the pole of strange stars of several masses as a function of rotational frequency. It is seen that t is a monotonically increasing function of Ω. Because of the stronger gravitational field, the impact of rotation on t is the smaller the more massive the strange star model of a sequence. The nuclear crust thickness on very light strange stars can be many kilometers thick as known from figure 18.22, growing to several thousand kilometers if the mass of the strange-matter core is on the order of planetary size or less, as will be discussed in section 18.5. However, for the range of stellar masses shown in figure 18.28, $0.7\,M_\odot \leq M \leq 1.5\,M_\odot$, the crust is less than 1 km thick even for rotation at the highest theoretically possible rotational frequency, given by the Kepler frequency $\Omega_{\rm K}$.

18.4.4 Moment of inertia

The mathematical expression for the moment of inertia of the hadronic crust, $I_{\rm crust}$, that can be carried by the rotating strange star schematically illustrated in figure 18.18 follows from equation (15.51) by means of substituting $R_{\rm crust}$ as the lower boundary of the radial integral there. One then gets

$$I_{\rm crust}(\Omega) = 2\,\pi \int_0^{\pi} {\rm d}\theta \int_{R_{\rm crust}(\theta)}^{R(\theta)} {\rm d}r\; {\rm e}^{\lambda+\mu+\nu+\psi}\, \frac{\epsilon + P(\epsilon)}{{\rm e}^{2\nu+2\psi} - (\omega - \Omega)^2}\, \frac{\Omega - \omega}{\Omega}\,. \tag{18.52}$$

Equation (18.52) enables us to calculate $I_{\rm crust}$ as a function of star mass, M, for a sample of selected rotational frequencies, for which we choose $\Omega = \Omega_{\rm K}$, $\Omega_{\rm K}/2$ and $\Omega_{\rm K} = 0~{\rm s}^{-1}$. The result of such a calculation is shown in figure 18.29. Because of the relatively small crust mass that can be

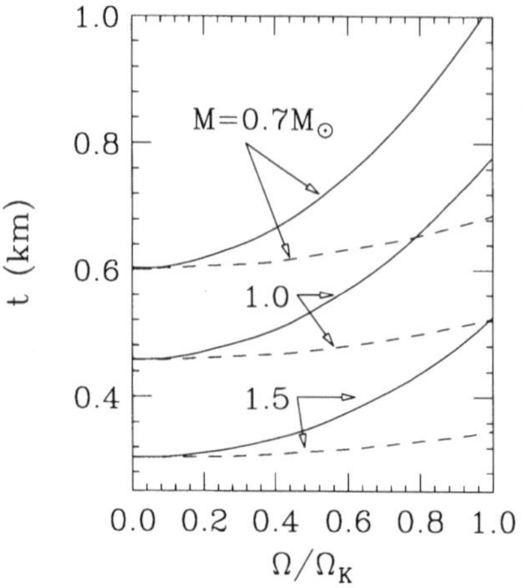

Figure 18.28. Crust thickness versus rotational frequency, Ω, for several star masses [148]. The solid and dashed lines refer to the thickness at the star's equator and pole. The bag constant is $B^{1/4} = 145$ MeV. (Reprinted courtesy of *Astrophys. J.*)

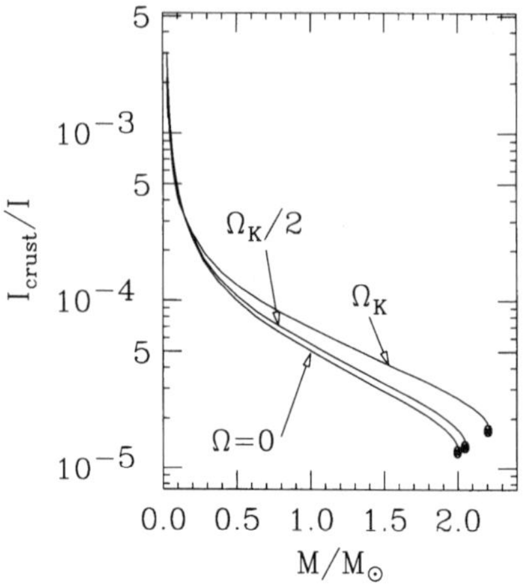

Figure 18.29. Ratio I_{crust}/I versus star mass, for several rotational frequencies [148]. The solid dots refer to the limiting-mass models. The bag constant is $B^{1/4} = 145$ MeV. (Reprinted courtesy of *Astrophys. J.*)

carried by the limiting-mass model (marked by solid dots in figure 18.29) of each stellar sequence, which is known from figure 18.24, the ratio I_{crust}/I is smallest for these stars. The less massive a strange star, the larger its radius (cf. figure 18.22) and therefore the larger I_{crust} as well as the total moment of inertia, I, the dependence of I_{crust}/I on M being such that the ratio I_{crust}/I is a monotonically decreasing function of M. One sees that there is only a slight difference between I_{crust} for $\Omega = 0$ and $\Omega = \Omega_K/2$. This confirms that equation (15.41), which defines the moment of inertia of a slowly rotating star, is a good approximation for I as long as the star rotates below about $\Omega_K/2$. In contrast to this, the full expression for I, given in equations (15.51) and (18.52), is to be employed for $\Omega > \Omega_K/2$, except for the very light stars of such sequences for which the general relativistic corrections of I become insignificantly small.

18.4.5 Compatibility of strange-matter hypothesis with pulsar glitches

As pointed out by Glendenning *et al* [148, 188], the amount of crust mass that can be carried by rotating strange stars is of considerable relevance to the question of whether or not this class of stars can exhibit glitches (cf.

figure 1.2), which, as described in section 3.6, are observed in the rotational frequencies of many pulsars. If strange stars should turn out to be incapable of producing pulsar glitches [173], the interpretation of pulsars as rotating strange stars would fail, rendering the strange matter hypothesis explained in section 2.3.1 rather questionable.

From figure 18.29 one reads off that $I_{\rm crust}/I$ varies between 10^{-3} for the lightest stars and about 10^{-5} at the upper mass limit, and differs by about a factor of 2 to 3 depending on the bag constant. If the angular momentum of the pulsar is conserved in the quake, then the relative frequency change and moment of inertia change are equal and can be written in the form [148]

$$\frac{\Delta\Omega}{\Omega} = \frac{|\Delta I|}{I_0} > \frac{|\Delta I|}{I} \equiv \frac{f\, I_{\rm crust}}{I} \quad \text{with} \quad 0 < f < 1\,, \tag{18.53}$$

where I_0 is the moment of inertia of that part of the star whose frequency is changed in the quake [148]. It might be that of the crust only, or some fraction thereof or all of the star, depending on how strongly the crust and core are coupled. In any case $I_0 < I$. Reading off from figure 18.29 that $10^{-5} \lesssim I_{\rm crust}/I \lesssim 10^{-3}$, we obtain from (18.53) the estimate

$$\frac{\Delta\Omega}{\Omega} \simeq \left(10^{-5} \text{ to } 10^{-3}\right) f\,. \tag{18.54}$$

The factor f in the above relations represents the fraction of the crustal moment of inertia that is altered in the quake, $|\Delta I| = f \times I_{\rm crust}$. Since observed glitches have relative frequency changes of $\Delta\Omega/\Omega \simeq 10^{-9}$ to 10^{-6}, a change in the crustal moment of inertia by less than 10% would cause a giant glitch even in the least favorable case. Of course there remains the question of whether there can be a sufficient build-up of stress and also of the recoupling of crust and core which involves the healing of the pulsar period. This is probably a very complicated process that does not simply involve the recoupling of two homogeneous substances. The variation of the pressure through the core can be as large as five orders of magnitude (cf. figure 18.21) over which the properties of quark matter may vary considerably. The pressure variation of the crust from inner edge to the surface of the star is even greater.

We note that whereas in the crust quake model of neutron stars, which is schematically illustrated in figure 18.30, the solid crust and an assumed almost purely neutron core are supposed to be weakly coupled by the magnetic field, in strange stars all particles – quarks and electrons – carry electric charge, so the crust and core or parts thereof should be rather strongly coupled. It is also worth noting that glitch phenomena, in their magnitude, frequency of occurrence, and healing time, vary greatly from one pulsar to another. The rate of glitching of PSR 1737–30, for instance, is nearly an order of magnitude greater than for any other known pulsar

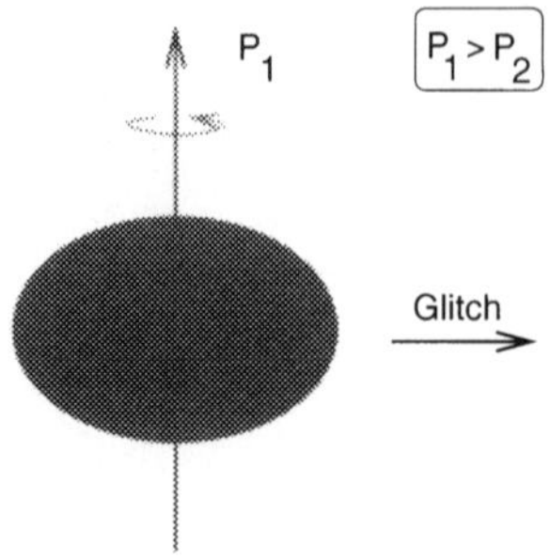

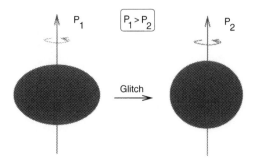

Figure 18.30. Pulsar glitch in the stellar quake model. A neutron (possibly strange?) star spins up (rotational periods $P_2 < P_1$) upon contracting toward a more spherically symmetric configuration. For illustration purposes, the change in shape deformation is greatly exaggerated.

[638]. The possibility of a highly individualistic behavior appears possible for crust quakes on strange stars because of the variation of the thickness and mass of the crust with the mass and frequency of the star, not to mention the temperature dependence of the shear modulus, the magnetic coupling between crust and core with a possibly varying structural nature of the quark core with depth as well as the varying nature of the nuclear crystalline crust with height above the core. Of course many of these remarks pertain also to conventional neutron stars [639], as pointed out by Glendenning and Weber [148].

The recoupling of crust and strange quark core is likely to have a long time scale, not unlike that expected for neutron stars (days to years [12]). Strange stars with crusts are likely to have a relatively short time scale coupling mechanism in addition, namely the abrasion of part of the inner surface of the crust by the core following a quake. Recall that the crust is suspended above the core by a few hundred fermis by the strong electric dipole layer on the core. If the relative change in the moment of inertia is attributed, for the sake of a rough order of magnitude estimate, to a change in the radius of the crust, then the relation $\Delta I/I = 2\Delta R/R$ holds. For the above quoted range of relative frequency glitches, $\Delta\Omega/\Omega = 10^{-9}$ to 10^{-6}, we have $\Delta R \sim 10^{-3}$ cm to 1 cm. So we expect a quake in the crust to cause the crust to momentarily come into physical contact with the core, a contact or bouncing that is unlikely to be symmetrical, and in any case that will transfer matter and therefore angular momentum to the core. However this abrasion of parts of the inner surface of the crust will be a momentary perturbation. The abraded material will be dissolved into the quark core. So, superimposed on the frequency change caused by the crust quake we suggest that there may be a smaller series of glitches associated with this transient mechanism [148].

Pointing out a possible discrepancy with the observed pulsar glitches, Alpar argues against the existence of strange stars [173]. His argument is twofold. One has to do with the assumption that strange quark stars have no internal structural differentiation that could give rise to pulsar glitches so that a bare strange star could not glitch. Our opinion on this is that we simply do not know enough about quark matter yet to make such a ruling and have alluded to at least some of the other uncertainties above [148]. Secondly, Alpar argues that even a strange star with a nuclear crust cannot glitch. This is based on assuming the near equality of $I_{\rm crust}/I$ and $\Delta\dot{\Omega}/\dot{\Omega}$ with the first ratio $\sim 10^{-5}$ by computation, and the second $\sim 10^{-3}$ to 10^{-2} by observation, yielding an apparent contradiction. However, with the aid of equation (18.53), we may write [148],

$$\frac{\Delta\dot{\Omega}}{\dot{\Omega}} \simeq \frac{\Delta\dot{\Omega}/\dot{\Omega}}{\Delta\Omega/\Omega}\frac{|\Delta I|}{I_0} > \frac{\Delta\dot{\Omega}/\dot{\Omega}}{\Delta\Omega/\Omega}\frac{f I_{\rm crust}}{I_0} > (10^{-1} \text{ to } 10)\, f\,, \tag{18.55}$$

yielding a small f value as before, namely $f < 10^{-4}$ to 10^{-1}. We have used measured values of the ratio $(\Delta\Omega/\Omega)/(\Delta\dot{\Omega}/\dot{\Omega}) \sim 10^{-6}$ to 10^{-4} for the Crab and Vela pulsars respectively [12]. So the observed range of the fractional change in $\dot{\Omega}$ *is consistent* with the crust having the small moment of inertia calculated and the quake involving only a small fraction, f, of that, just as in equation (18.54). Nevertheless, without undertaking a study of whether the nuclear solid crust on strange stars could sustain a sufficient build-up of stress before cracking to account for such a sudden change in relative moment of inertia, or whether the healing time and intervals between glitches can be understood, one cannot say definitely that strange stars with a nuclear solid crust can account for any complete set of glitch observations for a particular pulsar. However we have laid part of the groundwork for such an assessment. And we have shown that quite plausible fractional changes in the crustal moment of inertia can account for the magnitudes of glitches.

18.5 Complete sequences of strange stars with nuclear crusts

18.5.1 Strange stars versus neutron stars and white dwarfs

Figure 18.31 compares the mass–radius relationship of ordinary neutron stars, constructed for the relativistic Hartree–Fock equation of state HFV of table 12.4, and white dwarfs (BPS equation of state) with the one of strange-matter stars based on equation of state (18.50). For the latter, the relationship ranges from the compact strange stars (SS), which constitute the strange counterparts of neutron stars (NS), to the strange dwarfs (sd), which may be viewed as the strange counterparts of ordinary white dwarfs

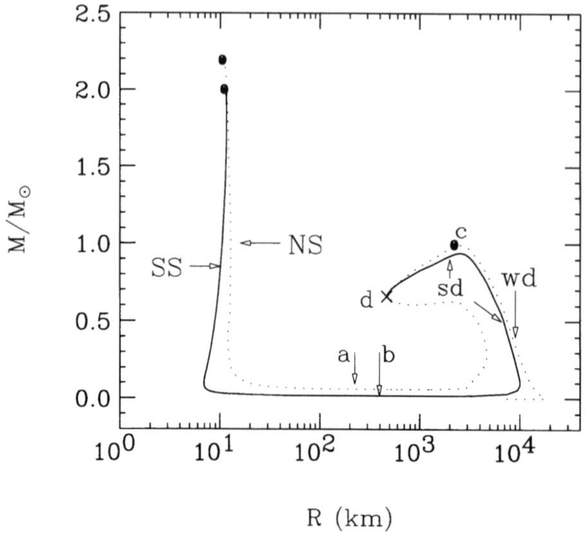

Figure 18.31. Mass versus radius of strange-star configurations with nuclear crust ($\epsilon_{\rm crust} = \epsilon_{\rm drip}$) and gravitationally bound stars [188]. The branches of neutron stars, strange stars, white dwarfs and strange dwarfs are denoted by NS, SS, wd, and sd, respectively. The solid dots in the upper left corner mark the heaviest NS and SS, while c marks the most massive white-dwarf star. The strange-dwarf sequence terminates at d. The arrows labeled a and b show the location of the lightest NS and SS, respectively. (Reprinted courtesy of *Astrophys. J.*)

(wd). From the maximum-mass star (solid dots), the central density decreases monotonically through the sequence in each case. The stable neutron stars and white dwarfs have radii of $\lesssim 200$ km and $\gtrsim 3000$ km, respectively. All other stars of the neutron star–white dwarf sequence are unstable against radial oscillations. The two sequences of figure 18.31 have certain resemblances, including the radius and mass of the limiting compact star of each sequence, which vary between $\sim$ 1.5 and 2.2 $M_\odot$ for non-rotating stars, and $\sim$ 1.8 to 2.5 $M_\odot$ for rotation at the Kepler frequency, depending on the underlying model for the equation of state. Table 18.3 shows the limiting masses of strange stars computed for deeply bound strange quark matter. Larger values for the bag constant correspond to less deeply bound strange matter and to smaller limiting masses.

Despite these similarities, however, there are several striking differences between both classes of stars that will be noticed next. For instance, the white dwarfs terminate at low mass in planetary objects while the strange stars terminate at a white dwarf (located at d for the example

Table 18.3. Properties of maximum-mass star configurations [188]. The properties listed are: star's radius, R; radius of strange core, R_{core}; total mass, M; crust mass, M_{crust}; central density, ϵ_c (in units of the density of normal nuclear matter, $\epsilon_0 = 2.5\times10^{14}$ g cm^{-3}); total number of baryons in the star, A; number of baryons in star's crust (core), A_{crust} (A_{core}); gravitational redshift, z.

Quantity	Neutron star	Strange stars ($B^{1/4} = 145$ MeV)	
	HFV †	$\epsilon_{\text{crust}} = 4\times10^{11}$ g cm^{-3}	$\epsilon_{\text{crust}} = 10^9$ g cm^{-3}
R (km)	10.70	11.143	11.052
R_{core} (km)	–	10.969	11.039
$M/M_\odot$	2.198	2.005	2.046
$M_{\text{crust}}/M_\odot$	–	1.3×10^{-5}	3.4×10^{-4}
ϵ_c/ϵ_0	9.64	7.81	7.86
$\log_{10} A$	57.5253	57.5178	57.5202
$\log_{10} A_{\text{crust}}$	–	53.7724	53.8369
$\log_{10} A_{\text{core}}$	–	57.5177	57.5201
z	0.6143	0.4611	0.4857

† Details of the HFV EOS are given in table 12.4.

shown in figure 18.31), when the strange core has shrunk to zero radius. Secondly, in the dwarf region, the direction of decreasing central density is in different senses on the strange dwarf and white dwarf sequences. We shall come back to this later in section 18.6 when performing a stability analysis of strange stars against radial oscillations. Here we have selected, for illustration, an inner nuclear crust density for the strange-star sequence of $\epsilon_{\text{crust}} = \epsilon_{\text{drip}}$ [174, 188]. The behavior of a sequence with an arbitrarily chosen, smaller value of $\epsilon_{\text{crust}} = 10^8$ g cm^{-3}, is shown in figure 18.32. It may serve to demonstrate the influence of less dense crusts on the mass–radius relationship of strange stars with crust. An enlargement of the left portion of this illustration is shown in figure 18.33. Obviously, there is some residue of the $R_c \propto M^{1/3}$c law that holds for low-mass bare strange stars, especially for low nuclear crust densities. However, since the nuclear crust on a strange star is bound by gravity and not by confinement, as for the strange core or a bare strange star, the mass–radius relationship ultimately resembles the one of gravitationally bound stars, that is, the radius becomes large as the mass approaches a small value, of about $10^{-1}\,M_\odot$ or less, as can be seen from figure 18.33. The minimum-mass configuration of the strange-star sequence with maximal inner crust density, $\epsilon_{\text{crust}} = 4\times10^{11}$ g cm^{-3}, in figure 18.32 has a mass of about $\sim 0.017\,M_\odot$, which corresponds to about 17 Jupiter masses. The stellar mass drops by two additional orders of magnitude if ϵ_{crust} is reduced to 10^8 g cm^{-3}. More than that, for inner crust densities smaller than 10^8 g cm^{-3} we find strange-matter stars that

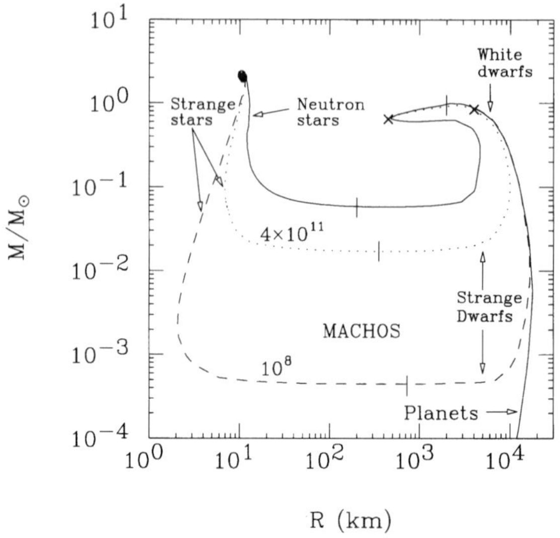

Figure 18.32. Neutron star–white dwarf sequence (solid line) [195]. The dotted and dashed curves refer to compact strange star–strange dwarf sequences with inner crust densities as indicated (in g cm^{-3}). Vertical bars mark minimum-mass stars, crosses mark the termination points of the strange-star sequences. (Reprinted courtesy of American Institute of Physics.)

can be even by orders of magnitude lighter than Jupiter. The properties of selected light stars are put together in table 18.4. So, if the strange quark-matter hypothesis is correct, there could exist stellar strange matter configurations with masses similar to those of brown dwarfs or ordinary planets but dramatically different radii, which could be even as small as those of neutron stars. Astrophysically, all such low-mass objects may be of considerable importance for the baryonic dark matter problem, since they may be difficult to detect and therefore may effectively hide baryonic matter. A gravitational lensing method of detection like that proposed for the massive astrophysical halo objects (MACHOs) may be feasible [640]. The detection, however, appears to be hampered by the fact that the strange low-mass objects may be indistinguishable from their non-strange counterparts by gravitational microlensing detection. Moreover, these light strange stars could only be seen by microlensing if they are abundant enough in our Galaxy.

In this connection we recall that microlensing [17, 641, 642, 643, 644] searches can give definitive results on dark objects in the mass range from 10^{-5} to $10^2\,M_\odot$ [641, 642], the lower bound of which overlaps with the mass range of the light planetary-like strange-matter objects. This would

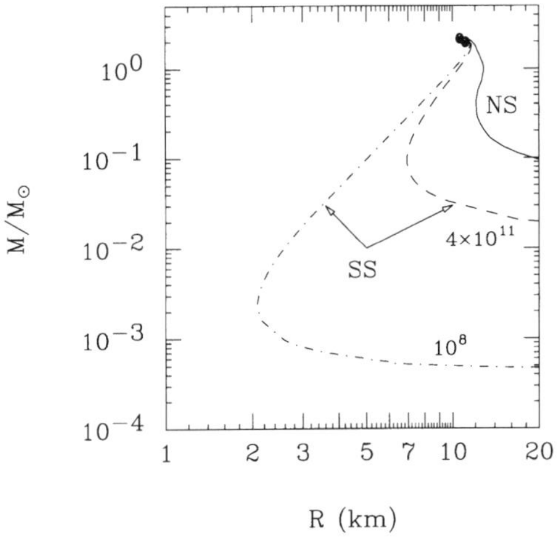

Figure 18.33. Mass versus radius of strange stars (SS) and neutron stars (NS) for inner crust densities $\epsilon_{\rm crust} = 4\times10^{11}$ g cm^{-3} and 10^8 g cm^{-3} [188]. (Reprinted courtesy of *Astrophys. J.*)

Table 18.4. Properties of minimum-mass star configurations [188]. The properties listed are explained in table 18.3.

Quantity	Neutron star	Strange stars ($B^{1/4} = 145$ MeV)	
	BPS †	$\epsilon_{\rm crust} = 4\times10^{11}$ g cm^{-3}	$\epsilon_{\rm crust} = 10^8$ g cm^{-3}
R (km)	224	395	820
$R_{\rm core}$ (km)	–	2.69	0.77
$M/M_\odot$	0.092	0.017	4×10^{-4}
$M_{\rm crust}/M_\odot$	–	1.8×10^{-4}	2.2×10^{-6}
ϵ_c/ϵ_0	0.575	1.667	1.646
$\log_{10} A$	56.0395	55.3637	53.7298
$\log_{10} A_{\rm crust}$	–	53.3149	51.4246
$\log_{10} A_{\rm core}$	–	55.3598	53.7276
z	6.220×10^{-4}	6.729×10^{-5}	

† Details of the BPS EOS are given in table 4.1.

be even more the case if the currently performed searches should prove to be sensitive to the entire theoretically possible range of baryonic dark matter, from 10^{-9} to $10^6\ M_\odot$ [645]. Therefore, if the dark baryonic matter consists of a certain fraction of strange objects, these experiments may have a good chance to detect them. The observation of the effects of an invisible object at least as massive as the planet Jupiter has been reported independently by the MACHO and French collaboration [646] and [647], respectively. 27 microlensing events in the Galactic Bulge have been reported in [648].

Strange stars located to the right of the minimum-mass configuration of each sequence consist of small strange cores ($\lesssim 3$ km) surrounded by a thick nuclear crust made up of white dwarf material. We thus call such objects strange dwarfs. Their cores have shrunk to zero at the crossed points. What is left is an ordinary white dwarf with a central density equal to the inner crust density of the former strange dwarf [174, 188]. A detailed stability analysis of strange stars against radial oscillations, to be performed in section 18.6, shows that all those strange-dwarf sequences that terminate at stable ordinary white dwarfs are stable against radial oscillations. Strange stars that are located to the left of the mass peak of ordinary white dwarfs, however, are unstable against oscillations and thus cannot exist stably in nature. So, in sharp contrast to neutron stars and white dwarfs, the branches of strange stars and strange dwarfs are *stably connected* with each other [174, 188]. Finally the strange dwarfs with 10^9 g cm$^{-3} < \epsilon_{\rm crust} < 4\times10^{11}$ g cm^{-3} form entire new classes of stars that contain nuclear material up to $\sim 4\times10^4$ times denser than in ordinary white dwarfs of average mass, $M \sim 0.6\,M_\odot$ (central density $\sim 10^7$ g cm^{-3}). The entire family of such strange stars owes its stability to the strange core. Without the core they would be placed into the unstable region between ordinary white dwarfs and neutron stars [174, 188, 193, 194].

18.5.2 Limiting rotational periods of strange stars

From the mass radius relationships exhibited in figures 18.31 to 18.33 it is clear that strange stars with crusts have *higher* mass-shedding frequencies, because of their smaller radii, than those of the gravitationally bound neutron stars [20, 149, 190]. This is already indicated by the classical expression (15.26) for $\Omega_{\rm K}$ and confirmed by the general relativistic expression (16.7). Furthermore, not only the strange star at the termination point can sustain very rapid rotation, but *all* stars down to rather small masses. In fact, as can be seen from figure 18.34, we find that even a certain part of the sequence of the low-mass strange stars can have rotational frequencies that rival even the maximal possible frequency of heavy neutron stars! The neutron star sequence is constructed for the same particular choice of equation of state as in section 18.5.1. It appears, however, that this trend seems to hold for any realistic nuclear equation of state and, thus, is not a specific feature of the model used here. Of course bare strange stars, possessing no nuclear crusts, can rotate faster than those with gravitationally bound crusts. This is particularly evident for the lighter strange stars of the sequences shown in figure 18.34. Strange-matter stars with typical pulsar masses of $\sim 1.45\,M_\odot$ can rotate at Kepler periods in the

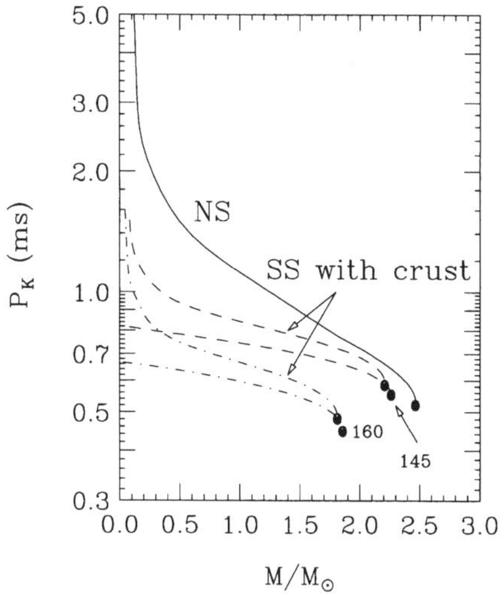

Figure 18.34. General relativistic Kepler period, $P_K \equiv 2\pi/\Omega_K$, versus rotational star mass for neutron stars (NS) and strange stars (SS) with ($\epsilon_{\rm crust} = \epsilon_{\rm drip}$) and without nuclear crust [188]. The numbers denote the value of the bag constant (in MeV). (Reprinted courtesy of *Astrophys. J.*)

range of 0.55 ms $\lesssim P_K \lesssim$ 0.8 ms, depending on the value of the bag constant and the existence of a crust on their surfaces. The situation is different for neutron stars of the same mass, for which a limiting rotational Kepler period of about 1 ms has been established [110, 112, 114, 187, 649]. Parenthetically, we recall that the gravity-wave instability limits the stable rotational periods of hot neutron stars more severely (to several milliseconds, as shown in figure 17.38) than mass shedding does (cf. figure 17.40). Since this instability does not pertain to hot strange stars, as demonstrated by Madsen [622], strange stars appear not to be prevented from rotating at periods $P \simeq P_K$.

18.5.3 Strange dwarfs versus white dwarfs

As we know from figure 18.31, the strange-matter stars with central densities smaller than the lightest strange star of the sequence (located at b possess masses and radii similar to those of ordinary white dwarfs. It is also known that the strange-matter cores are monotonically shrinking in the direction from b to d, since this is the direction in which the central star density decreases. The change in central density from b to d, strikingly, is less than 2%. The cross at d denotes that particular strange star whose core has shrunk to zero. What is left over at d is a star made up of the

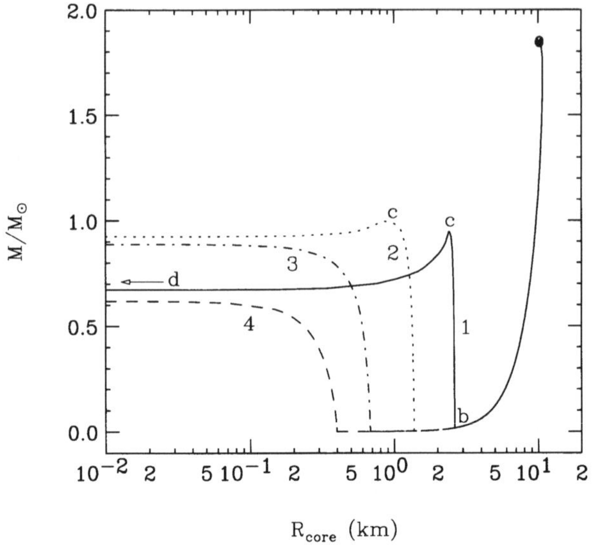

Figure 18.35. Mass versus strange-core radius for selected strange-star sequences with inner crust densities of: (1) 4×10^{11} g cm^{-3}, (2) 5×10^{9}, (3) 10^{8}, and (4) 10^{7} [188]. The mass–radius diagram of sequences 1 and 3 are shown in figures 18.31 and 18.32. The label c denotes maximum-mass strange dwarfs (cf. figure 18.31), which do not exist for sequences with $\epsilon_{\rm crust} \lesssim 10^{9}$ g cm^{-3} (3 and 4). Stars to the left of c are unstable against radial oscillations. The minimum-mass star models possess core radii 0.4 km $\lesssim R_{\rm core} \lesssim$ 3 km. (Reprinted courtesy of *Astrophys. J.*)

crust matter of the former strange dwarf, that is, a white dwarf. Its central density is equal to the inner crust density of the former strange dwarf, which, in the present case, is the neutron drip density, 4×10^{11} g cm^{-3}. Hence $\epsilon_c = 4 \times 10^{11}$ g cm^{-3} for the white dwarf at d in figure 18.31. A comparison with figure 18.35 (solid curve there) reveals that the interiors of the strange-matter stars between b and d consist of very small strange cores relative to the equatorial radii. The nuclear envelopes of these dwarfs are up to a few thousand kilometers thick. As the strange-matter core shrinks along any sequence corresponding to a chosen inner crust density, the final configuration will be a white dwarf. Thus each of the strange-matter sequences shown in figures 18.31 and 18.32 terminates on the white dwarf sequence. The crosses mark the termination points.

In contrast to the global properties of hypothetical strange dwarfs which do not differ significantly from their non-strange counterparts, their interior structures are drastically different from each other. A first

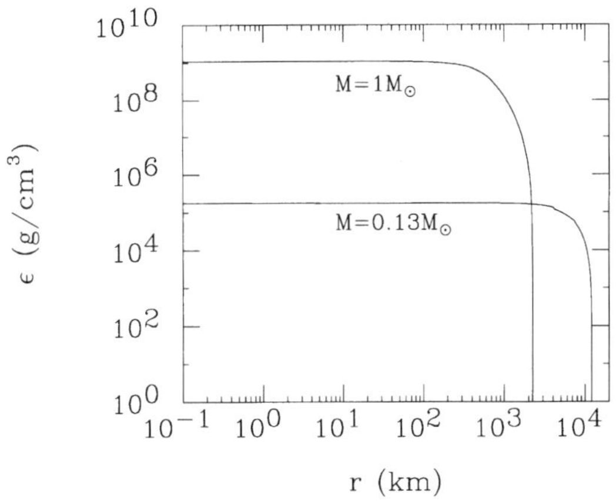

Figure 18.36. Mass density versus radius of a heavy and a light stable white dwarf [188]. (Reprinted courtesy of *Astrophys. J.*)

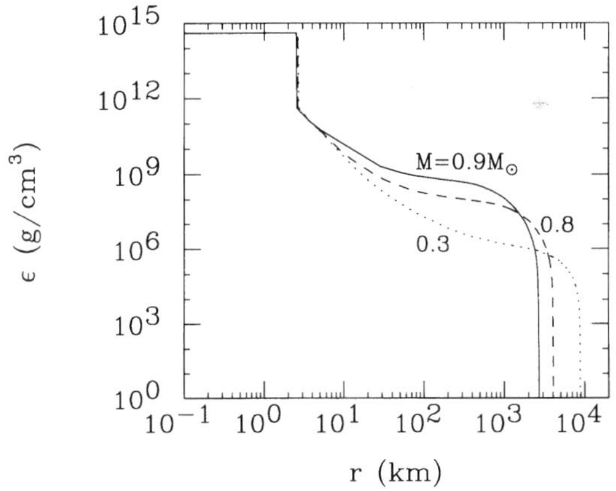

Figure 18.37. Mass density versus radius of a few selected stable strange dwarf models ($\epsilon_{\rm crust} = \epsilon_{\rm drip}$) [188]. (Reprinted courtesy of *Astrophys. J.*)

impression of this is given in figures 18.36 and 18.37. In the latter figure, the inner crust density of nuclear matter that overlays the strange core is equal to the neutron drip density. Due to the fact that the central density in stable dwarfs with strange matter cores changes only very little with mass (see, for example, the stars in figure 18.35 along sequence 1 between points b and c), the radii of their strange cores are nearly independent of star mass, as can be seen from figure 18.37. Such cores, because of their high mass density, still

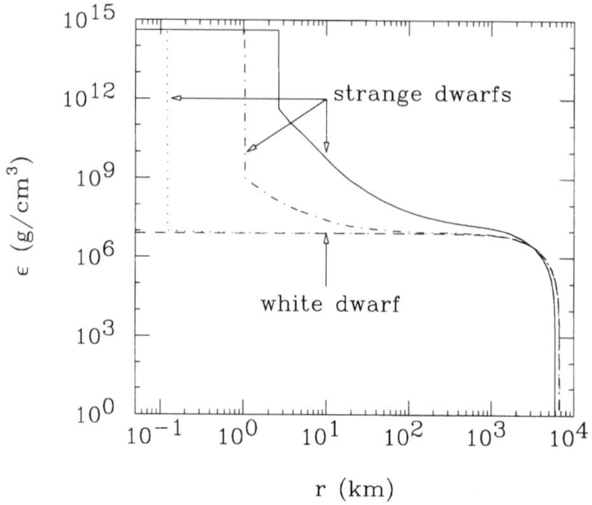

Figure 18.38. Mass density profiles of a white dwarf and several strange dwarfs of the same mass, $M = 0.6\,M_\odot$ [188]. The inner crust densities of the strange dwarfs are $\epsilon_{\rm crust} = 4\times 10^{11}$ g cm^{-3} (solid curve), 10^9 g cm^{-3} (dot-dashed), and 10^7 g cm^{-3} (dotted). (Reprinted courtesy of *Astrophys. J.*)

exert a non-negligible influence on the hydrostatic equilibrium of strange dwarfs, as follows from the profiles shown in figure 18.37. Indeed, as will be pointed out below (section 18.5.4), strange cores with $R_{\rm core} \sim 3$ km are sufficient to stabilize a new type of white-dwarf-like star that is much denser than ordinary white dwarfs [174].

The mass energy profiles of strange dwarfs constructed for several different inner crust densities *smaller* than the limiting drip density are compared to a white dwarf profile in figure 18.38. All profiles correspond to stars with $M = 0.6\,M_\odot$, the average mass value of observed white dwarfs [650]. The radii of the strange cores in these models decrease from about 3 km, obtained for $\epsilon_{\rm crust} = \epsilon_{\rm drip}$, to about 10^{-1} km if $\epsilon_{\rm crust} = 10^7$ g cm^{-3} (figure 18.35). Due to the core's much weaker gravitational impact in the latter case, the density profile of this strange dwarf resembles that of an ordinary white dwarf of the same mass.

18.5.4 Possible new class of dense white dwarfs

Glendenning *et al* [148, 174, 188] have pointed out a generic distinction among what we have heretofore referred to as strange dwarfs [188]. It has to do with the fact that there are no stable white dwarfs with central densities higher than $\sim 10^9$ g cm^{-3} [70] (i.e. the stellar model c in figure 18.31). If

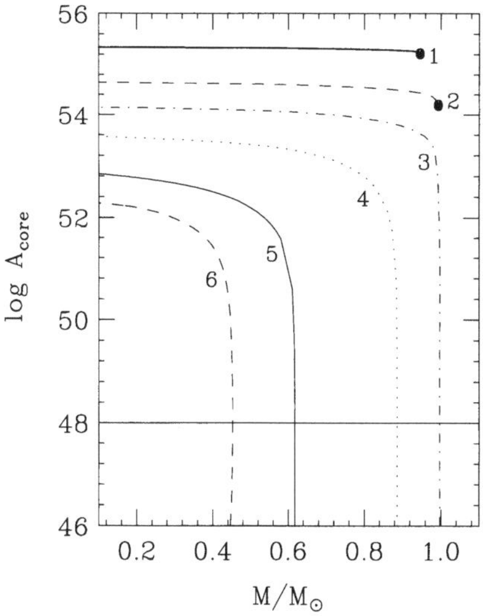

Figure 18.39. Baryon number of strange cores, A_{core}, in stable strange dwarfs versus star mass [188]. The inner crust densities (in units of g cm^{-3}) are: (1) neutron drip, 4×10^{11}, (2) 10^{10}, (3) 10^9, (4) 10^8, (5) 10^7, and (6) 10^6. The hydrostatically stable parts of sequences 1 and 2 terminate at the solid dots, which denote maximum-mass strange dwarfs labeled c in figure 18.35. (Reprinted courtesy of *Astrophys. J.*)

the inner density of the white dwarf material enveloping strange matter lies below $\sim 10^9$ g cm^{-3}, the star that would remain, were the strange core shrunk to zero, would be a stable ordinary white dwarf. Such strange stars may be referred to as 'trivial' strange dwarfs [174, 188]. The presence of the strange core does not disturb the configuration of the star appreciably, as can be verified by comparing the two strange dwarf configurations that lie close to the white dwarf in figure 18.38. Since there are no stable white dwarfs with central densities higher than about 10^9 g cm^{-3}, the strange dwarfs with inner crust densities of the nuclear material that lie in the range of 10^9 g cm^{-3} $< \epsilon_{\mathrm{crust}} \leq 4 \times 10^{11}$ g cm^{-3} are entirely new objects that owe their stability solely to the presence of the strange core and would be unstable without it [174, 188], for they would fall into the unstable region between the maximum-mass white dwarf and minimum-mass compact star (at c and a respectively in figure 18.31). Most strikingly, such new dwarfs have nuclear material that is 400 times more dense than the central density of the heaviest white dwarf, and 4×10^4 that of the typical $0.6\,M_{\odot}$ white dwarf. As such, provided that they are stable to radial oscillations, they would constitute a new class of white dwarfs, under the strange-matter hypothesis. To make a definitive statement as to whether such strange dwarfs can indeed exist in nature, we need to go beyond the construction of hydrostatic equilibrium configurations and investigate their stability against radial oscillations, which will be performed in section 18.6.

The baryon numbers of the strange cores in stable strange dwarfs are shown in figure 18.39. We recall that the trivial dwarf sequences ($\epsilon_{\rm crust} \le 10^9$ g cm^{-3}, curves 3 to 6) are stable down to the termination points, where the strange cores have shrunk to zero. Therefore the core baryon numbers in the trivial dwarfs can range from at most $\sim 10^{54}$, determined by hydrostatic equilibrium, down to zero. This is different for the dense strange-dwarf sequences ($\epsilon_{\rm crust} > 10^9$ g cm^{-3}, curves 1 and 2) because these terminate at the unstable dwarf region, beyond their mass peaks where all dense sequences become unstable against radial oscillations (section 18.6). Hence the first stable strange dwarfs (in the direction of growing strange cores) of the dense category, located right at the mass peaks, already possess strange cores of certain finite sizes depending on $\epsilon_{\rm crust}$ and star mass. Two examples are shown in figure 18.39. The solid dots refer to the cores' baryon number of stars at the mass peak of each respective sequence. For all dwarfs, $A_{\rm core}$ increases with decreasing star mass since this is the direction in which the central star density ϵ_c increases.

18.6 Stability of strange stars against radial oscillations

18.6.1 Eigenfrequencies of strange stars and strange dwarfs

Having explored the complete sequences of strange stars that are predicted to exist by virtue of the principle of hydrostatic equilibrium, we cannot make definitive statements yet about the real possible existence of such objects, since hydrostatic equilibrium alone is not sufficient to guarantee a star's stability. The still missing ingredient is a stability analysis against radial oscillations (i.e. compressional modes), as described in section 14.3, which we shall perform next.

The analysis is carried out by solving the Sturm–Liouville eigenequation (14.97) for the equation of state of strange stars with fixed inner crust density [630], which is given in (18.50). The result for the four lowest-lying eigenfrequencies of the complete sequence of strange-matter stars, i.e. from the compact strange stars to the strange dwarfs shown in figures 18.31 and 18.32, are exhibited in figure 18.40. An enlargement of the low-density portion of this figure is given in figure 18.41. One sees that the higher-lying modes form a discrete sequence, that is, $\omega_1^2 < \omega_2^2 < \omega_3^2$, as enforced by the mathematical structure of the eigenequation (14.97). Specifically, the eigenfrequencies of strange stars in the vicinity of the minimum-mass configuration at b of figure 18.41, which exhibit a dip-like behavior in that density region, fulfill this condition. This can barely be seen from fig-

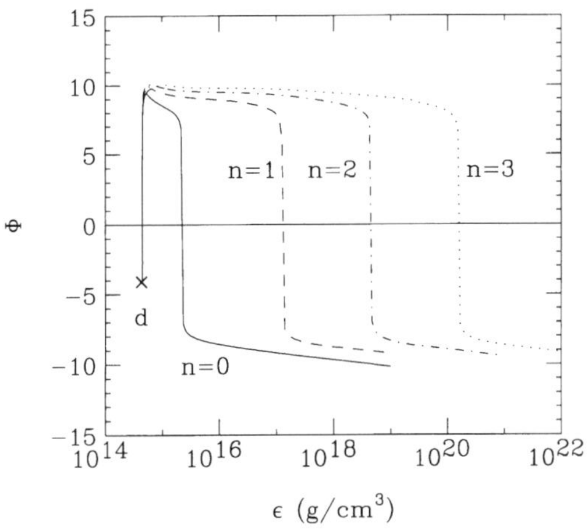

Figure 18.40. Oscillation frequencies of the lowest four ($n = 0, 1, 2, 3$) normal radial modes of strange-matter stars with $\epsilon_{\text{crust}} = \epsilon_{\text{drip}}$, measured by $\Phi(x) \equiv \text{sign}(x)\log_{10}(1 + |x|)$ with $x \equiv (\omega_n/\text{s}^{-1})^2$, as a function of central star density [188]. The cross at d refers to the termination point of the strange dwarf sequence shown in figure 18.31. For $\Phi < 0$ the eigenmodes are unstable. (Reprinted courtesy of *Astrophys. J.*)

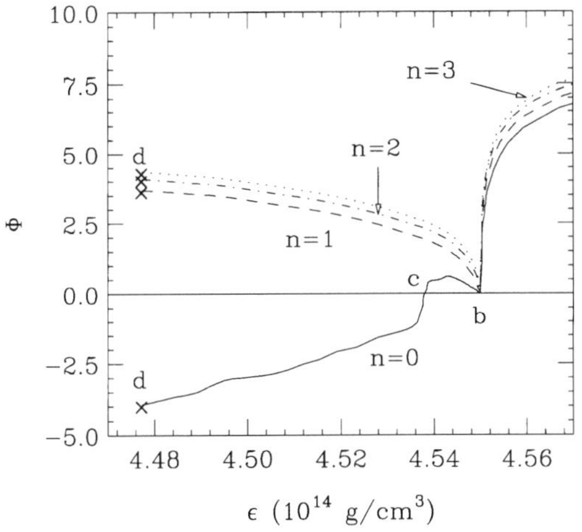

Figure 18.41. Same as figure 18.40, but for strange dwarfs in the vicinity of the termination point, d, of the sequence [188]. The labels b and c refer to the lightest and heaviest star, respectively, marked in figure 18.31. (Reprinted courtesy of *Astrophys. J.*)

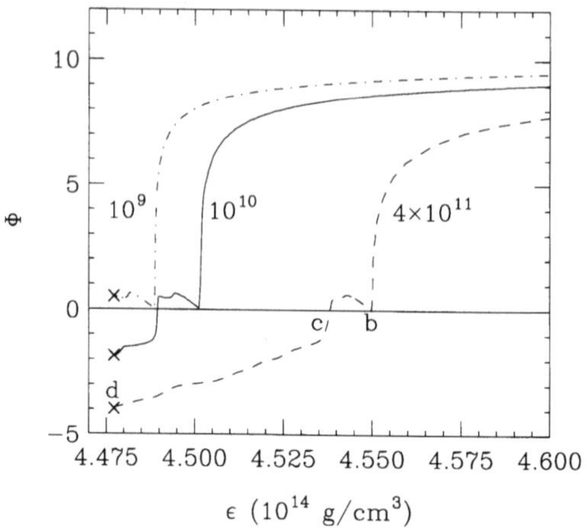

Figure 18.42. Eigenfrequencies of strange-dwarf sequences with $\epsilon_{\rm crust} < \epsilon_{\rm drip}$ [188]. The labels b, c and d have the same meaning as in figures 18.31 and 18.41. The numbers refer to the value of $\epsilon_{\rm crust}$. Strange dwarfs along sequences with $\epsilon_{\rm crust} \lesssim 10^9$ g cm^{-3} obey $\Phi > 0$ (cf. figure 18.40) and, thus, are all stable against radial oscillations. (Reprinted courtesy of *Astrophys. J.*)

ure 18.41 but has been confirmed numerically. These figures also show that strange stars possess a characteristic mode of vibration of zero frequency ($\omega_n^2 = 0$) when and only when $\mathrm{d}M/\mathrm{d}\epsilon_c = 0$, that is, only when the star's mass attains an extremum. Such a behavior is consistent with what has been established for neutron stars and ordinary white dwarfs [69]. Most interestingly, the lowest-lying eigenmode ($n = 0$) passes through zero at c of figure 18.41 and *remains* negative for all densities down to the central density of the strange dwarf at the termination point d. Therefore all members along this particular section are *unstable* against radial oscillations. Those between b and c are stable against radial oscillations because ω_0^2 (and thus all higher-lying modes of oscillation) is positive. So we conclude that the strange-dwarf sequence with $\epsilon_{\rm crust} = \epsilon_{\rm drip}$ has a segment whose stars are stable against radial oscillation modes, which is of decisive importance for their stable existence. These stars range in mass from exceedingly light ones, having masses typical of ordinary planets, to about one solar mass.

The eigenfrequencies of strange-dwarf sequences show a behavior qualitatively similar to those computed for the $\epsilon_{\rm crust} = \epsilon_{\rm drip}$ sequence as long as the inner crust density lies in the range of 10^9 g cm$^{-3} < \epsilon_{\rm crust} \le \epsilon_{\rm drip}$, which is illustrated in figure 18.42. This is different for sequences with $\epsilon_{\rm crust} \lesssim 10^9$ g cm^{-3} because these do not possess a maximum-mass dwarf star where ω_0^2 turns negative. Therefore the *complete* sequences of strange-matter stars with $\epsilon_{\rm crust} < 10^9$ g cm^{-3} – from the compact strange stars to the white dwarf at the termination point – are stable. However, since they are essentially white dwarfs which are only mildly perturbed by the strange core, they appear less interesting than the stable strange dwarfs with $\epsilon_{\rm crust} > 10^9$ g cm^{-3}. Nevertheless there is one aspect in which they important. Their stability demonstrates that the strange matter hypothesis is not inconsistent with observation of white dwarfs. For if the hypothesis is correct, then almost certainly all stars contain some strange matter.

For the rest of this section, we turn our interest to the eigenfrequencies and locations of the zero-frequency oscillation, ω_0^2, of the denser strange stars with crust, possessing central densities up to those at which even charm-quark states become populated in their dense interiors (charm stars). The second zero point of the oscillation frequencies ω_n^2 in figure 18.40 in the direction of increasing density is located at that density at which the strange stars attain their maximum mass (solid dots in figure 18.31). Since ω_0^2 remains negative at all densities greater than this one, it follows that no quark matter stars can exist in nature that are more compact than the hypothetical strange stars [30, 26, 194]. This rules out the possible existence of *charm stars*. In fact, as one sees from figure 18.40, going to higher and higher central star densities leads to the successive excitation of more and more unstable modes ($\omega_n^2 < 0$, $n = 2, 3, \ldots$). This situation is analogous to that of hydrostatic equilibrium configurations in the neutron star sequence with central densities above that of the maximum-mass neutron star [69, 536].

18.6.2 Schematic arguments for the stability of strange dwarfs against radial modes

To establish a connection with the more familiar stability arguments known for gravitationally bound stars [69], it is illustrative to recall the following schematic arguments about the stability of strange dwarfs against radial modes, as pointed out in [188]. Let us consider the strange dwarfs between points b and c in figures 18.31, 18.41, and 18.42 which obey $\mathrm{d}M/\mathrm{d}\epsilon_c < 0$, since the mass of these stars increases with decreasing central density. In the case of neutron stars and ordinary white dwarfs, $\mathrm{d}M/\mathrm{d}\epsilon_c < 0$ implies instability against at least the lowest-lying eigenmode of oscillation, $\omega_0^2 > 0$. This however is not the case for these strange dwarfs since their eigenfrequencies, as discussed above, are all positive. So these stars are stable against vibrational modes. The physical reason for this is that the acoustic modes in them propagate mostly in the compressible white dwarf material which comprises most of these stars and not in the much higher-density incompressible quark core of at most a few kilometers in radius for which the local adiabatic index, $\Gamma(\epsilon) \equiv [(\epsilon + P)/P]\mathrm{d}P/\mathrm{d}\epsilon$, is very large, as can be understood in reference to figure 18.19.

Due to the strict ordering of the eigenfrequencies, a star is stable against radial oscillations if it is stable against the $n = 0$ eigenmode. This mode corresponds to an oscillation with no nodes inside the star (pure compressional mode). Thus any star oscillating in this mode shrinks and expands as a whole. A strange dwarf with its highly incompressible strange core in its center can only oscillate in such a mode by means of varying its inner crust density, rather than the central density as is the case for neutron stars and ordinary white dwarfs. This leads one to reason that $\mathrm{d}M/\mathrm{d}\epsilon_{\mathrm{crust}} > 0$, rather than $\mathrm{d}M/\mathrm{d}\epsilon_c > 0$, constitutes the relevant condition necessary for stability against compressional vibrations. That this is indeed the case can be verified in close analogy to gravitationally bound stars [69]. As an example, any oscillating strange dwarf of the segment b–c with a given fixed strange core in its center contracts ($\epsilon_{\mathrm{crust}}$ increases) toward hydrostatic neighboring configurations that possess larger strange cores by remembering that the strange cores are monotonically growing in the clockwise direction and thus a stronger gravitational binding force. So the oscillating strange dwarf possessing too small a strange core expands back to its initial state. Conversely, any strange dwarf of the segment expands ($\epsilon_{\mathrm{crust}}$ decreases) during its oscillatory motion toward neighboring equilibrium configurations with cores that are smaller, and thus gravity pulls the star back. So any oscillating strange dwarf between points b and c is stable against compressional oscillations. The strict ordering of the eigenmodes implies stability against all higher-lying oscillation modes too.

18.7 On the making of strange dwarfs

Aside from the general remarks made in section 18.3.2, we have not said much about how strange dwarfs may have been made in the course of the evolution of the Universe. Following Glendenning, Kettner, and Weber [148, 174, 188], this will be done next. Of course, there may be other possibilities than those described below, and the history of astronomical discovery cautions one that the laws of nature are explored by the Universe in many unanticipated ways.

The crust on a compact strange star may be acquired either by accretion from the interstellar medium onto an initially, possibly primordial, bare strange star [651], as discussed in section 18.3.2, or during its creation in a supernova whose progenitor has already accreted strange-matter nuggets. The situation is different for the strange dwarfs. Such objects could be given birth in the collapse of *main-sequence stars* that were being bathed over their lifetime by a flux of strange nuggets. The existence of this flux is probably an inevitable consequence if the strange-matter hypothesis is correct. The strange-nugget flux would contaminate every object it came into contact with, i.e. planets, neutron stars, white dwarfs, and main-sequence stars. Naturally, due to the large radii of main-sequence stars, they are ideal large-surface long-integration-time detectors for the strange-matter flux [142]. In contrast to neutron stars and ordinary white dwarfs, whose material is characterized by a large structural constant so that the strange nuggets never reach their cores,[5] the nuggets accreted onto main-sequence stars can indeed gravitate to their centers, accumulate there and form a strange-matter core that grows with time until the star's demise as a main-sequence star. Those whose masses lies in the approximate mass range of $1 - 5\,M_\odot$ would then give birth to a strange dwarf [174, 195, 198]. As well, it has been discussed how *primordial strange-matter bodies* of masses between 10^{-2} and $1\,M_\odot$ can be formed in the early universe and survive to the present epoch [651]. Such objects will occasionally be captured by a main-sequence star and form a significant core in a single and singular event. The core's baryon number, however, cannot be significantly larger than $\sim 5 \times 10^{31}\ (M/M_\odot)^{-1.8}$ where M is the star's mass. Otherwise a main-sequence star is not capable of capturing the strange-matter core [142]. Finally we mention that in the very early evolution of the universe *lumps of hot strange matter* will evaporate nucleons which are plausibly gravitationally bound to the lump. The evaporation will continue until the quark matter has cooled sufficiently. Depending on the original baryon number of the quark lump, a strange star or dwarf, both with nuclear crusts, will have been formed.

18.8 Features of strange stars surveyed

As demonstrated in this book, if strange quark matter is more stable than nuclear matter, then not only must our comprehension of the composition of microscopically small objects be revised completely, but that of the interior structure of stellar bodies too [174, 188, 193, 194, 195, 652, 653]. This becomes very obvious by comparing figures 18.43 and 18.44 with each other. Most strikingly, the huge nuclear desert in figure 18.43, characteristic if nuclear matter is the true ground state of the strong interaction, disappears completely if strange matter is more stable than nuclear matter, giving way to a variety of different stable strange-matter objects that range from strangelets at the small baryon number end, $A \sim 10^2$, to strange stars at the high end, $A \sim 10^{57}$, where gravity terminates the sequence of the stable compact strange stars. The strange counterparts of ordinary atomic nuclei are the strange nuggets, those of neutron stars (pulsars) are the compact strange stars. Besides the very light planetary-like strange-matter stars, there should exist an expansive range of strange stellar objects whose masses range from that of Jupiter, $\sim 10^{-3}\,M_\odot$, to brown dwarfs, $\sim 10^{-1}\,M_\odot$, to those of white dwarfs, $10^{-1}\,M_\odot \lesssim M \lesssim 1.5\,M_\odot$. Such objects may be referred to as strange massive astrophysical compact halo

[5] This excludes the making of strange dwarfs from collapsing ordinary white dwarfs, which have reached the Chandrasekhar mass limit because of mass accretion from a companion.

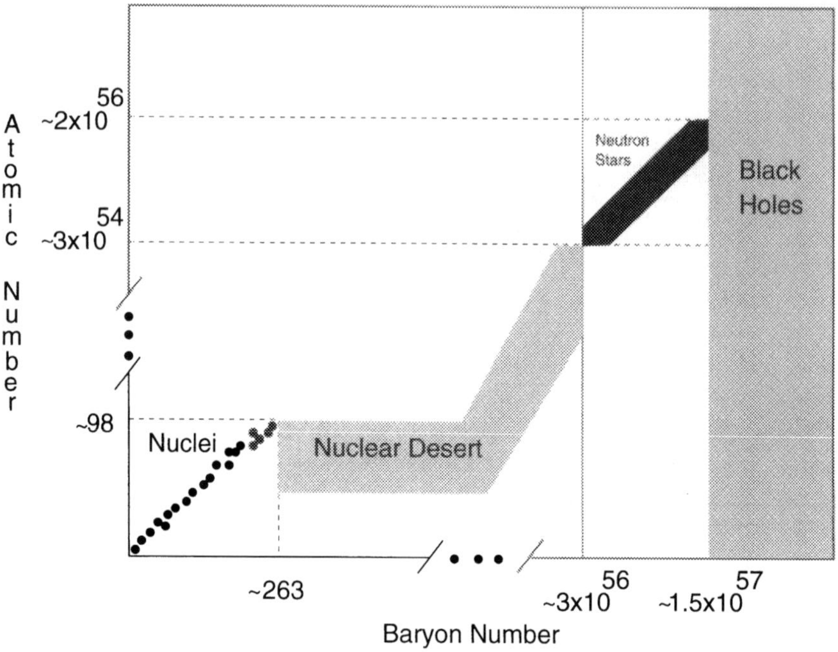

Figure 18.43. Graphical illustration of all possible stable nuclear objects, from nuclei to stellar configurations, if *nuclear matter* (that is, ^{56}Fe) is the most stable form of matter. In this case, the huge baryon number range of $\sim 3 \times 10^2$ to $\sim 3 \times 10^{56}$, referred to as the nuclear desert, is void of nuclear objects.

objects (strange MACHOs), in analogy to their non-strange counterparts known as MACHOs. The important astrophysical implication of the existence of strange MACHOs would be that they occur as natural stellar candidates which effectively hide baryonic matter, linking both light as well as compact strange stars to the fundamental dark baryonic matter problem in our Galaxy. Observationally, strange MACHOs should be seen by the gravitational microlensing experiments, provided such objects are sufficiently abundant. Experiments searching for microscopically small strange particles – which should be present in cosmic rays, terrestrial matter (causing Rutherford backscattering if an incident particle), and as being plausibly produced in relativistic heavy-ion collisions (HICs) – were summarized in section 2.3. Finally, we note the remarkable property of compact strange stars of withstanding stable rapid rotation down to periods considerably les than 1 ms (section 18.5.2), which appears hardly to be the case for neutron stars made up of confined hadronic matter (table 17.9), independent of whether part of their matter exists as deconfined quark

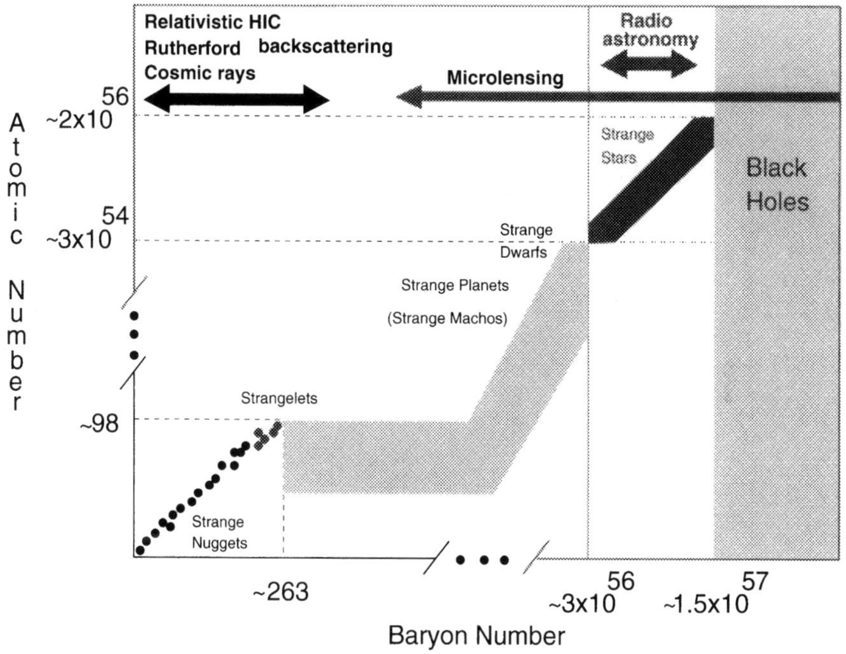

Figure 18.44. Same as figure 18.43 but for *strange matter* as the most stable configuration of matter. In sharp contrast to figure 18.43, now the nuclear desert no longer exists but is filled with a variety of different stable strange-matter objects (see text). Observational accessibilities are indicated at the top of the figure.

matter or is densely packed because of the condensation of mesons.[6] This feature offers a particular challenge for present and future radio astronomical searches for millisecond pulsars, with new equipment more sensitive to pulsars with periods of a few milliseconds or less where the sensitivity of previous pulsar searches dropped off [21, 22, 23]. This and other distinguishing features of strange-matter stars are compiled in table 18.5. Confirming or refuting them plays a key role in figuring out whether or not the confined hadronic phase of nucleons and nuclei is only metastable with respect to 3-flavor quark matter. Indeed the claim has been made by Alpar that the hypothesis does fail because strange stars are not capable of producing pulsar glitches [173]. As shown in

[6] This finding is based on the broad collection of competing models for the equation of state of superdense matter studied in section 12.6 rather than a study that is independent of the model adopted for the equation of state. This prevents us from making final conclusions about the smallest possible rotational period accessible to neutron stars, of course.

Table 18.5. Overview of the features of strange quark matter stars.

Features of strange quark matter stars	observable	unequivocal
• Strange stars:		
rotational periods $P < 1$ ms	yes †	possibly
• Planetary-like objects	yes	unlikely
• Strange white dwarfs	yes	unknown
• Cooling behavior	yes	possibly ‡
• Glitches	yes	unknown
• Post-glitch behavior	yes	unknown

† Until recently rotational periods of $P \sim 1$ ms set the borderline of detectability.
‡ Depending on star's age (see chapter 19).

section 18.4.5, however, this conclusion appears to be presumptuous. We find that the nuclear crust enveloping strange matter can have a moment of inertia sufficiently large that a fractional change can account for both the magnitude of the observed relative frequency changes $\Delta\Omega/\Omega$ of pulsars (glitches), even giant glitches, as well as the observed relative changes of the spin-down rates $\Delta\dot{\Omega}/\dot{\Omega}$ [148]. The situation seems to be different, however, for hot strange-matter stars with temperatures typical for stellar remnants newly formed in supernovae. Such objects cannot be enveloped in a nuclear crust [30], since high temperatures reduce the electric field at the surface of hot strange matter by means of which nuclear crust matter comes in contact with strange matter and becomes converted by hypothesis into the true ground state. In this case Alpar's claim may not be refutable. The last item to be mentioned when recounting the features of stable quark stars is that the density in them never grows high enough for the charm quarks to become populated. Stars containing c quarks, the so-called charm stars, form a new branch of hydrostatically stable stars which, however, turn out to be unstable against radial oscillations. Thus, they cannot exist stably in the Universe [30].

The white-dwarf-like strange stars, which we christened strange dwarfs, can be divided into two generically different categories. The dwarfs of the first category possess strange cores with radii in the range of $R_{\rm core} \sim 1$–3 km. Such cores compress the surrounding nuclear material, which can be several thousand kilometers thick, up to densities that are about 400 times higher than those in the most massive ordinary white dwarfs ($\epsilon_c \sim 10^9$ g cm^{-3}) and $\sim 4{\times}10^4$ times that of the typical $0.6\,M_\odot$ white dwarf ($\epsilon_c \sim 10^7$ g cm^{-3}). The entire sequences of such strange dwarfs owe their stability solely to the strange-matter cores, which possess baryon numbers of $\log_{10} A \lesssim 54 - 55$, compared to $\log_{10} A \sim 56 - 57$ for the baryon number of neutron stars or ordinary white dwarfs in their observed mass range. Their masses fall in the approximate range $10^{-4}\, M_\odot$ to $1.5\, M_\odot$, depending on the equation of state adopted for the stellar matter [174, 195].

The second, trivial, category of strange dwarfs consists of ordinary white dwarfs which envelop small quark cores, $R_{\rm core} \lesssim 0.7$ km ($\log_{10} A_{\rm core} < 55$). Being essentially ordinary white dwarfs, such configurations would be stable with or without the strange cores. This statement proves the compatibility of the strange matter hypothesis with the existence of white dwarfs.

Chapter 19

Cooling of neutron and strange stars

The predominant cooling mechanism of hot (temperatures $T \gtrsim 10^{10}$ K) newly formed neutron stars immediately after formation is neutrino emission, with an initial cooling time scale of seconds [224]. An overview of the relevant neutrino emission processes is compiled in table 19.1. Already a few minutes after birth, the internal neutron star temperature drops to $\sim 10^9$ K [654]. Photon emission overtakes neutrinos only when the internal temperature has fallen to $\sim 10^8$ K, with a corresponding surface temperature roughly two orders of magnitude smaller. Neutrino cooling dominates for at least the first thousand years, and typically for much longer in standard cooling calculations. Being sensitive to the adopted nuclear equation of state, the neutron star mass, the assumed magnetic field strength, the possible existence of superfluidity, meson condensates, quark matter, etc, theoretical cooling calculations (i.e. surface temperature as a function of time) provide most valuable information about the interior hadronic matter and neutron star structure. Below we shall study stellar cooling tracks that are computed for the broad collection of equations of state of table 12.4 which account for all these effects. Knowing the thermal evolution of a neutron star also yields information about such temperature-sensitive properties as transport coefficients, transition to superfluid states, crust solidification, internal pulsar heating mechanisms such as frictional dissipation at the crust–superfluid interfaces, and so on [224].

In order to study the thermal history of neutron stars, the equations of energy balance and energy transport will be derived in the framework of general relativity next [80].

19.1 Equation of thermal equilibrium (energy balance)

The energy radiated away by a star is supported from its rest mass by nuclear burning, from its gravitational and internal energy by quasi-static

Table 19.1. Overview of neutrino emitting processes relevant for neutron star cooling. The neutrino emissivities, ϵ_ν, associated with these processes are listed in table 19.4.

Name	Particle processes		
Modified Urca	$n + n$	$\longrightarrow$	$n + p + e^- + \bar{\nu}_e$
	$n + p + e^-$	$\longrightarrow$	$n + n + \nu_e$
Direct Urca	n	$\longrightarrow$	$p + e^- + \bar{\nu}_e$
	$p + e^-$	$\longrightarrow$	$n + \nu_e$
Quark modified Urca	$d + u + e^-$	$\longrightarrow$	$d + d + \nu_e$
	$u + u + e^-$	$\longrightarrow$	$u + d + \nu_e$
	$d + u + e^-$	$\longrightarrow$	$d + s + \nu_e$
	$u + u + e^-$	$\longrightarrow$	$u + s + \nu_e$
Quark direct Urca	d	$\longrightarrow$	$u + e^- + \bar{\nu}_e$
	$u + e^-$	$\longrightarrow$	$d + \nu_e$
	s	$\longrightarrow$	$u + e^- + \bar{\nu}_e$
	$u + e^-$	$\longrightarrow$	$s + \nu_e$
π^- condensate	$n + \langle \pi^- \rangle$	$\longrightarrow$	$n + e^- + \bar{\nu}_e$
K^- condensate	$n + \langle K^- \rangle$	$\longrightarrow$	$n + e^- + \bar{\nu}_e$
Quark bremsstrahlung	$Q_1 + Q_2$	$\longrightarrow$	$Q_1 + Q_2 + \nu + \bar{\nu}$
Core bremsstrahlung	$n + n$	$\longrightarrow$	$n + n + \nu_e + \bar{\nu}_e$
	$n + p$	$\longrightarrow$	$n + p + \nu_e + \bar{\nu}_e$
	$e^- + p$	$\longrightarrow$	$e^- + p + \nu_e + \bar{\nu}_e$
Crust bremsstrahlung	$e^- + (A, Z)$	$\longrightarrow$	$e^- + (A, Z) + \nu_e + \bar{\nu}_e$

contraction, or from both sources by both processes. The equation of thermal equilibrium, which will be derived next, expresses the energy balance which occurs during this conversion of mass energy from one form to another.

To derive this equation, let us consider a spherical mass shell of the star inside of which there are A baryons and which itself contains δA baryons, as illustrated in figure 19.1. Following Thorne [80], during a coordinate time interval $\mathrm{d}t$ the internal energy of this shell changes by

$$\begin{aligned} \mathrm{d}(\text{internal energy}) = &\ (\text{rest mass–energy converted to internal energy} \\ &\quad \text{by nuclear reactions}) \\ &+ (\text{work done on shell by gravitational forces to} \\ &\quad \text{change its volume during quasi-static contraction}) \\ &- (\text{energy radiated away, conducted away, or} \\ &\quad \text{convected away})\,. \end{aligned} \tag{19.1}$$

Next we define the rate per baryon, q, of thermonuclear energy generation as measured by an observer at rest in the star, at which rest mass-energy

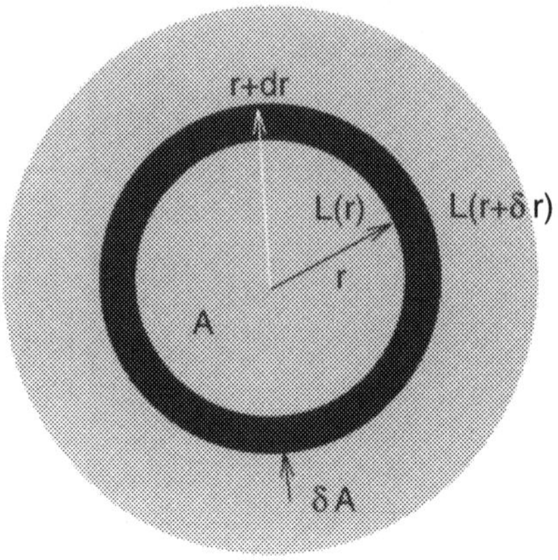

Figure 19.1. Schematic illustration of mass shell (solid black ring) of thickness δr inside a star, introduced to derive the equation of thermal stellar equilibrium. A denotes baryon number, L is the luminosity.

is converted into internal energy by thermonuclear reactions,

$$q = -\frac{\mathrm{d}\bar{m}_B}{\mathrm{d(proper\ time)}} = -\frac{\mathrm{d}\bar{m}_B}{\mathrm{e}^{\Phi}\,\mathrm{d}t}\,. \tag{19.2}$$

By virtue of (19.2) the amount of rest mass converted to internal energy is $q\,\delta A\,\mathrm{e}^{\Phi}\,\mathrm{d}t$. During quasi-static contraction the shell under consideration changes its volume by an amount

$$\mathrm{d}V = \mathrm{d}\Big(\frac{V}{\delta A}\,\delta A\Big) = \mathrm{d}\Big(\frac{1}{\rho}\,\delta A\Big)\,, \tag{19.3}$$

and the work done on the shell to produce this change is given by

$$-P\,\mathrm{d}V = -P\,\mathrm{d}\Big(\frac{1}{\rho}\,\delta A\Big)\,. \tag{19.4}$$

The rate at which the shell radiates, conducts, and convects away energy, is expressed in terms of the radial luminosity, L_r. It comprises the total mass energy carried by photons, by neutrinos, by conduction, and by convection outward across a sphere of coordinate radius r in unit time, as measured in the proper frame of an observer located at r and at rest with respect to the star (i.e. a clock fixed in the shell), and is mathematically given by

$$L_r(r+\delta r)\mathrm{e}^{2\Phi(r+\delta r)-2\Phi(r)} - L_r(r) = \Big(\frac{\mathrm{d}L_r}{\mathrm{d}r} + 2L_r\frac{\mathrm{d}\Phi}{\mathrm{d}r}\Big)\,\delta r\,. \tag{19.5}$$

The right-hand side of this equation follows by expanding the exponential function as

$$e^{2\Phi(r+\delta r)-2\Phi(r)} \approx 1 + 2\left[\Phi(r+\delta r) - \Phi(r)\right] = 1 + 2\,\frac{d\Phi}{dr}\,\delta r\,, \qquad (19.6)$$

and keeping only terms of order $\mathcal{O}(\delta r)$. One of the factors of $e^{2\Phi(r+\delta r)-2\Phi(r)}$ in (19.5) accounts for the gravitational redshift which the transported energy undergoes as it crosses the shell, while the other accounts for time dilation between the inner and outer surface of the shell. The total energy carried away from the shell in coordinate time dt then follows from (19.5) in the form

$$\Big(L_r(r+\delta r)e^{2\Phi(r+\delta r)-2\Phi(r)} - L_r(r)\Big)\,dt\,e^{\Phi} = \Big(\frac{dL_r}{dr} + 2\,L_r\,\frac{d\Phi}{dr}\Big)\,\delta r\,e^{\Phi}\,dt\,. \qquad (19.7)$$

With the aid of these relations, the equation of energy balance (19.1) can be written as

$$d\Big(\frac{\mathcal{E}^{\rm int}}{\rho}\delta A\Big) = q\,\delta A\,e^{\Phi}\,dt - P d\Big(\frac{1}{\rho}\delta A\Big) - \Big(\frac{dL_r}{dr} + 2\,L_r\,\frac{d\Phi}{dr}\Big)\,\delta r\,e^{\Phi}\,dt\,. \qquad (19.8)$$

To simplify this relation, note that

$$\Big(\frac{dL_r}{dr} + 2\,L_r\,\frac{d\Phi}{dr}\Big)\,e^{\Phi} = \frac{d}{dr}\Big(L_r\,e^{2\Phi}\Big)\,. \qquad (19.9)$$

Substituting (19.9) into (19.8) gives

$$e^{\Phi}\,d\Big(\frac{\mathcal{E}^{\rm int}}{\rho}\delta A\Big) = q\,\delta A\,e^{2\Phi}\,dt - P\,e^{\Phi}\,d\Big(\frac{1}{\rho}\delta A\Big) - \Big[\frac{d}{dr}\Big(L_r\,e^{2\Phi}\Big)\Big]\,\delta r\,dt\,, \qquad (19.10)$$

which leads for the gradient of L_r to

$$\begin{aligned}\frac{d}{dr}\left(L_r\,e^{2\Phi}\right) &= \frac{1}{\delta r\,dt}\Big\{q\,\delta A\,e^{2\Phi}\,dt - P\,e^{\Phi}\,d\Big(\frac{1}{\rho}\,\delta A\Big) - e^{\Phi}\,d\Big(\frac{\mathcal{E}^{\rm int}}{\rho}\delta A\Big)\Big\}\\ &= \frac{\delta A\,e^{2\Phi}}{\delta r}\Big\{q + \frac{P\,e^{-\Phi}}{\rho^2}\,\frac{d\rho}{dt} - e^{-\Phi}\,\frac{d}{dt}\,\frac{\mathcal{E}^{\rm int}}{\rho}\Big\}\,. \end{aligned} \qquad (19.11)$$

To arrive at the second relation in (19.11), we have used that the number of baryons δA in the mass shell is constant, in which case $d(\delta A/\rho) = -\delta A d\rho/\rho^2$ and $d(\mathcal{E}^{\rm int}\delta A/\rho) = \delta A d(\mathcal{E}^{\rm int}/\rho)$. To rearrange equation (19.11) further note that from (14.90),

$$\delta A = 4\pi\,r^2\,\rho\,\frac{1}{\sqrt{1-\frac{2m}{r}}}\,\delta r\,. \qquad (19.12)$$

The energy-balance equation of a star in thermal equilibrium thus takes the final form [655],

$$\frac{\mathrm{d}}{\mathrm{d}r}\left(L_r\,\mathrm{e}^{2\Phi}\right) = 4\pi\,\frac{\rho\,r^2\,\mathrm{e}^{2\Phi}}{\sqrt{1-\frac{2m}{r}}}\left\{q - \left[\mathrm{e}^{-\Phi}\,\frac{\mathrm{d}}{\mathrm{d}t}\frac{\mathcal{E}^{\mathrm{int}}}{\rho} - \frac{P}{\rho^2}\,\mathrm{e}^{-\Phi}\,\frac{\mathrm{d}\rho}{\mathrm{d}t}\right]_{\alpha=\mathrm{const}}\right\}. \tag{19.13}$$

It determines the energy flux inside a given stellar configuration. Inherent in (19.13) is that the rate of the fractional nuclear abundances, α_k, defined as

$$\alpha_k = \frac{\mathrm{d}Y_k}{\mathrm{d}(\mathrm{e}^{\Phi}\,t)} = \frac{\mathrm{d}Y_k}{\mathrm{e}^{\Phi}\,dt}, \tag{19.14}$$

is kept constant. From the above derivation it is evident that the time derivative in the equation of thermal equilibrium must be taken with baryon number A held fixed, *not* with radius held fixed. Finally we mention that (19.13) can be derived directly from the general relativity law of local energy conservation $\sum_{\mu\nu} u_\mu\, T^{\mu\nu}{}_{;\,\mu} = 0$ [656].

An alternative form of (19.13) follows by rewriting the term in curly brackets as

$$q - \left[\mathrm{e}^{-\Phi}\,\frac{\mathrm{d}}{\mathrm{d}t}\frac{\mathcal{E}^{\mathrm{int}}}{\rho} - \frac{P}{\rho^2}\,\mathrm{e}^{-\Phi}\,\frac{\mathrm{d}\rho}{\mathrm{d}t}\right] = q - \mathrm{e}^{-\Phi}\,\frac{\mathrm{d}}{\mathrm{d}t}\frac{\mathcal{E}^{\mathrm{int}}}{\rho} - P\,\mathrm{e}^{-\Phi}\,\frac{\mathrm{d}}{\mathrm{d}t}\frac{1}{\rho}. \tag{19.15}$$

The ratio $\mathcal{E}^{\mathrm{int}}/\rho$ can be replaced with $\mathcal{E}^{\mathrm{int}}/\rho = \epsilon/\rho - \bar{m}_B$, known from equation (12.107), which gives for the right-hand side of (19.15)

$$q + \mathrm{e}^{-\Phi}\,\frac{\mathrm{d}\bar{m}_B}{\mathrm{d}t} - \mathrm{e}^{-\Phi}\,\frac{\mathrm{d}}{\mathrm{d}t}\frac{\epsilon}{\rho} - P\,\mathrm{e}^{-\Phi}\,\frac{\mathrm{d}}{\mathrm{d}t}\frac{1}{\rho}. \tag{19.16}$$

Inspection of the expression of the rate of thermonuclear energy generation, $q = -\mathrm{d}\bar{m}_B/(\mathrm{d}t\,\mathrm{e}^{\Phi})$, introduced in equation (19.2), reveals that the first two terms on the left of (19.16) cancel each other, leaving us with

$$-\mathrm{e}^{-\Phi}\left[\frac{\mathrm{d}}{\mathrm{d}t}\frac{\epsilon}{\rho} + P\,\frac{\mathrm{d}}{\mathrm{d}t}\frac{1}{\rho}\right] = -\frac{1}{\rho}\,\mathrm{e}^{-\Phi}\left[\frac{\mathrm{d}\epsilon}{\mathrm{d}t} - \left(\frac{P+\epsilon}{\rho}\,\frac{\mathrm{d}\rho}{\mathrm{d}t}\right)\right] \tag{19.17}$$

$$= -\,\mathrm{e}^{-\Phi}\left[T\,\frac{\mathrm{d}s}{\mathrm{d}t} + \sum_k \mu^k\,\frac{\mathrm{d}Y_k}{\mathrm{d}t}\right]. \tag{19.18}$$

To get from equation (19.17) to (19.18) use of the first law of thermodynamics (cf. equation (14.67)),

$$\mathrm{d}\epsilon = \frac{P+\epsilon}{\rho}\,\mathrm{d}\rho + T\,\rho\,\mathrm{d}s + \sum_k \mu^k\,\rho\,\mathrm{d}Y_k \tag{19.19}$$

was made. Here, the relative nuclear abundances Y_k ($\equiv \rho^k/\rho$) denote the fraction of all baryons in a given sample of matter which are in the form

k, i.e. $Y_{\rm H}$, $Y_{\rm He}$, Y_n, Y_Λ, etc. Because of $\sum_k \rho^k = \rho$, the relative nuclear abundances must satisfy

$$\sum_{k={\rm H,He},n,\Lambda,\ldots} Y_k = 1\,, \quad \text{and} \quad \sum_{k={\rm H,He},n,\Lambda,\ldots} m_k\, Y_k = \bar{m}_B\,, \tag{19.20}$$

where m_k is the rest mass per baryon of the nuclear species k, and $\bar{m}_B$ the average baryon rest mass. Equation (19.19) differs from its Newtonian counterpart only with respect to changes in the rest mass-energies of baryons, which must be allowed for in relativistic theory because of the equivalence of mass and energy. One does not need to consider other special relativistic effects as long as the thermodynamic analysis is carried out in reference frames that comove with the matter.

With the aid of equations (19.15) through (19.18), the expression of energy balance (19.13) can be written as

$$\frac{\mathrm{d}}{\mathrm{d}r}\left(L_r\, \mathrm{e}^{2\Phi}\right) = -4\pi \frac{\rho\, r^2\, \mathrm{e}^{2\Phi}}{\sqrt{1-\frac{2m}{r}}}\left\{T\left(\frac{\mathrm{d}s}{\mathrm{d}t}\right)_{A=\mathrm{const}} + \sum_k \mu^k \left(\frac{\mathrm{d}Y_k}{\mathrm{d}t}\right)_{A=\mathrm{const}}\right\}. \tag{19.21}$$

With the help of $\mathrm{d}/\mathrm{d}r = (\mathrm{d}A/\mathrm{d}r)\,\mathrm{d}/\mathrm{d}A$ we introduce the baryon number A rather than r as the independent radial coordinate. This leads for (19.21) to

$$\frac{\mathrm{d}}{\mathrm{d}A}\left(L_r\, \mathrm{e}^{2\Phi}\right) = -\mathrm{e}^{\Phi}\left\{T\,\frac{\mathrm{d}s}{\mathrm{d}t} + \sum_k \mu^k\, \frac{\mathrm{d}Y_k}{\mathrm{d}t}\right\}, \tag{19.22}$$

with the result for $\mathrm{d}A/\mathrm{d}r$ given in equation (19.12). Obviously, the $\mathrm{d}Y_k/\mathrm{d}t$ terms vanish for systems whose nuclear abundances Y_k are constant in time, $\mathrm{d}Y_k/\mathrm{d}t = 0$.

After this excursion, let us now turn to the case of energy balance inside a star due to neutrino escape. Once a neutrino is produced by one of the particle processes listed in table 19.1, it escapes freely from the equilibrium configuration without ever being converted into any other form of energy. The rate of energy release into neutrinos, $q^{(\nu)}$, is the rate per baryon at which internal energy is converted into outgoing neutrinos. The corresponding neutrino emissivity, ϵ_ν, is connected with the rate of energy release as $q^{(\nu)} = -\epsilon_\nu/\rho$. Subject to these modifications, the equation of thermal equilibrium (19.13) then takes on the form

$$\frac{\mathrm{d}}{\mathrm{d}r}\left(L_r^{(\nu)}\, \mathrm{e}^{2\Phi}\right) = -\,4\,\pi\, \frac{r^2\, \mathrm{e}^{2\Phi}}{\sqrt{1-\frac{2m}{r}}}\, \epsilon_\nu\,, \tag{19.23}$$

which upon using A as the independent variable reads

$$\frac{\mathrm{d}}{\mathrm{d}A}\left(L_r^{(\nu)}\, \mathrm{e}^{2\Phi}\right) = -\,\frac{\epsilon_\nu}{\rho}\,. \tag{19.24}$$

Combining expression (19.24) for the neutrino luminosity with (19.22) for the one of photons leads for systems with constant nuclear abundances, $\mathrm{d}Y_k/\mathrm{d}t = 0$, to

$$\frac{\mathrm{d}}{\mathrm{d}A}\left[\left(L_r + L_r^{(\nu)}\right)\mathrm{e}^{2\Phi}\right] = -\,\mathrm{e}^{2\Phi}\,\frac{\epsilon_\nu}{\rho} - \mathrm{e}^{\Phi}\,T\,\frac{\mathrm{d}s}{\mathrm{d}t}\,. \qquad (19.25)$$

The heat capacity per particle, $\tilde{c}_\mathrm{V}$, for a system whose volume is constant is given by

$$\tilde{c}_\mathrm{V} = \left.\frac{\mathrm{d}q}{\mathrm{d}T}\right|_{V,A} = \left.\frac{\partial \mathcal{E}^{\mathrm{int}}/\rho}{\partial T}\right|_{V,A} = T\left.\frac{\mathrm{d}s}{\mathrm{d}T}\right|_{V,A}\,. \qquad (19.26)$$

Writing the last term in this relation as

$$T\,\frac{\mathrm{d}s}{\mathrm{d}t} = T\,\frac{\mathrm{d}s}{\mathrm{d}T}\,\frac{\mathrm{d}T}{\mathrm{d}t} = \tilde{c}_\mathrm{V}\,\frac{\mathrm{d}T}{\mathrm{d}t} \qquad (19.27)$$

and noticing that the heat capacity per unit volume, c_V, is connected with $\tilde{c}_\mathrm{V}$ as $\tilde{c}_\mathrm{V} = c_\mathrm{V}/\rho$, leads for the equation of energy balance (also referred to as the thermal equilibrium equation, given in (19.25), to the final representation

$$\frac{\mathrm{d}}{\mathrm{d}A}\left(L\,\mathrm{e}^{2\Phi}\right) = -\,\frac{\mathrm{e}^{2\Phi}}{\rho}\left(\epsilon_\nu + \mathrm{e}^{-\Phi}\,c_\mathrm{V}\,\frac{\mathrm{d}T}{\mathrm{d}t}\right)\,. \qquad (19.28)$$

The quantity L is the total stellar luminosity, which, in our case, originates from escaping photons and neutrinos, $L \equiv \sum_{\lambda=\gamma,\nu} L_r^{(\lambda)}$. As already noted, this is an equation for the gradient of the energy flux. The still missing ingredient needed to study the thermal history of compact stars is the equation of energy transport, which is an expression for the temperature gradient. This equation will be derived next.

19.2 Equation of energy transport

In general, energy is transported from the hot interior of a star toward its cool surface by a combination of diffusing photons, escaping neutrinos, heat conduction in the stellar material, and convective motions of the stellar material. The analysis of energy transport for stellar models in which conduction and radiation contributes roughly the same energy flux as convection is very difficult. Fortunately, for neutron stars convection is negligible compared to conduction and radiation, which eases the following analysis considerably.

19.2.1 Radiative energy transport by photon diffusion

The theory of radiative transport in general relativity has been developed in great detail by Linquist [657] plus several other authors [80]. In the following we shall follow Thorne [80] again and derive the equation of radiative energy transport on more physical grounds without going into the details of transport theory. Let us consider a stellar environment where photon transport is diffusive. At any point inside the star the electromagnetic radiation consists of two parts, namely a large isotropic part and a very much smaller, purely radial part which accounts for the photon luminosity, $L_r^{(\mathrm{R})}$. Consider the gradient in the radial component of the almost isotropic radiation pressure. It arises from two sources, firstly the gravitational attraction of the photon gas toward the center of the star, and, secondly the interaction of the radiation with matter. The gravitational attraction must be balanced by a pressure gradient [80],

$$\frac{\mathrm{d(radiation\ pressure)}}{\mathrm{d(proper\ radial\ distance)}} = -\,\mathrm{(internal\ mass\ per\ unit\ volume\ of\ photon\ gas)} \times \frac{\mathrm{d}\Phi}{\mathrm{d(proper\ radial\ distance)}}\,, \tag{19.29}$$

or, equivalently,

$$\left(\frac{\mathrm{d}P_{\mathrm{R}}}{\mathrm{d}r}\right)_{\mathrm{grav}} = -\,(\epsilon_{\mathrm{R}} + P_{\mathrm{R}})\,\frac{\mathrm{d}\Phi}{\mathrm{d}r}\,. \tag{19.30}$$

Of all interactions between matter and radiation, only absorption of the excess radial component, $L_r^{(\mathrm{R})}$, is anisotropic and can thus contribute to the radiation pressure gradient. The drop in radiation pressure over a coordinate interval $\mathrm{d}r$ due to absorption is the mass energy absorbed from the radial beam per unit proper time and per unit area by the matter in the shell of thickness $\mathrm{d}r$, i.e.

$$(\mathrm{d}P_{\mathrm{R}})_{\mathrm{absorption}} = -\,\kappa_{\mathrm{R}}\,\epsilon\,\frac{L_r^{(\mathrm{R})}}{4\pi r^2}\,\frac{1}{\sqrt{1-\frac{2m}{r}}}\,\mathrm{d}r\,, \tag{19.31}$$

where κ_{R} denotes the radiative absorption coefficient. The total gradient of the radiation pressure is given as the sum of equations (19.30) and (19.31),

$$\left(\frac{\mathrm{d}P_{\mathrm{R}}}{\mathrm{d}r}\right) = -\,(\epsilon_{\mathrm{R}} + P_{\mathrm{R}})\,\frac{\mathrm{d}\Phi}{\mathrm{d}r} - \kappa_{\mathrm{R}}\,\epsilon\,\frac{L_r^{(\mathrm{R})}}{4\pi r^2}\,\frac{1}{\sqrt{1-\frac{2m}{r}}}\,. \tag{19.32}$$

The very nearly isotropic radiation pressure, P_{R}, and the density of radiation energy, ϵ_{R}, are related to the temperature of the matter by

$$\epsilon_{\mathrm{R}} = 3\,P_{\mathrm{R}} = 4\,\sigma\,T^4\,, \tag{19.33}$$

where σ is the Stefan–Boltzmann constant. By combining equations (19.32) and (19.33), one obtains

$$\frac{\mathrm{d}}{\mathrm{d}r}\Big(\frac{4}{3}\,\sigma\,T^4\Big) = -\,\frac{16}{3}\,\sigma T^4\,\frac{\mathrm{d}\Phi}{\mathrm{d}r} - \kappa_{\mathrm{R}}\,\epsilon\,\frac{L_r^{(\mathrm{R})}}{4\pi r^2}\,\frac{1}{\sqrt{1-\frac{2m}{r}}}\,, \tag{19.34}$$

which, upon performing several straightforward manipulations, leads to an equation of radiative energy transport of the form

$$\frac{\mathrm{d}}{\mathrm{d}r}\Big(T\,\mathrm{e}^{\Phi}\Big) = -\,\frac{3}{16}\,\frac{\kappa_{\mathrm{R}}\,\epsilon}{\sigma\,T^3}\,\frac{L_r^{(\mathrm{R})}}{4\pi r^2}\,\frac{\mathrm{e}^{\Phi}}{\sqrt{1-\frac{2m}{r}}}\,, \tag{19.35}$$

applicable to stars with no mass motions inside (negligible convection). Changing to A as the independent variable leads for (19.35) to

$$\frac{\mathrm{d}}{\mathrm{d}A}\Big(T\,\mathrm{e}^{\Phi}\Big) = -\,\frac{3}{256}\,\frac{\kappa_{\mathrm{R}}\,\epsilon}{\sigma\,T^3}\,\frac{L_r^{(\mathrm{R})}\,\mathrm{e}^{\Phi}}{\pi^2 r^2 \rho}\,. \tag{19.36}$$

19.2.2 Conductive energy transport

The portion $L_r^{(\mathrm{c})}$ of the total luminosity L_r which is due solely to heat conduction, driven by relativistic electrons in the interior region of neutron stars and photons of the photosphere, is related to the temperature gradient by the equation of conductive energy transport, which will be derived next. In the Newtonian approximation the conductive luminosity is related to the temperature gradient by the equation of conductive energy transport [80],

$$\frac{\mathrm{d}T}{\mathrm{d}r} = -\kappa_{\mathrm{T}}^{-1} \times (\text{energy flux}) = -\,\frac{1}{4\,\pi\,r^2\,\kappa_{\mathrm{T}}}\,L_r^{(\mathrm{c})}\,, \tag{19.37}$$

where κ_{T} stands for the thermal conductivity. The relativistic generalization of this equation must be obtainable by insertion of the general relativistic correction factors e^{Φ} and $\sqrt{1-2m/r}$ in appropriate places. The latter factor is always used to convert differential radial coordinate intervals, $\mathrm{d}r$, to proper radial distances, $\mathrm{d}r/\sqrt{1-2m/r}$. Hence equation (19.37), when corrected for the effects of $\sqrt{1-2m/r}$, reads [80]

$$\frac{\mathrm{d}T}{\mathrm{d(proper\ radial\ distance)}} \equiv \frac{\mathrm{d}T}{\mathrm{d}r/\sqrt{1-2m/r}} = -\,\frac{1}{4\,\pi\,r^2\,\kappa_{\mathrm{T}}}\,L_r^{(\mathrm{c})}\,. \tag{19.38}$$

The second correction factor, e^{Φ}, must be inserted in such a manner as to account for the redshift of the energy which is being conducted upward through the star. A factor of e^{Φ} to some power, k, will appear inside the radial derivative; and another factor, $e^{k'\Phi}$, will appear outside the radial derivative, leaving us with

$$\sqrt{1-\frac{2m}{r}}\, e^{k'\Phi}\, \frac{d}{dr}\left(T\, e^{k\Phi}\right) = -\frac{1}{4\,\pi\, r^2\, \kappa_T}\, L_r^{(c)}\,. \tag{19.39}$$

The values of k and k' are determined (for details, see Thorne [80]) by the demand that the equation of conductive transport, (19.39), be invariant under the transformation $\Phi \to \Phi + \text{const}$ which leads to $k' = k$. Its correct numerical value must be $k' = k = -1$. Otherwise, the heat flow inside an insulated hot star would violate the second law of thermodynamics. Thus, by setting $k = k' = -1$ in (19.39) and rearranging terms, we arrive at the general relativistic equation for heat conduction in a star, also known as the general relativistic equaton of diffusion, given by

$$\frac{d}{dr}\left(T\, e^{\Phi}\right) = -\frac{1}{\kappa_T}\,\frac{L_r^{(c)}}{4\pi r^2}\,\frac{e^{\Phi}}{\sqrt{1-\frac{2m}{r}}}\,. \tag{19.40}$$

It is applicable to stars with no mass motions (negligible convection) inside, like the neutron stars. If there is more than one process contributing to heat conduction, then the total thermal conductivity, κ, is given by the sum over the individual contributions.

19.2.3 Photon diffusion and conduction combined

The equations of energy transport by photon diffusion, (19.35), and of conductive energy transport, (19.40), can be combined into a single equation. For this purpose, it is convenient to write equation (19.40) in a form that is formally identical to the equation of photon diffusion, namely

$$\frac{d}{dr}\left(T\, e^{\Phi}\right) = -\frac{3}{16}\,\frac{\kappa_c\,\epsilon}{\sigma T^3}\,\frac{L_r^{(c)}}{4\pi r^2}\,\frac{e^{\Phi}}{\sqrt{1-\frac{2m}{r}}}\,. \tag{19.41}$$

The quantity κ_c in (19.41) denotes the conductive absorption coefficient, which is connected with the thermal conductivity κ_T as

$$\frac{1}{\kappa_T} \equiv \frac{3}{16}\,\frac{\kappa_c\,\epsilon}{\sigma\, T^3} \quad\Rightarrow\quad \kappa_c = \frac{16\,\sigma\, T^3}{3\,\epsilon\,\kappa_T}\,. \tag{19.42}$$

Noticing that the total luminosity is determined by photon radiation, conduction, and neutrino emission, $L_r = L_r^{(R)} + L_r^{(c)} + L_r^{(\nu)}$ (convection is negligible for neutron stars), the conduction equation (19.41) may be combined with the equation of radiative energy transport (19.35) to give

$$\frac{d}{dr}\left(T\, e^{\Phi}\right) = -\frac{3}{16}\,\frac{\kappa^*\,\epsilon}{\sigma\, T^3}\,\frac{L_r - L_r^{(\nu)}}{4\pi r^2}\,\frac{e^{\Phi}}{\sqrt{1-\frac{2m}{r}}}\,, \tag{19.43}$$

with κ^* given by

$$\frac{1}{\kappa^*} = \frac{1}{\kappa_\mathrm{R}} + \frac{1}{\kappa_\mathrm{c}}, \qquad (19.44)$$

and κ_c defined in (19.42). Picking A as the independent variable rather than r leads for (19.43) to

$$\frac{\mathrm{d}}{\mathrm{d}A}\Big(T\,\mathrm{e}^{\Phi}\Big) = -\,\frac{3}{16}\,\kappa^*\,\frac{\epsilon}{\sigma\,T^3}\,\frac{\big(L_r - L_r^{(\nu)}\big)\mathrm{e}^{\Phi}}{16\pi^2 r^4 \rho}\,. \qquad (19.45)$$

If the thermal conductitivy is significantly larger than the radiation conductivity, i.e. $\kappa_\mathrm{c} \gg \kappa_\mathrm{R}$, the prefactor $\kappa^* \simeq \kappa_\mathrm{R}$. Conversely, if $\kappa_\mathrm{R} \gg \kappa_\mathrm{c}$, the prefactor $\kappa^* \simeq \kappa_\mathrm{c}$.

19.3 Summary of stellar cooling equations

The cooling calculations which we shall present below are performed for spherically symmetric, non-rotating neutron star models, which are constructed in the framework for general relativity theory. The line element of such stellar configurations was given in (14.14).

The equations that need to be solved to obtain the thermal evolution of a star can be divided into two distinct categories. The first category consists of those equations which determine the star's *global* structure. These equations were already derived in chapter 14. Here, however, we choose the number of baryons inside a radius r, $A = A(r)$, as the independent radial coordinate, rather than r itself. It is a very useful parameter for identifying shells of matter in successive configurations of an evolutionary sequence. The first category of equations is then given as follows:

(1) Baryon number equation,

$$\frac{\mathrm{d}r}{\mathrm{d}A} = \frac{1}{4\,\pi\,r^2\,\rho}\,\sqrt{1-\frac{2m}{r}}\,, \qquad r(0) = 0\,, \qquad (19.46)$$

(2) Mass equation,

$$\frac{\mathrm{d}m}{\mathrm{d}A} = \frac{\epsilon}{\rho}\,\sqrt{1-\frac{2m}{r}}\,, \qquad m(0) = 0\,, \qquad (19.47)$$

(3) Source equation for Φ,

$$\frac{\mathrm{d}\Phi}{\mathrm{d}A} = \frac{(4\,\pi\,r^3\,P + m)}{4\,\pi\,r^4\,\rho}\,\frac{1}{\sqrt{1-\frac{2m}{r}}}\,, \qquad \Phi(A_\mathrm{m}) = \frac{1}{2}\ln\!\left(1 - \frac{2m(A_\mathrm{m})}{r(A_\mathrm{m})}\right), \qquad (19.48)$$

(4) TOV equation of hydrostatic equilibrium,

$$\frac{\mathrm{d}P}{\mathrm{d}A} = -(P+\epsilon)\,\frac{\mathrm{d}\Phi}{\mathrm{d}A}\,, \qquad \epsilon(A_\mathrm{m}) = \epsilon_\mathrm{m}\,. \qquad (19.49)$$

The quantity $P(A)$ denotes pressure as a function of baryon number, $\epsilon(A) = \epsilon(P(A))$ is the total energy density, and $\rho(A) = \rho(P(A))$ the baryon number density. The quantities $\epsilon(A)$ and $P(A)$ are related with each other by the equation of state. The boundary conditions of (19.46) through (19.49) are listed to the right of each equation. They differ somewhat from the standard choice for equations (19.48) and (19.49), which fix the values of Φ and ϵ for a given baryon number $A_{\rm m}$ (the subscript 'm' refers to mantle) inside the star. The baryon number $A_{\rm m}$ is defined by $m(A_{\rm m}) = M_{\rm m}$, where $M_{\rm m}$ denotes a given mass.

The photosphere of the star is treated separately from the structure equations (19.46) through (19.49) for two reasons. Firstly, the equation of state of the photosphere depends much more sensitively on temperature than the equation of state of the star's core, which can be treated at zero temperature. Secondly, because of a scaling behavior [658, 659], the structure equations of the photosphere need to be solved only once, for a properly chosen star mass. The interface between crust and photosphere is located at $\epsilon_{\rm m}$, whose value will be chosen to be $\epsilon_{\rm m} = 10^{10}$ g cm^{-3} [659].

Since the central star temperature drops down to less than 10^9 K within a few minutes after birth [654], the effects of finite temperatures on the equation of state can be neglected to a very good approximation. Consequently, the stellar structure equations (19.46) through (19.49) do not depend on time and thus need to be solved only once, plausibly at the beginning of the numerical cooling simulation. The time dependence enters through the second category of equations – the equations of thermal equilibrium and energy transport derived in (19.28) and (19.45). For neutron stars, these take on the form:

(5) Equation of energy balance,

$$\frac{\mathrm{d}}{\mathrm{d}A}(L\,\mathrm{e}^{2\Phi}) = -\frac{1}{\rho}\left(\epsilon_\nu\,\mathrm{e}^{2\Phi} + c_{\rm v}\,\frac{\mathrm{d}}{\mathrm{d}t}(T\,\mathrm{e}^{\Phi})\right), \tag{19.50}$$

(6) Equation of thermal energy transport,

$$\frac{\mathrm{d}}{\mathrm{d}A}(T\,\mathrm{e}^{\Phi}) = -\frac{1}{\kappa}\,\frac{L\,\mathrm{e}^{\Phi}}{16\,\pi^2\,r^4\,\rho}, \tag{19.51}$$

where $\kappa^{-1} \equiv 3\kappa^*\epsilon/(16\sigma T^3)$ is the total thermal conductivity, $\kappa = \kappa(A,T)$. The other microphysical input quantities entering equations (19.50) and (19.51) are the equation of state, $P(\epsilon)$, the neutrino emissivity per unit volume, $\epsilon_\nu(A,T)$, and the heat capacity per unit volume, $c_{\rm v}(A,T)$. The boundary conditions of (19.50) and (19.51) read

$$L(A=0)=0\,, \quad \text{and} \quad T(A=A_{\rm m}) = T_{\rm m}(r_{\rm m}, A_{\rm m}, M_{\rm m})\,, \tag{19.52}$$

where the temperature of the mantle, $T_{\rm m}(r_{\rm m}, A_{\rm m}, M_{\rm m})$, is fixed by the properties of the photosphere at $r = r_{\rm m}$ [658, 659]. The star's initial

Table 19.2. Input quantities entering the cooling equations of compact stars.

Parameter	Symbol	References
Equation of state:	$P(\epsilon)$	
crust		[70, 298]
core		see table 12.4
Superfluidity		see table 19.3
Heat capacity	c_V	[224, 660]
Thermal conductivity:	κ	
crust		[393, 661, 662]
core		[663, 664]
Neutrino emissivity of crust:	ϵ_ν	
pair, photon, plasma processes		[665]
bremsstrahlung		[666, 667]
Neutrino emissivity of core:	ϵ_ν	see table 19.4
Photosphere (no magnetic field)		[659]

temperature can be taken, without loss of generality, to be $T(r) \equiv 10^{11}$ K, because the star's thermal evolution at times greater than say a few months does not depend on the choice of the initial temperature profile. Equations (19.50) and (19.51) can be solved numerically by means of a Newton–Raphson-like algorithm. Table 19.2 summarizes the microphysical input that enters the thermal structure equations (19.46) through (19.51).

19.4 Superfluidity

When the thermal energy $\propto k_B T$ is smaller than the latent heat, Δ, associated with the phase transition to a paired state, neutrons and protons may undergo pairing transitions inside neutron stars (see, for instance, [668]). One expects to find neutron superfluidity in the crust and the dense interior of neutron stars. Moreover, the protons in the superdense interior of neutron stars may be paired, forming a superconducting fluid in the cores of such objects. It is unlikely that the electrons will be superconducting, since the electron–phonon coupling is too weak in neutron star matter.

The pairing energy is typically less than about *one percent* of the total interaction energy and so it makes only very little difference to the pressure versus density relation, which, in turn, implies that pairing has only a very little effect on the *gross* properties of neutron stars. Some of the physical consequences of superfluidity that are important for neutron stars include (a) thermal, (b) magnetic, and (c) hydrodynamic effects. Thermal effects arise from the heat capacity of a superfluid, which is by a factor $\propto \exp(-\Delta/k_{\rm B}T)$ lower than that of a normal degenerate gas. Consequently less thermal energy can be stored in a superfluid star which causes them to cool more rapidly. This effect is counterbalanced however by the lower neutrino-emission rate from the superfluid, which delays cooling. As to item (b), the superfluid region inside a pulsar may rotate faster than the pulsar's outer regime, which slows down due to the emission of electromagnetic dipole radiation. This is likely to lead to weak frictional forces between the normal outer crust and the superfluid neutron interior which couple the two components and convert some rotational energy into frictional heating. Finally, with respect to (c), we recall that a superfluid has zero viscosity below a critical rotational velocity, but at higher velocities it develops a non-zero macroscopic viscosity. For example, superfluid helium placed in a pail rotating with a wall velocity just above the critical velocity quickly develops the parabolic meniscus characteristic of rigid-body rotation. The explanation of the spin-up lies in the creation of vortex sheets due to differential rotation at the interface between helium and the pail. Vortex sheets are created which subsequently break up into vortex lines that become tangled. Associated with this turbulent state is an eddy viscosity. Hence, macroscopically, the fluid threaded by a tangled array of vortices behaves *normally*. Therefore one unique consequence of neutron superfluidity in a rotating neutron star is that, because of the star's rotation, the superfluid will contain a tangled array of vortices which make the matter viscous. It is subject to Ekman pumping and turbulent mixing, too, and its average rotational velocity satisfies the condition for uniform rotation, $\boldsymbol{\Delta} \times \langle \boldsymbol{v} \rangle = 2\,\boldsymbol{\Omega}$. So if one considers $\boldsymbol{v}$, then the superfluid can be treated as rigidly rotating.

The Ekman mechanism describes the flow of matter of a differentially rotating star, that is, Ω decreases monotonically in the radial outward direction, from its interior radially outward to the surface [563]. The flow is driven by the difference in centrifugal acceleration between different adjacent layers of the star. The pressure gradient, on the other hand, acting on these mass shells is almost the same as before the differential rotation was established and therefore causes the fluid, because of mass conservation, to flow radially outward in the equatorial region, replacing low-angular-momentum fluid with high-angular-momentum fluid. It has been shown in [563] that differential rotation initially present in a hot, newly formed neutron star ($T \sim$ MeV) will be damped out via Ekman pumping (turbulent mixing) within a few days. In cold neutron stars, superfluid eddy viscosity, viscosity arising from electron–vortex scattering, or degenerate particle viscosity will mix vortices by means of which rigid-body rotation is achieved for such objects too.

The roots of the different types of pairing in neutron star matter are as follows. In the *inner crust* of a cool neutron star, that is, at densities between 4.3×10^{11} g cm^{-3} (neutron drip) and about nuclear matter density $\epsilon_0 = 2.5\times10^{14}$ g cm^{-3} (or $\rho_0 = 0.16$ fm^{-3}), a neutron gas coexists with a Coulomb lattice of neutron-rich nuclei and a sea of relativistic electrons (chapter 4). In this regime S-wave collisions predominate and the interacting gas-phase neutrons of the inner crust sample mainly the long-range attraction of the mutual $^1S_0$ potential. Hence the corresponding pairing interaction at the Fermi surface is negative, favoring isotropic pairing of neutrons in the $^1S_0$ two-body state, which should extend roughly to the depth of the crust–core boundary at about $(2–3)\times10^{14}$ g cm^{-3}. The effect peaks at a density of about one-tenth the saturation density ϵ_0 of symmetric nuclear matter. As the density approaches ϵ_0, the colliding neutrons see more and more of the strong short-range repulsion present in the nucleon–nucleon force, the pairing interaction at the Fermi surface turns positive, and the $^1S_0$ neutron gap closes. An analogous situation holds for proton pairing in the *quantum-fluid interior*, or baryonic region, at densities $\epsilon \gtrsim \epsilon_0$ where the dilute proton component reaches a partial density similar to that of the pair-condensed neutron gas of the inner crust. Hence the protons may pair in the $^1S_0$ state. Since the protons are charged, the proton superfluid is a superconductor. The strength of proton pairing appears to peak at densities somewhat beyond ϵ_0. Finally, it is expected that neutron pairing in the $^3P_2$–$^3F_2$ partial wave may be favored in roughly the same density range as the quantum-fluid interior, $\epsilon \gtrsim \epsilon_0$. The energetic advantages of neutron pairing in the $^3P_2$–$^3F_2$ state may be ascribed to the nature of the tensor and spin–orbit forces in this channel.

The different superfluid regimes can be summarized as follows [669, 670, 671]:

- 7×10^{11} g cm^{-3} to 2×10^{14} g cm^{-3}: neutrons in $^1S_0$-pair state,
- 2×10^{14} g cm^{-3} to 4×10^{14} g cm^{-3}: protons in $^1S_0$-pair state,
- 2×10^{14} g cm^{-3} to 5×10^{16} g cm^{-3}: neutrons in $^3P_2$-pair state.

These values depend on the composition of neutron star matter and, thus, vary somewhat from one equation of state to another. It is also known that very strong interior magnetic fields, on the order of $B \sim 10^{16}$ to 10^{17} G, can modify the gaps significanly [672]. The above density regimes, computed for the HV equation of state of table 12.4 assuming negligible interior magnetic fields, constitute representative estimates, which suffice to demonstrate the impact of superfluidity on neutron star cooling. Parenthetically we also mention that neutrons and protons may pair in the $^3D_2$, depending on the proton fraction [673], and that hyperons and quarks may form superfluids too [674, 675]. Until now, however, no detailed calculations for this last option exist, such that one is left with parameter studies (cf. section 19.7.5).

As we have seen in chapter 7, both single-nucleon properties and the two-body interaction between nucleons are modified in a dense baryonic

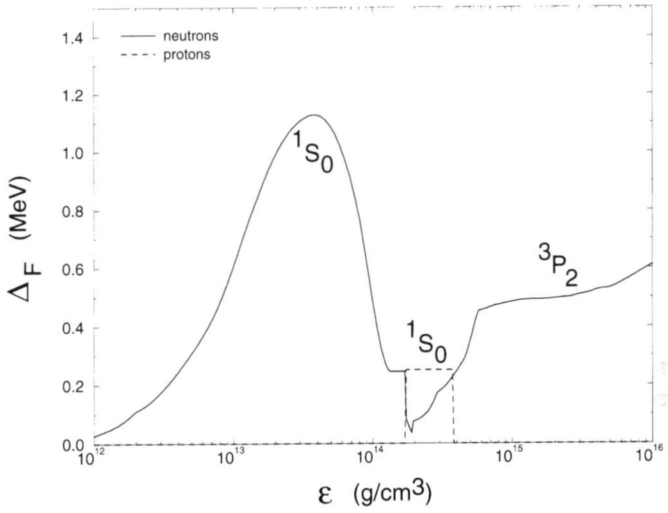

Figure 19.2. Superfluid energy gap versus mass density computed for the HV EOS of table 12.4 [543]. (Reprinted courtesy of *Nucl. Phys.*)

medium, relative to free space. The above picture of the superfluid content of neutron stars is predicted on the assumption that these modifications are not extreme. The simplest effect is that the medium is dispersive, leading to an effective baryon mass smaller than the bare mass, and, correspondingly, a somewhat reduced value for the pairing gap, $\Delta_F = \Delta_F(m^*)$. This effect is small at the very low densities relevant for $^1S_0$ neutron pairing, but becomes significant at the higher densities where $^1S_0$ proton pairing and $^3P_2$–$^3F_2$ neutron pairing is predicted to occur. Further uncertainties in the determination of Δ_F originate for instance from polarization effects [676]. Consequently, the $^1S_0$ proton pairing gaps as well as the $^3P_2$–$^3F_2$ neutron pairing gaps are not very well known. For the numerical simulations, we shall adopt the $^1S_0$ proton gaps of Wambach *et al* [671], and the $^1S_0$ and $^3P_2$ neutron gaps of Ainswort *et al* [669] and Amundsen and Østgaard [670], respectively. An overview of the gap energies and critical temperatures calculated for the HV equation of state is given in table 19.3. As a representative example, we show in figure 19.2 the density dependence of Δ_F for one of our model equations of state.

19.5 Neutrino emissivities

The neutrino emission processes can be divided into slow and fast ones. The main feature which distinguishes fast from slow processes is the existence

Table 19.3. Maximal gap energies, critical temperatures, and density range $\epsilon_1 < \epsilon < \epsilon_2$ for which the superfluid gap $\Delta_F(\epsilon) > 0$ in neutron star matter.

	Neutron $^1S_0$	Neutron $^3P_2$	Proton $^1S_0$
Reference	[669]	[670]	[671]
Δ_F^{max} (MeV)	1.13	0.62	0.25
T_c^{max} (10^9 K)	7.4	0.8	1.6
ϵ_1 (g cm^{-3})	7×10^{11}	2×10^{14}	2×10^{14}
ϵ_2 (g cm^{-3})	2×10^{14}	5×10^{16}	4×10^{14}

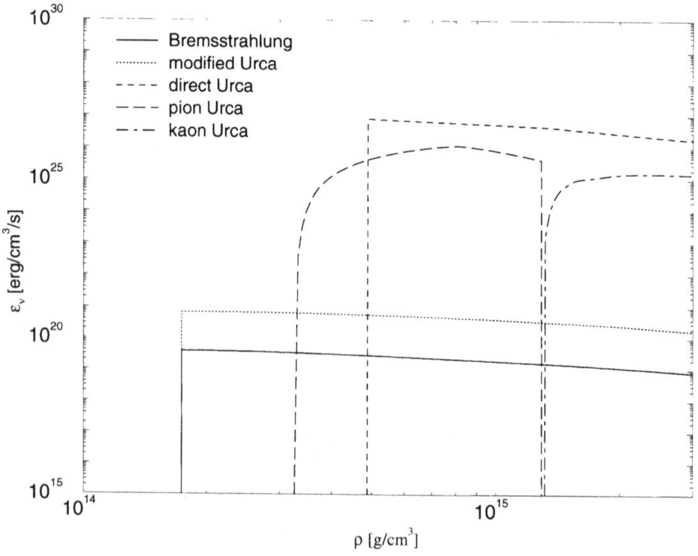

Figure 19.3. Neutrino emissivities of different neutrino-emitting processes as a function of energy density, for $T = 10^9$ K [543]. The particle composition is computed for G^{π}_{300} of table 12.4. (Reprinted courtesy of *Nucl. Phys.*)

of a critical threshold density for the fast processes. Below the thresholds, listed in table 19.4, the concentration of a certain particle species, $Y_i \equiv \rho^i/\rho$ ($i = p, e$) is either too small or entirely zero, as for the meson condensates, for the fast processes to occur. The density dependence of the neutrino emissivities associated with the physical processes of table 19.4 is graphically illustrated in figure 19.3.

Table 19.4. Neutrino emitting processes, their associated emissivities, ϵ_ν, and constraints (if any) which render some of these processes inoperative. The quantities ρ_0, Y_p, and Y_e denote the density of normal nuclear matter, and the relative proton and electron fractions in stellar matter.

Process	ϵ_ν (erg s^{-1})	Constraint	Rapidity	References
		Neutron stars		
Nucleon bremsstrahlung	$\sim 10^{19} \times T_9^8$	none	slow	[677, 678]
Modified Urca	$\sim 10^{21} \times T_9^8$	none	slow	[678]
Direct Urca	$\sim 10^{27} \times T_9^6$	$Y_p \gtrsim 0.11$	fast	[679]
π^- condensate	$\sim 10^{26} \times T_9^6$	$\rho \gtrsim 2\rho_0$	fast	[680]
K^- condensate	$\sim 10^{25} \times T_9^6$	$\rho \gtrsim 5\rho_0$	fast	[425, 681]
		Quark matter stars		
Modified Urca	$\sim 10^{20} \times T_9^8$	$Y_e > 0$	slow	[682]
Direct Urca	$\sim 10^{24} \times T_9^6$	$Y_e > 0$	fast	[683]
Quark bremsstrahlung	$\sim 10^{19} \times T_9^8$	none	slow	[684]

19.5.1 Neutrino pair bremsstrahlung

The most important neutrino process in the crust is neutrino–antineutrino pair production by electron bremsstrahlung,

$$e^- + (A, Z) \longrightarrow e^- + (A, Z) + \nu + \bar{\nu}\,. \tag{19.53}$$

In early estimates it was found that the contribution of this process to neutron star cooling is competitive with standard cooling via the modified Urca process. However, recent investigations have shown that the mass of the ionic crust is reduced by a factor of two to three if one takes the recently established transition value $\epsilon_{\rm tr} \approx 1.7 \times 10^{14}$ g cm^{-3} as boundary between crust and core [685] instead of the older value $\epsilon_{\rm tr} \approx 2.8 \times 10^{14}$ g cm^{-3} [70]. In addition, considerations by Pethick and Thorsson [667] indicate that the standard perturbative treatment adopted to determine ϵ_ν is not sufficiently accurate below a temperature of 10^{10} K. Furthermore, taking the electronic band structure into account can reduce the energy emission rates exponentially for temperatures below 10^{10} K. We include these corrections to the emissivities derived by Itoh *et al* [686, 687]. Additionally to the above process, pair, photon, and plasmon processes may also be important in the low-density regime. Their impact on ϵ_ν was estimated too by Itoh *et al* [665]. Neutrino–antineutrino pair bremsstrahlung production can also occur in the core according to

$$n + n \rightarrow n + n + \nu + \bar{\nu}\,, \qquad n + p \rightarrow n + p + \nu + \bar{\nu}\,,$$

$$e^- + p \to e^- + p + \nu + \bar{\nu} . \tag{19.54}$$

We note that the emissivity of the last bremsstrahlung process of (19.54) becomes very important if the neutrons become superfluid. In this case this process dominates over all other processes. To implement the processes (19.54) in the numerical calculation, we adopt the emissivities as calculated by Friman and Maxwell [677, 678].

19.5.2 Modified and direct Urca processes

The beta decay and electron capture processes,

$$n \to p + e^- + \bar{\nu}_e , \qquad p + e^- \to n + \nu_e , \tag{19.55}$$

also known as the direct Urca processes, were not included in cooling calculations [565, 609, 688, 689, 690] until the beginning of the 1990s, since energy and momentum conservation can only be fulfilled if the proton fraction exceeds a critical value (cf. section 19.5.2). Most interestingly, many of the older, non-relativistic models for the equations of state of neutron star matter indeed do not predict sufficiently high proton fractions, Y_p. Consequently, the so-called modified Urca process, which is characterized by the occurrence of a bystander particle necessary to conserve both energy and momentum of the baryons that are being scattered, that is,

$$n + n \to n + p + e^- + \bar{\nu}_e , \qquad n + p + e^- \to n + n + \nu_e , \tag{19.56}$$

was considered to be the dominant cooling mechanism of neutron stars. The emission rates associated with the processes (19.56) were calculated by Friman and Maxwell [678]. They treated the π meson exchanged among the nucleons as a free particle. Such an approximation neglects the influence of the medium on the propagating pion. Furthermore, the exchanged pions also undergo weak transitions which are associated with the exchange-current effect. An estimate of these effects on the neutrino emissivity, performed by Voskresenskiĭ and Senatorov [691, 692], indicates that these may increase by up to several orders of magnitude, which was shown to effect the cooling of heavier neutron stars [693].

As outlined just above, the direct Urca process (19.55) is only possible if the proton fraction exceeds a certain critical value [679, 694]. For a neutron star assumed to be made up of only neutrons, protons, and electrons, this critical proton fraction amounts about 11%. It is instructive to recapitulate the derivation of this figure. Let us begin with the condition of energy conservation, which constrains the chemical potentials (i.e. the Fermi energies) associated with the process (19.55) to $\mu^n = \mu^p + \mu^e$ (the neutrino is ignored). Secondly, momentum conservation implies that the particle Fermi-momenta obey $\boldsymbol{k}_{F_n} = \boldsymbol{k}_{F_p} + \boldsymbol{k}_{F_e}$. The triangle inequality then implies for the magnitudes of the particle momenta $|\boldsymbol{k}_{F_n}| \le |\boldsymbol{k}_{F_p}| + |\boldsymbol{k}_{F_e}|$. Thirdly, electric charge neutrality, $\sum_{i=e^-,p} q_i^{\rm el} \rho^i = 0$, constrains the Fermi momenta according to ${k_{F_p}}^3/(3\pi^2) - {k_{F_e}}^3/(3\pi^2) = 0 \Rightarrow k_{F_p} = k_{F_e}$. Substituting $k_{F_p} = k_{F_e}$ into the triangle inequality tells us that $k_{F_n} \le 2k_{F_p}$, or, in other words, $\rho^n \le 8\rho^p$. Expressed as a fraction of the system's total baryon number, $\rho \equiv \rho^p + \rho^n$, we arrive at $\rho^p/\rho = Y_p > 1/9 \simeq 0.11$! That is, the process (19.55) can take place without the need for a bystander particle if the proton fraction in the stellar matter exceeds about 11%. Medium effects and interactions among the particles will not modify this figure, since the triangle inequality depends only upon the momenta and charge conservation. The appearance of additional charge components however, such as hyperons, quarks, or bosons, will modify Y_p correspondingly. Prakash *et al* have shown that the hyperons increase the critical Y_p value slightly, from 11% to 13% [695]. Moreover, the direct Urca reactions involving the hyperons decrease the actual Urca threshold somewhat.

In contrast to older, non-relativistic equations of state which predict proton fractions that lie below the critical Y_p value, modern non-relativistic as well as relativistic equations of state lead to large proton fractions up to 20%, as can be seen from figure 19.4, such that the direct Urca process becomes easily possible. The proton fraction depends crucially on the *symmetry energy*, $a_{\rm sym}$, which unfortunately is not known at higher densities. Its value around nuclear matter density is generally accepted to be 32 MeV [409]. The collection of equations of state studied in this text (table 12.7) is characterized by symmetry energies in the relatively broad range of 26 MeV $\lesssim a_{\rm sym} \lesssim$ 37 MeV, with the symmetry energies of the relativistic equations of state in the narrower range 30 MeV $\lesssim a_{\rm sym} \lesssim$ 37 MeV. Larger $a_{\rm sym}$ values would increase the energy per baryon of neutron star matter, which can be compensated however by a closer equality in the number of protons and neutrons, i.e. a more isospin symmetric behavior of the matter. For that reason the proton fraction is generally higher for the relativistic equations of state than for the non-relativistic ones. This trend can be seen from figure 19.4, where the proton fractions are shown for several equations of state of our collection. The arrows indicate the critical densities beyond which the direct Urca process becomes possible for a given equation of state. Because of the different asymmetry behaviors of these equations of state, the critical densities vary considerably from one equation of state to another, linking the onset of the direct Urca process to a neutron star's mass. For the emissivities of the direct Urca process we shall use those calculated by Lattimer *et al* [679].

So far we have considered the direct Urca process in neutron star matter made up of protons and neutrons only. Strange baryons and Δ isobars, however, which may exist in such matter rather abundantly too,

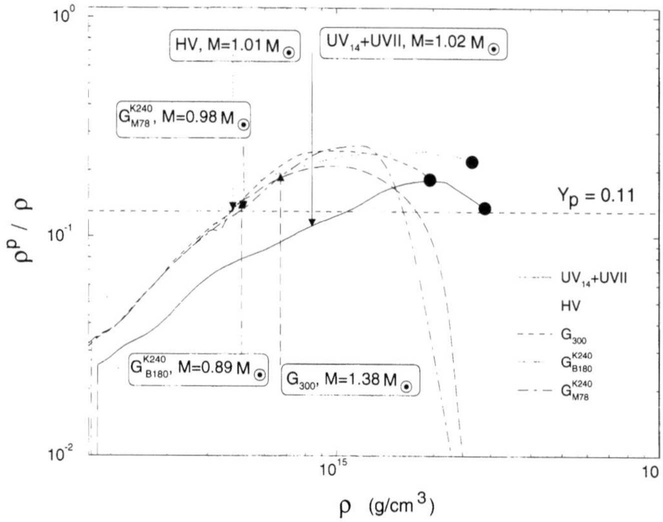

Figure 19.4. Proton fraction in neutron star matter as a function of density for different EOSs of table 12.4 [543]. The horizontal line shows the approximate critical proton fraction, $Y_p \simeq 0.11$, derived in the text, beyond which the direct Urca process becomes possible in chemically equilibrated proton–neutron matter. The exact critical densities are marked with arrows. The corresponding stellar masses (in units of $M_\odot$) are shown too. (Reprinted courtesy of *Nucl. Phys.*)

also produce neutrinos via direct Urca processes of the following type,

$$\Sigma^- \to \Lambda + e^- + \bar{\nu}_e\,, \quad \text{and} \quad \Lambda + e^- \to \Sigma^- + \nu_e\,. \tag{19.57}$$

Such processes were studied first by Prakash *et al* [695], and the implications for cooling of neutron stars were explored in [696]. We have included the above processes in cooling simulations elsewhere [697], where it was found that cooling driven by direct Urca among only nucleons clearly dominates over the energy loss associated with the processes (19.57).

19.5.3 Quasi-particle processes in boson-condensed matter

In the charged pion-condensed phase, the quasi-particle Urca processes

$$\eta(\pi) \to \eta(\pi) + e^- + \bar{\nu}_e\,, \quad \text{and} \quad \eta(\pi) + e^- \to \eta(\pi) + \nu_e \tag{19.58}$$

are possible. The quasi-particle state, η, is given as a superposition of proton and neutron states according to

$$|\eta_\pm\rangle = \cos\theta\,|n\rangle \mp \mathrm{i}\sin\theta\,|p\rangle\,, \tag{19.59}$$

with θ the mixing angle between proton and neutron states, $|p\rangle$ and $|n\rangle$. The emissivity for this process was calculated by Maxwell *et al* [680].

The critical density beyond which K^- mesons may form a condensate in neutron star matter is uncertain (cf. section 7.9). Calculations predict this to happen at densities in the range of 3–5 ρ_0 [420, 425, 698]. The emissivities of the quasi-particle direct Urca processes,

$$n(K) \to p(K) + e^- + \bar{\nu}_e \,, \qquad p(K) + e^- \to n(K) + \nu_e \,, \tag{19.60}$$

are reduced relative to the uncondensed case by a factor of $\cos^2(\theta/2)$ [681]. Additionally to (19.60), there are two more processes

$$\begin{aligned} n(K) &\to n(K) + e^- + \bar{\nu}_e \,, \qquad & n(K) + e^- &\to n(K) + \nu_e \,; \\ p(K) &\to p(K) + e^- + \bar{\nu}_e \,, \qquad & p(K) + e^- &\to p(K) + \nu_e \,, \end{aligned} \tag{19.61}$$

whose emissivities were calculated too by Thorsson *et al* [681]. The density dependence of the mixing angle, $\theta(\rho)$, and the value of the critical density, $4.18\rho_0$, are taken from [425].

19.5.4 Quark processes

The possible existence of quark matter in the cores of neutron stars has been investigated in chapter 8 and section 17.3.3, where it was pointed out that the density for onset of quark deconfinement may be as small as $\sim 2\rho$. Densities that small are even reached in light neutron stars, not to mention the heavy ones of course. The neutrino emission processes in quark matter can be divided into slow and fast ones, as listed in tables 19.1 and 19.4.

In complete analogy to the nucleon direct Urca process discussed in section 19.5.2 just above, the fast quark direct Urca processes,

$$d \to u + e^- + \bar{\nu}_e \,, \qquad u + e^- \to d + \nu_e \,, \tag{19.62}$$

$$s \to u + e^- + \bar{\nu}_e \,, \qquad u + e^- \to s + \nu_e \,, \tag{19.63}$$

are only possible if the Fermi momenta of quarks and electrons p_{F_i}, where $i = u, d, s, e^-$, obey the triangle inequality (e.g. $p_{F_d} < p_{F_u} + p_{F_e}$ for process (19.62)). This relation is the counterpart to the triangle inequality obeyed by nucleons and electrons in the nuclear matter case, section 19.5.2. In the latter case, individual neutrons and protons are free to scatter via the nucleon direct Urca process as soon as the proton fraction has grown sufficiently high in density [679, 694].

If the electron Fermi momentum p_{F_e} becomes too small, that is, Y_e becomes too little, then the triangle inequality of the processes (19.62) and (19.63) can no longer be fulfilled, and a bystander quark is needed to ensure energy and momentum conservation in the scattering process. This process, which constitutes the counterpart to the nucleon modified Urca process, is consequently called the quark modified Urca process. Its emissivity is considerably smaller than the emissivities of the direct Urca processes (19.62) and (19.63), because of the different phase spaces associated with two-quark scattering and quark decay, respectively. If the electron fraction vanishes entirely in quark matter, $Y_e = 0$ (cf. figure 19.15), both the quark direct and the quark modified Urca processes become unimportant. The neutrino emission is then dominated by bremsstrahlung processes only,

$$Q_1 + Q_2 \longrightarrow Q_1 + Q_2 + \nu + \bar{\nu}\,, \tag{19.64}$$

where Q_1, Q_2 denote any pair of quark flavors. In this case, stellar cooling would proceed rather slowly, as will be shown in section 19.8.2. Regarding the emissivities associated with the quark direct Urca, quark modified Urca, and quark bremsstrahlung processes, we refer to [682, 683, 684].

It has been suggested that the quarks eventually may form Cooper pairs [674, 699]. This would suppress, as in the nuclear matter case, the neutrino emissivities by an exponential factor of $\exp(-\Delta_{\rm F}/k_B T)$, where $\Delta_{\rm F}$ is the gap energy, k_B Boltzmann's constant, and T the temperature. Unfortunately, up to now there exists neither a reliable experimentally nor theoretically determined value for the gap energy. In order to provide a feeling for the influence of a possibly superfluid behavior of the quarks, we choose a value of $\Delta_{\rm F} = 0.4$ MeV estimated in the work of Bailin and Love [699]. We note that this value is not too different from the nuclear-matter case, where the proton $^1{\rm S}_0$ gap, for instance, is ~ 0.2–1.0 MeV [671, 700], depending on the nucleon–nucleon interaction and the microscopic many-body model.

19.6 Observed data

Among the soft X-ray observations of the two dozen sources which were identified as pulsars, the ROSAT and ASCA observations of PSRs 0002+62, 0833–45 (Vela), 0656+14, 0630+18 (Geminga), and 1055–52, listed in table 19.5, achieved a sufficiently high photon flux such that the effective surface temperatures of these pulsars could be extracted by two- or three-component spectral fits [37]. The effective surface temperatures obtained depend crucially on whether a hydrogen atmosphere is used or not. Since the photon flux measured solely in the X-ray energy band does not allow one to determine what kind of atmosphere one should use, we consider both the black-body model and the hydrogen-atmosphere model. The kind of atmosphere of individual pulsars may be determined by considering multi-wavelength observations [707]. All error bars shown in the figures represent typically a 3σ error range due to the small photon fluxes.

Except for PSRs 0833–45 (Vela) and 0002+62, all pulsar ages in table 19.5 are estimated by their spin-down age $\tau = P/2\dot{P}$. This relation

Table 19.5. Surface temperatures measured at infinity, T^∞, and spin-down ages, τ, of observed pulsars.

Pulsar	$\log \tau$ (years)	Model atmosphere	$\log\ T^\infty$ (K)	Reference
0002+62	~ 4 †	black body	$6.20^{+0.09}_{-0.40}$	[701]
0531+21 (Crab)	2.97	black body	$6.18^{+0.19}_{-0.06}$	[286]
0833–45 (Vela)	4.4 ± 0.1 †	black body	6.24 ± 0.08	[37]
		magnetic H-atmosphere	5.88 ± 0.06	[702]
0656+14	5.05	black body	$5.89^{+0.08}_{-0.33}$	[703]
		magnetic H-atmosphere	$5.72^{+0.06}_{-0.03}$	[704]
0630+18 (Geminga)	5.53	black body	$5.75^{+0.05}_{-0.08}$	[705]
		H-atmosphere	$5.42^{+0.12}_{-0.04}$	[706]
1055–52	5.73	black body	$5.90^{+0.09}_{-0.21}$	[703]

† Estimated true age instead of spin-down age (for details, see text).

implies however that both the moment of inertia and the magnetic surface field are constant with time, and that the braking index, n, is equal to its canonical value 3 (angular momentum loss due to pure magnetic dipole radiation). The true ages may therefore be quite different from the spin-down estimates. The age of Vela has been recently determined by Lyne *et al* [259], and the approximate age of PSR 0002+62 is given by an estimate of the age of the related supernova remnant G 117.7+06.

19.7 Comparison with observation

19.7.1 Standard cooling calculations

Figure 19.5 shows the cooling history of a neutron star with $M = 1.4\,M_\odot$ that cools down via the standard scenario only, that is, fast cooling processes are not allowed. The different curves correspond to different models for the equation of state. One sees that the standard cooling behavior depends only weakly on the equation of state. The UV_{14}+UVII model cools somewhat faster at later times than the other two models. This is due to the rather soft behavior of the UV_{14}+UVII equation of state at intermediate nuclear densities (figure 12.11), which leads to neutron stars with relatively small radii (see figures 14.9 to 14.11). Obviously, most of the observed data are not in agreement with the assumption of standard cooling. Specifically Vela's (0833–45) luminosity lies considerably below

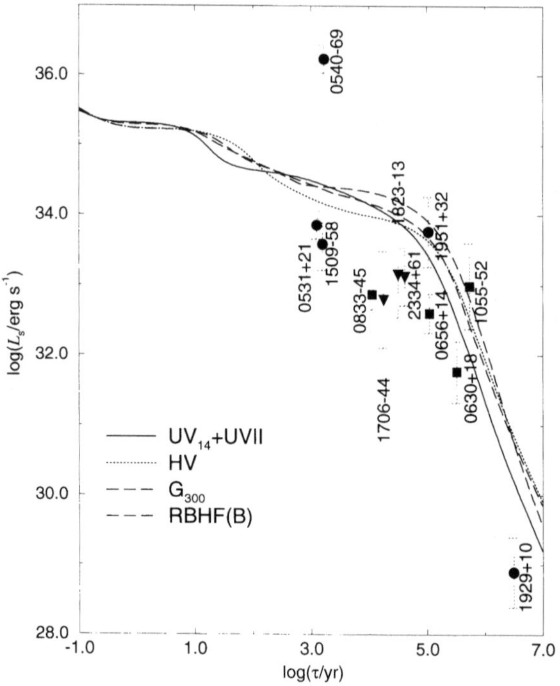

Figure 19.5. Standard cooling behavior of a neutron star with mass $M = 1.4\,M_\odot$ for several different models for the EOS [543]. (Reprinted courtesy of *Nucl. Phys.*)

the computed curves. The data of PSR 0630+18 (Geminga) and a few others are in relatively good agreement, however. We note again that there is considerable uncertainty in the ages of pulsars derived from the braking index, n, which, due to lack of information, usually is taken to be constant during a pulsar's spin-down. Moreover a canonical value of $n = 3$ [224], which corresponds to slow-down by emission of magnetic dipole radiation only, is assumed. This too constitutes a rather crude approximation.

19.7.2 Enhanced cooling via direct Urca process

As mentioned above, one way by means of which neutron stars can cool very efficiently is the direct Urca process [708, 709, 710]. This is demonstrated graphically in figures 19.6 to 19.8 for the equations of state UV$_{14}$+UVII, HV, and G$_{300}$. The proton fraction in the stars with $M = 1.0\,M_\odot$ is for all three cases below the threshold density for the direct Urca process. Hence these stars cool via the standard mechanism, the modified Urca process. The cooling behavior of all other stars at early stages is dominated by the direct

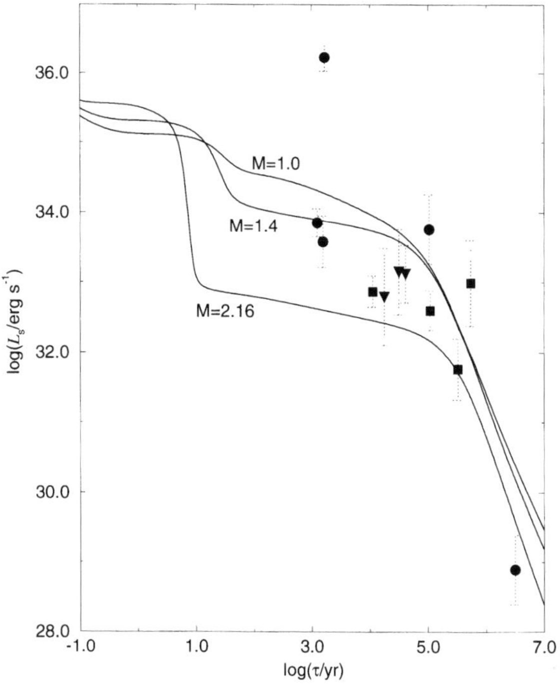

Figure 19.6. Enhanced cooling of neutron stars driven by direct Urca process [543]. The underlying EOS is UV_{14}+UVII. The observed data are labeled in figure 19.5. (Reprinted courtesy of *Nucl. Phys.*)

Urca process. The sudden drop in luminosity at ages of several hundred years has its origin in the fact that the core cools down much faster through the enhanced cooling process than the crust. This causes a temperature inversion in young stars. After several hundred years, depending on the thickness of the crust, the '*cooling wave*', built up by the temperature gradient, reaches the star's surface which leads to the sudden luminosity drop. (An estimate of the time needed by the cooling wave to reach the surface can be found in [711].) The drop of luminosity increases with mass, since the region inside the core of a star where the direct Urca process is possible grows with density. The drop is most strongly pronounced for UV_{14}+UVII. This has its origin in the $^3P_2$ neutron-superfluid which does not extent into the central core.

Changes in the hyperon coupling strengths in matter influence the cooling curves only very little, as can be seen in figure 19.9 for neutron stars of several different masses. This, however, may be drastically different if the direct Urca process among hyperons occurs in dense stellar matter

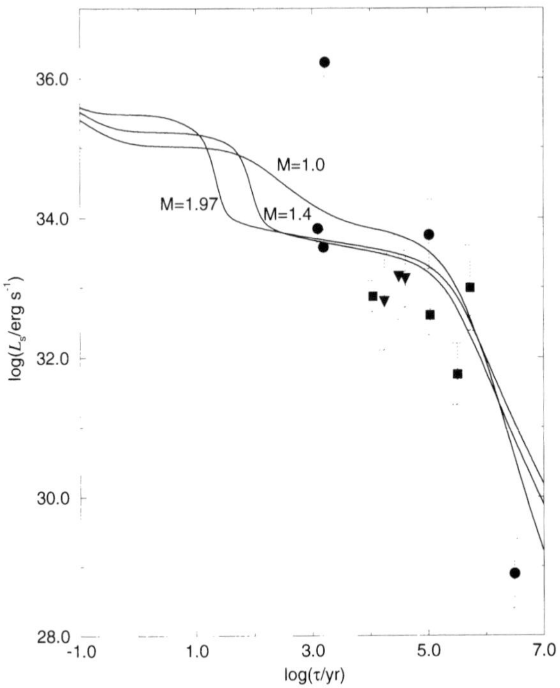

Figure 19.7. Same as figure 19.6, but for HV [543]. The observed data are labeled in figure 19.5. (Reprinted courtesy of *Nucl. Phys.*)

[695, 696]. The star with $M = 0.7\, M_\odot$ does not contain hyperons, since its core is not dense enough to reach the threshold density for hyperon population. Hence there is only one cooling curve in this case. The situation is different for the heavier stars shown in this figure, for which the hyperon concentration increases with star mass [61]. The hyperons are coupled to the meson fields either with the same strengths as the nucleons, that is, $x_\sigma = x_\omega = x_\rho = 1$, or are chosen such that mutual compatibility between the Λ binding energy in saturated nuclear matter, neutron star masses, and experimental data on hypernuclear levels is achieved (cf. section 7.4). As was the case for the other cooling tracks shown in this section, the direct Urca process among the nucleons is included, and no superfluidity is taken into account.

19.7.3 Enhanced cooling via pion condensation

The impact of pion condensation on the cooling of neutron stars [710] is demonstrated in figure 19.10. A comparison with figure 19.8 shows that

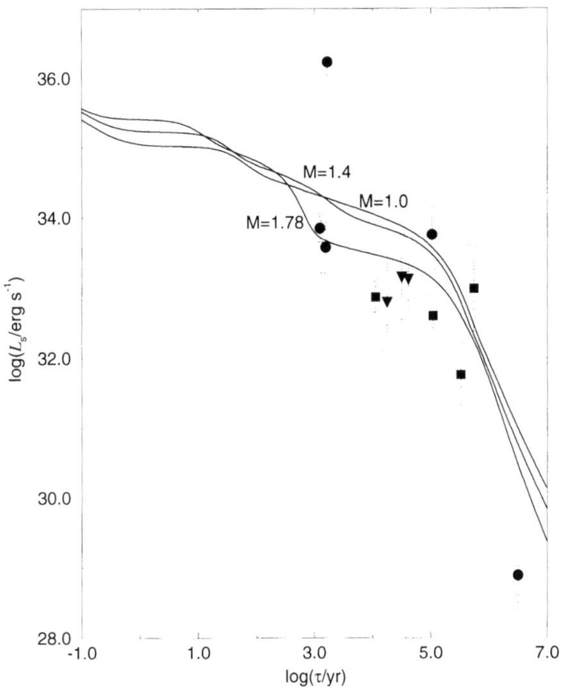

Figure 19.8. Same as figure 19.6, but for G_{300} [543]. The observed data are labeled in figure 19.5. (Reprinted courtesy of *Nucl. Phys.*)

the cooling behavior of such stars is qualitatively similar to the one of stars which cool via the direct Urca process. The threshold density of the pion condensate lies at 3.6×10^{14} g cm^{-3}, just slightly above normal nuclear matter saturation density. Hence all stars displayed in figure 19.10, with the exception of $M = 0.5\, M_\odot$, possess a pion condensate and therefore show enhanced cooling triggered by the condensate.

19.7.4 Enhanced cooling via kaon condensation

The cooling behavior of kaon-condensed stars [424, 675, 710, 712, 713] is qualitatively similar to those of pion-condensed stars. This is graphically illustrated in figure 19.11 for the $G_{300}^{K^-}$ equation of state. There is however a significant quantitative difference between both types of condensate which concerns the different threshold densities. In the present example, kaons go into a Bose condensate around $4.8\rho_0$ [425] which is by a factor of ~ 2.5 larger than the threshold density established for pion condensation, $\sim 1.3\,\rho_0$ [61]. For that reason, a K^- condensate exists only in the most massive stellar

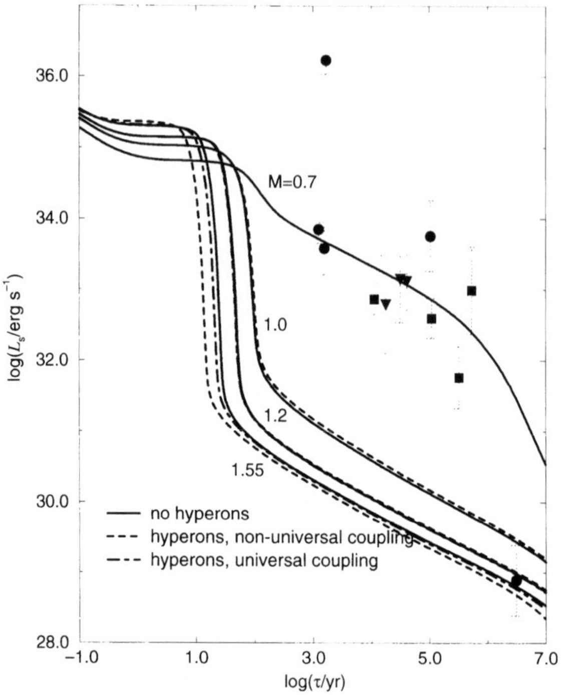

Figure 19.9. Enhanced cooling of neutron stars driven by direct Urca among nucleons [543]. The hyperon coupling strength is varied (see text). The underlying EOS is G_{M78}^{K240}. The observed data are labeled in figure 19.5. (Reprinted courtesy of *Nucl. Phys.*)

model, $M = 1.78\,M_\odot$, shown in figure 19.11. The other two models cool via the modified Urca process only. For the purpose of comparison, the central density of the $M = 1.4\,M_\odot$ neutron star is $2.4\rho_0$ which lies almost by a factor of two below the kaon threshold. The situation would be different for softer models for the equation of state, which lead to higher central star densities.

19.7.5 Enhanced cooling driven by quark matter

As discussed before, modern investigations of the possible transition of confined baryonic matter into quark matter predict transition densities less than twice normal nuclear matter density. Such densities are easily reached in neutron stars with masses $M \approx 1.4\,M_\odot$ [62]. To simulate the cooling history of such stars we apply Glendenning's G_{B180}^{K240} equation of state, whose specific properties have been discussed in connection with table 12.4. The

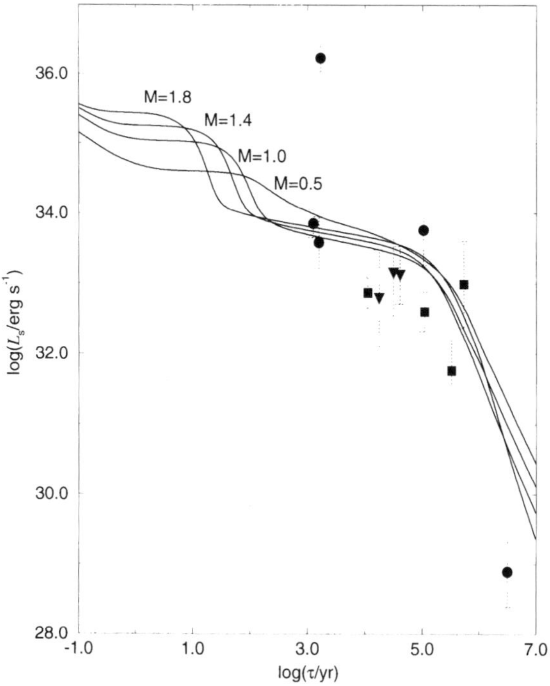

Figure 19.10. Enhanced cooling of neutron stars driven by π^- condensate [543]. The underlying EOS is G^{π}_{300}. The observed data are labeled in figure 19.5. (Reprinted courtesy of *Nucl. Phys.*)

weak reaction processes of the quarks are those given in equations (19.62) and (19.63). They speed up the energy loss of a star considerably, as is illustrated in figure 19.12, where the solid curves correspond to enhanced cooling caused by the quarks [660]. The dashed curves refer to stars with no quarks in their cores, which therefore cool more slowly. Note that the density in stars lighter than about $1\,M_{\odot}$ is not high enough to reach the transition density at which hadronic matter begins to dissolve into quark matter in our model. Enhanced quark cooling is therefore not possible for such light neutron stars. Finally, if the quarks form a superfluid, the emission of neutrinos is suppressed and cooling proceeds at a much slower pace (dotted curves). Bailin and Love have stated that quarks might become superfluid [674]. Unfortunately, there exists – as far as we know – no estimation of the gap energy of superfluid quark matter. Therefore, in order to simulated this effect, we simply assumed a plausible value of $\Delta_{\rm F} = 0.1$ MeV for the $^1{\rm S}_0$ pairing state of quarks, which was taken to be constant over the whole density range.

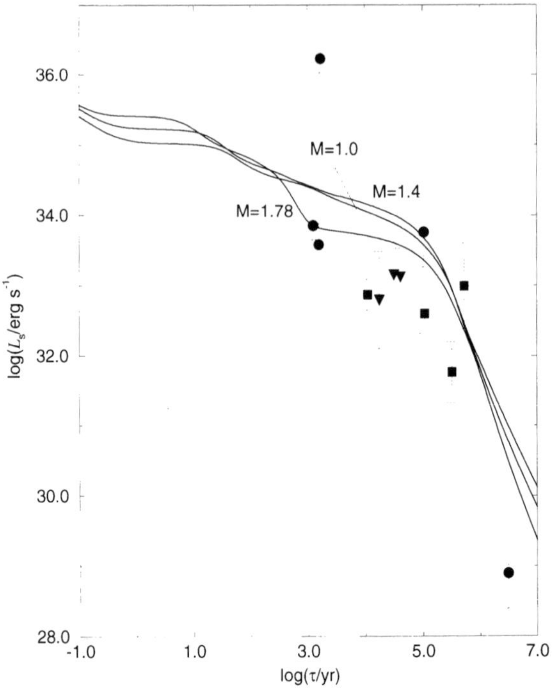

Figure 19.11. Enhanced cooling of neutron stars driven by K^- condensate [543]. The underlying EOS is $\mathrm{G}_{300}^{K^-}$. The observed data are labeled in figure 19.5. (Reprinted courtesy of *Nucl. Phys.*)

19.8 Strange quark matter stars

19.8.1 Enhanced cooling of strange stars

We restrict ourself to the determination of the cooling behavior of strange stars which are either bare or possess the densest possible hadronic crust that can be carried by such objects (i.e. inner crust density equal to neutron drip density). Figure 19.13 reveals that, no matter how thick the crust, such stars cool very rapidly [714]. Among the body of observed luminosities, there is at present one data point that would be consistent with the assumption that the underlying object is a strange star. This interpretation is not stringent of course, because some of the enhanced cooling mechanisms studied above might do as well. If one treats the quarks as superfluid, neutrino production ceases and therefore cooling is considerably slowed down. To simulate this effect qualitatively we assumed a density-independent energy gap of $\Delta_\mathrm{F} = 0.1$ MeV. The numerical outcome of such a simulation is exhibited in figure 19.14. It is seen

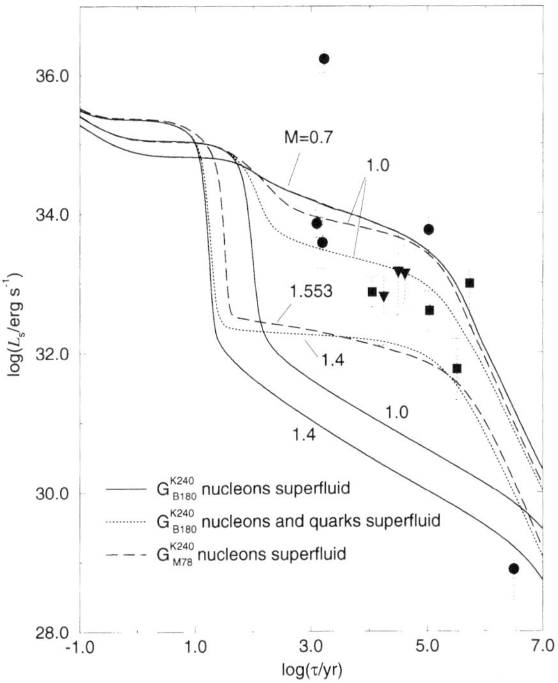

Figure 19.12. Enhanced cooling of neutron stars driven by an admixture of quark matter, computed for the G^{K240}_{B180} EOS [543]. Curves corresponding to G^{K240}_{M78} are without quarks. Enhanced cooling via direct Urca is taken into account too. The observed data are labeled in figure 19.5. (Reprinted courtesy of *Nucl. Phys.*)

that in this case the cooling behavior of strange stars is similar to the enhanced cooling behavior of neutron stars, shown in figures 19.6 to 19.12. Hence, with the exception of PSR 0540–69 practically all observed pulsar luminosities can now be reproduced too. A quantitative clarification of the question as to whether or not quark matter indeed becomes superfluid, and if so, at which densities, attains particular interest from astrophysics too. There is also the possibility that the quark flavors become superfluid at different densities. Consequently the heat capacity would be somewhat smaller. To simulate this possibility we reduce the heat capacity by an (arbitrary) factor of three. The consequences for the cooling behavior are shown by the dotted curves in 19.14.

In closing this section, we mention that one of the main differences between the cooling behavior of neutron stars and strange quark matter stars concerns the time needed for the cooling wave to reach the surface. This time is much shorter for strange quark matter stars because their

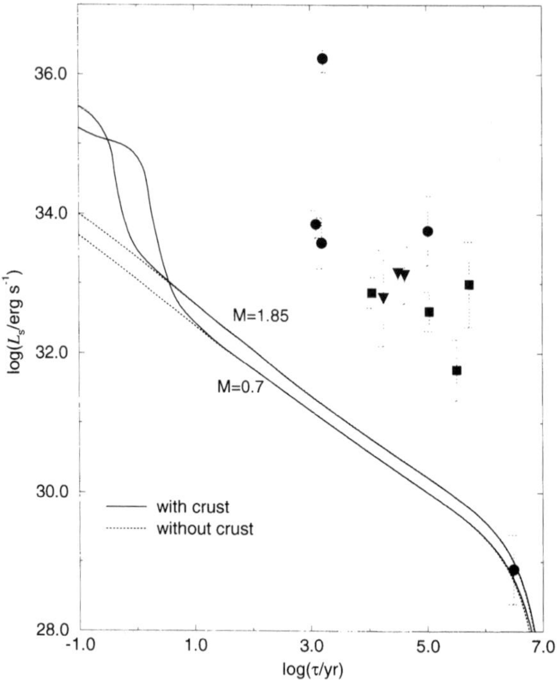

Figure 19.13. Cooling behavior of strange stars of masses $M = 0.7\,M_\odot$ and $M = 1.85\,M_\odot$, with and without crust [543]. The quarks are treated as non-superfluid. The observed data are labeled in figure 19.5. (Reprinted courtesy of *Nucl. Phys.*)

crusts have a thickness of only ~ 0.2 km as opposed to ~ 2 km for stars with $M \sim 1.4\,M_\odot$.

19.8.2 Slowly cooling strange stars

In the previous section we have seen that strange stars can cool more rapidly than ordinary neutron stars. This finding, however, does not hold in general, since the direct Urca process can be forbidden not only in neutron stars but in the strange stars too (cf. section 19.5.4). If so, strange stars would be slowly cooling and their surface temperatures could be more or less indistinguishable from those of slowly cooling neutron stars [715, 716].

To demonstrate this quantitatively, we model strange matter in accordance with the formalism developed in chapter 18, now however by accounting for $\mathcal{O}(\alpha_s)$ corrections too [623, 628]. The equation of state and the quark–lepton composition, governed by the conditions of chemical

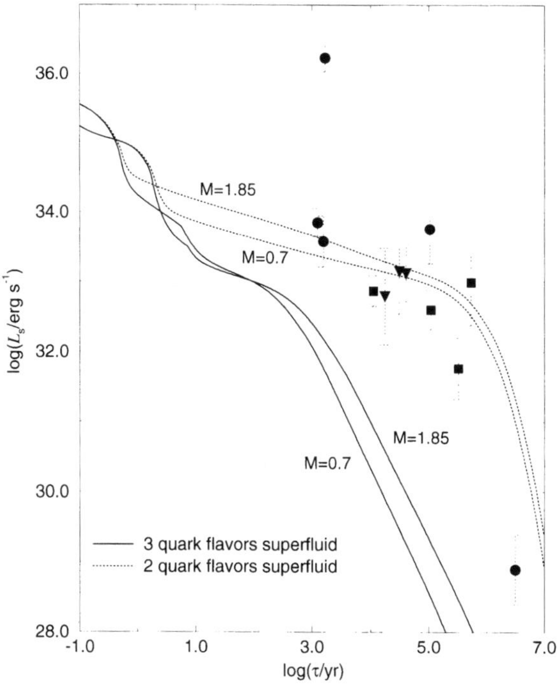

Figure 19.14. Cooling behavior of strange stars with crust [543]. The masses are $M = 0.7\,M_\odot$ and $M = 1.85\,M_\odot$. The quarks are treated as superfluid particles. The heat capacity is reduced by an exponential factor in one model, while in the other model it is kept unchanged. The observed data are labeled in figure 19.5. (Reprinted courtesy of *Nucl. Phys.*)

equilibrium and electric charge neutrality, is derived for that range of model parameters – that is, bag constant $B^{1/4}$, strange quark mass m_s, and strong coupling constant α_s – for which strange matter is absolutely stable with respect to nuclear matter.

In the limiting case of vanishing quark masses, electric charge neutrality can be achieved among the quarks themselves, and therefore electrons are not needed. In the more realistic case of a finite strange quark mass, the density of electrons can nevertheless vanish too above a certain density, depending on the value of the strong coupling constant α_s. Figure 19.15 shows the allowed parameter space of α_s and $B^{1/4}$ for a fixed strange quark mass of $m_s = 100$ MeV. This space is limited by two constraints. Firstly, the energy per baryon of 3-flavor quark matter has to be less than the one of iron (930 MeV), secondly, the energy per baryon of 2-flavor (u, d) quark matter has to be above the one of nucleons (938 MeV)

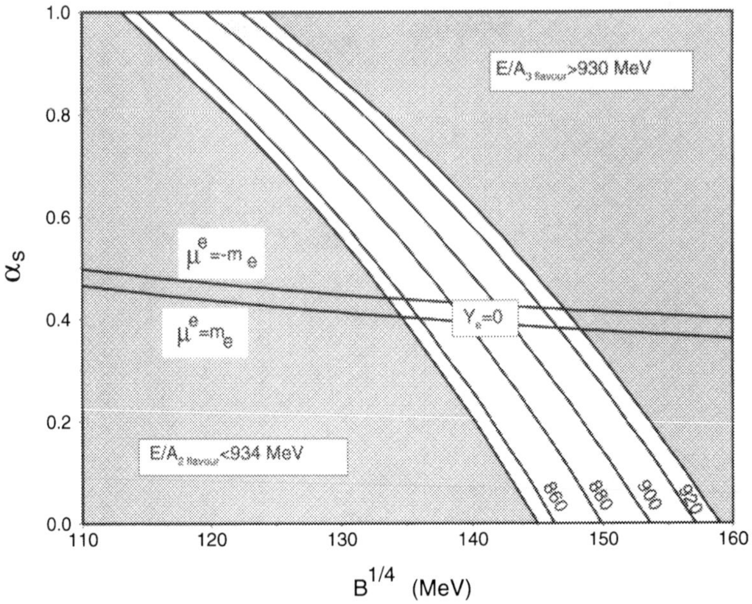

Figure 19.15. Parameter space of strange matter [716]. Shaded regions are excluded since they represent areas where 3-flavor quark matter is either not absolutely stable (i.e. $E/A > 930$ MeV), or nucleons would be unstable against decay into quark matter ($E/A < 934$ MeV). Absolutely stable strange matter exists only in the broad white band, extending from top left to bottom right, where four sample contours of constant energy per baryon (labels denote E/A in MeV) are shown. The area between the two lines labeled $\mu^e = \pm m_e$ is free of electrons, i.e. $Y_e = 0$. (Reprinted courtesy of *J. Phys. G: Nucl. Part. Phys.*)

minus a surface energy correction ($\sim$ 4 MeV) [628]. The horizontal lines labeled $\mu^e = \pm m_e$ represent the parameter sets for which the chemical potential of electrons and positrons are equal to their rest masses (see also section 18.1.1 where we considered the case $\alpha_s = 0$). Between these two lines, electrons and positrons disappear even for a non-vanishing strange quark mass. The behavior of the electron chemical potential depends on the chosen renormalization point ρ_{R}. We follow Duncan *et al* [683] by renormalizing on shell ($\rho_{\mathrm{R}} = m_s$). The renormalization $\rho_{\mathrm{R}} = 300$ MeV ($\approx \mu^s$) suggested by Farhi and Jaffe [628] reduces the strangeness fraction and thus enhances the electron chemical potential. The region for which the electron fraction $Y_e = 0$ is therefore shifted to higher α_s values. Since the MIT bag model is only phenomenological, it would be presumptuous to draw definitive conclusions for both cases.

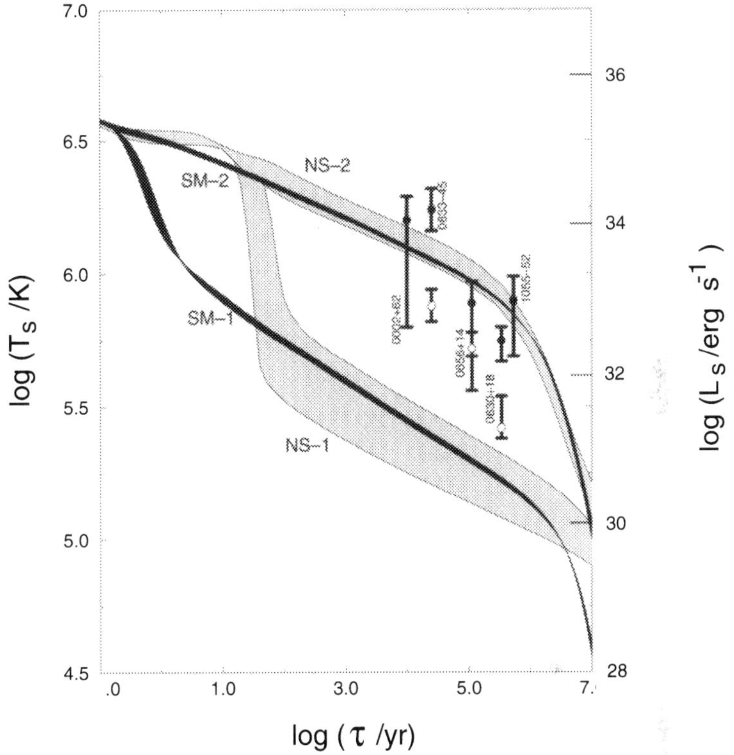

Figure 19.16. Cooling behavior of non-superfluid strange matter stars SM-1 (lower solid band) and SM-2 (upper solid band) and non-superfluid neutron stars NS-1 (lower shaded band) and NS-2 (upper shaded band) [715]. The surface temperatures of several observed neutron stars obtained for a black-body (magnetic hydrogen) atmosphere are marked with error bars with solid (hollow) circles. (Reprinted courtesy of *Astrophys. J.*)

As pointed out by Duncan *et al* [683] (see also [28, 717]), the neutrino emissivity of strange matter depends strongly on its electron fraction Y_e. For that reason we introduce two different complementary parameter sets denoted SM-1 and SM-2 which correspond to strange matter that contains a relatively high electron fraction or is entirely free ($Y_e = 0$) of electrons, respectively.

We have see in the above sections that neutron stars lose energy either via standard cooling or enhanced cooling. Both scenarios may be delayed if part of their matter is in the superfluid phase. Consequently all four options are taken into account in the cooling simulations of neutron stars

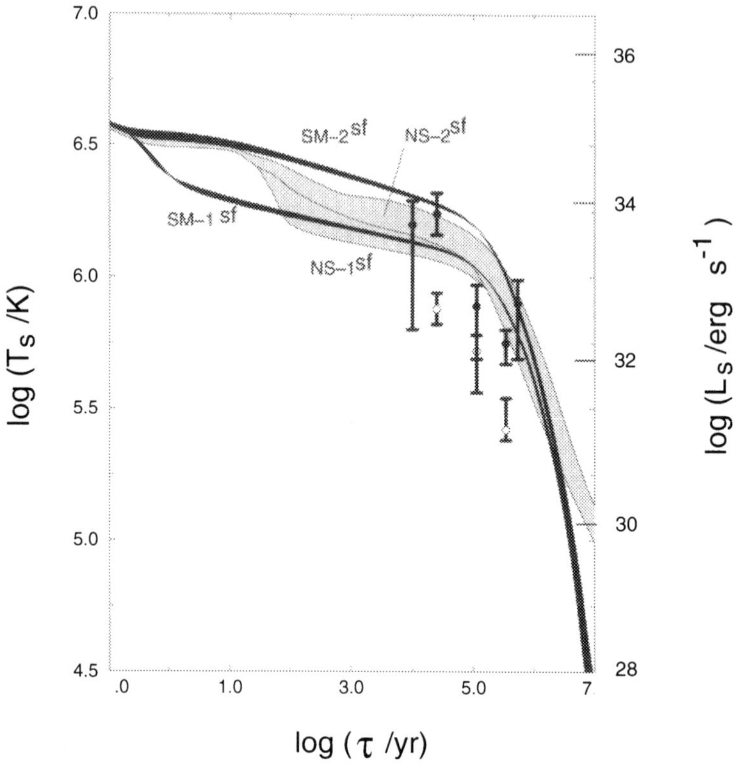

Figure 19.17. Same as figure 19.16, but for superfluid stellar models [715]. (Reprinted courtesy of *Astrophys. J.*)

and strange stars shown in figures 19.16 and 19.17. There the labels NS-1 and NS-2 refer to respectively enhanced cooling and standard cooling of normal neutron star matter, while NS-1^{sf} and NS-2^{sf} denote delayed cooling for superfluid neutron star matter, simulated for the parameters listed in table 19.3. In analogy to this, the corresponding strange-star cooling curves are SM-1 (enhanced cooling) and SM-2 (standard cooling) for normal strange quark matter, and SM-1^{sf} and SM-2^{sf} (delayed cooling) for superfluid quark matter.

All calculations are performed for a gravitational stellar mass of $M = 1.4\, M_{\odot}$, about which the observed pulsar masses tend to cluster. The band-like structure of the cooling curves reflects the uncertainties inherent in the equation of state of neutron-star and strange-star matter. These have their origin, in the case of neutron stars (gray bands), in the different many-body techniques used to solve the nuclear many-body problem. In the latter case, strange-star matter, the solid bands refer to the range of allowed bag values $B^{1/4}$, as indicated in figure 19.15, for which strange quark matter is absolutely stable. One might suspect that the large gap between the cooling tracks of SM-1 and SM-2 in figure 19.17 can be bridged steadily by varying the strong coupling constant α_s. However, it turns out that this gap can be filled only for α_s values that cover an extremely small range. This is caused by the sensitive functional relationship between α_s and the neutrino luminosity L_ν, which is rather steep around those α_s values for which the electrons vanish from the star's quark core. All other values of α_s give cooling tracks which are close to the upper or lower bands. This behavior resembles the case of neutron stars, where the neutrino luminosity depends sensitively on the star's mass.

Besides that, one sees from figures 19.16 and 19.17 that, except for the first ~ 30 years of the lifetime of a newly born pulsar, both neutron stars and strange stars may show more or less the *same* cooling behavior, provided both types of star are made up of either normal matter or superfluid matter. (We will come back to this issue immediately below.) This is made possible by the fact that both standard cooling (NS-2) as well as enhanced cooling (NS-1) in neutron stars has its counterpart in strange stars too (SM-2 and SM-1, respectively). The instant in time at which the surface temperature of a strange star suddenly drops depends on the thickness of the nuclear crust that may envelop the strange matter core and thermally insulate it from the surface [543]. In the present calculation, strange stars possess the densest possible nuclear crust, which is about 0.2 km thick. Thinner crusts would lead to temperature drops at even earlier times. Figures 19.16 and 19.17 indicate that the currently existing body of cooling data of observed pulsars does not allow one to decide about the true nature of the underlying collapsed star, that is, as to whether it is a strange star or a conventional neutron star. This could abruptly change with the observation of a very young pulsar shortly after its formation in a supernova explosion. In this case a prompt drop of the pulsar's temperature, say within the first 30 years after its formation, could offer a good signature of a strange star [146, 634]. This feature, provided it withstands a rigorous future analysis of the microscopic properties of quark matter, could become particularly interesting if continued observation of SN 1987A were to reveal the temperature of the possibly existing pulsar at its center.

Finally, what if only neutron stars are made up of superfluid matter but not strange stars? In this case one has to compare models SM-1 and SM-2 of figure 19.16 with models NS-$1^{\rm sf}$ and NS-$2^{\rm sf}$ of figure 19.17. One sees that this possibility yields to an overall different cooling history between neutron stars and enhanced-cooling strange stars (SM-1). Therefore, the standard argument pointed out quite frequently in the literature that strange stars cool much more rapidly than neutron stars applies only to this special case.

19.9 Summary and concluding remarks about cooling

For the reasons pointed out in sections 3.10 and 19.6, the observed cooling data of pulsars marked with squares in our figures appear to be the most significant ones. It is these data points that should be primarily reproduced by theoretical cooling calculations.

The observed luminosities of pulsars 1951+32 and 1055–52 are compatible with the assumption that these stars cool via the *standard* procedure (section 19.7.1) only. None of the extra cooling mechanisms, from the direct Urca process to quark matter degrees of freedom, is required for them. This may not be the case for pulsars like Geminga, Vela (PSR 0833–45), PSRs 0656+14, 2334+61, 1706–44, 1929+10 (cf. figure 19.5), whose observed luminosities are smaller than those of the pulsars quoted just above. Their low observed luminosities seem to require that (one or more) enhanced cooling processes are at work which, however, are somewhat weakened by a counter agitating physical effect, of which superfluidity fits most comfortably with our understanding of the behavior of matter inside neutron stars. This brings the theoretical cooling tracks computed for the broad range of competing equations of state (table 12.4), denoted as 'intermediate' in figure 19.18, closest to the observed data. A final decision on this issue – standard versus enhanced cooling [37] – is currently hampered by the uncertainties in a number of quantities, including the modified Urca neutrino-emission rate, superfluid gaps, equation of state, and mass and age of the star. If neutron stars should undergo enhanced cooling, it may register itself in their spin-up (lasting for several years) driven by thermal contraction in the earliest epoch of neutron star evolution, when the effect dominates over spin-down by magnetic dipole radiation [718]. The enormous progress that is being made at present in both the exploration of the properties of superdense matter [2, 3] and observational X-ray astronomy, makes one feel confident that a distinction between both types of cooling scenario will be possible in the foreseeable future. Clearly, knowledge of mass, age, and luminosity of one and the same pulsar would be a major step toward constraining the properties of superdense matter from astrophysics.

Although hampered by the large error bars of the data regarding the cooling curves, all equations of state of our collection appear to have problems in explaining more than just a few data points. This makes one wonder whether the observed data are possibly associated with stars of different masses. We recall that the canonical neutron star mass is about $1.4\,M_\odot$, and the observed data [31] seem to scatter about this value. Nevertheless, the existence of pulsars with considerably higher masses, as indicated by the analysis of quasi-periodic oscillations of neutron stars in binary systems, appears quite possible.

Figure 19.19 compares the influence of different *enhanced* neutrino

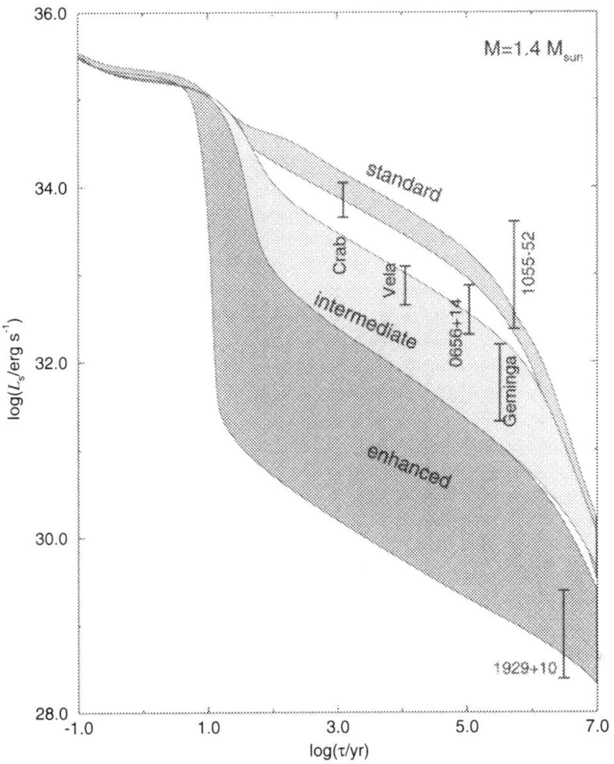

Figure 19.18. Overview of cooling calculations performed for different, competing assumptions about the behavior of superdense neutron star matter. Depending on the cooling processes that are taken into account, three distinct cooling scenarios arise referred to as 'standard', 'intermediate', and 'enhanced' (for details, see text). The band-like structure associated with each scenario reflects the uncertainties inherent in the EOS of superdense matter. All calculations are for a stellar mass of $M = 1.4\,M_\odot$.

emission processes – that is, direct Urca, pion and kaon condensation, and quark Urca – on the cooling behavior of neutron stars of mass $M = 1.4\,M_\odot$. The central density of the $1.4\,M_\odot$ models is not high enough to overcome the threshold density for kaon condensation. Only stars heavier than $1.78\,M_\odot$ provide the sufficiently high-pressure environments in their centers required (for this particular K^- condensation model (cf. section 19.5.3)) for the reaction $e^- \rightarrow K^- + \nu_e$ to take place. The nucleons and quarks are treated as non-superfluid particles. One sees that the inclusion of any of these enhanced cooling processes reduces the star temperature too quickly to achieve agreement with the observed data points. The only exception is

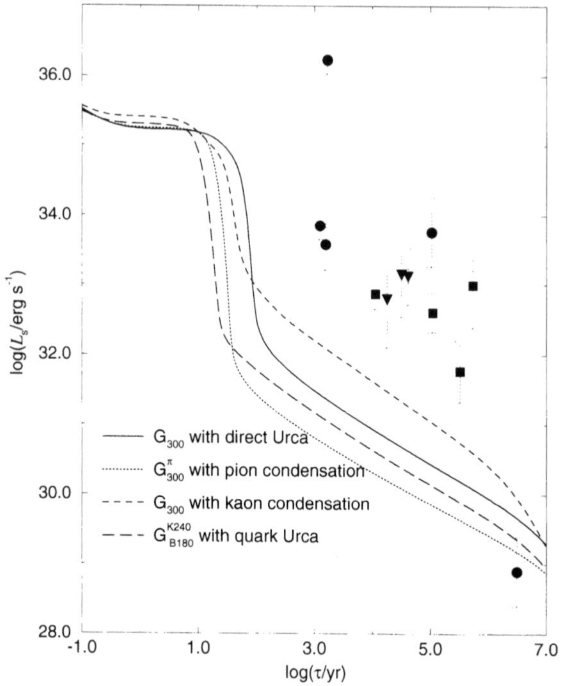

Figure 19.19. Influence of different enhanced neutrino-emitting processes on the cooling of neutron stars [543]. The mass in each case is $M = 1.4\,M_{\odot}$, except for the kaon-condensed model whose mass is $M = 1.78\,M_{\odot}$. The observed data are labeled in figure 19.5. (Reprinted courtesy of *Nucl. Phys.*)

PSR 1929+10. The most significant data points (squares), however, lie considerably above the enhanced cooling curves.

The enhanced cooling scenarios can be *slowed down* if one assumes that the neutrons in the cores of neutron stars are superfluid, as was the case in figures 19.6 to 19.12. Then the only problem left concerns the horizontal 'plateaus' in these figures at stellar ages between 10^2 and 10^5 years which tend to lie somewhat too high for stars with canonical mass, $M = 1.4\,M_{\odot}$, to be consistent with the data. Note, however, that one obtains agreement with some of the data points if the star mass is assumed to be different from the canonical value. Another possibility that would lead to an agreement with observation consists in varying the gap energy, Δ_F, which, as mentioned in section 19.4, is not very accurately known. This is particularly the case for the high-density $^3P_2$ superfluid. A reduction of this gap by a factor of two, for instance, shifts the cooling curves up right into the region where the observed data are concentrated, as can be seen

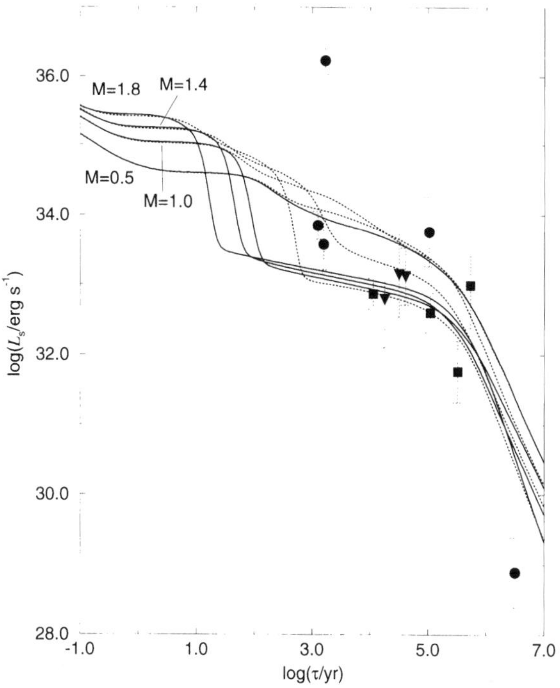

Figure 19.20. Influence of uncertainties in the superfluid $^3P_2$ gap on the cooling of neutron stars constructed for G_{300} [543]. The enhanced cooling processes are direct Urca in pion condensed (solid curves) and kaon condensed (dotted curves) matter. The observed data are labeled in figure 19.5. (Reprinted courtesy of *Nucl. Phys.*)

from figure 19.20. Of course there are other possibilities, besides reducing the gap energy, by means of which agreement with the observed data may be achieved. For example, the superfluid phase of neutrons may not reach the center of very massive neutron stars, as is the case for stars constructed for UV_{14}+UVII (see figure 19.6). This is due to the smaller proton fraction and the resulting higher Fermi momenta of neutrons. Another possibility is an additional cooling process not suppressed by superfluidity, as for example, the superfluid pair-braking [691, 693], or so-called internal heating of neutron stars [719, 720, 721, 722, 723]. The latter could greatly enhance the surface temperatures of middle aged and old pulsars (10^4 to 10^5 years of age).

Our last point concerns the cooling of the hypothetical strange stars. In the early years after the proposition of their possible existence, it was believed that strange stars should be easily distinguishable from their non-strange counterparts because of their much more rapid cooling. Unfortunately, nature seems not to be that kind, the reason being that the direct Urca process, which would be driving rapid strange-star cooling, can be forbidden in such stars, depending on the electron fraction of strange matter. If forbidden, strange stars are slowly cooling, and their surface temperatures would be more or less indistinguishable from those of their slowly cooling non-strange counterparts, the neutron stars. A clear distinction between both classes of stars is possible only under the proviso that the quark direct Urca process is effective in strange stars, leading to a drastic temperature reduction during the first ~ 30 years after the birth of such objects. This feature could become particularly interesting if continued observation of SN 1987A were to reveal the temperature of the possibly existing pulsar at its center.

Appendix A

Notation

A.1 Four-vectors in flat spacetime

The contravariant four-vectors in flat spacetime are denoted as $x^\mu = (x^0, x^1, x^2, x^3) \equiv (x^0, \boldsymbol{x})$, where a boldface letter denotes a three-vector or the three-dimensional part of a contravariant four-vector. The only exception concerns the three-dimensional gradient,

$$\boldsymbol{\nabla} = \left(\frac{\partial}{\partial x^1}, \frac{\partial}{\partial x^2}, \frac{\partial}{\partial x^3} \right) . \tag{A.1}$$

Conversion of a contravariant four-vector to a covariant one is accomplished as $x_\mu = g_{\mu\nu}\, x^\nu$, with the metric tensor of flat spacetime $g^{\mu\nu}$ given by

$$(g_{\mu\nu}) = (g^{\mu\nu}) = \begin{pmatrix} 1 & 0 & 0 & 0 \\ 0 & -1 & 0 & 0 \\ 0 & 0 & -1 & 0 \\ 0 & 0 & 0 & -1 \end{pmatrix} . \tag{A.2}$$

The column and row labels μ and ν are running from $0, 1, 2, 3$, or alternatively from t, r, θ, ϕ. From (A.2) it follows that $x_0 = x^0$ and $x_i = -x^i$.

Summation over repeated Lorentz (Greek) or space (Latin) indices is understood unless explicitly stated otherwise. For example

$$x^\mu\, x_\mu \equiv x^0\, x_0 + x^i\, x_i = (x^0)^2 - x^i\, x^i \equiv (x^0)^2 - \boldsymbol{x} \cdot \boldsymbol{x} \, . \tag{A.3}$$

Derivatives with respect to contravariant (x^μ) and covariant (x_μ) coordinates are defined as

$$\partial_\mu \equiv \frac{\partial}{\partial x^\mu} = \left(\frac{\partial}{\partial x^0}, \frac{\partial}{\partial x^1}, \frac{\partial}{\partial x^2}, \frac{\partial}{\partial x^3} \right) \equiv \left(\frac{\partial}{\partial x^0}, \boldsymbol{\nabla} \right) , \tag{A.4}$$

and

$$\partial^\mu \equiv \frac{\partial}{\partial x_\mu} = \left(\frac{\partial}{\partial x_0}, \frac{\partial}{\partial x_1}, \frac{\partial}{\partial x_2}, \frac{\partial}{\partial x_3} \right) \equiv \left(\frac{\partial}{\partial x_0}, -\boldsymbol{\nabla} \right) . \tag{A.5}$$

In terms of the Laplace operator $\Delta = \boldsymbol{\nabla}^2$, the d'Alembertian operator reads

$$\partial^\mu \, \partial_\mu = \frac{\partial^2}{\partial t^2} - \Delta \, . \tag{A.6}$$

The four-momentum operator is given by

$$p^\mu \equiv k^\mu = \mathrm{i} \, \partial^\mu = (\mathrm{i} \, \partial^0, -\mathrm{i} \, \boldsymbol{\nabla}) \, . \tag{A.7}$$

A.2 Dirac matrices

For the contravariant components of the γ matrices we have $\gamma^\mu \equiv (\gamma^0, \gamma^1, \gamma^2, \gamma^3) \equiv (\gamma^0, \boldsymbol{\gamma})$. The covariant components are obtained as $\gamma_\mu = g_{\mu\nu} \, \gamma^\nu$, which, for the metric tensor (A.2), leads to $\gamma_0 = \gamma^0$ and $\gamma_i = -\gamma^i$. The γ matrices satisfy

$$\gamma^\mu \, \gamma^\nu + \gamma^\nu \, \gamma^\mu = 2 \, g^{\mu\nu} \cdot \mathbf{1} \, , \tag{A.8}$$

with γ^0 Hermitian and γ^i anti-Hermitian, that is,

$$\gamma^{0\dagger} = \gamma^0, \quad \gamma^{i\dagger} = -\gamma^i \quad \Longleftrightarrow \quad \gamma^{\mu\dagger} = \gamma^0 \, \gamma^\mu \, \gamma^0 . \tag{A.9}$$

Equation (A.9) implies that $\gamma^\dagger_\mu = \gamma^\mu$, and from equation (A.8) it follows that $(\gamma^i)^2 = -1$ (no summation over i). The matrices γ^0 and γ^i are related to the β and $\boldsymbol{\alpha}$ matrices through

$$\gamma^0 = \beta, \qquad \boldsymbol{\gamma} = \beta \, \boldsymbol{\alpha} \, . \tag{A.10}$$

Next we introduce the Pauli matrices, which are given by

$$\sigma^1 = \begin{pmatrix} 0 & 1 \\ 1 & 0 \end{pmatrix}, \quad \sigma^2 = \begin{pmatrix} 0 & -\mathrm{i} \\ \mathrm{i} & 0 \end{pmatrix}, \quad \sigma^3 = \begin{pmatrix} 1 & 0 \\ 0 & -1 \end{pmatrix} . \tag{A.11}$$

Note that we reserved the symbol $\boldsymbol{\tau}$ to denote the Pauli matrices in isospin space (isospin operators). In contrast to this, the symbol $\boldsymbol{\sigma}$ is used for the Pauli matrices acting on the spin components of neutrons and protons. This notation complies with the standard notation widely used in the literature. Making use of (A.11), the Dirac representation of the γ matrices can be expressed as

$$\gamma^0 \equiv \beta = \sigma^3 \otimes \mathbf{1} = \begin{pmatrix} 1 & 0 \\ 0 & -1 \end{pmatrix} \otimes \mathbf{1} = \begin{pmatrix} \mathbf{1} & 0 \\ 0 & -\mathbf{1} \end{pmatrix} , \tag{A.12}$$

$$\gamma^i = \mathrm{i}\,\sigma^2 \otimes \sigma^i = \begin{pmatrix} 0 & 1 \\ -1 & 0 \end{pmatrix} \otimes \sigma^i = \begin{pmatrix} 0 & \sigma^i \\ -\sigma^i & 0 \end{pmatrix}, \qquad \text{(A.13)}$$

$$\alpha^i \equiv \gamma^0\,\gamma^i = \sigma^1 \otimes \sigma^i = \begin{pmatrix} 0 & 1 \\ 1 & 0 \end{pmatrix} \otimes \sigma^i = \begin{pmatrix} 0 & \sigma^i \\ \sigma^i & 0 \end{pmatrix}, \qquad \text{(A.14)}$$

$$\gamma^5 = \sigma^i \otimes \mathbf{1} = \begin{pmatrix} 0 & 1 \\ 1 & 0 \end{pmatrix} \otimes \mathbf{1} = \begin{pmatrix} 0 & \mathbf{1} \\ \mathbf{1} & 0 \end{pmatrix}. \qquad \text{(A.15)}$$

The γ^5 matrix can also be written as $\gamma^5 \equiv \mathrm{i}\gamma^0\,\gamma^1\,\gamma^2\,\gamma^3$, and $\gamma_5 \equiv \mathrm{i}\gamma_0\,\gamma_1\,\gamma_2\,\gamma_3$. One thus finds that $\gamma_5 = \gamma^5 = \gamma_5^\dagger$, for the anti-commutator $\{\gamma_5, \gamma^\mu\} = 0$, and $(\gamma^5)^2 = (\gamma_5)^2 = \mathbf{1}$. Moreover, one verifies the following relations:

$$\gamma^\mu\,\gamma_\mu = (\gamma^0)^2 - \gamma^i\,\gamma^i = \mathbf{1} - \boldsymbol{\gamma}\cdot\boldsymbol{\gamma} = 4\cdot\mathbf{1}, \qquad \text{(A.16)}$$

$$\gamma^\mu\,\gamma^\nu\,\gamma_\mu = -\,2\,\gamma^\nu, \qquad k^\mu\,\gamma_\mu\,k^\nu\,\gamma_\nu = k^2, \qquad \text{(A.17)}$$

$$\not{k}\,\not{k}' = k^\mu\,\gamma_\mu\,k'^\nu\,\gamma_\nu = k_\mu\,k'^\mu - \mathrm{i}\,\sigma_{\mu\nu}\,k^\mu\,k'^\nu, \qquad \text{(A.18)}$$

$$\gamma_\mu\,k^\nu\,\gamma_\nu\,k'^\kappa\,\gamma_\kappa\,\gamma^\mu = 4\,k\,k'\mathbf{1}, \qquad \text{(A.19)}$$

$$[\gamma_\mu, \sigma_\nu{}^\mu] = -\,6\,\mathrm{i}\,\gamma_\nu, \qquad k^\lambda\,k^\kappa\,\sigma_{\lambda\mu}\,\sigma_\kappa{}^\mu = 3\,k^2\,\mathbf{1}, \qquad \text{(A.20)}$$

$$k^\lambda\,k^\kappa\,\sigma_{\lambda\mu}\,\gamma^\nu\,\sigma_\kappa{}^\mu = -4\,k^\nu\,k^\mu\,\gamma_\mu + k^2\,\gamma^\nu, \qquad \text{(A.21)}$$

$$\{k^\nu\,\gamma_\nu, \gamma^\mu\} = 2\,k^\mu\,\mathbf{1}, \qquad \gamma^0\,(\gamma^\mu)^\dagger\,\gamma^0 = \gamma^\mu, \qquad \text{(A.22)}$$

$$\gamma^0\,(\sigma^{\mu\nu})^\dagger\,\gamma^0 = \sigma^{\mu\nu}, \qquad \gamma^0\,(\mathrm{i}\,\gamma_5)^\dagger\,\gamma^0 = \mathrm{i}\,\gamma_5. \qquad \text{(A.23)}$$

A numerator of the form $A_1\mathbf{1} + A_2\hat{\boldsymbol{k}}\cdot\boldsymbol{\gamma} + A_3\gamma^0$, where A_1, A_2 and A_3 denote any scalar functions, can be written as

$$\left(A_1\,\mathbf{1} + A_2\,\hat{\boldsymbol{k}}\cdot\boldsymbol{\gamma} + A_3\,\gamma^0\right)^{-1} = \frac{A_1\,\mathbf{1} - A_2\,\hat{\boldsymbol{k}}\cdot\boldsymbol{\gamma} - A_3\,\gamma^0}{A_1^2 + A_2^2 - A_3^2}. \qquad \text{(A.24)}$$

A.3 Traces

The trace of an ordinary product of two matrices is defined as

$$\mathrm{Tr}\,(\mathbf{A}\,\mathbf{B}) \equiv \sum_{\varsigma\varsigma'} (\mathbf{A})_{\varsigma\varsigma'}\,(\mathbf{B})_{\varsigma'\varsigma}\,, \tag{A.25}$$

while for the trace of a direct product

$$\mathrm{Tr}\,(\mathbf{A}\otimes\mathbf{B}) \equiv \sum_{\varsigma} (\mathbf{A}\otimes\mathbf{B})_{\varsigma\varsigma} = \sum_{\varsigma} (\mathbf{A})_{\varsigma\varsigma}\,(\mathbf{B})_{\varsigma\varsigma} = \mathrm{Tr}\,\mathbf{A}\;\mathrm{Tr}\,\mathbf{B}\,. \tag{A.26}$$

This leads to the following relations, which are encountered in chapters 7 and 12, for instance:

$$\mathrm{Tr}\,(\gamma_\nu\,\mathbf{1}) \equiv \mathrm{Tr}\,(\gamma_\nu\otimes\mathbf{1}) = \mathrm{Tr}\,(\gamma_\nu)\;\mathrm{Tr}\,(\mathbf{1}) = 0\,, \tag{A.27}$$

$$\begin{aligned}
\mathrm{Tr}\,(\gamma_\nu\,\boldsymbol{\gamma}\cdot\hat{\boldsymbol{k}}) &= \mathrm{Tr}\,(\gamma_\nu\,\gamma^i\,\hat{k}^i) \\
&= \hat{k}^i\;\mathrm{Tr}\,(\gamma_\nu\,\gamma^i) \\
&= \hat{k}^i\;\mathrm{Tr}\,(g_{\nu\mu}\,\gamma^\mu\,\gamma^i) \\
&= \hat{k}^i\;\mathrm{Tr}\,\left\{g_{\nu\mu}(2\,g^{\mu i}\,\mathbf{1}\;-\gamma^i\,\gamma^\mu)\right\} \\
&= \hat{k}^i\;\mathrm{Tr}\,(2\,\delta_\nu{}^i\,\mathbf{1}\;-\gamma^i\,\gamma_\nu) \\
&= 2\;\mathrm{Tr}\,(\hat{k}^i\,\delta_\nu{}^i\,\mathbf{1}) - \mathrm{Tr}\,(\boldsymbol{\gamma}\cdot\hat{\boldsymbol{k}}\,\gamma_\nu)\,, \\
\Longrightarrow \qquad \mathrm{Tr}\,(\gamma_\nu\,\boldsymbol{\gamma}\cdot\hat{\boldsymbol{k}}) &= 4\,\hat{k}^i\,\delta_\nu{}^i\,,
\end{aligned} \tag{A.28}$$

and, finally,

$$\begin{aligned}
\mathrm{Tr}\,(\gamma_\nu\,\gamma^0) &= \mathrm{Tr}\,(g_{\nu\mu}\,\gamma^\mu\,\gamma^0) \\
&= \mathrm{Tr}\,\left\{g_{\nu\mu}(2\,g^{\mu 0}\,\mathbf{1} - \gamma^0\,\gamma^\mu)\right\} \\
&= \mathrm{Tr}\,(2\,g_{\nu\mu}\,g^{\mu 0}\,\mathbf{1}) - \mathrm{Tr}\,(\gamma_\nu\,\gamma^0)\,, \\
\Longrightarrow \qquad \mathrm{Tr}\,(\gamma_\nu\,\gamma^0) &= 4\,\delta_\nu{}^0\,.
\end{aligned} \tag{A.29}$$

Appendix B

Useful mathematical relationships

B.1 Contour integrals

In chapter 6 we performed a spectral decomposition of the two-point baryon function, $\boldsymbol{g}_1^B$, which culminated in its spectral representation given in equations (6.48) and (6.71),

$$\boldsymbol{g}^B(q^0, \boldsymbol{q}) = \int_{-\infty}^{+\infty} \mathrm{d}\omega \, \frac{\Xi^B(\omega, \boldsymbol{q})}{\omega - (q^0 - \mu^B)(1 + \mathrm{i}\eta)} - 2\mathrm{i}\pi \, \mathrm{sign}(q^0 - \mu^B) \, f(|q^0 - \mu^B|) \, \Xi^B(q^0 - \mu^B, \boldsymbol{q}) \, . \quad \text{(B.1)}$$

With the aid of this representation the integrations over the energy variable in numerous expressions, from baryon self-energies to baryon densities to the equation of state, can then be performed analytically, by means of which $\boldsymbol{g}^B(q)$ becomes replaced by the associated spectral function Ξ^B. In the actual numerical evaluation of the many-body equations, one is thus left with the self-consistent determination of Ξ^B rather than the full $\boldsymbol{g}^B$ function, which renders the problem considerably simpler.

In the following we list a number of examples of such integrals, which are repeatedly encountered in the text. With the help of (B.1), we can write for the simplest type of such integrals

$$\begin{aligned} I_1 &\equiv \int \frac{\mathrm{d}^4 q}{(2\pi)^4} \, \mathrm{e}^{\mathrm{i}\eta q^0} \, \boldsymbol{g}^B(q) \\ &= \int \frac{\mathrm{d}^3 \boldsymbol{q}}{(2\pi)^3} \int_{-\infty}^{+\infty} \frac{\mathrm{d}q^0}{2\pi} \, \mathrm{e}^{\mathrm{i}\eta q^0} \left\{ \int_{-\infty}^{+\infty} \mathrm{d}\omega \, \frac{\Xi^B(\omega, \boldsymbol{q})}{\omega - (q^0 - \mu^B)(1 + \mathrm{i}\eta)} \right. \\ &\quad \left. - 2\mathrm{i}\pi \, \mathrm{sign}(q^0 - \mu^B) \, f(|q^0 - \mu^B|) \, \Xi^B(q^0 - \mu^B, \boldsymbol{q}) \right\} . \quad \text{(B.2)} \end{aligned}$$

Interchanging integrals gives

$$I_1 = \int_{-\infty}^{+\infty} \frac{d^3\boldsymbol{q}}{(2\pi)^3} \int_{-\infty}^{+\infty} d\omega \, \boldsymbol{\Xi}^B(\omega, \boldsymbol{q}) \int_{-\infty}^{+\infty} \frac{dq^0}{2\pi} \, e^{i\eta q^0} \frac{1}{\omega - (q^0 - \mu^B)(1 + i\eta)}$$
$$- i \int \frac{d^3\boldsymbol{q}}{(2\pi)^3} \int_{-\infty}^{+\infty} dq^0 \, e^{i\eta q^0} \, \mathrm{sign}(q^0 - \mu^B) \, f(|q^0 - \mu^B|) \, \boldsymbol{\Xi}^B(q^0 - \mu^B, \boldsymbol{q}) \,. \tag{B.3}$$

Residuum integration in the first term of (B.3) leads to

$$\int_{-\infty}^{+\infty} \frac{dq^0}{2\pi} \, e^{\pm i\eta q^0} \, \frac{1}{\omega - (q^0 - \mu)(1 + i\eta)} = \mp i \, \Theta(-\omega) \,. \tag{B.4}$$

As we know from equations (6.117) and (6.142), the spectral function is given by

$$\boldsymbol{\Xi}^B(\omega, \boldsymbol{q}) = \delta\big(\omega + \mu^B - \omega^B(\boldsymbol{q})\big) \, \boldsymbol{\Xi}^B(\boldsymbol{q}) \,, \tag{B.5}$$

so that upon rescaling the energy argument as $\omega \rightarrow \omega - \mu^B$, one has

$$\boldsymbol{\Xi}(\omega - \mu^B, \boldsymbol{q}) = \delta\big(\omega - \omega^B(\boldsymbol{q})\big) \, \boldsymbol{\Xi}^B(\boldsymbol{q}) \,. \tag{B.6}$$

Substituting (B.4) and (B.5) into (B.3) and making use of the identity

$$\Theta(\mu^B - \omega^B(\boldsymbol{q})) + \mathrm{sign}(\omega^B(\boldsymbol{q}) - \mu^B) \, f(|\omega^B(\boldsymbol{q}) - \mu^B|) = f(\omega^B(\boldsymbol{q}) - \mu^B) \tag{B.7}$$

then leads to

$$\int \frac{d^4q}{(2\pi)^4} \, e^{i\eta q^0} \, \boldsymbol{g}^B(q) \; = \; - i \int \frac{d^3\boldsymbol{q}}{(2\pi)^3} \, \boldsymbol{\Xi}^B(\boldsymbol{q}) \, f(\omega^B(\boldsymbol{q}) - \mu^B) \tag{B.8}$$

$$\stackrel{T \to 0}{\longrightarrow} - i \int \frac{d^3\boldsymbol{q}}{(2\pi)^3} \, \boldsymbol{\Xi}^B(\boldsymbol{q}) \, \Theta(q_{F_B} - |\boldsymbol{q}|) \,, \tag{B.9}$$

where the latter expression displays the zero-temperature limit. Similarly, one calculates for the antiparticle integral ($\mu^B > 0$)

$$\int \frac{d^4q}{(2\pi)^4} \, e^{-i\eta q^0} \, \boldsymbol{g}^B(q) \; = \; i \int \frac{d^3\boldsymbol{q}}{(2\pi)^3} \, \bar{\boldsymbol{\Xi}}^B(\boldsymbol{q}) \, f(-(\bar{\omega}^B(\boldsymbol{q}) - \mu^B)) \tag{B.10}$$

$$\stackrel{T \to 0}{\longrightarrow} 0 \,, \tag{B.11}$$

where the analog of (B.7) is given by

$$- \Theta(\bar{\omega}^B(\boldsymbol{q}) - \mu^B) + \mathrm{sign}(\bar{\omega}^B(\boldsymbol{q}) - \mu^B) \, f(|\bar{\omega}^B(\boldsymbol{q}) - \mu^B|)$$
$$= - f(-(\bar{\omega}^B(\boldsymbol{q}) - \mu^B)) \,. \tag{B.12}$$

The second type of integral has the structure

$$\begin{aligned} I_2 &\equiv \int \frac{\mathrm{d}^4 q}{(2\pi)^4}\, \mathrm{e}^{\mathrm{i}\eta q^0}\, q^0\, \boldsymbol{g}^B(q) \\ &= \int \frac{\mathrm{d}^3 \boldsymbol{q}}{(2\pi)^3} \int\limits_{-\infty}^{+\infty} \frac{\mathrm{d}q^0}{2\pi} \mathrm{e}^{\mathrm{i}\eta q^0}\, q^0 \Bigg\{ \int\limits_{-\infty}^{+\infty} \mathrm{d}\omega\, \frac{\boldsymbol{\Xi}^B(\omega, \boldsymbol{q})}{\omega - (q^0 - \mu)(1 + \mathrm{i}\eta)} \\ &\qquad - 2\mathrm{i}\pi\, \mathrm{sign}(q^0 - \mu^B)\, f(|q^0 - \mu^B|)\, \boldsymbol{\Xi}^B(q^0 - \mu^B, \boldsymbol{q}) \Bigg\}, \end{aligned} \tag{B.13}$$

which we write as

$$\begin{aligned} I_2 = &\int\limits_{-\infty}^{+\infty} \frac{\mathrm{d}^3 \boldsymbol{q}}{(2\pi)^3} \int\limits_{-\infty}^{+\infty} \mathrm{d}\omega\, \boldsymbol{\Xi}^B(\omega, \boldsymbol{q}) \int\limits_{-\infty}^{+\infty} \frac{\mathrm{d}q^0}{2\pi}\, \mathrm{e}^{\mathrm{i}\eta q^0} \frac{q^0}{\omega - (q^0 - \mu)(1 + \mathrm{i}\eta)} \\ &- \mathrm{i} \int \frac{\mathrm{d}^3 \boldsymbol{q}}{(2\pi)^3} \int\limits_{-\infty}^{+\infty} \frac{\mathrm{d}q^0}{2\pi}\, \mathrm{e}^{\mathrm{i}\eta q^0}\, q^0\, \mathrm{sign}(q^0 - \mu^B)\, f(|q^0 - \mu^B|)\, \boldsymbol{\Xi}^B(q^0 - \mu^B, \boldsymbol{q}) \,. \end{aligned} \tag{B.14}$$

Contour integration gives

$$\int\limits_{-\infty}^{+\infty} \frac{\mathrm{d}q^0}{2\pi}\, \mathrm{e}^{\pm\mathrm{i}\eta q^0} \frac{q^0}{\omega - (q^0 - \mu)(1 + \mathrm{i}\eta)} = \mp\mathrm{i}\, (\omega + \mu^B)\, \Theta(-\omega), \tag{B.15}$$

and therefore for the particles

$$\int \frac{\mathrm{d}^4 q}{(2\pi)^4}\, \mathrm{e}^{\mathrm{i}\eta q^0}\, q^0\, \boldsymbol{g}^B(q) = -\mathrm{i} \int \frac{\mathrm{d}^3 \boldsymbol{q}}{(2\pi)^3}\, \boldsymbol{\Xi}^B(\boldsymbol{q})\, \omega^B(\boldsymbol{q})\, f(\omega^B(\boldsymbol{q}) - \mu^B), \tag{B.16}$$

which for the thermally excited antiparticles gives

$$\int \frac{\mathrm{d}^4 q}{(2\pi)^4}\, \mathrm{e}^{-\mathrm{i}\eta q^0}\, q^0\, \boldsymbol{g}^B(q) = \mathrm{i} \int \frac{\mathrm{d}^3 \boldsymbol{q}}{(2\pi)^3}\, \bar{\boldsymbol{\Xi}}^B(\boldsymbol{q})\, \bar{\omega}^B(\boldsymbol{q})\, f(-(\bar{\omega}^B(\boldsymbol{q}) - \mu^B)) \,. \tag{B.17}$$

To evaluate the last type of integral, we note that

$$\int\limits_{-\infty}^{+\infty} \frac{\mathrm{d}q^0}{2\pi}\, \mathrm{e}^{\pm\mathrm{i}\eta q^0} \frac{\boldsymbol{\Sigma}^B(q^0, \boldsymbol{q})}{\omega - (q^0 - \mu)(1 + \mathrm{i}\eta)} = \mp\mathrm{i}\, \boldsymbol{\Sigma}^B(\omega + \mu^B, \boldsymbol{q})\, \Theta(-\omega) \,. \tag{B.18}$$

Therefore

$$\int \frac{d^4q}{(2\pi)^4}\, e^{i\eta q^0}\, \boldsymbol{\Sigma}^B(q)\, \boldsymbol{g}^B(q) = -i \int \frac{d^3\boldsymbol{q}}{(2\pi)^3}\, \boldsymbol{\Xi}^B(\boldsymbol{q})\, \boldsymbol{\Sigma}^B(\omega^B(\boldsymbol{q}),\boldsymbol{q})\, f(\omega^B(\boldsymbol{q}) - \mu^B)\,, \tag{B.19}$$

and

$$\int \frac{d^4q}{(2\pi)^4}\, e^{-i\eta q^0}\, \boldsymbol{\Sigma}^B(q)\, \boldsymbol{g}^B(q) = i \int \frac{d^3\boldsymbol{q}}{(2\pi)^3}\, \bar{\boldsymbol{\Xi}}^B(\boldsymbol{q})\, \boldsymbol{\Sigma}^B(\bar{\omega}^B(\boldsymbol{q}),\boldsymbol{q})\, f(-(\bar{\omega}^B(\boldsymbol{q}) - \mu^B))\,. \tag{B.20}$$

B.2 Fourier transformations and related formulae

For the Fourier transformations, we adopt the following conventions,

$$g(k) = \int d^4x\, e^{ikx}\, g(x)\,, \qquad g(x) = \int \frac{d^4k}{(2\pi)^4}\, e^{-ikx}\, g(k)\,, \tag{B.21}$$

$$\delta^4(k) = \int \frac{d^4x}{(2\pi)^4}\, e^{ikx}\,, \qquad \delta^4(x) = \int \frac{d^4k}{(2\pi)^4}\, e^{-ikx}\,. \tag{B.22}$$

It thus follows that $\partial^\mu \equiv \partial/\partial x_\mu = -ik^\mu$. Moreover, with the aid of (B.21) and (B.22), the following Fourier representations are readily verified:

$$\int d^4x'\, \Delta^{0\sigma}(x_1 - x')\, \boldsymbol{g}^B(x_1, x_1')\, \boldsymbol{g}^{B'}(x', x'^+) = \int \frac{d^4k}{(2\pi)^4} \int \frac{d^4q}{(2\pi)^4}\, e^{i\eta k^0}\, e^{-iq(x_1 - x_1')}\, \Delta^{0\sigma}(0)\, \boldsymbol{g}^{B'}(k)\, \boldsymbol{g}^B(q)\,, \tag{B.23}$$

$$\int d^4x'\, \Delta^{0\sigma}(x_1 - x')\, \boldsymbol{g}^B(x_1, x_1'^+)\, \boldsymbol{g}^{B'}(x', x_1') = \int \frac{d^4k}{(2\pi)^4} \int \frac{d^4q}{(2\pi)^4}\, e^{i\eta k^0}\, e^{-iq(x_1 - x_1')}\, \Delta^{0\sigma}(q-k)\, \boldsymbol{g}^B(k)\, \boldsymbol{g}^B(q)\,, \tag{B.24}$$

$$\lim_{x' \to x^+} \boldsymbol{g}^B(x, x') = \lim_{x' \to x^+} \int \frac{d^4k}{(2\pi)^4}\, e^{-ik(x-x')}\, \boldsymbol{g}^B(k) \equiv \int \frac{d^4k}{(2\pi)^4}\, e^{i\eta k^0}\, \boldsymbol{g}^B(k)\,, \tag{B.25}$$

and

$$\lim_{x' \to x^+} \partial_\nu\, \boldsymbol{g}^B(x, x') = -i \int \frac{d^4k}{(2\pi)^4}\, e^{i\eta k^0}\, k_\nu\, \boldsymbol{g}^B(k)\,. \tag{B.26}$$

The convergence factor,

$$\lim_{x' \to x^+} \mathrm{e}^{-\mathrm{i}k(x-x')} \equiv \lim_{\eta \to 0^+} \mathrm{e}^{\mathrm{i}\eta k^0} , \tag{B.27}$$

in the above relations and the related ones defines the respective path of integration in the complex k^0 plane. Throughout this book the limit $\eta \to 0^+$ is implicit whenever such a factor appears.

B.3 Momentum integrals

The following momentum integrals occur in the calculation of baryon self-energies, baryon densities, and the equation of state in the relativistic Hartree approximation:

$$\int_0^{k_F} \mathrm{d}k \, \frac{\boldsymbol{k}^2}{\sqrt{(m^*)^2 + \boldsymbol{k}^2}} = \frac{1}{2} \, k_F \sqrt{(m^*)^2 + k_F^2} - \frac{1}{2} \, (m^*)^2 \, \ln \left(\frac{k_F + \sqrt{(m^*)^2 + k_F^2}}{m^*} \right) , \tag{B.28}$$

$$\int_0^{k_F} \mathrm{d}k \, \frac{\boldsymbol{k}^4}{\sqrt{(m^*)^2 + \boldsymbol{k}^2}} = \frac{1}{4} \, k_F \left\{ k_F^2 - \frac{3}{2} \, (m^*)^2 \right\} \sqrt{(m^*)^2 + k_F^2} + \frac{3}{8} \, (m^*)^4 \, \ln \left(\frac{k_F + \sqrt{(m^*)^2 + k_F^2}}{m^*} \right) , \tag{B.29}$$

and

$$\int_0^{k_F} \mathrm{d}k \, \boldsymbol{k}^2 \, \sqrt{(m^*)^2 + \boldsymbol{k}^2} = \frac{1}{8} \left\{ k_F \, (2 \, k_F^2 + (m^*)^2) \, \sqrt{(m^*)^2 + k_F^2} - (m^*)^4 \, \ln \left(\frac{k_F + \sqrt{(m^*)^2 + k_F^2}}{m^*} \right) \right\} . \tag{B.30}$$

Appendix C

Hartree–Fock self-energies at zero temperature

In chapter 7 we have discussed at great length the derivation of the Hartree–Fock self-energy expressions which originate from the exchange of σ and ω mesons among the baryons, B. The contributions of the still missing mesons, ρ and π, to $\boldsymbol{\Sigma}^{\mathrm{HF},B}$ are listed below. The ρ meson contributes to $\boldsymbol{\Sigma}^{B}$ as

$$\boldsymbol{\Sigma}^{\mathrm{H},B}_{\zeta_1\zeta_1'}(q_{F_{B'}})\Big|_{\rho} = 2\,\gamma^0_{\zeta_1\zeta_1'}\left(\frac{g_{\rho B}}{m_\rho}\right)^2 \sum_{B'} (2J_{B'}+1)\, I_{3B'}\left(\frac{g_{\rho B'}}{g_{\rho B}}\right) \times \int \frac{\mathrm{d}^3\boldsymbol{q}}{(2\pi)^3}\, \Xi^{B'}_0(\boldsymbol{q})\,\Theta(q_{F_{B'}}-|\boldsymbol{q}|)\,, \tag{C.1}$$

with the quantum numbers of spin, J_B, and isospin, I_B, given in table 5.1.

The Fock terms to the total baryon self-energy, $\boldsymbol{\Sigma}^{B}$, are given by

$$\begin{aligned}\boldsymbol{\Sigma}^{\mathrm{F},B}_{\zeta_1\zeta_1'}(k)\Big|_{\pi} = &\left(\frac{f_{\pi B}}{m_\pi}\right)^2 \int \frac{\mathrm{d}^3\boldsymbol{q}}{(2\pi)^3}\Big\{\delta_{\zeta_1\zeta_1'}\Big[-(k^0-\omega^B(\boldsymbol{q}))^2+(\boldsymbol{k}-\boldsymbol{q})^2\Big]\\
&\times \Xi^B_S(\omega^B(\boldsymbol{q}),\boldsymbol{q}) + (\gamma\cdot\hat{\boldsymbol{k}})_{\zeta_1\zeta_1'}\Big[\Big(2(|\boldsymbol{q}|\,\hat{\boldsymbol{k}}\cdot\hat{\boldsymbol{q}}-|\boldsymbol{k}|)\,(|\boldsymbol{k}|\hat{\boldsymbol{k}}\cdot\hat{\boldsymbol{q}}-|\boldsymbol{q}|)\\
&-\Big((k^0-\omega^B(\boldsymbol{q}))^2-(\boldsymbol{k}-\boldsymbol{q})^2\Big)\,\hat{\boldsymbol{k}}\cdot\hat{\boldsymbol{q}}\Big)\,\Xi^B_V(\omega^B(\boldsymbol{q}),\boldsymbol{q})\\
&+2(|\boldsymbol{q}|\hat{\boldsymbol{k}}\cdot\hat{\boldsymbol{q}}-|\boldsymbol{k}|)\,(k^0-\omega^B(\boldsymbol{q}))\,\Xi^B_0(\omega^B(\boldsymbol{q}),\boldsymbol{q})\Big]\\
&+\gamma^0_{\zeta_1\zeta_1'}\Big[2(|\boldsymbol{k}|\,\hat{\boldsymbol{k}}\cdot\hat{\boldsymbol{q}}-|\boldsymbol{q}|)\,(k^0-\omega^B(\boldsymbol{q}))\,\Xi^B_V(\omega^B(\boldsymbol{q}),\boldsymbol{q})\\
&+\Big((k^0-\omega^B(\boldsymbol{q}))^2+(\boldsymbol{k}-\boldsymbol{q})^2\Big)\,\Xi^B_0(\omega^B(\boldsymbol{q}),\boldsymbol{q})\Big]\Big\}\\
&\times\,\Delta^{0\pi}(k^0-\omega^B(\boldsymbol{q}),\boldsymbol{k}-\boldsymbol{q})\,\Theta(q_{F_B}-|\boldsymbol{q}|)\,,\end{aligned} \tag{C.2}$$

and

$$
\begin{aligned}
\Sigma^{\mathrm{F},B}_{\zeta_1\zeta_1'}(k)\Big|_{\rho} = g^2_{\rho B} \int \frac{\mathrm{d}^3\boldsymbol{q}}{(2\pi)^3} \Big\{ \delta_{\zeta_1\zeta_1'} \Big[\Big(-4 + \mathcal{X}^B_1 \Big((k^0 - \omega^B(\boldsymbol{q}))^2 - (\boldsymbol{k}-\boldsymbol{q})^2 \Big) \Big) \\
\times \Xi^B_S(\omega^B(\boldsymbol{q}),\boldsymbol{q}) + \frac{3 f_{\rho B}}{m_B g_{\rho B}} \Big(\hat{\boldsymbol{q}} \cdot (\boldsymbol{k}-\boldsymbol{q})\, \Xi^B_V(\omega^B(\boldsymbol{q}),\boldsymbol{q}) \\
+ (k^0 - \omega^B(\boldsymbol{q}))\, \Xi^B_0(\omega^B(\boldsymbol{q}),\boldsymbol{q}) \Big) \Big] \\
+ (\boldsymbol{\gamma} \cdot \hat{\boldsymbol{k}})_{\zeta_1\zeta_1'} \Big[\frac{3 f_{\rho B}}{m_B g_{\rho B}} \left(|\boldsymbol{k}| - |\boldsymbol{q}|\, \hat{\boldsymbol{k}} \cdot \hat{\boldsymbol{q}} \right) \Xi^B_S(\omega^B(\boldsymbol{q}),\boldsymbol{q}) \\
+ \Big(2\, \hat{\boldsymbol{k}} \cdot \hat{\boldsymbol{q}} + |\boldsymbol{k}|\, |\boldsymbol{q}|\, (\hat{\boldsymbol{k}} \cdot \hat{\boldsymbol{q}})^2 \, (\mathcal{X}^B_2 - 2\mathcal{X}^B_3) + |\boldsymbol{k}|\, |\boldsymbol{q}|\, \mathcal{X}^B_2 \\
- (\boldsymbol{k}^2 + \boldsymbol{q}^2)\, (\hat{\boldsymbol{k}} \cdot \hat{\boldsymbol{q}})\, (\mathcal{X}^B_2 - \mathcal{X}^B_3) - \mathcal{X}^B_3\, (k^0 - \omega^B(\boldsymbol{q}))^2\, \hat{\boldsymbol{k}} \cdot \hat{\boldsymbol{q}} \Big)\, \Xi^B_V(\omega^B(\boldsymbol{q}),\boldsymbol{q}) \\
+ \mathcal{X}^B_2\, (k^0 - \omega^B(\boldsymbol{q}))\, (|\boldsymbol{q}| \hat{\boldsymbol{k}} \cdot \hat{\boldsymbol{q}} - |\boldsymbol{k}|)\, \Xi^B_0(\omega^B(\boldsymbol{q}),\boldsymbol{q}) \Big] \\
+ \gamma^0_{\zeta_1\zeta_1'} \Big[-\frac{3 f_{\rho B}}{m_B g_{\rho B}} (k^0 - \omega^B(\boldsymbol{q}))\, \Xi^B_S(\omega^B(\boldsymbol{q}),\boldsymbol{q}) + \mathcal{X}^B_2\, (k^0 - \omega^B(\boldsymbol{q})) \\
\times \boldsymbol{q} \cdot (\boldsymbol{k}-\boldsymbol{q})\, \Xi^B_V(\omega^B(\boldsymbol{q}),\boldsymbol{q}) \\
+ \Big((\mathcal{X}^B_2 - \mathcal{X}^B_3)\, (k^0 - \omega^B(\boldsymbol{q}))^2 + \mathcal{X}^B_3\, (\boldsymbol{k}-\boldsymbol{q})^2 + 2 \Big)\, \Xi^B_0(\omega^B(\boldsymbol{q}),\boldsymbol{q}) \Big] \Big\} \\
\times \Delta^{0\rho}(k^0 - \omega^B(\boldsymbol{q}), \boldsymbol{k}-\boldsymbol{q})\, \Theta(q_{F_B} - |\boldsymbol{q}|)\,. \qquad \text{(C.3)}
\end{aligned}
$$

The auxiliary quantities $\mathcal{X}^B_1$, $\mathcal{X}^B_2$, and $\mathcal{X}^B_3$ are defined in equations (D.16) to (D.18).

Appendix D

Hartree–Fock self-energies at finite temperature

Below we summarize the expressions for the baryon's self-energies, decomposed into Hartree (H) and Fock (F) terms, for the relativistic Hartree–Fock approximation at finite temperature. The self-energies are obtained from equations (7.2) and (7.3) with $\langle 12|\mathbf{V}^{BB'}|34\rangle$ given in (5.151). Adopting the notation introduced in section 5.4 and chapter 7, the Hartree–Fock self-energies originating from σ-meson exchange among the baryons read

$$\boldsymbol{\Sigma}^{\mathrm{H},B}_{\zeta_1\zeta_1'}\Big|_\sigma = -\mathrm{i}\,\delta_{\zeta_1\zeta_1'}\,\Delta^{0\sigma}(0)\,g_{\sigma B} \times \sum_{B'} g_{\sigma B'}\int\frac{\mathrm{d}^4q}{(2\pi)^4}\left(\mathrm{e}^{\mathrm{i}\eta q^0}+\mathrm{e}^{-\mathrm{i}\eta q^0}\right)\,\mathrm{Tr}\,\boldsymbol{g}^{B'}(q)\,, \tag{D.1}$$

$$\boldsymbol{\Sigma}^{\mathrm{F},B}_{\zeta_1\zeta_1'}(k)\Big|_\sigma = \mathrm{i}\,g^2_{\sigma B}\int\frac{\mathrm{d}^4q}{(2\pi)^4}\left(\mathrm{e}^{\mathrm{i}\eta q^0}+\mathrm{e}^{-\mathrm{i}\eta q^0}\right)\Delta^{0\sigma}(k-q)\big(\mathbf{1}\otimes\boldsymbol{g}^B(q)\otimes\mathbf{1}\big)_{\zeta_1\zeta_1'}\,. \tag{D.2}$$

The baryon self-energies due to ω-meson exchange read

$$\boldsymbol{\Sigma}^{\mathrm{H},B}_{\zeta_1\zeta_1'}(k)\Big|_\omega = \mathrm{i}\,\gamma^\mu_{\zeta_1\zeta_1'}\,\mathcal{D}^{0\omega}_{\mu\nu}(0)\,g_{\omega B}\sum_{B'} g_{\omega B'} \times\int\frac{\mathrm{d}^4q}{(2\pi)^4}\left(\mathrm{e}^{\mathrm{i}\eta q^0}+\mathrm{e}^{-\mathrm{i}\eta q^0}\right)\,\mathrm{Tr}\left(\gamma^\nu\boldsymbol{g}^{B'}(q)\right)\,, \tag{D.3}$$

$$\boldsymbol{\Sigma}^{\mathrm{F},B}_{\zeta_1\zeta_1'}(k)\Big|_\omega = -\mathrm{i}\,g^2_{\omega B}\int\frac{\mathrm{d}^4q}{(2\pi)^4}\left(\mathrm{e}^{\mathrm{i}\eta q^0}+\mathrm{e}^{-\mathrm{i}\eta q^0}\right)\gamma^\mu_{\zeta_1\zeta_3}\,\gamma^\nu_{\zeta_2\zeta_1'} \times\mathcal{D}^{0\omega}_{\mu\nu}(k-q)\,\boldsymbol{g}^B_{\zeta_3\zeta_2}(q)\,. \tag{D.4}$$

The π-meson contribution to the baryon self-energy vanishes, that is,

$$\boldsymbol{\Sigma}^{\mathrm{H},B}_{\zeta_1\zeta_1'}\Big|_{\pi} = 0\,, \tag{D.5}$$

since the derivatives in $\Gamma^{\pi B}$ (cf. equation (5.154)) lead to contributions in momentum space of the form $k^\mu \delta(k^\mu)$. The self-energy exchange term is not touched by this and is given by

$$\boldsymbol{\Sigma}^{\mathrm{F},B}_{\zeta_1\zeta_1'}(k)\Big|_{\pi} = \mathrm{i}\left(\frac{f_{\pi B}}{m_\pi}\right)^2 \int \frac{\mathrm{d}^4 q}{(2\pi)^4}\left(\mathrm{e}^{\mathrm{i}\eta q^0} + \mathrm{e}^{-\mathrm{i}\eta q^0}\right)(\gamma_5\gamma_\mu)_{\zeta_1\zeta_3}\left(\gamma_5\gamma_\nu\right)_{\zeta_2\zeta_1'} \\ \times (k-q)^\mu\,(k-q)^\nu\,\Delta^{0\pi}(k-q)\,\boldsymbol{g}^B_{\zeta_3\zeta_2}(q)\,. \tag{D.6}$$

Finally, the baryon self-energies originating from ρ-meson exchange have the form

$$\boldsymbol{\Sigma}^{\mathrm{H},B}_{\zeta_1\zeta_1'}(k)\Big|_{\rho} = \mathrm{i}\sum_{B'}\int \frac{\mathrm{d}^4 q}{(2\pi)^4}\left(\mathrm{e}^{\mathrm{i}\eta q^0} + \mathrm{e}^{-\mathrm{i}\eta q^0}\right) \\ \times \left[\left(g_{\rho B}\gamma_\mu - \mathrm{i}\,\frac{f_{\rho B}}{2m_B}\,(k-q)^\lambda\,\sigma_{\lambda\mu}\right)\otimes\boldsymbol{\tau}\right]_{\zeta_1\zeta_1'} \\ \times \mathcal{D}^{0\rho}(0)^{\mu\nu}\,\mathrm{Tr}\left\{\left[\left(g_{\rho B'}\gamma_\nu + \mathrm{i}\,\frac{f_{\rho B'}}{2m_{B'}}\,(k-q)^\kappa\,\sigma_{\kappa\nu}\right)\otimes\boldsymbol{\tau}\right]\boldsymbol{g}^{B'}(q)\right\}, \tag{D.7}$$

and

$$\boldsymbol{\Sigma}^{\mathrm{F},B}_{\zeta_1\zeta_1'}(k)\Big|_{\rho} = -\,\mathrm{i}\int \frac{\mathrm{d}^4 q}{(2\pi)^4}\left(\mathrm{e}^{\mathrm{i}\eta q^0} + \mathrm{e}^{-\mathrm{i}\eta q^0}\right) \\ \times \left[\left(g_{\rho B}\gamma_\mu - \mathrm{i}\,\frac{f_{\rho B}}{2m_B}\,(k-q)^\lambda\,\sigma_{\lambda\mu}\right)\otimes\boldsymbol{\tau}\right]_{\zeta_1\zeta_3}\mathcal{D}^{0\rho}(k-q)^{\mu\nu} \\ \times \left[\left(g_{\rho B}\gamma_\nu + \mathrm{i}\,\frac{f_{\rho B}}{2m_B}\,(k-q)^\kappa\,\sigma_{\kappa\nu}\right)\otimes\boldsymbol{\tau}\right]_{\zeta_2\zeta_1'}\boldsymbol{g}^B_{\zeta_3\zeta_2}(q)\,. \tag{D.8}$$

The meson propagators Δ^{0M} $(M = \sigma, \omega, \pi)$ and $\mathcal{D}^{0\rho}$ are defined in section 5.3. The contours of integration in equations (D.2) through (D.8) are such that for particles (antiparticles) the path is to be closed in the upper (lower) half of the complex energy plane, as graphically illustrated in figure D.1. Thereby one encounters expressions for $\boldsymbol{\Sigma}^B$ which (a) arise due to the propagation of virtual positive and negative-energy (quasi) baryons, and (b) terms that allow for both holes inside the Fermi sea of particles and antibaryon states (antiholes) inside the Fermi sea of antiparticles, as illustrated in figure 6.2. The contributions obtained in the former case (a), leading to infinities in the theory, can be viewed as being part of the (infinite) energy arising from the Dirac sea of occupied negative-energy states (antiparticle states of the quasi-baryons) shown in figure 9.3.

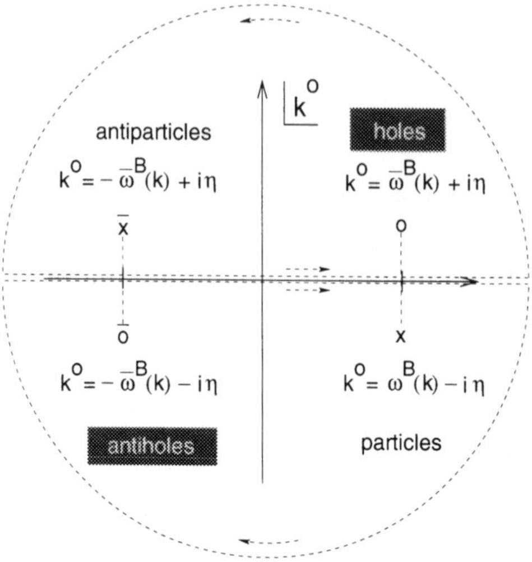

Figure D.1. Contours of integration encountered in the calculation of the baryon self-energies, $\mathbf{\Sigma}^B$, and the equation of state, $P(\epsilon)$. Contributions from the thermal Fermi sea of baryons (holes 'o') and antibaryons (antiholes '$\bar{\text{o}}$') are kept while those of virtual positive and negative-energy state baryons are dropped. Details are given in connection with figures 6.2 and 6.1.

The truncation of this expression leads for the Hartree approximation at absolute zero to Walecka's mean-field theory, introduced in section 7.2. We proceed in the same fashion for the Hartree–Fock theory at non-zero temperature. The truncation procedure can thus be briefly summarized by saying that both (thermal) Fermi sea contributions due to quasi-baryons of positive energy as well as those contributions arising from antibaryon states in the corresponding Fermi sea of thermal antiparticles are taken into account.

Upon performing the contour integrations in equations (D.2) through (D.8), followed by some algebraic manipulations regarding the γ matrices (section A.2), the self-energies from above take the, in some cases lengthy, forms:

$$\mathbf{\Sigma}^{\mathrm{H},B}_{\zeta_1\zeta_1'}\Big|_{\sigma} = -\,2\,\delta_{\zeta_1\zeta_1'}\left(\frac{g_{\sigma B}}{m_\sigma}\right)^2 \sum_{B'} \nu_{B'}\left(\frac{g_{\sigma B'}}{g_{\sigma B}}\right) \times \int \frac{\mathrm{d}^3\boldsymbol{q}}{(2\pi)^3}\left\{\Xi_S^{B'}(\boldsymbol{q})\, f^{B'}(\boldsymbol{q}) - \bar{\Xi}_S^{B'}(\boldsymbol{q})\, \bar{f}^{B'}(\boldsymbol{q})\right\}, \qquad \text{(D.9)}$$

$$\mathbf{\Sigma}^{\mathrm{H},B}_{\zeta_1\zeta_1'}\Big|_{\omega} = 2\,\gamma^0_{\zeta_1\zeta_1'}\left(\frac{g_{\omega B}}{m_\omega}\right)^2 \sum_{B'} \nu_{B'}\left(\frac{g_{\omega B'}}{g_{\omega B}}\right)$$

$$\times \int \frac{\mathrm{d}^3\boldsymbol{q}}{(2\pi)^3} \left\{ \Xi_0^{B'}(\boldsymbol{q})\, f^{B'}(\boldsymbol{q}) - \bar{\Xi}_0^{B'}(\boldsymbol{q})\, \bar{f}^{B'}(\boldsymbol{q}) \right\}, \quad \text{(D.10)}$$

$$\boldsymbol{\Sigma}^{\mathrm{H},B}_{\zeta_1\zeta_1'}\Big|_{\rho} = 2\,\gamma^0_{\zeta_1\zeta_1'} \left(\frac{g_{\rho B}}{m_\rho}\right)^2 \sum_{B'} \nu_{B'}\, I_{3B'} \left(\frac{g_{\rho B'}}{g_{\rho B}}\right)$$
$$\times \int \frac{\mathrm{d}^3\boldsymbol{q}}{(2\pi)^3} \left\{ \Xi_0^{B'}(\boldsymbol{q})\, f^{B'}(\boldsymbol{q}) - \bar{\Xi}_0^{B'}(\boldsymbol{q})\, \bar{f}^{B'}(\boldsymbol{q}) \right\}, \quad \text{(D.11)}$$

$$\boldsymbol{\Sigma}^{\mathrm{F},B}_{\zeta_1\zeta_1'}(k)\Big|_{\sigma} = g^2_{\sigma B} \int \frac{\mathrm{d}^3\boldsymbol{q}}{(2\pi)^3} \Big\{ \Big(\delta_{\zeta_1\zeta_1'}\, \Xi_S^B(\boldsymbol{q}) + (\gamma\cdot\hat{\boldsymbol{k}})_{\zeta_1\zeta_1'}\, \hat{\boldsymbol{k}}\cdot\hat{\boldsymbol{q}}\, \Xi_V^B(\boldsymbol{q})$$
$$+ \gamma^0_{\zeta_1\zeta_1'}\, \Xi_0^B(\boldsymbol{q}) \Big)\, \Delta^{0\sigma}(k^0 - \omega^B(\boldsymbol{q}), \boldsymbol{k}-\boldsymbol{q})\, f^B(\boldsymbol{q})$$
$$- \Big(\delta_{\zeta_1\zeta_1'}\, \bar{\Xi}_S^B(\boldsymbol{q}) \;+\; (\gamma\cdot\hat{\boldsymbol{k}})_{\zeta_1\zeta_1'}\, \hat{\boldsymbol{k}}\cdot\hat{\boldsymbol{q}}\, \bar{\Xi}_V^B(\boldsymbol{q}) \;+\; \gamma^0_{\zeta_1\zeta_1'}\, \bar{\Xi}_0^B(\boldsymbol{q}) \Big)$$
$$\times\, \Delta^{0\sigma}(k^0 - \bar{\omega}^B(\boldsymbol{q}), \boldsymbol{k}-\boldsymbol{q})\, \bar{f}^B(\boldsymbol{q}) \Big\}, \quad \text{(D.12)}$$

$$\boldsymbol{\Sigma}^{\mathrm{F},B}_{\zeta_1\zeta_1'}(k)\Big|_{\omega} = g^2_{\omega B} \int \frac{\mathrm{d}^3\boldsymbol{q}}{(2\pi)^3} \Big\{ \Big\{ \Big[-\delta_{\zeta_1\zeta_1'} \Big(4 - \frac{(k^0-\omega^B(\boldsymbol{q}))^2 - (\boldsymbol{k}-\boldsymbol{q})^2}{m_\omega^2} \Big) \Big]$$
$$\times\, \Xi_S^B(\boldsymbol{q}) + (\gamma\cdot\hat{\boldsymbol{k}})_{\zeta_1\zeta_1'} \Big[\Big(\hat{\boldsymbol{k}}\cdot\hat{\boldsymbol{q}} \Big(2 - \frac{\boldsymbol{k}^2 + \boldsymbol{q}^2 \,+\, (k^0-\omega^B(\boldsymbol{q}))^2}{m_\omega^2} \Big)$$
$$+ \frac{2\ |\,\boldsymbol{k}\,|\,|\,\boldsymbol{q}\,|}{m_\omega^2} \Big)\, \Xi_V^B(\boldsymbol{q}) \;-\; \frac{2}{m_\omega^2}\, \hat{\boldsymbol{k}} \,\cdot\, (\boldsymbol{k}-\boldsymbol{q})\, (k^0-\omega^B(\boldsymbol{q}))\, \Xi_0^B(\boldsymbol{q}) \Big]$$
$$+ \gamma^0_{\zeta_1\zeta_1'} \Big[\frac{2}{m_\omega^2}\, \hat{\boldsymbol{q}} \,\cdot\, (\boldsymbol{k}-\boldsymbol{q})\, (k^0-\omega^B(\boldsymbol{q}))\, \Xi_V^B(\boldsymbol{q})$$
$$+ \Big(2 + \frac{(k^0-\omega^B(\boldsymbol{q}))^2 + (\boldsymbol{k}-\boldsymbol{q})^2}{m_\omega^2} \Big)\, \Xi_0^B(\boldsymbol{q}) \Big] \Big\}$$
$$\times\, \Delta^{0\omega}(k^0-\omega^B(\boldsymbol{q}), \boldsymbol{k}-\boldsymbol{q})\, f^B(\boldsymbol{q})$$
$$- \Big\{ \Big[-\delta_{\zeta_1\zeta_1'} \Big(4 - \frac{(k^0-\bar{\omega}^B(\boldsymbol{q}))^2 - (\boldsymbol{k}-\boldsymbol{q})^2}{m_\omega^2} \Big) \Big]\, \bar{\Xi}_S^B(\boldsymbol{q})$$
$$+ (\gamma\cdot\hat{\boldsymbol{k}})_{\zeta_1\zeta_1'} \Big[\Big(\hat{\boldsymbol{k}}\cdot\hat{\boldsymbol{q}} \Big(2 - \frac{\boldsymbol{k}^2 + \boldsymbol{q}^2 \,+\, (k^0-\bar{\omega}^B(\boldsymbol{q}))^2}{m_\omega^2} \Big)$$
$$+ \frac{2\ |\,\boldsymbol{k}\,|\,|\,\boldsymbol{q}\,|}{m_\omega^2} \Big)\, \bar{\Xi}_V^B(\boldsymbol{q}) - \frac{2}{m_\omega^2}\, \hat{\boldsymbol{k}} \,\cdot\, (\boldsymbol{k}-\boldsymbol{q})\, (k^0-\bar{\omega}^B(\boldsymbol{q}))\, \bar{\Xi}_0^B(\boldsymbol{q}) \Big]$$
$$+ \gamma^0_{\zeta_1\zeta_1'} \Big[\frac{2}{m_\omega^2}\, \hat{q} \,\cdot\, (\boldsymbol{k}-\boldsymbol{q})\, (k^0-\bar{\omega}^B(\boldsymbol{q}))\, \bar{\Xi}_V^B(\boldsymbol{q})$$
$$+ \Big(2 + \frac{(k^0-\bar{\omega}^B(\boldsymbol{q}))^2 + (\boldsymbol{k}-\boldsymbol{q})^2}{m_\omega^2} \Big)\, \bar{\Xi}_0^B(\boldsymbol{q}) \Big]$$
$$\times\, \Delta^{0\omega}(k^0-\omega^{\bar{B}}(\boldsymbol{q}), \boldsymbol{k}-\boldsymbol{q})\, \bar{f}^B(\boldsymbol{q}) \Big\} \Big\}, \quad \text{(D.13)}$$

$$\Sigma^{\mathrm{F},B}_{\zeta_1\zeta_1'}(k)\Big|_\pi = 3\left(\frac{f_{\pi B}}{m_\pi}\right)^2 \int \frac{\mathrm{d}^3\boldsymbol{q}}{(2\pi)^3}\{\delta_{\zeta_1\zeta_1'}\,[[-(k^0-\omega^B(\boldsymbol{q}))^2+(\boldsymbol{k}-\boldsymbol{q})^2]\,\Xi^B_S(\boldsymbol{q})$$
$$\times\,\Delta^{0\pi}(k^0-\omega^B(\boldsymbol{q}),\boldsymbol{k}-\boldsymbol{q})\,f^B(\boldsymbol{q})$$
$$-\left[-(k^0-\bar{\omega}^B(\boldsymbol{q}))^2+(\boldsymbol{k}-\boldsymbol{q})^2\right]\bar{\Xi}^B_S(\boldsymbol{q})\,\Delta^0\pi(k^0-\bar{\omega}^B(\boldsymbol{q}),\boldsymbol{k}-\boldsymbol{q})\,\bar{f}^B(\boldsymbol{q})]$$
$$+(\gamma\cdot\hat{\boldsymbol{k}})_{\zeta_1\zeta_1'}[[(2(\mid\boldsymbol{q}\mid\ \hat{\boldsymbol{k}}\cdot\hat{\boldsymbol{q}}-\mid\boldsymbol{k}\mid)\,(\mid\boldsymbol{k}\mid\hat{\boldsymbol{k}}\cdot\hat{\boldsymbol{q}}-\mid\boldsymbol{q}\mid)$$
$$-((k^0-\omega^B(\boldsymbol{q}))^2-(\boldsymbol{k}-\boldsymbol{q})^2)\,\hat{\boldsymbol{k}}\cdot\hat{\boldsymbol{q}})\,\Xi^B_V(\boldsymbol{q})$$
$$+2(\mid\boldsymbol{q}\mid\hat{\boldsymbol{k}}\cdot\hat{\boldsymbol{q}}-\mid\boldsymbol{k}\mid)\,(k^0-\omega^B(\boldsymbol{q}))\,\Xi^B_0(\boldsymbol{q})]\,\Delta^{0\pi}(k^0-\omega^B(\boldsymbol{q}),\boldsymbol{k}-\boldsymbol{q})\,f^B(\boldsymbol{q})$$
$$-[\ B\ \rightarrow\ \bar{B}\ ]\,\Delta^{0\pi}(k^0-\bar{\omega}^B(\boldsymbol{q}),\boldsymbol{k}-\boldsymbol{q})\,\bar{f}^B(\boldsymbol{q})]$$
$$+\gamma^0_{\zeta_1\zeta_1'}\ [[2(\mid\boldsymbol{k}\mid\ \hat{\boldsymbol{k}}\cdot\hat{\boldsymbol{q}}-\mid\boldsymbol{q}\mid)\,(k^0-\omega^B(\boldsymbol{q}))\,\Xi^B_V(\boldsymbol{q})$$
$$+((k^0-\omega^B(\boldsymbol{q}))^2+(\boldsymbol{k}-\boldsymbol{q})^2)\,\Xi^B_0(\boldsymbol{q})]\,\Delta^{0\pi}(k^0-\omega^B(\boldsymbol{q}),\boldsymbol{k}-\boldsymbol{q})\,f^B(\boldsymbol{q})$$
$$-[\ B\ \rightarrow\ \bar{B}\ ]\,\Delta^{0\pi}(k^0-\bar{\omega}^B(\boldsymbol{q}),\boldsymbol{k}-\boldsymbol{q})\,\bar{f}^B(\boldsymbol{q})]\}\,, \tag{D.14}$$

and

$$\Sigma^{\mathrm{F},B}_{\zeta_1\zeta_1'}(k)\Big|_\rho = 3\,g^2_{\rho B}\int\frac{\mathrm{d}^3\boldsymbol{q}}{(2\pi)^3}\Big\{\delta_{\zeta_1\zeta_1'}\ \Big[\Big[(-4+\mathcal{X}^B_1((k^0-\omega^B(\boldsymbol{q}))^2$$
$$-(\boldsymbol{k}-\boldsymbol{q})^2))\,\Xi^B_S(\boldsymbol{q})+\frac{3f_{\rho B}}{m_B g_{\rho B}}\,(\hat{\boldsymbol{q}}\cdot(\boldsymbol{k}-\boldsymbol{q})\,\Xi^B_V(\boldsymbol{q})+(k^0-\omega^B(\boldsymbol{q}))\,\Xi^B_0(\boldsymbol{q}))\Big]$$
$$\times\,\Delta^{0\rho}(k^0-\omega^B(\boldsymbol{q}),\boldsymbol{k}-\boldsymbol{q})\,f^B(\boldsymbol{q})$$
$$-[\ B\ \rightarrow\ \bar{B}\ ]\,\Delta^{0\rho}(k^0-\omega^{\bar{B}}(\boldsymbol{q}),\boldsymbol{k}-\boldsymbol{q})\,\bar{f}^B(\boldsymbol{q})\Big]$$
$$+\ (\gamma\cdot\hat{\boldsymbol{k}})_{\zeta_1\zeta_1'}\Big[\Big[\frac{3f_{\rho B}}{m_B g_{\rho B}}\,(\mid\boldsymbol{k}\mid-\mid\boldsymbol{q}\mid\ \hat{\boldsymbol{k}}\cdot\hat{\boldsymbol{q}})\,\Xi^B_S(\boldsymbol{q})$$
$$+(2\hat{\boldsymbol{k}}\cdot\boldsymbol{q}+\mid\boldsymbol{k}\mid\mid\boldsymbol{q}\mid\ (\hat{\boldsymbol{k}}\cdot\hat{\boldsymbol{q}})^2\,(\mathcal{X}^B_2-2\mathcal{X}^B_3)+\mid\boldsymbol{k}\mid\mid\boldsymbol{q}\mid\mathcal{X}^B_2$$
$$-(\boldsymbol{k}^2+\boldsymbol{q}^2)\,(\hat{\boldsymbol{k}}\cdot\hat{\boldsymbol{q}})\,(\mathcal{X}^B_2-\mathcal{X}^B_3)-\mathcal{X}^B_3\,(k^0-\omega^B(\boldsymbol{q}))^2\,\hat{\boldsymbol{k}}\cdot\hat{\boldsymbol{q}})\,\Xi^B_V(\boldsymbol{q})$$
$$+\mathcal{X}^B_2\,(k^0-\omega^B(\boldsymbol{q}))\,(\mid\boldsymbol{q}\mid\hat{\boldsymbol{k}}\cdot\hat{\boldsymbol{q}}-\mid\boldsymbol{k}\mid)\,\Xi^B_0(\boldsymbol{q})\Big]$$
$$\times\,\Delta^{0\rho}(k^0-\omega^B(\boldsymbol{q}),\boldsymbol{k}-\boldsymbol{q})\,f^B(\boldsymbol{q})$$
$$-[\ B\ \rightarrow\ \bar{B}\ \Big]\,\Delta^{0\rho}(k^0-\omega^{\bar{B}}(\boldsymbol{q}),\boldsymbol{k}-\boldsymbol{q})\,\bar{f}^B(\boldsymbol{q})\Big]$$
$$+\gamma^0_{\zeta_1\zeta_1'}\ \Big[\Big[-\frac{3f_{\rho B}}{m_B g_{\rho B}}\,(k^0-\omega^B(\boldsymbol{q}))\,\Xi^B_S(\boldsymbol{q})$$
$$+\mathcal{X}^B_2\,(k^0-\omega^B(\boldsymbol{q}))\,\boldsymbol{q}\ \cdot(\boldsymbol{k}-\boldsymbol{q})\,\Xi^B_V(\boldsymbol{q})$$
$$+((\mathcal{X}^B_2-\mathcal{X}^B_3)\,(k^0-\omega^B(\boldsymbol{q}))^2+\mathcal{X}^B_3\,(\boldsymbol{k}-\boldsymbol{q})^2+2)\,\Xi^B_0(\boldsymbol{q})\Big]$$
$$\times\,\Delta^{0\rho}(k^0-\omega^B(\boldsymbol{q}),\boldsymbol{k}-\boldsymbol{q})\,f^B(\boldsymbol{q})$$
$$-[\ B\ \rightarrow\ \bar{B}\ ]\,\Delta^{0\rho}(k^0-\bar{\omega}^B(\boldsymbol{q}),\boldsymbol{k}-\boldsymbol{q})\,\bar{f}^B(\boldsymbol{q})\Big]\Big\}\,. \tag{D.15}$$

As everywhere else in this text, the factor $\nu_B=(2J_B+1)(2I_B+1)$ accounts for the spin–isospin degeneracy, and f^B and $\bar{f}^B$, defined in (6.67), are shorthand notations for the Fermi–Dirac distribution functions of baryons and antibaryons, respectively. The brackets $[B \to \bar{B}]$ in equations (D.14) and (D.15) denote the self-energy contributions of antibaryons, which are obtained by replacing Ξ_i^B, ω^B, and f^B with $\bar{\Xi}_i^B$, $\bar{\omega}^B$, and $\bar{f}^B$ in the respective expressions. Moreover, the following auxiliary functions were defined,

$$\mathcal{X}_1^B \equiv \frac{1}{m_\rho^2} - \frac{3}{(2m_B)^2}\left(\frac{f_{\rho B}}{g_{\rho B}}\right)^2 , \tag{D.16}$$

$$\mathcal{X}_2^B \equiv \frac{2}{m_\rho^2} + \frac{1}{m_B^2}\left(\frac{f_{\rho B}}{g_{\rho B}}\right)^2 , \tag{D.17}$$

$$\mathcal{X}_3^B \equiv \frac{1}{m_\rho^2} + \frac{1}{(2m_B)^2}\left(\frac{f_{\rho B}}{g_{\rho B}}\right)^2 . \tag{D.18}$$

Replacing the Hartree–Fock spectral functions Ξ_i^B and $\bar{\Xi}_i^B$, given in equations (6.139) through (6.141), with their Hartree approximated counterparts given in (7.29), which, as outlined in chapter 12, is a very good approximation because of the rather weak energy and momentum dependence of the self-energies Σ_S^B and Σ_0^B and the smallness of Σ_V^B [84, 125, 314], enables us to carry the calculation of expressions (D.12) and (D.13) one step further. After some algebra, one arrives for the total Fock contribution of σ and ω mesons (the σ–ω model) to the baryon self-energy at finite temperature at

$$\begin{aligned}\Sigma_S^{\mathrm{F},B}(|\boldsymbol{k}|) &= \frac{g_{\sigma B}^2}{4\pi^2}\int_0^\infty dq\, q^2\, \mathcal{Y}_1(|\boldsymbol{k}|,|\boldsymbol{q}|;m_\sigma)\,\Xi_S^B(\boldsymbol{q})\,\left(f^B(\boldsymbol{q}) + \bar{f}^B(\boldsymbol{q})\right) \\ &+ \frac{g_{\omega B}^2}{4\pi^2}\int_0^\infty dq\, q^2 \Big\{ -\Big[4 + \frac{\boldsymbol{k}^2+\boldsymbol{q}^2}{m_\omega^2}\Big]\mathcal{Y}_1(|\boldsymbol{k}|,|\boldsymbol{q}|;m_\omega) \\ &+ \frac{2|\boldsymbol{k}|\,|\boldsymbol{q}|}{m_\omega^2}\mathcal{Y}_2(|\boldsymbol{k}|,|\boldsymbol{q}|;m_\omega)\Big\}\,\Xi_S^B(\boldsymbol{q})\,\left(f^B(\boldsymbol{q}) + \bar{f}^B(\boldsymbol{q})\right) ,\end{aligned} \tag{D.19}$$

and

$$\begin{aligned}\Sigma_0^{\mathrm{F},B}(|\boldsymbol{k}|) &= \frac{g_{\sigma B}^2}{4\pi^2}\int_0^\infty dq\, q^2\, \mathcal{Y}_1(|\boldsymbol{k}|,|\boldsymbol{q}|;m_\sigma)\,\Xi_0^B(\boldsymbol{q})\,\left(f^B(\boldsymbol{q}) - \bar{f}^B(\boldsymbol{q})\right) \\ &+ \frac{g_{\omega B}^2}{4\pi^2}\int_0^\infty dq\, q^2 \Big\{ \Big[2 + \frac{\boldsymbol{k}^2+\boldsymbol{q}^2}{m_\omega^2}\Big]\mathcal{Y}_1(|\boldsymbol{k}|,|\boldsymbol{q}|;m_\omega) \\ &- \frac{2|\boldsymbol{k}|\,|\boldsymbol{q}|}{m_\omega^2}\mathcal{Y}_2(|\boldsymbol{k}|,|\boldsymbol{q}|;m_\omega)\Big\}\,\Xi_0^B(\boldsymbol{q})\,\left(f^B(\boldsymbol{q}) - \bar{f}^B(\boldsymbol{q})\right) .\end{aligned} \tag{D.20}$$

The functions $\mathcal{Y}_1$ and $\mathcal{Y}_2$ in (D.19) and (D.20), defined as

$$\mathcal{Y}_1(x, y; m_M) \equiv \frac{1}{2xy}\ln\frac{(x+y)^2 + m_M^2}{(x-y)^2 + m_M^2} ,$$

$$\mathcal{Y}_2(x, y; m_M) \equiv -\frac{1}{xy} + \frac{x^2 + y^2 + m_M^2}{2xy}\, \mathcal{Y}_1(x, y; m_M)\,, \qquad \text{(D.21)}$$

are auxiliary functions, where m_M denotes the meson mass ($M = \sigma, \omega$). The contributions of π and ρ mesons to $\mathbf{\Sigma}^{\mathrm{F},B}$, which we skip here for the sake of brevity, follow straightforwardly from expressions (D.14) and (D.15). We recall that the Hartree (direct) contributions, $\mathbf{\Sigma}^{\mathrm{H},B}$, to the total Hartree–Fock self-energy, $\mathbf{\Sigma}^{\mathrm{HF},B}$, can be read off immediately from equations (D.9) and (D.10). Their mathematical structure was discussed in great detail in chapter 7.

Appendix E

Helicity-state matrix elements of boson-exchange interactions

The baryon–baryon matrix elements of a given one-boson-exchange interaction for the case of scalar meson exchange are given in section 9.2. In the following we list the still missing matrix elements associated with the exchange of pseudovector and vector mesons (table 5.2) among the baryons. The notation is the same as in the section quoted just above (see also figure 9.1). To keep the notation to a minimum, henceforth we drop the indices 'B' carried by the baryon spinors, Φ^B, effective masses, m_B^*, and energy–momentum relations, ϵ^B. For the case of pseudovector coupling, the boson-exchange matrix elements in the self-consistent baryon basis then read,

$$\langle \Phi_{l'_1}(k'), \Phi_{l'_2}(k') | \mathbf{V}_{\rm pv} | \Phi_{l_1}(k), \Phi_{l_2}(k) \rangle = \Big(\frac{f_{\rm pv}}{m_{\rm pv}} \Big)^2 \tilde{\Delta}^0_{\rm pv}(k', k) \, \frac{\epsilon' \, \epsilon}{4 \, m'^* \, m^*}$$
$$\times \Big[(k'^0 - k^0) \Big(\frac{2l'_1 k'^*}{\epsilon'} + \frac{2l_1 k^*}{\epsilon} \Big) - (2l'_1 |\boldsymbol{k}'| - 2l_1 |\boldsymbol{k}|) \Big(1 + \frac{4l'_1 l_1 k'^* k^*}{\epsilon' \epsilon} \Big) \Big]$$
$$\times \Big[(k'^0 - k^0) \Big(\frac{2l'_2 k'^*}{\epsilon'} + \frac{2l_2 k^*}{\epsilon} \Big) - (2l'_2 |\boldsymbol{k}'| - 2l_2 |\boldsymbol{k}|) \Big(1 + \frac{4l'_2 l_2 k'^* k^*}{\epsilon' \epsilon} \Big) \Big]$$
$$\times \langle l'_1 \, l'_2 \, | \, 1 \, | \, l_1 \, l_2 \rangle \,, \qquad \text{(E.1)}$$

$$\langle \Theta_{l'_1}(k'), \Phi_{l'_2}(k') | \mathbf{V}_{\rm pv} | \Phi_{l_1}(k), \Phi_{l_2}(k) \rangle = \Big(\frac{f_{\rm pv}}{m_{\rm pv}} \Big)^2 \tilde{\Delta}^0_{\rm pv}(k', k) \, \frac{\epsilon' \, \epsilon}{4 \, m'^* \, m^*}$$
$$\times \Big[(k'^0 - k^0) \Big(1 + \frac{4l'_1 l_1 k'^* k^*}{\epsilon' \epsilon} \Big) - (2l'_1 |\boldsymbol{k}'| - 2l_1 |\boldsymbol{k}|) \Big(\frac{2l'_1 k'^*}{\epsilon'} + \frac{2l_1 k^*}{\epsilon} \Big) \Big]$$
$$\times \Big[(k'^0 - k^0) \Big(\frac{2l'_2 k'^*}{\epsilon'} + \frac{2l_2 k^*}{\epsilon} \Big) - (2l'_2 |\boldsymbol{k}'| - 2l_2 |\boldsymbol{k}|) \Big(1 + \frac{4l'_2 l_2 k'^* k^*}{\epsilon' \epsilon} \Big) \Big]$$
$$\times \langle l'_1 \, l'_2 \, | \, 1 \, | \, l_1 \, l_2 \rangle \,, \qquad \text{(E.2)}$$

$$\langle \Theta_{l_1'}(k'), \Phi_{l_2'}(k')|\mathbf{V}_{\rm pv}|\Phi_{l_1}(k), \Theta_{l_2}(k)\rangle = \Big(\frac{f_{\rm pv}}{m_{\rm pv}}\Big)^2 \tilde{\Delta}^0_{\rm pv}(k',k)\,\frac{\epsilon'\,\epsilon}{4\,m'^*\,m^*}$$
$$\times \Big[(k'^0 - k^0)\Big(1 + \frac{4l_1'l_1k'^*k^*}{\epsilon'\epsilon}\Big) - (2l_1'|\boldsymbol{k}'| - 2l_1|\boldsymbol{k}|)\Big(\frac{2l_1'k'^*}{\epsilon'} + \frac{2l_1k^*}{\epsilon}\Big)\Big]$$
$$\times \Big[(k'^0 - k^0)\Big(1 + \frac{4l_2'l_2k'^*k^*}{\epsilon'\epsilon}\Big) - (2l_2'|\boldsymbol{k}'| - 2l_2|\boldsymbol{k}|)\Big(\frac{2l_2'k'^*}{\epsilon'} + \frac{2l_2k^*}{\epsilon}\Big)\Big]$$
$$\times \langle l_1'\,l_2'\,|\,1\,|\,l_1\,l_2\rangle\,, \tag{E.3}$$

and

$$\langle \Theta_{l_1'}(k'), \Phi_{l_2'}(k')|\mathbf{V}_{\rm pv}|\Theta_{l_1}(k), \Phi_{l_2}(k)\rangle$$
$$= \langle \Phi_{l_1'}(k'), \Phi_{l_2'}(k')|\mathbf{V}_{\rm pv}|\Phi_{l_1}(k), \Phi_{l_2}(k)\rangle\,. \tag{E.4}$$

In case of the vector coupling, the expressions are rather lengthy. Because the vector bosons couple to the nucleon via Dirac ($g_{\rm v}$) plus tensor (also known as Pauli) coupling ($f_{\rm v}$) , one arrives at boson-exchange matrix elements that contain terms proportional to $g_{\rm v}^2$, $f_{\rm v}^2$, and $g_{\rm v}f_{\rm v}$. In listing the corresponding matrix elements, it is convenient to introduce the following decomposition,

$$\langle \Phi_{l_1'}(k'), \Phi_{l_2'}(k')|\mathbf{V}_{\rm v}|\Phi_{l_1}(k), \Phi_{l_2}(k)\rangle$$
$$\equiv \langle \Phi_{l_1'}(k'), \Phi_{l_2'}(k')|\mathbf{V}_{\rm vv}|\Phi_{l_1}(k), \Phi_{l_2}(k)\rangle$$
$$+ \langle \Phi_{l_1'}(k'), \Phi_{l_2'}(k')|\mathbf{V}_{\rm vt}|\Phi_{l_1}(k), \Phi_{l_2}(k)\rangle$$
$$+ \langle \Phi_{l_1'}(k'), \Phi_{l_2'}(k')|\mathbf{V}_{\rm tt}|\Phi_{l_1}(k), \Phi_{l_2}(k)\rangle\,, \tag{E.5}$$

with

$$\langle \Phi_{l_1'}(k'), \Phi_{l_2'}(k')|\mathbf{V}_{\rm vv}|\Phi_{l_1}(k), \Phi_{l_2}(k)\rangle = g_{\rm v}^2\,\tilde{\Delta}^0_{\rm v}(k',k)\,\frac{\epsilon'\,\epsilon}{4\,m'^*\,m^*}$$
$$\times \Big\{\Big[\Big(1 + \frac{4l_1'l_1k'^*k^*}{\epsilon'\epsilon}\Big)\Big(1 + \frac{4l_2'l_2k'^*k^*}{\epsilon'\epsilon}\Big) + \frac{1}{m_{\rm v}^2}\Big((k'^0 - k^0)\Big(1 + \frac{4l_1'l_1k'^*k^*}{\epsilon'\epsilon}\Big)$$
$$- (2l_1'|\boldsymbol{k}'| - 2l_1|\boldsymbol{k}|)\Big(\frac{2l_1'k'^*}{\epsilon'} + \frac{2l_1k^*}{\epsilon}\Big)\Big)\Big((k'^0 - k^0)\Big(1 + \frac{4l_2'l_2k'^*k^*}{\epsilon'\epsilon}\Big)$$
$$- (2l_2'|\boldsymbol{k}'| - 2l_2|\boldsymbol{k}|)\Big(\frac{2l_2'k'^*}{\epsilon'} + \frac{2l_2k^*}{\epsilon}\Big)\Big)\Big]\,\langle l_1'\,l_2'\,|\,1\,|\,l_1\,l_2\rangle$$
$$- \Big(\frac{2l_1'k'^*}{\epsilon'} + \frac{2l_1k^*}{\epsilon}\Big)\Big(\frac{2l_2'k'^*}{\epsilon'} + \frac{2l_2k^*}{\epsilon}\Big)\,\langle l_1'\,l_2'\,|\,\boldsymbol{\sigma}_1\cdot\boldsymbol{\sigma}_2\,|\,l_1\,l_2\rangle\Big\}\,, \tag{E.6}$$

$$\langle \Theta_{l_1'}(k'), \Phi_{l_2'}(k')|\mathbf{V}_{\rm vv}|\Phi_{l_1}(k), \Phi_{l_2}(k)\rangle = g_{\rm v}^2\,\tilde{\Delta}^0_{\rm v}(k',k)\,\frac{\epsilon'\,\epsilon}{4\,m'^*\,m^*}$$
$$\times \Big\{\Big[\Big(\frac{2l_1'k'^*}{\epsilon'} + \frac{2l_1k^*}{\epsilon}\Big)\Big(1 + \frac{4l_2'l_2k'^*k^*}{\epsilon'\epsilon}\Big) + \frac{1}{m_{\rm v}^2}\Big((k'^0 - k^0)\Big(\frac{2l_1'k'^*}{\epsilon'} + \frac{2l_1k^*}{\epsilon}\Big)$$

$$- (2l_1'|\boldsymbol{k}'| - 2l_1|\boldsymbol{k}|)\Big(1 + \frac{4l_1'l_1k'^*k^*}{\epsilon'\epsilon}\Big)\Big)\Big((k'^0 - k^0)\Big(1 + \frac{4l_2'l_2k'^*k^*}{\epsilon'\epsilon}\Big)$$
$$- (2l_2'|\boldsymbol{k}'| - 2l_2|\boldsymbol{k}|)\Big(\frac{2l_2'k'^*}{\epsilon'} + \frac{2l_2k^*}{\epsilon}\Big)\Big)\Big] \, \langle l_1'\, l_2' \,|\, 1 \,|\, l_1\, l_2\rangle$$
$$- \Big(1 + \frac{4l_1'l_1k'^*k^*}{\epsilon'\epsilon}\Big)\Big(\frac{2l_2'k'^*}{\epsilon'} + \frac{2l_2k^*}{\epsilon}\Big) \, \langle l_1'\, l_2' \,|\, \boldsymbol{\sigma}_1 \cdot \boldsymbol{\sigma}_2 \,|\, l_1\, l_2\rangle \Big\}, \qquad \text{(E.7)}$$

$$\langle \Theta_{l_1'}(k'), \Phi_{l_2'}(k') | \mathbf{V}_{\text{vv}} | \Phi_{l_1}(k), \Theta_{l_2}(k)\rangle = g_{\text{v}}^2 \, \tilde{\Delta}_{\text{v}}^0(k', k) \, \frac{\epsilon' \, \epsilon}{4\, m'^* \, m^*}$$
$$\times \Big\{\Big[\Big(\frac{2l_1'k'^*}{\epsilon'} + \frac{2l_1k^*}{\epsilon}\Big)\Big(\frac{2l_2'k'^*}{\epsilon'} + \frac{2l_2k^*}{\epsilon}\Big) + \frac{1}{m_{\text{v}}^2}\Big((k'^0 - k^0)\Big(\frac{2l_1'k'^*}{\epsilon'} + \frac{2l_1k^*}{\epsilon}\Big)$$
$$- (2l_1'|\boldsymbol{k}'| - 2l_1|\boldsymbol{k}|)\Big(1 + \frac{4l_1'l_1k'^*k^*}{\epsilon'\epsilon}\Big)\Big)\Big((k'^0 - k^0)\Big(\frac{2l_2'k'^*}{\epsilon'} + \frac{2l_2k^*}{\epsilon}\Big)$$
$$- (2l_2'|\boldsymbol{k}'| - 2l_2|\boldsymbol{k}|)\Big(1 + \frac{4l_2'l_2k'^*k^*}{\epsilon'\epsilon}\Big)\Big)\Big] \, \langle l_1'\, l_2' \,|\, 1 \,|\, l_1\, l_2\rangle$$
$$- \Big(1 + \frac{4l_1'l_1k'^*k^*}{\epsilon'\epsilon}\Big)\Big(1 + \frac{4l_2'l_2k'^*k^*}{\epsilon'\epsilon}\Big) \, \langle l_1'\, l_2' \,|\, \boldsymbol{\sigma}_1 \cdot \boldsymbol{\sigma}_2 \,|\, l_1\, l_2\rangle \Big\}, \qquad \text{(E.8)}$$

and

$$\langle \Theta_{l_1'}(k'), \Phi_{l_2'}(k') | \mathbf{V}_{\text{vv}} | \Theta_{l_1}(k), \Phi_{l_2}(k)\rangle$$
$$= \langle \Phi_{l_1'}(k'), \Phi_{l_2'}(k') | \mathbf{V}_{\text{vv}} | \Phi_{l_1}(k), \Phi_{l_2}(k)\rangle . \qquad \text{(E.9)}$$

The mixed matrix elements (second term in equation (E.5)) read,

$$\langle \Phi_{l_1'}(k'), \Phi_{l_2'}(k') | \mathbf{V}_{\text{vt}} | \Phi_{l_1}(k), \Phi_{l_2}(k)\rangle = \Big(\frac{f_{\text{v}}\, g_{\text{v}}}{2\, m_N}\Big) \tilde{\Delta}_{\text{v}}^0(k', k) \, \frac{\epsilon' \, \epsilon}{4\, m'^* \, m^*}$$
$$\times \Big\{\Big[-\Big(1 + \frac{4l_1'l_1k'^*k^*}{\epsilon'\epsilon}\Big)(2l_2'|\boldsymbol{k}'| - 2l_2|\boldsymbol{k}|)\Big(\frac{2l_2'k'^*}{\epsilon'} - \frac{2l_2k^*}{\epsilon}\Big)$$
$$- (2l_1'|\boldsymbol{k}'| + 2l_1|\boldsymbol{k}|)\Big(\frac{2l_1'k'^*}{\epsilon'} + \frac{2l_1k^*}{\epsilon}\Big)\Big(1 - \frac{4l_2'l_2k'^*k^*}{\epsilon'\epsilon}\Big)$$
$$- (2l_1'|\boldsymbol{k}'| - 2l_1|\boldsymbol{k}|)\Big(\frac{2l_1'k'^*}{\epsilon'} - \frac{2l_1k^*}{\epsilon}\Big)\Big(1 + \frac{4l_2'l_2k'^*k^*}{\epsilon'\epsilon}\Big)$$
$$- \Big(1 - \frac{4l_1'l_1k'^*k^*}{\epsilon'\epsilon}\Big)(2l_2'|\boldsymbol{k}'| + 2l_2|\boldsymbol{k}|)\Big(\frac{2l_2'k'^*}{\epsilon'} + \frac{2l_2k^*}{\epsilon}\Big)\Big] \, \langle l_1'\, l_2' \,|\, 1 \,|\, l_1\, l_2\rangle$$
$$+ \Big[\Big(\frac{2l_1'k'^*}{\epsilon'} + \frac{2l_1k^*}{\epsilon}\Big)\Big((k'^0 - k^0)\Big(\frac{2l_2'k'^*}{\epsilon'} - \frac{2l_2k^*}{\epsilon}\Big) - (2l_2'|\boldsymbol{k}'| + 2l_2|\boldsymbol{k}|)$$
$$\times \Big(1 - \frac{4l_2'l_2k'^*k^*}{\epsilon'\epsilon}\Big)\Big) + \Big((k'^0 - k^0)\Big(\frac{2l_1'k'^*}{\epsilon'} - \frac{2l_1k^*}{\epsilon}\Big) - (2l_1'|\boldsymbol{k}'| + 2l_1|\boldsymbol{k}|)$$
$$\times \Big(1 - \frac{4l_1'l_1k'^*k^*}{\epsilon'\epsilon}\Big)\Big)\Big(\frac{2l_2'k'^*}{\epsilon'} + \frac{2l_2k^*}{\epsilon}\Big)\Big] \, \langle l_1'\, l_2' \,|\, \boldsymbol{\sigma}_1 \cdot \boldsymbol{\sigma}_2 \,|\, l_1\, l_2\rangle \Big\}, \qquad \text{(E.10)}$$

$$\langle \Theta_{l_1'}(k'), \Phi_{l_2'}(k') | \mathbf{V}_{\text{vt}} | \Phi_{l_1}(k), \Phi_{l_2}(k)\rangle = \left(\frac{f_{\text{v}}\, g_{\text{v}}}{2\, m_N}\right) \tilde{\Delta}_{\text{v}}^0(k', k) \, \frac{\epsilon' \, \epsilon}{4\, m'^* \, m^*}$$

$$
\begin{aligned}
&\times \Big\{\Big[-\Big(\frac{2l_1'k'^*}{\epsilon'}+\frac{2l_1k^*}{\epsilon}\Big)(2l_2'|\boldsymbol{k}'|-2l_2|\boldsymbol{k}|)\Big(\frac{2l_2'k'^*}{\epsilon'}-\frac{2l_2k^*}{\epsilon}\Big)\\
&-\Big(1-\frac{4l_1'l_1k'^*k^*}{\epsilon'\epsilon}\Big)(2l_1'|\boldsymbol{k}'|-2l_1|\boldsymbol{k}|)\Big(1+\frac{4l_2'l_2k'^*k^*}{\epsilon'\epsilon}\Big)\\
&-\Big(1+\frac{4l_1'l_1k'^*k^*}{\epsilon'\epsilon}\Big)(2l_1'|\boldsymbol{k}'|+2l_1|\boldsymbol{k}|)\Big(1-\frac{4l_2'l_2k'^*k^*}{\epsilon'\epsilon}\Big)\\
&-\Big(\frac{2l_1'k'^*}{\epsilon'}-\frac{2l_1k^*}{\epsilon}\Big)(2l_2'|\boldsymbol{k}'|+2l_2|\boldsymbol{k}|)\Big(\frac{2l_2'k'^*}{\epsilon'}+\frac{2l_2k^*}{\epsilon}\Big)\Big]\,\langle l_1'\,l_2'\,|\,1\,|\,l_1\,l_2\rangle\\
&+\Big[(k'^0-k^0)\Big(\Big(1+\frac{4l_1'l_1k'^*k^*}{\epsilon'\epsilon}\Big)\Big(\frac{2l_2'k'^*}{\epsilon'}-\frac{2l_2k^*}{\epsilon}\Big)+\Big(1-\frac{4l_1'l_1k'^*k^*}{\epsilon'\epsilon}\Big)\\
&\times\Big(\frac{2l_2'k'^*}{\epsilon'}+\frac{2l_2k^*}{\epsilon}\Big)\Big)-\Big(1+\frac{4l_1'l_1k'^*k^*}{\epsilon'\epsilon}\Big)(2l_2'|\boldsymbol{k}'|+2l_2|\boldsymbol{k}|)\Big(1-\frac{4l_2'l_2k'^*k^*}{\epsilon'\epsilon}\Big)\\
&-(2l_1'|\boldsymbol{k}'|+2l_1|\boldsymbol{k}|)\Big(\frac{2l_1'k'^*}{\epsilon'}-\frac{2l_1k^*}{\epsilon}\Big)\Big(\frac{2l_2'k'^*}{\epsilon'}+\frac{2l_2k^*}{\epsilon}\Big)\Big]\\
&\times\ \langle l_1'\,l_2'\,|\,\boldsymbol{\sigma}_1\cdot\boldsymbol{\sigma}_2\,|\,l_1\,l_2\rangle\Big\},
\end{aligned}
\tag{E.11}
$$

$$
\begin{aligned}
&\langle\Theta_{l_1'}(k'),\Phi_{l_2'}(k')|\mathbf{V}_{\rm vt}|\Phi_{l_1}(k),\Theta_{l_2}(k)\rangle = \Big(\frac{f_{\rm v}\,g_{\rm v}}{2\,m_N}\Big)\tilde{\Delta}^0_{\rm v}(k',k)\,\frac{\epsilon'\,\epsilon}{4\,m'^*\,m^*}\\
&\times\Big\{\Big[\Big(\frac{2l_1'k'^*}{\epsilon'}+\frac{2l_1k^*}{\epsilon}\Big)(2l_2'|\boldsymbol{k}'|-2l_2|\boldsymbol{k}|)\Big(1-\frac{4l_2'l_2k'^*k^*}{\epsilon'\epsilon}\Big)\\
&+\Big(1+\frac{4l_1'l_1k'^*k^*}{\epsilon'\epsilon}\Big)(2l_1'|\boldsymbol{k}'|+2l_1|\boldsymbol{k}|)\Big(\frac{2l_2'k'^*}{\epsilon'}-\frac{2l_2k^*}{\epsilon}\Big)\\
&-\Big(1-\frac{4l_1'l_1k'^*k^*}{\epsilon'\epsilon}\Big)(2l_1'|\boldsymbol{k}'|-2l_1|\boldsymbol{k}|)\Big(\frac{2l_2'k'^*}{\epsilon'}+\frac{2l_2k^*}{\epsilon}\Big)\\
&-\Big(\frac{2l_1'k'^*}{\epsilon'}-\frac{2l_1k^*}{\epsilon}\Big)(2l_2'|\boldsymbol{k}'|+2l_2|\boldsymbol{k}|)\Big(1+\frac{4l_2'l_2k'^*k^*}{\epsilon'\epsilon}\Big)\Big]\,\langle l_1'\,l_2'\,|\,1\,|\,l_1\,l_2\rangle\\
&+\Big[(k'^0-k^0)\Big(-\Big(1+\frac{4l_1'l_1k'^*k^*}{\epsilon'\epsilon}\Big)\Big(1-\frac{4l_2'l_2k'^*k^*}{\epsilon'\epsilon}\Big)+\Big(1-\frac{4l_1'l_1k'^*k^*}{\epsilon'\epsilon}\Big)\\
&\times\Big(1+\frac{4l_2'l_2k'^*k^*}{\epsilon'\epsilon}\Big)\Big)+\Big(1+\frac{4l_1'l_1k'^*k^*}{\epsilon'\epsilon}\Big)(2l_2'|\boldsymbol{k}'|+2l_2|\boldsymbol{k}|)\Big(\frac{2l_2'k'^*}{\epsilon'}-\frac{2l_2k^*}{\epsilon}\Big)\\
&-\Big(\frac{2l_1'k'^*}{\epsilon'}-\frac{2l_1k^*}{\epsilon}\Big)(2l_1'|\boldsymbol{k}'|+2l_1|\boldsymbol{k}|)\Big(1+\frac{4l_2'l_2k'^*k^*}{\epsilon'\epsilon}\Big)\Big]\\
&\times\ \langle l_1'\,l_2'\,|\,\boldsymbol{\sigma}_1\cdot\boldsymbol{\sigma}_2\,|\,l_1\,l_2\rangle\Big\},
\end{aligned}
\tag{E.12}
$$

and

$$
\begin{aligned}
&\langle\Theta_{l_1'}(k'),\Phi_{l_2'}(k')|\mathbf{V}_{\rm vt}|\Theta_{l_1}(k),\Phi_{l_2}(k)\rangle = \Big(\frac{f_{\rm v}\,g_{\rm v}}{2\,m_N}\Big)\tilde{\Delta}^0_{\rm v}(k',k)\,\frac{\epsilon'\,\epsilon}{4\,m'^*\,m^*}\\
&\times\Big\{\Big[-\Big(1+\frac{4l_1'l_1k'^*k^*}{\epsilon'\epsilon}\Big)(2l_2'|\boldsymbol{k}'|-2l_2|\boldsymbol{k}|)\Big(\frac{2l_2'k'^*}{\epsilon'}-\frac{2l_2k^*}{\epsilon}\Big)\\
&-(2l_1'|\boldsymbol{k}'|+2l_1|\boldsymbol{k}|)\Big(\frac{2l_1'k'^*}{\epsilon'}+\frac{2l_1k^*}{\epsilon}\Big)\Big(1-\frac{4l_2'l_2k'^*k^*}{\epsilon'\epsilon}\Big)
\end{aligned}
$$

$$+ (2l_1'|\boldsymbol{k}'| - 2l_1|\boldsymbol{k}|)\Big(\frac{2l_1'k'^*}{\epsilon'} - \frac{2l_1k^*}{\epsilon}\Big)\Big(1 + \frac{4l_2'l_2k'^*k^*}{\epsilon'\epsilon}\Big)$$
$$+ \Big(1 - \frac{4l_1'l_1k'^*k^*}{\epsilon'\epsilon}\Big)(2l_2'|\boldsymbol{k}'| + 2l_2|\boldsymbol{k}|)\Big(\frac{2l_2'k'^*}{\epsilon'} + \frac{2l_2k^*}{\epsilon}\Big)\Big] \langle l_1'\, l_2' \,|\, 1 \,|\, l_1\, l_2\rangle$$
$$+ \Big[\Big(\frac{2l_1'k'^*}{\epsilon'} + \frac{2l_1k^*}{\epsilon}\Big)\Big((k'^0 - k^0)\Big(\frac{2l_2'k'^*}{\epsilon'} - \frac{2l_2k^*}{\epsilon}\Big) - (2l_2'|\boldsymbol{k}'| + 2l_2|\boldsymbol{k}|)$$
$$\times \Big(1 - \frac{4l_2'l_2k'^*k^*}{\epsilon'\epsilon}\Big)\Big) + \Big(-(k'^0 - k^0)\Big(\frac{2l_1'k'^*}{\epsilon'} - \frac{2l_1k^*}{\epsilon}\Big) + (2l_1'|\boldsymbol{k}'| + 2l_1|\boldsymbol{k}|)$$
$$\times \Big(1 - \frac{4l_1'l_1k'^*k^*}{\epsilon'\epsilon}\Big)\Big)\Big(\frac{2l_2'k'^*}{\epsilon'} + \frac{2l_2k^*}{\epsilon}\Big)\Big] \langle l_1'\, l_2' \,|\, \boldsymbol{\sigma}_1 \cdot \boldsymbol{\sigma}_2 \,|\, l_1\, l_2\rangle \Big\}. \tag{E.13}$$

Finally the matrix elements of the third term in equation (E.5)) are given as follows,

$$\langle \Phi_{l_1'}(k'), \Phi_{l_2'}(k')|\mathbf{V}_{\mathrm{tt}}|\Phi_{l_1}(k), \Phi_{l_2}(k)\rangle = \Big(\frac{f_{\mathrm{v}}}{2\,m_N}\Big)^2 \tilde{\Delta}_{\mathrm{v}}^0(k', k)\, \frac{\epsilon'\,\epsilon}{4\,m'^*\,m^*}$$
$$\times \Big\{\Big[\Big(\frac{2l_1'k'^*}{\epsilon'} - \frac{2l_1k^*}{\epsilon}\Big)\Big(\frac{2l_2'k'^*}{\epsilon'} - \frac{2l_2k^*}{\epsilon}\Big)(2l_1'|\boldsymbol{k}'| - 2l_1|\boldsymbol{k}|)(2l_2'|\boldsymbol{k}'| - 2l_2|\boldsymbol{k}|)$$
$$+ \Big(1 - \frac{4l_1'l_1k'^*k^*}{\epsilon'\epsilon}\Big) + \Big(1 - \frac{4l_2'l_2k'^*k^*}{\epsilon'\epsilon}\Big)$$
$$\times (-|\boldsymbol{k}'|^2 - |\boldsymbol{k}|^2 - 2\,(4l_1'l_1 + 4l_2'l_2)|\boldsymbol{k}'|\,|\boldsymbol{k}|)$$
$$+ (k'^0 - k^0)\,(2l_1'|\boldsymbol{k}'| + 2l_1|\boldsymbol{k}|)\Big(\frac{2l_1'k'^*}{\epsilon'} - \frac{2l_1k^*}{\epsilon}\Big)\Big(1 - \frac{4l_2'l_2k'^*k^*}{\epsilon'\epsilon}\Big)$$
$$+ (k'^0 - k^0)\,(2l_2'|\boldsymbol{k}'| + 2l_2|\boldsymbol{k}|)\Big(\frac{2l_2'k'^*}{\epsilon'} - \frac{2l_2k^*}{\epsilon}\Big)\Big(1 - \frac{4l_1'l_1k'^*k^*}{\epsilon'\epsilon}\Big)\Big]$$
$$\times \;\langle l_1'\, l_2' \,|\, 1 \,|\, l_1\, l_2\rangle$$
$$+ \Big[-(k'^0 - k^0)^2\Big(\frac{2l_1'k'^*}{\epsilon'} - \frac{2l_1k^*}{\epsilon}\Big)\Big(\frac{2l_2'k'^*}{\epsilon'} - \frac{2l_2k^*}{\epsilon}\Big)$$
$$- (2l_1'|\boldsymbol{k}'| + 2l_1|\boldsymbol{k}|)(2l_2'|\boldsymbol{k}'| + 2l_2|\boldsymbol{k}|)\Big(1 - \frac{4l_1'l_1k'^*k^*}{\epsilon'\epsilon}\Big)\Big(1 - \frac{4l_2'l_2k'^*k^*}{\epsilon'\epsilon}\Big)$$
$$+ (k'^0 - k^0)\Big(\frac{2l_1'k'^*}{\epsilon'} - \frac{2l_1k^*}{\epsilon}\Big)(2l_2'|\boldsymbol{k}'| + 2l_2|\boldsymbol{k}|)\Big(1 - \frac{4l_2'l_2k'^*k^*}{\epsilon'\epsilon}\Big)$$
$$+ (k'^0 - k^0)\Big(\frac{2l_2'k'^*}{\epsilon'} - \frac{2l_2k^*}{\epsilon}\Big)(2l_1'|\boldsymbol{k}'| + 2l_1|\boldsymbol{k}|)\Big(1 - \frac{4l_1'l_1k'^*k^*}{\epsilon'\epsilon}\Big)\Big]$$
$$\times \;\langle l_1'\, l_2' \,|\, \boldsymbol{\sigma}_1 \cdot \boldsymbol{\sigma}_2 \,|\, l_1\, l_2\rangle$$
$$+ \Big(1 - \frac{4l_1'l_1k'^*k^*}{\epsilon'\epsilon}\Big)\Big(1 - \frac{4l_2'l_2k'^*k^*}{\epsilon'\epsilon}\Big)(2|\boldsymbol{k}'|\,|\boldsymbol{k}|)\cos\vartheta\, \langle l_1'\, l_2' \,|\, 1 \,|\, l_1\, l_2\rangle \Big\}, \tag{E.14}$$

$$\langle \Theta_{l_1'}(k'), \Phi_{l_2'}(k')|\mathbf{V}_{\mathrm{tt}}|\Phi_{l_1}(k), \Phi_{l_2}(k)\rangle = \Big(\frac{f_{\mathrm{v}}}{2\,m_N}\Big)^2 \tilde{\Delta}_{\mathrm{v}}^0(k', k)\, \frac{\epsilon'\,\epsilon}{4\,m'^*\,m^*}$$
$$\times \Big\{\Big[(2l_1'|\boldsymbol{k}'| - 2l_1|\boldsymbol{k}|)\Big(1 - \frac{4l_1'l_1k'^*k^*}{\epsilon'\epsilon}\Big)(2l_2'|\boldsymbol{k}'| - 2l_2|\boldsymbol{k}|)\Big(\frac{2l_2'k'^*}{\epsilon'} - \frac{2l_2k^*}{\epsilon}\Big)$$

$$
\begin{aligned}
&+ (|\boldsymbol{k}'|^2 + |\boldsymbol{k}|^2)\Big(\frac{2l_1' k'^*}{\epsilon'} - \frac{2l_1 k^*}{\epsilon}\Big)\Big(1 - \frac{4l_2' l_2 k'^* k^*}{\epsilon' \epsilon}\Big) + \Big(\frac{2l_1' k'^*}{\epsilon'} - \frac{2l_1 k^*}{\epsilon}\Big) \\
&\times \Big(1 - \frac{4l_2' l_2 k'^* k^*}{\epsilon' \epsilon}\Big) + \Big(\frac{2l_1' k'^*}{\epsilon'} - \frac{2l_1 k^*}{\epsilon}\Big)(2l_2' |\boldsymbol{k}'| - 2l_2 |\boldsymbol{k}|) \\
&\times \Big((k'^0 - k^0)\Big(\frac{2l_2' k'^*}{\epsilon'} - \frac{2l_2 k^*}{\epsilon}\Big) - (2l_2' |\boldsymbol{k}'| + 2l_2 |\boldsymbol{k}|)\Big(1 - \frac{4l_2' l_2 k'^* k^*}{\epsilon' \epsilon}\Big)\Big) \\
&+ \Big((k'^0 - k^0)\Big(1 - \frac{4l_1' l_1 k'^* k^*}{\epsilon' \epsilon}\Big) - (2l_1' |\boldsymbol{k}'| + 2l_1 |\boldsymbol{k}|)\Big(\frac{2l_1' k'^*}{\epsilon'} - \frac{2l_1 k^*}{\epsilon}\Big)\Big) \\
&\times (2l_1' |\boldsymbol{k}'| + 2l_1 |\boldsymbol{k}|)\Big(1 - \frac{4l_2' l_2 k'^* k^*}{\epsilon' \epsilon}\Big)\Big] \quad \langle l_1'\, l_2' \,|\, 1 \,|\, l_1\, l_2 \rangle \\
&+ \Big[\Big(-(k'^0 - k^0)\Big(1 - \frac{4l_1' l_1 k'^* k^*}{\epsilon' \epsilon}\Big) + (2l_1' |\boldsymbol{k}'| + 2l_1 |\boldsymbol{k}|)\Big(\frac{2l_1' k'^*}{\epsilon'} - \frac{2l_1 k^*}{\epsilon}\Big)\Big) \\
&\times \Big((k'^0 - k^0)\Big(\frac{2l_2' k'^*}{\epsilon'} - \frac{2l_2 k^*}{\epsilon}\Big) - (2l_2' |\boldsymbol{k}'| + 2l_2 |\boldsymbol{k}|)\Big(1 - \frac{4l_2' l_2 k'^* k^*}{\epsilon' \epsilon}\Big)\Big)\Big] \\
&\times \ \langle l_1'\, l_2' \,|\, \boldsymbol{\sigma}_1 \cdot \boldsymbol{\sigma}_2 \,|\, l_1\, l_2 \rangle \ + \ (2|\boldsymbol{k}'|\,|\boldsymbol{k}|)\Big(\frac{2l_1' k'^*}{\epsilon'} - \frac{2l_1 k^*}{\epsilon}\Big)\Big(1 - \frac{4l_2' l_2 k'^* k^*}{\epsilon' \epsilon}\Big) \\
&\times \ \cos\vartheta \ \langle l_1'\, l_2' \,|\, 1 \,|\, l_1\, l_2 \rangle \Big\}, \qquad \text{(E.15)}
\end{aligned}
$$

$$
\begin{aligned}
&\langle \Theta_{l_1'}(k'), \Phi_{l_2'}(k') | \mathbf{V}_{\rm tt} | \Phi_{l_1}(k), \Theta_{l_2}(k) \rangle = \Big(\frac{f_{\rm v}}{2\, m_N}\Big)^2 \tilde{\Delta}_{\rm v}^0(k', k)\, \frac{\epsilon'\, \epsilon}{4\, m'^*\, m^*} \\
&\times \Big\{\Big[-(2l_1' |\boldsymbol{k}'| - 2l_1 |\boldsymbol{k}|)\Big(1 - \frac{4l_1' l_1 k'^* k^*}{\epsilon' \epsilon}\Big)(2l_2' |\boldsymbol{k}'| - 2l_2 |\boldsymbol{k}|)\Big(1 - \frac{4l_2' l_2 k'^* k^*}{\epsilon' \epsilon}\Big) \\
&- (|\boldsymbol{k}'|^2 + |\boldsymbol{k}|^2)\Big(\frac{2l_1' k'^*}{\epsilon'} - \frac{2l_1 k^*}{\epsilon}\Big)\Big(\frac{2l_2' k'^*}{\epsilon'} - \frac{2l_2 k^*}{\epsilon}\Big) + \Big(\frac{2l_1' k'^*}{\epsilon'} - \frac{2l_1 k^*}{\epsilon}\Big) \\
&\times (2l_2' |\boldsymbol{k}'| + 2l_2 |\boldsymbol{k}|)\Big(-(k'^0 - k^0)\Big(1 - \frac{4l_2' l_2 k'^* k^*}{\epsilon' \epsilon}\Big) + (2l_2' |\boldsymbol{k}'| + 2l_2 |\boldsymbol{k}|) \\
&\times \Big(\frac{2l_2' k'^*}{\epsilon'} - \frac{2l_2 k^*}{\epsilon}\Big)\Big) + \Big(-(k'^0 - k^0)\Big(1 - \frac{4l_1' l_1 k'^* k^*}{\epsilon' \epsilon}\Big) + (2l_1' |\boldsymbol{k}'| + 2l_1 |\boldsymbol{k}|) \\
&\times \Big(\frac{2l_1' k'^*}{\epsilon'} - \frac{2l_1 k^*}{\epsilon}\Big)\Big)(2l_1' |\boldsymbol{k}'| + 2l_1 |\boldsymbol{k}|)\Big(\frac{2l_2' k'^*}{\epsilon'} - \frac{2l_2 k^*}{\epsilon}\Big)\Big] \ \langle l_1'\, l_2' \,|\, 1 \,|\, l_1\, l_2 \rangle \\
&+ \Big[\Big(-(k'^0 - k^0)\Big(1 - \frac{4l_1' l_1 k'^* k^*}{\epsilon' \epsilon}\Big) + (2l_1' |\boldsymbol{k}'| + 2l_1 |\boldsymbol{k}|)\Big(\frac{2l_1' k'^*}{\epsilon'} - \frac{2l_1 k^*}{\epsilon}\Big)\Big) \\
&\times \Big(-(k'^0 - k^0)\Big(1 - \frac{4l_2' l_2 k'^* k^*}{\epsilon' \epsilon}\Big) + (2l_2' |\boldsymbol{k}'| + 2l_2 |\boldsymbol{k}|)\Big(\frac{2l_2' k'^*}{\epsilon'} - \frac{2l_2 k^*}{\epsilon}\Big)\Big)\Big] \\
&\times \ \langle l_1'\, l_2' \,|\, \boldsymbol{\sigma}_1 \cdot \boldsymbol{\sigma}_2 \,|\, l_1\, l_2 \rangle - (2|\boldsymbol{k}'|\,|\boldsymbol{k}|)\Big(\frac{2l_1' k'^*}{\epsilon'} - \frac{2l_1 k^*}{\epsilon}\Big) \\
&\times \Big(\frac{2l_2' k'^*}{\epsilon'} - \frac{2l_2 k^*}{\epsilon}\Big) \cos\vartheta \, \langle l_1'\, l_2' \,|\, 1 \,|\, l_1\, l_2 \rangle \Big\}, \qquad \text{(E.16)}
\end{aligned}
$$

and

$$
\begin{aligned}
&\langle \Theta_{l_1'}(k'), \Phi_{l_2'}(k') | \mathbf{V}_{\rm tt} | \Theta_{l_1}(k), \Phi_{l_2}(k) \rangle \\
&\qquad = -\langle \Phi_{l_1'}(k'), \Phi_{l_2'}(k') | \mathbf{V}_{\rm tt} | \Phi_{l_1}(k), \Phi_{l_2}(k) \rangle\,. \qquad \text{(E.17)}
\end{aligned}
$$

Appendix F

Partial-wave expansion of boson-exchange interactions

In chapter 10 we performed the partial-wave expansion of the one-boson-exchange matrix elements, such as, for instance, the nucleon–nucleon matrix elements $\langle \Phi_{l'_1}(k'), \Phi_{l'_2}(k') | \mathbf{V} | \Phi_{l_1}(k), \Phi_{l_2}(k) \rangle$, into angular momentum states $|jml_1l_2\rangle$ for mesons that possess a scalar coupling to the baryons. The still missing angular momentum state amplitudes associated with *pseudovector* and *vector* meson exchange (cf. table 5.2) among baryons are listed below [440]. To keep the notation to a minimum, we largely drop the functions' subscripts as well as their arguments. The explicit forms are easily recovered in reference to their defining equations. These are given for m^*, k^*, and W, for instance, in equations (9.2) and (9.3). Primes attached to these functions indicate $m'^* \equiv m'(k')$, $k'^* \equiv k'^*(k')$, and $W' \equiv W'(k')$, as in equations (9.36) and (9.42), where $k \equiv k^\mu$. The functions $\tilde{Q}_j$, and $\tilde{Q}_j^{(1)}$ through $\tilde{Q}_j^{(6)}$ are defined in equations (10.31), and (10.33) through (10.39) respectively.

F.1 Amplitudes associated with pseudovector meson exchange

The angular momentum state amplitudes associated with pseudovector meson exchange are given by,

$$
\begin{aligned}
{}^0\mathsf{V}^j_{\mathrm{pv}}(k',k) &= \left(\frac{f_{\mathrm{pv}}}{m_{\mathrm{pv}}}\right)^2 \frac{\pi}{m'^*\,m^*} \Big\{ \big[(k'^0 - k^0)^2 (W'W - m'^* m^*) + 2(k'^0 - k^0) \\
&\times (k^* |\boldsymbol{k}| W' - k'^* |\boldsymbol{k}'| W) + (\boldsymbol{k}'^2 + \boldsymbol{k}^2)(W'W + m'^* m^*) \\
&- 2|\boldsymbol{k}'||\boldsymbol{k}| k'^* k^* \big] \frac{\tilde{Q}_j}{|\boldsymbol{k}'|\,|\boldsymbol{k}|} + \big[(k'^0 - k^0)^2 k'^* k^* + 2(k'^0 - k^0) \\
&\times (k'^* |\boldsymbol{k}| W - k^* |\boldsymbol{k}'| W') + (\boldsymbol{k}'^2 + \boldsymbol{k}^2) k'^* k^*
\end{aligned}
$$

$$- 2|\boldsymbol{k}'||\boldsymbol{k}|(W'W + m'^*m^*)\Big] \frac{\tilde{Q}_j^{(1)}}{|\boldsymbol{k}'|\,|\boldsymbol{k}|}\Big\}, \tag{F.1}$$

$$\begin{aligned} {}^{12}\mathsf{V}^j_{\rm pv}(k',k) &= \left(\frac{f_{\rm pv}}{m_{\rm pv}}\right)^2 \frac{\pi}{m'^*\,m^*} \Big\{ \Big[(k'^0 - k^0)^2 k'^* k^* \\ &+ 2(k'^0 - k^0)(k'^*|\boldsymbol{k}|W - k^*|\boldsymbol{k}'|W') \\ &+ (\boldsymbol{k}'^2 + \boldsymbol{k}^2)k'^*k^* - 2|\boldsymbol{k}'||\boldsymbol{k}|(W'W + m'^*m^*)\Big] \frac{\tilde{Q}_j}{|\boldsymbol{k}'|\,|\boldsymbol{k}|} \\ &+ \Big[(k'^0 - k^0)^2(W'W - m'^*m^*) + 2(k'^0 - k^0)(k^*|\boldsymbol{k}|W' - k'^*|\boldsymbol{k}'|W) \\ &+ (\boldsymbol{k}'^2 + \boldsymbol{k}^2)(W'W + m'^*m^*) - 2|\boldsymbol{k}'||\boldsymbol{k}|k'^*k^*\Big] \frac{\tilde{Q}_j^{(1)}}{|\boldsymbol{k}'|\,|\boldsymbol{k}|}\Big\}, \end{aligned} \tag{F.2}$$

$$\begin{aligned} {}^{1}\mathsf{V}^j_{\rm pv}(k',k) &= \left(\frac{f_{\rm pv}}{m_{\rm pv}}\right)^2 \frac{\pi}{m'^*\,m^*} \Big\{ \Big[-(k'^0 - k^0)^2(W'W - m'^*m^*) \\ &- 2(k'^0 - k^0)(k^*|\boldsymbol{k}|W' - k'^*|\boldsymbol{k}'|W) - (\boldsymbol{k}'^2 + \boldsymbol{k}^2)(W'W + m'^*m^*) \\ &+ 2|\boldsymbol{k}'||\boldsymbol{k}|k'^*k^*\Big] \frac{\tilde{Q}_j}{|\boldsymbol{k}'|\,|\boldsymbol{k}|} + \Big[-(k'^0 - k^0)^2 k'^*k^* - 2(k'^0 - k^0) \\ &\times (k'^*|\boldsymbol{k}|W - k^*|\boldsymbol{k}'|W') - (\boldsymbol{k}'^2 + \boldsymbol{k}^2)k'^*k^* \\ &+ 2|\boldsymbol{k}'||\boldsymbol{k}|(W'W + m'^*m^*)\Big] \frac{\tilde{Q}_j^{(2)}}{|\boldsymbol{k}'|\,|\boldsymbol{k}|}\Big\}, \end{aligned} \tag{F.3}$$

$$\begin{aligned} {}^{34}\mathsf{V}^j_{\rm pv}(k',k) &= \left(\frac{f_{\rm pv}}{m_{\rm pv}}\right)^2 \frac{\pi}{m'^*\,m^*} \Big\{ \Big[-(k'^0 - k^0)^2 k'^* k^* \\ &- 2(k'^0 - k^0)(k'^*|\boldsymbol{k}|W - k^*|\boldsymbol{k}'|W') \\ &- (\boldsymbol{k}'^2 + \boldsymbol{k}^2)k'^*k^* + 2|\boldsymbol{k}'||\boldsymbol{k}|(W'W + m'^*m^*)\Big] \frac{\tilde{Q}_j}{|\boldsymbol{k}'|\,|\boldsymbol{k}|} \\ &+ \Big[-(k'^0 - k^0)^2(W'W - m'^*m^*) - 2(k'^0 - k^0)(k'^*|\boldsymbol{k}|W' - k'^*|\boldsymbol{k}'|W) \\ &- (\boldsymbol{k}'^2 + \boldsymbol{k}^2)(W'W + m'^*m^*) + 2|\boldsymbol{k}'||\boldsymbol{k}|k'^*k^*\Big] \frac{\tilde{Q}_j^{(2)}}{|\boldsymbol{k}'|\,|\boldsymbol{k}|}\Big\}, \end{aligned} \tag{F.4}$$

$$\begin{aligned} {}^{5}\mathsf{V}^j_{\rm pv}(k',k) &= \left(\frac{f_{\rm pv}}{m_{\rm pv}}\right)^2 \frac{\pi}{m'^*\,m^*} \Big\{ \Big[(k'^0 - k^0)^2(m'^*W - m^*W') \\ &+ 2(k'^0 - k^0)(m^*|\boldsymbol{k}'|k'^* + m'^*|\boldsymbol{k}|k^*) + (\boldsymbol{k}^2 - \boldsymbol{k}'^2)(m^*W' + m'^*W)\Big] \\ &\times \frac{\tilde{Q}_j^{(3)}}{|\boldsymbol{k}'|\,|\boldsymbol{k}|}\Big\}, \end{aligned} \tag{F.5}$$

$$ {}^{6}\mathsf{V}^{j}_{\rm pv}(k',k) = -\,{}^{5}\mathsf{V}^{j}_{\rm pv}(k',k)\,, \tag{F.6} $$

$$ \begin{aligned} {}^{0}\mathsf{U}^{j}_{\rm pv}(k',k) &= \left(\frac{f_{\rm pv}}{m_{\rm pv}}\right)^{2} \frac{\pi}{m'^{*}\, m^{*}} \Big\{ \big[(k'^{0}-k^{0})^{2} k'^{*} W \\ &+ 2(k'^{0}-k^{0})(|\boldsymbol{k}| k'^{*} k^{*} - |\boldsymbol{k}'| W' W) \\ &+ (\boldsymbol{k}'^{2}+\boldsymbol{k}^{2}) k'^{*} W - 2|\boldsymbol{k}'||\boldsymbol{k}| k^{*} W'\big] \frac{\tilde{Q}_{j}}{|\boldsymbol{k}'|\,|\boldsymbol{k}|} \\ &+ \big[(k'^{0}-k^{0})^{2}(k^{*} W' - 2(k'^{0}-k^{0})(|\boldsymbol{k}'| k'^{*} k^{*} - |\boldsymbol{k}| W' W) \\ &+ (\boldsymbol{k}'^{2}+\boldsymbol{k}^{2}) k^{*} W' - 2|\boldsymbol{k}'||\boldsymbol{k}| k'^{*} W\big] \frac{\tilde{Q}^{(1)}_{j}}{|\boldsymbol{k}'|\,|\boldsymbol{k}|} \Big\}, \end{aligned} \tag{F.7} $$

$$ \begin{aligned} {}^{12}\mathsf{U}^{j}_{\rm pv}(k',k) &= \left(\frac{f_{\rm pv}}{m_{\rm pv}}\right)^{2} \frac{\pi}{m'^{*}\, m^{*}} \Big\{ \big[(k'^{0}-k^{0})^{2} k^{*} W' \\ &- 2(k'^{0}-k^{0})(|\boldsymbol{k}'| k'^{*} k^{*} - |\boldsymbol{k}| W' W) \\ &+ (\boldsymbol{k}'^{2}+\boldsymbol{k}^{2}) k^{*} W' - 2|\boldsymbol{k}'||\boldsymbol{k}| k'^{*} W\big] \frac{\tilde{Q}_{j}}{|\boldsymbol{k}'|\,|\boldsymbol{k}|} \\ &+ \big[(k'^{0}-k^{0})^{2}(k'^{*} W + 2(k'^{0}-k^{0})(|\boldsymbol{k}| k'^{*} k^{*} - |\boldsymbol{k}'| W' W) \\ &+ (\boldsymbol{k}'^{2}+\boldsymbol{k}^{2}) k'^{*} W - 2|\boldsymbol{k}'||\boldsymbol{k}| k^{*} W'\big] \frac{\tilde{Q}^{(1)}_{j}}{|\boldsymbol{k}'|\,|\boldsymbol{k}|} \Big\}, \end{aligned} \tag{F.8} $$

$$ \begin{aligned} {}^{1}\mathsf{U}^{j}_{\rm pv}(k',k) &= \left(\frac{f_{\rm pv}}{m_{\rm pv}}\right)^{2} \frac{\pi}{m'^{*}\, m^{*}} \Big\{ -\big[(k'^{0}-k^{0})^{2} k'^{*} W \\ &+ 2(k'^{0}-k^{0})(|\boldsymbol{k}| k'^{*} k^{*} - |\boldsymbol{k}'| W' W) \\ &+ (\boldsymbol{k}'^{2}+\boldsymbol{k}^{2}) k'^{*} W - 2|\boldsymbol{k}'||\boldsymbol{k}| k^{*} W'\big] \frac{\tilde{Q}_{j}}{|\boldsymbol{k}'|\,|\boldsymbol{k}|} \\ &- \big[(k'^{0}-k^{0})^{2}(k^{*} W' - 2(k'^{0}-k^{0})(|\boldsymbol{k}'| k'^{*} k^{*} - |\boldsymbol{k}| W' W) \\ &+ (\boldsymbol{k}'^{2}+\boldsymbol{k}^{2}) k^{*} W' - 2|\boldsymbol{k}'||\boldsymbol{k}| k'^{*} W\big] \frac{\tilde{Q}^{(2)}_{j}}{|\boldsymbol{k}'|\,|\boldsymbol{k}|} \Big\}, \end{aligned} \tag{F.9} $$

$$ \begin{aligned} {}^{34}\mathsf{U}^{j}_{\rm pv}(k',k) &= \left(\frac{f_{\rm pv}}{m_{\rm pv}}\right)^{2} \frac{\pi}{m'^{*}\, m^{*}} \Big\{ -\big[(k'^{0}-k^{0})^{2} k^{*} W' \\ &- 2(k'^{0}-k^{0})(|\boldsymbol{k}'| k'^{*} k^{*} - |\boldsymbol{k}| W' W) \\ &+ (\boldsymbol{k}'^{2}+\boldsymbol{k}^{2}) k^{*} W' - 2|\boldsymbol{k}'||\boldsymbol{k}| k'^{*} W\big] \frac{\tilde{Q}_{j}}{|\boldsymbol{k}'|\,|\boldsymbol{k}|} \\ &- \big[(k'^{0}-k^{0})^{2}(k'^{*} W + 2(k'^{0}-k^{0})(|\boldsymbol{k}| k'^{*} k^{*} - |\boldsymbol{k}'| W' W) \\ &+ (\boldsymbol{k}'^{2}+\boldsymbol{k}^{2}) k'^{*} W - 2|\boldsymbol{k}'||\boldsymbol{k}| k^{*} W'\big] \frac{\tilde{Q}^{(2)}_{j}}{|\boldsymbol{k}'|\,|\boldsymbol{k}|} \Big\}, \end{aligned} \tag{F.10} $$

$$
\begin{aligned}
{}^{3}\mathsf{U}^{j}_{\mathrm{pv}}(k',k) &= \left(\frac{f_{\mathrm{pv}}}{m_{\mathrm{pv}}}\right)^2 \frac{\pi}{m'^{*}\, m^{*}} \{-2(k'^0-k^0)^2 m'^* k^* - 4(k'^0-k^0) \\
&\times m'^*|\boldsymbol{k}|W - 2(\boldsymbol{k}'^2+\boldsymbol{k}^2)m'^*k^* + 4m'^*|\boldsymbol{k}|\boldsymbol{k}'^2\} \frac{\tilde{Q}^{(3)}_j}{|\boldsymbol{k}'|\,|\boldsymbol{k}|}\,,
\end{aligned}
\tag{F.11}
$$

$$
\begin{aligned}
{}^{56}\mathsf{U}^{j}_{\mathrm{pv}}(k',k) &= \left(\frac{f_{\mathrm{pv}}}{m_{\mathrm{pv}}}\right)^2 \frac{\pi}{m'^{*}\, m^{*}} \{2(k'^0-k^0)^2 m^* k'^* - 4(k'^0-k^0) \\
&\times m^*|\boldsymbol{k}'|W' - 2(\boldsymbol{k}'^2+\boldsymbol{k}^2)m^*k'^* + 4m^*k'^*\boldsymbol{k}'^2\} \frac{\tilde{Q}^{(3)}_j}{|\boldsymbol{k}'|\,|\boldsymbol{k}|}\,,
\end{aligned}
\tag{F.12}
$$

$$
{}^{5}\mathsf{U}^{j}_{\mathrm{pv}}(k',k) = -\ {}^{3}\mathsf{U}^{j}_{\mathrm{pv}}(k',k)\,,
\tag{F.13}
$$

$$
{}^{78}\mathsf{U}^{j}_{\mathrm{pv}}(k',k) = -\ {}^{56}\mathsf{U}^{j}_{\mathrm{pv}}(k',k)\,,
\tag{F.14}
$$

$$
\begin{aligned}
{}^{0}\mathsf{W}^{j}_{\mathrm{pv}}(k',k) &= \left(\frac{f_{\mathrm{pv}}}{m_{\mathrm{pv}}}\right)^2 \frac{\pi}{m'^{*}\, m^{*}} \Big\{ \big[(k'^0-k^0)^2(W'W+m'^*m^*) \\
&+ 2(k'^0-k^0)(k^*|\boldsymbol{k}|W' - k'^*|\boldsymbol{k}'|W) + (\boldsymbol{k}'^2+\boldsymbol{k}^2)(W'W-m'^*m^*) \\
&- 2|\boldsymbol{k}'||\boldsymbol{k}|k'^*k^*\big] \frac{\tilde{Q}_j}{|\boldsymbol{k}'|\,|\boldsymbol{k}|} + \big[(k'^0-k^0)^2k'^*k^* + 2(k'^0-k^0) \\
&\times (k'^*|\boldsymbol{k}|W - k^*|\boldsymbol{k}'|W') + (\boldsymbol{k}'^2+\boldsymbol{k}^2)k'^*k^* \\
&- 2|\boldsymbol{k}'||\boldsymbol{k}|(W'W-m'^*m^*)\big] \frac{\tilde{Q}^{(1)}_j}{|\boldsymbol{k}'|\,|\boldsymbol{k}|} \Big\}\,,
\end{aligned}
\tag{F.15}
$$

$$
\begin{aligned}
{}^{12}\mathsf{W}^{j}_{\mathrm{pv}}(k',k) &= \left(\frac{f_{\mathrm{pv}}}{m_{\mathrm{pv}}}\right)^2 \frac{\pi}{m'^{*}\, m^{*}} \Big\{ \big[(k'^0-k^0)^2k'^*k^* \\
&+ 2(k'^0-k^0)(k'^*|\boldsymbol{k}|W - k^*|\boldsymbol{k}'|W') \\
&+ (\boldsymbol{k}'^2+\boldsymbol{k}^2)k'^*k^* - 2|\boldsymbol{k}'||\boldsymbol{k}|(W'W-m'^*m^*)\big] \frac{\tilde{Q}_j}{|\boldsymbol{k}'|\,|\boldsymbol{k}|} \\
&+ \big[(k'^0-k^0)^2(W'W+m'^*m^*) + 2(k'^0-k^0)(k^*|\boldsymbol{k}|W' - k'^*|\boldsymbol{k}'|W) \\
&+ (\boldsymbol{k}'^2+\boldsymbol{k}^2)(W'W-m'^*m^*) - 2|\boldsymbol{k}'||\boldsymbol{k}|k'^*k^*\big] \frac{\tilde{Q}^{(1)}_j}{|\boldsymbol{k}'|\,|\boldsymbol{k}|} \Big\}\,,
\end{aligned}
\tag{F.16}
$$

$$
\begin{aligned}
{}^{1}\mathsf{W}^{j}_{\mathrm{pv}}(k',k) &= \left(\frac{f_{\mathrm{pv}}}{m_{\mathrm{pv}}}\right)^2 \frac{\pi}{m'^{*}\, m^{*}} \Big\{ \big[(k'^0-k^0)^2(W'W+m'^*m^*) \\
&+ 2(k'^0-k^0)(k^*|\boldsymbol{k}|W' - k'^*|\boldsymbol{k}'|W) + (\boldsymbol{k}'^2+\boldsymbol{k}^2)(W'W-m'^*m^*)
\end{aligned}
$$

$$- 2|\boldsymbol{k}'||\boldsymbol{k}|k'^* k^*\big] \frac{\tilde{Q}_j}{|\boldsymbol{k}'|\,|\boldsymbol{k}|} + \big[(k'^0 - k^0)^2 k'^* k^* + 2(k'^0 - k^0)$$
$$\times (k'^* |\boldsymbol{k}| W - k^* |\boldsymbol{k}'| W') + (\boldsymbol{k}'^2 + \boldsymbol{k}^2) k'^* k^*$$
$$- 2|\boldsymbol{k}'||\boldsymbol{k}|(W'W - m'^* m^*)\big] \frac{\tilde{Q}_j^{(2)}}{|\boldsymbol{k}'|\,|\boldsymbol{k}|}\Big\}, \qquad \text{(F.17)}$$

$$^{34}\mathsf{W}^j_{\mathrm{pv}}(k', k) = \left(\frac{f_{\mathrm{pv}}}{m_{\mathrm{pv}}}\right)^2 \frac{\pi}{m'^*\, m^*} \Big\{\big[(k'^0 - k^0)^2 k'^* k^*$$
$$+ 2(k'^0 - k^0)(k'^* |\boldsymbol{k}| W - k^* |\boldsymbol{k}'| W')$$
$$+ (\boldsymbol{k}'^2 + \boldsymbol{k}^2) k'^* k^* - 2|\boldsymbol{k}'||\boldsymbol{k}|(W'W - m'^* m^*)\big] \frac{\tilde{Q}_j}{|\boldsymbol{k}'|\,|\boldsymbol{k}|}$$
$$+ \big[(k'^0 - k^0)^2 (W'W + m'^* m^*) + 2(k'^0 - k^0)(k^* |\boldsymbol{k}| W' - k'^* |\boldsymbol{k}'| W)$$
$$+ (\boldsymbol{k}'^2 + \boldsymbol{k}^2)(W'W - m'^* m^*) - 2|\boldsymbol{k}'||\boldsymbol{k}| k'^* k^*\big] \frac{\tilde{Q}_j^{(2)}}{|\boldsymbol{k}'|\,|\boldsymbol{k}|}\Big\}, \qquad \text{(F.18)}$$

$$^{3}\mathsf{W}^j_{\mathrm{pv}}(k', k) = 0\,, \qquad \text{(F.19)}$$

$$^{56}\mathsf{W}^j_{\mathrm{pv}}(k', k) = \left(\frac{f_{\mathrm{pv}}}{m_{\mathrm{pv}}}\right)^2 \frac{\pi}{m'^*\, m^*} \Big\{\big[-2(k'^0 - k^0)^2 (m'^* W + m^* W')$$
$$+ 4(k'^0 - k^0)(m^* |\boldsymbol{k}'| k'^* - m'^* k k^*) + 2(\boldsymbol{k}^2 + \boldsymbol{k}'^2)(m^* W' - m'^* W)$$
$$+ 4\boldsymbol{k}'^2 (m'^* W - m^* W')\big] \frac{\tilde{Q}_j^{(3)}}{|\boldsymbol{k}'|\,|\boldsymbol{k}|}\Big\}, \qquad \text{(F.20)}$$

$$^{5}\mathsf{W}^j_{\mathrm{pv}}(k', k) = 0\,, \qquad \text{(F.21)}$$

and

$$^{78}\mathsf{W}^j_{\mathrm{pv}}(k', k) = {}^{56}\mathsf{W}^j_{\mathrm{pv}}(k', k)\,. \qquad \text{(F.22)}$$

F.2 Amplitudes associated with vector meson exchange

The angular momentum state amplitudes associated with vector meson exchange are given by,

$$^{0}\mathsf{V}^j_{\mathrm{vv}}(k', k) = g_{\mathrm{v}}^2 \frac{\pi}{m'^*\, m^*} \Big\{\Big[(4W'W - 2m'^* m^*) + \frac{1}{m_{\mathrm{v}}^2}\big((k'^0 - k^0)^2$$
$$\times (W'W + m'^* m^*) + 2(k'^0 - k^0)(k^* |\boldsymbol{k}| W' - k'^* |\boldsymbol{k}'| W) + (\boldsymbol{k}'^2 + \boldsymbol{k}^2)$$
$$\times (W'W - m'^* m^*) - 2|\boldsymbol{k}'||\boldsymbol{k}| k'^* k^*\big)\Big] \frac{\tilde{Q}_j}{|\boldsymbol{k}'|\,|\boldsymbol{k}|}$$

$$+\frac{1}{m_{\mathrm{v}}^2}\big[(k'^0-k^0)^2k'^*k^*+2(k'^0-k^0)(k'^*|\boldsymbol{k}|W-k^*|\boldsymbol{k}'|W')$$

$$+(\boldsymbol{k}'^2+\boldsymbol{k}^2)k'^*k^*-2|\boldsymbol{k}'||\boldsymbol{k}|(W'W-m'^*m^*)\big]\frac{\tilde{Q}_j^{(1)}}{|\boldsymbol{k}'|\,|\boldsymbol{k}|}\Big\}, \tag{F.23}$$

$$^{12}\mathsf{V}^j_{\mathrm{vv}}(k',k)=g_{\mathrm{v}}^2\frac{\pi}{m'^*\,m^*}\Big\{\big[4k'^*k^*+\frac{1}{m_{\mathrm{v}}^2}\big((k'^0-k^0)^2k'^*k^*$$

$$+2(k'^0-k^0)(k'^*|\boldsymbol{k}|W-k^*|\boldsymbol{k}'|W')$$

$$+(\boldsymbol{k}'^2+\boldsymbol{k}^2)k'^*k^*-2|\boldsymbol{k}'||\boldsymbol{k}|(W'W-m'^*m^*)\big)\big]\frac{\tilde{Q}_j}{|\boldsymbol{k}'|\,|\boldsymbol{k}|}$$

$$+\big[2m'^*m^*+\frac{1}{m_{\mathrm{v}}^2}\big((k'^0-k^0)^2(W'W+m'^*m^*)$$

$$+2(k'^0-k^0)(k^*|\boldsymbol{k}|W'-k'^*|\boldsymbol{k}'|W)$$

$$+(\boldsymbol{k}'^2+\boldsymbol{k}^2)(W'W-m'^*m^*)-2|\boldsymbol{k}'||\boldsymbol{k}|k'^*k^*\big)\big]\frac{\tilde{Q}_j^{(1)}}{|\boldsymbol{k}'|\,|\boldsymbol{k}|}\Big\}, \tag{F.24}$$

$$^{1}\mathsf{V}^j_{\mathrm{vv}}(k',k)=g_{\mathrm{v}}^2\frac{\pi}{m'^*\,m^*}\Big\{\big[2W'W+\frac{1}{m_{\mathrm{v}}^2}\big((k'^0-k^0)^2(W'W+m'^*m^*)$$

$$+2(k'^0-k^0)(k^*|\boldsymbol{k}|W'-k'^*|\boldsymbol{k}'|W)+(\boldsymbol{k}'^2+\boldsymbol{k}^2)(W'W-m'^*m^*)$$

$$-2|\boldsymbol{k}'||\boldsymbol{k}|k'^*k^*\big)\big]\frac{\tilde{Q}_j}{|\boldsymbol{k}'|\,|\boldsymbol{k}|}+\big[2k'^*k^*+\frac{1}{m_{\mathrm{v}}^2}(k'^0-k^0)^2k'^*k^*$$

$$+2(k'^0-k^0)(k'^*|\boldsymbol{k}|W-k^*|\boldsymbol{k}'|W')$$

$$+(\boldsymbol{k}'^2+\boldsymbol{k}^2)k'^*k^*-2|\boldsymbol{k}'||\boldsymbol{k}|(W'W-m'^*m^*)\big)\big]\frac{\tilde{Q}_j^{(2)}}{|\boldsymbol{k}'|\,|\boldsymbol{k}|}\Big\}, \tag{F.25}$$

$$^{34}\mathsf{V}^j_{\mathrm{vv}}(k',k)=g_{\mathrm{v}}^2\frac{\pi}{m'^*\,m^*}\Big\{\big[2k'^*k^*+\frac{1}{m_{\mathrm{v}}^2}\big((k'^0-k^0)^2k'^*k^*$$

$$+2(k'^0-k^0)(k'^*|\boldsymbol{k}|W-k^*|\boldsymbol{k}'|W')$$

$$+(\boldsymbol{k}'^2+\boldsymbol{k}^2)k'^*k^*-2|\boldsymbol{k}'||\boldsymbol{k}|(W'W-m'^*m^*)\big)\big]\frac{\tilde{Q}_j}{|\boldsymbol{k}'|\,|\boldsymbol{k}|}$$

$$+\big[2W'W+\frac{1}{m_{\mathrm{v}}^2}\big((k'^0-k^0)^2(W'W+m'^*m^*)$$

$$+2(k'^0-k^0)(k^*|\boldsymbol{k}|W'-k'^*|\boldsymbol{k}'|W)$$

$$+(\boldsymbol{k}'^2+\boldsymbol{k}^2)(W'W-m'^*m^*)-2|\boldsymbol{k}'||\boldsymbol{k}|k'^*k^*\big]\frac{\tilde{Q}_j^{(2)}}{|\boldsymbol{k}'|\,|\boldsymbol{k}|}\Big\}, \tag{F.26}$$

$$^{5}\mathsf{V}^j_{\mathrm{vv}}(k',k)=g_{\mathrm{v}}^2\frac{\pi}{m'^*\,m^*}\Big\{\big[-2m'^*W+\frac{1}{m_{\mathrm{v}}^2}\big(-(k'^0-k^0)^2(m'^*W+m^*W')$$

$$+ 2(k'^0 - k^0)(m^*|\boldsymbol{k}'|k'^* - m'^*|\boldsymbol{k}|k^*) + (\boldsymbol{k}^2 - \boldsymbol{k}'^2)(m^*W' - m'^*W))\big]$$
$$\times \frac{\tilde{Q}_j^{(3)}}{|\boldsymbol{k}'|\,|\boldsymbol{k}|}\Big\}, \tag{F.27}$$

$${}^6\mathsf{V}^j_{\mathrm{vv}}(k',k) = g_{\mathrm{v}}^2 \frac{\pi}{m'^*\, m^*}\Big\{\big[-2m^*W' + \frac{1}{m_{\mathrm{v}}^2}(-(k'^0 - k^0)^2(m^*W' + m'^*W)$$
$$+ 2(k'^0 - k^0)(m^*|\boldsymbol{k}'|k'^* - m'^*|\boldsymbol{k}|k^*) + (\boldsymbol{k}^2 - \boldsymbol{k}'^2)(m^*W' - m'^*W))\big]$$
$$\times \frac{\tilde{Q}_j^{(3)}}{|\boldsymbol{k}'|\,|\boldsymbol{k}|}\Big\}, \tag{F.28}$$

$${}^0\mathsf{U}^j_{\mathrm{vv}}(k',k) = g_{\mathrm{v}}^2 \frac{\pi}{m'^*\, m^*}\Big\{\big[4k'^*W + \frac{1}{m_{\mathrm{v}}^2}((k'^0 - k^0)^2 k'^*W + 2(k'^0 - k^0)$$
$$\times (|\boldsymbol{k}|k'^*k^* - |\boldsymbol{k}'|W'W) + (\boldsymbol{k}'^2 + \boldsymbol{k}^2)k'^*W - 2|\boldsymbol{k}'||\boldsymbol{k}|k^*W')\big]\frac{\tilde{Q}_j}{|\boldsymbol{k}'|\,|\boldsymbol{k}|}$$
$$+ \frac{1}{m_{\mathrm{v}}^2}\big[(k'^0 - k^0)^2 k^*W' - 2(k'^0 - k^0)(|\boldsymbol{k}'|k'^*k^* - |\boldsymbol{k}|W'W)$$
$$+ (\boldsymbol{k}'^2 + \boldsymbol{k}^2)k^*W' - 2|\boldsymbol{k}'||\boldsymbol{k}|k'^*W\big]\frac{\tilde{Q}_j^{(1)}}{|\boldsymbol{k}'|\,|\boldsymbol{k}|}\Big\}, \tag{F.29}$$

$${}^{12}\mathsf{U}^j_{\mathrm{vv}}(k',k) = g_{\mathrm{v}}^2 \frac{\pi}{m'^*\, m^*}\Big\{\big[4k^*W' + \frac{1}{m_{\mathrm{v}}^2}((k'^0 - k^0)^2 k^*W'$$
$$- 2(k'^0 - k^0)(|\boldsymbol{k}'|k'^*k^* - |\boldsymbol{k}|W'W)$$
$$+ (\boldsymbol{k}'^2 + \boldsymbol{k}^2)k^*W' - 2|\boldsymbol{k}'||\boldsymbol{k}|k'^*W))\big]\frac{\tilde{Q}_j}{|\boldsymbol{k}'|\,|\boldsymbol{k}|}$$
$$+ \big[\frac{1}{m_{\mathrm{v}}^2}((k'^0 - k^0)^2 k'^*W + 2(k'^0 - k^0)(|\boldsymbol{k}|k'^*k^* - |\boldsymbol{k}'|W'W)$$
$$+ (\boldsymbol{k}'^2 + \boldsymbol{k}^2)k'^*W - 2|\boldsymbol{k}'||\boldsymbol{k}|k^*W')\big]\frac{\tilde{Q}_j^{(1)}}{|\boldsymbol{k}'|\,|\boldsymbol{k}|}\Big\}, \tag{F.30}$$

$${}^1\mathsf{U}^j_{\mathrm{vv}}(k',k) = g_{\mathrm{v}}^2 \frac{\pi}{m'^*\, m^*}\Big\{\big[2k'^*W + \frac{1}{m_{\mathrm{v}}^2}((k'^0 - k^0)^2 k'^*W$$
$$+ 2(k'^0 - k^0)(|\boldsymbol{k}|k'^*k^* - |\boldsymbol{k}'|W'W) + (\boldsymbol{k}'^2 + \boldsymbol{k}^2)k'^*W$$
$$- 2|\boldsymbol{k}'||\boldsymbol{k}|k^*W')\big]\frac{\tilde{Q}_j}{|\boldsymbol{k}'|\,|\boldsymbol{k}|} + \big[2k^*W' + \frac{1}{m_{\mathrm{v}}^2}((k'^0 - k^0)^2 k^*W'$$
$$+ 2(k'^0 - k^0)(|\boldsymbol{k}'|k'^*k^* - |\boldsymbol{k}|W'W)$$
$$+ (\boldsymbol{k}'^2 + \boldsymbol{k}^2)k^*W' - 2|\boldsymbol{k}'||\boldsymbol{k}|k'^*W)\big]\frac{\tilde{Q}_j^{(2)}}{|\boldsymbol{k}'|\,|\boldsymbol{k}|}\Big\}, \tag{F.31}$$

$$
\begin{aligned}
{}^{34}\mathsf{U}^j_{\text{pv}}(k',k) &= g_{\text{v}}^2 \frac{\pi}{m'^*\,m^*}\Big\{\big[2k^*W' + \frac{1}{m_{\text{v}}^2}\big((k'^0-k^0)^2 k^* W' \\
&- 2(k'^0-k^0)(|\boldsymbol{k}'|k'^*k^* - |\boldsymbol{k}|W'W) + (\boldsymbol{k}'^2+\boldsymbol{k}^2)k^*W' \\
&- 2|\boldsymbol{k}'||\boldsymbol{k}|k'^*W\big)\big]\frac{\tilde{Q}_j}{|\boldsymbol{k}'|\,|\boldsymbol{k}|} + \big[2k'^*W + \frac{1}{m_{\text{v}}^2}\big((k'^0-k^0)^2 k'^* W \\
&+ 2(k'^0-k^0)(|\boldsymbol{k}|k'^*k^* - |\boldsymbol{k}'|W'W) + (\boldsymbol{k}'^2+\boldsymbol{k}^2)k'^*W \\
&- 2|\boldsymbol{k}'||\boldsymbol{k}|k^*W'\big)\big]\frac{\tilde{Q}^{(2)}_j}{|\boldsymbol{k}'|\,|\boldsymbol{k}|}\Big\}, \qquad \text{(F.32)}
\end{aligned}
$$

$$
\begin{aligned}
{}^{3}\mathsf{U}^j_{\text{vv}}(k',k) &= g_{\text{v}}^2 \frac{\pi}{m'^*\,m^*}\Big\{\Big(\frac{2}{m_{\text{v}}^2}\big[-(k'^0-k^0)^2 m'^*k^* - 2(k'^0-k^0) \\
&\times m'^*|\boldsymbol{k}|W + (\boldsymbol{k}'^2-\boldsymbol{k}^2)m'^*k^*\big]\Big)\frac{\tilde{Q}^{(3)}_j}{|\boldsymbol{k}'|\,|\boldsymbol{k}|}\Big\}, \qquad \text{(F.33)}
\end{aligned}
$$

$$
\begin{aligned}
{}^{56}\mathsf{U}^j_{\text{vv}}(k',k) &= g_{\text{v}}^2 \frac{\pi}{m'^*\,m^*}\Big\{\Big(-4m^*k'^* + \frac{2}{m_{\text{v}}^2}\big[-(k'^0-k^0)^2 m^*k'^* \\
&+ 2(k'^0-k^0)m^*|\boldsymbol{k}'|W' - (\boldsymbol{k}'^2-\boldsymbol{k}^2)m^*k'^*\big]\Big)\frac{\tilde{Q}^{(3)}_j}{|\boldsymbol{k}'|\,|\boldsymbol{k}|}\Big\}, \qquad \text{(F.34)}
\end{aligned}
$$

$$
\begin{aligned}
{}^{5}\mathsf{U}^j_{\text{vv}}(k',k) &= g_{\text{v}}^2 \frac{\pi}{m'^*\,m^*}\Big\{\Big(-4m'^*k^* + \frac{2}{m_{\text{v}}^2}\big[-(k'^0-k^0)^2 m'^*k^* \\
&- 2(k'^0-k^0)m'^*|\boldsymbol{k}|W + (\boldsymbol{k}'^2-\boldsymbol{k}^2)m'^*k^*\big]\Big)\frac{\tilde{Q}^{(3)}_j}{|\boldsymbol{k}'|\,|\boldsymbol{k}|}\Big\}, \qquad \text{(F.35)}
\end{aligned}
$$

$$
\begin{aligned}
{}^{78}\mathsf{U}^j_{\text{vv}}(k',k) &= g_{\text{v}}^2 \frac{\pi}{m'^*\,m^*}\Big\{\Big(\frac{2}{m_{\text{v}}^2}\big[-(k'^0-k^0)^2 m^*k'^* + 2(k'^0-k^0) \\
&\times m^*|\boldsymbol{k}'|W' - (\boldsymbol{k}'^2-\boldsymbol{k}^2)m^*k'^*\big]\Big)\frac{\tilde{Q}^{(3)}_j}{|\boldsymbol{k}'|\,|\boldsymbol{k}|}\Big\}, \qquad \text{(F.36)}
\end{aligned}
$$

$$
\begin{aligned}
{}^{0}\mathsf{W}^j_{\text{vv}}(k',k) &= g_{\text{v}}^2 \frac{\pi}{m'^*\,m^*}\Big\{\big[(4W'W + 2m'^*m^*) + \frac{1}{m_{\text{v}}^2}\big((k'^0-k^0)^2 \\
&\times (W'W - m'^*m^*) + 2(k'^0-k^0)(k^*|\boldsymbol{k}|W' - k'^*|\boldsymbol{k}'|W) \\
&+ (\boldsymbol{k}'^2+\boldsymbol{k}^2)(W'W + m'^*m^*) - 2|\boldsymbol{k}'||\boldsymbol{k}|k'^*k^*\big)\big]\frac{\tilde{Q}_j}{|\boldsymbol{k}'|\,|\boldsymbol{k}|} \\
&+ \frac{1}{m_{\text{v}}^2}\big((k'^0-k^0)^2 k'^*k^* + 2(k'^0-k^0)(k'^*|\boldsymbol{k}|W - k^*|\boldsymbol{k}'|W') \\
&+ (\boldsymbol{k}'^2+\boldsymbol{k}^2)k'^*k^* - 2|\boldsymbol{k}'||\boldsymbol{k}|(W'W + m'^*m^*)\big)\frac{\tilde{Q}^{(1)}_j}{|\boldsymbol{k}'|\,|\boldsymbol{k}|}\Big\}, \qquad \text{(F.37)}
\end{aligned}
$$

$$
\begin{aligned}
{}^{12}\mathsf{W}^{j}_{\mathrm{vv}}(k',k) &= g_{\mathrm{v}}^2 \frac{\pi}{m'^*\, m^*} \Big\{ \big[4k'^* k^* + \frac{1}{m_{\mathrm{v}}^2} \big((k'^0 - k^0)^2 k'^* k^* \\
&+ 2(k'^0 - k^0)(k'^* |\boldsymbol{k}| W - k^* |\boldsymbol{k}'| W') + (\boldsymbol{k}'^2 + \boldsymbol{k}^2) k'^* k^* \\
&- 2|\boldsymbol{k}'||\boldsymbol{k}|(W'W + m'^* m^*) \big) \big] \frac{\tilde{Q}_j}{|\boldsymbol{k}'|\,|\boldsymbol{k}|} \\
&+ \big[-2m'^* m^* + \frac{1}{m_{\mathrm{v}}^2} \big((k'^0 - k^0)^2 (W'W - m'^* m^*) \\
&+ 2(k'^0 - k^0)(k^* |\boldsymbol{k}| W' - k'^* |\boldsymbol{k}'| W) \\
&+ (\boldsymbol{k}'^2 + \boldsymbol{k}^2)(W'W + m'^* m^*) - 2|\boldsymbol{k}'||\boldsymbol{k}| k'^* k^* \big] \frac{\tilde{Q}_j^{(1)}}{|\boldsymbol{k}'|\,|\boldsymbol{k}|} \Big\} ,
\end{aligned} \tag{F.38}
$$

$$
\begin{aligned}
{}^{1}\mathsf{W}^{j}_{\mathrm{vv}}(k',k) &= g_{\mathrm{v}}^2 \frac{\pi}{m'^*\, m^*} \Big\{ \big[-2W'W + \frac{1}{m_{\mathrm{v}}^2} - \big((k'^0 - k^0)^2 (W'W - m'^* m^*) \\
&- 2(k'^0 - k^0)(k^* |\boldsymbol{k}| W' - k'^* |\boldsymbol{k}'| W) - (\boldsymbol{k}'^2 + \boldsymbol{k}^2)(W'W + m'^* m^*) \\
&+ 2|\boldsymbol{k}'||\boldsymbol{k}| k'^* k^* \big) \big] \frac{\tilde{Q}_j}{|\boldsymbol{k}'|\,|\boldsymbol{k}|} + \big[-2k'^* k^* + \frac{1}{m_{\mathrm{v}}^2} \big(-(k'^0 - k^0)^2 k'^* k^* \\
&- 2(k'^0 - k^0)(k'^* |\boldsymbol{k}| W - k^* |\boldsymbol{k}'| W') - (\boldsymbol{k}'^2 + \boldsymbol{k}^2) k'^* k^* \\
&+ 2|\boldsymbol{k}'||\boldsymbol{k}|(W'W + m'^* m^*) \big) \big] \frac{\tilde{Q}_j^{(2)}}{|\boldsymbol{k}'|\,|\boldsymbol{k}|} \Big\} ,
\end{aligned} \tag{F.39}
$$

$$
\begin{aligned}
{}^{34}\mathsf{W}^{j}_{\mathrm{pv}}(k',k) &= g_{\mathrm{v}}^2 \frac{\pi}{m'^*\, m^*} \Big\{ \big[-2k'^* k^* + \frac{1}{m_{\mathrm{v}}^2} \big(-(k'^0 - k^0)^2 k'^* k^* \\
&- 2(k'^0 - k^0)(k'^* |\boldsymbol{k}| W - k^* |\boldsymbol{k}'| W') - (\boldsymbol{k}'^2 + \boldsymbol{k}^2) k'^* k^* \\
&+ 2|\boldsymbol{k}'||\boldsymbol{k}|(W'W + m'^* m^*) \big) \big] \frac{\tilde{Q}_j}{|\boldsymbol{k}'|\,|\boldsymbol{k}|} + \big[-2W'W + \frac{1}{m_{\mathrm{v}}^2} \big(-(k'^0 - k^0)^2 \\
&\times (W'W - m'^* m^*) - 2(k'^0 - k^0)(k^* |\boldsymbol{k}| W' - k'^* |\boldsymbol{k}'| W) - (\boldsymbol{k}'^2 + \boldsymbol{k}^2) \\
&\times (W'W + m'^* m^*) + 2|\boldsymbol{k}'||\boldsymbol{k}| k'^* k^* \big) \big] \frac{\tilde{Q}_j^{(2)}}{|\boldsymbol{k}'|\,|\boldsymbol{k}|} \Big\} ,
\end{aligned} \tag{F.40}
$$

$$
{}^{3}\mathsf{W}^{j}_{\mathrm{pv}}(k',k) = 0 , \tag{F.41}
$$

$$
\begin{aligned}
{}^{56}\mathsf{W}^{j}_{\mathrm{vv}}(k',k) &= g_{\mathrm{v}}^2 \frac{\pi}{m'^*\, m^*} \Big\{ \big[4m^* W' + \frac{2}{m_{\mathrm{v}}^2} \big((k'^0 - k^0)^2 (m^* W' - m'^* W) \\
&- 2(k'^0 - k^0)(m^* |\boldsymbol{k}'| k'^* + m'^* |\boldsymbol{k}| k^*) - (\boldsymbol{k}^2 - \boldsymbol{k}'^2)(m^* W' + m'^* W) \big) \big] \\
&\times \frac{\tilde{Q}_j^{(3)}}{|\boldsymbol{k}'|\,|\boldsymbol{k}|} \Big\} ,
\end{aligned} \tag{F.42}
$$

$$^{5}\mathsf{W}^{j}_{\mathrm{pv}}(k',k) = 0\,, \tag{F.43}$$

$$^{78}\mathsf{W}^{j}_{\mathrm{vv}}(k',k) = g_{\mathrm{v}}^{2}\frac{\pi}{m'^{*}\,m^{*}}\Big\{\big[4m'^{*}W + \frac{2}{m_{\mathrm{v}}^{2}}(-(k'^{0}-k^{0})^{2}(m^{*}W'-m'^{*}W)$$
$$+\,2(k'^{0}-k^{0})(m^{*}|\boldsymbol{k}'|k'^{*}+m'^{*}|\boldsymbol{k}|k^{*})+(\boldsymbol{k}^{2}-\boldsymbol{k}'^{2})(m^{*}W'+m'^{*}W))\big]$$
$$\times\,\frac{\tilde{Q}^{(3)}_{j}}{|\boldsymbol{k}'|\,|\boldsymbol{k}|}\Big\}, \tag{F.44}$$

$$^{0}\mathsf{V}^{j}_{\mathrm{tt}}(k',k) = \Big(\frac{f_{\mathrm{v}}}{2\,m_{B}}\Big)^{2}\frac{\pi}{m'^{*}\,m^{*}}\Big\{\big[3(k'^{0}-k^{0})^{2}(W'W-m'^{*}m^{*})$$
$$+\,4(k'^{0}-k^{0})(k^{*}|\boldsymbol{k}|W'-k'^{*}|\boldsymbol{k}'|W)+(\boldsymbol{k}'^{2}+\boldsymbol{k}^{2})(m'^{*}m^{*}+3W'W)\big]$$
$$\times\,\frac{\tilde{Q}_{j}}{|\boldsymbol{k}'|\,|\boldsymbol{k}|}+\big[(k'^{0}-k^{0})^{2}k'^{*}k^{*}+4(k'^{0}-k^{0})(k'^{*}|\boldsymbol{k}|W-k^{*}|\boldsymbol{k}'|W')$$
$$+\,(\boldsymbol{k}'^{2}+\boldsymbol{k}^{2})k'^{*}k^{*}-2|\boldsymbol{k}'||\boldsymbol{k}|(m'^{*}m^{*}+3W'W)\big]\frac{\tilde{Q}^{(1)}_{j}}{|\boldsymbol{k}'|\,|\boldsymbol{k}|}$$
$$-\,2|\boldsymbol{k}'||\boldsymbol{k}|k'^{*}k^{*}\frac{\tilde{Q}^{(4)}_{j}}{|\boldsymbol{k}'|\,|\boldsymbol{k}|}\Big\}, \tag{F.45}$$

$$^{12}\mathsf{V}^{j}_{\mathrm{tt}}(k',k) = \Big(\frac{f_{\mathrm{v}}}{2\,m_{B}}\Big)^{2}\frac{\pi}{m'^{*}\,m^{*}}\Big\{\big[-3(k'^{0}-k^{0})^{2}k'^{*}k^{*}-4(k'^{0}-k^{0})$$
$$\times\,(k'^{*}|\boldsymbol{k}|W-k^{*}|\boldsymbol{k}'|W')-3(\boldsymbol{k}'^{2}+\boldsymbol{k}^{2})k'^{*}k^{*}+4|\boldsymbol{k}'||\boldsymbol{k}|m'^{*}m^{*}\big]$$
$$\times\,\frac{\tilde{Q}_{j}}{|\boldsymbol{k}'|\,|\boldsymbol{k}|}+\big[-(k'^{0}-k^{0})^{2}(W'W-m'^{*}m^{*})-4(k'^{0}-k^{0})(k^{*}|\boldsymbol{k}|W'$$
$$-\,k'^{*}|\boldsymbol{k}'|W)-(\boldsymbol{k}'^{2}+\boldsymbol{k}^{2})(W'W+3m'^{*}m^{*})+6|\boldsymbol{k}'||\boldsymbol{k}|k'^{*}k^{*}\big]\frac{\tilde{Q}^{(1)}_{j}}{|\boldsymbol{k}'|\,|\boldsymbol{k}|}$$
$$+\,2|\boldsymbol{k}'||\boldsymbol{k}|(W'W+m'^{*}m^{*})\frac{\tilde{Q}^{(4)}_{j}}{|\boldsymbol{k}'|\,|\boldsymbol{k}|}\Big\}, \tag{F.46}$$

$$^{1}\mathsf{V}^{j}_{\mathrm{tt}}(k',k) = \Big(\frac{f_{\mathrm{v}}}{2\,m_{B}}\Big)^{2}\frac{\pi}{m'^{*}\,m^{*}}\Big\{\big[(k'^{0}-k^{0})^{2}(W'W-m'^{*}m^{*})$$
$$+\,(\boldsymbol{k}'^{2}+\boldsymbol{k}^{2})(W'W-m'^{*}m^{*})+4|\boldsymbol{k}'||\boldsymbol{k}|k'^{*}k^{*}\big]\frac{\tilde{Q}_{j}}{|\boldsymbol{k}'|\,|\boldsymbol{k}|}$$
$$+\,2|\boldsymbol{k}'||\boldsymbol{k}|(W'W+m'^{*}m^{*})\frac{\tilde{Q}^{(1)}_{j}}{|\boldsymbol{k}'|\,|\boldsymbol{k}|}+\big[-(k'^{0}-k^{0})^{2}k'^{*}k^{*}-(\boldsymbol{k}'^{2}+\boldsymbol{k}^{2})$$
$$\times\,k'^{*}k^{*}-4|\boldsymbol{k}'||\boldsymbol{k}|W'W\big]\frac{\tilde{Q}^{(2)}_{j}}{|\boldsymbol{k}'|\,|\boldsymbol{k}|}-2|\boldsymbol{k}'||\boldsymbol{k}|k'^{*}k^{*}\frac{\tilde{Q}^{(5)}_{j}}{|\boldsymbol{k}'|\,|\boldsymbol{k}|}\Big\}, \tag{F.47}$$

$$
\begin{aligned}
{}^{34}\mathsf{V}^{j}_{\rm tt}(k',k) = \Big(\frac{f_{\rm v}}{2\,m_B}\Big)^2 \frac{\pi}{m'^*\,m^*}\Big\{\big[-(k'^0-k^0)^2 k'^* k^* \\
- (\boldsymbol{k}'^2+\boldsymbol{k}^2)k'^*k^* - 4|\boldsymbol{k}'||\boldsymbol{k}|W'W\big]\frac{\tilde{Q}_j}{|\boldsymbol{k}'|\,|\boldsymbol{k}|} \\
- 2|\boldsymbol{k}'||\boldsymbol{k}|k'^*k^*\frac{\tilde{Q}^{(1)}_j}{|\boldsymbol{k}'|\,|\boldsymbol{k}|} + \big[(k'^0-k^0)^2(W'W-m'^*m^*) \\
+ (\boldsymbol{k}'^2+\boldsymbol{k}^2)(W'W-m'^*m^*) + 4|\boldsymbol{k}'||\boldsymbol{k}|k'^*k^*\big] \\
\times \frac{\tilde{Q}^{(2)}_j}{|\boldsymbol{k}'|\,|\boldsymbol{k}|} + 2|\boldsymbol{k}'||\boldsymbol{k}|(W'W+m'^*m^*)\frac{\tilde{Q}^{(5)}_j}{|\boldsymbol{k}'|\,|\boldsymbol{k}|}\Big\},
\end{aligned} \tag{F.48}
$$

$$
\begin{aligned}
{}^{5}\mathsf{V}^{j}_{\rm tt}(k',k) = \Big(\frac{f_{\rm v}}{2\,m_B}\Big)^2 \frac{\pi}{m'^*\,m^*}\Big\{\big[(k'^0-k^0)^2(m^*W'-m'^*W) \\
- 4(k'^0-k^0)m^*k'^*|\boldsymbol{k}'| + (\boldsymbol{k}'^2+\boldsymbol{k}^2)(m^*W'-m'^*W) + 4m'^*\boldsymbol{k}'^2W\big] \\
\times \frac{\tilde{Q}^{(3)}_j}{|\boldsymbol{k}'|\,|\boldsymbol{k}|} - 2|\boldsymbol{k}'||\boldsymbol{k}|(m^*W'+m'^*W)\frac{\tilde{Q}^{(6)}_j}{|\boldsymbol{k}'|\,|\boldsymbol{k}|}\Big\},
\end{aligned} \tag{F.49}
$$

$$
\begin{aligned}
{}^{6}\mathsf{V}^{j}_{\rm tt}(k',k) = \Big(\frac{f_{\rm v}}{2\,m_B}\Big)^2 \frac{\pi}{m'^*\,m^*}\Big\{\big[-(k'^0-k^0)^2(m^*W'-m'^*W) \\
+ 4(k'^0-k^0)m'^*k^*|\boldsymbol{k}| + (\boldsymbol{k}'^2+\boldsymbol{k}^2)(3m^*W'+m'^*W) - 4m^*\boldsymbol{k}'^2W'\big] \\
\times \frac{\tilde{Q}^{(3)}_j}{|\boldsymbol{k}'|\,|\boldsymbol{k}|} - 2|\boldsymbol{k}'||\boldsymbol{k}|(m^*W'+m'^*W)\frac{\tilde{Q}^{(6)}_j}{|\boldsymbol{k}'|\,|\boldsymbol{k}|}\Big\},
\end{aligned} \tag{F.50}
$$

$$
\begin{aligned}
{}^{0}\mathsf{U}^{j}_{\rm tt}(k',k) = \Big(\frac{f_{\rm v}}{2\,m_B}\Big)^2 \frac{\pi}{m'^*\,m^*}\Big\{\big[3(k'^0-k^0)^2k'^*W + 4(k'^0-k^0) \\
\times (|\boldsymbol{k}|k'^*k^* - |\boldsymbol{k}'|W'W) + 3(\boldsymbol{k}'^2+\boldsymbol{k}^2)k'^*W\big]\frac{\tilde{Q}_j}{|\boldsymbol{k}'|\,|\boldsymbol{k}|} \\
+ \big[(k'^0-k^0)^2k^*W' - 4(k'^0-k^0)(|\boldsymbol{k}'|k'^*k^* - |\boldsymbol{k}|W'W) \\
+ (\boldsymbol{k}'^2+\boldsymbol{k}^2)k^*W' - 6|\boldsymbol{k}'||\boldsymbol{k}|k'^*W\big]\frac{\tilde{Q}^{(1)}_j}{|\boldsymbol{k}'|\,|\boldsymbol{k}|} - 2|\boldsymbol{k}'||\boldsymbol{k}|k^*W'\frac{\tilde{Q}^{(4)}_j}{|\boldsymbol{k}'|\,|\boldsymbol{k}|}\Big\},
\end{aligned} \tag{F.51}
$$

$$
\begin{aligned}
{}^{12}\mathsf{U}^{j}_{\rm tt}(k',k) = \Big(\frac{f_{\rm v}}{2\,m_B}\Big)^2 \frac{\pi}{m'^*\,m^*}\Big\{\big[-3(k'^0-k^0)^2k^*W' + 4(k'^0-k^0) \\
\times (|\boldsymbol{k}'|k'^*k^* - |\boldsymbol{k}|W'W) - 3(\boldsymbol{k}'^2+\boldsymbol{k}^2)k^*W'\big]\frac{\tilde{Q}_j}{|\boldsymbol{k}'|\,|\boldsymbol{k}|} \\
+ \big[-(k'^0-k^0)^2k'^*W - 4(k'^0-k^0)(|\boldsymbol{k}|k'^*k^* - |\boldsymbol{k}'|W'W)
\end{aligned}
$$

$$- (\boldsymbol{k}'^2 + \boldsymbol{k}^2)k'^*W + 6|\boldsymbol{k}'||\boldsymbol{k}|k^*W'\Big]\frac{\tilde{Q}_j^{(1)}}{|\boldsymbol{k}'|\,|\boldsymbol{k}|} + 2|\boldsymbol{k}'||\boldsymbol{k}|k'^*W\frac{\tilde{Q}_j^{(4)}}{|\boldsymbol{k}'|\,|\boldsymbol{k}|}\Big\}, \tag{F.52}$$

$$\begin{aligned}{}^{1}\mathsf{U}^j_{\mathrm{tt}}(k',k) &= \Big(\frac{f_{\mathrm{v}}}{2m_B}\Big)^2\frac{\pi}{m'^*\,m^*}\Big\{\Big[(k'^0-k^0)^2k'^*W + (\boldsymbol{k}'^2+\boldsymbol{k}^2)k'^*W \\ &+ 4(|\boldsymbol{k}'||\boldsymbol{k}|k^*W'\Big]\frac{\tilde{Q}_j}{|\boldsymbol{k}'|\,|\boldsymbol{k}|} + 2(|\boldsymbol{k}'||\boldsymbol{k}|k'^*W\frac{\tilde{Q}_j^{(1)}}{|\boldsymbol{k}'|\,|\boldsymbol{k}|} + \Big[-(k'^0-k^0)^2k^*W' \\ &- (\boldsymbol{k}'^2+\boldsymbol{k}^2)k^*W' - 4|\boldsymbol{k}'||\boldsymbol{k}|k'^*W\Big]\frac{\tilde{Q}_j^{(2)}}{|\boldsymbol{k}'|\,|\boldsymbol{k}|} - 2|\boldsymbol{k}'||\boldsymbol{k}|k^*W')\frac{\tilde{Q}_j^{(5)}}{|\boldsymbol{k}'|\,|\boldsymbol{k}|}\Big\},\end{aligned} \tag{F.53}$$

$$\begin{aligned}{}^{34}\mathsf{U}^j_{\mathrm{tt}}(k',k) &= \Big(\frac{f_{\mathrm{v}}}{2m_B}\Big)^2\frac{\pi}{m'^*\,m^*}\Big\{\Big[-(k'^0-k^0)^2k^*W' - (\boldsymbol{k}'^2+\boldsymbol{k}^2)k^*W' \\ &- 4|\boldsymbol{k}'||\boldsymbol{k}|k'^*W\Big]\frac{\tilde{Q}_j}{|\boldsymbol{k}'|\,|\boldsymbol{k}|} - 2|\boldsymbol{k}'||\boldsymbol{k}|k^*W'\frac{\tilde{Q}_j^{(1)}}{|\boldsymbol{k}'|\,|\boldsymbol{k}|} + \Big[(k'^0-k^0)^2k'^*W \\ &+ (\boldsymbol{k}'^2+\boldsymbol{k}^2)k'^*W + 4|\boldsymbol{k}'||\boldsymbol{k}|k^*W'\Big]\frac{\tilde{Q}_j^{(2)}}{|\boldsymbol{k}'|\,|\boldsymbol{k}|} + 2|\boldsymbol{k}'||\boldsymbol{k}|k'^*W)\frac{\tilde{Q}_j^{(5)}}{|\boldsymbol{k}'|\,|\boldsymbol{k}|}\Big\},\end{aligned} \tag{F.54}$$

$$\begin{aligned}{}^{3}\mathsf{U}^j_{\mathrm{tt}}(k',k) &= \Big(\frac{f_{\mathrm{v}}}{2m_B}\Big)^2\frac{\pi}{m'^*\,m^*}\Big\{\Big[-2(k'^0-k^0)^2m'^*k^* - 8(k'^0-k^0)m'^*|\boldsymbol{k}|W \\ &- 2(\boldsymbol{k}'^2+\boldsymbol{k}^2)m'^*k^*\Big]\frac{\tilde{Q}_j^{(3)}}{|\boldsymbol{k}'|\,|\boldsymbol{k}|} + 4|\boldsymbol{k}'||\boldsymbol{k}|m'^*k^*\frac{\tilde{Q}_j^{(6)}}{|\boldsymbol{k}'|\,|\boldsymbol{k}|}\Big\},\end{aligned} \tag{F.55}$$

$$\begin{aligned}{}^{56}\mathsf{U}^j_{\mathrm{tt}}(k',k) &= \Big(\frac{f_{\mathrm{v}}}{2m_B}\Big)^2\frac{\pi}{m'^*\,m^*}\Big\{\Big[-2(k'^0-k^0)^2m^*k'^* + 6(k'^0-k^0) \\ &\times m^*k'^* - 8m^*\boldsymbol{k}'^2k'^*\Big]\frac{\tilde{Q}_j^{(3)}}{|\boldsymbol{k}'|\,|\boldsymbol{k}|} - 4|\boldsymbol{k}'||\boldsymbol{k}|m^*k'^*\frac{\tilde{Q}_j^{(6)}}{|\boldsymbol{k}'|\,|\boldsymbol{k}|}\Big\},\end{aligned} \tag{F.56}$$

$$\begin{aligned}{}^{5}\mathsf{U}^j_{\mathrm{tt}}(k',k) &= \Big(\frac{f_{\mathrm{v}}}{2m_B}\Big)^2\frac{\pi}{m'^*\,m^*}\Big\{\Big[2(k'^0-k^0)^2m'^*k^* + 2(\boldsymbol{k}'^2-\boldsymbol{k}^2)m'^*k^* \\ &- 8m'^*\boldsymbol{k}'^2k'^*\Big]\frac{\tilde{Q}_j^{(3)}}{|\boldsymbol{k}'|\,|\boldsymbol{k}|} + 2|\boldsymbol{k}'||\boldsymbol{k}|m'^*k^*\frac{\tilde{Q}_j^{(6)}}{|\boldsymbol{k}'|\,|\boldsymbol{k}|}\Big\},\end{aligned} \tag{F.57}$$

$$\begin{aligned}{}^{78}\mathsf{U}^j_{\mathrm{tt}}(k',k) &= \Big(\frac{f_{\mathrm{v}}}{2m_B}\Big)^2\frac{\pi}{m'^*\,m^*}\Big\{\Big[2(k'^0-k^0)^2m^*k'^* - 8(k'^0-k^0)m^* \\ &\times \boldsymbol{k}'W' + 2(\boldsymbol{k}'^2+\boldsymbol{k}^2)m^*k'^*\Big]\frac{\tilde{Q}_j^{(3)}}{|\boldsymbol{k}'|\,|\boldsymbol{k}|} - 4|\boldsymbol{k}'||\boldsymbol{k}|m^*k'^*\frac{\tilde{Q}_j^{(6)}}{|\boldsymbol{k}'|\,|\boldsymbol{k}|}\Big\},\end{aligned} \tag{F.58}$$

$$
\begin{aligned}
{}^{0}\mathsf{W}^{j}_{\rm tt}(k',k) &= \Big(\frac{f_{\rm V}}{2\,m_B}\Big)^2 \frac{\pi}{m'^*\,m^*}\Big\{\big[-3(k'^0-k^0)^2(W'W+m'^*m^*)\\
&- 4(k'^0-k^0)(k^*|\boldsymbol{k}|W'-k'^*|\boldsymbol{k}'|W)+(\boldsymbol{k}'^2+\boldsymbol{k}^2)(m'^*m^*-3W'W)\big]\\
&\times \frac{\tilde{Q}_j}{|\boldsymbol{k}'|\,|\boldsymbol{k}|}+\big[-(k'^0-k^0)^2k'^*k^*-4(k'^0-k^0)(k'^*|\boldsymbol{k}|W-k^*|\boldsymbol{k}'|W')\\
&-(\boldsymbol{k}'^2+\boldsymbol{k}^2)k'^*k^*+2|\boldsymbol{k}'||\boldsymbol{k}|(3W'W-m'^*m^*)\big]\frac{\tilde{Q}_j^{(1)}}{|\boldsymbol{k}'|\,|\boldsymbol{k}|}\\
&+2|\boldsymbol{k}'||\boldsymbol{k}|k'^*k^*\frac{\tilde{Q}_j^{(4)}}{|\boldsymbol{k}'|\,|\boldsymbol{k}|}\Big\}, \qquad \text{(F.59)}
\end{aligned}
$$

$$
\begin{aligned}
{}^{12}\mathsf{W}^{j}_{\rm tt}(k',k) &= \Big(\frac{f_{\rm V}}{2\,m_B}\Big)^2 \frac{\pi}{m'^*\,m^*}\Big\{\big[3(k'^0-k^0)^2k'^*k^*+4(k'^0-k^0)\\
&\times(k'^*|\boldsymbol{k}|W-k^*|\boldsymbol{k}'|W')+3(\boldsymbol{k}'^2+\boldsymbol{k}^2)k'^*k^*+4|\boldsymbol{k}'||\boldsymbol{k}|m'^*m^*\big]\frac{\tilde{Q}_j}{|\boldsymbol{k}'|\,|\boldsymbol{k}|}\\
&+\big[(k'^0-k^0)^2(W'W+m'^*m^*)+4(k'^0-k^0)(k'^*|\boldsymbol{k}|W'-k'^*|\boldsymbol{k}'|W)\\
&+(\boldsymbol{k}'^2+\boldsymbol{k}^2)(W'W-3m'^*m^*)-6|\boldsymbol{k}'||\boldsymbol{k}|k'^*k^*\big]\frac{\tilde{Q}_j^{(4)}}{|\boldsymbol{k}'|\,|\boldsymbol{k}|}\Big\}, \qquad \text{(F.60)}
\end{aligned}
$$

$$
\begin{aligned}
{}^{1}\mathsf{W}^{j}_{\rm tt}(k',k) &= \Big(\frac{f_{\rm V}}{2\,m_B}\Big)^2 \frac{\pi}{m'^*\,m^*}\Big\{\big[(k'^0-k^0)^2(W'W+m'^*m^*)+(\boldsymbol{k}'^2+\boldsymbol{k}^2)\\
&\times(W'W+m'^*m^*)+4|\boldsymbol{k}'||\boldsymbol{k}|k'^*k^*\big]\frac{\tilde{Q}_j}{|\boldsymbol{k}'|\,|\boldsymbol{k}|}+2|\boldsymbol{k}'||\boldsymbol{k}|(W'W-m'^*m^*)\\
&\times\frac{\tilde{Q}_j^{(1)}}{|\boldsymbol{k}'|\,|\boldsymbol{k}|}+\big[-(k'^0-k^0)^2k'^*k^*-(\boldsymbol{k}'^2+\boldsymbol{k}^2)k'^*k^*-4|\boldsymbol{k}'||\boldsymbol{k}|W'W\big]\frac{\tilde{Q}_j^{(2)}}{|\boldsymbol{k}'|\,|\boldsymbol{k}|}\\
&-2|\boldsymbol{k}'||\boldsymbol{k}|k'^*k^*\frac{\tilde{Q}_j^{(5)}}{|\boldsymbol{k}'|\,|\boldsymbol{k}|}\Big\}, \qquad \text{(F.61)}
\end{aligned}
$$

$$
\begin{aligned}
{}^{34}\mathsf{W}^{j}_{\rm tt}(k',k) &= \Big(\frac{f_{\rm V}}{2\,m_B}\Big)^2 \frac{\pi}{m'^*\,m^*}\Big\{\big[-(k'^0-k^0)^2k'^*k^*-(\boldsymbol{k}'^2+\boldsymbol{k}^2)\\
&\times k'^*k^*-4|\boldsymbol{k}'||\boldsymbol{k}|W'W\big]\frac{\tilde{Q}_j}{|\boldsymbol{k}'|\,|\boldsymbol{k}|}-2|\boldsymbol{k}'||\boldsymbol{k}|k'^*k^*\frac{\tilde{Q}_j^{(1)}}{|\boldsymbol{k}'|\,|\boldsymbol{k}|}\\
&+\big[(k'^0-k^0)^2(W'W+m'^*m^*)+(\boldsymbol{k}'^2+\boldsymbol{k}^2)(W'W+m'^*m^*)\\
&+4|\boldsymbol{k}'||\boldsymbol{k}|k'^*k^*\big]\frac{\tilde{Q}_j^{(2)}}{|\boldsymbol{k}'|\,|\boldsymbol{k}|}+2|\boldsymbol{k}'||\boldsymbol{k}|(W'W-m'^*m^*)\frac{\tilde{Q}_j^{(5)}}{|\boldsymbol{k}'|\,|\boldsymbol{k}|}\Big\}, \qquad \text{(F.62)}
\end{aligned}
$$

$$
{}^{3}\mathsf{W}^{j}_{\rm tt}(k',k) = 0\,, \qquad \text{(F.63)}
$$

$$
\begin{aligned}
{}^{56}\mathsf{W}^{j}_{\mathrm{tt}}(k',k) &= \left(\frac{f_{\mathrm{V}}}{2\,m_B}\right)^2 \frac{\pi}{m'^*\,m^*}\Big\{\big[-2(k'^0-k^0)^2(m^*W'+m'^*W) \\
&- 8(k'^0-k^0)m'^*(k^*|\boldsymbol{k}| + 2(\boldsymbol{k}'^2+\boldsymbol{k}^2)(3m^*W'-m'^*W) \\
&- 8m^*\boldsymbol{k}'^2W'\big]\frac{\tilde{Q}^{(3)}_j}{|\boldsymbol{k}'|\,|\boldsymbol{k}|} - 4|\boldsymbol{k}'||\boldsymbol{k}|(m^*W'-m'^*W)\frac{\tilde{Q}^{(6)}_j}{|\boldsymbol{k}'|\,|\boldsymbol{k}|}\Big\}, \qquad \text{(F.64)}
\end{aligned}
$$

$$
{}^{5}\mathsf{W}^{j}_{\mathrm{tt}}(k',k) = 0\,, \qquad \text{(F.65)}
$$

$$
\begin{aligned}
{}^{78}\mathsf{W}^{j}_{\mathrm{tt}}(k',k) &= \left(\frac{f_{\mathrm{V}}}{2\,m_B}\right)^2 \frac{\pi}{m'^*\,m^*}\Big\{\big[-2(k'^0-k^0)^2(m^*W'+m'^*W) \\
&+ 8(k'^0-k^0)m^*(k'^*|\boldsymbol{k}'| - 2(\boldsymbol{k}'^2+\boldsymbol{k}^2)(m^*W'+m'^*W) \\
&+ 8m'^*\boldsymbol{k}'^2W\big]\frac{\tilde{Q}^{(3)}_j}{|\boldsymbol{k}'|\,|\boldsymbol{k}|} + 4|\boldsymbol{k}'||\boldsymbol{k}|(m^*W'-m'^*W)\frac{\tilde{Q}^{(6)}_j}{|\boldsymbol{k}'|\,|\boldsymbol{k}|}\Big\}, \qquad \text{(F.66)}
\end{aligned}
$$

$$
\begin{aligned}
{}^{0}\mathsf{V}^{j}_{\mathrm{vt}}(k',k) &= \left(\frac{f_{\mathrm{V}}\,g_{\mathrm{V}}}{2\,m_B}\right)\frac{\pi}{m'^*\,m^*}\Big\{\big[2(m^*k'^*|\boldsymbol{k}'| + m'^*k^*|\boldsymbol{k}|) - 6(k'^0-k^0) \\
&\times (m^*W'-m'^*W)\big]\frac{\tilde{Q}_j}{|\boldsymbol{k}'|\,|\boldsymbol{k}|} - 2(m^*k'^*|\boldsymbol{k}| + m'^*k^*|\boldsymbol{k}'|)\frac{\tilde{Q}^{(1)}_j}{|\boldsymbol{k}'|\,|\boldsymbol{k}|}\Big\},
\end{aligned}
$$
(F.67)

$$
\begin{aligned}
{}^{12}\mathsf{V}^{j}_{\mathrm{vt}}(k',k) &= \left(\frac{f_{\mathrm{V}}\,g_{\mathrm{V}}}{2\,m_B}\right)\frac{\pi}{m'^*\,m^*}\Big\{\big[6(m^*k'^*|\boldsymbol{k}| + m'^*k^*|\boldsymbol{k}'|)\big]\frac{\tilde{Q}_j}{|\boldsymbol{k}'|\,|\boldsymbol{k}|} \\
&+ \big[-6(m^*k'^*|\boldsymbol{k}'| + m'^*k^*|\boldsymbol{k}|) + 2(k'^0-k^0)(m^*W'-m'^*W)\big]\frac{\tilde{Q}^{(1)}_j}{|\boldsymbol{k}'|\,|\boldsymbol{k}|}\Big\},
\end{aligned}
$$
(F.68)

$$
\begin{aligned}
{}^{1}V^{j}_{\mathrm{vt}}(k',k) &= \left(\frac{f_{\mathrm{V}}\,g_{\mathrm{V}}}{2\,m_B}\right)\frac{\pi}{m'^*\,m^*}\Big\{\big[-2(m^*k'^*|\boldsymbol{k}'| + m'^*k^*|\boldsymbol{k}|) - 2(k'^0-k^0) \\
&\times (m^*W'-m'^*W)\big]\frac{\tilde{Q}_j}{|\boldsymbol{k}'|\,|\boldsymbol{k}|} + 2(m^*k'^*|\boldsymbol{k}| + m'^*k^*|\boldsymbol{k}'|)\frac{\tilde{Q}^{(2)}_j}{|\boldsymbol{k}'|\,|\boldsymbol{k}|}\Big\}, \qquad \text{(F.69)}
\end{aligned}
$$

$$
\begin{aligned}
{}^{34}\mathsf{V}^{j}_{\mathrm{vt}}(k',k) &= \left(\frac{f_{\mathrm{V}}\,g_{\mathrm{V}}}{2\,m_B}\right)\frac{\pi}{m'^*\,m^*}\Big\{\big[2(m^*k'^*|\boldsymbol{k}| + m'^*k^*|\boldsymbol{k}'|)\big]\frac{\tilde{Q}_j}{|\boldsymbol{k}'|\,|\boldsymbol{k}|} \\
&+ \big[-2(m^*k'^*|\boldsymbol{k}'| + m'^*k^*|\boldsymbol{k}|) - 2(k'^0-k^0)(m^*W'-m'^*W)\big]\frac{\tilde{Q}^{(2)}_j}{|\boldsymbol{k}'|\,|\boldsymbol{k}|}\Big\},
\end{aligned}
$$
(F.70)

$$
{}^{5}\mathsf{V}^{j}_{\mathrm{vt}}(k',k) = \left(\frac{f_{\mathrm{v}}\,g_{\mathrm{v}}}{2\,m_B}\right)\frac{\pi}{m'^*\,m^*}\{6|\boldsymbol{k}'|k'^*W + 2|\boldsymbol{k}|k^*W') - 2(k'^0 - k^0)
$$

$$
\times\,(W'W - m'^*m^*)\}\,\frac{\tilde{Q}^{(3)}_j}{|\boldsymbol{k}'|\,|\boldsymbol{k}|}\,, \tag{F.71}
$$

$$
{}^{6}\mathsf{V}^{j}_{\mathrm{vt}}(k',k) = \left(\frac{f_{\mathrm{v}}\,g_{\mathrm{v}}}{2\,m_B}\right)\frac{\pi}{m'^*\,m^*}\{2|\boldsymbol{k}'|k'^*W + 6|\boldsymbol{k}|k^*W') + 2(k'^0 - k^0)
$$

$$
\times\,(W'W - m'^*m^*)\}\,\frac{\tilde{Q}^{(3)}_j}{|\boldsymbol{k}'|\,|\boldsymbol{k}|}\,, \tag{F.72}
$$

$$
{}^{0}\mathsf{Z}^{j}_{\mathrm{vt}}(k',k) = {}^{12}\mathsf{Z}^{j}_{\mathrm{vt}}(k',k) = {}^{1}\mathsf{Z}^{j}_{\mathrm{vt}}(k',k) = {}^{34}\mathsf{Z}^{j}_{\mathrm{vt}}(k',k) = 0\,, \tag{F.73}
$$

$$
{}^{5}\mathsf{Z}^{j}_{\mathrm{vt}}(k',k) = \left(\frac{f_{\mathrm{v}}\,g_{\mathrm{v}}}{2\,m_B}\right)\frac{\pi}{m'^*\,m^*}\Big\{\big[-2(k'^0 - k^0)k'^*k^* + 2|\boldsymbol{k}|k'^*W
$$

$$
+\,6|\boldsymbol{k}'|k^*W'\big]\,\frac{\tilde{Q}^{(3)}_j}{|\boldsymbol{k}'|\,|\boldsymbol{k}|}\Big\}\,, \tag{F.74}
$$

$$
{}^{6}\mathsf{Z}^{j}_{\mathrm{vt}}(k',k) = \left(\frac{f_{\mathrm{v}}\,g_{\mathrm{v}}}{2\,m_B}\right)\frac{\pi}{m'^*\,m^*}\Big\{\big[2(k'^0 - k^0)k'^*k^* + 6|\boldsymbol{k}|k'^*W
$$

$$
+\,2|\boldsymbol{k}'|k^*W'\big]\,\frac{\tilde{Q}^{(3)}_j}{|\boldsymbol{k}'|\,|\boldsymbol{k}|}\Big\}\,, \tag{F.75}
$$

$$
{}^{0}\mathsf{U}^{j}_{\mathrm{vt}}(k',k) = \left(\frac{f_{\mathrm{v}}\,g_{\mathrm{v}}}{2\,m_B}\right)\frac{\pi}{m'^*\,m^*}\Big\{\big[-6(k'^0 - k^0)m^*k'^* + 2m^*|\boldsymbol{k}'|W'\big]
$$

$$
\times\,\frac{\tilde{Q}_j}{|\boldsymbol{k}'|\,|\boldsymbol{k}|} - 2m^*|\boldsymbol{k}|W'\,\frac{\tilde{Q}^{(1)}_j}{|\boldsymbol{k}'|\,|\boldsymbol{k}|}\Big\}\,, \tag{F.76}
$$

$$
{}^{12}\mathsf{U}^{j}_{\mathrm{vt}}(k',k) = \left(\frac{f_{\mathrm{v}}\,g_{\mathrm{v}}}{2\,m_B}\right)\frac{\pi}{m'^*\,m^*}\Big\{6m^*|\boldsymbol{k}|W'\,\frac{\tilde{Q}_j}{|\boldsymbol{k}'|\,|\boldsymbol{k}|}
$$

$$
+\,\big[2(k'^0 - k^0)m^*k'^* - 6m^*|\boldsymbol{k}'|W'\big]\,\frac{\tilde{Q}^{(1)}_j}{|\boldsymbol{k}'|\,|\boldsymbol{k}|}\Big\}\,, \tag{F.77}
$$

$$
{}^{1}\mathsf{U}^{j}_{\mathrm{vt}}(k',k) = \left(\frac{f_{\mathrm{v}}\,g_{\mathrm{v}}}{2\,m_B}\right)\frac{\pi}{m'^*\,m^*}\Big\{\big[-2(k'^0 - k^0)m^*k'^* - 2m^*|\boldsymbol{k}'|W'\big]
$$

$$
\times\,\frac{\tilde{Q}_j}{|\boldsymbol{k}'|\,|\boldsymbol{k}|} + 2m^*|\boldsymbol{k}|W'\,\frac{\tilde{Q}^{(2)}_j}{|\boldsymbol{k}'|\,|\boldsymbol{k}|}\Big\}\,, \tag{F.78}
$$

$$
\begin{aligned}
{}^{34}\mathsf{U}^{j}_{\mathrm{vt}}(k',k) = \left(\frac{f_{\mathrm{V}}\,g_{\mathrm{V}}}{2\,m_B}\right)\frac{\pi}{m'^*\,m^*}\Big\{2m^*|\boldsymbol{k}|W'\frac{\tilde{Q}_j}{|\boldsymbol{k}'|\,|\boldsymbol{k}|} \\
+ \left[-2(k'^0 - k^0)m^*k'^* - 2m^*|\boldsymbol{k}'|W'\right]\frac{\tilde{Q}^{(2)}_j}{|\boldsymbol{k}'|\,|\boldsymbol{k}|}\Big\}\,, \qquad \text{(F.79)}
\end{aligned}
$$

$$
{}^{3}\mathsf{U}^{j}_{\mathrm{vt}}(k',k) = \left(\frac{f_{\mathrm{V}}\,g_{\mathrm{V}}}{2\,m_B}\right)\frac{\pi}{m'^*\,m^*}\left(4m'^*m^*|\boldsymbol{k}|\right)\frac{\tilde{Q}^{(3)}_j}{|\boldsymbol{k}'|\,|\boldsymbol{k}|}\,, \qquad \text{(F.80)}
$$

$$
\begin{aligned}
{}^{56}\mathsf{U}^{j}_{\mathrm{vt}}(k',k) = \left(\frac{f_{\mathrm{V}}\,g_{\mathrm{V}}}{2\,m_B}\right)\frac{\pi}{m'^*\,m^*}\Big\{\big[4(k'^0 - k^0)k'^*W \\
+ 4(3|\boldsymbol{k}|k'^*k^* + |\boldsymbol{k}'|W'W)\big]\frac{\tilde{Q}^{(3)}_j}{|\boldsymbol{k}'|\,|\boldsymbol{k}|}\Big\}\,, \qquad \text{(F.81)}
\end{aligned}
$$

$$
{}^{5}\mathsf{U}^{j}_{\mathrm{vt}}(k',k) = \left(\frac{f_{\mathrm{V}}\,g_{\mathrm{V}}}{2\,m_B}\right)\frac{\pi}{m'^*\,m^*}\left(-4m'^*m^*|\boldsymbol{k}|\right)\frac{\tilde{Q}^{(3)}_j}{|\boldsymbol{k}'|\,|\boldsymbol{k}|}\,, \qquad \text{(F.82)}
$$

$$
\begin{aligned}
{}^{78}\mathsf{U}^{j}_{\mathrm{vt}}(k',k) = \left(\frac{f_{\mathrm{V}}\,g_{\mathrm{V}}}{2\,m_B}\right)\frac{\pi}{m'^*\,m^*}\Big\{\big[-4(k'^0 - k^0)k'^*W \\
+ 4(|\boldsymbol{k}|k'^*k^* + 3|\boldsymbol{k}'|W'W)\big]\frac{\tilde{Q}^{(3)}_j}{|\boldsymbol{k}'|\,|\boldsymbol{k}|}\Big\}\,, \qquad \text{(F.83)}
\end{aligned}
$$

$$
{}^{0}\mathsf{W}^{j}_{\mathrm{vt}}(k',k) = {}^{12}\mathsf{W}^{j}_{\mathrm{vt}}(k',k) = {}^{1}\mathsf{W}^{j}_{\mathrm{vt}}(k',k) = {}^{34}\mathsf{W}^{j}_{\mathrm{vt}}(k',k) = 0\,, \qquad \text{(F.84)}
$$

$$
\begin{aligned}
{}^{3}\mathsf{W}^{j}_{\mathrm{vt}}(k',k) = \left(\frac{f_{\mathrm{V}}\,g_{\mathrm{V}}}{2\,m_B}\right)\frac{\pi}{m'^*\,m^*}\Big\{\big[4(k'^0 - k^0)k'^*k^* \\
+ 4(3|\boldsymbol{k}|k'^*W + |\boldsymbol{k}'|k^*W')\big]\frac{\tilde{Q}^{(3)}_j}{|\boldsymbol{k}'|\,|\boldsymbol{k}|}\Big\}\,, \qquad \text{(F.85)}
\end{aligned}
$$

$$
{}^{56}\mathsf{W}^{j}_{\mathrm{vt}}(k',k) = 0\,, \qquad \text{(F.86)}
$$

$$
\begin{aligned}
{}^{5}\mathsf{W}^{j}_{\mathrm{vt}}(k',k) = \left(\frac{f_{\mathrm{V}}\,g_{\mathrm{V}}}{2\,m_B}\right)\frac{\pi}{m'^*\,m^*}\Big\{\big[4(k'^0 - k^0)k'^*k^* \\
- 4(|\boldsymbol{k}|k'^*W + 3|\boldsymbol{k}'|k^*W')\big]\frac{\tilde{Q}^{(3)}_j}{|\boldsymbol{k}'|\,|\boldsymbol{k}|}\Big\}\,, \qquad \text{(F.87)}
\end{aligned}
$$

and

$$
{}^{78}\mathsf{W}^{j}_{\mathrm{vt}}(k',k) = 0\,. \qquad \text{(F.88)}
$$

Appendix G

Rotating stars in general relativity

As we know from chapters 14 and 15, the metric of rotating neutron stars is more complicated than the one of non-rotating neutron stars, which has its origin in the additional dependence of the metric on the polar angle and the frame dragging frequency. The latter is a function of radial coordinate and polar angle too, like all metric functions. This seemingly innocuous complication, however, renders the structure of Einstein's equations considerably more complicated. We begin with the Riemann–Christoffel tensor, $R^{\tau}{}_{\mu\nu\sigma}$, whose non-vanishing components will be given first. This is followed by the expressions for the Ricci tensor $R^{\mu\nu}$, the Ricci scalar R, the Einstein tensor $G^{\mu\nu}$, and finally the energy–momentum tensor $T^{\mu\nu}$. Several covariant derivatives of the latter are given at the end of this chapter. For the definitions of these quantities, we refer to chapter 13.

Throughout this appendix, the absence of a + or − sign implies multiplication.

G.1 Riemann–Christoffel tensor

The components of the Riemann–Christoffel curvature tensor, $R^{\tau}{}_{\mu\nu\sigma}$, are given by,

$$R^{t}{}_{tr\theta} = \frac{1}{4}\Big\{\omega(r,\theta)\Big[-\Big(\frac{\partial}{\partial\theta}\nu(r,\theta)\Big)\frac{\partial}{\partial r}\omega(r,\theta) + \Big(\frac{\partial}{\partial\theta}\omega(r,\theta)\Big)\frac{\partial}{\partial r}\nu(r,\theta) - \Big(\frac{\partial}{\partial r}\psi(r,\theta)\Big)\frac{\partial}{\partial\theta}\omega(r,\theta) + \Big(\frac{\partial}{\partial\theta}\psi(r,\theta)\Big)\frac{\partial}{\partial r}\omega(r,\theta)\Big] e^{2\,\psi(r,\theta)-2\,\nu(r,\theta)}\Big\}, \tag{G.1}$$

$$R^{t}{}_{rtr} = \frac{1}{4}\Big\{-2\Big(\frac{\partial}{\partial r}\lambda(r,\theta)\Big)\omega(r,\theta)\Big(\frac{\partial}{\partial r}\omega(r,\theta)\Big)e^{2\,\mu(r,\theta)+2\,\psi(r,\theta)}$$

$$- 4\Big(\frac{\partial}{\partial\theta}\lambda(r,\theta)\Big)\Big(\frac{\partial}{\partial\theta}\nu(r,\theta)\Big)e^{2\,\lambda(r,\theta)+2\,\nu(r,\theta)} - 4\Big(\frac{\partial^2}{\partial r^2}\nu(r,\theta)\Big)$$
$$e^{2\,\mu(r,\theta)+2\,\nu(r,\theta)} + 6\Big(\frac{\partial}{\partial r}\psi(r,\theta)\Big)\omega(r,\theta)\Big(\frac{\partial}{\partial r}\omega(r,\theta)\Big)e^{2\,\mu(r,\theta)+2\,\psi(r,\theta)}$$
$$+ 2\Big(\frac{\partial}{\partial\theta}\lambda(r,\theta)\Big)\omega(r,\theta)\Big(\frac{\partial}{\partial\theta}\omega(r,\theta)\Big)e^{2\,\lambda(r,\theta)+2\,\psi(r,\theta)}$$
$$- 4\Big(\frac{\partial}{\partial r}\nu(r,\theta)\Big)^2 e^{2\,\mu(r,\theta)+2\,\nu(r,\theta)} - 2\Big(\frac{\partial}{\partial r}\nu(r,\theta)\Big)\omega(r,\theta)$$
$$\Big(\frac{\partial}{\partial r}\omega(r,\theta)\Big)e^{2\,\mu(r,\theta)+2\,\psi(r,\theta)} + 2\,\omega(r,\theta)\Big(\frac{\partial^2}{\partial r^2}\omega(r,\theta)\Big)$$
$$e^{2\,\mu(r,\theta)+2\,\psi(r,\theta)} + 3\Big(\frac{\partial}{\partial r}\omega(r,\theta)\Big)^2 e^{2\,\mu(r,\theta)+2\,\psi(r,\theta)} + 4\Big(\frac{\partial}{\partial r}\lambda(r,\theta)\Big)$$
$$\Big(\frac{\partial}{\partial r}\nu(r,\theta)\Big)e^{2\,\mu(r,\theta)+2\,\nu(r,\theta)}\Big\}\, e^{-2\,\mu(r,\theta)-2\,\nu(r,\theta)}\,, \tag{G.2}$$

$$R^t{}_{rt\theta} = \frac{1}{4}\Big\{3\,e^{2\,\psi(r,\theta)}\Big(\frac{\partial}{\partial\theta}\omega(r,\theta)\Big)\frac{\partial}{\partial r}\omega(r,\theta) + 2\,e^{2\,\psi(r,\theta)}\omega(r,\theta)$$
$$\frac{\partial^2}{\partial\theta\partial r}\omega(r,\theta) + 4\Big(\frac{\partial}{\partial\theta}\psi(r,\theta)\Big)e^{2\,\psi(r,\theta)}\omega(r,\theta)\frac{\partial}{\partial r}\omega(r,\theta)$$
$$- 4\Big(\frac{\partial}{\partial r}\nu(r,\theta)\Big)\Big(\frac{\partial}{\partial\theta}\nu(r,\theta)\Big)e^{2\,\nu(r,\theta)} + 4\Big(\frac{\partial}{\partial r}\mu(r,\theta)\Big)e^{2\,\nu(r,\theta)}$$
$$\frac{\partial}{\partial\theta}\nu(r,\theta) + 4\Big(\frac{\partial}{\partial\theta}\lambda(r,\theta)\Big)e^{2\,\nu(r,\theta)}\frac{\partial}{\partial r}\nu(r,\theta) - 4\Big(\frac{\partial^2}{\partial\theta\partial r}\nu(r,\theta)\Big)$$
$$e^{2\,\nu(r,\theta)} + 2\Big(\frac{\partial}{\partial r}\psi(r,\theta)\Big)e^{2\,\psi(r,\theta)}\omega(r,\theta)\frac{\partial}{\partial\theta}\omega(r,\theta)$$
$$- 2\Big(\frac{\partial}{\partial r}\mu(r,\theta)\Big)e^{2\,\psi(r,\theta)}\omega(r,\theta)\frac{\partial}{\partial\theta}\omega(r,\theta) - 2\Big(\frac{\partial}{\partial\theta}\lambda(r,\theta)\Big)$$
$$e^{2\,\psi(r,\theta)}\omega(r,\theta)\frac{\partial}{\partial r}\omega(r,\theta) - 2\Big(\frac{\partial}{\partial\theta}\nu(r,\theta)\Big)e^{2\,\psi(r,\theta)}\omega(r,\theta)$$
$$\frac{\partial}{\partial r}\omega(r,\theta)\Big\}\, e^{-2\,\nu(r,\theta)}\,, \tag{G.3}$$

$$R^t{}_{rr\phi} = -\frac{1}{2}\Big\{-3\Big(\frac{\partial}{\partial r}\psi(r,\theta)\Big)\Big(\frac{\partial}{\partial r}\omega(r,\theta)\Big)e^{2\,\mu(r,\theta)} - \Big(\frac{\partial^2}{\partial r^2}\omega(r,\theta)\Big)$$
$$e^{2\,\mu(r,\theta)} + e^{2\,\mu(r,\theta)}\Big(\frac{\partial}{\partial r}\nu(r,\theta)\Big)\frac{\partial}{\partial r}\omega(r,\theta) + \Big(\frac{\partial}{\partial r}\lambda(r,\theta)\Big)$$
$$e^{2\,\mu(r,\theta)}\frac{\partial}{\partial r}\omega(r,\theta) - \Big(\frac{\partial}{\partial\theta}\lambda(r,\theta)\Big)e^{2\,\lambda(r,\theta)}\frac{\partial}{\partial\theta}\omega(r,\theta)\Big\}$$
$$e^{-2\,\nu(r,\theta)-2\,\mu(r,\theta)+2\,\psi(r,\theta)}\,, \tag{G.4}$$

$$R^t{}_{r\theta\phi} = \frac{1}{2}\Big\{-\Big(\frac{\partial}{\partial\theta}\lambda(r,\theta)\Big)\frac{\partial}{\partial r}\omega(r,\theta) - \Big(\frac{\partial}{\partial\theta}\nu(r,\theta)\Big)\frac{\partial}{\partial r}\omega(r,\theta)$$

$$+\left(\frac{\partial}{\partial r}\psi(r,\theta)\right)\frac{\partial}{\partial\theta}\omega(r,\theta)+2\left(\frac{\partial}{\partial\theta}\psi(r,\theta)\right)\frac{\partial}{\partial r}\omega(r,\theta)$$
$$+\frac{\partial^2}{\partial\theta\partial r}\omega(r,\theta)-\left(\frac{\partial}{\partial r}\mu(r,\theta)\right)\frac{\partial}{\partial\theta}\omega(r,\theta)\Big\}\,\mathrm{e}^{2\,\psi(r,\theta)-2\,\nu(r,\theta)}\,, \tag{G.5}$$

$$R^t{}_{\theta tr}=\frac{1}{4}\,\mathrm{e}^{-2\,\nu(r,\theta)}\Big\{3\,\mathrm{e}^{2\,\psi(r,\theta)}\left(\frac{\partial}{\partial\theta}\omega(r,\theta)\right)\frac{\partial}{\partial r}\omega(r,\theta)+2\,\mathrm{e}^{2\,\psi(r,\theta)}\omega(r,\theta)$$
$$\frac{\partial^2}{\partial\theta\partial r}\omega(r,\theta)+2\left(\frac{\partial}{\partial\theta}\psi(r,\theta)\right)\mathrm{e}^{2\,\psi(r,\theta)}\omega(r,\theta)\frac{\partial}{\partial r}\omega(r,\theta)$$
$$-4\left(\frac{\partial}{\partial r}\nu(r,\theta)\right)\left(\frac{\partial}{\partial\theta}\nu(r,\theta)\right)\mathrm{e}^{2\,\nu(r,\theta)}+4\left(\frac{\partial}{\partial r}\mu(r,\theta)\right)\mathrm{e}^{2\,\nu(r,\theta)}$$
$$\frac{\partial}{\partial\theta}\nu(r,\theta)+4\left(\frac{\partial}{\partial\theta}\lambda(r,\theta)\right)\mathrm{e}^{2\,\nu(r,\theta)}\frac{\partial}{\partial r}\nu(r,\theta)$$
$$-4\left(\frac{\partial^2}{\partial\theta\partial r}\nu(r,\theta)\right)\mathrm{e}^{2\,\nu(r,\theta)}+4\left(\frac{\partial}{\partial r}\psi(r,\theta)\right)\mathrm{e}^{2\,\psi(r,\theta)}\omega(r,\theta)$$
$$\frac{\partial}{\partial\theta}\omega(r,\theta)-2\left(\frac{\partial}{\partial r}\mu(r,\theta)\right)\mathrm{e}^{2\,\psi(r,\theta)}\omega(r,\theta)\frac{\partial}{\partial\theta}\omega(r,\theta)$$
$$-2\left(\frac{\partial}{\partial\theta}\lambda(r,\theta)\right)\mathrm{e}^{2\,\psi(r,\theta)}\omega(r,\theta)\frac{\partial}{\partial r}\omega(r,\theta)-2\,\mathrm{e}^{2\,\psi(r,\theta)}\omega(r,\theta)$$
$$\left(\frac{\partial}{\partial\theta}\omega(r,\theta)\right)\frac{\partial}{\partial r}\nu(r,\theta)\Big\}\,, \tag{G.6}$$

$$R^t{}_{\theta t\theta}=-\frac{1}{4}\Big\{-2\left(\frac{\partial}{\partial r}\mu(r,\theta)\right)\omega(r,\theta)\left(\frac{\partial}{\partial r}\omega(r,\theta)\right)\mathrm{e}^{2\,\mu(r,\theta)+2\,\psi(r,\theta)}$$
$$-2\,\omega(r,\theta)\left(\frac{\partial^2}{\partial\theta^2}\omega(r,\theta)\right)\mathrm{e}^{2\,\lambda(r,\theta)+2\,\psi(r,\theta)}+2\left(\frac{\partial}{\partial\theta}\nu(r,\theta)\right)$$
$$\omega(r,\theta)\left(\frac{\partial}{\partial\theta}\omega(r,\theta)\right)\mathrm{e}^{2\,\lambda(r,\theta)+2\,\psi(r,\theta)}-6\left(\frac{\partial}{\partial\theta}\psi(r,\theta)\right)\omega(r,\theta)$$
$$\left(\frac{\partial}{\partial\theta}\omega(r,\theta)\right)\mathrm{e}^{2\,\lambda(r,\theta)+2\,\psi(r,\theta)}+4\left(\frac{\partial}{\partial r}\mu(r,\theta)\right)\left(\frac{\partial}{\partial r}\nu(r,\theta)\right)$$
$$\mathrm{e}^{2\,\nu(r,\theta)+2\,\nu(r,\theta)}-4\left(\frac{\partial}{\partial\theta}\mu(r,\theta)\right)\left(\frac{\partial}{\partial\theta}\nu(r,\theta)\right)\mathrm{e}^{2\,\lambda(r,\theta)+2\,\nu(r,\theta)}$$
$$+2\left(\frac{\partial}{\partial\theta}\mu(r,\theta)\right)\omega(r,\theta)\left(\frac{\partial}{\partial\theta}\omega(r,\theta)\right)\mathrm{e}^{2\,\lambda(r,\theta)+2\,\psi(r,\theta)}$$
$$-3\left(\frac{\partial}{\partial\theta}\omega(r,\theta)\right)^2\mathrm{e}^{2\,\lambda(r,\theta)+2\,\psi(r,\theta)}+4\left(\frac{\partial}{\partial\theta}\nu(r,\theta)\right)^2\mathrm{e}^{2\,\lambda(r,\theta)+2\,\nu(r,\theta)}$$
$$+4\left(\frac{\partial^2}{\partial\theta^2}\nu(r,\theta)\right)\mathrm{e}^{2\,\lambda(r,\theta)+2\,\nu(r,\theta)}\Big\}\,\mathrm{e}^{-2\,\lambda(r,\theta)-2\,\nu(r,\theta)}\,, \tag{G.7}$$

$$R^t{}_{\theta r\phi}=\frac{1}{2}\Big\{-\left(\frac{\partial}{\partial\theta}\lambda(r,\theta)\right)\frac{\partial}{\partial r}\omega(r,\theta)-\left(\frac{\partial}{\partial\theta}\omega(r,\theta)\right)\frac{\partial}{\partial r}\nu(r,\theta)$$

$$+ 2\Big(\frac{\partial}{\partial r}\psi(r,\theta)\Big)\frac{\partial}{\partial\theta}\omega(r,\theta) + \Big(\frac{\partial}{\partial\theta}\psi(r,\theta)\Big)\frac{\partial}{\partial r}\omega(r,\theta)$$
$$+ \frac{\partial^2}{\partial\theta\partial r}\omega(r,\theta) - \Big(\frac{\partial}{\partial r}\mu(r,\theta)\Big)\frac{\partial}{\partial\theta}\omega(r,\theta)\Big\}\, \mathrm{e}^{2\,\psi(r,\theta)-2\,\nu(r,\theta)}\,, \tag{G.8}$$

$$R^t{}_{\theta\theta\phi} = \frac{1}{2}\Big\{3\Big(\frac{\partial}{\partial\theta}\psi(r,\theta)\Big)\Big(\frac{\partial}{\partial\theta}\omega(r,\theta)\Big)\mathrm{e}^{2\,\lambda(r,\theta)} + \Big(\frac{\partial^2}{\partial\theta^2}\omega(r,\theta)\Big)$$
$$\mathrm{e}^{2\,\lambda(r,\theta)} - \mathrm{e}^{2\,\lambda(r,\theta)}\Big(\frac{\partial}{\partial\theta}\nu(r,\theta)\Big)\frac{\partial}{\partial\theta}\omega(r,\theta) + \Big(\frac{\partial}{\partial r}\mu(r,\theta)\Big)$$
$$\mathrm{e}^{2\,\mu(r,\theta)}\frac{\partial}{\partial r}\omega(r,\theta) - \Big(\frac{\partial}{\partial\theta}\mu(r,\theta)\Big)\mathrm{e}^{2\,\lambda(r,\theta)}\frac{\partial}{\partial\theta}\omega(r,\theta)\Big\}$$
$$\mathrm{e}^{-2\,\nu(r,\theta)-2\,\lambda(r,\theta)+2\,\psi(r,\theta)}\,, \tag{G.9}$$

$$R^t{}_{\phi t\phi} = -\frac{1}{4}\Big\{\Big(\frac{\partial}{\partial r}\omega(r,\theta)\Big)^2\mathrm{e}^{2\,\mu(r,\theta)+2\,\psi(r,\theta)} + 4\Big(\frac{\partial}{\partial r}\psi(r,\theta)\Big)\Big(\frac{\partial}{\partial r}\nu(r,\theta)\Big)$$
$$\mathrm{e}^{2\,\mu(r,\theta)+2\,\nu(r,\theta)} + \Big(\frac{\partial}{\partial\theta}\omega(r,\theta)\Big)^2\mathrm{e}^{2\,\lambda(r,\theta)+2\,\psi(r,\theta)} + 4\Big(\frac{\partial}{\partial\theta}\psi(r,\theta)\Big)$$
$$\Big(\frac{\partial}{\partial\theta}\nu(r,\theta)\Big)\mathrm{e}^{2\,\lambda(r,\theta)+2\,\nu(r,\theta)}\Big\}\mathrm{e}^{-2\,\nu(r,\theta)-2\,\lambda(r,\theta)+2\,\psi(r,\theta)-2\,\mu(r,\theta)}\,, \tag{G.10}$$

$$R^t{}_{\phi r\theta} = \frac{1}{2}\Big\{\Big(\frac{\partial}{\partial\theta}\nu(r,\theta)\Big)\frac{\partial}{\partial r}\omega(r,\theta) - \Big(\frac{\partial}{\partial\theta}\omega(r,\theta)\Big)\frac{\partial}{\partial r}\nu(r,\theta)$$
$$+ \Big(\frac{\partial}{\partial r}\psi(r,\theta)\Big)\frac{\partial}{\partial\theta}\omega(r,\theta) - \Big(\frac{\partial}{\partial\theta}\psi(r,\theta)\Big)\frac{\partial}{\partial r}\omega(r,\theta)\Big\}$$
$$\mathrm{e}^{2\,\psi(r,\theta)-2\,\nu(r,\theta)}\,, \tag{G.11}$$

$$R^r{}_{ttr} = -\frac{1}{4}\Big\{-4\Big(\frac{\partial}{\partial r}\psi(r,\theta)\Big)^2\omega^2(r,\theta)\mathrm{e}^{2\,\mu(r,\theta)+2\,\psi(r,\theta)+2\,\nu(r,\theta)}$$
$$+ 4\Big(\frac{\partial}{\partial r}\nu(r,\theta)\Big)\omega(r,\theta)\Big(\frac{\partial}{\partial r}\omega(r,\theta)\Big)\mathrm{e}^{2\,\mu(r,\theta)+2\,\psi(r,\theta)+2\,\nu(r,\theta)}$$
$$- \omega^2(r,\theta)\Big(\frac{\partial}{\partial r}\omega(r,\theta)\Big)^2\mathrm{e}^{2\,\mu(r,\theta)+4\,\psi(r,\theta)} - 4\Big(\frac{\partial}{\partial r}\lambda(r,\theta)\Big)$$
$$\Big(\frac{\partial}{\partial r}\nu(r,\theta)\Big)\mathrm{e}^{2\,\mu(r,\theta)+4\,\nu(r,\theta)} + 4\Big(\frac{\partial}{\partial r}\lambda(r,\theta)\Big)\Big(\frac{\partial}{\partial r}\psi(r,\theta)\Big)$$
$$\omega^2(r,\theta)\mathrm{e}^{2\,\mu(r,\theta)+2\,\psi(r,\theta)+2\,\nu(r,\theta)} + 4\Big(\frac{\partial}{\partial r}\lambda(r,\theta)\Big)\omega(r,\theta)\Big(\frac{\partial}{\partial r}\omega(r,\theta)\Big)$$
$$\mathrm{e}^{2\,\mu(r,\theta)+2\,\psi(r,\theta)+2\,\nu(r,\theta)} + 4\Big(\frac{\partial}{\partial\theta}\lambda(r,\theta)\Big)\Big(\frac{\partial}{\partial\theta}\nu(r,\theta)\Big)\mathrm{e}^{2\,\lambda(r,\theta)+4\,\nu(r,\theta)}$$

$$
\begin{aligned}
&- 4\Big(\frac{\partial}{\partial\theta}\lambda(r,\theta)\Big)\Big(\frac{\partial}{\partial\theta}\psi(r,\theta)\Big)\omega^2(r,\theta)\mathrm{e}^{2\,\lambda(r,\theta)+2\,\nu(r,\theta)+2\,\psi(r,\theta)}\\
&- 4\Big(\frac{\partial}{\partial\theta}\lambda(r,\theta)\Big)\omega(r,\theta)\Big(\frac{\partial}{\partial\theta}\omega(r,\theta)\Big)\mathrm{e}^{2\,\lambda(r,\theta)+2\,\nu(r,\theta)+2\,\psi(r,\theta)}\\
&- 12\Big(\frac{\partial}{\partial r}\psi(r,\theta)\Big)\omega(r,\theta)\Big(\frac{\partial}{\partial r}\omega(r,\theta)\Big)\mathrm{e}^{2\,\mu(r,\theta)+2\,\psi(r,\theta)+2\,\nu(r,\theta)}\\
&- 3\Big(\frac{\partial}{\partial r}\omega(r,\theta)\Big)^2 e^{2\,\mu(r,\theta)+2\,\psi(r,\theta)+2\,\nu(r,\theta)} - 4\,\omega(r,\theta)\Big(\frac{\partial^2}{\partial r^2}\omega(r,\theta)\Big)\\
&\mathrm{e}^{2\,\mu(r,\theta)+2\,\psi(r,\theta)+2\,\nu(r,\theta)} + 4\Big(\frac{\partial^2}{\partial r^2}\nu(r,\theta)\Big)\mathrm{e}^{2\,\mu(r,\theta)+4\,\nu(r,\theta)}\\
&+ 4\Big(\frac{\partial}{\partial r}\nu(r,\theta)\Big)^2\mathrm{e}^{2\,\mu(r,\theta)+4\,\nu(r,\theta)} - 4\Big(\frac{\partial^2}{\partial r^2}\psi(r,\theta)\Big)\omega^2(r,\theta)\\
&\mathrm{e}^{2\,\mu(r,\theta)+2\,\psi(r,\theta)+2\,\nu(r,\theta)}\Big\}\ \mathrm{e}^{-2\,\nu(r,\theta)-2\,\lambda(r,\theta)-2\,\mu(r,\theta)}\,, \qquad \text{(G.12)}
\end{aligned}
$$

$$
\begin{aligned}
R^{r}{}_{tt\theta} = &-\frac{1}{4}\Big\{4\Big(\frac{\partial^2}{\partial\theta\partial r}\nu(r,\theta)\Big)\mathrm{e}^{4\,\nu(r,\theta)} - 4\Big(\frac{\partial}{\partial\theta}\lambda(r,\theta)\Big)\mathrm{e}^{4\,\nu(r,\theta)}\frac{\partial}{\partial r}\nu(r,\theta)\\
&- 4\Big(\frac{\partial}{\partial r}\mu(r,\theta)\Big)\mathrm{e}^{4\,\nu(r,\theta)}\frac{\partial}{\partial\theta}\nu(r,\theta) + 4\Big(\frac{\partial}{\partial r}\nu(r,\theta)\Big)\Big(\frac{\partial}{\partial\theta}\nu(r,\theta)\Big)\\
&\mathrm{e}^{4\,\nu(r,\theta)} - 4\Big(\frac{\partial^2}{\partial\theta\partial r}\psi(r,\theta)\Big)\omega^2(r,\theta)\mathrm{e}^{2\,\psi(r,\theta)+2\,\nu(r,\theta)}\\
&- 4\Big(\frac{\partial}{\partial r}\psi(r,\theta)\Big)\Big(\frac{\partial}{\partial\theta}\psi(r,\theta)\Big)\omega^2(r,\theta)\mathrm{e}^{2\,\psi(r,\theta)+2\,\nu(r,\theta)}\\
&- 6\Big(\frac{\partial}{\partial r}\psi(r,\theta)\Big)\omega(r,\theta)\Big(\frac{\partial}{\partial\theta}\omega(r,\theta)\Big)\mathrm{e}^{2\,\psi(r,\theta)+2\,\nu(r,\theta)}\\
&- 6\Big(\frac{\partial}{\partial\theta}\psi(r,\theta)\Big)\omega(r,\theta)\Big(\frac{\partial}{\partial r}\omega(r,\theta)\Big)\mathrm{e}^{2\,\psi(r,\theta)+2\,\nu(r,\theta)}\\
&- 3\Big(\frac{\partial}{\partial\theta}\omega(r,\theta)\Big)\Big(\frac{\partial}{\partial r}\omega(r,\theta)\Big)\mathrm{e}^{2\,\psi(r,\theta)+2\,\nu(r,\theta)} - 4\,\omega(r,\theta)\\
&\Big(\frac{\partial^2}{\partial\theta\partial r}\omega(r,\theta)\Big)\mathrm{e}^{2\,\psi(r,\theta)+2\,\nu(r,\theta)} + 2\Big(\frac{\partial}{\partial\theta}\nu(r,\theta)\Big)\omega(r,\theta)\\
&\Big(\frac{\partial}{\partial r}\omega(r,\theta)\Big)\mathrm{e}^{2\,\psi(r,\theta)+2\,\nu(r,\theta)} + 2\,\omega(r,\theta)\Big(\frac{\partial}{\partial\theta}\omega(r,\theta)\Big)\Big(\frac{\partial}{\partial r}\nu(r,\theta)\Big)\\
&\mathrm{e}^{2\,\psi(r,\theta)+2\,\nu(r,\theta)} - \mathrm{e}^{4\,\psi(r,\theta)}\omega^2(r,\theta)\Big(\frac{\partial}{\partial\theta}\omega(r,\theta)\Big)\frac{\partial}{\partial r}\omega(r,\theta)\\
&+ 4\Big(\frac{\partial}{\partial\theta}\lambda(r,\theta)\Big)\Big(\frac{\partial}{\partial r}\psi(r,\theta)\Big)\omega^2(r,\theta)\mathrm{e}^{2\,\psi(r,\theta)+2\,\nu(r,\theta)}\\
&+ 4\Big(\frac{\partial}{\partial\theta}\lambda(r,\theta)\Big)\omega(r,\theta)\Big(\frac{\partial}{\partial r}\omega(r,\theta)\Big)\mathrm{e}^{2\,\psi(r,\theta)+2\,\nu(r,\theta)}\\
&+ 4\Big(\frac{\partial}{\partial r}\mu(r,\theta)\Big)\Big(\frac{\partial}{\partial\theta}\psi(r,\theta)\Big)\omega^2(r,\theta)\mathrm{e}^{2\,\psi(r,\theta)+2\,\nu(r,\theta)}\\
&+ 4\Big(\frac{\partial}{\partial r}\mu(r,\theta)\Big)\omega(r,\theta)\Big(\frac{\partial}{\partial\theta}\omega(r,\theta)\Big)\mathrm{e}^{2\,\psi(r,\theta)+2\,\nu(r,\theta)}\Big\}
\end{aligned}
$$

$$e^{-2\lambda(r,\theta)-2\nu(r,\theta)}, \tag{G.13}$$

$$\begin{aligned}
R^r{}_{tr\phi} = -\frac{1}{4}\Big\{&-4\Big(\frac{\partial^2}{\partial r^2}\psi(r,\theta)\Big)\omega(r,\theta)e^{2\mu(r,\theta)+2\nu(r,\theta)} - 4\Big(\frac{\partial}{\partial r}\psi(r,\theta)\Big)^2 \\
&\omega(r,\theta)e^{2\mu(r,\theta)+2\nu(r,\theta)} - 6\Big(\frac{\partial}{\partial r}\psi(r,\theta)\Big)\Big(\frac{\partial}{\partial r}\omega(r,\theta)\Big) \\
&e^{2\mu(r,\theta)+2\nu(r,\theta)} - 2\Big(\frac{\partial^2}{\partial r^2}\omega(r,\theta)\Big)e^{2\mu(r,\theta)+2\nu(r,\theta)} \\
&+ 2\Big(\frac{\partial}{\partial r}\nu(r,\theta)\Big)\Big(\frac{\partial}{\partial r}\omega(r,\theta)\Big)e^{2\mu(r,\theta)+2\nu(r,\theta)} - \omega(r,\theta) \\
&\Big(\frac{\partial}{\partial r}\omega(r,\theta)\Big)^2 e^{2\mu(r,\theta)+2\psi(r,\theta)} + 4\Big(\frac{\partial}{\partial r}\lambda(r,\theta)\Big)\Big(\frac{\partial}{\partial r}\psi(r,\theta)\Big) \\
&\omega(r,\theta)e^{2\mu(r,\theta)+2\nu(r,\theta)} + 2\Big(\frac{\partial}{\partial r}\lambda(r,\theta)\Big)\Big(\frac{\partial}{\partial r}\omega(r,\theta)\Big) \\
&e^{2\mu(r,\theta)+2\nu(r,\theta)} - 4\Big(\frac{\partial}{\partial\theta}\lambda(r,\theta)\Big)\Big(\frac{\partial}{\partial\theta}\psi(r,\theta)\Big)\omega(r,\theta) \\
&e^{2\lambda(r,\theta)+2\nu(r,\theta)} - 2\Big(\frac{\partial}{\partial\theta}\lambda(r,\theta)\Big)\Big(\frac{\partial}{\partial\theta}\omega(r,\theta)\Big)e^{2\lambda(r,\theta)+2\nu(r,\theta)}\Big\} \\
&e^{-2\nu(r,\theta)-2\lambda(r,\theta)+2\psi(r,\theta)-2\mu(r,\theta)},
\end{aligned} \tag{G.14}$$

$$\begin{aligned}
R^r{}_{t\theta\phi} = \frac{1}{4}\Big\{&-2\Big(\frac{\partial}{\partial\theta}\lambda(r,\theta)\Big)e^{2\nu(r,\theta)}\frac{\partial}{\partial r}\omega(r,\theta) - 2\Big(\frac{\partial}{\partial\theta}\nu(r,\theta)\Big) \\
&e^{2\nu(r,\theta)}\frac{\partial}{\partial r}\omega(r,\theta) + e^{2\psi(r,\theta)}\omega(r,\theta)\Big(\frac{\partial}{\partial\theta}\omega(r,\theta)\Big)\frac{\partial}{\partial r}\omega(r,\theta) \\
&- 4\Big(\frac{\partial}{\partial\theta}\lambda(r,\theta)\Big)e^{2\nu(r,\theta)}\Big(\frac{\partial}{\partial r}\psi(r,\theta)\Big)\omega(r,\theta) - 4\Big(\frac{\partial}{\partial r}\mu(r,\theta)\Big) \\
&e^{2\nu(r,\theta)}\Big(\frac{\partial}{\partial\theta}\psi(r,\theta)\Big)\omega(r,\theta) + 4\Big(\frac{\partial^2}{\partial\theta\partial r}\psi(r,\theta)\Big)\omega(r,\theta)e^{2\nu(r,\theta)} \\
&+ 4\Big(\frac{\partial}{\partial r}\psi(r,\theta)\Big)\Big(\frac{\partial}{\partial\theta}\psi(r,\theta)\Big)\omega(r,\theta)e^{2\nu(r,\theta)} + 2\Big(\frac{\partial}{\partial r}\psi(r,\theta)\Big) \\
&\Big(\frac{\partial}{\partial\theta}\omega(r,\theta)\Big)e^{2\nu(r,\theta)} + 4\Big(\frac{\partial}{\partial\theta}\psi(r,\theta)\Big)\Big(\frac{\partial}{\partial r}\omega(r,\theta)\Big)e^{2\nu(r,\theta)} \\
&+ 2\Big(\frac{\partial^2}{\partial\theta\partial r}\omega(r,\theta)\Big)e^{2\nu(r,\theta)} - 2\Big(\frac{\partial}{\partial r}\mu(r,\theta)\Big)e^{2\nu(r,\theta)}\frac{\partial}{\partial\theta}\omega(r,\theta)\Big\} \\
&e^{-2\nu(r,\theta)-2\lambda(r,\theta)+2\psi(r,\theta)},
\end{aligned} \tag{G.15}$$

$$R^r{}_{\theta t\phi} = -\frac{1}{2}\Big\{\Big(\frac{\partial}{\partial\theta}\nu(r,\theta)\Big)\frac{\partial}{\partial r}\omega(r,\theta) - \Big(\frac{\partial}{\partial\theta}\omega(r,\theta)\Big)\frac{\partial}{\partial r}\nu(r,\theta)$$

$$+\left(\frac{\partial}{\partial r}\psi(r,\theta)\right)\frac{\partial}{\partial\theta}\omega(r,\theta)-\left(\frac{\partial}{\partial\theta}\psi(r,\theta)\right)\frac{\partial}{\partial r}\omega(r,\theta)\Big\}$$
$$e^{2\,\psi(r,\theta)-2\,\lambda(r,\theta)}\,, \tag{G.16}$$

$$\begin{aligned}R^{r}{}_{\theta r\theta} = & -\Big\{\left(\frac{\partial^2}{\partial\theta^2}\lambda(r,\theta)\right)e^{2\,\lambda(r,\theta)}+\left(\frac{\partial}{\partial\theta}\lambda(r,\theta)\right)^2 e^{2\,\lambda(r,\theta)}\\ & +\left(\frac{\partial^2}{\partial r^2}\mu(r,\theta)\right)e^{2\,\mu(r,\theta)}+\left(\frac{\partial}{\partial r}\mu(r,\theta)\right)^2 e^{2\,\mu(r,\theta)}-\left(\frac{\partial}{\partial r}\lambda(r,\theta)\right)\\ & \left(\frac{\partial}{\partial r}\mu(r,\theta)\right)e^{2\,\mu(r,\theta)}-\left(\frac{\partial}{\partial\theta}\lambda(r,\theta)\right)e^{2\,\lambda(r,\theta)}\frac{\partial}{\partial\theta}\mu(r,\theta)\Big\}\,e^{-2\,\lambda(r,\theta)}\,,\end{aligned} \tag{G.17}$$

$$\begin{aligned}R^{r}{}_{\phi tr} = & -\frac{1}{4}\Big\{4\left(\frac{\partial^2}{\partial r^2}\psi(r,\theta)\right)\omega(r,\theta)e^{2\,\mu(r,\theta)+2\,\nu(r,\theta)}+4\left(\frac{\partial}{\partial r}\psi(r,\theta)\right)^2\\ & \omega(r,\theta)e^{2\,\mu(r,\theta)+2\,\nu(r,\theta)}+6\left(\frac{\partial}{\partial r}\psi(r,\theta)\right)\left(\frac{\partial}{\partial r}\omega(r,\theta)\right)\\ & e^{2\,\mu(r,\theta)+2\,\nu(r,\theta)}+2\left(\frac{\partial^2}{\partial r^2}\omega(r,\theta)\right)e^{2\,\mu(r,\theta)+2\,\nu(r,\theta)}-2\left(\frac{\partial}{\partial r}\nu(r,\theta)\right)\\ & \left(\frac{\partial}{\partial r}\omega(r,\theta)\right)e^{2\,\mu(r,\theta)+2\,\nu(r,\theta)}+\omega(r,\theta)\left(\frac{\partial}{\partial r}\omega(r,\theta)\right)^2 e^{2\,\mu(r,\theta)+2\,\psi(r,\theta)}\\ & -4\left(\frac{\partial}{\partial r}\lambda(r,\theta)\right)\left(\frac{\partial}{\partial r}\psi(r,\theta)\right)\omega(r,\theta)e^{2\,\mu(r,\theta)+2\,\nu(r,\theta)}\\ & -2\left(\frac{\partial}{\partial r}\lambda(r,\theta)\right)\left(\frac{\partial}{\partial r}\omega(r,\theta)\right)e^{2\,\mu(r,\theta)+2\,\nu(r,\theta)}+4\left(\frac{\partial}{\partial\theta}\lambda(r,\theta)\right)\\ & \left(\frac{\partial}{\partial\theta}\psi(r,\theta)\right)\omega(r,\theta)e^{2\,\lambda(r,\theta)+2\,\nu(r,\theta)}+2\left(\frac{\partial}{\partial\theta}\lambda(r,\theta)\right)\left(\frac{\partial}{\partial\theta}\omega(r,\theta)\right)\\ & e^{2\,\lambda(r,\theta)+2\,\nu(r,\theta)}\Big\}\,e^{-2\,\nu(r,\theta)-2\,\lambda(r,\theta)+2\,\psi(r,\theta)-2\,\mu(r,\theta)}\,,\end{aligned} \tag{G.18}$$

$$\begin{aligned}R^{r}{}_{\phi t\theta} = & -\frac{1}{4}\Big\{-2\left(\frac{\partial}{\partial\theta}\lambda(r,\theta)\right)e^{2\,\nu(r,\theta)}\frac{\partial}{\partial r}\omega(r,\theta)-2\left(\frac{\partial}{\partial\theta}\omega(r,\theta)\right)\\ & \left(\frac{\partial}{\partial r}\nu(r,\theta)\right)e^{2\,\nu(r,\theta)}+e^{2\,\psi(r,\theta)}\omega(r,\theta)\left(\frac{\partial}{\partial\theta}\omega(r,\theta)\right)\frac{\partial}{\partial r}\omega(r,\theta)\\ & -4\left(\frac{\partial}{\partial\theta}\lambda(r,\theta)\right)e^{2\,\nu(r,\theta)}\left(\frac{\partial}{\partial r}\psi(r,\theta)\right)\omega(r,\theta)-4\left(\frac{\partial}{\partial r}\mu(r,\theta)\right)\\ & e^{2\,\nu(r,\theta)}\left(\frac{\partial}{\partial\theta}\psi(r,\theta)\right)\omega(r,\theta)+4\left(\frac{\partial^2}{\partial\theta\partial r}\psi(r,\theta)\right)\omega(r,\theta)e^{2\,\nu(r,\theta)}\\ & +4\left(\frac{\partial}{\partial r}\psi(r,\theta)\right)\left(\frac{\partial}{\partial\theta}\psi(r,\theta)\right)\omega(r,\theta)e^{2\,\nu(r,\theta)}+4\left(\frac{\partial}{\partial r}\psi(r,\theta)\right)\\ & \left(\frac{\partial}{\partial\theta}\omega(r,\theta)\right)e^{2\,\nu(r,\theta)}+2\left(\frac{\partial}{\partial\theta}\psi(r,\theta)\right)\left(\frac{\partial}{\partial r}\omega(r,\theta)\right)e^{2\,\nu(r,\theta)}\end{aligned}$$

$$+ 2\Big(\frac{\partial^2}{\partial\theta\partial r}\omega(r,\theta)\Big)e^{2\,\nu(r,\theta)} - 2\Big(\frac{\partial}{\partial r}\mu(r,\theta)\Big)e^{2\,\nu(r,\theta)}\frac{\partial}{\partial\theta}\omega(r,\theta)\Big\}$$

$$e^{-2\,\nu(r,\theta)-2\,\lambda(r,\theta)+2\,\psi(r,\theta)}\,, \tag{G.19}$$

$$R^{r}{}_{\phi r\phi} = -\frac{1}{4}\Big\{4\Big(\frac{\partial^2}{\partial r^2}\psi(r,\theta)\Big)e^{2\,\mu(r,\theta)+2\,\nu(r,\theta)} + 4\Big(\frac{\partial}{\partial r}\psi(r,\theta)\Big)^2$$
$$e^{2\,\mu(r,\theta)+2\,\nu(r,\theta)} + \Big(\frac{\partial}{\partial r}\omega(r,\theta)\Big)^2 e^{2\,\mu(r,\theta)+2\,\psi(r,\theta)} - 4\Big(\frac{\partial}{\partial r}\lambda(r,\theta)\Big)$$
$$\Big(\frac{\partial}{\partial r}\psi(r,\theta)\Big)e^{2\,\mu(r,\theta)+2\,\nu(r,\theta)} + 4\Big(\frac{\partial}{\partial\theta}\lambda(r,\theta)\Big)\Big(\frac{\partial}{\partial\theta}\psi(r,\theta)\Big)$$
$$e^{2\,\lambda(r,\theta)+2\,\nu(r,\theta)}\Big\}\,e^{-2\,\nu(r,\theta)-2\,\lambda(r,\theta)+2\,\psi(r,\theta)-2\,\mu(r,\theta)}\,, \tag{G.20}$$

$$R^{r}{}_{\phi\theta\phi} = -\frac{1}{4}\Big\{4\Big(\frac{\partial^2}{\partial\theta\partial r}\psi(r,\theta)\Big)e^{2\,\nu(r,\theta)} + 4\Big(\frac{\partial}{\partial r}\psi(r,\theta)\Big)\Big(\frac{\partial}{\partial\theta}\psi(r,\theta)\Big)$$
$$e^{2\,\nu(r,\theta)} + e^{2\,\psi(r,\theta)}\Big(\frac{\partial}{\partial\theta}\omega(r,\theta)\Big)\frac{\partial}{\partial r}\omega(r,\theta) - 4\Big(\frac{\partial}{\partial r}\psi(r,\theta)\Big)$$
$$\Big(\frac{\partial}{\partial\theta}\lambda(r,\theta)\Big)e^{2\,\nu(r,\theta)} - 4\Big(\frac{\partial}{\partial r}\mu(r,\theta)\Big)\Big(\frac{\partial}{\partial\theta}\psi(r,\theta)\Big)e^{2\,\nu(r,\theta)}\Big\}$$
$$e^{-2\,\nu(r,\theta)-2\,\lambda(r,\theta)+2\,\psi(r,\theta)}\,, \tag{G.21}$$

$$R^{\theta}{}_{tt} = \frac{1}{4}\Big\{-4\Big(\frac{\partial^2}{\partial\theta\partial r}\nu(r,\theta)\Big)e^{4\,\nu(r,\theta)} + 4\Big(\frac{\partial}{\partial\theta}\lambda(r,\theta)\Big)e^{4\,\nu(r,\theta)}$$
$$\frac{\partial}{\partial r}\nu(r,\theta) + 4\Big(\frac{\partial}{\partial r}\mu(r,\theta)\Big)e^{4\,\nu(r,\theta)}\frac{\partial}{\partial\theta}\nu(r,\theta) - 4\Big(\frac{\partial}{\partial r}\nu(r,\theta)\Big)$$
$$\Big(\frac{\partial}{\partial\theta}\nu(r,\theta)\Big)e^{4\,\nu(r,\theta)} + 4\Big(\frac{\partial^2}{\partial\theta\partial r}\psi(r,\theta)\Big)\omega^2(r,\theta)e^{2\,\psi(r,\theta)+2\,\nu(r,\theta)}$$
$$+ 4\Big(\frac{\partial}{\partial r}\psi(r,\theta)\Big)\Big(\frac{\partial}{\partial\theta}\psi(r,\theta)\Big)\omega^2(r,\theta)e^{2\,\psi(r,\theta)+2\,\nu(r,\theta)}$$
$$+ 6\Big(\frac{\partial}{\partial r}\psi(r,\theta)\Big)\omega(r,\theta)\Big(\frac{\partial}{\partial\theta}\omega(r,\theta)\Big)e^{2\,\psi(r,\theta)+2\,\nu(r,\theta)}$$
$$+ 6\Big(\frac{\partial}{\partial\theta}\psi(r,\theta)\Big)\omega(r,\theta)\Big(\frac{\partial}{\partial r}\omega(r,\theta)\Big)e^{2\,\psi(r,\theta)+2\,\nu(r,\theta)}$$
$$+ 3\Big(\frac{\partial}{\partial\theta}\omega(r,\theta)\Big)\Big(\frac{\partial}{\partial r}\omega(r,\theta)\Big)e^{2\,\psi(r,\theta)+2\,\nu(r,\theta)} + 4\,\omega(r,\theta)$$
$$\Big(\frac{\partial^2}{\partial\theta\partial r}\omega(r,\theta)\Big)e^{2\,\psi(r,\theta)+2\,\nu(r,\theta)} - 2\Big(\frac{\partial}{\partial\theta}\nu(r,\theta)\Big)\omega(r,\theta)$$
$$\Big(\frac{\partial}{\partial r}\omega(r,\theta)\Big)e^{2\,\psi(r,\theta)+2\,\nu(r,\theta)} - 2\,\omega(r,\theta)\Big(\frac{\partial}{\partial\theta}\omega(r,\theta)\Big)\Big(\frac{\partial}{\partial r}\nu(r,\theta)\Big)$$

$$e^{2\psi(r,\theta)+2\nu(r,\theta)} + e^{4\psi(r,\theta)}\omega^2(r,\theta)\Big(\frac{\partial}{\partial\theta}\omega(r,\theta)\Big)\frac{\partial}{\partial r}\omega(r,\theta)$$
$$- 4\Big(\frac{\partial}{\partial\theta}\lambda(r,\theta)\Big)\Big(\frac{\partial}{\partial r}\psi(r,\theta)\Big)\omega^2(r,\theta)e^{2\psi(r,\theta)+2\nu(r,\theta)}$$
$$- 4\Big(\frac{\partial}{\partial\theta}\lambda(r,\theta)\Big)\omega(r,\theta)\Big(\frac{\partial}{\partial r}\omega(r,\theta)\Big)e^{2\psi(r,\theta)+2\nu(r,\theta)}$$
$$- 4\Big(\frac{\partial}{\partial r}\mu(r,\theta)\Big)\Big(\frac{\partial}{\partial\theta}\psi(r,\theta)\Big)\omega^2(r,\theta)e^{2\psi(r,\theta)+2\nu(r,\theta)}$$
$$- 4\Big(\frac{\partial}{\partial r}\mu(r,\theta)\Big)\omega(r,\theta)\Big(\frac{\partial}{\partial\theta}\omega(r,\theta)\Big)e^{2\psi(r,\theta)+2\nu(r,\theta)}\Big\}$$
$$e^{-2\mu(r,\theta)-2\nu(r,\theta)}, \tag{G.22}$$

$$R^\theta{}_{tt\theta} = \frac{1}{4}\Big\{-4\Big(\frac{\partial^2}{\partial\theta^2}\nu(r,\theta)\Big)e^{2\lambda(r,\theta)+4\nu(r,\theta)} - 4\Big(\frac{\partial}{\partial\theta}\nu(r,\theta)\Big)^2$$
$$e^{2\lambda(r,\theta)+4\nu(r,\theta)} + 4\Big(\frac{\partial^2}{\partial\theta^2}\psi(r,\theta)\Big)\omega^2(r,\theta)$$
$$e^{2\lambda(r,\theta)+2\nu(r,\theta)+2\psi(r,\theta)} + 4\Big(\frac{\partial}{\partial\theta}\psi(r,\theta)\Big)^2\omega^2(r,\theta)$$
$$e^{2\lambda(r,\theta)+2\nu(r,\theta)+2\psi(r,\theta)} + 12\Big(\frac{\partial}{\partial\theta}\psi(r,\theta)\Big)\omega(r,\theta)\Big(\frac{\partial}{\partial\theta}\omega(r,\theta)\Big)$$
$$e^{2\lambda(r,\theta)+2\nu(r,\theta)+2\psi(r,\theta)} + 3\Big(\frac{\partial}{\partial\theta}\omega(r,\theta)\Big)^2e^{2\lambda(r,\theta)+2\nu(r,\theta)+2\psi(r,\theta)}$$
$$+ 4\omega(r,\theta)\Big(\frac{\partial^2}{\partial\theta^2}\omega(r,\theta)\Big)e^{2\lambda(r,\theta)+2\nu(r,\theta)+2\psi(r,\theta)} - 4\Big(\frac{\partial}{\partial\theta}\nu(r,\theta)\Big)$$
$$\omega(r,\theta)\Big(\frac{\partial}{\partial\theta}\omega(r,\theta)\Big)e^{2\lambda(r,\theta)+2\nu(r,\theta)+2\psi(r,\theta)} + \omega^2(r,\theta)$$
$$\Big(\frac{\partial}{\partial\theta}\omega(r,\theta)\Big)^2e^{2\lambda(r,\theta)+4\psi(r,\theta)} - 4\Big(\frac{\partial}{\partial r}\mu(r,\theta)\Big)\Big(\frac{\partial}{\partial r}\nu(r,\theta)\Big)$$
$$e^{2\mu(r,\theta)+4\nu(r,\theta)} + 4\Big(\frac{\partial}{\partial r}\mu(r,\theta)\Big)\Big(\frac{\partial}{\partial r}\psi(r,\theta)\Big)\omega^2(r,\theta)$$
$$e^{2\mu(r,\theta)+2\psi(r,\theta)+2\nu(r,\theta)} + 4\Big(\frac{\partial}{\partial r}\mu(r,\theta)\Big)\omega(r,\theta)\Big(\frac{\partial}{\partial r}\omega(r,\theta)\Big)$$
$$e^{2\mu(r,\theta)+2\psi(r,\theta)+2\nu(r,\theta)} + 4\Big(\frac{\partial}{\partial\theta}\mu(r,\theta)\Big)\Big(\frac{\partial}{\partial\theta}\nu(r,\theta)\Big)e^{2\lambda(r,\theta)+4\nu(r,\theta)}$$
$$- 4\Big(\frac{\partial}{\partial\theta}\mu(r,\theta)\Big)\Big(\frac{\partial}{\partial\theta}\psi(r,\theta)\Big)\Big(\omega(r,\theta)\Big)^2e^{2\lambda(r,\theta)+2\nu(r,\theta)+2\psi(r,\theta)}$$
$$- 4\Big(\frac{\partial}{\partial\theta}\mu(r,\theta)\Big)\omega(r,\theta)\Big(\frac{\partial}{\partial\theta}\omega(r,\theta)\Big)e^{2\lambda(r,\theta)+2\nu(r,\theta)+2\psi(r,\theta)}\Big\}$$
$$e^{-2\nu(r,\theta)-2\lambda(r,\theta)-2\mu(r,\theta)}, \tag{G.23}$$

$$R^\theta{}_{tr\phi} = \frac{1}{4}\Big\{-2\Big(\frac{\partial}{\partial\theta}\lambda(r,\theta)\Big)e^{2\nu(r,\theta)}\frac{\partial}{\partial r}\omega(r,\theta) - 2\Big(\frac{\partial}{\partial\theta}\omega(r,\theta)\Big)$$

$$\left(\frac{\partial}{\partial r}\nu(r,\theta)\right)e^{2\,\nu(r,\theta)} + e^{2\,\psi(r,\theta)}\omega(r,\theta)\left(\frac{\partial}{\partial\theta}\omega(r,\theta)\right)\frac{\partial}{\partial r}\omega(r,\theta)$$
$$- 4\left(\frac{\partial}{\partial\theta}\lambda(r,\theta)\right)e^{2\,\nu(r,\theta)}\left(\frac{\partial}{\partial r}\psi(r,\theta)\right)\omega(r,\theta) - 4\left(\frac{\partial}{\partial r}\mu(r,\theta)\right)$$
$$e^{2\,\nu(r,\theta)}\left(\frac{\partial}{\partial\theta}\psi(r,\theta)\right)\omega(r,\theta) + 4\left(\frac{\partial^2}{\partial\theta\partial r}\psi(r,\theta)\right)\omega(r,\theta)e^{2\,\nu(r,\theta)}$$
$$+ 4\left(\frac{\partial}{\partial r}\psi(r,\theta)\right)\left(\frac{\partial}{\partial\theta}\psi(r,\theta)\right)\omega(r,\theta)e^{2\,\nu(r,\theta)} + 4\left(\frac{\partial}{\partial r}\psi(r,\theta)\right)$$
$$\left(\frac{\partial}{\partial\theta}\omega(r,\theta)\right)e^{2\,\nu(r,\theta)} + 2\left(\frac{\partial}{\partial\theta}\psi(r,\theta)\right)\left(\frac{\partial}{\partial r}\omega(r,\theta)\right)e^{2\,\nu(r,\theta)}$$
$$+ 2\left(\frac{\partial^2}{\partial\theta\partial r}\omega(r,\theta)\right)e^{2\,\nu(r,\theta)} - 2\left(\frac{\partial}{\partial r}\mu(r,\theta)\right)e^{2\,\nu(r,\theta)}\frac{\partial}{\partial\theta}\omega(r,\theta)\Big\}$$
$$e^{-2\,\nu(r,\theta)-2\,\mu(r,\theta)+2\,\psi(r,\theta)}\,, \tag{G.24}$$

$$R^{\theta}{}_{t\theta\phi} = \frac{1}{4}\Big\{4\left(\frac{\partial^2}{\partial\theta^2}\psi(r,\theta)\right)\omega(r,\theta)e^{2\,\lambda(r,\theta)+2\,\nu(r,\theta)} + 4\left(\frac{\partial}{\partial\theta}\psi(r,\theta)\right)^2\omega(r,\theta)$$
$$e^{2\,\lambda(r,\theta)+2\,\nu(r,\theta)} + 6\left(\frac{\partial}{\partial\theta}\psi(r,\theta)\right)\left(\frac{\partial}{\partial\theta}\omega(r,\theta)\right)e^{2\,\lambda(r,\theta)+2\,\nu(r,\theta)}$$
$$+ 2\left(\frac{\partial^2}{\partial\theta^2}\omega(r,\theta)\right)e^{2\,\lambda(r,\theta)+2\,\nu(r,\theta)} - 2\left(\frac{\partial}{\partial\theta}\nu(r,\theta)\right)\left(\frac{\partial}{\partial\theta}\omega(r,\theta)\right)$$
$$e^{2\,\lambda(r,\theta)+2\,\nu(r,\theta)} + \omega(r,\theta)\left(\frac{\partial}{\partial\theta}\omega(r,\theta)\right)^2 e^{2\,\lambda(r,\theta)+2\,\psi(r,\theta)}$$
$$+ 4\left(\frac{\partial}{\partial r}\mu(r,\theta)\right)\left(\frac{\partial}{\partial r}\psi(r,\theta)\right)\omega(r,\theta)e^{2\,\mu(r,\theta)+2\,\nu(r,\theta)} + 2\left(\frac{\partial}{\partial r}\mu(r,\theta)\right)$$
$$\left(\frac{\partial}{\partial r}\omega(r,\theta)\right)e^{2\,\mu(r,\theta)+2\,\nu(r,\theta)} - 4\left(\frac{\partial}{\partial\theta}\mu(r,\theta)\right)\left(\frac{\partial}{\partial\theta}\psi(r,\theta)\right)\omega(r,\theta)$$
$$e^{2\,\lambda(r,\theta)+2\,\nu(r,\theta)} - 2\left(\frac{\partial}{\partial\theta}\mu(r,\theta)\right)\left(\frac{\partial}{\partial\theta}\omega(r,\theta)\right)e^{2\,\lambda(r,\theta)+2\,\nu(r,\theta)}\Big\}$$
$$e^{-2\,\nu(r,\theta)-2\,\lambda(r,\theta)+2\,\psi(r,\theta)-2\,\mu(r,\theta)}\,, \tag{G.25}$$

$$R^{\theta}{}_{rt\phi} = \frac{1}{2}\Big\{\left(\frac{\partial}{\partial\theta}\nu(r,\theta)\right)\frac{\partial}{\partial r}\omega(r,\theta) - \left(\frac{\partial}{\partial\theta}\omega(r,\theta)\right)\frac{\partial}{\partial r}\nu(r,\theta)$$
$$+ \left(\frac{\partial}{\partial r}\psi(r,\theta)\right)\frac{\partial}{\partial\theta}\omega(r,\theta) - \left(\frac{\partial}{\partial\theta}\psi(r,\theta)\right)\frac{\partial}{\partial r}\omega(r,\theta)\Big\}$$
$$e^{2\,\psi(r,\theta)-2\,\mu(r,\theta)}\,, \tag{G.26}$$

$$R^{\theta}{}_{rr\theta} = e^{-2\,\mu(r,\theta)}\Big\{\left(\frac{\partial^2}{\partial\theta^2}\lambda(r,\theta)\right)e^{2\,\lambda(r,\theta)} + \left(\frac{\partial}{\partial\theta}\lambda(r,\theta)\right)^2 e^{2\,\lambda(r,\theta)}$$
$$+ \left(\frac{\partial^2}{\partial r^2}\mu(r,\theta)\right)e^{2\,\mu(r,\theta)} + \left(\frac{\partial}{\partial r}\mu(r,\theta)\right)^2 e^{2\,\mu(r,\theta)} - \left(\frac{\partial}{\partial r}\lambda(r,\theta)\right)$$

$$\left(\frac{\partial}{\partial r}\mu(r,\theta)\right)\mathrm{e}^{2\,\mu(r,\theta)} - \left(\frac{\partial}{\partial\theta}\lambda(r,\theta)\right)\mathrm{e}^{2\,\lambda(r,\theta)}\frac{\partial}{\partial\theta}\mu(r,\theta)\Big\}, \qquad \text{(G.27)}$$

$$R^{\theta}{}_{\phi tr} = -\frac{1}{4}\Big\{-2\left(\frac{\partial}{\partial\theta}\lambda(r,\theta)\right)\mathrm{e}^{2\,\nu(r,\theta)}\frac{\partial}{\partial r}\omega(r,\theta) - 2\left(\frac{\partial}{\partial\theta}\nu(r,\theta)\right)\mathrm{e}^{2\,\nu(r,\theta)}$$
$$\frac{\partial}{\partial r}\omega(r,\theta) + \mathrm{e}^{2\,\psi(r,\theta)}\omega(r,\theta)\left(\frac{\partial}{\partial\theta}\omega(r,\theta)\right)\frac{\partial}{\partial r}\omega(r,\theta)$$
$$- 4\left(\frac{\partial}{\partial\theta}\lambda(r,\theta)\right)\mathrm{e}^{2\,\nu(r,\theta)}\left(\frac{\partial}{\partial r}\psi(r,\theta)\right)\omega(r,\theta) - 4\left(\frac{\partial}{\partial r}\mu(r,\theta)\right)$$
$$\mathrm{e}^{2\,\nu(r,\theta)}\left(\frac{\partial}{\partial\theta}\psi(r,\theta)\right)\omega(r,\theta) + 4\left(\frac{\partial^2}{\partial\theta\partial r}\psi(r,\theta)\right)\omega(r,\theta)\mathrm{e}^{2\,\nu(r,\theta)}$$
$$+ 4\left(\frac{\partial}{\partial r}\psi(r,\theta)\right)\left(\frac{\partial}{\partial\theta}\psi(r,\theta)\right)\omega(r,\theta)\mathrm{e}^{2\,\nu(r,\theta)} + 2\left(\frac{\partial}{\partial r}\psi(r,\theta)\right)$$
$$\left(\frac{\partial}{\partial\theta}\omega(r,\theta)\right)\mathrm{e}^{2\,\nu(r,\theta)} + 4\left(\frac{\partial}{\partial\theta}\psi(r,\theta)\right)\left(\frac{\partial}{\partial r}\omega(r,\theta)\right)\mathrm{e}^{2\,\nu(r,\theta)}$$
$$+ 2\left(\frac{\partial^2}{\partial\theta\partial r}\omega(r,\theta)\right)\mathrm{e}^{2\,\nu(r,\theta)} - 2\left(\frac{\partial}{\partial r}\mu(r,\theta)\right)\mathrm{e}^{2\,\nu(r,\theta)}\frac{\partial}{\partial\theta}\omega(r,\theta)\Big\}$$
$$\mathrm{e}^{-2\,\nu(r,\theta)-2\,\mu(r,\theta)+2\,\psi(r,\theta)}, \qquad \text{(G.28)}$$

$$R^{\theta}{}_{\phi t\theta} = -\frac{1}{4}\Big\{4\left(\frac{\partial^2}{\partial\theta^2}\psi(r,\theta)\right)\omega(r,\theta)\mathrm{e}^{2\,\lambda(r,\theta)+2\,\nu(r,\theta)} + 4\left(\frac{\partial}{\partial\theta}\psi(r,\theta)\right)^2$$
$$\omega(r,\theta)\mathrm{e}^{2\,\lambda(r,\theta)+2\,\nu(r,\theta)} + 6\left(\frac{\partial}{\partial\theta}\psi(r,\theta)\right)\left(\frac{\partial}{\partial\theta}\omega(r,\theta)\right)$$
$$\mathrm{e}^{2\,\lambda(r,\theta)+2\,\nu(r,\theta)} + 2\left(\frac{\partial^2}{\partial\theta^2}\omega(r,\theta)\right)\mathrm{e}^{2\,\lambda(r,\theta)+2\,\nu(r,\theta)} - 2\left(\frac{\partial}{\partial\theta}\nu(r,\theta)\right)$$
$$\left(\frac{\partial}{\partial\theta}\omega(r,\theta)\right)\mathrm{e}^{2\,\lambda(r,\theta)+2\,\nu(r,\theta)} + \omega(r,\theta)\left(\frac{\partial}{\partial\theta}\omega(r,\theta)\right)^2$$
$$\mathrm{e}^{2\,\lambda(r,\theta)+2\,\psi(r,\theta)} + 4\left(\frac{\partial}{\partial r}\mu(r,\theta)\right)\left(\frac{\partial}{\partial r}\psi(r,\theta)\right)\omega(r,\theta)\mathrm{e}^{2\,\mu(r,\theta)+2\,\nu(r,\theta)}$$
$$+ 2\left(\frac{\partial}{\partial r}\mu(r,\theta)\right)\left(\frac{\partial}{\partial r}\omega(r,\theta)\right)\mathrm{e}^{2\,\mu(r,\theta)+2\,\nu(r,\theta)} - 4\left(\frac{\partial}{\partial\theta}\mu(r,\theta)\right)$$
$$\left(\frac{\partial}{\partial\theta}\psi(r,\theta)\right)\omega(r,\theta)\mathrm{e}^{2\,\lambda(r,\theta)+2\,\nu(r,\theta)} - 2\left(\frac{\partial}{\partial\theta}\mu(r,\theta)\right)\left(\frac{\partial}{\partial\theta}\omega(r,\theta)\right)$$
$$\mathrm{e}^{2\,\lambda(r,\theta)+2\,\nu(r,\theta)}\Big\}\,\mathrm{e}^{-2\,\nu(r,\theta)-2\,\lambda(r,\theta)+2\,\psi(r,\theta)-2\,\mu(r,\theta)}, \qquad \text{(G.29)}$$

$$R^{\theta}{}_{\phi r\phi} = -\frac{1}{4}\Big\{4\left(\frac{\partial^2}{\partial\theta\partial r}\psi(r,\theta)\right)\mathrm{e}^{2\,\nu(r,\theta)} + 4\left(\frac{\partial}{\partial r}\psi(r,\theta)\right)\left(\frac{\partial}{\partial\theta}\psi(r,\theta)\right)$$
$$\mathrm{e}^{2\,\nu(r,\theta)} + \mathrm{e}^{2\,\psi(r,\theta)}\left(\frac{\partial}{\partial\theta}\omega(r,\theta)\right)\frac{\partial}{\partial r}\omega(r,\theta) - 4\left(\frac{\partial}{\partial r}\psi(r,\theta)\right)$$

$$\left(\frac{\partial}{\partial\theta}\lambda(r,\theta)\right)\mathrm{e}^{2\,\nu(r,\theta)} - 4\left(\frac{\partial}{\partial r}\mu(r,\theta)\right)\left(\frac{\partial}{\partial\theta}\psi(r,\theta)\right)\mathrm{e}^{2\,\nu(r,\theta)}\Big\}$$
$$\mathrm{e}^{-2\,\nu(r,\theta)-2\,\mu(r,\theta)+2\,\psi(r,\theta)}\,, \tag{G.30}$$

$$R^{\theta}{}_{\phi\theta\phi} = -\frac{1}{4}\Big\{4\left(\frac{\partial^2}{\partial\theta^2}\psi(r,\theta)\right)\mathrm{e}^{2\,\lambda(r,\theta)+2\,\nu(r,\theta)} + 4\left(\frac{\partial}{\partial\theta}\psi(r,\theta)\right)^2$$
$$\mathrm{e}^{2\,\lambda(r,\theta)+2\,\nu(r,\theta)} + \left(\frac{\partial}{\partial\theta}\omega(r,\theta)\right)^2\mathrm{e}^{2\,\lambda(r,\theta)+2\,\psi(r,\theta)} + 4\left(\frac{\partial}{\partial r}\psi(r,\theta)\right)$$
$$\left(\frac{\partial}{\partial r}\mu(r,\theta)\right)\mathrm{e}^{2\,\mu(r,\theta)+2\,\nu(r,\theta)} - 4\left(\frac{\partial}{\partial\theta}\mu(r,\theta)\right)\left(\frac{\partial}{\partial\theta}\psi(r,\theta)\right)$$
$$\mathrm{e}^{2\,\lambda(r,\theta)+2\,\nu(r,\theta)}\Big\}\;\mathrm{e}^{-2\,\nu(r,\theta)-2\,\lambda(r,\theta)+2\,\psi(r,\theta)-2\,\mu(r,\theta)}\,, \tag{G.31}$$

$$R^{\phi}{}_{tt\phi} = \frac{1}{4}\Big\{(-\mathrm{e}^{2\,\nu(r,\theta)} + \mathrm{e}^{2\,\psi(r,\theta)}\omega^2(r,\theta))\left(\left(\frac{\partial}{\partial r}\omega(r,\theta)\right)^2\mathrm{e}^{2\,\mu(r,\theta)+2\,\psi(r,\theta)}\right.$$
$$+\,4\left(\frac{\partial}{\partial r}\psi(r,\theta)\right)\left(\frac{\partial}{\partial r}\nu(r,\theta)\right)\mathrm{e}^{2\,\mu(r,\theta)+2\,\nu(r,\theta)} + \left(\frac{\partial}{\partial\theta}\omega(r,\theta)\right)^2$$
$$\left.\mathrm{e}^{2\,\lambda(r,\theta)+2\,\psi(r,\theta)} + 4\left(\frac{\partial}{\partial\theta}\psi(r,\theta)\right)\left(\frac{\partial}{\partial\theta}\nu(r,\theta)\right)\mathrm{e}^{2\,\lambda(r,\theta)+2\,\nu(r,\theta)}\right)$$
$$\mathrm{e}^{-2\,\nu(r,\theta)-2\,\lambda(r,\theta)-2\,\mu(r,\theta)}\Big\}, \tag{G.32}$$

$$R^{\phi}{}_{tr\theta} = -\frac{1}{2}\,(-\mathrm{e}^{2\,\nu(r,\theta)} + \mathrm{e}^{2\,\psi(r,\theta)}\omega^2(r,\theta))\Big\{\left(\frac{\partial}{\partial\theta}\nu(r,\theta)\right)\frac{\partial}{\partial r}\omega(r,\theta)$$
$$-\left(\frac{\partial}{\partial\theta}\omega(r,\theta)\right)\frac{\partial}{\partial r}\nu(r,\theta) + \left(\frac{\partial}{\partial r}\psi(r,\theta)\right)\frac{\partial}{\partial\theta}\omega(r,\theta)$$
$$-\left(\frac{\partial}{\partial\theta}\psi(r,\theta)\right)\frac{\partial}{\partial r}\omega(r,\theta)\Big\}\;\mathrm{e}^{-2\,\nu(r,\theta)}\,, \tag{G.33}$$

$$R^{\phi}{}_{rtr} = \frac{1}{2}\Big\{2\,\omega(r,\theta)\left(\frac{\partial}{\partial r}\lambda(r,\theta)\right)\left(\frac{\partial}{\partial r}\nu(r,\theta)\right)\mathrm{e}^{2\,\mu(r,\theta)+2\,\nu(r,\theta)}$$
$$+\left(\frac{\partial^2}{\partial r^2}\omega(r,\theta)\right)\mathrm{e}^{2\,\mu(r,\theta)+2\,\nu(r,\theta)} - 2\,\omega(r,\theta)\left(\frac{\partial}{\partial\theta}\lambda(r,\theta)\right)$$
$$\left(\frac{\partial}{\partial\theta}\nu(r,\theta)\right)\mathrm{e}^{2\,\lambda(r,\theta)+2\,\nu(r,\theta)} + 2\,\omega(r,\theta)\left(\frac{\partial}{\partial r}\omega(r,\theta)\right)^2$$
$$\mathrm{e}^{2\,\mu(r,\theta)+2\,\psi(r,\theta)} - 2\,\omega(r,\theta)\left(\frac{\partial}{\partial r}\nu(r,\theta)\right)^2\mathrm{e}^{2\,\mu(r,\theta)+2\,\nu(r,\theta)}$$
$$+\,2\left(\frac{\partial^2}{\partial r^2}\psi(r,\theta)\right)\omega(r,\theta)\mathrm{e}^{2\,\mu(r,\theta)+2\,\nu(r,\theta)} + 2\left(\frac{\partial}{\partial r}\psi(r,\theta)\right)^2\omega(r,\theta)$$
$$\mathrm{e}^{2\,\mu(r,\theta)+2\,\nu(r,\theta)} + 3\left(\frac{\partial}{\partial r}\psi(r,\theta)\right)\left(\frac{\partial}{\partial r}\omega(r,\theta)\right)\mathrm{e}^{2\,\mu(r,\theta)+2\,\nu(r,\theta)}$$

$$
\begin{aligned}
&- \Big(\frac{\partial}{\partial r}\nu(r,\theta)\Big)\Big(\frac{\partial}{\partial r}\omega(r,\theta)\Big)e^{2\,\mu(r,\theta)+2\,\nu(r,\theta)} - 2\Big(\frac{\partial}{\partial r}\lambda(r,\theta)\Big)\\
&\Big(\frac{\partial}{\partial r}\psi(r,\theta)\Big)\omega(r,\theta)e^{2\,\mu(r,\theta)+2\,\nu(r,\theta)} - \Big(\frac{\partial}{\partial r}\lambda(r,\theta)\Big)\Big(\frac{\partial}{\partial r}\omega(r,\theta)\Big)\\
&e^{2\,\mu(r,\theta)+2\,\nu(r,\theta)} + 2\Big(\frac{\partial}{\partial\theta}\lambda(r,\theta)\Big)\Big(\frac{\partial}{\partial\theta}\psi(r,\theta)\Big)\omega(r,\theta)e^{2\,\lambda(r,\theta)+2\,\nu(r,\theta)}\\
&+ \Big(\frac{\partial}{\partial\theta}\lambda(r,\theta)\Big)\Big(\frac{\partial}{\partial\theta}\omega(r,\theta)\Big)e^{2\,\lambda(r,\theta)+2\,\nu(r,\theta)} - \Big(\frac{\partial}{\partial r}\nu(r,\theta)\Big)\\
&\omega^2(r,\theta)\Big(\frac{\partial}{\partial r}\omega(r,\theta)\Big)e^{2\,\mu(r,\theta)+2\,\psi(r,\theta)} - \Big(\frac{\partial}{\partial r}\lambda(r,\theta)\Big)\omega^2(r,\theta)\\
&\Big(\frac{\partial}{\partial r}\omega(r,\theta)\Big)e^{2\,\mu(r,\theta)+2\,\psi(r,\theta)} + \Big(\frac{\partial}{\partial\theta}\lambda(r,\theta)\Big)\omega^2(r,\theta)\\
&\Big(\frac{\partial}{\partial\theta}\omega(r,\theta)\Big)e^{2\,\lambda(r,\theta)+2\,\psi(r,\theta)} + 3\Big(\frac{\partial}{\partial r}\psi(r,\theta)\Big)\omega^2(r,\theta)\\
&\Big(\frac{\partial}{\partial r}\omega(r,\theta)\Big)e^{2\,\mu(r,\theta)+2\,\psi(r,\theta)} + \omega^2(r,\theta)\Big(\frac{\partial^2}{\partial r^2}\omega(r,\theta)\Big)\\
&e^{2\,\mu(r,\theta)+2\,\psi(r,\theta)} - 2\,\omega(r,\theta)\Big(\frac{\partial^2}{\partial r^2}\nu(r,\theta)\Big)e^{2\,\mu(r,\theta)+2\,\nu(r,\theta)}\Big\}\\
&e^{-2\,\mu(r,\theta)-2\,\nu(r,\theta)}, \qquad \text{(G.34)}
\end{aligned}
$$

$$
\begin{aligned}
R^{\phi}{}_{rt\theta} = &-\frac{1}{4}\Big\{-\Big(\frac{\partial}{\partial r}\psi(r,\theta)\Big)e^{2\,\psi(r,\theta)}\omega^2(r,\theta)\frac{\partial}{\partial\theta}\omega(r,\theta)\\
&- 2\Big(\frac{\partial}{\partial\theta}\psi(r,\theta)\Big)e^{2\,\psi(r,\theta)}\omega^2(r,\theta)\frac{\partial}{\partial r}\omega(r,\theta) - e^{2\,\psi(r,\theta)}\omega^2(r,\theta)\\
&\frac{\partial^2}{\partial\theta\partial r}\omega(r,\theta) + \Big(\frac{\partial}{\partial\theta}\nu(r,\theta)\Big)e^{2\,\psi(r,\theta)}\omega^2(r,\theta)\frac{\partial}{\partial r}\omega(r,\theta)\\
&+ \Big(\frac{\partial}{\partial\theta}\lambda(r,\theta)\Big)e^{2\,\psi(r,\theta)}\omega^2(r,\theta)\frac{\partial}{\partial r}\omega(r,\theta) + \Big(\frac{\partial}{\partial r}\mu(r,\theta)\Big)\\
&e^{2\,\psi(r,\theta)}\omega^2(r,\theta)\frac{\partial}{\partial\theta}\omega(r,\theta) - 2\,\omega(r,\theta)\Big(\frac{\partial}{\partial\theta}\lambda(r,\theta)\Big)e^{2\,\nu(r,\theta)}\frac{\partial}{\partial r}\nu(r,\theta)\\
&- 2\,\omega(r,\theta)\Big(\frac{\partial}{\partial r}\mu(r,\theta)\Big)e^{2\,\nu(r,\theta)}\frac{\partial}{\partial\theta}\nu(r,\theta) + 2\,\omega(r,\theta)\Big(\frac{\partial}{\partial r}\nu(r,\theta)\Big)\\
&\Big(\frac{\partial}{\partial\theta}\nu(r,\theta)\Big)e^{2\,\nu(r,\theta)} - 2\,e^{2\,\psi(r,\theta)}\omega(r,\theta)\Big(\frac{\partial}{\partial\theta}\omega(r,\theta)\Big)\frac{\partial}{\partial r}\omega(r,\theta)\\
&+ 2\Big(\frac{\partial}{\partial\theta}\lambda(r,\theta)\Big)e^{2\,\nu(r,\theta)}\Big(\frac{\partial}{\partial r}\psi(r,\theta)\Big)\omega(r,\theta) + 2\Big(\frac{\partial}{\partial r}\mu(r,\theta)\Big)\\
&e^{2\,\nu(r,\theta)}\Big(\frac{\partial}{\partial\theta}\psi(r,\theta)\Big)\omega(r,\theta) - 2\Big(\frac{\partial}{\partial r}\psi(r,\theta)\Big)\Big(\frac{\partial}{\partial\theta}\psi(r,\theta)\Big)\omega(r,\theta)\\
&e^{2\,\nu(r,\theta)} + 2\,\omega(r,\theta)\Big(\frac{\partial^2}{\partial\theta\partial r}\nu(r,\theta)\Big)e^{2\,\nu(r,\theta)} + \Big(\frac{\partial}{\partial\theta}\lambda(r,\theta)\Big)e^{2\,\nu(r,\theta)}\\
&\frac{\partial}{\partial r}\omega(r,\theta) + \Big(\frac{\partial}{\partial\theta}\omega(r,\theta)\Big)\Big(\frac{\partial}{\partial r}\nu(r,\theta)\Big)e^{2\,\nu(r,\theta)} - 2\Big(\frac{\partial^2}{\partial\theta\partial r}\psi(r,\theta)\Big)
\end{aligned}
$$

$$\omega(r,\theta)\mathrm{e}^{2\,\nu(r,\theta)} - 2\Big(\frac{\partial}{\partial r}\psi(r,\theta)\Big)\Big(\frac{\partial}{\partial \theta}\omega(r,\theta)\Big)\mathrm{e}^{2\,\nu(r,\theta)}$$
$$- \Big(\frac{\partial}{\partial \theta}\psi(r,\theta)\Big)\Big(\frac{\partial}{\partial r}\omega(r,\theta)\Big)\mathrm{e}^{2\,\nu(r,\theta)} + \Big(\frac{\partial}{\partial r}\mu(r,\theta)\Big)\mathrm{e}^{2\,\nu(r,\theta)}$$
$$\frac{\partial}{\partial \theta}\omega(r,\theta) - \Big(\frac{\partial^2}{\partial \theta \partial r}\omega(r,\theta)\Big)\mathrm{e}^{2\,\nu(r,\theta)}\Big\}\ \mathrm{e}^{-2\,\nu(r,\theta)}\,, \tag{G.35}$$

$$R^{\phi}{}_{rr\phi} = \frac{1}{4}\Big\{6\Big(\frac{\partial}{\partial r}\psi(r,\theta)\Big)\omega(r,\theta)\Big(\frac{\partial}{\partial r}\omega(r,\theta)\Big)\mathrm{e}^{2\,\mu(r,\theta)+2\,\psi(r,\theta)}$$
$$+ 2\,\omega(r,\theta)\Big(\frac{\partial^2}{\partial r^2}\omega(r,\theta)\Big)\mathrm{e}^{2\,\mu(r,\theta)+2\,\psi(r,\theta)} - 2\,\Big(\frac{\partial}{\partial r}\nu(r,\theta)\Big)\omega(r,\theta)$$
$$\Big(\frac{\partial}{\partial r}\omega(r,\theta)\Big)\mathrm{e}^{2\,\mu(r,\theta)+2\,\psi(r,\theta)} - 2\Big(\frac{\partial}{\partial r}\lambda(r,\theta)\Big)\omega(r,\theta)\Big(\frac{\partial}{\partial r}\omega(r,\theta)\Big)$$
$$\mathrm{e}^{2\,\mu(r,\theta)+2\,\psi(r,\theta)} + 2\Big(\frac{\partial}{\partial \theta}\lambda(r,\theta)\Big)\omega(r,\theta)\Big(\frac{\partial}{\partial \theta}\omega(r,\theta)\Big)\mathrm{e}^{2\,\lambda(r,\theta)+2\,\psi(r,\theta)}$$
$$+ 4\Big(\frac{\partial^2}{\partial r^2}\psi(r,\theta)\Big)\mathrm{e}^{2\,\mu(r,\theta)+2\,\nu(r,\theta)} + 4\Big(\frac{\partial}{\partial r}\psi(r,\theta)\Big)^2\mathrm{e}^{2\,\mu(r,\theta)+2\,\nu(r,\theta)}$$
$$+ \Big(\frac{\partial}{\partial r}\omega(r,\theta)\Big)^2\mathrm{e}^{2\,\mu(r,\theta)+2\,\psi(r,\theta)} - 4\Big(\frac{\partial}{\partial r}\lambda(r,\theta)\Big)\Big(\frac{\partial}{\partial r}\psi(r,\theta)\Big)$$
$$\mathrm{e}^{2\,\mu(r,\theta)+2\,\nu(r,\theta)} + 4\Big(\frac{\partial}{\partial \theta}\lambda(r,\theta)\Big)\Big(\frac{\partial}{\partial \theta}\psi(r,\theta)\Big)\mathrm{e}^{2\,\lambda(r,\theta)+2\,\nu(r,\theta)}\Big\}$$
$$\mathrm{e}^{-2\,\mu(r,\theta)-2\,\nu(r,\theta)}\,, \tag{G.36}$$

$$R^{\phi}{}_{r\theta\phi} = -\frac{1}{4}\Big\{2\Big(\frac{\partial}{\partial \theta}\lambda(r,\theta)\Big)\mathrm{e}^{2\,\psi(r,\theta)}\omega(r,\theta)\frac{\partial}{\partial r}\omega(r,\theta) + 2\Big(\frac{\partial}{\partial \theta}\nu(r,\theta)\Big)$$
$$\mathrm{e}^{2\,\psi(r,\theta)}\omega(r,\theta)\frac{\partial}{\partial r}\omega(r,\theta) - 2\,\Big(\frac{\partial}{\partial r}\psi(r,\theta)\Big)\mathrm{e}^{2\,\psi(r,\theta)}\omega(r,\theta)\frac{\partial}{\partial \theta}\omega(r,\theta)$$
$$- 4\Big(\frac{\partial}{\partial \theta}\psi(r,\theta)\Big)\mathrm{e}^{2\,\psi(r,\theta)}\omega(r,\theta)\frac{\partial}{\partial r}\omega(r,\theta) - 2\,\mathrm{e}^{2\,\psi(r,\theta)}\omega(r,\theta)$$
$$\frac{\partial^2}{\partial \theta \partial r}\omega(r,\theta) + 2\Big(\frac{\partial}{\partial r}\mu(r,\theta)\Big)\mathrm{e}^{2\,\psi(r,\theta)}\omega(r,\theta)\frac{\partial}{\partial \theta}\omega(r,\theta)$$
$$- 4\Big(\frac{\partial^2}{\partial \theta \partial r}\psi(r,\theta)\Big)\mathrm{e}^{2\,\nu(r,\theta)} - 4\Big(\frac{\partial}{\partial r}\psi(r,\theta)\Big)\Big(\frac{\partial}{\partial \theta}\psi(r,\theta)\Big)\mathrm{e}^{2\,\nu(r,\theta)}$$
$$- \mathrm{e}^{2\,\psi(r,\theta)}\Big(\frac{\partial}{\partial \theta}\omega(r,\theta)\Big)\frac{\partial}{\partial r}\omega(r,\theta) + 4\Big(\frac{\partial}{\partial r}\psi(r,\theta)\Big)\Big(\frac{\partial}{\partial \theta}\lambda(r,\theta)\Big)$$
$$\mathrm{e}^{2\,\nu(r,\theta)} + 4\Big(\frac{\partial}{\partial r}\mu(r,\theta)\Big)\Big(\frac{\partial}{\partial \theta}\psi(r,\theta)\Big)\mathrm{e}^{2\,\nu(r,\theta)}\Big\}\ \mathrm{e}^{-2\,\nu(r,\theta)}\,, \tag{G.37}$$

$$R^{\phi}{}_{\theta tr} = -\frac{1}{2}\Big\{-2\Big(\frac{\partial}{\partial r}\psi(r,\theta)\Big)\mathrm{e}^{2\,\psi(r,\theta)}\omega^2(r,\theta)\frac{\partial}{\partial \theta}\omega(r,\theta) - \Big(\frac{\partial}{\partial \theta}\psi(r,\theta)\Big)$$
$$\mathrm{e}^{2\,\psi(r,\theta)}\omega^2(r,\theta)\frac{\partial}{\partial r}\omega(r,\theta) - \mathrm{e}^{2\,\psi(r,\theta)}\Big(\omega(r,\theta)\Big)^2\frac{\partial^2}{\partial \theta \partial r}\omega(r,\theta)$$

$$
\begin{aligned}
&+ \mathrm{e}^{2\,\psi(r,\theta)} \Big(\omega(r,\theta)\Big)^2 \Big(\frac{\partial}{\partial\theta}\omega(r,\theta)\Big) \frac{\partial}{\partial r}\nu(r,\theta) + \Big(\frac{\partial}{\partial\theta}\lambda(r,\theta)\Big) \\
&\mathrm{e}^{2\,\psi(r,\theta)} \omega^2(r,\theta) \frac{\partial}{\partial r}\omega(r,\theta) + \Big(\frac{\partial}{\partial r}\mu(r,\theta)\Big) \mathrm{e}^{2\,\psi(r,\theta)} \omega^2(r,\theta) \\
&\frac{\partial}{\partial\theta}\omega(r,\theta) - 2\,\omega(r,\theta)\Big(\frac{\partial}{\partial\theta}\lambda(r,\theta)\Big)\mathrm{e}^{2\,\nu(r,\theta)}\frac{\partial}{\partial r}\nu(r,\theta) - 2\,\omega(r,\theta) \\
&\Big(\frac{\partial}{\partial r}\mu(r,\theta)\Big)\mathrm{e}^{2\,\nu(r,\theta)}\frac{\partial}{\partial\theta}\nu(r,\theta) + 2\,\omega(r,\theta)\Big(\frac{\partial}{\partial r}\nu(r,\theta)\Big)\Big(\frac{\partial}{\partial\theta}\nu(r,\theta)\Big) \\
&\mathrm{e}^{2\,\nu(r,\theta)} - 2\,\mathrm{e}^{2\,\psi(r,\theta)}\omega(r,\theta)\Big(\frac{\partial}{\partial\theta}\omega(r,\theta)\Big)\frac{\partial}{\partial r}\omega(r,\theta) + 2\Big(\frac{\partial}{\partial\theta}\lambda(r,\theta)\Big) \\
&\mathrm{e}^{2\,\nu(r,\theta)}\Big(\frac{\partial}{\partial r}\psi(r,\theta)\Big)\omega(r,\theta) + 2\Big(\frac{\partial}{\partial r}\mu(r,\theta)\Big)\mathrm{e}^{2\,\nu(r,\theta)}\Big(\frac{\partial}{\partial\theta}\psi(r,\theta)\Big) \\
&\omega(r,\theta) - 2\Big(\frac{\partial}{\partial r}\psi(r,\theta)\Big)\Big(\frac{\partial}{\partial\theta}\psi(r,\theta)\Big)\omega(r,\theta)\mathrm{e}^{2\,\nu(r,\theta)} + 2\,\omega(r,\theta) \\
&\Big(\frac{\partial^2}{\partial\theta\partial r}\nu(r,\theta)\Big)\mathrm{e}^{2\,\nu(r,\theta)} + \Big(\frac{\partial}{\partial\theta}\lambda(r,\theta)\Big)\mathrm{e}^{2\,\nu(r,\theta)}\frac{\partial}{\partial r}\omega(r,\theta) \\
&- 2\Big(\frac{\partial^2}{\partial\theta\partial r}\psi(r,\theta)\Big)\omega(r,\theta)\mathrm{e}^{2\,\nu(r,\theta)} - \Big(\frac{\partial}{\partial r}\psi(r,\theta)\Big)\Big(\frac{\partial}{\partial\theta}\omega(r,\theta)\Big) \\
&\mathrm{e}^{2\,\nu(r,\theta)} - 2\Big(\frac{\partial}{\partial\theta}\psi(r,\theta)\Big)\Big(\frac{\partial}{\partial r}\omega(r,\theta)\Big)\mathrm{e}^{2\,\nu(r,\theta)} + \Big(\frac{\partial}{\partial r}\mu(r,\theta)\Big) \\
&\mathrm{e}^{2\,\nu(r,\theta)}\frac{\partial}{\partial\theta}\omega(r,\theta) - \Big(\frac{\partial^2}{\partial\theta\partial r}\omega(r,\theta)\Big)\mathrm{e}^{2\,\nu(r,\theta)} + \Big(\frac{\partial}{\partial\theta}\nu(r,\theta)\Big)\mathrm{e}^{2\,\nu(r,\theta)} \\
&\frac{\partial}{\partial r}\omega(r,\theta)\Big\}\ \mathrm{e}^{-2\,\nu(r,\theta)}\,, \qquad \text{(G.38)}
\end{aligned}
$$

$$
\begin{aligned}
R^{\phi}{}_{\theta t\theta} = \frac{1}{2}\Big\{ &\Big(\frac{\partial^2}{\partial\theta^2}\omega(r,\theta)\Big)\mathrm{e}^{2\,\lambda(r,\theta)+2\,\nu(r,\theta)} - \Big(\frac{\partial}{\partial\theta}\nu(r,\theta)\Big)\omega^2(r,\theta) \\
&\Big(\frac{\partial}{\partial\theta}\omega(r,\theta)\Big)\mathrm{e}^{2\,\lambda(r,\theta)+2\,\psi(r,\theta)} + \Big(\frac{\partial}{\partial r}\mu(r,\theta)\Big)\omega^2(r,\theta)\Big(\frac{\partial}{\partial r}\omega(r,\theta)\Big) \\
&\mathrm{e}^{2\,\mu(r,\theta)+2\,\psi(r,\theta)} - \Big(\frac{\partial}{\partial\theta}\mu(r,\theta)\Big)\omega^2(r,\theta)\Big(\frac{\partial}{\partial\theta}\omega(r,\theta)\Big) \\
&\mathrm{e}^{2\,\lambda(r,\theta)+2\,\psi(r,\theta)} - 2\,\omega(r,\theta)\Big(\frac{\partial^2}{\partial\theta^2}\nu(r,\theta)\Big)\mathrm{e}^{2\,\lambda(r,\theta)+2\,\nu(r,\theta)} \\
&- 2\,\omega(r,\theta)\Big(\frac{\partial}{\partial\theta}\nu(r,\theta)\Big)^2\mathrm{e}^{2\,\lambda(r,\theta)+2\,\nu(r,\theta)} + 2\,\omega(r,\theta)\Big(\frac{\partial}{\partial\theta}\omega(r,\theta)\Big)^2 \\
&\mathrm{e}^{2\,\lambda(r,\theta)+2\,\psi(r,\theta)} - 2\,\omega(r,\theta)\Big(\frac{\partial}{\partial r}\mu(r,\theta)\Big)\Big(\frac{\partial}{\partial r}\nu(r,\theta)\Big)\mathrm{e}^{2\,\mu(r,\theta)+2\,\nu(r,\theta)} \\
&+ 2\,\omega(r,\theta)\Big(\frac{\partial}{\partial\theta}\mu(r,\theta)\Big)\Big(\frac{\partial}{\partial\theta}\nu(r,\theta)\Big)\mathrm{e}^{2\,\lambda(r,\theta)+2\,\nu(r,\theta)} \\
&+ 2\Big(\frac{\partial^2}{\partial\theta^2}\psi(r,\theta)\Big)\omega(r,\theta)\mathrm{e}^{2\,\lambda(r,\theta)+2\,\nu(r,\theta)} + 2\Big(\frac{\partial}{\partial\theta}\psi(r,\theta)\Big)^2\omega(r,\theta) \\
&\mathrm{e}^{2\,\lambda(r,\theta)+2\,\nu(r,\theta)} + 3\Big(\frac{\partial}{\partial\theta}\psi(r,\theta)\Big)\Big(\frac{\partial}{\partial\theta}\omega(r,\theta)\Big)\mathrm{e}^{2\,\lambda(r,\theta)+2\,\nu(r,\theta)}
\end{aligned}
$$

$$
\begin{aligned}
&-\Big(\frac{\partial}{\partial\theta}\nu(r,\theta)\Big)\Big(\frac{\partial}{\partial\theta}\omega(r,\theta)\Big)\mathrm{e}^{2\,\lambda(r,\theta)+2\,\nu(r,\theta)}+2\Big(\frac{\partial}{\partial r}\mu(r,\theta)\Big)\\
&\Big(\frac{\partial}{\partial r}\psi(r,\theta)\Big)\omega(r,\theta)\mathrm{e}^{2\,\mu(r,\theta)+2\,\nu(r,\theta)}+\Big(\frac{\partial}{\partial r}\mu(r,\theta)\Big)\Big(\frac{\partial}{\partial r}\omega(r,\theta)\Big)\\
&\mathrm{e}^{2\,\mu(r,\theta)+2\,\nu(r,\theta)}-2\Big(\frac{\partial}{\partial\theta}\mu(r,\theta)\Big)\Big(\frac{\partial}{\partial\theta}\psi(r,\theta)\Big)\omega(r,\theta)\mathrm{e}^{2\,\lambda(r,\theta)+2\,\nu(r,\theta)}\\
&-\Big(\frac{\partial}{\partial\theta}\mu(r,\theta)\Big)\Big(\frac{\partial}{\partial\theta}\omega(r,\theta)\Big)\mathrm{e}^{2\,\lambda(r,\theta)+2\,\nu(r,\theta)}+3\Big(\frac{\partial}{\partial\theta}\psi(r,\theta)\Big)\\
&\omega^2(r,\theta)\Big(\frac{\partial}{\partial\theta}\omega(r,\theta)\Big)\mathrm{e}^{2\,\lambda(r,\theta)+2\,\psi(r,\theta)}+\omega^2(r,\theta)\Big(\frac{\partial^2}{\partial\theta^2}\omega(r,\theta)\Big)\\
&\mathrm{e}^{2\,\lambda(r,\theta)+2\,\psi(r,\theta)}\Big\}\,\mathrm{e}^{-2\,\lambda(r,\theta)-2\,\nu(r,\theta)}\,,
\end{aligned}
\tag{G.39}
$$

$$
\begin{aligned}
R^{\phi}{}_{\theta r\phi}=&-\frac{1}{4}\Big\{2\Big(\frac{\partial}{\partial\theta}\lambda(r,\theta)\Big)\mathrm{e}^{2\,\psi(r,\theta)}\omega(r,\theta)\frac{\partial}{\partial r}\omega(r,\theta)+2\,\mathrm{e}^{2\,\psi(r,\theta)}\omega(r,\theta)\\
&\Big(\frac{\partial}{\partial\theta}\omega(r,\theta)\Big)\frac{\partial}{\partial r}\nu(r,\theta)-4\Big(\frac{\partial}{\partial r}\psi(r,\theta)\Big)\mathrm{e}^{2\,\psi(r,\theta)}\omega(r,\theta)\frac{\partial}{\partial\theta}\omega(r,\theta)\\
&-2\Big(\frac{\partial}{\partial\theta}\psi(r,\theta)\Big)\mathrm{e}^{2\,\psi(r,\theta)}\omega(r,\theta)\frac{\partial}{\partial r}\omega(r,\theta)-2\,\mathrm{e}^{2\,\psi(r,\theta)}\omega(r,\theta)\\
&\frac{\partial^2}{\partial\theta\partial r}\omega(r,\theta)+2\Big(\frac{\partial}{\partial r}\mu(r,\theta)\Big)\mathrm{e}^{2\,\psi(r,\theta)}\omega(r,\theta)\frac{\partial}{\partial\theta}\omega(r,\theta)\\
&-4\Big(\frac{\partial^2}{\partial\theta\partial r}\psi(r,\theta)\Big)\mathrm{e}^{2\,\nu(r,\theta)}-4\Big(\frac{\partial}{\partial r}\psi(r,\theta)\Big)\Big(\frac{\partial}{\partial\theta}\psi(r,\theta)\Big)\\
&\mathrm{e}^{2\,\nu(r,\theta)}-\mathrm{e}^{2\,\psi(r,\theta)}\Big(\frac{\partial}{\partial\theta}\omega(r,\theta)\Big)\frac{\partial}{\partial r}\omega(r,\theta)+4\Big(\frac{\partial}{\partial r}\psi(r,\theta)\Big)\\
&\Big(\frac{\partial}{\partial\theta}\lambda(r,\theta)\Big)\mathrm{e}^{2\,\nu(r,\theta)}+4\Big(\frac{\partial}{\partial r}\mu(r,\theta)\Big)\Big(\frac{\partial}{\partial\theta}\psi(r,\theta)\Big)\mathrm{e}^{2\,\nu(r,\theta)}\Big\}\\
&\mathrm{e}^{-2\,\nu(r,\theta)}\,,
\end{aligned}
\tag{G.40}
$$

$$
\begin{aligned}
R^{\phi}{}_{\theta\theta\phi}=&\frac{1}{4}\Big\{6\Big(\frac{\partial}{\partial\theta}\psi(r,\theta)\Big)\omega(r,\theta)\Big(\frac{\partial}{\partial\theta}\omega(r,\theta)\Big)\mathrm{e}^{2\,\lambda(r,\theta)+2\,\psi(r,\theta)}\\
&+2\,\omega(r,\theta)\Big(\frac{\partial^2}{\partial\theta^2}\omega(r,\theta)\Big)\mathrm{e}^{2\,\lambda(r,\theta)+2\,\psi(r,\theta)}-2\Big(\frac{\partial}{\partial\theta}\nu(r,\theta)\Big)\\
&\omega(r,\theta)\Big(\frac{\partial}{\partial\theta}\omega(r,\theta)\Big)\mathrm{e}^{2\,\lambda(r,\theta)+2\,\psi(r,\theta)}+2\Big(\frac{\partial}{\partial r}\mu(r,\theta)\Big)\omega(r,\theta)\\
&\Big(\frac{\partial}{\partial r}\omega(r,\theta)\Big)\mathrm{e}^{2\,\mu(r,\theta)+2\,\psi(r,\theta)}-2\Big(\frac{\partial}{\partial\theta}\mu(r,\theta)\Big)\omega(r,\theta)\\
&\Big(\frac{\partial}{\partial\theta}\omega(r,\theta)\Big)\mathrm{e}^{2\,\lambda(r,\theta)+2\,\psi(r,\theta)}+4\Big(\frac{\partial^2}{\partial\theta^2}\psi(r,\theta)\Big)\mathrm{e}^{2\,\lambda(r,\theta)+2\,\nu(r,\theta)}\\
&+4\Big(\frac{\partial}{\partial\theta}\psi(r,\theta)\Big)^2\mathrm{e}^{2\,\lambda(r,\theta)+2\,\nu(r,\theta)}+\Big(\frac{\partial}{\partial\theta}\omega(r,\theta)\Big)^2\mathrm{e}^{2\,\lambda(r,\theta)+2\,\psi(r,\theta)}
\end{aligned}
$$

$$+\,4\Big(\frac{\partial}{\partial r}\psi(r,\theta)\Big)\Big(\frac{\partial}{\partial r}\mu(r,\theta)\Big)\mathrm{e}^{2\,\mu(r,\theta)+2\,\nu(r,\theta)} - 4\Big(\frac{\partial}{\partial\theta}\mu(r,\theta)\Big)$$
$$\Big(\frac{\partial}{\partial\theta}\psi(r,\theta)\Big)\mathrm{e}^{2\,\lambda(r,\theta)+2\,\nu(r,\theta)}\Big\}\ \mathrm{e}^{-2\,\lambda(r,\theta)-2\,\nu(r,\theta)}\,, \tag{G.41}$$

$$R^{\phi}{}_{\phi t\phi} = -\frac{1}{4}\Big\{\omega(r,\theta)\Big(\Big(\frac{\partial}{\partial r}\omega(r,\theta)\Big)^2\mathrm{e}^{2\,\mu(r,\theta)+2\,\psi(r,\theta)} + 4\Big(\frac{\partial}{\partial r}\psi(r,\theta)\Big)$$
$$\Big(\frac{\partial}{\partial r}\nu(r,\theta)\Big)\mathrm{e}^{2\,\mu(r,\theta)+2\,\nu(r,\theta)} + \Big(\frac{\partial}{\partial\theta}\omega(r,\theta)\Big)^2\mathrm{e}^{2\,\lambda(r,\theta)+2\,\psi(r,\theta)}$$
$$+\,4\Big(\frac{\partial}{\partial\theta}\psi(r,\theta)\Big)\Big(\frac{\partial}{\partial\theta}\nu(r,\theta)\Big)\mathrm{e}^{2\,\lambda(r,\theta)+2\,\nu(r,\theta)}\Big)$$
$$\mathrm{e}^{-2\,\nu(r,\theta)-2\,\lambda(r,\theta)+2\,\psi(r,\theta)-2\,\mu(r,\theta)}\Big\}\,, \tag{G.42}$$

$$R^{\phi}{}_{\phi r\theta} = \frac{1}{2}\Big\{\omega(r,\theta)\Big(\Big(\frac{\partial}{\partial\theta}\nu(r,\theta)\Big)\frac{\partial}{\partial r}\omega(r,\theta) - \Big(\frac{\partial}{\partial\theta}\omega(r,\theta)\Big)\frac{\partial}{\partial r}\nu(r,\theta)$$
$$+\,\Big(\frac{\partial}{\partial r}\psi(r,\theta)\Big)\frac{\partial}{\partial\theta}\omega(r,\theta) - \Big(\frac{\partial}{\partial\theta}\psi(r,\theta)\Big)\frac{\partial}{\partial r}\omega(r,\theta)\Big)$$
$$\mathrm{e}^{2\,\psi(r,\theta)-2\,\nu(r,\theta)}\Big\}\,. \tag{G.43}$$

G.2 Ricci tensor

The Ricci tensor is obtained from the Riemann tensor by contraction, $R_{\mu\nu} = R^{\tau}{}_{\mu\sigma\nu}\, g^{\sigma}{}_{\tau} = R^{\tau}{}_{\mu\tau\nu}$ (cf. equation (13.45)). Its individual components are as follows,

$$R_{tt} = -\frac{1}{2}\Big\{2\Big(\frac{\partial^2}{\partial\theta^2}\psi(r,\theta)\Big)\omega^2(r,\theta)\mathrm{e}^{2\,\lambda(r,\theta)+2\,\nu(r,\theta)+2\,\psi(r,\theta)} + 2\Big(\frac{\partial}{\partial r}\psi(r,\theta)\Big)^2$$
$$\omega^2(r,\theta)\mathrm{e}^{2\,\mu(r,\theta)+2\,\psi(r,\theta)+2\,\nu(r,\theta)} + 2\Big(\frac{\partial}{\partial\theta}\psi(r,\theta)\Big)\omega^2(r,\theta)\Big(\frac{\partial}{\partial\theta}\nu(r,\theta)\Big)$$
$$\mathrm{e}^{2\,\lambda(r,\theta)+2\,\nu(r,\theta)+2\,\psi(r,\theta)} - 2\Big(\frac{\partial}{\partial\theta}\nu(r,\theta)\Big)\Big(\frac{\partial}{\partial\theta}\psi(r,\theta)\Big)\mathrm{e}^{2\,\lambda(r,\theta)+4\,\nu(r,\theta)}$$
$$-\,2\Big(\frac{\partial}{\partial r}\nu(r,\theta)\Big)\Big(\frac{\partial}{\partial r}\psi(r,\theta)\Big)\mathrm{e}^{2\,\mu(r,\theta)+4\,\nu(r,\theta)} - 2\Big(\frac{\partial}{\partial\theta}\lambda(r,\theta)\Big)\Big(\frac{\partial}{\partial\theta}\nu(r,\theta)\Big)$$
$$\mathrm{e}^{2\,\lambda(r,\theta)+4\,\nu(r,\theta)} + 2\Big(\frac{\partial}{\partial r}\nu(r,\theta)\Big)\Big(\frac{\partial}{\partial r}\psi(r,\theta)\Big)\omega^2(r,\theta)\mathrm{e}^{2\,\mu(r,\theta)+2\,\psi(r,\theta)+2\,\nu(r,\theta)}$$
$$+\,2\Big(\frac{\partial}{\partial r}\lambda(r,\theta)\Big)\Big(\frac{\partial}{\partial r}\nu(r,\theta)\Big)\mathrm{e}^{2\,\mu(r,\theta)+4\,\nu(r,\theta)} + \omega^2(r,\theta)\Big(\frac{\partial}{\partial r}\omega(r,\theta)\Big)^2$$
$$\mathrm{e}^{2\,\mu(r,\theta)+4\,\psi(r,\theta)} + \Big(\frac{\partial}{\partial r}\omega(r,\theta)\Big)^2\mathrm{e}^{2\,\mu(r,\theta)+2\,\psi(r,\theta)+2\,\nu(r,\theta)} - 2\Big(\frac{\partial}{\partial r}\nu(r,\theta)\Big)$$
$$\omega(r,\theta)\Big(\frac{\partial}{\partial r}\omega(r,\theta)\Big)\mathrm{e}^{2\,\mu(r,\theta)+2\,\psi(r,\theta)+2\,\nu(r,\theta)} + 2\Big(\frac{\partial^2}{\partial r^2}\psi(r,\theta)\Big)\omega^2(r,\theta)$$

$$
\begin{aligned}
&e^{2\mu(r,\theta)+2\psi(r,\theta)+2\nu(r,\theta)} + 2\,\omega(r,\theta)\Big(\frac{\partial^2}{\partial r^2}\omega(r,\theta)\Big)e^{2\mu(r,\theta)+2\psi(r,\theta)+2\nu(r,\theta)}\\
&-2\Big(\frac{\partial}{\partial r}\lambda(r,\theta)\Big)\omega(r,\theta)\Big(\frac{\partial}{\partial r}\omega(r,\theta)\Big)e^{2\mu(r,\theta)+2\psi(r,\theta)+2\nu(r,\theta)} - 2\Big(\frac{\partial}{\partial r}\lambda(r,\theta)\Big)\\
&\Big(\frac{\partial}{\partial r}\psi(r,\theta)\Big)\omega^2(r,\theta)e^{2\mu(r,\theta)+2\psi(r,\theta)+2\nu(r,\theta)} + 6\Big(\frac{\partial}{\partial r}\psi(r,\theta)\Big)\omega(r,\theta)\\
&\Big(\frac{\partial}{\partial r}\omega(r,\theta)\Big)e^{2\mu(r,\theta)+2\psi(r,\theta)+2\nu(r,\theta)} + 2\Big(\frac{\partial}{\partial\theta}\lambda(r,\theta)\Big)\omega(r,\theta)\Big(\frac{\partial}{\partial\theta}\omega(r,\theta)\Big)\\
&e^{2\lambda(r,\theta)+2\nu(r,\theta)+2\psi(r,\theta)} + 2\,\Big(\frac{\partial}{\partial\theta}\lambda(r,\theta)\Big)\Big(\frac{\partial}{\partial\theta}\psi(r,\theta)\Big)\omega(r,\theta)\\
&e^{2\lambda(r,\theta)+2\nu(r,\theta)+2\psi(r,\theta)} - 2\Big(\frac{\partial^2}{\partial r^2}\nu(r,\theta)\Big)e^{2\mu(r,\theta)+4\nu(r,\theta)} - 2\Big(\frac{\partial}{\partial r}\nu(r,\theta)\Big)^2\\
&e^{2\mu(r,\theta)+4\nu(r,\theta)} - 2\Big(\frac{\partial^2}{\partial\theta^2}\nu(r,\theta)\Big)e^{2\lambda(r,\theta)+4\nu(r,\theta)} - 2\Big(\frac{\partial}{\partial\theta}\nu(r,\theta)\Big)^2\\
&e^{2\lambda(r,\theta)+4\nu(r,\theta)} + \Big(\frac{\partial}{\partial\theta}\omega(r,\theta)\Big)^2 e^{2\lambda(r,\theta)+2\nu(r,\theta)+2\psi(r,\theta)}\\
&+2\Big(\frac{\partial}{\partial\theta}\psi(r,\theta)\Big)^2\omega^2(r,\theta)e^{2\lambda(r,\theta)+2\nu(r,\theta)+2\psi(r,\theta)} + 2\,\omega(r,\theta)\Big(\frac{\partial^2}{\partial\theta^2}\omega(r,\theta)\Big)\\
&e^{2\lambda(r,\theta)+2\nu(r,\theta)+2\psi(r,\theta)} + \omega^2(r,\theta)\Big(\frac{\partial}{\partial\theta}\omega(r,\theta)\Big)^2 e^{2\lambda(r,\theta)+4\psi(r,\theta)}\\
&-2\Big(\frac{\partial}{\partial r}\mu(r,\theta)\Big)\Big(\frac{\partial}{\partial r}\nu(r,\theta)\Big)e^{2\mu(r,\theta)+4\nu(r,\theta)} + 2\Big(\frac{\partial}{\partial r}\mu(r,\theta)\Big)\\
&\Big(\frac{\partial}{\partial r}\psi(r,\theta)\Big)\omega^2(r,\theta)e^{2\mu(r,\theta)+2\psi(r,\theta)+2\nu(r,\theta)} - 2\Big(\frac{\partial}{\partial\theta}\nu(r,\theta)\Big)\\
&\omega(r,\theta)\Big(\frac{\partial}{\partial\theta}\omega(r,\theta)\Big)e^{2\lambda(r,\theta)+2\nu(r,\theta)+2\psi(r,\theta)} + 2\Big(\frac{\partial}{\partial\theta}\mu(r,\theta)\Big)\Big(\frac{\partial}{\partial\theta}\nu(r,\theta)\Big)\\
&e^{2\lambda(r,\theta)+4\nu(r,\theta)} + 2\Big(\frac{\partial}{\partial r}\mu(r,\theta)\Big)\omega(r,\theta)\Big(\frac{\partial}{\partial r}\omega(r,\theta)\Big)e^{2\mu(r,\theta)+2\psi(r,\theta)+2\nu(r,\theta)}\\
&+6\Big(\frac{\partial}{\partial\theta}\psi(r,\theta)\Big)\omega(r,\theta)\Big(\frac{\partial}{\partial\theta}\omega(r,\theta)\Big)e^{2\lambda(r,\theta)+2\nu(r,\theta)+2\psi(r,\theta)} - 2\Big(\frac{\partial}{\partial\theta}\mu(r,\theta)\Big)\\
&\Big(\frac{\partial}{\partial\theta}\psi(r,\theta)\Big)\omega^2(r,\theta)e^{2\lambda(r,\theta)+2\nu(r,\theta)+2\psi(r,\theta)} - 2\Big(\frac{\partial}{\partial\theta}\mu(r,\theta)\Big)\omega(r,\theta)\\
&\Big(\frac{\partial}{\partial\theta}\omega(r,\theta)\Big)e^{2\lambda(r,\theta)+2\nu(r,\theta)+2\psi(r,\theta)}\Big\}\; e^{-2\nu(r,\theta)-2\lambda(r,\theta)-2\mu(r,\theta)}\,, \qquad \text{(G.44)}
\end{aligned}
$$

$$
\begin{aligned}
R_{t\phi} = &\frac{1}{2}\Big\{2\Big(\frac{\partial}{\partial\theta}\psi(r,\theta)\Big)^2\omega(r,\theta)e^{2\lambda(r,\theta)+2\nu(r,\theta)} + 3\Big(\frac{\partial}{\partial\theta}\psi(r,\theta)\Big)\Big(\frac{\partial}{\partial\theta}\omega(r,\theta)\Big)\\
&e^{2\lambda(r,\theta)+2\nu(r,\theta)} + \Big(\frac{\partial^2}{\partial\theta^2}\omega(r,\theta)\Big)e^{2\lambda(r,\theta)+2\nu(r,\theta)} + 2\Big(\frac{\partial^2}{\partial\theta^2}\psi(r,\theta)\Big)\omega(r,\theta)\\
&e^{2\lambda(r,\theta)+2\nu(r,\theta)} - \Big(\frac{\partial}{\partial\theta}\nu(r,\theta)\Big)\Big(\frac{\partial}{\partial\theta}\omega(r,\theta)\Big)e^{2\lambda(r,\theta)+2\nu(r,\theta)} + \omega(r,\theta)\\
&\Big(\frac{\partial}{\partial\theta}\omega(r,\theta)\Big)^2 e^{2\lambda(r,\theta)+2\psi(r,\theta)} + 2\Big(\frac{\partial}{\partial r}\mu(r,\theta)\Big)\Big(\frac{\partial}{\partial r}\psi(r,\theta)\Big)\omega(r,\theta)
\end{aligned}
$$

$$e^{2\,\mu(r,\theta)+2\,\nu(r,\theta)} + \left(\frac{\partial}{\partial r}\mu(r,\theta)\right)\left(\frac{\partial}{\partial r}\omega(r,\theta)\right)e^{2\,\mu(r,\theta)+2\,\nu(r,\theta)} - 2\left(\frac{\partial}{\partial \theta}\mu(r,\theta)\right)$$
$$\left(\frac{\partial}{\partial \theta}\psi(r,\theta)\right)\omega(r,\theta)e^{2\,\lambda(r,\theta)+2\,\nu(r,\theta)} - \left(\frac{\partial}{\partial \theta}\mu(r,\theta)\right)\left(\frac{\partial}{\partial \theta}\omega(r,\theta)\right)$$
$$e^{2\,\lambda(r,\theta)+2\,\nu(r,\theta)} + \left(\frac{\partial^2}{\partial r^2}\omega(r,\theta)\right)e^{2\,\mu(r,\theta)+2\,\nu(r,\theta)} + \omega(r,\theta)\left(\frac{\partial}{\partial r}\omega(r,\theta)\right)^2$$
$$e^{2\,\mu(r,\theta)+2\,\psi(r,\theta)} + 2\left(\frac{\partial}{\partial r}\psi(r,\theta)\right)\omega(r,\theta)\left(\frac{\partial}{\partial r}\nu(r,\theta)\right)e^{2\,\mu(r,\theta)+2\,\nu(r,\theta)}$$
$$+ 2\left(\frac{\partial}{\partial \theta}\psi(r,\theta)\right)\omega(r,\theta)\left(\frac{\partial}{\partial \theta}\nu(r,\theta)\right)e^{2\,\lambda(r,\theta)+2\,\nu(r,\theta)} + 2\left(\frac{\partial}{\partial \theta}\lambda(r,\theta)\right)$$
$$\left(\frac{\partial}{\partial \theta}\psi(r,\theta)\right)\omega(r,\theta)e^{2\,\lambda(r,\theta)+2\,\nu(r,\theta)} + \left(\frac{\partial}{\partial \theta}\lambda(r,\theta)\right)\left(\frac{\partial}{\partial \theta}\omega(r,\theta)\right)$$
$$e^{2\,\lambda(r,\theta)+2\,\nu(r,\theta)} - \left(\frac{\partial}{\partial r}\lambda(r,\theta)\right)\left(\frac{\partial}{\partial r}\omega(r,\theta)\right)e^{2\,\mu(r,\theta)+2\,\nu(r,\theta)}$$
$$+ 3\left(\frac{\partial}{\partial r}\psi(r,\theta)\right)\left(\frac{\partial}{\partial r}\omega(r,\theta)\right)e^{2\,\mu(r,\theta)+2\,\nu(r,\theta)} - 2\left(\frac{\partial}{\partial r}\lambda(r,\theta)\right)\left(\frac{\partial}{\partial r}\psi(r,\theta)\right)$$
$$\omega(r,\theta)e^{2\,\mu(r,\theta)+2\,\nu(r,\theta)} - \left(\frac{\partial}{\partial r}\nu(r,\theta)\right)\left(\frac{\partial}{\partial r}\omega(r,\theta)\right)e^{2\,\mu(r,\theta)+2\,\nu(r,\theta)}$$
$$+ 2\left(\frac{\partial}{\partial r}\psi(r,\theta)\right)^2\omega(r,\theta)e^{2\,\mu(r,\theta)+2\,\nu(r,\theta)} + 2\left(\frac{\partial^2}{\partial r^2}\psi(r,\theta)\right)\omega(r,\theta)$$
$$e^{2\,\mu(r,\theta)+2\,\nu(r,\theta)}\Big\}\, e^{-2\,\nu(r,\theta)-2\,\lambda(r,\theta)+2\,\psi(r,\theta)-2\,\mu(r,\theta)}\,, \qquad \text{(G.45)}$$

$$R_{rr} = -\frac{1}{2}\Big\{2\left(\frac{\partial}{\partial \theta}\lambda(r,\theta)\right)\left(\frac{\partial}{\partial \theta}\nu(r,\theta)\right)e^{2\,\lambda(r,\theta)+2\,\nu(r,\theta)} + 2\left(\frac{\partial^2}{\partial r^2}\nu(r,\theta)\right)$$
$$e^{2\,\mu(r,\theta)+2\,\nu(r,\theta)} + 2\left(\frac{\partial}{\partial r}\nu(r,\theta)\right)^2 e^{2\,\mu(r,\theta)+2\,\nu(r,\theta)} + 2\left(\frac{\partial^2}{\partial r^2}\mu(r,\theta)\right)$$
$$e^{2\,\mu(r,\theta)+2\,\nu(r,\theta)} - 2\left(\frac{\partial}{\partial r}\lambda(r,\theta)\right)\left(\frac{\partial}{\partial r}\psi(r,\theta)\right)e^{2\,\mu(r,\theta)+2\,\nu(r,\theta)}$$
$$+ 2\left(\frac{\partial}{\partial \theta}\lambda(r,\theta)\right)\left(\frac{\partial}{\partial \theta}\psi(r,\theta)\right)e^{2\,\lambda(r,\theta)+2\,\nu(r,\theta)} + 2\left(\frac{\partial^2}{\partial r^2}\psi(r,\theta)\right)$$
$$e^{2\,\mu(r,\theta)+2\,\nu(r,\theta)} + 2\left(\frac{\partial}{\partial r}\psi(r,\theta)\right)^2 e^{2\,\mu(r,\theta)+2\,\nu(r,\theta)} - \left(\frac{\partial}{\partial r}\omega(r,\theta)\right)^2$$
$$e^{2\,\mu(r,\theta)+2\,\psi(r,\theta)} - 2\left(\frac{\partial}{\partial r}\lambda(r,\theta)\right)\left(\frac{\partial}{\partial r}\nu(r,\theta)\right)e^{2\,\mu(r,\theta)+2\,\nu(r,\theta)}$$
$$+ 2\left(\frac{\partial}{\partial r}\mu(r,\theta)\right)^2 e^{2\,\mu(r,\theta)+2\,\nu(r,\theta)} - 2\left(\frac{\partial}{\partial r}\lambda(r,\theta)\right)\left(\frac{\partial}{\partial r}\mu(r,\theta)\right)$$
$$e^{2\,\mu(r,\theta)+2\,\nu(r,\theta)} - 2\left(\frac{\partial}{\partial \theta}\lambda(r,\theta)\right)\left(\frac{\partial}{\partial \theta}\mu(r,\theta)\right)e^{2\,\lambda(r,\theta)+2\,\nu(r,\theta)}$$
$$+ 2\left(\frac{\partial^2}{\partial \theta^2}\lambda(r,\theta)\right)e^{2\,\lambda(r,\theta)+2\,\nu(r,\theta)} + 2\left(\frac{\partial}{\partial \theta}\lambda(r,\theta)\right)^2 e^{2\,\lambda(r,\theta)+2\,\nu(r,\theta)}\Big\}$$

$$e^{-2\,\mu(r,\theta)-2\,\nu(r,\theta)}\,, \tag{G.46}$$

$$R_{r\theta} = \frac{1}{2}\Big\{e^{2\,\psi(r,\theta)}\Big(\frac{\partial}{\partial\theta}\omega(r,\theta)\Big)\frac{\partial}{\partial r}\omega(r,\theta) - 2\Big(\frac{\partial}{\partial r}\nu(r,\theta)\Big)\Big(\frac{\partial}{\partial\theta}\nu(r,\theta)\Big)e^{2\,\nu(r,\theta)}$$
$$+ 2\Big(\frac{\partial}{\partial r}\mu(r,\theta)\Big)e^{2\,\nu(r,\theta)}\frac{\partial}{\partial\theta}\nu(r,\theta) + 2\,\Big(\frac{\partial}{\partial\theta}\lambda(r,\theta)\Big)e^{2\,\nu(r,\theta)}\frac{\partial}{\partial r}\nu(r,\theta)$$
$$- 2\Big(\frac{\partial^2}{\partial\theta\partial r}\nu(r,\theta)\Big)e^{2\,\nu(r,\theta)} - 2\Big(\frac{\partial^2}{\partial\theta\partial r}\psi(r,\theta)\Big)e^{2\,\nu(r,\theta)} - 2\Big(\frac{\partial}{\partial r}\psi(r,\theta)\Big)$$
$$\Big(\frac{\partial}{\partial\theta}\psi(r,\theta)\Big)e^{2\,\nu(r,\theta)} + 2\Big(\frac{\partial}{\partial r}\psi(r,\theta)\Big)\Big(\frac{\partial}{\partial\theta}\lambda(r,\theta)\Big)e^{2\,\nu(r,\theta)} + 2\Big(\frac{\partial}{\partial r}\mu(r,\theta)\Big)$$
$$\Big(\frac{\partial}{\partial\theta}\psi(r,\theta)\Big)e^{2\,\nu(r,\theta)}\Big\}\,e^{-2\,\nu(r,\theta)}\,, \tag{G.47}$$

$$R_{\theta\theta} = -\frac{1}{2}\Big\{2\Big(\frac{\partial}{\partial r}\mu(r,\theta)\Big)\Big(\frac{\partial}{\partial r}\nu(r,\theta)\Big)e^{2\,\mu(r,\theta)+2\,\nu(r,\theta)} - 2\,\Big(\frac{\partial}{\partial\theta}\mu(r,\theta)\Big)$$
$$\Big(\frac{\partial}{\partial\theta}\nu(r,\theta)\Big)e^{2\,\lambda(r,\theta)+2\,\nu(r,\theta)} - \Big(\frac{\partial}{\partial\theta}\omega(r,\theta)\Big)^2 e^{2\,\lambda(r,\theta)+2\,\psi(r,\theta)}$$
$$+ 2\Big(\frac{\partial}{\partial\theta}\nu(r,\theta)\Big)^2 e^{2\,\lambda(r,\theta)+2\,\nu(r,\theta)} + 2\Big(\frac{\partial^2}{\partial\theta^2}\nu(r,\theta)\Big)e^{2\,\lambda(r,\theta)+2\,\nu(r,\theta)}$$
$$+ 2\Big(\frac{\partial^2}{\partial r^2}\mu(r,\theta)\Big)e^{2\,\mu(r,\theta)+2\,\nu(r,\theta)} - 2\Big(\frac{\partial}{\partial\theta}\mu(r,\theta)\Big)\Big(\frac{\partial}{\partial\theta}\psi(r,\theta)\Big)$$
$$e^{2\,\lambda(r,\theta)+2\,\nu(r,\theta)} + 2\Big(\frac{\partial}{\partial r}\psi(r,\theta)\Big)\Big(\frac{\partial}{\partial r}\mu(r,\theta)\Big)e^{2\,\mu(r,\theta)+2\,\nu(r,\theta)}$$
$$+ 2\Big(\frac{\partial^2}{\partial\theta^2}\psi(r,\theta)\Big)e^{2\,\lambda(r,\theta)+2\,\nu(r,\theta)} + 2\Big(\frac{\partial}{\partial\theta}\psi(r,\theta)\Big)^2 e^{2\,\lambda(r,\theta)+2\,\nu(r,\theta)}$$
$$+ 2\Big(\frac{\partial}{\partial r}\mu(r,\theta)\Big)^2 e^{2\,\mu(r,\theta)+2\,\nu(r,\theta)} - 2\,\Big(\frac{\partial}{\partial r}\lambda(r,\theta)\Big)\Big(\frac{\partial}{\partial r}\mu(r,\theta)\Big)e^{2\,\mu(r,\theta)+\,\nu(r,\theta)}$$
$$- 2\Big(\frac{\partial}{\partial\theta}\lambda(r,\theta)\Big)\Big(\frac{\partial}{\partial\theta}\mu(r,\theta)\Big)e^{2\,\lambda(r,\theta)+2\,\nu(r,\theta)} + 2\Big(\frac{\partial^2}{\partial\theta^2}\lambda(r,\theta)\Big)e^{2\,\lambda(r,\theta)+2\,\nu(r,\theta)}$$
$$+ 2\Big(\frac{\partial}{\partial\theta}\lambda(r,\theta)\Big)^2 e^{2\,\lambda(r,\theta)+2\,\nu(r,\theta)}\Big\}\,e^{-2\,\lambda(r,\theta)-2\,\nu(r,\theta)}\,, \tag{G.48}$$

$$R_{\phi\phi} = -\frac{1}{2}\Big\{2\Big(\frac{\partial}{\partial r}\psi(r,\theta)\Big)\Big(\frac{\partial}{\partial r}\nu(r,\theta)\Big)e^{2\,\mu(r,\theta)+2\,\nu(r,\theta)} + \Big(\frac{\partial}{\partial r}\omega(r,\theta)\Big)^2$$
$$e^{2\,\mu(r,\theta)+2\,\psi(r,\theta)} + 2\Big(\frac{\partial}{\partial\theta}\psi(r,\theta)\Big)\Big(\frac{\partial}{\partial\theta}\nu(r,\theta)\Big)e^{2\,\lambda(r,\theta)+2\,\nu(r,\theta)} + \Big(\frac{\partial}{\partial\theta}\omega(r,\theta)\Big)^2$$
$$e^{2\,\lambda(r,\theta)+2\,\psi(r,\theta)} - 2\Big(\frac{\partial}{\partial r}\lambda(r,\theta)\Big)\Big(\frac{\partial}{\partial r}\psi(r,\theta)\Big)e^{2\,\mu(r,\theta)+2\,\nu(r,\theta)} + 2\Big(\frac{\partial^2}{\partial r^2}\psi(r,\theta)\Big)$$
$$e^{2\,\mu(r,\theta)+2\,\nu(r,\theta)} + 2\Big(\frac{\partial}{\partial r}\psi(r,\theta)\Big)^2 e^{2\,\mu(r,\theta)+2\,\nu(r,\theta)} + 2\Big(\frac{\partial}{\partial\theta}\lambda(r,\theta)\Big)\Big(\frac{\partial}{\partial\theta}\psi(r,\theta)\Big)$$
$$e^{2\,\lambda(r,\theta)+2\,\nu(r,\theta)} - 2\Big(\frac{\partial}{\partial\theta}\mu(r,\theta)\Big)\Big(\frac{\partial}{\partial\theta}\psi(r,\theta)\Big)e^{2\,\lambda(r,\theta)+2\,\nu(r,\theta)} + 2\Big(\frac{\partial^2}{\partial\theta^2}\psi(r,\theta)\Big)$$

$$e^{2\lambda(r,\theta)+2\nu(r,\theta)} + 2\Big(\frac{\partial}{\partial\theta}\psi(r,\theta)\Big)^2 e^{2\lambda(r,\theta)+2\nu(r,\theta)} + 2\Big(\frac{\partial}{\partial r}\psi(r,\theta)\Big)\Big(\frac{\partial}{\partial r}\mu(r,\theta)\Big)$$
$$e^{2\mu(r,\theta)+2\nu(r,\theta)}\Big\}\, e^{-2\nu(r,\theta)-2\lambda(r,\theta)+2\psi(r,\theta)-2\mu(r,\theta)}\,. \tag{G.49}$$

G.3 Ricci scalar

The Ricci scalar follows from the Ricci tensor as $R = R_{\mu\nu}\, g^{\mu\nu}$ (cf. equation (13.47)). One arrives at

$$\begin{aligned}
R = &-\frac{1}{2}\Big[4\Big(\frac{\partial}{\partial\theta}\lambda(r,\theta)\Big)^2 e^{2\lambda(r,\theta)+2\nu(r,\theta)} + 4\Big(\frac{\partial}{\partial\theta}\lambda(r,\theta)\Big)\Big(\frac{\partial}{\partial\theta}\psi(r,\theta)\Big)\\
&e^{2\lambda(r,\theta)+2\nu(r,\theta)} - 4\Big(\frac{\partial}{\partial\theta}\mu(r,\theta)\Big)\Big(\frac{\partial}{\partial\theta}\psi(r,\theta)\Big)e^{2\lambda(r,\theta)+2\nu(r,\theta)}\\
&+4\Big(\frac{\partial}{\partial\theta}\lambda(r,\theta)\Big)\Big(\frac{\partial}{\partial\theta}\nu(r,\theta)\Big)e^{2\lambda(r,\theta)+2\nu(r,\theta)} - 4\Big(\frac{\partial}{\partial\theta}\mu(r,\theta)\Big)\\
&\Big(\frac{\partial}{\partial\theta}\nu(r,\theta)\Big)e^{2\lambda(r,\theta)+2\nu(r,\theta)} + 4\Big(\frac{\partial}{\partial\theta}\psi(r,\theta)\Big)\Big(\frac{\partial}{\partial\theta}\nu(r,\theta)\Big)\\
&e^{2\lambda(r,\theta)+2\nu(r,\theta)} + 4\Big(\frac{\partial}{\partial r}\psi(r,\theta)\Big)\Big(\frac{\partial}{\partial r}\nu(r,\theta)\Big)e^{2\mu(r,\theta)+2\nu(r,\theta)}\\
&-4\Big(\frac{\partial}{\partial r}\lambda(r,\theta)\Big)\Big(\frac{\partial}{\partial r}\nu(r,\theta)\Big)e^{2\mu(r,\theta)+2\nu(r,\theta)} + 4\Big(\frac{\partial}{\partial r}\mu(r,\theta)\Big)\\
&\Big(\frac{\partial}{\partial r}\nu(r,\theta)\Big)e^{2\mu(r,\theta)+2\nu(r,\theta)} - 4\Big(\frac{\partial}{\partial r}\lambda(r,\theta)\Big)\Big(\frac{\partial}{\partial r}\psi(r,\theta)\Big)\\
&e^{2\mu(r,\theta)+2\nu(r,\theta)} - 4\Big(\frac{\partial}{\partial\theta}\lambda(r,\theta)\Big)\Big(\frac{\partial}{\partial\theta}\mu(r,\theta)\Big)e^{2\lambda(r,\theta)+2\nu(r,\theta)}\\
&+4\Big(\frac{\partial}{\partial r}\psi(r,\theta)\Big)\Big(\frac{\partial}{\partial r}\mu(r,\theta)\Big)e^{2\mu(r,\theta)+2\nu(r,\theta)} - 4\Big(\frac{\partial}{\partial r}\lambda(r,\theta)\Big)\\
&\Big(\frac{\partial}{\partial r}\mu(r,\theta)\Big)e^{2\mu(r,\theta)+2\nu(r,\theta)} + 4\Big(\frac{\partial}{\partial r}\psi(r,\theta)\Big)^2 e^{2\mu(r,\theta)+2\nu(r,\theta)}\\
&+4\Big(\frac{\partial}{\partial r}\mu(r,\theta)\Big)^2 e^{2\mu(r,\theta)+2\nu(r,\theta)} + 4\Big(\frac{\partial^2}{\partial r^2}\nu(r,\theta)\Big)e^{2\mu(r,\theta)+2\nu(r,\theta)}\\
&+4\Big(\frac{\partial}{\partial\theta}\psi(r,\theta)\Big)^2 e^{2\lambda(r,\theta)+2\nu(r,\theta)} + 4\Big(\frac{\partial^2}{\partial r^2}\psi(r,\theta)\Big)e^{2\mu(r,\theta)+2\nu(r,\theta)}\\
&+4\Big(\frac{\partial^2}{\partial\theta^2}\nu(r,\theta)\Big)e^{2\lambda(r,\theta)+2\nu(r,\theta)} + 4\Big(\frac{\partial}{\partial\theta}\nu(r,\theta)\Big)^2 e^{2\lambda(r,\theta)+2\nu(r,\theta)}\\
&+4\Big(\frac{\partial^2}{\partial\theta^2}\psi(r,\theta)\Big)e^{2\lambda(r,\theta)+2\nu(r,\theta)} + 4\Big(\frac{\partial}{\partial r}\nu(r,\theta)\Big)^2 e^{2\mu(r,\theta)+2\nu(r,\theta)}\\
&+4\Big(\frac{\partial^2}{\partial\theta^2}\lambda(r,\theta)\Big)e^{2\lambda(r,\theta)+2\nu(r,\theta)} + 4\Big(\frac{\partial^2}{\partial r^2}\mu(r,\theta)\Big)e^{2\mu(r,\theta)+2\nu(r,\theta)}\\
&-\Big(\frac{\partial}{\partial\theta}\omega(r,\theta)\Big)^2 e^{2\lambda(r,\theta)+2\psi(r,\theta)} - \Big(\frac{\partial}{\partial r}\omega(r,\theta)\Big)^2 e^{2\mu(r,\theta)+2\psi(r,\theta)}\Big]\\
&e^{-2\nu(r,\theta)-2\lambda(r,\theta)-2\mu(r,\theta)}\,.
\end{aligned} \tag{G.50}$$

G.4 Einstein tensor

The covariant components of the Einstein tensor (13.43) are found to be,

$$
\begin{aligned}
G_{tt} = \frac{1}{4}\Big\{ & -3\,\omega^2(r,\theta)\Big(\frac{\partial}{\partial r}\omega(r,\theta)\Big)e^{2\,\mu(r,\theta)+4\,\psi(r,\theta)} - \Big(\frac{\partial}{\partial r}\omega(r,\theta)\Big)^2 \\
& e^{2\,\mu(r,\theta)+2\,\psi(r,\theta)+2\,\nu(r,\theta)} + 4\Big(\frac{\partial}{\partial r}\nu(r,\theta)\Big)\omega(r,\theta)\Big(\frac{\partial}{\partial r}\omega(r,\theta)\Big) \\
& e^{2\,\mu(r,\theta)+2\,\psi(r,\theta)+2\,\nu(r,\theta)} - 4\,\omega(r,\theta)\Big(\frac{\partial^2}{\partial r^2}\omega(r,\theta)\Big)e^{2\,\mu(r,\theta)+2\,\psi(r,\theta)+2\,\nu(r,\theta)} \\
& + 4\Big(\frac{\partial}{\partial r}\lambda(r,\theta)\Big)\omega(r,\theta)\Big(\frac{\partial}{\partial r}\omega(r,\theta)\Big)e^{2\,\mu(r,\theta)+2\,\psi(r,\theta)+2\,\nu(r,\theta)} \\
& - 12\Big(\frac{\partial}{\partial r}\psi(r,\theta)\Big)\omega(r,\theta)\Big(\frac{\partial}{\partial r}\omega(r,\theta)\Big)e^{2\,\mu(r,\theta)+2\,\psi(r,\theta)+2\,\nu(r,\theta)} \\
& - 4\Big(\frac{\partial}{\partial \theta}\lambda(r,\theta)\Big)\omega(r,\theta)\Big(\frac{\partial}{\partial \theta}\omega(r,\theta)\Big)e^{2\,\lambda(r,\theta)+2\,\nu(r,\theta)+2\,\psi(r,\theta)} \\
& - 4\Big(\frac{\partial^2}{\partial r^2}\psi(r,\theta)\Big)e^{2\,\mu(r,\theta)+4\,\nu(r,\theta)} - 4\Big(\frac{\partial}{\partial r}\psi(r,\theta)\Big)^2 e^{2\,\mu(r,\theta)+4\,\nu(r,\theta)} \\
& - 4\Big(\frac{\partial^2}{\partial r^2}\mu(r,\theta)\Big)e^{2\,\mu(r,\theta)+4\,\nu(r,\theta)} - \Big(\frac{\partial}{\partial \theta}\omega(r,\theta)\Big)^2 e^{2\,\lambda(r,\theta)+2\,\nu(r,\theta)+2\,\psi(r,\theta)} \\
& + 4\,\omega^2(r,\theta)\Big(\frac{\partial^2}{\partial r^2}\mu(r,\theta)\Big)e^{2\,\mu(r,\theta)+2\,\psi(r,\theta)+2\,\nu(r,\theta)} \\
& + 4\,\omega^2(r,\theta)\Big(\frac{\partial^2}{\partial \theta^2}\lambda(r,\theta)\Big)e^{2\,\lambda(r,\theta)+2\,\nu(r,\theta)+2\,\psi(r,\theta)} + 4\,\omega^2(r,\theta) \\
& \Big(\frac{\partial}{\partial \theta}\lambda(r,\theta)\Big)^2 e^{2\,\lambda(r,\theta)+2\,\nu(r,\theta)+2\,\psi(r,\theta)} + 4\,\omega^2(r,\theta)\Big(\frac{\partial}{\partial r}\mu(r,\theta)\Big)^2 \\
& e^{2\,\mu(r,\theta)+2\,\psi(r,\theta)+2\,\nu(r,\theta)} - 4\,\omega(r,\theta)\Big(\frac{\partial^2}{\partial \theta^2}\omega(r,\theta)\Big)e^{2\,\lambda(r,\theta)+2\,\nu(r,\theta)+2\,\psi(r,\theta)} \\
& - 3\,\omega^2(r,\theta)\Big(\frac{\partial}{\partial \theta}\omega(r,\theta)\Big)^2 e^{2\,\lambda(r,\theta)+4\,\psi(r,\theta)} - 4\Big(\frac{\partial}{\partial r}\lambda(r,\theta)\Big)\omega^2(r,\theta) \\
& \Big(\frac{\partial}{\partial r}\nu(r,\theta)\Big)e^{2\,\mu(r,\theta)+2\,\psi(r,\theta)+2\,\nu(r,\theta)} + 4\Big(\frac{\partial}{\partial \theta}\nu(r,\theta)\Big)\omega(r,\theta)\Big(\frac{\partial}{\partial \theta}\omega(r,\theta)\Big) \\
& e^{2\,\lambda(r,\theta)+2\,\nu(r,\theta)+2\,\psi(r,\theta)} - 4\Big(\frac{\partial}{\partial r}\mu(r,\theta)\Big)\omega(r,\theta)\Big(\frac{\partial}{\partial r}\omega(r,\theta)\Big) \\
& e^{2\,\mu(r,\theta)+2\,\psi(r,\theta)+2\,\nu(r,\theta)} - 4\Big(\frac{\partial}{\partial \theta}\lambda(r,\theta)\Big)\omega^2(r,\theta)\Big(\frac{\partial}{\partial \theta}\mu(r,\theta)\Big) \\
& e^{2\,\lambda(r,\theta)+2\,\nu(r,\theta)+2\,\psi(r,\theta)} + 4\Big(\frac{\partial}{\partial r}\lambda(r,\theta)\Big)\Big(\frac{\partial}{\partial r}\psi(r,\theta)\Big)e^{2\,\mu(r,\theta)+4\,\nu(r,\theta)} \\
& - 4\Big(\frac{\partial}{\partial \theta}\lambda(r,\theta)\Big)\Big(\frac{\partial}{\partial \theta}\psi(r,\theta)\Big)e^{2\,\lambda(r,\theta)+4\,\nu(r,\theta)} \\
& - 4\Big(\frac{\partial}{\partial r}\lambda(r,\theta)\Big)\omega^2(r,\theta)\Big(\frac{\partial}{\partial r}\mu(r,\theta)\Big)e^{2\,\mu(r,\theta)+2\,\psi(r,\theta)+2\,\nu(r,\theta)}
\end{aligned}
$$

$$- 12\Big(\frac{\partial}{\partial\theta}\psi(r,\theta)\Big)\omega(r,\theta)\Big(\frac{\partial}{\partial\theta}\omega(r,\theta)\Big)e^{2\,\lambda(r,\theta)+2\,\nu(r,\theta)+2\,\psi(r,\theta)}$$
$$+ 4\Big(\frac{\partial}{\partial\theta}\lambda(r,\theta)\Big)\omega(r,\theta)\Big(\frac{\partial}{\partial\theta}\nu(r,\theta)\Big)e^{2\,\lambda(r,\theta)+2\,\nu(r,\theta)+2\,\psi(r,\theta)}$$
$$+ 4\Big(\frac{\partial}{\partial\theta}\mu(r,\theta)\Big)\omega(r,\theta)\Big(\frac{\partial}{\partial\theta}\omega(r,\theta)\Big)e^{2\,\lambda(r,\theta)+2\,\nu(r,\theta)+2\,\psi(r,\theta)}$$
$$+ 4\Big(\frac{\partial}{\partial r}\lambda(r,\theta)\Big)\Big(\frac{\partial}{\partial r}\mu(r,\theta)\Big)e^{2\,\mu(r,\theta)+4\,\nu(r,\theta)} + 4\Big(\frac{\partial}{\partial\theta}\lambda(r,\theta)\Big)$$
$$\Big(\frac{\partial}{\partial\theta}\mu(r,\theta)\Big)e^{2\,\lambda(r,\theta)+4\,\nu(r,\theta)} + 4\Big(\frac{\partial}{\partial\theta}\mu(r,\theta)\Big)\Big(\frac{\partial}{\partial\theta}\psi(r,\theta)\Big)$$
$$e^{2\,\lambda(r,\theta)+4\,\nu(r,\theta)} - 4\Big(\frac{\partial}{\partial r}\mu(r,\theta)\Big)\Big(\frac{\partial}{\partial r}\psi(r,\theta)\Big)e^{2\,\mu(r,\theta)+4\,\nu(r,\theta)}$$
$$- 4\Big(\frac{\partial}{\partial r}\mu(r,\theta)\Big)^2 e^{2\,\mu(r,\theta)+4\,\nu(r,\theta)} - 4\Big(\frac{\partial^2}{\partial\theta^2}\psi(r,\theta)\Big)e^{2\,\lambda(r,\theta)+4\,\nu(r,\theta)}$$
$$- 4\Big(\frac{\partial}{\partial\theta}\psi(r,\theta)\Big)^2 e^{2\,\lambda(r,\theta)+4\,\nu(r,\theta)} + 4\,\omega^2(r,\theta)\Big(\frac{\partial^2}{\partial r^2}\nu(r,\theta)\Big)$$
$$e^{2\,\mu(r,\theta)+2\,\psi(r,\theta)+2\,\nu(r,\theta)} + 4\,\omega^2(r,\theta)\Big)\Big(\frac{\partial^2}{\partial\theta^2}\nu(r,\theta)\Big)$$
$$e^{2\,\lambda(r,\theta)+2\,\nu(r,\theta)+2\,\psi(r,\theta)} + 4\omega^2(r,\theta)\Big(\frac{\partial}{\partial r}\nu(r,\theta)\Big)^2 e^{2\,\mu(r,\theta)+2\,\psi(r,\theta)+2\,\nu(r,\theta)}$$
$$- 4\Big(\frac{\partial}{\partial\theta}\mu(r,\theta)\Big)\omega^2(r,\theta)\Big(\frac{\partial}{\partial\theta}\nu(r,\theta)\Big)e^{2\,\lambda(r,\theta)+2\,\nu(r,\theta)+2\,\psi(r,\theta)} + 4\,\omega^2(r,\theta)$$
$$\Big(\frac{\partial}{\partial\theta}\nu(r,\theta)\Big)^2 e^{2\,\lambda(r,\theta)+2\,\nu(r,\theta)+2\,\psi(r,\theta)} + 4\Big(\frac{\partial}{\partial r}\mu(r,\theta)\Big)\omega^2(r,\theta)\Big(\frac{\partial}{\partial r}\nu(r,\theta)\Big)$$
$$e^{2\,\mu(r,\theta)+2\,\psi(r,\theta)+2\,\nu(r,\theta)} - 4\Big(\frac{\partial^2}{\partial\theta^2}\lambda(r,\theta)\Big)e^{2\,\lambda(r,\theta)+4\,\nu(r,\theta)} - 4\Big(\frac{\partial}{\partial\theta}\lambda(r,\theta)\Big)^2$$
$$e^{2\,\lambda(r,\theta)+4\,\nu(r,\theta)}\Big\}\,e^{-2\,\nu(r,\theta)-2\,\lambda(r,\theta)-2\,\mu(r,\theta)}\,, \tag{G.51}$$

$$G_{t\phi} = -\frac{1}{4}\Big\{4\,\omega(r,\theta)\Big(\frac{\partial}{\partial\theta}\lambda(r,\theta)\Big)^2 e^{2\,\lambda(r,\theta)+2\,\nu(r,\theta)} + 4\,\omega(r,\theta)\Big(\frac{\partial^2}{\partial r^2}\mu(r,\theta)\Big)$$
$$e^{2\,\mu(r,\theta)+2\,\nu(r,\theta)} + 4\,\omega(r,\theta)\Big(\frac{\partial}{\partial r}\mu(r,\theta)\Big)^2 e^{2\,\mu(r,\theta)+2\,\nu(r,\theta)}$$
$$+ 4\,\omega(r,\theta)\Big(\frac{\partial}{\partial r}\nu(r,\theta)\Big)^2 e^{2\,\mu(r,\theta)+2\,\nu(r,\theta)} + 4\,\omega(r,\theta)\Big(\frac{\partial^2}{\partial r^2}\nu(r,\theta)\Big)$$
$$e^{2\,\mu(r,\theta)+2\,\nu(r,\theta)} + 4\,\omega(r,\theta)\Big(\frac{\partial}{\partial\theta}\lambda(r,\theta)\Big)\Big(\frac{\partial}{\partial\theta}\nu(r,\theta)\Big)$$
$$e^{2\,\lambda(r,\theta)+2\,\nu(r,\theta)} - 4\,\omega(r,\theta)\Big(\frac{\partial}{\partial r}\lambda(r,\theta)\Big)\Big(\frac{\partial}{\partial r}\nu(r,\theta)\Big)e^{2\,\mu(r,\theta)+2\,\nu(r,\theta)}$$
$$- 3\,\omega(r,\theta)\Big(\frac{\partial}{\partial r}\omega(r,\theta)\Big)^2 e^{2\,\mu(r,\theta)+2\,\psi(r,\theta)} - 6\Big(\frac{\partial}{\partial r}\psi(r,\theta)\Big)\Big(\frac{\partial}{\partial r}\omega(r,\theta)\Big)$$

$$e^{2\mu(r,\theta)+2\nu(r,\theta)} + 2\left(\frac{\partial}{\partial r}\nu(r,\theta)\right)\left(\frac{\partial}{\partial r}\omega(r,\theta)\right)e^{2\mu(r,\theta)+2\nu(r,\theta)} - 2\left(\frac{\partial^2}{\partial r^2}\omega(r,\theta)\right)$$
$$e^{2\mu(r,\theta)+2\nu(r,\theta)} + 2\left(\frac{\partial}{\partial r}\lambda(r,\theta)\right)\left(\frac{\partial}{\partial r}\omega(r,\theta)\right)e^{2\mu(r,\theta)+2\nu(r,\theta)} - 2\left(\frac{\partial}{\partial\theta}\lambda(r,\theta)\right)$$
$$\left(\frac{\partial}{\partial\theta}\omega(r,\theta)\right)e^{2\lambda(r,\theta)+2\nu(r,\theta)} - 6\left(\frac{\partial}{\partial\theta}\psi(r,\theta)\right)\left(\frac{\partial}{\partial\theta}\omega(r,\theta)\right)e^{2\lambda(r,\theta)+2\nu(r,\theta)}$$
$$- 3\,\omega(r,\theta)\left(\frac{\partial}{\partial\theta}\omega(r,\theta)\right)^2 e^{2\lambda(r,\theta)+2\psi(r,\theta)} + 2\left(\frac{\partial}{\partial\theta}\nu(r,\theta)\right)\left(\frac{\partial}{\partial\theta}\omega(r,\theta)\right)$$
$$e^{2\lambda(r,\theta)+2\nu(r,\theta)} + 2\left(\frac{\partial}{\partial\theta}\mu(r,\theta)\right)\left(\frac{\partial}{\partial\theta}\omega(r,\theta)\right)e^{2\lambda(r,\theta)+2\nu(r,\theta)} - 2\left(\frac{\partial}{\partial r}\mu(r,\theta)\right)$$
$$\left(\frac{\partial}{\partial r}\omega(r,\theta)\right)e^{2\mu(r,\theta)+2\nu(r,\theta)} - 2\left(\frac{\partial^2}{\partial\theta^2}\omega(r,\theta)\right)e^{2\lambda(r,\theta)+2\nu(r,\theta)} - 4\,\omega(r,\theta)$$
$$\left(\frac{\partial}{\partial\theta}\mu(r,\theta)\right)\left(\frac{\partial}{\partial\theta}\nu(r,\theta)\right)e^{2\lambda(r,\theta)+2\nu(r,\theta)} + 4\,\omega(r,\theta)\left(\frac{\partial}{\partial r}\mu(r,\theta)\right)\left(\frac{\partial}{\partial r}\nu(r,\theta)\right)$$
$$e^{2\mu(r,\theta)+2\nu(r,\theta)} + 4\,\omega(r,\theta)\left(\frac{\partial^2}{\partial\theta^2}\nu(r,\theta)\right)e^{2\lambda(r,\theta)+2\nu(r,\theta)} + 4\,\omega(r,\theta)$$
$$\left(\frac{\partial}{\partial\theta}\nu(r,\theta)\right)^2 e^{2\lambda(r,\theta)+2\nu(r,\theta)} - 4\,\omega(r,\theta)\left(\frac{\partial}{\partial r}\lambda(r,\theta)\right)\left(\frac{\partial}{\partial r}\mu(r,\theta)\right)$$
$$e^{2\mu(r,\theta)+2\nu(r,\theta)} - 4\,\omega(r,\theta)\left(\frac{\partial}{\partial\theta}\lambda(r,\theta)\right)\left(\frac{\partial}{\partial\theta}\mu(r,\theta)\right)$$
$$e^{2\lambda(r,\theta)+2\nu(r,\theta)} + 4\,\omega(r,\theta)\left(\frac{\partial^2}{\partial\theta^2}\lambda(r,\theta)\right)e^{2\lambda(r,\theta)+2\nu(r,\theta)}\Big\}$$
$$e^{-2\nu(r,\theta)-2\lambda(r,\theta)+2\psi(r,\theta)-2\mu(r,\theta)}, \tag{G.52}$$

$$G_{rr} = \frac{1}{4}\Big\{\left(\frac{\partial}{\partial r}\omega(r,\theta)\right)^2 e^{2\mu(r,\theta)+2\psi(r,\theta)} + 4\left(\frac{\partial}{\partial r}\psi(r,\theta)\right)\left(\frac{\partial}{\partial r}\nu(r,\theta)\right)$$
$$e^{2\mu(r,\theta)+2\nu(r,\theta)} + 4\left(\frac{\partial}{\partial\theta}\psi(r,\theta)\right)\left(\frac{\partial}{\partial\theta}\nu(r,\theta)\right)e^{2\lambda(r,\theta)+2\nu(r,\theta)}$$
$$+ 4\left(\frac{\partial}{\partial r}\mu(r,\theta)\right)\left(\frac{\partial}{\partial r}\nu(r,\theta)\right)e^{2\mu(r,\theta)+2\nu(r,\theta)} - 4\left(\frac{\partial}{\partial\theta}\mu(r,\theta)\right)\left(\frac{\partial}{\partial\theta}\nu(r,\theta)\right)$$
$$e^{2\lambda(r,\theta)+2\nu(r,\theta)} + 4\left(\frac{\partial}{\partial\theta}\nu(r,\theta)\right)^2 e^{2\lambda(r,\theta)+2\nu(r,\theta)} + 4\left(\frac{\partial^2}{\partial\theta^2}\nu(r,\theta)\right)$$
$$e^{2\lambda(r,\theta)+2\nu(r,\theta)} - \left(\frac{\partial}{\partial\theta}\omega(r,\theta)\right)^2 e^{2\lambda(r,\theta)+2\psi(r,\theta)} + 4\left(\frac{\partial^2}{\partial\theta^2}\psi(r,\theta)\right)$$
$$e^{2\lambda(r,\theta)+2\nu(r,\theta)} + 4\left(\frac{\partial}{\partial\theta}\psi(r,\theta)\right)^2 e^{2\lambda(r,\theta)+2\nu(r,\theta)}$$
$$+ 4\left(\frac{\partial}{\partial r}\psi(r,\theta)\right)\left(\frac{\partial}{\partial r}\mu(r,\theta)\right)e^{2\mu(r,\theta)+2\nu(r,\theta)} - 4\left(\frac{\partial}{\partial\theta}\mu(r,\theta)\right)\left(\frac{\partial}{\partial\theta}\psi(r,\theta)\right)$$
$$e^{2\lambda(r,\theta)+2\nu(r,\theta)}\Big\}\, e^{-2\mu(r,\theta)-2\nu(r,\theta)}, \tag{G.53}$$

$$G_{r\theta} = \frac{1}{2}\Big\{ e^{2\,\psi(r,\theta)} \Big(\frac{\partial}{\partial\theta}\omega(r,\theta)\Big) \frac{\partial}{\partial r}\omega(r,\theta) - 2\Big(\frac{\partial}{\partial r}\nu(r,\theta)\Big)\Big(\frac{\partial}{\partial\theta}\nu(r,\theta)\Big) e^{2\,\nu(r,\theta)}$$
$$+ 2\Big(\frac{\partial}{\partial r}\mu(r,\theta)\Big) e^{2\,\nu(r,\theta)} \frac{\partial}{\partial\theta}\nu(r,\theta) + 2\Big(\frac{\partial}{\partial\theta}\lambda(r,\theta)\Big) e^{2\,\nu(r,\theta)} \frac{\partial}{\partial r}\nu(r,\theta)$$
$$- 2\Big(\frac{\partial^2}{\partial\theta\partial r}\nu(r,\theta)\Big) e^{2\,\nu(r,\theta)} - 2\Big(\frac{\partial^2}{\partial\theta\partial r}\psi(r,\theta)\Big) e^{2\,\nu(r,\theta)} - 2\Big(\frac{\partial}{\partial r}\psi(r,\theta)\Big)$$
$$\Big(\frac{\partial}{\partial\theta}\psi(r,\theta)\Big) e^{2\,\nu(r,\theta)} + 2\Big(\frac{\partial}{\partial r}\psi(r,\theta)\Big)\Big(\frac{\partial}{\partial\theta}\lambda(r,\theta)\Big) e^{2\,\nu(r,\theta)} + 2\Big(\frac{\partial}{\partial r}\mu(r,\theta)\Big)$$
$$\Big(\frac{\partial}{\partial\theta}\psi(r,\theta)\Big) e^{2\,\nu(r,\theta)} \Big\} e^{-2\,\nu(r,\theta)}, \tag{G.54}$$

$$G_{\theta\theta} = \frac{1}{4}\Big\{ -\Big(\frac{\partial}{\partial r}\omega(r,\theta)\Big)^2 e^{2\,\mu(r,\theta)+2\,\psi(r,\theta)} + 4\Big(\frac{\partial^2}{\partial r^2}\psi(r,\theta)\Big) e^{2\,\mu(r,\theta)+2\,\nu(r,\theta)}$$
$$+ 4\Big(\frac{\partial}{\partial r}\psi(r,\theta)\Big)\Big(\frac{\partial}{\partial r}\nu(r,\theta)\Big) e^{2\,\mu(r,\theta)+2\,\nu(r,\theta)} + 4\Big(\frac{\partial}{\partial\theta}\psi(r,\theta)\Big)\Big(\frac{\partial}{\partial\theta}\nu(r,\theta)\Big)$$
$$e^{2\,\lambda(r,\theta)+2\,\nu(r,\theta)} + 4\Big(\frac{\partial}{\partial\theta}\lambda(r,\theta)\Big)\Big(\frac{\partial}{\partial\theta}\nu(r,\theta)\Big) e^{2\,\lambda(r,\theta)+2\,\nu(r,\theta)} + 4\Big(\frac{\partial}{\partial r}\nu(r,\theta)\Big)^2$$
$$e^{2\,\mu(r,\theta)+2\,\nu(r,\theta)} + 4\Big(\frac{\partial^2}{\partial r^2}\nu(r,\theta)\Big) e^{2\,\mu(r,\theta)+2\,\nu(r,\theta)} + 4\Big(\frac{\partial}{\partial r}\psi(r,\theta)\Big)^2$$
$$e^{2\,\mu(r,\theta)+2\,\nu(r,\theta)} + \Big(\frac{\partial}{\partial\theta}\omega(r,\theta)\Big)^2 e^{2\,\lambda(r,\theta)+2\,\psi(r,\theta)} - 4\Big(\frac{\partial}{\partial r}\lambda(r,\theta)\Big)$$
$$\Big(\frac{\partial}{\partial r}\psi(r,\theta)\Big) e^{2\,\mu(r,\theta)+2\,\nu(r,\theta)} + 4\Big(\frac{\partial}{\partial\theta}\lambda(r,\theta)\Big)\Big(\frac{\partial}{\partial\theta}\psi(r,\theta)\Big) e^{2\,\lambda(r,\theta)+2\,\nu(r,\theta)}$$
$$- 4\Big(\frac{\partial}{\partial r}\lambda(r,\theta)\Big)\Big(\frac{\partial}{\partial r}\nu(r,\theta)\Big) e^{2\,\mu(r,\theta)+2\,\nu(r,\theta)} \Big\} e^{-2\,\lambda(r,\theta)-2\,\nu(r,\theta)}, \tag{G.55}$$

$$G_{\phi\phi} = \frac{1}{4}\Big\{ -3\Big(\frac{\partial}{\partial r}\omega(r,\theta)\Big)^2 e^{2\,\mu(r,\theta)+2\,\psi(r,\theta)} + 4\Big(\frac{\partial}{\partial\theta}\lambda(r,\theta)\Big)\Big(\frac{\partial}{\partial\theta}\nu(r,\theta)\Big)$$
$$e^{2\,\lambda(r,\theta)+2\,\nu(r,\theta)} + 4\Big(\frac{\partial}{\partial r}\mu(r,\theta)\Big)\Big(\frac{\partial}{\partial r}\nu(r,\theta)\Big) e^{2\,\mu(r,\theta)+2\,\nu(r,\theta)}$$
$$- 4\Big(\frac{\partial}{\partial\theta}\mu(r,\theta)\Big)\Big(\frac{\partial}{\partial\theta}\nu(r,\theta)\Big) e^{2\,\lambda(r,\theta)+2\,\nu(r,\theta)}$$
$$+ 4\Big(\frac{\partial}{\partial\theta}\nu(r,\theta)\Big)^2 e^{2\,\lambda(r,\theta)+2\,\nu(r,\theta)} + 4\Big(\frac{\partial}{\partial r}\nu(r,\theta)\Big)^2 e^{2\,\mu(r,\theta)+2\,\nu(r,\theta)}$$
$$+ 4\Big(\frac{\partial^2}{\partial r^2}\nu(r,\theta)\Big) e^{2\,\mu(r,\theta)+2\,\nu(r,\theta)} + 4\Big(\frac{\partial^2}{\partial\theta^2}\nu(r,\theta)\Big) e^{2\,\lambda(r,\theta)+2\,\nu(r,\theta)}$$
$$- 3\Big(\frac{\partial}{\partial\theta}\omega(r,\theta)\Big)^2 e^{2\,\lambda(r,\theta)+2\,\psi(r,\theta)} + 4\Big(\frac{\partial^2}{\partial r^2}\mu(r,\theta)\Big) e^{2\,\mu(r,\theta)+2\,\nu(r,\theta)}$$
$$+ 4\Big(\frac{\partial}{\partial r}\mu(r,\theta)\Big)^2 e^{2\,\mu(r,\theta)+2\,\nu(r,\theta)} + 4\Big(\frac{\partial^2}{\partial\theta^2}\lambda(r,\theta)\Big) e^{2\,\lambda(r,\theta)+2\,\nu(r,\theta)}$$
$$+ 4\Big(\frac{\partial}{\partial\theta}\lambda(r,\theta)\Big)^2 e^{2\,\lambda(r,\theta)+2\,\nu(r,\theta)} - 4\Big(\frac{\partial}{\partial r}\lambda(r,\theta)\Big)\Big(\frac{\partial}{\partial r}\mu(r,\theta)\Big) e^{2\,\mu(r,\theta)+2\,\nu(r,\theta)}$$

$$- 4\Big(\frac{\partial}{\partial\theta}\lambda(r,\theta)\Big)\Big(\frac{\partial}{\partial\theta}\mu(r,\theta)\Big)e^{2\,\lambda(r,\theta)+2\,\nu(r,\theta)} - 4\Big(\frac{\partial}{\partial r}\lambda(r,\theta)\Big)\Big(\frac{\partial}{\partial r}\nu(r,\theta)\Big)$$

$$e^{2\,\mu(r,\theta)+2\,\nu(r,\theta)}\Big\}\,e^{-2\,\nu(r,\theta)-2\,\lambda(r,\theta)+2\,\psi(r,\theta)-2\,\mu(r,\theta)}\,. \tag{G.56}$$

The mixed components of the Einstein tensor are found from $G^{\mu}{}_{\nu} = g^{\mu\kappa}\, G_{\kappa\nu}$,

$$G^{t}{}_{t} = \frac{1}{4}\Big\{4\Big(\frac{\partial}{\partial\theta}\lambda(r,\theta)\Big)^2 e^{2\,\lambda(r,\theta)+2\,\nu(r,\theta)} + 4\Big(\frac{\partial}{\partial\theta}\lambda(r,\theta)\Big)\Big(\frac{\partial}{\partial\theta}\psi(r,\theta)\Big)$$

$$e^{2\,\lambda(r,\theta)+2\,\nu(r,\theta)} - 4\Big(\frac{\partial}{\partial\theta}\mu(r,\theta)\Big)\Big(\frac{\partial}{\partial\theta}\psi(r,\theta)\Big)e^{2\,\lambda(r,\theta)+2\,\nu(r,\theta)}$$

$$- 2\Big(\frac{\partial}{\partial\theta}\mu(r,\theta)\Big)\omega(r,\theta)\Big(\frac{\partial}{\partial\theta}\omega(r,\theta)\Big)e^{2\,\lambda(r,\theta)+2\,\psi(r,\theta)}$$

$$+ 6\Big(\frac{\partial}{\partial\theta}\psi(r,\theta)\Big)\omega(r,\theta)\Big(\frac{\partial}{\partial\theta}\omega(r,\theta)\Big)e^{2\,\lambda(r,\theta)+2\,\psi(r,\theta)} - 2\Big(\frac{\partial}{\partial\theta}\nu(r,\theta)\Big)$$

$$\omega(r,\theta)\Big(\frac{\partial}{\partial\theta}\omega(r,\theta)\Big)e^{2\,\lambda(r,\theta)+2\,\psi(r,\theta)} - 4\Big(\frac{\partial}{\partial r}\lambda(r,\theta)\Big)\Big(\frac{\partial}{\partial r}\psi(r,\theta)\Big)$$

$$e^{2\,\mu(r,\theta)+2\,\nu(r,\theta)} - 4\Big(\frac{\partial}{\partial\theta}\lambda(r,\theta)\Big)\Big(\frac{\partial}{\partial\theta}\mu(r,\theta)\Big)e^{2\,\lambda(r,\theta)+2\,\nu(r,\theta)}$$

$$+ 4\Big(\frac{\partial}{\partial r}\psi(r,\theta)\Big)\Big(\frac{\partial}{\partial r}\mu(r,\theta)\Big)e^{2\,\mu(r,\theta)+2\,\nu(r,\theta)} + 2\Big(\frac{\partial}{\partial r}\mu(r,\theta)\Big)$$

$$\omega(r,\theta)\Big(\frac{\partial}{\partial r}\omega(r,\theta)\Big)e^{2\,\mu(r,\theta)+2\,\psi(r,\theta)} - 4\Big(\frac{\partial}{\partial r}\lambda(r,\theta)\Big)\Big(\frac{\partial}{\partial r}\mu(r,\theta)\Big)$$

$$e^{2\,\mu(r,\theta)+2\,\nu(r,\theta)} + 4\Big(\frac{\partial}{\partial r}\psi(r,\theta)\Big)^2 e^{2\,\mu(r,\theta)+2\,\nu(r,\theta)} + 4\Big(\frac{\partial}{\partial r}\mu(r,\theta)\Big)^2$$

$$e^{2\,\mu(r,\theta)+2\,\nu(r,\theta)} + 4\Big(\frac{\partial}{\partial\theta}\psi(r,\theta)\Big)^2 e^{2\,\lambda(r,\theta)+2\,\nu(r,\theta)} + 4\Big(\frac{\partial^2}{\partial r^2}\psi(r,\theta)\Big)$$

$$e^{2\,\mu(r,\theta)+2\,\nu(r,\theta)} + 4\Big(\frac{\partial^2}{\partial\theta^2}\psi(r,\theta)\Big)e^{2\,\lambda(r,\theta)+2\,\nu(r,\theta)} + 4\Big(\frac{\partial^2}{\partial\theta^2}\lambda(r,\theta)\Big)$$

$$e^{2\,\lambda(r,\theta)+2\,\nu(r,\theta)} + 4\Big(\frac{\partial^2}{\partial r^2}\mu(r,\theta)\Big)e^{2\,\mu(r,\theta)+2\,\nu(r,\theta)} + \Big(\frac{\partial}{\partial\theta}\omega(r,\theta)\Big)^2$$

$$e^{2\,\lambda(r,\theta)+2\,\psi(r,\theta)} + \Big(\frac{\partial}{\partial r}\omega(r,\theta)\Big)^2 e^{2\,\mu(r,\theta)+2\,\psi(r,\theta)} + 2\,\omega(r,\theta)\Big(\frac{\partial^2}{\partial\theta^2}\omega(r,\theta)\Big)$$

$$e^{2\,\lambda(r,\theta)+2\,\psi(r,\theta)} - 2\Big(\frac{\partial}{\partial r}\nu(r,\theta)\Big)\omega(r,\theta)\Big(\frac{\partial}{\partial r}\omega(r,\theta)\Big)e^{2\,\mu(r,\theta)+2\,\psi(r,\theta)}$$

$$+ 2\,\omega(r,\theta)\Big(\frac{\partial^2}{\partial r^2}\omega(r,\theta)\Big)e^{2\,\mu(r,\theta)+2\,\psi(r,\theta)} + 2\Big(\frac{\partial}{\partial\theta}\lambda(r,\theta)\Big)\omega(r,\theta)$$

$$\Big(\frac{\partial}{\partial\theta}\omega(r,\theta)\Big)e^{2\,\lambda(r,\theta)+2\,\psi(r,\theta)} + 6\Big(\frac{\partial}{\partial r}\psi(r,\theta)\Big)\omega(r,\theta)\Big(\frac{\partial}{\partial r}\omega(r,\theta)\Big)$$

$$e^{2\,\mu(r,\theta)+2\,\psi(r,\theta)} - 2\Big(\frac{\partial}{\partial r}\lambda(r,\theta)\Big)\omega(r,\theta)\Big(\frac{\partial}{\partial r}\omega(r,\theta)\Big)e^{2\,\mu(r,\theta)+2\,\psi(r,\theta)}\Big\}$$

$$e^{-2\,\nu(r,\theta)-2\,\lambda(r,\theta)-2\,\mu(r,\theta)}\,, \tag{G.57}$$

$$G^t{}_\phi = -\frac{1}{2}\Big\{\Big(\frac{\partial^2}{\partial r^2}\omega(r,\theta)\Big)e^{2\,\mu(r,\theta)} + \Big(\frac{\partial^2}{\partial \theta^2}\omega(r,\theta)\Big)e^{2\,\lambda(r,\theta)} + \Big(\frac{\partial}{\partial \theta}\lambda(r,\theta)\Big)$$
$$e^{2\,\lambda(r,\theta)}\frac{\partial}{\partial \theta}\omega(r,\theta) + \Big(\frac{\partial}{\partial r}\mu(r,\theta)\Big)e^{2\,\mu(r,\theta)}\frac{\partial}{\partial r}\omega(r,\theta) + 3\Big(\frac{\partial}{\partial r}\psi(r,\theta)\Big)$$
$$\Big(\frac{\partial}{\partial r}\omega(r,\theta)\Big)e^{2\,\mu(r,\theta)} - e^{2\,\mu(r,\theta)}\Big(\frac{\partial}{\partial r}\nu(r,\theta)\Big)\frac{\partial}{\partial r}\omega(r,\theta) - \Big(\frac{\partial}{\partial r}\lambda(r,\theta)\Big)$$
$$e^{2\,\mu(r,\theta)}\frac{\partial}{\partial r}\omega(r,\theta) - \Big(\frac{\partial}{\partial \theta}\mu(r,\theta)\Big)e^{2\,\lambda(r,\theta)}\frac{\partial}{\partial \theta}\omega(r,\theta) - e^{2\,\lambda(r,\theta)}\Big(\frac{\partial}{\partial \theta}\nu(r,\theta)\Big)$$
$$\frac{\partial}{\partial \theta}\omega(r,\theta) + 3\Big(\frac{\partial}{\partial \theta}\psi(r,\theta)\Big)\Big(\frac{\partial}{\partial \theta}\omega(r,\theta)\Big)e^{2\,\lambda(r,\theta)}\Big\}$$
$$e^{-2\,\nu(r,\theta)-2\,\lambda(r,\theta)+2\,\psi(r,\theta)-2\,\mu(r,\theta)} \tag{G.58}$$

$$G^r{}_r = \frac{1}{4}\Big\{\Big(\frac{\partial}{\partial r}\omega(r,\theta)\Big)^2 e^{2\,\mu(r,\theta)+2\,\psi(r,\theta)} + 4\Big(\frac{\partial}{\partial r}\psi(r,\theta)\Big)\Big(\frac{\partial}{\partial r}\nu(r,\theta)\Big)$$
$$e^{2\,\mu(r,\theta)+2\,\nu(r,\theta)} + 4\Big(\frac{\partial}{\partial \theta}\psi(r,\theta)\Big)\Big(\frac{\partial}{\partial \theta}\nu(r,\theta)\Big)e^{2\,\lambda(r,\theta)+2\,\nu(r,\theta)}$$
$$+ 4\Big(\frac{\partial}{\partial r}\mu(r,\theta)\Big)\Big(\frac{\partial}{\partial r}\nu(r,\theta)\Big)e^{2\,\mu(r,\theta)+2\,\nu(r,\theta)} - 4\Big(\frac{\partial}{\partial \theta}\mu(r,\theta)\Big)$$
$$\Big(\frac{\partial}{\partial \theta}\nu(r,\theta)\Big)e^{2\,\lambda(r,\theta)+2\,\nu(r,\theta)} + 4\Big(\frac{\partial}{\partial \theta}\nu(r,\theta)\Big)^2 e^{2\,\lambda(r,\theta)+2\,\nu(r,\theta)}$$
$$+ 4\Big(\frac{\partial^2}{\partial \theta^2}\nu(r,\theta)\Big)e^{2\,\lambda(r,\theta)+2\,\nu(r,\theta)} - \Big(\frac{\partial}{\partial \theta}\omega(r,\theta)\Big)^2 e^{2\,\lambda(r,\theta)+2\,\psi(r,\theta)}$$
$$+ 4\Big(\frac{\partial^2}{\partial \theta^2}\psi(r,\theta)\Big)e^{2\,\lambda(r,\theta)+2\,\nu(r,\theta)} + 4\Big(\frac{\partial}{\partial \theta}\psi(r,\theta)\Big)^2 e^{2\,\lambda(r,\theta)+2\,\nu(r,\theta)}$$
$$+ 4\Big(\frac{\partial}{\partial r}\psi(r,\theta)\Big)\Big(\frac{\partial}{\partial r}\mu(r,\theta)\Big)e^{2\,\mu(r,\theta)+2\,\nu(r,\theta)} - 4\Big(\frac{\partial}{\partial \theta}\mu(r,\theta)\Big)$$
$$\Big(\frac{\partial}{\partial \theta}\psi(r,\theta)\Big)e^{2\,\lambda(r,\theta)+2\,\nu(r,\theta)}\Big\}e^{-2\,\nu(r,\theta)-2\,\lambda(r,\theta)-2\,\mu(r,\theta)}, \tag{G.59}$$

$$G^r{}_\theta = \frac{1}{2}\Big\{e^{2\,\psi(r,\theta)}\Big(\frac{\partial}{\partial \theta}\omega(r,\theta)\Big)\frac{\partial}{\partial r}\omega(r,\theta) - 2\Big(\frac{\partial}{\partial r}\nu(r,\theta)\Big)\Big(\frac{\partial}{\partial \theta}\nu(r,\theta)\Big)e^{2\,\nu(r,\theta)}$$
$$+ 2\Big(\frac{\partial}{\partial r}\mu(r,\theta)\Big)e^{2\,\nu(r,\theta)}\frac{\partial}{\partial \theta}\nu(r,\theta) + 2\,\Big(\frac{\partial}{\partial \theta}\lambda(r,\theta)\Big)e^{2\,\nu(r,\theta)}\frac{\partial}{\partial r}\nu(r,\theta)$$
$$- 2\Big(\frac{\partial^2}{\partial \theta\partial r}\nu(r,\theta)\Big)e^{2\,\nu(r,\theta)} - 2\Big(\frac{\partial^2}{\partial \theta\partial r}\psi(r,\theta)\Big)e^{2\,\nu(r,\theta)} - 2\Big(\frac{\partial}{\partial r}\psi(r,\theta)\Big)$$
$$\Big(\frac{\partial}{\partial \theta}\psi(r,\theta)\Big)e^{2\,\nu(r,\theta)} + 2\Big(\frac{\partial}{\partial r}\psi(r,\theta)\Big)\Big(\frac{\partial}{\partial \theta}\lambda(r,\theta)\Big)e^{2\,\nu(r,\theta)}$$
$$+ 2\Big(\frac{\partial}{\partial r}\mu(r,\theta)\Big)\Big(\frac{\partial}{\partial \theta}\psi(r,\theta)\Big)e^{2\,\nu(r,\theta)}\Big\}\,e^{-2\,\lambda(r,\theta)-2\,\nu(r,\theta)}, \tag{G.60}$$

$$G^\theta{}_r = \frac{1}{2}\Big\{e^{2\,\psi(r,\theta)}\Big(\frac{\partial}{\partial \theta}\omega(r,\theta)\Big)\frac{\partial}{\partial r}\omega(r,\theta) - 2\Big(\frac{\partial}{\partial r}\nu(r,\theta)\Big)\Big(\frac{\partial}{\partial \theta}\nu(r,\theta)\Big)e^{2\,\nu(r,\theta)}$$

$$+ 2\Big(\frac{\partial}{\partial r}\mu(r,\theta)\Big)e^{2\,\nu(r,\theta)}\frac{\partial}{\partial\theta}\nu(r,\theta) + 2\,\Big(\frac{\partial}{\partial\theta}\lambda(r,\theta)\Big)e^{2\,\nu(r,\theta)}\frac{\partial}{\partial r}\nu(r,\theta)$$
$$- 2\Big(\frac{\partial^2}{\partial\theta\partial r}\nu(r,\theta)\Big)e^{2\,\nu(r,\theta)} - 2\Big(\frac{\partial^2}{\partial\theta\partial r}\psi(r,\theta)\Big)e^{2\,\nu(r,\theta)} - 2\Big(\frac{\partial}{\partial r}\psi(r,\theta)\Big)$$
$$\Big(\frac{\partial}{\partial\theta}\psi(r,\theta)\Big)e^{2\,\nu(r,\theta)} + 2\Big(\frac{\partial}{\partial r}\psi(r,\theta)\Big)\Big(\frac{\partial}{\partial\theta}\lambda(r,\theta)\Big)e^{2\,\nu(r,\theta)}$$
$$+ 2\Big(\frac{\partial}{\partial r}\mu(r,\theta)\Big)\Big(\frac{\partial}{\partial\theta}\psi(r,\theta)\Big)e^{2\,\nu(r,\theta)}\Big\}\,e^{-2\,\mu(r,\theta)-2\,\nu(r,\theta)}\,, \tag{G.61}$$

$$G^{\theta}{}_{\theta} = \frac{1}{4}\Big\{-\Big(\frac{\partial}{\partial r}\omega(r,\theta)\Big)^2 e^{2\,\mu(r,\theta)+2\,\psi(r,\theta)} + 4\Big(\frac{\partial^2}{\partial r^2}\psi(r,\theta)\Big)e^{2\,\mu(r,\theta)+2\nu(r,\theta)}$$
$$+ 4\Big(\frac{\partial}{\partial r}\psi(r,\theta)\Big)\Big(\frac{\partial}{\partial r}\nu(r,\theta)\Big)e^{2\,\mu(r,\theta)+2\,\nu(r,\theta)} + 4\Big(\frac{\partial}{\partial\theta}\psi(r,\theta)\Big)$$
$$\Big(\frac{\partial}{\partial\theta}\nu(r,\theta)\Big)e^{2\,\lambda(r,\theta)+2\,\nu(r,\theta)} + 4\Big(\frac{\partial}{\partial\theta}\lambda(r,\theta)\Big)\Big(\frac{\partial}{\partial\theta}\nu(r,\theta)\Big)e^{2\,\lambda(r,\theta)+2\,\nu(r,\theta)}$$
$$+ 4\Big(\frac{\partial}{\partial r}\nu(r,\theta)\Big)^2 e^{2\,\mu(r,\theta)+2\,\nu(r,\theta)} + 4\Big(\frac{\partial^2}{\partial r^2}\nu(r,\theta)\Big)e^{2\,\mu(r,\theta)+2\,\nu(r,\theta)}$$
$$+ 4\Big(\frac{\partial}{\partial r}\psi(r,\theta)\Big)^2 e^{2\,\mu(r,\theta)+2\,\nu(r,\theta)} + \Big(\frac{\partial}{\partial\theta}\omega(r,\theta)\Big)^2 e^{2\,\lambda(r,\theta)+2\,\psi(r,\theta)}$$
$$- 4\Big(\frac{\partial}{\partial r}\lambda(r,\theta)\Big)\Big(\frac{\partial}{\partial r}\psi(r,\theta)\Big)e^{2\,\mu(r,\theta)+2\,\nu(r,\theta)} + 4\Big(\frac{\partial}{\partial\theta}\lambda(r,\theta)\Big)$$
$$\Big(\frac{\partial}{\partial\theta}\psi(r,\theta)\Big)e^{2\,\lambda(r,\theta)+2\,\nu(r,\theta)} - 4\Big(\frac{\partial}{\partial r}\lambda(r,\theta)\Big)\Big(\frac{\partial}{\partial r}\nu(r,\theta)\Big)e^{2\,\mu(r,\theta)+2\,\nu(r,\theta)}\Big\}$$
$$e^{-2\,\nu(r,\theta)-2\,\lambda(r,\theta)-2\,\mu(r,\theta)}\,, \tag{G.62}$$

$$G^{\phi}{}_{t} = -\frac{1}{2}\Big\{-2\Big(\frac{\partial^2}{\partial\theta^2}\psi(r,\theta)\Big)\omega(r,\theta)e^{2\,\lambda(r,\theta)+2\,\nu(r,\theta)} + 2\,\omega(r,\theta)\Big(\frac{\partial}{\partial r}\nu(r,\theta)\Big)^2$$
$$e^{2\,\mu(r,\theta)+2\,\nu(r,\theta)} - 2\Big(\frac{\partial}{\partial r}\mu(r,\theta)\Big)\Big(\frac{\partial}{\partial r}\psi(r,\theta)\Big)\omega(r,\theta)e^{2\,\mu(r,\theta)+2\,\nu(r,\theta)}$$
$$- 2\Big(\frac{\partial^2}{\partial r^2}\psi(r,\theta)\Big)\omega(r,\theta)e^{2\,\mu(r,\theta)+2\,\nu(r,\theta)} + 2\,\omega(r,\theta)\Big(\frac{\partial^2}{\partial r^2}\nu(r,\theta)\Big)$$
$$e^{2\,\mu(r,\theta)+2\,\nu(r,\theta)} - \omega^2(r,\theta)\Big(\frac{\partial^2}{\partial r^2}\omega(r,\theta)\Big)e^{2\,\mu(r,\theta)+2\,\psi(r,\theta)}$$
$$+ 2\Big(\frac{\partial}{\partial\theta}\mu(r,\theta)\Big)\Big(\frac{\partial}{\partial\theta}\psi(r,\theta)\Big)\omega(r,\theta)e^{2\,\lambda(r,\theta)+2\,\nu(r,\theta)} + 2\,\omega(r,\theta)$$
$$\Big(\frac{\partial}{\partial\theta}\lambda(r,\theta)\Big)\Big(\frac{\partial}{\partial\theta}\nu(r,\theta)\Big)e^{2\,\lambda(r,\theta)+2\,\nu(r,\theta)} - 2\,\omega(r,\theta)$$
$$\Big(\frac{\partial}{\partial r}\lambda(r,\theta)\Big)\Big(\frac{\partial}{\partial r}\nu(r,\theta)\Big)e^{2\,\mu(r,\theta)+2\,\nu(r,\theta)} - 2\,\omega(r,\theta)\Big(\frac{\partial}{\partial r}\omega(r,\theta)\Big)^2$$
$$e^{2\,\mu(r,\theta)+2\,\psi(r,\theta)} - 3\Big(\frac{\partial}{\partial r}\psi(r,\theta)\Big)\Big(\frac{\partial}{\partial r}\omega(r,\theta)\Big)e^{2\,\mu(r,\theta)+2\,\nu(r,\theta)}$$

$$+\left(\frac{\partial}{\partial r}\nu(r,\theta)\right)\omega^2(r,\theta)\left(\frac{\partial}{\partial r}\omega(r,\theta)\right)e^{2\,\mu(r,\theta)+2\,\psi(r,\theta)}+\left(\frac{\partial}{\partial r}\nu(r,\theta)\right)$$
$$\left(\frac{\partial}{\partial r}\omega(r,\theta)\right)e^{2\,\mu(r,\theta)+2\,\nu(r,\theta)}-3\left(\frac{\partial}{\partial r}\psi(r,\theta)\right)\omega^2(r,\theta)\left(\frac{\partial}{\partial r}\omega(r,\theta)\right)$$
$$e^{2\,\mu(r,\theta)+2\,\psi(r,\theta)}-\left(\frac{\partial}{\partial\theta}\lambda(r,\theta)\right)\omega^2(r,\theta)\left(\frac{\partial}{\partial\theta}\omega(r,\theta)\right)e^{2\,\lambda(r,\theta)+2\,\psi(r,\theta)}$$
$$+\left(\frac{\partial}{\partial r}\lambda(r,\theta)\right)\omega^2(r,\theta)\left(\frac{\partial}{\partial r}\omega(r,\theta)\right)e^{2\,\mu(r,\theta)+2\,\psi(r,\theta)}-2\left(\frac{\partial}{\partial r}\psi(r,\theta)\right)^2$$
$$\omega(r,\theta)e^{2\,\mu(r,\theta)+2\,\nu(r,\theta)}-\left(\frac{\partial^2}{\partial r^2}\omega(r,\theta)\right)e^{2\,\mu(r,\theta)+2\,\nu(r,\theta)}+\left(\frac{\partial}{\partial r}\lambda(r,\theta)\right)$$
$$\left(\frac{\partial}{\partial r}\omega(r,\theta)\right)e^{2\,\mu(r,\theta)+2\,\nu(r,\theta)}-\left(\frac{\partial}{\partial\theta}\lambda(r,\theta)\right)\left(\frac{\partial}{\partial\theta}\omega(r,\theta)\right)e^{2\,\lambda(r,\theta)+2\,\nu(r,\theta)}$$
$$+2\left(\frac{\partial}{\partial r}\lambda(r,\theta)\right)\left(\frac{\partial}{\partial r}\psi(r,\theta)\right)\omega(r,\theta)e^{2\,\mu(r,\theta)+2\,\nu(r,\theta)}-2\left(\frac{\partial}{\partial\theta}\psi(r,\theta)\right)^2$$
$$\omega(r,\theta)e^{2\,\lambda(r,\theta)+2\,\nu(r,\theta)}-2\left(\frac{\partial}{\partial\theta}\lambda(r,\theta)\right)\left(\frac{\partial}{\partial\theta}\psi(r,\theta)\right)\omega(r,\theta)e^{2\,\lambda(r,\theta)+2\,\nu(r,\theta)}$$
$$-3\left(\frac{\partial}{\partial\theta}\psi(r,\theta)\right)\left(\frac{\partial}{\partial\theta}\omega(r,\theta)\right)e^{2\,\lambda(r,\theta)+2\,\nu(r,\theta)}-2\,\omega(r,\theta)\left(\frac{\partial}{\partial\theta}\omega(r,\theta)\right)^2$$
$$e^{2\,\lambda(r,\theta)+2\,\psi(r,\theta)}+\left(\frac{\partial}{\partial\theta}\nu(r,\theta)\right)\left(\frac{\partial}{\partial\theta}\omega(r,\theta)\right)e^{2\,\lambda(r,\theta)+2\,\nu(r,\theta)}$$
$$+\left(\frac{\partial}{\partial\theta}\mu(r,\theta)\right)\left(\frac{\partial}{\partial\theta}\omega(r,\theta)\right)e^{2\,\lambda(r,\theta)+2\,\nu(r,\theta)}-\left(\frac{\partial}{\partial r}\mu(r,\theta)\right)\left(\frac{\partial}{\partial r}\omega(r,\theta)\right)$$
$$e^{2\,\mu(r,\theta)+2\,\nu(r,\theta)}-\left(\frac{\partial^2}{\partial\theta^2}\omega(r,\theta)\right)e^{2\,\lambda(r,\theta)+2\,\nu(r,\theta)}-3\left(\frac{\partial}{\partial\theta}\psi(r,\theta)\right)\omega^2(r,\theta)$$
$$\left(\frac{\partial}{\partial\theta}\omega(r,\theta)\right)e^{2\,\lambda(r,\theta)+2\,\psi(r,\theta)}-2\,\omega(r,\theta)\left(\frac{\partial}{\partial\theta}\mu(r,\theta)\right)\left(\frac{\partial}{\partial\theta}\nu(r,\theta)\right)$$
$$e^{2\,\lambda(r,\theta)+2\,\nu(r,\theta)}+2\,\omega(r,\theta)\left(\frac{\partial}{\partial r}\mu(r,\theta)\right)\left(\frac{\partial}{\partial r}\nu(r,\theta)\right)e^{2\,\mu(r,\theta)+2\,\nu(r,\theta)}$$
$$+\left(\frac{\partial}{\partial\theta}\mu(r,\theta)\right)\omega^2(r,\theta)\left(\frac{\partial}{\partial\theta}\omega(r,\theta)\right)e^{2\,\lambda(r,\theta)+2\,\psi(r,\theta)}+\left(\frac{\partial}{\partial\theta}\nu(r,\theta)\right)$$
$$\omega^2(r,\theta)\left(\frac{\partial}{\partial\theta}\omega(r,\theta)\right)e^{2\,\lambda(r,\theta)+2\,\psi(r,\theta)}-\left(\frac{\partial}{\partial r}\mu(r,\theta)\right)\omega(r,\theta)\left(\frac{\partial}{\partial r}\omega(r,\theta)\right)$$
$$e^{2\,\mu(r,\theta)+2\,\psi(r,\theta)}+2\,\omega(r,\theta)\left(\frac{\partial^2}{\partial\theta^2}\nu(r,\theta)\right)e^{2\,\lambda(r,\theta)+2\,\nu(r,\theta)}+2\,\omega(r,\theta)$$
$$\left(\frac{\partial}{\partial\theta}\nu(r,\theta)\right)^2e^{2\,\lambda(r,\theta)+2\,\nu(r,\theta)}-\omega^2(r,\theta)\left(\frac{\partial^2}{\partial\theta^2}\omega(r,\theta)\right)e^{2\,\lambda(r,\theta)+2\,\psi(r,\theta)}\Big\}$$
$$e^{-2\,\nu(r,\theta)-2\,\lambda(r,\theta)-2\,\mu(r,\theta)},\tag{G.63}$$

$$G^{\phi}{}_{\phi}=\frac{1}{4}\Big\{4\left(\frac{\partial}{\partial\theta}\lambda(r,\theta)\right)^2e^{2\,\lambda(r,\theta)+2\,\nu(r,\theta)}+4\left(\frac{\partial}{\partial\theta}\lambda(r,\theta)\right)\left(\frac{\partial}{\partial\theta}\nu(r,\theta)\right)$$
$$e^{2\,\lambda(r,\theta)+2\,\nu(r,\theta)}-4\left(\frac{\partial}{\partial\theta}\mu(r,\theta)\right)\left(\frac{\partial}{\partial\theta}\nu(r,\theta)\right)e^{2\,\lambda(r,\theta)+2\,\nu(r,\theta)}$$

$$- 4\Big(\frac{\partial}{\partial r}\lambda(r,\theta)\Big)\Big(\frac{\partial}{\partial r}\nu(r,\theta)\Big)\mathrm{e}^{2\,\mu(r,\theta)+2\,\nu(r,\theta)} + 4\Big(\frac{\partial}{\partial r}\mu(r,\theta)\Big)$$
$$\Big(\frac{\partial}{\partial r}\nu(r,\theta)\Big)\mathrm{e}^{2\,\mu(r,\theta)+2\,\nu(r,\theta)} + 2\Big(\frac{\partial}{\partial \theta}\mu(r,\theta)\Big)\omega(r,\theta)\Big(\frac{\partial}{\partial \theta}\omega(r,\theta)\Big)$$
$$\mathrm{e}^{2\,\lambda(r,\theta)+2\,\psi(r,\theta)} - 6\Big(\frac{\partial}{\partial \theta}\psi(r,\theta)\Big)\omega(r,\theta)\Big(\frac{\partial}{\partial \theta}\omega(r,\theta)\Big)\mathrm{e}^{2\,\lambda(r,\theta)+2\,\psi(r,\theta)}$$
$$+ 2\Big(\frac{\partial}{\partial \theta}\nu(r,\theta)\Big)\omega(r,\theta)\Big(\frac{\partial}{\partial \theta}\omega(r,\theta)\Big)\mathrm{e}^{2\,\lambda(r,\theta)+2\,\psi(r,\theta)} - 4\Big(\frac{\partial}{\partial \theta}\lambda(r,\theta)\Big)$$
$$\Big(\frac{\partial}{\partial \theta}\mu(r,\theta)\Big)\mathrm{e}^{2\,\lambda(r,\theta)+2\,\nu(r,\theta)} - 2\Big(\frac{\partial}{\partial r}\mu(r,\theta)\Big)\omega(r,\theta)\Big(\frac{\partial}{\partial r}\omega(r,\theta)\Big)$$
$$\mathrm{e}^{2\,\mu(r,\theta)+2\,\psi(r,\theta)} - 4\Big(\frac{\partial}{\partial r}\lambda(r,\theta)\Big)\Big(\frac{\partial}{\partial r}\mu(r,\theta)\Big)\mathrm{e}^{2\,\mu(r,\theta)+2\,\nu(r,\theta)}$$
$$+ 4\Big(\frac{\partial}{\partial r}\mu(r,\theta)\Big)^2\mathrm{e}^{2\,\mu(r,\theta)+2\,\nu(r,\theta)} + 4\Big(\frac{\partial^2}{\partial r^2}\nu(r,\theta)\Big)\mathrm{e}^{2\,\mu(r,\theta)+2\,\nu(r,\theta)}$$
$$+ 4\Big(\frac{\partial^2}{\partial \theta^2}\nu(r,\theta)\Big)\mathrm{e}^{2\,\lambda(r,\theta)+2\,\nu(r,\theta)} + 4\Big(\frac{\partial}{\partial \theta}\nu(r,\theta)\Big)^2\mathrm{e}^{2\,\lambda(r,\theta)+2\,\nu(r,\theta)}$$
$$+ 4\Big(\frac{\partial}{\partial r}\nu(r,\theta)\Big)^2\mathrm{e}^{2\,\mu(r,\theta)+2\,\nu(r,\theta)} + 4\Big(\frac{\partial^2}{\partial \theta^2}\lambda(r,\theta)\Big)\mathrm{e}^{2\,\lambda(r,\theta)+2\,\nu(r,\theta)}$$
$$+ 4\Big(\frac{\partial^2}{\partial r^2}\mu(r,\theta)\Big)\mathrm{e}^{2\,\mu(r,\theta)+2\,\nu(r,\theta)} - 3\Big(\frac{\partial}{\partial \theta}\omega(r,\theta)\Big)^2\mathrm{e}^{2\,\lambda(r,\theta)+2\,\psi(r,\theta)}$$
$$- 3\Big(\frac{\partial}{\partial r}\omega(r,\theta)\Big)^2\mathrm{e}^{2\,\mu(r,\theta)+2\,\psi(r,\theta)} - 2\,\omega(r,\theta)\Big(\frac{\partial^2}{\partial \theta^2}\omega(r,\theta)\Big)\mathrm{e}^{2\,\lambda(r,\theta)+2\,\psi(r,\theta)}$$
$$+ 2\Big(\frac{\partial}{\partial r}\nu(r,\theta)\Big)\omega(r,\theta)\Big(\frac{\partial}{\partial r}\omega(r,\theta)\Big)\mathrm{e}^{2\,\mu(r,\theta)+2\,\psi(r,\theta)} - 2\,\omega(r,\theta)$$
$$\Big(\frac{\partial^2}{\partial r^2}\omega(r,\theta)\Big)\mathrm{e}^{2\,\mu(r,\theta)+2\,\psi(r,\theta)} - 2\Big(\frac{\partial}{\partial \theta}\lambda(r,\theta)\Big)\omega(r,\theta)\Big(\frac{\partial}{\partial \theta}\omega(r,\theta)\Big)$$
$$\mathrm{e}^{2\,\lambda(r,\theta)+2\,\psi(r,\theta)} - 6\Big(\frac{\partial}{\partial r}\psi(r,\theta)\Big)\omega(r,\theta)\Big(\frac{\partial}{\partial r}\omega(r,\theta)\Big)\mathrm{e}^{2\,\mu(r,\theta)+2\,\psi(r,\theta)}$$
$$+ 2\Big(\frac{\partial}{\partial r}\lambda(r,\theta)\Big)\omega(r,\theta)\Big(\frac{\partial}{\partial r}\omega(r,\theta)\Big)\mathrm{e}^{2\,\mu(r,\theta)+2\,\psi(r,\theta)}\Big\}$$
$$\mathrm{e}^{-2\,\nu(r,\theta)-2\,\lambda(r,\theta)-2\,\mu(r,\theta)}\,. \tag{G.64}$$

G.5 Energy–momentum tensor

The contravariant components of the energy–momentum tensor of a rotationally deformed body made up of a perfect fluid follow from (13.20),

$$T_{tt} = \big\{-\epsilon(r)\mathrm{e}^{4\,\nu(r,\theta)} + 2\,\epsilon(r)\omega^2(r,\theta)\mathrm{e}^{2\,\nu(r,\theta)+2\,\psi(r,\theta)} - 2\,\epsilon(r)\omega(r,\theta)\Omega$$
$$\mathrm{e}^{2\,\nu(r,\theta)+2\,\psi(r,\theta)} - \epsilon(r)\mathrm{e}^{4\,\psi(r,\theta)}\omega^4(r,\theta) + 2\,\epsilon(r)\mathrm{e}^{4\,\psi(r,\theta)}\omega^3(r,\theta)\Omega$$
$$- \epsilon(r)\mathrm{e}^{4\,\psi(r,\theta)}\omega^2(r,\theta)\Omega^2 + P(r)\omega^2(r,\theta)\mathrm{e}^{2\,\nu(r,\theta)+2\,\psi(r,\theta)} - 2\,P(r)$$
$$\omega(r,\theta)\Omega\,\mathrm{e}^{2\,\nu(r,\theta)+2\,\psi(r,\theta)} - P(r)\mathrm{e}^{4\,\psi(r,\theta)}\omega^4(r,\theta) + 2\,P(r)\mathrm{e}^{4\,\psi(r,\theta)}$$

$$
\omega^3(r,\theta)\Omega - P(r)\mathrm{e}^{4\,\psi(r,\theta)}\omega^2(r,\theta)\Omega^2 - \Omega^2 P(r)\mathrm{e}^{2\,\nu(r,\theta)+2\,\psi(r,\theta)}
+ 2\,P(r)\omega(r,\theta)\Omega\,\mathrm{e}^{2\,\nu(r,\theta)+2\,\psi(r,\theta)} - P(r)\omega^2(r,\theta)\mathrm{e}^{2\,\nu(r,\theta)+2\,\psi(r,\theta)}
+ P(r)\omega^2(r,\theta)\Omega^2\mathrm{e}^{2\,\psi(r,\theta)+2\,\psi(r,\theta)} - 2\,P(r)\omega^3(r,\theta)\Omega\,\mathrm{e}^{2\,\psi(r,\theta)+2\,\psi(r,\theta)}
+ P(r)\omega^4(r,\theta)\mathrm{e}^{2\,\psi(r,\theta)+2\,\psi(r,\theta)}\big\}
\big\{-\mathrm{e}^{2\,\nu(r,\theta)} + [\Omega - \omega(r,\theta)]^2\,\mathrm{e}^{2\,\psi(r,\theta)}\big\}^{-1}, \tag{G.65}
$$

$$
T_{t\phi} = -\big\{\mathrm{e}^{2\,\psi(r,\theta)}\big[\epsilon(r)\omega(r,\theta)\mathrm{e}^{2\,\nu(r,\theta)} - \epsilon(r)\mathrm{e}^{2\,\nu(r,\theta)}\Omega - \epsilon(r)\mathrm{e}^{2\,\psi(r,\theta)}
\omega^3(r,\theta) + 2\,\epsilon(r)\mathrm{e}^{2\,\psi(r,\theta)}\omega^2(r,\theta)\Omega - \epsilon(r)\mathrm{e}^{2\,\psi(r,\theta)}\omega(r,\theta)\Omega^2 - P(r)
\mathrm{e}^{2\,\nu(r,\theta)}\Omega - P(r)\mathrm{e}^{2\,\psi(r,\theta)}\omega^3(r,\theta) + 2\,P(r)\mathrm{e}^{2\,\psi(r,\theta)}\omega^2(r,\theta)\Omega
- P(r)\mathrm{e}^{2\,\psi(r,\theta)}\omega(r,\theta)\Omega^2 + P(r)\omega(r,\theta)\mathrm{e}^{2\,\psi(r,\theta)}\Omega^2 - 2\,P(r)\omega^2(r,\theta)
\mathrm{e}^{2\,\psi(r,\theta)}\Omega + P(r)\omega^3(r,\theta)\mathrm{e}^{2\,\psi(r,\theta)}\big]\big\}
\big\{-\mathrm{e}^{2\,\nu(r,\theta)} + [\Omega - \omega(r,\theta)]^2\,\mathrm{e}^{2\,\psi(r,\theta)}\big\}^{-1}, \tag{G.66}
$$

$$
T_{rr} = P(r)\,\mathrm{e}^{2\,\lambda(r,\theta)}, \tag{G.67}
$$

$$
T_{\theta\theta} = P(r)\,\mathrm{e}^{2\,\mu(r,\theta)}, \tag{G.68}
$$

$$
T_{\phi t} = -\big\{\mathrm{e}^{2\,\psi(r,\theta)}\big[\epsilon(r)\omega(r,\theta)\mathrm{e}^{2\,\nu(r,\theta)} - \epsilon(r)\mathrm{e}^{2\,\nu(r,\theta)}\Omega - \epsilon(r)\mathrm{e}^{2\,\psi(r,\theta)}
\omega^3(r,\theta) + 2\,\epsilon(r)\mathrm{e}^{2\,\psi(r,\theta)}\omega^2(r,\theta)\Omega - \epsilon(r)\mathrm{e}^{2\,\psi(r,\theta)}\omega(r,\theta)\Omega^2
- P(r)\mathrm{e}^{2\,\nu(r,\theta)}\Omega - P(r)\mathrm{e}^{2\,\psi(r,\theta)}\omega^3(r,\theta) + 2\,P(r)\mathrm{e}^{2\,\psi(r,\theta)}
\omega(r,\theta)\Omega - P(r)\mathrm{e}^{2\,\psi(r,\theta)}\omega(r,\theta)\Omega^2 + P(r)\omega(r,\theta)\mathrm{e}^{2\,\psi(r,\theta)}\Omega^2
- 2\,P(r)\omega^2(r,\theta)\mathrm{e}^{2\,\psi(r,\theta)}\Omega + P(r)\omega^3(r,\theta)\mathrm{e}^{2\,\psi(r,\theta)}\big]\big\}
\big\{-\mathrm{e}^{2\,\nu(r,\theta)} + [\Omega - \omega(r,\theta)]^2\mathrm{e}^{2\,\psi(r,\theta)}\big\}^{-1}, \tag{G.69}
$$

$$
T_{\phi\phi} = \big\{\mathrm{e}^{2\,\psi(r,\theta)}\big[-\epsilon(r)\mathrm{e}^{2\,\psi(r,\theta)}\omega^2(r,\theta) + 2\,\epsilon(r)\mathrm{e}^{2\,\psi(r,\theta)}\omega(r,\theta)\Omega
- \epsilon(r)\mathrm{e}^{2\,\psi(r,\theta)}\Omega^2 - P(r)\mathrm{e}^{2\,\psi(r,\theta)}\omega^2(r,\theta) + 2\,P(r)\mathrm{e}^{2\,\psi(r,\theta)}
\omega(r,\theta)\Omega - P(r)\mathrm{e}^{2\,\psi(r,\theta)}\Omega^2 - P(r)\mathrm{e}^{2\,\nu(r,\theta)} + P(r)\mathrm{e}^{2\,\psi(r,\theta)}\Omega^2
- 2\,P(r)\mathrm{e}^{2\,\psi(r,\theta)}\omega(r,\theta)\Omega + P(r)\omega^2(r,\theta)\mathrm{e}^{2\,\psi(r,\theta)}\big]\big\}
\big\{-\mathrm{e}^{2\,\nu(r,\theta)} + [\Omega - \omega(r,\theta)]^2\,\mathrm{e}^{2\,\psi(r,\theta)}\big\}^{-1}. \tag{G.70}
$$

The mixed components of the energy–momentum tensor read,

$$
T_t{}^t = \big\{P(r)\mathrm{e}^{2\,\psi(r,\theta)}\Omega^2 - P(r)\mathrm{e}^{2\,\psi(r,\theta)}\omega(r,\theta)\Omega
+ \epsilon(r)\mathrm{e}^{2\,\psi(r,\theta)}\omega(r,\theta)\Omega - \epsilon(r)\mathrm{e}^{2\,\psi(r,\theta)}\omega^2(r,\theta) + \epsilon(r)\mathrm{e}^{2\,\nu(r,\theta)}\big\}
\big\{-\mathrm{e}^{2\,\nu(r,\theta)} + [\Omega - \omega(r,\theta)]^2\,\mathrm{e}^{2\,\psi(r,\theta)}\big\}^{-1}, \tag{G.71}
$$

$$T_r{}^r = P(r)\,, \tag{G.72}$$

$$T_\phi{}^t = -\frac{\mathrm{e}^{2\,\psi(r,\theta)}[\epsilon(r)+P(r)]\,[\Omega-\omega(r,\theta)]}{-\mathrm{e}^{2\,\nu(r,\theta)}+[\Omega-\omega(r,\theta)]^2\,\mathrm{e}^{2\,\psi(r,\theta)}}\,. \tag{G.73}$$

G.6 Covariant derivatives of the energy–momentum tensor

The covariant derivatives of the energy–momentum tensor, introduced in equation (13.29), are given by

$$T^\kappa{}_{t;\,\kappa} = 0\,, \tag{G.74}$$

$$\begin{aligned} T^\kappa{}_{r;\,\kappa} = \Big\{ & -\epsilon(r)\Big(\frac{\partial}{\partial r}\nu(r,\theta)\Big)\mathrm{e}^{2\,\nu(r,\theta)} - P(r)\Big(\frac{\partial}{\partial r}\nu(r,\theta)\Big)\mathrm{e}^{2\,\nu(r,\theta)} \\ & + \epsilon(r)\Big(\frac{\partial}{\partial r}\psi(r,\theta)\Big)\mathrm{e}^{2\,\psi(r,\theta)}\omega^2(r,\theta) + \epsilon(r)\mathrm{e}^{2\,\psi(r,\theta)}\omega(r,\theta) \\ & \frac{\partial}{\partial r}\omega(r,\theta) + P(r)\Big(\frac{\partial}{\partial r}\psi(r,\theta)\Big)\mathrm{e}^{2\,\psi(r,\theta)}\omega^2(r,\theta) + P(r) \\ & \mathrm{e}^{2\,\psi(r,\theta)}\omega(r,\theta)\frac{\partial}{\partial r}\omega(r,\theta) - 2\,\epsilon(r)\Big(\frac{\partial}{\partial r}\psi(r,\theta)\Big)\mathrm{e}^{2\,\psi(r,\theta)}\omega(r,\theta)\Omega \\ & - \epsilon(r)\mathrm{e}^{2\,\psi(r,\theta)}\Big(\frac{\partial}{\partial r}\omega(r,\theta)\Big)\Omega - 2\,P(r)\Big(\frac{\partial}{\partial r}\psi(r,\theta)\Big)\mathrm{e}^{2\,\psi(r,\theta)} \\ & \omega(r,\theta)\Omega - P(r)\mathrm{e}^{2\,\psi(r,\theta)}\Big(\frac{\partial}{\partial r}\omega(r,\theta)\Big)\Omega - \Big(\frac{\mathrm{d}}{\mathrm{d}r}P(r)\Big)\mathrm{e}^{2\,\nu(r,\theta)} \\ & + \Big(\frac{\mathrm{d}}{\mathrm{d}r}P(r)\Big)\mathrm{e}^{2\,\psi(r,\theta)}\Omega^2 - 2\Big(\frac{\mathrm{d}}{\mathrm{d}r}P(r)\Big)\mathrm{e}^{2\,\psi(r,\theta)}\omega(r,\theta)\Omega \\ & + \Big(\frac{\mathrm{d}}{\mathrm{d}r}P(r)\Big)\omega^2(r,\theta)\mathrm{e}^{2\,\psi(r,\theta)} + \epsilon(r)\Big(\frac{\partial}{\partial r}\psi(r,\theta)\Big)\mathrm{e}^{2\,\psi(r,\theta)}\Omega^2 \\ & + P(r)\Big(\frac{\partial}{\partial r}\psi(r,\theta)\Big)\mathrm{e}^{2\,\psi(r,\theta)}\Omega^2\Big\} \\ & \big\{-\mathrm{e}^{2\,\nu(r,\theta)} + [\Omega-\omega(r,\theta)]^2\mathrm{e}^{2\,\psi(r,\theta)}\big\}^{-1}\,, \end{aligned} \tag{G.75}$$

$$\begin{aligned} T^\kappa{}_{\theta;\,\kappa} = \Big\{ & -2\,P(r)\Big(\frac{\partial}{\partial\theta}\psi(r,\theta)\Big)\mathrm{e}^{2\,\psi(r,\theta)}\omega(r,\theta)\Omega - P(r)\mathrm{e}^{2\,\psi(r,\theta)} \\ & \Big(\frac{\partial}{\partial\theta}\omega(r,\theta)\Big)\Omega - 2\,\epsilon(r)\Big(\frac{\partial}{\partial\theta}\psi(r,\theta)\Big)\mathrm{e}^{2\,\psi(r,\theta)}\omega(r,\theta)\Omega \\ & - \epsilon(r)\mathrm{e}^{2\,\psi(r,\theta)}\Big(\frac{\partial}{\partial\theta}\omega(r,\theta)\Big)\Omega - \epsilon(r)\Big(\frac{\partial}{\partial\theta}\nu(r,\theta)\Big)\mathrm{e}^{2\,\nu(r,\theta)} \\ & - P(r)\Big(\frac{\partial}{\partial\theta}\nu(r,\theta)\Big)\mathrm{e}^{2\,\nu(r,\theta)} + \epsilon(r)\mathrm{e}^{2\,\psi(r,\theta)}\omega(r,\theta)\frac{\partial}{\partial\theta}\omega(r,\theta) \\ & + \epsilon(r)\Big(\frac{\partial}{\partial\theta}\psi(r,\theta)\Big)\mathrm{e}^{2\,\psi(r,\theta)}\omega^2(r,\theta) + P(r)\mathrm{e}^{2\,\psi(r,\theta)}\omega(r,\theta) \end{aligned}$$

$$\frac{\partial}{\partial\theta}\omega(r,\theta) + P(r)\Big(\frac{\partial}{\partial\theta}\psi(r,\theta)\Big)e^{2\,\psi(r,\theta)}\omega^2(r,\theta)$$
$$+\,\Omega^2 e^{2\,\psi(r,\theta)}\epsilon(r)\frac{\partial}{\partial\theta}\psi(r,\theta) + \Omega^2 e^{2\,\psi(r,\theta)}P(r)\frac{\partial}{\partial\theta}\psi(r,\theta)\Big\}$$
$$\big\{-e^{2\,\nu(r,\theta)} + [\Omega-\omega(r,\theta)]^2 e^{2\,\psi(r,\theta)}\big\}^{-1}, \qquad \text{(G.76)}$$

$$T^{\kappa}{}_{\phi;\,\kappa} = 0\,. \qquad \text{(G.77)}$$

Appendix H

Quark matter at finite temperature

The coefficients occurring in the expansion of pressure, equation (18.30), are obtained from equations (18.4) and (18.5) as follows,

$$\frac{\partial P^i}{\partial \frac{\Delta\mu}{\mu_0}} = -\frac{\nu_i \mu_0 \eta_i}{2\pi^2} \int_{m_i}^{\infty} \mathrm{d}E \; E \; \sqrt{E^2 - m_i^2} \; f^i(E) \tag{H.1}$$

$$\overset{T\to 0}{\longrightarrow} -\frac{\nu_i \mu_0 \eta_i}{2\pi^2} \int_{m_i}^{\mu\eta_i} \mathrm{d}E \; E \; \sqrt{E^2 - m_i^2} \tag{H.2}$$

$$= -\frac{\nu_i \mu_0 \eta_i}{2\pi^2} \frac{1}{3} \left((\mu\, \eta_i)^2 - m_i^2 \right)^{3/2} . \tag{H.3}$$

At finite temperatures the quantity η_i is given by (cf. equation (18.13))

$$\eta_i = \begin{cases} 1 - (x_0 - \Delta x) & \text{if } i = u, c\,, \\ 1 & \text{if } i = d, s\,, \\ x_0 + \Delta x & \text{if } i = e^-, \mu^- \,. \end{cases} \tag{H.4}$$

All those quantities not defined here are explained in chapter 18. Using equation (18.23), one derives from (H.3)

$$\left. \frac{\partial P^i}{\partial \frac{\Delta\mu}{\mu}} \right|_{\mu_0, x_0, T=0} = -\frac{\nu_i}{6\pi^2} \, \mu_0^4 \, (\rho^i)^4 \, (1 - z_i^2)^{3/2} . \tag{H.5}$$

The second coefficient reads

$$\frac{\partial P^i}{\partial \Delta x} = \frac{\nu_i c_i \mu}{2\pi^2} \int_{m_i}^{\infty} \mathrm{d}E \; E \; \sqrt{E^2 - m_i^2} \; f^i(E) \tag{H.6}$$

$$\overset{T\to 0}{\longrightarrow} \frac{\nu_i c_i \mu_0}{2\pi^2} \int_{m_i}^{\mu_0 \eta_i} \mathrm{d}E \; E \; \sqrt{E^2 - m_i^2} \,, \tag{H.7}$$

and thus

$$\left.\frac{\partial P^i}{\partial \Delta x}\right|_{\mu_0, x_0, T=0} = \frac{\nu_i c_i}{6\pi^2}\,\mu_0^4\,\eta_0^3\,(1-z_i)^{3/2}\,. \tag{H.8}$$

The third coefficient is obtained from

$$\frac{\partial P^i}{\partial \frac{T^2}{\mu_0^2}} = \frac{\mu_0^2 \nu_i}{12\pi^2}\left\{\int\limits_{-\infty}^{\infty} \mathrm{d}\xi\;\xi\;(\xi T + \mu\eta_i)^3 \frac{\mathrm{e}^\xi}{(1+\mathrm{e}^\xi)^2}\frac{1}{T}\right.$$
$$\left. -\frac{3}{2}m_i^2 \int\limits_{-\infty}^{\infty} \mathrm{d}\xi\;\xi\;(\xi T + \mu\eta_i)\frac{\mathrm{e}^\xi}{(1+\mathrm{e}^\xi)^2}\frac{1}{T}\right\}, \tag{H.9}$$

where the transformation $\xi \equiv (E-\mu\eta_i)/T$ has been introduced. For the zero-temperature coefficient one obtains

$$\left.\frac{\partial P^i}{\partial \frac{T^2}{\mu_0^2}}\right|_{\mu_0, x_0, T=0} = \frac{\mu_0^2 \nu_i}{12\pi^2}\left(3(\mu_0\eta_i)^2\frac{\pi^2}{3} - \frac{3}{2}m_i^2\frac{\pi^2}{3}\right) \tag{H.10}$$

$$= \frac{\nu_i}{12}\,\mu_0^4\,\eta_i^2\left(1-\frac{1}{2}z_i^2\right)\,. \tag{H.11}$$

The coefficients occurring in the expansion of particle density are obtained from equation (18.6), and are given by

$$\frac{\partial \rho^i}{\partial \frac{\Delta\mu}{\mu_0}} = -\frac{\nu_i \mu_0 \eta_i}{2\pi^2}\int\limits_{(m_i-\mu\eta_i)/T}^{\infty} \mathrm{d}\xi\;(\xi T + \mu\eta_i)\;\sqrt{(\xi T+\mu\eta_i)^2 - m_i^2}\;\frac{\mathrm{e}^\xi}{(1+\mathrm{e}^\xi)^2}\,, \tag{H.12}$$

$$\left.\frac{\partial \rho^i}{\partial \frac{\Delta\mu}{\mu_0}}\right|_{\mu_0, x_0, T=0} = -\,\frac{\nu_i \mu_0 \eta_i}{2\pi^2}\,\mu_0\,\eta_i\,\sqrt{(\mu\eta_i)^2 - m_i^2}\int\limits_{-\infty}^{\infty}\mathrm{d}\xi\;\frac{\mathrm{e}^\xi}{(1+\mathrm{e}^\xi)^2} \tag{H.13}$$

$$= -\,\frac{\nu_i}{2\pi^2}\,\mu_0^3\,\eta_i^3\,\sqrt{1-z_i^2}\,, \tag{H.14}$$

and

$$\frac{\partial \rho^i}{\partial \Delta x} = \frac{\nu_i \mu c_i}{2\pi^2}\int\limits_{(m_i-\mu\eta_i)/T}^{\infty} \mathrm{d}\xi\;(\xi T + \mu\eta_i)\;\sqrt{(\xi T+\mu\eta_i)^2 - m_i^2}\;\frac{\mathrm{e}^\xi}{(1+\mathrm{e}^\xi)^2}\,, \tag{H.15}$$

which has the same mathematical structure as equation (H.12). At zero temperature it reduces to

$$\frac{\partial \rho^i}{\partial \Delta x}\bigg|_{\mu_0, x_0, T=0} = \frac{\nu_i\, c_i}{2\pi^2}\, \mu_0^3\, \eta_i^2\, \sqrt{1 - z_i^2}\,. \tag{H.16}$$

Finally, from

$$\frac{\partial \rho^i}{\partial \frac{T^2}{\mu_0^2}} = \frac{\nu_i \mu_0^2}{4\pi^2} \int\limits_{(m_i - \mu\eta_i)/T}^{\infty} d\xi\, \frac{\xi}{T}\, (\xi T + \mu\eta_i) \sqrt{(\xi T + \mu\eta_i)^2 - m_i^2}\, \frac{e^{\xi}}{(1 + e^{\xi})^2} \tag{H.17}$$

$$= \frac{\nu_i \mu_0^2}{4\pi^2} \left\{ \int\limits_{-\infty}^{+\infty} d\xi\, \frac{\xi}{T}\, (\xi T + \mu\eta_i)^2\, \frac{e^{\xi}}{(1 + e^{\xi})^2} - \frac{1}{2} \int\limits_{-\infty}^{+\infty} d\xi\, \frac{\xi}{T}\, m_i^2\, \frac{e^{\xi}}{(1 + e^{\xi})^2} \right\} \tag{H.18}$$

one obtains

$$\frac{\partial \rho^i}{\partial \frac{T^2}{\mu_0^2}}\bigg|_{\mu_0, x_0, T=0} = \frac{\nu_i \mu_0^2}{4\pi^2}\, 2\mu_0\, \eta_i \frac{\pi^2}{3} = \frac{\nu_i}{6}\, \mu_0^3\, \eta_i\,. \tag{H.19}$$

The expansion coefficients of particle density enter in the condition of electric charge neutrality, equation (18.3), written at finite temperature,

$$0 = \sum_i q_i^{\rm el} \left\{ \rho_0^i + \frac{\partial \rho^i}{\partial \frac{\Delta\mu}{\mu_0}}\bigg|_{\mu_0, x_0, T_0} \frac{\Delta\mu}{\mu_0} + \frac{\partial \rho^i}{\partial \Delta x}\bigg|_{\mu_0, x_0, T_0} \Delta x + \frac{\partial \rho^i}{\partial \frac{T^2}{\mu_0^2}}\bigg|_{\mu_0, x_0, T_0} \frac{T^2}{\mu_0^2} \right\}. \tag{H.20}$$

Now, equation (H.20) holds at all temperatures $T \geq 0$. In the special case of $T = 0$ one has $\Delta\mu = \Delta x = 0$, and thus it follows that $\sum_i q_i^{\rm el}\, \rho_0^i = 0$, in agreement with equation (18.3). Since the sum of the remaining three terms on the right-hand side of equation (H.20) must be equal to zero at any temperature different from zero, all three must possess the same functional temperature dependence. We thus make the ansatz

$$\frac{\Delta\mu}{\mu_0} = a\, \pi^2 \left(\frac{T}{\mu_0}\right)^2 \quad \text{and} \quad \Delta x = b\, \pi^2 \left(\frac{T}{\mu_0}\right)^2. \tag{H.21}$$

Inserting equation (H.21) in equation (H.20) leads to an equation of the form

$$\mathcal{J}_1\, a + \mathcal{J}_2\, b \; - \; \mathcal{J}_3 = 0\,, \tag{H.22}$$

where

$$\mathcal{J}_1 \equiv \pi^2 \sum_i q_i \frac{\partial \rho^i}{\partial \frac{\Delta\mu}{\mu_0}}\bigg|_{\mu_0, x_0, T=0}$$
$$= 1 - \sqrt{1 - z_s^2} - 3x_0(x_0^2 - 2x_0 + 2) + 2(1 - x_0)^3\sqrt{1 - z_c^2} - x_0^3\sqrt{1 - z_\mu^2}\,, \tag{H.23}$$

$$\mathcal{J}_2 \equiv \pi^2 \sum_i q_i \frac{\partial \rho^i}{\partial \Delta x}\bigg|_{\mu_0, x_0, T=0}$$
$$= 2 - x_0(4 - 3x_0) + 2(1 - x_0)^2\sqrt{1 - z_c^2} + x_0^2\sqrt{1 - z_\mu^2}\,, \tag{H.24}$$

$$\mathcal{J}_3 \equiv -\sum_i q_i \frac{\partial \rho^i}{\partial \frac{T^2}{\mu_0^2}}\bigg|_{\mu_0, x_0, T=0} = \frac{1}{3}(1 - 5x_0)\,. \tag{H.25}$$

To find the second linear equation that supplements (H.22), we turn to the pressure–energy density relation of (18.1), from which one derives

$$P + B = \sum_i \Bigg[P_0^i + \Bigg\{ \frac{\partial P^i}{\partial \frac{\Delta\mu}{\mu_0}}\bigg|_{\mu_0, x_0, T_0} \frac{\Delta\mu}{\mu_0} + \frac{\partial P^i}{\partial \Delta x}\bigg|_{\mu_0, x_0, T_0} \Delta x + \frac{\partial P^i}{\partial \frac{T^2}{\mu_0^2}}\bigg|_{\mu_0, x_0, T_0} \frac{T^2}{\mu_0^2} \Bigg\} \Bigg]. \tag{H.26}$$

Since at zero temperature $P + B = \sum_i P_0^i$, it follows that the expansion in curly brackets must vanish identically. By means of equation (H.21) one obtains, in analogy to equation (H.22),

$$\mathcal{J}_4\, a + \mathcal{J}_5\, b - \mathcal{J}_6 = 0\,, \tag{H.27}$$

where

$$\mathcal{J}_4 \equiv \pi^2 \sum_i \frac{\partial P^i}{\partial \frac{\Delta\mu}{\mu_0}}\bigg|_{\mu_0, x_0, T=0}$$
$$= (1 - x_0)^4 + \frac{1}{3}x_0^4 + 1 + (1 - z_s^2)^{3/2} + (1 - x_0)^4(1 - z_c^2)^{3/2} + \frac{1}{3}x_0^4(1 - z_\mu^2)^{3/2}\,, \tag{H.28}$$

$$\mathcal{J}_5 \equiv \pi^2 \sum_i \frac{\partial P^i}{\partial \Delta x}\bigg|_{\mu_0, x_0, T=0}$$
$$= (1 - x_0)^3 - \frac{1}{3}x_0^3 + (1 - x_0)^3(1 - z_c^2)^{3/2} - \frac{1}{3}x_0^3(1 - z_\mu^2)^{3/2}\,, \tag{H.29}$$

$$\mathcal{J}_6 \equiv - \sum_i \frac{\partial P^i}{\partial \frac{T^2}{\mu_0^2}}\bigg|_{\mu_0, x_0, T=0}$$
$$= \frac{1}{2}\Big[(1-x_0)^2 + 1 + \Big(1 - \frac{1}{2}z_s^2\Big) + \frac{1}{3}x_0^2 + (1-x_0)^2\Big(1 - \frac{1}{2}z_c^2\Big)$$
$$+ \frac{1}{3}x_0^2\Big(1 - \frac{1}{2}z_\mu^2\Big)\Big] . \tag{H.30}$$

Muons and charm quarks make their appearance only at densities larger than 3.8×10^{17} g cm^{-3} and 1.1×10^{17} g cm^{-3}, respectively. Quark matter at lower densities consists of only u, d, s quarks and electrons. For such matter the above coefficients, $\mathcal{J}_1$ through $\mathcal{J}_6$, are given by the simpler relations

$$\mathcal{J}_1 = 1 - \sqrt{1 - z_s^2} - 3x_0(x_0^2 - 2x_0 + 2) , \tag{H.31}$$
$$\mathcal{J}_2 = 2 - x_0(4 - 3x_0) , \tag{H.32}$$
$$\mathcal{J}_3 = - \ x_0 , \tag{H.33}$$
$$\mathcal{J}_4 = (1-x_0)^4 + \frac{1}{3}x_0^4 + 1 + (1 - z_s^2)^{3/2} , \tag{H.34}$$
$$\mathcal{J}_5 = (1-x_0)^3 - \frac{1}{3}x_0^3 , \tag{H.35}$$
$$\mathcal{J}_6 = \frac{1}{2}\Big[(1-x_0)^2 + 1 + \Big(1 - \frac{1}{2}z_s^2\Big) + \frac{1}{3}x_0^2\Big] . \tag{H.36}$$

Equations (H.22) and (H.27) constitute two relations for the two unknowns a and b. Solving for them and inserting the results into equations (18.30) and (H.21) leads to the expressions for P^i and ρ^i given in equations (18.31) and (18.32), respectively. Finally, the expression for the energy density is obtained from equations (18.1) and (18.2) which, for this purpose, are combined to

$$\epsilon = 3\,P + 4\,B + \sum_{i=c,s,\mu^-} (\epsilon^i \ - \ 3\,P^i) . \tag{H.37}$$

Appendix I

Models of rotating relativistic neutron stars of selected masses

The properties of models of neutron stars of selected gravitational masses are listed in the tables on the following pages. Each model star rotates at its respective general relativistic Kepler period, $P_K = 2\pi/\Omega_K$, computed self-consistently from equations (16.5) and (16.7).

Table I.1. Properties of rotating neutron stars of gravitational mass $M = 1.30\,M_\odot$. The underlying EOSs are explained in table 12.4. The listed properties are: central energy density, ϵ_c; Kepler period, P_K (in 10^{-3} s); baryon mass, M_A; equatorial radius, R_{eq} (in km); ratio of rotational to gravitational energy, T/W; angular momentum, J; polar redshift, z_p; equatorial redshifts in backward and forward direction, z_B and z_F, respectively.

EOS	$\frac{\epsilon_c}{\epsilon_0}$ †	P_K	$\frac{M_A}{M_\odot}$ ‡	R_{eq}	$\frac{T}{W}$	$\frac{cJ}{GM^2}$	z_p	z_B	z_F
Relativistic field-theoretical equations of state									
1	1.98	1.136	1.40	16.3	0.093	0.449	0.23	0.58	−0.17
2	1.89	1.235	1.35	17.3	0.086	0.442	0.21	0.55	−0.17
3	2.18	1.140	1.40	16.1	0.088	0.398	0.24	0.60	−0.18
4	2.57	1.070	1.41	15.9	0.090	0.462	0.24	0.61	−0.18
5	2.25	1.102	1.40	15.9	0.092	0.435	0.24	0.60	−0.18
6	3.71	0.944	–	14.7	0.084	0.445	0.26	0.68	−0.19
7	1.91	1.031	–	15.3	0.099	0.460	0.26	0.62	−0.18
8	3.29	0.967	1.52	15.2	0.092	0.455	0.28	0.69	−0.19
9	3.14	1.020	1.41	15.3	0.090	0.449	0.25	0.63	−0.18
10	2.43	1.044	–	15.5	0.096	0.458	0.25	0.61	−0.18
11	2.69	0.979	–	14.9	0.097	0.465	0.26	0.65	−0.18
Non-relativistic potential-model equations of state									
12	3.04	1.028	–	15.6	0.082	0.449	0.24	0.64	−0.19
13	3.61	0.780	1.43	12.7	0.103	0.459	0.34	0.78	−0.19
14	3.94	0.806	–	13.1	0.093	0.451	0.31	0.76	−0.20
15	3.39	0.798	1.42	12.9	0.101	0.452	0.33	0.77	−0.19
16	4.03	0.713	1.44	11.9	0.103	0.448	0.37	0.85	−0.20

† The quantity ϵ_0 is defined in table 1.3.

‡ The baryon mass has not been calculated for every stellar model.

Table I.2. Properties of rotating neutron stars of gravitational mass $M = 1.44\,M_\odot$. The underlying EOSs are explained in table 12.4. The listed properties are: central energy density, ϵ_c; Kepler period, P_K (in 10^{-3} s); baryon mass, M_A; equatorial radius, R_{eq} (in km); ratio of rotational to gravitational energy, T/W; angular momentum, J; polar redshift, z_p; equatorial redshifts in backward and forward direction, z_B and z_F, respectively.

EOS	$\frac{\epsilon_c}{\epsilon_0}$ †	P_K	$\frac{M_A}{M_\odot}$ ‡	R_{eq}	$\frac{T}{W}$	$\frac{cJ}{GM^2}$	z_p	z_B	z_F
Relativistic field-theoretical equations of state									
1	2.21	1.068	1.57	16.2	0.096	0.450	0.27	0.66	−0.18
2	2.07	1.160	1.51	17.2	0.088	0.437	0.24	0.62	−0.18
3	2.57	1.060	1.57	15.9	0.091	0.398	0.28	0.69	−0.19
4	3.36	0.977	1.58	15.9	0.092	0.457	0.28	0.70	−0.19
5	2.56	1.018	1.58	15.7	0.094	0.441	0.28	0.69	−0.19
6	4.57	0.845	–	11.9	0.086	0.441	0.32	0.80	−0.20
7	2.11	0.997	–	15.3	0.102	0.456	0.29	0.68	−0.18
8	3.63	0.932	1.58	15.0	0.092	0.455	0.30	0.73	−0.19
9	4.57	0.877	1.59	14.4	0.089	0.443	0.32	0.78	−0.20
10	2.86	0.966	–	15.2	0.097	0.454	0.30	0.71	−0.19
11	2.96	0.915	–	14.7	0.099	0.460	0.31	0.74	−0.19
Non-relativistic potential-model equations of state									
12	3.43	0.947	–	15.2	0.085	0.445	0.28	0.73	−0.20
13	3.95	0.739	1.61	12.7	0.105	0.456	0.40	0.89	−0.20
14	4.70	0.742	–	12.9	0.096	0.451	0.38	0.89	−0.21
15	3.65	0.758	1.60	12.9	0.104	0.452	0.38	0.87	−0.20
16	4.34	0.679	1.61	12.0	0.106	0.450	0.44	0.97	−0.21

† The quantity ϵ_0 is defined in table 1.3.

‡ The baryon mass has not been calculated for every stellar model.

Table I.3. Properties of rotating neutron stars of gravitational mass $M = 1.60 M_{\odot}$. The underlying EOSs are explained in table 12.4. The listed properties are: central energy density, ϵ_c; Kepler period, P_K (in 10^{-3} s); baryon mass, M_A; equatorial radius, R_{eq} (in km); ratio of rotational to gravitational energy, T/W; angular momentum, J; polar redshift, z_p; equatorial redshifts in backward and forward direction, z_B and z_F, respectively.

EOS	$\frac{\epsilon_c}{\epsilon_0}$ †	P_K	$\frac{M_A}{M_{\odot}}$ ‡	R_{eq}	$\frac{T}{W}$	$\frac{cJ}{GM^2}$	z_p	z_B	z_F
Relativistic field-theoretical equations of state									
1	2.54	1.008	1.76	16.1	0.098	0.440	0.32	0.75	−0.19
2	2.36	1.077	1.71	17.0	0.092	0.434	0.29	0.72	−0.19
3	3.57	0.954	1.77	15.4	0.092	0.402	0.34	0.81	−0.20
4	5.07	0.830	1.77	14.4	0.091	0.450	0.37	0.88	−0.21
5	3.02	0.939	1.76	15.5	0.096	0.451	0.33	0.79	−0.20
6	6.43	0.729	–	13.2	0.088	0.443	0.41	0.99	−0.22
7	2.46	0.948	–	15.5	0.104	0.456	0.35	0.78	−0.19
8	5.21	0.801	1.77	14.0	0.092	0.452	0.38	0.91	−0.21
9	7.03	0.706	1.78	12.9	0.089	0.445	0.43	1.03	−0.22
10	3.21	0.885	–	14.9	0.099	0.449	0.36	0.84	−0.20
11	3.29	0.850	–	14.5	0.101	0.457	0.37	0.86	−0.20
Non-relativistic potential-model equations of state									
12	3.98	0.858	–	14.7	0.088	0.444	0.40	0.86	−0.21
13	4.46	0.694	1.81	12.6	0.107	0.458	0.48	1.04	−0.21
14	5.29	0.671	–	12.4	0.099	0.456	0.48	1.07	−0.22
15	3.92	0.720	1.79	12.9	0.106	0.450	0.46	1.00	−0.21
16	4.61	0.646	1.81	11.9	0.109	0.450	0.53	1.12	−0.22

† The quantity ϵ_0 is defined in table 1.3.
‡ The baryon mass has not been calculated for every stellar model.

Table I.4. Properties of rotating neutron stars of gravitational mass $M = 1.80 M_{\odot}$. The underlying EOSs are explained in table 12.4. The listed properties are: central energy density, ϵ_c; Kepler period, P_K (in 10^{-3} s); baryon mass, M_A; equatorial radius, R_{eq} (in km); ratio of rotational to gravitational energy, T/W; angular momentum, J; polar redshift, z_p; equatorial redshifts in backward and forward direction, z_B and z_F, respectively.

EOS§	$\frac{\epsilon_c}{\epsilon_0}$ †	P_K	$\frac{M_A}{M_{\odot}}$ ‡	R_{eq}	$\frac{T}{W}$	$\frac{cJ}{GM^2}$	z_p	z_B	z_F
Relativistic field-theoretical equations of state									
1	3.29	0.913	2.01	15.7	0.100	0.441	0.40	0.90	−0.21
2	2.87	0.986	1.95	16.6	0.094	0.434	0.36	0.85	−0.20
3	5.93	0.749	2.02	13.8	0.091	0.418	0.48	1.11	−0.23
5	3.64	0.854	2.01	15.1	0.099	0.443	0.42	0.96	−0.21
7	3.04	0.880	–	15.4	0.106	0.459	0.42	0.92	−0.20
10	3.58	0.811	–	14.6	0.101	0.450	0.45	1.00	−0.21
11	3.64	0.781	–	14.2	0.103	0.456	0.47	1.03	−0.21
Non-relativistic potential-model equations of state									
12	4.97	0.757	–	14.1	0.092	0.449	0.46	1.06	−0.22
13	5.32	0.636	2.07	12.3	0.110	0.465	0.62	1.27	−0.22
14	5.81	0.605	–	12.0	0.103	0.465	0.63	1.34	−0.23
15	4.43	0.670	2.06	12.8	0.110	0.458	0.58	1.21	−0.22
16	5.11	0.602	2.09	11.9	0.112	0.459	0.68	1.37	−0.23

§ EOSs not listed here do not support a rotating neutron star of this mass.
† The quantity ϵ_0 is defined in table 1.3.
‡ The baryon mass has not been calculated for every stellar model.

Table I.5. Properties of rotating neutron stars of gravitational mass $M = 2.00M_\odot$. The underlying EOSs are explained in table 12.4. The listed properties are: central energy density, ϵ_c; Kepler period, P_K (in 10^{-3} s); baryon mass, M_A; equatorial radius, R_{eq} (in km); ratio of rotational to gravitational energy, T/W; angular momentum, J; polar redshift, z_p; equatorial redshifts in backward and forward direction, z_B and z_F, respectively.

EOS§	$\frac{\epsilon_c}{\epsilon_0}$ †	P_K	$\frac{M_A}{M_\odot}$ ‡	R_{eq}	$\frac{T}{W}$	$\frac{cJ}{GM^2}$	z_p	z_B	z_F
Relativistic field-theoretical equations of state									
1	5.03	0.774	2.27	14.6	0.100	0.446	0.53	1.17	−0.22
2	3.64	0.879	2.20	15.9	0.096	0.440	0.45	1.03	−0.22
5	5.29	0.732	2.28	14.1	0.101	0.456	0.56	1.23	−0.23
7	3.86	0.795	–	14.9	0.107	0.464	0.53	1.12	−0.21
10	4.14	0.731	–	12.1	0.105	0.457	0.59	1.25	−0.23
11	4.14	0.718	–	13.9	0.107	0.463	0.67	1.26	−0.22
Non-relativistic potential-model equations of state									
12	7.07	0.629	–	12.9	0.097	0.470	0.66	1.43	−0.23
13	7.14	0.556	2.4	11.8	0.113	0.491	0.86	1.66	−0.23
14	6.36	0.551	–	11.7	0.109	0.484	0.85	1.68	−0.24
15	4.99	0.621	2.34	12.6	0.113	0.472	0.75	1.48	−0.23
16	5.64	0.561	2.13	11.8	0.119	0.488	0.89	1.70	−0.24

§ EOSs not listed here do not support a rotating neutron star of this mass.
† The quantity ϵ_0 is defined in table 1.3.
‡ The baryon mass has not been calculated for every stellar model.

Appendix J

Equations of state in tabular form

The equations of state studied in this work can be downloaded from the following web site:
`http://www.physik.uni-muenchen.de/sektion/suessmann/astro`
and are available via email request from
`fweber@mfl.sue.physik.uni-muenchen.de` or `fweber@nta2.lbl.gov`.

Eventual changes of these addresses can be looked up at
`http://www.iop.org/cgi-bin/PEERS/main`.

References

[1] Greiner W and Stöcker H (ed) 1990 *Proc. NATO Advanced Study Institute on The Nuclear Equation of State (Peniscola)* NATO ASI Series B vol 216A&B (New York: Plenum Press)

[2] Greiner W, Stöcker H and Gallmann A (ed) 1994 *Proc. NATO Advanced Study Institute on Hot and Dense Nuclear Matter (Bodrum)* NATO ASI Series B vol 335 (New York: Plenum Press)

[3] Stöcker H, Gallmann A and Hamilton J H (ed) 1997 *Proc. Int. Conf. on Nuclear Physics at the Turn of the Millennium: Structure of Vacuum and Elementary Matter (Wilderness)* (Singapore: World Scientific)

[4] Gutbrod H and Stöcker H 1991 *Scientific American* November issue, p 32

[5] Timmes F X, Woosley S E and Weaver T A 1996 *Astrophys. J.* **457** 834

[6] Bethe H A and Brown G E 1985 *Scientific American* May issue, p 32

[7] Bethe H A 1990 *Rev. Mod. Phys.* **62** 801

[8] Müller E 1990 *J. Phys. G: Nucl. Part. Phys.* **16** 1571

[9] Woosley S E and Weaver T A 1992 Theory of neutron star formation, in: *The Structure and Evolution of Neutron Stars* ed D Pines, R Tamagaki and S Tsuruta (New York: Addison-Wesley) p 235

[10] A M Lenchek (ed) 1972 *The Physics of Pulsars* (New York: Gordon and Breach)

[11] Smith F G 1977 *Pulsars* (Cambridge: Cambridge University Press)

[12] Manchester R N and Taylor J H 1977 *Pulsars* (San Francisco: Freeman)

[13] Lyne A G and Graham-Smith F 1990 *Pulsar Astronomy* (Cambridge: Cambridge University Press)

[14] Blandford R D, Hewish A, Lyne A G and Mestel L 1992 Pulsars as Physics Laboratories *Phil. Trans. R. Soc.* A **341** 1–192

[15] Nagase F 1989 *Publ. Astron. Soc. Japan* **41** 1

[16] Wheeler J A 1966 *Ann. Rev. Astron. Astrophys.* **4** 393

[17] Blandford R D and Narayan R 1992 *Ann. Rev. Astron. Astrophys.* **30** 311

[18] Pines D and Alpar M A 1992 Inside neutron stars: a 1990 perspective, in: *The Structure and Evolution of Neutron Stars* ed D Pines, R Tamagaki and S Tsuruta (New York: Addison-Wesley) p 7

[19] Taylor J H and Stinebring D R 1986 *Ann. Rev. Astron. Astrophys.* **24** 285

[20] Glendenning N K 1990 *Mod. Phys. Lett.* A **5** 2197

[21] Lyne A G 1995 A review of galactic millisecond pulsar searches *IAU Symposium No 165: Compact Stars in Binaries (The Hague)* ed J van Paredijs, E P J van den Heuvel and E Kuulkers (Dordrecht: Kluwer)

[22] Lyne A G 1997 Millisecond pulsar searches, in: *Proc. 7th M Grossmann Meeting on General Relativity (Stanford)* ed R T Jantzen and G MacKeiser (Singapore: World Scientific)

[23] Camilo F 1995 Millisecond pulsar searches, in: *The Lives of Neutron Stars* ed M A Alpar, Ü Kiziloglu and J van Paradijs (Dordrecht: Kluwer) p 243

Camilo F 1996 Present and future pulsar searches, in: *Proc. High Sensitivity Radio Astronomy* (Cambridge: Cambridge University Press)

[24] Bhattacharya D and van den Heuvel E P J 1991 *Phys. Rep.* **203** 1

[25] Backer D C and Kulkarni S R 1990 *Phys. Today* **43** 26

[26] Glendenning N K 1991 Nuclear and particle astrophysics, in: *Proc. Int. Summer School on the Structure of Hadrons and Hadronic Matter (Dronten)* ed O Scholten and J H Koch (Singapore: World Scientific) p 275

[27] Witten E 1984 *Phys. Rev.* D **30** 272

[28] Alcock C, Farhi E and Olinto A V 1986 *Astrophys. J.* **310** 261

[29] Haensel P, Zdunik J L and Schaeffer R 1986 *Astron. Astrophys.* **160** 121

[30] Kettner Ch, Weber F, Weigel M K and Glendenning N K 1995 *Phys. Rev.* D **51** 1440

[31] Thorsett S E, Arzoumanian Z, McKinnon M M and Taylor J H 1993 *Astrophys. J.* **405** L29

[32] Nagase F 1992 Properties of neutron stars from X-ray pulsar observation, in: *The Structure and Evolution of Neutron Stars* ed D Pines, R Tamagaki and S Tsuruta (New York: Addison-Wesley) p 77

[33] Manchester R N 1992 Radio pulsar timing, in: *The Structure and Evolution of Neutron Stars* ed D Pines, R Tamagaki and S Tsuruta (New York: Addison-Wesley) p 32

[34] Trümper J 1992 *Mon. Notes R. Astron. Soc.* **33** 165

[35] Trümper J 1993 *Science* **260** 1769

[36] Iwasawa K, Koyama K and Halpern J P 1992 Cyclotron lines and the lulse leriod change of X-ray pulsar 1E2259+586, in: *The Structure and Evolution of Neutron Stars* ed D Pines, R Tamagaki and S Tsuruta (New York: Addison-Wesley) p 203

[37] Ögelman H 1995 X-Ray observations of cooling neutron stars, in: *The Lives of Neutron Stars* ed M A Alpar, Ü Kiziloglu and J van Paradijs (Dordrecht: Kluwer) p 101

[38] Trümper J, Pietsch W, Reppin C, Voges W, Staubert R and Kendziorra E 1978 *Astrophys. J.* **219** L105

[39] Voges W, Pietsch W, Reppin C, Trümper J, Kendziorra E and Staubert R 1982 *Astrophys. J.* **263** 803

[40] Wheaton W 1979 *Nature* **282** 240

[41] Makishima K 1992 Magnetic fields of binary X-ray pulsars, in: *The Structure and Evolution of Neutron Stars* ed D Pines, R Tamagaki and S Tsuruta (New York: Addison-Wesley) p 86

[42] Walter F M, Wolk S J and Neuhäuser R 1996 *Nature* **379** 233

[43] Prakash M 1997 *AIP Conf. Proceedings (Big Sky)* vol 412 ed T W Donnelly (New York: American Institute of Physics) p 1007

[44] Prakash M and Lattimer J M 1997 *The Radius of the Neutron Star RXJ185635–3754 and Implications for the Equation of State* (SUNY Preprint)

[45] Hoshi R 1992 Mass–radius relations in X-ray pulsars, in: *The Structure and Evolution of Neutron Stars* ed D Pines, R Tamagaki and S Tsuruta (New York: Addison-Wesley) p 98

[46] Liang E P 1986 *Astrophys. J.* **304** 682

[47] Hillebrandt W, Müller E and Mönchmeyer R 1990 Constraints on the nuclear equation of state from type II supernovae and newly born neutron stars, in: *The Nuclear Equation of State* NATO ASI Series B vol 216A, ed W Greiner and H Stöcker (New York: Plenum Press) p 689

[48] Backer D C, Kulkarni S R, Heiles C, Davis M M and Goss W M 1982 *Nature* **300** 615

[49] Woosley S E 1986 Nucleosynthesis and chemical evolution, in: *16th Advanced Course, Swiss Society of Astrophysics and Astronomy* ed B Hauck *et al* (Geneva: Geneva Observatory) p 1

[50] Woosley S E and Weaver T A 1986 *Ann. Rev. Astron. Astrophys.* **24** 205

[51] Hillebrandt W 1987, *NATO ASI on High Energy Phenomena Around Collapsed Stars* ed F Pacini (Dodrecht: Reidel) p 73

[52] Kahana S H 1989 *Ann. Rev. Nucl. Part. Sci.* **39** 231

[53] Cooperstein J and Baron E 1990 *Supernovae* ed A Petschek (Berlin: Springer)

[54] Bethe H A 1971 *Ann. Rev. Nucl. Sci.* **21** 93

[55] Ruderman M 1972 *Ann. Rev. Astron. Astrophys.* **10** 427

[56] Börner G 1972 *On the Properties of Matter in Neutron Stars, Habilitation Thesis* University of Munich

[57] Arnett W D and Bowers R L 1977 *Astrophys. J. Suppl.* **33** 415

[58] Baym G 1978 Neutron stars and the physics of matter at high density, in: *Nuclear Physics with Heavy Ions and Mesons (Les Houches)* session XXX vol 2 ed R Balian, M Rho and G Ripka (Amsterdam: North-Holland) p 745

[59] Baym G and Pethick Ch 1979 *Ann. Rev. Astron. Astrophys.* **17** 415

[60] Canuto V 1978 *Proc. Int. School Phys. Enrico Fermi, Course LXV* (New York: North-Holland) 448

[61] Glendenning N K 1985 *Astrophys. J.* **293** 470

[62] Weber F and Glendenning N K 1993 Hadronic matter and rotating relativistic neutron stars, in: *Proc. of Nankai Summer School Astrophysics and Neutrino Physics (Tianjin)* ed D H Feng, G Z He and X Q Li (Singapore: World Scientific) p 64–183

[63] Lamb F K 1991 Neutron stars and black holes, in: *Frontiers of Stellar Evolution* vol 20, ed D L Lambert (ASP Conference Series) p 299 (San Francisco: Astronomical Society of the Pacific)

[64] Pines D, Tamagaki R and Tsuruta S (ed) 1992 *Proc. Conf. on The Structure and Evolution of Neutron Stars (Kyoto)* (New York: Addison-Wesley)

[65] Muto T, Takatsuka T, Tamagaki R and Tatsumi T 1993 *Prog. Theor. Phys. Suppl.* **112** 221.

[66] Glendenning N K 1997 *Compact Stars, Nuclear Physics, Particle Physics, and General Relativity* (New York: Springer)

[67] Prakash M, Bombaci I, Prakash M, Ellis P J, Lattimer J M and Knorren R 1997 *Phys. Rep.* **280** 1
[68] Feynman R P, Metropolis N and Teller E 1949 *Phys. Rev.* **75** 1561
[69] Harrison B K and Wheeler J A 1965 *Gravitation Theory and Gravitational Collapse* ed B K Harrison, K S Thorne, M Wakano and J A Wheeler (Chicago: University of Chicago Press) pp 1–177
[70] Baym G, Pethick C and Sutherland P 1971 *Astrophys. J.* **170** 299
[71] Baym G, Bethe H A and Pethick C J 1971 *Astrophys. J.* **175** 225
[72] Negele J W and Vautherin D 1973 *Nucl. Phys.* A **178** 123
[73] Fritzsch H, Gell-Mann M and Leutwyler H 1973 *Phys. Lett.* B **47** 365
[74] Baym G and Chin S 1976 *Phys. Lett.* B **62** 241
[75] Keister B D and Kisslinger L S 1976 *Phys. Lett.* B **64** 117
[76] Chapline G and Nauenberg M 1976 *Nature* **264** 235; 1977 *Phys. Rev.* D **16** 450
[77] Fechner W B and Joss P C 1978 *Nature* **274** 347
[78] Glendenning N K 1982 *Phys. Lett.* B **114** 392; 1985 *Astrophys. J.* **293** 470; 1987 *Z. Phys.* A **326** 57; 1987 *Z. Phys.* A **327** 295
[79] Weber F and Weigel M K 1989 *Nucl. Phys.* A **505** 779
[80] Thorne K S 1966 *Proc. Int. School of Phys. Enrico Fermi, Course 35, High Energy Astrophysics* ed L Gratton (New York: Academic Press) p 166
[81] Bowers R L, Campbell J A and Zimmermann R L 1973 *Phys. Rev.* D **7** 2278; 2289
[82] Moszkowski S A 1974 *Phys. Rev.* D **9** 1613
[83] Leung Y C 1989 *Physics of Dense Matter* (Singapore: World Scientific)
[84] Weber F and Weigel M K 1989 *Nucl. Phys.* A **493** 549
[85] Weber F and Weigel M K 1989 *Nucl. Phys.* A **495** 363c
[86] Glendenning N K, Weber F and Moszkowski S A 1992 *Phys. Rev.* C **45** 844
[87] Glendenning N K 1991 *Nucl. Phys.* B *Proc. Suppl.* **24B** 110
[88] Glendenning N K 1992 *Phys. Rev.* D **46** 1274
[89] Glendenning N K and Pei S 1995 *Acta Phys. Hungarica (New Ser.), Heavy Ion Phys.* **1** 323
[90] Glendenning N K 1995 *Phys. Rep.* **264** 143
[91] Glendenning N K 1986 *Phys. Rev. Lett.* **57** 1120
[92] Serot B D and Walecka J D 1986 *Adv. Nucl. Phys.* **16** 1
[93] Ter Haar B and Malfliet R 1987 *Phys. Rep.* **149** 208
[94] Gambhir Y K, Ring P and Thimet A 1990 *Ann. Phys., NY* **198** 132
[95] Reinhard P G 1989 *Rep. Prog. Phys.* **55** 439
[96] Lalazissis G A and Sharma M M 1995 *Nucl. Phys.* A **586** 201
[97] Lalazissis G A, Sharma M M and Ring P 1996 *Nucl. Phys.* A **597** 35
[98] Bouyssy A, Marcos S, Mathiot J F and Van Giai N 1985 *Phys. Rev. Lett.* **55** 1731
[99] Bouyssy A, Mathiot J F and Van Giai N 1987 *Phys. Rev.* C **30** 380
[100] Jetter M, Weber F and Weigel M K 1991 *Europhys. Lett.* **14** 633
[101] Huber H, Weber F and Weigel M K 1993 *Phys. Lett.* B **317** 485
[102] Huber H, Weber F and Weigel M K 1996 *Nucl. Phys.* A **596** 684
[103] Nuppenau C, Lee Y J and Mac Kellar A D 1989 *Nucl. Phys.* A **504** 839
[104] Brockmann R and Machleidt R 1990 *Phys. Rev.* C **42** 1965
[105] Friedman J L, Ipser J R and Parker L 1984 *Nature* **312** 255

[106] Friedman J L, Ipser J R and Parker L 1986 *Astrophys. J.* **304** 115

[107] Friedman J L, Ipser J R and Parker L 1989 *Phys. Rev. Lett.* **62** 3015

[108] Lattimer J M, Prakash M, Masak D and Yahil A 1990 *Astrophys. J.* **355** 241

[109] Weber F, Glendenning N K and Weigel M K 1991 *Astrophys. J.* **373** 579

[110] Weber F and Glendenning N K 1991 *Z. Phys.* A **339** 211

[111] Weber F and Glendenning N K 1991 *Phys. Lett.* B **265** 1

[112] Weber F and Glendenning N K 1992 *Astrophys. J.* **390** 541

[113] Weber F, Glendenning N K and Weigel M K 1991 Rotating neutron stars and the equation of state of dense matter, in: *Progress in High Energy Physics* ed W-Y Pauchy Hwang, S-C Lee, C-E Lee and D J Ernst (New York: North-Holland) p 309

[114] Weber F and Glendenning N K 1992 Impact of the nuclear equation of state on models of rotating neutron stars *Proc. Int. Workshop on Unstable Nuclei in Astrophysics (Tokyo)* ed S Kubono and T Kajino (Singapore: World Scientific) p 307

[115] Hartle J B 1967 *Astrophys. J.* **150** 1005

[116] Hartle J B and Thorne K S 1968 *Astrophys. J.* **153** 807

[117] Poschenrieder P and Weigel M K 1988 *Phys. Lett.* B **200** 231

[118] Poschenrieder P and Weigel M K 1988 *Phys. Rev.* C **38** 471

[119] Machleidt R, Holinde K and Elster Ch 1987 *Phys. Rep.* **149** 1

[120] Holinde K, Erkelenz K and Alzetta R 1972 *Nucl. Phys.* A **194** 161; **198** 598

[121] Celenza L S and Shakin C M 1986 *Relativistic Nuclear Structure Physics* World Scientific Lecture Notes in Physics vol 2 (Singapore: World Scientific)

[122] Horowitz C J and Serot B D 1987 *Nucl. Phys.* A **464** 613

[123] Götz J, Ramschütz J, Weber F and Weigel M K 1989 *Phys. Lett.* B **226** 213

[124] Ramschütz J, Weber F and Weigel M K 1990 *J. Phys. G: Nucl. Part. Phys.* **16** 987

[125] Weber F and Weigel M K 1988 *Z. Phys.* A **330** 249

[126] Hewish A, Bell S J, Pilkington J D H, Scott P F and Collins R A 1968 *Nature* **217** 709

[127] Pacini F 1967 *Nature* **216** 567

[128] Pacini F 1968 *Nature* **219** 146

[129] Gold T 1968 *Nature* **218** 731; 1969 **221** 25

[130] Gunn J E and Ostriker J P 1969 *Nature* **221** 455

[131] Ostriker J P and Gunn J E 1969 *Astrophys. J.* **157** 1395

[132] Gunn J E and Ostriker J P 1970 *Astrophys. J.* **160** 979

[133] Hulse R A and Taylor J H 1975 *Astrophys. J.* **195** L51

[134] Brown G E and Bethe H A 1994 *Astrophys. J.* **423** 659

[135] Ellis P J, Lattimer J M and Prakash M 1996 *Comments Nucl. Part. Phys.* **22** (no 2) 63

[136] Ter Haar D 1972 *Phys. Rep.* **3** 57

[137] Michel F C 1982 *Rev. Mod. Phys.* **54** 1

[138] Trimble V 1968 *Astron. J.* **73** 535

[139] Brown G E 1996 Supernova explosions, black holes and nucleon stars, in: *Proc. Nuclear Physics Conf. INPC '95 (Beijing)* ed S Zuxun and X Jincheng (Singapore: World Scientific) p 623

[140] Glendenning N K 1982 *Phys. Lett.* B **114** 392; 1985 *Astrophys. J.* **293** 470; 1987 *Z. Phys.* A **326** 57; **327** 295; 1989 *Nucl. Phys.* A **493** 521

[141] Madsen J 1988 *Phys. Rev. Lett.* **61** 2909

[142] Madsen J 1994 Physics and astrophysics of strange quark matter, in: *Proc. 2nd Int. Conf. on Physics and Astrophysics of Quark–Gluon Plasma (Calcutta)* ed B Sinha, Y P Viyogi and S Raha (Singapore: World Scientific) p 186

[143] Friedman J L 1990 How fast can pulsars spin?, in: *General Relativity and Gravitation* ed N Ashby, D F Bartlett and W Wyss (Cambridge: Cambridge University Press)

[144] Caldwell R R and Friedman J L 1991 *Phys. Lett.* B **264** 143

[145] Madsen J 1997 *AIP Conf. Proc. 412 (Big Sky)* ed T W Donnelly (New York: American Institute of Physics) p 999

[146] Alcock C and Olinto A V 1988 *Ann. Rev. Nucl. Part. Sci.* **38** 161

[147] Glendenning N K 1990 *Mod. Phys. Lett.* **5** 713

[148] Glendenning N K and Weber F 1992 *Astrophys. J.* **400** 647

[149] Glendenning N K 1990 Strange quark matter stars, in: *Proc. Int. Workshop on Relativistic Aspects of Nuclear Physics (Rio de Janeiro)* ed T Kodama, K C Chung, S J B Duarte and M C Nemes (Singapore: World Scientific)

[150] Bodmer A R 1971 *Phys. Rev.* D **4** 1601

[151] Terazawa H 1979 *INS Report 338* University of Tokyo; 1989 *J. Phys. Soc. Japan* **58** 3555; 4388; 1990 **59** 1199

[152] Madsen J and Haensel P (ed) 1991 *Proc. Int. Workshop on Strange Quark Matter in Physics and Astrophysics (Aarhus)*, *Nucl. Phys.* B *Proc. Suppl.* **24B** 1

[153] Liu H C and Shaw G L 1984 *Phys. Rev.* D **30** 1137

[154] Greiner C, Koch P and Stöcker H 1987 *Phys. Rev. Lett.* **58** 1825

[155] Greiner C, Rischke D H, Stöcker H and Koch P 1988 *Phys. Rev.* D **38** 2797

[156] Shaw G L, Shin M, Dalitz R H and Desai M 1989 *Nature* **337** 436

[157] Greiner C and Stöcker H 1991 *Phys. Rev.* D **44** 3517

[158] Barz H W, Friman B L, Knoll J and Schulz H 1991 *Proc. Int. Workshop on Strange Quark Matter in Physics and Astrophysics (Aarhus)* ed J Madsen and P Haensel *Nucl. Phys.* B *Proc. Suppl.* **24B** 211

[159] Crawford H J, Desai M S and Shaw G L 1992 *Phys. Rev.* D **45** 857

[160] Usov V V 1998 *Phys. Rev. Lett.* **80** 230

[161] McLerran L 1989 Strange matter: a short review, in: *The Nuclear Equation of State (Part B)* NATO ASI Series vol 216B ed W Greiner and H Stöcker (New York: Plenum Press) p 155

[162] Madsen J 1993 *Phys. Rev. Lett.* **70** 391

[163] Heiselberg H, Pethick C J and Staubo E F 1993 *Phys. Rev. Lett.* **70** 1355

[164] Alcock C and Farhi E 1985 *Phys. Rev.* D **32** 1273

[165] Applegate J and Hogan C 1985 *Phys. Rev.* D **31** 3037

[166] Madsen J, Heiselberg H and Riisager K 1986 *Phys. Rev.* D **34** 2947

[167] Madsen J and Olesen M L 1991 *Phys. Rev.* D **43** 1069; **44** 566 (erratum)

[168] Ruderman M A 1969 *Nature* **223** 597

[169] Baym G and Pines D 1971 *Ann. Phys., NY* **66** 816
[170] Olinto A V 1987 *Phys. Lett.* B **192** 71
[171] Frieman J A and Olinto A V 1989 *Nature* **341** 633
[172] Horvath J E and Benvenuto O G 1988 *Phys. Lett.* B **213** 516
[173] Alpar M A 1987 *Phys. Rev. Lett.* **58** 2152
[174] Glendenning N K, Kettner Ch and Weber F 1995 *Phys. Rev. Lett.* **74** 3519
[175] De Rújula A and Glashow S L 1984 *Nature* **312** 734
[176] Price P B 1988 *Phys. Rev.* D **38** 3813
[177] Bjorken J D and McLerran L 1979 *Phys. Rev.* D **20** 2353
[178] Chin S A and Kerman A K 1979 *Phys. Rev. Lett.* **43** 1292
[179] Chinellato J A *et al* (Brazil–Japan Collaboration) 1990 *Proc. 21st Int. Cosmic Ray Conf. (Adelaide)* vol 8, ed R J Protheroe (Northfield, South Australia: Graphic Services) p 259
[180] Jaffe R L 1977 *Phys. Lett.* **38** 195
[181] Baym G, Kolb E W, McLerran L, Walker T P and Jaffe R L 1985 *Phys. Lett.* B **160** 181
[182] Terazawa H 1991 *J. Phys. Soc. Japan* **60** 1848
[183] Terazawa H 1993 *J. Phys. Soc. Japan* **62** 1415
[184] Michel F C 1988 *Phys. Rev. Lett.* **60** 677
[185] Benvenuto O G, Horvath J E and Vucetich H 1989 *Int. J. Mod. Phys.* A **4** 257
Benvenuto O G and Horvath J E 1989 *Phys. Rev. Lett.* **63** 716
[186] Horvath J E, Benvenuto O G and Vucetich H 1992 *Phys. Rev.* D **45** 3865
[187] Weber F and Glendenning N K 1993 Neutron stars, strange stars, and the nuclear equation of state, in: *Proc. 1st Symp. on Nuclear Physics in the Universe (Oak Ridge)* ed M W Guidry and M R Strayer (Bristol: Institute of Physics Publishing) p 127
[188] Glendenning N K, Kettner Ch and Weber F 1995 *Astrophys. J.* **450** 253
[189] Haensel P and Zdunik J L 1989 *Nature* **340** 617
[190] Glendenning N K 1989 *Phys. Rev. Lett.* **63** 2629
[191] Glendenning N K 1990 Supernovae, compact stars and nuclear physics, in: *Proc. 1989 Int. Nuclear Physics Conf. (Sao Paulo)* vol 2, ed M S Hussein *et al* (Singapore: World Scientific)
[192] Sawyer R F 1989 *Phys. Lett.* B **233** 412
[193] Weber F, Kettner Ch, Weigel M K and Glendenning N K 1995 Strange matter stars, in: *Proc. Int. Symp. on Strangeness and Quark Matter* ed G Vassiliadis, A D Panagiotou, B S Kumar and J Madsen (Singapore: World Scientific) p 308
[194] Kettner Ch, Weber F, Weigel M K and Glendenning N K 1995 Stability of Strange quark stars with nuclear crust against radial oscillations, in: *Proc. Int. Symp. on Strangeness and Quark Matter* ed G Vassiliadis, A D Panagiotou, B S Kumar and J Madsen (Singapore: World Scientific) p 333
[195] Glendenning N K, Kettner Ch and Weber F 1995 Possible new class of dense white dwarfs, in: *Proc. Int. Conf. on Strangeness in Hadronic Matter* AIP 340 ed J Rafelski (New York: American Institute of Physics) p 46

[196] Romanelli P F C 1986 Strange Star Crusts *BS Thesis* (Massachusetts Institute of Technology, USA)

[197] Benvenuto O G and Althaus L G 1996 *Astrophys. J.* **462** 364; *Phys. Rev.* D **53** 635

[198] Weber F, Schaab Ch, Weigel M K and Glendenning N K 1997 From quark matter to strange MACHOs, in: *Frontier Objects in Astrophysics and Particle Physics (Vulcano)* ed F Giovannelli and G Mannocchi (Bologna: Editrice Compositori) p 87

[199] Ghosh S K, Phatak S C, Sahu P K 1996 *Nucl. Phys.* A **596** 670
Dai Z, Lu T and Peng Q 1993 *Phys. Lett.* B **319** 199

[200] Horvath J E, Vucetich H and Benvenuto O G 1993 *Mon. Not. R. Astron. Soc.* **262** 506

[201] Cho S J, Lee K S and Heinz U 1994 *Phys. Rev.* D **50** 4771

[202] Kumar B S 1995 Strange quark matter: past, present, and future, in: *Proc. Int. Symp. on Strangeness and Quark Matter* ed G Vassiliadis, A D Panagiotou, B S Kumar and J Madsen (Singapore: World Scientific) p 318

[203] Pretzl K (NA52 collaboration) 1997 (Private communication)

[204] Wilk G and Wlodarczyk Z 1996 *J. Phys. G: Nucl. Part. Phys.* **22** L105

[205] Saito T, Hatano Y, Fukuda Y and Oda H 1990 *Phys. Rev. Lett.* **65** 2094

[206] Saito T 1995 Test of the CRASH experiment counters with heavy ions, in: *Proc. Int. Symp. on Strangeness and Quark Matter* ed G Vassiliadis, A D Panagiotou, B S Kumar and J Madsen (Singapore: World Scientific) p 259

[207] Ichimura M *et al* 1993 *Nuovo Cimento* A **36** 843

[208] Shaulov S B 1996 *Acta Phys. Hungarica (New Ser.), Heavy Ion Phys.* **4** 403

[209] MACRO collaboration 1992 *Phys. Rev. Lett.* **69** 1860

[210] De Rújula A, Glashow S L, Wilson R R and Charpak G 1983 *Phys. Rep.* **99** 341

[211] Price P B 1984 *Phys. Rev. Lett.* **52** 1265

[212] Brügger M, Lützenkirchen K, Polikanov S, Herrmann G, Overbeck M, Trautmann N, Breskin A, Chechik R, Fraenkel Z and Smilansky U 1989 *Nature* **337** 434

[213] Thomas J and Jacobs P 1995 *A Guide to the High Energy Heavy Ion Experiments* (Lawrence Livermore National Laboratory, UCRL-ID-119181)

[214] Rusek A *et al* (E886 collaboration) 1996 *Phys. Rev.* C **54** R15

[215] Belz J *et al* (E888 collaboration) 1996 *Phys. Rev.* D **53** R3487

[216] Belz J *et al* (E888 collaboration) 1996 *Phys. Rev. Lett.* **76** 3277

[217] Pretzl K *et al* (NA52 collaboration) 1995 Search for strange quark matter in relativistic heavy ion collisions at CERN, in: *Proc. Int. Symp. on Strangeness and Quark Matter* ed G Vasiliadis, A D Panagiotou, B S Kumar and J Madsen (Singapore: World Scientific) p 230

[218] Dittus F *et al* (NA52 collaboration) 1995 First look at NA52 Data on Pb–Pb interactions at 158 A GeV/c, in: *Proc. Int. Conf. on Strangeness in Hadronic Matter* AIP Conf. Ser. 340, ed J Rafelski (New York: American Institute of Physics) p 24

[219] Pretzl K *et al* (NA52 collaboration) 1997 Search for strangelets and production of antinuclei in Pb–Pb collisions at CERN, in: *Proc. Int. Conf. on Nucl. Phys. at the Turn of the Millennium: Structure of Vacuum and Elementary Matter* ed H Stöcker, A Gallmann and J H Hamilton (Singapore: World Scientific) p 379

[220] Price P B 1978 *Phys. Rev.* D **18** 1382

[221] Saito T 1995 *Proc. 24th Int. Cosmic Ray Conf. (Rome)* vol 1, p 898

[222] Miyamura O 1995 *Proc. 24th Int. Cosmic Ray Conf. (Rome)* vol 1, p 890

[223] Capdevielle J N 1995 *Proc. 24th Int. Cosmic Ray Conf. (Rome)* vol 1, p 910

[224] Shapiro S L and Teukolsky S A 1983 *Black Holes, White Dwarfs, and Neutron Stars* (New York: Wiley)

[225] Meszaros P 1992 *High Energy Radiation from Magnetized Neutron Stars* (Chicago: University of Chicago Press)

[226] Hartmann D 1994 *High Energy Astrophysics* ed J M Matthews (Singapore: World Scientific)

[227] Fishman G J and Meegan Ch A 1995 *Ann. Rev. Astron. Astrophys.* **33** 415

[228] Piran T 1997 *Unsolved Problems in Astrophysics* ed J N Bahcall and J P Ostriker (Princeton: Princeton University Press)

[229] Haensel P, Paczynski B and Amsterdamski P 1991 *Astrophys. J.* **375** 209

[230] Gladysz-Dzioaduś E and Panagiotou A D 1995 Strangelet frmation in Centauro cosmic ray events, in: *Proc. Int. Symp. on Strangeness and Quark Matter* ed G Vasiliadis, A D Panagiotou, B S Kumar and J Madsen (Singapore: World Scientific) p 265

[231] Bahcall J N 1978 *Ann. Rev. Astron. Astrophys.* **16** 241

[232] Taylor J H and Weisberg J M 1989 *Astrophys. J.* **345** 434

[233] Thielemann F K, Nomoto K and Hashimoto M 1992 Neutron star masses from supernova explosions, in: *The Structure and Evolution of Neutron Stars* ed D Pines, R Tamagaki and S Tsuruta (New York: Addison-Wesley) p 298

[234] Thielemann F K, Hashimoto M, Nomoto K 1996 *Astrophys. J.* **460** 408

[235] Rappaport S A and Joss P C 1983 *Accretion Driven Stellar X-Ray Sources* ed W H G Lewin and E P J van den Heuvel (Cambridge: Cambridge University Press)

[236] Joss P C and Rappaport S A 1984 *Ann. Rev. Astron. Astrophys.* **22** 537

[237] Bildsten L and Cutler C 1995 *Astrophys. J.* **449** 800

[238] Kaaret P, Ford E and Chen K 1997 *Astrophys. J.* **480** L27

[239] Alpar M A and Shaham J 1985 *Nature* **316** 239

[240] Lamb F K, Shibazaki N, Alpar M A and Shaham J 1985 *Nature* **317** 681

[241] Epstein R I, Lamb F K and Piedhorsky W C 1986 X-Ray Variability in Astrophysics *Los Alamos Science* vol 13, ed N G Cooper (Los Alamos: LANL) p 34

[242] Shapiro I I 1964 *Phys. Rev. Lett.* **13** 789

[243] Weinberg S 1972 *Gravitation and Cosmology* (New York: Wiley)

[244] Rhoades C E and Ruffini R 1974 *Phys. Rev. Lett.* **32** 324

[245] Sabbadini A G and Hartle J B 1977 *Ann. Phys., NY* **104** 95

[246] Hartle J B 1978 *Phys. Rep.* **46** 201

[247] Glendenning N K 1998 Pulsar signal of deconfinement, in: *Proc. Quark Matter 1997 (Tsukuba), Nucl. Phys.* A **638** 239c

[248] Shibazaki N, Murakami T, Shaham J and Nomoto K 1989 *Nature* **342** 656
[249] Bell J F, Kulkarni S R, Bailes M, Leitch E M and Lyne A G 1995 *Astrophys. J.* **452** L121
[250] Van Paradijs J 1978 *Astrophys. J.* **234** 609
[251] Fujimoto M Y and Taam R E 1986 *Astrophys. J.* **305** 246
[252] Yancopoulos S D, Hamilton T T and Helfand D J 1994 *Astrophys. J.* **429** 832
[253] Walter F M, Lattimer J M, Prakash M, Matthews L, An P and Neuhäuser R 1997 *HST Cycle 7 General Observer Proposal*
[254] Brown G E 1997 *Phys. Blätter* **53** 671
[255] Li G Q, Lee C H and Brown G E 1997 *Nucl. Phys.* A **625** 372
[256] Li G Q, Lee C H and Brown G E 1997 *Phys. Rev. Lett.* **79** 5214
[257] Wasserman I and Shapiro S L 1983 *Astrophys. J.* **265** 1036
[258] Ruderman M A 1987 *High Energy Phenomena around Collapsed Stars* ed F Pacini (Dodrecht: Reidel)
[259] Lyne A G, Pritchard R S, Graham-Smith F and Camilo F 1996 *Nature* **381** 497
[260] Lyne A G, Pritchard R S and Graham-Smith F 1993 *Mon. Not. R. Astron. Soc.* **265** 1003
[261] Kaspi V M, Manchester R N, Siegman B, Johnston S and Lyne A G 1994 *Astrophys. J.* **422** L83
[262] Boyd P T *et al* 1995 *Astrophys. J.* **448** 365
[263] Shemar S L and Lyne A G 1996 *Mon. Not. R. Astron. Soc.* **282** 667
[264] Lyne A G, Kaspi V M, Bailes M, Manchester R N, Taylor H and Arzoumanian Z 1996 *Mon. Not. R. Astron. Soc.* **281** L14
[265] Anderson P W and Itoh N 1975 *Nature* **256** 25
[266] Pines D and Alpar M A 1985 *Nature* **316** 27
[267] Alpar M A 1992 Some topics in neutron star superfluid dynamics, in: *The Structure and Evolution of Neutron Stars* ed D Pines, R Tamagaki and S Tsuruta (New York: Addison-Wesley) p 148
[268] Clark J W, Davé R D and Chen J M C 1992 Microscopic calculations of superfluid gaps, in: *The Structure and Evolution of Neutron Stars* ed D Pines, R Tamagaki and S Tsuruta (New York: Addison-Wesley) p 134
[269] Epstein R I, Link B and Baym G 1992 Superfluid dynamics in the inner crust of neutron stars, in: *The Structure and Evolution of Neutron Stars* ed D Pines, R Tamagaki and S Tsuruta (New York: Addison-Wesley) p 156
[270] Alpar M A, Cheng K S and Pines D 1989 *Astrophys. J.* **346** 823
[271] Sedrakian A 1995 *Mon. Not. R. Astron. Soc.* **277** 225
[272] Sedrakian A and Sedrakian D M 1995 *Astrophys. J.* **447** 305
[273] Sedrakian A, Sedrakian D M, Cordes J and Terzian Y 1995 *Astrophys. J.* **447** 324
[274] Baym G and Pethick C 1975 *Ann. Rev. Nucl. Sci.* **25** 27
[275] Trimble V and Rees M 1970 *Astrophys. Lett.* **5** 93
[276] Börner G and Cohen J M 1973 *Astrophys. J.* **185** 959
[277] Stella L and Vietri M 1998 *Astrophys. J.* **492** L59
[278] Brecher K and Burrows A 1980 *Astrophys. J.* **240** 642
[279] Ramaty R and Meszaros P 1981 *Astrophys. J.* **250** 384

[280] Ruderman M 1992 Neutron star crust braking and magnetic field evolution, in: *The Structure and Evolution of Neutron Stars* ed D Pines, R Tamagaki and S Tsuruta (New York: Addison-Wesley) p 353

[281] Sedrakyan D M 1970 *Nature* **228** 1074; *Astrophysics* **6** 339

[282] Chakrabarty S, Bandyopadhyay D and Pal S 1997 *Phys. Rev. Lett.* **78** 2898

[283] Becker W, Predehl P, Trümper J and Ögelman H 1992 *IAU Circular* **5554** 1

[284] Finley J and Ögelman H 1993 *IAU Circular* **5787**

[285] Becker W 1993 *IAU Circular* **5805**

[286] Becker W and Aschenbach B 1995 *The Lives of Neutron Stars* ed M A Alpar, Ü Kiziloglu and J van Paradijs (Dordrecht: Kluwer) p 47

[287] Seward F, Harnden F, Murdin P and Clark D 1983 *Astrophys. J.* **267** 698

[288] Trussoni E, Brinkmann W, Ögelman H, Hasinger G and Aschenbach B 1990 *Astron. Astrophys.* **234** 403

[289] Finley J P Ögelman H, Hasinger G and Trümper J 1993 *Astrophys. J.* **410** 323

[290] Safi-Harb S and Ögelman H 1995 *The Lives of Neutron Stars* ed M A Alpar, Ü Kiziloglu and J van Paradijs (Dordrecht: Kluwer) p 53

[291] Yancopoulos S D, Hamilton T D and Helfland D J 1993 *Bull. Am. Astron. Soc.* **25** 912

[292] Seward F and Wang Z R 1988 *Astrophys. J.* **332** 199

[293] Becker W and Trümper J 1993 *Nature* **365** 528

[294] Ögelman H, Finley J and Zimmermann H 1993 *Nature* **361** 136

[295] Finley J P Ögelman H and Kiziloglu Ü 1992 *Astrophys. J.* **394** L21

[296] Halpern J and Ruderman M 1993 *Astrophys. J.* **415** 286

[297] Ögelman H and Finley J 1993 *Astrophys. J.* **413** L31

[298] Negele J W and Vautherin D 1973 *Nucl. Phys.* A **207** 298

[299] Abrahams A M and Shapiro S L 1991 *Astrophys. J.* **374** 652

[300] Friedman E, Gal A and Batty C J 1994 *Nucl. Phys.* A **579** 518

[301] Barth R *et al* 1997 *Phys. Rev. Lett.* **78** 4027

[302] Cassing W, Bratkovskaya E L, Mosel U, Teis S and Sibirtsev A 1997 *Nucl. Phys.* A **614** 415
Bratkovskaya E L Cassing W and Mosel U 1997 *Nucl. Phys.* A **622** 593

[303] Weise W 1998 Quarks, hadrons and dense nuclear matter, in: *Trends in Nuclear Physics – 100 Years Later (Les Houches)* (Amsterdam: Elsevier Science) in press (see also: *Preprint* Technical University of Munich TUM/T39–97–1)

[304] Dashen R F and Rajaraman R 1974 *Phys. Rev.* D **10** 696; 708

[305] Rajaraman R 1979 Elementarity of baryon resonances in nuclear matter, in: *Mesons in Nuclei* vol III ed M Rho and D Wilkinson (Amsterdam: North-Holland) p 197

[306] Particle Data Group 1984 *Rev. Mod. Phys.* **56** S1

[307] Holzenkamp B, Holinde K and Speth J 1989 *Nucl. Phys.* A **500** 485

[308] Holinde K 1981 *Phys. Rep.* **68** 121

[309] Boguta J and Bodmer A R 1977 *Nucl. Phys.* A **292** 413

[310] Boguta J and Rafelski J 1977 *Phys. Lett.* B **71** 22

[311] Brown G E and Rho M 1991 *Phys. Rev. Lett.* **66** 2720

[312] Guichon P A M, Saito K, Rodionov E and Thomas A W 1996 *Nucl. Phys.* A **601** 349
[313] Saito K, Tsushima K and Thomas A W 1997 *Phys. Rev.* C **55** 2637
[314] Weber F and Weigel M K 1989 *J. Phys. G: Nucl. Part. Phys.* **15** 765
[315] Wilets L 1979 Green's functions method for the relativistic field theory many-body problem, in: *Mesons in Nuclei* vol III ed M Rho and D Wilkinson (Amsterdam: North-Holland) p 791
[316] Weber F and Weigel M K 1990 *Nucl. Phys.* A **519** 303c
[317] Weigel M K and Wegmann G 1971 *Fortschr. Phys.* **19** 451
[318] Müther H, Prakash M and Ainsworth T L 1987 *Phys. Lett.* B **199** 469
[319] Martin P C and Schwinger J 1959 *Phys. Rev.* **115** 1342
[320] Erkelenz K 1974 *Phys. Rev.* C **13** 191
[321] Machleidt R 1989 *Adv. Nucl. Phys.* **19** 188
[322] Iachello F, Jackson A D and Landé A 1973 *Phys. Lett.* B **43** 191
[323] Höhler G and Pietarinen E 1975 *Nucl. Phys.* B **95** 210
[324] Grein W 1977 *Nucl. Phys.* B **131** 255
[325] Weber F and Weigel M K 1985 *Phys. Rev.* C **32** 2141
[326] Jackson A D 1983 *Ann. Rev. Nucl. Part. Sci.* **33** 105
[327] Haug P K 1969 *Dissertation* University of Frankfurt
[328] Joachain Ch J 1975 *Quantum Collision Theory* (Amsterdam: North-Holland)
[329] Salpeter E and Bethe H A 1951 *Phys. Rev.* **84** 1232
[330] Nakanishi N 1969 *Prog. Theor. Phys. Suppl.* **43** 131
[331] Woloshyn R M and Jackson A D 1973 *Nucl. Phys.* B **64** 269
[332] Zuilhof M J and Tjon J A 1982 *Phys. Rev.* C **26** 1277
[333] Blankenbecler R and Sugar R 1966 *Phys. Rev.* **142** 1051
[334] Reiner A S 1964 *Phys. Rev.* **133** 1105
[335] Barton G 1965 *Introduction to Dispersion Techniques in Field Theory* (New York: Benjamin)
[336] Arfken G 1985 *Mathematical Methods for Physicists* (Orlando: Academic Press)
[337] Weber F and Weigel M K 1990 *Europhys. Lett.* **12** 603
[338] Dolan L and Jackiw R 1974 *Phys. Rev.* D **9** 3320
[339] Kadanoff L P and Baym G 1962 *Quantum Statistical Mechanics* (New York: Benjamin)
[340] Fetter A L and Walecka J D 1971 *Quantum Theory of Many-Particle Systems* (New York: McGraw-Hill)
[341] Bernard C W 1974 *Phys. Rev.* D **9** 3312
[342] Shuryak E V 1980 *Phys. Rep.* **61** 71
[343] Gross D J, Pisarski R D and Yaffe L G 1981 *Rev. Mod. Phys.* **53** 43
[344] Puff R D 1961 *Ann. Phys., NY* **13** 317
[345] Bowers R L and Zimmerman R L 1973 *Phys. Rev.* D **7** 296
[346] Bowers R L, Gleeson A M and Pedigo R D 1975 *Phys. Rev.* D **12** 3043; 3056
[347] Bjorken J D and Drell S D 1965 *Relativistic Quantum Fields* (New York: McGraw-Hill)
[348] Johnson M H and Teller E 1955 *Phys. Rev.* **98** 783
[349] Duerr H P 1956 *Phys. Rev.* **103** 469

[350] Walecka J D 1974 *Ann. Phys., NY* **83** 491
[351] Zimanyi J and Moszkowski S A 1990 *Phys. Rev.* C **42** 1416
[352] Glendenning N K, Banerjee B and Gyulassy M 1983 *Ann. Phys., NY* **149** 1
[353] Particle Data Group 1976 *Rev. Mod. Phys.* **48** 30
[354] Sakurai J J 1966 *Phys. Rev. Lett.* **17** 1021
[355] Glendenning N K 1989 *Nucl. Phys.* A **493** 521
[356] Horowitz C J and Serot B D 1983 *Nucl. Phys.* A **399** 529
[357] Boguta J 1982 *Phys. Lett.* B **109** 251
[358] Waldhauser B M, Maruhn J A, Stöcker H and Greiner W 1987 *Z. Phys.* A **328** 19
[359] Waldhauser B M, Theis J, Maruhn J A, Stöcker H and Greiner W 1987 *Phys. Rev.* C **36** 1019
[360] Pannert W, Ring P and Boguta J 1987 *Phys. Rev. Lett.* **59** 2420
[361] Glendenning N K 1987 *Phys. Lett.* B **185** 275; *Nucl. Phys.* A **469** 600
[362] Glendenning N K and Moszkowski S A 1991 *Phys. Rev. Lett.* **67** 2414
[363] Boguta J and Bohrmann S 1981 *Phys. Lett.* B **102** 93
[364] Walker G E 1986 *Nucl. Phys.* A **450** 287c.
[365] Rufa M, Schaffner J, Marhun J, Stöcker H, Greiner W and Reinhard P G 1990 *Phys. Rev.* C **42** 2469
[366] Millener D J, Dover C B and Gal A 1988 *Phys. Rev.* C **38** 2700
[367] Weisskopf V F 1957 *Rev. Mod. Phys.* **29** 174
[368] Hugenholtz N M and Van Hove L 1958 *Physica* **24** 363
[369] Banerjee B, Glendenning N K and Gyulassy M 1981 *Nucl. Phys.* A **361** 326
[370] Glendenning N K, Hecking P and Ruck V 1983 *Ann. Phys., NY* **149** 22
[371] Campbell D K 1977 Field theory, chiral symmetry, and pion–nucleus interaction, in: *Nuclear Physics with Heavy Ions and Mesons, Les Houches* vol 2 (Amsterdam: North-Holland) p 549
[372] Kaplan D and Nelson A 1986 *Phys. Lett.* B **175** 57
[373] Brown G E, Kubodera K and Rho M 1987 *Phys. Lett.* B **192** 273
[374] Lee C H and Rho M 1995 Kaon condensation in dense stellar matter, in: *Proc. Int. Symp. on Strangeness and Quark Matter* ed G Vasiliadis, A Panagiotou, B S Kumar and J Madsen (Singapore: World Scientific) p 283
[375] Ainsworth T L, Baron E, Brown G E, Cooperstein J and Prakash M 1987 *Nucl. Phys.* A **464** 740
[376] Kruse H, Jacak B and Stöcker H 1985 *Phys. Rev. Lett.* **54** 289
[377] Molitoris J J and Stöcker H 1985 *Phys. Rev.* C **32** 346
[378] Kruse H *et al* 1985 *Phys. Rev.* C **31** 1770
[379] Stock R *et al* 1982 *Phys. Rev. Lett.* **49** 1236
[380] Harris J W *et al* 1987 *Phys. Rev. Lett.* **58** 463
[381] Renfordt R E 1984 *Phys. Rev. Lett.* **53** 763
[382] Glendenning N K 1988 *Phys. Rev.* C **37** 2733
[383] Waldhauser B M, Maruhn J A, Stöcker H and Greiner W 1988 *Phys. Rev.* C **38** 1003
[384] Friedman B and Pandharipande V R 1981 *Nucl. Phys.* A **361** 502
[385] Harris J W *et al* 1985 *Phys. Lett.* B **153** 377; 1984 *Phys. Rev. Lett.* **52** 289
[386] Ter Haar B and Malfliet R 1986 *Phys. Rev. Lett.* **56** 1237

[387] Sauer G, Chandra H and Mosel U 1976 *Nucl. Phys.* A **264** 221
[388] Freedman R A 1977 *Phys. Lett.* B **71** 369
[389] Schmidt K E and Pandharipande V R 1979 *Phys. Lett.* B **87** 11
[390] Lattimer J M 1981 *Ann. Rev. Nucl. Part. Sci.* **31** 337; 1992 The equation of state in neutron star matter, in: *The Structure and Evolution of Neutron Stars* ed D Pines, R Tamagaki and S Tsuruta (New York: Addison-Wesley) p 50
[391] Huber H, Weber F, Weigel M K and Schaab Ch 1998 *Int. J. Mod. Phys.* E **7** (no 3) 301
[392] Garpman S I A, Glendenning N K and Karant Y J 1979 *Nucl. Phys.* A **322** 382
[393] Itoh N, Mitake S, Iyetomi H and Ichimaru S 1983 *Astrophys. J.* **273** 774
[394] Itoh N, Hayashi H and Kohyama Y 1993 *Astrophys. J.* **418** 405
Itoh N and Kohyama Y 1993 *Astrophys. J.* **404** 268
[395] Østgaard E 1992 Electrical conductivity in neutron stars, in: *Proc. 1st Symp. on Nuclear Physics in the Universe (Oak Ridge)* ed M W Guidry and M R Strayer (Bristol: Institute of Physics Publishing) p 471
[396] Itoh N and Hiraki K 1994 *Astrophys. J.* **435** 784
Itoh N Kotouda T and Hiraki K 1995 *Astrophys. J.* **455** 244
[397] Keil W and Janka H T 1995 *Astron. Astrophys.* **296** 145
[398] Schröter A *et al* 1994 *Z. Phys.* A **350** 101
[399] Senger P 1996 *Acta Phys. Hungarica (New Ser.), Heavy Ion Phys.* **4** 317
[400] Waas T, Rho M and Weise W 1997 *Nucl. Phys.* A **617** 449
[401] Schaffner J and Mishustin I N 1996 *Phys. Rev.* C **53** 1416
[402] Tamagaki R 1991 *Prog. Theor. Phys.* **85** 321
[403] Sakai T, Mori J, Buchmann A J, Shimizu K and Yazaki K 1997 *Nucl. Phys.* A **625** 192
[404] Faessler A, Buchmann A J, Krivoruchenko M I and Martemyanov B V 1997 *Phys. Lett.* B **391** 255
[405] Faessler A, Buchmann A J and Krivoruchenko M I 1997 *Phys. Rev.* C **56** 1576
[406] Glendenning N K and Schaffner-Bielich J 1998 *Phys. Rev.* C **58** 1298
[407] Pandharipande V R 1971 *Nucl. Phys.* A **178** 123
[408] Bethe H A and Johnson M 1974 *Nucl. Phys.* A **230** 1
[409] Myers W D and Swiatecki W J 1996 *Nucl. Phys.* A **601** 141
[410] Glendenning N K 1990 Neutron stars, fast pulsars, supernovae and the equation of state of dense matter, in: *Proc. NATO Advanced Study Institute on The Nuclear Equation of State (Peniscola)* NATO ASI Series B vol 216A, ed W Greiner and H Stöcker (New York: Plenum Press) p 751
[411] Kapusta J I and Olive K A 1990 *Phys. Rev. Lett.* **64** 13
[412] Baym G 1992 Ultrarelativistic heavy-ion collisions and neutron stars, in: *The Structure and Evolution of Neutron Stars* ed D Pines, R Tamagaki and S Tsuruta (New York: Addison-Wesley) p 188
[413] Brown G E and Weise W 1976 *Phys. Rep.* **27** 1
[414] Migdal A B 1978 *Rev. Mod. Phys.* **50** 107
[415] Baym G and Campbell D K 1979 Chiral symmetry and pion condensation, in: *Mesons in Nuclei* vol III ed M Rho and D Wilkinson (Amsterdam: North-Holland) p 1031

[416] Muto T and Tatsumi T 1987 *Prog. Theor. Phys.* **78** 1405
[417] Takatsuka T, Tamiya K, Tatsumi T and Tamagaki R 1978 *Prog. Theor. Phys.* **59** 1933
[418] Bäckman S O and Weise W 1979 Pion condensation and the pion–nuclear interaction, in: *Mesons in Nuclei* vol III, ed M Rho and D Wilkinson (Amsterdam: North-Holland) p 1095
[419] Barshay S and Brown G E 1973 *Phys. Lett.* B **47** 107
[420] Brown G E, Kubodera K, Page D and Pizzochero P 1988 *Phys. Rev.* D **37** 2042
[421] Tatsumi T and Muto T 1991 *Nuclei in the Cosmos* ed H Oberhummer and C Rolfs (Berlin: Springer)
[422] Umeda H, Nomoto K, Tsuruta S, Muto T and Tatsumi T 1992 Neutron star cooling and pion condensation, in: *The Structure and Evolution of Neutron Stars* ed D Pines, R Tamagaki and S Tsuruta (New York: Addison-Wesley) p 406
[423] Tatsumi T 1988 *Prog. Theor. Phys.* **80** 22
[424] Page D and Baron E 1990 *Astrophys. J.* **354** (1990) L17
[425] Thorsson V, Prakash M and Lattimer J M 1994 *Nucl. Phys.* A **572** 693
[426] Glendenning N K, Pei S and Weber F 1997 *Phys. Rev. Lett.* **79** 1603
[427] Chapline G and Nauenberg M 1977 *Ann. NY Acad. Sci.* **302** 191
[428] Gentile N A, Aufderheide M B, Mathews G J, Swesty F D and Fuller G M 1993 *Astrophys. J.* **414** 701
[429] Prakash M, Cooke J R and Lattimer J M 1995 *Phys. Rev.* D **52** 661
[430] Hasenfratz A and Hasenfratz P 1985 *Ann. Rev. Nucl. Part. Sci.* **35** 559
[431] Ellis J, Kapusta J I and Olive K A 1991 *Nucl. Phys.* B **348** 345
[432] Bethe H A, Brown G E and Cooperstein J 1987 *Nucl. Phys.* A **462** 791
[433] Serot B D and Uechi H 1987 *Ann. Phys., NY* **179** 272
[434] Ravenhall D G, Pethick C J and Wilson J R 1983 *Phys. Rev. Lett.* **50** 2066
[435] Ravenhall D G, Pethick C J and Lattimer J M 1983 *Nucl. Phys.* A **407** 571
[436] Williams R D and Koonin S E 1985 *Nucl. Phys.* A **435** 844
[437] Hermann B 1996 *Master Thesis* University of Munich
[438] Glendenning N K and Pei S 1995 *Phys. Rev.* C **52** 2250
[439] Heiselberg H 1995 Quark matter structure in neutron stars, in: *Proc. Int. Symp. on Strangeness and Quark Matter* ed G Vasiliadis, A D Panagiotou, B S Kumar and J Madsen (Singapore: World Scientific) p 298
[440] Poschenrieder P 1987 *Dissertation* University of Munich
[441] Rose M E 1957 *Elementary Theory of Angular Momentum* (New York: Wiley)
[442] Erkelenz K, Alzetta R and Holinde K 1971 *Nucl. Phys.* A **176** 413
[443] Ho-Kim Q and Khanna F C 1974 *Ann. Phys., NY* **86** 233
[444] Wegmann G 1968 *Z. Phys.* **211** 235
[445] Wegmann G and Weigel M K 1968 *Phys. Lett.* B **26** 245
[446] Jacob M and Wick G C 1959 *Ann. Phys., NY* **7** 404
[447] Weber F 1992 Hadron physics and neutron star properties *Habilitation Thesis* University of Munich
[448] Huber H 1993 *Master Thesis* University of Munich
[449] De Jonga F and Malfliet R 1991 *Phys. Rev.* C **44** 998

[450] Huber H, Weber F and Weigel M K 1995 *Phys. Rev.* C **51** 1790
[451] Brockmann R and Toki H 1992 *Phys. Rev. Lett.* **68** 3408
[452] Haddad S and Weigel M K 1993 *Phys. Rev.* C **48** 2740
[453] Haddad S and Weigel M K 1994 *Nucl. Phys.* A **578** 471
[454] Von-Eiff D, Stocker W and Weigel M K 1994 *Phys. Rev.* C **50** 1436
[455] Engvik L, Hjorth-Jensen M, Osnes E, Bao G and Østgaard E 1994 *Phys. Rev. Lett.* **73** 2650
[456] Bao G, Engvik L, Hjorth-Jensen M, Osnes E and Østgaard E 1994 *Nucl. Phys.* A **575** 707
[457] Bao G Østgaard E and Dybvik B 1994 *Int. J. Mod. Phys.* D **3** 813
[458] Engvik L, Osnes E, Hjorth-Jensen M, Bao G and Østgaard E 1996 *Astrophys. J.* **469** 794
[459] Huber H, Weigel M K and Weber F 1994 The relativistic treatment of symmetric and asymmetric nuclear matter, in: *Proc of NATO Advanced Study Institute on Hot and Dense Nuclear Matter* Series B vol 335 ed W Greiner, H Stöcker and A Gallmann (New York: Plenum Press) p 799
[460] Huber H, Weber F and Weigel M K 1994 *Phys. Rev.* C **50** R1287
[461] Lopez-Quelle M, Marcos S, Niembro R, Bouyssy A and Van Giai N 1988 *Nucl. Phys.* A **483** 479
[462] Bernardos P, Fomenko V N, Van Giai N, Quelle M L, Marcos S, Niembro R and Savushkin L N 1983 *Phys. Rev.* C **48** 2665
[463] Zhang J K, Jin Y and Onley D S 1993 *Phys. Rev.* C **48** 2697
[464] Wiringa R B, Fiks V and Fabrocini A 1988 *Phys. Rev.* C **38** 1010
[465] Bombaci I and Lombardo U 1991 *Phys. Rev.* C **44** 1892
[466] Rapp R, Durso J W and Wambach J 1997 *Nucl. Phys.* A **651** 501
[467] Glendenning N K 1987 *Nucl. Phys.* A **469** 600
[468] Anastasio M R, Celenza L S and Shakin C M 1981 *Phys. Rev.* C **23** 2258
[469] Sehn L, Fuchs C and Faessler A 1997 *Phys. Rev.* C **56** 216
[470] McNeil J A, Ray L and Wallace S J 1983 *Phys. Rev.* C **27** 2123
[471] Ter Haar B and Malfliet R 1987 *Phys. Rev. Lett.* **59** 1652
[472] Strobel K, Weber F, Weigel M K and Schaab Ch 1997 *Int. J. Mod. Phys.* E **6** (no 4) 669
[473] Baldo M, Bombaci I and Burgio G F 1998 *Astron. Astrophys.* in press (see also: astro-ph/9707277)
[474] Lee C H, Kuo T T S, Li G Q and Brown G E 1997 *Phys. Lett.* B **412** 235
[475] Von-Eiff D, Pearson J M, Stocker W and Weigel M K 1994 *Phys. Lett.* B **324** 279
[476] Farine M, Coté J and Pearson J M 1981 *Phys. Rev.* C **24** 303
[477] Von-Eiff D, Pearson J M, Stocker W and Weigel M K 1994 *Phys. Rev.* C **50** 831
[478] Backer D C 1990 (Private communication)
[479] Sprung D W L 1972 *Adv. Nucl. Phys.* **5** 225
[480] Day B D *Rev. Mod. Phys.* **51** 821
[481] Pandharipande V R and Wiringa R B 1979 *Rev. Mod. Phys.* **51** 821
[482] Myers W D and Swiatecki W J 1990 *Ann. Phys., NY* **204** 401
[483] Myers W D and Swiatecki W J 1991 *Ann. Phys., NY* **211** 292

[484] Myers W D and Swiatecki W J 1995 The nuclear Thomas–Fermi model, in: *Proc. XXIX Zakopane School of Physics (Zakopane)* ed R Broda and W Meczynski *Acta Phys. Polonica* B **26** 111 (*see also* 1996 B **27** 99; 1997 B **28** 9; 1998 B **29** 313)

Myers W D and Swiatecki W J 1998 *Phys. Rev.* C **57** 3020

[485] Myers W D and Swiatecki W J 1994 Thomas–Fermi treatment of nuclear masses, deformations and density distributions, in: *Proc. 4th KINR Int. School on Nuclear Physics (Kiev)* ed F A Ivanyuk (Kiev: National Academy of Sciences of the Ukraine) p 40

[486] Seyler R G and Blanchard C H 1961 *Phys. Rev.* **124** 227; 1963 **131** 355

[487] Myers W D 1977 *Droplet Model of Atomic Nuclei* (New York: McGraw-Hill)

Myers W D and Swiatecki W 1969 *Ann. Phys., NY* **55** 395

Myers W D and Schmidt K H 1983 *Nucl. Phys.* A **410** 61

[488] Hartmann D, El Eid M F and Barranco M 1984 *Astron. Astrophys.* **131** 249

[489] Randrup J and Medeiros E L 1991 *Nucl. Phys.* A **529** 115

[490] Randrup J and Medeiros E L 1992 *Phys. Rev.* C **45** 372

[491] Glendenning N K 1992 *Phys. Rev.* D **46** 4161

[492] Zel'dovich Ya B 1962 *Sov. Phys.-JETP* **14** 1143

[493] Zel'dovich Ya B and Novikov I D 1971 *Relativistic Astrophysics* vol I II (Chicago, University of Chicago Press); 1983 vol II

[494] Ambartsumyan V A and Saakyan G S 1960 *Sov. Astrophys.-AJ* **4** 187; 1963 *Astron. Zh.* **37** 193

[495] Langer W D and Rosen L C 1970 *Astrophys. Space Sci.* **6** 217

[496] Libby L M and Thomas F J 1969 *Phys. Lett.* B **30** 88

[497] Pandharipande V R and Garde V K 1972 *Phys. Lett.* B **39** 608

[498] Banerjee P K and Sprung D W L 1971 *Can. J. Phys.* **49** 1871

[499] Mahaux C 1981 *Brueckner Theory of Infinite Fermi Systems* Lecture Notes in Physics vol 138 (Berlin: Springer)

[500] Fiset E O and Foster T C 1972 *Nucl. Phys.* A **184** 588

[501] Ho-Kim Q and Khanna F C 1974 *Ann. Phys., NY* **86** 233

[502] Coester F, Cohen S, Day B and Vincent C M 1970 *Phys. Rev.* C **1** 769

[503] Wong C W 1972 *Ann. Phys., NY* **72** 107

[504] Celenza L S and Shakin C M 1981 *Phys. Rev.* C **24** 2704

[505] Green A M 1976 *Rep. Prog. Phys.* **39** 1109

[506] Thirring H 1918 *Phys. Z.* **19** 33

[507] Hönl H and Soergel-Fabricius Ch 1961 *Z. Phys.* **163** 571

[508] Hönl H and Dehnen H 1962 *Z. Phys.* **166** 544

[509] Brill D R and Cohen J M 1966 *Phys. Rev.* **143** 1011

[510] Misner Ch W, Thorne K S and Wheeler J A 1973 *Gravitation* (San Francisco: Freeman)

[511] Lightman A P, Press W H, Price R H, Teukolsky S A 1975 *Problem Book in Relativity and Gravitation* (Princeton: Princeton University Press)

[512] Tolman R C 1987 *Relativity, Thermodynamics and Cosmology* (New York: Dover)

[513] Schwarzschild K 1916 *Sitzber. Deut. Akad. Wiss. Berlin, Kl. Math.-Phys. Tech.* pp 424–434

[514] Schutz B F 1990 *A First Course in General Relativity* (Cambridge: Cambridge University Press)

[515] Oppenheimer J R and Volkoff G M 1939 *Phys. Rev.* **55** 374

[516] Misner C W and Sharp D H 1964 *Phys. Rev.* **136** 571

[517] Ledoux P and Warlaven Th 1958 *Handbuch der Physik* vol 51 ed S Flügge (Berlin: Springer) p 353
Ledoux P *Handbuch der Physik* vol 51, ed S Flügge (Berlin: Springer) p 635

[518] Unno W, Osaki Y, Ando H and Shibahashi H 1979 *Nonradial Oscillations of Stars* (Tokyo: University Press)

[519] Cox J P 1980 *Theory of Stellar Pulsations* (Princeton: Princeton University Press)

[520] Thorne K S and Campolattaro A 1967 *Astrophys. J.* **149** 591
Price R and Thorne K S 1969 *Astrophys. J.* **155** 163
Thorne K S 1969 *Astrophys. J.* **158** 1; 997
Campolattaro A and Thorne K S 1970 *Astrophys. J.* **159** 847

[521] Ferrari V 1992 Non-radial Oscillations of Stars in General Relativity: a Scattering Problem *Phil. Trans. R. Soc.* A **340** 423

[522] Bastrukov S, Molodtsova I, Papoyan V and Weber F 1996 *J. Phys. G: Nucl. Part. Phys.* **22** L33

[523] Bastrukov S, Molodtsova I, Papoyan V and Weber F 1996 *Proc. VII Int. Conf. on Symmetry Methods in Physics (Dubna)* vol 1, ed A N Sissakian and G S Pogosyan (Dubna: JINR) p 42

[524] Bastrukov S and Podgainy D 1997 *Astron. J.* **74** 39
Bastrukov S Molodtsova I Papoyan V and Podgainy D 1997 *Astrophysics* **40** 77

[525] Bastrukov S, Weber F and Podgainy D 1999 *J. Phys. G: Nucl. Part. Phys.* **25** 1

[526] Bertsch G F 1974 *Ann. Phys., NY* **86** 138
Bertsch G F 1978 *Nuclear Physics with Heavy Ions and Mesons* vol 1, ed R Balian, M Rho and G Ripka (Amsterdam: North-Holland) p 175

[527] Nix J R and Sierk A J 1980 *Phys. Rev.* C **21** 396
Bastrukov S I and Molodtsova I V 1995 *Phys. Part. Nucl.* **26** 180
Bastrukov S I, Libert J and Molodtsova I V 1978 *Int. J. Mod. Phys.* E **6** (no 1) 89

[528] Nörenberg W 1986 *New Vistas in Nuclear Dynamics* ed P J Brussard and J H Koch (New York: Plenum Press)

[529] Bastrukov S I, Podgainy D V, Molodtsova I V and Kosenko G 1998 *J. Phys. G: Nucl. Part. Phys.* **24** L1
Bastrukov S I Podgainy D V 1998 *Physica* A **250** (no 3, 4) 435

[530] Koester D and Chanmugam G 1990 *Rep. Prog. Phys.* **53** 837

[531] Hansen C J and Van Horn H M 1979 *Astrophys. J.* **233** 253

[532] Hansen C J 1980 *Nonradial and Nonlinear Stellar Pulsations* Lecture Notes in Physics, ed H A Hill and W A Dziembowski (Berlin: Springer) p 445

[533] McDermott P N, Van Horn H M and Hansen C J 1988 *Astrophys. J.* **325** 725

[534] Saakian G S 1995 *Physics of Neutron Stars* (Dubna: JINR)

[535] Chandrasekhar S 1964 *Phys. Rev. Lett.* **12** 114

[536] Bardeen J M, Thorne K S and Meltzer D W 1966 *Astrophys. J.* **145** 505

[537] Friedman J L and Ipser J R 1992 *Phil. Trans. R. Soc.* A **340** 391

[538] Bardeen J M 1972 Rapidly rotating stars, disks and black holes, in: *Black Holes* ed C DeWitt and B S DeWitt (New York: Gordon and Breach)

[539] Datta B, Sahu P K, Anand J D and Goyal A 1992 *Phys. Lett.* B **283** 313

[540] Vät H M and Chanmugam G 1992 *Astron. Astrophys.* **260** 250

[541] Hillebrandt W and Steinmetz K O 1976 *Astron. Astrophys.* **53** 283

[542] Lattimer J 1998 (Private communication)

[543] Schaab Ch Weber F, Weigel M K and Glendenning N K 1996 *Nucl. Phys.* A **605** 531

[544] Ruffini R 1978 *Physics and Astrophysics of Neutron Stars and Black Holes* (Amsterdam: North-Holland) p 287

[545] Lattimer J M and Yahil A 1989 *Astrophys. J.* **340** 426

[546] Ruderman M A and Fowler W A 1971 Elementary particles, in: *Science, Technology and Society* ed L C L Yuan (New York: Academic Press) p 72

[547] Alonso J D and Ibanez-Cabanell J H 1985 *Astrophys. J.* **291** 308

[548] Wilson J R 1972 *Astrophys. J.* **176** 195

[549] Abramowicz M A and Wagoner R V 1976 *Astrophys. J.* **204** 896

[550] Lightman A P and Shapiro S L 1976 *Astrophys. J.* **207** 263

[551] Butterworth E M and Ipser J R 1975 *Astrophys. J.* **200** L103; **204** 200

[552] Butterworth E M and Ipser J R 1976 *Astrophys. J.* **204** 200

[553] Butterworth E M 1979 *Astrophys. J.* **231** 219

[554] Butterworth E M 1976 *Astrophys. J.* **204** 561

[555] Datta B 1988 *Fundamentals Cosmic Phys.* **12** 151

[556] Eriguchi Y 1993 *Rotating Objects and Relativistic Physics* ed F J Chinea and L M González-Romero (Berlin: Springer)

[557] Eriguchi Y, Hachisu I and Nomoto K 1994 *Mon. Not. R. Astron. Soc.* **266** 179

[558] Neugebauer G and Herold H 1992 Gravitational fields of rapidly rotating neutron stars, in: *Relativistic Gravity Research* Lecture Notes in Physics vol 410, ed J Ehlers and G Schäfer (Berlin: Springer) p 305; p 319

[559] Salgado M, Bonazzola S, Gourgoulhon E and Haensel P 1994 *Astron. Astrophys.* **291** 155

[560] Cook G B, Shapiro S L, Teukolsky S A 1994 *Astrophys. J.* **422** 227

[561] Cook G B, Shapiro S L, Teukolsky S A 1994 *Astrophys. J.* **424** 823

[562] Hartle J B and Sharp D H 1967 *Astrophys. J.* **147** 317

[563] Hegyi D J 1977 *Astrophys. J.* **217** 244

[564] Cutler C, Lindblom L and Splinter R J 1990 *Astrophys. J.* **363** 603

[565] Nomoto K and Tsuruta S 1987 *Astrophys. J.* **312** 711

[566] Eisenhart L 1926 *Differential Geometry* (Princeton: Princeton University Press)

[567] Irvine J M 1978 *Neutron Stars* (Oxford: Clarendon Press)

[568] Kapoor R C and Datta B 1984 *Mon. Not. R. Astron. Soc.* **209** 895

[569] Buchdahl H A 1959 *Phys. Rev.* **116** 1027

[570] Bondi H 1964 *Proc. R. Soc.* A **284** 303

[571] Hartle J B 1973 *Astrophys.Space Sci.* **24** 385

[572] Ray A and Datta B 1984 *Astrophys. J.* **282** 542

[573] Abramowitz M and Stegun I A 1964 *Handbook of Mathematical Functions* (Washington, DC: Government Printing Office)

[574] Sedrakian D M, Papoian V V and Chubarian E V 1970 *Mon. Not. R. Astron. Soc.* **149** 25
[575] Arutyunyan G G, Sedrakian D M and Chubaryan É V 1971 *Astrofizika* **7** (no 3) 467
[576] Friedman J L and Schutz B F 1978 *Astrophys. J.* **222** 281
[577] Friedman J L 1983 *Phys. Rev. Lett.* **51** 11
[578] Chandrasekhar S 1969 *Ellipsoidal Figures of Equilibrium* (New Haven: Yale University Press)
[579] Detweiler S L and Lindblom L 1977 *Astrophys. J.* **213** 193
[580] Wu X Müther H, Soffel M, Herold H and Ruder H 1991 *Astron. Astrophys.* **246** 411
[581] Datta B and Ray A 1983 *Mon. Not. R. Astron. Soc.* **204** 75
[582] Shapiro S L, Teukolsky S A and Wasserman I 1983 *Astrophys. J.* **272** 702
[583] Bardeen J M and Wagoner R V 1971 *Astrophys. J.* **167** 359
[584] Abramowicz M A and Prasanna A R 1990 *Mon. Not. R. Astron. Soc.* **245** 720
[585] Abramowicz M A 1990 *Mon. Not. R. Astron. Soc.* **245** 733
[586] Lindblom L 1986 *Astrophys. J.* **303** 146
[587] Chandrasekhar S 1970 *Phys. Rev. Lett.* **24** 611
[588] Friedman J L 1978 *Commun. Math. Phys.* **62** 247
[589] Lindblom L 1992 Instabilities in rotating neutron stars, in: *The Structure and Evolution of Neutron Stars* ed D Pines, R Tamagaki and S Tsuruta (New York: Addison-Wesley) p 122
[590] Lindblom L and Detweiler S L 1977 *Astrophys. J.* **211** 565
[591] Lindblom L and Hiscock W A 1983 *Astrophys. J.* **267** 384
[592] Ipser J R and Managan R A 1985 *Astrophys. J.* **292** 517
[593] Managan R A 1985 *Astrophys. J.* **294** 463
[594] Ipser J R and Lindbolm L 1989 *Phys. Rev. Lett.* **62** 2777
[595] Ipser J R and Lindblom L 1990 *Astrophys. J.* **355** 226
[596] Ipser J R and Lindblom L 1991 *Astrophys. J.* **373** 213
[597] Ipser J R and Lindblom L 1991 *Astrophys. J.* **379** 285
[598] Cutler C and Lindblom L 1991 *Ann. NY Acad. Sci.* **631** 97
[599] Sawyer R F 1989 *Phys. Rev.* D **39** 3804
[600] Flowers E and Itoh N 1976 *Astrophys. J.* **206** 218
[601] Flowers E and Itoh N 1979 *Astrophys. J.* **230** 847
[602] Mendell G 1991 *Astrophys. J.* **380** 515
[603] Lindblom L and Mendell G 1992 Superfluid effects on the stability of rotating neutron stars, in: *The Structure and Evolution of Neutron Stars* ed D Pines, R Tamagaki and S Tsuruta (New York: Addison-Wesley) p 188
[604] Baumgart D and Friedman J L 1986 *Proc. R. Soc.* A **405** 65
[605] Cutler C and Lindblom L 1987 *Astrophys. J.* **314** 234
[606] Managan R A 1986 *Astrophys. J.* **309** 598
[607] Detweiler S L 1975 *Astrophys. J.* **197** 203
[608] Lamb H 1881 *Proc. London Math. Soc.* **13** 51
[609] Tsuruta S 1979 *Phys. Rep.* **56** 237
[610] Takatsuka T 1996 *Prog. Theor. Phys. Suppl.* **95** 901
[611] Lindblom L 1987 *Astrophys. J.* **317** 325

[612] Hartle J B and Munn M W 1975 *Astrophys. J.* **198** 467
[613] Friedman J L, Ipser J R and Sorkin R D 1988 *Astrophys. J.* **325** 722
[614] Straumann N 1992 Fermion and boson stars, in: *Relativistic Gravity Research* Lecture Notes in Physics vol 410 ed J Ehlers and G Schäfer (Berlin: Springer) p 267
[615] Imamura J, Durisen R, Friedman J L 1985 *Astrophys. J.* **294** 474
[616] Weber F, Glendenning N K and Pei S 1997 Signal for the quark–hadron phase transition in rotating hybrid stars, in: *Proc. 3rd Int. Conf. on the Physics and Astrophysics of Quark–Gluon Plasma (Jaipur)* ed D C Sinha, D K Srivastava and Y P Viyogi (New Delhi: Narosa) p 237
[617] Weber F and Glendenning N K 1997 (unpublished calculations)
[618] Mottelson B R and Valatin J G 1960 *Phys. Rev. Lett.* **5** 511
[619] Stephens F S and Simon R S 1972 *Nucl. Phys.* A **183** 257
[620] Johnson A, Ryde H and Hjorth S A 1972 *Nucl. Phys.* A **179** 753
[621] Sato K and Suzuki H 1989 *Prog. Theor. Phys.* **81** 997
[622] Madsen J 1998 *Phys. Rev. Lett.* **81** 3311
[623] Chodos A, Jaffe R L, Johnson K, Thorne C B and Weisskopf V F 1974 *Phys. Rev.* D **9** 3471
[624] Shuryak E V 1988 *The QCD Vacuum, Hadrons, and the Superdense Matter* (Singapore: World Scientific)
[625] Chmaj T and Slominiski W 1988 *Phys. Rev.* D **40** 165
[626] Freedman B A and McLerran L D 1977 *Phys. Rev.* D **16** 1130; 1147; 1169
[627] Freedman B A and McLerran L D 1978 *Phys. Rev.* D **17** 1109
[628] Farhi E and Jaffe R L 1984 *Phys. Rev.* D **30** 2379
[629] Burrows A 1988 *Astrophys. J.* **335** 891
[630] Kettner Ch 1994 *Master Thesis* University of Munich
[631] Misner C W and Zapolsky H S 1964 *Phys. Rev. Lett.* **12** 635
[632] Madsen J 1992 *Phys. Rev.* D **46** 3290
[633] Haensel P and Zdunik J L 1991 *Nucl. Phys.* B *Proc. Suppl.* **24B** 139
[634] Pizzochero P 1991 *Phys. Rev. Lett.* **66** 2425
[635] Schertler K, Greiner C and Thoma M 1997 *Nucl. Phys.* A **616** 659
[636] Wang Ch and Wang W 1991 *Commun. Theor. Phys.* **15** 347
[637] Schaab Ch 1998 *Dissertation* University of Munich
[638] McKenna J and Lyne A G 1990 *Nature* **343** 349
[639] Canuto V and Chitre S M 1974 *Phys. Rev.* D **9** 1587
[640] Alcock C *et al* 1991 *Robotic Telescopes in the 1990s* ed A V Filipenko (San Francisco: Astronomical Society of the Pacific)
[641] Paczynski B 1986 *Astrophys. J.* **304** 1
[642] Griest K 1991 *Astrophys. J.* **366** 412
[643] Griest K 1993 The search for the dark matter: WIMPs and MACHOs, in: *Texas/Pascos '92 (Berkeley)* ed C W Akerlof and M A Srednicki *Ann. NY Acad. Sci.* **688**
[644] Hewitt J N 1993 Gravitational lenses, in: *Texas/Pascos '92 (Berkeley)* ed C W Akerlof and M A Srednicki *Ann. NY Acad. Sci.* **688**
[645] Gould A 1992 *Astrophys. J.* **392** 234
[646] Bennett D P *et al* 1993 The first data from the MACHO experiment, in: *Texas/Pascos '92 (Berkeley)* ed C W Akerlof and M A Srednicki *Ann. NY Acad. Sci.* **688**

[647] Magneville C 1993 Status report on the French search for MACHOs in the Galactic halo, in: *(Texas/Pascos '92 (Berkeley))* ed C W Akerlof and M A Srednicki *Ann. NY Acad. Sci.* **688**

[648] Sutherland W *et al* 1995 *Nucl. Phys. Proc. Suppl.* **38** 379

[649] Weber F and Glendenning N K 1993 Interpretation of rapidly rotating pulsars, in: *Proc. Conf. on Nuclei in the Cosmos (Karlsruhe)* ed F Käppeler and K Wisshak (Bristol: Institute of Physics Publishing) p 399

[650] Lang K R 1992 *Astrophysical Data, Planets and Stars* (New York: Springer)

[651] Cottingham W N, Kalafatis D and Vinh Mau R 1994 *Phys. Rev. Lett.* **73** 1328

[652] Weber F and Glendenning N K 1994 Fast pulsars, compact stars, and the strange matter hypothesis, in: *Proc. 2nd Int. Conf. on Physics and Astrophysics of Quark–Gluon Plasma (Calcutta)* ed B Sinha, Y P Viyog and S Raha (Singapore: World Scientific) p 343

[653] Weber F and Glendenning N K 1995 Fast pulsars, strange stars, and strange dwarfs, in: *Proc 7th Texas Symposium on Relativistic Astrophysics (Munich)* ed H Böhringer, G E Morfill and J Trümper *Ann. NY Acad. Sci.* **759** 303

[654] Burrows A and Lattimer J M 1986 *Astrophys. J.* **307** 178

[655] Thorne K 1977 *Astrophys. J.* **212** 825

[656] Misner C W and Sharp D H 1965 *Phys. Rev. Lett.* **15** 279

[657] Linquist R W 1966 *Ann. Phys., NY* **37** 487

[658] Gudmundson E, Pethick C and Epstein R 1983 *Astrophys. J.* **272** 286

[659] Van Riper K A 1988 *Astrophys. J.* **329** 339

[660] Van Riper K A 1991 *Astrophys. J. Suppl.* **75** 449

[661] Itoh N, Kohyama Y, Matsumoto N and Seki M 1984 *Astrophys. J.* **285** 758

[662] Mitake S, Ichimaru S and Itoh N 1984 *Astrophys. J.* **277** 375

[663] Gnedin O Y and Yakovlelv D 1995 *Nucl. Phys.* A **582** 697

[664] Haensel P 1991 Transport properties of strange matter, in: *Proc. Int. Workshop on Strange Quark Matter in Physics and Astrophysics (Aarhus)* ed J Madsen and P Haensel, *Nucl. Phys.* B *Proc. Suppl.* **24B** p 23

[665] Itoh N, Tomoo A, Nakagawa M and Kohyama Y 1989 *Astrophys. J.* **339** 354

[666] Itoh N and Kohyama Y 1983 *Astrophys. J.* **275** 859

[667] Pethick C and Thorsson V 1994 *Phys. Rev. Lett.* **72** 1964

[668] Takatsuka T and Tamagaki R 1993 *Prog. Theor. Phys. Suppl.* **112** 27

[669] Ainsworth T, Wambach J and Pines D 1989 *Phys. Lett.* B **222** 173
Wambach J Ainsworth T L and Pines D 1993 *Nucl. Phys.* A **555** 128

[670] Amundsen L and Østgaard E 1985 *Nucl. Phys.* A **442** 163

[671] Wambach J, Ainsworth T and Pines D 1991 *Neutron Stars: Theory and Observation* ed J Ventura and D Pines (Dordrecht: Kluwer) pp 37–48

[672] Schaab Ch Weber F and Weigel M K 1998 *Astron. Astrophys.* **335** 596

[673] Alm Th Röpke G, Sedrakian A and Weber F 1996 *Nucl. Phys.* A **604** 491

[674] Bailin D and Love A 1984 *Phys. Rep.* **107** 325

[675] Page D 1995 *Rev. Mex. Fis.* **41** *Suppl. 1* 178

[676] Chen J, Clark J, Davé R and Khodel V 1993 *Nucl. Phys.* A **555** 59

[677] Maxwell O 1979 *Astrophys. J.* **231** 201

[678] Friman B and Maxwell O 1979 *Astrophys. J.* **232** 541

[679] Lattimer J M, Pethick C J, Prakash M and Haensel P 1991 *Phys. Rev. Lett.* **66** 2701

[680] Maxwell O, Brown G, Campbell D, Dashen R and Manassah J 1977 *Astrophys. J.* **216** 77

[681] Thorsson V, Prakash M, Tatsumi T and Pethick C 1995 *Phys. Rev.* D **52** 3739

[682] Price C 1980 *Phys. Rev.* D **22** 1910

[683] Duncan R C, Shapiro S L and Wasserman I 1983 *Astrophys. J.* **267** 358

[684] Iwamoto N 1980 *Phys. Rev. Lett.* **44** 1637; 1982 *Ann. Phys., NY* **141** 1

[685] Pethick C, Ravenhall P and Lorenz C 1995 *Nucl. Phys.* A **584** 675

[686] Itoh N, Kohyama Y, Matsumoto N and Seki M 1984 *Astrophys. J.* **280** 787

[687] Itoh N, Kohyama Y, Matsumoto N and Seki M 1984 *Astrophys. J.* **285** 304

[688] Tsuruta S and Cameron A G W 1966 *Can. J. Phys.* **44** 1863

[689] Glen G and Sutherland P 1980 *Astrophys. J.* **239** 671

[690] Richardson M, Van Horn K R and Malone R 1982 *Astrophys. J.* **255** 624

[691] Voskresenskiĭ D and Senatorov A 1896 *Zh. Eksp. Teor. Fiz.* **90** 1505

[692] Voskresenskiĭ D and Senatorov A 1987 *Sov. J. Nucl. Phys.* **45** 411

[693] Schaab C, Voskresenskiĭ D, Sedrakian A, Weber F and Weigel M K 1997 *Astron. Astrophys.* **321** 591

[694] Boguta J 1981 *Phys. Lett.* B **106** 255

[695] Prakash M, Prakash M, Lattimer J M and Pethick C J 1992 *Astrophys. J.* **390** L77

[696] Haensel P and Gnedin O Y 1994 *Astron. Astrophys.* **290** 458

[697] Schaab Ch 1995 *Master Thesis* University of Munich

[698] Muto T and Tatsumi T 1992 *Phys. Lett.* B **283** 165

[699] Bailin D and Love A 1979 *J. Phys.* A **12** L283

[700] Elgarøy Ø, Engvik L, Hjorth-Jensen M and Osnes E 1996 *Phys. Rev. Lett.* **77** 1429

[701] Hailey C J and Craig W W 1995 *Astrophys. J.* **455** L151

[702] Page D, Shibanov Y A and Zavlin V E Röntgenstrahlung from the Universe *Proc. Int. Conf. on X-Ray Astronomy and Astrophysics* MPE Report 263 Garching, p 103

[703] Greiveldinger C *et al* 1996 *Astrophys. J.* **465** L35

[704] Anderson S, Cordova F, Pavlov G, Robinson C and Thompson R 1993 *Astrophys. J.* **414** 867

[705] Halpern J P and Wang F Y H 1997 *Astrophys. J.* **477** 905

[706] Meyer R, Parlov G and Mészáros P 1994 *Astrophys. J.* **433** 265

[707] Pavlov G, Zavlin V, Trümper J and Neuhäuser R 1996 *Astrophys. J.* **472** L33

[708] Page D and Applegate J H 1992 *Astrophys. J.* **394** L17

[709] Gnedin O Y and Yakovlev D G 1993 *Astron. Lett.* **19** 104

[710] Umeda H, Nomoto K, Tsuruta S, Muto T and Tatsumi T 1994 *Astrophys. J.* **431** 309

Umeda H Tsuruta S and Nomoto K 1994 *Astrophys. J.* **433** 256

[711] Lattimer J M, Van Riper K A, Prakash M and Prakash M 1994 *Astrophys. J.* **425** 802

[712] Page D 1993 The Geminga neutron star: evidence for nucleon superfluidity at very high density, in: *Proc. 1st Symp. on Nuclear Physics in the Universe (Oak Ridge)* ed M R Strayer and M W Guidry (Bristol: Institute of Physics Publishing) p 151

[713] Page D 1994 *Astrophys. J.* **428** 250

[714] Page D 1992 *High Energy Phenomenology* ed R Huerta and M A Perez (Singapore: World Scientific) p 347

[715] Schaab Ch, Hermann B, Weber F and Weigel M K 1997 *Astrophys. J.* **480** L111

[716] Schaab Ch, Hermann B, Weber F and Weigel M K 1997 *J. Phys. G: Nucl. Part. Phys.* **23** 1

[717] Pethick C 1992 *Rev. Mod. Phys.* **64** 1133

[718] Alpar M A and Ögelman H 1990 *Astrophys. J.* **349** L55

[719] Shibazaki N and Lamb F 1989 *Astrophys. J.* **346** 808

[720] Umeda H, Shibazaki N, Tsuruta S and Nomoto K 1993 *Astrophys. J.* **408** 186

[721] Sedrakian A and Sedrakian D 1993 *Astrophys. J.* **413** 658

[722] Reisenegger A 1995 *Astrophys. J.* **442** 749

[723] Van Riper K A, Link B and Epstein R I 1995 *Astrophys. J.* **448** 294

Index

9 780750 303323

Printed and bound by CPI Group (UK) Ltd, Croydon, CR0 4YY
17/10/2024
01775685-0020